Racing for the Surface

Bingyun Li • Thomas Fintan Moriarty
Thomas Webster • Malcolm Xing
Editors

Racing for the Surface

Antimicrobial and Interface Tissue Engineering

Springer

Editors
Bingyun Li
Department of Orthopaedics
School of Medicine
West Virginia University
Morgantown, WV, USA

Thomas Webster
Department of Chemical Engineering
Northeastern University
Boston, MA, USA

Thomas Fintan Moriarty
AO Research Institute Davos
Davos Platz, Graubünden, Switzerland

Malcolm Xing
Department of Mechanical Engineering
University of Manitoba
Winnipeg, MB, Canada

ISBN 978-3-030-34470-2 ISBN 978-3-030-34471-9 (eBook)
https://doi.org/10.1007/978-3-030-34471-9

This Springer imprint is published by the registered company Springer Nature Switzerland AG
The registered company address is: Gewerbestrasse 11, 6330 Cham, Switzerland

Foreword

Orthopedic and trauma surgery has been among the most innovative and successful areas of medicine in recent decades. Survival from major trauma and bone cancer has increased exponentially, and many older people can now retain their mobility and activities well beyond the expectations of past generations. However, we face major challenges with the complications of injury and surgery. Among these complications, infection and bone loss cause great disability for our patients. When they occur together, they produce chronic ill health, recurring need for medical intervention, social isolation, loss of income and employment, and a high risk of depression and mental disorder.

The problem of infected bone defects is now much more common than previously. Prior to 1990, many patients with severe head or chest trauma did not survive the initial resuscitation, so no reconstruction of their associated limb injuries was required. Similarly, primary bone cancer of the limbs was often treated with amputation. Now we need better reconstructive options to restore function for the survivors of major trauma and cancer.

We have made major strides in understanding the mechanisms behind fracture healing, bone regeneration, bacterial invasion, adhesion, biofilm formation, and clinical infection. This has allowed us to develop strategies to prevent bacterial colonisation, for the treatment of established infection and restoration of bone defects. However, our current treatments could not be regarded as 'patient-friendly'. They often rely on prolonged hospital treatment with staged surgery, invasive bone grafting, use of toxic antimicrobials with frequent adverse effects, and bone stabilisation using cumbersome external fixation. In the best units, there are still failures with recurrence of infection in up to 20% of cases and late amputation rates of 10%.

This book focusses our attention on an attractive range of options for enhanced tissue engineering. The use of interactive biomaterials has been an aspiration of many surgeons and scientists for at least a century. It is fascinating to read that in 1932, Alexander Fleming noted the ability of tellurium to kill bacteria which were resistant to penicillin. Now we are investigating a plethora of materials, ions, and elements, which can interfere with bacterial physiology and help us with the rapidly advancing problem of antimicrobial resistance.

Initially, we asked little of our orthopedic materials. They needed to be biocompatible or at least bio-tolerant, with certain mechanical properties. But this evolved into the concept of bioactivity and biodegradability. Now we look for materials which can provide mechanical support, drug delivery, inhibition of certain cells or pathogenic organisms, and then allow remodelling into normal human tissue by induction of complex cellular pathways. This task is equivalent to the moon landing of 50 years ago!

The following chapters introduce novel ideas which will be fascinating for surgeons. The book is inspiring in the range of possibilities which await future clinicians. 'Anti-biofouling materials' or 'bone-like multifunctional coatings' are certainly engaging ideas. Better understanding of the natural mechanisms which prevent bacterial infection and promote tissue healing may allow us to exploit those areas of overlap which achieve both goals. The concept of using endogenous human responses to heal defects (in antibody-mediated osseous regeneration) promises to allow improved healing without the toxic effects of exogenous therapy. In general, a move from systemic therapy to local therapy within bones has already seen major improvements in clinical outcomes with reduced adverse effects.

The next phase of research in orthopedic biomaterials will be critical in establishing what is useful in clinical practice. It is interesting to show that materials can exhibit osteoinductive behaviour or antimicrobial properties under ideal laboratory conditions. This now needs to be tested in the more taxing clinical environment. Patients with multiple comorbid conditions, taking several disease modifying drugs, and being non-compliant with therapy will fully expose any weaknesses in the concepts. It is essential that scientists and surgeons learn to work together to direct the studies which will be required. Our patients need us to step up to this challenge.

Martin McNally
Oxford Bone Infection Unit, Oxford University Hospitals
Oxford, UK
2019

Preface

Half a Century and Billions of Dollars Later, Is the Charnley Hip Implant Still the Best We Have?

From around 1947 to now, the Charnley hip implant has revolutionised medicine, returning motor function to millions of patients suffering from hip fractures. It is hard to imagine life before what is now a common surgical procedure to restore motor function to patients suffering from hip fractures as a result of accidents, disease, cancer, and so much more. But, it was not a pretty time. Consider if you broke your hip and then came to the quick realisation that there is no treatment for you to restore motor function, or even walk to the store to get groceries ever again. Although implant failures were reported (and still are) with the original Charnley hip implant, it has certainly given hope to patients worldwide that they can once again walk on their own after a debilitating hip fracture.

Of course, as with any medical advance, over the decades since the Charnley hip implant was first introduced to medicine, numerous researchers and companies have tried to improve its original design from changing chemistry, geometry, surface texture, to even getting rid of the whole idea by using injectable or biodegradable chemistries. One can now find alterations to the original Charnley hip implant in all of today's commercially available hip implants. Yet, some fundamental design parameters from the original Charnley hip have remained the same—the general geometry (with a stem inserted into the femur marrow space and the acetabular cup into the hip), strong metal components to help support physiological loads, components that can be easily imaged for diagnosis using X-rays, and other features.

Some of the more 'revolutionary' design enhancements to the original Charnley hip have been new surface textures, which provide a more beneficial three-dimensional surface that promotes bone growth. But is this enough? Aren't these just baby steps in the scheme of improving orthopedic implant functionality? Weren't these modifications easy? Would anyone call them a *revolutionary* change in orthopedic implants? Don't problems still exist? Yes, and here are some: orthopedic implant lifetimes less than the lifetime of the patient, lack of diagnosis of bacteria presence at an early enough time to do anything about it, infection, orthopedic implants that cannot be used in children to grow with the child, chronic inflammation, wear debris causing bone loss surrounding the implant, lack of return to a fully active lifestyle for high-end athletes after hip implant surgery, clinical outcomes dependent on following proper rehabilitation schemes. These are just some of the problems that 70 years after the first Charnley implant we are yet to solve.

This book offers some solutions. It is focused on the *big* steps, not the *small* steps commonplace today, necessary for a *revolutionary* advancement in hip (or any orthopedic) implant design to meet such persistent problems. Chapters cover new strategies such as immunomodulation to aid in bone growth surrounding implants, peptide-functionalized biomaterials and injectable peptide approaches, nanoparticles, gasotransmitters, 3D nanofibre tissue engineering matrices, nanotechnology, antibody approaches to growing bone, bioprinting, new osteoconductive and osteoinductive materials, additive manufacturing, new bone biomaterial chemistries, antimicrobial approaches, nanocoatings, drug delivery coatings, decellularized extracellular matrices, and so much more. It also emphasises common problems encountered in developing new orthopedic biomaterials, such as novel animal models that more accurately assess bone biomaterials, strategies for infection control, and more.

This book *pushes* the envelope and encourages all of us to think out of the box for new orthopedic biomaterials, not just the *incremental* changes that have occurred for advancing the Charnley implant over the past three quarters of a century. We need such *revolutionary* changes to meet the growing demands placed on today's orthopedic implants, which are inserted into more patients than ever before to finally move beyond the Charnley hip for improved patient health.

Morgantown, WV, USA — Bingyun Li
Graubünden, Switzerland — Thomas Fintan Moriarty
Boston, MA, USA — Thomas Webster
Winnipeg, MB, Canada — Malcolm Xing

Contents

Part I
Innovative Antimicrobial and Osteoinductive Therapeutics

Advances in Antimicrobial and Osteoinductive Biomaterials

Samson Afewerki, Nicole Bassous, Samarah Harb, Carlos Palo-Nieto, Guillermo U. Ruiz-Esparza, Fernanda R. Marciano, Thomas Webster, and Anderson Oliveira Lobo

Abstract The enormous growing problem with antibiotic resistance in pathogenic microbes is one of the greatest threats we are facing today. In the context of orthopedic applications, infections also lead to the limited healing ability of infected and defected bone. Generally, these problems are treated with a load of antibiotics or surgical intervention. Therefore, having antibacterial properties integrated with a biomaterial would reduce the time of healing and treatment, amount of antibiotic needed, and total cost. Currently, there exists several strategies and materials with the potential of tackling these challenges. Some materials with antibacterial properties currently employed are silver nanoparticles (AgNPs), cerium oxide nanoparticles (CeO_2NPs), selenium nanoparticles (SeNPs), copper nanoparticles (CuNPs), antimicrobial peptides (AMPs), biopolymers (such as chitosan), and carbon nanostructures. On the other hand, osteoinductive and osteoconductive materials are important to promote bone healing and regeneration. Within this framework, materials which have been employed widely are bioactive glasses (BG), calcium phosphates

S. Afewerki (✉) · G. U. Ruiz-Esparza
Division of Engineering in Medicine, Department of Medicine,
Harvard Medical School, Brigham & Women's Hospital, Cambridge, MA, USA

Harvard-MIT Division of Health Science and Technology,
Massachusetts Institute of Technology, MIT, Cambridge, MA, USA

N. Bassous · T. Webster
Nanomedicine Laboratory, Department of Chemical Engineering, Northeastern University, Boston, MA, USA

S. Harb
Institute of Chemistry, São Paulo State University, Araraquara, SP, Brazil

C. Palo-Nieto
Department of Medicinal Chemistry, BMC, Uppsala University, Uppsala, Sweden

F. R. Marciano
Department of Physics, UFPI—Federal University of Piauí, Teresina, PI, Brazil

A. O. Lobo (✉)
LIMAV—Interdisciplinary Laboratory for Advanced Materials, Department of Materials Engineering, UFPI—Federal University of Piauí, Teresina, PI, Brazil
e-mail: lobo@ufpi.edu.br

B. Li et al. (eds.), *Racing for the Surface*,
https://doi.org/10.1007/978-3-030-34471-9_1

(CaPs) (e.g., hydroxyapatite (HA), tricalcium β-phosphate (β-TCP), and biphasic calcium phosphate (BCP)), peptides, growth factors, and other elements (e.g., magnesium (Mg), zinc (Zn), strontium (Sr), silicon (Si), selenium (Se), and Cu, to name a few). Some of the current technological solutions that have been employed are, for instance, the use of a co-delivery system, where both the antibacterial and the osteo-inducing agents are delivered from the same delivery system. However, this approach requires overcoming challenges with local delivery in a sustained and prolonged way, thus avoiding tissue toxicity. To address these challenges and promote novel biomaterials with dual action, sophisticated thinking and approaches have to be employed. For this, it is of the utmost importance to have a solid fundamental understanding of current technologies, bacteria behavior and response to treatments, and also a correlation between the material of use, the host tissue and bacteria. We hope by highlighting these aspects, we will promote the invention of the next generation of smart biomaterials with dual action ability to both inhibit infection and promote tissue growth.

Keywords Antibacterial · Osteoinduction · Osteoconduction · Biomaterials Orthopedic treatment · Tissue engineering · Defect · Infection · Antibiotic resistant Dentistry

Introduction

Currently, there is no doubt that the grand challenging problem with the prevalence of multi-antibiotic resistant microbes due to the overuse of antibiotics is among the greatest threats to society and the healthcare system. With respect to orthopedic challenges, infection also plays an important role in negatively impacting the treatment and healing process significantly [1]. For instance, defected or damaged bone can be treated with osteoinductive biomaterials in order to promote healing and regeneration; however, these materials does not prevent infection. There are several challenges in orthopedic problems associated with infections, foremost, they could be difficult to detect at an early stage, as vide supra stated an increased challenge to treat multidrug-resistant organisms, and persistence and recurrence of infection, particularly associated with implants [2, 3]. In the context of implants associated with a risk of microbial infection, the general approaches are, for instance, implant replacement, or in worst case amputation or mortality [4]. Hence, integrating antimicrobial properties with the implant would provide huge advantages [4]. Furthermore, a great solution to the vide supra mentioned challenges would be the development of dual functional biomaterials with the ability to promote the healing of the bone by displaying osteoinductive properties, and simultaneously inherent antibacterial properties, without the use of antibiotics [5]. This could enable the treatment or prevention of future conceivable infections [6, 7]. Here, over the years, a plethora of biomaterials with antibacterial or osteoinductive properties have been reported. Examples of the latter, in particularly, in their nanoparticle forms are silver

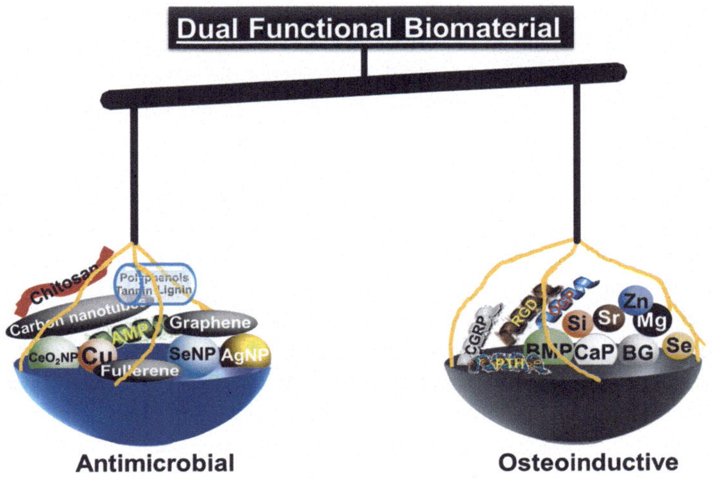

Fig. 1 Examples of biomaterials with antimicrobial and osteoinductive properties widely employed in orthopedic applications

nanoparticles (AgNPs) [8], cerium oxide nanoparticles (CeO_2NPs) [9] selenium nanoparticles (SeNPs) [10], and polymers or materials such as carbon nanostructures [11], chitosan [12], natural-based polyphenols [13–15], and antimicrobial peptides (AMPs) [16] (Fig. 1). Examples of biomaterials with osteoinductive properties are those which include osteogenic growth factors (OGF), fibroblast growth factor (FGF), vascular endothelial growth factor (VEGF) and epidermal growth factor (EGF), bioactive glass (BG), bone morphogenic proteins (BMP), hydroxyapatite (HA), elements (e.g., magnesium (Mg), zinc (Zn), silicon (Si), selenium (Se), and copper (Cu)), and peptides such as those in parathyroid hormone (PTH) and arginine-glycine-aspartic acid (RGD) [17–21] (Fig. 1). In this chapter, the current challenges with bone repair/regeneration and antibacterial infection will be highlighted. Furthermore, fundamental aspects of antimicrobial and osteoinductive properties will be discussed providing the reader an essential platform information within this topic and then some examples of antimicrobial and osteoinductive biomaterials. Subsequently biomaterials displaying dual functions or dual delivery systems with both antimicrobial and osteoinductive properties in various orthopedic applications will be presented.

Current Challenges

Reviewing the historical development of bone tissue engineering and its advancements, it is clear to see that a large impact has been mainly made due to the integration of multidisciplinary fields such as biology, material science and engineering, and clinical avenues [22]. Starting from the first example of prosthetics employed in humans in the 1960s, and then further through a more sophisticated design and development of biomaterials by 1984 [23]. Here, the dogma transformed from designing a bioinert tissue responsive material to a bioactive one, which more resembles the host bone and also with similar mechanical properties. These types of biomaterials mainly consisted of ceramics, glasses, and glass-ceramics, and were thought to have better performance due to their ability to promote cellular functions such as colonization, proliferation, and differentiation within the surrounding environment of the implant. Therefore, this class of materials was further implemented into various orthopedics and dentistry applications [24]. The incorporation of various bioactive components (such as HA) onto implants and prostheses improved their performance and osteoinduction properties [25, 26]. Further, important features of this new class of biomaterials, besides resembling the native bone, both structurally and mechanically, were that they were also resorbable [27]. This property allowed for the chemical breakdown of the material, thus, eventually being completely replaced by newly formed tissue. In this regard, an important study by Hench on the impact of time on the resorption of the Dexon sutures in vivo promoted the employment of resorbable polymers as implants [27]. To date, a wide range of implant products have been approved by the Food and Drug Administration (FDA) for a wide range of applications and are available for clinical use, displaying properties such as bioinertness, bioactivity, and resorption [22, 23]. Nevertheless, great research efforts have been devoted to further improve biomaterial properties and performance, thus overcoming some current limitations leading to imperfect implant function and survival (Fig. 2a). One of the greatest challenges to be addressed is the precise control of the biomaterial features such as material composition, surface chemistry, pore size, porosity, morphology, degradation rate, and mechanical performance (Fig. 2b) [28]. In order to tackle these problems, a fundamental understanding of the vital processes such as osteogenesis, biomaterial and bone interactions, a mechanistic interfacial interaction between the host tissue and the biomaterial impact at the cellular level are important aspects. All of these features would highly promote the invention of the next generation of materials. Furthermore, the more sophisticated challenge of designing a material which displays both osteoconductive and antimicrobial properties functioning for a wide range of applications is an important research topic. This would allow for the design of biomaterials that are resistant to infections, prevent drug resistance and at the same time promote bone healing and regeneration. To pursue this vision, the right optimization between its fabrication (sustainable and eco-friendly technologies), safety, and performance need to be included, without impeding one another [29]. As vide supra mentioned, an optimal biomaterial for promoting the healing and regeneration of bone defects should not only possess the right mechanical and degradation properties, but also the right surface chemistry to endorse cellular processes such as cell attachment, proliferation, migration, differentiation, and remodeling leading to vascularization and eventually the formation of new bone tissue [30]. Some of the central challenges and desired properties in such a devised biomaterial for orthopedic engineering are highlighted in Fig. 2.

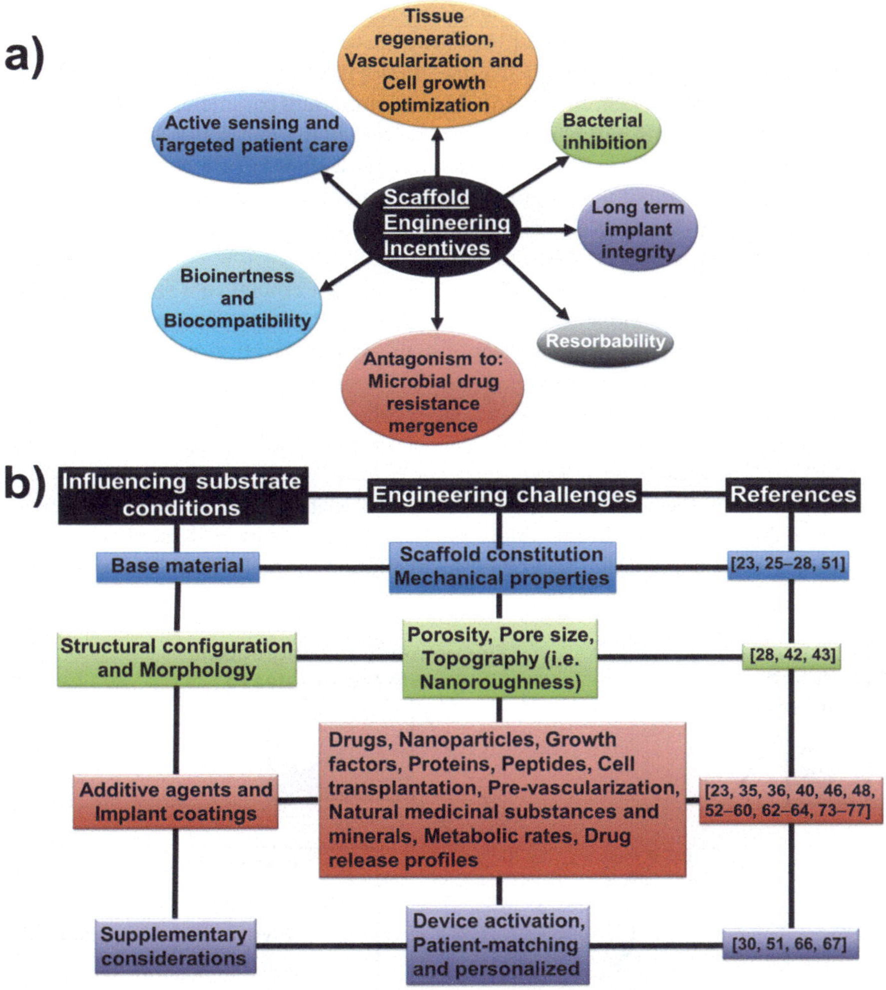

Fig. 2 Demonstrating (**a**) scaffold engineering incentives and (**b**) their influencing factors and challenges

Nowadays, the concept of taking biomaterial features to the next level is paramount, where it should not only function as a replacement, but rather regenerate the damaged/defected tissue [23]. This paradigm shift is highly dependent on the biomaterial design, hence due to the general lack of synthetic biomaterials responding to physiological stimuli [31], naturally or biologically derived materials can be employed. For instance, decellularization of biological structures has proven to be able to function as vascularized scaffolds [32]. The aim here is to promote vascularization, allowing the transport of oxygen and nutrients to cells and simultaneously removal of waste products [33]. Biomaterials for bone tissue regeneration are generally more challenging to design due to the lack of proximal blood supply, thus, less access to existing blood supply comparable for instance to cartilage tissues [34]. However, several strategies have been developed to address these challenges, such as the employment of prevascularized scaffolds and/or the use of growth factors [35, 36]. Another important parameter of the scaffolds is the porosity, since this will play an important role in the interaction with the localized blood vessels, which indirectly impact vascularization [37]. Here, the design of the scaffolds which smoothly integrate with the host vasculature is another challenge and necessity [38]. To date, several biomaterials with various properties have been developed for improving vascularization (Fig. 3) [39].

Furthermore, features on the implant surface play a crucial role since they have the ability to direct protein adsorption or cellular attachments. This mechanism can start a cascade reaction where it promotes the vascularization and subsequently endorses the proliferation of osteoblasts. In this context, Bielby et al. demonstrated the differentiation of murine embryonic stem cells into osteogenic cells which further enhanced proliferation through the incorporation of soluble ions onto the scaffold [40, 41]. Several reports have disclosed various strategies for improving the interface between the implant and host bone, for instance through the employment of interconnected porous biomaterials [42], which also can be loaded with cells [43]. Here, in the context of direct cell transplantation in implants, it is vital that the

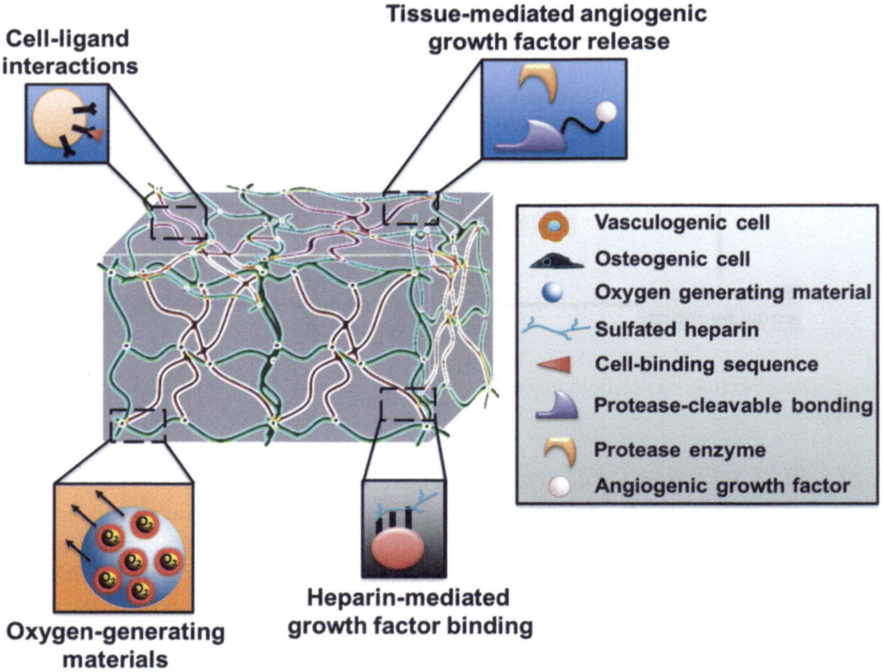

Fig. 3 Biomaterial design strategies for improving vascularization post-implantation in the region of tissue defects

right cell type is employed, considering, its accessibility, generated in high yield, and efficiently to promote the repair of the tissue and with a high survival rate [44]. These will overcome the limitations due to poor adjacent vascularization and cell–cell interactions, causing cell death due to insufficient nutrient and oxygen uptake. The overall goal is to obtain a smooth host-tissue cell scaffold interphase, which eventually will allow for incorporation into the surrounding host bone and endure a normal bone remodeling processes [38, 45]. Recently, we have seen blossom advances and interest in the employment of stem cells in various regenerative and tissue engineering applications. However, despite the great promise of stem cell technology and their potential, several contemplations have to be made such as developing a solid controlled approach for stem cell differentiation to the desired phenotype, acceptable purities and negligible carcinogenic latent [23, 46]. Over the years, several types of stem cells have been distinguished starting from the earliest embryonically derived stem cells to stem cells from the bone marrow, gut, liver, brain, and the circulatory system [47, 48]. In the context of orthopedic applications, the induced pluripotent stem cells (iPS cells) have been shown to be good candidates [46]. However, despite the improvement of implant scaffolds through surface tailoring or the addition of cells into the scaffolds, other challenges remain, in order to devise a material with high performance and with no limitations. Here, orthopedic implants can also promote bacterial adhesion and growth ensuing a negative impact on clinical outcomes and increase healthcare expense [2]. This is also one of the major factors leading to orthopedic implant failure. There are several mechanisms triggering this failure, for instance, lapses in surgical hygiene, contact with microbial flora, or the invasion of microorganisms due to implant failure [49]. Moreover, the bacterial adhesion on the implant causes several problems, firstly it promotes colonization leading to biofilm formation and this in turn can hinder tissue integration and thereby block various cellular functions and regeneration processes [50]. Furthermore, these also result in a prodigious negative impact on the patient leading to pain, surgical intervention for removal or replacement of the implant, and continuation of antibiotic treatment. Nevertheless, we have witnessed the problem with the frequent use of antibiotics promoting drug-resistant bacteria such as methicillin-resistant *Staphylococcus aureus* (MRSA) [51]. Over the years, several strategies and technologies have been invented to overcome these grand challenges, such as implant coatings with/without releasing bactericidal agents, but with the ability to prevent or reduce bacterial adhesion [52–54]. In this regard, a bactericidal agent frequently employed within such a coating is silver or their respective nanoparticle system (AgNPs) [55]. Nevertheless, they have to be engineered to provide a sustained, controlled and prolonged release preferable for at least 1 year [51, 55]. Interestingly, common food ingredients or natural medicinal components, such as garlic extracts and Aloe Vera, have been successfully demonstrated for their incorporation with implants promoting bone repair and preventing bacterial infections [56–58]. Additionally, bactericidal agents or strategies that have been employed to reduce bacteria adhesion and prevent plausible biofilm formation are the employment of nitric oxide [59] or the use of self-assembled monolayers (SAMs) to block bacterial attachment [60]. Furthermore, these strategies can be designed to provide both bacterial protective and at the same time promote bone healing/repair processes through the incorporation of osteoinductive components, such as BMP [61] and transforming growth factor beta (TGF-β) [62]. There are several challenges encountered when designing a system and materials with dual functions, such as the precise control of the delivery of each component, longstanding over the desired time frame, early and long-term osseointegration, and controlled resorbability [28]. Here resorbability, allowing the material to degrade into non-toxic components, is an important feature of the biomaterials; however, it requires a sophisticated design, in particular when it comes to the precise control of the in vivo degradation rate [63, 64]. Figure 4 highlights some of the most common bone grafting materials, including their resorptions mechanism [65]. Ideally, the material should degrade at the same rate as the tissue ingrowth and healing process [51]. However, this is highly

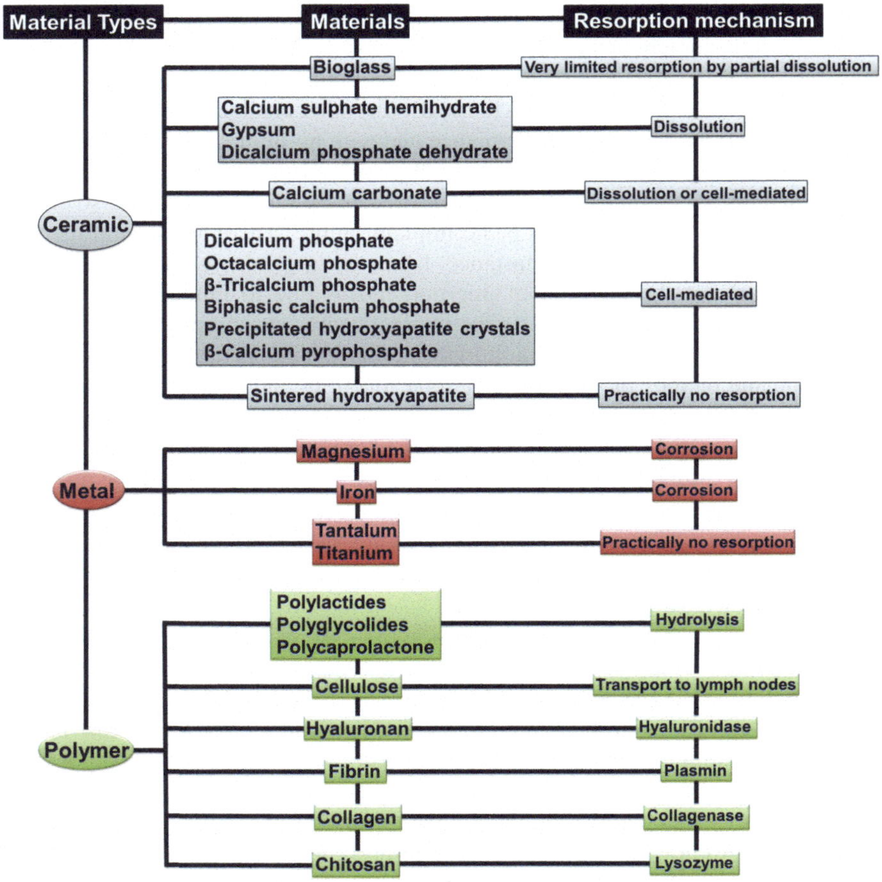

Fig. 4 Selected bone graft substitutes and their resorption mechanisms. Reproduced with permission [65]. Copyright 2010, Elsevier Ltd. (CC BY) license

influenced and dependent on several factors such as the site of implantation, the in vivo conditions, and the nature and degree of the infected/defected site, which make it a great challenge in devising biomaterials with the desired resorbability.

Despite all the current challenges, the vast advancement and innovations in the field of biomaterial technology will, most probably, successively promote inventions of grand solutions. For instance, the advancement of wireless technologies has the potential of providing tools to monitor or remotely control the healing process, the delivery of drugs in situ, or even indirectly stimulate the formation of bone tissue [51]. In this context, a great complementary would be the development of personalized medicine, this would allow overcoming current limitations, such as mechanical variations arising from the biomaterial and the treated person, and irregularities in various procedures [66, 67].

The Fundamental Basics of Antimicrobial and Osteoinductive Properties

One of the ideal approaches solving the problems posed by the grand challenges is not only to identify the fundamental mechanism leading to the problem, but also the fundamentals behind the solution. In the context of orthopedics, the biomaterial employed should display an ability to support the adhesion and localization of proteins, osseous cells, and growth factors in the region of the bone defect in order to promote the repair and regeneration of bone [68]. The process of successful bone formation promoted through biomaterials is depicted in Fig. 5. In step 1, osteogenesis starts, where mature osteoblasts are differentiated into progenitor cells, followed by osteoconduction, where bone starts growing on the surface of the biomaterial and simultaneously osseointegration takes place, meaning the direct contact between bone and biomaterial [69, 70]. Next, the osteoinduction process takes place, where cells are developed into bone forming cells (osteoprogenitor cells), and the progression of of osteogenesis is induced. Subsequently, the process of angiogenesis is promoted and cells are recruited and afterward the bone is fruitfully formed (Fig. 5) [71].

Additionally, alongside with promoting bone healing and repair, a secondary fundamental pursuit in biomaterial design is the prevention of bacterial infections and biofilm formation on the biomaterial employed [70]. In this regard, what makes it difficult is to predict any possible infection due to challenges of early detection. Here, when biomaterials, for instance titanium, are implanted in vivo, several mechanisms are triggered in the process of integration with the microenvironment promoting osteointegration and the prevention of bacterial infections [72]. Initially (Phase I), negatively charged biomolecules are adsorbed onto the positively charged titanium surface and further interact through various non-covalent forces such as hydrophobic, electrostatic, hydrogen bonding, and Van der Waals forces. This process proceeds within seconds. Next (Phase II), cells and bacterial attachment starts (minutes), followed by (Phase III) non-specific cellular adhesion resulting in fixing the cells with the aid of extracellular matrix (ECM) attaching proteins to the

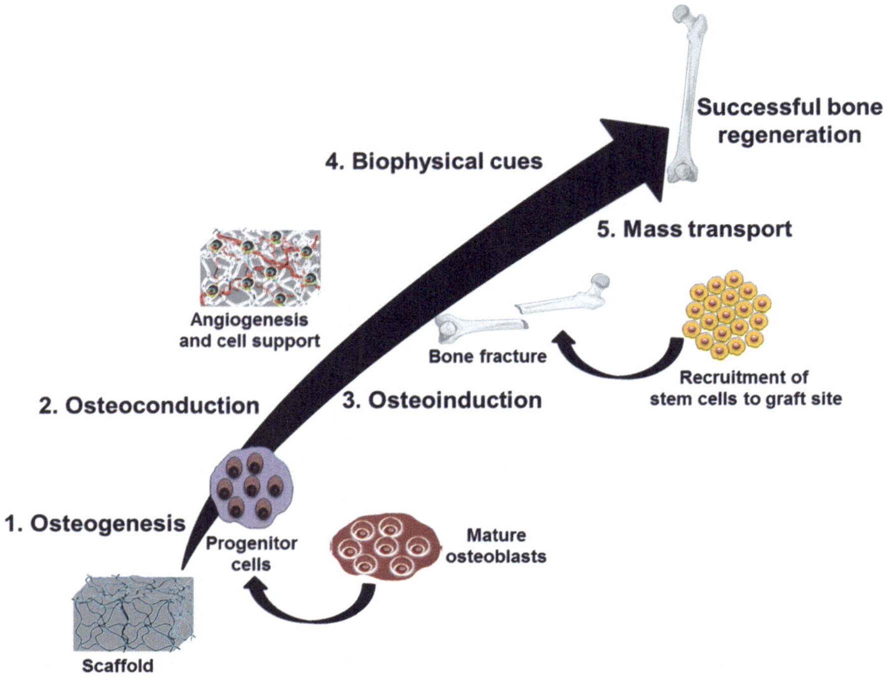

Fig. 5 The mechanism of a scaffold's promotion of bone repair and regeneration through the endorsing of osteogenic, osteoinductive, and osteoconductive processes in vivo

surface (can take hours). In the next stage (Phase IV) (days to weeks) migration, proliferation and differentiation proceeds, and lasts (V) the entire process of mineralization and bone remodeling which starts (Weeks) (Fig. 6) [72]. Therefore, tailoring the surface of the biomaterial is an important strategy [73]. For instance, introducing nanotopographies on the surface has shown to successfully promote the detachment of bacteria [74–76]. Other approaches regarding surface fabrication are coating of the surface to provide different properties or altering the surface roughness and surface energy [73, 77, 78]. Figure 7 presents the various antibacterial arsenals and therapeutics available to combat against infections and biofilm formation.

Antimicrobial Biomaterials

To date, we are entering an era where antimicrobial diseases are on the rise and predicted to cause enormous of death, even more than all cancers together [80]. Therefore, the development of novel materials with antimicrobial properties would provide an alternative to traditional antibiotics for various biomedical applications [81]. In this

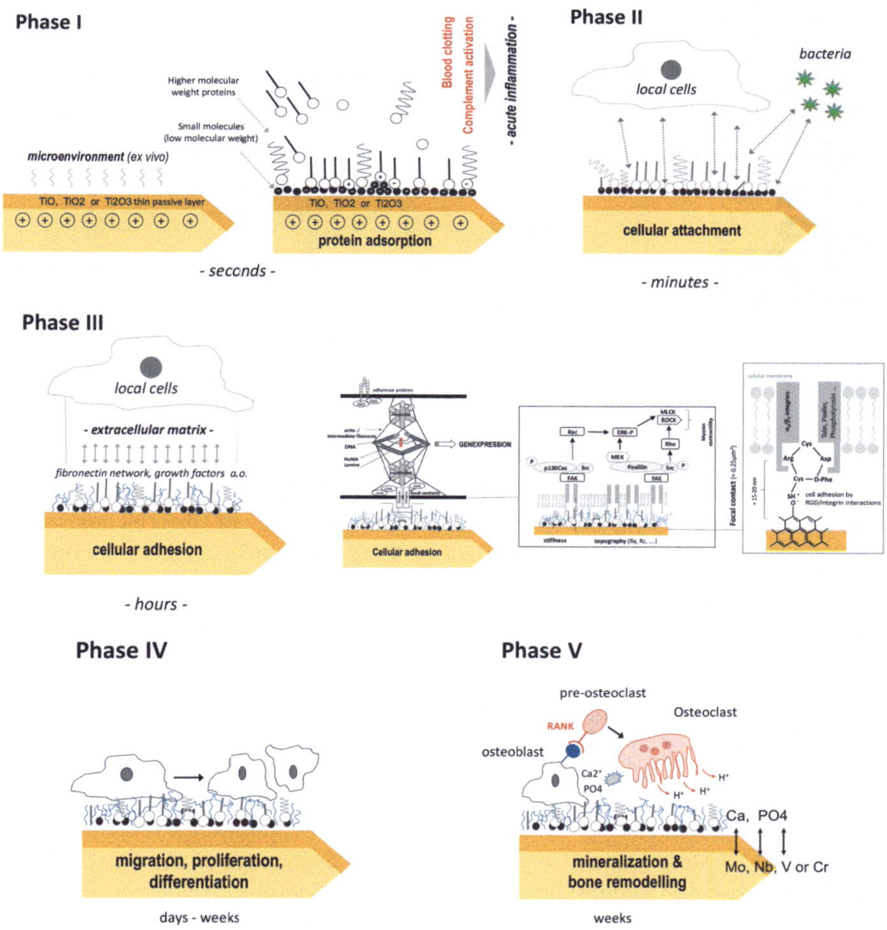

Fig. 6 The proposed in vivo local reaction onto the surface of a titanium-based implant. Reproduced with permission [72, 79]. Copyright 2017, Basel, Switzerland (CC BY) license

context, nanotechnology presents a potential approach to antimicrobial resistance, which could stimulate innovation and create a new generation of antibiotic treatment for future medicines [82]. Within this topic, AgNPs are some of the most employed antimicrobial agents in the biomedical field due to its wide antibacterial activity [83]. Additionally, other materials and elements, and their respective nanoparticle (such as Se, cerium (Ce), gold (Au), titanium (Ti), Cu, iron (Fe), carbon (carbon nanotubes (CNT), fullerene, graphene, etc.)) have been proven to display antimicrobial properties [84]. Noteworthy, while some of these materials, for instance, Ag and Cu, are intrinsically antibacterial even in their bulk state, others such as iron oxide need to be transformed to their respective nanoform in order to display antibacterial properties. The success of these strategies and nanotechnologies have led to several commercial-

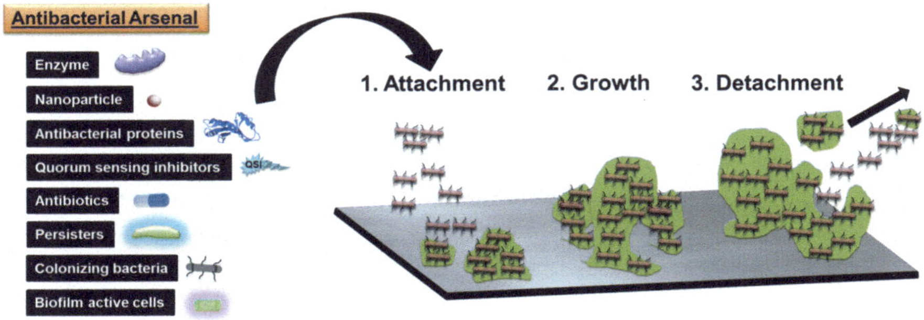

Fig. 7 Examples of various antibacterial arsenals and therapeutics for combatting bacterial biofilms and the mechanism of biofilm formation

ized products for applications in bacterial diagnosis, antibiotic delivery, and medical devices [83]. Several strategies have been developed in order to enhance the antibacterial properties of these nanomaterials. Some of these strategies are: their incorporation into biomaterials thus controlling the release, protection from aggregation, improved solubility, and engineering their size and shape. All of these aspects are also important for providing low toxicity for in vivo applications [83, 85].

Elements

In this section, various elements with antibacterial properties demonstrated in the literature will be presented (Fig. 8a). In this context, AgNPs constitute a very promising approach for the development of new antimicrobial technologies [86]. Nanoparticle formulations can add significant improvements to the antibacterial activity of elements through specific actions, such as improved adsorption at the bacterial surface [8, 87, 88]. AgNPs have attracted increasing interest due to their chemical stability, catalytic activity, localized surface plasma resonance, and high conductivity. In addition, previous reports have shown that the reactive oxygen species (ROS) formed at the surface of the AgNPs, or the release of free silver ions under certain conditions may induce cell death of either mammalian cells or microbial cells, which endows the AgNPs with unique antibacterial and antifungal effects [89, 90]. Based on these effects, AgNPs hold great potential in preventing wound inflammation and hence promoting wound healing in the form of topical administration. Here, for topical use, skin penetration ability and safety of AgNPs should be assessed [88, 91]. Small silver particles (e.g., 4–122 nm) with lower toxicity to humans have been developed, but must be released in a controlled manner to minimize side effects and maximize antimicrobial activity [84, 88]. To date, a vast number of reports have been disclosed demonstrating AgNPs and combinations as an efficient antimicrobial agent for a wide range of bacteria [8, 92, 93]. Prominently, despite the promising potential of AgNPs, some reports have demonstrated bacterial resistance against silver [94].

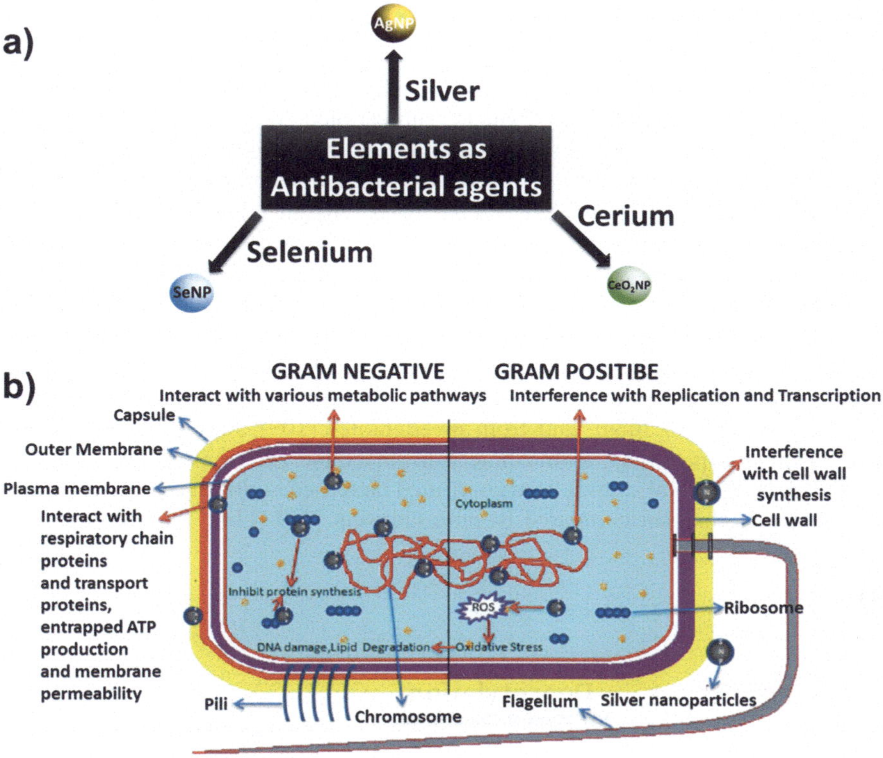

Fig. 8 (**a**) Various elements with antibacterial activity. (**b**) The proposed mechanism of the antibacterial activity of AgNPs. Reproduced with permission [79]. Copyright 2014, Springer Science Business Media New York

However, which the high antibacterial effect of AgNPs has been widely described the exact mechanism of their action has yet to be fully elucidated. In fact, their potent antibacterial and broad-spectrum activity against morphologically and metabolically different microorganisms seems to be correlated with a multifaceted mechanism by which the nanoparticles interact with the microbes. As depicted in Fig. 8b, the mechanism may proceed via various pathways, such as an interaction with the bacterial cell wall to cause leakage, interaction with various metabolic pathways, inhibited protein synthesis and promoted ROS triggering DNA damage and degradation [8, 79, 95, 96]. Furthermore, elemental Se has also shown to be a good antimicrobial candidate, particularly in its nanoform (SeNPs) [10, 97]. Interestingly, despite its broad and high antibacterial performance, an in vitro study demonstrated a safe toxicity profile [10, 98]. Several groups have demonstrated the successful employment of SeNPs and its derivative against various bacteria [99–101]. For instance, they have been employed as a medical device coating for preventing biofilm formation [98, 102]. Moreover, Ce in its oxide form (CeO_2NPs) also displays antibacterial properties through a similar mechanism as the AgNPs [9, 83, 103].

Polymers and Miscellaneous

Polymers have also been shown to have characteristic antimicrobial activity, but here it is well established that polymers with cationic components interact better with the negatively charge bacteria membrane and promote damage and cell lysis [104]. A good example within the subject is nitrogen containing compounds such as chitosan, poly-ε-lysine, polyethyleneimine, and polyguanidines [105]. A careful material design will allow for the tailoring of the antibacterial properties of the polymers, for instance, addition of a hydrophobic group will endorse infiltration into the hydrophobic bacterial membrane [85, 106, 107]. Here, chitosan is probably one of the most known and widely employed nitrogen containing polymer with antimicrobial properties [108]. As depicted in Fig. 9a, chitosan is obtained after deacetylation from chitin, a polysaccharide extracted from the exoskeletons from insects, cell walls of fungi, and from invertebrates [109]. It is well known that materials with a quaternary amine moiety displayed increased antibacterial properties [110]; here several studies have confirmed increased antibacterial performance of quaternary chitosan compared to pure chitosan [111–114]. Interestingly, the

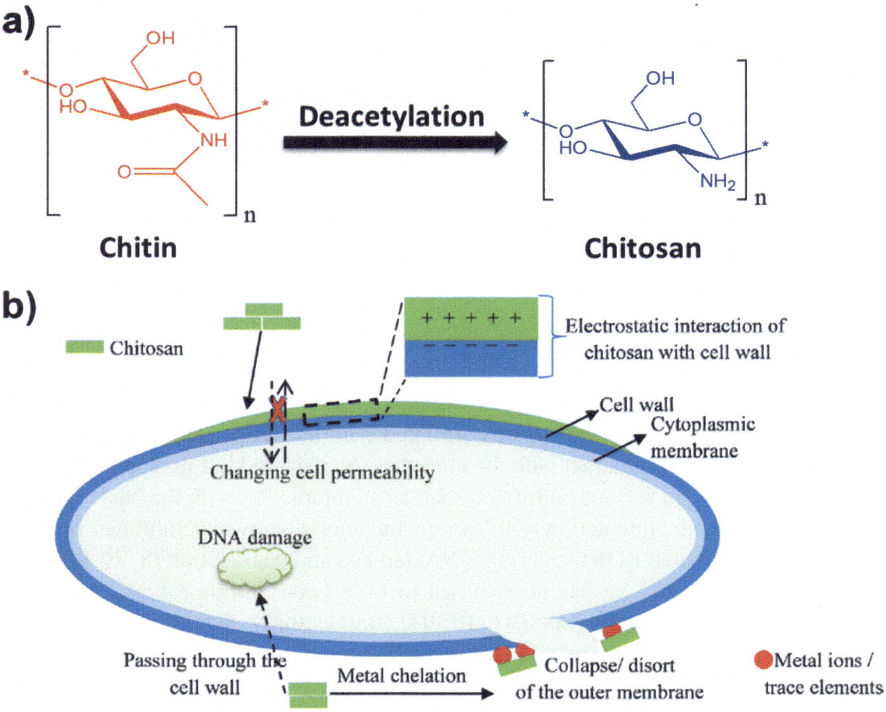

Fig. 9 (**a**) The generation of chitosan through a deacetylation step from chitin and their chemical structures. (**b**) The plausible antibacterial mechanisms of chitosan. Reproduced with permission [108]. Copyright 2019, Elsevier B.V

antibacterial property of chitosan can be improved through several strategies, by the length of the alkyl moiety on the amine generating quaternary amine, where generally an increased alkyl chain promotes higher performance, molecular weight, and degree of acetylation, where lower molecular weight and lower degree of acetylation result in improved performance [115]. The plausible antibacterial mechanism of action of chitosan is depicted in Fig. 9b, which proceeds through several pathways [108]. As vide supra mentioned, the positive charge interferes with the negatively charge bacteria surface, or through inhibition of the mRNA and protein synthesis, chelation of important metals and nutrients and thereby changing cell permeability or preventing nutrients from entering the cell through electrostatic interaction with the cell wall [115–117].

Furthermore, antimicrobial peptides (AMPs), widely found in nature, have also shown the potential for being a source of antibacterial material for a wide range of microbials [16, 85, 118]. These AMPs can be categorized based on their structure as following: α-helical AMPs, cysteine-rich AMPs, β-sheet AMPs, AMPs rich in regular amino acids, and AMPs with rare modified amino acids [16]. Moreover, some modes of action of these AMPs are, for instance, bacteria membrane disruption and ion channel formation leading to leakage of potassium ions and other components [16]. Going further, carbon nanostructure such as fullerene [11, 119], carbon nanotubes (CNTs), and graphene all have inherent antibacterial properties and proceed with a wide range of mode of actions [11, 120–123]. Some of the antibacterial mechanisms are, for instance, reduction of biofilm formation and cell attachment, generation of oxidative stress and ROS, and promoted the loss of the cellular integrity [11, 124, 125]. Nevertheless, the antibacterial efficiency and mode of action depend on several factors such as the composition of the material, size, type of microbe, etc., and can also be tailored through surface modification [11, 125]. Here, in 2007, Kang et al. disclosed a seminal work demonstrating the evidence of the antimicrobial activity of single-walled carbon nanotubes (SWCNT). The authors concluded that the SWCNT promoted membrane damage causing cell inactivation and bacterial cell death. Considering the high cost of a pure carbon nanostructure material, a good alternative could be the merging with other materials. In this regard, Aslan et al. incorporated the polymer poly(lactic-*co*-glycolic acid) (PLGA) with SWCNT and employed the constructs against *Escherichia coli* (*E. coli*) and *Staphylococcus epidermidis* (*S. epidermidis*), providing up to 98% bacteria death [124]. Furthermore, graphene in various forms, such as graphene oxide (GO) [126], graphene oxide nanoribbons (O-GNR) [127], and graphene-wrapped silver nanowires (AgNWs) [128] have all been reported to be good candidates for eliminating or reducing various bacterial types. There have also been reports where several types of carbon nanostructures have been merged in order to provide for higher antibacterial performance [129]. Nevertheless, in order for the carbon materials to find translational applications, limitations such as high cost and also in some cases relatively low solubility or insolubility in water must not be overlooked. Furthermore, another class of natural based omnipresent materials with antimicrobial properties are polyphenols [130] such as lignin [131–133] and tannin [134, 135], which are widely found in nature [13, 136]. Besides displaying antimicrobial properties, these components have several additional advantages such as being cheap, readily available and renewable, and valorization of these products is of great interest and importance, and therefore they are good candidates, particularly in the quest for fighting microbial resistance challenges [137–139]. It is believed that the hydroxyl (OH) groups in the polyphenol structure are inherent for its antibacterial properties [140]. However, the unique structure of polyphenols allows for a wide range of interaction possibilities such as covalent and physical interactions, e.g., hydrogen bonding, metal coordination, hydrophobic, imine and amine formation through a Schiff base reaction, and ionic based interaction [141]. All these interaction possibilities will promote the interaction with the bacterial cell wall and membrane, inhibition of biofilm formation, inhibition of bacterial enzymes and substrate deprivation, protein regulation, and metal iron deprivation due to chelating ability [140].

Osteoinductive Biomaterials

Biomaterials employed for various orthopedic applications should possess the ability to function as a scaffold and induced new bone formation (osteoinductive) [142]. Nevertheless, little is known about the exact mechanism for how these processes proceed, despite that several materials identified and implemented with an osteoinductive ability [143, 144]. One good strategy for the invention of a new biomaterial for bone repair and regeneration could be to mimic the composition of native bone, where it mainly consists of collagen (type 1) fibers combined with inorganic minerals such as HA and other important materials such as osteogenic factors [143, 145, 146]. In this context, materials such as calcium phosphates (CaPs) have shown osteoinductive and osteoconductive properties due to their resembling of the minerals in native bone [17, 147]. Examples of CaPs are HA, β-tricalcium phosphate (β-TCP) (both started to be employed in 1980), and biphasic calcium phosphates (BCP) (started being employed in 1990) [143]. Some differences between these materials are their mechanical properties, solubility, and resorbability [17, 148, 149]. Therefore, it is important in selecting the appropriate material with suitable properties for their intended application without compromising any other properties. For instance, employing a material with high resistance could also mean increased brittleness [149]. Alternative approaches surmounting these limitations could be through merging with other materials such as polymeric based biomaterials, which also would provide an ECM like composition [150–155]. Enduring inorganic materials, silica-based material bioglasses with an ability to easily form bonds with bone and stimulate new bond formation are other types of materials widely employed for various orthopedic applications [19, 25]. The conventional BG (45S5Bioglass®) is comprised of SiO_2 (45 wt%), Na_2O (24.5 wt%), CaO (6 wt%), and P_2O_5 (6 wt%) [148, 156]. Advantages with BG except being non-toxic, biocompatible, osteoconductive, and osteoinductive are their fast reaction with tissue and the ability to bind both hard tissues like bone and also soft tissues [155, 157]. Lately, BG based on borate and borosilicate have demonstrated an improved performance compared to the silica-based materials, due to their controllable degradation; however, one limitation is the concern of toxicity due to boron released [148].

Moreover, the mechanism of BG bioactivity has been well studied and starts with the formation of silanol (SiOH) bonds and the release of silicic acid ($Si(OH)_4$), and then a polycondensation step of the SiOH takes place generating hydrated silica gel. Subsequently, adsorption of an amorphous formed carbonate film composed of $CaO\text{-}P_2O_5$, and then crystallization of hydroxyl carbonate apatite (HCA). Afterwards, adsorption of biological moieties in the HCA layer occurs, then BG reacts with macrophages, osteoblast stem cells attach, and differentiation and proliferation of osteoblasts leading to matrix formation ensues. Next, crystallization of the matrix and growth of bone ensues, and finally bone formation takes place [148].

To date, several different BG have been developed and even commercialized; for more detailed information within this subject, the readers are referred to other beautiful reviews and articles found in the literature [148, 158–162]. Beside BGs, other inorganic biomaterials with osteoinductive properties employed in orthopedics are for instance elements such as Mg, Zn, Sr, Si, Se and Cu [20], and also the silicate nanoplatelets Laponite® [163]. Interestingly, Si as an element has also proven to be an important agent promoting various fundamental processes such as metabolism, formation and calcification of bone tissue, increasing bone mineral density and stimulating the formation of collagen and osteoblastic differentiation [164–166]. Here, each element contributes differently, for instance, Mg plays a vital role in the structure, density, and mechanical properties of bone. However, it also promotes ECM interactions and the activation of alkaline phosphatase (ALP) and integrins [167]. On the other hand, Zn is important for various cellular processes such as formation, mineralization, development, and maintenance of healthy bones [17, 148, 168]. Moreover, the element Sr not only stimulates bone formation, but also inhibits bone resorption and promotes the death of osteoclasts [17, 169–171]. Several reports have demonstrated the use of Sr as an additive in combination with other osteoinductive biomaterials [172–174]. Different from the other elements, Se functions as a protecting agent and aids in immune defenses, and antioxidant protection against ROS, reactive nitrogen species (RNS) and oxidative damage. However, it also promotes collagen expression, calcium (Ca) deposition, and osteoblastic differentiation [175–177]. Lastly, the element Cu has proven several functions, such as promoting the synthesis of bone and connective tissues, inhibiting bone resorption, and enhancing angiogenesis through functioning as a hypoxia-mimicking material [158, 178–180].

Other strategies stimulating bone production could be through the addition of OGF with the potential of stimulating osteoinduction, osteoconduction, and osseointegration. Some examples are FGF, TGF, VEGF, BMPs [181, 182], and EGF [18, 21]. De facto, these materials have been employed by native bone during bone formation (osteogenesis). In order to stimulate several processes simultaneously, within this framework, Zhan et al. disclosed the combination of both VEGF and BMP-2 in a silk scaffold. Here, VEGF promoted angiogenesis and BMP-2 enhanced bone formation [18]. Moreover, in addition to OGF, certain type of peptides have also shown important applications in orthopedic challenges, such as RGD peptides (promoting the osseointegration), PTH [183], thrombin peptide 508 (TP508),

PepGen (P-15), calcitonin gene-related peptides (CGRP) [21], osteogenic growth peptide (OGP), and ECM-derived peptides. All these agents play different roles in the final quest of improving regeneration and repair of bone tissue. For instance, the PTH influences the regulation of calcium phosphate metabolism and activates osteoblasts through several processes such as promoting osteoblast proliferation and differentiation, and reducing osteoblast apoptosis and peroxisome activator receptor [18, 21, 184]. On the other hand, OGP plays an important role to increase bone formation through promoting ALP activity, and regulation of osteoprogenitor cell proliferation, differentiation, osteocalcin secretion, and collagen and matrix mineralization [21].

Dual Functional Biomaterials

Biomaterial-associated infection in various orthopedic applications is a great challenge and an increasing problem. Prominently, since the current golden standard of treatment includes high doses of antibiotics, or in some cases, additional surgical debridement of the infected tissue, such advances are badly needed [185]. Consequently, advances in the development of novel technologies for this overwhelming problem are of great interest. There are several strategies disclosed to address this problem, which will be highlighted in this section. One approach could be through the use of a co-delivery system, where both osteoinductive and antibacterial agents are delivered from a scaffold. However, there are several challenges with this strategy. Here, some of the challenges with a co-delivery system are difficulties with precise control of the delivery rate (sustained), sequential or simultaneously delivery, site (locally), and avoiding interference between the two agents [6, 186]. Therefore, the next generation of materials in this context would be a material that possesses dual function inherently without the need of adding any antibacterial or osteoinductive agents [6]. Nevertheless, designing and inventing these kinds of materials requires a sophisticated strategy including rational design. Moreover, the material also needs to display other vital properties, such as biocompatibility and adequate mechanical and degradation properties [31]. From the perspective of dual functional biomaterials, Lobo and coworkers very recently disclosed the fabrication and generation of nanofibers based on the combination of polycaprolactone (PCL), polyethylene glycol (PEG), and gelatin methacryloyl (GelMA) with the potential to promote bone regeneration and repair through stimulating ALP activity and Ca deposition [187]. Interestingly, the same material also showed antibacterial activity against *S. aureus*, *P. aeruginosa* and MRSA [188]. Based on these studies, the material could be a potential candidate with dual function for orthopedic applications. Very recently, Wang and coauthors designed core shell nanofibers as a co-delivery system, where the shell was comprised of PCL and the core of gelatin [189]. The nanofibers were designed as a bone regeneration and anti-infective membrane; therefore, the core was loaded with the antibiotic metronidazole, while the shell with nano-HA. A prolonged release of the antibiotic was observed for more than

20 days and showed significant improvement compared to nanofibers without core shell structure. There are still several limitations with the presented technology, besides employing an antibiotic, with the potential of promoting antibiotic resistance; for example, about 55% of the drug was already released at just day 1. Moreover, Shi et al. employed different strategies for inducing both antibacterial and osteoinductive properties through the design of a bio-interface consisting of the cationic polymer (polyhexamethylene biguanide (PHMB)) [190]. Primarily, the material was coated with polydopamine and then further through cation–π interactions with the PHMB bioelectrical environment that could be generated. Surface modification or coating is a powerful and facile approach allowing for tailoring of the properties of biomaterials and providing osteogenic and antibacterial properties [191]. This strategy has been employed by Kumar et al. for the addition of amine and carboxylic functionalities onto PCL and multiwall CNT (MWCNT) [192]. The material did not only exhibit osteoinductive but also antibacterial properties, nevertheless it also improved mechanical properties (increased tensile strength and elastic modulus) and polymer crystallinity. Other groups have also employed the surface coating strategy in order to induce dual functionality on titanium implants [193, 194].

Moreover, chitosan has been widely employed in various orthopedic applications due to its favorable antimicrobial property [114, 195]. In the context of dual functional biomaterials for orthopedic applications, it has been merged with BG-poly(3-hydroxybutyrate-*co*-3-hydroxyvalerate) (PHBV) for the generation of microspheres [196]. The membrane was designed as a multidrug delivery scaffold against bacterial infection and osteoporosis for periodontal repair, therefore, the antibiotic drug tetracycline hydrochloride and the antiosteoporosis agent daidzein were selected as the model drugs. Nevertheless, no bacterial test was demonstrated, which could be interesting to see if the antibacterial properties could be inherent from the chitosan within the membrane, thus avoiding the use of antibiotics [196]. Overall, the material displayed multifunctionality, besides antibacterial and antiosteoporosis properties, it also displayed enhanced osteoblast activity, increased surface roughness, improved hydrophilicity, decreased swelling ratio, and decreased degradation. Moreover, the quaternary chitosan hydroxypropyltrimethyl ammonium chloride chitosan (HACC) has been integrated with BMP-2 for inducing dual function within zein-based materials [197]. Here, the BMP-2 was incorporated into the porous silica material SBA-15 in order to provide sustained and localized release. The multicomponent scaffold displayed prolonged antibacterial activity for up to 5 days and the BMP-2 could be released for more than 27 days. The performance of the material was demonstrated using an in vivo rabbit model of a critical-sized radius bone defect, which showed efficient bone formation [197].

Alternative to the fibers for periodontal applications, recently, dual functional PCL based electrospun materials loaded with zinc oxide nanoparticles (ZnO-NPs) have been disclosed [198]. The authors demonstrated the translational application of their devised material through an in vivo experiment by implanting the material in a rat periodontal defect model. The successful performance of the material could be observed in the decreased distance between the cementoenamel junction (CEJ) and the bone crest [198]. Moreover, a bone implant composed of calcium silicate-gelatin

(CSG) coated with chitosan or chitosan oligosaccharide has been demonstrated as a dual functional biomaterial [199]. The authors demonstrate that having 0.2% of chitosan or 0.4% of chitosan oligosaccharide displayed comparable antibacterial properties against *S. aureus* and *E. coli* as to Ag coating. Nevertheless, the Ag showed significant toxicity even at a low concentration (0.004%), while the chitosan-based material did not show any cytotoxicity at <0.4% concentration. In parallel, osteogenic properties were also demonstrated showing an ability to promote cell attachment, proliferation, ALP activity, and osteocalcin and Ca deposition.

Furthermore, Bari et al. designed a Cu containing mesoporous BG (Cu-MBG) nanoparticle composed of SiO_2-CaO as a multifunctional biomaterial for bone regeneration. As stated above, Cu has the ability of inducing both antibacterial and osteoinductive properties through various mechanisms [200, 201]. The biomaterial released copper ions (Cu^{2+}) in a sustained manner for up to 7 days. However, no proper biological study was performed on its osteogenic ability, rather through analyzing the HA-forming ability of the biomaterial. The biomaterial showed improved antibacterial properties tested against *E. coli*, *S. aureus,* and *S. epidermidis* compared to the material without Cu (e.g., a 50% reduction against *S. epidermidis* at day 3, while in the absence of Cu no reduction was observed). Moreover, the osteoinductive component BMP-2 was merged with AgNPs within a scaffold made from PLGA and was successfully demonstrated for the repair of rat femoral infected segmental defects [202]. A similar combination with PLGA as the base polymer was also demonstrated by other research groups [203, 204]. Additionally, Sun et al. employed the same combinations; however, the scaffold employed in this study was a collagen composite [205]. Several reports have demonstrated the use of BMP-2 in combination with vancomycin employing different scaffolds such as a silica calcium phosphate nanocomposite [206], a calcium sulfate composite [207], poly(2-hydroxyethyl methacrylate)-nanocrystalline HA (pHEMA-nHA) [208], and polyurethane [209]. Additionally, an interesting strategy and dual combination was fruitfully demonstrated for bone regeneration using Sr and Ag combination within NTs [210, 211]. Furthermore, Fig. 10 depicts additional antimicrobial and osteoinductive biomaterials, their delivery methods and approaches employed for the incorporation of the agents [212].

Future Perspective and Remarks

A wide range of biomaterials and technologies displaying both antimicrobial and osteoinductive functionalities have been presented and developed over the years. Despite this fruitful progress, no ideal biomaterials providing dual functionalities with high efficiency, controllable triggering mechanisms, avoiding the use of antibiotic and long-lasting have been invented. All these should also be integrated with facile, scalable and sustainable fabrication and preparation technologies. Organic chemistry can play a crucial role in the quest of designing novel dual functional biomaterials allowing for the employment of green chemistry parameters such as

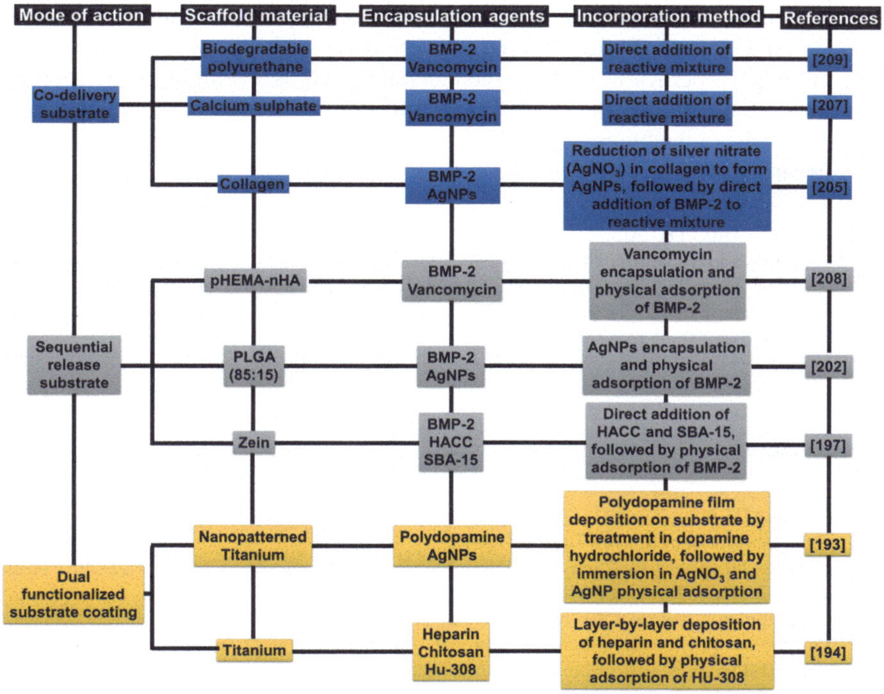

Fig. 10 Antibacterial and osteogenic scaffold constituents and delivery methods

atom economy, waste reduction, reduced toxicity, and the use of renewable resources [213, 214]. This would also allow tailoring the surface of the biomaterials and in that way introducing antimicrobial and osteoinductive functionalities [215]. Therefore, further advancements within this field are important for not only future perspectives, but also the fundamental understanding of various cellular mechanism, material and tissue interactions, long-term performance of such materials, etc. These will provide for the development and invention of a more solid technological platform which will be easier to translate into the clinic and find useful and suitable applications. A co-delivery system is one strategy extensively employed allowing the delivery of both agents consecutively without interfering with one and other, nevertheless, some of the limitations and challenges that need to be overcome in the future are, for instance, controlled local and sustained delivery, thus avoiding systemic toxicity, delivery over the desired time point and avoiding the use of antibiotics, thus circumventing the risk and promotion of the development of antibiotic resistance organisms. Moreover, having biomaterials that prevent future infection will also promote the healing and repair of bone since infections have the ability to decline or hinder the healing process. Besides being dual functional biomaterials, they should also display vital properties such as biocompatibility, biodegradability, support tissue attachment, regeneration, proliferation, optimal mechanical properties, and good integration with the host tissue. Another option for future advanced biomaterial within the discussed topic could be the employment of smart materials [31]. Adding properties to existing biomaterials, such as the ability of triggering itself in case of any treatment or future damages or infections, is progressing. These are some of the future materials we hope will boost this endeavor and advance the current topic.

Acknowledgements Dr. Afewerki gratefully acknowledges the financial support from the Sweden-America Foundation (The Family Mix Entrepreneur Foundation) and the Olle Engkvist Byggmästare Foundation. Professor Lobo and Professor Marciano acknowledge the National Council for Scientific and Technological Development for support (CNPq, #303752/2017-3 to AOL and #304133/2017-5 to FRM).

References

1. Thomas MV, Puleo DA (2011) Infection, inflammation, and bone regeneration: a paradoxical relationship. J Dent Res 90(9):1052–1061
2. Moriarty TF, Kuehl R, Coenye T, Metsemakers W-J, Morgenstern M, Schwarz EM et al (2016) Orthopaedic device-related infection: current and future interventions for improved prevention and treatment. EFORT Open Rev 1(4):89–99
3. Winkler H (2017) Treatment of chronic orthopaedic infection. EFORT Open Rev 2(5):110–116
4. Ribeiro M, Monteiro FJ, Ferraz MP (2012) Infection of orthopedic implants with emphasis on bacterial adhesion process and techniques used in studying bacterial-material interactions. Biomatter 2(4):176–194
5. Zhang K, Wang S, Zhou C, Cheng L, Gao X, Xie X et al (2018) Advanced smart biomaterials and constructs for hard tissue engineering and regeneration. Bone Res 6(1):31
6. Lu H, Liu Y, Guo J, Wu H, Wang J, Wu G (2016) Biomaterials with antibacterial and osteoinductive properties to repair infected bone defects. Int J Mol Sci 17(3):334
7. Raphel J, Holodniy M, Goodman SB, Heilshorn SC (2016) Multifunctional coatings to simultaneously promote osseointegration and prevent infection of orthopaedic implants. Biomaterials 84:301–314
8. Franci G, Falanga A, Galdiero S, Palomba L, Rai M, Morelli G et al (2015) Silver nanoparticles as potential antibacterial agents. Molecules 20(5):8856–8874
9. Pelletier DA, Suresh AK, Holton GA, McKeown CK, Wang W, Gu B et al (2010) Effects of engineered cerium oxide nanoparticles on bacterial growth and viability. Appl Environ Microbiol 76(24):7981–7989
10. Guisbiers G, Wang Q, Khachatryan E, Mimun L, Mendoza-Cruz R, Larese-Casanova P et al (2016) Inhibition of E. coli and S. aureus with selenium nanoparticles synthesized by pulsed laser ablation in deionized water. Int J Nanomedicine 11:3731–3736
11. Al-Jumaili A, Alancherry S, Bazaka K, Jacob MV (2017) Review on the antimicrobial properties of carbon nanostructures. Materials 10(9):1066
12. Tan H, Ma R, Lin C, Liu Z, Tang T (2013) Quaternized chitosan as an antimicrobial agent: antimicrobial activity, mechanism of action and biomedical applications in orthopedics. Int J Mol Sci 14(1):1854–1869
13. Coppo E, Marchese A (2014) Antibacterial activity of polyphenols. Curr Pharm Biotechnol 15(4):380–390
14. Lee E, Song Y, Lee S (2014) Antimicrobial property and biodegradability of lignin nanofibers. Master's thesis, Yonsei University, Republic of Korea
15. Scalbert A (1991) Antimicrobial properties of tannins. Phytochemistry 30(12):3875–3883

16. Reddy K, Yedery R, Aranha C (2004) Antimicrobial peptides: premises and promises. Int J Antimicrob Agents 24(6):536–547
17. Habibovic P, Barralet J (2011) Bioinorganics and biomaterials: bone repair. Acta Biomater 7(8):3013–3026
18. Zhang W, Zhu C, Wu Y, Ye D, Wang S, Zou D et al (2014) VEGF and BMP-2 promote bone regeneration by facilitating bone marrow stem cell homing and differentiation. Eur Cell Mater 27(12):1–11
19. Hench LL, Splinter RJ, Allen W, Greenlee T (1971) Bonding mechanisms at the interface of ceramic prosthetic materials. J Biomed Mater Res 5(6):117–141
20. Tomás H, Alves CS, Rodrigues J (2018) Laponite®: a key nanoplatform for biomedical applications? Nanomedicine 14(7):2407–2420
21. Pountos I, Panteli M, Lampropoulos A, Jones E, Calori GM, Giannoudis PV (2016) The role of peptides in bone healing and regeneration: a systematic review. BMC Med 14(1):103
22. Ratner BD, Hoffman AS, Schoen FJ, Lemons JE (2004) Biomaterials science—an introduction to materials in medicine, 2nd edn. Elsevier, New York
23. Hench LL, Thompson I (2010) Twenty-first century challenges for biomaterials. J R Soc Interface 7(Suppl 4):S379–SS91
24. Campana V, Milano G, Pagano E, Barba M, Cicione C, Salonna G et al (2014) Bone substitutes in orthopaedic surgery: from basic science to clinical practice. J Mater Sci Mater Med 25(10):2445–2461
25. Hench LL (1998) Bioceram J Am Ceram Soc 81(7):1705–1728
26. Rea S, Bonfield W (2004) Biocomposites for medical applications. J Australas Ceram Soc 40(1):43–57
27. Hench L (1980) Biomaterials. Science 208(4446):826–831
28. Ratner BD, Hoffman AS, Yaszemski MJ, Lemons JE, Schoen FJ (2012) Biomaterials science : an introduction to materials in medicine. Elsevier Science & Technology, San Diego
29. Uludağ H (2014) Grand challenges in biomaterials. Front Bioeng Biotechnol 2:43
30. Muschler GF, Nakamoto C, Griffith LG (2004) Engineering principles of clinical cell-based tissue engineering. J Bone Joint Surg Am 86(7):1541–1558
31. Kowalski PS, Bhattacharya C, Afewerki S, Langer R (2018) Smart biomaterials: recent advances and future directions. ACS Biomater Sci Eng 4(11):3809–3817
32. Novosel EC, Kleinhans C, Kluger PJ (2011) Vascularization is the key challenge in tissue engineering. Adv Drug Deliv Rev 63(4):300–311
33. Soker S, Machado M, Atala A (2000) Systems for therapeutic angiogenesis in tissue engineering. World J Urol 18(1):10–18
34. Rouwkema J, Rivron NC, van Blitterswijk CA (2008) Vascularization in tissue engineering. Trends Biotechnol 26(8):434–441
35. Mikos AG, Sarakinos G, Lyman MD, Ingber DE, Vacanti JP, Langer R (1993) Prevascularization of porous biodegradable polymers. Biotechnol Bioeng 42(6):716–723
36. Patel ZS, Young S, Tabata Y, Jansen JA, Wong ME, Mikos AG (2008) Dual delivery of an angiogenic and an osteogenic growth factor for bone regeneration in a critical size defect model. Bone 43(5):931–940
37. Whang K, Healy K, Elenz D, Nam E, Tsai D, Thomas C et al (1999) Engineering bone regeneration with bioabsorbable scaffolds with novel microarchitecture. Tissue Eng 5(1):35–51
38. Petite H, Viateau V, Bensaid W, Meunier A, de Pollak C, Bourguignon M et al (2000) Tissue-engineered bone regeneration. Nat Biotechnol 18(9):959–963
39. García JR, García AJ (2016) Biomaterial-mediated strategies targeting vascularization for bone repair. Drug Delivery Transl Res 6(2):77–95
40. Bielby RC, Christodoulou IS, Pryce RS, Radford WJ, Hench LL, Polak JM (2004) Time- and concentration-dependent effects of dissolution products of 58S Sol–gel bioactive glass on proliferation and differentiation of murine and human osteoblasts. Tissue Eng 10:1018–1026
41. Bielby RC, Pryce RS, Hench LL, Polak JM (2005) Enhanced derivation of osteogenic cells from murine embryonic stem cells after treatment with ionic dissolution products of 58S bioactive sol–gel glass. Tissue Eng 11(3–4):479–488

42. Gao J, Dennis JE, Solchaga LA, Awadallah AS, Goldberg VM, Caplan AI (2001) Tissue-engineered fabrication of an osteochondral composite graft using rat bone marrow-derived mesenchymal stem cells. Tissue Eng 7(4):363–371
43. Karp JM, Shoichet MS, Davies JE (2003) Bone formation on two-dimensional poly (DL-lactide-co-glycolide)(PLGA) films and three-dimensional PLGA tissue engineering scaffolds in vitro. J Biomed Mater Res A 64(2):388–396
44. Grande DA, Breitbart AS, Mason J, Paulino C, Laser J, Schwartz RE (1999) Cartilage tissue engineering: current limitations and solutions. Clin Orthop Relat Res 367:S176–SS85
45. Hutmacher DW, Sittinger M (2003) Periosteal cells in bone tissue engineering. Tissue Eng 9(Suppl 1):S45–S64
46. De Miguel MP, Fuentes-Julián S, Alcaina Y (2010) Pluripotent stem cells: origin, maintenance and induction. Stem Cell Rev 6(4):633–649
47. Thomson JA, Itskovitz-Eldor J, Shapiro SS, Waknitz MA, Swiergiel JJ, Marshall VS et al (1998) Embryonic stem cell lines derived from human blastocysts. Science 282(5391):1145–1147
48. Hubbell JA, Thomas SN, Swartz MA (2009) Materials engineering for immunomodulation. Nature 462(7272):449–460
49. Quaile A (2012) Infections associated with spinal implants. Int Orthop 36(2):451–456
50. Arciola CR, Campoccia D, Montanaro L (2018) Implant infections: adhesion, biofilm formation and immune evasion. Nat Rev Microbiol 16(7):397–409
51. Bose S, Bandyopadhyay A (2017) Materials and devices for bone disorders, 1st edn. Elsevier, Amsterdam, pp 1–560
52. Romanò CL, Scarponi S, Gallazzi E, Romanò D, Drago L (2015) Antibacterial coating of implants in orthopaedics and trauma: a classification proposal in an evolving panorama. J Orthop Surg Res 10(1):157
53. Aranya AK, Pushalkar S, Zhao M, LeGeros RZ, Zhang Y, Saxena D (2017) Antibacterial and bioactive coatings on titanium implant surfaces. J Biomed Mater Res A 105(8):2218–2227
54. Orapiriyakul W, Young PS, Damiati L, Tsimbouri PM (2018) Antibacterial surface modification of titanium implants in orthopaedics. J Tissue Eng 9:2041731418789838
55. Kim JS, Kuk E, Yu KN, Kim J-H, Park SJ, Lee HJ et al (2007) Antimicrobial effects of silver nanoparticles. Nanomedicine 3(1):95–101
56. Rahman S, Carter P, Bhattarai N (2017) Aloe vera for tissue engineering applications. J Funct Biomater 8(1):6
57. Bansal SS, Kausar H, Vadhanam MV, Ravoori S, Pan J, Rai SN et al (2014) Curcumin implants, not curcumin diet, inhibit estrogen-induced mammary carcinogenesis in ACI rats. Cancer Prev Res (Phila) 7(4):456–465
58. Pakdel F, Ghasemi S, Babaloo A, Javadzadeh Y, Momeni R, Ghanizadeh M et al (2017) Antibacterial effects of garlic extracts and ziziphora essential oil on bacteria associated with peri-implantitis. J Clin Diagn Res 11(4):ZC16–ZC19
59. Nichols SP, Schoenfisch MH (2013) Nitric oxide flux-dependent bacterial adhesion and viability at fibrinogen-coated surfaces. Biomater Sci 1(11):1151–1159
60. Freitas SC, Correa-Uribe A, Cristina L, Martins M, Pelaez-Vargas A (2018) Self-assembled monolayers for dental implants. Int J Dent 2018:4395460
61. Bessa PC, Casal M, Reis RL (2008) Bone morphogenetic proteins in tissue engineering: the road from laboratory to clinic, part II (BMP delivery). J Tissue Eng Regen Med 2(2–3):81–96
62. Simmons CA, Alsberg E, Hsiong S, Kim WJ, Mooney DJ (2004) Dual growth factor delivery and controlled scaffold degradation enhance in vivo bone formation by transplanted bone marrow stromal cells. Bone 35(2):562–569
63. Zhao L, Chu PK, Zhang Y, Wu Z (2009) Antibacterial coatings on titanium implants. J Biomed Mater Res B Appl Biomater 91(1):470–480
64. Modjarrad K, Ebnesajjad S (2013) Handbook of polymer applications in medicine and medical devices. Elsevier, Amsterdam, pp 1–354
65. Bohner M (2010) Resorbable biomaterials as bone graft substitutes. Mater Today 13(1):24–30
66. Wilhelmi M, Haverich A (2013) Functionalized medical implants in the era of personalized medicine. Clin Pract 10(2):119–121

67. Qin M, Liu Y, Wang L, Li D, Jin Z, Liu Y et al (2017) Laser metal direct forming of the customized titanium implants. Rare Metal Mater Eng 46(2017):1924–1928
68. Bosetti M, Fusaro L, Nicolì E, Borrone A, Aprile S, Cannas M (2014) Poly-L-lactide acid-modified scaffolds for osteoinduction and osteoconduction. J Biomed Mater Res Part A 102(10):3531–3539
69. Goonoo N, Bhaw-Luximon A (2018) Regenerative medicine: induced pluripotent stem cells and their benefits on accelerated bone tissue reconstruction using scaffolds. J Mater Res 33(11):1573–1591
70. Algburi A, Comito N, Kashtanov D, Dicks LMT, Chikindas ML (2017) Control of biofilm formation: antibiotics and beyond. Appl Environ Microbiol 83(3):e02508–e02516
71. Albrektsson T, Johansson C (2001) Osteoinduction, osteoconduction and osseointegration. Eur Spine J 10(Suppl 2):S96–S101
72. Jäger M, Jennissen HP, Dittrich F, Fischer A, Köhling HL (2017) Antimicrobial and osseointegration properties of nanostructured titanium orthopaedic implants. Materials (Basel) 10(11):1302
73. Hatton BD (2015) Antimicrobial coatings for metallic biomaterials. In: Wen C (ed) Surface coating and modification of metallic biomaterials. Woodhead Publishing, Sawston, pp 379–391
74. Betancourt T, Brannon-Peppas L (2006) Micro- and nanofabrication methods in nanotechnological medical and pharmaceutical devices. Int J Nanomed 1(4):483–495
75. Besinis A, Hadi SD, Le HR, Tredwin C, Handy RD (2017) Antibacterial activity and biofilm inhibition by surface modified titanium alloy medical implants following application of silver, titanium dioxide and hydroxyapatite nanocoatings. Nanotoxicology 11(3):327–338
76. Tripathy A, Pahal S, Mudakavi RJ, Raichur AM, Varma MM, Sen P (2018) Impact of bioinspired nanotopography on the antibacterial and antibiofilm efficacy of chitosan. Biomacromolecules 19(4):1340–1346
77. Ercan B, Khang D, Carpenter J, Webster TJ (2013) Using mathematical models to understand the effect of nanoscale roughness on protein adsorption for improving medical devices. Int J Nanomedicine 8(Suppl 1):75–81
78. Slavin YN, Asnis J, Häfeli UO, Bach H (2017) Metal nanoparticles: understanding the mechanisms behind antibacterial activity. J Nanobiotechnol 15(1):65
79. Pandey JK, Swarnkar R, Soumya K, Dwivedi P, Singh MK, Sundaram S et al (2014) Silver nanoparticles synthesized by pulsed laser ablation: as a potent antibacterial agent for human enteropathogenic gram-positive and gram-negative bacterial strains. Appl Biochem Biotechnol 174(3):1021–1031
80. O'Neill J (2016) The review on antimicrobial resistance. Tackling drug-resistant infections globally: final report and recommendations
81. Gupta A, Mumtaz S, Li C-H, Hussain I, Rotello VM (2019) Combatting antibiotic-resistant bacteria using nanomaterials. Chem Soc Rev 48(2):415–427
82. Kumar M, Curtis A, Hoskins C (2018) Application of nanoparticle technologies in the combat against anti-microbial resistance. Pharmaceutics 10(1):11
83. Alpaslan E, Geilich BM, Yazici H, Webster TJ (2017) pH-controlled cerium oxide nanoparticle inhibition of both gram-positive and gram-negative bacteria growth. Sci Rep 7:45859
84. Zhang BG, Myers DE, Wallace GG, Brandt M, Choong PF (2014) Bioactive coatings for orthopaedic implants-recent trends in development of implant coatings. Int J Mol Sci 15(7):11878–11921
85. Gupta A, Landis RF, Li C-H, Schnurr M, Das R, Lee Y-W et al (2018) Engineered polymer nanoparticles with unprecedented antimicrobial efficacy and therapeutic indices against multidrug-resistant bacteria and biofilms. J Am Chem Soc 140(38):12137–12143
86. Rai M, Yadav A, Gade A (2009) Silver nanoparticles as a new generation of antimicrobials. Biotechnol Adv 27(1):76–83
87. Le Ouay B, Stellacci F (2015) Antibacterial activity of silver nanoparticles: a surface science insight. Nano Today 10(3):339–354

88. Chaloupka K, Malam Y, Seifalian AM (2010) Nanosilver as a new generation of nanoproduct in biomedical applications. Trends Biotechnol 28(11):580–588
89. Qin H, Zhu C, An Z, Jiang Y, Zhao Y, Wang J et al (2014) Silver nanoparticles promote osteogenic differentiation of human urine-derived stem cells at noncytotoxic concentrations. Int J Nanomedicine 9:2469–2478
90. Mahmood M, Li Z, Casciano D, Khodakovskaya MV, Chen T, Karmakar A et al (2011) Nanostructural materials increase mineralization in bone cells and affect gene expression through miRNA regulation. J Cell Mol Med 15(11):2297–2306
91. Sportelli MC, Izzi M, Volpe A, Clemente M, Picca RA, Ancona A et al (2018) The pros and cons of the use of laser ablation synthesis for the production of silver nano-antimicrobials. Antibiotics (Basel) 7(3):67
92. Loo YY, Rukayadi Y, Nor-Khaizura M-A-R, Kuan CH, Chieng BW, Nishibuchi M, Radu S (2018) In vitro antimicrobial activity of green synthesized silver nanoparticles against selected gram-negative foodborne pathogens. Front Microbiol 9:1555
93. Huma Z-e, Gupta A, Javed I, Das R, Hussain SZ, Mumtaz S et al (2018) Cationic silver nanoclusters as potent antimicrobials against multidrug-resistant bacteria. ACS Omega 3(12):16721–16727
94. Panáček A, Kvítek L, Smékalová M, Večeřová R, Kolář M, Röderová M et al (2018) Bacterial resistance to silver nanoparticles and how to overcome it. Nat Nanotechnol 13(1):65–71
95. Pareek V, Gupta R, Panwar J (2018) Do physico-chemical properties of silver nanoparticles decide their interaction with biological media and bactericidal action? A review. Mater Sci Eng C 90(11):739–749
96. Patil MP, Kim G-D (2017) Eco-friendly approach for nanoparticles synthesis and mechanism behind antibacterial activity of silver and anticancer activity of gold nanoparticles. Appl Microbiol Biotechnol 101(1):79–92
97. Hosnedlova B, Kepinska M, Skalickova S, Fernandez C, Ruttkay-Nedecky B, Peng Q et al (2018) Nano-selenium and its nanomedicine applications: a critical review. Int J Nanomedicine 13:2107–2128
98. Stolzoff M, Wang S, Webster T (eds) (2016) Efficacy and mechanism of selenium nanoparticles as antibacterial agents. In: Front. bioeng. biotechnol. conference abstract: 10th world biomaterials congress. https://doi.org/10.3389/conf.FBIOE.2016.01.01826
99. Srivastava N, Mukhopadhyay M (2015) Green synthesis and structural characterization of selenium nanoparticles and assessment of their antimicrobial property. Bioprocess Biosyst Eng 38(9):1723–1730
100. Shoeibi S, Mashreghi M (2017) Biosynthesis of selenium nanoparticles using Enterococcus faecalis and evaluation of their antibacterial activities. J Trace Elem Med Biol 39:135–139
101. Huang X, Chen X, Chen Q, Yu Q, Sun D, Liu J (2016) Investigation of functional selenium nanoparticles as potent antimicrobial agents against superbugs. Acta Biomater 30:397–407
102. Wang Q, Webster TJ (2012) Nanostructured selenium for preventing biofilm formation on polycarbonate medical devices. J Biomed Mater Res Part A 100(12):3205–3210
103. Farias IAP, dos Santos CCL, Sampaio FC (2018) Antimicrobial activity of cerium oxide nanoparticles on opportunistic microorganisms: a systematic review. Biomed Res Int 2018:1
104. Li S, Dong S, Xu W, Tu S, Yan L, Zhao C et al (2018) Antibacterial hydrogels. Adv Sci 5(5):1700527
105. Jain A, Duvvuri LS, Farah S, Beyth N, Domb AJ, Khan W (2014) Antimicrobial polymers. Adv Healthc Mater 3(12):1969–1985
106. Yang Y, Cai Z, Huang Z, Tang X, Zhang X (2018) Antimicrobial cationic polymers: from structural design to functional control. Polym J 50(1):33–44
107. Du H, Wang Y, Yao X, Luo Q, Zhu W, Li X et al (2016) Injectable cationic hydrogels with high antibacterial activity and low toxicity. Polym Chem 7(36):5620–5624
108. Hosseinnejad M, Jafari SM (2016) Evaluation of different factors affecting antimicrobial properties of chitosan. Int J Biol Macromol 85:467–475
109. Rinaudo M (2006) Chitin and chitosan: properties and applications. Prog Polym Sci 31(7):603–632

110. Muñoz-Bonilla A, Fernández-García M (2012) Polymeric materials with antimicrobial activity. Prog Polym Sci 37(2):281–339
111. Cheah WY, Show P-L, Ng IS, Lin G-Y, Chiu C-Y, Chang Y-K (2019) Antibacterial activity of quaternized chitosan modified nanofiber membrane. Int J Biol Macromol 126:569–577
112. Ignatova M, Starbova K, Markova N, Manolova N, Rashkov I (2006) Electrospun nanofibre mats with antibacterial properties from quaternised chitosan and poly(vinyl alcohol). Carbohydr Res 341(12):2098–2107
113. Sajomsang W, Gonil P, Tantayanon S (2009) Antibacterial activity of quaternary ammonium chitosan containing mono or disaccharide moieties: preparation and characterization. Int J Biol Macromol 44(5):419–427
114. Wang D (2016) Osteoinductive and antibacterial biomaterials for bone tissue engineering. Dissertation, Vrije Universiteit Amsterdam
115. Goy RC, Britto D, Assis OBG (2009) A review of the antimicrobial activity of chitosan. Polímeros 19(3):241–247
116. Rabea EI, Badawy MET, Stevens CV, Smagghe G, Steurbaut W (2003) Chitosan as antimicrobial agent: applications and mode of action. Biomacromolecules 4(6):1457–1465
117. Raafat D, Sahl HG (2009) Chitosan and its antimicrobial potential—a critical literature survey. Microb Biotechnol 2(2):186–201
118. Eckhard LH, Sol A, Abtew E, Shai Y, Domb AJ, Bachrach G et al (2014) Biohybrid polymer-antimicrobial peptide medium against Enterococcus faecalis. PLoS One 9(10):e109413
119. Chen Q, Ma Z, Liu G, Wei H, Xie X (2016) Antibacterial activity of cationic cyclen-functionalized fullerene derivatives: membrane stress. Dig J Nanomater Bios 11:753–761
120. Medeiros SJ, Oliveira AM, de Carvalho JO, Ricci R, Martins MCC, Rodrigues BV et al (2018) Nanohydroxyapatite/graphene nanoribbons nanocomposites induce in vitro osteogenesis and promote in vivo bone neoformation. ACS Biomater Sci Eng 4(5):1580–1590
121. Siqueira IA, Corat MAF, Cavalcanti BN, Neto WAR, Martin AA, Bretas RES et al (2015) In vitro and in vivo studies of novel poly (D, L-lactic acid), superhydrophilic carbon nanotubes, and nanohydroxyapatite scaffolds for bone regeneration. ACS Appl Mater Interfaces 7(18):9385–9398
122. Van Noorden R (2011) Chemistry: the trials of new carbon. Nature 469(7328):14–16
123. Mocan T, Matea CT, Pop T, Mosteanu O, Buzoianu AD, Suciu S et al (2017) Carbon nanotubes as anti-bacterial agents. Cell Mol Life Sci 74(19):3467–3479
124. Aslan S, Loebick CZ, Kang S, Elimelech M, Pfefferle LD, Van Tassel PR (2010) Antimicrobial biomaterials based on carbon nanotubes dispersed in poly (lactic-co-glycolic acid). Nanoscale 2(9):1789–1794
125. Dizaj SM, Mennati A, Jafari S, Khezri K, Adibkia K (2015) Antimicrobial activity of carbon-based nanoparticles. Adv Pharm Bull 5(1):19–23
126. Gurunathan S, Han JW, Dayem AA, Eppakayala V, Kim J-H (2012) Oxidative stress-mediated antibacterial activity of graphene oxide and reduced graphene oxide in Pseudomonas aeruginosa. Int J Nanomed 7:5901–5914
127. Ricci R, Leite N, Da-Silva N, Pacheco-Soares C, Canevari R, Marciano F et al (2017) Graphene oxide nanoribbons as nanomaterial for bone regeneration: effects on cytotoxicity, gene expression and bactericidal effect. Mater Sci Eng C 78:341–348
128. Zhao C, Deng B, Chen G, Lei B, Hua H, Peng H et al (2016) Large-area chemical vapor deposition-grown monolayer graphene-wrapped silver nanowires for broad-spectrum and robust antimicrobial coating. Nano Res 9(4):963–973
129. Rodrigues BV, Leite NC, das Neves Cavalcanti B, da Silva NS, Marciano FR, Corat EJ et al (2016) Graphene oxide/multi-walled carbon nanotubes as nanofeatured scaffolds for the assisted deposition of nanohydroxyapatite: characterization and biological evaluation. Int J Nanomedicine 11:2569–2585
130. Lochab B, Shukla S, Varma IK (2014) Naturally occurring phenolic sources: monomers and polymers. RSC Adv 4(42):21712–21752
131. Upton BM, Kasko AM (2016) Strategies for the conversion of lignin to high-value polymeric materials: review and perspective. Chem Rev 116(4):2275–2306

132. Dong X, Dong M, Lu Y, Turley A, Jin T, Wu C (2011) Antimicrobial and antioxidant activities of lignin from residue of corn stover to ethanol production. Ind Crop Prod 34(3):1629–1634
133. Erakovic S, Jankovic A, Tsui GCP, Tang C-Y, Miskovic-Stankovic V, Stevanovic T (2014) Novel bioactive antimicrobial lignin containing coatings on titanium obtained by electrophoretic deposition. Int J Mol Sci 15(7):12294–12322
134. Chung K-T, Wong TY, Wei C-I, Huang Y-W, Lin Y (1998) Tannins and human health: a review. Crit Rev Food Sci Nutr 38(6):421–464
135. Park JH, Choi S, Moon HC, Seo H, Kim JY, Hong S-P et al (2017) Antimicrobial spray nanocoating of supramolecular Fe(III)-tannic acid metal-organic coordination complex: applications to shoe insoles and fruits. Sci Rep 7(1):6980
136. Sarjit A, Wang Y, Dykes GA (2015) Antimicrobial activity of gallic acid against thermophilic campylobacter is strain specific and associated with a loss of calcium ions. Food Microbiol 46:227–233
137. Arbenz A, Averous L (2015) Chemical modification of tannins to elaborate aromatic biobased macromolecular architectures. Green Chem 17(5):2626–2646
138. Redondo LM, Chacana PA, Dominguez JE, Fernandez Miyakawa ME (2014) Perspectives in the use of tannins as alternative to antimicrobial growth promoter factors in poultry. Front Microbiol 5:118
139. Daglia M (2012) Polyphenols as antimicrobial agents. Curr Opin Biotechnol 23(2):174–181
140. Papuc C, Goran GV, Predescu CN, Nicorescu V, Stefan G (2017) Plant polyphenols as antioxidant and antibacterial agents for shelf-life extension of meat and meat products: classification, structures, sources, and action mechanisms. Compr Rev Food Sci Food Saf 16(6):1243–1268
141. Ahn BK (2017) Perspectives on mussel-inspired wet adhesion. J Am Chem Soc 139(30):10166–10171
142. Habibovic P, de Groot K (2007) Osteoinductive biomaterials—properties and relevance in bone repair. J Tissue Eng Regen Med 1(1):25–32
143. Zhu Y, Zhang K, Zhao R, Ye X, Chen X, Xiao Z et al (2017) Bone regeneration with micro/nano hybrid-structured biphasic calcium phosphate bioceramics at segmental bone defect and the induced immunoregulation of MSCs. Biomaterials 147:133–144
144. Barradas A, Yuan H, van Blitterswijk CA, Habibovic P (2011) Osteoinductive biomaterials: current knowledge of properties, experimental models and biological mechanisms. Eur Cell Mater 21:407–429
145. Holzwarth JM, Ma PX (2011) Biomimetic nanofibrous scaffolds for bone tissue engineering. Biomaterials 32(36):9622–9629
146. Webster TJ, Massa-Schlueter EA, Smith JL, Slamovich EB (2004) Osteoblast response to hydroxyapatite doped with divalent and trivalent cations. Biomaterials 25(11):2111–2121
147. LeGeros RZ (2008) Calcium phosphate-based osteoinductive materials. Chem Rev 108(11):4742–4753
148. Rahaman MN, Day DE, Bal BS, Fu Q, Jung SB, Bonewald LF et al (2011) Bioactive glass in tissue engineering. Acta Biomater 7(6):2355–2373
149. Roseti L, Parisi V, Petretta M, Cavallo C, Desando G, Bartolotti I et al (2017) Scaffolds for bone tissue engineering: state of the art and new perspectives. Mater Sci Eng C 78:1246–1262
150. Caballero SSR, Saiz E, Montembault A, Tadier S, Maire E, David L et al (2019) 3-D printing of chitosan-calcium phosphate inks: rheology, interactions and characterization. J Mater Sci Mater Med 30(1):6
151. Iviglia G, Morra M, Cassinelli C, Torre E, Rodriguez Y, Baena R (2018) New collagen-coated calcium phosphate synthetic bone filler (Synergoss®): a comparative surface analysis. Int J Appl Ceram Technol 15(4):910–920
152. Inzana JA, Olvera D, Fuller SM, Kelly JP, Graeve OA, Schwarz EM et al (2014) 3D printing of composite calcium phosphate and collagen scaffolds for bone regeneration. Biomaterials 35(13):4026–4034

153. Gao X, Song J, Ji P, Zhang X, Li X, Xu X et al (2016) Polydopamine-templated hydroxyapatite reinforced polycaprolactone composite nanofibers with enhanced cytocompatibility and osteogenesis for bone tissue engineering. ACS Appl Mater Interfaces 8(5):3499–3515
154. Li H, Chang J (2005) pH-compensation effect of bioactive inorganic fillers on the degradation of PLGA. Compos Sci Technol 65(14):2226–2232
155. Stevanović M, Filipović N, Djurdjević J, Lukić M, Milenković M, Boccaccini A (2015) 45S5Bioglass®-based scaffolds coated with selenium nanoparticles or with poly (lactide-co-glycolide)/selenium particles: processing, evaluation and antibacterial activity. Colloids Surf B Biointerfaces 132:208–215
156. Li A, Ren H, Cui Y, Wang C, Zhou X, Lin H et al (2017) Detailed structure of a new bioactive glass composition for the design of bone repair materials. J Non-Cryst Solids 475:10–14
157. Sarin S, Rekhi A (2016) Bioactive glass: a potential next generation biomaterial. SRM J Res Dent Sci 7(1):27–32
158. El-Rashidy AA, Roether JA, Harhaus L, Kneser U, Boccaccini AR (2017) Regenerating bone with bioactive glass scaffolds: a review of in vivo studies in bone defect models. Acta Biomater 62:1–28
159. Jones JR (2013) Review of bioactive glass: from Hench to hybrids. Acta Biomater 9(1):4457–4486
160. Chen QZ, Thompson ID, Boccaccini AR (2006) 45S5 Bioglass®-derived glass–ceramic scaffolds for bone tissue engineering. Biomaterials 27(11):2414–2425
161. Kargozar S, Baino F, Hamzehlou S, Hill RG, Mozafari M (2018) Bioactive glasses: sprouting angiogenesis in tissue engineering. Trends Biotechnol 36(4):430–444
162. Jones JR, Brauer DS, Hupa L, Greenspan DC (2016) Bioglass and bioactive glasses and their impact on healthcare. Int J Appl Glas Sci 7(4):423–434
163. Sheikhi A, Afewerki S, Oklu R, Gaharwar AK, Khademhosseini A (2018) Effect of ionic strength on shear-thinning nanoclay–polymer composite hydrogels. Biomater Sci 6:2073–2083
164. Hoppe A, Güldal NS, Boccaccini AR (2011) A review of the biological response to ionic dissolution products from bioactive glasses and glass-ceramics. Biomaterials 32(11):2757–2774
165. Valliant EM, Romer F, Wang D, McPhail DS, Smith ME, Hanna JV et al (2013) Bioactivity in silica/poly (γ-glutamic acid) sol–gel hybrids through calcium chelation. Acta Biomater 9(8):7662–7671
166. Catauro M, Bollino F, Papale F (2018) Surface modifications of titanium implants by coating with bioactive and biocompatible poly (ε-caprolactone)/SiO2 hybrids synthesized via sol–gel. Arab J Chem 11(7):1126–1133
167. Hickey DJ, Ercan B, Sun L, Webster TJ (2015) Adding MgO nanoparticles to hydroxyapatite–PLLA nanocomposites for improved bone tissue engineering applications. Acta Biomater 14:175–184
168. Webster TJ, Ergun C, Doremus RH, Bizios R (2002) Hydroxylapatite with substituted magnesium, zinc, cadmium, and yttrium. II. Mechanisms of osteoblast adhesion. J Biomed Mater Res 59(2):312–317
169. Querido W, Rossi AL, Farina M (2016) The effects of strontium on bone mineral: a review on current knowledge and microanalytical approaches. Micron 80:122–134
170. Yang F, Yang D, Tu J, Zheng Q, Cai L, Wang L (2011) Strontium enhances osteogenic differentiation of mesenchymal stem cells and in vivo bone formation by activating Wnt/catenin signaling. Stem Cells 29(6):981–991
171. Lemaire-Hurtel A-S, Mentaverri R, Caudrillier A, Cournarie F, Wattel A, Kamel S et al (2009) The calcium-sensing receptor is involved in strontium ranelate-induced osteoclast apoptosis. New insights into the associated signaling pathways. J Biol Chem 284(1):575–584
172. Fiorilli S, Molino G, Pontremoli C, Iviglia G, Torre E, Cassinelli C et al (2018) The incorporation of strontium to improve bone-regeneration ability of mesoporous bioactive glasses. Materials 11(5):678

173. Kannan S, Pina S, Ferreira JMF (2006) Formation of strontium-stabilized β-tricalcium phosphate from calcium-deficient apatite. J Am Ceram Soc 89(10):3277–3280
174. Oryan A, Baghaban Eslaminejad M, Kamali A, Hosseini S, Sayahpour FA, Baharvand H (2019) Synergistic effect of strontium, bioactive glass and nano-hydroxyapatite promotes bone regeneration of critical-sized radial bone defects. J Biomed Mater Res B Appl Biomater 107(1):50–64
175. Zeng H, Cao J (2013) Selenium in bone health: roles in antioxidant protection and cell proliferation. Nutrients 5:97–110
176. Ebert R, Ulmer M, Zeck S, Meissner-Weigl J, Schneider D, Stopper H et al (2006) Selenium supplementation restores the antioxidative capacity and prevents cell damage in bone marrow stromal cells in vitro. Stem Cells 24(5):1226–1235
177. Liu H, Bian W, Liu S, Huang K (2012) Selenium protects bone marrow stromal cells against hydrogen peroxide-induced inhibition of osteoblastic differentiation by suppressing oxidative stress and ERK signaling pathway. Biol Trace Elem Res 150(1):441–450
178. Dollwet HHA, Sorenson JRJ (1988) Roles of copper in bone maintenance and healing. Biol Trace Elem Res 18(1):39–48
179. Shi M, Chen Z, Farnaghi S, Friis T, Mao X, Xiao Y et al (2016) Copper-doped mesoporous silica nanospheres, a promising immunomodulatory agent for inducing osteogenesis. Acta Biomater 30:334–344
180. Li X, He Q, Shi J (2014) Global gene expression analysis of cellular death mechanisms induced by mesoporous silica nanoparticle-based drug delivery system. ACS Nano 8(2):1309–1320
181. Vukicevic S, Sampath KT (2002) Bone morphogenetic proteins: from laboratory to clinical practice, 1st edn. Birkhäuser, Basel
182. Ho-Shui-Ling A, Bolander J, Rustom LE, Johnson AW, Luyten FP, Picart C (2018) Bone regeneration strategies: engineered scaffolds, bioactive molecules and stem cells current stage and future perspectives. Biomaterials 180:143–162
183. Lombardi G, Di CS, Rubino M, Faggiano A, Vuolo L, Guerra E et al (2011) The roles of parathyroid hormone in bone remodeling: prospects for novel therapeutics. J Endocrinol Investig 34(7 Suppl):18–22
184. Jung RE, Hämmerle CH, Kokovic V, Weber FE (2007) Bone regeneration using a synthetic matrix containing a parathyroid hormone peptide combined with a grafting material. Int J Oral Maxillofac Implants 22(2):258–266
185. Johnson CT, García AJ (2015) Scaffold-based anti-infection strategies in bone repair. Ann Biomed Eng 43(3):515–528
186. Salles GN, Calió ML, Afewerki S, Pacheco-Soares C, Porcionatto M, Hölscher C et al (2018) Prolonged drug-releasing fibers attenuate Alzheimer's disease-like pathogenesis. ACS Appl Mater Interfaces 10(43):36693–36702
187. Lobo AO, Afewerki S, de Paula MMM, Ghannadian P, Marciano FR, Zhang YS et al (2018) Electrospun nanofiber blend with improved mechanical and biological performance. Int J Nanomedicine 13:7891–7903
188. De Paula MMM, Bassous NJ, Afewerki S, Harb SV, Ghannadian P, Marciano FR et al (2018) Understanding the impact of crosslinked PCL/PEG/GelMA electrospun nanofibers on bactericidal activity. PLoS One 13(12):e0209386
189. Wang Y, Jiang Y, Zhang Y, Wen S, Wang Y, Zhang H (2019) Dual functional electrospun core-shell nanofibers for anti-infective guided bone regeneration membranes. Mater Sci Eng C 98:134–139
190. Shi L, Zhang W, Yang K, Shi H, Li D, Liu J et al (2015) Antibacterial and osteoinductive capability of orthopedic materials via cation–π interaction mediated positive charge. J Mater Chem B 3(5):733–737
191. Yang G, Yang H, Shi L, Wang T, Zhou W, Zhou T et al (2018) Enhancing corrosion resistance, osteoinduction, and antibacterial properties by Zn/Sr additional surface modification of magnesium alloy. ACS Biomater Sci Eng 4(12):4289–4298

192. Kumar S, Bose S, Chatterjee K (2014) Amine-functionalized multiwall carbon nanotubes impart osteoinductive and bactericidal properties in poly(ε-caprolactone) composites. RSC Adv 4(37):19086–19098
193. Zhang Y, Dong C, Yang S, Chiu T-W, Wu J, Xiao K et al (2018) Enhanced silver loaded antibacterial titanium implant coating with novel hierarchical effect. J Biomater Appl 32(9):1289–1299
194. Qian X, Qing F, Jun O, Hong S (2014) Construction of drug-loaded titanium implants via layer-by-layer electrostatic self-assembly. West China J Stomatol 32:537–541
195. Xu C, Lei C, Meng L, Wang C, Song Y (2012) Chitosan as a barrier membrane material in periodontal tissue regeneration. J Biomed Mater Res B Appl Biomater 100(5):1435–1443
196. Li W, Ding Y, Yu S, Yao Q, Boccaccini AR (2015) Multifunctional chitosan-45S5 bioactive glass-poly(3-hydroxybutyrate-co-3-hydroxyvalerate) microsphere composite membranes for guided tissue/bone regeneration. ACS Appl Mater Interfaces 7(37):20845–20854
197. Zhou P, Xia Y, Cheng X, Wang P, Xie Y, Xu S (2014) Enhanced bone tissue regeneration by antibacterial and osteoinductive silica-HACC-zein composite scaffolds loaded with rhBMP-2. Biomaterials 35(38):10033–10045
198. Nasajpour A, Ansari S, Rinoldi C, Shahrokhi Rad A, Aghaloo T, Ryon Shin S et al (2018) A multifunctional polymeric periodontal membrane with osteogenic and antibacterial characteristics. Adv Funct Mater 28:1703437
199. Wei C-K, Ding S-J (2017) Dual-functional bone implants with antibacterial ability and osteogenic activity. J Mater Chem B 5(10):1943–1953
200. Bergemann C, Zaatreh S, Wegner K, Arndt K, Podbielski A, Bader R et al (2017) Copper as an alternative antimicrobial coating for implants—an in vitro study. World J Transplant 7(3):193–202
201. Lemire JA, Harrison JJ, Turner RJ (2013) Antimicrobial activity of metals: mechanisms, molecular targets and applications. Nat Rev Microbiol 11(6):371–384
202. Zheng Z, Yin W, Zara JN, Li W, Kwak J, Mamidi R et al (2010) The use of BMP-2 coupled—nanosilver-PLGA composite grafts to induce bone repair in grossly infected segmental defects. Biomaterials 31(35):9293–9300
203. Liu Y, Zheng Z, Zara JN, Hsu C, Soofer DE, Lee KS et al (2012) The antimicrobial and osteoinductive properties of silver nanoparticle/poly (DL-lactic-co-glycolic acid)-coated stainless steel. Biomaterials 33(34):8745–8756
204. Stevanović M, Uskoković V, Filipović M, Škapin SD, Uskoković D (2013) Composite PLGA/AgNpPGA/AscH nanospheres with combined osteoinductive, antioxidative, and antimicrobial activities. ACS Appl Mater Interfaces 5(18):9034–9042
205. Sun CY, Che YJ, Lu SJ (2015) Preparation and application of collagen scaffold-encapsulated silver nanoparticles and bone morphogenetic protein 2 for enhancing the repair of infected bone. Biotechnol Lett 37(2):467–473
206. Pacheco H, Vedantham K, Aniket, Young A, Marriott I, El-Ghannam A (2014) Tissue engineering scaffold for sequential release of vancomycin and rhBMP2 to treat bone infections. J Biomed Mater Res A 102(12):4213–4223
207. Wang Y, Wang X, Li H, Xue D, Shi Z, Qi Y et al (2011) Assessing the character of the rhBMP-2- and vancomycin-loaded calcium sulphate composites in vitro and in vivo. Arch Orthop Trauma Surg 131(7):991–1001
208. Li X, Xu J, Filion TM, Ayers DC, Song J (2013) pHEMA-nHA encapsulation and delivery of vancomycin and rhBMP-2 enhances its role as a bone graft substitute. Clin Orthop Relat Res 471(8):2540–2547
209. Guelcher SA, Brown KV, Li B, Guda T, Lee B-H, Wenke JC (2011) Dual-purpose bone grafts improve healing and reduce infection. J Orthop Trauma 25(8):477–482
210. Neoh KG, Hu X, Zheng D, Kang ET (2012) Balancing osteoblast functions and bacterial adhesion on functionalized titanium surfaces. Biomaterials 33(10):2813–2822
211. Cheng H, Xiong W, Fang Z, Guan H, Wu W, Li Y et al (2016) Strontium (Sr) and silver (Ag) loaded nanotubular structures with combined osteoinductive and antimicrobial activities. Acta Biomater 31:388–400

212. Bose S, Banerjee D, Bandyopadhyay A (2017) Chapter 1—Introduction to biomaterials and devices for bone disorders. In: Bose S, Bandyopadhyay A (eds) Materials for bone disorders. Academic, New York, pp 1–27
213. Anastas PT, Warner JC (1998) Green chemistry: theory and practice. Oxford University Press, Oxford
214. Anastas P, Eghbali N (2010) Green chemistry: principles and practice. Chem Soc Rev 39(1):301–312
215. Kiran A, Kumar T, Sanghavi R, Doble M, Ramakrishna S (2018) Antibacterial and bioactive surface modifications of titanium implants by PCL/TiO2 nanocomposite coatings. Nanomaterials. 8(10):860

Recent Advances in Controlled Release Technologies for the Co-delivery of Antimicrobial and Osteoconductive Therapeutics

Chukwuazam Nwasike, Kyle Reeser, Yizhong Liu, Jaspreet Singh Nagi, Erin Purr, Chendong Han, and Amber L. Doiron

Abstract Bone defects are a significant cause of morbidity in the fields of orthopedics, maxillofacial surgery, and oral implantology, yet their treatment currently faces many challenges including the defect size and location, underlying disease, and microbial infection. Bacteria may be introduced to healing bone through several routes including colonization during open-wound trauma, introduction during surgery, from blood-borne bacteria, or infection of a medical device such as a bone screw. Unfortunately, current treatment strategies are often inadequate and lead to severe and costly consequences. To tackle the problem of infection during bone healing, novel biomaterials such as scaffolds, cements, surface-modified implants, and particles have been developed that comprise both antimicrobial and osteoconductive properties. The antimicrobial properties of these biomaterials typically stem from the addition of antimicrobial agents like antibiotics and silver nanoparticles to the composite material, while osteoconductive properties are conveyed by biomolecules such as growth factors or hydroxyapatite. By controlling modes of delivery and/or release kinetics, these antibacterial and osteoconductive therapeutic constructs are potentially capable of significantly improving bone healing. Recent findings have shown very promising results in the application of these constructs with dual functions in treating infected bone defects. Here, we summarize the advances within the last decade in particle technologies, implant coatings, tissue engineering, and bone cements with both antimicrobial and osteoconductive activity with an

C. Nwasike · K. Reeser · E. Purr · C. Han
Department of Biomedical Engineering, Binghamton University (SUNY), Binghamton, NY, USA

Y. Liu
Interdisciplinary Faculty of Toxicology and Department of Veterinary Integrative Biosciences, Texas A&M University, College Station, TX, USA

J. S. Nagi · A. L. Doiron (✉)
Department of Electrical and Biomedical Engineering, University of Vermont, Burlington, VT, USA
e-mail: Amber.Doiron@uvm.edu

B. Li et al. (eds.), *Racing for the Surface*,
https://doi.org/10.1007/978-3-030-34471-9_2

emphasis on fabrication and the performance of constructs in various in vitro and in vivo models.

Keywords Drug delivery · Bone tissue engineering · Nanoparticles Microparticles · Antibacterial · Bactericidal · Antibiotic · Silver · Titanium · Wound healing · Regenerative medicine · Medical implant infection · Biofilm · Scaffold Cement

Abbreviations

Ag-HA	Silver-containing hydroxyapatite
AgNP	Silver nanoparticle
AgNP/GS	Silver nanoparticle gentamicin
AgNP-BHAC	Silver nanoparticle-doped hydroxyapatite coatings with oriented block arrays
AgNPPGA	Poly(L-glutamic acid)-capped silver nanoparticles
ALP	Alkaline phosphatase
Bbr	Berberine
BMP	Bone morphogenetic protein
BMSC	Bone marrow-derived mesenchymal stem/stromal cells
CAP	Calcium phosphate
CD	Zero-dimensional carbon dot
CFU	Colony forming unit
CL	Clindamycin phosphate
CMCS	O-carboxymethyl chitosan
Col I	Type I collagen/procollagen
CS	Chitosan
CT	Computed tomography
Cu	Copper
E. coli	*Escherichia coli*
F	Fluorine
Fe	Iron
GR-HA	Glass-reinforced hydroxyapatite
GS	Gentamicin sulfate
HA	Hydroxyapatite
IGF-1	Insulin-like growth factor 1
LbL	Layer-by-layer (deposition)
MCPM	Monocalcium phosphate monohydrate
MIC	Minimum inhibitory concentration
MRSA	Methicillin-resistant *Staphylococcus aureus*
MSC	Mesenchymal stem cell
nHA	Nanohydroxyapatite

NPs	Nanoparticles
NT	Nanotube
OCN	Osteocalcin
P. aeruginosa	*Pseudomonas aeruginosa*
PBS	Phosphate buffered saline
PCL	Polycaprolactone
PDLLA	Poly(D,L-lactide)
PLGA	Poly(lactic co-glycolic acid)
PM	PLGA Microparticles
QRT-PCR	Quantitative real-time polymerase chain reaction
RUNX2	Runt-related transcription factor 2
S. albus	*Staphylococcus albus*
S. aureus	*Staphylococcus aureus*
S. epidermidis	*Staphylococcus epidermidis*
SBA-15	Mesoporous silica nanoparticle
SEM	Scanning electron microscopy
SNPSA	Silver nanoparticle/poly(DL-lactic-co-glycolic acid)(PLGA)-coated stainless steel alloy
Sr	Strontium
SR	Strontium ranelate
TCP	Tricalcium phosphate
Ti	Titanium
TNT	Titania nanotubes
XRD	X-ray diffraction
Zn	Zinc
ZnO	Zinc oxide

Introduction

While it is one of the few human tissues that repairs itself through regeneration [1], bone healing can be limited in certain situations due to location or size of the defect, extensiveness of trauma, underlying disease, patient-specific factors such as smoking, or infection [2, 3]. The incidence of bone fracture is estimated to be 2704 per 100,000 person-years in the United States, with incidence increasing with age and being 49% greater among women [4]. Although bone repair typically results in a return of the tissue to pre-injury cellular, structural, and functional status, approximately 10% of fractures do not heal [1]. Genetic and age-related diseases or disorders such as congenital malformation, osteogenesis imperfecta, rheumatoid arthritis, or osteoporosis complicate the healing process and limit the bone's capacity for self-healing within a reasonable timeline. The most common clinical interventions to improve bone regeneration include human or bovine tissues and can include orthobiologics such as the "gold standard" bone autograft, bone allograft, bone graft substitutes, bone matrices, and various other isolated biological molecules [5]. While these treatments are viable in many situations, factors such as patient-specific limitations and morbidity associated with the harvest site for autografts and antigenicity, donor shortages, and low osteoinductivity of allografts motivate research towards future improved interventions.

Infections and poor osteoinductivity are the major challenges in bone fracture healing. Bone development, repair, and remodeling are dynamic processes, with the local environment playing a large role in the phenomena. Bone healing occurs through one of two mechanisms, direct or indirect healing. Direct healing requires anatomical reduction and internal fixation, which are most common through surgical intervention. These conditions allow for healing to immediately regenerate lamellar bone as opposed to the remodeling steps that precede this in the indirect healing mechanism. In non-stable conditions, indirect bone healing starts with blood clot formation providing the scaffold that becomes the site for inflammation, mesenchymal stem cell (MSC) recruitment, and creation of the primary cartilaginous callous. The primary callous is revascularized, calcified, and finally remodeled into normal bone [1, 6]. The presence of signaling molecules in the local environment is key; cytokines and growth factors are necessary to promote the actions of osteoprogenitor cells that produce extracellular matrix. Significant evidence suggests that bone morphogenetic proteins (BMP) and other factors play an important role in recruiting MSCs to the injury site [6]. Key markers of cellular osteoconductive activity include alkaline phosphatase (ALP), osteocalcin (OCN), type one procollagen (Col I), and runt-related transcription factor 2 (RUNX2), which are often used to indicate the differentiation of MSCs towards an osteoblastic phenotype and bone regeneration.

Infections present a unique challenge in orthopedic surgery and bone grafting, with bacterial contamination commonly associated with failure to heal despite antibiotic treatment. Osteomyelitis is an infection of the bone that can result from open fracture traumatic injuries, surgical interventions, or colonization from an infection spread through the bloodstream. Infections resulting from surgery present an interesting and pressing situation [7, 8] as the incidence of orthopedic and dental surgeries continues to rise with population aging and advances in hip replacements, knee replacements, and dental fixation that make these options more widely available. Infections at the surgical site can result from early or late infection, owing to the introduction of bacteria during surgery or the acute healing process compared to colonization of the implanted biomaterial device or surrounding bone weeks, months, or years after implantation [9]. Due to the adhesive nature of many materials used for bone grafting to encourage fixation to surrounding healthy bone, bone grafts and related products like cements can be a breeding ground for bacterial attachment and infection. Titanium (Ti) and Ti alloys are widely used in medical implants, but bacterial infections and poor osseointegration often occur at the early stage of implantation and lead to early orthopedic implant failures. While risk of infection after total joint replacement is considered low at 0.5–5% [7, 8, 10], the treatment of these infections is complicated and costly with severe cases involving implant removal, surgical debridement, long-term antibiotic treatment, and possible amputation [8, 11]. Current clinical approaches for preventing infection include minimizing contamination during surgery and administering peri-operative prophylactic antibiotics, but the continuing incidence of post-surgical infection indicates that this is an inadequate regimen. There exists a significant need for new strategies for the prevention of early and late infections during bone healing as a result of traumatic injury and orthopedic surgery.

Care following infection is complex and depends on the source of infection, location, and severity. *Staphylococcus* species (including MRSA), *Acinetobacter baumannii*, *Enterobacter* and *Klebsiella* species, and *Pseudomonas aeruginosa* are among the most commonly isolated bacterial strains from traumatic and open wounds [12–14]. Interestingly, gram-negative species are likely the initial colonizers during care, but gram-positive bacteria appear to be cultured more readily from recurrent infections [12, 13]. Common first-line treatments include antibiotics administered either orally, intravenously, or locally to the site of infection as well as wound debridement and irrigation [13, 15–19]. Local dosing at high concentrations of antibiotics is recommended to improve efficacy and avoid systemic side effects, but without a means of containing and releasing the pharmaceutical agent in a controlled manner, patient well-being can be compromised. Eliminating the need for repetitive dosing, enabling localized delivery and controlled kinetics of antibiotics, and increasing the efficacy of drugs are all advantages to the use of a drug delivery technology in the treatment of bone infections.

While many strategies detailed below include antibiotics in cements, particles, and scaffolds, other approaches highlight the limitations of antibiotics including their fragile and finite nature (i.e., controlled release cannot continue forever), the growing concern of antibiotic resistant bacterial strains, and complications associated with biofilm infections that may limit the use of traditional antibiotics in the next generation of tissue healing approaches. In fact, the majority of surgical site infections including implant-related infections are caused by biofilms, which are surface-associated bacteria that exist within a self-produced polysaccharide matrix. Biofilms are up to 1000-times more tolerant to antibiotics than their planktonic counterparts and wound debridement is insufficient [20–22]. Often, high potency antibiotics such as vancomycin and tobramycin must be employed, and even they are oftentimes unsuccessful in biofilm eradication [23]. Due to these limitations, researchers have alternatively investigated the use of metallic particles and ion inclusions into implant materials. One of the most popular non-antibiotic, antibacterial technologies is silver. The bactericidal activity of silver has been known for centuries, with more recent use being pervasive in orthopedic and dental applications [24, 25]. The antibacterial nature of silver is thought to require either direct contact between metallic nanoparticles and bacteria or ionization, with ions readily reacting with negatively charged molecules such as proteins, RNA, DNA, and chloride [26]. A notable limitation of silver is the protein-rich nature of infected or healing wounds, which may prompt a requirement that high concentrations of silver be used for effective bacterial killing. Delivery of silver to the infection site varies from research group to group, and many current strategies are detailed below.

Due to these unique challenges, many researchers have worked towards tissue engineering constructs and drug delivery technologies that not only improve rates of bone healing but also prevent or treat bacterial infections. Research efforts in tissue engineering to promote more rapid bone healing have focused on the introduction of scaffolds or constructs made of polymers and/or ceramics containing biomolecules to promote bone growth in addition to antibiotics to deter infection. Additionally, nanomaterials have been used to induce osteoconductive but infection-protective conditions through the delivery of biomolecules. Implant coatings and bone cements are also promising means of delivering osteoconductive, antibacterial agents. Of note, implant coatings and cements may hold promise because they are typically lower-cost than tissue engineering approaches, which would enable quicker and more widespread adoption in industry.

In this chapter, we summarize efforts over approximately the last decade to accomplish both improved bone healing and antimicrobial activity in various in vitro and in vivo animal studies through the use of micro- and nanoparticle technologies, implant coatings, tissue engineering, and bone cements. With a focus on study outcomes, technologies are summarized starting from fabrication or synthesis as well as any important characterization and cell and animal studies.

Nanoparticles, Microparticles, and Powders as a Means of Osteoconductive and Antimicrobial Therapy

Microparticles and nanoparticles (NPs) have been in wide use for several decades in the field of drug and gene delivery, starting from the delivery of small molecules and advancing to deliver complex biomolecules such as proteins and nucleic acids. Microparticles are most often locally delivered by such means as injection or implantation at the site of intended action, whereas nanoparticles are frequently designed to traverse physiological barriers via passive or active targeting. Both microparticles and nanoparticles can be made of various materials including polymers, ceramics, carbon, metals, or composites thereof. Metallic silver nanoparticles (AgNPs) have great significance in antimicrobial applications due to their ability to kill bacteria through an as-of-yet not fully understood mechanism that may involve release of silver ions or direct contact between AgNPs and bacteria [26]. Likewise, ceramic particles are of keen interest in this topic area due to the important role of calcium phosphate (CAP) and the natural form of CAP, hydroxyapatite (HA), in bone tissue. Therefore, the use of ceramic particles to recapitulate some of the natural properties of bone is widespread. Nanoparticles comprising CAP are frequently used to increase the mechanical strength and osteointegration of polymeric tissue engineering scaffolds. Finally, polymers are tunable materials that allow for the facile synthesis of particles or scaffolds with varied properties such as molecular weight, size, charge, and surface chemistry. Polymeric materials allow for control over release kinetics and are appropriate for a broad range of encapsulants from small to large molecules. Due to the different advantages of each of these materials,

Table 1 Antibacterial and osteoconductive particle-based technologies

Type of particle	Material of particle	Key properties	Antimicrobial used	Key properties	Bacteria strains tested against
Nanosphere [30]	PLGA	Biocompatible, biodegradable, nontoxic	Silver nanoparticles	Effective against both gram-positive and gram-negative bacteria	MRSA, *E. coli*
Microsphere [31]	PLGA	Biocompatible, biodegradable, nontoxic	Silver nanoparticles	Effective against a broad spectrum of bacteria and microorganisms, unlikely to encounter resistant strains	MRSA, *S. epidermis*
Nano-sized powder [32]	Calcium phosphate	Composed of two main inorganic materials that make up bone, nontoxic, biocompatible	Silver	Effective against most microorganisms, nontoxic, prevents proliferation of bacterial colonies	*S. aureus*, *P. aeruginosa*, *E. coli*, *C. albicans*
Microparticle [33]	Calcium phosphate	Composed of materials found in human bone, nontoxic	Clindamycin phosphate	Effective against a wide range of bacteria, commonly prescribed to treat osteomyelitis	*S. aureus*
Microsphere [34]	Chitosan	Derived from nontoxic chitin, positively charged, easy to assemble into nanospheres and nanofibers	Berberine chloride	Water soluble, antimicrobial, anti-inflammatory, and antifungal	*S. aureus*

much research has focused on composite approaches that involve the use of more than one class of material; recent reviews of nanoparticulate platforms for bone infection are highlighted in the field [27–29]. The particles used by the researchers summarized in this section are detailed in Table 1.

Bone Healing Particles That Treat or Prevent Infections Based on Silver

To create particles appropriate for wound dressings and prophylaxis, Stevanović et al. investigated combining a lyophilization method and a physicochemical solvent/nonsolvent approach in synthesizing multifunctional poly(lactic-co-glycolic acid) (PLGA) nanospheres encapsulating poly(L-glutamic acid)-capped silver nanoparticles (AgNPPGA) and ascorbic acid, which were designed to allow for both osteoinductive and antibacterial properties. Over the last several decades, lactic and glycolic acid-based polymeric particles have been widely used in pharmaceutics and tissue engineering because they are degradable and considered to be highly biocompatible in medical applications and in drug delivery systems. AgNPPGA were synthesized by adding PGA and silver nitrate ($AgNO_3$) to distilled water containing sodium hydroxide and glucose. Ascorbic acid was added to AgNPPGA before encapsulating the particles with PLGA dissolved in acetone. PLGA/AgNPPGA/ascorbic acid nanosphere synthesis was confirmed using X-ray diffraction (XRD), scanning electron microscopy (SEM), and dynamic light scattering with reported hydrodynamic sizes of 86 nm for 10% of particles, 50% of particles were smaller than 142 nm, and 90% of particles were smaller than 397 nm. XRD analysis showed no crystalline peaks of PLGA, suggesting that the polymer was amorphous in this formulation, while characteristic signals strongly suggested encapsulation of AgNPPGA by PLGA particles. Osteoinductivity was investigated in MC3T3-E1 osteoblastic cells in vitro by treating cells with PLGA encapsulated AgNPPGA ascorbic acid particles and using quantitative real-time polymerase chain reaction (qRT-PCR) to quantify the expression of two major osteogenic markers, OCN and Col I. Both markers were significantly upregulated in cells treated with ascorbic acid nanoparticles—OCN was over 60% higher than the control while Col I was over 80% higher than the control—compared to cells not treated with particles. The antibacterial activity of the particles was investigated in methicillin-resistant *Staphylococcus aureus* (MRSA) and *Escherichia coli* (*E. coli*) using a microdilution assay of the broth to obtain minimal inhibitory concentrations (MIC) in the bacterial cell lines. In *E. coli*, the MIC remained low and steady at about 0.09 μg/mL for over 50 days, while MRSA experienced a decrease in MIC from 0.09 μg/mL to approximately 0.03 μg/mL over the same time period, confirming antibacterial activity of the AgNPPGA with ascorbic acid. While further study is necessary, this paper adds to a wealth of knowledge contributed by a very active lab in bactericidal and osteoconductive technologies [30].

In a related approach using composite materials and PLGA, Mao et al. used a novel solid-in-oil nanosuspension method to fabricate strontium ranelate (SR)-loaded PLGA microspheres (PM) assembled with silver and hydroxyapatite nanoparticles (SR-PM-Ag-HA) for the treatment of bone infections and enhancement of bone regeneration. SR, which increases bone deposition by osteoblasts and reduces bone resorption by osteoclasts, is often used in the treatment of osteoporosis. The advantages of this combined formulation included the high biocompatibility of PLGA as the carrier, the osteoinductive properties of HA, the controlled release of SR, and the antimicrobial properties of AgNPs. In vitro SR release results showed ~90% release of the drug in the SR-PM, SR-PM-Ag, and SR-PM-Ag-HA samples after 22 days. The samples with NPs showed a steady release over time, whereas the SR-PM sample showed a slight burst release between 2 and 10 days. The four microsphere types—PM, SR-PM, SR-PM-Ag, and SR-PM-Ag-HA—were tested in the presence of two bacterial stains *Staphylococcus epidermidis* and MRSA. SR-PM-Ag and SR-PM-Ag-HA showed the best antimicrobial properties as a negligible number of surviving colony forming units (CFUs) were observed as compared to more than 350 CFUs seen for PM and SR-PM in both bacterial stains. Similar results were observed when biofilm formation of the same two strains on Ti disks using the four microspheres was tested by staining viable cells using BacLight Green. The cells treated with SR-PM-Ag and SR-PM-Ag-HA showed inhibition of biofilm formation; few viable cells were observed for either bacterial stain in the presence of these particles. The authors stated that human bone marrow mesenchymal stem cells (hBMSC) and an extra control sample such as SR-PM-HA would be tested to validate these results in the future, but it is clear that SR-PM-Ag-HA showed potential for combating infection and promoting bone regrowth in bone-related infections [31].

Bostancıoğlu et al. investigated the angiogenic, cytotoxic, and antibacterial properties of silver-doped, nano-sized, CAP-based inorganic powders. CAP is already well established as a therapeutic agent for bone defects, while the antibacterial properties of silver have also been well investigated. In this study, powders were synthesized using wet chemical manufacturing methods where both calcium hydroxide and silver nitrate were dissolved in ultra-filtered water and mixed by stirring for 2 h. The pH of the mixture was adjusted to 5.5 using ortho-phosphoric acid, and precipitates were collected and dried at 80 °C. XRD analysis showed characteristic peaks of nano-sized apatite crystals, while elemental analysis using energy-dispersive X-ray spectroscopy was used to confirm the presence of silver in the formulation. Ultrasonic homogenization was used to determine the size of the particles, approximately 60 nm. The antimicrobial activity of the silver-doped powders was investigated in *S. aureus*, *P. aeruginosa*, *E. coli*, and *C. albicans* using the halo test. Powders with higher concentrations of silver reduced bacterial colonies and fungi. In vitro toxicity was investigated in human umbilical vein endothelial cells, mouse embryo fibroblasts NIH-3T3, hamster lung fibroblasts V79 379A, and normal human fibroblasts TIG-114 using the MTT assay. The most significant toxicity result was observed in V79 379A cells, which showed an almost 90% decrease in viability after cells were treated with the highest tested concentration of silver-doped CAP powder (1 mg/mL). Human endothelial cells seeded on Matrigel substratum and treated with the synthesized powders showed high motility and differentiated into fine network-like structures in a dose-dependent manner. These powders were not tested for osteoinductive potential but have strong antibacterial and angiogenic activity, suggesting their future investigation as bone fillers or implant coatings [32].

Bone Healing Particles That Treat or Prevent Infections Based on Antibiotics

In order to locally administer antibiotics for the treatment of osteomyelitis, Makarov and colleagues designed a bone defect-filling material to promote healing and deliver high concentrations of vancomycin. Degradable materials were used in this approach in order to prevent the need for an additional surgery to remove the material and to avoid the introduction of a permanent biomaterial with surface area for bacterial colonization. Therefore, resorbable CAPs and degradable polymers were selected for the basis of the composite material. Composite beads were synthesized by admixing dicalcium phosphate–polycaprolactone (PCL) or solution-mixed beta-tricalcium phosphate (TCP)-PCL composite powder with either 1 or 4 wt% vancomycin. All composite matrices were composed of 40% PCL and 60% CAP, and they all had low porosity of less than 7%, which was important to sustain release of the antibiotic. All composites had slow release profiles, with the slowest releasing of 10% vancomycin within the first 24 h and 90% over 10 weeks. Importantly, the formulation process did not significantly diminish the activity of the vancomycin, which had a similar MIC to the reference vancomycin. Follow-up in vitro and in vivo studies are necessary to validate the material's ability to promote healing [35].

Uskoković and Desai added foundational knowledge to the field by studying the effect of five different previously synthesized types of CAPs along with the antibiotic clindamycin phosphate (CL) on bacterial and osteoblastic cell cultures to replace the conventional treatment materials for osteomyelitis. The authors aimed to eliminate the need for repetitive dosing as the CAP/CL materials would enable localized delivery and controlled release of antibiotics, increasing the efficacy of the drug. The optical density of the broth inoculated with *S. aureus* and incubated with different concentrations of the CAP/CL particles, monocalcium phosphate monohydrate (MCPM), dicalcium phosphate anhydrous, amorphous CAP, HA, and calcium pyrophosphate was measured. The MIC levels for all the particles were below 15 mg/mL and were the highest for MCPM and less than 1 mg/mL for amorphous CAP. Incubation of CAP/CL particles with osteoblastic cells in vitro showed minimal toxicity (<2%) for all of the particles except MCPM/CL. When the CAP/CL particles were exposed to MC3T3-E1 cells for gene expression analysis, each particle upregulated osteogenic genes, with dicalcium phosphate anhydrous showing the highest upregulation of osteopontin BSP-1, osteocalcin BGLAP, Col I, and RUNX2, which were used as markers for bone regeneration. Due to the combination of antimicrobial and osteogenic properties, these materials show potential to be used as alternatives to conventional osteomyelitis treatment [33].

Dual-function methods have been studied intensively, but most approaches exhibit an initial burst release of encapsulated agents, whereas Cai and colleagues have designed a material for the sustained release of both loaded agents. Cai et al. investigated the use of an injectable gel with dual regenerative and antimicrobial effects based on chitosan (CS) microspheres loaded with the osteoconductive growth factor BMP-2 and berberine (Bbr), a water-soluble isoquinoline alkaloid with antimicrobial characteristics, for bone healing. Positively charged CS microspheres were loaded with BMP-2 via swollen encapsulation, and negatively charged O-carboxymethyl chitosan (CMCS) microspheres were loaded with Bbr via physical adsorption. Then CS and CMCS microspheres were mixed together using a dispersion system through extrusion to form an injectable gel. The release of BMP-2 and Bbr from microspheres over time in phosphate buffered saline (PBS) was studied, showing a steady increase in accumulated concentration of both agents in the first 40 days, which then plateaued as the protein release was completed. The antibacterial activity of Bbr-loaded CMCS microspheres was evaluated using the disk diffusion method to observe the bacteria inhibition zone on *S. aureus* on an agar plate overnight. Diffused Bbr-loaded CMCS microspheres significantly inhibited bacterial growth on agar plates. Cell differentiation was assessed using BMSCs. Cell morphology and cell viability were evaluated with little deleterious effect from the microparticles after 4 days of incubation. Cells were cultured in medium containing BMP-2-loaded CS microspheres for 7 days and exhibited significantly higher ALP activity level (three times higher after 2 weeks) and OCN production (three times after 1 week) compared to BMSCs cultured in regular medium over 3 weeks of incubation. Furthermore, a femoral defect animal model was studied using New Zealand white rabbits. The defects were either left empty or filled with a microsphere gel, and regenerated bone volume was measured using reconstructed micro-computed tomography (CT) images. Five experimental conditions were evaluated: no filling material (control) or filled with (a) bare CS + CMCS microspheres, (b) Bbr-loaded CMCS microspheres, (c) BMP-2-loaded CS microspheres, or (d) BMP-2-loaded CS + Bbr-loaded CMCS microspheres. At 4 weeks, groups c and d significantly increased new bone volume. At 8 weeks, all groups showed increases in bone volume with group d having the highest volume, and at 12 weeks there was no difference among groups. This study developed an innovative approach with demonstrated results in enhancing bone tissue volume at short time periods after defect filling and inhibiting bacterial infections during the bone fracture healing process. The injectable gel form of the release vessel can be very versatile in future therapeutic applications [34].

While the above particle-based materials are stand-alone technologies, it is also important to realize the pervasiveness of particle technologies in this field that is evident in the remainder of this chapter where particles are used as part of many composite strategies for coatings or constructs to improve osteoconductive or antibacterial outcomes.

Modifying Medical Implants to Have Combinatorial Osteoconductive and Antimicrobial Properties

While metallic medical implants are widely used, bacterial infections associated with implant material surfaces lead to local inflammation and can eventually result in implant failures and deleterious effects on patients. Technologies to promote healing and prevent or treat infections detailed below come in the form of implant coatings. Coatings are a popular strategy due to their often lower-cost and potential for adaptability to allow for widespread industrial implementation. Implant coatings may or may not incorporate growth factors, antibiotics, silver, ceramics, polymers, and multiple layers. A ceramic material of significant importance in this field is hydroxyapatite, which has been used commercially on metallic implants since the

1980s. HA is a naturally occurring CAP in bone that gives the tissue its compressive strength, and a synthetic version of HA as well as synthetic CAP are widely used. Delamination from the surface and release of free HA particles have been widely discussed as potential failure mechanisms of these coatings, although recent data suggests that outcomes of HA-coated implants are similar to uncoated implants at 10- and 20-years post-implantation. HA continues to be a material of great focus for improved outcomes for orthopedic implants [36–42] (Table 2).

Implant Coatings Using Metallic Antimicrobial Strategies

Coatings Without Growth Factors

The surface properties of the implants play a very important role in initial bacterial adhesion, and the use of Ag to control infection in dental and orthopedic devices has been widely studied. In 2007, Chen et al. investigated the use of silver-containing hydroxyapatite (Ag-HA) thin film coatings created by a sol-gel method in order to control initial antibacterial adhesion and enhance osteoblast cell proliferation and differentiation. Passivated Ti substrate surfaces were coated with an Ag-doped HA sol. The film coating was designed to passively repel bacterial adhesion and increase tissue compatibility. All experiments were carried out on three surface coatings: HA, Ag-HA1.0, and Ag-HA1.5, which contained 0, 1, and 1.5 wt% $AgNO_3$ in doping steps, respectively. The antibacterial effects of the coatings were assessed in the presence of *S. epidermidis* or *S. aureus*, using the total number of CFU as an indicator of infection. Both silver-containing HA coatings showed a significant reduction in *S. aureus* CFU compared to the HA coating; Ag-HA1.5 showed an approximate 40% reduction while Ag-HA1.0 showed more than an 80% reduction. In the presence of *S. epidermidis*, Ag-HA1.5 showed a 25% reduction and Ag-HA1.0 showed more than a 80% reduction in CFU compared to the HA coating. Osteoblast cell proliferation and differentiation were investigated in human embryonic palatal mesenchyme cells using dsDNA quantification and ALP activity. No significant differences in dsDNA (cell proliferation) amount were observed among the 3 surfaces, and no differences in ALP activity were observed between HA and the Ag-HA1.0 surfaces; however, the Ag-HA1.5 surface exhibited significantly less ALP activity after 12 and 15 days of culture. Silver-containing HA coatings showed efficacy in controlling initial bacterial adhesion while maintaining similar osteoconductive activity as HA surfaces [43].

Similarly, Rameshbabu et al. created silver substituted nanosize hydroxyapatite (Ag-nHA) thin film coatings synthesized using the fast, simple, and efficient method of microwave processing in order to minimize the risk of bacterial contamination without compromising bioactivity or osseointegration. Calcium hydroxide and $AgNO_3$ mixed in water were subjected to microwave irradiation for approximately 30 min at 800 W power. Stainless steel disks were surface coated with 0.5Ag-nHA, 1.0Ag-nHA, or 1.5Ag-nHA, which contained 0.5, 1, and 1.5 wt%

Table 2 Surface modified implantable technologies

Type of implant	Coating factors	Growth factor (present or absent)	Properties	Bacteria strains tested against
Titanium [43]	Silver-containing hydroxyapatite (HA)	Absent	Effective against gram-positive bacteria, functions like hydroxyapatite in bone regeneration	*S. aureus*, *S. epidermis*
Stainless steel disk [44]	Silver substituted nanosize HA	Absent	Effective against both gram-negative and gram-positive bacteria, however, poor cell adhesion	*E. coli*, *S. aureus*
Titanium [45]	Silver-containing, strontium (Sr)-doped plasma sprayed HA	Absent	Effective against gram-negative bacteria, nontoxic, strongly promotes osteoconductivity	*P. aeruginosa*
Titanium [46]	Silver-containing HA	Absent	Effective against different types of bacteria, biocompatible	*E. coli*, *S. albus*
Alloy [47]	Silver nanoparticle-doped HA coatings with oriented block arrays	Absent	Biocompatible, potential application in dental implants for infection prophylaxis, effective against broad range bacteria	*S. aureus*, *E. coli*
Titania nanotube [48]	HA partially substituted with silver and fluorine ions	Absent	Strong antibacterial activity on gram-positive bacteria, nontoxic, promotes growth in osteoblasts	*S. aureus*
Titanium [49]	Strontium and silver loaded nanotubes	Absent	Strong inducer of osteogenesis, effective against a wide range of microorganisms including gram-positive, and gram-negative bacteria	MRSA, *E. coli*, methicillin-sensitive *S. aureus*
Titanium [50]	Silver nanoparticle gentamicin loaded silk fibroin	Absent	Promotes adhesion and proliferation of preosteoblasts, strong antibacterial activity	*S. aureus*
Stainless steel alloy [51]	Silver nanoparticle/PLGA	Present (BMP-2)	Increases ALP, and terminal differentiation. Strong antimicrobial activity	*P. aeruginosa*, *S. aureus*
Titanium [52]	Silver nanoparticles, BMP-2, HA	Present (BMP-2)	Promotes osteoblast differentiation, and osteocytes in rabbits	N/A
Titanium [53]	Multi-layer; PDLLA, GS, BMP-2, IGF-1	Present (IGF-1, BMP-2)	Causes very high initial metabolic activity in osteoblasts	N/A
Laponite clay [54, 55]	Multi-layer; poly(β-amino esters), poly(acrylic acid), BMP-2, GS	Present (BMP-2)	Effective against gram-positive bacteria, promotes ALP and host-implant integration	*S. aureus*

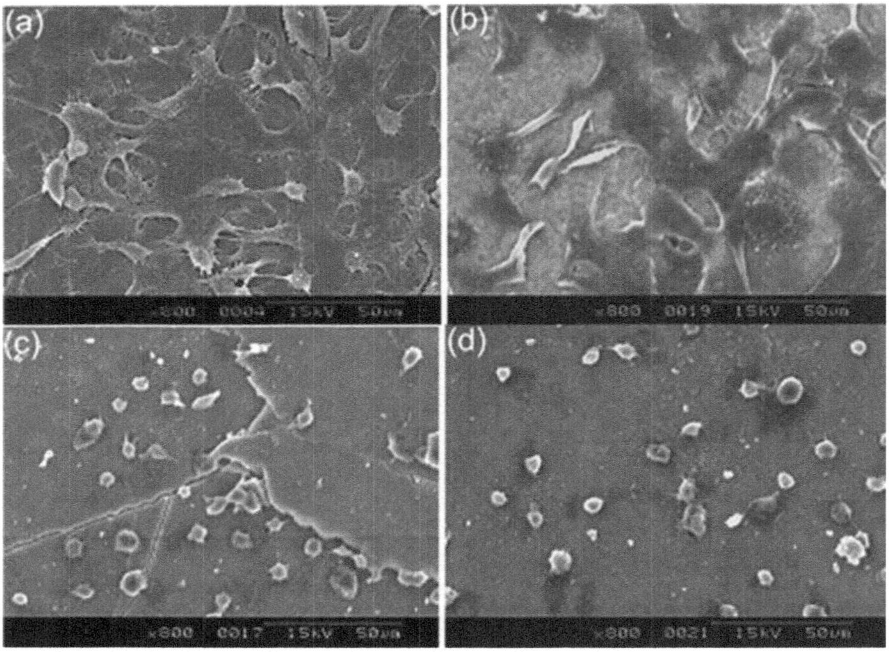

Fig. 1 SEM images of osteoblast cell attachment on tissue culture plate (**a**), 0.5Ag-nHA (**b**), 1.0Ag-nHA (**c**), and 1.5Ag-nHA (**d**) pellets [44]

$AgNO_3$, respectively. *E. coli* and *S. aureus* were cultured in the presence of the coated disks, and the spread plate method was used to quantify the bacterial colonies. At an exposure level of 10^5 bacterial cells/mL, all Ag-nHA coatings showed almost zero bacterial colonies after 48 h. At 10^8 cells/mL, all Ag-nHA coatings showed almost zero *S. aureus* colonies, while the *E. coli* colony numbers were the greatest for the 0.5Ag-HA surface and the least for the 1.5Ag-nHA. A cell adhesion test using human osteoblast cells and SEM imaging revealed that all Ag-nHA surfaces showed poor cell attachment except 0.5Ag-nHA (Fig. 1). While this paper developed an efficient and fast way to synthesize Ag-nHA particles for surface coatings and antibacterial effects were evident, poor cell adhesion presented a challenge [44].

While previous studies showed the antibacterial effects of Ag-HA coatings, Ag-HA was considered to have poor biocompatibility and cause cell death and nearly complete loss of ALP activity. Fielding et al. incorporated strontium in the Ag-HA coatings to offset the negative effects of Ag-HA coatings and improve their performance compared to pure HA coatings. The authors investigated the use of silver-containing, strontium (Sr)-doped plasma sprayed hydroxyapatite thin film coatings in order to improve the antimicrobial properties of implant materials and reduce the cytotoxic effects of the Ag-HA coatings. Strontium is a non-essential element that

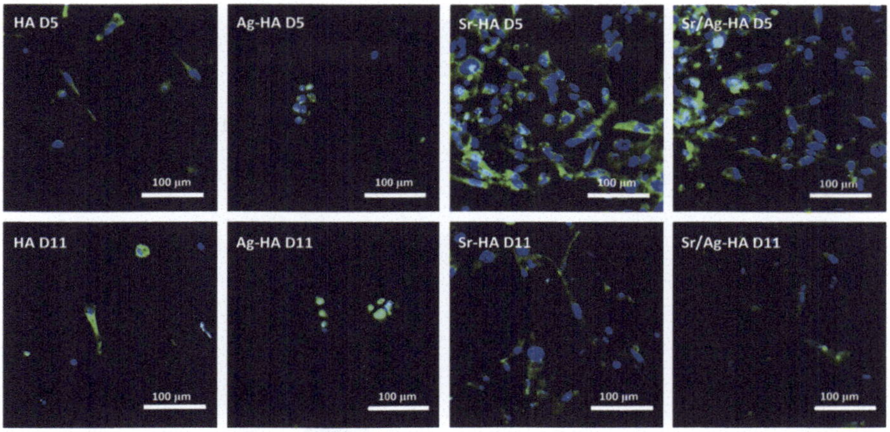

Fig. 2 Confocal micrographs showing ALP expression in human fetal osteoblast cells at 5 and 11 days. Green fluorescence indicates active ALP; blue fluorescence indicates the nuclei of cells [45]

displaces calcium in osteoblastic calcium-mediated processes and stimulates bone formation. HA was doped with 2 wt% Ag_2O and 1 wt% SrO to create Sr/Ag-HA-coated Ti substrates, which were then compared to HA alone, Sr-HA, and Ag-HA coatings. The antibacterial effects of the coatings were investigated in *P. aeruginosa* using live/dead fluorescent dyes. Both Ag-HA and Sr/Ag-HA coatings showed a strong antibacterial effect compared to HA and Sr-HA coatings, as judged by qualitative fluorescence microscopy. Osteoblast cell proliferation and osteoconductivity were investigated in human fetal osteoblast cells using MTT and ALP immunohistochemistry. Both Sr-HA and Sr/Ag-HA showed significantly higher cell viability compared to Ag-HA after 3 and 11 days of culture, suggesting a reduced cytotoxic effect by adding Sr to the Ag-HA coating. Both Ag-HA and Sr/Ag-HA coatings showed stronger ALP activity compared to HA and Sr-HA coatings at all-time points, suggesting improved osteoconductivity by the addition of Sr to the Ag-HA coatings (Fig. 2). This study showed that the thin film Sr/Ag-HA coating passively repelled bacterial adhesion and decreased the cytotoxic effects of the Ag-HA coating alone. These authors presented a method for enhancing the long-term antimicrobial properties of implant surface coatings while minimizing the negative cytotoxic effects of the coating to improve cell proliferation and osteoconductivity. This approach could potentially improve orthopedic device clinical outcomes [45].

Most methods for creating Ag-HA coatings on surfaces cannot be applied to porous surfaces, so Qu et al. designed a process for placing Ag-HA composite coatings on porous titanium surfaces using a sol-gel method in order to reduce the risk of bacterial contamination while maintaining the biocompatibility of pure HA surfaces. $Ca(NO_3)_2$ and P_2O_5 were mixed to create a Ca/P sol, which was then mixed with either 0.8 or 1.6 wt% $AgNO_3$ to create a Ag-HA gel that could be dip-coated onto porous Ti substrates (Fig. 3). The antibacterial effects of the coatings were

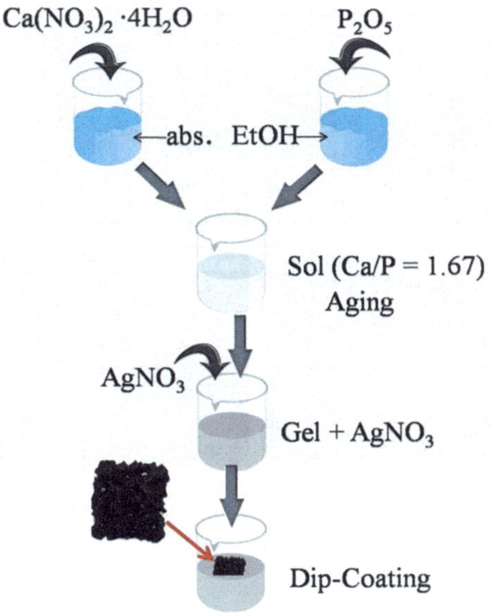

Fig. 3 The schematic drawing of the sol-gel process [46]

investigated in the presence of *E. coli* and *Staphylococcus albus* (*S. albus*) using the spread plate method. All Ag-HA coatings showed significantly fewer bacterial colonies when compared to pure HA coatings, and the coatings with higher silver concentration showed the strongest effect. Biocompatibility testing was done using SEM imaging and ALP quantification on samples seeded with osteoblasts derived from Sprague-Dawley rats. All Ag-HA surfaces showed cells adhering to the coating and spreading on the surfaces at all imaged time points from 1 to 7 days after seeding. Reduced ALP activity was shown in Ag-HA1.6 but not Ag-HA0.8. These authors developed an approach to modify porous Ti substrates with Ag-HA coatings, showing that these coating technologies have the potential for wide clinical applicability [46].

Tian and colleagues presented a method that could potentially be applied to a variety of metallic implant surfaces to improve their antimicrobial properties and reduce negative long-term outcomes by modifying the surface with silver nanoparticle-doped HA coatings with oriented block arrays (AgNP-BHAC). The authors showed that AgNP-BHAC improved the biocompatibility and osteoinductivity of Ti-6Al-4V Ti alloys while also exhibited antibacterial effects. The method used incorporated silver into HA coatings with block morphology, taking advantage of the antimicrobial effects of AgNPs and the excellent osteoconductivity and bioactivity of HA. The AgNP-BHAC coatings were applied in three steps beginning with the creation of a bioglass coating on Ti alloy substrates by dip coating into a mixed solution (calcium nitrate, tetraethyl orthosilicate, triethyl phosphate, ethanol, and nitric acid) followed by calcination. Next, the BHAC layer was added by hydrothermal treatment with a simulated body fluid solution (NaCl, $NaHCO_3$, KCl, K_2HPO_4-$3H_2O$, $MgCl_2$-$6H_2O$, $CaCl_2$, Na_2SO_4, and $(CH_2OH)_3CNH_2$) at 120 °C for

24 h. Finally, the AgNP-BHAC coatings were formed via facile Ag mirror reaction by dipping BHAC-coated alloy substrates in an $[Ag(NH_3)_2]OH$ solution for 20 s and drying. Standard physiochemical property characterization was performed including SEM, transmission electron microscopy, and Fourier-transform infrared spectroscopy. Bare alloy substrates, alloy substrates coated with BHAC, and alloy substrates coated with Ag-BHAC were exposed to *S. aureus* or *E. coli* cultured in suspension for 24 h. Bacterial colony formation was examined using confocal laser scanning microscopy and SEM imaging. The AgNP-BHAC-coated alloy substrates showed much less colony formation compared to other samples. Antibacterial activity was evaluated using the spread plate method: substrates incubated in bacterial suspensions were washed with PBS, sonicated to detach the adherent cells, and the detached cells were collected and cultured on soy agar plates. CFUs were quantified, with the AgNP-BHAC-coated alloy showing extremely strong antibacterial activity with near-zero CFU compared with the bare alloy and the BHAC-coated alloy, which formed multiple colonies. Furthermore, the osteoinductivity of the coating was studied using hBMSCs cultured on the coated and uncoated implant surfaces. A higher expression of integrins and better cell proliferation was shown in hBMSCs that were cultured on the AgNP-BHAC-coated alloy surface, and mRNA expression of osteogenic differentiation-related genes also showed significantly higher expression compared to controls. This new hydrothermal method for rapid synthesis of AgNP-BHAC coatings improved implant biocompatibility and osteoinductivity while demonstrating potential applications for surface modification on orthopedic and dental implants for infection prophylaxis [47].

Recently, another composite approach in the use of silver to mediate bacterial infections and HA to promote bone growth was investigated by Huang and colleagues, who investigated the use of a hybrid coating of nanostructured HA partially substituted with Ag^+ and fluorine (F^-) ions electrodeposited on titania nanotube (TNT) arrays on Ti substrates. The authors stated the need for a bifunctional coating that incorporates antimicrobial activity and osteoinductivity to overcome current challenges in the use of Ti and its alloys. The authors provided a simple and rapid method for creating a bifunctional coating via chemical bonding, incorporating silver into the HA-TiO_2 hybrid coating for improved antimicrobial properties with a second binary element fluorine to increase the structural stability of the coating on the Ti substrate. The release of silver was quantified by analyzing the silver ion concentration after incubation of the coated substrates in PBS; a steady release was observed for the entire 16-day experiment with 230 ppb cumulatively released. Three groups were studied for cytotoxicity via the MTT assay and for osteoinductivity using MC3T3-E1 mouse osteoblast precursor cells in vitro, including bare Ti substrates, Ti substrates with HA coating, and Ti substrates with a bilayer coating of HA partially substituted with F- and Ag+ electrodeposited on titania nanotubes (F-Ag-HA/TNT). On day 4 and day 7 but not on day 1, the HA coating and F-Ag-HA coating showed a significant increase in cell proliferation compared with the bare Ti substrate. The ALP expression level was significantly increased (~30%) in cells cultured on HA and F-Ag-HA coatings compared to control Ti, indicating improved differentiation of osteoblast cells. Moreover, Ti substrates with different coatings were placed in agar plates with pre-cultured *S. aureus* bacteria, and Ag-HA and F-Ag-HA showed a 100% inhibition after 24- and 36-h incubation, while HA and FHA showed less than a 20% inhibition. Follow-on animal studies are necessary, but this newly developed, simple and rapid method for creating a uniform, bifunctional nanostructure coating incorporating fluorine and silver on Ti implant surfaces holds promise [48].

Cheng et al. investigated the seeding of Sr- and Ag-loaded nanotubular structures on Ti surfaces for the repair of bone defects while providing antibacterial protection. The persistent release of Sr was intended to rapidly increase the filling of the bone defect by enhancing the repair of damaged cortical bone and increasing trabecular bone microarchitecture. The continuous and controlled release of Ag from the nanotubes was designed to provide anti-adherent and antibacterial properties against MRSA and *E. coli*. Ti foils (average size: 10.0 mm × 10.0 mm × 1.0 mm) and Ti rods (diameter: 1.0 mm; length: 12 mm) were anodized and treated hydrothermally for 3 h at 200 °C in 0.02 M $Sr(OH)_2$ solution to incorporate Sr into the Ti and create nanotubes (NT-Sr). NT-Sr was soaked in a 1.5 M or 2.0 M $AgNO_3$ solution for 10 min at room temperature to incorporate Ag into the system, overall resulting in Sr- and Ag-loaded nanotubular Ti (NT-AgSr). Initially, inductively coupled plasma-atomic emission spectrometry was used to monitor the release of Ag and Sr from NT-AgSr into PBS over time. Ag release from NT-AgSr decreased from 0.096 ppm on day 1 to 0.02 ppm on day 3, while Sr released from the composite also decreased from over 0.11 ppm on day 1 to just under 0.05 ppm on day 3. Release of both Ag and Sr remained constant for the next 11 days. The effectiveness of NT-AgSr to induce osteogenesis was investigated in vitro in MC3T3-E1 mouse preosteoblasts. Expression levels measured via qRT-PCR of osteogenesis-related genes ALP, OCN, RUNX2, and Col I were significantly higher (over 100%, 100%, 50%, and 70% higher, respectively) in cells treated with NT-AgSr compared to control cells treated with Ti only. Antibacterial activity was detected in samples treated with NT-AgSr, which resulted in a clear zone of growth inhibition on bacterial strains *E. coli*, methicillin-sensitive *S. aureus*, and MRSA, with the largest zone observed for all three bacterial strains on the plates containing the NT-AgSr with the highest concentration of silver. Unique among many similar reports of new materials, these authors also tested in vivo bone regeneration in 6-week-old Sprague-Dawley rats using histomorphometrical analysis. Rats treated with NT-AgSr showed a 15–20% increase in new bone tissue compared to animals treated with Ti without Sr and Ag. This formulation seems to be effective at inducing bone growth in vivo, while interrupting bacterial activity in vitro. Moving forward, the researchers plan to investigate the effectiveness of NT-AgSr in infected bone defects before progressing to clinical studies [49].

Zhou et al. have developed a multi-layered coating intended for use on Ti implants that exhibits increasing bactericidal activity with decreasing pH. This work is a follow-up to earlier research by the same group, honing the functionality of the previously developed silver nanoparticle gentamicin (AgNP/GS) loaded silk fibroin coating intended to decrease Ti implant-associated infection [50], by incorporating a pH-sensitive CS "nanovalve" layer (Fig. 4). Cleaned and polished Ti disks (10 mm × 10 mm × 0.5 mm) were first decorated with polydopamine and

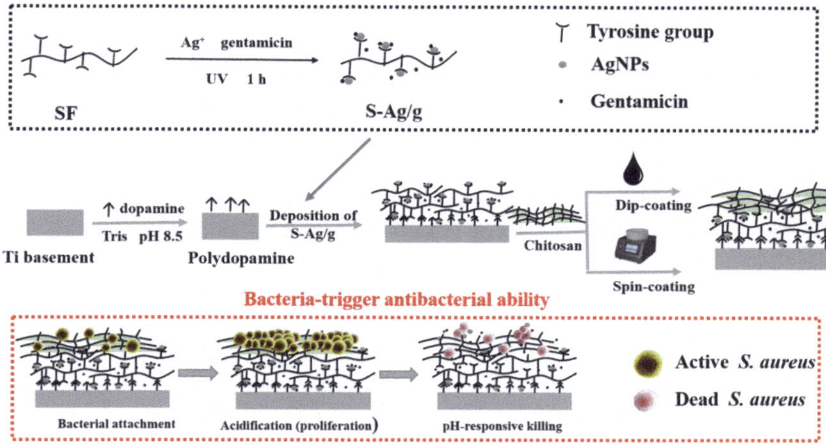

Fig. 4 Conceptual illustration of constructing multifunctional multi-layers through a facile depositing process. The depicted bacteria-responsive drug release mechanism is that attached bacteria proliferate and secrete lactic and acetic acid, which cause a local decrease in pH near the functional coating. The reduced local pH triggers the outmost CS barrier layer to swell and be corrupted, and increase the release of AgNPs/GS complexes loaded in the inner silk fibroin layer to inhibit bacterial attachment and kill planktonic bacteria [56]

dip-coated in an AgNP/GS-loaded silk fibroin solution as previously described [50]. The disks were then either dip-coated or spin-coated with CS [56]. Local acidification induced by bacterial growth on the multi-layered coating resulted in protonation of the amino groups of the CS, collapsing the CS layer and allowing the bactericidal release of Ag^{+}. This pH-dependent "nanovalve" was found to mitigate the cytotoxicity associated with the initial burst release followed by sustained, uncontrolled AgNP release seen previously [50]. Local acidification caused by *S. aureus* proliferation increased the bactericidal killing rate of an AgNP/GS-loaded silk fibroin coating on polydopamine-decorated Ti disks with dip-coated or spin-coated CS layers by 85.1% and 94.6%, respectively, as compared to a control without a pH-responsive CS layer. Preosteoblast MC3T3-E1 cells exhibited the greatest increase in attachment and proliferation on the spin-coated CS layer, with increased osteoblast differentiation potential compared to the control, as evidenced by ALP expression. The authors believe there is great clinical potential in their work, developing a multi-layered coating that simultaneously defends a Ti implant from bacterial invasion while promoting osteogenic activity [56].

While most studies use silver as the metallic antibacterial agent, this study used copper (Cu) and zinc (Zn). Huang et al. designed Cu/Zn co-substituted HA coated on pure titanium by electrodeposition in order to improve the antibacterial properties of pure HA while maintaining the material's biocompatibility. Since high copper content would induce cytotoxicity, the investigators also incorporated a secondary agent, Zn, to lessen the negative effects of Cu. Furthermore, they used electrodeposition, which is a cost-friendly method for coating both smooth and porous surfaces that establish a strong substrate-coating bond. ZnCuHA coatings showed significantly fewer *E. coli* bacterial colonies compared to pure HA coatings and bare substrates, with an over 90% decrease in colony number. Also, MC3T3-E1 osteoblast-like cells grown on the ZnCuHA surface appeared more flattened and evenly spread compared to the HA-coated surface. MTT assay results showed significantly higher viability of cells on the ZnCuHA surface than HA or bare substrates. This unique material surface coating shows potential as it outperforms HA coatings, but a direct comparison to Ag-HA coatings was not done [57].

Coatings with Growth Factors

Liu et al. investigated the use of an AgNP/PLGA-coated stainless steel alloy (SNPSA) as a potential implant material to target a broad spectrum of bacteria and reduce the risk of developing antibiotic resistance associated with the use of broad-spectrum antibiotics [51]. This study followed up on the work done by Zheng et al. on AgNP-based PLGA bone grafts loaded with BMP-2 that were shown to regenerate bone in vivo in rat femoral segmental defects infected with *S. aureus* [58]. Liu et al. soaked a 316 L stainless steel wire in a solution of 17.5% (w/v) PLGA and either 0%, 1%, or 2% AgNP for 30 s, air dried, and then soaked an additional two times to allow for a uniform coating of AgNP to form. The high surface energy of the nanoparticle coating encouraged the adsorption of BMP-2 growth factor, which aided in the promotion of bone growth [54–56, 58]. When treated with 10^5 CFU/mL *S. aureus* and *P. aeruginosa*, 2%-SNPSA demonstrated significant inhibition of initial adherence with less than 1% of cells remaining after a 24-h incubation. In in vitro culture with preosteoblastic MC3T3-E1 murine cells, these 2% SNPSA materials significantly increased the ALP activity by almost 50% per cell and significantly promoted the terminal differentiation of osteoblasts in cells that attached to the coating when compared to the 0%-SNPSA control. A hind femur rat model loaded with either *S. aureus* or *P. aeruginosa* was used to assess SNPSAs. After 8 weeks, 2%-SNPSA demonstrated significant bone formation despite a contamination level that could cause invasive tissue infection, and the samples showed no signs of osteolysis as confirmed by radiography and 3D micro-CT analysis. H&E staining showed no evidence of bacterial survival with minimal inflammatory cell infiltration in the surrounding tissue with 2%-SNPSA after 8 weeks, whereas both were present in the 0%-SNPSA control. The authors stated that further studies must be performed to test the interfacial adhesion of the silver/PLGA nanoparticle coating to different commonly used metal implant materials like Ti, Ti alloys, cobalt alloys, and different stainless steel alloys, but based on these findings there is potential for the use of the coating to improve biomaterial implants [51].

Similarly, Xie et al. investigated a method that incorporates AgNPs and BMP-2 into HA coatings on a Ti surface that could be used to modify metallic orthopedic and dental implants. The AgNPs were electrochemically deposited on the Ti surface along with CS, which was used for stabilization, uniform distribution, and reducing toxicity of the AgNPs. BMP-2 mixed with heparin was immobilized on the surface of the coating through an electrostatic reaction with CS/Ag/HA. The release studies indicated a sustained release of Ag from BMP/CS/Ag/HA and CS/Ag/HA coatings at 4.7 and 4.8 mg/L after soaking in PBS for 30 days. Similar results were observed for release studies of BMP-2, with 56% being from BMP/CS/Ag/HA and 86% being released from CS/Ag/HA coatings after 25 days. The ALP activity of BMSCs on BMP/CS/Ag/HA coatings was 11 U/gprot, which was higher than all the other coatings being tested and indicated the differentiation of BMSCs towards osteoblasts. A rabbit model was used to assess the osteoinductivity of the Ti implants with BMP/CS/Ag/HA; histological analysis after 12 weeks indicated linkage of the newly grown bone from the implant surface with a nearby host bone along with the presence of osteocytes, which indicated bone maturation [52].

Implant Coatings Containing Antibiotics

Direct Surface Modification with Antibiotics

An approach used infrequently but with possible wide-reaching impact involves the direct immobilization of an antibiotic onto the surface of a material intended for implantation. This approach is designed to decrease the chances of early infection after device implantation, which is key especially in instances of open fractures, but is more limited in long-term device infection abatement. Nie et al. investigated the use of bacitracin immobilized on a Ti surface in order to improve antibacterial activity and biocompatibility of bone implants, which they studied using a rat osteomyelitis model with femoral medullary cavity placement of Ti rods. Ti and Ti alloys are widely used in medical implants, but bacterial infections and poor osseointegration often occur soon after implantation and lead to early orthopedic implant failures. Ti is bio-inert; therefore, an approach that can promote osseointegration while preventing early stage infection around the implant at the same time is needed to overcome this challenge. As a follow-up in vivo study of their previous in vitro study [59], rods were machined to a 1.5 mm diameter and 20 mm height from Ti6Al4V titanium alloy. After rigorous cleaning with sodium hydroxide and water, the Ti rods were soaked in a dopamine solution to create a dopamine bioactive layer for drug immobilization, then they were soaked in bacitracin dissolved in ethanoic acid solution to allow for chemical bonding to the surface. Ti rods and Ti rods with bacitracin surface modification were exposed to a suspension of *S. aureus* and compared to control bare Ti rods in PBS. All three groups of rods were implanted in the femoral cavity of Sprague-Dawley rats without further treatment, and all mice were sacrificed 3 weeks later. Bacteria dwelling on the rods post-sacrifice were quantified via a spread plate method, and osteoregeneration was monitored using micro-CT on post-surgery day 1 and week 3. Rods in the bacitracin surface modification group and bare rods without bacterial incubation showed significantly higher regeneration of bone (twofold) and fewer bacteria CFU compared to bare rods with bacterial incubation. This study explored a simple, rapid method for immobilizing bacitracin on a Ti alloy surface with a demonstrated antibacterial and osteoinductivity-promoting effect. The simplicity of the approach might make it more translatable to industrial use and clinical application for Ti/Ti alloy implants [60].

Multi-layer Implant Coatings

Release rate is a key parameter in the effective delivery of therapeutics, with varied physiological conditions and encapsulants requiring different rates of release for optimal efficacy. With multiple components embedded within a single carrier, Strobel et al. investigated a sequential drug delivery system consisting of poly(D,L-lactide) (PDLLA) as the carrier, GS as the antibiotic, and insulin-like growth factor 1G (IGF-1) and BMP-2 as the two growth factors for osteoinduction. Ti wires were coated with either PDLLA, a suspension of GS and PDLLA, BMP-2 and PDLLA, or IGF-1 and PDLLA by dipping the wires in the suspension and allowing them to dry between coatings. All three active agents were released in a specific manner from the coating, starting with a burst release of the antibiotic, a burst release followed by a sustained release of IGF-1, and finally a slow release of BMP-2. Drug release data from the two-layer (GS/BMP-2) coated wire soaked in PBS showed that 44% of the GS was released on day 1, which slowed down and reached a maximum of 57% after 8 weeks, whereas 42% of the BMP-2 was released after 2 weeks, which slowed down even further reaching a maximum of only 49% after 8 weeks. The three-layer (GS/IGF-1/BMP-2) coated wire showed a burst release of GS releasing 38% in 3 days, burst release of IGF-1 releasing 26% in 3 days followed by a plateau at 47% until 8 weeks, and slow release of BMP-2 releasing only 6% in 8 weeks. Consequences of the BMP-2 and IGF-1 release were tested in vitro using human primary osteoblast-like cells; the three-layer coated wires showed the highest cell metabolic activity (130%) in the first 2 weeks as compared to the PDLLA control with a slight decrease in activity in week 3. On the other hand, single-layer coated wires of BMP-2 and IGF-1 showed 110% metabolic activity in the first 2 weeks, which then dropped and was comparable to the PDLLA control until 8 weeks. These results showed that the multiple coatings had a definite effect over the single implant coating, and the approach had potential to stimulate bone healing by controlling the release kinetics by varying the drug concentrations in the layers [53].

Min et al. investigated the tunable release of multiple therapeutics from layer-by-layer (LbL) coatings to prevent biofilm formation. A novel clay interlayer barrier was used to prevent interlayer diffusion. The designed coating was intended to release both the antibiotic GS, and the osteoinductive growth factor recombinant human BMP-2 in therapeutically relevant amounts with tunable rates of release, thus reducing the chance of infection or biofilm formation in new implants, and making the treatment of infected implants much simpler. The osteoinductive layer consisted of alternating layers of poly(β-amino esters), poly(acrylic acid), and BMP-2, while a similar antibacterial layer substituted GS for BMP-2. The poly(β-amino esters) of molecular weights 15 kDa and 20 kDa degrade at different rates under physiological conditions, thereby releasing GS and BMP-2 at different rates.

GS is an antibiotic effective against strains of *S. aureus* and is most efficacious with a localized delivery at low concentrations [61]. BMP-2 is a known osteoinductive growth factor that best delivers advantageous effects with a localized release at low rates [62–65]. Laponite clay has been shown to promote osteogenic differentiation of stem cells and has already been utilized in drug delivery systems because of its intercalation capacity [66–68]. The LbL coatings were created by dipping plasma-etched silicon substrates in cationic and anionic solutions of BMP-2 or GS until the desired thickness of the layers was achieved. A programmable spraying apparatus was used to deposit the polymer/clay barriers in between or on top of the layered films. In a Kirby–Bauer disk diffusion assay, the composite film consisting of BMP-2 with 80 iterations of application, laponite clay with 15 iterations of application, GS with 40 iterations of application and another layer of laponite clay with 15 iterations of application ($B_{80}L_{15}G_{40}L_{15}$), the composite film without laponite clay layers($B_{80}G_{40}$), as well as a film containing just GS (G_{40}) all showed clear zones of inhibition of *S. aureus* greater than the 15 mm benchmark, indicating that the laponite clay barrier was not interfering with the film's ability to deliver GS. The release of BMP-2 was modeled by measuring ALP activity in preosteoblast MC3T3-E1 cells cultured in vitro with a single-barrier film containing BMP-2 or composite films containing $B_{80}G_{40}$, or $B_{80}L_{15}G_{40}L_{15}$. Over the course of 5 weeks, BMP-2 released from the barrier-containing film maintained ALP activity. In comparison, a 50% reduction in the ALP activity was observed when a barrier layer was not included in the film. The activity of ALP reduced to zero over time when a single layer of BMP-2 was used. This indicated that the barrier composite film was capable of delivering a low, consistent rate of BMP-2 for a long period of time and reduced the occurrence of an initial burst release. Researchers concluded that the implementation of laponite clay barriers allowed for the separation and delivery of complex therapeutics in one film. These films have high potential to be tuned to achieve different release rates by varying layer separation and thickness [54].

In another study conducted by the same group, Min et al. developed a multilayered thin film implant coating to deliver both a broad-spectrum antibiotic and an osteoinductive growth factor, which followed controlled but disparate release kinetics. The authors suggest that such an implant coating would reduce revision arthroplasty to a single surgical step, decreasing the cost and timescale associated with the current two-stage standard procedure. The LbL deposition technique employing electrostatic multi-layer assembly was used to create a thin film incorporating BMP-2 releasing and GS releasing layers (Fig. 5). The osteoinductive layer consisted of alternating layers of poly(β-amino esters), poly(acrylic acid), and BMP-2, while a similar antibacterial layer substituted GS for BMP-2. As before, the poly(β-amino esters) of differing molecular weights degraded at different rates, thereby releasing GS and BMP-2 at different rates. A rat tibia model was used to assess the antibiotic and osteoinductive efficacy of the multi-layered implant coating on a radiolucent polyetheretherketone implant. A bare implant infected with bioluminescent *S. aureus* Xen 29 was removed 7 days post inoculation and replaced in a single-stage revision by an implant with the antibacterial coating, the osteoinductive coating, a multi-layered coating including both antibacterial and osteoinductive

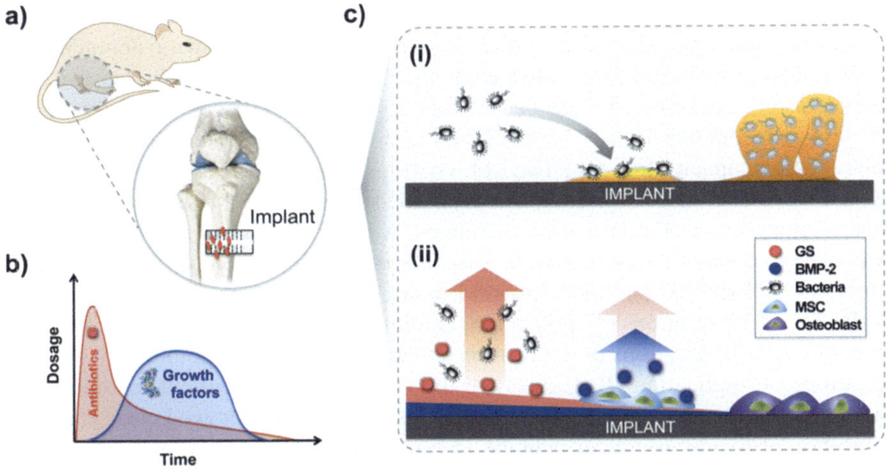

Fig. 5 Programmed sequential dual therapy delivery strategy to win the "Race to the Surface" against bacteria. (**a**) Illustration of a rat tibia model with induced osteomyelitis. (**b**) Desired release profile of an antibiotic and a growth factor and illustration of the top-down degradation of an LbL coating on an orthopedic implant. (**c**) Possible scenarios following in vivo application: (**ci**) In an uncoated implant, the residual bacteria in the defect and avascular tissue act as foreign bodies and can cause reinfection and form a biofilm (represented by the yellow area). (**cii**) In our dual therapy LbL coating, however, local delivery of an antibiotic (red circles) controls infection until the implant is vascularized and immune-competent. The subsequent release of a growth factor (blue circles) induces the osteogenic differentiation potential of endogenous precursor bone marrow stem cells, resulting in optimal bone healing and bone-implant integrity [55]

layers, or by a bare control implant. Excised implants that included both functional layers showed a bioluminescent signal comparable to the background at 8 weeks, indicating no further infection of the implant, compared to an obvious infection in the untreated controls. Micro-CT and histological analysis revealed that implants containing both functional layers induced bone coverage of up to 80% in 3 weeks. A complete host-implant integration of excised implants containing both functional layers was observed at 8 weeks compared to bone destruction, fracture, and ultimate failure seen in control. In terms of long-term antimicrobial efficacy, untreated control tibiae exhibited a strong bioluminescent signal, suggesting robust bacterial metabolic activity, while antibiotic treated tibiae showed no expression of bioluminescence. By rolling the implant on the blood agar, the surface bacterial growth was examined, and a three orders of magnitude of difference was observed between the uncoated implant and the coated one. Min et al. suggested that the multi-layered thin film coating could be explored for use with biomedical applications beyond bone implants, such as vascular grafts and artificial hearts [55].

Most dual-action antimicrobial systems describe a diffusion-based release of a biocide and contact-killing or bacteria repelling capabilities; however, Wong et al. designed a bifunctional film system that exhibited a bolus release of a biocide and a sustained release of an anti-inflammatory agent. Many implant failures are caused by an adverse foreign body response by the patient's immune system, which occurs in the days to weeks following implant placement and can be worsened by the presence of implant-related infection. Wong et al. designed a bifunctional LbL construct that incorporated antibiotic GS and anti-inflammatory drug diclofenac. This work was a follow-up study to previously demonstrated results with hydrolytically degradable polycation-based multi-layers for both the sustained and designed substantive bolus release of small-molecule drugs from surfaces [61]. The construct consisted of a permanent microbicidal polyelectrolyte multi-layered base film *N,N*-dodecyl,methyl-PEI/Poly(acrylic acid) with a hydrolytically degradable polyelectrolyte multi-layered top film. Two degradable film architectures were presented in this study: one for the burst release of GS to eradicate initial infection and one for the sustained release of diclofenac to suppress the inflammation at the implantation site. In vitro release experiments in PBS showed that GS was released within the first 6 h, while diclofenac was released sustainably over 16 days. Cytotoxicity of the film substrate and cell adhesion were assessed by culturing MC3T3-E1 cells and A549 epithelial cancer cells on the substrate; compared with cells cultured on bare glass substrate, no statistical difference was observed. The activity of GS after release from the film was visually confirmed using a Kirby–Bauer assay with *S. aureus*, and the activity of diclofenac released from the film was confirmed for up to 9 days by investigating the inhibition of cyclooxygenase. The bactericidal activity of the permanent base film after drug release and complete degradation of the top film was also confirmed using the Kirby–Bauer assay. The use of a permanent bactericidal base film with added tunable release of therapeutic agents potentially has broad applications for a variety of medical devices [69].

Tissue Engineering Scaffolds

Tissue engineering plays an increasingly important role in the future of clinical bone repair. The use of autologous and allogenic grafting has multiple limitations such as a lack of donors, additional risk of infection at donor sites, and immunological antigenicity; tissue engineering aims to circumvent these technologies. Tissue engineering most often employs a scaffold or construct material to house or recruit cells and provide necessary biochemical and biomechanical cues to recapitulate normal tissue. Important concerns in tissue engineering include being able to closely match the biochemical and mechanical properties of native bone and allowing for appropriate transport through the scaffold material for nutrient and waste exchange as well as oxygen delivery to cells in the scaffold. The use of tissue engineering for bone regeneration has been very widely researched, but here we focus on approaches that also include a means of discouraging bacterial infection (Table 3).

Limited graft availability and infection at the surgical sites cause serious risk in bone regeneration. However, the brittleness and the lack of an ideal implant shape are still key challenges in bone surgery; therefore, Schneider et al. designed flexible, osteoconductive scaffolds capable of delivering antimicrobials. Synthetic, flexible,

Table 3 Scaffold technologies

Type of scaffold	Material of scaffold	Key properties	Antimicrobial used	Key properties	Bacteria strains tested against
Polymer/ Mineral [70]	PLGA/calcium phosphate	Biodegradable, osteoconductive, capable of inducing bioactivity in bone defects	Silver	Potential antimicrobial activity owning to the presence of silver nanoparticles	N/A
Tissue-based [71]	Collagen	Capable of upregulating key osteogenic gene markers to initiate differentiation into osteoblast	Silver nanoparticle	Highly effective against gram-positive bacteria	*S. aureus*
Mineral, Polymer [72]	Chitosan, nano hydroxyapatite	Increases ALP activity and capable of inducing osteogenic markers	Near-infrared light	Effective against both gram-positive and gram-negative bacteria	*S. aureus*, *E. coli*
Polymer, Mineral [73]	Chitosan, tricalcium phosphate	Nontoxic, induces rapid cellular attachment and proliferation	Amoxicillin	Effective against a wide range of bacteria, commonly prescribed antibiotics	N/A but commercially available and FDA approved
Mineral, Polymer [74]	Calcium phosphate, chitosan	Capable of mimicking the extracellular matrix, shows bone-linking ability	Silver	Strong antibacterial activity on a broad range of bacteria	*Porphyromonas gingivalis*, *Staphylococcus mutans*

cotton wool-like nanocomposites consisting of a biodegradable PLGA matrix loaded with either CAP nanoparticle (PLGA/CAP) or silver-doped CAP nanoparticles (PLGA/Ag-CAP) in order to achieve bioactivity while providing additional antimicrobial properties have been developed. CAP and Ag-CAP were prepared by flame spray synthesis, and PLGA/CAP and PLGA/Ag-CAP scaffolds were fabricated by low-temperature electrospinning. A total number of 8 drill hole defects were created in two female sheep. The bone cavities were washed and then filled with PLGA/CAP or PLGA/Ag-CAP scaffolds, and the sheep were sacrificed 8 weeks after surgery. Bone regeneration was evaluated using micro X-ray and histological imaging. The average area of newly formed bone was not statistically different between the two samples, 20.5 ± 11.2% for PLGA/CAP and 22.5 ± 9.2% for PLGA/Ag-CAP treated defects (Fig. 6). This easily applicable system using a compressible material has great potential in applications to minimize invasive surgery or for specific indications in dental or orthopedic surgery [70].

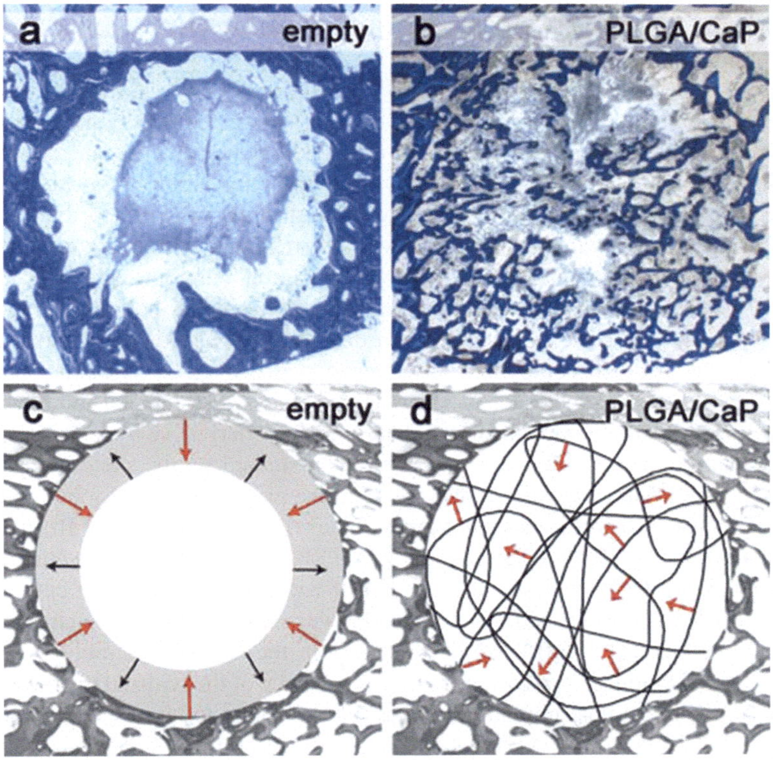

Fig. 6 Bone formation in defects left empty and for bone cotton wool treated defects. Histological ground sections for an (**a**) empty defect show the formation of an organized blood clot in the center of a defect, whereas the (**b**) PLGA/CAP treated defect bone formation (blue) is visible. Schematic description (**c, d**) of bone formation (red arrow) and bone resorption (black arrow) [70]

Sun et al. designed a novel collagen scaffold composite that encapsulated AgNP and BMP-2 to heal bone defects. This hybrid collagen scaffold combined porous structure, antibacterial activity, and osteoinductivity of collagen scaffolds with AgNP and BMP-2 to control bacterial infection and enhance bone regeneration, respectively. The absorbance of suspension-cultured vancomycin-resistant *S. aureus* strain Mu50 was monitored after exposure to varying concentrations of diluted collagen, Ag/NP/collagen, or BMP-2/AgNP/collagen scaffolds. Collagen scaffolds alone did not inhibit bacterial growth, but *S. aureus* proliferation was completely inhibited by Ag/NP/collagen and BMP-2/AgNP/collagen scaffolds at 10 μg mL^{-1} in vitro. Longer Ag^+ release periods enhanced the antibacterial activity of these hybrid composites; only 45% (AgNP/collagen) and 50.5% (BMP-2/AgNP/collagen) of Ag^+ was released in the first 24 h with continuing steady release over 300 h, whereas 89% of the Ag^+ was released from the control $AgNO_3$ sample in the first 24 h. The expression levels of RUNX2 showed upregulation upon analysis of BMSCs, which confirmed that the hybrid BMP-2/AgNP/collagen scaffold retained the ability to differentiate BMSCs towards osteoblasts. The authors stated that further study was warranted to understand the mechanical strength of these composites towards their use in bone repair, but these hybrid scaffolds show potential based on these preliminary findings [71].

The use of an osteoinductive scaffold material with controlled-release growth factors that additionally serves to mitigate implant-associated infection can be a key combination for the long-term efficacy of orthopedic bone defect repair. Zhou et al. have investigated the use of silica/hydroxypropyltrimethyl ammonium chloride CS/zein composite scaffolds loaded with recombinant human BMP-2 for bone tissue engineering. The growth factor BMP-2 was impregnated within the mesoporous SBA-15 nanoparticles, which acted to prevent the initial burst release of the growth factor, a common issue seen when therapeutic carriers are not incorporated in such scaffolds. The synthesis of SBA-15 nanoparticles was published in a previous study [75]; in brief, a block copolymer surfactant was dissolved in an acidic solution before the addition of silica source. The obtained gel was crystallized and filtered to remove the surfactant template. The water-soluble CS derivative hydroxypropyltrimethyl ammonium chloride chitosan, previously studied by the same group [76], was incorporated into a scaffold made of zein, the principle protein found in corn, for its demonstrated antibacterial properties. The relative release rate of BMP-2 was significantly improved by its impregnation into SBA-15. Compared to a cumulative protein release of 97% by day 11 in a control without the carrier, the cumulative release of BMP-2 was found to be reduced between 40 and 80%, depending on the relative weight ratio of SBA-15 nanoparticles in the scaffold. A rabbit forelimb segmental defect model was used to assess the efficacy of the scaffold in bone defect repair. At 12 weeks post-implant, the scaffold was found to have been completely resorbed and replaced with more extensive new bone growth compared to controls as evaluated with micro-CT analysis and histological staining. Zhou et al. demonstrate the utility of SBA-15 particles in facilitating the retention and controlled release of BMP-2 at the scaffold site, which may be incorporated into osteoinductive scaffolds or a myriad of additional biomedical applications [77].

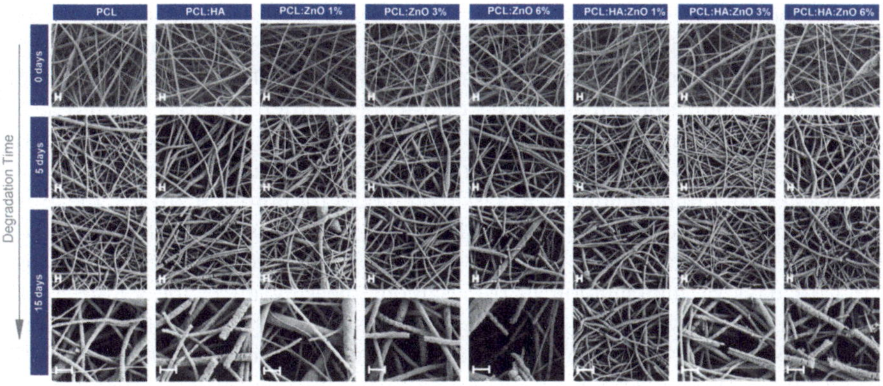

Fig. 7 SEM of scaffolds with different degradation times. Micrographs of scaffolds with different ZnO content. PCL, PCL:HA, PCL:ZnO 1%, PCL:ZnO 3%, PCL:ZnO 6%, PCL:HA:ZnO 1%, PCL:HA:ZnO 3%, and PCL:HA:ZnO 6%. Last row shows a zoom of samples after 15 days of degradation. Magnification bars are equal to 2 μm [81]

Felice et al. investigated the use of scaffolds containing PCL and zinc oxide (ZnO) nanostructures with or without HA, for their potential application in bone tissue engineering. ZnO has been shown to exhibit both osteoinductive and antibacterial effects, whereas HA has been studied for its osteoconductive properties. Due to the brittle nature of these components, incorporation of various concentrations of HA and ZnO into an electrospun, hydrolytically degradable nanofibrous PCL substrate allowed for their use in scaffolds (Fig. 7). While several mechanisms have been proposed [78–80], the authors inferred the antibacterial property associated with ZnO in this study might be ascribed to the action of the released Zn^{2+} ions, a mechanism which was explored elsewhere. ALP activity of human fetal osteoblasts cultured on PCL scaffolds containing 6% ZnO was found to be 141 times higher compared to the PCL-only control on day 14, a strong marker of the ability of the ZnO-containing scaffolds to initiate mineralization. The authors found that PCL exhibited a variable degradation rate with ZnO concentration, and therefore the proposed synthetic scaffold might be employed in engineered resorption systems of varied timescales [81].

Lu et al. created a novel zero-dimensional carbon dot (CD) doped chitosan/nanohydroxyapatite (CS/nHA/CD) scaffold to investigate the effects of CD on osteogenesis, osteosarcoma ablation, and the elimination of clinical bacteria. The fabricated scaffold would circumvent the current limitations in treating bone defects by promoting osteogenesis, eradicating common bacteria, and treating tumors and infections through photothermal therapy. CDs were synthesized by the traditional microwave method and were then doped onto the prepared CS/nHA scaffolds. CDs were shown to guide MSC osteogenesis and demonstrated photothermal effects under near-infrared radiation, making them a viable option for the photothermal treatment of tumors and infections [82–84]. Rat BMSCs were used to investigate the osteogenic effects of the scaffold. At 7 days, the rat BMSCs on the CS/nHA/CD scaffold showed statistically significantly higher ALP expression levels in comparison to the control CS/nHA scaffold. After 14 days, the CS/nHA/CD showed significantly higher levels of Col I and OCN than the CS/nHA scaffold ($p < .01$ and $< .05$, respectively), indicating greater levels of osteoblast differentiation. When treated with near-infrared light, the CS/nHA/CD scaffold showed an antibacterial rate of 99% with *S. aureus* and 97% with *E. coli*, compared to just 75% in the control. Four weeks after implantation in a classic rat model for porous scaffolds, the average bone mineral density in the CS/nHA/CD scaffold was 205.55 mg/cm^3 compared to 183.9 mg/cm^3 in the CS/nHA scaffold as measured by micro-CT scans. After 14 days of implantation in osteosarcoma-bearing mice, the CS/nHA/CD scaffold irradiated with light showed complete suppression of the subcutaneous osteosarcoma tumor as compared to an average tumor volume of 2500 mm^3 in CS/nHA/CD, CS/nHA, and CS/nHA groups. However, the authors stated that this test was limited by the fact that there were no existing in vivo models of osteosarcoma. However, they concluded that CD promoted osteogenesis within the CS/nHA scaffold and the CS/nHA/CD scaffold enhanced the osteogenic differentiation of rat BMSCs. Also, when irradiated with light, the scaffold had a high antibacterial rate against common clinical bacteria and was able to completely suppress an osteosarcoma tumor, making it a promising method for bone repair and treatment for tumors and bacterial infections [72].

Topsakal et al. fabricated nanofiber scaffolds loaded with the antibiotic amoxicillin and synthesized it with polyurethane, CS, and TCP that could induce bone tissue regeneration while providing antibacterial activity. The amoxicillin-loaded nanofiber was synthesized by mixing a 3 wt% CS solution in acetic acid with 5 wt% polyurethane, and varying levels of TCP (3–10 wt%). Amoxicillin (40 mg) was added to the resulting polymeric solution and was mixed for 1 h before electrospinning with flow rates between 0.8 and 2.3 mL/h. Amoxicillin-loaded nanofiber synthesis was confirmed using Fourier-transform infrared spectroscopy analysis which showed characteristic peaks of CS, polyurethane, and TCP. The morphology of the nanofiber was visualized using SEM, which showed macropores throughout the scaffolds, few micropores in the pore walls, and nanofibers mostly ranging from 87 to 112 nm. In vitro studies were conducted with L929 fibroblast cells to investigate the attachment, toxicity, and proliferation of cells in the presence of amoxicillin-loaded nanofibers. Nanofiber attachment occurred rapidly with about 75–80% cells attached to the fibers after 3 h. Toxicity was not observed, and cells treated with the nanofibers proliferated at a 30% faster rate than cells treated with TCP only. Loaded nanofibers released 60% of the amoxicillin within 24 h. Moving forward, the researchers plan to further investigate these loaded nanofibers for their potential in bone tissue engineering [73].

Jin et al. fabricated a biomimetic and bioactive silver ion-loaded CAP/CS membrane capable of mimicking the extracellular matrix to induce cell function naturally while providing antibacterial activity. This biomimetic process is classified as guided bone regeneration, which has become popular for reconstructing bone defects in oral and maxillofacial surgery in both orthopedic and dental applications. Guided bone regeneration makes use of a membrane that has two main functions: preventing fibrous ingrowth and directing growth of neo-bone tissue. In this study, the membrane was synthesized by dissolving CS in trifluoroacetic acid before Ag-CAP was added and mixed ultrasonically for 30 min. The mixture was electrospun at 26 kV with a distance to collector of 20 cm and a pump rate of 1 mL/h. The membrane was stabilized and crosslinked with 5% (w/v) vanillin dissolved in ethanol for 2 h at 50 °C. Synthesis of Ag-CAP/CS was confirmed using XRD analysis, which showed characteristic peaks of $CaHPO_4$, indicating that CAP was transformed into $CaHPO_4$ during electrospinning. Additionally, X-ray photoelectron spectroscopy was used to detect and confirm the presence of silver ions in the membrane. The bone-linking ability of Ag-CAP/CS was investigated by observing mineralization on the surface of membranes incubated in simulated body fluid. After 2 weeks, pure chitosan did not show any mineralization in contrast to the Ag-CAP/CS membrane that was covered with spherical minerals that increased as the immersion time increased. For cell studies, BMSC were used to investigate attachment and proliferation facilitated by Ag-CAP/CS. SEM images showed good attachment of cells to the Ag-CAP/CS membrane after 4 days. The membrane also induced 30% higher cellular proliferation than cells in the presence of chitosan membranes. Finally, antibacterial activity was also investigated using the direct contact test on *Porphyromonas gingivalis* and *Staphylococcus mutans* bacterial cells. In the presence of both bacterial cells, Ag-CAP/CS reduced growth by at least three times more than the control [74].

Bone Cements

CAP cement is commonly used as a therapeutic agent for bone defects, especially in infected defects. Cements have several benefits such as full degradability, low price, osteoconduction, and biocompatibility. The combinatorial delivery of osteoconductive and antibacterial factors from cements that are often employed anyway would be cost effective for the promotion of tissue healing (Table 4).

Table 4 Bone cement-based osteoconductive and antimicrobial technologies

Type of cement	Growth factor used	Antimicrobial used	Key properties	Bacteria strains tested against
Calcium phosphate [85]	BMP-2	Vancomycin	Sustained release over a long time period, causes mineralization of primary rat MSC, effective against a broad range of bacteria	N/A but commercially available and FDA approved
Calcium phosphate [86]	Absent	Iron	Induces expression of several osteogenic genes, strong antimicrobial activity especially effective against a broad range of bacteria	*E. coli*, *Salmonella enteritidis*, *P. aeruginosa*, *S. aureus*
Hydroxyapatite [87]	Absent	Cerium(III) oxide	Nontoxic, initiates bone repair rapidly, exhibits strong antimicrobial activity	*S. aureus*, *S. epidermidis*, *P. aeruginosa*

Calcium-based bone cement loaded with recombinant human BMP-2 and the antibiotic vancomycin holds potential to be effective at eradicating causative organisms while repairing bone defects simultaneously. Wang et al. created calcium sulfate cement constructs loaded with BMP-2 and vancomycin and studied their release kinetics and bioactivity in vitro and in vivo. BMP-2 and vancomycin-loaded calcium sulfate cement were fabricated in a mixing bowl containing osteoset at concentrations of 180 μg of BMP-2, 48 mg of vancomycin, and 960 mg of CAP. An ALP activity assay demonstrated the sustained release of BMP-2 for 29 days, while high-performance liquid chromatography detected vancomycin within 1 cm of the implant site for about 10 days post treatment, resulting in mineralization of primary rat MSCs in vitro. Detected vancomycin concentrations detected ranged from 293.3 ± 54.3 μg/g and 196.1 ± 92.2 μg/g at different locations on bone on day 1 post treatment to 8.7 ± 19.3 μg/g on day 7 post treatment, while BMP-2 detected in new bone tissue declined from over 100 ng/g detected on day 1 to 1.49 ng/g on day 28. Histomorphometrical analysis using adult male New Zealand white rabbits showed an almost 60% increase in new bone tissue after treatment with the BMP-2/vancomycin-loaded calcium sulfate cement compared to bone defects that were treated with calcium sulfate alone, which only produced about 20% new bone tissue 28 days post treatment. The composite initially released a bolus of vancomycin and BMP-2 to the bone before a continued release for over 14 and 21 days, respectively, showing extensive opportunity for translation to clinical studies [85].

Uskoković et al. enhanced self-hardening CAP cements with self-setting, iron (Fe)-doped bone-integrative cements to improve deficiencies of CAP in bone tissue substitutes. These deficiencies are expected to be improved by iron ion dopants that transform Fe-doped β-tricalcium phosphate (Fe-TCP) to nanocrystalline brushite. TCP has been shown to incorporate foreign ions; therefore, cationic substitution was hypothesized to not impose any significant lattice distortions in the material. Fe-TCP precursor was prepared by mixing $Ca(NO_3)_2$ with $Fe(NO_3)_3$ before adding $(NH_4)_2HPO_4$ dropwise at the rate of 20 mL/min. The precipitates were crystallized by heat treatment for 1-h at 900 °C. Using energy-dispersive XRD, the researchers observed that cement composite stabilized compositionally and mechanically 50-h after fabrication. The compressive strength of the composite doubled from 11.5 ± 0.5 to 24.5 ± 2.0 MPa after Fe dopant was added. The osteoinductivity of Fe-TCP was investigated in MC3T3-E1 cells by measuring mRNA expression levels of the osteogenic genes OCN and RUNX2. Gene expression was elevated for both genes; OCN was expressed over 90% more in cells treated with Fe-TCP compared to control cells treated with ascorbic acid and β-glycerophosphate. Antibacterial activity was observed against all four bacterial strains investigated: *E. coli*, *Salmonella enteritidis*, *P. aeruginosa*, and *S. aureus*. Among all the bacterial strains, the diameter of the bacteria inhibition zones was at least 30% higher than bacteria not treated with Fe-TCP. The only exception was *P. aeruginosa* treated with a low concentration of Fe-TCP, which showed very weak antibacterial activity. Moving forward, Fe-TCP performance should be investigated in animal models to discern the performance of the composite in vivo [86].

Morais et al. investigated the biological performance of three previously fabricated hydrogels based on CS, alginate, and HA to improve osteoconductive properties of glass-reinforced hydroxyapatite (GR-HA). GR-HA is a CAP-based bone substitute capable of chemically mimicking the inorganic phase of bone. Synthetic bone substitutes like GR-HA have been associated with hydrogels, which make the formulation injectable in clinical settings, aiding in minimally invasive placement into complex bone defect geometries. GR-HA was synthesized by adding 2.5% (w/w) of glass (which was composed of $15CaO$-$65P_2O_5$-$10Na_2O$–$10CaF_2$, mol%) to a mixture of HA and microcrystalline cellulose. Injectable bone substitutes were fabricated by mixing and aggregating GR-HA granules with hydrogels previously synthesized. Cellular viability using the MTT assay with MG63 human osteosarcoma cells showed HA injectable hydrogels with almost a 60% increase in cell growth 72 h post treatment. SEM analysis showed that cells treated with HA hydrogels developed more filopodia with more spreading just 1 day post treatment compared to cells treated with other types of hydrogels that exhibited lamellipodia. Hyaluronic acid hydrogels were chosen for animal studies and were subcutaneously implanted in Sprague-Dawley rats, resulting in histologically visible tissue repair with little inflammatory response. Antimicrobial activity was investigated by incorporating Ce(III) ions into the HA hydrogel using cross-linking solutions to target microbial activities of *S. aureus*, *S. epidermidis*, *Candida albicans*, and *P. aeruginosa*. The Ce(III) incorporated hydrogel showed over 70% antimicrobial activity in all microbes except for *P. aeruginosa*, which was just over 20%. The weaker Ce(III) effect on *P. aeruginosa* was attributed to the fact that gram-negative bacteria lack peptidoglycans on their cell wall [87].

Conclusion

Infection remains one of the main obstacles in the repair of bone defects. Current clinical methods are not effective enough to handle both bone repair and antimicrobial activity, so the development of biomaterials with both osteoconductive and antimicrobial properties could be a viable treatment option. As detailed herein, dual osteoconductive and antimicrobial constructs are capable of repairing bone defects in different environments while providing antimicrobial activity. Many of these constructs did not pose any cytotoxicity issue. Moreover, most silver-nanoparticle-containing technologies reported similar outcomes to antibiotic-loaded constructs suggesting versatility and room for the future optimization of constructs with silver as a broad-spectrum treatment. These new dual functional biomaterials show a very promising potential to lessen the burden of osteomyelitis and surgical site infections.

References

1. Einhorn TA, Gerstenfeld LC (2015) Fracture healing: mechanisms and interventions. Nat Rev Rheumatol 11:45–54. https://doi.org/10.1038/nrrheum.2014.164
2. Castillo RC, Bosse MJ, MacKenzie EJ et al (2005) Impact of smoking on fracture healing and risk of complications in limb-threatening open tibia fractures. J Orthop Trauma 19:151–157

3. Santolini E, West R, Giannoudis PV (2015) Risk factors for long bone fracture non-union: a stratification approach based on the level of the existing scientific evidence. Injury 46(Suppl 8):S8–S19. https://doi.org/10.1016/S0020-1383(15)30049-8
4. Amin S, Achenbach SJ, Atkinson EJ et al (2014) Trends in fracture incidence: a population-based study over 20 years. J Bone Miner Res 29:581–589. https://doi.org/10.1002/jbmr.2072
5. Calcei JG, Rodeo SA (2019) Orthobiologics for bone healing. Clin Sports Med 38:79–95. https://doi.org/10.1016/j.csm.2018.08.005
6. Marsell R, Einhorn TA (2011) The biology of fracture healing. Injury 42:551–555. https://doi.org/10.1016/j.injury.2011.03.031
7. Al-Mulhim FA, Baragbah MA, Sadat-Ali M et al (2014) Prevalence of surgical site infection in orthopedic surgery: a 5-year analysis. Int Surg 99:264–268. https://doi.org/10.9738/INTSURG-D-13-00251.1
8. Sodhi N, Anis HK, Garbarino LJ et al (2019) Have we actually reduced our 30-day short-term surgical site infection rates in primary total hip arthroplasty in the United States? J Arthroplast 34(9):2102–2106. https://doi.org/10.1016/j.arth.2019.04.045
9. Tsukayama DT, Estrada R, Gustilo RB (1996) Infection after total hip arthroplasty. A study of the treatment of one hundred and six infections. J Bone Joint Surg Am 78:512–523. https://doi.org/10.2106/00004623-199604000-00005
10. Campoccia D, Montanaro L, Arciola CR (2006) The significance of infection related to orthopedic devices and issues of antibiotic resistance. Biomaterials 27:2331–2339. https://doi.org/10.1016/j.biomaterials.2005.11.044
11. Berbari EF, Hanssen AD, Duffy MC et al (1998) Risk factors for prosthetic joint infection: case-control study. Clin Infect Dis 27:1247–1254. https://doi.org/10.1086/514991
12. Johnson EN, Burns TC, Hayda RA et al (2007) Infectious complications of open type III tibial fractures among combat casualties. Clin Infect Dis 45:409–415. https://doi.org/10.1086/520029
13. Murray CK, Obremskey WT, Hsu JR et al (2011) Prevention of infections associated with combat-related extremity injuries. J Trauma 71:S235–S257. https://doi.org/10.1097/TA.0b013e318227ac5f
14. Blanchette KA, Prabhakara R, Shirtliff ME, Wenke JC (2017) Inhibition of fracture healing in the presence of contamination by Staphylococcus aureus: effects of growth state and immune response. J Orthop Res 35:1845–1854. https://doi.org/10.1002/jor.23573
15. Gosselin RA, Roberts I, Gillespie WJ (2004) Antibiotics for preventing infection in open limb fractures. Cochrane Database Syst Rev (1):CD003764. https://doi.org/10.1002/14651858.CD003764.pub2
16. Brown KV, Walker JA, Cortez DS et al (2010) Earlier debridement and antibiotic administration decrease infection. J Surg Orthop Adv 19:18–22
17. Penn-Barwell JG, Murray CK, Wenke JC (2012) Early antibiotics and debridement independently reduce infection in an open fracture model. J Bone Joint Surg Br 94:107–112. https://doi.org/10.1302/0301-620X.94B1.27026
18. Lack WD, Karunakar MA, Angerame MR et al (2015) Type III open tibia fractures: immediate antibiotic prophylaxis minimizes infection. J Orthop Trauma 29:1–6. https://doi.org/10.1097/BOT.0000000000000262
19. Rand BCC, Penn-Barwell JG, Wenke JC (2015) Combined local and systemic antibiotic delivery improves eradication of wound contamination: an animal experimental model of contaminated fracture. Bone Joint J 97-B:1423–1427. https://doi.org/10.1302/0301-620X.97B10.35651
20. Costerton JW, Stewart PS, Greenberg EP (1999) Bacterial biofilms: a common cause of persistent infections. Science 284:1318–1322
21. Donlan RM, Costerton JW (2002) Biofilms: survival mechanisms of clinically relevant microorganisms. Clin Microbiol Rev 15:167–193. https://doi.org/10.1128/cmr.15.2.167-193.2002
22. Hall-Stoodley L, Stoodley P, Kathju S et al (2012) Towards diagnostic guidelines for biofilm-associated infections. FEMS Immunol Med Microbiol 65:127–145. https://doi.org/10.1111/j.1574-695X.2012.00968.x
23. Badha V, Moore R, Heffernan J et al (2019) Determination of tobramycin and vancomycin exposure required to eradicate biofilms on muscle and bone tissue in vitro. J Bone Jt Infect 4:1–9. https://doi.org/10.7150/jbji.29711

24. Hangst K, Eitenmüller J, Weltin R, Peters G (1987) Hydroxylapatite silver phosphate ceramics: production, analysis and biological testing of their antibacterial effectiveness. MRS Online Proc Lib Arch 110:269. https://doi.org/10.1557/PROC-110-269
25. Kim TN, Feng QL, Kim JO et al (1998) Antimicrobial effects of metal ions (Ag+, Cu2+, Zn2+) in hydroxyapatite. J Mater Sci Mater Med 9:129–134. https://doi.org/10.1023/A:1008811501734
26. Qing Y, Cheng L, Li R et al (2018) Potential antibacterial mechanism of silver nanoparticles and the optimization of orthopedic implants by advanced modification technologies. Int J Nanomedicine 13:3311–3327. https://doi.org/10.2147/IJN.S165125
27. Uskoković V, Desai TA (2014) Nanoparticulate drug delivery platforms for advancing bone infection therapies. Expert Opin Drug Deliv 11:1899–1912. https://doi.org/10.1517/17425247.2014.944860
28. Uskoković V (2015) Nanostructured platforms for the sustained and local delivery of antibiotics in the treatment of osteomyelitis. Crit Rev Ther Drug Carrier Syst 32:1–59
29. Uskoković V (2015) When 1+1>2: nanostructured composites for hard tissue engineering applications. Mater Sci Eng C Mater Biol Appl 57:434–451. https://doi.org/10.1016/j.msec.2015.07.050
30. Stevanović M, Uskoković V, Filipović M et al (2013) Composite PLGA/AgNpPGA/AscH nanospheres with combined osteoinductive, antioxidative, and antimicrobial activities. ACS Appl Mater Interfaces 5:9034–9042. https://doi.org/10.1021/am402237g
31. Mao Z, Li Y, Yang Y et al (2018) Osteoinductivity and antibacterial properties of strontium ranelate-loaded poly(lactic-co-glycolic acid) microspheres with assembled silver and hydroxyapatite nanoparticles. Front Pharmacol 9:368. https://doi.org/10.3389/fphar.2018.00368
32. Bostancıoğlu RB, Peksen C, Genc H et al (2015) Analyses of the modulatory effects of antibacterial silver doped calcium phosphate-based ceramic nano-powder on proliferation, survival, and angiogenic capacity of different mammalian cells in vitro. Biomed Mater 10:045024. https://doi.org/10.1088/1748-6041/10/4/045024
33. Uskoković V, Desai TA (2013) Phase composition control of calcium phosphate nanoparticles for tunable drug delivery kinetics and treatment of osteomyelitis. II. Antibacterial and osteoblastic response. J Biomed Mater Res A 101A:1427–1436. https://doi.org/10.1002/jbm.a.34437
34. Cai B, Zou Q, Zuo Y et al (2018) Injectable gel constructs with regenerative and anti-infective dual effects based on assembled chitosan microspheres. ACS Appl Mater Interfaces 10:25099–25112. https://doi.org/10.1021/acsami.8b06648
35. Makarov C, Cohen V, Raz-Pasteur A, Gotman I (2014) In vitro elution of vancomycin from biodegradable osteoconductive calcium phosphate-polycaprolactone composite beads for treatment of osteomyelitis. Eur J Pharm Sci 62:49–56. https://doi.org/10.1016/j.ejps.2014.05.008
36. Coathup MJ, Blunn GW, Flynn N et al (2001) A comparison of bone remodelling around hydroxyapatite-coated, porous-coated and grit-blasted hip replacements retrieved at post-mortem. J Bone Joint Surg Br 83:118–123
37. Hayakawa T, Yoshinari M, Kiba H et al (2002) Trabecular bone response to surface roughened and calcium phosphate (Ca-P) coated titanium implants. Biomaterials 23:1025–1031
38. Darimont GL, Cloots R, Heinen E et al (2002) In vivo behaviour of hydroxyapatite coatings on titanium implants: a quantitative study in the rabbit. Biomaterials 23:2569–2575
39. Yoshinari M, Oda Y, Inoue T et al (2002) Bone response to calcium phosphate-coated and bisphosphonate-immobilized titanium implants. Biomaterials 23:2879–2885
40. Cho J-H, Garino JP, Choo S-K et al (2010) Seven-year results of a tapered, titanium, hydroxyapatite-coated cementless femoral stem in primary total hip arthroplasty. Clin Orthop Surg 2:214–220. https://doi.org/10.4055/cios.2010.2.4.214
41. Vidalain J-P (2011) Twenty-year results of the cementless Corail stem. Int Orthop 35:189–194. https://doi.org/10.1007/s00264-010-1117-2
42. Lazarinis S, Kärrholm J, Hailer NP (2011) Effects of hydroxyapatite coating on survival of an uncemented femoral stem. A Swedish Hip Arthroplasty Register study on 4,772 hips. Acta Orthop 82:399–404. https://doi.org/10.3109/17453674.2011.597699

43. Chen W, Oh S, Ong AP et al (2007) Antibacterial and osteogenic properties of silver-containing hydroxyapatite coatings produced using a sol gel process. J Biomed Mater Res A 82:899–906. https://doi.org/10.1002/jbm.a.31197
44. Rameshbabu N, Sampath Kumar TS, Prabhakar TG et al (2007) Antibacterial nanosized silver substituted hydroxyapatite: synthesis and characterization. J Biomed Mater Res A 80:581–591. https://doi.org/10.1002/jbm.a.30958
45. Fielding GA, Roy M, Bandyopadhyay A, Bose S (2012) Antibacterial and biological characteristics of silver containing and strontium doped plasma sprayed hydroxyapatite coatings. Acta Biomater 8:3144–3152. https://doi.org/10.1016/j.actbio.2012.04.004
46. Qu J, Lu X, Li D et al (2011) Silver/hydroxyapatite composite coatings on porous titanium surfaces by sol-gel method. J Biomed Mater Res B Appl Biomater 97B:40–48. https://doi.org/10.1002/jbm.b.31784
47. Tian B, Chen W, Yu D et al (2016) Fabrication of silver nanoparticle-doped hydroxyapatite coatings with oriented block arrays for enhancing bactericidal effect and osteoinductivity. J Mech Behav Biomed Mater 61:345–359. https://doi.org/10.1016/j.jmbbm.2016.04.002
48. Huang Y, Song G, Chang X et al (2018) Nanostructured Ag+-substituted fluorhydroxyapatite-TiO2 coatings for enhanced bactericidal effects and osteoinductivity of Ti for biomedical applications. Int J Nanomedicine 13:2665–2684. https://doi.org/10.2147/IJN.S162558
49. Cheng H, Xiong W, Fang Z et al (2016) Strontium (Sr) and silver (Ag) loaded nanotubular structures with combined osteoinductive and antimicrobial activities. Acta Biomater 31:388–400. https://doi.org/10.1016/j.actbio.2015.11.046
50. Zhou W, Jia Z, Xiong P et al (2017) Bioinspired and biomimetic AgNPs/gentamicin-embedded silk fibroin coatings for robust antibacterial and osteogenetic applications. ACS Appl Mater Interfaces 9:25830–25846. https://doi.org/10.1021/acsami.7b06757
51. Liu Y, Zheng Z, Zara JN et al (2012) The antimicrobial and osteoinductive properties of silver nanoparticle/poly (dl-lactic-co-glycolic acid)-coated stainless steel. Biomaterials 33:8745–8756. https://doi.org/10.1016/j.biomaterials.2012.08.010
52. Xie C-M, Lu X, Wang K-F et al (2014) Silver nanoparticles and growth factors incorporated hydroxyapatite coatings on metallic implant surfaces for enhancement of osteoinductivity and antibacterial properties. ACS Appl Mater Interfaces 6:8580–8589. https://doi.org/10.1021/am501428e
53. Strobel C, Bormann N, Kadow-Romacker A et al (2011) Sequential release kinetics of two (gentamicin and BMP-2) or three (gentamicin, IGF-I and BMP-2) substances from a one-component polymeric coating on implants. J Control Release 156:37–45. https://doi.org/10.1016/j.jconrel.2011.07.006
54. Min J, Braatz RD, Hammond PT (2014) Tunable staged release of therapeutics from layer-by-layer coatings with clay interlayer barrier. Biomaterials 35:2507–2517. https://doi.org/10.1016/j.biomaterials.2013.12.009
55. Min J, Choi KY, Dreaden EC et al (2016) Designer dual therapy nanolayered implant coatings eradicate biofilms and accelerate bone tissue repair. ACS Nano 10:4441–4450. https://doi.org/10.1021/acsnano.6b00087
56. Zhou W, Li Y, Yan J et al (2018) Construction of self-defensive antibacterial and osteogenic AgNPs/gentamicin coatings with chitosan as nanovalves for controlled release. Sci Rep 8:13432. https://doi.org/10.1038/s41598-018-31843-2
57. Huang Y, Zhang X, Mao H et al (2015) Osteoblastic cell responses and antibacterial efficacy of Cu/Zn co-substituted hydroxyapatite coatings on pure titanium using electrodeposition method. RSC Adv 5:17076–17086. https://doi.org/10.1039/C4RA12118J
58. Zheng Z, Yin W, Zara JN et al (2010) The use of BMP-2 coupled—Nanosilver-PLGA composite grafts to induce bone repair in grossly infected segmental defects. Biomaterials 31:9293–9300. https://doi.org/10.1016/j.biomaterials.2010.08.041
59. Nie B, Ao H, Zhou J et al (2016) Biofunctionalization of titanium with bacitracin immobilization shows potential for anti-bacteria, osteogenesis and reduction of macrophage inflammation. Colloids Surf B Biointerfaces 145:728–739. https://doi.org/10.1016/j.colsurfb.2016.05.089

60. Nie B, Ao H, Long T et al (2017) Immobilizing bacitracin on titanium for prophylaxis of infections and for improving osteoinductivity: an in vivo study. Colloids Surf B Biointerfaces 150:183–191. https://doi.org/10.1016/j.colsurfb.2016.11.034
61. Moskowitz JS, Blaisse MR, Samuel RE et al (2010) The effectiveness of the controlled release of gentamicin from polyelectrolyte multilayers in the treatment of Staphylococcus aureus infection in a rabbit bone model. Biomaterials 31:6019–6030. https://doi.org/10.1016/j.biomaterials.2010.04.011
62. Groeneveld EH, Burger EH (2000) Bone morphogenetic proteins in human bone regeneration. Eur J Endocrinol 142:9–21
63. Jeon O, Song SJ, Yang HS et al (2008) Long-term delivery enhances in vivo osteogenic efficacy of bone morphogenetic protein-2 compared to short-term delivery. Biochem Biophys Res Commun 369:774–780. https://doi.org/10.1016/j.bbrc.2008.02.099
64. Zara JN, Siu RK, Zhang X et al (2011) High doses of bone morphogenetic protein 2 induce structurally abnormal bone and inflammation in vivo. Tissue Eng Part A 17:1389–1399. https://doi.org/10.1089/ten.TEA.2010.0555
65. James AW, LaChaud G, Shen J et al (2016) A review of the clinical side effects of bone morphogenetic protein-2. Tissue Eng Part B Rev 22:284–297. https://doi.org/10.1089/ten.TEB.2015.0357
66. Liu L, Grunlan JC (2007) Clay assisted dispersion of carbon nanotubes in conductive epoxy nanocomposites. Adv Funct Mater 17:2343–2348. https://doi.org/10.1002/adfm.200600785
67. Podsiadlo P, Kaushik AK, Arruda EM et al (2007) Ultrastrong and stiff layered polymer nanocomposites. Science 318:80–83. https://doi.org/10.1126/science.1143176
68. Wang C, Wang S, Li K et al (2014) Preparation of laponite bioceramics for potential bone tissue engineering applications. PLoS One 9:e99585. https://doi.org/10.1371/journal.pone.0099585
69. Wong SY, Moskowitz JS, Veselinovic J et al (2010) Dual functional polyelectrolyte multilayer coatings for implants: permanent microbicidal base with controlled release of therapeutic agents. J Am Chem Soc 132:17840–17848. https://doi.org/10.1021/ja106288c
70. Schneider OD, Mohn D, Fuhrer R et al (2011) Biocompatibility and bone formation of flexible, cotton wool-like PLGA/calcium phosphate nanocomposites in sheep. Open Orthop J 5:63–71. https://doi.org/10.2174/1874325001105010063
71. Sun C-Y, Che Y-J, Lu S-J (2015) Preparation and application of collagen scaffold-encapsulated silver nanoparticles and bone morphogenetic protein 2 for enhancing the repair of infected bone. Biotechnol Lett 37:467–473. https://doi.org/10.1007/s10529-014-1698-8
72. Lu Y, Li L, Li M et al (2018) Zero-dimensional carbon dots enhance bone regeneration, osteosarcoma ablation, and clinical bacterial eradication. Bioconjug Chem 29:2982–2993. https://doi.org/10.1021/acs.bioconjchem.8b00400
73. Topsakal A, Uzun M, Ugar G et al (2018) Development of amoxicillin-loaded electrospun polyurethane/chitosan/β-tricalcium phosphate scaffold for bone tissue regeneration. IEEE Trans Nanobioscience 17:321–328. https://doi.org/10.1109/TNB.2018.2844870
74. Jin S, Li J, Wang J et al (2018) Electrospun silver ion-loaded calcium phosphate/chitosan antibacterial composite fibrous membranes for guided bone regeneration. Int J Nanomedicine 13:4591–4605. https://doi.org/10.2147/IJN.S167793
75. Zhao D, Huo Q, Feng J et al (1998) Nonionic triblock and star diblock copolymer and oligomeric sufactant syntheses of highly ordered, hydrothermally stable, mesoporous silica structures. J Am Chem Soc 120:6024–6036. https://doi.org/10.1021/ja974025i
76. Zhou P, Xia Y, Wang J et al (2013) Antibacterial properties and bioactivity of HACC- and HACC–Zein-modified mesoporous bioactive glass scaffolds. J Mater Chem B 1:685–692. https://doi.org/10.1039/C2TB00102K
77. Zhou P, Xia Y, Cheng X et al (2014) Enhanced bone tissue regeneration by antibacterial and osteoinductive silica-HACC-zein composite scaffolds loaded with rhBMP-2. Biomaterials 35:10033–10045. https://doi.org/10.1016/j.biomaterials.2014.09.009
78. Pasquet J, Chevalier Y, Pelletier J et al (2014) The contribution of zinc ions to the antimicrobial activity of zinc oxide. Colloids Surf A Physicochem Eng Asp 457:263–274. https://doi.org/10.1016/j.colsurfa.2014.05.057

79. Sirelkhatim A, Mahmud S, Seeni A et al (2015) Review on zinc oxide nanoparticles: antibacterial activity and toxicity mechanism. Nanomicro Lett 7:219–242. https://doi.org/10.1007/s40820-015-0040-x
80. Joe A, Park S-H, Shim K-D et al (2017) Antibacterial mechanism of ZnO nanoparticles under dark conditions. J Ind Eng Chem 45:430–439. https://doi.org/10.1016/j.jiec.2016.10.013
81. Felice B, Sánchez MA, Socci MC et al (2018) Controlled degradability of PCL-ZnO nanofibrous scaffolds for bone tissue engineering and their antibacterial activity. Mater Sci Eng C Mater Biol Appl 93:724–738. https://doi.org/10.1016/j.msec.2018.08.009
82. Zhang M, Wang W, Cui Y et al (2018) Near-infrared light-mediated photodynamic/photothermal therapy nanoplatform by the assembly of Fe3O4 carbon dots with graphitic black phosphorus quantum dots. Int J Nanomedicine 13:2803–2819. https://doi.org/10.2147/IJN.S156434
83. Peng X, Wang R, Wang T et al (2018) Carbon Dots/Prussian Blue Satellite/Core Nanocomposites for optical imaging and photothermal therapy. ACS Appl Mater Interfaces 10:1084–1092. https://doi.org/10.1021/acsami.7b14972
84. Geng B, Qin H, Zheng F et al (2019) Carbon dot-sensitized MoS2 nanosheet heterojunctions as highly efficient NIR photothermal agents for complete tumor ablation at an ultralow laser exposure. Nanoscale 11:7209–7220. https://doi.org/10.1039/c8nr10445j
85. Wang Y, Wang X, Li H et al (2011) Assessing the character of the rhBMP-2- and vancomycin-loaded calcium sulphate composites in vitro and in vivo. Arch Orthop Trauma Surg 131:991–1001. https://doi.org/10.1007/s00402-011-1269-6
86. Uskoković V, Graziani V, Wu VM et al (2019) Gold is for the mistress, silver for the maid: enhanced mechanical properties, osteoinduction and antibacterial activity due to iron doping of tricalcium phosphate bone cements. Mater Sci Eng C 94:798–810. https://doi.org/10.1016/j.msec.2018.10.028
87. Morais DS, Rodrigues MA, Lopes MA et al (2013) Biological evaluation of alginate-based hydrogels, with antimicrobial features by Ce(III) incorporation, as vehicles for a bone substitute. J Mater Sci Mater Med 24:2145–2155. https://doi.org/10.1007/s10856-013-4971-9

Biofilm-inhibiting and Osseointegration-promoting Orthopedic Implants with Novel Nanocoatings

Meng Chen, Hongmin Sun, Hongjiao Ouyang, John E. Jones, Qingsong Yu, Yuanxi Xu, and Shankar Revu

Abstract Orthopedic implants are medical devices surgically placed into the body to replace a missing joint or bone or to reinforce a damaged structure. However, there is up to a 28% loosening rate on cementless implanted knee joint prostheses within a 4–10-year period after implant insertion, and a 2–5% infection rate for orthopedic implants (joint prostheses and fracture fixation devices). In the USA, total hip and knee arthroplasties currently account for over one million interventions each year. Due to the enormous size of the patient population with orthopedic implants, even a currently low risk of infection or failure has not only caused many patients to suffer, but it has also incurred huge costs for the associated health care system. Therefore, there is an urgent need to develop a novel dual-functional nano-coating technology with judiciously engineered physicochemical properties to address simultaneously the two critical issues long facing orthopedic implants: lack of integration with bone tissue and biofilm-caused infections for the enhanced success of implants.

M. Chen (✉) · J. E. Jones
Nanova, Columbia, MO, USA
e-mail: chenmeng@nanovamed.com; Jonesjohn@nanovamed.com

H. Sun
School of Medicine, University of Missouri, Columbia, MO, USA
e-mail: sunh@health.missouri.edu

H. Ouyang · S. Revu
College of Dentistry, Texas A&M University, Dallas, TX, USA
e-mail: ouyang@tamhsc.edu

Q. Yu
College of Engineering, University of Missouri, Columbia, MO, USA
e-mail: yuq@missouri.edu

Y. Xu
School of Medicine, University of Missouri, Columbia, MO, USA

Department of Clinical Development and Registration,
Protech Pharmaservices Corporation (PPC), Shanghai, China
e-mail: Yuanxi.xu@ppccro.com

B. Li et al. (eds.), *Racing for the Surface*,
https://doi.org/10.1007/978-3-030-34471-9_3

We have generated a nanocoating showing a very promising capability of inhibiting biofilm formation by *Staphylococcus aureus* and *Staphylococcus epidermidis*, two of the most common biofilm formers on orthopedic implants, and enhancing bone conductivity simultaneously. The dual-functional nanocoatings coming out of our research demonstrated the following unique features for orthopedic implants: (1) inhibit bacterial colonization and concomitantly promote osteoblast functions; (2) generate long-lasting functionalities for practical clinical applications because these nanocoatings are dense and highly cross-linked without substances of low molecular weight; (3) provide needed abrasion resistance for orthopedic implants and ensure strong coating adhesion to the surface; and (4) improve bone integration and reduce device-related infections in the long run.

Keywords Anti-biofilm · Nanocoating · Osseointegration · Bone conductivity *Staphylococcus aureus* · *Staphylococcus epidermidis* · Low temperature plasma deposition · Dual-function · Orthopedic implants · Coating adhesion · Abrasion resistance · Surface chemistry · Contact angle · Proliferation · Differentiation Infection

Introduction

Existing Problems with Metal-Based Orthopedic Implants

Orthopedic implants, mainly made from stainless steel and titanium alloys for strength, have been increasingly used to provide fixation of bone or to replace articulating surfaces of a joint to restore the function of fractured bone segments, impaired limbs, or affected joints. However, metal, as a foreign material with a very different chemical composition from the bone (a living tissue made of minerals and collagens, more like a polymer-ceramic composites), when implanted in the human body, inevitably has very different responses at the cellular level and tissue level compared to human bone. Some of the responses are detrimental clinically and might cause significant complications and even lead to painful revision. For example, slow or incomplete osseointegration between surrounding bone and orthopedic implants could lead to implant loosening [1], and the osseointegration is affected by the differentiation of osteoblastic cells on the implant surface [2]. As another example, infection of metal implants with a biofilm has also been an unsolved problem for orthopedic implants. Bacteria in biofilms are extremely resistant to antibiotics being protected from antimicrobial agents and from host defense mechanisms [1–5]. It is even worse considering that the infection can happen during the implantation or months or years later and infection may also exist after the implants are surgically removed [6]. *Staphylococcus aureus* (*S. aureus*) and *Staphylococcus epidermidis* (*S. epidermidis*) account for approximately two thirds of infections associated with surgical implants [7]. In the USA, total hip and knee arthroplasties currently account for almost 1.1 million interventions per year [8], and the number is expected to increase to four million annually by 2030 [9]. Around seven million Americans are living with a hip or knee replacement [10]. Approximately 10% of patients within a 10-year period require revision surgery with biofilm-related infection as the second major cause for implant failure, costing approximately $1.9 billion in the USA per year [7].

Therefore, novel strategies are urgently needed to simultaneously address the two critical issues of incomplete implant integration with bone tissue and biofilm-caused infections to further enhance the success of orthopedic implants, and to reduce associated health care costs.

Existing Strategies Modifying Implants to Prevent Biofilm Formation and Promote Osteoconductivity

Surface modification of implants has a distinguishing feature that it only modifies the top surfaces and does not alter the bulk properties. Therefore, a lot of effort has been made to modify the implant surface in order to achieve either an optimal biocompatibility, or anti-bacterial property, or both, while maintaining the excellent mechanical strength of metals. Antibiotics were successfully coated on the surface of implants to prevent biofilm formation [11–13]. However, coating a medical device with a bactericidal compound could potentially increase the risk of selecting for antibiotic-resistant pathogens in humans over time [14, 15]. Heavy metal silver was used as an anti-biofilm agent on the surfaces of biomaterials [16–18]. However, medical devices coated with silver ions or metallic silver have disappointing clinical outcomes, probably due to inactivation of metallic silver when the devices are contacting blood and the coating is wearing off [19]. Other bactericidal agents (such as furanones) have been coated on surfaces to inhibit biofilm formation [20, 21], but encountered with one critical shortcoming, that is, the surfaces could be covered by macromolecules and dead microorganisms, causing loss of their antimicrobial function [22]. The infection-resistant surface of implants was also developed by depositing a thin layer of anti-adhesion coating on implants to prevent attachment of pathogenic bacteria. Those approaches include coating of peptide-functionalized poly(L-lysine)-grafted-poly(ethylene glycol) copolymers [23], grafting of long-chain zwitterionic poly(sulfobetaine methacrylate) [24], superhydrophobic xerogel coating [25], and manufacturing of submicron-textured biomaterial surfaces [26]. However, no animal studies have been reported for those studies, and thus the in vivo anti-biofilm effectiveness remains unclear.

Rapid and complete integration between the bone and implant surface is of great importance for a successful outcome of orthopedic implantation procedures. As a result, surface modification of orthopedic implants that can improve osteoconductivity will be of great benefit to the patients receiving implants. Being the major mineral component of natural bone and structurally similar to the bone, hydroxyapatite (HA) has been used to coat orthopedic implants with the hope of achieving a high osteogenic activity (osteoconduction and osseointegration) [27–30]. However, such coatings have not yet led to successful clinical results and displayed additional failure modes, such as delamination of coatings and undesirable osteoconductivity. Studies have shown that porous titanium orthopedic implants coated with HA could incur more severe infection in a rabbit tibiae implant model [31, 32]. Currently, none of these coatings have received FDA approval for their use on orthopedic metal implants due to unproven clinical safety and efficacy.

To bring more benefits to patients, other surface modification methods have been used to simultaneously inhibit bacterial adhesion and promote osteoblast functions [33]. Titanium surfaces modified with poly(methacrylic acid) brushes and silk sericin have shown enhanced osteoblast adhesion, proliferation, and alkaline phosphatase (ALP) activity while concomitantly reducing the adhesion of *S. aureus* and *S. epidermidis* [34]. Titanium surfaces grafted with RGD-functionalized hydrophilic polymers have also been investigated [23, 35, 36] to take advantage of the tripeptide RGD (Arg-Gly-Asp) motif present in a number of extracellular host proteins (including fibronectin and fibrinogen) that interact specifically with the integrin receptors on host cells, but is not recognized by bacteria [35]. Titanium substrates functionalized with chitosan and subsequent modification by RGD also showed substantial reduction in adherent bacteria and significantly increased osteoblast proliferation and ALP activity [37, 38]. Another approach to achieve the dual purpose of inhibiting bacterial colonization and enhancing osteoblast functions is to immobilize bone morphogenetic protein-2 (BMP-2) on titanium surfaces with either an anti-adhesive polymer or bactericidal polymer as an intermediate layer [39, 40]. Vascular endothelial growth factor (VEGF) conjugated to either bactericidal carboxymethyl chitosan (CMCS) or anti-adhesive hyaluronic acid grafted on titanium was shown to achieve similar results [41]. However, those dual-function approaches are facing the following major challenges: (1) adverse effects of friction or handling during implantation on the surface moieties of implants functionalized with proteins; (2) unfavorable development prospects from positive in vitro studies to a similar clinical outcome; and (3) uncertainties regarding the long-term performance of the implant surface modified with functional polymer coatings and growth factors that will degrade over time.

Nanocoating with Tailored Functional Groups for Biomedical Applications

Our judiciously designed dual-functional and durable nanocoatings for orthopedic implants have demonstrated promising in vitro and in vivo efficacy in inhibiting biofilm formation and concomitantly promoting osseointegration. In addition, the nanocoating is conformal (not changing the surface topography of the implant surface), durable, and tenaciously adhered to implant surfaces, suitable for orthopedic implant applications. The novel dual-functional nanocoatings of 20–30 nm in thickness are deposited on stainless steel and titanium alloy, from which orthopedic implants are mainly made, through plasma deposition using silicon-based monomers, and its mixture with oxygen. Plasma deposition is a thin film forming process in a vacuum reactor, where thin films deposit on the surface of substrates under plasma conditions. In a plasma deposition process, monomers are introduced into a plasma reactor and get activated to produce a gaseous complex composed of highly energetic electrons, ions, free radicals and excited monomer molecules, known as the plasma state. In recent decades, plasma process has been widely used in the preparation of biomedical materials with unique performance and in the manufacturing of medical devices [42]. For instance, a new nitrogen-rich plasma-deposited biomaterial as an external coating for stent-grafts can promote healing around the implant after endovascular aneurysm repair [43]. Through plasma deposition, many appropriate functional groups, such as amine, hydroxyl, and carboxylic acid, useful for the immobilization of bioactive molecules, can be created in the deposited coatings. More importantly, these chemical groups can be put onto almost any material by choosing the right monomers and plasma processes.

Our research group has investigated plasma deposition processes extensively for various applications. One example of our work is the development of biocompatible coatings for improved thrombo-resistance and endothelialization for medical devices and implants [44–47]. It is hoped that by tailoring the chemistry and functional groups of the nanocoating, we can regulate the growth of cells and bacteria on an implant surface, thus achieving dual functions. Successful application of this nanocoating technology will help to achieve the desired outcome of rendering the implant surface more infection-resistant and more osseoconductive without the use of any antibiotics, peptides, or growth factors, as supported by the promising preliminary results of significantly less biofilm formation by *S. aureus* or *S. epidermidis* and enhanced osteoblast cell proliferation/alkaline phosphatase level on the nanocoated surfaces. Therefore, we anticipate that the novel nanocoatings could make significant impact on public health care and the area of orthopedic implants through benefiting millions of patients in the USA.

The novelty of our nanocoating technology stems from the following essential advantages: (1) it will not affect the underlying topography of the implant surface because of its nanoscale (20–30 nm) nature; (2) it is a sterile process, environmentally friendly, and cost-effective, unlike the wet chemistry processes [24–26], (3) it poses no risk for promoting antibiotic resistance because of its non-drug-based nature; and (4) it creates long-lasting functionalities due to the tenacious adhesion of the abrasion-resistant nanocoating to the implant surface through covalent chemical bonding. As demonstrated in our preliminary studies, this nanocoating approach has shown its great promise of translating positive in vitro results into in vivo efficacy and a future similar clinical outcome.

Other plasma-deposited coatings on titanium showed significantly reduced attachment of bacteria [48]. However, there was no animal study to demonstrate in vivo efficacy and no mention of osteoconductivity. The orthopedic implants with stable and durable dual functions of inhibiting biofilm and promoting osseointegration to come out of our research could become a high-impact innovation in medical implant procedures.

Experimental Setup and Methods

Fabrication of Nanocoatings with Tailored Coating Chemistry and Surface Properties

Through process optimization by means of changing plasma power level, working pressure, working gas, gas mass flow, and treatment/deposition time, the amount of surface functionalities, coating thickness and consequently the final surface properties can be adjusted and well controlled. Specifically, nanocoating optimization can be focused on balancing surface $-CH_3$ groups, which has been considered as the most important factor for reducing protein adsorption which in turn resulted in less biofilm formation as compared to a bare metal surface [49], and Si-O groups, which represent hard surfaces that could provide favorable conditions to osteoblast cells for improved proliferation even at a lower level of protein (e.g., fibrinogen) adsorption, by means of using different ratios of trimethylsilane (TMS) to O_2 in the coating process. It has been reported that a dense and rigid layer of nanoscale SiOx on the surface could promote osteogenesis of human mesenchymal stem cells (hMSCs) [50]. The nanocoating optimization strategy can be adjusted based on feedback from anti-biofilm activity and osteoconductivity studies in multiple rounds. Further work will be focused on keeping a good anti-biofilm performance already achieved with TMS nanocoatings while trying to maximize bone regrowth properties by producing denser, more rigid or harder coatings.

A bell-jar plasma reactor, which had been described in our previous work [49], was powered by a direct current (DC) power supply to generate a low temperature gas discharge plasma. TMS or its mixture with oxygen (O_2) was introduced into the plasma reactor for coating deposition. Plasma surface pretreatment using O_2 as a working gas would provide a clean and reproducible starting condition for further plasma coating deposition, by forming a well-controlled surface layer. Specifically, it was used to introduce oxygen-containing groups on the metal substrate surface for covalent chemical binding to the subsequent TMS plasma coating. The substrates included stainless steel (SS) and titanium alloy (Ti). The main operational parameters investigated included: mass flow rate of TMS (1–4 standard cubic centimeters per minute [sccm]), discharge power (5–10 W), working pressure (20–80 mTorr), and deposition time (0.2–2 min). The ratio of TMS to O_2 was varied from 1:0, 1:1, 1:2, 1:4, and 1:8 for desired coating properties by changing the surface chemistry, energy, or hardness.

It is worth mentioning that our nanocoatings are designed to be 20–30 nm thick, which could go up to 60 nm to be achieved with some combinations of process parameters for increased surface hardness, much thinner than other plasma-deposited coatings of 100–285 nm thick siloxane and fluorosiloxane on titanium [48], in which the internal stress usually inherent in thicker coatings could increase the risk of coating cracking or delamination. In our coating process, DC plasma was used, and metallic implants served as part of the cathode to ensue stronger coating adhesion and better pin-hole free coating due to positive ion bombardment to the implant surface removing loosely bound elements, advantageous to radio frequency (RF) plasma coating processes widely used in other similar approaches for coating deposition or surface modification. Furthermore, in this DC plasma process, due to the configuration of a metallic implant as part of the cathode, every single spot of the substrate surface exposed to the plasma environment is deposited with a uniform coating of desired abrasion-resistant strength.

Characterization of Coatings and Surfaces

Surface morphology, energy, and chemical composition are often investigated to better understand how they affect surface bioactivity such as biofilm formation, osteoblast functions, and how they are correlated.

Scanning Electron Microscopy (SEM) was used to analyze the surface morphology of the nanocoated surfaces. **Atomic Force Microscopy (AFM)** was used to characterize surface roughness of coatings on SS and Ti and measure coating thickness on Si wafers. **Contact angle analysis** was used to evaluate the surface energy of nanocoatings and how surface hydrophobicity or wettability could affect bacterial and osteoblast cell attachment. **X-ray Photoelectron Spectroscopy (XPS)** was used to analyze the chemical composition of the coatings and chemical bonding states of the elements contained. The change in elemental composition of carbon, silicone, and oxygen can be correlated with surface biological activities. **Attenuated Total Reflection (ATR) Fourier Transformed Infrared (FTIR) Spectroscopy** was used to characterize functional groups on nanocoatings such as $-CH_3$, Si-O and their changes over plasma coating conditions.

Evaluation of Durability of Bioactivity and Nanocoating Integrity

The durability or stability of the bioactivity created on the substrate surface is very critical to the successful clinical application of orthopedic implants. It has to provide a long-lasting (preferably longer than 2 years) bioactivity on the surface of orthopedic implants to make a successful product. The nanocoating also has to maintain its mechanical integrity since premature delamination or insufficient abrasion resistance will lessen its benefits in clinical applications.

Bioactivity durability test in wet condition was performed in an environment simulating the conditions of medical devices when inserted or implanted in patients where the surfaces will be in contact with human body fluid. SS and Ti substrates with optimized nanocoatings were immersed in wells of 24-well cell plates filled with simulated body fluid (SBF) [51] with 10% fetal bovine serum (FBS) and then placed in an incubator (37 °C, 5% CO_2, humidified) for 1 and 2 months. The medium

(SBF + 10% FBS) was replaced every other day. At the end of each incubation period, the specimens were removed, rinsed, and utilized in the biofilm assay and osteoblast cell culture test to determine the durability of the nanocoatings.

Accelerated adhesion test was conducted using 2″×2″ SS and Ti substrates (wafers) coated with optimized nanocoatings. The coated wafers were immersed into a 60 °C water bath for 1, 3, and 10 days. Following standard ASTM D 3359, wafers at different time-points were taken out and scribed with a cross-hatch to form 100 tiny squares, and then a tape pull test was performed. The tested surfaces were inspected by both visual and optical microscopy. A rating scale of 0–100, which was the number of squares that remained after the tape test, was used. Rating of over 96 after 10 days of immersion was considered as a pass.

Abrasion resistance test: Nanocoated SS and Ti bone screws were implanted into bovine femur obtained from a local slaughter house and then removed. This procedure was repeated twice. Then those screws were rinsed with a PBS solution, dried and examined under optical microscope for any possible cracking or delamination.

Assessment of Nanocoatings with In Vitro and In Vivo Models

In vitro biofilm assay: TMS nanocoatings with desired mechanical strength, durability, and biocompatibility, as determined by the aforementioned studies, were further tested for in vitro anti-biofilm property evaluation. Bacteria were cultured on wafers that were coated with 20% (v/v) human plasma coated [52] in 24-well flat-bottomed sterile microtiter plates overnight. Bacteria (*S. aureus* NRS234 and *S. epidermidis* RP62A) in biofilms were counted by a plate counting method following being dispersed by ultrasonication, as previously reported [49].

Small animal treatment: The mouse bone implant infection model allowed us to study the peri-implant bacterial infections of bone and soft tissues after femur intramedullary pin implantation [53–55]. SS pins were used in a mouse bone implant infection model adopted from Bernthal et al. [56] to characterize the effect of nanocoating on implant biofilm infections in vivo.

Study of Fibrinogen and Fibronectin's Roles in Mediating Anti-biofilm Activity of Nanocoatings

We have observed that TMS nanocoatings reduce host plasma protein adsorption of both fibronectin and fibrinogen onto the implant surface. Since these proteins have been shown to play an important role in biofilm formation via facilitating the bacterial adhesion [33], we hypothesize that the ability of the TMS nanocoating to reduce fibrinogen and fibronectin adsorption onto implant surface is a driving force underlying its inhibitory effect on biofilm formation. These studies not only can elucidate the mechanisms important for understanding how nanocoatings work, but also identify additional functional readouts such as protein adsorption to screen new materials for expanded and improved biological and clinical functions. The protein adsorption of fibrinogen, fibronectin, and albumin on SS coupons was measured by an ELISA approach [57].

Results and Discussion

Using low temperature plasma deposition technology, we fabricated nanoscale (20~30 nm) coatings of TMS or its mixture with oxygen on the surfaces of 316L SS and a Grade 5 Ti alloy of medical grade widely used in making orthopedic implants. This plasma process was performed using a DC plasma source. The silicon-containing monomer, trimethylsilane can be polymerized and deposited rapidly onto the metallic substrate surface with strong adhesion inside the plasma deposition reactor. Plasma-deposited organosilicon coatings exhibit not only a dense film as conventional plasma coatings do, but also provide a certain level of abrasion resistance for the surface of orthopedic implants due to its inorganic -Si-Si- and -Si-C-Si- backbone. The good adhesion is attributed to the formation of the -Si-O-Fe- or -Si-O-Ti- chemical bonds between the plasma-deposited layer and the oxide layer on the substrate surface. Because surface chemistry, topography, and wettability, among many other factors, affect cell attachment and interaction to the surface of biomaterials [58], we have investigated those surface properties in our preliminary studies.

Surface chemistry of coated surfaces: The surface chemistry of the coated SS and Ti substrates was analyzed with XPS. High-resolution scans of C1s were conducted for control SS, SS with TMS plasma coating, control Ti, and Ti with TMS plasma coating [49].

Compared to the uncoated SS, the surface with the TMS coating exhibited more components, with a binding energy of 284.5 eV, indicating a large amount of CH_3 formed on the surface, which could contribute to decreased protein adoption and bacterial attachment [49]. A similar phenomenon was also observed on Ti surfaces coated with a TMS coating, indicative of functional CH_3 groups generated at the surface regardless of the underlying bulk material.

Plasma nanocoatings of TMS mixed with oxygen at various ratios deposited on SS substrates were also analyzed and the elemental composition data are listed in Table 1. Increasing O_2 mass flow in the coating process resulted in decreased C percentage and elevated O while the Si percentage remained relatively stable, indicative of more Si-O formation on the coated surface, which could be one of the causes leading to reduced biofilm formation and improved osteoblast functions as described in the following subsections.

Water contact angle: As an indication of surface wettability affecting cell attachment, contact angle has been measured. The results (Fig. 1) demonstrated

Table 1 Elemental composition of SS as determined by XPS survey scan (atomic %)

Sample ID	C	Si	O	Zn	Ni	Fe	Cr	N	Mg
316L SS	27.71	1.37	47.72	0.31	0.45	15.84	4.39	0.33	1.87
TMS	50.34	25.09	24.57	0	0	0	0	0	0
TMS/O_2 1:1	33.24	24	42.76	0	0	0	0	0	0
TMS/O_2 1:2	23.13	23.49	53.38	0	0	0	0	0	0
TMS/O_2 1:4	12.07	26.61	61.32	0	0	0	0	0	0
TMS/O_2 1:8	6.87	28.09	65.04	0	0	0	0	0	0

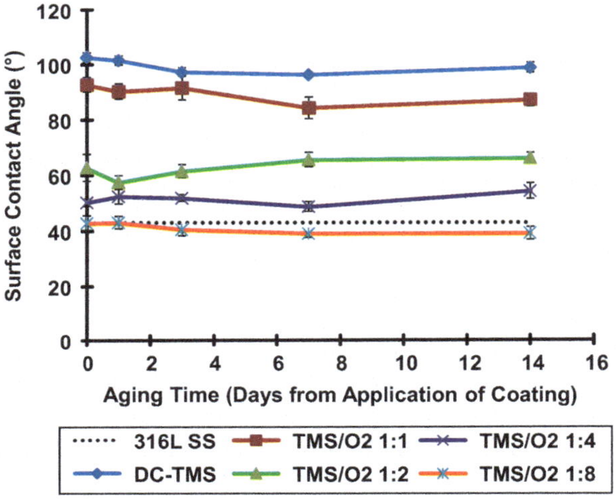

Fig. 1 Surface contact angle stability measurements over 14-day period following coating deposition

that the TMS plasma coating without oxygen rendered the surface of SS more hydrophobic (the contact angle for bare SS was about 40° as shown in the figure), whereas increasing the ratio of O_2 for the coating process tended to turn the TMS plasma coated surfaces more hydrophilic. On the other hand, the data also suggested that all the coated SS surfaces appeared to be very stable during the time period of 2 weeks after coating deposition. Similar contact angle results were also seen on Ti surfaces: 99° ± 6° for TMS coated, 43° ± 5° for TMS + O_2 (1:4), and 66° ± 7° for bare Ti ($n = 5$).

Plasma nanocoatings displayed strong adhesion to stainless steel surfaces: Robust adhesion of functional nanocoatings to implant surfaces is a fundamental feature required for clinical applications since premature delamination will lessen its benefit. Our initial test of a TMS coating to 2 × 2 inch SS wafers via standard ASTM D 3359 indicated that there was no coating coming off of the cross-hatched and surrounding area, indicative of strong adhesion to the underlying surface, which warrants the coating integrity on the surface of orthopedic implants during the clinical implantation procedure.

TMS coating displayed potent inhibition of staphylococcal biofilm formation: We found that there was significantly less biofilm formation on both SS and Ti coated with TMS plasma nanocoating by *S. epidermidis* (biofilm forming RP62A strain) than uncoated controls in vitro [49], with a reduction of 99.6 ± 0.6% on SS and 99.6 ± 0.2% on Ti. Only sporadic cells or cell clusters were observed on TMS coated surfaces while multilayer biofilms were formed on uncoated surfaces [49].

Further modification of the coating process was made to incorporate oxygen into the coating to inhibit *S. aureus* biofilm formation. A variety of mixtures of TMS with oxygen at molar ratio of 1:1, 1:2, 1:4, and 1:8 were used to deposit plasma coatings on SS surfaces, and it was found the ratio of TMS to O_2 at 1:4 could generate the highest inhibition of *S. aureus* (biofilm forming NRS234 strain) [59] biofilm formation on SS surfaces while also demonstrating significant inhibition of *S. epidermidis* biofilm (Fig. 2a). Similar anti-biofilm activity was also observed on silicone substrate [57]. Very recently we have completed an immersion test using SBF to study the long-term stability of stainless steel wafers with nanocoatings (test procedure described in Subsection "Evaluation of Durability of Bioactivity and Nanocoating Integrity"). The anti-biofilm activity of the TMS/O_2 coating was well preserved (~80%) against *S. aureus* (NRS234) after 8 weeks of immersion.

Immediately upon insertion into the host, the surface of implants adsorbs plasma and extracellular proteins, such as fibrinogen and fibronectin [60]. *S. aureus* and *S. epidermidis* display a number of bacterial surface proteins that specifically bind to fibronectin and fibrinogen [61], which may mediate bacterial attachment to biomaterials [61, 62]. Interestingly, as shown in Fig. 2a, the TMS/O_2 1:4 nanocoating significantly decreased fibrinogen deposition on SS surfaces ($p < 0.04$) with a trend of decreased fibronectin binding. This finding suggests that the TMS nanocoating prevents the adhesion of proteins that favor bacterial adhesion to inhibit biofilm formation.

TMS nanocoatings displayed anti-infection efficacy: To further analyze the in vivo anti-infection efficacy of the nanocoating, TMS/O_2 1:4 coated and uncoated SS pins were dipped into 10^8 CFU of methicillin-resistant *S. aureus* (MRSA) strain NRS 384 and implanted into the femoral intramedullary canal of 12-week-old male C57BL/6 mice following a bone implant model by Bernthal et al. [56]. The mice receiving TMS/O_2 1:4 SS pins demonstrated markedly improved symptoms than mice with uncoated pins at 4 weeks with less swollen knee joints and less inflammation around the implantation sites by visual examination (Fig. 2b). The results suggested that the TMS nanocoating could reduce infection and inflammation caused by *S. aureus* infection of bone implants.

TMS nanocoatings displayed osteoconductivity for bone regeneration: Optimal bone regeneration around implants is critical for long-term success of implants. Intriguingly, we observed that when the murine osteoblast MC3T3-E1 cells were cultured on the TMS/O_2 1:4 coated SS for 7 or 14 days in vitro, they displayed significantly increased cell numbers (Fig. 3a) and alkaline phosphatase (ALP) protein levels (Fig. 3b), compared with those growing on uncoated surface. These results demonstrated that TMS/O_2 1:4 stimulated proliferation and differentiation of murine osteoblasts in vitro. Importantly, the TMS nanocoating also stimulated the gene expression of osteoblast marker bone sialoprotein (BSP) (Fig. 3c), a cell surface adhesion molecule, and induced ALP protein expression (Fig. 3d) in

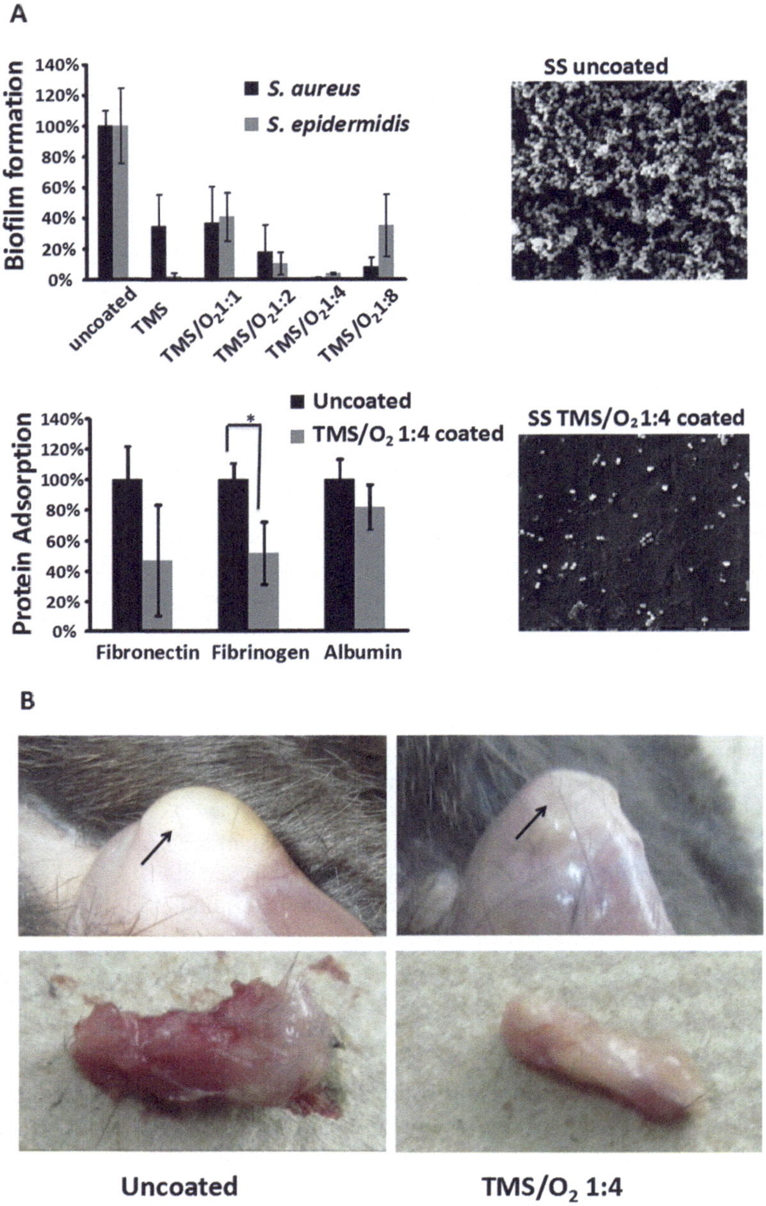

Fig. 2 Nanocoatings inhibited biofilm-related infections. (**A**) TMS/O_2 1:4 nanocoating inhibited *staphylococcal* biofilm formation and protein adsorption ($n = 3$). Student's t tests were performed. A p-value of <0.05 was considered to be statistically significant. Data was pooled from three samples and presented as mean ± standard deviation. * $p < 0.05$. (**B**) Mice implanted with MRSA-infected SS pins coated with TMS/O_2 1:4 nanocoating exhibited reduced infection and inflammation symptoms

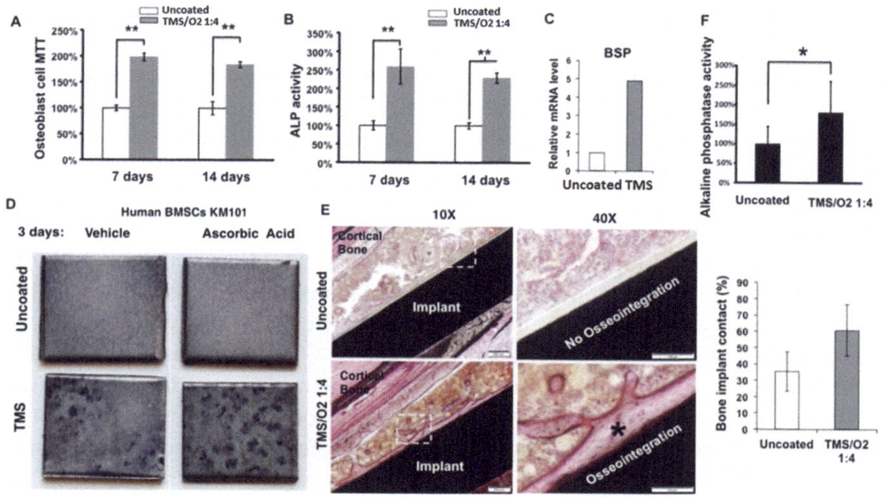

Fig. 3 TMS nanocoatings promoted osteogenesis. (**A, B**) TMS/O_2 1:4 coated SS surfaces promoted (**A**) murine osteoblast proliferation and (**B**) alkaline phosphatase levels, compared with uncoated surfaces ($n = 9$; $^{**}P < 0.01$). (**C, D**) TMS coating stimulated osteoblastic differentiation of human BMSCs, as reflected by increased human BSP mRNA levels, as shown by (**C**) qRT-PCR and (**D**) enhanced ALP staining, compared with uncoated surfaces. (**E**) H&E stain and quantification of bone-implant contact on the TMS/O_2 1:4 coated pins, compared with the uncoated pins, in vivo ($n = 3$) with ∗ indicating newly formed bone tissues. (**F**) Mice implanted with the TMS/O_2 1:4 coated pins demonstrated significantly increased serum levels of ALP activity, compared with the control mice with uncoated pins in the presence of MRSA ($n = 9$ in uncoated group, $n = 10$ in TMS/O_2 1:4 group)

human bone marrow stromal cells (hBMSCs) KM101. Taken together, these observations demonstrate that the TMS nanocoatings have a potent osteoconductive capacity for both murine and human osteoblastic cells in vitro.

Moreover, the TMS/O_2 1:4 coated SS pins, when implanted into the mouse femoral intramedullary canals, displayed markedly increased bone-implant contact (%) at 8 weeks post surgery, when compared with uncoated SS pins (Fig. 3e). Hematoxylin and Eosin (H&E) staining demonstrated that the TMS/O_2 1:4 coated SS pin surfaces were covered by well-organized bone tissues; in contrast, the infiltration of adipocytes, inflammatory cells, and fibrous tissues were often found associated with the uncoated SS (Fig. 3e). Furthermore, intriguingly, even in the presence of bacterial infection, the mice receiving TMS/O_2 1:4 pins displayed significantly increased serum levels of ALP, indicating increased bone formation in these mice, compared with those receiving uncoated pins (Fig. 3f). The former mice also displayed significantly reduced levels of bacterial infection and local tissue inflammation (Fig. 2b). Taken together, the results demonstrate the potent stimulatory effects of the TMS nanocoatings on osteogenesis both in vitro and in vivo, with or without bacterial infections.

Future research in this area will cover fundamental understanding of interaction of host and bacteria on foreign body implant, or study of mechanisms. Integration of an implant into bone, determining the long-term performance of the device, takes place largely at the interface of tissue and implant. Surface chemistry and surface topography of the implant, among other various factors, could impact the development of this interface [33]. Infections associated with orthopedic implants are manifested by bacterial colonization and biofilm formation on the implanted device and infection of the adjacent tissues. It is thus imperative to understand how implant surface chemistry and topography modulate host protein adsorption, bone cell and bacterial cell signaling. The novel dual-function nanocoating discussed here in this entry could also serve as a research tool to explore the cellular and molecular mechanisms underlying the interaction of host and bacteria on foreign body implant. By identifying critical host and bacterial genes and proteins that contribute to the dual anti-biofilm and osteoconductive properties of the nanocoatings, we would not only shed light on bone development and bacterial pathogenesis, but also identify novel therapeutic molecular targets for treating infectious and bone diseases.

Conclusions

Biofilm-inhibiting and osseointegration-promoting orthopedic implants are in urgent need to battle implant-related infections and lack of integration of implant with bone tissue. We have identified novel TMS nanocoatings with acceptable mechanical durability and unique dual properties of anti-biofilm formation and osteoconductivity, which would warrant further development of TMS coating technology for better mechanical duration, biological compatibility as well as more potent anti-biofilm and osteoconductivity. Our preliminary studies have demonstrated that $-CH_3$ and Si-O groups could be major surface factors that regulate the anti-biofilm function of the TMS nanocoatings. Further, we have found out that the molecular mechanisms underlying the dual functions of the TMS nanocoatings can be distinct from but intricately linked with and mutually beneficial to each other, thus making the optimization of the dual functions possible. Successful application of this nanocoating technology may not only lead to improved clinical efficacy, increased quality of life, and decreased health care costs to the patients, but also generate rich fundamental knowledge of the complex and highly inter-related events occurring at the implant surface after implantation.

Acknowledgments Some of the research results presented in this entry were generated from the project funded by the National Heart, Lung, and Blood Institute (NHLBI) of the NIH, grant R44HL097485 and NIH grant P01HL573461. The authors are grateful for the contributions of all colleagues and collaborators in this research area of nanocoating technology for orthopedic implant application. We are also thankful for the thoughtful and constructive comments and suggestions of the reviewers, which have improved the presentation.

Conflict of interest: Dr. Hongmin Sun owns stocks in Nanova, Inc. This does not detract from an author's objectivity in presentation of study results.

References

1. Davies D (2003) Understanding biofilm resistance to antibacterial agents. Nat Rev Drug Discov 2:114–122
2. Donlan RM (2001) Biofilms and device-associated infections. Emerg Infect Dis 7:277–281
3. Campoccia D, Montanaro L, Arciola CR (2006) The significance of infection related to orthopedic devices and issues of antibiotic resistance. Biomaterials 27(11):2331–2339
4. Zaborowska M, Tillander J, Brånemark R, Hagberg L, Thomsen P, Trobos M (2017) Biofilm formation and antimicrobial susceptibility of staphylococci and enterococci from osteomyelitis associated with percutaneous orthopaedic implants. J Biomed Mater Res B Appl Biomater 105(8):2630–2640
5. Li B, Webster TJ (2018) Bacteria antibiotic resistance: new challenges and opportunities for implant-associated orthopedic infections. J Orthop Res 36(1):22–32
6. Donlan RM, Costerton JW (2002) Biofilms: survival mechanisms of clinically relevant microorganisms. Clin Microbiol Rev 15:167–193
7. Darouiche RO (2004) Treatment of infections associated with surgical implants. N Engl J Med 350:1422–1429
8. Steiner C, Andrews R, Barrett M, Weiss A (2012) HCUP Projections: Mobility/Orthopedic Procedures 2003 to 2012. HCUP Projections Report # 2012-03. 2012 Sep 20. U.S. Agency for Healthcare Research and Quality. http://hcup-us.ahrq.gov/reports/projections/2012-03.pdf. Accessed 12 Apr 2019
9. Kurtz S, Ong K, Lau E, Mowat F, Halpern M (2007) Projections of primary and revision hip and knee arthroplasty in the United States from 2005 to 2030. J Bone Joint Surg Am 89:780–785
10. Kremers HM, Larson DR, Crowson CS, Kremers WK, Washington RE, Steiner CA, Jiranek WA, Berry DJ (2015) Prevalence of total hip and knee replacement in the United States. J Bone Joint Surg Am 97(17):1386–1397
11. Antoci V Jr et al (2008) The inhibition of Staphylococcus epidermidis biofilm formation by vancomycin-modified titanium alloy and implications for the treatment of periprosthetic infection. Biomaterials 29:4684–4690
12. Lucke M et al (2003) Gentamicin coating of metallic implants reduces implant-related osteomyelitis in rats. Bone 32:521–531
13. Popat KC, Eltgroth M, LaTempa TJ, Grimes CA, Desai TA (2007) Decreased Staphylococcus epidermis adhesion and increased osteoblast functionality on antibiotic-loaded titania nanotubes. Biomaterials 28:4880–4888
14. Sampath LA, Tambe SM, Modak SM (2001) In vitro and in vivo efficacy of catheters impregnated with antiseptics or antibiotics: evaluation of the risk of bacterial resistance to the antimicrobials in the catheters. Infect Control Hosp Epidemiol 22:640–646
15. Tambe SM, Sampath L, Modak SM (2001) In vitro evaluation of the risk of developing bacterial resistance to antiseptics and antibiotics used in medical devices. J Antimicrob Chemother 47:589–598
16. Jiang H, Manolache S, Wong ACL, Denes FS (2004) Plasma-enhanced deposition of silver nanoparticles onto polymer and metal surfaces for the generation of antimicrobial characteristics. J Appl Polym Sci 93:1411–1422
17. Stobie N et al (2008) Prevention of Staphylococcus epidermidis biofilm formation using a low-temperature processed silver-doped phenyltriethoxysilane sol-gel coating. Biomaterials 29:963–969
18. Zeng X, Xiong S, Zhuo S, Liu C, Miao J, Liu D, Wang H, Zhang Y, Zheng Z, Ting K, Wang C, Liu Y (2019) Nanosilver/poly (dl-lactic-co-glycolic acid) on titanium implant surfaces for the enhancement of antibacterial properties and osteoinductivity. Int J Nanomedicine 14:1849–1863
19. Rai M, Yadav A, Gade A (2009) Silver nanoparticles as a new generation of antimicrobials. Biotechnol Adv 27:76–83

20. Baveja JK et al (2004) Furanones as potential anti-bacterial coatings on biomaterials. Biomaterials 25:5003–5012
21. Hume EB et al (2004) The control of Staphylococcus epidermidis biofilm formation and in vivo infection rates by covalently bound furanones. Biomaterials 25:5023–5030
22. Klibanov AM (2007) Permanently microbicidal materials coatings. J Mater Chem 17:2479–2482
23. Harris LG, Tosatti S, Wieland M, Textor M, Richards RG (2004) Staphylococcus aureus adhesion to titanium oxide surfaces coated with non-functionalized and peptide-functionalized poly(L-lysine)-grafted-poly(ethylene glycol) copolymers. Biomaterials 25:4135–4148
24. Cheng G, Zhang Z, Chen S, Bryers JD, Jiang S (2007) Inhibition of bacterial adhesion and biofilm formation on zwitterionic surfaces. Biomaterials 28:4192–4199
25. Privett BJ et al (2011) Antibacterial fluorinated silica colloid superhydrophobic surfaces. Langmuir 27:9597–9601
26. Xu LC, Siedlecki CA (2012) Submicron-textured biomaterial surface reduces staphylococcal bacterial adhesion and biofilm formation. Acta Biomater 8:72–81
27. Chien CY, Liu TY, Kuo WH, Wang MJ, Tsai WB (2013) Dopamine-assisted immobilization of hydroxyapatite nanoparticles and RGD peptides to improve the osteoconductivity of titanium. J Biomed Mater Res A 101:740–747
28. Wang G, Zreiqat H (2010) Functional coatings or films for hard-tissue applications. Materials 3:3994–4050
29. Wang Y, Liu X, Fan T, Tan Z, Zhou Z, He D (2017) In vitro evaluation of hydroxyapatite coatings with (002) crystallographic texture deposited by micro-plasma spraying. Mater Sci Eng C Mater Biol Appl 75:596–601
30. Łukaszewska-Kuska M, Krawczyk P, Martyla A, Hędzelek W, Dorocka-Bobkowska B (2018) Hydroxyapatite coating on titanium endosseous implants for improved osseointegration: physical and chemical considerations. Adv Clin Exp Med 27(8):1055–1059
31. Oosterbos CJ et al (2002) Osseointegration of hydroxyapatite-coated and noncoated Ti6Al4V implants in the presence of local infection: a comparative histomorphometrical study in rabbits. J Biomed Mater Res 60:339–347
32. Vogely HC et al (2000) Effects of hydrosyapatite coating on Ti-6A1-4V implant-site infection in a rabbit tibial model. J Orthop Res 18:485–493
33. Neoh KG, Hu X, Zheng D, Kang ET (2012) Balancing osteoblast functions and bacterial adhesion on functionalized titanium surfaces. Biomaterials 33:2813–2822
34. Zhang F, Zhang Z, Zhu X, Kang ET, Neoh KG (2008) Silk-functionalized titanium surfaces for enhancing osteoblast functions and reducing bacterial adhesion. Biomaterials 29:4751–4759
35. Maddikeri RR, Tosatti S, Schuler M, Chessari S, Textor M, Richards RG et al (2008) Reduced medical infection related bacterial strains adhesion on bioactive RGD modified titanium surfaces: a first step toward cell selective surfaces. J Biomed Mater Res A 84:425–435
36. Subbiahdoss G et al (2010) Bacterial biofilm formation versus mammalian cell growth on titanium-based mono- and bi-functional coating. Eur Cell Mater 19:205–213
37. Chua PH, Neoh KG, Kang ET, Wang W (2008) Surface functionalization of titanium with hyaluronic acid/chitosan polyelectrolyte multilayers and RGD for promoting osteoblast functions and inhibiting bacterial adhesion. Biomaterials 29:1412–1421
38. Shi Z, Neoh KG, Kang ET, Poh C, Wang W (2008) Bacterial adhesion and osteoblast function on titanium with surface-grafted chitosan and immobilized RGD peptide. J Biomed Mater Res A 86:865–872
39. Shi Z, Neoh KG, Kang ET, Poh C, Wang W (2009) Titanium with surface-grafted dextran and immobilized bone morphogenetic protein-2 for inhibition of bacterial adhesion and enhancement of osteoblast functions. Tissue Eng Part A 15:417–426
40. Shi Z, Neoh KG, Kang ET, Poh CK, Wang W (2009) Surface functionalization of titanium with carboxymethyl chitosan and immobilized bone morphogenetic protein-2 for enhanced osseointegration. Biomacromolecules 10:1603–1611
41. Hu X et al (2010) An in vitro assessment of titanium functionalized with polysaccharides conjugated with vascular endothelial growth factor for enhanced osseointegration and inhibition of bacterial adhesion. Biomaterials 31:8854–8863

42. Ratner BD (1997) Plasma processing of polymers. In: d'Agostino R, Favia P, Fracassi F (eds) . Kluwer Academic Publishers, Dordrecht, The Netherlands
43. Lerouge S, Major A, Girault-Lauriault PL, Raymond MA, Laplante P, Soulez G et al (2007) Nitrogen-rich coatings for promoting healing around stent-grafts after endovascular aneurysm repair. Biomaterials 28:1209–1217
44. Chen M, Osaki S, Zamora PO, Potekhin M (2003) Effect of nitrogen and oxygen incorporated into TMSAA plasma coating on surface-bound heparin activity. J Appl Polym Sci 89:1875–1883
45. Shen Y et al (2009) Investigation of surface endothelialization on biomedical nitinol (NiTi) alloy: effects of surface micropatterning combined with plasma nanocoatings. Acta Biomater 5:3593–3604
46. Tang CJ et al (2010) A study on surface endothelialization of plasma coated intravascular stents. Surf Coat Technol 204:1487–1492
47. Jones JE, Yu Q, M Chen M (2017) A chemical stability study of trimethylsilane plasma nanocoatings for coronary stents. J Biomater Sci Polym Ed 28(1):15–32
48. Stallard CP, McDonnell KA, Onayemi OD, O'Gara JP, Dowling DP (2012) Evaluation of protein adsorption on atmospheric plasma deposited coatings exhibiting superhydrophilic to superhydrophobic properties. Biointerphases 7:31
49. Ma Y et al (2012) Inhibition of Staphylococcus epidermidis biofilm by trimethylsilane plasma coating. Antimicrob Agents Chemother 56:5923–5937
50. Yang Y, Kulangara K, Lam RT, Dharmawan R, Leong KW (2012) Effects of topographical and mechanical property alterations induced by oxygen plasma modification on stem cell behavior. ACS Nano 6:8591–8598
51. Lee JT et al (2011) Cell culture medium as an alternative to conventional simulated body fluid. Acta Biomater 7:2615–2622
52. Cassat JE, Lee CY, Smeltzer MS (2007) Investigation of biofilm formation in clinical isolates of Staphylococcus aureus. Methods Mol Biol 391:127–144
53. Niska JA et al (2012) Monitoring bacterial burden, inflammation and bone damage longitudinally using optical and muCT imaging in an orthopaedic implant infection in mice. PLoS One 7:e47397
54. Niska JA et al (2013) Vancomycin-rifampin combination therapy has enhanced efficacy against an experimental Staphylococcus aureus prosthetic joint infection. Antimicrob Agents Chemother 57:5080–5086
55. Niska JA et al (2012) Daptomycin and tigecycline have broader effective dose ranges than vancomycin as prophylaxis against a Staphylococcus aureus surgical implant infection in mice. Antimicrob Agents Chemother 56:2590–2597
56. Bernthal NM et al (2010) A mouse model of post-arthroplasty Staphylococcus aureus joint infection to evaluate in vivo the efficacy of antimicrobial implant coatings. PLoS One 5:e12580
57. Xu Y et al (2015) Nanoscale plasma coating inhibits formation of Staphylococcus aureus biofilm. Antimicrob Agents Chemother 59(12):7308–7315
58. Wu S et al (2011) Plasma-modified biomaterials for self-antimicrobial applications. ACS Appl Mater Interfaces 3:2851–2860
59. Ma Y et al (2012) Novel inhibitors of Staphylococcus aureus virulence gene expression and biofilm formation. PLoS One 7:e47255
60. Thevenot P et al (2008) Surface chemistry influence implant biocompatibility.Curr Top Med Chem 8(4): 270–280
61. Cheung AL, Fischetti VA (1990) The role of fibrinogen in staphylococcal adherence to catheters in vitro. J Infect Dis 161:1177–1186
62. Pei L, Flock JI (2001) Lack of fbe, the gene for a fibrinogen-binding protein from Staphylococcus epidermidis, reduces its adherence to fibrinogen coated surfaces. Microb Pathog 31:185–193

Three-Dimensional (3D) and Drug-Eluting Nanofiber Coating for Prosthetic Implants

Liang Chen and Weiping Ren

Abstract Failure of osseointegration and implant infection are the two main causes of implant failure and loosening. There is an urgent need for orthopedic implants that promote rapid osseointegration and prevent infection, particularly when placed in bone compromised by disease or physiology of the patients. This chapter reviews current and potential future use of biologic and drug-eluting coatings for orthopedic implants to facilitate osseointegration and prevent implant infection. The potential application of porous and drug-eluting coaxial nanofiber as a means of alternative implant surface coating was discussed.

Keywords Osseointegration · Infection · Periprosthetic implant · Implant coating Three-dimensional (3D) nanofibers · Automatic electrospun nanofibers collector Cellular activity · Electrospinning · Corona discharge · Porous structure · Coaxial Drug delivery

Abbreviations

AL	Aseptic loosening
AMP	Antimicrobial peptides
CS	Chondroitin sulfate
DAC	Defensive antibacterial coating
Doxy	Doxycycline
ECM	Extracellular matrices
HA	Hydroxyapatite
LBL	Layer-by-layer
MSCs	Mesenchymal stem cells

L. Chen
Department of Biomedical Engineering, Wayne State University/IBIO, Detroit, MI, USA
e-mail: ej7518@wayne.edu

W. Ren (✉)
Department of Biomedical Engineering, Wayne State University/IBIO, Detroit, MI, USA

John D. Dingell VA Medical Center, Detroit, MI, USA
e-mail: as7606@wayne.edu

B. Li et al. (eds.), *Racing for the Surface*,
https://doi.org/10.1007/978-3-030-34471-9_4

NFs	Nanofibers
PAA	Poly(acrylic acid)
PCL	Polycaprolactone
PEO	Poly(ethylene oxide)
PJI	Prosthetic joint infection
PLGA	Poly(lactic-*co*-glycolic acid)
PVP	Polyvinyl pyrrolidone
rhBMP-2	Recombinant human bone morphogenetic protein-2
rhBMP-4	Recombinant human bone morphogenetic protein-4
THA	Total hip arthroplasties
Ti	Titanium
TiColl	Type-I collagen-coated titanium
TJA	Total joint arthroplasty
TKA	Total knee arthroplasties
$VEGF_{165}$	Recombinant human vascular endothelial growth factor
β-TCP	β-tricalcium

Background

Introduction

Total joint arthroplasty (TJA) is a life-improving intervention for millions of people all over the world. It is for total replacement of hip, knee, ankle, elbow, and shoulder by orthopedic implant/prosthesis. In the USA alone, there were approximately 332,000 total hip arthroplasties (THA), and over 700,000 total knee arthroplasties (TKA) performed in 2010 [1, 2]. The number of TJA continues to increase significantly along with an aging population. It may reach 572,000 and 3,480,000 for hip and knee arthroplasties, respectively, by 2030 [2, 3]. Besides, the procedure for other joints, including ankles, elbow, and shoulders, are also available, and also increasing performed.

TJA is one of the most successful clinical procedures. It helps patients with functional restoration, pain and stiffness relief, thereby improving patients' quality of life. Several surveys reported that the patients who received TKA expressed 90–95% of satisfaction rate [4–9]. The survival rate of the implants within 10–15 years was greater than 90% [4–9], which indicates the reliability and durability of the implants.

A biocompatible and bioactive orthopedic implant is one of the keys that determine the success of the TJA. An implant should be sufficiently inert to avoid triggering systemic immune/inflammatory reactions; and in the meanwhile it should stimulate the integration of the implant to the surrounding tissues. The most widely used implant materials are titanium (Ti) and its alloys. It has notable biocompatibility and lower stress shielding comparing to other metallic materials. Therefore, the Ti alloy has become the gold standard in cementless implants [10]. Focusing on the orthopedic implants of TJA, this chapter will introduce the potential complications of TJA and current strategies on how to address issues by utilizing implant coatings (design, materials, and drugs).

Complications of Total Joint Arthroplasty: Osseointegration Insufficiency and Infection

The TJA is a well-established orthopedic procedure in both surgical technique and implant design. Nevertheless, a small portion of patients has poor outcomes, like implant loosening or infection. As a result, they may need a revision surgery. The revision rate, within the first 2 years after TKA, is approximately 3% [4, 11]. This percentage, although seems insignificant, would lead to considerable medical costs considering the total number of TKA procedures each year.

A successful TJA requires the orthopedic implant to be appropriately stabilized (primary stability) through mechanical press-fit during the surgery [10]. It is followed by native bone ingrowth to bridge the gaps between the implant surface and periprosthetic tissues. Subsequently, osteoblast-like cells deposit on the interface of the implant and surrounding bone and start active proliferation (first 10–12 days after implantation) [10]. Through a series of spatiotemporal cellular activities, the extracellular matrices (ECM) are finally mineralized to a mature ECM (28 days after implantation), which is composed of ~65% minerals, like hydroxyapatite (HA) and ~35% of organic components, mainly type-I collagen [12]. Thus, the implant could be further stabilized to the host bone, which is called secondary stability [10, 13]. The bone continues remodeling until it completely integrates with the implant to restore the function. It is a dynamic and long-term healing process. If any one of these steps is delayed or disrupted, it may lead to TJA failure. For instance, if the implant was not fixed perfectly at beginning, it would initiate micromotion and enlarge the gap between the implants and bones. Later, more and more wear debris may be accumulated in the gap, leading to macrophage-induced osteolysis [14–16].

The reasons of implant failure include aseptic loosening (AL) and septic loosening, also referred to as periprosthetic joint infection (PJI) [2]. AL is the leading cause of TJA failure and its incidence continues to increase [17]. Rapid and sufficient osseointegration can enhance implant stability and increase the implant life. In 1950s, Dr. Brånemark et al. first presented the concept of "osseointegration." It was based on the observation of the formation of a direct interface between a Ti implant surface and periprosthetic bone [10, 18]. The clinical definition of the osseointegration is that alloplastic materials are rigidly fixed to bone and maintained during load bearing [10]. The lack of sufficient osseointegration causes implants micromotion, instability, osteolysis, and loosening [14, 16]. Ryd et al. reported that early implant loosening in both hips and knees might result in implant failure [19]. Kärrholm et al. also concluded that the subsidence of the implant could increase the risk of AL

[20]. The physiochemical and bioactive characters of the implant surface have a big impact on the consequence of the osseointegration.

PJI is another leading cause of TJA failure. The annual incidence of PJI after TJA is more than 2% among the Medicare population [21]. This is a large number considering ~581,000 total knees and more than 193,000 total hips are performed each year in the USA alone. The direct cost/hospital cost to treat PJI in the American health care system was $566 million in 2009 [2, 22]. In general, the first 2 years has the highest risk of PJI in that roughly 60–70% PJI happens in this period [2, 21, 23, 24]. The incidence of implant infection is even higher after revision surgery than after primary surgery. The average cost of treatment PJI in revision surgery is 3–6 times higher than the primary implantation.

The PJI is mainly initiated through introducing bacteria like *Staphylococcus aureus*, a leading pathogen (60%) of PJI, during the surgery [25, 26]. The bacteria contaminate the orthopedic implant surface or periprosthetic tissue, and quickly colonize on the implant surface. Later, the infection spreads and progresses to adjacent tissues during early onset infections (within first few months after implantation). Hematogenous spreading is another path for PJI. The overall rate of hematogenous spreading is low however, the patient may remain at the risk of hematogenous infection throughout the life of the implant and is one of the causes of late-onset PJI (over 1 year after implantation) [27]. Biofilm is a complex community of one type or multiple types of microorganisms that forms on a surface of the implant. It may start to take shape at any time, including during the late-onset PJIs [23]. The formation of the biofilm makes the treatment of infection more difficult and complicated. The biofilm protects the cells from the treatment of antibiotics and the action of the immune system [23, 28] because of their low growth rate, antibiotic resistance property, and the protective extracellular matrix [29, 30].

The PJI treatment that aims to control the infection and restore the function of joints can be reached by many different medical and surgical strategies. It includes antibacterial treatment without surgery, debridement with implant retention, and resection of the implant without reimplantation or with reimplantation through one-stage arthroplasty or two-stage arthroplasty exchange, and amputation [21]. Prophylactic systematic administration of antibiotics is a routine treatment to prevent infection. However, long-term use of antibiotics may lead to drug resistance. In addition, the systematic antibiotic treatment would be less effective when the biofilm was formed on the implant surface. To prevent implant infection, various strategies have been attempted, either by implant surface fabrication or incorporation of antibiotics into the implant devices [31].

Recent Implant Coating Developments: Advantages and Disadvantages

The aim of the TJA is to restore or improve the pre-morbid function. Over the past 25 years, the orthopedic implant concept or design has progressed from the restoration of the mechanical functions of bone tissue to regenerative medicine. Researchers have no longer been satisfied with the Ti or its alloy implants for the lack of the bioactive property. An increasing number of scientists have been focusing on adding biological properties of the implants to enhance the bone healing. Implant surface coating is one of the strategies to modify the physiochemical properties of the implant surface, and to reach locally pharmacological treatments [10, 32]. Recent developments in implant surface coating technologies for the osseointegration enhancement and infection inhibition are summarized and discussed below.

Hydroxyapatite (HA) Coating

HA coating fabricated by plasma spray is a common coating in the clinic. It has been used clinically since 1987 [14]. The HA coating has similar component to the bone, which provides calcium and phosphate for new bone formation. Clinical studies indicated that HA coating not only bridges the interface of implant and bone, but also enhances the osseointegration of cementless metallic implants within bone [33, 34]. In a canine study, the formation of new bone was discovered at distance of 400 μm from the HA coated implant, which was inserted in the femoral condyles of mature dogs [35]. This finding proved the osteoconductive capability of HA coating [35]. However, there were also controversial reports stating that no differences were found between HA coated and non-coated implants for the long-term clinical outcomes [36, 37]. The HA coating may impair initial osseointegration because it lacks a physiological surface [38]; It's brittle in nature [39] and the poor adhesion strength [40] additionally effect the clinical outcomes.

Recently, the traditional HA coating has been used as a drug delivery device for the local delivery of growth factors, peptides, antibacterial drugs, and DNA [41–46]. He J. et al. [41] developed a porous HA coating infiltrated with collagen, RGD peptide, and recombinant human bone morphogenetic protein-2 (rhBMP-2) for Ti alloy implant. The collagen/rhBMP-2-modified HA coating increased the attachment, proliferation, and differentiation of mesenchymal stem cells (MSCs) in vitro. It also significantly accelerated bone growth rate after implantation into dog femora. Thus, the modification of the HA coating by embedding osteogenic factors becomes an effective method to enhance osseointegration at bone–implant surface [41].

With the aim to inhibit implant-associated infection, the HA coating was used to load with antibacterial drugs, such as silver [46, 47], antibiotics [48], and antimicrobial peptides (AMP) [49]. Silver has a broad antibacterial spectrum. The bactericidal effect of silver coating is through the interaction of silver with the membranes, proteins, and DNA of bacteria [50]. Moreover, silver can interrupt the formed biofilm [14]. Thus, silver is an effective bactericide, which has been applied to the HA coating. For example, Chen W. et al. used co-sputtering technology to create a silver-HA coating on Ti implant surface [47]. The silver-HA coating significantly reduced the attachment of *S. epidermidis* and *S. aureus* when compared to uncoated surfaces. Moreover, the silver-HA coating did not induce in vitro cytotoxicity. In another study, although the silver-HA coating fabricated by plasma spray inhibited bacterial colonization, it showed cytotoxic effect [46]. The silver-HA coating reduced the viability and osteogenic differentiation of human fetal osteoblast cells. Fortunately, adding strontium to the silver-HA coating offsets the negative effects, and even improved the performance when compared to pure HA coating [46]. Thus, the silver embedded HA coating is a multifunctional surface, which enhances osseointegration and inhibits infection. Like silver, other alternative inorganic antibacterial elements, including copper, zinc, nitrogen, and gold, can be applied to HA coating for infection inhibition in the future [51, 52].

Systemic administration of antibiotics is common clinical practice to prevent infection for TJA patients. However, the effectiveness may be reduced because of relatively low dose in the implant site, and the risk of antibiotic resistance occurred after long time use. Therefore, local delivery of antibiotics is expected to directly eliminate the bacteria on the implant surface and hence even more effective when combined with the systemic antibiotic treatment. HA coatings have embedded various types of antibiotics, such as gentamicin, tobramycin, and vancomycin [53–55]. The antibiotic-embedded HA coatings have shown effective antibacterial properties [53–55]. A biodegradable, poly (lactic-*co*-glycolic acid) (PLGA), gentamicin-embedded HA coating on cementless hip implant was developed by Neut D. et al. for the prevention of PJI [53]. The PLGA-gentamicin-HA coated pin reduced staphylococcal infection rate in a bacterially contaminated medullary canal of rabbit; and didn't impair the bone ingrowth rate through a condylar defects of Beagle dog model [53].

Although HA coatings embedded with osteogenic or antibacterial agents can enhance osseointegration and prevent infection, there are some unsolved issues regarding the efficacies of local drug delivery. The first issue is the methodology of embedding agents into HA coating. The plasma spray is a high temperature procedure and may lead to the inactivation of embedded drugs, such as growth factors and antibiotics during the procedure. Therefore, the types of drugs that can be incorporated within the HA coating is limited. The second issue is uncontrollability of drug release [56]. In HA coating, physical absorption is the mechanism of agent embedding. The weak bonding force results in a burst release of embedded agents. A study showed that most antibiotics were released from HA coating within 1-h incubation [57]. Besides the HA coating, many coating technologies have been developed to extend and control the release of embedded drugs to enhance osseointegration and inhibit infection. They can be classified into three categories: hydrogel coating, layer-by-layer (LBL) coatings, and immobilization [58], which are introduced in the following sections.

Hydrogel Coating

Hydrogel networks are generally obtained by chemical or physical cross-linking, ultraviolet (UV) irradiation, and electrochemical polymerization. Hydrogel coatings are usually achieved by simply immersing implants into a hydrogel solution and drying out afterward. Hydrogel coatings have been easily applied to many types of implants for stabilizing the implant through bridge of the bone–implant interface

[14]. In addition, a broad range of drugs can be easily added into hydrogel solution before coating. Many studies demonstrated that Ti implants coated with type-I collagen promoted osseointegration [59–62]. Sartori M. et al. developed a type-I collagen-coated titanium (TiColl) screw [60]. The TiColl screws increased bone–implant contact and bone ingrowth in the femoral condyles of healthy and osteopenic rats. The results proved that the TiColl coating enhanced osseointegration even in the physiologically compromised animals. Stadlinger B. et al. combined chondroitin sulfate (CS) and recombinant human bone morphogenetic protein-4 (rhBMP-4) to the type-I collagen coating on Ti implant [61]. The in vivo results showed that the highest bone–implant contact was formed on CS-collagen-coated implant, followed by collagen-coated implant and CS-rhBMP-4-collagen-coated implant [61].

Chitosan is a derivative of chitin, which is a popular polymer material in tissue engineering because of its good biocompatibility and antibacterial property. It has been reported that either chitosan alone [63] or combined with other polymers such as poly(acrylic acid) (PAA) [64] and polyvinyl pyrrolidone (PVP) [65] can form hydrogels to carry antibiotics. For example, a drug-eluting chitosan-vancomycin coating on Ti foil was biocompatible and bactericidal. It reduced the infection risk in antibacterial tests [63]. Recently, an antibiotic-loaded fast-resorbable hydrogel coating (defensive antibacterial coating, DAC) has been applied in an European clinical trial for THA and TKA [66]. The DAC composes of covalently linked hyaluronan and poly-D,L-lactide with antibiotics. In a clinical trial, the DAC was used by simply spreading on the hip/knee prosthesis surface during the surgery. The results showed that the DAC reduced the rate of early surgical site infection. There were no detectable side effects after THA and TKA with a cementless or hybrid implant [66].

The limitation of hydrogel coating as a drug-eluting device is the burst drug release. Like the DAC, it completely degraded within 72 h with 100% antibiotics released [66]. It can be applied as antibacterial coating to prevent early onset infection but is not suitable for the inhibition of the late-onset infection.

Layer-by-Layer (LBL) Coatings

Layer-by-layer (LBL) coating is by depositing layers of polyelectrolyte solutions with opposite charges in an alternating fashion on the implant surface, resulting in a thin film that can be used to load a variety of biomolecules [58]. The number of layers, concentration of molecules in the solution, and chemical properties of the polyelectrolyte solution can be modified to reach an optimal drug loading efficiency and release kinetics. Various growth factors have been deposited on the implant via LBL coating technology [44, 67, 68]. Shah NJ et al. [67] developed [poly (β-amin ester)/polyanion/growth factor/polyanion] LBL tetralayer coating on polycaprolactone/β-tricalcium (PCL/β-TCP) scaffolds. With the aim to mimic the healing process, this coating not only extended the release time of recombinant human vascular endothelial growth factor ($VEGF_{165}$) and BMP-2, but also delivered the two growth factors at different times. As a result, the implant coating facilitated blood vessel ingrowth and bone formation in vitro [67].

Bactericidal LBL coating has been developed by incorporation of polyelectrolyte multilayer films with silver [69], gentamicin [70], and vancomycin [71]. The LBL coating can control drug densities and release profiles. For example, it has been reported that polyelectrolyte multilayers on Ti implants showed a sustained release of bioactive gentamicin over 1 month [70]. The coating has been demonstrated to be bactericidal against *S. aureus* and biocompatible in vitro. An in vivo study has shown that the coated Ti implant successfully decreased the degree of infection in a rabbit *S. aureus* bone infection model [70].

Overall, the LBL coating is a promising technique for massive drug loading and controllable drug release. It can be applied to prevent late-stage infection and inhibit biofilm formation. However, the broad application of LBL techniques has been limited due to several technical challenges. Firstly, the fabrication of LBL coating is labor intensive and expensive. In order to reduce initial drug burst release and to extend releasing duration, usually a few hundreds layers would be required. Secondly, the LBL coating is performed in acidic solution, which may cause toxicity to tissue [58].

Immobilization of Drugs on the Implant Surface

An alternative strategy for long-term drug delivery is to immobilize drugs directly on the implant surface. Osteogenic peptides are the most common used agents that have been used for the improvement of osseointegration [58, 72–74]. Peptide GFOGER, derived from type-I collagen, was proven to promote osteogenic differentiation through binding to the $\alpha2\beta1$ integrin receptor on the surface of osteoblast-like cells [72]. Wojtowicz et al. immobilized the GFOFER peptides on the surface of PCL scaffolds via passive absorption [72]. The implantation of GROFER-coated PCL scaffolds was performed in rat femoral defects model. The results showed that the GROFER-coated PCL scaffold effectively promoted bone repair with significant bone volume increase after 12-week implantation [72]. The peptides can be easily absorbed on the polymer surface but is difficult to attach on the metallic surface via physical absorption. Therefore, a strong covalent bonding between drugs and metallic surface needs to be formed to immobilize the drugs. For example, in order to improve the cell adhesion on Ti implant, modified cyclic-RGD peptide with phosphonic acid anchors was developed. The phosphonic acid anchor can bond to titanium oxide and indirectly immobilize the RGD peptides to the metal surface [74].

To inhibit infection, immobilization of vancomycin (Vanc) on the Ti or Ti6Al4V(Vanc-Ti) implant surface through covalent bonding was reported [75–77]. In this way, the Vanc-Ti coating presented the antibiotic for a long period. It inhibited *S. aureus* colonization up to 11 months in vitro [75]; and even inhibited *S. epidermidis* biofilm formation [76]. Although the immobilization technology achieved long-term drug delivery, the immobilized antibiotics or peptides should remain function in their tethered form, which limits the application of many types agents.

Other Coatings

Other coatings focus on modifying the surface structure of the implant. Porous implant surface coatings have been used clinically on ceramic or metallic implants since 1970 to assist osteoconduction [14, 78]. Various types of porous structures have been developed and investigated, such as open pores and highly interconnected porous structure. The most commonly used porous coating is trabecular metal (Ti) [14]. The trabecular metal with high porosity (80%) allows rapid bone ingrowth and implant stabilization as reported in in vivo studies [79]. However, clinical revision rates of patients using these porous implants are not reduced [80]. Thus, more clinical investigations are needed to evaluate the long-term survival rate of the implants with porous coatings. Currently, several engineered implant surfaces with micro- or nanostructure have been developed [81, 82]. This micro- or nanoscale surface with increased porous structure enhances the cell adhesion and bony ingrowth. At the same time the rough surface increases the friction force and enhances the osseointegration. A cell-favored surface may also attract bacterial attachment. It is a challenge to balance promoting host cells growth and inhibiting bacterial growth.

Future Direction

The implant surface coatings for enhancing osteointegration and inhibiting infection have been closely related and stated as a "race for the surface" by Gristina [83]. The host cells and bacteria will be racing for the implant surface right after implantation. The ideal implant surface should promote strong osseointegration by facilitating host cells attachment to the surface and meanwhile inhibit the bacterial colonization. Thus, the strategies for osteointegration and anti-infection coating could be combined together to reach a multifunctional coating. In addition, the applicable coating should be simpler for preparation and economic for fabrication.

Electrospun Nanofibers (NFs) Coating to Enhance Osseointegration

Characters and Current Researches in Electrospinning

The native ECM of bone tissues is a nanofibrous collagen network. The fundamental unit of the bone is mineralized and highly ordered collagen I fibrils, only a few nanometers thick [84] with collagen. The collagen I fibrils are aligned and arranged to form a higher order structure seen in a mature bone matrix [85–87]. One of the promising technologies that can be used to mimic bone nanoscale ECM structure is electrospinning [88, 89].

Electrospinning, developed in the early 1930s, has been applied in various industrial products, such as highly efficient filters, lightweight and protective cloth, and battery cells, as well as tissue engineering and regenerative medicine [90]. It uses an electrical charge to exceed the surface tension of a charged polymer solution, resulting in the formation of micro- or nanoscale fibers [91]. With the expanding availability of nature, synthetic or combined polymer materials, electrospun NFs have been applied to tissue engineering because of its unique characteristics, such as high surface area and porosity [89]. Recent studies indicated that the attachment, proliferation, and differentiation of bone cells can be enhanced by the physiochemical and microstructural properties of electrospun NFs [92, 93]. The potential application of implants with NF coating for the enhancement of osseointegration is promising but often is overlooked [89]. More efforts are obviously required to better understand the dynamic interplay between the physiochemical and microstructural natures of NFs and the fate of bone cells [89].

Limitations (Dense and Compact Structure)

The porous structure of the NF scaffolds is critical for its application in tissue engineering and regenerative medicine. The nutrition and waste of cells should be transported through the pores [94]. Cells growth and differentiation demand a porous structure of the local environment. It is well known that different cells require different pore sizes [94]. For example, vascularization happens at pore sizes over 300 μm in the bone tissue [95]; while fibroblasts prefer a pore size of 6–20 μm [96]. Thus, ideal electrospun NFs scaffold should have three-dimensional shape and macroscale pores, which provide sufficient space for cell infiltration and differentiation [97]. From this aspect, one main limitation of current electrospinning technology is that the electrospun NFs are firmly packed that only provide a superficial porous structure due to the sheet-like assembly process [98]. This inevitable event impedes cell infiltration and growth throughout the NF mats [98]. There are no satisfactory resolutions of this technical barrier. The fabrication of loose, thick, and bulky scaffolds (3D scaffolds) with controllable microstructures remains a technical challenge [99].

Current 3D NFs Fabrication Techniques

Many efforts have been explored in past decades to fabricate 3D porous and looser NF scaffolds. The first strategy to form the 3D NF scaffold is by simply stacking, folding, or rolling multiple thin NF films [100]. For example, by layer-by-layer interval stacking, micro- and nanofiber membrane were formed into a sandwich-like 3D scaffold. The nanofiber layers assisted cell adhesion and proliferation; while the microfiber layers with larger pores helped cell infiltration [100]. Levorson et al.

found that this scaffold has increased the chondrogenic differentiation of MSCs in vitro [101]. However, the microstructure of the layer-by-layer stacking NFs is still dense and compact.

Adding porogens, including salt particles [98, 102], ice crystal, and washable polymers [103], is an alternative strategy to fabricate 3D porous NFs. The embedded porogens can quickly built-up the NFs volume during the electrospinning. The porogens will be then washed away after electrospinning, leaving numerous larger pores in the formed NFs. Salt leaching uses salt particles as the porogens. Nam et al. introduced NaCl crystal (diameter: 90–106 μm) to the Taylor Cone by a sheath surrounding the spinneret [98]. Formed 3D PCL NFs characterized a uniform porous structure with average pore size of 200 μm. The highly porous structure facilitated CFK2 cell infiltration to the depth of 4 mm [98]. The salt leaching method can be used to control the pore size; while the requirement of multiple steps and the modification of electrospun device together make fabrication far more complicated.

Cryogenic electrospinning, by embedding ice crystal as porogens, was first reported by Simonet et al. in 2007, and later was termed as cryogenic electrospinning in 2009 by Leong et al. [104, 105]. From condensing humidity, the ice crystals are formed simultaneously with NFs deposition by a low-temperature fiber collector device; and the crystals will be then removed by freeze-drying procedure [105]. The porosity of formed 3D NFs was four times higher than the traditional NFs [105]. Correctly balancing between the fiber and ice crystal formation is the key to achieve the 3D NFs by cryogenic electrospinning. Unlike the salt leaching, another limitation of this method is relatively smaller and uncontrollable pore size. Another removable polymer porogen is poly (ethylene oxide) (PEO) [106]. PEO is a water-soluble material electrospun with PCL polymer to achieve a combination of PEO NFs and PCL NFs. Later, the PEO NFs were washed away leaving pure and porous PCL NFs. The porosity could be adjusted by the ratio of PCL and PEO NFs. However, it is difficult to increase the pore size and scale up the formed NF volume through this approach.

It has been demonstrated that the fabrication of NF collector surface design is one of the most effective approaches to create 3D fibers with desired fibrous structure and patterns [102]. Some advanced NF collector modification techniques have been reported recently, such as rolling or stacking collectors [100, 107], liquid bath collector [108, 109], and micro-patterned collector. The liquid bath collector design increased the dispersion effect and decreased the fiber bonding through collecting NFs in various liquid solutions, such as water and organic solvents [103, 110]. Yang et al. collected 3D cotton-like poly(lactic-*co*-glycolic acid) (PLGA)/PCL NFs in ethanol bath for bone regeneration in vivo [109]. The super loose and uncompressed NFs scaffolds were found to enhance the chondrogenic differentiation both in vitro and in vivo. Yarn, a bundle of aligned nanofibers, is formed in a liquid bath as well. Usually, the NFs deposit on the surface of the liquid bath, and through water vortex, the NFs are pulled and twisted into a continuous yarn. The yarns could be further collected by a rotating collector to compose a 3D nanoyarn scaffold [111]. The nanoyarn scaffolds have been studied in tendon tissue regeneration [111] and cartilage tissue regeneration [112] for the improvement of cell penetration and vascularization. Overall, the NFs formed in the liquid bath collector have homogenous structure, but they are difficult to scale up.

Micro-patterned collector is another collection technique allowing the formation of highly porous NFs. Li and Xia arranged conductive and nonconductive void spaces to make a patterned collector [113]. NFs were aligned across a nonconductive void. These methods make it possible to form 3D NFs scaffolds in certain forms. However, the processing is relatively complex, slow, and difficult to control. In addition, this process cannot be used to create scalable, block scaffolds with an interconnected porous structure. It is obviously that a simple and one-step real-time technology for the preparation of controllable porous NF matrix is urgently needed for the application of tissue engineering of different tissues and/or organs.

The Technique of 3D NFs Collector (Mechanism, Device, Physiochemical Properties of NFs, and Cellular Behavior)

The working mechanism of electrospinning is that driven by high voltage, a charged polymer jet overcomes its surface tension and deposits onto low potential targets in the form of numerous NFs. Commonly, the NFs collected on the flat surface with equipotential density. We designed a NF collector mounted with multiple movable sharp and electric conductive needles. The corona discharge effect leads to continuous deposition of 3D NF matrices on the surface of the NF collector [114]. As a result, the local electric field around the needle tip creates strength much higher than the surrounding conductor, resulting in an acceleration of free electrons to a high velocity, which ionizes neutral air molecules [114]. Thus, the charged polymer jet prefers to deposit onto the sharp tip of the needle during the electrospinning (Fig. 1a) [114]. According to this mechanism, we have designed a 3D NF collector with numerous movable needles where electrospun NFs are gradually deposited to form 3D architectures (Fig. 1) [114]. Unlike conventional electrospinning that lays down a uniform deposition, the electric field vectors in the vicinity of the collector majorly target two fractions—the projecting points of needles (A) and the edging corner of the platform (B/C), which enforces the deposition of spinning nanofibers along the alignment of B-A-C and allows a triangle-shaped fiber sheet formation as shown in Fig. 1b-1 [114]. When two points are more prominent on the surface of collector such as points D and E, the spinning fibers are deposited to these points giving a wave-shaped fiber sheet formation (Fig. 1b) [114]. When the collector was fully covered by a deposited fiber sheet, the needles' positions were re-adjusted by gradually pushing those pierced needles forward. At the same time, a new fiber sheet would start depositing on the tips. After several rounds, 3D NFs architectures were gradually built on the surface of the collector by stacking multilayers of fiber sheet into bulk (Fig. 1). Thus, using the coronal charge effect provides a simple and one-step approach to develop the 3D nanofibers. In comparison to the 2D PCL nanofibers, the 3D PCL nanofibers have a looser microstructure and larger pore sizes via scanning electron microscopy (SEM) [114]. The pore sizes of 2D NFs was in the range of 0–1 μm^2; the

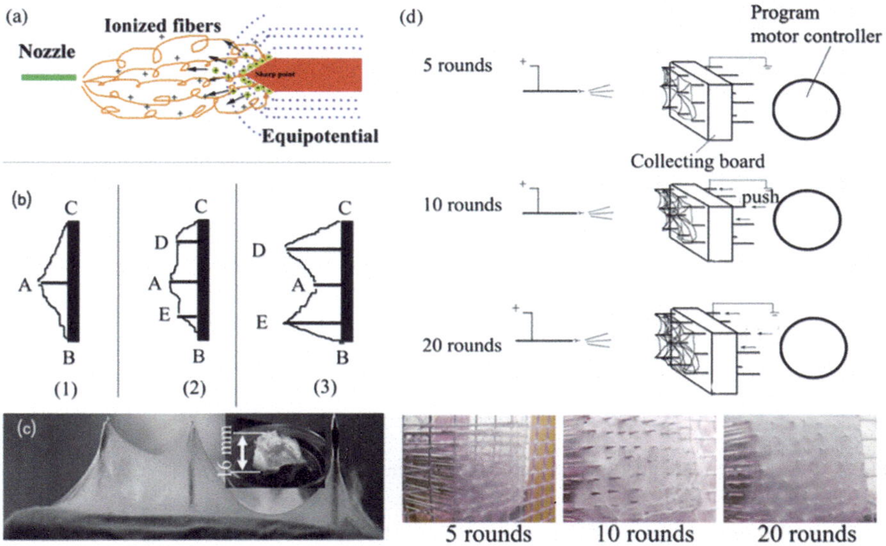

Fig. 1 The fabrication of PCL 3D nanofibers (NFs). (**a**) A diagram of the mechanism of NFs depositing on the needle tip by coronal discharge effect. (**b**) Illustration of a cross-sectional view of electrospun fibers built-up between the spinneret and needle-collector. (**c**) Photograph of collected fibers deposited along needles and platform during electrospinning. (**d**) Multiple rounds (5, 10, and 20) to form 3D nanofibers on needle collectors [114]

3D NFs had larger pores size were mainly in the range of 0.1–10 μm^2. In addition, 3D NFs with looser structure stimulated the infiltration, proliferation, and differentiation of murine pre-osteoblastic MC3T3-E1 cells [114]. The pre-osteoblast cells infiltrated the entire 3D PCL nanofibers, while they only spread on the surface of 2D nanofibers after 7 days culture. A significantly higher cellular proliferation was also discovered on 3D NFs at 7-day culture than that on the 2D NFs ($p < 0.01$). The looser structure further increased the differentiation level of the cells, which had a significantly higher alkaline phosphatase (ALP) concentration on 3D NFs ($p < 0.01$) [114]. However, one of the key limitations is that the microstructure and shape of the formed 3D NFs based on the corona discharge mechanism are neither controllable nor reproducible because of the manual movement of the mounted needles. Since both the macrostructure and microstructure of the NFs affect cell behavior, establishing reference NF scaffolds with well-characterized cell response is critical to advancing their use in the tissue engineering field. Further development of this novel coronal discharge-based porous NF fabrication technique requires standardization of the electrospinning process and characterization methods. Therefore, we developed a programmed electrospun 3D NF collector that can be used to fabricate 3D NFs with desired microstructure, such as pore size and porosity, by precisely controlling the moving speed of NF collector during electrospinning [115]. This device can be used to precisely control the needle collectors constantly moving forward via different

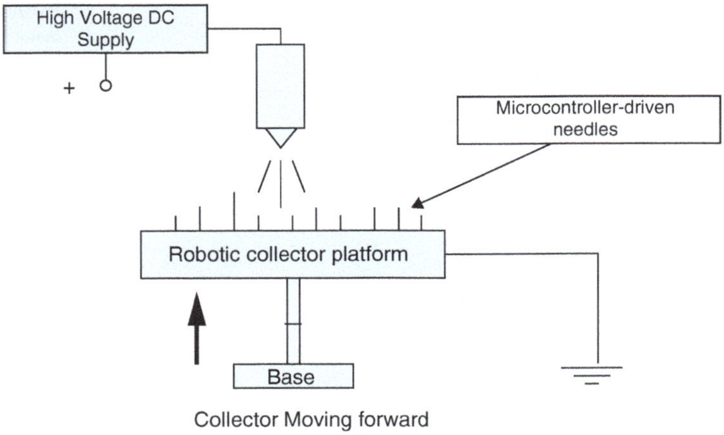

Fig. 2 Illustration of automatic 3D nanofibers collector (unpublished data)

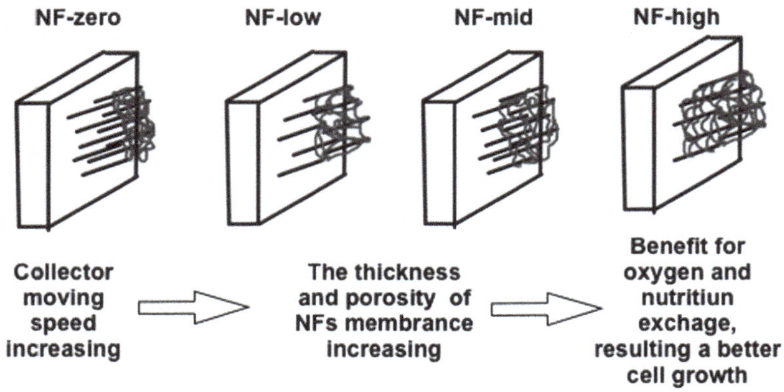

Fig. 3 Illustration of four types of NFs formed on collector with different moving speeds

moving speeds (0–0.232 mm/min) (Fig. 2). Four types of polycaprolactone (PCL) 3D NF matrices with different microstructures can be obtained concurrently on the NF collector surface by setting different forward moving speed of the NF collector device (from low to high) (Fig. 3). A linear increase in the NF sheet thickness was recorded with increasing NF collector moving speed with 1-h electrospinning. Scanning electron microscopy (SEM) measurement showed a looser microstructure and an increase of porosity with the increase of collector moving speed (Fig. 4). NFs prepared at high collector speed showed enhanced cell proliferation and differentiation (ALP expression) of pre-osteoblastic MC3T3 cells compared to NFs collected on a static collector. A programmable NF collector permits the fabrication of reproducible 3D NF scaffolds of variable size and adjustable microstructure. This simple, controllable, one-step 3D electrospun NFs fabrication may help move forwarding the clinical translation of electrospun NFs in regenerative medicine.

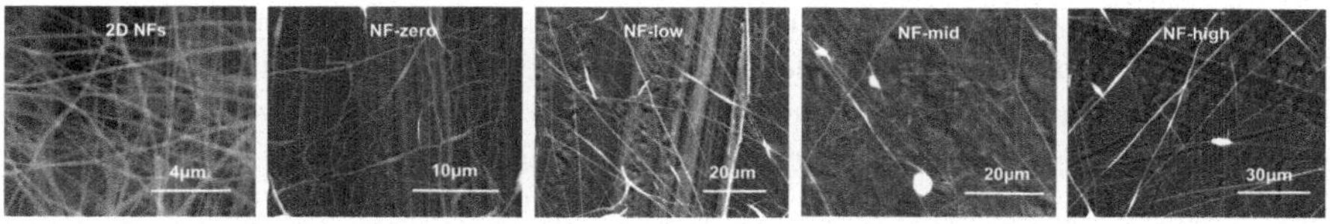

	2D-NFs	NF-zero	NF-low	NF-mid	NF-high
Porosity (%)	65.39±1.58	70.14±2.7	76.11±2.02	84.63±3.2	88.88±1.82

Fig. 4 The morphology of five types PCL NFs. Porosity was calculated by SEM images and Image J software. Experiments repeated three times with triplicate. 2D NF: NFs were collected in a flat collector surface

Nanofibers Coating as Drug-Eluting Device to Enhance Osteointegration and Treat Periprosthetic Infection (PJI)

Coaxial Nanofibrous Coating as a Controlled Drug-Eluting Device (Current Technology Development Status)

Electrospun NF can be used as a drug-eluting device by embedding drugs into the polymer solution before electrospinning. The drug release kinetics is determined by NFs structure and the degradation rate of NFs. Polyvinyl alcohol (PVA) has been extensively used in electrospinning because of its excellent biocompatibility and good electrospun NF-forming capability [116]. It has been reported that the morphology and chemical composition of electrospun PVA/HA NFs were similar to the basic architecture of bone [117]. We [116] developed electrospun PVA/Collagen/HA NFs and found that the inclusion of HA and collagen in the PVA NFs significantly increased the fiber stability and the mechanical strength. The encapsulated nano-HA crystals and collagen also enhanced the adhesion and proliferation of osteoblastic MC3T3 cells in vitro. However, these blended PVA/Collagen/HA NFs cannot be used as a desired drug release device because of their fast degradation rate (~10 days) [116].

Coaxial electrospinning has been used to prepare coaxial core-sheath NFs that can be used to control and extend the embedded drug release. We have used this technology to prepare coaxial NFs scaffolds as a local drug-eluting device to enhance osseointegration and prevent infection both in vitro and in vivo [89, 118]. During coaxial electrospinning, a spinneret is employed to trap a secondary fluid layer (containing labile drugs) within the core of the forming NFs [89]. The sheath solution acts as a guide and surrounds the core material. The sheath structure represents a physical barrier to reduce the initial burst release and protects the drugs in the core fiber. The concentration gradient inside the core fiber is the driving force for diffusion [89]. In a core-sheath system, a drug release rate is affected by both the concentration gradient and the degradation rate of the sheath barrier. Therefore, control of the drug release rate can be achieved by preparation of various formulations and thicknesses of slow/fast degradation sheath fibers and/or modification of physiochemical properties of NFs [89, 119]. We [89] developed coaxial electrospun PCL/PVA core-sheath NFs blended with both HA nanorods and type-I collagen (PCL^{Col}/PVA^{HA}). The incorporation of collagen into the PCL sheath (PCL^{col}) increased its hydrophilicity and provided numerous binding sites for cell adhesion. The incorporation of HA into the PVA core (PVA^{HA}) increased the surface roughness and mechanical strength of NFs. The PVA^{HA} core was used as a drug reservoir. This hybrid core-sheath NF scaffold takes advantage of the slow degradation nature of PCL and the bioactivity of PVA while minimizing the disadvantages of both. Doxycycline (Doxy) embedded in the PCL^{Col}/PVA^{HA} NFs showed more sustained release (~1 month) compared with the blended NFs (completely released within 48 h). Doxy released is stable and bactericidal as evidenced by a modified *S. aureus* growth inhibition assay [120]. We also found that PCL^{col}/PVA^{HA} NF coating enhanced osseointegration in vivo.

In the next section, we would like to introduce our two recent studies using implants with coaxial nanofiber coating from the aspects of osseointegration enhancement and infection inhibition.

Sustained Strontium Release from Coaxial NFs to Enhance Osseointegration

The use of NF coating needs careful understanding and coordination of its rate of degradation with the physiology of osseointegration. An early and sufficient osseointegration resulting in "the formation of a direct interface between an implant and bone without intervening soft tissue" is critical for the early implant stability (~1 month). Obviously PCL is not an appropriate NF material because of its much slower degradation rate both in vivo and in vitro [121] that has been verified in our previous pilot study. Therefore, we added PLGA to PCL as the sheath fiber in this study because of its faster and controllable degradation rate (~1 month) comparable to that of osseointegration physiology [122]. Another benefit of PLGA is its stronger binding to the Ti surface [123, 124] than that of PCL [89]. One reason for this is that PLGA has much higher ratio of oxygen atoms in its molecular structure than that of PCL, thus providing more electrostatic interaction on the Ti surface [125].

Strontium (Sr^{2+}) is a minor element that can be found in our body and daily diet [126]. Nearly 99% of Sr^{2+} ions are deposited in bone [126]. The Sr^{2+} ion has the similar cellular transport pathway as calcium ions, and has strong affinity for the incorporation in the bone matrix during mineralization [126, 127]. Sr^{2+} enhances bone formation and strength through the inhibition of osteoclasts and activation of osteoblasts [128–131]. There are few studies that investigated the role(s) of Sr^{2+} in the field of implant osseointegration [132, 133]. Park et al. found that Sr^{2+}-embedded Ti implants significantly enhanced implant osseointegration as compared with the control Ti implants in a rabbit tibia implantation model [134].

We [135] have developed a Sr^{2+}-doped coaxial PCL/PLGA-PVA NF coating to enhance the osseointegration. The sheath fiber formula PCL/PLGA (1:1, v/v) was optimized to match the NFs degradation rate to the implant osseointegration physiology, which is about 1 month. Although an initial Sr^{2+} burst release was observed, a sustained release of Sr^{2+} from the PCL/PLGA (1:1, v/v)-PVA coaxial NFs was detected for over 2 months. The Sr^{2+}-doped PCL/PLGA-PVA coaxial NFs were biocompatible and significantly enhanced the differentiation of murine pre-osteoblast MC3T3-E1 cells using both indirect and direct contact approaches in vitro. Taken together, Sr^{2+}-doped PCL/PLGA-PVA coaxial NFs are promising nanofabricated implant coatings to promote earlier and sufficient implant osseointegration.

Sustained Release of Doxycycline from Coaxial NFs to Prevent and Treat PJI (In Vitro and In Vivo Study)

For the prevention and treatment of PJI, we have developed a Doxy-doped coaxial electrospun PCL/PVA NFs as the Ti pin coating [118]. This hybrid core-sheath NF scaffold takes advantage of the slow degradation nature of PCL and the bioactivity of PVA while minimizing the disadvantages of both. The slow degradation of the PCL concomitantly reduced the Doxy diffusion from the PVA (core materials). Doxy embedded in the PCL/PVA NFs showed more sustained release (~1 month) compared with the blended NFs (completely released within 48 h). The Doxy-doped coaxial PCL/PVA NFs were directly deposited on the Ti pin surface during electrospinning with the aim to improve osseointegration and inhibition infection. The bone–implant surface (%) in the NFs-coated Ti pin groups was significantly higher ($p < 0.05$) than the non-coating groups after implantation 2, 4, and 8 weeks in a *S. aureus* infected rat tibia implantation model [118]. In addition, the Doxy-loaded NFs inhibited bacterial growth up to 8 weeks in vivo [118]. However, the bacteria grown back and formed biofilm after 16 weeks implantation. These two studies showed the great potential of NFs coating for the enhancement of implant osseointegration and infection inhibition. The results of the in vivo study revealed the two challenges in the application of nanofibers as implant coating. The first problem is the weak bonding of the nanofibers with the implant surface. The nanofibers may be separated from the Ti pin surface during implantation, which may cause larger gaps between the implant and the surrounding tissue. We have blended PLGA with the NFs to increase the bonding strength to Ti pin surface. More strategies could be developed to strengthen adhesion of nanofibers coatings to different metal implant surfaces. The second problem is the lack of a long-term infection inhibition.

Nanofiber is an ideal matrix for host cells and bacteria adhesion and growth. If any bacterial residues are left in the matrix, they may slowly colonize, and form a biofilm when the antibiotic dose was low or missing. Thus, the antibiotic dose should be high enough to kill the majority of bacteria without toxicity to the host tissue within 48 h after implantation. In addition, a sustained and controllable antibiotic dose at a sufficient level is needed to prevent the recurrence of infection. Multiple antibacterial drugs can be embedded within the same NF coating to increase the antibacterial efficiency, especially those that inhibit later infection or disrupt biofilm formation.

Summary and Conclusions

Orthopedic implants have been widely and successfully applied in TJA worldwide. As for the duration of the implants, although 10- to 15-year survival rate is higher than 90%, the amount of implant failures is a clinical challenge. It lays huge burden on patients, physically and mentally. This chapter generally introduced the two leading complications of TJA, that is, insufficient osseointegration and infection. The implant surface is the "racing arena" for host cells and bacteria after implantation. An ideal implant surface should benefit host cell growth and inhibit bacteria adhesion. For this reason, implant coatings have become a potential solution to promote TJA success. It has drawn much interest to enhance osseointegration and inhibit infection. Various current coating strategies for osseointegration and infection prevention are compared and summarized in the chapter. At the end, we discussed the possibility of using nanofibers as implant coating and briefly introduced our researches about nanofibers Ti implant coating. More efforts are needed to develop advanced implant coating technologies that are more "bone-like" and multifunctional to both enhance osseointegration and prevent infection.

References

1. Centers for Disease Control and Prevention (2013) National hospital discharge survey: 2010 table, procedures by selected patient characteristics. Centers for Disease Control and Prevention, Atlanta
2. Tande AJ, Patel R (2014) Prosthetic joint infection. Clin Microbiol Rev 27(2):302–345
3. Kurtz S et al (2007) Projections of primary and revision hip and knee arthroplasty in the United States from 2005 to 2030. J Bone Joint Surg Am 89(4):780–785
4. Sharkey PF et al (2002) Insall Award paper. Why are total knee arthroplasties failing today? Clin Orthop Relat Res (404):7–13
5. Colizza WA, Insall JN, Scuderi GR (1995) The posterior stabilized total knee prosthesis. Assessment of polyethylene damage and osteolysis after a ten-year-minimum follow-up. J Bone Joint Surg Am 77(11):1713–1720
6. Emmerson KP, Moran CG, Pinder IM (1996) Survivorship analysis of the Kinematic Stabilizer total knee replacement: a 10- to 14-year follow-up. J Bone Joint Surg Br 78(3):441–445

7. Ranawat CS, Luessenhop CP, Rodriguez JA (1997) The press-fit condylar modular total knee system. Four-to-six-year results with a posterior-cruciate-substituting design. J Bone Joint Surg Am 79(3):342–348
8. Font-Rodriguez DE, Scuderi GR, Insall JN (1997) Survivorship of cemented total knee arthroplasty. Clin Orthop Relat Res (345):79–86
9. Weir DJ, Moran CG, Pinder IM (1996) Kinematic condylar total knee arthroplasty. 14-year survivorship analysis of 208 consecutive cases. J Bone Joint Surg Br 78(6):907–911
10. Goriainov V et al (2014) Bone and metal: an orthopaedic perspective on osseointegration of metals. Acta Biomater 10(10):4043–4057
11. Heck DA et al (1998) Revision rates after knee replacement in the United States. Med Care 36(5):661–669
12. Baroli B (2009) From natural bone grafts to tissue engineering therapeutics: brainstorming on pharmaceutical formulative requirements and challenges. J Pharm Sci 98(4):1317–1375
13. Davies JE (2003) Understanding peri-implant endosseous healing. J Dent Educ 67(8):932–949
14. Raphel J et al (2016) Multifunctional coatings to simultaneously promote osseointegration and prevent infection of orthopaedic implants. Biomaterials 84:301–314
15. Sundfeldt M, Carlsson LV, Johansson CB, Thomsen P, Gretzer C (2006) Aseptic loosening, not only a question of wear: a review of different theories. Acta Orthop 77(2):177–197
16. Amstutz HC et al (1992) Mechanism and clinical significance of wear debris-induced osteolysis. Clin Orthop Relat Res (276):7–18
17. Sadoghi P, Liebensteiner M, Agreiter M, Leithner A, Böhler N, Labek G (2013) Revision surgery after total joint arthroplasty: a complication-based analysis using worldwide arthroplasty registers. J Arthroplast 28(8):1329–1332
18. Brånemark R et al (2001) Osseointegration in skeletal reconstruction and rehabilitation: a review. J Rehabil Res Dev 38(2):175–181
19. Ryd L (1992) Roentgen stereophotogrammetric analysis of prosthetic fixation in the hip and knee joint. Clin Orthop Relat Res (276):56–65
20. Kärrholm J, Borssén B, Löwenhielm G, Snorrason F (1994) Does early micromotion of femoral stem prostheses matter? 4-7-year stereoradiographic follow-up of 84 cemented prostheses. J Bone Joint Surg Br 76(6):912–917
21. Kurtz SM et al (2010) Prosthetic joint infection risk after TKA in the Medicare population. Clin Orthop Relat Res 468(1):52–56
22. Kurtz SM et al (2012) Economic burden of periprosthetic joint infection in the United States. J Arthroplasty 27(8 Suppl):61–5.e1
23. Pulido L et al (2008) Periprosthetic joint infection: the incidence, timing, and predisposing factors. Clin Orthop Relat Res 466(7):1710–1715
24. Ratto N, Arrigoni C, Rosso F, Bruzzone M, Dettoni F, Bonasia DE, Rossi R (2016) Total knee arthroplasty and infection: how surgeons can reduce the risks. EFORT Open Rev 1(9):339–344
25. Wisplinghoff H et al (2004) Nosocomial bloodstream infections in US hospitals: analysis of 24,179 cases from a prospective nationwide surveillance study. Clin Infect Dis 39(3):309–317
26. Friedman ND et al (2002) Health care—associated bloodstream infections in adults: a reason to change the accepted definition of community-acquired infections. Ann Intern Med 137(10):791–797
27. Uckay I et al (2009) Low incidence of haematogenous seeding to total hip and knee prostheses in patients with remote infections. J Infect 59(5):337–345
28. Donlan RM, Costerton JW (2002) Biofilms: survival mechanisms of clinically relevant microorganisms. Clin Microbiol Rev 15(2):167–193
29. Molina-Manso D et al (2013) In vitro susceptibility to antibiotics of staphylococci in biofilms isolated from orthopaedic infections. Int J Antimicrob Agents 41(6):521–523
30. del Pozo JL, Patel R (2007) The challenge of treating biofilm-associated bacterial infections. Clin Pharmacol Ther 82(2):204–209
31. Simchi A et al (2011) Recent progress in inorganic and composite coatings with bactericidal capability for orthopaedic applications. Nanomedicine 7(1):22–39

32. Niinomi M (2008) Metallic biomaterials. J Artif Organs 11(3):105–110
33. Rahbek O et al (2001) Sealing effect of hydroxyapatite coating on peri-implant migration of particles. An experimental study in dogs. J Bone Joint Surg Br 83(3):441–447
34. Geesink RG (2002) Osteoconductive coatings for total joint arthroplasty. Clin Orthop Relat Res 395:53–65
35. Soballe K (1993) Hydroxyapatite ceramic coating for bone implant fixation. Mechanical and histological studies in dogs. Acta Orthop Scand Suppl 255:1–58
36. Bauer TW (1995) Hydroxyapatite: coating controversies. Orthopedics 18(9):885–888
37. Bloebaum RD et al (1994) Complications with hydroxyapatite particulate separation in total hip arthroplasty. Clin Orthop Relat Res (298):19–26
38. Goosen JH, Kums AJ, Kollen BJ, Verheyen CC (2008) Porous-coated femoral components with or without hydroxyapatite in primary uncemented total hip arthroplasty: a systematic review of randomized controlled trials. Arch Orthop Trauma Surg 129(9):1165–1169
39. Song Y, Zhang S, Li J, Zhao C, Zhang X (2010) Electrodeposition of Ca-P coatings on biodegradable Mg alloy: in vitro biomineralization behavior. Acta Biomater 6(5):1736–1742
40. de Jonge LT et al (2008) Organic-inorganic surface modifications for titanium implant surfaces. Pharm Res 25(10):2357–2369
41. He J et al (2012) Collagen-infiltrated porous hydroxyapatite coating and its osteogenic properties: in vitro and in vivo study. J Biomed Mater Res A 100(7):1706–1715
42. Choi S, Murphy WL (2010) Sustained plasmid DNA release from dissolving mineral coatings. Acta Biomater 6(9):3426–3435
43. Saran N, Zhang R, Turcotte RE (2011) Osteogenic protein-1 delivered by hydroxyapatite-coated implants improves bone ingrowth in extracortical bone bridging. Clin Orthop Relat Res 469(5):1470–1478
44. Shah NJ et al (2012) Osteophilic multilayer coatings for accelerated bone tissue growth. Adv Mater 24(11):1445–1450
45. LeGeros RZ (2002) Properties of osteoconductive biomaterials: calcium phosphates. Clin Orthop Relat Res 395:81–98
46. Fielding GA et al (2012) Antibacterial and biological characteristics of silver containing and strontium doped plasma sprayed hydroxyapatite coatings. Acta Biomater 8(8):3144–3152
47. Chen W et al (2006) In vitro anti-bacterial and biological properties of magnetron co-sputtered silver-containing hydroxyapatite coating. Biomaterials 27(32):5512–5517
48. Pan CJ et al (2011) Enhancing the antibacterial activity of biomimetic HA coatings by incorporation of norvancomycin. J Orthop Sci 16(1):105–113
49. Kazemzadeh-Narbat M et al (2012) Drug release and bone growth studies of antimicrobial peptide-loaded calcium phosphate coating on titanium. J Biomed Mater Res B Appl Biomater 100(5):1344–1352
50. Cao H et al (2011) Biological actions of silver nanoparticles embedded in titanium controlled by micro-galvanic effects. Biomaterials 32(3):693–705
51. Huo K et al (2013) Osteogenic activity and antibacterial effects on titanium surfaces modified with Zn-incorporated nanotube arrays. Biomaterials 34(13):3467–3478
52. Svensson S et al (2013) Osseointegration of titanium with an antimicrobial nanostructured noble metal coating. Nanomedicine 9(7):1048–1056
53. Neut D et al (2015) A biodegradable gentamicin-hydroxyapatite-coating for infection prophylaxis in cementless hip prostheses. Eur Cell Mater 29:42–56
54. Stigter M, de Groot K, Layrolle P (2002) Incorporation of tobramycin into biomimetic hydroxyapatite coating on titanium. Biomaterials 23(20):4143–4153
55. Stigter M, Bezemer J, de Groot K, Layrolle P (2004) Incorporation of different antibiotics into carbonated hydroxyapatite coatings on titanium implants, release and antibiotic efficacy. J Control Release 99(1):127–137
56. Renwen Zhang DX, Tracy L, Carol T (2004) Ectopic bone formation using osteogenic protein-1 carried by a solution precipitated hydroxyapatite. J Biomed Mater Res 71A(3):412–418
57. Yamamura K, Iwata H, Yotsuyanagi T (1992) Synthesis of antibiotic-loaded hydroxyapatite beads and in vitro drug release testing. J Biomed Mater Res 26(8):1053–1064

58. Goodman SB et al (2013) The future of biologic coatings for orthopaedic implants. Biomaterials 34(13):3174–3183
59. Rammelt S et al (2004) Coating of titanium implants with type-I collagen. J Orthop Res 22(5):1025–1034
60. Sartori M et al (2015) Collagen type I coating stimulates bone regeneration and osteointegration of titanium implants in the osteopenic rat. Int Orthop 39(10):2041–2052
61. Stadlinger B et al (2008) Evaluation of osseointegration of dental implants coated with collagen, chondroitin sulphate and BMP-4: an animal study. Int J Oral Maxillofac Surg 37(1):54–59
62. Dupont KM et al (2012) Synthetic scaffold coating with adeno-associated virus encoding BMP2 to promote endogenous bone repair. Cell Tissue Res 347(3):575–588
63. Ordikhani F, Tamjid E, Simchi A (2014) Characterization and antibacterial performance of electrodeposited chitosan-vancomycin composite coatings for prevention of implant-associated infections. Mater Sci Eng C Mater Biol Appl 41:240–248
64. de la Torre PM, Enobakhare Y, Torrado G, Torrado S (2003) Release of amoxicillin from polyionic complexes of chitosan and poly(acrylic acid). Study of polymer/polymer and polymer/drug interactions within the network structure. Biomaterials 24(8):1499–1506
65. Risbud MV, Hardikar AA, Bhat SV, Bhonde RR (2000) pH-sensitive freeze-dried chitosan-polyvinyl pyrrolidone hydrogels as controlled release system for antibiotic delivery. J Control Release 68(1):23–30
66. Romanò CL, Malizos K, Capuano N, Mezzoprete R, D'Arienzo M, Van Der Straeten C, Scarponi S, Drago L (2016) Does an antibiotic-loaded hydrogel coating reduce early post-surgical infection after joint arthroplasty? J Bone Jt Infect 1:34–41
67. Shah NJ et al (2011) Tunable dual growth factor delivery from polyelectrolyte multilayer films. Biomaterials 32(26):6183–6193
68. Macdonald ML et al (2011) Tissue integration of growth factor-eluting layer-by-layer polyelectrolyte multilayer coated implants. Biomaterials 32(5):1446–1453
69. Malcher M et al (2008) Embedded silver ions-containing liposomes in polyelectrolyte multilayers: cargos films for antibacterial agents. Langmuir 24(18):10209–10215
70. Moskowitz JS et al (2010) The effectiveness of the controlled release of gentamicin from polyelectrolyte multilayers in the treatment of Staphylococcus aureus infection in a rabbit bone model. Biomaterials 31(23):6019–6030
71. Shukla A et al (2010) Tunable vancomycin releasing surfaces for biomedical applications. Small 6(21):2392–2404
72. Wojtowicz AM et al (2010) Coating of biomaterial scaffolds with the collagen-mimetic peptide GFOGER for bone defect repair. Biomaterials 31(9):2574–2582
73. Petrie TA et al (2010) Multivalent integrin-specific ligands enhance tissue healing and biomaterial integration. Sci Transl Med 2(45):45–60
74. Auernheimer J et al (2005) Titanium implant materials with improved biocompatibility through coating with phosphonate-anchored cyclic RGD peptides. Chembiochem 6(11):2034–2040
75. Antoci V Jr et al (2007) Covalently attached vancomycin provides a nanoscale antibacterial surface. Clin Orthop Relat Res 461:81–87
76. Antoci V Jr et al (2008) The inhibition of Staphylococcus epidermidis biofilm formation by vancomycin-modified titanium alloy and implications for the treatment of periprosthetic infection. Biomaterials 29(35):4684–4690
77. Hickok NJ, Shapiro IM (2012) Immobilized antibiotics to prevent orthopaedic implant infections. Adv Drug Deliv Rev 64(12):1165–1176
78. Engh CA, Bobyn JD, Glassman AH (1987) Porous-coated hip replacement. The factors governing bone ingrowth, stress shielding, and clinical results. J Bone Joint Surg Br 69(1):45–55
79. Black J (1994) Biological performance of tantalum. Clin Mater 16(3):167–173
80. Kawamura H et al (2001) The porous coated anatomic total hip replacement. A ten to fourteen-year follow-up study of a cementless total hip arthroplasty. J Bone Joint Surg Am 83-A(9):1333–1338

81. Harrison N et al (2013) Micromotion and friction evaluation of a novel surface architecture for improved primary fixation of cementless orthopaedic implants. J Mech Behav Biomed Mater 21:37–46
82. Shah NJ, Hong J, Hyder MN, Hammond PT (2012) Promoting bone mesenchymal stem cells and inhibiting bacterial adhesion of acid-etched nanostructured titanium by ultraviolet functionalization. Adv Mater 24:1445–1450
83. Gristina AG (1987) Biomaterial-centered infection: microbial adhesion versus tissue integration. Science 237(4822):1588–1595
84. Fratzl P, Groschner M, Vogl G, Plenk H Jr, Eschberger J, Fratzl-Zelman N, Koller K, Klaushofer K (1992) Mineral crystals in calcified tissues: a comparative study by SAXS. J Bone Miner Res 7(3):329–334
85. Zhang R, Ma PX (2000) Synthetic nano-fibrillar extracellular matrices with predesigned macroporous architectures. J Biomed Mater Res 52(2):430–438
86. Li WJ, Laurencin CT, Caterson EJ, Tuan RS, Ko FK (2002) Electrospun nanofibrous structure: a novel scaffold for tissue engineering. J Biomed Mater Res 60(3):613–621
87. White CA, Carsen S, Rasuli K, Feibel RJ, Kim PR, Beaulé PE (2012) High incidence of migration with poor initial fixation of the Accolade stem. Clin Orthop Relat Res 470(2):410–417
88. Baker BM, Gee AO, Metter RB, Nathan AS, Marklein RA, Burdick JA, Mauck RL (2008) The potential to improve cell infiltration in composite fiber-aligned electrospun scaffolds by the selective removal of sacrificial fibers. Biomaterials 29(15):2348–2358
89. Song W et al (2013) Coaxial PCL/PVA electrospun nanofibers: osseointegration enhancer and controlled drug release device. Biofabrication 5(3):035006
90. Katti DS et al (2004) Bioresorbable nanofiber-based systems for wound healing and drug delivery: optimization of fabrication parameters. J Biomed Mater Res B Appl Biomater 70(2):286–296
91. Reneker DH, Chun I (1996) Nanometre diameter fibres of polymer, produced by electrospinning. Nanotechnology 7:216–223
92. Kohgo T et al (2011) Bone regeneration with self-assembling peptide nanofiber scaffolds in tissue engineering for osseointegration of dental implants. Int J Periodontics Restorative Dent 31(4):e9–e16
93. Huang Z et al (2008) Effect of nanofiber-coated surfaces on the proliferation and differentiation of osteoprogenitors in vitro. Tissue Eng Part A 14(11):1853–1859
94. Rampichová M, Chvojka J, Buzgo M, Prosecká E, Mikeš P, Vysloužilová L, Tvrdík D, Kochová P, Gregor T, Lukáš D, Amler E (2013) Elastic three-dimensional poly (ε-caprolactone) nanofibre scaffold enhances migration, proliferation and osteogenic differentiation of mesenchymal stem cells. Cell Prolif 46(1):23–37
95. Karageorgiou V, Kaplan D (2005) Porosity of 3D biomaterial scaffolds and osteogenesis. Biomaterials 26(27):5474–5491
96. Lowery JL, Datta N, Rutledge GC (2010) Effect of fiber diameter, pore size and seeding method on growth of human dermal fibroblasts in electrospun poly(epsilon-caprolactone) fibrous mats. Biomaterials 31(3):491–504
97. Blakeney BA, Tambralli A, Anderson JM, Andukuri A, Lim DJ, Dean DR, Jun W (2011) Cell infiltration and growth in a low density, uncompressed three-dimensional electrospun nanofibrous scaffold. Biomaterials 32(6):1583–1590
98. Nam J et al (2007) Improved cellular infiltration in electrospun fiber via engineered porosity. Tissue Eng 13(9):2249–2257
99. Holzwarth JM, Ma PX (2011) Biomimetic nanofibrous scaffolds for bone tissue engineering. Biomaterials 32(36):9622–9629
100. Pham QP, Sharma U, Mikos AG (2006) Electrospun poly(epsilon-caprolactone) microfiber and multilayer nanofiber/microfiber scaffolds: characterization of scaffolds and measurement of cellular infiltration. Biomacromolecules 7(10):2796–2805
101. Levorson EJ et al (2013) Fabrication and characterization of multiscale electrospun scaffolds for cartilage regeneration. Biomed Mater 8(1):014103

102. Teo WE, Inai R, Ramakrishna S (2011) Technological advances in electrospinning of nanofibers. Sci Technol Adv Mater 12(1):013002
103. Wu J, Hong Y (2016) Enhancing cell infiltration of electrospun fibrous scaffolds in tissue regeneration. Bioact Mater 1(1):56–64
104. Leong MF et al (2009) In vitro cell infiltration and in vivo cell infiltration and vascularization in a fibrous, highly porous poly(D,L-lactide) scaffold fabricated by cryogenic electrospinning technique. J Biomed Mater Res A 91(1):231–240
105. Leong MF et al (2010) Fabrication and in vitro and in vivo cell infiltration study of a bilayered cryogenic electrospun poly(D,L-lactide) scaffold. J Biomed Mater Res A 94(4):1141–1149
106. Baker BM et al (2008) The potential to improve cell infiltration in composite fiber-aligned electrospun scaffolds by the selective removal of sacrificial fibers. Biomaterials 29(15):2348–2358
107. Thorvaldsson A et al (2008) Electrospinning of highly porous scaffolds for cartilage regeneration. Biomacromolecules 9(3):1044–1049
108. Ki CS et al (2008) Development of 3-D nanofibrous fibroin scaffold with high porosity by electrospinning: implications for bone regeneration. Biotechnol Lett 30(3):405–410
109. Yang W et al (2013) In vivo bone generation via the endochondral pathway on three-dimensional electrospun fibers. Acta Biomater 9(1):4505–4512
110. Smit E, Bűttner U, Sanderson RD (2005) Continuous yarns from electrospun fibers. Polymer 46(8):2419–2423
111. Wu J et al (2014) Cell infiltration and vascularization in porous nanoyarn scaffolds prepared by dynamic liquid electrospinning. J Biomed Nanotechnol 10(4):603–614
112. Xu Y, Wu J, Wang H, Li H, Di N, Song L, Li S, Li D, Xiang Y, Liu W, Mo X, Zhou Q (2013) Fabrication of electrospun poly(L-lactide-co-ε-caprolactone)/collagen nanoyarn network as a novel, three-dimensional, macroporous, aligned scaffold for tendon tissue engineering. Tissue Eng Part C Methods 19(12):925–936
113. Li DX, Xia YN (2004) Electrospinning of nanofibers: reinventing the wheel? Adv Mater 16(14):1151–1170
114. Song W, Chen L, Seta J, Markel DC, Yu X, Ren W (2018) Corona discharge: a novel approach to fabricate three-dimensional electrospun nanofibers for bone tissue engineering. ACS Biomater Sci Eng 4(10):3624
115. Chen L et al (2020) Preparation of electrospun nanofibers with desired microstructures using a programmed three-dimensional (3D) nanofiber collector. Mater Sci Eng C 106:110188
116. Song W et al (2012) Electrospun polyvinyl alcohol-collagen-hydroxyapatite nanofibers: a biomimetic extracellular matrix for osteoblastic cells. Nanotechnology 23(11):115101
117. Kim GM, Asran AS, Michler GH, Simon P, Kim JS (2008) Electrospun PVA/HAp nanocomposite nanofibers: biomimetics of mineralized hard tissues at a lower level of complexity. Bioinspir Biomim 3(4):046003
118. Song W et al (2017) Doxycycline-loaded coaxial nanofiber coating of titanium implants enhances osseointegration and inhibits Staphylococcus aureus infection. Biomed Mater 12(4):045008
119. Szentivanyi A et al (2011) Electrospun cellular microenvironments: understanding controlled release and scaffold structure. Adv Drug Deliv Rev 63(4):209–220
120. Song W et al (2011) A novel strontium-doped calcium polyphosphate/erythromycin/poly(vinyl alcohol) composite for bone tissue engineering. J Biomed Mater Res A 98(3):359–371
121. Fu SZ et al (2014) In vitro and in vivo degradation behavior of n-HA/PCL-Pluronic-PCL polyurethane composites. J Biomed Mater Res A 102(2):479–486
122. Lu L et al (2000) In vitro and in vivo degradation of porous poly(DL-lactic-co-glycolic acid) foams. Biomaterials 21(18):1837–1845
123. Xiao D et al (2014) Room-temperature attachment of PLGA microspheres to titanium surfaces for implant-based drug release. Appl Surf Sci 309:112–118
124. Abdal-hay A, Hwang M-G, Lim JK (2012) In vitro bioactivity of titanium implants coated with bicomponent hybrid biodegradable polymers. J Sol-Gel Sci Technol 64(3):756–764

125. Xu L, Yamamoto A (2012) Characteristics and cytocompatibility of biodegradable polymer film on magnesium by spin coating. Colloids Surf B Biointerfaces 93:67–74
126. Nielsen SP (2004) The biological role of strontium. Bone 35(3):583–588
127. Meka SRK, Jain S, Chatterjee K (2016) Strontium eluting nanofibers augment stem cell osteogenesis for bone tissue regeneration. Colloids Surf B: Biointerfaces 146:649–656
128. Marie P (2003) Optimizing bone metabolism in osteoporosis: insight into the pharmacologic profile of strontium ranelate. Osteoporos Int 14(3):9–12
129. Barbara A et al (2004) Normal matrix mineralization induced by strontium ranelate in MC3T3-E1 osteogenic cells. Metabolism 53(4):532–537
130. Yang F et al (2011) Strontium enhances osteogenic differentiation of mesenchymal stem cells and in vivo bone formation by activating Wnt/catenin signaling. Stem Cells 29(6):981–991
131. Schumacher M et al (2013) A novel strontium (II)-modified calcium phosphate bone cement stimulates human-bone-marrow-derived mesenchymal stem cell proliferation and osteogenic differentiation in vitro. Acta Biomater 9(12):9547–9557
132. Karrholm J et al (1994) Does early micromotion of femoral stem prostheses matter? 4-7-year stereoradiographic follow-up of 84 cemented prostheses. J Bone Joint Surg Br 76(6):912–917
133. Makadia HK, Siegel SJ (2011) Poly lactic-co-glycolic acid (PLGA) as biodegradable controlled drug delivery carrier. Polymers 3(3):1377–1397
134. Park JW (2011) Increased bone apposition on a titanium oxide surface incorporating phosphate and strontium. Clin Oral Implants Res 22(2):230–234
135. Chen L, Mazeh H, Guardia A, Song W, Begeman P, Markel DC, Ren W (2019) Sustained release of strontium (Sr2+) from polycaprolactone (PCL)/ poly (D,L-lactide-co-glycolide) (PLGA)-polyvinyl alcohol (PVA) coaxial nanofibers enhances osteoblastic differentiation. J Biomater Appl 34(4):533–545

Cationic Antimicrobial Coatings with Osteoinductive Properties

Qing Song, Yangyang Pei, Xiaoting Ye, Peng Li, and Wei Huang

Abstract Orthopedic implant-associated infections caused by pathogenic bacteria, especially the *Staphylococcus* genus, have been a medical and surgical challenge. The infections not only delay the healing process, and have patients suffer from severe pain and even be subjected to re-implantation, but also cause enormous economic losses. It is clear that both a reduction of bacterial infections and acceleration of bone healing are critical to improving the osseointegration of orthopedic implants. Recently, various antibacterial coatings have been employed for the surface modification of orthopedic implants to reduce the bacterial infections. Interestingly, it has been found that some antibacterial coatings, including polycations and metal cations, also possess osteoinductive properties, and thus effectively speed up the

Q. Song
MIIT Key Laboratory of Flexible Electronics & Shaanxi Key Laboratory of Flexible Electronics, Xi'an Key Laboratory of Flexible Electronics & Xi'an Key Laboratory of Biomedical Materials and Engineering, Xi'an Institute of Flexible Electronics (IFE) & Xi'an Institute of Biomedical Materials and Engineering (IBME), Northwestern Polytechnical University (NPU), Xi'an, China

Key Laboratory for Organic Electronics and Information Displays, Institute of Advanced Materials (IAM), Nanjing University of Posts & Telecommunications, Nanjing, China

Y. Pei · X. Ye
MIIT Key Laboratory of Flexible Electronics & Shaanxi Key Laboratory of Flexible Electronics, Xi'an Key Laboratory of Flexible Electronics & Xi'an Key Laboratory of Biomedical Materials and Engineering, Xi'an Institute of Flexible Electronics (IFE) & Xi'an Institute of Biomedical Materials and Engineering (IBME), Northwestern Polytechnical University (NPU), Xi'an, China

P. Li (✉)
MIIT Key Laboratory of Flexible Electronics & Shaanxi Key Laboratory of Flexible Electronics, Xi'an Key Laboratory of Flexible Electronics & Xi'an Key Laboratory of Biomedical Materials and Engineering, Xi'an Institute of Flexible Electronics (IFE) & Xi'an Institute of Biomedical Materials and Engineering (IBME), Northwestern Polytechnical University (NPU), Xi'an, China

Key Laboratory of Flexible Electronics (KLOFE) & Institute of Advanced Materials (IAM), Jiangsu National Synergetic Innovation Center for Advanced Materials (SICAM), Nanjing Tech University, Nanjing, China
e-mail: iampli@nwpu.edu.cn

B. Li et al. (eds.), *Racing for the Surface*,
https://doi.org/10.1007/978-3-030-34471-9_5

healing process. In this chapter, we will shed light on the antibacterial and osteogenic mechanisms of positively charged biomaterials and present some typical cationic antimicrobial coatings with osteoinductive properties in detail.

Keywords Polycations · Copper · Antibacterial · Osteoinductive · Angiogenesis

Introduction

Annually, more than $200 billion is spent on bone fracture treatment and over 500,000 bone grafting procedures are carried out in the United States. Because of the limitations of autografts (e.g., donor-site morbidity and supply) and allografts (e.g., immunogenic response to foreign tissue and the risk of disease transmission), the demands of bone graft substitutes (BGSs) are increasing. The market for BGS was valued at about $2.4 billion in 2015 and is estimated to double in the next decade [1]. The ideal BGSs should be osteoconductive and osteoinductive, and possess similar mechanical strength to real bones. Osteoinductive activity of biomaterials is an important consideration for orthopedic implants since it determines the degree of osteogenesis and bone defect healing process. When BGSs are implanted into patients, the adhesion of bacterial pathogens on their surfaces often causes infections and even osteomyelitis, which is a tough medical issue [2]. The *Staphylococcus* genus, especially *Staphylococcus aureus* (*S. aureus*) and *Staphylococcus epidermidis*, are the main bacterial pathogens in implant-related infections and their drug-resistant strains have even been termed superbugs [3]. The infections may negatively influence the osseointegration of bone implants, require the removal of implants, and prolong treatment time. Therefore, it is critical to develop antimicrobial biomaterials to control implant-associated pathogenic infections. Given the above concerns, some innovative biomaterials with antibacterial and osteoinductive properties, consisting of polycations and metal cations, have been designed and developed in order to meet the urgent requirements of orthopedic implants. This chapter will focus on these positively charged coatings and shed light on their functional mechanisms.

W. Huang
MIIT Key Laboratory of Flexible Electronics & Shaanxi Key Laboratory of Flexible Electronics, Xi'an Key Laboratory of Flexible Electronics & Xi'an Key Laboratory of Biomedical Materials and Engineering, Xi'an Institute of Flexible Electronics (IFE) & Xi'an Institute of Biomedical Materials and Engineering (IBME), Northwestern Polytechnical University (NPU), Xi'an, China

Key Laboratory for Organic Electronics and Information Displays, Institute of Advanced Materials (IAM), Nanjing University of Posts & Telecommunications, Nanjing, China

Key Laboratory of Flexible Electronics (KLOFE) & Institute of Advanced Materials (IAM), Jiangsu National Synergetic Innovation Center for Advanced Materials (SICAM), Nanjing Tech University, Nanjing, China

Antimicrobial and Osteoinductive Mechanisms of Cationic Coatings

Antibacterial Mechanisms of Positively Charged Coatings

Due to the growing global concerns about antibiotic-resistant pathogens, cationic antibacterial materials have attracted more and more attention from researchers because they kill bacteria usually through physical approaches that are not easy to generate bacterial drug-resistance. In nature, most bacteria are negatively charged because their cell wall components contain the negatively charged teichoic acids in Gram-positive bacteria and phospholipids and lipopolysaccharides in Gram-negative bacteria [4], and therefore, the positively charged materials are inclined to attract negatively charged bacteria and subsequently disrupt the bacterial wall/membrane, finally resulting in cell death. The positively charged synthesized polymers, natural antimicrobial peptides (AMPs), and metal cations (e.g., Ag^{+} and Cu^{2+}) have been utilized in the preparation of antibacterial coatings on orthopedic implants.

Polycations

The synthesized bactericidal polycations, including the polymers derived from quaternary ammonium compounds, polyethylenimine derivatives, and chitosan derivatives, are capable of interrupting the integrity of bacterial cell walls/membranes via the mechanisms of "phospholipid sponge effect" or "polymeric spacer effect," or inactivating the bacterial enzymes to kill bacteria upon contact. However, the majority of cationic polymers have poor biocompatibility and are prone to provoke inflammation or hemolysis. On the contrary, AMPs and their mimics normally have good biocompatibility, and they are able to exchange the divalent cations (e.g., Mg^{2+} and Ca^{2+}) on the bacterial membrane via an "ion-exchange mechanism" to destabilize the cytoplasmic membrane, leading to bacterial death. Additionally, through electrostatic attraction, the positively charged AMPs attach onto the negatively charged cytoplasmic membrane that is comprised by phospholipid bilayers, and then insert into the membrane to disturb the membrane integrity, finally resulting in the leakage of cytoplasm and bacterial death [5].

Metal Cations

Silver is an effective broad-spectrum antibacterial metal, and various silver coatings have been utilized to reduce orthopedic implant-associated infections [6]. However, the highly dose-dependent cytotoxicity of silver toward osteoblast and mesenchymal stem cells (MSCs) has a negative influence on implant osseointegration, while copper has better biocompatibility compared with other antimicrobial metal cations such as Ag^{+}, Zn^{2+}, Hg^{2+}, and Co^{2+} as presented in Table 1. Based on the results of

Table 1 LD_{50} concentrations of different metal cations for L929 fibroblasts and corresponding bacterial reduction rates

Metal ions	LD_{50} (L929) (mmol/L)	Bacterial reduction rates at LD_{50}
Ag^+	3.5×10^{-3}	0.93
Zn^{2+}	3.6×10^{-3}	1.11
Hg^{2+}	4.2×10^{-3}	7.58
Cu^{2+}	2.3×10^{-1}	2.5×10^4
Co^{2+}	3.4×10^{-2}	1.42
Al^{3+}	1.8	0.46

Source: Adapted from Heidenau F, *et al.*, *J Mater Sci Mater Med*, 16 (10):883–888, 2005

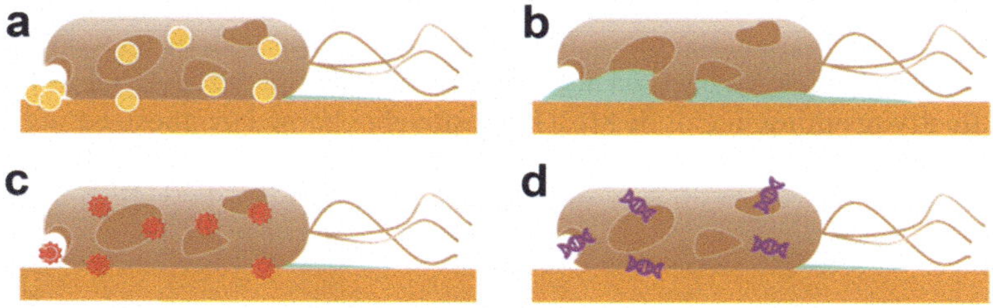

Fig. 1 Antibacterial mechanisms of copper. (**a**) Copper cations influx into bacteria and damage cytoplasmic membrane integrity. (**b**) Bacterial lysis caused by copper. (**c**) Copper cations generate reactive oxygen species which cause further bacterial damage. (**d**) Nucleic acids are degraded [8]

growth inhibition tests with L929 fibroblasts, the lethal doses 50 (LD_{50}, the specific concentration of metal cations to reduce cell growth by 50%) and its corresponding biocidal efficiency showed that cells tolerated Cu^{2+} at relatively high concentrations (LD_{50}: 2.3×10^{-1} mmol/L) and Cu^{2+} showed the highest antibacterial effectiveness (reduction rate: 2.5×10^4) at the corresponding LD_{50}, which indicated that the antibacterial Cu^{2+} had better biocompatibility compared with other metal cations [7]. In this chapter, we focus on the functional activities of copper coatings.

The antibacterial mechanisms of copper are summarized in Fig. 1. Copper is capable of entering into bacteria and generating reactive oxygen species, which leads to damage of the cytoplasmic membrane integrity and degradation of DNA and RNA, resulting in cell lysis, cytoplasm leakage, and bacterial death. It is interesting to find that bacteria have copper resistance systems which prolong bacterial survival, but the system cannot completely protect bacteria from the antibacterial performance and thus cannot threaten the final antibacterial effects of copper [8].

Osteoinduction and Signaling Pathways

Osteoinductivity is another important property of BGSs. The osteoinduction phenomenon can be divided into three principles: (1) osteoinductive materials should be able to recruit mesenchymal cells to bone graft surfaces, (2) the materials should

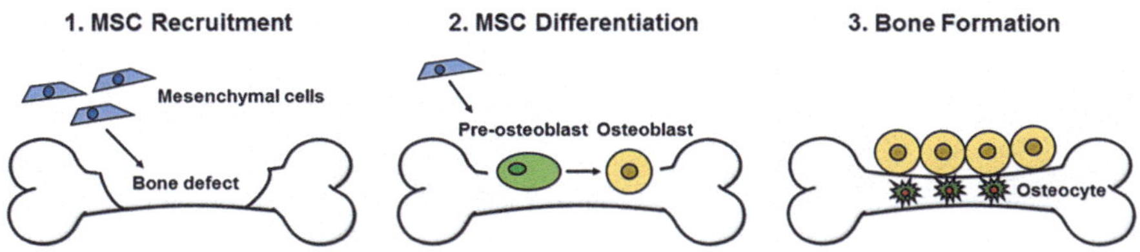

Fig. 2 Principles of osteoinductive materials [9]

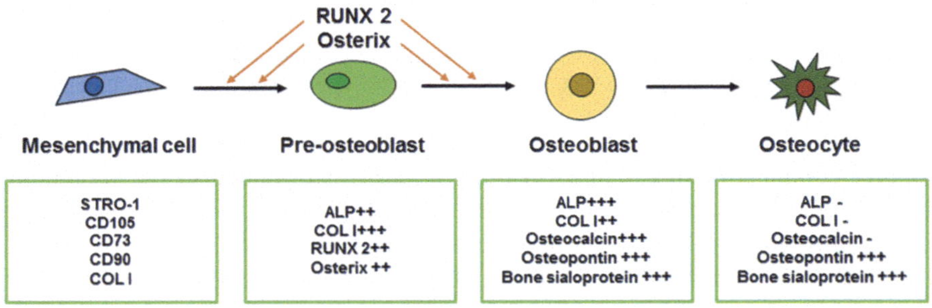

Fig. 3 Osteoblast lineage cells and the differentiation markers for MSCs, pre-osteoblasts, osteoblasts, and osteocytes [9]

promote mesenchymal differentiation into bone-forming osteoblasts, and (3) osteoblasts should form ectopic bone in vivo (Fig. 2) [9].

The multipotent mesenchymal stem cells (MSCs) are capable of differentiating into osteoblasts, adipocytes, chondrocytes, myocytes, and fibroblasts, and play a critical role in cell-based therapy [10]. Osteoblast lineage cells are a group of cells that include mesenchymal cells, pre-osteoblasts, osteoblasts, bone-lining cells, and osteocytes [11]. Osteoblasts originate from mesenchymal progenitors and then via the proliferation and differentiation of pre-osteoblasts to form matured osteoblasts. The functional osteoblasts secrete an extracellular matrix (ECM) and then deposit hydroxyapatite on collagen to form bones. At the final stage of bone development, the osteoblasts either become bone-lining cells, which cover the surface of bone to prevent certain ions, or mature into osteocytes entrapped in the bone matrix to maintain a matrix and keep calcium homeostasis [12]. During the differentiation from MSCs into osteoblasts, there are various phenotypic markers such as Runt-related transcription factor 2 (RUNX2), osteoblast-specific transcription factor (osterix or OSX), alkaline phosphatase (ALP), type I collagen (COL I), osteocalcin (OCN), osteopontin (OPN), and bone sialoprotein (BSP) as shown in Fig. 3. RUNX2 and OSX are the essential and sufficient transcription factors during mesenchymal differentiation into osteoblasts. RUNX2 is the key regulator for determining cell fate and is needed for the expression of non-collagenous proteins such as OCN and BSP. OSX, as a downstream factor of RUNX2, ensures the full differentiation to osteoblasts [13]. ALP is an early marker of osteoblast differentiation and participates in the initiation of mineralization through producing Pi to form CaP crystals

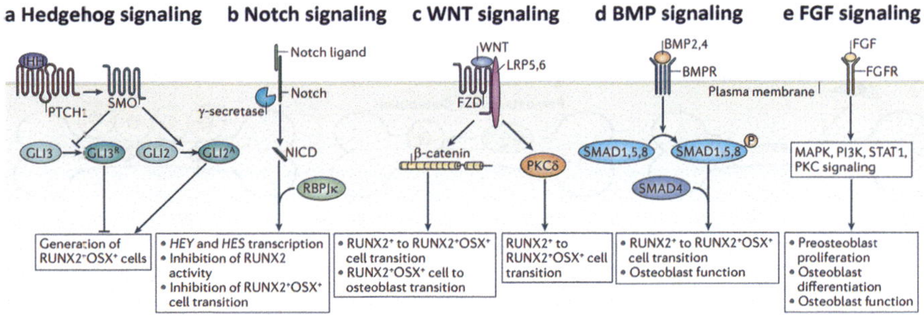

Fig. 4 Main signaling pathways of osteoblast differentiation (Adapted from Long F, *Nat Rev Mol Cell Biol*, 13 (1):27–38, 2012). (**a**) Hedgehog signaling, (**b**) Notch signaling, (**c**) WNT signaling, (**d**) BMP signaling, (**e**) FGF signaling

with the combination of Ca^{2+}. COL I is the primary extracellular matrix protein in bone. Both non-collagenous proteins BSP and OCN play a crucial role in hydroxyapatite formation on COL I and bone turnover [9]. The abovementioned factors have been widely used as biomarkers for investigating osteoinductivity of biomaterials at different osteo-differentiation stages (Fig. 3).

Osteoblast differentiation is regulated by various signaling pathways, and the main pathways are shown in Fig. 4. In the hedgehog signaling pathway, Indian hedgehog (IHH) binds to the receptor Patched homologue 1 (PTCH1) and activates signaling through Smoothened (SMO), thus impeding the production of the proteolytically cleaved GLI3 repressor ($GLI3^R$) and increasing the production of the GLI2 activator ($GLI2^A$). While derepression of $GLI3^R$ is able to produce $RUNX2^+$cells, both derepression of $GLI3^R$ and activation of $GLI2^A$ are required for generation of the $RUNX2^+$ and OSX^+ cells. The Notch signaling pathway suppresses osteoblast differentiation. In the WNT signaling pathway, wingless-type (WNT) ligands are important in the development and homeostasis of bone metabolism through both canonical β-catenin dependent and non-canonical protein kinase Cδ (PKCδ) signaling pathways. During β-catenin dependent WNT signaling, WNT protein binds to its receptors Frizzled (FZD) and lipoprotein receptor-related protein 5 (LRP5) or LRP6, leading to the stabilization and translocation of cytosolic β-catenin into the nucleus where it binds transcription factors including members of the T-cell factor (TCF) and lymphoid enhancer binding factor (LEF-1) family, and thus upregulates the transcription of *RUNX2* and *OSX*. In the BMP signaling pathway, bone morphogenetic proteins (BMPs) bind to their receptors and subsequently phosphorylate SMAD1, SMAD5, or SMAD8, which are capable of forming a complex with SMAD4, and then enter the nucleus to regulate gene expression, finally stimulating the transcription of *RUNX2/OSX* and osteoblast differentiation. In the FGF signaling pathway, fibroblast growth factors (FGFs) bind to cell surface Tyr kinase FGF receptors (FGFRs), resulting in the activation of multiple signaling modules and thus regulating pre-osteoblast proliferation and differentiation [11].

Biomaterials interplay with cells via an ECM. Biomaterial–ECM–cell interactions involve integrins that are a family of transmembrane αβ heterodimer and

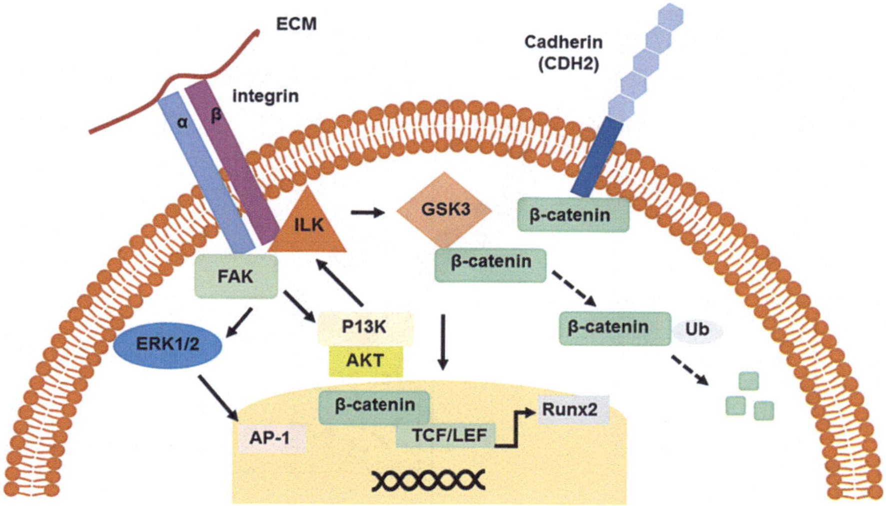

Fig. 5 Osteogenic signaling mediated by integrins [14]

convey the outside signals into the cells through a cascade of signaling pathways (Fig. 5). The interaction between the ligands in the ECM and integrins across the cell membrane induces the recruitment and phosphorylation of focal adhesion kinase (FAK) and activation of phosphatidylinositol 3-kinase (PI3K) and mitogen-activated protein kinase (MAPK) ERK1/2, resulting in stimulation of *RUNX2* transcription and osteogenic differentiation. In another way, integrins recruit and activate the integrin-linked kinase (ILK) and PI3K/AKT, leading to the inhibition of glycogen synthase kinase-3b (GSK3-b) and ubiquitin (Ub)-mediated β-catenin degradation. Then, the accumulated β-catenin proteins translocate into the nucleus and stimulate osteogenic differentiation via the WNT/β-catenin signaling pathway [14].

Angiogenesis and Osteogenesis

Blood supply is a principal issue in bone reconstruction when using orthopedic implants. Vascularization plays an important role in the bone defect healing process by providing cells with nutrients, and thus angiogenesis is believed to play an essential role in the osteogenesis of the implants [15]. The interaction between osteogenesis and angiogenesis is regulated by the factors produced by osteoblasts and endothelial cells and their precursors. The most important angiogenic factor, vascular endothelial growth factor (VEGF), also promotes osteogenesis. As shown in Fig. 6, at the normal physiological levels, VEGF maintains bone homeostasis, but too little VEGF disturbs osteogenic differentiation and too much VEGF upregulates osteoclast recruitment, resulting in bone resorption. During bone restoration,

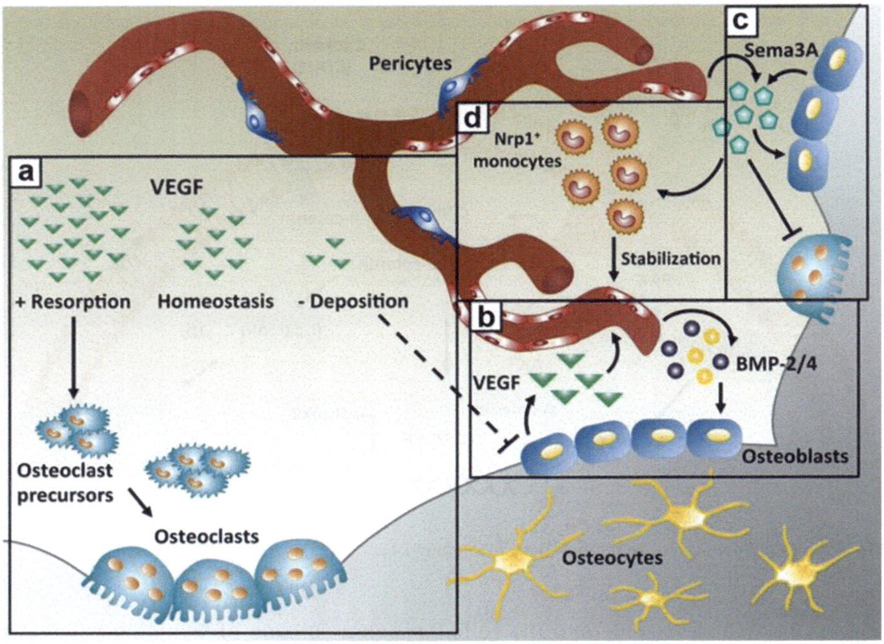

Fig. 6 Interplay between angiogenesis and osteogenesis (Adapted from Grosso A, et al., *Front Bioeng Biotechnol*, 5:68, 2017)

VEGF is generated by osteoblasts, and increases migration and proliferation of endothelial cells. In sequence, endothelial cells secrete osteogenic factors including BMP-2 and BMP-4 to promote osteogenic differentiation. Besides, VEGF controls Sema3A expression of endothelial cells in a dose-dependent manner and Sema3A from different sources inhibits osteoclast differentiation, thus promoting bone deposition. Moreover, Sema3A is able to recruit neuropilin 1-expressing ($Nrp1^+$) monocytes, which supports vessel stabilization [16].

Positively Charged Polymer Coatings with Antimicrobial and Osteoinductive Properties

Various antimicrobial polycations are capable of controlling orthopedic implant-associated infections, and thus prevent inflammatory osteolysis, reduce loosening of orthopedic implants, and improve their osseointegration. Furthermore, in order to improve the osteoinductivity of implants and speed up the bone healing process, some polycations with both antimicrobial and osteoinductive properties have been investigated. Shi et al. reported that a coating consisting of cationic polyhexamethylene biguanidine (PHMB) and polydopamine had excellent antibacterial efficacy and upregulated the expression of osteogenic marker genes including *Runx2*, *Alp*, *Opn*, *Ocn*, and *Bsp* in bone MSCs from rats [17]. Schaer et al. reported that the hydrophobic polycation *N*,*N*-dodecyl,methyl-polyethylenimine coatings on the surfaces of titanium and stainless steel resisted *S. aureus* biofilm formation and improved the healing of orthopedic fractures in a sheep trauma model [18]. Makihira et al. found that a cationic antimicrobial peptide (JH8194) immobilized on titanium inhibited biofilm formation from *Porphyromonas gingivalis* and enhanced the differentiation of mouse pre-osteoblast MC3T3-E1 cells. The mRNA expression of *Runx2* and *Opn*, and ALP activity obviously increased compared with the control titanium [19]. Tripathi et al. added positively charged arginine residues to a KLD peptide that possessed three positive and three negative charges with a net charge of zero, and thus obtained a series of positively charged KLD-R/2R/3R/4R peptides that significantly improved the antimicrobial activity of KLD and did not show any cytotoxicity [20]. In addition, KLD and KLD-2R/3R enhanced mineralized nodule formation and stimulated the expressions of *Runx2* and *Col-1* in bone marrow stromal cells. It was interesting to find that KLD-2R improved bone regeneration in adult Sprague Dawley rats in a dose-dependent manner and KLD-2R at 250 μg/rat exhibited higher activity than KLD in vivo [20]. Lee et al. studied the in vivo effects of the antibacterial human beta-defensin-2 (hBD2) and -3 (hBD3) on bone healing. The defensin-overexpressing rat bone marrow stromal cells were implanted into the *S. aureus*-contaminated rat calvarial defect model, and the results revealed that hBD2 and hBD3 significantly reduced viable *S. aureus* in vivo and remarkably enhanced new bone formation to improve bone healing [21]. Choe et al. reported that the innate defense regulator peptide IDR-1018 promoted the recruitment of macrophages to the site of infections, improved *S. aureus* clearance, inhibited excess cytokine production, and preserved osseointegration [22] (Table 2).

The antimicrobial mechanisms of the positively charged polymers have been explained in "Polycations" section, while their osteoinductive mechanisms still remain to be elucidated. The surface chemistry of the polycations may modulate the structure of fibronectin or other ligands such as OPN and COL I, which conveys signals into cells through αβ integrins, as illustrated in Fig. 5, to induce osteogenesis through a cascade of signaling pathways [14, 23]. Additionally, the positive charge may influence the NO synthase (NOS) system that has a critical role in maintaining the bone homeostasis [24, 25]. Besides the direct osteoinductive effects of the polycations, for the bacteria-contaminated model, especially *S. aureus* infected implants, the antimicrobial property of polycations also contributed to the inhibition of the negative effects of *S. aureus* on osteoblasts. During bone homeostasis, osteoblasts and osteoclasts are responsible for the formation and resorption of the bone ECM, respectively. Once infection sits in, *S. aureus* can be internalized and survived in osteoblasts. Through the interaction with the extracellular/intracellular receptors, *S. aureus* or its components may induce over-production of cytokines, chemokines, and growth factors to increase inflammation, recruit inflammatory cells to stimulate osteoclastogenesis and thus increase bone resorption, lead to osteoblast death through apoptosis or necrosis, and eventually result in bone loss or failure of bone regeneration [26]. Therefore, through killing bacteria, the antimicrobial polymers

Table 2 Antimicrobial and osteoinductive polycations

Polycations	Antimicrobial effects	Osteoinductive characterizations	Cell/animal models
Polyhexamethylene biguanidine and polydopamine [17]	Killing *S. aureus*	Upregulating expressions of *Runx2*, *Alp*, *Opn*, *Ocn*, and *Bsp*	Rat BMSCs
N,N-dodecyl,methyl-polyethylenimine [18]	Resisting biofilm formation from *S. aureus*	Promoting bone healing	Sheep trauma model
Peptide JH8194 [19]	Inhibiting biofilm formation from *Porphyromonas gingivalis*	Enhancing ALP activity and expressions of *Runx2* and *Opn*	Mouse pre-osteoblast MC3T3-E1 cells
Arginine modified KLD-12 peptide [20]	Inhibiting growth of *E. coli*, *S. aureus*, *B. Subtilis*, *P. aeruginosa*	Increasing ALP activity and expressions of *Runx2*, *Col 1*, and *BMP-2*	Rat calvarial osteoblasts
Human beta-defensin-2 and -3 [21]	Decreasing viable *S. aureus*	Promoting bone healing in *S. aureus*-contaminated bone defects	Rat calvarial defect model
Peptide IDR-1018 [22]	Improving *S. aureus* clearance	Preserving osseointegration	Murine model of implant infection

are capable of preventing the disruption of *S. aureus* on bone homeostasis. Additionally, although some polycations have not been fabricated as functional coatings, they are able to be coated on the orthopedic implants to endow the surfaces with antimicrobial and osteoinductive properties.

Cationic Copper with Antimicrobial and Osteoinductive Activities

Due to the excellent biocompatibility of copper as presented in "Metal cations" section, herein, we focus on the copper-bearing orthopedic implants that exhibit antimicrobial and osteoinductive activities. Wu et al. found that copper (5% Cu)-containing mesoporous bioactive glass (Cu-MBG) scaffolds significantly inhibited the bacterial viability of *Escherichia coli* compared to MBG without Cu in vitro. Cu-MBG scaffolds stimulated HIF-1 and VEGF expression in human bone marrow stromal cells, and they significantly increased the expressions of *ALP*, *OPN*, and *OCN*, and thus promoted the osteogenic differentiation of the bone marrow stromal cells [27]. Burghardt et al. galvanically deposited copper to a titanium surface, and found the released Cu^{2+} at concentrations from 0.3 to 1.75 mM inhibited the growth of *S. aureus* and cleared the adherent bacteria from the implant surface within 24 h in vitro. The released copper at a concentration around 0.1 mM promoted the proliferation of human bone marrow derived MSCs, increased ALP activity, enhanced the expression of *COL I*, *OPN*, *osteoprotegerin*, and stimulated mineralization, revealing that the coating stimulated osteogenic differentiation [28]. Wu et al. found that Cu-incorporated TiO_2 coatings had potent antimicrobial ability against *Escherichia coli*, with an antibacterial ratio of above 90% at Cu concentrations of 10 and 100 nM in vitro. The incorporated copper ions significantly enhanced the expression of the angiogenesis-associated *Vegf* and promoted the expressions of osteogenesis-related *Opn*, *Bmp-2* and *Col I* as well as ALP activity of rat bone marrow stem cells, compared with the pristine coating, which revealed that copper stimulated the coupling activities of angiogenesis and osteogenesis [29].

The above in vitro investigations clearly demonstrated that Cu^{2+} possesses antimicrobial and osteoinductive properties. Furthermore, in vivo animal tests were employed to observe the bone restoration on the orthopedic implants. Prinz et al. reported that copper-coated nails, with an average copper load of 1 mg/mm^2, prevented bacterial infection of *S. aureus*. The copper-coated nails were used to repair the fractured tibia of female white New Zealand rabbits and promoted new bone formation through a stimulation of callus formation, which might be attributed to the enhanced angiogenesis activity stimulated by Cu^{2+} [30]. Ren et al. implanted the 317L-Cu stainless steel (Cu-SS) into 2-month-old female Sprague Dawley rats and found more new bone tissue formed around the Cu-SS implant compared with the control 317L SS without Cu. The bone volume on the Cu-SS sample increased by 10% at day 3 post-operation, while the bone volume on SS without Cu reduced by 10% [31] (Table 3).

Cationic copper is able to kill bacteria via the mechanisms illustrated in Fig. 1. Through the generation of reactive oxygen species, copper damages the cell

Table 3 Antimicrobial and osteoinductive Cu-containing materials

Materials	Antimicrobial effects	Osteoinductive characterizations	Cell/animal models
Cu-containing mesoporous bioactive glass scaffolds [27]	Inhibiting bacterial viability of *E. coli*	Increasing expressions of *ALP*, *OPN*, and *OCN*; stimulating angiogenesis-associated HIF-1α and VEGF expression	Human bone marrow stromal cells
Cu-containing titanium surfaces [28]	Inhibiting growth of *S. aureus*	Increasing ALP activity, enhancing expressions of *COL I*, *OPN*, *osteoprotegerin*, and stimulating mineralization	Human bone marrow derived MSC
Cu-incorporated TiO_2 coatings[29]	Killing *E. coli*	Promoting expressions of *Opn*, *Bmp-2*, and *Col I* as well as ALP activity; enhancing expressions of VEGF and HIF-1α	Rat bone marrow stem cells
Cu-coated nails [30]	Preventing *S. aureus* attachment and biofilm formation	Stimulating bone formation	Female white New Zealand rabbits
317L-Cu stainless steel [31, 32]	Killing *S. aureus* and *E. coli*	Stimulating APL activity and expressions of *Runx2*, *Col I*, and *Opn*	Mouse MC3T3-E1 pre-osteoblasts

membrane, enzymes [29], and nucleic acids, finally leading to bacterial death [8]. Copper not only has good biocompatibility [7], but also exhibits an osteoinductive property which might be attributed to the results that copper increased the expression of VEGF [29] and subsequently promoted bone regeneration through VEGF-associated angiogenesis and osteogenesis [16] as illustrated in Fig. 6.

Final Comment and Future Directions

Orthopedic implant-associated infections, especially caused by *S. aureus*, have been a serious clinical problem. Recently, the positively charged biomaterials have been explored and the research reveals that some cationic biomaterials possess both antibacterial and osteoinductive properties and are capable of reducing bacterial infections and improving the osseointegration of the bone implants, simultaneously. In this chapter, we summarized the antibacterial and osteoinductive mechanisms of positively charged biomaterials and presented some studies about cationic antibacterial coatings with osteoinductive properties, which will offer some new clues for the design of novel antimicrobial materials and benefit the development of orthopedic biomaterials.

With the in-depth research on mechanisms of anti-bacteria and osteogenesis and the development of various new biomaterials, in the future, novel composite biomaterials consisting of antibacterial and osteoinductive cationic components will be applied to orthopedic biomaterials to prevent biofilm formation and increase osseointegration of implants. In addition, blood supply is of great importance in bone healing, and therefore the effect of cationic antimicrobial materials on acceleration of angiogenesis and osteogenesis will be a key concern on the development of advanced orthopedic biomaterials.

Acknowledgments This work was financially supported by the National Key R&D Program of China (2018YFC1105402 and 2017YFA0207202), the National Natural Science Foundation of China (21706222 and 21875189), Key R&D Program of Jiangsu Province (BE2017740), the open research fund of Key Laboratory for Organic Electronics and Information Displays, and the Fundamental Research Funds for the Central Universities.

References

1. Lobb DC, DeGeorge BR, Chhabra AB (2019) Bone graft substitutes: current concepts and future expectations. J Hand Surg Am 44(6):497-505.e2
2. Lu H, Liu Y, Guo J, Wu H, Wang J, Wu G (2016) Biomaterials with antibacterial and osteoinductive properties to repair infected bone defects. Int J Mol Sci 17(3):334
3. Campoccia D, Montanaro L, Arciola CR (2006) The significance of infection related to orthopedic devices and issues of antibiotic resistance. Biomaterials 27(11):2331–2339
4. Poortinga AT, Bos R, Norde W, Busscher HJ (2002) Electric double layer interactions in bacterial adhesion to surfaces. Surf Sci Rep 47(1):1–32

5. Ding X, Duan S, Ding X, Liu R, Xu F-J (2018) Versatile antibacterial materials: an emerging arsenal for combatting bacterial pathogens. Adv Funct Mater 28(40):1802140
6. Croes M, Bakhshandeh S, van Hengel IAJ, Lietaert K, van Kessel KPM, Pouran B, van der Wal BCH, Vogely HC, Van Hecke W, Fluit AC, Boel CHE, Alblas J, Zadpoor AA, Weinans H, Amin Yavari S (2018) Antibacterial and immunogenic behavior of silver coatings on additively manufactured porous titanium. Acta Biomater 81:315–327
7. Heidenau F, Mittelmeier W, Detsch R, Haenle M, Stenzel F, Ziegler G, Gollwitzer H (2005) A novel antibacterial titania coating: metal ion toxicity and in vitro surface colonization. J Mater Sci Mater Med 16(10):883–888
8. Grass G, Rensing C, Solioz M (2011) Metallic copper as an antimicrobial surface. Appl Environ Microbiol 77(5):1541–1547
9. Miron RJ, Zhang YF (2012) Osteoinduction: a review of old concepts with new standards. J Dent Res 91(8):736–744
10. Hu L, Yin C, Zhao F, Ali A, Ma J, Qian A (2018) Mesenchymal stem cells: cell fate decision to osteoblast or adipocyte and application in osteoporosis treatment. Int J Mol Sci 19(2):360
11. Long F (2012) Building strong bones: molecular regulation of the osteoblast lineage. Nat Rev Mol Cell Biol 13(1):27–38
12. Krishnan V, Dhurjati R, Vogler EA, Mastro AM (2010) Osteogenesis in vitro: from pre-osteoblasts to osteocytes. In Vitro Cell Dev Biol-Animal 46(1):28–35
13. Shahi M, Peymani A, Sahmani M (2017) Regulation of bone metabolism. Rep Biochem Mol Biol 5(2):73-82
14. Marie PJ, Haÿ E, Saidak Z (2014) Integrin and cadherin signaling in bone: role and potential therapeutic targets. Trends Endocrinol Metab 25(11):567–575
15. Götz W, Reichert C, Canullo L, Jäger A, Heinemann F (2012) Coupling of osteogenesis and angiogenesis in bone substitute healing—a brief overview. Ann Anat 194(2):171–173
16. Grosso A, Burger MG, Lunger A, Schaefer DJ, Banfi A, Maggio ND (2017) It takes two to tango: coupling of angiogenesis and osteogenesis for bone regeneration. Front Bioeng Biotechnol 5:68
17. Shi L, Zhang W, Yang K, Shi H, Li D, Liu J, Ji J, Chu PK (2015) Antibacterial and osteoinductive capability of orthopedic materials via cation–π interaction mediated positive charge. J Mater Chem B 3(5):733–737
18. Schaer TP, Stewart S, Hsu BB, Klibanov AM (2012) Hydrophobic polycationic coatings that inhibit biofilms and support bone healing during infection. Biomaterials 33(5):1245–1254
19. Makihira S, Shuto T, Nikawa H, Okamoto K, Mine Y, Takamoto Y, Ohara M, Tsuji K (2010) Titanium immobilized with an antimicrobial peptide derived from histatin accelerates the differentiation of osteoblastic cell line, MC3T3-E1. Int J Mol Sci 11(4):1458–1470
20. Tripathi JK, Pal S, Awasthi B, Kumar A, Tandon A, Mitra K, Chattopadhyay N, Ghosh JK (2015) Variants of self-assembling peptide, KLD-12 that show both rapid fracture healing and antimicrobial properties. Biomaterials 56:92–103
21. Lee PH, Chen MY, Lai YL, Lee SY, Chen HL (2018) Human beta-defensin-2 and -3 mitigate the negative effects of bacterial contamination on bone healing in rat calvarial defect. Tissue Eng A 24(7–8):653–661
22. Choe H, Narayanan AS, Gandhi DA, Weinberg A, Marcus RE, Lee Z, Bonomo RA, Greenfield EM (2015) Immunomodulatory peptide IDR-1018 decreases implant infection and preserves osseointegration. Clin Orthop Relat Res 473(9):2898–2907
23. Keselowsky BG, Collard DM, García AJ (2005) Integrin binding specificity regulates biomaterial surface chemistry effects on cell differentiation. Proc Natl Acad Sci U S A 102(17):5953–5957
24. Sabanai K, Tsutsui M, Sakai A, Hirasawa H, Tanaka S, Nakamura E, Tanimoto A, Sasaguri Y, Ito M, Shimokawa H, Nakamura T, Yanagihara N (2008) Genetic disruption of all NO synthase isoforms enhances BMD and bone turnover in mice in vivo: involvement of the renin-angiotensin system. J Bone Miner Res 23(5):633–643
25. Zhang W, Liu J, Shi H, Yang K, Wang P, Wang G, Liu N, Wang H, Ji J, Chu PK (2016) Communication between nitric oxide synthase and positively-charged surface and bone formation promotion. Colloids Surf B: Biointerfaces 148:354–362

26. Josse J, Velard F, Gangloff SC (2015) *Staphylococcus aureus* vs. osteoblast: relationship and consequences in osteomyelitis. Front Cell Infect Microbiol 5:85
27. Wu C, Zhou Y, Xu M, Han P, Chen L, Chang J, Xiao Y (2013) Copper-containing mesoporous bioactive glass scaffolds with multifunctional properties of angiogenesis capacity, osteostimulation and antibacterial activity. Biomaterials 34(2):422–433
28. Burghardt I, Lüthen F, Prinz C, Kreikemeyer B, Zietz C, Neumann H-G, Rychly J (2015) A dual function of copper in designing regenerative implants. Biomaterials 44:36–44
29. Wu Q, Li J, Zhang W, Qian H, She W, Pan H, Wen J, Zhang X, Liu X, Jiang X (2014) Antibacterial property, angiogenic and osteogenic activity of Cu-incorporated TiO2 coating. J Mater Chem B 2(39):6738–6748
30. Prinz C, Elhensheri M, Rychly J, Neumann H-G (2017) Antimicrobial and bone-forming activity of a copper coated implant in a rabbit model. J Biomater Appl 32(2):139–149
31. Ren L, Wong HM, Yan CH, Yeung KW, Yang K (2015) Osteogenic ability of Cu-bearing stainless steel. J Biomed Mater Res Part B Appl Biomater 103(7):1433–1444
32. Ren L, Yang K, Guo L, Chai H-w (2012) Preliminary study of anti-infective function of a copper-bearing stainless steel. Mater Sci Eng C 32(5):1204–1209

Peptide-functionalized Biomaterials with Osteoinductive or Anti-biofilm Activity

Jennifer Patterson

Abstract Peptides are short sequences of amino acids. Peptides with biological functionality can be derived from the active domain of proteins or determined from peptide screening experiments. Combined with modern chemical techniques to facilitate peptide synthesis, this leads to peptide modification as an interesting approach to render synthetic biomaterials bioactive. Peptides have been used to functionalize implant surfaces as well as bulk biomaterials, and they can be incorporated within controlled release systems. This chapter considers both osteoinductive peptides and anti-biofilm peptides with the goals to improve bone regeneration and reduce implant-associated infection, respectively.

Keywords Peptide synthesis · Phage display · Surface modification · Hydrogels Nanofibers · Titanium · Controlled release · Osteoinductive peptides · Osteogenic differentiation · Bone morphogenetic protein · Osteogenic growth peptide Hydroxyapatite · Antimicrobial peptides · Anti-biofilm · Bone tissue engineering Dentistry

Introduction

Proteins are the functional building blocks of cells and tissues. They are made up of differing combinations of 20 natural amino acids (Table 1), which have an amino group on one end, a carboxylic acid group on the other end, and a variable side chain coming off the central carbon atom. Within a protein or polypeptide, the amino acid residues are joined together by the formation of an amide bond between the carboxylic acid of one amino acid and the amino group of the next. The sequences of natural proteins are determined from the genetic code whereby DNA is first tran-

J. Patterson (✉)
Independent Consultant, Leuven, Belgium

Department of Imaging and Pathology, KU Leuven, Leuven, Belgium
e-mail: pattersn@alumni.princeton.edu

B. Li et al. (eds.), *Racing for the Surface*,
https://doi.org/10.1007/978-3-030-34471-9_6

Table 1 The 20 standard amino acids, their abbreviations, and types of side chains

Amino acid	Three-letter abbreviation	One-letter abbreviation	Type of side chain
Alanine	Ala	A	Hydrophobic
Cysteine	Cys	C	Thiol
Aspartic acid	Asp	D	Negatively charged
Glutamic acid	Glu	E	Negatively charged
Phenylalanine	Phe	F	Hydrophobic
Glycine	Gly	G	None (hydrogen)
Histidine	His	H	Positively charged
Isoleucine	Ile	I	Hydrophobic
Lysine	Lys	K	Positively charged
Leucine	Leu	L	Hydrophobic
Methionine	Met	M	Hydrophobic
Asparagine	Asn	N	Polar uncharged
Proline	Pro	P	Special
Glutamine	Gln	Q	Polar uncharged
Arginine	Arg	R	Positively charged
Serine	Ser	S	Polar uncharged
Threonine	Thr	T	Polar uncharged
Valine	Val	V	Hydrophobic
Tryptophan	Trp	W	Hydrophobic
Tyrosine	Tyr	Y	Hydrophobic

scribed to messenger RNA and then translated to protein with each three-base-long codon specifying an amino acid. Similar to proteins, peptides are shorter chains of amino acids, also linked by amide bonds, that are typically less than 50 amino acids in length. With advances in chemical and molecular biology techniques, peptides can be chemically synthesized and modified, and de novo sequences can be designed. As a treatment strategy, peptides offer many of the advantages of protein therapeutics while addressing some of their limitations.

This chapter focuses on the use of peptides in biomaterial-based approaches for bone regeneration and anti-biofilm applications. First, a perspective is given on peptide design with two main approaches being the derivation of amino acid sequences from natural proteins and the screening of de novo libraries of peptides for specific functions. Then, relevant peptide synthesis techniques are presented along with chemical methods that allow functionalization of surfaces and bulk materials with peptides. The main focus of the chapter is the introduction of specific peptides from two main classes: osteoinductive peptides and anti-biofilm peptides. In both cases, in vitro assays to assess peptide activity as well as in vivo studies to demonstrate efficacy are presented. To date, however, few studies have combined osteoinductive peptides with antimicrobial peptides to stimulate bone formation while simultaneously reducing infection. The chapter concludes with some future perspectives on the topic.

Peptide Design

Peptides Derived from Proteins

The primary structure of a protein is its chemical composition determined by the specific sequence of amino acids. Protein secondary structure, which includes units such as α-helices and β-sheets, arises from the organization of portions of the protein by weak intramolecular bonds, such as hydrogen bonding and ionic interactions. Hydrogen bonding can occur between atoms on the protein backbone, but in other cases, these interactions are driven by the chemical composition of the side chains of the amino acids. With some exceptions, these side chains can be either positively or negatively charged, polar uncharged, or hydrophobic (Table 1). In de novo protein design, specifying the pattern of hydrophobic and hydrophilic amino acids in the protein sequence is in many cases sufficient to drive protein folding into α-helices [1] or β-sheets [2], as measured by circular dichroism. As with proteins, peptides with specific patterns of amino acids can self-assemble to form these structural units, such as RADA16 [RADARADARADARADA [3]; single-letter amino acid abbreviations used throughout this chapter, see Table 1] and Q11 [QQKFQFQFEQQ [4]] that form β-sheets or LD_6 [LIVAGD [5]] that forms α-helical intermediates. These self-assembling peptides have also been used to generate cytocompatible hydrogels for biomedical applications, but further discussion of their use is outside the scope of the current chapter.

Likewise, these interactions can occur intermolecularly, that is, between two proteins, leading to receptor–ligand pairing or enzyme–substrate docking for example. As the amino acid sequences that are involved in these interactions are usually specific, they can serve as a starting point for the design of bioactive peptides that mimic these active domains of the protein. One of the most simple and commonly used peptide mimetics is the cell adhesive RGD sequence. RGD is found in several proteins including fibronectin, laminin, collagen, vitronectin, and others and mediates cell attachment through various integrins on the surface of cells [6]. Cell attachment to RGD is sequence specific, as evidenced by reduced binding to scrambled peptides such as RDG, and exposure to soluble RGD causes cells to detach from their substrate [6]. Functionalization of an otherwise nonadhesive biomaterial with RGD will allow cells to attach to and spread on or within the material. Patterning of RGD can also restrict cell attachment to specific areas, as has been shown by observing encapsulated cell spreading in inert poly(ethylene glycol) (PEG) hydrogels compared to PEG hydrogels with uniform functionalization with RGD and with spatially patterned RGD (Fig. 1) [7].

Besides RGD, there are a number of other integrin binding peptides that can be used as cell adhesion ligands, as summarized in Table 2. The use of integrin-binding peptides is a generic strategy in tissue engineering approaches, as most cell types are adherent and thus will attach, spread, proliferate, and migrate on the peptide-functionalized materials. Knowledge of the surface receptor profile of different cells can aid in the selection of which ligand or ligands to choose to be able to bind a

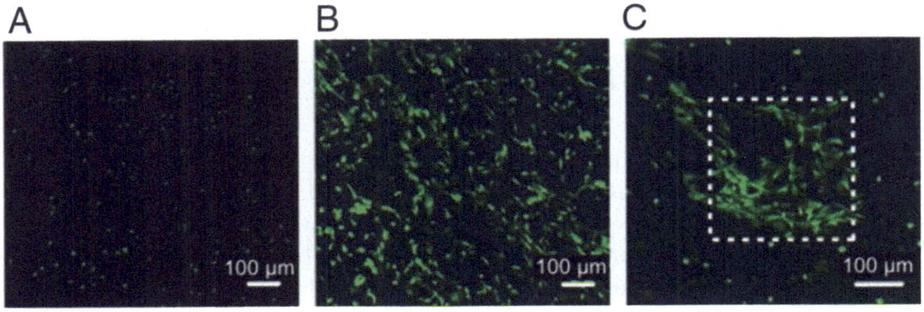

Fig. 1 Effect of the addition of RGD uniformly throughout compared to spatially patterned within PEG hydrogels that were formed by click chemistry on 3 T3 fibroblast spreading. (**A**) No RGD. (**B**) Uniform RGD. (**C**) Patterned RGD. The scale bars are 100 μm. (Reprinted by permission from Springer Nature, *Nature Materials*, ©2009 [7])

Table 2 Cell adhesion (integrin binding) peptides

Peptide sequence	Binding partners	Source and/or function	References
RGD	Several integrins	Fibronectin and others	[6]
YIGSR	Integrins $\alpha_1\beta_1$, $\alpha_2\beta_1$	Laminin; binds endothelial cells	[8–10]
VAPG	Galactoside-binding protein	Elastin; binds smooth muscle cells	[8, 11]
SIKVAV	Integrins α_3, α_6, β_1; $\alpha_V\beta_3$	Laminin; binds endothelial cells	[12, 13]
RNIAEIIKDI	Integrins $\alpha_3\beta_1$, $\alpha_6\beta_1$, $\alpha_6\beta_4$, $\alpha_7\beta_1$	Laminin	[9]
EILDVPST	Integrin $\alpha_4\beta_1$	Fibronectin; binds lymphoid and tumor cells	[14]
KLDAPT	Integrins α_4, β_1, β_7	Fibronectin; binds lymphoid cells	[15]
WQPPRARI	Heparin-binding	Fibronectin; binds endothelial cells and fibroblasts	[16]
PHSRN	Integrin $\alpha_5\beta_1$	Fibronectin	[17]
AEIDGIEL	Integrin $\alpha_9\beta_1$	Tenascin-C	[18, 19]
REDV	Integrin $\alpha_4\beta_1$	Fibronectin; binds endothelial cells	[20]

specific cell type. These peptides can then be used to functionalize biomaterials in both 2D and 3D. For example, nanofibrous matrices were formed from the self-assembly of peptide amphiphiles functionalized with either YIGSR or VAPG. The materials presenting YIGSR enhanced the attachment of endothelial cells, whereas the matrices containing VAPG promoted greater spreading of smooth muscle cells [8]. Hydrogels formed from photo-cross-linked PEG diacrylate combined with VAPG-functionalized PEG monoacrylate were also shown to preferentially support attachment of smooth muscle cells over fibroblasts, endothelial cells, and platelets [11]. Further, SIKVAV-conjugated chitosan hydrogels were demonstrated to promote angiogenesis and re-epithelialization in vivo [12]. These cell adhesive peptides can potentially be combined with peptides targeting other receptors, and the design of osteoinductive and anti-biofilm peptides that are based on naturally occurring domains within proteins are further described in sections "Osteoinductive Peptides" and "Anti-biofilm Peptides" of this chapter.

Moving beyond the direct use of peptides derived from natural proteins, these mimetics can be varied, often in a systematic way, by changing individual amino acids within the peptide sequence to increase or decrease the activity of the peptide. One good example of this is the use of collagen-derived peptides to explore the sequence specificity of matrix metalloproteinases (MMPs), which are a family of enzymes that are active during normal tissue remodeling processes and also in disease states. Starting from an eight-amino-acid-long MMP substrate sequence found in the α1 chain of type I collagen, GPQGIAGQ, Nagase and Fields systematically varied the amino acids in each position and found that the substitution of A with W led to increases in the Michaelis-Menten parameter k_{cat}/K_M and the relative rate of hydrolysis by MMP-1, MMP-8, and MMP-3 [21]. These peptide sequences were then used in innovative enzymatically degradable synthetic hydrogels formed by the Michael-type addition reaction of end-functionalized multi-arm PEG macromers with cross-linker peptides containing the MMP substrate sequences [22, 23]. Indeed, the hydrogels containing the modified GPQGIWGQ substrate degraded faster than the hydrogels containing the collagen-derived GPQGIAGQ sequence, and this allowed for faster remodeling of the matrix and migration of embedded fibroblasts (Fig. 2) [22]. These degradable hydrogels were also found to support bone formation in combination with the osteoinductive molecule, bone morphogenetic protein (BMP)-2 [23]. In a rat calvarial defect model, hydrogels that were not degradable but loaded with BMP-2 led to similar low amounts of bone formation as degradable hydrogels without BMP-2 (approximately 20% bone coverage); however, the

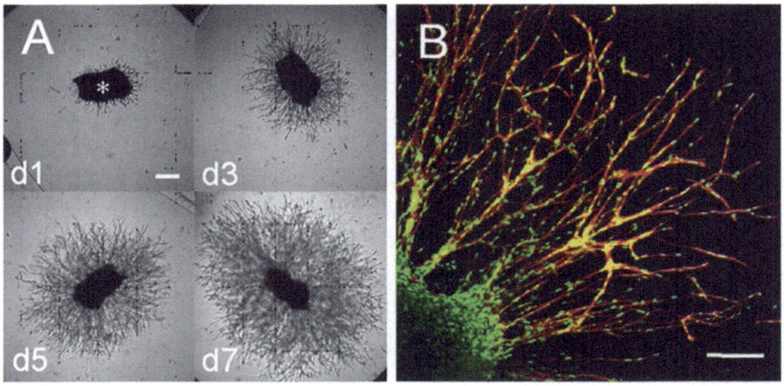

Fig. 2 Fibroblast migration into PEG hydrogels prepared through a Michael-type addition reaction with the MMP-sensitive GPQGIWGQ peptide cross-linker. (**A**) Phase contrast images after 1, 3, 5, and 7 days with scale bar = 250 μm. (**B**) Confocal image showing cell membranes and nuclei with scale bar = 150 μm. (Reprinted with permission from [22]. ©2003 National Academy of Sciences, USA)

hydrogels that were degradable and loaded with BMP-2 resulted in similar amounts of bone formation as the standardly used collagen sponge with BMP-2 (approximately 90% bone coverage) [23].

Library Screening Approaches

It is also possible to design peptides completely from scratch, and several methodologies have been developed to screen libraries of peptides for specific biological activity. In fact, half of the 2018 Nobel Prize in Chemistry was jointly awarded to George P. Smith and Sir Gregory P. Winter for developing the technique of phage display of peptides [24] and antibodies [25]. Phage display utilizes gene editing of the coat proteins of bacteriophage to incorporate peptide sequences that are displayed as part of the coat proteins. These phage can then be screened against target molecules for an interaction of interest and further amplified. The DNA within the selected phage can finally be sequenced to determine the relevant peptide sequence. Phage display can be useful in determining substrate specificity of enzymes. For example, to determine mutant substrates that are not cleaved by an enzyme, the coat proteins are modified with the library of peptide sequences, each with a tag on the terminus to allow for separation using antibody-coated beads (Fig. 3). After incubation with the enzyme, the phage that are still attached to the beads can be separated, washed, amplified, treated with enzyme, and sorted again. After several rounds of sorting and amplification, the individual phage can be isolated and sequenced. This approach has been used to identify substrates that are less susceptible to cleavage by the metalloproteinase ADAMTS13 [26], and by changing the method to select the phage of interest, one can identify enzyme substrates that are preferentially cleaved, peptides that bind to specific proteins or other molecules, etc. [24].

An alternative to phage display for peptide design is mixture-based oriented peptide libraries [27]. The mixture-based oriented peptide library approach screens the peptides in solution and relies on peptide sequencing methods (Edman degradation) to determine the sequence. It has been applied for determining enzyme substrates and protein interaction domains [28–31]. For example, to determine an enzyme substrate, in a first round, a totally degenerate peptide library with the amino (N-) termini acetylated is treated with the protease. The fragments that are generated have a free amine group on the N-terminus and are subjected to N-terminal sequencing. The amino acids that are most abundant in each position are then fixed in a second-generation library, which have the N-termini as free amines and the carboxy (C-) termini tagged with biotin. After cleavage with the protease, the C-terminal fragments and uncleaved peptides can be separated using immobilized avidin, and the N-terminal fragments can be sequenced. This approach has been used to generate substrates for a variety of MMPs [27]. Another approach to screen for enzyme substrates is a technique called cellular libraries of peptides substrates (CLiPS) [32]. A library of peptides containing potential substrates tagged with a ligand for a fluorescent probe are displayed on the surface of *E. coli*.

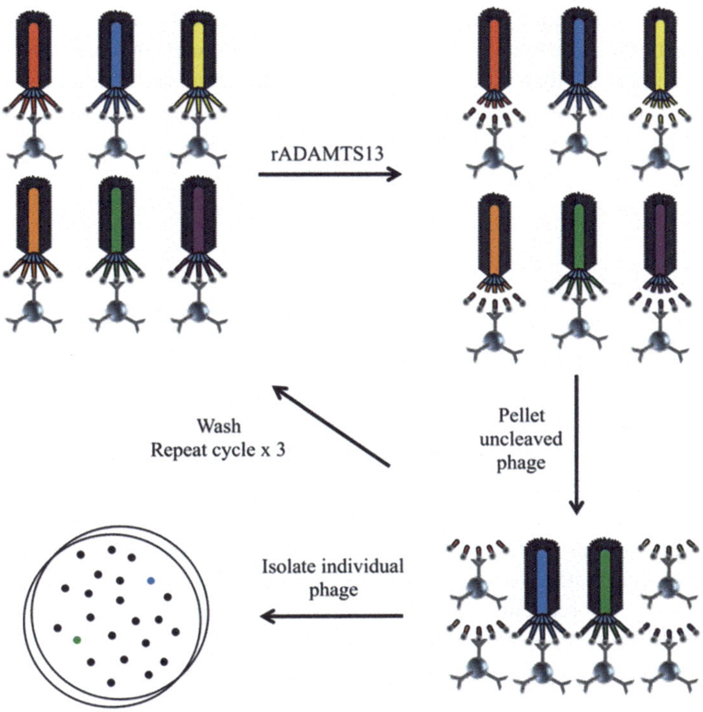

Fig. 3 Working principle of phage display. In this case, the substrates that are not cleaved by the enzyme ADAMTS13 are selected and amplified. (Reprinted under the terms of the Creative Commons Attribution License from Desch et al. ©2015 [26])

After treating the cells with the protease of interest, the intact peptides are able to be detected using fluorescence-activated cell sorting (FACS), and the clones with hydrolyzed substrates can be enriched by repeating this cycle several times. Compared to phage display and soluble peptide libraries, up to 10^4 copies of the substrate are displayed on a single cell, which allows quantitation of substrate conversion for an individual clone [32].

These techniques are commonly used to identify peptides that bind to target receptors or to natural ligands, for epitope mapping or mimicking, in drug discovery to find new enzyme inhibitors and receptor agonists and antagonists, for epitope discovery in vaccine development and diagnostics, and selection of DNA-binding motifs, among others [24]. Of relevance to biomaterials development, they have also been used to identify new substrates for various MMPs, such as MMP-2 [33], MMP-3 or -7 [34], MMP-11 [35], MMP-13 [36], and MT1-MMP [37]. These peptide substrates were evaluated with the PEG hydrogel system described above, and sequences resulting in hydrogels with increased susceptibility to MMP-1, MMP-2, or plasmin were identified [38, 39]. The faster degrading hydrogels supported increased proliferation of encapsulated fibroblasts as well as sprouting of endothelial cells in a chick aortic ring outgrowth assay [38, 39]. Similar to the cell adhesion peptides already described, the functionalization of scaffolds with enzyme substrate peptides also provides a generic strategy to support the matrix remodeling, proliferation, and migration of cells within the matrix. More specific peptides that are osteoinductive or anti-biofilm, which have been found by library screening approaches, are detailed in sections "Osteoinductive Peptides" and "Anti-biofilm Peptides" of this chapter.

Chemical Methods

Peptide Synthesis

During protein translation from messenger RNA, the nascent protein chain is synthesized from N- to C- terminus in the ribosome via the action of transfer RNA, which recognizes the codon in the RNA strand. A new amide bond is formed to attach the amino acid to the growing polypeptide chain. Specific codons initiate and terminate the synthesis of the protein. On the other hand, peptides are typically synthesized chemically using a technique called solid-phase peptide synthesis, which allows excess reagents and by-products of the reactions to be easily removed by washing and filtration after each step. They can also be synthesized in solution, although this requires additional work to separate the peptide intermediate after each step. In the case of solid-phase peptide synthesis, the peptide is synthesized on a solid support from the C- to N-terminus. Cleavable protecting groups typically block reactive functionalities on the side chains and amino group of the amino acid. The latter protecting group is often an Fmoc (base-labile) or Boc (acid-labile) group, which can be removed before the addition of the next amino acid. Merrifield was the first to describe the solid-phase peptide synthesis technique in 1963 [40], and the Fmoc/*t*-Bu orthogonal protecting group strategy has become the most common approach applied today. Solid-phase peptide synthesis typically proceeds by coupling of the first amino acid to the supporting resin followed by a deprotection step to remove the Fmoc group. The next amino acid is added, followed by deprotection, and so on until the last amino acid or other N-terminal group is added. Then, the peptide is cleaved from the resin, and the protecting groups are also removed from the side chains. The peptide is typically purified by reversed-phase HPLC-MS, with the MS used to confirm the molecular mass of the synthesized peptide.

Important considerations when planning to synthesize a peptide include the general synthesis strategy, the choice of protecting groups for the side chains, the resin used as solid support as well as how it is linked to the C-terminus of the peptide, and the choice of coupling reagents [41]. The speed and quality of solid-phase peptide synthesis can be improved by the application of microwave irradiation, which raises the temperature and helps to break up chain aggregation of the peptide intermediates [42]. The microwave energy reduces the reaction time for both the coupling of amino acids and also the removal of the Fmoc protecting group [42, 43]. Nevertheless, synthesis of longer peptides tends to result in lower yields, and some amino acids are more difficult to couple than others. As a result, chemical ligation strategies can be used to join two shorter peptides together in solution phase. These methods result in the formation of amide or other chemical bonds between the two peptide segments, and typically chemoselective functional groups are chosen to allow ligation in the presence of the unprotected side chains on the peptides. Commonly used reactions include thioester ligation, thioether ligation, and imine ligation [44]. Enzyme-catalyzed methods, which have been primarily developed for ligation of proteins, can also be applied to peptides. For example, sortase A derived from *Staphylococcus aureus* recognizes and cleaves the LPxTG motif, which can be tagged on one peptide. The resulting thioester is then ligated to the N-terminus of a second peptide that starts with a G residue [45, 46]. The enzyme subtiligase, an engineered mutant of subtilisin, can also catalyze the reaction between a peptide thioester or peptide ester and the N-terminus of a second peptide [47, 48].

An advantage of solid-phase peptide synthesis is that it is not limited to the 20 natural amino acids, and in fact any compounds with appropriate reactive groups can be added to the peptide chain. Some examples are provided in Table 3. The functional group on the C-terminus of the peptide is usually determined by the choice of starting resin, and the N-terminus is often acetylated to block the reactivity of the amine group. Fluorescent labels or dyes can be incorporated into the peptides, including them as modifications on side chains or directly within the peptide chain if proper reactive groups are available. Examples of molecules used in this approach are summarized in Table 4. Fluorescent dyes such as fluorescein isothiocyanate (FITC), 5-carboxyfluorescein (5-FAM), Dansyl, 5- (and 6-) carboxytetramethylrhodamine [5(6)-TAMRA],

Table 3 Potential modifications to peptides prepared by solid-phase peptide synthesis

Modification	Molecules
N-terminus	Free amine, acetylation, Fmoc
C-terminus	Free acid, amidation
D-form amino acids	D-Ala, D-Cys, D-Asp, D-Glu, D-Phe, D-His, D-Ile, D-Lys, D-Leu, D-Met, D-Asn, D-Pro, D-Gln, D-Arg, D-Ser, D-Thr, D-Val, D-Trp, D-Tyr
Phosphorylation	pSer, pTyr, pThr
PEGylation	PEG_X
Cyclization	Disulfide bridge

Table 4 Fluorescent tags for incorporation into peptides, their excitation (λ_{ex}) and emission (λ_{em}) wavelengths, and their function

Fluorophore	λ_{ex}	λ_{em}	Function
Biotin	NA	NA	Binding-tagged streptavidin
FITC	492 nm	515 nm	Fluorophore, donor for FRET
5-FAM	492 nm	518 nm	Fluorophore
Dansyl	333 nm	518 nm	Fluorophore, FRET
TAMRA	546 nm	575 nm	Fluorophore, acceptor for FRET
Dnp	350 nm	NA	Acceptor for FRET (quencher)
MCA	322 nm	390 nm	Fluorophore, FRET with Dnp
5(6)-CR6G	522 nm	550 nm	Fluorophore
DABCYL	428 nm	NA	Acceptor for FRET (quencher)

7-methoxycoumarin-4-acetic acid (MCA), and 5- (and 6-) carboxyrhodamine 6G [5(6)-CR6G] have different excitation and emission wavelengths, and they can be selected to avoid overlap with tissue autofluorescence or with other fluorescent molecules used in a biomaterial formulation. Biotin can be incorporated for later tagging with fluorophores conjugated to streptavidin, and other molecules such as dinitrophenol (Dnp) or DABCYL can be used as part of Forster energy resonance transfer (FRET) pairs. Some of these molecules can be used as an N-terminal modification by reaction of the molecule with the amine group at the end of the peptide as the last step during peptide synthesis. This reaction can be achieved through a carboxylic acid group already present in the molecule or by first modifying the molecule with a succinimidyl ester group, which can then react with the terminal amine group. Alternatively, the fluorescent molecule can be added to the side chain of a lysine residue, and then this modified lysine can be incorporated anywhere in the peptide using standard coupling and deprotection reactions.

Labeling of peptides with fluorescent dyes can be a useful strategy for examining the loading and release of the peptide from a biomaterial carrier. For example, FITC-labeled peptides were used to track release from electrospun nanofibrous membranes and aerogels of poly(lactide-co-glycolide) (PLGA)/collagen/gelatin [49, 50]. A fluorescence plate reader can directly measure the concentration of the released peptide over time without the need for more complicated assays, such as ELISA, for the detection of the peptide. FRET pairs and dye quenching can also be used in peptide design to track spatial information about the peptides or peptide fragments. Fluorescent dyes and molecules that quench their fluorescence when in close proximity, such as the combination of fluorescein with DABCYL, can be incorporated on the opposite ends of enzyme substrate peptides similar to the ones described above. When the peptide is intact, the dye and quencher are close enough that there is no fluorescence; however, once the substrate is cleaved, the two fragments can diffuse away from each other, and the fluorescence is restored. In addition to being useful for determining kinetic parameters of the enzyme substrate interaction in solution [21], this approach has been used to track the local degradation of a PEG hydrogel incorporating such peptides (Fig. 4) [51].

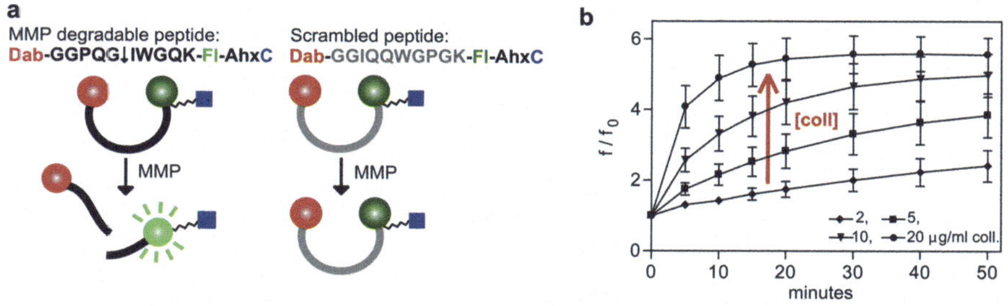

Fig. 4 Design and characterization of quenched fluorescent peptides as enzyme substrates. (**a**) Sequence and design of cleavable peptide and non-degradable control. (**b**) Degradation kinetics tracked by monitoring fluorescence over time. (Reprinted from *Biomaterials*, 34, Leight JL et al., Direct measurement of matrix metalloproteinase activity in 3D cellular microenvironments using a fluorogenic peptide substrate, 7344–52, ©2013 with permission from Elsevier [51])

Peptide Functionalization

In addition to synthesis of the desired peptide, it must be able to be chemically bound to or mixed in with the biomaterial while maintaining its biological activity. This can also influence the design of the peptide. For covalent binding of the peptide, an appropriate reactive group or additional amino acids must be added to allow the reaction to occur. One example of this is the use of a peptide substrate, which can be covalently cross-linked through the action of an enzyme. One such substrate is NQEQVSP(L), which is derived from α_2-plasmin inhibitor and is a substrate for the transglutaminase enzyme factor XIIIa. This peptide allows covalent cross-linking with some natural biomaterials, such as fibrin [52, 53], or with synthetic biomaterials that have a complementary substrate (FKGG) recognized by the enzyme [54]. For covalently bound peptides that act as a ligand, spacer amino acids may need to be added within the peptide to prevent steric hindrance interfering with receptor recognition. Cyclization of the peptide to constrain its presentation may also help to improve ligand–receptor interactions, as has been demonstrated with the RGD peptide [55]. For peptides that are active in soluble form, an MMP substrate can be added in the peptide sequence between the covalent attachment point and the peptide of interest to allow enzymatic cleavage and release of the peptide. For example, this has been done by adding the MMP substrate sequence PVGLIG to the N-terminus of osteogenic growth peptide (OGP) and then using carbodiimide chemistry to link the peptide to a partially oxidized form of alginate. Release of the peptide from the alginate matrix through the action of MMP-2 was confirmed [56].

In the PEG hydrogels prepared by a Michael-type addition reaction described above, a covalent bond was formed between the polymer and the peptides through the functionalization of the PEG with vinyl sulfone groups and the incorporation of cysteine residues, which have a reactive thiol group, in the peptides. This allowed both the functionalization of the PEG with the cell adhesive RGD peptide and the cross-linking of the PEG into an enzymatically degradable network using the substrate peptides [22, 23, 38, 39]. Similar conjugation reactions can also be performed between maleimide groups and thiol groups [57]. This strategy has been used to functionalize surfaces with peptides, for example, with layer-by-layer films of peptide-grafted poly(allylamine hydrochloride) (PAH) [58] and maleimide self-assembling monolayers on quartz substrates [59].

Depending on the reactive groups present on the material, peptides can also be covalently attached via an amide bond using 1-ethyl-3-(3-dimethylaminopropyl) carbodiimide (EDC) and *N*-hydroxysuccinimide (NHS) as coupling reagents to materials such as maleic anhydride-modified poly(lactic acid) (PLA) [60] or the carboxyl groups of partially oxidized alginate [56]. Typically, this approach relies on the reaction of an amine group coming from the N-terminus or a lysine side chain of the peptide with a carboxyl group on the material. Further, copper-catalyzed azide alkyne cycloaddition reactions can be used with azide-functionalized peptides and, for example, propargyl functionalized L-phenylalanine-based poly(ester urea)s [61]. Alternatively, the reactive groups can be switched between peptide and polymer, as in the case of azidopropyl hyperbranched poly(arylene oxindole) functionalization with *5-Hexynoic*-RGDS [62].

For noncovalent binding of peptides, the peptides may be allowed to adsorb to the material or be mixed in during processing, so that they are entrapped in the matrix. Alternatively, the peptide can be modified with an affinity domain that promotes noncovalent interactions with the material. For example, the addition of a highly negatively charged E7 domain, consisting of seven glutamic acid residues in a row, enables peptides to bind to calcium-based materials [63–66] or to premineralized materials, such as PLGA/collagen/gelatin nanofibers [49]. For an anorganic bovine bone graft material, coupling of a BMP-2-derived peptide to E7 led to increased loading and greater retention, even after 8 weeks of implantation in vivo [63]. Alternatively, the hydroxyapatite binding domain of statherin (N15 domain) was used to bind either PGRGDS or PDGEA (a cell adhesive peptide derived from type I collagen) to hydroxyapatite surfaces. The peptide-coated materials bound MC3T3-E1 osteoblasts via the $\alpha_V\beta_3$ integrin (for PGRGDS) and $\alpha_2\beta_1$ integrin (for PDGEA) [67].

Other linkers, such as polydopamine, can be used to mediate peptide attachment to surfaces [68–72]. In one approach, the surface can be coated with polydopamine, and then the peptide can be bound to the polydopamine coating [68]. Polydopamine coatings can be formed by dip-coating of a material in an aqueous solution of dopamine, which self-polymerizes, and they allow for further functionalization via secondary reactions [73]. This methodology is inspired by the adhesive proteins in mussels, which contain 3,4-dihydroxy-L-phenylalanine and lysine amino acids, and the covalent and noncovalent interactions that can occur with catechol compounds, and it is particularly interesting because it is compatible with a number of different inorganic and organic materials, such as metals, oxides, polymers, and ceramics [73]. Alternatively, peptides can be functionalized with a short polydopamine tag that can then interact with surfaces via the catechol groups (Fig. 5) [70]. This approach has been used to functionalize otherwise inert titanium (Ti) surfaces with RGDS and OGP, which enhanced the attachment, proliferation, and osteogenic differentiation of bone marrow-derived mesenchymal stem cells (BM-MSCs) in vitro as well as improved osteogenesis and mechanical stability of peptide-coated screws implanted in the femoral condyles of New Zealand White rabbits in vivo [70].

Osteoinductive Peptides

Peptide-functionalized materials provide an interesting approach to stimulate bone regeneration. In the fields of orthopedics and dentistry, peptides are relevant in two major ways. The first is to provide surface functionalization of the frequently Ti-based implants that are used as permanent implants. The use and characterization of Ti implants [74] as well as their combination with biopolymers [75] have been previously reviewed. The second is to use peptides in tissue engineering approaches. Typically, tissue engineering aims to combine cells, biomaterials, and biological

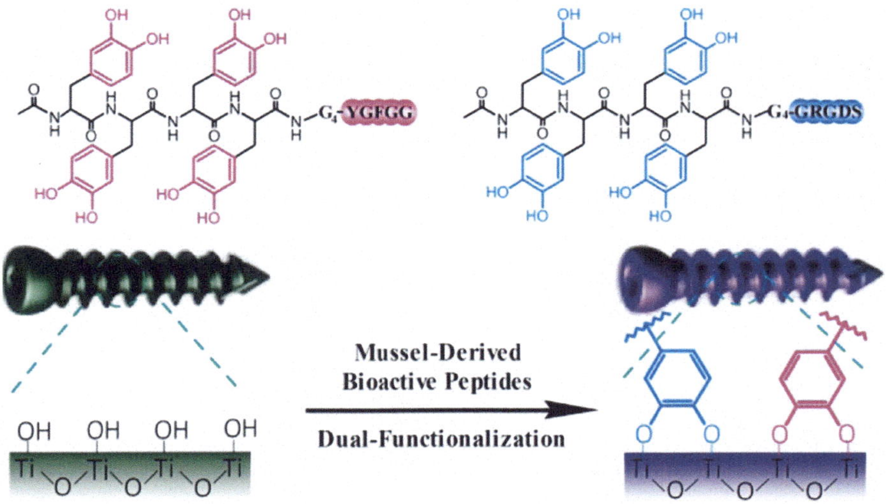

Fig. 5 Cell adhesive $(DOPA)_4$-G_4-GRGDS and osteoinductive $(DOPA)_4$-G_4-YGFGG peptides that mimic mussel-derived proteins with catechol groups to enable coordination between the catechol groups and titanium oxide for surface functionalization of Ti cortical bone screws. (Reprinted with permission from [70]. ©2016 American Chemical Society)

factors to create living implants that can repair or regenerate tissue function [76], and peptides are one means of providing these biological signals. In this case, tissue engineering scaffolds of a variety of materials have been modified and examined for their effects both in vitro and in vivo.

This section focuses on the application of peptides with osteoinductive capabilities, that is, peptides that can promote the osteogenic differentiation of progenitor cells in vitro and can lead to de novo bone formation in vivo. While the cell adhesive and enzymatically degradable peptides discussed above as well as other peptides, such as pro-angiogenic peptides, can support tissue remodeling and repair more generally and thus are interesting to consider as part of treatments to regenerate bone, as considered in recent reviews [77, 78], in this section, peptides that influence bone progenitor cells are specifically considered. These peptides are derived from a number of sources including identification of active domains in BMP-2, BMP-7, and other osteogenic proteins. A number of osteogenic peptides, their sequences and variants, and their effects are briefly summarized in Table 5.

OGP Peptides

Osteogenic growth peptide (OGP; ALKRQGRTLYGFGG) is a 14 amino acid peptide derived from the C-terminus of histone H4. It is found naturally in soluble form in serum and has been shown to influence osteoblast proliferation and differentiation [79]. OGP increased alkaline phosphatase (ALP) activity of MC3T3-E1 pre-

Table 5 Osteoinductive/osteogenic peptides and their effects

Peptide	Sequence	Effects	Reference
OGP	ALKRQGRTLYGFGG YGFGG (shorter variant)	Osteogenic differentiation in vitro; bone formation in vivo	[70, 79, 80]
BMP-2 derived	NSVNSKIPKACCVPTELSAI	Ectopic bone formation in vivo	[81]
	KIPKASSVPTELSAISTLYL	Osteogenic differentiation in vitro; ectopic bone formation in vivo	[82]
	DWIVA	Osteogenic differentiation in vitro; increased bone formation in vivo	[83]
BMP-7 derived	TVPKPSSAPTQLNAISTLYF	Osteogenic differentiation in vitro	[84]
	GQGFSYPYKAVFSTQ (BFP-1)	Osteogenic differentiation in vitro	[85]
	ETLDGQSINPKLAGL (BFP-3)	Osteogenic differentiation in vitro	[86]
BMP-4 derived	RKKNPNCRRH	Osteogenic differentiation in vitro (ERK1/2 activation); bone formation in vivo	[87]
BMP-9 derived	KVGKACCVPTKLSPISVLYK	Osteogenic differentiation in vitro; ectopic bone formation in vivo	[88, 89]
$PTH_{1\text{-}34}$	SVSEIQLMHNLGKHLNS-MERVEWLRKKLQDVHNF	Orthotopic bone formation in vivo	[90]
Others	GTPGPQGIAGQRGVV (P-15)	Proliferation in vitro; bone formation in vivo	[91]
	GLRSKSKKFRRPDIQY-PDATDEDITSHM (CBM)	Mineralization in vitro and in vivo	[92]
	PFSSTKT (BMHP1)	Osteogenic differentiation in vitro	[59]

osteoblasts in vitro and improved bone remodeling in vivo in a rat mandibular condyle model when delivered systemically [79]. In addition, systemic administration of OGP has been shown to improve the structural and mechanical properties of the fracture callus in a rat model [93].

OGP has also been used with several biomaterials, both as a 2D surface modification and for incorporation in a 3D scaffold [94, 95]. When OGP and its shorter variant, OGP(10–14) (YGFGG) [80], were tethered to 2D surfaces via click chemistry between an azide-functionalized peptide and an alkyne-terminated self-assembled monolayer (SAM), they led to increases in MC3T3-E1 attachment and proliferation in vitro [96]. A similar approach has been used with azide-functionalized OGP and poly(ester urea) scaffolds via propargyl groups on the surface of the scaffolds. This functionalization increased the differentiation of human mesenchymal stem cells (hMSCs), as demonstrated by increases in ALP activity, calcium deposition, and expression of osteogenic genes [61]. Carbodiimide chemistry has been utilized to link OGP(10–14) to maleic anhydride-modified PLA scaffolds, improving the proliferation, differentiation, and mineralization of neonatal rat calvarial osteoblasts [60]. Moreover, OGP has been used to functionalize the surface of Ti implants through a polydomamine linker. In combination with the cell adhesive peptide RGD, the OGP-functionalized surfaces promoted

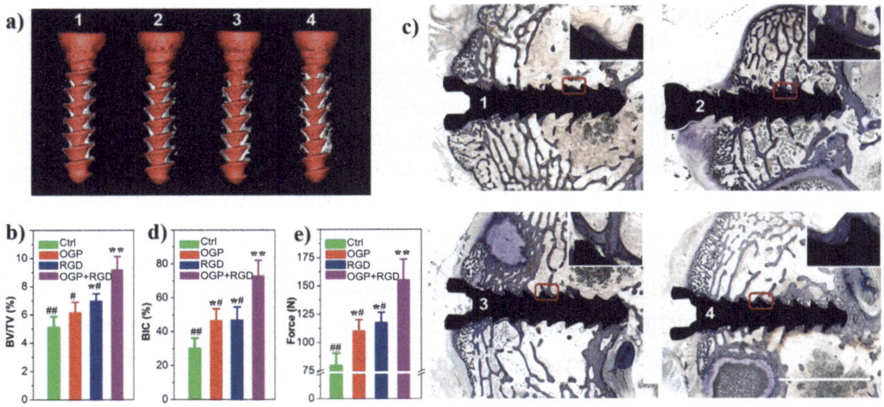

Fig. 6 Effect of OGP surface treatment of Ti bone screws on osteogenesis and mechanical stability in a rabbit femoral condyle. (**a**) Reconstructed microCT images. (**b**) Quantitation of bone volume (BV)/total volume (TV) from microCT images. (**c**) Representative histological images stained with toluidine blue with 1 = untreated control, 2 = OGP, 3 = RGD, and 4 = OGP/RGD (3,1). (**d**) Quantification of bone-implant contact (BIC) from histological images. (**e**) Biomechanical pull-out testing. (Reprinted with permission from [70]. ©2016 American Chemical Society)

cell attachment and mineralization in vitro, and functionalized Ti screws led to increased osseointegration, as shown by increased bone volume, bone-implant contact, and pull-out force (Fig. 6) [70].

Additionally, hydrogels have been formed using OGP(10–14) tethered to alginate matrices by a protease-sensitive linker. Exposure to enzymatic treatment led to the release of the peptide and osteogenic differentiation of treated hMSCs. The MSC-laden, OGP-functionalized alginate hydrogels also resulted in ectopic bone formation in vivo [56]. Moreover, functionalization with OGP increased MC3T3-E1 proliferation and alkaline phosphatase activity for cells cultured on hydrogels formed from the self-assembling peptide RADA16 [97].

Peptides Derived from BMP-2

As BMP-2 is one of the most potent osteoinductive growth factors [98], peptides derived from BMP-2 have also been studied for their ability to stimulate osteogenic differentiation in vitro and bone formation in vivo. One of the first peptides derived from BMP-2 was a 20-amino-acid-long sequence (NSVNSK-IPKACCVPTELSAI), which led to ectopic bone formation in the calf muscle of rats when linked to an alginate hydrogel [81]. A peptide containing the motif DWIVA has also been demonstrated to promote proliferation and osteogenic differentiation of MC3T3-E1 cells in vitro and to increase bone growth when conjugated to the surface of Ti implants used in mandibular bone defects in beagle dogs [83].

The peptide KIPKASSVPTELSAISTLYL, derived from residues 73–92 of the knuckle epitope of BMP-2, has promoted ALP activity and osteocalcin gene expression by C3H10T1/2 cells in culture and, when combined with an alginate hydrogel, has led to ectopic bone formation in rat muscle tissue [82, 99] and accelerated bone formation in a rat tibial defect [100]. Entrapping the peptide in chitosan microspheres that were then embedded in nano-hydroxyapatite/collagen/poly(lactic acid) (PLA) scaffolds provided controlled release of the peptide, which still retained its activity [101]. Further, grafting the peptide to self-assembled monolayers in combination with the cell adhesive RGD peptide led to the upregulation of bone sialoprotein expression as well as promotion of mineralization, even in the absence of other osteogenic supplements [102]. As has been shown with OGP, polydopamine has also been used to link the BMP-2-derived peptide to PLGA scaffolds [69, 71]. For example, the modified scaffolds promoted osteogenic differentiation of human adipose-derived stem cells (hADSCs) in culture and resulted in bone formation in calvarial defects in vivo [69].

The BMP-2-derived KIPKASSVPTELSAISTLYL has also been incorporated into surface coatings using electrostatic interactions. One way this has been achieved is by forming layer-by-layer films of PAH and poly(sodium 4-styrenesulfonate), with the peptide grafted on the PAH. The films were coated on electrospun membranes of nano-hydroxyapatite and PLGA, and they promoted alkaline phosphatase activity by MSCs in vitro and bone formation in a rat calvarial defect in vivo [58]. A second method to immobilize the peptide is through the E7 calcium binding domain described earlier. E7-modified BMP-2-derived peptides have been bound to anorganic bovine bone [63], α-tricalcium phosphate (α-TCP) scaffolds [64], and even nanofibrous membranes of PLGA/collagen/gelatin that were pre-mineralized [49]. Nanofibrous aerogels of electropsun PLGA/collagen/gelatin and Sr-Cu-doped bioactive glass fibers were also functionalized with E7-BMP-2 peptides. The addition of the peptide increased bone formation in vivo in rat calvarial bone defects (Fig. 7) [50].

Peptides Derived from BMP-7

Bone forming peptide-1 (BFP-1; GQGFSYPYKAVFSTQ) is derived from residues 100–114 of BMP-7. Treatment with soluble BFP-1 led to increased ALP activity and calcium deposition by bone marrow stromal cells, and further these BFP-1-treated cells increased bone formation when injected subcutaneously in mice [85]. Likewise, hMSCs also showed increased ALP activity, calcium deposition, and osteogenic gene expression when cultured on nanofibrous scaffolds consisting of polycaprolactone (PCL) functionalized with the peptide via a polydopamine linker. This enhancement in osteogenic differentiation was also seen when the cells were cultured in medium lacking other osteoinductive factors [68]. Further, human-induced pluripotent stem cells (hiPSCs) also increased ALP activity, calcium deposition, and osteogenic gene expression when cultured on 2D surfaces that had been functionalized with BFP-1 and carboxymethyl chitosan [72]. A second osteoinductive peptide,

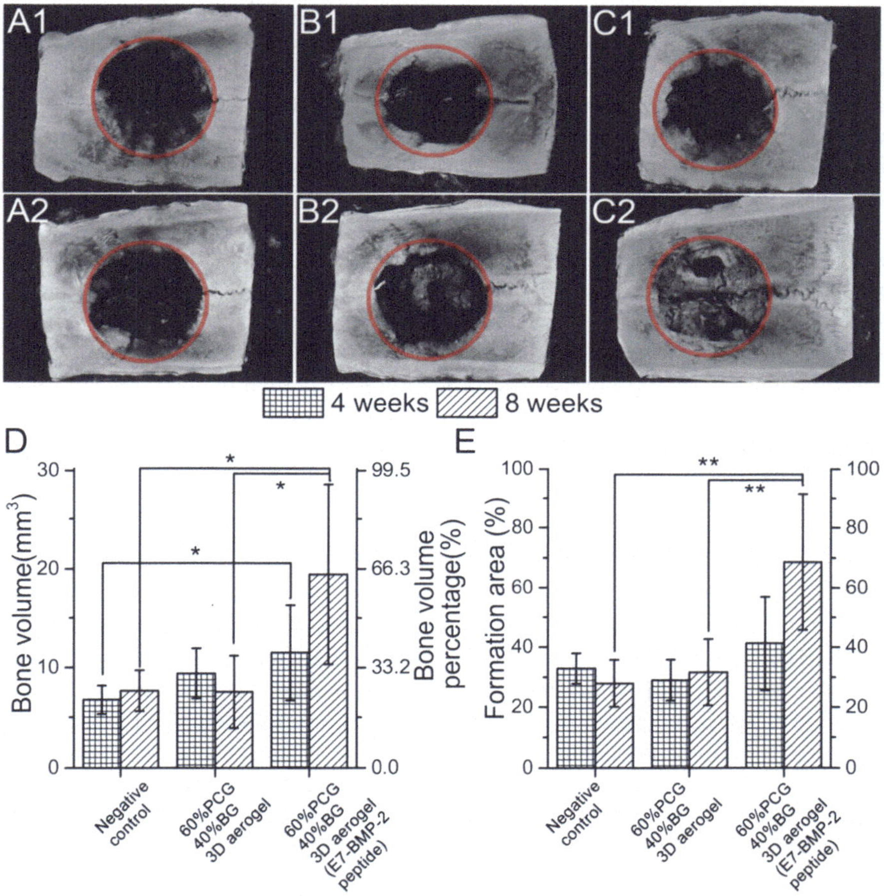

Fig. 7 Calvarial bone regeneration induced by E7-BMP-2 peptide loaded aerogels. Radiographs of (**A**) untreated controls, (**B**) aerogel only control, (**C**) E7-BMP-2 peptide loaded aerogel group. Quantification of (**D**) regenerated bone volume and (**E**) bone formation area. (Reprinted with permission from [50]. ©2018 John Wiley and Sons)

BFP-3, was also determined from BMP-7 residues 250–265 (ETLDGQSI-NPKLAGL). It increased in vitro osteogenic differentiation of bone marrow stromal cells by increasing ERK1/2 and Smad1/5/8 phosphorylation [86].

Peptides Derived from BMP-4 and BMP-9

Osteoinductive peptides have also been identified within BMP-9, including from the knuckle epitope, residues 68–87 (KVGKACCVPTKLSPISVLYK) [103]. In vitro, this peptide-induced gene expression of Runx2, osterix, type I collagen, and osteocalcin by MC3T3-E1 pre-osteoblast cells, and it activated the Smad pathway [89]. The phosphorylation of Smad1/5/8 was further enhanced when this peptide was presented in combination with a fibronectin-derived peptide containing both PHSRN and RGD ligands and functionalized on films of PCL [104]. When combined with a chitosan-based delivery system, the peptide induced ectopic bone formation in the quadriceps of mice. Interestingly, a collagen-based delivery system abrogated this response [88].

A heparin binding domain from residues 15–24 of BMP-4, RKKNPNCRRH, has been shown to have osteogenic activity. Administered in soluble form, the peptide promoted the osteogenic differentiation of hMSCs, as shown by increased matrix mineralization as well as upregulation of the osteogenic genes, ALP, osteopontin, and osteonectin. Phosphorylation of Smad 1/5/8 and MAPK was demonstrated by Western blotting and confirmed the activation of the ERK1/2 pathway. Further, when combined with an alginate hydrogel matrix, the peptide was able to stimulate in vivo bone formation in a rabbit calvarial defect model [87].

Peptides Derived from Parathyroid Hormone

A peptide derived from parathyroid hormone residues 1–34 (PTH_{1-34}; SVSEIQLMHNLGKHLNSMERVEWLRKKLQDVHNF) has been used clinically as treatment for osteoporosis [105]. For stimulating bone regeneration, it has been covalently bound to an RGD-functionalized PEG hydrogel and been shown to lead to increases in in vivo bone formation, at levels similar to treatment with autologous bone, when implanted into alveolar bone defects surrounding standard Ti implants in a dog model [90]. Similarly, PTH_{1-34} was covalently bound to a fibrin matrix via an enzymatically degradable peptide linker, and this led to increased bone formation in defects of the femur and humerus of sheep [106]. The soluble form of PTH_{1-34} was also administered via injection to determine dosing amount and frequency to improve bone healing in a mouse femoral allograft model [107]. Moreover, it was shown to protect against radiotherapy-induced trabecular bone loss in a rat model [108].

Other Peptides

A number of other peptides have shown osteoinductive activity. Bone marrow homing peptide 1 (BMHP1; PFSSTKT), which was discovered from screening a phage display library, led to osteogenic differentiation of MSCs cultured on quartz substrates functionalized with the peptide [59]. The peptide P-15 derived from residues 766–780 of the type I collagen α chain (GTPGPQGIAGQRG-VV) accelerated bone formation in a rabbit calvarial defect model when it was coated on deproteinized bovine bone [91]. A collagen-binding motif (CBM) from osteopontin (GLRSKSKKFRRPDIQYPDATDEDITSHM) in combination with a type I collagen

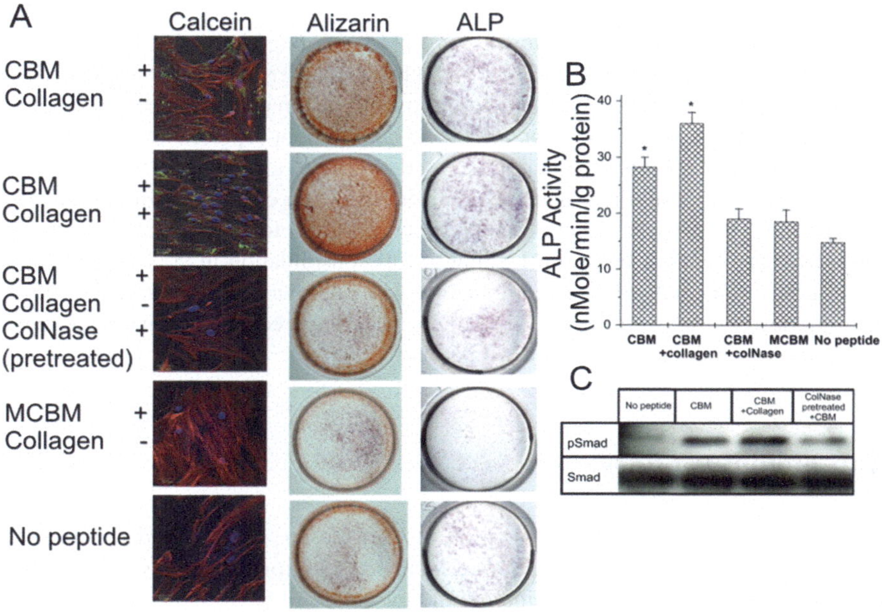

Fig. 8 Effect of CBM on bone marrow stromal cells. (**A**) Micrographs of calcein staining (left panel), Alizarin Red S staining (middle panel), and ALP expression (right panel). (**B**) Quantification of ALP activity. (**C**) Western blot analysis for phosphorylated-Smad expression. (Reprinted from *Biomaterials*, 28, Lee J-Y et al., Assembly of collagen-binding peptide with collagen as a bioactive scaffold for osteogenesis in vitro and in vivo, 4257–67, ©2007 with permission from Elsevier [92])

matrix increased mineralization by cultured bone marrow stromal cells, as shown by calcein, Alizarin Red S, and ALP staining, as well as phosphorylation of Smad1/5/8 (Fig. 8). The peptide functionalized matrices also led to increased bone formation in rabbit calvarial defects [92].

Use of Osteoinductive Peptides Clinically

The majority of peptides used to stimulate bone healing and regeneration have only been evaluated in vitro or in preclinical animal studies [109, 110]. Of the osteoinductive peptides discussed in this chapter, two have seen more extensive clinical use, PTH_{1-34} and P-15. In addition, chrysalin, also known as thrombin peptide 508 (TP508), has been evaluated in Phase I/II clinical trials to treat distal radial fractures, showing reduced time to healing [111]. PTH_{1-34} has been approved for the prevention and treatment of osteoporosis clinically [105]. In a clinical study for the treatment of distal radial fractures, PTH_{1-34} shortened the time to healing with the lower of two doses tested [112] and appeared to improve early callus formation [113]. The use of

PTH_{1-34} in bone healing has also been documented in several case reports, including for sternal nonunion [114] and hip fracture [115], among others [110]. P-15 was combined with anorganic bone mineral and a hydrogel carrier material and tested in a clinical study for anterior cervical discectomy and fusion in comparison with autograft. The radiographic fusion rates, clinical and patient-reported outcomes, and safety profile of the P-15 formulation were all similar to autograft bone [116]. In a first example of its use for orthopedic applications, again combined with anorganic bone mineral, P-15 has been tested in a pilot clinical trial to treat patients with malunion or delayed union fractures, leading to full consolidation in 90% of the cases [117]. In the oral cavity, P-15 in combination with anorganic bone mineral has been shown to lead to improved defect fill results when compared to anorganic bone mineral alone in a multicenter trial with 33 patients [118].

Anti-biofilm Peptides

Biofilms are formed by the colonization of bacteria on surfaces and represent a leading cause of chronic and implant-associated infections clinically [119, 120]. They are characterized by the formation of aggregates of bacteria that encapsulate themselves in a dense extracellular matrix composed of polysaccharides, extracellular DNA, proteins, and lipids [121]. As a result, bacteria in biofilms are 10- to 1000-fold more resistant to antibiotics than planktonic (motile) bacteria [122, 123]. Of relevance to biomaterials, biofilms can form on the surface of implanted medical devices, such as catheters, valves, stents, and orthopedic prostheses, and even of contact lenses [124]. In dentistry, the plaque that develops on the surface of teeth is also a biofilm [125]. Peptide-based treatments for preventing or eliminating biofilms are becoming interesting due to the development of antimicrobial peptides, which demonstrate broad-spectrum activity against planktonic Gram-positive and Gram-negative bacteria. A subset of these antimicrobial peptides have also demonstrated activity against bacteria in biofilms, both on their own and in combination with other antibiotics [126–132], and a number of examples are summarized in Table 6.

Antimicrobial peptides, a subset host defense peptides, are naturally produced by organisms including animals, fungi, plants, and bacteria, resulting in the identification of more than 2000 such peptides [152]. Antimicrobial peptides typically act rapidly and have a broad-spectrum antimicrobial activity against planktonic cells, while host defense peptides also include those peptides that act as innate immune modulators through anti-infective, anti-inflammatory, wound healing, and/or anti-biofilm activities [153, 154]. Thus, these natural peptides are interesting to explore as potential therapeutics against biofilms. Alternatively, peptide screening methods can be used to identify peptides with antibacterial activity [155]. As the literature on antimicrobial peptides is vast, this chapter will focus on a few key examples of antimicrobial peptides that are naturally derived or identified from peptide screening studies and have shown anti-biofilm activity. Additional emphasis will be placed on

Table 6 Anti-biofilm antimicrobial peptides and their in vitro and in vivo effects

Peptide sequence	Bacteria	Effect	Reference
3002 ILVRWIRWRIQW-NH_2	MRSA	Mouse cutaneous abscess model in vivo	[133]
AS10 KLKKIAQKIKNFFQKLVP	*C. albicans* *P. aeruginosa* *E. coli*	Inhibition in vitro	[134]
Battacin 4-Methylhexanoyl-D-Dab-Dab-Dab-L-D-F-Dab-Dab-L-NH_2	*P. aeruginosa* *P. syringae* pv. *actinidiae* *S. aureus*	Inhibition/eradication in vitro	[135]
BMAP-27-melittin KFKKLFKKLSPVIGAVLKVLT	*S. aureus* *P. aeruginosa*	Inhibition/eradication in vitro	[136]
BMAP-28 GGLRSLGRKILRAWKKYGPIIVPIIRI-NH_2	*S. aureus* *E. faecalis*	Rat ureteral stent infection model in vivo	[137]
CAMA KWKLFKKIGIGKFLQSAKKF-NH_2	MRSA	Inhibition in vitro	[138]
Cyclic lipopeptide 3 12-Aminododecanoic-Dap-D-L-L-D-I-N-D-K	*S. aureus* *P. aeruginosa*	Inhibition/eradication in vitro	[139]
DJK-5 vqwrairvrvir (D-amino acids) DJK-6 vqwrrirvwvir (D-amino acids)	*P. aeruginosa* *E. coli* *A. baumannii* *K. pneumoniae* *S. enterica*	Inhibition/eradication in vitro *C. elegans* nematodes in vivo *G. mellonella* larvae in vivo Mouse cutaneous abscess model in vivo	[140, 141]
D-RR4 wlrrikawlrrika-NH_2 (D-amino acids)	*P. aeruginosa* *A. baumannii*	*C. elegans* model in vivo	[142]
hep20 ICIFCCGCCHRSHCGMCCKT	*S. epidermis*	Inhibition in vitro	[143]

(continued)

Table 6 (continued)

Peptide sequence	Bacteria	Effect	Reference
IDR-1018 VRLIVAVRIWRR-NH_2	*P. aeruginosa* *E. coli* *A. baumannii* *K. pneumoniae* *S. enterica* MRSA	Inhibition/eradication in vitro	[133, 144]
KT2 NGVQPKYKWWKWWKKWW-NH_2	*E. coli*	Inhibition/eradication in vitro	[145]
LL-37 LLGDFFRKSKEKIGKEFKRIVQRIKDFLRNLVPRTES	*P. aeruginosa*	Inhibition in vitro	[146]
P10 LAREYKKIVEKLKRWLRQVLRTLR	MDR *S. aureus*	Inhibition/eradication in vitro	[147]
RT2 NGVQPKYRWWRWWRRWW-NH_2	*E. coli*	Inhibition/eradication in vitro	[145]
SAAP-148 LKRVWKRVFKLLKRYWRQLKKPVR	*A. baumannii* MRSA	Mouse skin wound model in vivo	[148]
Tachyplesin III KWCFRVCYRGICYRKCR-NH_2	*P. aeruginosa*	Inhibition in vitro and rat ureteral stent infection model in vivo	[149]
UP-5 RBRBR (B = biphenylalanine)	MRSA	Inhibition in vitro	[150]
WRL3 WLRAFRRLVRRLARGLRR-NH_2	MRSA	Mouse infected burn wound in vivo	[151]

peptides that have shown anti-biofilm activity in vivo in animal models or that have been used to modify biomaterial surfaces. Anti-biofilm activity is typically demonstrated in vitro by the ability to kill multiple species of bacteria that can be present in biofilms, including *Enterococcus faecium*, *Staphylococcus aureus*, *Klebsiella pneumoniae*, *Acinetobacter baumannii*, *Pseudomonas aeruginosa*, and *Enterobacter* species [131]. In vivo activity has primarily been tested in mouse skin wound or cutaneous abscess models as well as rat ureteral catheterization models. In addition to killing planktonic bacteria that can form or detach from biofilms as well as killing embedded bacteria, antimicrobial peptides can also act to interfere with bacterial adhesion and gene expression as well as influence the host response to the biofilm [127]. These latter activities can occur at concentrations much lower than the minimum inhibitory concentration (MIC), which is usually used to characterize the effects of antimicrobial agents on planktonic bacteria [127] and provides a measurement of anti-biofilm activity.

LL-37, P10, AS10, and hep20 Peptides (Naturally Derived)

Of the many antimicrobial peptides, cathelicidin LL-37 (LLGDFFRKSKEKIGKEFKRIVQRIKDFLRNLVPRTES) was one of the first human-derived peptides to demonstrate anti-biofilm activity. It can be found at mucosal surfaces and in the granules of phagocytes, and its concentration at sites of chronic inflammation is higher than baseline levels in most bodily fluids [146]. It is able to prevent biofilm formation in vitro at concentrations well below its MIC, and it can also reduce existing *P. aeruginosa* biofilms [146]. Shorter peptides derived from LL-37, such as P10 (LAREYKKIVEKLKRWLRQVLRTL-R), have also shown antimicrobial activity. P10 was more effective at killing the methicillin-resistant *S. aureus* (MRSA) strain LUH14616 than LL-37, and it could eradicate MRSA strains LUH14616 and LUH15051 from thermally wounded human skin equivalents in vitro without showing toxicity towards the fibroblasts and keratinocytes within the tissue models [147].

Likewise, the mouse cathelicidin-related antimicrobial peptide (CRAMP) has been shown to inhibit fungal biofilm formation [156]. Shorter peptides based on CRAMP, such as AS10 (KLKKIAQKIKNFFQKLVP), were also able to inhibit *Candida albicans* biofilm growth on Ti disks as well as formation of biofilms from other bacteria [134]. The study also tested the effect of AS10 on human osteoblasts, mesenchymal stromal cells, and endothelial cells in vitro and demonstrated no negative effects on cell viability. Further, AS10 did not interfere with calcium deposition by the osteoblasts or mesenchymal stromal cells or with tube formation by the endothelial cells, which suggests that AS10 could be used as an anti-biofilm coating on implants without affecting cells in the surrounding tissues [134].

A second human-derived antimicrobial peptide is hepcidin 20 (hep20; ICIFCCGCCHRSHCGMCCKT), a 20-amino-acid peptide that is found in the liver. The related hepcidin 25 (hep25) is involved in the regulation of plasma iron levels by binding ferroportin on macrophages, enterocytes, and hepatocytes [157]. Both hep25 and hep20 have demonstrated antimicrobial activity against several bacterial strains [158] with hep20 also showing antifungal activity [159]. Hep20 inhibited biofilm formation of polysaccharide intercellular adhesion (PIA)-positive and PIA-negative clinical isolates of *Staphylococcus epidermis*, most likely by inhibiting the accumulation of extracellular matrix within the biofilm [143].

IDR-1018 and 3002 Peptides (from Peptide Library Screenings)

Innate defense regulator peptide 1018 (IDR-1018) and 3002 are two examples of peptides that have been optimized by peptide library screening methods. Starting from the bovine bactenecin derivative Bac2a (RLARIVVIRVAR-NH_2), a library of over 100 peptides, each 12 amino acids long, was generated by performing point mutations, scrambling, and deletion of amino acids [160]. The immunomodulatory potential of the peptides from the library was evaluated by measuring the ex vivo induction of monocyte chemotactic protein 1 (MCP-1) and MCP-3 by human peripheral blood mononuclear cells, and IDR-1018 (VRLI-VAVRIWRR-NH_2) resulted in a >50-fold increase compared to Bac2a [160]. IDR-1018 also exhibited broad-spectrum anti-biofilm activity towards *P. aeruginosa*, *E. coli*, *A. baumannii*, *K. pneumoniae*, MRSA, *S. typhimurium*, and *B. cenocepacia* when tested with sub-MIC concentrations [161]. It could also inhibit or eliminate biofilms, as demonstrated by growing *P. aeruginosa* biofilms in continuous-culture flow cells in a minimal medium. The medium flowing through the cells was supplemented with IDR-1018 either during or after biofilm establishment, and in both cases, the treatment led to thinner biofilms that lacked the structural features of mature biofilms or eliminated them entirely (Fig. 9) [161].

The sequence specificity of IDR-1018 was further explored by generating a second peptide library that contained 96 peptides with single amino acid substitutions [133]. This library was SPOT-synthesized on cellulose arrays, and the ability to prevent MRSA biofilms was evaluated. The resulting data were fed into a quantitative structure–activity relationship (QSAR) model, which predicted new peptides with anti-biofilm activity from an in silico library of 100,000 peptide sequences. The identified peptide 3002 (ILVRWIRWRIQW-NH_2) was more effective than IDR-1018 in inhibiting biofilm growth and in eradicating pre-formed biofilms. Further, peptide 3002 was able to reduce abscess size in a mouse cutaneous model of high-density bacterial infection, an MRSA chronic infection model [133].

DJK-5, DJK-6, and D-RR4 Peptides (D-Enantiomeric Peptides)

While peptides represent a potentially powerful therapeutic option, they have some drawbacks for use in vivo including degradation by proteases and inactivation in bodily fluids. The incorporation of non-natural amino acids into synthetic peptides

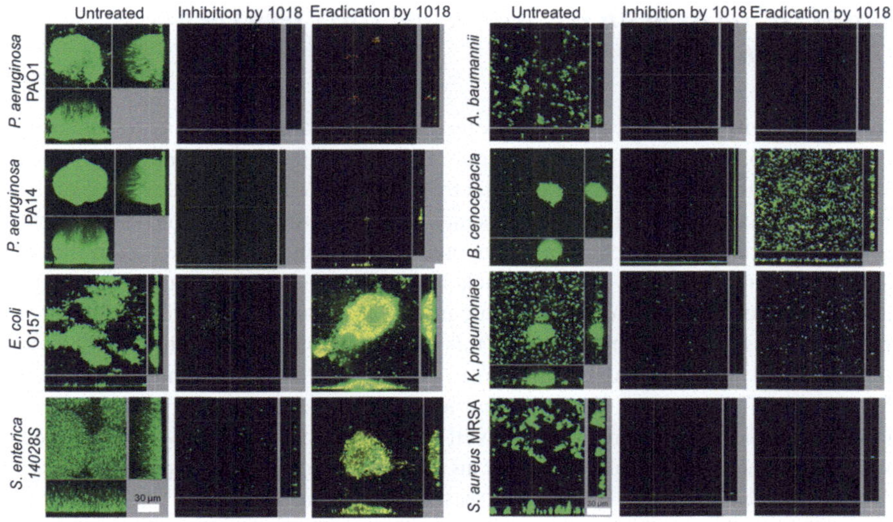

Fig. 9 Inhibition and eradication/reduction of biofilms of different bacterial strains by sub-inhibitory concentrations of peptide IDR-1018, as demonstrated in a flow cell assay. The scale bars are 30 μm. (Reprinted under the terms of the Creative Commons Attribution License from de la Fuente-Núñez et al. ©2014 [161])

can address some of these limitations by providing more stable variants, which may lead to increased activity [140–142, 162]. The synthetic antimicrobial peptides DJK-5, DJK-6, and D-RR4 have been developed including the D-enantiomeric form of amino acids and have shown strong anti-biofilm activity both in vitro and in vivo [140–142]. DJK-5 (vqwrairvrvir; D-amino acids) and DJK-6 (vqwrrirvw-vir; D-amino acids) were identified from screening a library of 12-amino-acid-long peptides with the following design constraints: D-enantiomeric forms of only 9 amino acids (A, F, I, K, L, Q, R, V, W), 4 charged residues, 7–8 hydrophobic residues, and 0–1 Q residues. DJK-5 and DJK-6 were identified as being able to inhibit biofilm formation by several common bacterial strains, at concentrations below their MIC, and to eradicate established biofilms [140]. Further, DJK-5 was shown to decrease abscess size and reduce bacterial burden in a mouse cutaneous abscess model induced by *P. aeruginosa* [141].

The peptide D-RR4 (wlrrikawlrrika-NH_2; D-amino acids) was developed to be resistant to proteolytic digestion, as confirmed by stability in the presence of both bacterial (proteinase K) and mammalian (trypsin) proteases, and it remained active in the presence of physiologic concentrations of salt, serum proteins, and acidic pH [142]. It also showed reduced toxicity towards macrophages and keratinocytes when compared to the L-enantiomeric form of the peptide. The in vivo antibacterial activity of D-RR4 was demonstrated by increased survival and decreased bacterial burden of *Caenorhabditis elegans* worms infected with *P. aeruginosa* or *A. baumannii*, after treatment with the peptide [142].

BMAP28, Tachyplesin III, WRL3, and DRGN-1 Peptides (Evaluation In Vivo)

In vitro activity to inhibit or kill relevant bacterial strains as well as to inhibit or eradicate biofilms helps to demonstrate the potential of anti-biofilm peptides. However, ultimately, they must be effective in vivo, and thus the development of relevant animal models is important. As mentioned previously, the survival of *C. elegans* worms and the reduction of cutaneous abscesses in mice are two examples of such in vivo models. However, they lack some of the complex features of implant-associated biofilm formation and chronic infection. A more challenging model is an infected mouse burn wound model. In this case, a region of skin is first scalded and then infected with MRSA. In addition to the size of the wound and the bacterial burden, the effects on host response [interleukin (IL)-6, IL-10, tumor necrosis factor (TNF)-α, and MCP-1 production] as well as angiogenesis can be evaluated. Using this model, the antimicrobial effects of an engineered amphiphathic peptide, WRL3 (WLRAFRRLVRRLARGLRR-NH_2), was demonstrated (Fig. 10). This confirmed its potential as an anti-biofilm agent as also shown by antimicrobial activity against MRSA and inhibition of biofilm activity in vitro [151]. Likewise, DRGN-1 (PSKKTKPVKPKKVA) enhanced healing in a mouse skin wound model that was infected with a biofilm containing both *P. aeruginosa* and *S. aureus* [163].

As catheters frequently develop biofilms, in vivo models have also been developed to study the effects of antimicrobial peptides on ureteral stent infection. These rat models have involved the implantation of ureteral stents either subcutaneously or in the bladder. The subcutaneous model was used to assess the effectiveness of stents coated with Tachyplesin III (KWCFRVCYRGICYRKCR-NH_2), which is derived from horseshoe crabs and exhibits antimicrobial activity against *P. aeruginosa*. After implantation of the Tachyplesin III-coated stent, *P. aeruginosa* was injected onto the implant surface. The Tachyplesin-III-treated group showed a reduction in bacterial count compared to uncoated stents, and this effect was further enhanced by co-treatment with intraperitoneal injection of piperacillin-tazobactam (TZP) [149]. On the other hand, the bladder implantation model was used with BMAP-28 (GGLRSLGR KILRAWKKYGPIIVPIIRI-NH_2), a cathelicidin peptide. After implantation of the BMAP-28-coated stent, *S. aureus* or *E. faecalis* was inoculated into the bladder. The BMAP-28 coating reduced the bacterial counts for both strains compared to uncoated controls, and this level was further reduced by coadministration of vancomycin intraperitoneally [137].

Immobilization of Antimicrobial Peptides

A final consideration in the development of anti-biofilm peptides is their delivery mechanism. As biofilms tend to form on the surface of implanted devices, peptide immobilization techniques become relevant [164]. Antimicrobial peptides can simply be adsorbed on the surface of materials by soaking them in solutions of the peptides

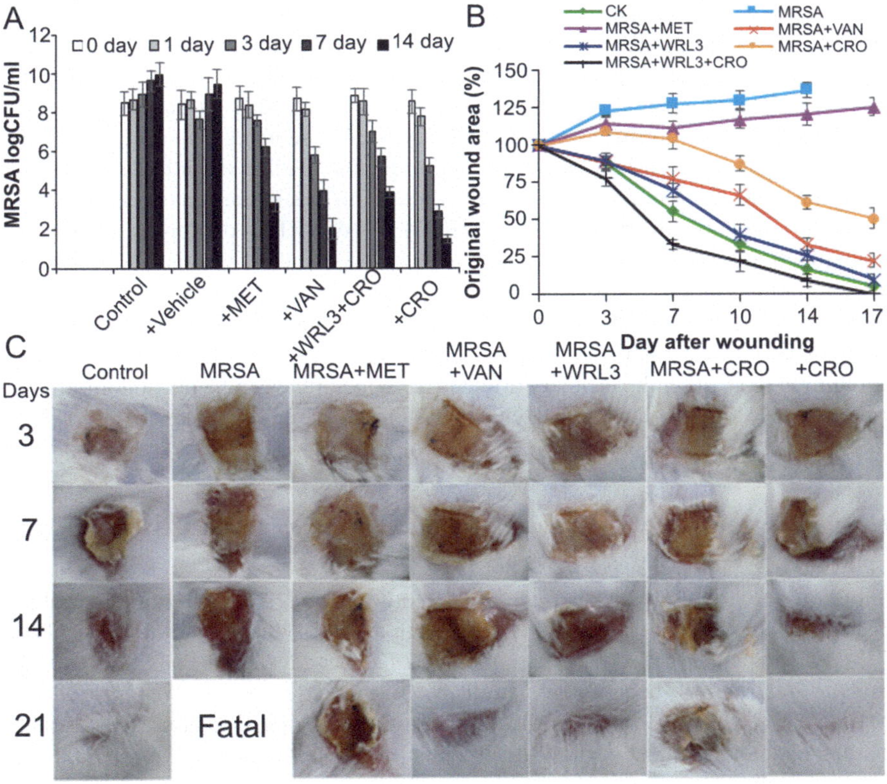

Fig. 10 In vivo antimicrobial activity of WRL3 as demonstrated in a mouse infected burn wound model. (**A**) Viable bacterial counts in colony-forming units (CFU). (**B**) Wound area measurements. (**C**) Images of wounded regions. (Reprinted with permission from [151]. ©2016 American Chemical Society)

[137, 149, 165]. Layer-by-layer films provide a next step in complexity of preparation and controlled release of the peptide. These films are formed from alternating deposition of cationic and anionic molecules (polymers or peptides) [166, 167]. For example, the thickness and stability of the coating, amount of peptide loaded, and the release rates of the antimicrobial peptide ponericin G1 (GWKDWA KKAGGWLKKKGPGMAKAALKAAMQ-NH_2) were affected by the choice of polyanion used, and the film composition also influenced resistance to attachment of *S. aureus* [167].

Polymer brushes provide a means for covalent conjugation of antimicrobial peptides. Ti and quartz surfaces have been functionalized by surface-initiated polymerization of *N*,*N*-dimethylacrylamide and aminopropyl methacrylamide hydrochloride, resulting in a primary amine functionality. These amine groups were modified to maleimide groups, which then allowed the reaction with cysteine-containing peptides [168, 169]. The presentation of antimicrobial peptide

Tet-20 (KRWRIRVRVIRKC) inhibited *P. aeruginosa* and *S. aureus* growth in vitro and *S. aureus* adherence in vivo on subcutaneous implants [169]. Similarly, an allyl glycidyl ether polymer brush with PEG-maleimide spacer was used to bind cysteine-containing peptides to polydimethylsiloxane (PDMS) surfaces [170].

As reduction in biofilm formation and improved osseointegration are important considerations for Ti orthopedic and dental implants, loading of antimicrobial peptides in porous calcium phosphate (CaP) coatings on Ti surfaces has been explored. Coatings containing the peptide Tet213 (KRWWKWWRRC) demonstrated antimicrobial activity against *S. aureus* and *P. aeruginosa* bacterial strains while remaining cytocompatible towards MG-63 osteoblast-like cells [171]. A similar coating loaded with HHC36 (KRW-WKWWRR) did not interfere with bone growth around press-fit grafts in a rabbit femoral defect model [172]. A more complex coating combining the CaP with vertically oriented TiO_2 nanotubes and a phospholipid film were also able to provide sustained release and antimicrobial activity of HHC36 [173].

Use of Anti-biofilm and Antimicrobial Peptides Clinically

Similar to the osteoinductive peptides, anti-biofilm peptides as well as antimicrobial peptides more generally have not yet resulted in many products for clinical use [174, 175]. Of the peptides discussed herein, LL-37 has been tested in Phase I/II clinical trials to treat hard-to-heal venous leg ulcers. In this case, the LL-37 improved the healing rate constants [176], but the application was a topical treatment for wound healing and not treatment of biofilms. Polymyxin B and Polymyxin E (colistin) have been used clinically to treat drug-resistant infections resulting from Gram-negative bacteria [177–179]. However, the majority of antimicrobial peptides tested clinically have been used topically in clinical trials to treat infected diabetic foot ulcers, catheter infections, and skin and fungal infections [175] and not for treatment and/or prevention of biofilms. While several anti-biofilm peptides show promise, clinical translation of antimicrobial peptides has met a number of regulatory hurdles [174].

Conclusion/Summary and Future Perspectives

This chapter started out by introducing basic concepts in peptide design, from learning from nature and deriving peptides from active domains within proteins to utilizing peptide library screening methods to identify peptides with a desired response. Standard methods for peptide synthesis, most commonly solid-phase peptide synthesis, and various techniques for functionalizing biomaterials with peptides were summarized. These chemical methods for peptide synthesis bring about additional opportunities in peptide design, as non-natural amino acids and fluorescent tags, among other molecules, can be incorporated during the synthesis. This allows for the development of peptides that may be more stable or have higher activity than their naturally derived counterparts, and the possibility of adding dyes further provides simplified methods for detecting the peptides. These so-called peptidomimetics have started to be explored for their antimicrobial activities [126, 132, 180], and it would be interesting to explore further osteoinductive peptidomimetics. Covalent and non-covalent binding of peptides to biomaterials also provides a means to localize the peptide and in some cases provide controlled release. It would also be interesting to explore osteoinductive and antimicrobial peptides in the context of other controlled release systems. For example, research is rapidly advancing in the development of stimuli-responsive systems that release their cargo in response to changes in temperature, pH, light, etc., and conjugation to polymers or encapsulation in micro-/nano-particles also provide a means for extended delivery of compounds [181].

Finally, this chapter focused on the development and application of two classes of peptides, those with osteoinductive and anti-biofilm activity. The osteoinductive peptides have been derived for a large part from the active domains of the different BMPs. Their activities have been demonstrated in vitro by stimulating osteogenic differentiation of different types of progenitor cells and in vivo by promoting bone growth, ectopically or orthopically. These peptides show promise to support osseointegration when functionalized on the surface of Ti or other inert implant materials. Further work to combine these peptides with newly developed biomaterial scaffolds [182, 183] would help to advance the field of bone tissue engineering. The anti-biofilm peptides presented have been developed from naturally occurring antimicrobial peptides as well as from screening of peptide libraries. Candidate peptides have shown activity against several strains of bacteria that are involved in biofilm formation, and they have also been shown to reduce bacterial burden or wound size in in vivo models. Development of more complex animal models that mimic the clinical conditions of biofilm infection would help to further identify and characterize peptides with anti-biofilm activity. Last but not least, the combination of osteoinductive and anti-biofilm peptides would allow for the development of materials that stimulate bone formation while simultaneously protecting against chronic infection. Anti-biofilm peptides have clearly been intended for orthopedic [184] and dental applications, as they have been studied while functionalized on Ti surfaces. Full-length BMPs have been combined with antibacterial agents, such as vancomycin and silver (Ag^+) [185], so the combination of osteoinductive and anti-biofilm peptides, functionalized on the surface of permanent implants or implemented in tissue engineering scaffolds, is a logical next step.

Acknowledgments This work in the research group of the author has been partially supported by the Research Foundation Flanders (FWO), grant number G.0B39.14, and the special research fund of the KU Leuven, grant numbers CREA/13/017 and IDO/13/016. The author also gratefully acknowledges the interesting discussions about peptide-functionalized biomaterials and protein engineering over the years with the members of her research group, particularly Dr. Al Halifa Soultan, Dr. Susanna Piluso, Dr. Abhijith Kudva, Burak Toprakhisar, and Christian Garcia Abrego, as well as her former mentors Prof. Jeffrey Hubbell, Prof. Patrick Stayton, and Prof. Michael Hecht.

References

1. Kamtekar S, Schiffer JM, Xiong H, Babik JM, Hecht MH (1993) Protein design by binary patterning of polar and nonpolar amino acids. Science 262:1680. https://doi.org/10.1126/science.8259512
2. West MW, Wang W, Patterson J, Mancias JD, Beasley JR, Hecht MH (1999) *De novo* amyloid proteins from designed combinatorial libraries. Proc Natl Acad Sci U S A 96:11211–11216. https://doi.org/10.1073/pnas.96.20.11211
3. Zhang S, Holmes TC, DiPersio CM, Hynes RO, Su X, Rich A (1995) Self-complementary oligopeptide matrices support mammalian cell attachment. Biomaterials 16:1385–1393. https://doi.org/10.1016/0142-9612(95)96874-Y
4. Collier JH, Messersmith PB (2004) Self-assembling polymer–peptide conjugates: nanostructural tailoring. Adv Mater 16:907–910. https://doi.org/10.1002/adma.200306379
5. Hauser CAE et al (2011) Natural tri- to hexapeptides self-assemble in water to amyloid β-type fiber aggregates by unexpected α-helical intermediate structures. Proc Natl Acad Sci U S A 108:1361. https://doi.org/10.1073/pnas.1014796108
6. Ruoslahti E (1996) RGD and other recognition sequences for integrins. Annu Rev Cell Dev Biol 12:697–715. https://doi.org/10.1146/annurev.cellbio.12.1.697
7. DeForest CA, Polizzotti BD, Anseth KS (2009) Sequential click reactions for synthesizing and patterning three-dimensional cell microenvironments. Nat Mater 8:659. https://doi.org/10.1038/nmat2473
8. Andukuri A, Minor WP, Kushwaha M, Anderson JM, Jun H-W (2010) Effect of endothelium mimicking self-assembled nanomatrices on cell adhesion and spreading of human endothelial cells and smooth muscle cells. Nanomedicine 6:289–297. https://doi.org/10.1016/j.nano.2009.09.004
9. Dalet-Fumeron V, Boudjennah L, Pagano M (1998) Binding of the cysteine proteinases papain and cathepsin B-like to coated laminin: use of synthetic peptides from laminin and from the laminin binding region of the β1Integrin subunit to characterize the binding site. Arch Biochem Biophys 358:283–290. https://doi.org/10.1006/abbi.1998.0868
10. Massia SP, Hubbell JA (1991) Human endothelial cell interactions with surface-coupled adhesion peptides on a nonadhesive glass substrate and two polymeric biomaterials. J Biomed Mater Res 25:223–242
11. Gobin AS, West JL (2003) Val-Ala-Pro-Gly, an elastin-derived non-integrin ligand: smooth muscle cell adhesion and specificity. J Biomed Mater Res A 67A:255–259. https://doi.org/10.1002/jbm.a.10110
12. Chen S et al (2015b) A laminin mimetic peptide SIKVAV-conjugated chitosan hydrogel promoting wound healing by enhancing angiogenesis, re-epithelialization and collagen deposition. J Mater Chem B 3:6798–6804. https://doi.org/10.1039/C5TB00842E
13. Davel LE, Puricelli LI, Del Carmen M, Vidal C, De Lorenzo MS, Sacerdote de Lustig E, Bal de Kier Joffe ED (1999) Soluble factors from the target organ enhance tumor cell angiogenesis: role of laminin SIKVAV sequence. Oncol Rep 6:907–918
14. Maeda T, Oyama R, Titani K, Sekiguchi K (1993) Engineering of artificial cell-adhesive proteins by grafting EILDVPST sequence derived from fibronectin the. J Biochem 113:29–35
15. Moyano JV et al (1997) Fibronectin type III5 repeat contains a novel cell adhesion sequence, KLDAPT, which binds activated α4β1 and α4β7 integrins. J Biol Chem 272:24832–24836
16. Woods A, McCarthy JB, Furcht LT, Couchman JR (1993) A synthetic peptide from the COOH-terminal heparin-binding domain of fibronectin promotes focal adhesion formation. Mol Biol Cell 4:605–613
17. Feng Y, Mrksich M (2004) The synergy peptide PHSRN and the adhesion peptide RGD mediate cell adhesion through a common mechanism. Biochemistry 43:15811–15821. https://doi.org/10.1021/bi049174+

18. Lee ST et al (2010) Engineering integrin signaling for promoting embryonic stem cell self-renewal in a precisely defined niche. Biomaterials 31:1219–1226. https://doi.org/10.1016/j.biomaterials.2009.10.054
19. Yokosaki Y et al (1998) Identification of the ligand binding site for the integrin $\alpha9\beta1$ in the third fibronectin type III repeat of tenascin-C. J Biol Chem 273:11423–11428
20. Massia SP, Hubbell JA (1992) Vascular endothelial cell adhesion and spreading promoted by the peptide REDV of the IIICS region of plasma fibronectin is mediated by integrin alpha 4 beta 1. J Biol Chem 267:14019–14026
21. Nagase H, Fields GB (1996) Human matrix metalloproteinase specificity studies using collagen sequence-based synthetic peptides. Biopolymers 40:399–416. https://doi.org/10.1002/(SICI)1097-0282(1996)40:4<399::AID-BIP5>3.0.CO;2-R
22. Lutolf MP, Lauer-Fields JL, Schmoekel HG, Metters AT, Weber FE, Fields GB, Hubbell JA (2003a) Synthetic matrix metalloproteinase-sensitive hydrogels for the conduction of tissue regeneration: engineering cell-invasion characteristics. Proc Natl Acad Sci U S A 100:5413. https://doi.org/10.1073/pnas.0737381100
23. Lutolf MP, Weber FE, Schmoekel HG, Schense JC, Kohler T, Müller R, Hubbell JA (2003b) Repair of bone defects using synthetic mimetics of collagenous extracellular matrices. Nat Biotechnol 21:513. https://doi.org/10.1038/nbt818
24. Smith GP, Petrenko VA (1997) Phage display. Chem Rev 97:391–410. https://doi.org/10.1021/cr960065d
25. Winter GP, James K, Potter G (1989) Antibody engineering. Philos Trans R Soc Lond B Biol Sci 324:537–547. https://doi.org/10.1098/rstb.1989.0066
26. Desch KC et al (2015) Probing ADAMTS13 substrate specificity using phage display. PLoS One 10:e0122931. https://doi.org/10.1371/journal.pone.0122931
27. Turk BE, Huang LL, Piro ET, Cantley LC (2001) Determination of protease cleavage site motifs using mixture-based oriented peptide libraries. Nat Biotechnol 19:661–667. https://doi.org/10.1038/90273
28. Songyang Z, Blechner S, Hoagland N, Hoekstra MF, Piwnica-Worms H, Cantley LC (1994) Use of an oriented peptide library to determine the optimal substrates of protein kinases. Curr Biol 4:973–982. https://doi.org/10.1016/S0960-9822(00)00221-9
29. Songyang Z et al (1997) Recognition of unique carboxyl-terminal motifs by distinct PDZ domains. Science 275:73–77
30. Songyang Z et al (1993) SH2 domains recognize specific phosphopeptide sequences. Cell 72:767–778. https://doi.org/10.1016/0092-8674(93)90404-E
31. Yaffe MB et al (1997) The structural basis for 14-3-3:phosphopeptide binding specificity. Cell 91:961–971. https://doi.org/10.1016/S0092-8674(00)80487-0
32. Boulware KT, Daugherty PS (2006) Protease specificity determination by using cellular libraries of peptide substrates (CLiPS). Proc Natl Acad Sci U S A 103:7583. https://doi.org/10.1073/pnas.0511108103
33. Chen EI, Kridel SJ, Howard EW, Li W, Godzik A, Smith JW (2002) A unique substrate recognition profile for matrix metalloproteinase-2. J Biol Chem 277:4485–4491. https://doi.org/10.1074/jbc.M109469200
34. Smith MM, Shi L, Navre M (1995) Rapid identification of highly active and selective substrates for Stromelysin and Matrilysin using bacteriophage peptide display libraries. J Biol Chem 270:6440–6449
35. Pan W, Arnone M, Kendall M, Grafstrom RH, Seitz SP, Wasserman ZR, Albright CF (2003) Identification of peptide substrates for human MMP-11 (Stromelysin-3) using phage display. J Biol Chem 278:27820–27827
36. Deng S-J et al (2000) Substrate specificity of human collagenase 3 assessed using a phage-displayed peptide library. J Biol Chem 275:31422–31427. https://doi.org/10.1074/jbc.M004538200

37. Shuichi O, Kazutaka M, Yoshikazu S, Ken-ichi M, Konstanty W, Yuji Y (2001) Substrate phage as a tool to identify novel substrate sequences of proteases. Comb Chem High Throughput Screen 4:573–583. https://doi.org/10.2174/1386207013330788
38. Patterson J, Hubbell JA (2010) Enhanced proteolytic degradation of molecularly engineered PEG hydrogels in response to MMP-1 and MMP-2. Biomaterials 31:7836–7845. https://doi.org/10.1016/j.biomaterials.2010.06.061
39. Patterson J, Hubbell JA (2011) SPARC-derived protease substrates to enhance the plasmin sensitivity of molecularly engineered PEG hydrogels. Biomaterials 32:1301–1310. https://doi.org/10.1016/j.biomaterials.2010.10.016
40. Merrifield RB (1963) Solid phase peptide synthesis. I. The synthesis of a tetrapeptide. J Am Chem Soc 85:2149–2154. https://doi.org/10.1021/ja00897a025
41. Mäde V, Els-Heindl S, Beck-Sickinger AG (2014) Automated solid-phase peptide synthesis to obtain therapeutic peptides. Beilstein J Org Chem 10:1197–1212. https://doi.org/10.3762/bjoc.10.118
42. Vanier GS (2013) Microwave-assisted solid-phase peptide synthesis based on the Fmoc protecting group strategy (CEM). In: Jensen KJ, Tofteng Shelton P, Pedersen SL (eds) Peptide synthesis and applications. Humana Press, Totowa, pp 235–249. https://doi.org/10.1007/978-1-62703-544-6_17
43. Ramesh S, de la Torre BG, Albericio F, Kruger HG, Govender T (2017) Microwave-assisted synthesis of antimicrobial peptides. In: Hansen PR (ed) Antimicrobial peptides: methods and protocols. Springer, New York, pp 51–59. https://doi.org/10.1007/978-1-4939-6737-7_4
44. Chen M, Heimer P, Imhof D (2015a) Synthetic strategies for polypeptides and proteins by chemical ligation. Amino Acids 47:1283–1299. https://doi.org/10.1007/s00726-015-1982-5
45. Clancy KW, Melvin JA, McCafferty DG (2010) Sortase transpeptidases: insights into mechanism, substrate specificity, and inhibition. Pept Sci 94:385–396. https://doi.org/10.1002/bip.21472
46. Schmohl L, Schwarzer D (2014) Chemo-enzymatic three-fragment assembly of semisynthetic proteins. J Pept Sci 20:145–151. https://doi.org/10.1002/psc.2600
47. Chang TK, Jackson DY, Burnier JP, Wells JA (1994) Subtiligase: a tool for semisynthesis of proteins. Proc Natl Acad Sci U S A 91:12544. https://doi.org/10.1073/pnas.91.26.12544
48. Tan X, Yang R, Liu C-F (2018) Facilitating Subtiligase-catalyzed peptide ligation reactions by using peptide thioester substrates. Org Lett 20:6691–6694. https://doi.org/10.1021/acs.orglett.8b02747
49. Boda SK et al (2019) Mineralized nanofiber segments coupled with calcium-binding BMP-2 peptides for alveolar bone regeneration. Acta Biomater 85:282–293. https://doi.org/10.1016/j.actbio.2018.12.051
50. Weng L, Boda SK, Wang H, Teusink MJ, Shuler FD, Xie J (2018) Novel 3D hybrid nanofiber aerogels coupled with BMP-2 peptides for cranial bone regeneration. Adv Healthc Mater 7:e1701415. https://doi.org/10.1002/adhm.201701415
51. Leight JL, Alge DL, Maier AJ, Anseth KS (2013) Direct measurement of matrix metalloproteinase activity in 3D cellular microenvironments using a fluorogenic peptide substrate. Biomaterials 34:7344–7352. https://doi.org/10.1016/j.biomaterials.2013.06.023
52. Sakiyama SE, Schense JC, Hubbell JA (1999) Incorporation of heparin-binding peptides into fibrin gels enhances neurite extension: an example of designer matrices in tissue engineering. FASEB J 13:2214–2224
53. Schense JC, Hubbell JA (1999) Cross-linking exogenous bifunctional peptides into fibrin gels with factor XIIIa. Bioconjug Chem 10:75–81. https://doi.org/10.1021/bc9800769
54. Ehrbar M, Rizzi SC, Schoenmakers RG, San Miguel B, Hubbell JA, Weber FE, Lutolf MP (2007) Biomolecular hydrogels formed and degraded via site-specific enzymatic reactions. Biomacromolecules 8:3000–3007. https://doi.org/10.1021/bm070228f
55. Zhu J, Tang C, Kottke-Marchant K, Marchant RE (2009) Design and synthesis of biomimetic hydrogel scaffolds with controlled organization of cyclic RGD peptides. Bioconjug Chem 20:333–339. https://doi.org/10.1021/bc800441v

56. Maia FR, Barbosa M, Gomes DB, Vale N, Gomes P, Granja PL, Barrias CC (2014) Hydrogel depots for local co-delivery of osteoinductive peptides and mesenchymal stem cells. J Control Release 189:158–168. https://doi.org/10.1016/j.jconrel.2014.06.030
57. Phelps EA et al (2012) Maleimide cross-linked bioactive PEG hydrogel exhibits improved reaction kinetics and cross-linking for cell encapsulation and in situ delivery. Adv Mater 24:64–70. https://doi.org/10.1002/adma.201103574
58. Gentile P, Ferreira AM, Callaghan JT, Miller CA, Atkinson J, Freeman C, Hatton PV (2017) Multilayer nanoscale encapsulation of biofunctional peptides to enhance bone tissue regeneration in vivo. Adv Healthc Mater 6:1601182. https://doi.org/10.1002/adhm.201601182
59. Cao F-Y, Yin W-N, Fan J-X, Zhuo R-X, Zhang X-Z (2015) A novel function of BMHP1 and cBMHP1 peptides to induce the osteogenic differentiation of mesenchymal stem cells. Biomater Sci 3:345–351. https://doi.org/10.1039/C4BM00300D
60. Hou R et al (2018) Novel osteogenic growth peptide C-terminal pentapeptide grafted poly(d,l-lactic acid) improves the proliferation and differentiation of osteoblasts: the potential bone regenerative biomaterial. Int J Biol Macromol 119:874–881. https://doi.org/10.1016/j.ijbiomac.2018.08.010
61. Li S, Xu Y, Yu J, Becker ML (2017) Enhanced osteogenic activity of poly(ester urea) scaffolds using facile post-3D printing peptide functionalization strategies. Biomaterials 141:176–187. https://doi.org/10.1016/j.biomaterials.2017.06.038
62. Soultan AH, Verheyen T, Smet M, De Borggraeve WM, Patterson J (2018) Synthesis and peptide functionalization of hyperbranched poly(arylene oxindole) towards versatile biomaterials. Polym Chem 9:2775–2784. https://doi.org/10.1039/C8PY00139A
63. Bain JL, Bonvallet PP, Abou-Arraj RV, Schupbach P, Reddy MS, Bellis SL (2015) Enhancement of the regenerative potential of anorganic bovine bone graft utilizing a polyglutamate-modified BMP2 peptide with improved binding to calcium-containing materials. Tissue Eng Part A 21:2426–2436. https://doi.org/10.1089/ten.tea.2015.0160
64. Cao Q, He Z, Sun WQ, Fan G, Zhao J, Bao N, Ye T (2019) Improvement of calcium phosphate scaffold osteogenesis in vitro via combination of glutamate-modified BMP-2 peptides. Mater Sci Eng C Mater Biol Appl 96:412–418. https://doi.org/10.1016/j.msec.2018.11.048
65. Culpepper BK, Webb WM, Bonvallet PP, Bellis SL (2014) Tunable delivery of bioactive peptides from hydroxyapatite biomaterials and allograft bone using variable-length polyglutamate domains. J Biomed Mater Res A 120A:1008–1016. https://doi.org/10.1002/jbm.a.34766
66. Sawyer AA, Weeks DM, Kelpke SS, McCracken MS, Bellis SL (2005) The effect of the addition of a polyglutamate motif to RGD on peptide tethering to hydroxyapatite and the promotion of mesenchymal stem cell adhesion. Biomaterials 26:7046–7056. https://doi.org/10.1016/j.biomaterials.2005.05.006
67. Gilbert M, Giachelli CM, Stayton PS (2003) Biomimetic peptides that engage specific integrin-dependent signaling pathways and bind to calcium phosphate surfaces. J Biomed Mater Res A 67A:69–77. https://doi.org/10.1002/jbm.a.10053
68. Gao X et al (2015) Osteoinductive peptide-functionalized nanofibers with highly ordered structure as biomimetic scaffolds for bone tissue engineering. Int J Nanomedicine 10:7109–7128
69. Ko E, Yang K, Shin J, Cho S-W (2013) Polydopamine-assisted osteoinductive peptide immobilization of polymer scaffolds for enhanced bone regeneration by human adipose-derived stem cells. Biomacromolecules 14:3202–3213. https://doi.org/10.1021/bm4008343
70. Pan G et al (2016) Biomimetic design of mussel-derived bioactive peptides for dual-functionalization of titanium-based biomaterials. J Am Chem Soc 138:15078–15086. https://doi.org/10.1021/jacs.6b09770
71. Pan H, Zheng Q, Yang S, Guo X (2014) Effects of functionalization of PLGA-[Asp-PEG]n copolymer surfaces with Arg-Gly-Asp peptides, hydroxyapatite nanoparticles, and BMP-2-derived peptides on cell behavior in vitro. J Biomed Mater Res A 102:4526–4535. https://doi.org/10.1002/jbm.a.35129
72. Wang M et al (2015) In vitro culture and directed osteogenic differentiation of human pluripotent stem cells on peptides-decorated two-dimensional microenvironment. ACS Appl Mater Interfaces 7:4560–4572. https://doi.org/10.1021/acsami.5b00188

73. Lee H, Dellatore SM, Miller WM, Messersmith PB (2007a) Mussel-inspired surface chemistry for multifunctional coatings. Science 318:426. https://doi.org/10.1126/science.1147241
74. Geetha M, Singh AK, Asokamani R, Gogia AK (2009) Ti based biomaterials, the ultimate choice for orthopaedic implants—a review. Prog Mater Sci 54:397–425. https://doi.org/10.1016/j.pmatsci.2008.06.004
75. Tejero R, Anitua E, Orive G (2014) Toward the biomimetic implant surface: biopolymers on titanium-based implants for bone regeneration. Prog Polym Sci 39:1406–1447. https://doi.org/10.1016/j.progpolymsci.2014.01.001
76. Langer R, Vacanti JP (1993) Tissue engineering. Science 260:920. https://doi.org/10.1126/science.8493529
77. Moeinzadeh S, Jabbari E (2015) Morphogenic peptides in regeneration of load bearing tissues. In: Bertassoni LE, Coelho PG (eds) Engineering mineralized and load bearing tissues. Springer, Cham, pp 95–110. https://doi.org/10.1007/978-3-319-22345-2_6
78. Visser R, Rico-Llanos GA, Pulkkinen H, Becerra J (2016) Peptides for bone tissue engineering. J Control Release 244:122–135. https://doi.org/10.1016/j.jconrel.2016.10.024
79. Bab I et al (1992) Histone H4-related osteogenic growth peptide (OGP): a novel circulating stimulator of osteoblastic activity. EMBO J 11:1867–1873
80. Gabarin N et al (2001) Mitogenic Gi protein-MAP kinase signaling cascade in MC3T3-E1 osteogenic cells: activation by C-terminal pentapeptide of osteogenic growth peptide [OGP(10–14)] and attenuation of activation by cAMP. J Cell Biochem 81:594–603. https://doi.org/10.1002/jcb.1083
81. Suzuki Y, Tanihara M, Suzuki K, Saitou A, Sufan W, Nishimura Y (2000) Alginate hydrogel linked with synthetic oligopeptide derived from BMP-2 allows ectopic osteoinduction in vivo. J Biomed Mater Res 50:405–409. https://doi.org/10.1002/(SICI)1097-4636(20000605)50:3<405::AID-JBM15>3.0.CO;2-Z
82. Saito A, Suzuki Y, Ogata S-i, Ohtsuki C, Tanihara M (2003) Activation of osteo-progenitor cells by a novel synthetic peptide derived from the bone morphogenetic protein-2 knuckle epitope. Biochim Biophys Acta 1651:60–67. https://doi.org/10.1016/S1570-9639(03)00235-8
83. Seol Y-J et al (2006) Enhanced osteogenic promotion around dental implants with synthetic binding motif mimicking bone morphogenetic protein (BMP)-2. J Biomed Mater Res A 77A:599–607. https://doi.org/10.1002/jbm.a.30639
84. Zouani OF, Chollet C, Guillotin B, Durrieu M-C (2010) Differentiation of pre-osteoblast cells on poly(ethylene terephthalate) grafted with RGD and/or BMPs mimetic peptides. Biomaterials 31:8245–8253. https://doi.org/10.1016/j.biomaterials.2010.07.042
85. Kim HK et al (2012) Osteogenesis induced by a bone forming peptide from the prodomain region of BMP-7. Biomaterials 33:7057–7063. https://doi.org/10.1016/j.biomaterials.2012.06.036
86. Lee JS, Kim ME, Seon JK, Kang JY, Yoon TR, Park Y-D, Kim HK (2018) Bone-forming peptide-3 induces osteogenic differentiation of bone marrow stromal cells via regulation of the ERK1/2 and Smad1/5/8 pathways. Stem Cell Res 26:28–35. https://doi.org/10.1016/j.scr.2017.11.016
87. Choi YJ et al (2010) The identification of a heparin binding domain peptide from bone morphogenetic protein-4 and its role on osteogenesis. Biomaterials 31:7226–7238. https://doi.org/10.1016/j.biomaterials.2010.05.022
88. Bergeron E, Leblanc E, Drevelle O, Giguère R, Beauvais S, Grenier G, Faucheux N (2011) The evaluation of ectopic bone formation induced by delivery systems for bone morphogenetic protein-9 or its derived peptide. Tissue Eng Part A 18:342–352. https://doi.org/10.1089/ten.tea.2011.0008
89. Bergeron E, Senta H, Mailloux A, Park H, Lord E, Faucheux N (2009) Murine preosteoblast differentiation induced by a peptide derived from bone morphogenetic Proteins-9. Tissue Eng Part A 15:3341–3349. https://doi.org/10.1089/ten.tea.2009.0189
90. Jung RE, Cochran DL, Domken O, Seibl R, Jones AA, Buser D, Hammerle CHF (2007) The effect of matrix bound parathyroid hormone on bone regeneration. Clin Oral Implants Res 18:319–325. https://doi.org/10.1111/j.1600-0501.2007.01342.x

91. Park J-B et al (2007) Osteopromotion with synthetic oligopeptide–coated bovine bone mineral in vivo. J Periodontol 78:157–163. https://doi.org/10.1902/jop.2007.060200
92. Lee J-Y et al (2007b) Assembly of collagen-binding peptide with collagen as a bioactive scaffold for osteogenesis in vitro and in vivo. Biomaterials 28:4257–4267. https://doi.org/10.1016/j.biomaterials.2007.05.040
93. Gabet Y et al (2004) Osteogenic growth peptide modulates fracture callus structural and mechanical properties. Bone 35:65–73. https://doi.org/10.1016/j.bone.2004.03.025
94. Pigossi SC, Medeiros MC, Saska S, Cirelli JA, Scarel-Caminaga RM (2016) Role of osteogenic growth peptide (OGP) and OGP(10-14) in bone regeneration: a review. Int J Mol Sci 17:1885. https://doi.org/10.3390/ijms17111885
95. Policastro GM, Becker ML (2016) Osteogenic growth peptide and its use as a bio-conjugate in regenerative medicine applications. Wiley Interdiscip Rev Nanomed Nanobiotechnol 8:449–464. https://doi.org/10.1002/wnan.1376
96. Moore NM, Lin NJ, Gallant ND, Becker ML (2010) The use of immobilized osteogenic growth peptide on gradient substrates synthesized via click chemistry to enhance MC3T3-E1 osteoblast proliferation. Biomaterials 31:1604–1611. https://doi.org/10.1016/j.biomaterials.2009.11.011
97. Horii A, Wang X, Gelain F, Zhang S (2007) Biological designer self-assembling peptide nanofiber scaffolds significantly enhance osteoblast proliferation, differentiation and 3-D migration. PLoS One 2:e190. https://doi.org/10.1371/journal.pone.0000190
98. Wozney JM (1989) Bone morphogenetic proteins. Prog Growth Factor Res 1:267–280. https://doi.org/10.1016/0955-2235(89)90015-X
99. Saito A, Suzuki Y, Ogata S-I, Ohtsuki C, Tanihara M (2004) Prolonged ectopic calcification induced by BMP-2–derived synthetic peptide. J Biomed Mater Res A 70A:115–121. https://doi.org/10.1002/jbm.a.30071
100. Saito A, Suzuki Y, Ogata S-I, Ohtsuki C, Tanihara M (2005) Accelerated bone repair with the use of a synthetic BMP-2-derived peptide and bone-marrow stromal cells. J Biomed Mater Res A 72A:77–82. https://doi.org/10.1002/jbm.a.30208
101. Niu X, Feng Q, Wang M, Guo X, Zheng Q (2009) Porous nano-HA/collagen/PLLA scaffold containing chitosan microspheres for controlled delivery of synthetic peptide derived from BMP-2. J Control Release 134:111–117. https://doi.org/10.1016/j.jconrel.2008.11.020
102. Moore NM, Lin NJ, Gallant ND, Becker ML (2011) Synergistic enhancement of human bone marrow stromal cell proliferation and osteogenic differentiation on BMP-2-derived and RGD peptide concentration gradients. Acta Biomater 7:2091–2100. https://doi.org/10.1016/j.actbio.2011.01.019
103. Bergeron E, Marquis ME, Chrétien I, Faucheux N (2007) Differentiation of preosteoblasts using a delivery system with BMPs and bioactive glass microspheres. J Mater Sci Mater Med 18:255–263. https://doi.org/10.1007/s10856-006-0687-4
104. Beauvais S, Drevelle O, Lauzon M-A, Daviau A, Faucheux N (2016) Modulation of MAPK signalling by immobilized adhesive peptides: effect on stem cell response to BMP-9-derived peptides. Acta Biomater 31:241–251. https://doi.org/10.1016/j.actbio.2015.12.005
105. Neer RM et al (2001) Effect of parathyroid hormone (1-34) on fractures and bone mineral density in postmenopausal women with osteoporosis. N Engl J Med 344:1434–1441. https://doi.org/10.1056/NEJM200105103441904
106. Arrighi I, Mark S, Alvisi M, von Rechenberg B, Hubbell JA, Schense JC (2009) Bone healing induced by local delivery of an engineered parathyroid hormone prodrug. Biomaterials 30:1763–1771. https://doi.org/10.1016/j.biomaterials.2008.12.023
107. Takahata M, Schwarz EM, Chen T, O'Keefe RJ, Awad HA (2012) Delayed short-course treatment with teriparatide (PTH1–34) improves femoral allograft healing by enhancing intramembranous bone formation at the graft–host junction. J Bone Miner Res 27:26–37. https://doi.org/10.1002/jbmr.518
108. Chandra A et al (2014) PTH1–34 alleviates radiotherapy-induced local bone loss by improving osteoblast and osteocyte survival. Bone 67:33–40. https://doi.org/10.1016/j.bone.2014.06.030

109. Dent-Acosta RE, Storm N, Steiner RS, San Martin J (2012) The tactics of modern-day regulatory trials. JBJS 94:39–44. https://doi.org/10.2106/jbjs.l.00194
110. Pountos I, Panteli M, Lampropoulos A, Jones E, Calori GM, Giannoudis PV (2016) The role of peptides in bone healing and regeneration: a systematic review. BMC Med 14:103. https://doi.org/10.1186/s12916-016-0646-y
111. Ryaby JT, Sheller MR, Levine BP, Bramlet DG, Ladd AL, Carney DH (2006) Thrombin peptide TP508 stimulates cellular events leading to angiogenesis, revascularization, and repair of dermal and musculoskeletal tissues. JBJS 88:132–139. https://doi.org/10.2106/jbjs.f.00892
112. Aspenberg P et al (2010) Teriparatide for acceleration of fracture repair in humans: a prospective, randomized, double-blind study of 102 postmenopausal women with distal radial fractures. J Bone Miner Res 25:404–414. https://doi.org/10.1359/jbmr.090731
113. Aspenberg P, Johansson T (2010) Teriparatide improves early callus formation in distal radial fractures. Acta Orthop 81:234–236. https://doi.org/10.3109/17453671003761946
114. Chintamaneni S, Finzel K, Gruber BL (2010) Successful treatment of sternal fracture nonunion with teriparatide. Osteoporos Int 21:1059–1063. https://doi.org/10.1007/s00198-009-1061-4
115. Yu C-T, Chang C-C, Chen C-L, Wei JC-C, Wu J-K (2008) Early callus formation in human hip fracture treated with internal fixation and teriparatide. J Rheumatol 35:2082–2083
116. Arnold PM et al (2016) Efficacy of i-factor bone graft versus autograft in anterior cervical discectomy and fusion: results of the prospective, randomized, single-blinded Food and Drug Administration investigational device exemption study. Spine 41:1075–1083. https://doi.org/10.1097/brs.0000000000001466
117. Gomar F, Orozco R, Villar JL, Arrizabalaga F (2007) P-15 small peptide bone graft substitute in the treatment of non-unions and delayed union. A pilot clinical trial. Int Orthop 31:93–99. https://doi.org/10.1007/s00264-006-0087-x
118. Yukna RA, Krauser JT, Callan DP, Evans GH, Cruz R, Martin M (2000) Multi-center clinical comparison of combination anorganic bovine-derived hydroxyapatite matrix (ABM)/cell binding peptide (P-15) and ABM in human periodontal osseous defects. 6-month results. J Periodontol 71:1671–1679. https://doi.org/10.1902/jop.2000.71.11.1671
119. Bjarnsholt T (2013) The role of bacterial biofilms in chronic infections. APMIS 121:1–58. https://doi.org/10.1111/apm.12099
120. Darouiche RO (2004) Treatment of infections associated with surgical implants. N Engl J Med 350:1422–1429. https://doi.org/10.1056/NEJMra035415
121. Sutherland IW (2001) The biofilm matrix—an immobilized but dynamic microbial environment. Trends Microbiol 9:222–227. https://doi.org/10.1016/S0966-842X(01)02012-1
122. Davies D (2003) Understanding biofilm resistance to antibacterial agents. Nat Rev Drug Discov 2:114. https://doi.org/10.1038/nrd1008
123. Jefferson KK (2004) What drives bacteria to produce a biofilm? FEMS Microbiol Lett 236:163–173. https://doi.org/10.1111/j.1574-6968.2004.tb09643.x
124. Costerton JW, Stewart PS, Greenberg EP (1999) Bacterial biofilms: a common cause of persistent infections. Science 284:1318. https://doi.org/10.1126/science.284.5418.1318
125. Karygianni L, Al-Ahmad A, Argyropoulou A, Hellwig E, Anderson AC, Skaltsounis AL (2016) Natural antimicrobials and oral microorganisms: a systematic review on herbal interventions for the eradication of multispecies oral biofilms. Front Microbiol 6:1529. https://doi.org/10.3389/fmicb.2015.01529
126. Andrea A, Molchanova N, Jenssen H (2018) Antibiofilm peptides and peptidomimetics with focus on surface immobilization. Biomol Ther 8:27. https://doi.org/10.3390/biom8020027
127. Batoni G, Maisetta G, Esin S (2016) Antimicrobial peptides and their interaction with biofilms of medically relevant bacteria. Biochim Biophys Acta 1858:1044–1060. https://doi.org/10.1016/j.bbamem.2015.10.013
128. de la Fuente-Núñez C, Cardoso MH, de Souza Cândido E, Franco OL, Hancock REW (2016) Synthetic antibiofilm peptides. Biochim Biophys Acta 1858:1061–1069. https://doi.org/10.1016/j.bbamem.2015.12.015
129. Dostert M, Belanger CR, Hancock REW (2018) Design and assessment of anti-biofilm peptides: steps toward clinical application. J Innate Immun 11:193. https://doi.org/10.1159/000491497

130. Pletzer D, Coleman SR, Hancock REW (2016) Anti-biofilm peptides as a new weapon in antimicrobial warfare. Curr Opin Microbiol 33:35–40. https://doi.org/10.1016/j.mib.2016.05.016
131. Pletzer D, Hancock REW (2016) Antibiofilm peptides: potential as broad-Spectrum agents. J Bacteriol 198:2572. https://doi.org/10.1128/JB.00017-16
132. Strempel N, Strehmel J, Overhage J (2015) Potential application of antimicrobial peptides in the treatment of bacterial biofilm infections. Curr Pharm Des 21:67–84. https://doi.org/10.2174/1381612820666140905124312
133. Haney EF, Brito-Sánchez Y, Trimble MJ, Mansour SC, Cherkasov A, Hancock REW (2018) Computer-aided discovery of peptides that specifically attack bacterial biofilms. Sci Rep 8:1871. https://doi.org/10.1038/s41598-018-19669-4
134. De Brucker K et al (2014) Derivatives of the mouse cathelicidin-related antimicrobial peptide (CRAMP) inhibit fungal and bacterial biofilm formation. Antimicrob Agents Chemother 58:5395. https://doi.org/10.1128/AAC.03045-14
135. De Zoysa GH, Cameron AJ, Hegde VV, Raghothama S, Sarojini V (2015) Antimicrobial peptides with potential for biofilm eradication: synthesis and structure activity relationship studies of battacin peptides. J Med Chem 58:625–639. https://doi.org/10.1021/jm501084q
136. Almaaytah A, Tarazi S, Al-Fandi M, Abuilhaija A, Al-shar'i N, Al-Balas Q, Abu-Awad A (2015) The design and functional characterization of the antimicrobial and antibiofilm activities of BMAP27-melittin, a rationally designed hybrid peptide. Int J Pept Res Ther 21:165–177. https://doi.org/10.1007/s10989-014-9444-6
137. Orlando F et al (2008) BMAP-28 improves the efficacy of vancomycin in rat models of gram-positive cocci ureteral stent infection. Peptides 29:1118–1123. https://doi.org/10.1016/j.peptides.2008.03.005
138. Mataraci E, Dosler S (2012) *In vitro* activities of antibiotics and antimicrobial cationic peptides alone and in combination against methicillin-resistant *Staphylococcus aureus* biofilms. Antimicrob Agents Chemother 56:6366. https://doi.org/10.1128/AAC.01180-12
139. Bionda N et al (2016) Identification of novel cyclic lipopeptides from a positional scanning combinatorial library with enhanced antibacterial and antibiofilm activities. Eur J Med Chem 108:354–363. https://doi.org/10.1016/j.ejmech.2015.11.032
140. de la Fuente-Núñez C et al (2015) D-Enantiomeric peptides that eradicate wild-type and multidrug-resistant biofilms and protect against lethal Pseudomonas aeruginosa infections. Chem Biol 22:1280–1282. https://doi.org/10.1016/j.chembiol.2015.09.004
141. Pletzer D, Wolfmeier H, Bains M, Hancock REW (2017) Synthetic peptides to target stringent response-controlled virulence in a Pseudomonas aeruginosa murine cutaneous infection model. Front Microbiol 8:1867–1867. https://doi.org/10.3389/fmicb.2017.01867
142. Mohamed MF, Brezden A, Mohammad H, Chmielewski J, Seleem MN (2017) A short D-enantiomeric antimicrobial peptide with potent immunomodulatory and antibiofilm activity against multidrug-resistant Pseudomonas aeruginosa and Acinetobacter baumannii. Sci Rep 7:6953. https://doi.org/10.1038/s41598-017-07440-0
143. Brancatisano FL et al (2014) Inhibitory effect of the human liver-derived antimicrobial peptide hepcidin 20 on biofilms of polysaccharide intercellular adhesin (PIA)-positive and PIA-negative strains of Staphylococcus epidermidis. Biofouling 30:435–446. https://doi.org/10.1080/08927014.2014.888062
144. Mansour SC, de la Fuente-Núñez C, Hancock REW (2015) Peptide IDR-1018: modulating the immune system and targeting bacterial biofilms to treat antibiotic-resistant bacterial infections. J Pept Sci 21:323–329. https://doi.org/10.1002/psc.2708
145. Anunthawan T, de la Fuente-Núñez C, Hancock REW, Klaynongsruang S (2015) Cationic amphipathic peptides KT2 and RT2 are taken up into bacterial cells and kill planktonic and biofilm bacteria. Biochim Biophys Acta 1848:1352–1358. https://doi.org/10.1016/j.bbamem.2015.02.021
146. Overhage J, Campisano A, Bains M, Torfs ECW, Rehm BHA, Hancock REW (2008) Human host defense peptide LL-37 prevents bacterial biofilm formation. Infect Immun 76:4176–4182. https://doi.org/10.1128/IAI.00318-08

147. Haisma EM et al (2014) LL-37-derived peptides eradicate multidrug-resistant *Staphylococcus aureus* from thermally wounded human skin equivalents. Antimicrob Agents Chemother 58:4411. https://doi.org/10.1128/AAC.02554-14
148. de Breij A et al (2018) The antimicrobial peptide SAAP-148 combats drug-resistant bacteria and biofilms. Sci Transl Med 10:eaan4044. https://doi.org/10.1126/scitranslmed.aan4044
149. Minardi D et al (2007) The antimicrobial peptide Tachyplesin III coated alone and in combination with intraperitoneal piperacillin-tazobactam prevents ureteral stent Pseudomonas infection in a rat subcutaneous pouch model. Peptides 28:2293–2298. https://doi.org/10.1016/j.peptides.2007.10.001
150. Almaaytah A, Qaoud MT, Khalil Mohammed G, Abualhaijaa A, Knappe D, Hoffmann R, Al-Balas Q (2018) Antimicrobial and antibiofilm activity of UP-5, an ultrashort antimicrobial peptide designed using only arginine and biphenylalanine. Pharmaceuticals (Basel, Switzerland) 11(1). https://doi.org/10.3390/ph11010003
151. Ma Z et al (2017) Membrane-active amphipathic peptide WRL3 with in vitro antibiofilm capability and in vivo efficacy in treating methicillin-resistant Staphylococcus aureus burn wound infections. ACS Infect Dis 3:820–832. https://doi.org/10.1021/acsinfecdis.7b00100
152. Haney EF, Mansour SC, Hancock REW (2017) Antimicrobial peptides: an introduction. In: Hansen PR (ed) Antimicrobial peptides: methods and protocols. Springer, New York, pp 3–22. https://doi.org/10.1007/978-1-4939-6737-7_1
153. Hancock REW, Sahl H-G (2006) Antimicrobial and host-defense peptides as new anti-infective therapeutic strategies. Nat Biotechnol 24:1551–1557. https://doi.org/10.1038/nbt1267
154. Haney EF, Straus SK, Hancock REW (2019) Reassessing the host defense peptide landscape. Front Chem 7:43. https://doi.org/10.3389/fchem.2019.00043
155. Hilpert K, Volkmer-Engert R, Walter T, Hancock REW (2005) High-throughput generation of small antibacterial peptides with improved activity. Nat Biotechnol 23:1008. https://doi.org/10.1038/nbt1113
156. Butts A, Krysan DJ (2012) Antifungal drug discovery: something old and something new. PLoS Pathog 8:e1002870
157. Ganz T, Nemeth E (2012) Hepcidin and iron homeostasis. Biochim Biophys Acta 1823:1434–1443. https://doi.org/10.1016/j.bbamcr.2012.01.014
158. Maisetta G, Petruzzelli R, Brancatisano FL, Esin S, Vitali A, Campa M, Batoni G (2010) Antimicrobial activity of human hepcidin 20 and 25 against clinically relevant bacterial strains: effect of copper and acidic pH. Peptides 31:1995–2002. https://doi.org/10.1016/j.peptides.2010.08.007
159. Tavanti A, Maisetta G, Del Gaudio G, Petruzzelli R, Sanguinetti M, Batoni G, Senesi S (2011) Fungicidal activity of the human peptide hepcidin 20 alone or in combination with other antifungals against Candida glabrata isolates. Peptides 32:2484–2487. https://doi.org/10.1016/j.peptides.2011.10.012
160. Wieczorek M et al (2010) Structural studies of a peptide with immune modulating and direct antimicrobial activity. Chem Biol 17:970–980. https://doi.org/10.1016/j.chembiol.2010.07.007
161. de la Fuente-Núñez C, Reffuveille F, Haney EF, Straus SK, Hancock REW (2014) Broad-spectrum anti-biofilm peptide that targets a cellular stress response. PLoS Pathog 10:e1004152. https://doi.org/10.1371/journal.ppat.1004152
162. Hamamoto K, Kida Y, Zhang Y, Shimizu T, Kuwano K (2002) Antimicrobial activity and stability to proteolysis of small linear cationic peptides with D-amino acid substitutions. Microbiol Immunol 46:741–749. https://doi.org/10.1111/j.1348-0421.2002.tb02759.x
163. Chung EMC, Dean SN, Propst CN, Bishop BM, van Hoek ML (2017) Komodo dragon-inspired synthetic peptide DRGN-1 promotes wound-healing of a mixed-biofilm infected wound NPJ biofilms and microbiomes. NPJ Biofilms Microbiomes 3:9. https://doi.org/10.1038/s41522-017-0017-2
164. Onaizi SA, Leong SSJ (2011) Tethering antimicrobial peptides: current status and potential challenges. Biotechnol Adv 29:67–74. https://doi.org/10.1016/j.biotechadv.2010.08.012

165. Forbes S, McBain AJ, Felton-Smith S, Jowitt TA, Birchenough HL, Dobson CB (2013) Comparative surface antimicrobial properties of synthetic biocides and novel human apolipoprotein E derived antimicrobial peptides. Biomaterials 34:5453–5464. https://doi.org/10.1016/j.biomaterials.2013.03.087
166. Etienne O et al (2004) Multilayer polyelectrolyte films functionalized by insertion of Defensin: a new approach to protection of implants from bacterial colonization. Antimicrob Agents Chemother 48:3662. https://doi.org/10.1128/AAC.48.10.3662-3669.2004
167. Shukla A, Fleming KE, Chuang HF, Chau TM, Loose CR, Stephanopoulos GN, Hammond PT (2010) Controlling the release of peptide antimicrobial agents from surfaces. Biomaterials 31:2348–2357. https://doi.org/10.1016/j.biomaterials.2009.11.082
168. Gao G, Cheng John TJ, Kindrachuk J, Hancock Robert EW, Straus Suzana K, Kizhakkedathu Jayachandran N (2012) Biomembrane interactions reveal the mechanism of action of surface-immobilized host defense IDR-1010 peptide. Chem Biol 19:199–209. https://doi.org/10.1016/j.chembiol.2011.12.015
169. Gao G et al (2011) The biocompatibility and biofilm resistance of implant coatings based on hydrophilic polymer brushes conjugated with antimicrobial peptides. Biomaterials 32:3899–3909. https://doi.org/10.1016/j.biomaterials.2011.02.013
170. Lim K et al (2013) Immobilization studies of an engineered arginine–tryptophan-rich peptide on a silicone surface with antimicrobial and antibiofilm activity. ACS Appl Mater Interfaces 5:6412–6422. https://doi.org/10.1021/am401629p
171. Kazemzadeh-Narbat M, Kindrachuk J, Duan K, Jenssen H, Hancock REW, Wang R (2010) Antimicrobial peptides on calcium phosphate-coated titanium for the prevention of implant-associated infections. Biomaterials 31:9519–9526. https://doi.org/10.1016/j.biomaterials.2010.08.035
172. Kazemzadeh-Narbat M, Noordin S, Masri BA, Garbuz DS, Duncan CP, Hancock REW, Wang R (2012) Drug release and bone growth studies of antimicrobial peptide-loaded calcium phosphate coating on titanium. J Biomed Mater Res B Appl Biomater 100B:1344–1352. https://doi.org/10.1002/jbm.b.32701
173. Kazemzadeh-Narbat M, Lai BFL, Ding C, Kizhakkedathu JN, Hancock REW, Wang R (2013) Multilayered coating on titanium for controlled release of antimicrobial peptides for the prevention of implant-associated infections. Biomaterials 34:5969–5977. https://doi.org/10.1016/j.biomaterials.2013.04.036
174. Fox JL (2013) Antimicrobial peptides stage a comeback. Nat Biotechnol 31:379. https://doi.org/10.1038/nbt.2572
175. Mahlapuu M, Håkansson J, Ringstad L, Björn C (2016) Antimicrobial peptides: an emerging category of therapeutic agents. Front Cell Infect Microbiol 6:194–194. https://doi.org/10.3389/fcimb.2016.00194
176. Grönberg A, Mahlapuu M, Ståhle M, Whately-Smith C, Rollman O (2014) Treatment with LL-37 is safe and effective in enhancing healing of hard-to-heal venous leg ulcers: a randomized, placebo-controlled clinical trial. Wound Repair Regen 22:613–621. https://doi.org/10.1111/wrr.12211
177. Falagas ME, Kasiakou SK, Saravolatz LD (2005) Colistin: the revival of polymyxins for the management of multidrug-resistant gram-negative bacterial infections. Clin Infect Dis 40:1333–1341. https://doi.org/10.1086/429323
178. Landman D, Georgescu C, Martin DA, Quale J (2008) Polymyxins revisited. Clin Microbiol Rev 21:449. https://doi.org/10.1128/CMR.00006-08
179. Zavascki AP, Goldani LZ, Li J, Nation RL (2007) Polymyxin B for the treatment of multidrug-resistant pathogens: a critical review. J Antimicrob Chemother 60:1206–1215. https://doi.org/10.1093/jac/dkm357
180. Rotem S, Mor A (2009) Antimicrobial peptide mimics for improved therapeutic properties. Biochim Biophys Acta 1788:1582–1592. https://doi.org/10.1016/j.bbamem.2008.10.020
181. Piluso S, Soultan AH, Patterson J (2017) Molecularly engineered polymer-based Systems in drug delivery and regenerative medicine. Curr Pharm Des 23:281–294. https://doi.org/10.2174/1381612822666161021104239

182. Moreira Teixeira LS, Patterson J, Luyten FP (2014) Skeletal tissue regeneration: where can hydrogels play a role? Int Orthop 38:1861–1876. https://doi.org/10.1007/s00264-014-2402-2
183. Van den Broeck L, Piluso S, Soultan AH, De Volder M, Patterson J (2019) Cytocompatible carbon nanotube reinforced polyethylene glycol composite hydrogels for tissue engineering. Mater Sci Eng C Mater Biol Appl 98:1133–1144. https://doi.org/10.1016/j.msec.2019.01.020
184. Romanò CL, Toscano M, Romanò D, Drago L (2013) Antibiofilm agents and implant-related infections in orthopaedics: where are we? J Chemother 25:67–80. https://doi.org/10.1179/1973947812Y.0000000045
185. Lu H, Liu Y, Guo J, Wu H, Wang J, Wu G (2016) Biomaterials with antibacterial and osteo-inductive properties to repair infected bone defects. Int J Mol Sci 17:334–334. https://doi.org/10.3390/ijms17030334

Construction of Bio-functionalized ZnO Coatings on Titanium Implants with Both Self-Antibacterial and Osteoinductive Properties

Lei Tan, Xiangmei Liu, and Shuilin Wu

Abstract Bacterial infection and lack of osteoinductive ability are the major concerns of titanium-based bone implants. This chapter describes the construction of bio-functionalized ZnO coatings with both self-antibacterial and osteoinductive properties on titanium implants. To obtain ZnO-modified coatings with strong binding forces with substrates, two preparation methods are presented including atomic layer deposition and laser cladding techniques. Next, the antibacterial and osteoinductive properties of ZnO-based coatings are described including Ag/ZnO/hydroxyapatite-Ti, Ti-ZnO/polydopamine/arginine-glycine-aspartic acid-cysteine and poly(lactic-co-glycolic acid)/Ag/ZnO-Ti. We summarize the balancing strategies of reducing cytotoxicity of ZnO to bone cells and enhancing its toxicity towards bacteria.

Keywords Titanium implant · ZnO · Self-antibacterial · Osteoinductive · Atomic layer deposition · Laser cladding · Controlled release · Balancing

Introduction

It is well known that zinc ions possess both antibacterial and osteoinductive properties. The appropriate amount of zinc incorporated on the surface of implants can improve bone tissue integration and inhibit bone resorption [1, 2]. However, these biological functions are associated with release behavior because the excess release of zinc ions and related reactive oxygen species (ROS) can produce high cytotoxicity and thus inhibit the functions of improving bone formation [1]. On the contrary, high concentrations of zinc ion are helpful to its antibacterial effect. To address this

L. Tan · X. Liu
Ministry-of-Education Key Laboratory for the Green Preparation and Application of Functional Materials, Hubei Key Laboratory of Polymer Materials, School of Materials Science and Engineering, Hubei University, Wuhan, China

S. Wu (✉)
School of Materials Science and Engineering, Tianjin University, Tianjin, China
e-mail: shuilinwu@tju.edu.cn

B. Li et al. (eds.), *Racing for the Surface*,
https://doi.org/10.1007/978-3-030-34471-9_7

problem, the key point is to weaken the cytotoxicity of zinc ions towards bone cells and to enhance its toxicity towards bacteria to achieve both antibacterial and osteoinductive properties simultaneously.

As a promising zinc source, ZnO can be incorporated into and onto the coating of implants. From our previous investigations, we mainly chose two strategies to incorporate ZnO in the coating of Ti implants including atomic layer deposition (ALD) and laser cladding techniques [3, 4]. The ALD technique was first applied by Suntola et al. in 1970s [5]. Through two successive and cyclic sequentially self-limiting half-reactions, a controllable film can be prepared by a layer-by-layer mode on the surface of a substrate. Different from other methods such as vapor deposition, the thickness of the deposited film can be adjusted precisely through ALD at a monolayer level. Moreover, ALD can be operated under low deposition temperatures (even below ambient temperature), which makes it practicable to modify some thermal-sensitive substrates such as polymers [6]. As for the laser cladding method, it can increase the specific surface area, change the surface roughness, and increase the wettability of implants to influence the attachment and differentiation of bone cells [7]. Moreover, some ceramic powders such as hydroxyapatite (HA) can be fixed tightly on the surface of implants through laser melting [4, 8].

ZnO possesses good antibacterial ability through releasing Zn^{2+} and ROS or contact-killing. But the obvious cytotoxicity induced by a high concentration of Zn^{2+} and ROS can induce cell apoptosis and inhibit osteogenesis [1]. As for decreasing cytotoxicity of zinc ions to bone cells, we summarize three strategies according to our previous studies as follows: (1) chelating Zn^{2+} through coordination with a biocompatible coating; (2) reducing the produced ROS using an antioxidant; and (3) controlling the release behavior of Zn^{2+} and extending the release period [1, 4, 9, 10]. In order to balance the cytotoxicity and antibacterial activity of ZnO, the antibacterial performance of ZnO needs additional enhancement after the above treatments. Generally, Ag nanoparticles can be added to improve the antibacterial efficacy. Due to the synergistic antibacterial effect of Ag and Zn, only a small amount of Zn^{2+} and Ag nanoparticles is needed to kill bacteria efficiently [4, 9]. Besides, the additional antibacterial agents are not limited to Ag [10, 11]. Herein, we introduce two strategies of fabricating ZnO-decorated coatings on Ti implants and describe their antibacterial and osteoinductive behavior.

Fabrication and Characteristics of ZnO-Decorated Coatings on Ti Implants

The atomic layer deposition (ALD) technique is proven to be a promising strategy to fabricate a ZnO film on the surface of Ti implants. During deposition, diethyl zinc and H_2O are used as Zn and O precursors, respectively. Since the thickness of ZnO films is dependent on the cycle times, the content of ZnO on Ti implants can be well adjusted to control their biological functions. Based on this method, ZnO nanofilms could be deposited on the surface of carbon nanotubes/chitosan (CNT/CS) or titania nanotubes (TNTs) coatings [3]. Figure 1a shows the modified CNT/CS coating on

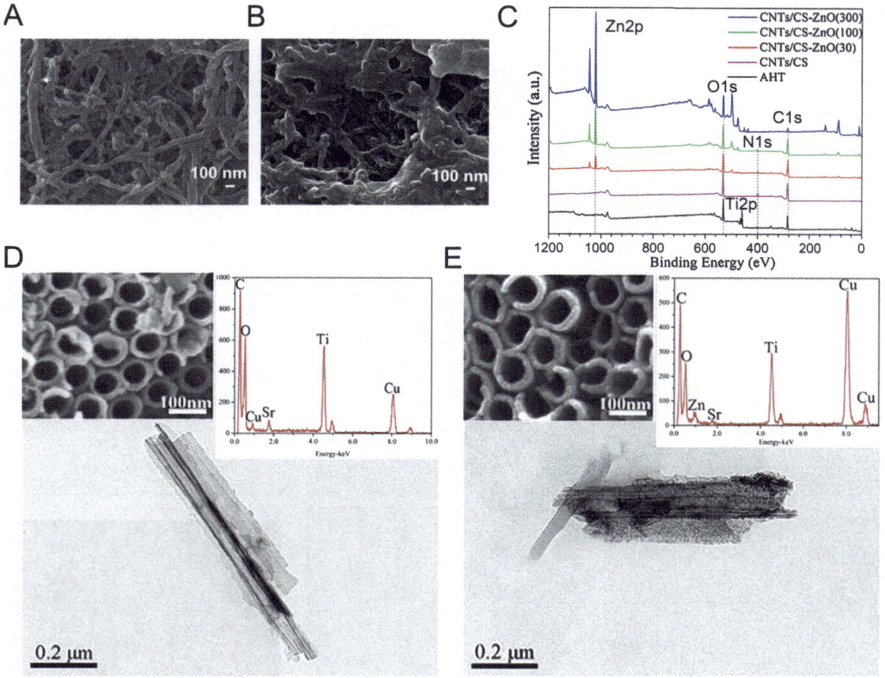

Fig. 1 FE-SEM images of (**A**) CNTs/CS and (**B**) CNTs/CS-ZnO (300 cycles). (**C**) XPS spectra of alkali-heat-treated Ti (AHT), CNTs/CS, CNTs/CS-ZnO (30 cycles), CNTs/CS-ZnO (100 cycles), and CNTs/CS-ZnO (300 cycles) [3]. (Copyright 2016. Adapted with permission from the Elsevier and Copyright Clearance Center.) FE-SEM (insert), EDS, and TEM images of (**D**) TNTs-Sr and (**E**) TNTs-Sr/ZnO [12]. (Copyright 2017. Adapted with permission from the Elsevier and Copyright Clearance Center)

the Ti implant via electrophoretic deposition. After depositing ZnO through the ALD technique, the dense nanoparticles distributed on CNT/CS were observed (Fig. 1b). From the X-ray photoelectron spectroscopy (XPS) results (Fig. 1c), the detected Zn element further proved the successful modification of ZnO nanofilms. The content of Zn increased with cycle times indicating that the content of ZnO could be regulated. Figure 1d shows the TNTs-Sr array on a Ti implant. It can be observed that the orifices of TNTs-Sr became rough after ZnO deposition (Fig. 1e). The signal of the Zn element was also detected by an energy dispersive spectrometer (EDS) [12]. Obviously, the ZnO nanofilms can be deposited uniformly on the surface of organic or inorganic coatings via the ALD technique.

Moreover, ZnO nanorods (NRs) arrays could grow from the ZnO seed layer (ZnOs) obtained by the ALD through hydrothermal method. Arginine-glycine-aspartic acid-cysteine (RGDC) peptide was covalently modified with Ti-ZnO/polydopamine (PDA) to improve osteogenesis (Fig. 2a). As shown in Fig. 2b, the size of ZnOs on Ti plate was around 20 nm. The hexagonal ZnO NRs with a 100 nm diameter were observed (Fig. 2c). Compared with ZnOs that is the

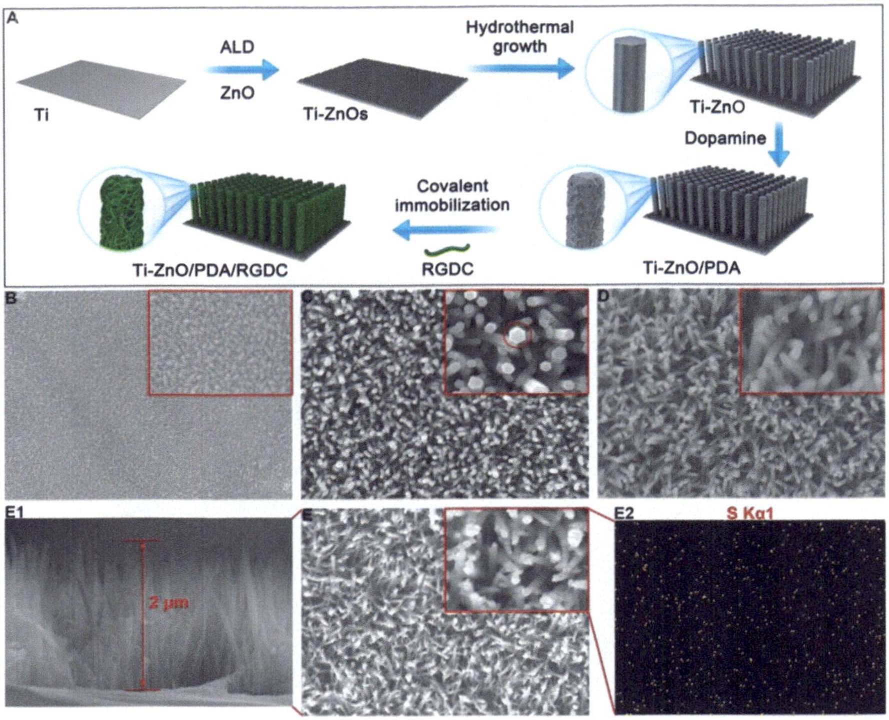

Fig. 2 (**A**) Schematic illustration of the fabrication process of the hybrid ZnO/PDA/RGDC NR arrays on Ti. FE-SEM images of (**B**) Ti-ZnOs (scale bars = 100 nm), (**C**) Ti-ZnO, (**D**) Ti-ZnO/PDA, and (**E**) Ti-ZnO/PDA/RGDC. (**E1**) Cross-sectional image of Ti-ZnO/PDA/RGDC. (**E2**) Elemental mapping of Ti-ZnO/PDA/RGDC, (scale bar = 100 nm (inset figure = 1 μm)) [1]. (Copyright 2017. Adapted with permission from the American Chemical Society)

ZnO seed layer (ZnOs) on Ti substrate obtained by ALD, ZnO NRs on Ti plate possessed a higher specific surface area and thus provided more active sites to improve the antibacterial performance and to increase the content of the modified RGDC. After the modification of PDA or RGDC, the surface of ZnO NRs became rougher (Fig. 2d, e). From the SEM image section (Fig. 2e1), the length of ZnO/PDA/RGDC NRs was approximately 2 μm. The detected S element proved the successful modification of RGDC (Fig. 2e2) [1].

In addition to the above methods of preparing ZnO films, our previous investigation shows that ZnO also can be incorporated with hydroxyapatite (HA) and Ag nanoparticles onto Ti6Al4V (Ti6) implants through laser cladding. The prefabricated Ag/HA and ZnO nanoparticles were first mixed and dispersed in deionized (DI) water. Then AgxZnOyHA dispersion was added to the surface of Ti6 and dried in a vacuum prior to laser cladding. The content of each component can be adjusted in the mixed powders for further antibacterial and osteoinductive investigation. The AgxZnOyHA modified Ti6 by laser cladding is shown in Fig. 3a. A regular texture

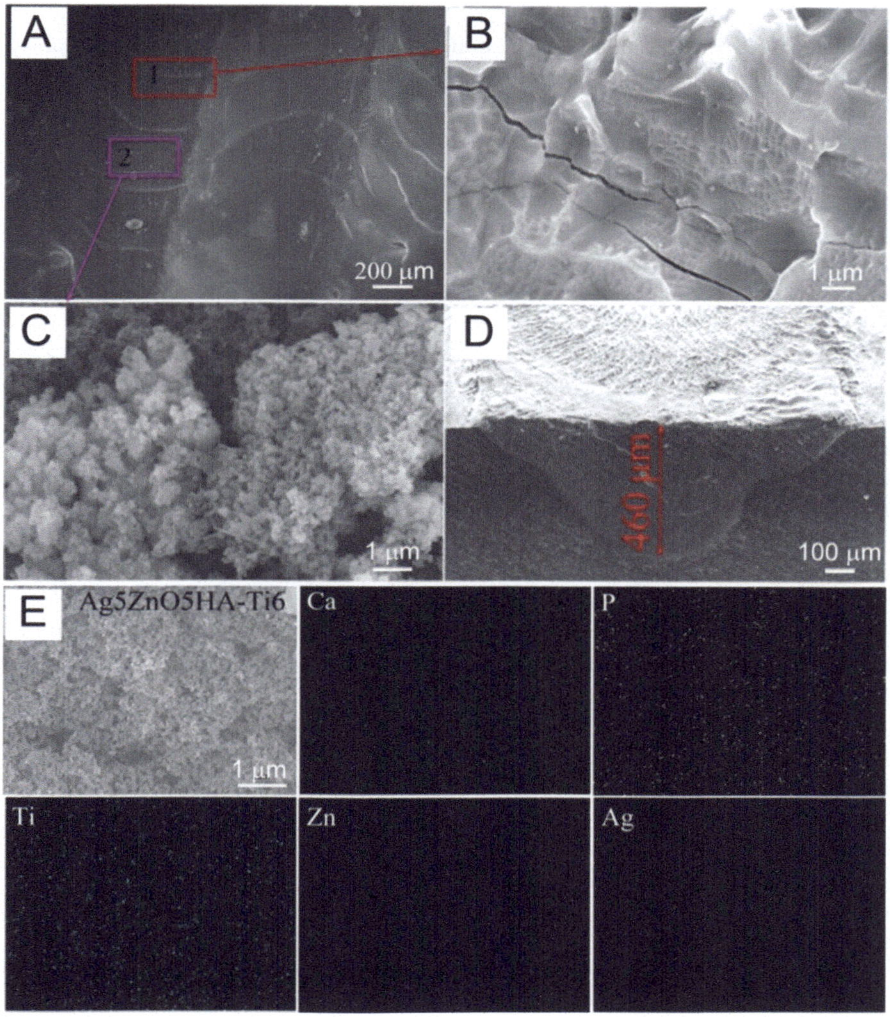

Fig. 3 (**A**) SEM image showing the surface morphology of the Ag5ZnO5HA-Ti6 (5:5:90, wt%) composite coating. (**B**) SEM image of bulging area (zone 1). (**C**) SEM image of the flat area (zone 2). (**D**) Cross-sectional SEM micrograph. (**E**) Elemental distribution by EDS [4]. (Copyright 2018. Adapted with permission from the American Chemical Society)

with parallel ridges was observed due to the back and forth movement of the laser spot which led to rapid melting and freezing of the composite on Ti6. The bulging areas of zone 1 (Fig. 3b) were attributed to the molten peaks while the flat areas of zone 2 (Fig. 3c) exhibited a loccular morphology, suggesting that the laser cladding can change the surface roughness of Ti6. From the section of the SEM image of AgxZnOyHA-Ti6 (Fig. 3d), the cladding with a 460 μm depth can be observed.

For the sample of the Ag5ZnO5HA-Ti6 sample, the EDS images showed that the elements of Ca, P, Ti, Zn, and Ag were uniformly distributed in the coating, indicating that these biofunctional elements were successfully introduced (Fig. 3e) [4].

Antibacterial Behavior Investigation

As we know, Zn^{2+} has antibacterial activity which is highly related to its content. However, high content or fast release can result in high cytotoxicity and cannot improve osteogenesis effectively. Benefiting from the laser melting of ZnO with coatings, the ZnO had a strong binding force with the coating. Therefore, the Zn^{2+} could be released slowly for 20 days (Fig. 4a). Whereas, the slowly release Zn^{2+} could not kill *Escherichia coli* (*E. coli*) or *Staphylococcus aureus* (*S. aureus*) efficiently. From Fig. 4b, c, the addition of Ag significantly increased antibacterial efficiency. On the contrary, the ZnO also could reduce the usage of Ag to lower its toxicity [4].

Due to the high specific surface area of ZnO NRs arrays on Ti (Ti-ZnO), the antibacterial performance of different modified ZnO NRs arrays were also investigated. Before that, the release behavior of Zn ions and ROS were first studied. As shown in Fig. 5a, a1, the release amount of Zn^{2+} in each group was similar, but the slight increased release of Zn^{2+} in Ti-ZnO/PDA-RGDC was due to that of the oxidation product of PDA which decreased the pH value (Fig. 5b, e). The decreased absorbance and faded color of PDA in the presence of Ti-ZnO demonstrated the antioxidant effect of PDA (Fig. 5c, d). Moreover, the electron spin resonance (ESR) results showed that Ti-ZnO generated an obvious more hydroxyl radical compared with Ti-ZnOs due to its higher specific surface area. However, the hydroxyl radical was decreased when PDA was introduced, which might weaken the antibacterial performance of Ti-ZnO (Fig. 5f). Besides, the released Zn^{2+} from Ti-ZnO/PDA could be chelated with the pyrocatechol of PDA (Fig. 5g, h), which thus might reduce the cytotoxicity [1].

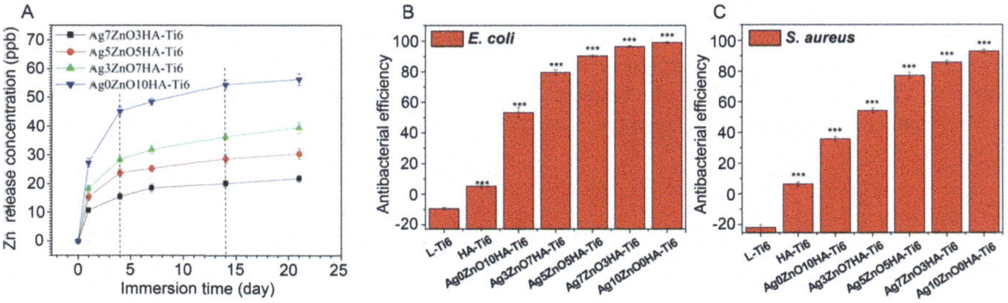

Fig. 4 (**A**) Release of Zn^{2+} from different ZnO nanoparticles (NPs). (**B**) Antibacterial efficiency of the samples compared to Ti6 for *E. coli*. (**C**) Antibacterial efficiency of the samples for *S. aureus*. The error bars indicate mean ± standard deviations: $***P < 0.001$ vs. L-Ti6 (*t* test) [4]. (Copyright 2018. Adapted with permission from the American Chemical Society)

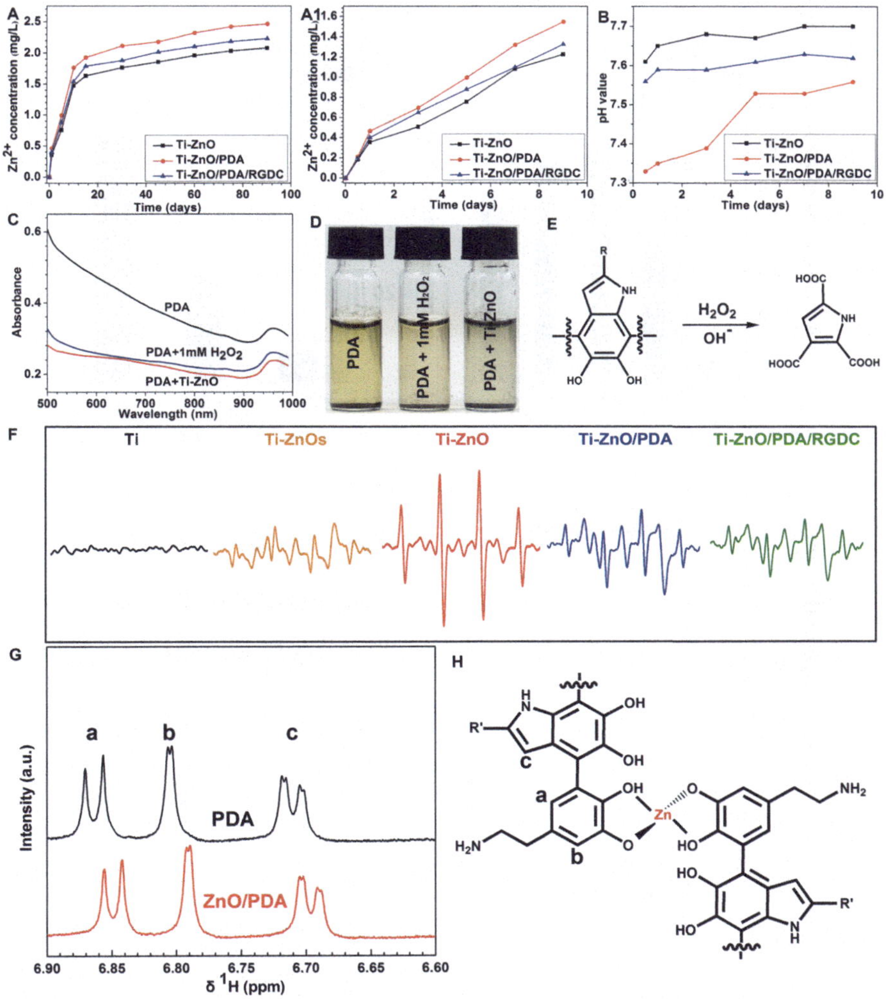

Fig. 5 (**A**) Cumulative zinc ion release curves of Ti-ZnO, Ti-ZnO/PDA, and Ti-ZnO/PDA/RGDC after immersion at 37 °C for 90 days. (**A1**) Short-term release; (**B**) pH values of the liquids corresponding to (**A1**). Degradation of PDA: (**C**) UV-vis absorption spectra, (**D**) corresponding photograph, (**E**) chemical degradation pathway of PDA by H_2O_2. (**F**) ESR spectra of Ti, Ti-ZnOs, Ti-ZnO, Ti-ZnO/PDA, and Ti-ZnO/PDA/RGDC. (**G**) HNMR spectra of PDA and PDA/ZnO [1]. (**H**) Structure of the Zn-PDA complex [1]. (Copyright 2017. Adapted with permission from the American Chemical Society)

The antibacterial behavior of samples was investigated by the spread plate method. As shown in Fig. 6a, the antibacterial rate of Ti-ZnO was much higher than Ti-ZnOs, indicating that ZnO NRs possessed better antibacterial ability. After modifying with PDA and RGDC, the antibacterial rates decreased but were still higher than 72%. From Fig. 6b, the antibacterial rate of Ti-ZnO/PDA-RGDC was much higher than Ti-ZnOs/PDA-RGDC [1].

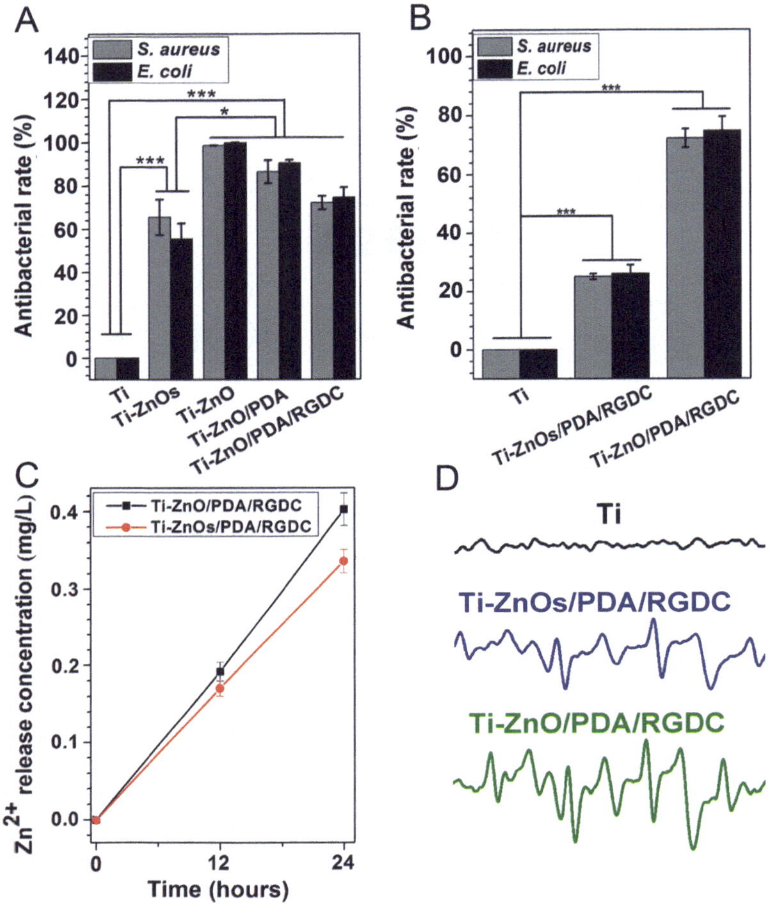

Fig. 6 (**A**, **B**) Antibacterial activity of the different samples. (**C**) Cumulative zinc ion release curves of Ti-ZnOs/PDA/RGDC and Ti-ZnO/PDA/RGDC after immersion at 37 °C for 1 day. (**D**) ESR spectra of Ti, Ti-ZnOs/PDA/RGDC and Ti-ZnO/PDA/RGDC. The error bars indicate mean ± standard deviations: $*P < 0.05$ and $***P < 0.001$ (t test) [1]. (Copyright 2017. Adapted with permission from the American Chemical Society)

Since the Ti-ZnOs/PDA-RGDC and Ti-ZnO/PDA-RGDC exhibited similar Zn^{2+} and hydroxyl radical release behavior, we speculated that the bacteria were physically punctured by Ti-ZnO/PDA/RGDC, which was proved by the TEM images of bacteria. As shown in Fig. 7, the bacteria treated by Ti-ZnO/PDA/RGDC showed obvious damages and were punctured by the needle-like NRs. Besides, the Zn^{2+} content was also higher than the Ti-ZnOs/PDA-RGDC group. These results showed that the physical puncture behavior and Zn^{2+} release of NRs had a synergistic antibacterial effect [1].

We also utilized poly(lactic-co-glycolic acid) (PLGA) to modify the ZnO NRs arrays because PLGA not only can control the release behavior of Zn^{2+} but can also

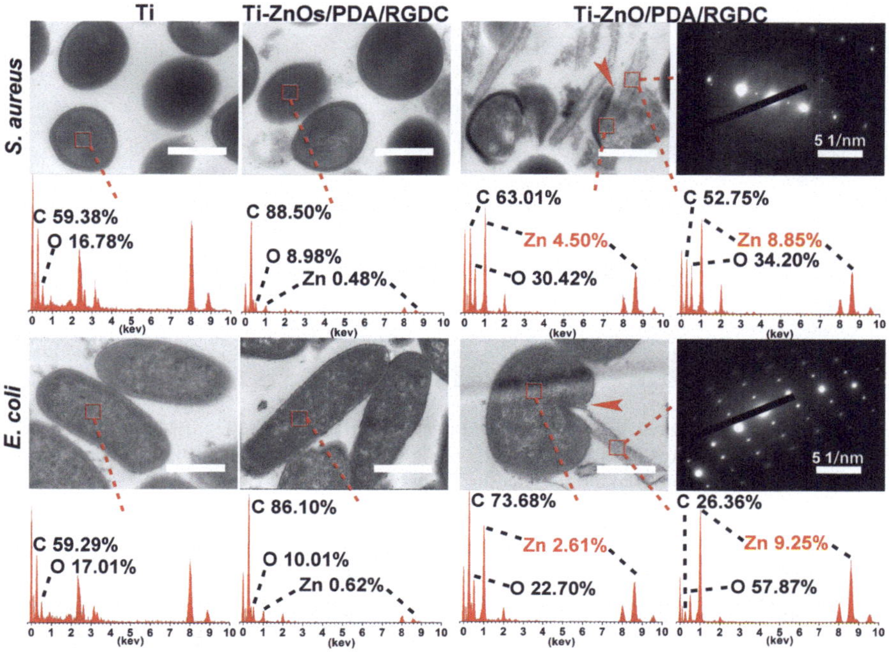

Fig. 7 TEM images of ultrathin section (about 70 nm) and the corresponding EDS of *S. aureus* and *E. coli* treated with Ti, Ti-ZnOs/PDA/RGDC, and Ti-ZnO/PDA/RGDC (scale bars = 500 nm) [1]. (Copyright 2017. Adapted with permission from the American Chemical Society)

reduce the cytotoxicity of ZnO NRs. The ZnO NRs surface showed a hydrophobic property due to its special nanostructure. The introduction of PLGA could increase the wettability of implant surfaces and thus be a benefit to the bone cell adhesion. However, the antibacterial ability of ZnO NRs was also weakened due to the coverage of PLGA. Our previous work showed that the antibacterial rate decreased from 94.1% to 70.3% against *S. aureus* after the modification of PLGA (PLGA/ZnO-Ti). Figure 8a, b shows that the surface of ZnO NRs was almost covered completely by a thick film after spinning PLGA. Unlike Ti-ZnOs/PDA-RGDC, the NRs structure was not maintained on the surface of PLGA/Ag/ZnO-Ti. The antibacterial performance of PLGA/ZnO-Ti is mainly attributed to the released ROS and Zn^{2+} but not the physical puncture effect. Similar to the abovementioned results, the introduction of Ag nanoparticles can further increase the antibacterial rate [9].

Osteoinductive Behavior Investigation

In addition to self-antibacterial property, the cytotoxicity and osteoinductive properties of Zn^{2+} cannot be ignored. The osteoinductive property of ZnO is attributed to the fact that the released Zn^{2+} can promote bone formation. However, high concentration

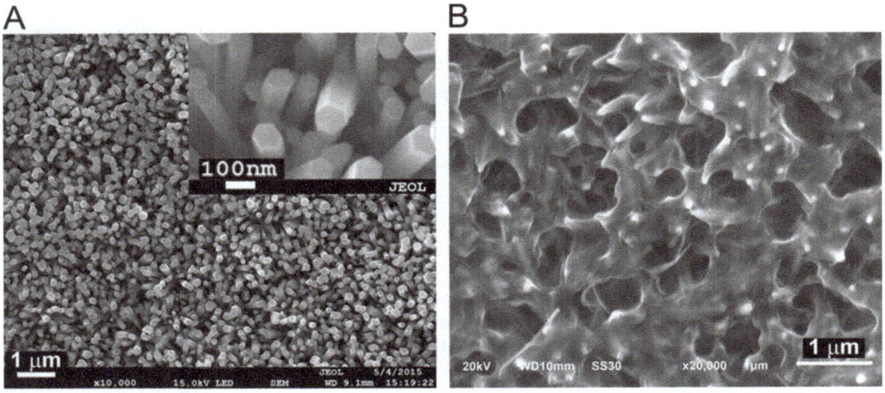

Fig. 8 (**A**) Morphological and microstructural characterization by the SEM image of ZnO nanorods. (**B**) SEM image of PLGA/3Ag/ZnO nanorods composite coating [9]. (Copyright 2017. Adapted with permission from the Elsevier and Copyright Clearance Center)

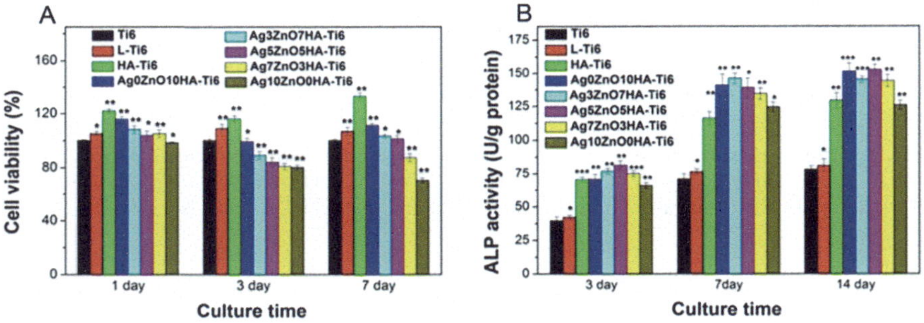

Fig. 9 (**A**) Cell viability assessed by MTT after culturing for 1, 3, and 7 days. (**B**) Cell osteogenesis evaluated by the ALP activity assay for 3, 7, and 14 days. The error bars indicate mean ± standard deviations: $*P < 0.05$, $**P < 0.05$, and $***P < 0.001$ vs. Ti6 (*t* test) [4]. (Copyright 2018. Adapted with permission from the American Chemical Society)

of Zn^{2+} can induce cytotoxicity and inhibit cell growth and bone tissue formation. We know that high concentration of Zn^{2+} possesses strong antibacterial ability. Therefore, how to use Zn^{2+} appropriately should be of concern. In the case of laser cladding method, the Zn^{2+} from Ag0ZnO10HA-Ti6 released quickly with cumulative concentration of 1.91 μg/L on day 4 because of some decorated ZnO on the surface, which was still lower than the safe concentration of 3 mg/L provided by the World Health Organization. Besides, quick release of Zn^{2+} is helpful to early infection prevention. After 4 days, the Zn^{2+} released slowly and steadily due to the strong fixation of ZnO in the composite coating (Fig. 4a). Benefiting from the release behavior of Zn^{2+}, the Ag0ZnO10HA-Ti6 exhibited high cell viability after 1, 3, or 7 days culture (Fig. 9a). It also showed that the alkaline phosphatase (ALP) activity of Ag0ZnO10HA-Ti6 was higher than the untreated Ti6, laser-treated Ti6,

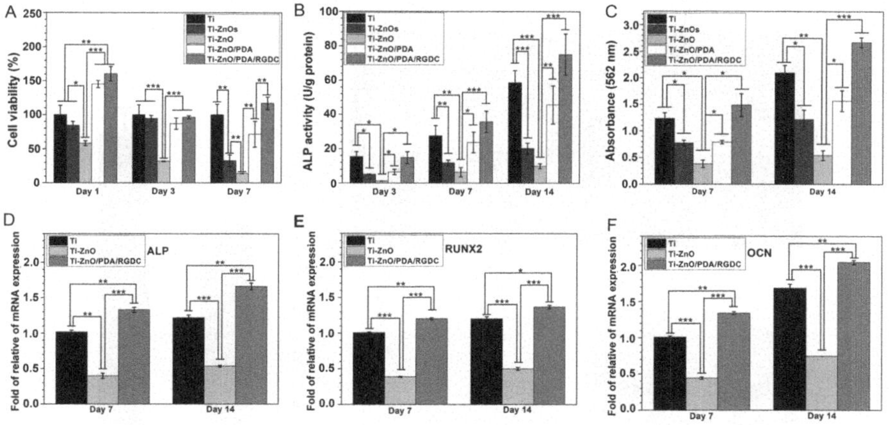

Fig. 10 (**A**) MTT assay of cell viabilities cultured in the medium with different samples after 1, 3, and 7 days. (**B**) Specific ALP activities of MC3T3-El osteoblasts cultured in the medium with different samples after 1, 3, and 7 days. (**C**) Quantitative measurement of the Alizarin Red staining after 7 and 14 days. Quantification of the osteoblast-related gene expressions of (**D**) ALP, (**E**) RUNX2, (**F**) OCN on Ti, Ti-ZnO, and Ti-ZnO/PDA/RGDC using normalization against a β-actin reference on the 7th and 14th day. The error bars indicate mean ± standard deviations: $*P < 0.05$, $**P < 0.05$, and $***P < 0.001$ (t test) [1]. (Copyright 2017. Adapted with permission from the American Chemical Society)

and HA-treated Ti6 on day 7 and day 14, suggesting that the osteogenesis ability of coating was obviously improved by the laser fixed ZnO. The appropriate addition of Ag nanoparticles not only did not influence the osteogenesis ability (Fig. 9b) but also showed efficient antibacterial performance [4].

As for ZnO NRs (Ti-ZnO), they exhibited high cytotoxicity after 1, 3, and 7 days towards MC3T3-El osteoblasts, which led to the lowest ALP expression. Although they could release Zn^{2+}, the osteoinductive behavior was mainly dominated by its high cytotoxicity. The cytotoxicity of ZnO was significantly decreased by the modification of PDA (Fig. 10a). We found that the ZnO NRs arrays could puncture bacteria but not osteoblasts, which could be explained by the fact that the size of osteoblasts is larger than bacteria and the bacteria were selectively punctured by ZnO NRs. After that, the ALP activity of Ti-ZnO/PDA increased compared to Ti-ZnO due to a decreased cytotoxicity and released Zn^{2+} from ZnO NRs (Fig. 10b). The modified RGDC further increased the ALP activity because it could promote cell attachment and improve osteogenic ability. The results of Alizarin Red staining and osteoblast-related gene expression further proved the great osteoinductive property of Ti-ZnO/PDA/RGDC (Fig. 10c–f) [1].

A bone implantation experiment in vivo was also conducted and we studied the osteoinductive property of Ti-ZnO/PDA/RGDC after 4 weeks. The results of micro-CT and Van Gieson's picro fuchsin staining showed that the newly formed bone tissues on the surface of implants obviously increased compared to Ti, suggesting that the decorated RGDC and released Zn^{2+} from ZnO NRs could accelerate osseointegration of the implants in vivo (Fig. 11) [1].

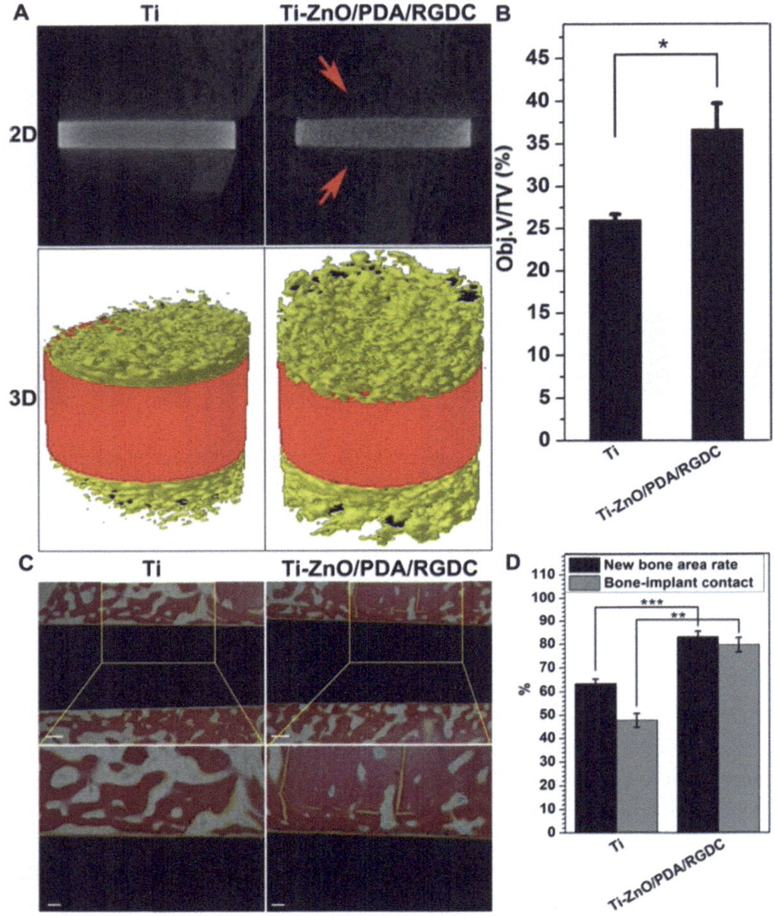

Fig. 11 (**A**) Micro-CT 2D and 3D images of new bone formation around the Ti and Ti-ZnO/PDA/RGDC implants in the rabbit femur 4 weeks after implantation. (**B**) Quantitative measurement of micro-CT 3D images of bone remodeling. (**C**) Histological characteristics at the bone-implant interfaces stained with Van Gieson's picro fuchsin (scale bars = 200 μm and 100 μm). (**D**) New bone area rate and bone-implant contact from the histomorphometric measurements. The error bars indicate mean ± standard deviations: $*P < 0.05$, $**P < 0.05$, and $***P < 0.001$ (t test) [1]. (Copyright 2017. Adapted with permission from the American Chemical Society)

Conclusion

This chapter focuses on the construction of bio-functionalized ZnO coatings with both self-antibacterial and osteoinductive properties on titanium implants. A laser cladding method fixed the ZnO tightly on the surface of substrates, which controlled the release of Zn^{2+} and thus showed great biocompatibility and improved the osteoinductive property. The additional antibacterial agent such as Ag nanoparticles enhanced the antibacterial performance of ZnO. The ALD technique provided a simple and mild method for preparing ZnO coatings with controlled thickness. Polymers such as PLGA and PDA reduced the cytotoxicity of ZnO NRs and the controlled release of Zn^{2+} can accelerate bone tissue integration. We believe that the successful balance of self-antibacterial and osteoinductive properties of ZnO can make it a promising coating for implants.

Future Directions

At present, the development of the bio-functionalized ZnO coatings still focuses on the modification of ZnO to control the release of Zn ions to improve the biocompatibility and osteoinductive ability. The antibacterial property of ZnO should be improved by additional antibacterial strategies through synergistic effect. All-in-one bio-functionalized ZnO coating is the future development direction. Moreover, our previous work reported that the bacteria on the surface of implants could be eliminated through near-infrared light [13]. Unlike the self-antibacterial behavior of endogenous antibacterial coatings, the bacteria can be killed rapidly and efficiently using exogenous irradiation. As for the photo-induced antibacterial application of ZnO, we prepared an antibacterial surface using red phosphorus/ZnO heterointerface to realize a light-activated rapid disinfection through the generation of ROS and heat (Fig. 12) [14]. However, the light source was solar light which cannot penetrate into

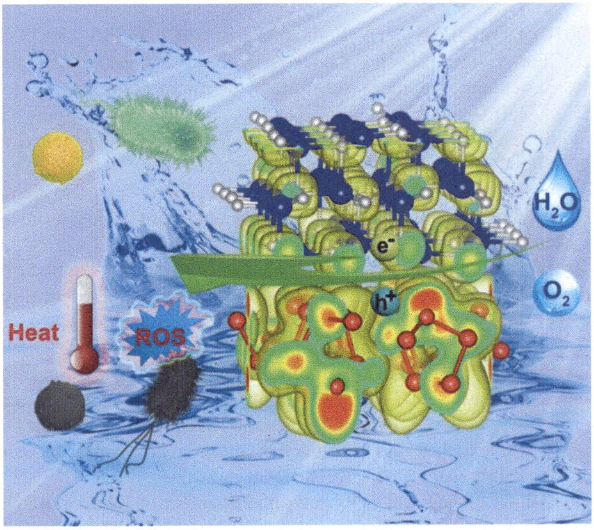

Fig. 12 Light-activated ultrafast bacterial inactivation due to the synergetic ROS and heat by RP/ZnO heterojunction [14]. (Copyright 2019. Adapted with permission from the John Wiley and Sons and Copyright Clearance Center)

the deep site of tissue. In the future, the key point is how to produce ROS or heat from ZnO-based coatings on implants using NIR light. So, the ZnO-based heterointerface or up-conversion nanoparticles can be incorporated into the coatings to realize NIR light induced disinfection on the surface of implant.

Acknowledgements This work was partially supported by the National Natural Science Foundation of China, Nos. 51801056, 51671081, 51871162, and 51422102, and the National Key Research and Development Program of China No. 2016YFC1100600 (subproject 2016YFC1100604), and the Natural Science Fund of Hubei Province, 2018CFA064. The authors also gratefully acknowledge the helpful comments and suggestions of the reviewers, which have improved the presentation.

References

1. Li J, Tan L, Liu X et al (2017) Balancing bacteria–osteoblast competition through selective physical puncture and biofunctionalization of ZnO/polydopamine/arginine-glycine-aspartic acid-cysteine nanorods. ACS Nano 11:11250–11263
2. Jung Kyu P, Yong-Jin K, Junseok Y et al (2010) The topographic effect of zinc oxide nanoflowers on osteoblast growth and osseointegration. Adv Mater 22:4857–4861
3. Zhu Y, Liu X, Yeung KWK, Chu PK, Wu S (2016) Biofunctionalization of carbon nanotubes/chitosan hybrids on Ti implants by atom layer deposited ZnO nanostructures. Appl Surf Sci 400:14–23
4. Zhang Y, Liu X, Li Z et al (2017) Nano Ag/ZnO incorporated hydroxyapatite composite coatings: highly effective infection prevention and excellent osteointegration. ACS Appl Mater Interfaces 10:1266–1277
5. Suntola T, Antson J (1977) Method for producing compound thin films: U.S. Patent 4,058,430 [P]. 11–15
6. George SM (2010) Atomic layer deposition: an overview. Chem Rev 110:111–131
7. Park JW, Kim Y, Park C et al (2009) Enhanced osteoblast response to an equal channel angular pressing-processed pure titanium substrate with microrough surface topography. Acta Biomater 5:3272–3280
8. Comesaña R, Quintero F, Lusquiños F et al (2010) Laser cladding of bioactive glass coatings. Acta Biomater 6:953–961
9. Xiang Y, Li J, Liu X et al (2017) Construction of poly(lactic-co-glycolic acid)/ZnO nanorods/Ag nanoparticles hybrid coating on Ti implants for enhanced antibacterial activity and biocompatibility. Mater Sci Eng C 79:629–637
10. Wang T, Liu X, Zhu Y et al (2017) Metal ion coordination polymer-capped pH-triggered drug release system on titania nanotubes for enhancing self-antibacterial capability of Ti implants. ACS Biomater Sci Eng 3:816–825
11. Li Y, Liu X, Tan L et al (2017) Construction of N-halamine labeled silica/zinc oxide hybrid nanoparticles for enhancing antibacterial ability of Ti implants. Mater Sci Eng C 76:50–58
12. Zhang K, Zhu Y, Liu X et al (2017) Sr/ZnO doped titania nanotube array: an effective surface system with excellent osteoinductivity and self-antibacterial activity. Mater Des 130:403–412
13. Tan L, Li J, Liu X et al (2018) Rapid biofilm eradication on bone implants using red phosphorus and near-infrared light. Adv Mater 30:1801808
14. Li J, Liu X, Tan L et al (2019) Light-activated rapid disinfection by accelerated charge transfer in red phosphorus/ZnO heterointerface. Small Methods 3:1900048

Gasotransmitters: Antimicrobial Properties and Impact on Cell Growth for Tissue Engineering

Kenyatta S. Washington and Chris A. Bashur

Abstract Several clinical situations including birth defects, trauma, and fracture nonunions often result in critical-sized defects that require a graft that can remodel and integrate with the existing bone as well as mitigate the risk of infectious complications. Delivery of gasotransmitters from tissue engineering scaffolds is a potential option to provide antibacterial properties while simultaneously promoting osteogenesis and tissue vascularization. Gasotransmitters, such as nitric oxide, carbon monoxide, and hydrogen sulfide, are inorganic gases that have an important role in cell signaling, and supplemental doses have also been shown to provide bactericidal properties. This chapter reviews the importance of understanding the complex and dose-dependent impacts of different gasotransmitters on both bacterial and mammalian cells. The current research into the selectivity of a gasotransmitter dose for killing bacterial cells compared to mammalian cells is a particular focus. The chapter also discusses the applications of gasotransmitters to engineered tissues, with a focus on bone and microvasculature, as well as the current limitations for incorporating gasotransmitters within scaffolds that need to be addressed.

Keywords Gasotransmitters · Antimicrobial agents · Tissue engineering Scaffolds · Bone regeneration · Carbon monoxide · Nitric oxide · Hydrogen sulfide Tissue vascularization · Drug delivery

Introduction

Gasotransmitters are small, inorganic gases that have an important role in cell signaling and represent a major area of research. Perhaps the most commonly investigated gasotransmitter is nitric oxide (NO). For example, cell-produced NO has been

K. S. Washington · C. A. Bashur (✉)
Department of Biomedical and Chemical Engineering and Sciences, Florida Institute of Technology, Melbourne, FL, USA
e-mail: Kwashington2015@my.fit.edu; cbashur@fit.edu

B. Li et al. (eds.), *Racing for the Surface*,
https://doi.org/10.1007/978-3-030-34471-9_8

shown to be involved in bone marrow-derived mesenchymal stem cell (BMSC) differentiation into osteoblasts for bone tissue engineering [1], and supplemental delivery of NO, through NO donors with a controlled release rate, has been shown to improve engineering strategies for bone and other tissues [2]. However, NO donors have alternatively been used to provide antibacterial properties for bone tissue engineering scaffolds [3]. This elicits the possibility that a gasotransmitter can simultaneously provide antibacterial properties and promote cell remodeling and integration of a tissue engineered graft. Thus, this chapter reviews the importance of understanding the complex and dose-dependent impacts of gasotransmitters on both bacterial and mammalian cells. The gasotransmitters of NO, carbon monoxide (CO), and hydrogen sulfide (H_2S) are discussed. This chapter also discusses the applications of gasotransmitters to engineered tissues, with a focus on bone and microvasculature, as well as the current limitations for incorporating gasotransmitters within scaffolds that need to be addressed.

Need for Antimicrobial Engineered Grafts

Bone frequently heals with minimal scarring, but critical-sized defects fail to heal naturally [4, 5]. This can occur due to several clinical situations including birth defects, trauma, infection, fracture non-unions, and tumor removal and will require some type of intervention. Current treatment options include permanent implants, such as metallic implants, and implants used to promote regeneration such as bone substitutes and grafts. However, there is still an inherently high risk of infectious complications with permanent implants for patients and challenges for physicians [6]. For oral disease or maxillofacial trauma, implants are exposed to an environment that is colonized with a wide array of microorganisms, biofilms, and inflammatory mediators that can contribute to infection and tissue destruction in some patients [7]. Further, infections occurring in relation to implants can inhibit bone remodeling and integration. A variety of antimicrobial agents and delivery strategies for bone tissue engineering have been developed and are discussed in detail in the other chapters of this book. An important consideration for these strategies is the ability to both prevent infection and allow efficient bone regeneration. The impact of the antimicrobial agent on the inflammatory response is an important consideration for effectively regenerating bone because some pro-inflammatory mediators are required for bone regeneration, but an extended inflammatory response interferes with bone healing [8]. In addition, tissue vascularization through angiogenesis or vasculogenesis within grafts is required for healing of these critical-sized bone defects. However, the lack of vascularization is a concern for current bone tissue engineering strategies. The delivery of gasotransmitters from scaffolds presents promise for bone regeneration by potentially providing antimicrobial properties and simultaneously promoting osteogenesis and tissue vascularization.

There are also many other types of engineered tissues (e.g., small diameter vascular grafts) where gasotransmitters may offer the ability to provide both antimicrobial and pro-tissue healing responses. Some applications in addition to bone tissue engineering have an especially high risk of infection. One example of this is an arterio-venous shunt for hemodialysis access from the arm, where repeated puncture will introduce exposure of the graft to bacteria and fungi on the skin. However, for any implanted graft, there is some risk of infection. This can occur at the time of surgery or even years afterward, as discussed in recent review articles [9]. A biofilm can develop on biomaterials after an infection and it is very hard to treat. It is also possible that biomaterial properties can have an influence on the long-term infection risk [10]. Degradable biomaterials appear to reduce the risk of infection compared to non-degradable options [11]. In addition, scaffolds with pores of a particular size may allow bacteria to enter but prevent the immune cells from entering and targeting the bacteria [12]. For these reasons, as well as traditional drug delivery considerations, it is important to consider both the antimicrobial agent and the delivery strategy from the scaffold. This will be considered in the upcoming sections.

Gasotransmitters in Mammalian Cells

Gasotransmitters are gaseous molecules that are endogenously produced in the body and are crucial to a wide range of biological applications. These include NO, H_2S, and CO, which are often more commonly known only for their toxicity at high concentrations. However, there has been a growing interest in gasotransmitters because of their cell signaling function, antibacterial properties, and overall potential for biomedical use. It is important to note that controlling the dose level of gasotransmitters is critical for the desired response and can determine if it will be beneficial or toxic. This is illustrated in the situations when gasotransmitters are naturally produced in the body at different levels.

Nitric Oxide

NO was the earliest discovered gasotransmitter. It is also the most widely researched for therapeutic applications, and it is in clinical use for treating pulmonary fibrosis. It has been used because of its ability to promote vasodilation, reduce platelet aggregation, and regulate a large range of other physiological functions in mammalian cells. For example, osteoclasts release NO in response to mechanical loading, and this can result in repair of microfractures and bone strengthening [13]. These physiological functions also include maintaining redox homeostasis within cells and acting as an effector molecule during immune responses [14]. During the immune response to bacteria or other pathogens, NO is released by macrophages to kill pathogens that have been engulfed [15]. Thus, NO has multiple roles in maintaining homeostasis within the body.

NO is a reactive species that is produced endogenously through the L-arginine/nitric oxide synthase (NOS) pathway. There are specific isoforms of mammalian NOS based upon the common type of cell that they are expressed in, such as endothelial (eNOS) and neuron (nNOS), which are expressed at constant levels. There is also induced nitric oxide synthase (iNOS) that exhibits an elevated expression with inflammation and contributes to NO fluctuations within the body. In chronic diseased conditions, a much higher level of NO is often found [16]. This variation in the expression levels in the body results in differences that are also important to consider when using NO as part of a strategy to engineer a graft. For example, overproduction of NO has been connected to bone loss in some inflammatory conditions, but NO at lower levels has been shown to mediate the beneficial effects of estrogen on bone repair [17, 18]. One of the main concerns with high NO concentrations is that it can react to generate peroxynitrites that result in DNA damage and other cytotoxic impacts [19].

Hydrogen Sulfide

H_2S has more recently been investigated as a gasotransmitter and for therapeutic use [20–23]. H_2S is synthesized from L-cysteine and is highly expressed in intestines and in the brain. It has been shown to regulate neurotransmission, neuromodulation, brain development, and insulin secretion as well as be involved in pathologies [24–28]. H_2S has also shown specific effects in bone and vascular tissue which include hypoxia sensing and bone remodeling as well as vasodilation [22]. Like NO and CO, the dose level of H_2S is also very important. H_2S at high concentrations can cause toxicity by inhibiting cytochrome C oxidase in the nucleus, and it has five times higher toxicity than CO [29]. There are still questions about the role that endogenous H_2S has on mammalian host–pathogen interactions [30].

Carbon Monoxide

Investigations into the benefits of CO as a gasotransmitter and antimicrobial agent are also more recent, and there are fewer studies than with NO. There are many similarities between the impacts of CO and NO. For example, carbon monoxide is also an endogenously produced signaling molecule within mammalian tissues. In addition, one of the pathways that CO acts through to provide responses desirable for tissue engineering (e.g., endothelial cell proliferation) involves binding to NOS molecules to induce NO production [31]. Several detailed reviews highlight the anti-inflammatory and tissue protective activity of CO in specific conditions, such as acute gastrointestinal inflammation [32] and preeclampsia [33]. There are also different responses of CO at varying levels. For the endogenous level found in the body, this has been shown to provide anti-inflammatory properties, reduce cell apoptosis, and to be necessary for vascular function. However, at higher levels, CO is pro-inflammatory, and it is fatal at even higher concentrations. A detailed review of the mechanisms of cell response to CO is provided elsewhere [34].

There are also important differences between CO and the other gasotransmitters. Table 1 shows examples of different gasotransmitter that have been tested for their bactericidal properties, with some of these studies directly comparing to the corresponding impact on mammalian cells. Many studies that have investigated the benefits of these gasotransmitters on mammalian cells are not a focus of this table since this has been described in detail in several review articles [45]. While CO can also provide antibacterial properties and modulate the inflammatory response when added as a pharmacological agent, it is not thought to be directly involved in the body's natural defense to bacterial infection. In addition, CO is more stable than NO, so it is less affected by the reactive oxygen species and other components of the inflammatory response found in many disease conditions. It has been proposed that the antibacterial impacts of CO is different than NO and H_2S [46], likely due to several of these different properties. A potential reason for this is described in more detail in the next section where the role of gasotransmitters produced by bacterial cells is considered.

Gasotransmitters in Bacterial Cells

In addition to gasotransmitters released by mammals killing bacterial cells, bacterial cells themselves produce all three of these gasotransmitters. This is described in detail in a review article by Tinajero-Trejo et al., which included a description of signaling pathways as well as the roles that these gasotransmitters have within bacteria [46]. The roles for each of these gasotransmitters vary. For example, CO is a carbon and energy source in bacteria and H_2S is a metabolic intermediate during anoxic sulfate respiration [47]. NO has been shown to promote biofilm formation in the Gram-negative bacteria *Azospirillum brasilense* [48], but adding NO donors has also been shown to have the opposite effect on some bacteria (i.e., the Gram-negative *Pseudomonas aeruginosa* that cause infections in hospitals) by reducing biofilms that have already formed [49]. The expression of these gasotransmitters within bacteria can have direct implications on their ability to be used as antibacterial agents.

Role in Protecting Bacteria

Interestingly, NO has been shown to protect microbes in some studies, which is the opposite of their typical biomedical purpose. In these examples, NO has been able to protect bacteria against oxidative stress [39, 50], which can be produced by the host's immune system in response to a pathogen. There are several reasons for this protective effect, including detoxification of NO by converting it to nitrate. In *E. coli* and other bacteria, flavohemoglobins have been shown to play a role in this detoxification process [51, 52]. There are also other methods of protection that have been identified, including a reversal of *S*-nitrosothiol formation that would prevent the release of NO [53]. At the least, these protective mechanisms have the potential

Table 1 Examples of gasotransmitters delivered as microbial agents

Molecule	Delivery method	Cell type	Outcome	References
Carbon monoxide	Carbon monoxide-releasing molecules (CORMs) – Seven CORMs tested – Release with water	*Bacterial cells* – *Escherichia coli* (*E. Coli*) – *Staphylococcus aureus* *Eukaryotic cells* – Murine macrophages – Porcine kidney epithelial cells – Human hepatoma cells	– Toxic to *E. coli* and non-toxic to eukaryotic cells – CO into the intercellular space did not enter *E. coli* cells – Released CO at high concentration (1 mM) became effective as bactericide	Nobre et al. [35]
	CORM-3-Ru(CO)3Cl(glycinate) – Release with water	*E. coli*	– Killed bacterial cells – Increased membrane permeability – Inhibition of terminal oxidases	Wilson et al. [36]
	CORM-3, CORM-2, and CORM-371 – Release with water	*Pseudomonas aeruginosa (PAO1)*	– CO carriers inhibit bacterial growth and O_2 consumption in vitro	Desmard et al. [37]
Nitric oxide	Endogenous NO in bacteria	*Bacillus anthracis*	– Bacteria produces NO – Provides resistance to macrophage killing of the bacteria	Deupree and Schoenfisch [38]; Shatalin et al. [39]
	NO released from a surface flux (xerogel coating) and bolus (PROLI/NO)	*Pseudomonas aeruginosa* and *E. coli*	– High levels of NO released over short durations are more damaging to bacteria than sustained lower-level surface fluxes	Deupree and Schoenfisch [38]
	Gaseous NO	*P. aeruginosa* *S. aureus* Human dermal fibroblast	– Lower cell numbers in bacteria at higher doses – NO was not toxic to human dermal fibroblast	Ghaffari et al. [40]
	NO-releasing molecules – Several compounds tested – Release with water	*Periodontal bacteria* *P. gingivalis and A. israelii* Human gingival fibroblasts	– Reached minimum bactericidal concentrations – Reduction in cell viability for HGF-1 cells	Shim et al. [41]
Hydrogen sulfide	Endogenous H_2S in bacteria	*Bacillus anthracis, P. aeruginosa, Staphylococcus aureus, and Escherichia coli*	– Cytoprotective mechanism on bacteria – Antioxidant effect on bacteria – Increased bacteria resistance to oxidative stress and antibiotics by a dual mechanism	Shatalin et al. [42]
	H_2S donor (NaHS)	Hepatocellular carcinoma	– Cell proliferation – Promoted apoptosis at higher doses	Wu et al. [43]
	H_2S donor (NaHS)	*E. coli*	– Varying dose changes selectivity of H_2S for killing bacteria	Fu et al. [44]

to reduce the effectiveness of NO delivery as antimicrobial agent and need to be considered. In addition, there has been an antioxidant effect of bacteria-produced NO that has been proposed to be involved in bacterial resistance to some antibiotics [54]. However, there is still some debate about this [46].

Similar results have been observed with H_2S. This is produced in large amounts by bacteria within the intestines. It is also produced by *E. coli* and other bacteria. There are a variety of different pathways due to the significant diversity of bacteria. Similar to mammalian cells, there is also less known about the role of H_2S for protecting bacteria. One study has shown that H_2S protects bacteria from several different antibiotics through a reduction in oxidative stress, similar to what has been shown with NO [42]. However, other studies have shown the opposite (e.g., H_2S promoting hydrogen peroxide generation to kill *E. coli*). Thus, further study is needed. It is likely that the dose of the H_2S and maybe also the species of bacteria are important considerations for the effect of this gasotransmitter on bacterial cells.

Unlike NO and H_2S, there does not appear to be a protective effect of CO on bacteria, even though it is also produced by bacteria [46]. CO is an inhibitor of mitochondrial respiration in bacterial and mammalian cells. Like the impact of CO on mammalian cells, it is likely that the impact on bacterial cells is dose specific. Low levels are present within bacteria as an energy source, yet higher levels can lead to bacterial cell death [55].

Gasotransmitter Dose

An important consideration for the effect of gasotransmitters on mammalian cells is the effective vs. toxic dose because they exhibit bimodal effects in these cells. Doses can periodically be found in the literature, although many of these doses are from in vitro tests without a biomaterial and those from in vivo measurements do not consider the biodistribution into the appropriate tissue. However, these offer a starting point to design gasotransmitter delivery systems from a biomaterial carrier. The importance of specific doses on positive vs. negative effects for mammalian cells is reviewed in detail for CO elsewhere [34]. Dose has also been shown to have an impact with NO delivery [2]. Bone-specific effects include osteoblast proliferation at low levels and bone resorption at high levels. Elevated levels of NO have also been implicated in cell fusion of pre-osteoclasts that may help to explain the increase in bone resorption rates [56]. Most of the NO delivery strategies involve NO donor molecules to provide more controlled release since free NO has a short half-life of less than 1 s in the presence of oxygen and hemoglobin [2]. Most NO donors (e.g., diazeniumdiolates and *S*-nitrosothiols) release over time when in contact with biological fluids. The half-life for breakdown of the molecule and release of NO is an important consideration for NO donor molecule design. A study by Mancini et al. showed in vitro that NO released from an NO donor (i.e., sodium nitroprusside or SNP) led to osteoblast proliferation at the NO donor concentration of 10 μM, but 50 μM and higher induced osteoblast apoptosis [57]. In addition to dose, this study also suggested that the rate of NO release is important. NO release from the slower releasing NO donor *N*-diazeniumdiolate (DETA) did not reduce osteoblast proliferation after 3 days until a concentration of 100 μM. As a further complication, studies have indicated that the growth phase of the cell can impact the NO at the same dose. For example, osteoclast activity was shown to not be affected at the proliferation phase of culture (1–3 days), but it was impacted at later times that corresponded to their differentiation (4–6 days) [2, 58]. H_2S is also typically delivered from donor molecules and like other gasotransmitters exhibits a biphasic dose effect [59]. For example, growth of cancer cells with the H_2S donor sodium hydrosulfide (NaHS) was observed at 10–50 μM, but proliferation was inhibited at a dose of 1000 μM in cancerous cells [60]. This result was observed in vitro. Overall, these examples indicate the importance of understanding the rate of release and duration since it can have a major impact on the response of mammalian cells. The balance between the effect of a particular dose on mammalian vs. bacterial cells is also important and can be used as a measure of selectivity for particular types of gasotransmitters.

Gasotransmitter Selectivity

Data regarding the selectivity of a gasotransmitter in its ability to kill bacteria but not mammalian cells is an important parameter to consider when choosing a particular gasotransmitter for an antibiotic purpose. However, this data is even harder to find because there are few studies that do a detailed comparison of the impacts on both prokaryotic and eukaryotic cells. This direct comparison is especially important because of the complexity of the dose and its impact on cells (e.g., release rate and duration) as well as complications in interpreting the results when different reaction products from gasotransmitter donor molecules are also present. For example, the antibacterial impact varies for different CO donors, independent of the amount of CO that could be released [37].

A study by Nobre et al. provided a rare detailed analysis of this selectivity using CO delivered from CO-releasing molecules (CORMs) [35]. There are different CORMs, some of which are more or less toxic to bacterial cells by themselves without CO release, that were tested in this study at the same CORM doses. The different CORMs gave different levels of bactericidal properties for *E. coli*, and similar trends were observed for ROS generation within mammalian cells. This shows that certain CORMs in particular can be toxic to both bacterial and mammalian cells. However, they also showed that CORM levels that were toxic to bacteria were generally not toxic to a variety of mammalian cells (e.g., macrophages, kidney epithelial cells, and a liver cell line). In particular, CORM compounds with minimum inhibitory concentrations of 350–650 μM and minimum bactericidal concentrations of 500–750 μM were tested with mammalian cells at doses from 50 to 500 μM. They found that for most CORMs, there was negligible toxicity to kidney epithelial cells. They also found that some cell types were more sensitive. For example, certain CORMs did cause a significant reduction in macrophage viability by 250 μM, but others continued to exhibit close to 100% viability even at 500 μM. The authors mention that the ability of the CORM to enter the cell instead of releasing into the media is necessary for bactericidal activity. One of the challenges with interpreting these results is that certain CORM molecules have different levels of toxicity, which makes it hard to differentiate between the CO and the byproducts that often contain heavy metals. Overall, CORMs were less toxic to mammalian than bacterial cells, but there was variability with different types of CORMs that were used. CORMs that had very good bactericidal properties but limited impact on most mammalian cells at the same concentration were found.

A similar comparison has been performed for gaseous NO delivered as part of a custom incubator, and its transfer to cell culture media was characterized by the concentration of stable NO products (NO_2^-) [40]. In this study, lower bacterial cell numbers were observed with higher NO doses as expected. *P. aeruginosa* started to exhibit decreased numbers by 40 parts per million (ppm), and *S. aureus* started to exhibit decreased numbers by 80 ppm. By 160 ppm, more than 84% cell death was observed for different types of bacterial cells. However, NO did not exhibit any toxicity for human dermal fibroblasts up to 200 ppm, and it was at 400 ppm that toxicity was observed (i.e., less than 80% viability relative to the control). This provided a well-controlled experiment to demonstrate that NO can provide selective in vitro toxicity for bacterial and not mammalian cells when no nitric oxide donor molecules are present. Interestingly, it also showed that NO gas uptake depended on the type of aqueous media, with cell culture media formulations leading to ~2–4 times higher NO uptake than a saline solution. This demonstrates complexity in controlling the dose available to the cells, even in this well-controlled system. For the 200 ppm NO gas that was not toxic for the human dermal fibroblasts, the concentration of nitrate/nitrite in the media rose to 4 mM concentrations.

The dose and selectivity of NO that leads to cell toxicity is even more complicated when traditional NO donors or nanoparticles are introduced. These donors are easier to include within a therapeutic approach and will typically be necessary for tissue engineering applications. For example, a 50 μM dose of the NO donor DETA has been shown to lead to osteoblast apoptosis [57]. However, a separate study has shown for several NO donors that cytotoxic effects were not observed in fibroblasts and keratinocytes until greater than 250 μM and 500 μM, respectively [61]. This could be due to a combination of different types of cells, different NO donors, and other differences in the experiments that were performed and requires further investigation. It also appears that, like CO-releasing molecules, the NO donors are more toxic than the pure gasotransmitter itself. This is suggested by the comparison with the toxic concentrations observed in the separate study with gaseous NO. Further, this may have an impact on the selective toxicity for bacterial vs. mammalian cells at the same dose. A study by Shim et al. directly compared the antibacterial activity of several NO-releasing compounds against periodontal bacteria and their cytotoxicity for mammalian cells [41]. The minimum bactericidal concentration was observed at 1–25 mM for a variety of NO donors (e.g., DETA-NO) for *P. gingivalis* and *A. israelii* (Fig. 1). The NO donors were also provided to human gingival fibroblasts (HGF-1) at the minimum bactericidal concentration level found for each NO donor (Fig. 2).

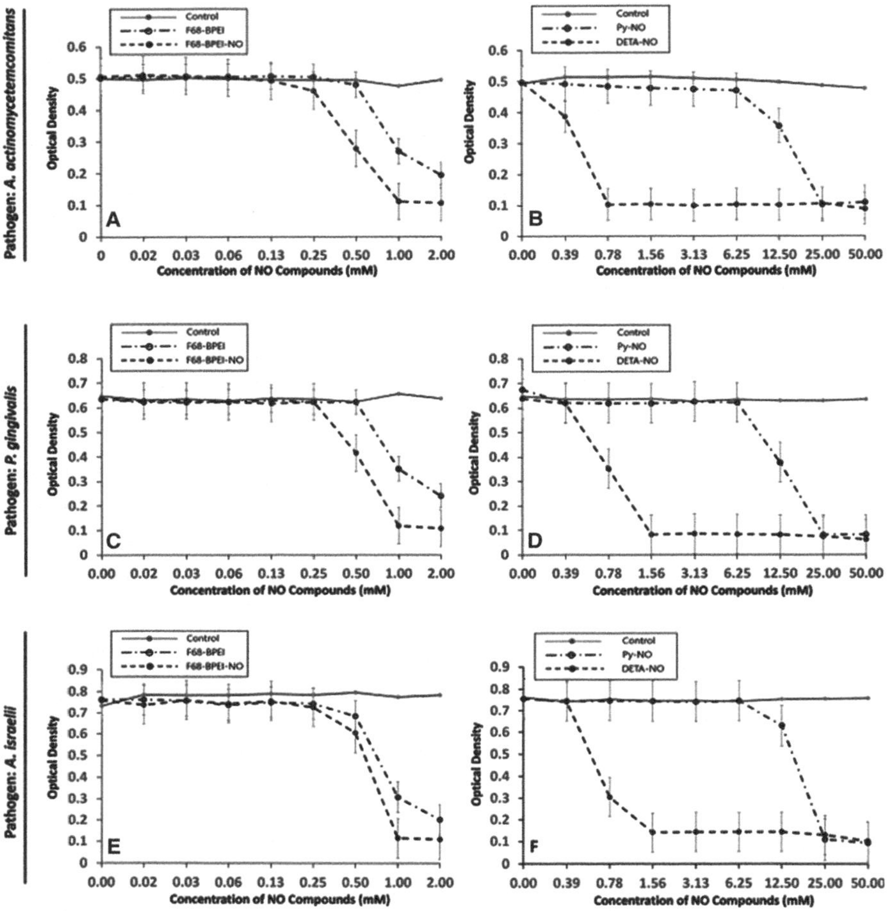

Fig. 1 An example comparison between toxicity of different gasotransmitters to bacterial cells. (**A–F**) Bactericidal efficacy of NO-releasing compounds. (Reprinted from an open access article from, *PLoS One,* Shim JS et al., 2018, copyright 2018, Creative Common Attribution License)

However, this led to reductions in cell viability ranging from 44 to 70% for the HGF-1 cells. This is a potential concern, and it stresses that further study is needed to develop NO delivery systems that make sure that the selectivity toxicity for bacteria seen with NO gas is not lost if it were to be delivered in the body.

There is less literature available on the selective toxicity of H_2S to bacterial compared to mammalian cells. It has been reported that low doses of the H_2S donor sodium hydrosulfide (NaHS) (10–100 μM) promoted hepatocellular carcinoma cell proliferation, but higher doses ≥400 μM promoted apoptosis of the same cell population [43]. This demonstrated that there are different toxic doses for different types of mammalian cells, but these studies did not directly compare with the bactericidal properties. In a separate study, antibacterial effects of the same H_2S donor (i.e., NaHS) on *E. coli* cells have been reported starting at 0.250 mM, with the minimum bactericidal concentration at 2.5 mM [44]. This dose raises concerns

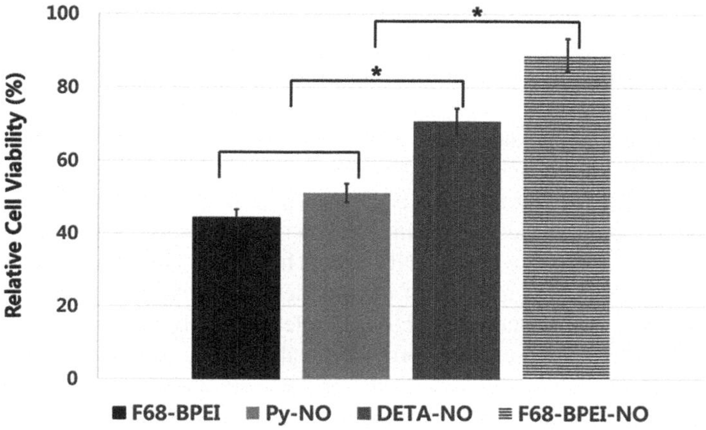

Fig. 2 The different response observed for gasotransmitter impact on mammalian cells. Cytotoxic effects of NO-releasing compounds on HGF-1 cells. Cell viability in each group was calculated by comparing the viability with that of an untreated control (100%). Asterisks indicate significant differences ($P < 0.05$) between groups. (Reprinted from an open access article from, *PLoS One*, Shim JS et al., 2018, copyright 2018, Creative Common Attribution License)

about the selectivity of H_2S for killing bacteria vs. mammalian cells, at least for these particular species. The fact that *E. coli* have been shown to use H_2S to protect against oxidative stress may play a role in what appears to be resistance of this cell type to the antibacterial effects. However, more research is needed with different types of pathogenic bacterial cells and side-by-side comparisons with mammalian cells to determine the effects. This is especially true for H_2S, but it is also needed for other gasotransmitters.

Gasotransmitter Inclusion into Scaffolds

When gasotransmitters have been delivered in vivo, they most often have been delivered systemically. The main example has been using inhaled gasotransmitters for treating pulmonary fibrosis, which has progressed to clinical trials and clinical use [62]. In addition, gasotransmitter molecules have been injected clinically (e.g., sodium nitroprusside) and in animal models [63]. However, systemic delivery for regenerative medicine or tissue engineering would result in low levels reaching the tissue of interest due to limitations with each specific gasotransmitter. For NO, this includes its high reactivity, and for CO, this includes its propensity to bind to hemoglobin. Thus, incorporating these strategies within a scaffold is advantageous.

The most research for incorporating a gasotransmitter within a scaffold has been performed for NO. Much of this research has involved vascular applications, including modifying traditional ePTFE grafts [64] and incorporating as part of a drug-eluting stent [65]. However, NO-containing scaffolds have also been used to target a variety of types of tissues. For skin applications, gaseous NO delivery has been included as part of a patch to provide antibacterial properties [66]. Separate studies used electrospun NO-loaded materials that provided controlled release of NO over the course of 14 days as a wound dressing [67], with the focus more on promoting wound healing than on the antibacterial properties. With both vascular and skin pathologies (e.g., diabetic ulcers), underexpression of NO is a concern, and one possible reason why studies have shown promise for NO delivery to be used to treat these conditions [68]. NO has also been incorporated within a scaffold for bone tissue engineering, as described in the section "Gasotransmitters for Bone Applications" [3]. For these studies, NO donors have been incorporated within scaffolds and as coating with a range of chemistries, including poly(ε-caprolactone)/keratin mats and an injectable nitric oxide releasing poly(ethylene) glycol-fibrin adhesive [69, 70]. These scaffolds provide local delivery of the encapsulated NO donors, and they also typically modify the NO release rate.

There are fewer studies that have incorporated H_2S within scaffolds, and many of these are more broadly targeted instead of for a particular tissue type. For example, Feng et al. developed an electrospun microfiber containing an H_2S donor, *N*-(benzoylthio)benzamide (NSHD1), that releases hydrogen sulfide in the presence of biological thiols [71]. This study illustrates the important impact that including a gasotransmitter donor within a scaffold can have on the release. The fibers slowed the rate of H_2S release 10–20 times compared to free NSHD1 within a phosphate-buffered saline solution, and H_2S released from the fibers was able to reduce cell damage from hydrogen peroxide added to the system. A recent study has also investigated the incorporation of another H_2S donor, GYY4137, within a silk fibroin salt-leached scaffold that would provide local delivery. This study was specifically targeted for engineering bone tissue [72]. The scaffold took ~30 min to fully hydrate, and a burst release of H_2S occurred over 90 min before the release started to plateau. However, for this scaffold/gasotransmitter donor system, the rate of release compared to free donor not contained within a scaffold was not investigated. Since this same H_2S donor was reported elsewhere to provide a more controlled, extended release out to more than 1 day when it was not included within a scaffold [73], it is not clear what role incorporating the H_2S donor within the scaffold had on controlled release in this study.

There are also limited studies that have incorporated carbon monoxide-releasing molecules within a scaffold, and to our knowledge, none have been used for bone tissue engineering applications. It has, however, been used for both antibacterial properties and vascular tissue engineering. Bohlender et al. incorporated a photoactivatable manganese-based CORM with a high CO yield within a poly(L-lactide-*co*-D/L-lactide) scaffold, with a stated goal to provide antibacterial properties that would be tested in future studies [74]. With the high amount of CO released from these scaffolds, there was some toxicity observed for 3T3 mouse fibroblast cells. One of the goals of incorporating the CORM within the scaffold was to contain the potentially toxic metal byproducts within the fibers, although the study was not designed with necessary controls to test this effect. Another study by Michael et al. incorporated a different photo CORM within an electrospun PCL scaffold for a different reason [75]. This organic CORM had demonstrated low toxicity to mammalian cells, but the reaction and CO release would not occur in the presence of water, so it needed to be contained within a scaffold (Fig. 3). Further comparison between the impact of CORMs contained within

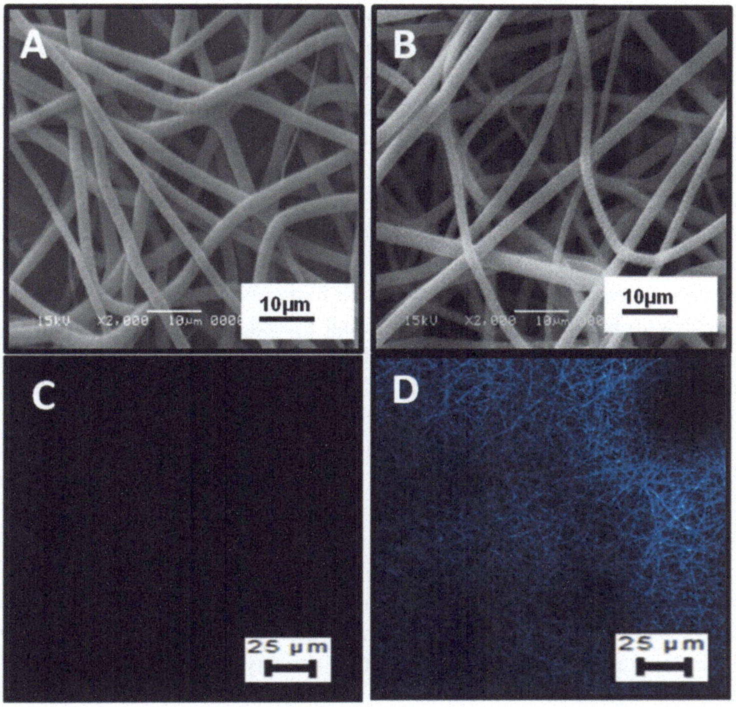

Fig. 3 An example of a gasotransmitter donor incorporated within a scaffold. SEM images of electrospun scaffolds produced from (**A**) pure PCL and (**B**) PCL/CORM solutions. The solution concentrations were 16% and 19%, respectively. Confocal images of PCL/CORM scaffolds taken with a DAPI filter (**C**) before and (**D**) after activation are also shown. (Reprinted with permission and license from IOP Publishing, Biomedical Materials, Michael et al., 2016, copyright 2016 IOP Publishing Ltd)

scaffolds and free CORMs within a solution that are able to be endocytosed within cells is needed to determine their antibacterial impacts vs. their effects on healthy cells. For all types of gasotransmitters, the scaffold-gasotransmitter donor needs to be carefully investigated to determine how they impact the gasotransmitter dose, release rate, and the corresponding impact on the cells.

Gasotransmitters for Bone Applications

For bone applications, most of the studies have focused on elucidating the important role of cell-produced NO in bone healing. As mentioned earlier, inducible NOS (iNOS) leads to much higher levels of NO during inflammation, and properly controlled inflammatory processes involving NO are necessary for bone healing. For example, the delivery of a drug to induce cells to synthesize iNOS in rat models with a femoral fracture was shown to result in 46% higher maximum strength for the remodeled bone compared to conditions without the drug [76]. It has also been shown that induction of iNOS by lipopolysaccharide (LPS) resulted in MMP-1 induction in vitro [77], which is necessary for bone remodeling and osteointegration, but can be a concern if it is not balanced with the corresponding deposition of new bone tissue. They also tested a rat model of dental pulp infection that linked iNOS upregulation to pathological bone resorption through the use of an iNOS inhibitor. Separately, nitric oxide has been shown to promote osteoclast cell fusion, which could also help to explain bone resorption in this diseased model [56]. Overall, these results for cell-produced NO indicate several challenges with understanding the impact of different NO levels in native bone tissue due to its biphasic effects (e.g., with healthy vs. diseased animal models). Similar potential for bone healing has been found for endogenous hydrogen sulfide. One study found that endogenous H_2S increased nuclear RUNX2 accumulation, likely contributing to osteogenesis. Further, overexpression of H_2S in the tissue promoted bone healing in a bone—fracture rat model [78]. These studies make it clear that gasotransmitters have an important role in native bone healing and suggest that adding supplemental doses may also be beneficial.

Studies have also investigated the impact of supplemental doses of gasotransmitters from donor molecules on bone cells. Osteoblasts, for instance, have been shown to increase the expression of the bone-specific marker osteocalcin in response to NO donors at the right doses, but these donors surprisingly also appeared to decrease the alkaline phosphatase activity [79]. This is explained in the article as due to a pathway not related to NO signaling. NO has also been shown to promote the differentiation of stem cells to osteoblasts. For example, rat dental pulp stem cells exposed to an NO donor increased the expression of RUNX2 and alkaline phosphatase as well as deposited higher levels of mineral deposits [80]. At the right doses, NO donors have been shown to also reduce osteoclast formation [81]. As mentioned previously, the dose is very important as negative responses will be observed if the dose of NO or other gasotransmitters is too high. Appropriate doses of hydrogen sulfide have also been shown to provide beneficial properties for bone cells. This includes protecting osteoblasts against damage from oxidative stress [82]. Further, a hydrogen sulfide donor was shown to lead to bone marrow cells differentiate in culture to form colonies of alkaline phosphate-positive cells that also exhibited increased mRNA expression of osteogenic genes (e.g., RUNX2, osterix, and osteocalcin) [83]. The impact of exogenous CO on cells for bone tissue engineering is less clear.

There have been few studies that have developed constructs for bone tissue engineering that included gasotransmitter delivery. There is one study that assessed H_2S delivery for engineering bone [72]. Their system included the H_2S donor within a silk fibroin scaffold with a well-characterized release profile (Fig. 4). In this study,

Fig. 4 (continued) γSF_1%, containing 1% of GYY4137; γSF_5%, containing 5% of GYY4137; γSF_10%, containing 10% of GYY4137) (Reprinted with permission from Raggio, Rosasilvia, et al. "Silk fibroin porous scaffolds loaded with a slow-releasing hydrogen sulfide agent (GYY4137) for applications of tissue engineering." *ACS Biomaterials Science & Engineering* 4.8 (2018): 2956–2966. Copyright 2018 American Chemical Society Publications, ACS Biomaterials Science and Engineering)

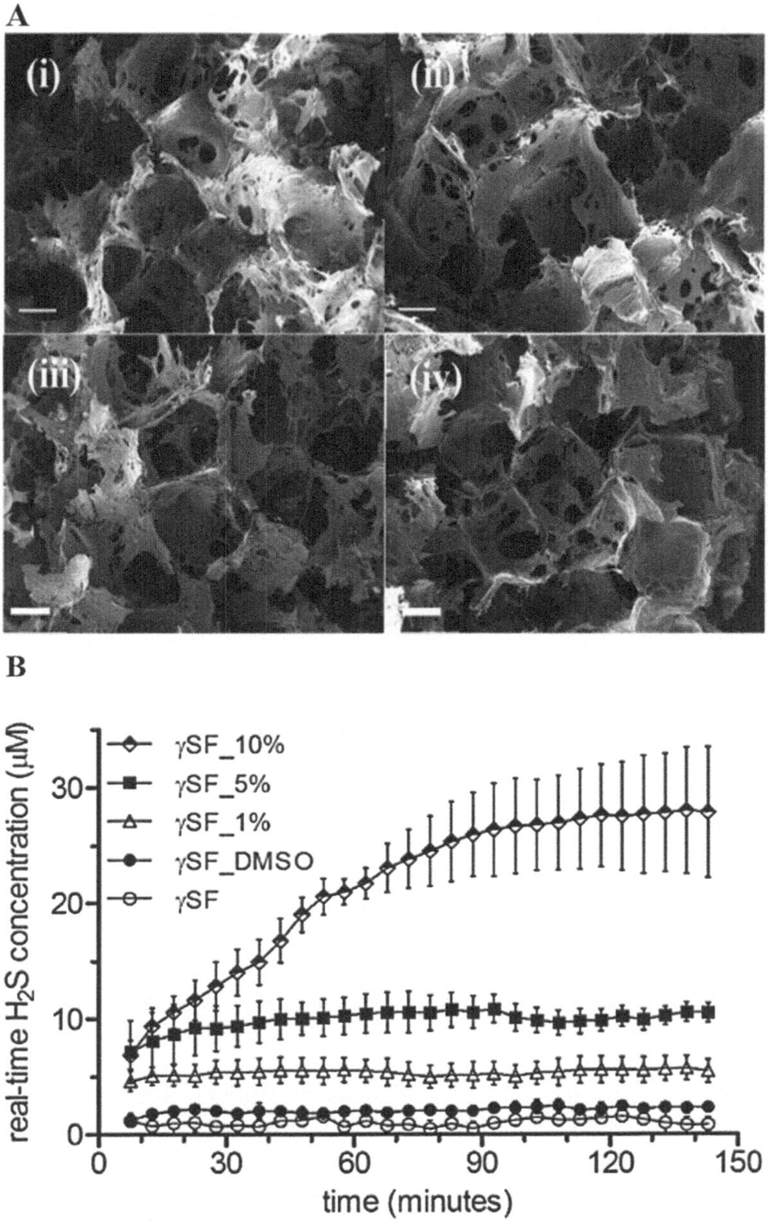

Fig. 4 Bone tissue engineering application for the gasotransmitter H_2S. (**A**) FESEM images of the internal porous structure of dry γ-irradiated SF scaffolds: (i) γSF, untreated scaffold; (ii) γSF_1%, scaffold containing 1% GYY4137; (iii) γSF_5%, scaffold containing 5% GYY4137; (iv) γSF_10%, scaffold containing 10% GYY4137; scale bar 200 μm. (**B**) Release of H_2S monitored by H_2S-selective microelectrode for SF scaffolds incubated in PBS pH 7.4 with L-cysteine 4 mM, at room temperature (mean with SEM, $n = 4$); error bars are not visible when lying within dimensions of symbols. Cytotoxicity of the different groups of scaffolds (γSF_DMSO, treated with DMSO;

cell culture was performed with NIH/3T3 fibroblasts and human MSCs, and no toxicity was observed when the H_2S donor was loaded from 1 to 10% w/w within the scaffold. However, the analysis was limited to cytotoxicity testing, and markers of the osteoblast phenotype were not tested. In addition, this study did not determine if the released H_2S could provide antibacterial properties at the doses tested. It is difficult to correlate the doses provided within this scaffold systems to free H_2S donor tested in other studies, so the osteogenic properties of these systems need to be carefully investigated.

A study by Pant et al. performed an analysis of both the mammalian and bacterial cell response to their 3D scaffold containing the NO donor, *S*-nitroso-*N*-acetyl-penicillamine (SNAP) [3]. In this molecule, penicillamine is a metabolite of penicillin but does not have any antibacterial properties itself. This NO donor was added to a mixture of nano-hydroxyapatite, starch, and either alginate or chitosan, and this mixture was processed with freeze-drying to produce a porous cross-linked scaffold. The NO donor was added at 10% w/w to the nanoparticles, and this resulted in a burst release of the NO for 30 min and then a slower release after that. A 99% reduction in *S. aureus* and *P. aeruginosa* was observed with the NO donor compared to controls without it, demonstrating bactericidal properties. Negligible toxicity was observed for NIH-3T3 fibroblasts, but the cytotoxicity tests were performed from conditioned media instead of cells cultured within the same scaffold. Overall, the studies using gasotransmitter donors for bone tissue engineering are currently limited and in the early stages.

Gasotransmitters for Tissue Vascularization

Another important challenge for bone tissue engineering is the formation of microvasculature to allow for enough oxygen and nutrient diffusion for the generation of viable thick bone grafts [84]. While current studies do not explicitly investigate gasotransmitters for promoting angiogenesis in bone grafts, the literature suggests that they can be beneficial. Gasotransmitters are known to control vascular tone, and they have also been shown to be involved in the formation of microvasculature in native tissue. CO, NO, and H_2S are known to be needed for endothelial cell proliferation and capillary function [85, 86]. The role of NO on angiogenic sprout formation is less clear. It has recently been shown that NO is involved in microvessel sprouting, with ~35% more NO in the tip cells that initiate the sprout [87]. However, an inhibition of NO in tumor tissue led to improved angiogenesis, with longer vessels forming along with the formation of sprouts [88]. Further, dose-specific effects have been observed with the delivery of an H_2S donor, with no pro-angiogenic effect observed at a high dose [89]. This is consistent with the bimodal effects observed for other cells and physiological processes, such as inflammation. Inflammation is necessary for bone healing. For example, it was shown in a Ccr2 −/− mouse model of decreased macrophage recruitment that disruption of Ccr2 signaling impaired vascularization, decreased formation of callus, and delayed maturation of cartilage and bone remodeling up to 21 days after injury

[90, 91]. However, levels and duration of inflammation that are too high can have the opposite effect and can interfere with bone healing and regeneration [8]. Thus, the dose-specific effects of gasotransmitters on inflammation likely correlate with their ability to promote angiogenesis. With appropriate doses, gasotransmitters can likely aid tissue vascularization and other aspects of remodeling for bone grafts.

Conclusions and Future Challenges

Gasotransmitters represent a promising approach to provide both antibacterial properties and remodeling for engineering of bone as well as other types of tissues. Decades of research illustrate that gasotransmitters play a key role in physiological function of biological systems, and there have been promising studies demonstrating the antibacterial effects of NO as well as other gasotransmitters. There have also been promising studies demonstrating the ability of these gasotransmitters to promote tissue remodeling and integration. However, few studies have directly compared the doses of specific gasotransmitter delivery strategies needed to achieve both of these results within that same tissue engineered graft. Previous studies suggest that there are differences in selectivity in killing bacterial cells while avoiding mammalian cell toxicity between the different gasotransmitters (i.e., NO, H_2S, and CO) that should be considered. These studies also suggest that other variables, such as the type of gasotransmitter donor molecule and if it is included within a scaffold, can have a major impact on the results that are obtained. Further studies with direct comparisons between parameters of interest are needed to reliably and consistently provide gasotransmitter strategies for tissue engineering of bone or other tissue types. This direct comparison is particularly important because of the challenges with controlling and predicting the dose of these gaseous molecules available to cells and the variability that will be found between studies. For bone tissue engineering in particular, there is evidence from native tissue that gasotransmitters play an important role in tissue maintenance and healing, but these studies have been limited. Further, most of the works that consider both bacterial and mammalian cell responses have been either basic cell culture or regenerative medicine studies without a scaffold component. One potential challenge with getting many of these strategies to the clinic is that beneficial effects seen in vitro may not be seen when they are tested in the more complex environment found in animal models.

In addition to the need to perform further research into the selectivity of different gasotransmitter doses, there are other areas that will likely be important to move this field forward. This includes developing new systems to provide long-term controlled release that is not possible with most current delivery strategies. Controlled release is critical because these gaseous molecules quickly leave the system or, in the case of NO and H_2S, react quickly. In addition, a way to translate findings for appropriate doses from in vitro cell culture to inside the body is needed to develop safe and effective delivery systems. This is one area where computational modeling will be very useful. With further research, gasotransmitter delivery systems offer a promising approach for engineering bone and other tissues.

Acknowledgment *Funding*: Efforts were supported by the National Science Foundation under Grant No. CBET 1510003.

References

1. Damoulis PD, Drakos DE, Gagari E, Kaplan DL (2007) Osteogenic differentiation of human mesenchymal bone marrow cells in silk scaffolds is regulated by nitric oxide. Ann N Y Acad Sci 1117:367–376. https://doi.org/10.1196/annals.1402.038. NIH Public Access
2. Nichols SP, Storm WL, Koh A, Schoenfisch MH (2012) Local delivery of nitric oxide: targeted delivery of therapeutics to bone and connective tissues. Adv Drug Deliv Rev 64(12):1177–1188. https://doi.org/10.1016/J.ADDR.2012.03.002. Elsevier
3. Pant J, Sundaram J, Goudie MJ, Nguyen DT, Handa H (2018) Antibacterial 3D bone scaffolds for tissue engineering application. J Biomed Mater Res B Appl Biomater 107:1068–1078. https://doi.org/10.1002/jbm.b.34199
4. Einhorn TA, Lee CA (2001) Bone regeneration: new findings and potential clinical applications. J Am Acad Orthop Surg 9(3):157–165. https://doi.org/10.5435/00124635-200105000-00002
5. Keramaris NC, Calori GM, Nikolaou VS, Schemitsch EH, Giannoudis PV (2008) Fracture vascularity and bone healing: a systematic review of the role of VEGF. Injury 39(Suppl 2):S45–S57. https://doi.org/10.1016/S0020-1383(08)70015-9. Elsevier
6. Darouiche RO (2003) Antimicrobial approaches for preventing infections associated with surgical implants. Clin Infect Dis 36(10):1284–1289. https://doi.org/10.1086/374842
7. Dentino A, Lee S, Mailhot J, Hefti AF (2013) Principles of periodontology. Periodontology 2000 61(1):16–53. https://doi.org/10.1111/j.1600-0757.2011.00397.x. Wiley
8. Thomas MV, Puleo DA (2011) Infection, inflammation, and bone regeneration: a paradoxical relationship. J Dent Res 90(9):1052–1061. https://doi.org/10.1177/0022034510393967. International Association for Dental Research
9. Johnson CT, García AJ (2015) Scaffold-based anti-infection strategies in bone repair. Ann Biomed Eng 43(3):515–528. https://doi.org/10.1007/s10439-014-1205-3. NIH Public Access
10. Mortimer CJ, Widdowson JP, Wright CJ (2018) Electrospinning of functional nanofibers for regenerative medicine: from bench to commercial scale. In: Novel aspects of nanofibers. InTech. https://doi.org/10.5772/intechopen.73677
11. Daghighi S, Sjollema J, van der Mei HC, Busscher HJ, Rochford ETJ (2013) Infection resistance of degradable versus non-degradable biomaterials: an assessment of the potential mechanisms. Biomaterials 34(33):8013–8017. https://doi.org/10.1016/J.BIOMATERIALS.2013.07.044. Elsevier
12. Kim J, Li WA, Sands W, Mooney DJ (2014) Effect of pore structure of macroporous poly(lactide-*co*-glycolide) scaffolds on the *in vivo* enrichment of dendritic cells. ACS Appl Mater Interfaces 6(11):8505–8512. https://doi.org/10.1021/am501376n. American Chemical Society
13. Klein-Nulend J, van Oers RFM, Bakker AD, Bacabac RG (2013) Nitric oxide signaling in mechanical adaptation of bone. Osteoporos Int 25(5):1427–1437. https://doi.org/10.1007/s00198-013-2590-4
14. Chinta KC, Saini V, Glasgow JN, Mazorodze JH, Rahman MA, Reddy D, Lancaster JR, Steyn AJC, Steyn AJC (2016) The emerging role of gasotransmitters in the pathogenesis of tuberculosis. Nitric Oxide 59:28–41. https://doi.org/10.1016/j.niox.2016.06.009. NIH Public Access
15. Fang FC (2004) Antimicrobial reactive oxygen and nitrogen species: concepts and controversies. Nat Rev Microbiol 2(10):820–832. https://doi.org/10.1038/nrmicro1004

16. Arkenau H-T, Stichtenoth DO, Frölich JC, Manns MP, Böker K-HW (2002) Elevated nitric oxide levels in patients with chronic liver disease and cirrhosis correlate with disease stage and parameters of hyperdynamic circulation. Z Gastroenterol 40(11):907–913. https://doi.org/10.1055/s-2002-35413. © Karl Demeter Verlag im Georg Thieme Verlag Stuttgart, New York
17. Wimalawansa SJ (2010) Nitric oxide and bone. Ann N Y Acad Sci 1192(1):391–403. https://doi.org/10.1111/j.1749-6632.2009.05230.x
18. Wimalawansa SJ (2008) Nitric oxide: novel therapy for osteoporosis. Expert Opin Pharmacother 9(17):3025–3044. https://doi.org/10.1517/14656560802197162
19. Pryor WA, Squadrito GL (1995) The chemistry of peroxynitrite: a product from the reaction of nitric oxide with superoxide. Am J Physiol 268(5 Pt 1):L699–L722. https://doi.org/10.1152/ajplung.1995.268.5.L699. American Physiological Society, Bethesda, MD
20. Carter JM, Qian Y, Foster JC, Matson JB (2015) Peptide-based hydrogen sulphide-releasing gels. Chem Commun 51(66):13131–13134. https://doi.org/10.1039/C5CC04883D. Royal Society of Chemistry
21. George TJ, Arnaoutakis GJ, Beaty CA, Jandu SK, Santhanam L, Berkowitz DE, Shah AS (2012) Inhaled hydrogen sulfide improves graft function in an experimental model of lung transplantation. J Surg Res 178(2):593–600. https://doi.org/10.1016/j.jss.2012.06.037. NIH Public Access
22. Pu H, Hua Y (2017) Hydrogen sulfide regulates bone remodeling and promotes orthodontic tooth movement. Mol Med Rep 16(6):9415–9422. https://doi.org/10.3892/mmr.2017.7813. Spandidos Publications
23. Zhang L, Wang Y, Li Y, Li L, Xu S, Feng X, Liu S (2018) Hydrogen sulfide (H2S)-releasing compounds: therapeutic potential in cardiovascular diseases. Front Pharmacol 9:1066. https://doi.org/10.3389/fphar.2018.01066. Frontiers Media SA
24. Bhatia M (2005) Hydrogen sulfide as a vasodilator. IUBMB Life 57(9):603–606. https://doi.org/10.1080/15216540500217875. Wiley
25. Kimura H (2002) Hydrogen sulfide as a neuromodulator. Mol Neurobiol 26(1):013–020. https://doi.org/10.1385/MN:26:1:013. Humana Press
26. Łowicka E, Bełtowski J (2007) Hydrogen sulfide (H2S)—the third gas of interest for pharmacologists. Pharmacol Rep 59(1):4–24. http://www.ncbi.nlm.nih.gov/pubmed/17377202
27. Wang R (2003) The gasotransmitter role of hydrogen sulfide. Antioxid Redox Signal 5(4):493–501. https://doi.org/10.1089/152308603768295249. Mary Ann Liebert, Inc.
28. WANG RUI (2002) Two's company, three's a crowd: can H2S be the third endogenous gaseous transmitter? FASEB J 16(13):1792–1798. https://doi.org/10.1096/fj.02-0211hyp. Federation of American Societies for Experimental Biology
29. Gadalla MM, Snyder SH (2010) Hydrogen sulfide as a gasotransmitter. J Neurochem 113(1):14–26. https://doi.org/10.1111/j.1471-4159.2010.06580.x. NIH Public Access
30. Pal VK, Bandyopadhyay P, Singh A (2018) Hydrogen sulfide in physiology and pathogenesis of bacteria and viruses. IUBMB Life 70(5):393–410. https://doi.org/10.1002/iub.1740. Europe PMC Funders
31. Wegiel B, Gallo DJ, Raman KG, Karlsson JM, Ozanich B, Chin BY, Tzeng E et al (2010) Nitric oxide–dependent bone marrow progenitor mobilization by carbon monoxide enhances endothelial repair after vascular injury. Circulation 121(4):537–548. https://doi.org/10.1161/CIRCULATIONAHA.109.887695
32. Babu D, Motterlini R, Lefebvre RA (2015) CO and CO-releasing molecules (CO-RMs) in acute gastrointestinal inflammation. Br J Pharmacol 172(6):1557–1573. https://doi.org/10.1111/bph.12632
33. Ahmed A, Ramma W (2015) Unravelling the theories of pre-eclampsia: are the protective pathways the new paradigm? Br J Pharmacol 172(6):1574–1586. https://doi.org/10.1111/bph.12977
34. Washington KS, Bashur CA (2017) Delivery of antioxidant and anti-inflammatory agents for tissue engineered vascular grafts. Front Pharmacol 8:659. https://doi.org/10.3389/fphar.2017.00659. Frontiers

35. Nobre LS, Jeremias H, Romão CC, Saraiva LM, Schenk WA, Benz R, Zimmermann U et al (2016) Examining the antimicrobial activity and toxicity to animal cells of different types of CO-releasing molecules. Dalton Trans 45(4):1455–1466. https://doi.org/10.1039/C5DT02238J. The Royal Society of Chemistry
36. Wilson JL, Jesse HE, Hughes B, Lund V, Naylor K, Davidge KS, Cook GM, Mann BE, Poole RK (2013) Ru(CO)$_3$ Cl(glycinate) (CORM-3): a carbon monoxide–releasing molecule with broad-spectrum antimicrobial and photosensitive activities against respiration and cation transport in *Escherichia Coli*. Antioxid Redox Signal 19(5):497–509. https://doi.org/10.1089/ars.2012.4784
37. Desmard M, Foresti R, Morin D, Dagouassat M, Berdeaux A, Denamur E, Crook SH et al (2012) Differential antibacterial activity against *Pseudomonas aeruginosa* by carbon monoxide-releasing molecules. Antioxid Redox Signal 16(2):153–163. https://doi.org/10.1089/ars.2011.3959
38. Deupree SM, Schoenfisch MH (2009) Morphological analysis of the antimicrobial action of nitric oxide on gram-negative pathogens using atomic force microscopy. Acta Biomater 5(5):1405–1415. https://doi.org/10.1016/J.ACTBIO.2009.01.025. Elsevier
39. Shatalin K, Gusarov I, Avetissova E, Shatalina Y, McQuade LE, Lippard SJ, Nudler E (2008) Bacillus anthracis-derived nitric oxide is essential for pathogen virulence and survival in macrophages. Proc Natl Acad Sci 105(3):1009–1013. https://doi.org/10.1073/pnas.0710950105
40. Ghaffari A, Neil DH, Ardakani A, Road J, Ghahary A, Miller CC (2005) A direct nitric oxide gas delivery system for bacterial and mammalian cell cultures. Nitric Oxide 12(3):129–140. https://doi.org/10.1016/J.NIOX.2005.01.006. Academic
41. Shim JS, Park D-s, Baek D-H, Jha N, In Park S, Yun HJ, Kim WJ, Ryu JJ (2018) Antimicrobial activity of NO-releasing compounds against periodontal pathogens. PLoS One 13(10):e0199998. https://doi.org/10.1371/journal.pone.0199998. Edited by Salomon Amar. Public Library of Science
42. Shatalin K, Shatalina E, Mironov A, Nudler E (2011) H2S: a universal defense against antibiotics in bacteria. Science 334(6058):986–990. https://doi.org/10.1126/science.1209855
43. Wu D, Li M, Tian W, Wang S, Cui L, Li H, Wang H, Ji A, Li Y (2017) Hydrogen sulfide acts as a double-edged sword in human hepatocellular carcinoma cells through EGFR/ERK/MMP-2 and PTEN/AKT signaling pathways. Sci Rep 7(1):5134. https://doi.org/10.1038/s41598-017-05457-z. Nature Publishing Group
44. Fu L-H, Wei Z-Z, Hu K-D, Hu L-Y, Li Y-H, Chen X-Y, Han Z, Yao G-F, Zhang H (2018) Hydrogen sulfide inhibits the growth of Escherichia coli through oxidative damage. J Microbiol 56(4):238–245. https://doi.org/10.1007/s12275-018-7537-1. The Microbiological Society of Korea
45. Motterlini R, Otterbein LE (2010) The therapeutic potential of carbon monoxide. Nat Rev Drug Discov 9(9):728–743. https://doi.org/10.1038/nrd3228
46. Tinajero-Trejo M, Jesse HE, Poole RK (2013) Gasotransmitters, poisons, and antimicrobials: it's a gas, gas, gas! F1000Prime Rep 5:28. https://doi.org/10.12703/P5-28. Faculty of 1000 Ltd
47. Slonczewski J, Foster JW (2009) Microbiology: an evolving science, 1st edn. W.W. Norton & Co, New York. https://www.worldcat.org/title/microbiology-an-evolving-science/oclc/185042615
48. Arruebarrena Di Palma A, Pereyra CM, Moreno Ramirez L, Xiqui Vázquez ML, Baca BE, Pereyra MA, Lamattina L, Creus CM (2013) Denitrification-derived nitric oxide modulates biofilm formation in *Azospirillum brasilense*. FEMS Microbiol Lett 338(1):77–85. https://doi.org/10.1111/1574-6968.12030
49. Barnes RJ, Bandi RR, Wong WS, Barraud N, McDougald D, Fane A, Kjelleberg S, Rice SA (2013) Optimal dosing regimen of nitric oxide donor compounds for the reduction of *Pseudomonas aeruginosa* biofilm and isolates from wastewater membranes. Biofouling 29(2):203–212. https://doi.org/10.1080/08927014.2012.760069
50. Gusarov I, Nudler E (2005) NO-mediated cytoprotection: instant adaptation to oxidative stress in bacteria. Proc Natl Acad Sci 102(39):13855–13860. https://doi.org/10.1073/pnas.0504307102

51. Gardner PR, Gardner AM, Martin LA, Salzman AL (1998) Nitric oxide dioxygenase: an enzymic function for flavohemoglobin. Proc Natl Acad Sci U S A 95(18):10378–10383. http://www.ncbi.nlm.nih.gov/pubmed/9724711
52. Wisecaver JH, Alexander WG, King SB, Todd Hittinger C, Rokas A (2016) Dynamic evolution of nitric oxide detoxifying flavohemoglobins, a family of single-protein metabolic modules in bacteria and eukaryotes. Mol Biol Evol 33(8):1979–1987. https://doi.org/10.1093/molbev/msw073
53. Laver JR, McLean S, Bowman LAH, Harrison LJ, Read RC, Poole RK (2013) Nitrosothiols in bacterial pathogens and pathogenesis. Antioxid Redox Signal 18(3):309–322. https://doi.org/10.1089/ars.2012.4767
54. Gusarov I, Shatalin K, Starodubtseva M, Nudler E (2009) Endogenous nitric oxide protects bacteria against a wide spectrum of antibiotics. Science 325(5946):1380–1384. https://doi.org/10.1126/science.1175439
55. Nakahira K, Choi AMK (2015) Carbon monoxide in the treatment of sepsis. Am J Physiol Lung Cell Mol Physiol 309(12):L1387–L1393. https://doi.org/10.1152/ajplung.00311.2015. American Physiological Society
56. Nilforoushan D, Gramoun A, Glogauer M, Manolson MF (2009) Nitric oxide enhances osteoclastogenesis possibly by mediating cell fusion. Nitric Oxide 21(1):27–36. https://doi.org/10.1016/J.NIOX.2009.04.002. Academic
57. Mancini L, Moradi-Bidhendi N, Becherini L, Martineti V, MacIntyre I (2000) The biphasic effects of nitric oxide in primary rat osteoblasts are CGMP dependent. Biochem Biophys Res Commun 274(2):477–481. https://doi.org/10.1006/bbrc.2000.3164
58. Holliday LS, Dean AD, Lin RH, Greenwald JE, Gluck SL (1997) Low NO concentrations inhibit osteoclast formation in mouse marrow cultures by CGMP-dependent mechanism. Am J Physiol 272(3):F283–F291. https://doi.org/10.1152/ajprenal.1997.272.3.F283
59. Guo F-F, Yu T-C, Hong J, Fang J-Y (2016) Emerging roles of hydrogen sulfide in inflammatory and neoplastic colonic diseases. Front Physiol 7:156. https://doi.org/10.3389/fphys.2016.00156. Frontiers Media SA
60. Cai W-J, Wang M-J, Ju L-H, Wang C, Zhu Y-C (2010) Hydrogen sulfide induces human colon cancer cell proliferation: role of Akt, ERK and P21. Cell Biol Int 34(6):565–572. https://doi.org/10.1042/CBI20090368
61. Krischel V, Bruch-Gerharz D, Suschek C, Kröncke K-D, Ruzicka T, Kolb-Bachofen V (1998) Biphasic effect of exogenous nitric oxide on proliferation and differentiation in skin derived keratinocytes but not fibroblasts. J Investig Dermatol 111(2):286–291. https://doi.org/10.1046/j.1523-1747.1998.00268.x
62. Lowson SM (2004) Alternatives to nitric oxide. Br Med Bull 70:119–131. https://doi.org/10.1093/bmb/ldh028
63. Ziesche S, Franciosa JA (1977) Clinical application of sodium nitroprusside. Heart Lung 6(1):99–103. http://www.ncbi.nlm.nih.gov/pubmed/583902
64. Gregory EK, Vavra AK, Moreira ES, Havelka GE, Jiang Q, Lee VR, Van Lith R, Ameer GA, Kibbe MR (2011) Antioxidants modulate the antiproliferative effects of nitric oxide on vascular smooth muscle cells and adventitial fibroblasts by regulating oxidative stress. Am J Surg 202(5):536–540. https://doi.org/10.1016/j.amjsurg.2011.06.018
65. Elnaggar MA, Seo SH, Gobaa S, Lim KS, Bae I-H, Jeong MH, Han DK, Joung YK (2016) Nitric oxide releasing coronary stent: a new approach using layer-by-layer coating and liposomal encapsulation. Small 12(43):6012–6023. https://doi.org/10.1002/smll.201600337
66. Jones ML, Ganopolsky JG, Labbé A, Prakash S (2010) A novel nitric oxide producing probiotic patch and its antimicrobial efficacy: preparation and in vitro analysis. Appl Microbiol Biotechnol 87(2):509–516. https://doi.org/10.1007/s00253-010-2490-x
67. Lowe A, Bills J, Verma R, Lavery L, Davis K, Balkus KJ (2015) Electrospun nitric oxide releasing bandage with enhanced wound healing. Acta Biomater 13:121–130. https://doi.org/10.1016/J.ACTBIO.2014.11.032. Elsevier
68. Frank S, Kämpfer H, Wetzler C, Pfeilschifter J (2002) Nitric oxide drives skin repair: novel functions of an established mediator. Kidney Int 61(3):882–888. https://doi.org/10.1046/J.1523-1755.2002.00237.X. Elsevier

69. Joseph CA, McCarthy CW, Tyo AG, Hubbard KR, Fisher HC, Altscheffel JA, He W et al (2019) Development of an injectable nitric oxide releasing poly(ethylene) glycol-fibrin adhesive hydrogel. ACS Biomater Sci Eng 5(2):959–969. https://doi.org/10.1021/acsbiomaterials.8b01331. American Chemical Society
70. Wan X, Wang Y, Jin X, Li P, Yuan J, Shen J (2018) Heparinized PCL/keratin mats for vascular tissue engineering scaffold with potential of catalytic nitric oxide generation. J Biomater Sci Polym Ed 29(14):1785–1798. https://doi.org/10.1080/09205063.2018.1504192. Taylor & Francis
71. Feng S, Zhao Y, Xian M, Wang Q (2015) Biological thiols-triggered hydrogen sulfide releasing microfibers for tissue engineering applications. Acta Biomater 27:205–213. https://doi.org/10.1016/j.actbio.2015.09.010. NIH Public Access
72. Raggio R, Bonani W, Callone E, Dirè S, Gambari L, Grassi F, Motta A (2018) Silk fibroin porous scaffolds loaded with a slow-releasing hydrogen sulfide agent (GYY4137) for applications of tissue engineering. ACS Biomater Sci Eng 4(8):2956–2966. https://doi.org/10.1021/acsbiomaterials.8b00212. American Chemical Society
73. Lee ZW, Zhou J, Chen C-S, Zhao Y, Tan C-H, Li L, Moore PK, Deng L-W (2011) The slow-releasing hydrogen sulfide donor, GYY4137, exhibits novel anti-cancer effects in vitro and in vivo. PLoS One 6(6):e21077. https://doi.org/10.1371/journal.pone.0021077. Edited by Joseph Alan Bauer. Public Library of Science
74. Bohlender C, Gläser S, Klein M, Weisser J, Thein S, Neugebauer U, Popp J, Wyrwa R, Schiller A (2014) Light-triggered CO release from nanoporous non-wovens. J Mater Chem B 2(11):1454–1463. https://doi.org/10.1039/C3TB21649G. Royal Society of Chemistry
75. Michael E, Abeyrathna N, Patel AV, Liao Y, Bashur CA (2016) Incorporation of photo-carbon monoxide releasing materials into electrospun scaffolds for vascular tissue engineering. Biomed Mater (Bristol, England) 11(2):025009. https://doi.org/10.1088/1748-6041/11/2/025009
76. Rajfer RA, Kilic A, Neviaser AS, Schulte LM, Hlaing SM, Landeros J, Ferrini MG, Ebramzadeh E, Park S-H (2017) Enhancement of fracture healing in the rat, modulated by compounds that stimulate inducible nitric oxide synthase: acceleration of fracture healing via inducible nitric oxide synthase. Bone Joint Res 6(2):90–97. https://doi.org/10.1302/2046-3758.62.BJR-2016-0164.R2. British Editorial Society of Bone and Joint Surgery
77. Lin S-K, Kok S-H, Kuo MY-P, Lee M-S, Wang C-C, Lan W-H, Hsiao M, Goldring SR, Hong C-Y (2003) Nitric oxide promotes infectious bone resorption by enhancing cytokine-stimulated interstitial collagenase synthesis in osteoblasts. J Bone Miner Res 18(1):39–46. https://doi.org/10.1359/jbmr.2003.18.1.39
78. Zheng Y, Liao F, Lin X, Zheng F, Fan J, Cui Q, Yang J, Geng B, Cai J (2017) Cystathionine γ-lyase-hydrogen sulfide induces runt-related transcription factor 2 sulfhydration, thereby increasing osteoblast activity to promote bone fracture healing. Antioxid Redox Signal 27(11):742–753. https://doi.org/10.1089/ars.2016.6826
79. Hikiji H, Shin WS, Oida S, Takato T, Koizumi T, Toyo-oka T (1997) Direct action of nitric oxide on osteoblastic differentiation. FEBS Lett 410(2–3):238–242. https://doi.org/10.1016/S0014-5793(97)00597-8. No longer published by Elsevier
80. Sonoda S, Mei Y-f, Atsuta I, Danjo A, Yamaza H, Hama S, Nishida K et al (2018) Exogenous nitric oxide stimulates the odontogenic differentiation of rat dental pulp stem cells. Sci Rep 8(1):3419. https://doi.org/10.1038/s41598-018-21183-6. Nature Publishing Group
81. Kalyanaraman H, Ramdani G, Joshua J, Schall N, Boss GR, Cory E, Sah RL, Casteel DE, Pilz RB (2017) A novel, direct NO donor regulates osteoblast and osteoclast functions and increases bone mass in ovariectomized mice. J Bone Miner Res 32(1):46–59. https://doi.org/10.1002/jbmr.2909. Wiley
82. Xu Z-S, Wang X-Y, Xiao D-M, Hu L-F, Lu M, Wu Z-Y, Bian J-S (2011) Hydrogen sulfide protects MC3T3-E1 osteoblastic cells against H2O2-induced oxidative damage—implications for the treatment of osteoporosis. Free Radic Biol Med 50(10):1314–1323. https://doi.org/10.1016/j.freeradbiomed.2011.02.016

83. Grassi F, Tyagi AM, Calvert JW, Gambari L, Walker LD, Yu M, Robinson J et al (2016) Hydrogen sulfide is a novel regulator of bone formation implicated in the bone loss induced by estrogen deficiency. J Bone Miner Res 31(5):949–963. https://doi.org/10.1002/jbmr.2757. NIH Public Access
84. Almubarak S, Nethercott H, Freeberg M, Beaudon C, Jha A, Jackson W, Marcucio R, Miclau T, Healy K, Bahney C (2016) Tissue engineering strategies for promoting vascularized bone regeneration. Bone 83:197–209. https://doi.org/10.1016/j.bone.2015.11.011. NIH Public Access
85. Volti GL, Sacerdoti D, Sangras B, Vanella A, Mezentsev A, Scapagnini G, Falck JR, Abraham NG (2005) Carbon monoxide signaling in promoting angiogenesis in human microvessel endothelial cells. Antioxid Redox Signal 7(5–6):704–710. https://doi.org/10.1089/ars.2005.7.704
86. Ziche M, Morbidelli L, Masini E, Amerini S, Granger HJ, Maggi CA, Geppetti P, Ledda F (1994) Nitric oxide mediates angiogenesis in vivo and endothelial cell growth and migration in vitro promoted by substance P. J Clin Invest 94(5):2036–2044. https://doi.org/10.1172/JCI117557
87. Priya MK, Sahu G, Soto-Pantoja DR, Goldy N, Sundaresan AM, Jadhav V, Barathkumar TR et al (2015) Tipping off endothelial tubes: nitric oxide drives tip cells. Angiogenesis 18(2):175–189. https://doi.org/10.1007/s10456-014-9455-0
88. Phillips PG, Birnby LM, Narendran A, Milonovich WL (2001) Nitric oxide modulates capillary formation at the endothelial cell-tumor cell interface. Am J Physiol Lung Cell Mol Physiol 281(1):L278–L290. https://doi.org/10.1152/ajplung.2001.281.1.L278. American Physiological Society, Bethesda, MD
89. Wang M-J, Cai W-J, Zhu Y-C (2010) Mechanisms of angiogenesis: role of hydrogen sulphide. Clin Exp Pharmacol Physiol 37(7):764–771. https://doi.org/10.1111/j.1440-1681.2010.05371.x. Wiley
90. Shantz S, Alan J, Yu Y-Y, Andres W, Miclau T, Marcucio R (2014) Modulation of macrophage activity during fracture repair has differential effects in young adult and elderly mice. J Orthop Trauma 28:S10–S14. https://doi.org/10.1097/BOT.0000000000000062
91. Xing Z, Lu C, Hu D, Yu Y-y, Wang X, Colnot C, Nakamura M, Wu Y, Miclau T, Marcucio RS (2010) Multiple roles for CCR2 during fracture healing. Dis Model Mech 3(7–8):451–458. https://doi.org/10.1242/dmm.003186

Carbon Nanotubes: Their Antimicrobial Properties and Applications in Bone Tissue Regeneration

Pei Wang, Malcolm Xing, and Bingyun Li

Abstract Carbon nanotubes (CNTs) possess unique physicochemical properties that are suitable for a variety of uses in manufacturing, environmental, and biomedical applications, as well as in the food and agriculture industries. The uses and manufacturing of CNTs are increasing due to advances in the large-scale production of CNTs and the development of various CNT-added composites. The antimicrobial activities and potential applications of CNTs and CNT-based materials as antimicrobials have recently attracted great interest. In this mini review, we first introduce the commonly seen nanotechnology-based antimicrobial materials, especially carbon-based nanomaterials such as CNTs, fullerenes, and graphene. Next, the history of the antimicrobial studies of CNTs is briefly described, followed by an overview of CNT toxicity. The antimicrobial properties of single-walled and multi-walled CNTs and the antimicrobial mechanisms of CNTs are summarized and discussed. The future challenges and potential applications of CNTs in bone tissue engineering and antimicrobial applications are provided.

Keywords Carbon nanotube · Antimicrobial · Antibacterial · Antibiotic · Infection Tissue engineering · Nanomaterials · Nanotechnology · Toxicity

---

P. Wang
Department of Materials Science and Engineering, Dalian Maritime University, Dalian, China

M. Xing
Department of Mechanical Engineering, University of Manitoba, Winnipeg, MB, Canada

B. Li (✉)
Nanomedicine Laboratory, Department of Orthopaedics, School of Medicine,
West Virginia University, Morgantown, WV, USA
e-mail: bili@hsc.wvu.edu; https://directory.hsc.wvu.edu/Profile/30485

B. Li et al. (eds.), *Racing for the Surface*,
https://doi.org/10.1007/978-3-030-34471-9_9

Introduction

Nanomaterials are defined as materials with at least one dimension in the size range of approximately 1–100 nm. In 1991, carbon nanotubes (CNTs) were reported by Sumio Iijima, whose work has stimulated an intense research interest in the area of nanomaterials [1]. Since then, CNTs have been found to exhibit not only high specific surface areas but also unique chemical, physical, and electronic properties, which make them potentially useful in many applications in the fields of energy, environment, electronics, and medicine, to name a few.

CNTs are structurally described as cylindrical nanostructures rolled with sheets of six-membered carbon atom rings (i.e., graphene). CNTs with a single cylinder of only one layer are known as single-walled CNTs (SWCNTs), and those with two or more layers are referred to as multi-walled CNTs (MWCNTs). In the biomaterial and biomedical fields, CNTs have been extensively studied as biomaterials for the diagnosis of diseases (e.g., in vivo imaging) and as platforms for disease treatment by immobilizing therapeutic agents (e.g., anticancer drugs and antimicrobial agents). One of the potential benefits of using CNTs is to make treatment and imaging more accurate, more feasible, and less invasive to living organisms compared to the use of conventional materials. Another benefit of CNT-based biomaterials is the ease of modifying the surfaces of CNTs with a broad scope of molecules due to the high reactivities of CNT surfaces [2–4].

With the rapid development of new implants and biomedical devices, developing antimicrobial surfaces on various implants and biomedical devices has attracted significant research interest. Antimicrobial surfaces could be designed to kill or at least inhibit the growth of microbial species (e.g., bacteria). Studies using CNTs have indicated that CNTs have antimicrobial properties; these properties may depend on their physicochemical properties. The antimicrobial properties can be tuned by their size, oxidation level, surface functionalization, and electronic structures [5, 6]. Besides CNTs, there are other types of nanomaterials that have also shown antimicrobial properties. These nanomaterials may include graphene, fullerene, dendrimer, nanocomposite, and metallic nanoparticles (e.g., nanogold, nanosilver, zinc oxide). The antimicrobial activities and potential adverse effects of these nanomaterials are listed and compared in Table 1. One can see that various nanomaterials have presented antimicrobial properties against a variety of pathogens including bacteria and fungi (Table 1). However, clinical applications of such nanomaterials are still very limited, and nanomaterials including CNTs have not yet been used clinically as broadly as was hoped (despite the dramatically increasing amount of research into biomaterial applications worldwide). Their limited success is likely due to the potential safety concerns of nanomaterials. The impacts of nanomaterials on human health and the environment should always be considered; however, these impacts have not been thoroughly studied [32–35]. Better understanding of the characteristics of nanomaterials, from their basic physicochemical properties to their therapeutic effectiveness within biological systems, is much needed. In this mini review, we mainly focus on CNTs, which are one of the most widely studied and applied nanomaterials. We not only emphasize the antimicrobial properties and toxicity of SWCNTs and MWCNTs but also discuss their antimicrobial mechanisms.

Table 1 Major nanomaterials as antimicrobial biomaterials

Nanomate-rial	Scheme	Antimicrobial property	Toxicity	References
SWCNT	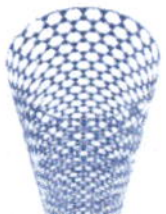	Presented antimicrobial activities against *S. aureus*, *E. coli*, *Candida tropicalis*, *K. pneumoniae*, and *Salmonella* species	Longer SWCNTs caused more inflammation, epithelioid granulomas, fibrosis in lungs, pulmonary blockage, and asphyxiation	[7–10]
MWCNT		Showed antimicrobial activities against *E. coli*, *S. aureus*, and *B. subtilis*, etc. However, MWCNTs presented fewer antimicrobial properties compared to SWCNTs	Thinner and shorter MWCNTs induced relatively higher toxicity	[11, 12]
Fullerene		Presented spectrum and light-activated antimicrobial properties against Gram-positive (e.g., *S. aureus*) and Gram-negative (e.g., *E. coli*) bacteria and fungi (e.g., *C. albicans*)	Fullerenes were toxic to human dermal fibroblasts, lung epithelial cells, and astrocytes	[13–16]
Graphene	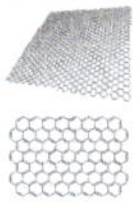	Presented antimicrobial properties against both Gram-positive (e.g., *S. aureus*) and Gram-negative (e.g., *E. coli*) bacteria	May elicit toxic effects both in vitro and in vivo toward human cells	[17, 18]
Dendrimer	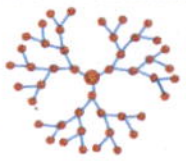	Phloroglucinol succinic acid dendrimers showed high antimicrobial properties against Gram-positive bacteria, and dendrimers like SB056 scaffold presented high antimicrobial activities against Gram-negative bacteria	Generated neuroinflammatory reactions and induced both cytotoxicity and utophagic flux in glioma cell lines	[19–21]
Nanocomposite		Nanocomposites like CNT/Ag presented antimicrobial activities against *S. aureus*, *K. pneumoniae*, *P. aeruginosa*, *A. baumannii*, and *P. mirabilis*	CNT/Ag nanocomposite caused damage in plasma membranes and mitochondria	[22, 23]
Metal nanoparticle		Ag, Cu, ZnO, Fe_2O_3, and TiO_2 presented antimicrobial properties. For instance, Ag nanoparticles had strong antimicrobial properties against *E. coli*, *S. aureus*, *E. faecalis*, *T. parceramosum*, *A. fumigatus*, *F. oxysporum*, *C. albicans*, *T. mentagrophytes*, and *P. variotii*	Metallic nanoparticles like Ag, Cu, etc. presented relatively high toxicity toward human cells	[24–31]

A. baumannii, *Acinetobacter baumannii*; *A. fumigatus*, *Aspergillus fumigatus*; *B. subtilis*, *Bacillus subtilis*; *C. albicans*, *Candida albicans*; *E. coli*, *Escherichia coli*; *E. faecalis*, *Enterococcus faecalis*; *F. oxysporum*, *Fusarium oxysporum*; *K. Pneumoniae*, *Klebsiella pneumoniae*; *P. aeruginosa*, *Pseudomonas aeruginosa*; *P. mirabilis*, *Proteus mirabilis*; *S. aureus*, *Staphylococcus aureus*; *T. mentagrophytes*, *Trichophyton mentagrophytes*; *T. parceramosum*, *Trichoderma parceramosum*; *P. variotii*, *Paecilomyces variotii*

Brief History of Antimicrobial Studies of CNTs

CNTs have been studied in many fields of science and technology since the first report of the helical microtubules of graphitic carbon in 1991 by Sumio Iijima [1]. As to their antimicrobial properties, one of the first studies came from Kang et al. of Yale University, who demonstrated that SWCNTs in aqueous solution had strong antimicrobial activities against *E. coli* [36]. The antimicrobial mechanism of CNTs was then examined in subsequent studies in 2008 [5], where they showed that *E. coli* underwent severe membrane damage from contact with SWCNTs and subsequently lost their viability. They believed that oxidative stress was one of the direct causes of the membrane damage observed, and the physical interaction of SWCNTs with the bacterial cells was believed to play a major role in antimicrobial mechanisms. The size of CNTs was a vital factor affecting their antimicrobial properties. They also found that SWCNTs had stronger antimicrobial properties against *E. coli* compared to MWCNTs [5].

Taking advantage of the ease in surface modification of CNTs, researchers have further produced various CNT-based antimicrobial nanomaterials. Researchers have explored the antimicrobial activities of surface functionalized MWCNTs and SWCNTs against various pathogens including Gram-positive and Gram-negative bacteria. For instance, CNT-based nanocomposites with metal and metal oxides have been developed, including CNT/Ag, CNT/TiO_2, CNT/SiO_2, CNT/Cu, CNT/ZnO, and CNT/SnO_2 nanocomposites; some have shown significant antimicrobial activities [37–41]. As an example, CNT/Ag nanocomposites have shown excellent antimicrobial properties against *S. aureus* and *E. coli* [39]. Overall, the studies of CNTs as an antimicrobial biomaterial have attracted increased attention; some examples are listed in Table 2.

Antimicrobial Properties of SWCNTs

SWCNTs were shown to present strong antimicrobial activities first by Kang et al. in 2007 [36]. In their study, the interaction of SWCNTs with *E. coli* K12 in an isotonic aqueous solution was examined, and it was found that the antimicrobial activities of SWCNTs were dependent on their interaction time with bacteria. After incubating for 30, 60, and 120 min, the percentage of bacterial cell viability loss on SWCNTs were $73.1 \pm 5.4\%$, $79.9 \pm 9.8\%$, and $87.6 \pm 4.7\%$, respectively. It was also found that direct contact between *E. coli* and SWCNTs was necessary for the inactivation of *E. coli*. Significantly, morphological changes were observed in SWCNT treated bacteria; *E. coli* incubated with SWCNTs lost cellular integrity while the controls maintained their outer membrane structure (Fig. 1).

Yang et al. investigated the effect of the length of SWCNTs on their antimicrobial activities [10]. Three different lengths of SWCNTs (<1 μm, 1–5 μm, and ~5 μm) were tested against *S. typhimurium*. They found that longer SWCNTs presented

Table 2 Examples of antimicrobial studies of CNTs

Year	CNT type	Bacteria studied	Key finding	References
2007	SWCNTs	*E. coli*	Highly purified SWCNTs exhibit strong antimicrobial activity against *E. coli*	[36]
2008	SWCNTs and MWCNTs	*E. coli*	Diameter of CNTs is a key factor of antimicrobial properties, cell membrane damage by direct contact is a main antimicrobial mechanism, and SWCNTs are much more toxic to bacteria than MWCNTs	[5]
2010	MWCNTs	*S. aureus*, *S. pyogenes*, *E. coli*, *S. typhimurium*	CNT/Ag nanocomposites present significant antimicrobial properties	[37]
2010	SWCNTs	*S. typhimurium*	At the same weight concentration, longer CNTs presented higher antimicrobial properties	[10]
2013	MWCNTs	*L. acidophilus*, *B. adolescentis*, *E. coli*, *E. faecalis*, *S. aureus*	CNTs have shown broad-spectrum antibacterial activities to human gut bacteria	[42]
2014	SWCNTs	*S. aureus*, *E. coli*, *C. tropicalis*	CNT/chitosan composite hydrogel exhibited increasing antimicrobial activities with increasing CNT concentration	[7]
2014	MWCNTs	*P. aeruginosa*, *S. typhimurium*, *K. pneumonia*, *E. coli*, *S. pneumonia*, *B. subtilis*, *B. cereus*, *S. aureus*	Functionalization of CNT surfaces changes the antimicrobial properties	[43]
2015	MWCNTs	*K. pneumonia*, *E. coli*	CNTs increased the antimicrobial activities of gelatin	[8]
2016	MWCNTs	*S. aureus*, *E. coli*	CNT/Ag composites presented excellent antibacterial activity, cell compatibility, and long-term stability	[39]
2016	MWCNTs	*E. coli*, *S. aureus*	With the addition of unsaturated phospholipids, MWCNTs may present antibacterial effects comparable to SWCNTs	[44]
2018	MWCNTs	*E. coli*	Conjugation of MWCNTs led to efficient photodynamic inactivation in *E. coli*	[45]
2019	MWCNTs	*S. aureus*, *S. mutans*, *C. albicans*	Significant anti-adhesive effects against oral microbial species without cytotoxicity to oral keratinocytes were observed for the 1% CNT group compared to the PMMA control group	[46]

B. adolescentis, *Bifidobacterium adolescentis*; *B. cereus*, *Bacillus cereus*; *C. tropicalis*, *Candida tropicalis*; *L. acidophilus*, *Lactobacillus acidophilus*; *PMMA* poly(methyl methacrylate); *S. mutans*, *Streptococcus mutans*; *S. pneumonia*, *Streptococcus pneumonia*; *S. pyogenes*, *Streptococcus pyogenes*; *S. typhimurium*, *Salmonella enterica serovar typhimurium*

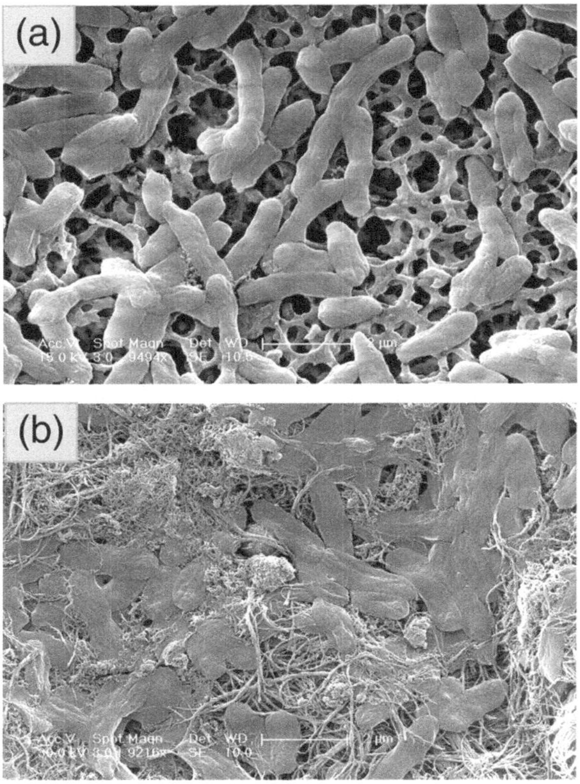

Fig. 1 Scanning electron microscopy (SEM) images of *E. coli*. (**a**) Cells incubated without SWCNTs for 60 min. Cells were filtered and observed via SEM on the filter. (**b**) Cells incubated with SWCNTs for 60 min. (Reprinted with permission from [36]. Copyright (2007) American Chemical Society)

stronger antimicrobial activities compared to shorter SWCNTs at the same weight concentration, and longer SWCNTs tended to aggregate more with bacteria while shorter SWCNTs aggregated less with bacterial cells (Fig. 2). Furthermore, compared to shorter SWCNTs, the antimicrobial activities of longer SWCNTs were more concentration- and treatment time-dependent [10].

Dong et al. also assessed the antimicrobial activities of SWCNTs and found that SWCNTs exhibited increasing antimicrobial activities against both *Salmonella enteric* (*S. enteric*) and *E. coli* with increasing SWCNT concentration. Overall, SWCNTs showed poor antimicrobial activities against *S. enteric* [47].

More recently, much attention has been paid to CNT derivatives and how to increase the antimicrobial activities of SWCNTs. Venkatesan et al. prepared chitosan-conjugated CNT hydrogels which showed strong antimicrobial activities against *S. aureus* and *Candida tropicalis* [7]. Chaudhari et al. found that pegylated Ag/SWCNT nanocomposites had good antimicrobial activities against *Salmonella* species based on findings from their growth curve assay and minimum inhibitory concentration (MIC) studies. These nanocomposites inhibited *S. typhimurium* at 62.5 μg/mL and were non-toxic to human cells [9]. In their further studies, Chaudhari et al. reported a novel method for covalently functionalizing SWCNT/Ag with an antimicrobial peptide (i.e., TP359—a lipidated cationic oligopeptide of myristoyl-

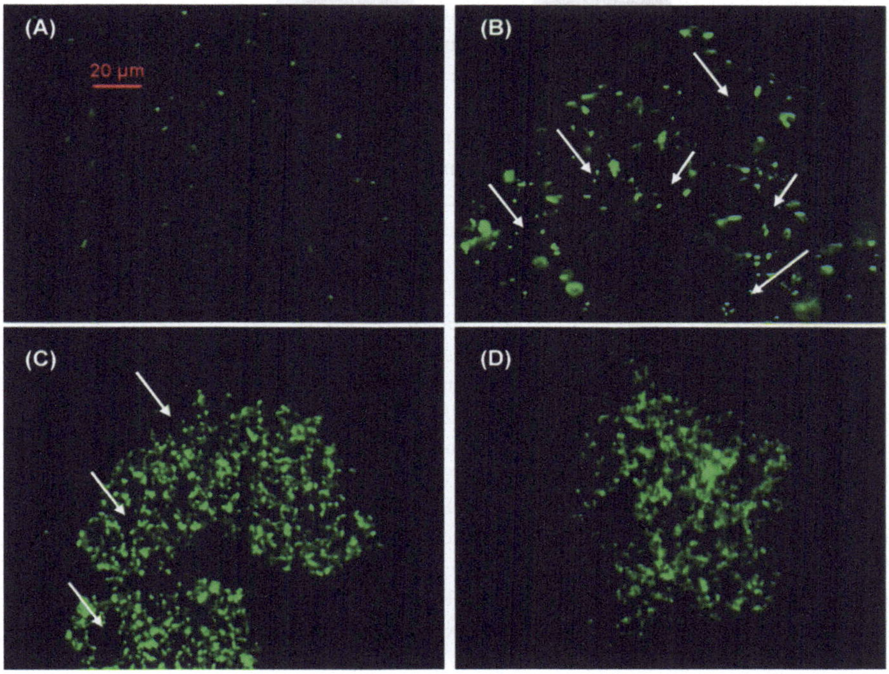

Fig. 2 Representative images of *Salmonella* cells in deionized water suspension (**A**) without SWCNTs, and the aggregates formed by cells and SWCNTs of different lengths (**B**) <1 μm, (**C**) 1–5 μm, and (**D**) ~5 μm. Cells were stained with green fluorescence. Black spots in the aggregates (indicated by arrows in B and C) were clusters of SWCNTs. (Reprinted with permission from [10]. Copyright (2010) American Chemical Society)

KKALKd amide). The new conjugate exhibited additive antimicrobial activities not only against two Gram-positive bacteria (i.e., *S. aureus* and *S. pyogenes*) but also against two Gram-negative bacteria (i.e., *S. typhimurium* and *E. coli*) [48]. Their data showed that the MIC of the new conjugate was lowered to 7.8–3.9 μg/mL with an IC50 of ~4–5 μg/mL from an MIC of 62.6 μg/mL and an IC50 of ~31–35 μg/mL for SWCNT/Ag. The new conjugate also showed reduced cellular toxicity toward murine macrophages and lung carcinoma cells [48].

Overall, SWCNTs and their composites have presented strong antimicrobial properties; those antimicrobial properties may vary with diameter, interaction time, etc.

Antimicrobial Properties of MWCNTs

MWCNTs have also been demonstrated to have antimicrobial properties. Surface functionalization has been widely utilized to improve their antimicrobial properties. For instance, Lohan et al. reported that nonionic surfactants like polysorbates (Tween 20, Tween 40, and Tween 80) could enhance the solubility of MWCNTs, and MWCNTs presented comparable antimicrobial efficacy as SWCNTs against *E. coli* and *S. aureus* upon the addition of unsaturated phospholipids [44]. Booshehri et al. synthesized a hollow fiber membrane by depositing Ag nanoparticle-conjugated MWCNTs on the external surfaces, and the resultant nanocomposite membranes exhibited significantly improved antimicrobial and antifouling properties [49]. In another study, MWCNT surfaces were functionalized with mono-, di-, and triethanolamine (MEA, DEA, and TEA) using a microwave method [43]. The antimicrobial performance of pristine and functionalized MWCNTs were tested with multiple bacterial species including four Gram-negative bacteria (*P. aeruginosa, S. typhimurium, K. pneumonia,* and *E. coli*) as well as four Gram-positive bacteria (*Streptococcus pneumonia, B. subtilis, Bacillus cereus,* and *S. aureus*). The data based on MIC and radial diffusion studies demonstrated that the type of functional group had a direct influences on their antimicrobial properties. The antimicrobial results were: MWCNT-TEA > MWCNT-DEA > MWCNT-MEA > pristine MWCNT. It was evident that the functional groups played a key role in antimicrobial activities of MWCNTs [43].

It was also shown that the length of MWCNTs may be important for their antimicrobial properties. The antimicrobial activities of MWCNTs were tested against *L. acidophilus, B. adolescentis, E. coli, E. faecalis,* and *S. aureus* in the human digestive system, and it was found that the shorter MWCNTs (s-MWCNTs) and hydroxyl modified MWCNTs (s-MWCNT-OH) were more toxic toward bacteria compared to the longer MWCNTs (l-MWCNTs) and s-MWCNT-COOH [42].

Spizzirria et al. also synthesized bioconjugates using gelatin, MWCNTs, and fluoroquinolones and demonstrated that the hybrid nanomaterials had greatly enhanced antimicrobial activities against *K. pneumoniae* and *E. coli* [8].

Therefore, it seems that both SWCNTs and MWCNTs have antimicrobial properties; the thin and rigid SWCNTs may have higher antimicrobial effectiveness compared to MWCNTs, probably due to their ability to better pierce the wall/membrane of spherical bacteria. However, MWCNTs may offer better biocompatibility and lower toxicity by functionalization or chemical modification [50, 51].

Toxicity of SWCNTs and MWCNTs Against Eukaryotic Cells

CNTs have emerged as one of the most widely used nanomaterials for biomedical applications, such as bioimaging and drug delivery carriers. The safety and biocompatibility of CNT products are therefore essential for their biomedical applications. In general, the physical interaction of CNTs with cells could cause membrane damage and cell death. Significant cytotoxicity of SWCNTs was observed in macrophages after 6 h of in vitro exposure, and SWCNTs significantly reduced the phagocytosis of macrophages at a dose as low as 0.38 $\mu g/cm^2$. Among the three types of carbon materials tested, the cytotoxicity followed a sequence order on a mass basis: SWCNT > MWCNT > fullerene (C60) [52]. In vitro studies have also shown that CNTs appeared to present much less cytotoxicity toward eukaryotic cells (e.g., human epithelial cells, T-cells) compared to prokaryotic cells [53, 54]. The in vivo toxicity studies of CNTs, currently mainly focused on skin-contacted, oral, and inhalational exposures, which are the likely routes for the exposure of CNTs to human beings, have shown that exposures to CNTs may lead to toxicity. Some studies suggested that the toxicity observed was probably attributed to the residual metals (from the synthesis process) on CNTs, which might result in elevated oxidative stress, thereby causing inflammation and toxicity, since there was no obvious increase of dermal cell number and skin thickening in mice treated with highly pure and clean CNTs [55, 56]. It was suggested that purification might be a good way to reduce the potential toxicity of CNTs.

Antimicrobial Mechanisms of CNTs

The antimicrobial effect of nanomaterials has been widely studied, and some mechanisms of the observed antimicrobial activities have been reported [57]. However, accurate antimicrobial mechanisms behind each nanomaterial are still not well-understood [58]. The antimicrobial mechanisms of CNTs are found to be associated with the physicochemical and structural characteristics (diameter, surface functional group, length, residual catalyst contamination, electronic structure, and electronic transfer) of CNTs and the shape of the bacteria. The diameter of CNTs is found to be a vital factor governing their antimicrobial capacities, and small-diameter SWCNTs (length ranging at 1–3 μm) displayed the strongest bacterial cytotoxicity compared to other CNTs with similar length. The antimicrobial mechanism is diameter-dependent piercing that leads to lysis of the cell wall and membrane. The length of the CNTs is another critical factor contributing to the antimicrobial activities of CNTs. In general, shorter CNTs were found to be more toxic to bacteria than longer CNTs, while long CNTs (>50 μm) may wrap around the surface of bacteria, thereby increasing contact area with the bacterial wall and promoting osmotic lysis of bacteria and cause lysis of the cell wall and membrane [42]. The antimicrobial activities of CNTs are therefore mainly associated with diameter-dependent piercing and length-dependent wrapping on the lysis of microbial membranes and subsequent release of intracellular components (e.g., DNA and RNA) and loss of bacterial membrane potential. Moreover, the high specific surface and chemical groups on the surfaces of CNTs may facilitate chemical and physical interactions, such as hydrogen-binding or electrostatic absorption, with the microbial membranes, disintegrate the microbial membranes, and eventually result in cell death. In addition, the shape of the bacteria may contribute to the antimicrobial activities of CNTs; CNTs were found to have less antimicrobial activities against rod-like bacteria as opposed to spherical ones [42].

The antimicrobial mechanisms of CNTs were summarized in three main phases, since it was demonstrated that the number of damaged bacteria and the extent of

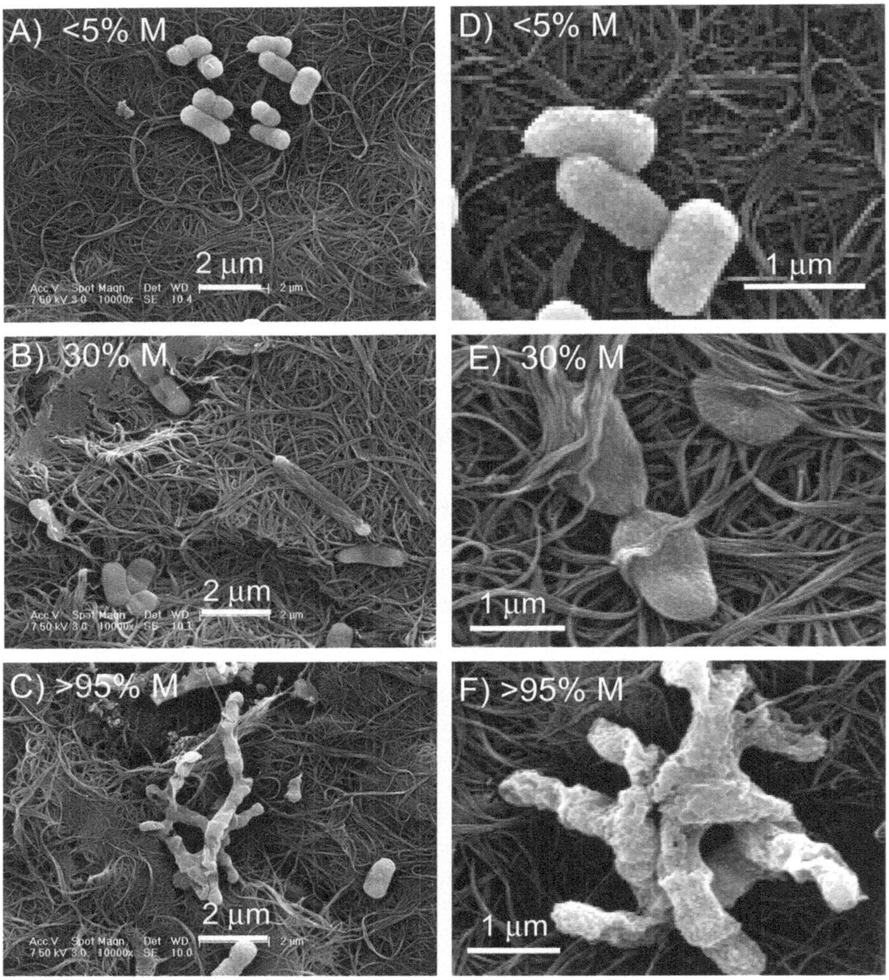

Fig. 3 Representative SEM images of *E. coli* deposited on SWCNT filters. *E. coli* was deposited on the SWCNT filter, incubated for 45 min in isotonic saline, and fixed with glutaraldehyde and osmium tetroxide prior to SEM imaging: (**A**, **D**) <5% metallic, (**B**, **E**) 30% metallic, and (**C**, **F**) >95% metallic. Note the differences in cell membrane hydration, structure, and roughness between the three samples. (Reprinted with permission from [6]. Copyright (2010) American Chemical Society)

damage to the bacteria depend on the electronic properties of CNTs (Fig. 3) [6]. The first phase includes physical contact and interaction between CNTs and bacteria. The second phase is the bacterial membrane perturbation due to the physical contact and interaction. The third phase is electronic structure-dependent bacterial oxidation. The direct cell contact with CNTs may cause cell membrane damage and subsequent cell inactivation and death [6].

Perspectives and Summary

Comparing CNTs with conventional antibiotics The mechanism of many conventional antibiotics, such as ofloxacin and ceftazidime, acts on specific targets within the microbes, for instance, causing the breakage of double-stranded DNA or disturbance of protein synthesis and blockage of cell division [59]. As a result, the morphology of the bacteria may be preserved, and consequently, the bacteria may develop resistance. Compared to these conventional antibiotics, the antimicrobial activities of CNTs are mainly mediated by their direct interactions and perturbation of cell walls and membranes; such physical damage to the cell walls and membranes may be less likely to induce the development of microbial resistance. As antibiotic resistance continues to spread worldwide [60], antimicrobial materials like CNTs may find unique applications as coatings or components of devices to reduce implant-associated infections.

Applying CNTs in bone tissue regeneration CNTs have not only been shown to have antimicrobial properties, they may also promote osteogenesis. CNTs were found to promote bone tissue formation in vivo [61] and suppress the differentiation of osteoclasts as well as expression of the transcription factor nuclear factor-kappa B (NFκB) in osteoclasts [29]. CNTs also served as seed materials for the crystallization of hydroxyapatite, the major component of bone, and CNTs could attract Ca ions and activate osteoblasts. Composites like CNT/polylactic acid could promote osteoblast proliferation, and CNT/polycarbonate urethane and CNT/poly lactic-co-glycolic acid enhanced the adhesion of osteoblasts [62–65]. Therefore, CNTs could promote osteogenesis and thereby the process of bone tissue regeneration. Considering both the antimicrobial and osteogenic properties, CNTs and CNT-based composites may become innovative materials to reduce two of the major complications (i.e., delayed bone healing and infection) associated with major injuries like trauma which involves implantation and potential contamination. In addition, composites of CNT/polyethylene, CNT/polyether ether ketone (PEEK), and CNT/ceramic may be developed as artificial joints. For instance, MWCNT-conjugated ultra-high-molecular-weight polyethylene (UHMWPE) is suitable as a sliding material for artificial joints with high wear resistance and low breakability. Combining CNTs with ceramics may increase fracture toughness and can transform ceramics into ideal, wear-free, and antifracture sliding parts for artificial joints.

Impacting human health and the environment Like most nanomaterials, CNTs are considered to have potential mobility and therefore may impact in and across air, water, soil, and biota. Once CNTs are released into the environment, they may induce unforeseen environmental impacts not only to human beings but also to ecosystems. For instance, SWCNTs were detected in fecal material collected from the digestive tract of fathead minnows, and clumps of SWCNTs were found on the gills [66]. MWCNTs were found to cause immune responses and presented reproductive toxicity [67]. CNTs may stay in the digestive tract of bottom consumers in the ecological pyramid, and could move up through the food chain as these bottom consumers (e.g., worms) are further consumed by benthivores [68, 69]. As for plants or plant cells, CNTs were found to inhibit the root elongation in tomatoes and to enhance root elongation of onions and cucumbers, and upon functionalization, CNTs inhibited root elongation in lettuce while the root growth of cabbages and carrots were not affected by CNTs (either functionalized or nonfunctionalized). Exposing tomato seeds to CNTs increased the germination percentage and enhanced the growth of seedlings; it was hypothesized that CNTs penetrated the seed coat and enhanced the amount of water uptake of seeds during the germination period. Due to the complexity of ecosystems, the microenvironments that CNTs are released into likely are more complicated than most lab experimental settings. The temperature, pH, and interactions with natural substances will likely change the surface chemistry. Therefore, the original surface chemistry of CNTs before entering ecosystems may not be important. In addition, it is important to standardize the protocols for analyzing the potential impacts of CNTs in ecosystems.

In summary, CNTs have displayed antimicrobial properties, which are associated with the physicochemical and structural characteristics of CNTs and the shape of the microbes. Their antimicrobial properties are mainly through diameter dependent piercing and length dependent wrapping that leads to lysis of the cell wall and membrane, and may be less likely to induce microbial resistance. Meanwhile, great effort and progress have been made to examine the in vitro and in vivo toxicity of CNTs. However, due to the complexity of ecosystems and expanding applications of nanomaterials, more should be done to investigate and understand the toxicological and environmental effects of direct and indirect exposures and uses of CNTs. Standardized experimental protocols and techniques are much needed to determine the impacts of CNTs (may be applied to other nanomaterials as well) on individual cells, organs, entire organisms, and ecosystems.

Acknowledgments This work is supported by the Office of the Assistant Secretary of Defense for Health Affairs, through the Peer Reviewed Medical Research Program, Discovery Awards under Award Nos. W81XWH1710603 and W81XWH1810203. We also acknowledge the financial support from WVU PSCoR and WVCTSI. Opinions, interpretations, conclusions, and recommendations are those of the authors and are not necessarily endorsed by the funding agencies. We thank Suzanne Danley for proofreading.

References

1. Iijima S (1991) Helical microtubules of graphitic carbon. Nature 354(6348):56–58
2. Rodriguez-Fernandez L, Valiente R, Gonzalez J, Villegas JC, Fanarraga ML (2012) Multiwalled carbon nanotubes display microtubule biomimetic properties in vivo, enhancing microtubule assembly and stabilization. ACS Nano 6(8):6614–6625
3. Lacerda L, Bianco A, Prato M, Kostarelos K (2006) Carbon nanotubes as nanomedicines: from toxicology to pharmacology. Adv Drug Deliv Rev 58(14):1460–1470
4. Pagona G, Tagmatarchis N (2006) Carbon nanotubes: materials for medicinal chemistry and biotechnological applications. Curr Med Chem 13(15):1789–1798
5. Kang S, Herzberg M, Rodrigues DF, Elimelech M (2008) Antibacterial effects of carbon nanotubes: size does matter! Langmuir 24(13):6409–6413

6. Vecitis CD, Zodrow KR, Kang S, Elimelech M (2010) Electronic-structure-dependent bacterial cytotoxicity of single-walled carbon nanotubes. ACS Nano 4(9):5471–5479
7. Venkatesan J, Jayakumar R, Mohandas A, Bhatnagar I, Kim SK (2014) Antimicrobial activity of chitosan-carbon nanotube hydrogels. Materials (Basel) 7(5):3946–3955
8. Spizzirri UG, Hampel S, Cirillo G, Mauro MV, Vittorio O, Cavalcanti P, Giraldi C, Curcio M, Picci N, Iemma F (2015) Functional gelatin-carbon nanotubes nanohybrids with enhanced antibacterial activity. Int J Polym Mater Polym Biomater 64(9):439–447
9. Chaudhari AA, Jasper SL, Dosunmu E, Miller ME, Arnold RD, Singh SR, Pillai S (2015) Novel pegylated silver coated carbon nanotubes kill Salmonella but they are non-toxic to eukaryotic cells. J Nanobiotechnology 13:23
10. Yang C, Mamouni J, Tang Y, Yang L (2010) Antimicrobial activity of single-walled carbon nanotubes: length effect. Langmuir 26(20):16013–16019
11. Azizian J, Hekmati M, Dadras OG (2014) Functionalization of carboxylated multiwall nanotubes with dapsone derivatives and study of their antibacterial activities against E. coli and S. aureus. Orient J Chem 30(2):667–673
12. Shahriary L, Nair R, Sabharwal S, Athawale AA (2015) One-step synthesis of Ag–reduced graphene oxide–multiwalled carbon nanotubes for enhanced antibacterial activities. New J Chem 39(6):4583–4590
13. Yang X, Ebrahimi A, Li J, Cui Q (2014) Fullerene-biomolecule conjugates and their biomedicinal applications. Int J Nanomedicine 9:77–92
14. Mizuno K, Zhiyentayev T, Huang L, Khalil S, Nasim F, Tegos GP, Gali H, Jahnke A, Wharton T, Hamblin MR (2011) Antimicrobial photodynamic therapy with functionalized fullerenes: quantitative structure-activity relationships. J Nanomed Nanotechnol 2(2):1–9
15. Mizuno T, Masuda Y, Irie K (2015) The Saccharomyces cerevisiae AMPK, Snf1, negatively regulates the Hog1 MAPK pathway in ER stress response. PLoS Genet 11(9):e1005491
16. Tegos GP, Demidova TN, Arcila-Lopez D, Lee H, Wharton T, Gali H, Hamblin MR (2005) Cationic fullerenes are effective and selective antimicrobial photosensitizers. Chem Biol 12(10):1127–1135
17. Maleki Dizaj S, Mennati A, Jafari S, Khezri K, Adibkia K (2015) Antimicrobial activity of carbon-based nanoparticles. Adv Pharm Bull 5(1):19–23
18. Guo X, Mei N (2014) Assessment of the toxic potential of graphene family nanomaterials. J Food Drug Anal 22(1):105–115
19. Scorciapino MA, Pirri G, Vargiu AV, Ruggerone P, Giuliani A, Casu M, Buerck J, Wadhwani P, Ulrich AS, Rinaldi AC (2012) A novel dendrimeric peptide with antimicrobial properties: structure-function analysis of SB056. Biophys J 102(5):1039–1048
20. Bertero A, Boni A, Gemmi M, Gagliardi M, Bifone A, Bardi G (2014) Surface functionalisation regulates polyamidoamine dendrimer toxicity on blood–brain barrier cells and the modulation of key inflammatory receptors on microglia. Nanotoxicology 8(2):158–168
21. Kumar MS, Karthikeyan S, Ramprasad C, Aruna PR, Mathivanan N, Velmurugan D, Ganesan S (2015) Investigation of phloroglucinol succinic acid dendrimer as antimicrobial agent against Staphylococcus Aureus, Escherichia Coli and Candida Albicans. Nano Biomed Eng 7(2):62–74
22. Yun H, Kim JD, Choi HC, Lee CW (2013) Antibacterial activity of CNT-Ag and GO-Ag nanocomposites against gram-negative and gram-positive bacteria. Bull Korean Chem Soc 34(11):3261–3264
23. Shankar S, Teng X, Rhim JW (2014) Properties and characterization of agar/CuNP bionanocomposite films prepared with different copper salts and reducing agents. Carbohydr Polym 114:484–492
24. Rai M, Yadav A, Gade A (2009) Silver nanoparticles as a new generation of antimicrobials. Biotechnol Adv 27(1):76–83
25. Chatterjee AK, Chakraborty R, Basu T (2014) Mechanism of antibacterial activity of copper nanoparticles. Nanotechnology 25(13):135101
26. Besinis A, De Peralta T, Handy RD (2014) The antibacterial effects of silver, titanium dioxide and silica dioxide nanoparticles compared to the dental disinfectant chlorhexidine on Streptococcus mutans using a suite of bioassays. Nanotoxicology 8(1):1–16

27. Azam A, Ahmed AS, Oves M, Khan MS, Habib SS, Memic A (2012) Antimicrobial activity of metal oxide nanoparticles against gram-positive and gram-negative bacteria: a comparative study. Int J Nanomedicine 7:6003–6009
28. Sui M, Zhang L, Sheng L, Huang S, She L (2013) Synthesis of ZnO coated multi-walled carbon nanotubes and their antibacterial activities. Sci Total Environ 452-453:148–154
29. Tran N, Mir A, Mallik D, Sinha A, Nayar S, Webster TJ (2010) Bactericidal effect of iron oxide nanoparticles on Staphylococcus aureus. Int J Nanomedicine 5:277–283
30. Brunet L, Lyon DY, Hotze EM, Alvarez PJ, Wiesner MR (2009) Comparative photoactivity and antibacterial properties of C60 fullerenes and titanium dioxide nanoparticles. Environ Sci Technol 43(12):4355–4360
31. Maness PC, Smolinski S, Blake DM, Huang Z, Wolfrum EJ, Jacoby WA (1999) Bactericidal activity of photocatalytic TiO(2) reaction: toward an understanding of its killing mechanism. Appl Environ Microbiol 65(9):4094–4098
32. Soto K, Garza KM, Murr LE (2007) Cytotoxic effects of aggregated nanomaterials. Acta Biomater 3(3):351–358
33. Wick P, Manser P, Limbach LK, Dettlaff-Weglikowska U, Krumeich F, Roth S, Stark WJ, Bruinink A (2007) The degree and kind of agglomeration affect carbon nanotube cytotoxicity. Toxicol Lett 168(2):121–131
34. Fraczek A, Menaszek E, Paluszkiewicz C, Blazewicz M (2008) Comparative in vivo biocompatibility study of single- and multi-wall carbon nanotubes. Acta Biomater 4(6):1593–1602
35. Jain KK (2012) Advances in use of functionalized carbon nanotubes for drug design and discovery. Expert Opin Drug Discov 7(11):1029–1037
36. Kang S, Pinault M, Pfefferle LD, Elimelech M (2007) Single-walled carbon nanotubes exhibit strong antimicrobial activity. Langmuir 23(17):8670–8673
37. Rangari VK, Mohammad GM, Jeelani S, Hundley A, Vig K, Singh SR, Pillai S (2010) Synthesis of Ag/CNT hybrid nanoparticles and fabrication of their nylon-6 polymer nanocomposite fibers for antimicrobial applications. Nanotechnology 21(9):095102
38. Afzal MA, Kalmodia S, Kesarwani P, Basu B, Balani K (2013) Bactericidal effect of silver-reinforced carbon nanotube and hydroxyapatite composites. J Biomater Appl 27(8):967–978
39. Nie C, Cheng C, Ma L, Deng J, Zhao C (2016) Mussel-inspired antibacterial and biocompatible silver-carbon nanotube composites: green and universal nanointerfacial functionalization. Langmuir 32(23):5955–5965
40. Sivaraj D, Vijayalakshmi K (2017) Preferential killing of bacterial cells by hybrid carbon nanotube-MnO2 nanocomposite synthesized by novel microwave assisted processing. Mater Sci Eng C Mater Biol Appl 81:469–477
41. Pandiyan R, Mahalingam S, Ahn YH (2019) Antibacterial and photocatalytic activity of hydrothermally synthesized SnO2 doped GO and CNT under visible light irradiation. J Photochem Photobiol B 191:18–25
42. Chen H, Wang B, Gao D, Guan M, Zheng L, Ouyang H, Chai Z, Zhao Y, Feng W (2013) Broad-spectrum antibacterial activity of carbon nanotubes to human gut bacteria. Small 9(16):2735–2746
43. Zardini HZ, Davarpanah M, Shanbedi M, Amiri A, Maghrebi M, Ebrahimi L (2014) Microbial toxicity of ethanolamines—multiwalled carbon nanotubes. J Biomed Mater Res A 102(6):1774–1781
44. Lohan S, Raza K, Singla S, Chhibber S, Wadhwa S, Katare OP, Kumar P, Singh B (2016) Studies on enhancement of anti-microbial activity of pristine MWCNTs against pathogens. AAPS PharmSciTech 17(5):1042–1048
45. Vt A, Paramanantham P, Sb SL, Sharan A, Alsaedi MH, Dawoud TMS, Asad S, Busi S (2018) Antimicrobial photodynamic activity of rose Bengal conjugated multi walled carbon nanotubes against planktonic cells and biofilm of Escherichia coli. Photodiagn Photodyn Ther 24:300–310
46. Kim KI, Kim DA, Patel KD, Shin US, Kim HW, Lee JH, Lee HH (2019) Carbon nanotube incorporation in PMMA to prevent microbial adhesion. Sci Rep 9(1):4921
47. Dong L, Henderson A, Field C (2012) Antimicrobial activity of single-walled carbon nanotubes suspended in different surfactants. J Nanotechnology 2012:7

48. Chaudhari AA, Ashmore D, Nath SD, Kate K, Dennis V, Singh SR, Owen DR, Palazzo C, Arnold RD, Miller ME, Pillai SR (2016) A novel covalent approach to bio-conjugate silver coated single walled carbon nanotubes with antimicrobial peptide. J Nanobiotechnology 14(1):58
49. Booshehri AY, Wang R, Xu R (2013) The effect of re-generable silver nanoparticles/multi-walled carbon nanotubes coating on the antibacterial performance of hollow fiber membrane. Chem Eng J 230:251–259
50. Zardini HZ, Amiri A, Shanbedi M, Maghrebi M, Baniadam M (2012) Enhanced antibacterial activity of amino acids-functionalized multi walled carbon nanotubes by a simple method. Colloids Surf B Biointerfaces 92:196–202
51. Qi XB, Gunawan P, Xu R, Chang MW (2012) Cefalexin-immobilized multi-walled carbon nanotubes show strong antimicrobial and anti-adhesion properties. Chem Eng Sci 84:552–556
52. Jia G, Wang H, Yan L, Wang X, Pei R, Yan T, Zhao Y, Guo X (2005) Cytotoxicity of carbon nanomaterials: single-wall nanotube, multi-wall nanotube, and fullerene. Environ Sci Technol 39(5):1378–1383
53. Alpatova AL, Shan W, Babica P, Upham BL, Rogensues AR, Masten SJ, Drown E, Mohanty AK, Alocilja EC, Tarabara VV (2010) Single-walled carbon nanotubes dispersed in aqueous media via non-covalent functionalization: effect of dispersant on the stability, cytotoxicity, and epigenetic toxicity of nanotube suspensions. Water Res 44(2):505–520
54. Fadel TR, Steenblock ER, Stern E, Li N, Wang X, Haller GL, Pfefferle LD, Fahmy TM (2008) Enhanced cellular activation with single walled carbon nanotube bundles presenting antibody stimuli. Nano Lett 8(7):2070–2076
55. Koyama S, Kim YA, Hayashi T, Takeuchi K, Fujii C, Kuroiwa N, Koyama H, Tsukahara T, Endo M (2009) In vivo immunological toxicity in mice of carbon nanotubes with impurities. Carbon 47(5):1365–1372
56. Murray AR, Kisin E, Leonard SS, Young SH, Kommineni C, Kagan VE, Castranova V, Shvedova AA (2009) Oxidative stress and inflammatory response in dermal toxicity of single-walled carbon nanotubes. Toxicology 257(3):161–171
57. Baranwal A, Srivastava A, Kumar P, Bajpai VK, Maurya PK, Chandra P (2018) Prospects of nanostructure materials and their composites as antimicrobial agents. Front Microbiol 9:422
58. Beyth N, Houri-Haddad Y, Domb A, Khan W, Hazan R (2015) Alternative antimicrobial approach: nano-antimicrobial materials. Evid Based Complement Alternat Med 2015:246012
59. Nederberg F, Zhang Y, Tan JP, Xu K, Wang H, Yang C, Gao S, Guo XD, Fukushima K, Li L, Hedrick JL, Yang YY (2011) Biodegradable nanostructures with selective lysis of microbial membranes. Nat Chem 3(5):409–414
60. Li B, Webster TJ (2018) Bacteria antibiotic resistance: new challenges and opportunities for implant-associated orthopedic infections. J Orthop Res 36(1):22–32
61. Saito N, Usui Y, Aoki K, Narita N, Shimizu M, Ogiwara N, Nakamura K, Ishigaki N, Kato H, Taruta S, Endo M (2008) Carbon nanotubes for biomaterials in contact with bone. Curr Med Chem 15(5):523–527
62. Supronowicz PR, Ajayan PM, Ullmann KR, Arulanandam BP, Metzger DW, Bizios R (2002) Novel current-conducting composite substrates for exposing osteoblasts to alternating current stimulation. J Biomed Mater Res 59(3):499–506
63. Bajaj P, Khang D, Webster TJ (2006) Control of spatial cell attachment on carbon nanofiber patterns on polycarbonate urethane. Int J Nanomedicine 1(3):361–365
64. Shi X, Hudson JL, Spicer PP, Tour JM, Krishnamoorti R, Mikos AG (2006) Injectable nanocomposites of single-walled carbon nanotubes and biodegradable polymers for bone tissue engineering. Biomacromolecules 7(7):2237–2242
65. Lin C, Wang Y, Lai Y, Yang W, Jiao F, Zhang H, Ye S, Zhang Q (2011) Incorporation of carboxylation multiwalled carbon nanotubes into biodegradable poly(lactic-co-glycolic acid) for bone tissue engineering. Colloids Surf B Biointerfaces 83(2):367–375
66. Helland A, Wick P, Koehler A, Schmid K, Som C (2007) Reviewing the environmental and human health knowledge base of carbon nanotubes. Environ Health Perspect 115(8):1125–1131

67. Cheng J, Chan CM, Veca LM, Poon WL, Chan PK, Qu L, Sun YP, Cheng SH (2009) Acute and long-term effects after single loading of functionalized multi-walled carbon nanotubes into zebrafish (Danio rerio). Toxicol Appl Pharmacol 235(2):216–225
68. Roberts AP, Mount AS, Seda B, Souther J, Qiao R, Lin S, Ke PC, Rao AM, Klaine SJ (2007) In vivo biomodification of lipid-coated carbon nanotubes by Daphnia magna. Environ Sci Technol 41(8):3025–3029
69. Ghafari P, St-Denis CH, Power ME, Jin X, Tsou V, Mandal HS, Bols NC, Tang XS (2008) Impact of carbon nanotubes on the ingestion and digestion of bacteria by ciliated protozoa. Nat Nanotechnol 3(6):347–351

Part II
Interface Tissue Engineering and Advanced Material for Scaffolds

Fracture Healing and Progress Towards Successful Repair

William A. Lackington and Keith Thompson

Abstract Despite the intrinsic healing capacity of bone and advancements in orthopedic technologies, well-established interventions, including autologous bone grafting, have had a relatively limited impact on easing the burden of a proportion of the 5–20% of long bone fracture patients who suffer from delayed healing or nonunion. In this chapter, we describe how the biology of bone development and bone homeostasis are recapitulated in bone healing, and how immunological and mechanical factors regulate healing. We present the current barriers faced clinically, outlining some of the main risk factors associated with the development of delayed healing and nonunion, with a focus on bone infection, and how it hijacks the bone healing process, ultimately leading to bone destruction. We conclude by depicting the outlook on fracture healing, outlining the progress to-date and the biggest challenges we face, while highlighting how our increasing understanding of the immunomodulation of bone healing can potentially be harnessed to develop innovative strategies for patient benefit.

Keywords Fracture healing · Mechanical factor · Immunological factor · Delayed healing · Nonunion · Risk factor

Introduction

Bone is a dynamic and highly vascularized tissue, which has the rare capacity to heal without the formation of a fibrotic scar [1]. Advancements in orthopedic technologies and methods of fracture fixation have led to high standards in the treatment and care of patients with fractures [2]. However, despite its intrinsic healing capacity and modern orthopedic fixation methods, a proportion of fractures exhibit delayed healing or result in nonunion. In the case of large bone defects, interventions such as

---

W. A. Lackington (✉) · K. Thompson
AO Research Institute Davos, Davos, Switzerland
e-mail: William.Lackington@aofoundation.org

B. Li et al. (eds.), *Racing for the Surface*,
https://doi.org/10.1007/978-3-030-34471-9_10

bone grafting are used to replace damaged and lost bone tissue, which remains the second most transplanted tissue after blood [2] with over 2.5 million bone grafting procedures taking place worldwide annually [3]. Complications including disturbed vascularization, soft-tissue damage, lack of adequate mechanical stability, and bacterial infections have all been identified as likely causative factors for impaired healing, although their specific contributions remain to be adequately addressed, while the observed rates of delayed healing or nonunion (10–20% of all cases) highlight that it remains a major clinical problem [4, 5]. Autologous bone grafting remains the clinical gold standard for the treatment of complex long bone defects despite the known constraints of donor site morbidity, limited tissue availability, and a reduction in the regenerative capacity of donor tissue with increasing donor age [3, 6–8]. Grafting alternatives have emerged in the form of tissue-engineered osteoinductive and osteoconductive biomaterials, paving the way for combinatorial treatment strategies that utilize modern methods of fracture fixation together with biomaterials to support bone healing. Furthermore, the crosstalk between immune cells and the biology of bone healing is now better understood than ever before [9]. Combining these developments may enable innovative solutions in the form of "immuno-informed" tissue-engineered biomaterials and fracture fixation technologies, which might elicit favorable immune responses upon implantation and thereby complement the intrinsic healing capacity of bone. In this chapter, we review our understanding of the process of fracture healing with a focus on the role of immunology, outlining our progress towards overcoming the barriers towards successful repair, which are currently faced clinically.

Origin of Bone

The processes of embryonic bone development and bone homeostasis in the adult are recapitulated, at least in part, in the process of bone healing after fracture. The intrinsic healing capacity of bone has evolved in parallel with the functionality of bone tissue [10]. Bone mechanically supports soft tissue, is a lever for the action of muscles, protects the central nervous system from trauma, regulates calcium levels in extracellular fluid, and houses and supports hematopoiesis [11]. Bone begins to form during the sixth to seventh week of embryonic development via two osteogenic pathways, namely intramembranous ossification, which gives rise to the flat bones of the cranial vault, including the cranial suture lines, some facial bones, and parts of the clavicle and mandible, and endochondral ossification, which gives rise to long bones and bones at the base of the skull [12, 13].

During early limb development, endochondral ossification (Fig. 1) is initiated at the limb bud with the condensation of mesenchymal stromal cells (MSCs) expressing collagen type II [14], which forms an anlage for individual bones in the endochondral skeleton [15]. MSCs undergo chondrogenic differentiation into chondrocytes while the mesenchyme located on the periphery forms the perichondrium [16]. Chondrocytes in the center of the cartilaginous template undergo hypertrophy and

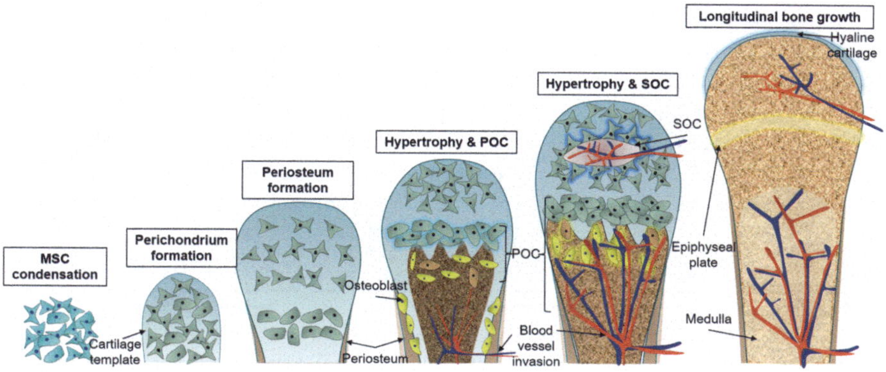

Fig. 1 Endochondral ossification (e.g., developing long bone). Schematic illustrating the phases of endochondral ossification, beginning with the condensation of MSCs and their chondrogenic differentiation to form an early cartilage template and perichondrium. Cells on the periphery undergo direct osteoblastic differentiation to form the perichondrium, while cells in the center proliferate rapidly and undergo hypertrophy, initiating mineralization of the cartilaginous matrix, which is then invaded by blood vessels, forming the primary ossification center (POC). The epiphyses are then invaded by blood vessels, forming a secondary ossification center (SOC), while the periphery maintains a stable cartilage phenotype, resulting in the formation of hyaline cartilage

begin to produce collagen type X, while cells in the periphery undergo direct osteoblastic differentiation to form an encircling bone collar [17, 18]. Hypertrophic cells then initiate bone synthesis by mineralizing the transient cartilaginous template, and the hypertrophic zone is invaded by blood vessels and an influx of cells, forming the primary ossification center [19]. The mineralized cartilage template is remodeled by osteoclasts, while osteoprogenitors differentiate into osteoblasts and lay down the osteoid of new bone. The developing epiphyses are then invaded by blood vessels, forming a secondary ossification center, while the periphery maintains a stable cartilage phenotype, resulting in hyaline articular cartilage surfaces seen within joints. The growth plate persists between primary and secondary ossification centers, propagating longitudinally to allow long bone growth before ossifying in early adulthood [12, 20].

Intramembranous ossification (Fig. 2) involves cells originating from the neural crest [21], which begins with the condensation of MSCs to form an ossification center, where they undergo direct osteoblastic differentiation [21]. These cells produce and secrete osteoid, which subsequently becomes calcified. Some osteoblasts become entrapped in this calcified matrix to become osteocytes. Bony spicules radiate out from the primary ossification center, while the entire region becomes surrounded by a compact layer of MSCs to form the periosteum. Cells on the inner surface of the periosteum also undergo osteoblastic differentiation and repeat the process, so that many layers of bone are formed. While fracture healing predominantly recapitulates endochondral ossification, intramembranous ossification also occurs subperiosteally, in both distal and proximal ends of the fracture to generate a hard callus from the periphery of the fracture towards the center of the fracture gap [1, 22]. The bridging

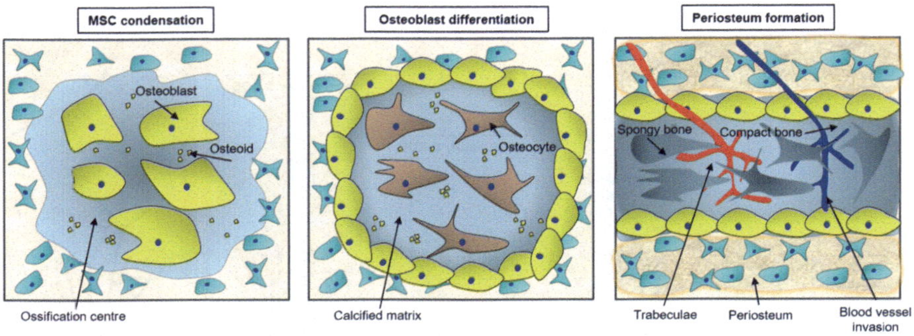

Fig. 2 Intramembranous ossification (e.g., developing calvaria). Schematic illustrating the phases of intramembranous ossification, which begins with the condensation of MSCs and the formation of ossification centers, where osteoblasts become entrapped in newly formed calcified matrix and become osteocytes. Subsequently, blood vessel invasion promotes surrounding osteoid to become calcified and the formation of trabeculae, while compact layers of MSCs on the surface of spongy bone become the periosteum, which in turn facilitates the formation of compact bone superficial to trabeculae

of this periosteal hard callus, which is a product of both endochondral and intramembranous ossification, ultimately provides fractures with a rigid structure to allow weight bearing—and is as such a hallmark of healing [23].

After fracture, long bones primarily heal following the route of endochondral ossification [1], which has led to an increase in the development of strategies aimed at recapitulating endochondral ossification in the field of tissue engineering using, for example, engineered cartilage as a template to promote bone formation [12, 24, 25].

Bone Healing: An Interplay Between Immunological and Mechanical Factors

Bone healing also recapitulates the process of bone remodeling. Remodeling is the process responsible for maintaining the general health and mechanical properties of bone tissue throughout the lifetime of an adult. Old or damaged bone is removed by bone resorbing osteoclasts while new bone matrix is produced by osteoblasts, allowing bone to withstand dynamic stress while repairing developing fatigue fractures [26]. In the healthy skeleton, a homeostatic balance between bone resorption and formation exists to maintain the function of bone throughout the lifetime of adults [11]. Homeostasis in bone remodeling is maintained by both immunological and mechanical factors.

Macrophages, an integral part of the innate immune system, have been shown to be a key facilitator of maintaining homeostasis in bone remodeling [27], not only by serving as the precursor to osteoclasts but also by coordinating osteoclast–osteoblast coupling and by serving as a cellular canopy over bone remodeling sites [28, 29]. Key signaling molecules responsible for mediating osteoclast–osteoblast coupling include the receptor activator of nuclear factor κB (RANK), RANK ligand (RANKL), and osteoprotegerin (OPG), which collectively form what is referred to as the RANK/RANKL/OPG axis [30]. Osteoblasts produce the transmembrane protein RANKL, which is responsible for inducing fusion of osteoclast progenitors into mature osteoclasts via binding to its receptor RANK on the surface of osteoclast progenitors [30]. OPG is a soluble decoy receptor, secreted by osteoblasts, responsible for maintaining a balance between bone resorption and formation. Adaptive immune cells, including B and T lymphocytes, can both positively and negatively influence bone homeostasis. For instance, T helper 17 (Th17) cells indirectly stimulate bone resorption through the production of interleukin 17 (IL-17), which stimulates RANKL expression on osteoblasts and stromal cells, and the synthesis of matrix-degrading enzymes [31]. Conversely, T helper 1 (Th1) and 2 (Th2) subsets of T lymphocytes have the capacity to inhibit osteoclastogenesis through their secretion of interferon gamma (IFN-γ) and interleukin 4 (IL-4), respectively [32]. Similarly, B lymphocytes can regulate bone homeostasis by producing OPG [30]. Taken together, it is clear that bone homeostasis is dependent on a complex interplay of factors produced by immune cells, which may, in part, be responsible for the increased risk of patients with chronic immune disorders, such as type 1 diabetes, to develop delayed healing and nonunion after fracture [33].

Mechanical factors are also important influencers of bone remodeling: for example, increased loading increases bone formation and decreases resorption, decreased loading decreases formation and increases resorption, while absolute immobilization stimulates resorption and halts formation [34]. The importance of mechanical regulation of bone remodeling is highlighted by Wolff's law, which states that the structure of bone will adapt to its mechanical usage [35] and can be aptly demonstrated in astronauts who lose bone mass after spending time in weightless environments due to reduced loading [36] and in tennis players who gain bone mass in their playing arm due to increased loading compared to their non-playing arm [37]. Given the importance of such immunological and mechanical factors in bone homeostasis, it is perhaps not surprising that these factors also play important roles during fracture healing. There are two main types of bone healing, namely primary and secondary fracture healing, which are dependent on the distance between the fractured bone ends, in addition to the mechanical stabilization of the fracture environment [38].

Primary Fracture Healing

Primary fracture healing, which seldom occurs, is characterized by minimal fracture gap and inter-fragmentary movement and can ensue either via contact healing or gap healing. Contact healing resembles bone remodeling, whereby macrophages play a key role in establishing osteoclast–osteoblast coupling to allow for resorption and subsequent ossification. In this process, osteoclasts generate longitudinal cavities perpendicular to the long axis, which are later filled by osteoblasts, resulting in bone formation in the correct axial direction [39]. However, contact healing typically occurs only if the displacement between bone ends is less than 0.01 mm and the interfragmentary strain is less than 2% [1, 40]. Gap healing takes place when similar inter-fragmentary stabilization is achieved; however, the fracture gap is larger than in contact healing, but typically less than 1 mm [1]. In this process, the gap is first filled with an intermediate of lamellar bone oriented perpendicular to the long axis, which is later remodeled by a process resembling contact healing [41].

Secondary Fracture Healing

Secondary fracture healing, which is more clinically relevant and applicable to large defects, follows well-defined, histologically and mechanically distinct phases, namely hematoma formation associated with an initial proinflammatory phase, followed by the formation of granulation tissue, callus formation and remodeling, which ultimately results in bone formation via endochondral and intramembranous ossification [40, 42, 43] (Fig. 3). It has been suggested that the goal of secondary

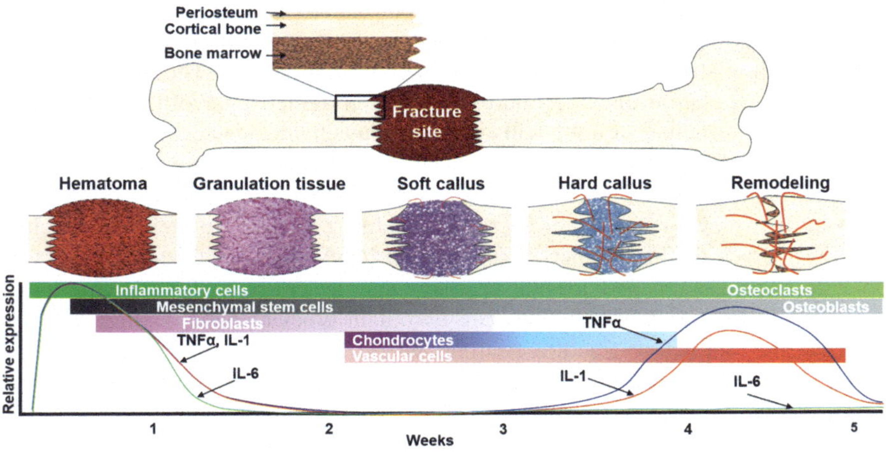

Fig. 3 Immunomodulation of fracture healing. Schematic illustrating the transient phases of fracture healing, which are progressively transformed from a proinflammatory hematoma to remodeled bone via a fibrocartilaginous intermediate, in part recapitulating both endochondral and intramembranous ossification. Temporal immunomodulation facilitates the smooth transition between phases of the healing cascade, orchestrating the influx of key cell types highlighted here with the relative expression pattern of some proinflammatory cytokines (TNFα, IL-1 and IL-6). Duality in cytokine functionality is also depicted with TNFα and IL-1, which are proinflammatory mediators initially, but later promote bone remodeling in the latter phase of bone healing. Remodeling can take significantly longer than shown here, particularly with larger injuries [44, 45]

fracture healing is to replace soft transient templates of bone tissue with more stable and rigid structures that allow weight bearing [41, 42].

The disruption of bone vasculature after fracture leads to the formation of a hematoma between bone fragments through activation of a plasma coagulation cascade and exposure of platelets to the extravascular environment, marking the beginning of a transient proinflammatory phase [40]. Rising importance has been given to the inflammatory phase of fracture healing as we shed more light on how the hematoma serves as the site where inflammatory cells can dock and control the expression of a temporally regulated cytokine pattern, which directs cell recruitment for subsequent stages of bone healing. As such, removal of the hematoma from fractures dramatically impacts on fracture healing, resulting in delayed healing. For example, in an ovine open tibial fracture model where the hematoma is removed in the first week post-injury, the quality of bone formation formed after 2 weeks is significantly reduced in comparison to undisturbed controls [46].

The fibrin-rich hematoma formed after fracture serves as the first transient matrix and docking site for the influx of inflammatory cells, mesenchymal cells, and endothelial cells which are attracted by resident macrophages, platelet-derived factors, complement fragments, and danger signals released from necrotic cells [26]. Among the inflammatory cells are neutrophils, which are the first responders to the fracture site [47]. While the complete role of neutrophils in bone healing has not been fully elucidated, it has been shown that neutrophils are responsible for recruiting a second wave of inflammatory cells, namely macrophages and T lymphocytes, through the secretion of proinflammatory and chemotactic mediators including IL-6 and monocyte chemoattractant protein-1 (MCP-1) [48]. Taken together, this group of inflammatory cells are responsible for initiating the subsequent stages in bone healing through the temporal regulation of cytokine patterns, which in many cases have bimodal functionality [49]. For example, TNFα (tumor necrosis factor α) is a potent proinflammatory cytokine, which is first produced by recruited inflammatory cells and resident macrophages. TNFα is now well regarded as a primary mediator of the proinflammatory phase within the hematoma, with its concentration peaking shortly after fracture (1–3 days) to promote MSC infiltration and proliferation [45, 50]. However, thereafter the concentration of TNFα drops for subsequent stages in bone healing until the remodeling phase where the level of TNFα is elevated again to facilitate osteoclast differentiation [45]. Consequently, cases where TNFα expression patterns are disturbed, particularly in the inflammatory phase, are those which are typically associated with delayed bone healing or nonunion [51]. Duality in cytokine functionality is not just specific to TNFα, for instance interleukin-1β (IL-1β) has a very similar bimodal expression pattern to TNFα [52]. Other examples of cytokine duality include IL-17, which is produced by Th17 cells and has both catabolic effects, by enhancing osteoclast-mediated bone resorption, and anabolic effects, by enhancing osteoblast-mediated bone formation [53]. While the initial proinflammatory fracture hematoma is critical for establishing the correct cytokine pattern to facilitate subsequent phases of bone healing, the effective "switching off" of the proinflammatory phase via anti-inflammatory mediators, such as IL-1 receptor antagonist (IL-1Ra) and IL-10, appears to be equally important to facilitate healing.

When the acute inflammation is cleared, a transient granulation tissue (7–14 days post-fracture) develops, whereby cells within the fracture hematoma gradually change the extracellular matrix into a proteoglycan and collagen-rich intermediate, while capillaries grow into the fracture site from endosteal circulation [54]. Facilitating angiogenesis is crucial during the formation of granulation tissue and later phases of bone healing. For instance, rats with femoral fractures treated with angiogenesis inhibitors fail to develop granulation tissue and exhibit minimal bone formation compared to control animals, which follow the typical healing process [55]. During the granulation tissue phase, recruited MSCs and fibroblasts are actively proliferating to prepare for the subsequent stages of healing where they will need to differentiate.

Soft callus formation is marked by chondrogenic differentiation of MSCs at the fracture site (2–3 weeks post-fracture). Chondrogenic differentiation is promoted by a combination of mechanical signals derived from micromotion provided by relative stability fixation techniques [56], the hypoxic microenvironment due to disrupted vasculature [57, 58], and macrophage-derived signals. Chondrogenic differentiation is induced and maintained by the coordinated expression of growth factors including transforming growth factor-β2 and -β3 (TGF-β2 and -β3), platelet-derived growth factor (PDGF), fibroblast growth factor-1 (FGF-1), and insulin-like growth factor (IGF) [57, 58]. Chondrocytes form a cartilaginous matrix rich in collagen type II and collagen type X, which serves as a scaffold for endochondral bone formation. As the soft callus develops with the help of fibroblasts to help pull the wound together and give it structure, intramembranous bone formation begins to take place in local areas that have improved blood supply, namely subperiosteally where periosteal stem cells differentiate directly into osteoblasts and form woven bone in both the distal and proximal ends of the fracture while advancing towards the fracture gap [22, 53]. The advancing bone front ultimately surrounds the external surface of the cartilaginous matrix, providing some degree of mechanical stability to the soft callus [59]. Initially, the soft callus matrix remains largely avascular to promote enough cartilaginous template for endochondral ossification [60]; however, as healing proceeds, the callus is invaded by endothelial cells, which promote vascularization into the fracture site [61]. Vascularization, stimulated by pro-angiogenic factors including vascular endothelial growth factor (VEGF), bone morphogenic proteins (BMPs), FGF-1, and TGF-β [62], promotes hypertrophy and the mineralization of the cartilaginous matrix, marking the end of the soft callus phase and beginning of the hard callus phase [63].

Hard callus formation recapitulates the events that occur in the secondary ossification center during long bone development whereby chondrocytes undergo hypertrophy and begin to calcify the cartilaginous matrix [64]. Concomitant with revascularization of the fracture site, osteoprogenitor cells, stimulated by osteogenic factors including BMPs secreted by MSCs [65], differentiate into osteoblasts, which facilitate the transition of the cartilaginous scaffold into a transient woven bone matrix. The exact source of osteoprogenitor cells remains ambiguous. Periosteal stem cells have recently been identified as the cell niche responsible for mediating intramembranous ossification subperiosteally [53], while bone marrow MSCs have been known to contribute only to a limited amount of direct osteoblastic differentiation [66]. The hypothesis that osteoprogenitors originate from multiple sources including vasculature and surrounding local tissue stem cell niches [67, 68] is supported since a hard callus can also form, albeit to a limited extent, in the absence of MSCs and periosteum. The recent discovery of the periosteal stem cell [53] suggests that bone contains multiple resident stem cell niches, each with individual specialized functions.

In the final phase of bone healing, the irregular woven bone in the hard callus is remodeled into cortical and trabecular bone in a process that can take several months or even years to complete. Osteoclasts adhere to mineralized surfaces and, using a combination of proteinases and acid, are capable of degrading organic components such as collagen and demineralizing the matrix [69]. Bone resorption creates pits known as Howship's lacunae, which can be identified histologically, where osteoblasts, guided by macrophages, are able to deposit new bone [63]. Together with the aforementioned production of RANKL [70, 71], osteoblasts may also regulate osteoclast function via the production of macrophage colony-stimulating factor (M-CSF), which stimulates the differentiation of hematopoietic stem cells into osteoclast precursors [72]. Ultimately, remodeling can restore the original structure and function of the bone, completing the process of fracture healing.

Current Barriers to Successful Bone Healing

Given the multifactorial pathophysiology of fractures, a multitude of risk factors make it more likely that delayed healing and nonunions might develop. The United States Food and Drug Administration (FDA) defines a nonunion as a fracture that has not healed within 9 months of injury and shows poor progression of healing radiographically between months 6 and 9 [73]. However, the variable pathophysiology of fractures has also made it difficult to select the criteria that define a nonunion clinically, with citations ranging between 2 and 12 months [74]. Risk factors, including patient-related, fracture-related, and trauma-related, pose barriers to successful bone healing, which need to be overcome using innovative therapies that complement the intrinsic healing capacity of bone.

Patient-Related Risk Factors

Three of the most prevalent patient-related risk factors for impaired bone healing are diabetes mellitus, nonsteroidal anti-inflammatory drugs (NSAIDs), and smoking. Other patient-related risk factors including vitamin D deficiency [75], thyroid imbalance [76], hyperparathyroidism [77], and increasing age [78] are not covered in this review.

Diabetes mellitus was classically thought as a metabolic disease with high blood glucose levels [79], resulting from deficits in insulin production (type 1) caused by the autoimmune-mediated destruction of insulin-producing β-cells in the pancreas [80] or by a resistance to insulin (type 2) [81]. More recently, however, type 1 diabetes is increasingly being considered as an inflammatory disease characterized by dysregulated inflammation [82]. During type 1 diabetes, proinflammatory cytokines including IL-1β, IL-6, IL-18, and TNFα are significantly upregulated [83], and this inflammatory state appears resilient towards attempts to downregulate this inflammation once it has been induced [84]. Therefore, it is perhaps no surprise that the bone healing process, which is heavily influenced by proinflammatory mediators, is perturbed in patients with diabetes. Specifically, enhanced inflammation, and the inability to successfully resolve it due to diabetes, increases osteoclastogenesis during fracture healing, significantly increasing the likelihood of nonunion or delayed healing [84, 85].

NSAIDs, which inhibit the enzymes cyclo-oxygenase 1 and 2 (COX)-1/2 to varying extents depending on their chemical structure, are widely used drugs typically used to treat pain after surgery, including after fracture repair. However, usage of NSAIDs, including readily available drugs such as ibuprofen and aspirin, has been shown to be associated with an increased likelihood of developing fracture healing complications [86]. Prostaglandin E2 (PGE-2) is the most abundant prostaglandin in bone and plays a role via binding to its receptor, E prostanoid receptor 2 (EP2R), in the stimulation of bone formation, and in bone resorption via binding to EP4R [87]. NSAIDs can also lead to PGE-2 inhibition [88], impairing endochondral ossification, specifically limiting hypertrophy and bone deposition in both in vitro and ex vivo models [89]. However, the current clinical evidence is not sufficient to warrant discontinuation of all NSAIDs in all contexts of bone fracture and rehabilitation protocols but will certainly benefit from a greater number of randomized trials. For example, the association between nonunion after long bone fracture and duration of NSAID usage was recently assessed in several studies, but only one of these used a randomized controlled trial design [90–92]. Having used NSAIDs for 90 days postoperatively, the findings from the clinical studies suggest that NSAIDs have a detrimental effect on fracture healing. While the effects of NSAIDs are beneficial for pain management, it seems that their detrimental effect on fracture healing might be dependent on their relative use [86, 93]. Thus, further prospective randomized studies are required to fully elucidate the effects of short-term and long-term NSAID use, as well as cumulative doses, on fracture healing, and perhaps to find a balance between benefitting from the pain management aspect of NSAIDs without significantly impairing fracture healing.

A further important patient-related risk factor is smoking status. A recent meta-analysis, which sampled 40 studies incorporating over 8000 adults identified that smokers take 27.7 days longer (14.2–21.3) for union to occur after fracture and that smokers have greater than double (1.9–2.6) the risk of developing nonunion compared to non-smokers [94]. Nicotine and carbon monoxide are two constituents that particularly affect fracture healing. Nicotine decreases blood flow to the extremities due to increased peripheral vasoconstriction [95], reduces microvascular perfusion [96], and increases blood viscosity and fibrinogen levels, which in turn increases the potential risk of microvascular clotting [97]. Additionally, nicotine directly damages osteoblasts and macrophages [98]. Carbon monoxide, with its 200-fold greater affinity for hemoglobin binding than oxygen, greatly reduces oxygen tension in tissues [99], exacerbating the nicotine-induced inhibitory effects on perfusion. Taken together, it is perhaps no surprise why smoking is such a significant risk factor for bone-healing complications. To minimize this risk, smoking cessation perioperatively is highly recommended [100]. The data here is categorically undebatable, with bone healing rates increasing in patients who give up smoking, particularly those who give up smoking for longer than 6 months postoperatively [101]. However, while these benefits are dependent on the length of smoking cessation, they also likely depend on lifetime smoking duration of the patient.

Fracture-Related Risk Factors

Fracture-related factors are dependent on the characteristics of the injury, including the location of the fracture, the extent of bone loss, the pattern of bone injury, and the condition of the soft tissue envelope surrounding the fracture. Several anatomical positions have been reported to have increased risk of nonunion. For example, comminution and poor interfragmentary cortical apposition in clavicle fractures have been associated with increased risk of nonunion [102]. Some locations within a single bone might also have a higher risk of nonunion. The poor blood supply associated with the distal tibia, the metadiaphyseal region of the fifth metatarsal, the tarsal navicular body and the scaphoid waist put these regions at higher risk of nonunion compared to other parts within the same bone [103–105]. Even though the exact quantity of bone loss required to develop nonunion has not been defined, the concept of a critical sized defect is often used, and thus, the extent of bone loss is also a significant risk factor for the development of nonunion. While these risk factors for the development of nonunion are inherently inevitable, there are other fracture-related risk factors that arise from fracture management and can thus be addressed.

The risk of nonunion might also be elevated through inadequate fracture management despite the high standards set by modern fracture stabilization techniques. When fixation strategies are used, excessive stripping of the periosteum might compromise native periosteal stem cell niches and dampen the fracture-healing capacity. Fractures that are not stabilized appropriately might also develop atrophic or hypertrophic nonunion. Inappropriately rigid stabilization with insufficient interfragmentary movement might inhibit bone growth leading to atrophic nonunion, while too much micromotion and interfragmentary strain can lead to large amounts of connective tissue being formed, resulting in a hypertrophic nonunion [106].

Trauma-Related Risk Factors

Concomitant with the severity of the fracture is the extent of the trauma-induced damage to the surrounding soft tissue. Maintaining the health of the surrounding tissue envelope aims to preserve the blood supply for fracture healing while extensive damage might limit revascularization during the bone healing process. Another trauma-related risk factor for nonunion is infection, presenting one of the biggest clinical challenges of the twenty-first century modern trauma medicine.

Osteomyelitis is an infectious disease that triggers inflammation, caused primarily by *Staphylococcus aureus* and *Staphylococcus epidermidis*, which often leads to bone destruction and bone loss [107, 108]. Infection is predominantly caused via open fractures, where there is a breach of the skin during the injury itself, permitting microorganisms to enter the wound and to colonize the bone tissue. A much reduced, although not insignificant, risk of infection occurs during surgical procedures themselves, for example, with prosthetic joint replacements or implantation of fracture fixation devices, where the surface of implants themselves are at potential risk of bacterial colonization.

In the general population, the incidence of bone infection after fracture can vary between 1.8 and 27% depending on the fracture type (closed vs. open) and location; however, with lower extremity open fractures, e.g., the tibia, demonstrating the highest incidence and being most affected [109–111]. Osteomyelitis also has an incidence rate of up to 2.4% in total hip arthroplasties and up to 3% for total knee arthroplasties [112–114]. In subpopulations with predispositions to infection, including patients with underlying disease such as diabetes or peripheral vascular disease, the incidence of osteomyelitis can be significantly greater [115]. The source of infection can be either contiguous, where osteomyelitis originates from trauma, direct inoculation during surgery, and surrounding infectious tissues, or hematogenous, where osteomyelitis arises from existing infection in another part of the body and is facilitated access to the fracture site via the circulating blood. In adults, 80% of osteomyelitis cases are contiguous, while in children the source is predominantly hematogenous [116, 117].

The pathogenesis of osteomyelitis follows targeting of bone healing processes and is mediated, in part, by microbial surface components recognizing adhesive matrix molecules (MSCRAMMs) and, in part, by the toxins they produce. Infection begins with colonization, the attachment of *Staphylococcus* to the surface bone or the surface of implants, once coated with host plasma proteins. This 'race for the surface' is mediated by the presence of MSCRAMMs, such as protein A (SpA) or fibronectin and collagen binding protein (FnBP A/B) which interact with bone cells, the extracellular matrix (ECM) and plasma proteins. Attachment of staphylococci to the surface of bone or implant facilitates biofilm formation, which are colonies of microorganisms enveloped in ECM that allow the infection to persist during treatment [118]. MSCRAMMs may also be secreted; for example, SpA can bind directly to osteoblasts, mediate cell death, and inhibit bone formation [119–121]. FnBPs can mediate internalization via the osteoblast integrin receptor $\alpha5\beta1$ (the fibronectin receptor) [122, 123], which can lead to apoptosis of the cell via binding to TNF-related apoptosis-inducing ligand (TRAIL), and activation of IL-6, IL-12, and CSF, which further exacerbate bone loss by enhancing inflammation (or impairing healing) [124, 125]. *Staphylococcus* can also persist intracellularly to evade the immune system [126] and even reside internally within hematopoietic cells, hijacking osteoclastogenesis to further the effects of bone resorption [127, 128]. During osteomyelitis, many toxins are also produced which negatively impact the bone healing process. For instance, *S. aureus* produces toxic shock syndrome toxin 1 (TSST-1), coagulase, Panton-Valentine leucocidin (PVL), hemolysins (Hla), and phenol-soluble modulins (PSMs) [129]. Through an unknown mechanism, TSST-1 mediates immune evasion and is also a mediator of osteoclast activation while not being directly cytotoxic towards them, resulting in increased bone resorption [130]. Hla, which lyses red blood cells, typically serves as an antigen for the innate immune system to detect; however, in osteomyelitis it is downregulated, contributing to the quiescence of bone infection, allowing the infection to evade the immune system [127]. The production of coagulase, which converts fibrinogen to fibrin, provides *S. aureus* with a physical shield against the innate immune system [131]. In mouse models, PVL has been shown to be responsible for the spreading of osteomyelitis [132], while PSMs contribute to the severity of infection. Taken together, these mechanisms allow *Staphylococcus*-induced osteomyelitis to prolong infection and evade the immune system while the natural processes of bone healing are hijacked, leading to bone destruction and bone loss.

Conclusion

Bone attempts to self-heal in response to injury by recapitulating the biology of bone development and bone homeostasis. Specifically, an acute proinflammatory hematoma is established for the docking of immunomodulatory cells, which set up highly regulated transient cytokine patterns to facilitate the transformation of the fracture hematoma to remodeled bone via a fibrocartilaginous intermediate. Despite its intrinsic healing capacity and modern orthopedic fixation methods to provide mechanical stability, large bone defects do not always heal successfully, which might result in delayed healing and nonunion. Barriers to successful healing, which arise from patient-, fracture-, and trauma-related risk factors, can be minimized to increase the likelihood of healing. However, bone infection is still a major clinical burden, exasperating patients with fractures due to its capacity to hijack and impact the bone healing process, and an alarming clinical concern due to the emerging prevalence of antibiotic resistance. Concomitant with our increasing understanding of the immunomodulation of bone healing, the development of novel biomaterials to serve in place of bone autografts may also permit the local delivery of immunomodulators and/or antibiotics, thus paving the way for innovative therapeutic strategies aimed at restoring a pro-healing environment in patient populations at increased risk of healing complications.

References

1. Marsell R, Einhorn TA (2011) The biology of fracture healing. Injury 42:551–555
2. Shegarfi H, Reikeras O (2009) Review article: bone transplantation and immune response. J Orthop Surg (Hong Kong) 17:206–211
3. Wang W, Yeung KWK (2017) Bone grafts and biomaterials substitutes for bone defect repair: a review. Bioact Mater 2:224–247
4. Haas NP (2000) Callus modulation—fiction or reality? Chirurg 71:987–988
5. Winkler T, Sass FA, Duda GN, Schmidt-Bleek K (2018) A review of biomaterials in bone defect healing, remaining shortcomings and future opportunities for bone tissue engineering: the unsolved challenge. Bone Joint Res 7:232–243
6. Grabowski G, Cornett CA (2013) Bone graft and bone graft substitutes in spine surgery: current concepts and controversies. J Am Acad Orthop Surg 21:51–60
7. Flierl MA, Smith WR, Mauffrey C, Irgit K, Williams AE, Ross E, Peacher G, Hak DJ, Stahel PF (2013) Outcomes and complication rates of different bone grafting modalities in long bone fracture nonunions: a retrospective cohort study in 182 patients. J Orthop Surg Res 8:33
8. Betz RR (2002) Limitations of autograft and allograft: new synthetic solutions. Orthopedics 25:s561–s570
9. Sridharan R, Reilly RB, Buckley CT (2015) Decellularized grafts with axially aligned channels for peripheral nerve regeneration. J Mech Behav Biomed Mater 41:124–135
10. Hirasawa T, Kuratani S (2015) Evolution of the vertebrate skeleton: morphology, embryology, and development. Zoological Lett 1:2. eCollection 2015
11. Rodan GA (1998) Bone homeostasis. Proc Natl Acad Sci U S A 95:13361–13362
12. Thompson EM, Matsiko A, Farrell E, Kelly DJ, O'Brien FJ (2015) Recapitulating endochondral ossification: a promising route to in vivo bone regeneration. J Tissue Eng Regen Med 9:889–902
13. Karaplis AC (2008) Chapter 3—Embryonic development of bone and regulation of intramembranous and endochondral bone formation. In: Bilezikian J, Raisz LG, Martin TJ (eds) Principles of bone biology. Academic Press, San Diego, pp 53–84
14. Nah HD, Rodgers BJ, Kulyk WM, Kream BE, Kosher RA, Upholt WB (1988) In situ hybridization analysis of the expression of the type II collagen gene in the developing chicken limb bud. Coll Relat Res 8:277–294
15. Hall BK, Miyake T (1992) The membranous skeleton: the role of cell condensations in vertebrate skeletogenesis. Anat Embryol (Berl) 186:107–124
16. Kosher RA, Kulyk WM, Gay SW (1986) Collagen gene expression during limb cartilage differentiation. J Cell Biol 102:1151–1156
17. Sila-Asna M, Bunyaratvej A, Maeda S, Kitaguchi H, Bunyaratavej N (2007) Osteoblast differentiation and bone formation gene expression in strontium-inducing bone marrow mesenchymal stem cell. Kobe J Med Sci 53:25–35
18. Schmid TM, Linsenmayer TF (1985) Immunohistochemical localization of short chain cartilage collagen (type X) in avian tissues. J Cell Biol 100:598–605
19. Liu Z, Xu J, Colvin JS, Ornitz DM (2002) Coordination of chondrogenesis and osteogenesis by fibroblast growth factor 18. Genes Dev 16:859–869
20. Ornitz DM, Marie PJ (2002) FGF signaling pathways in endochondral and intramembranous bone development and human genetic disease. Genes Dev 16:1446–1465
21. Couly GF, Coltey PM, Le Douarin NM (1993) The triple origin of skull in higher vertebrates: a study in quail-chick chimeras. Development 117:409–429
22. Perren SM (1991) The concept of biological plating using the limited contact-dynamic compression plate (LC-DCP). Scientific background, design and application. Injury 22 Suppl 1:1–41
23. Uthgenannt BA, Kramer MH, Hwu JA, Wopenka B, Silva MJ (2007) Skeletal self-repair: stress fracture healing by rapid formation and densification of woven bone. J Bone Miner Res 22:1548–1556

24. Matsiko A, Thompson EM, Lloyd-Griffith C, Cunniffe GM, Vinardell T, Gleeson JP, Kelly DJ, O'Brien FJ (2018) An endochondral ossification approach to early stage bone repair: use of tissue-engineered hypertrophic cartilage constructs as primordial templates for weight-bearing bone repair. J Tissue Eng Regen Med 12:e2147–e2150
25. Sheehy EJ, Mesallati T, Kelly L, Vinardell T, Buckley CT, Kelly DJ (2015) Tissue engineering whole bones through endochondral ossification: regenerating the distal phalanx. Biores Open Access 4:229–241
26. Loi F, Cordova LA, Pajarinen J, Lin TH, Yao Z, Goodman SB (2016) Inflammation, fracture and bone repair. Bone 86:119–130
27. Fujiwara N, Kobayashi K (2005) Macrophages in inflammation. Curr Drug Targets Inflamm Allergy 4:281–286
28. Cho SW (2015) Role of osteal macrophages in bone metabolism. J Pathol Transl Med 49:102–104
29. Lassus J, Salo J, Jiranek WA, Santavirta S, Nevalainen J, Matucci-Cerinic M, Horak P, Konttinen Y (1998) Macrophage activation results in bone resorption. Clin Orthop Relat Res 352:7–15
30. Pacifici R (2013) Osteoimmunology and its implications for transplantation. Am J Transplant 13:2245–2254
31. J C-L, H C, J.E F (2009) Osteoimmunology—the hidden immune regulation of bone. Autoimmun Rev 8:250–255
32. Mori G, D'Amelio P, Faccio R, Brunetti G (2013) The interplay between the bone and the immune system. Clin Dev Immunol 2013:720504
33. Retzepi M, Donos N (2010) The effect of diabetes mellitus on osseous healing. Clin Oral Implants Res 21:673–681
34. Rodan GA (1997) Bone mass homeostasis and bisphosphonate action. Bone 20:1–4
35. Frost HM (1994) Wolff's law and bone's structural adaptations to mechanical usage: an overview for clinicians. Angle Orthod 64:175–188
36. Robling AG, Turner CH (2009) Mechanical signaling for bone modeling and remodeling. Crit Rev Eukaryot Gene Expr 19:319–338
37. Jones HH, Priest JD, Hayes WC, Tichenor CC, Nagel DA (1977) Humeral hypertrophy in response to exercise. J Bone Joint Surg Am 59:204–208
38. Jagodzinski M, Krettek C (2007) Effect of mechanical stability on fracture healing—an update. Injury 38(Suppl 1):S3–S10
39. Kaderly RE (1991) Primary bone healing. Semin Vet Med Surg 6:21–25
40. Claes L, Recknagel S, Ignatius A (2012) Fracture healing under healthy and inflammatory conditions. Nat Rev Rheumatol 8:133–143
41. Shapiro F (1988) Cortical bone repair. The relationship of the lacunar-canalicular system and intercellular gap junctions to the repair process. J Bone Joint Surg Am 70:1067–1081
42. Perren SM (2002) Evolution of the internal fixation of long bone fractures. The scientific basis of biological internal fixation: choosing a new balance between stability and biology. J Bone Joint Surg Br 84:1093–1110
43. Morshed S (2014) Current options for determining fracture union. Adv Med 2014:708574
44. Cho TJ, Gerstenfeld LC, Einhorn TA (2002) Differential temporal expression of members of the transforming growth factor beta superfamily during murine fracture healing. J Bone Miner Res 17:513–520
45. Kon T, Cho TJ, Aizawa T, Yamazaki M, Nooh N, Graves D, Gerstenfeld LC, Einhorn TA (2001) Expression of osteoprotegerin, receptor activator of NF-kappaB ligand (osteoprotegerin ligand) and related proinflammatory cytokines during fracture healing. J Bone Miner Res 16:1004–1014
46. Schell H, Duda GN, Peters A, Tsitsilonis S, Johnson KA, Schmidt-Bleek K (2017) The haematoma and its role in bone healing. J Exp Orthop 4:5. Epub 2017 Feb 7
47. Bastian O, Pillay J, Alblas J, Leenen L, Koenderman L, Blokhuis T (2011) Systemic inflammation and fracture healing. J Leukoc Biol 89:669–673

48. Xing Z, Lu C, Hu D, Yu YY, Wang X, Colnot C, Nakamura M, Wu Y, Miclau T, Marcucio RS (2010) Multiple roles for CCR2 during fracture healing. Dis Model Mech 3:451–458
49. Hoff P, Maschmeyer P, Gaber T, Schutze T, Raue T, Schmidt-Bleek K, Dziurla R, Schellmann S, Lohanatha FL, Rohner E, Ode A, Burmester GR, Duda GN, Perka C, Buttgereit F (2013) Human immune cells' behavior and survival under bioenergetically restricted conditions in an in vitro fracture hematoma model. Cell Mol Immunol 10:151–158
50. Lu Z, Wang G, Dunstan CR, Zreiqat H (2012) Short-term exposure to tumor necrosis factor-alpha enables human osteoblasts to direct adipose tissue-derived mesenchymal stem cells into osteogenic differentiation. Stem Cells Dev 21:2420–2429
51. Karnes JM, Daffner SD, Watkins CM (2015) Multiple roles of tumor necrosis factor-alpha in fracture healing. Bone 78:87–93
52. Einhorn TA, Majeska RJ, Rush EB, Levine PM, Horowitz MC (1995) The expression of cytokine activity by fracture callus. J Bone Miner Res 10:1272–1281
53. Debnath S, Yallowitz AR, McCormick J, Lalani S, Zhang T, Xu R, Li N, Liu Y, Yang YS, Eiseman M, Shim J, Hameed M, Healey JH, Bostrom MP, Landau DA, Greenblatt MB (2018) Discovery of a periosteal stem cell mediating intramembranous bone formation. Nature 562:133–139
54. Singer AJ, Clark RA (1999) Cutaneous wound healing. N Engl J Med 341:738–746
55. Hausman MR, Schaffler MB, Majeska RJ (2001) Prevention of fracture healing in rats by an inhibitor of angiogenesis. Bone 29:560–564
56. Ghiasi MS, Chen J, Vaziri A, Rodriguez EK, Nazarian A (2017) Bone fracture healing in mechanobiological modeling: a review of principles and methods. Bone Rep 6:87–100
57. Melnyk M, Henke T, Claes L, Augat P (2008) Revascularisation during fracture healing with soft tissue injury. Arch Orthop Trauma Surg 128:1159–1165
58. Grundnes O, Reikeras O (1992) Blood flow and mechanical properties of healing bone. Femoral osteotomies studied in rats. Acta Orthop Scand 63:487–491
59. Bielby R, Jones E, McGonagle D (2007) The role of mesenchymal stem cells in maintenance and repair of bone. Injury 38(Suppl 1):S26–S32
60. Petersen A, Princ A, Korus G, Ellinghaus A, Leemhuis H, Herrera A, Klaumunzer A, Schreivogel S, Woloszyk A, Schmidt-Bleek K, Geissler S, Heschel I, Duda GN (2018) A biomaterial with a channel-like pore architecture induces endochondral healing of bone defects. Nat Commun 9:4430
61. Carano RA, Filvaroff EH (2003) Angiogenesis and bone repair. Drug Discov Today 8:980–989
62. Deckers MM, van Bezooijen RL, van der Horst G, Hoogendam J, van Der Bent C, Papapoulos SE, Lowik CW (2002) Bone morphogenetic proteins stimulate angiogenesis through osteoblast-derived vascular endothelial growth factor A. Endocrinology 143:1545–1553
63. Schindeler A, McDonald MM, Bokko P, Little DG (2008) Bone remodeling during fracture repair: the cellular picture. Semin Cell Dev Biol 19:459–466
64. Tsiridis E, Upadhyay N, Giannoudis P (2007) Molecular aspects of fracture healing: which are the important molecules? Injury 38(Suppl 1):S11–S25
65. Clines GA (2010) Prospects for osteoprogenitor stem cells in fracture repair and osteoporosis. Curr Opin Organ Transplant 15:73–78
66. Colnot C, Huang S, Helms J (2006) Analyzing the cellular contribution of bone marrow to fracture healing using bone marrow transplantation in mice. Biochem Biophys Res Commun 350:557–561
67. Rumi MN, Deol GS, Singapuri KP, Pellegrini VD Jr (2005) The origin of osteoprogenitor cells responsible for heterotopic ossification following hip surgery: an animal model in the rabbit. J Orthop Res 23:34–40
68. Collett GD, Canfield AE (2005) Angiogenesis and pericytes in the initiation of ectopic calcification. Circ Res 96:930–938
69. Feng X, Teitelbaum SL (2013) Osteoclasts: new insights. Bone Res 1:11–26
70. Boyce BF, Xing L (2008) Functions of RANKL/RANK/OPG in bone modeling and remodeling. Arch Biochem Biophys 473:139–146

71. Teitelbaum SL (2000) Bone resorption by osteoclasts. Science 289:1504–1508
72. Kodama H, Nose M, Niida S, Yamasaki A (1991) Essential role of macrophage colony-stimulating factor in the osteoclast differentiation supported by stromal cells. J Exp Med 173:1291–1294
73. Bishop JA, Palanca AA, Bellino MJ, Lowenberg DW (2012) Assessment of compromised fracture healing. J Am Acad Orthop Surg 20:273–282
74. Bhandari M, Guyatt GH, Swiontkowski MF, Tornetta P 3rd, Sprague S, Schemitsch EH (2002) A lack of consensus in the assessment of fracture healing among orthopaedic surgeons. J Orthop Trauma 16:562–566
75. Priemel M, von Domarus C, Klatte TO, Kessler S, Schlie J, Meier S, Proksch N, Pastor F, Netter C, Streichert T, Puschel K, Amling M (2010) Bone mineralization defects and vitamin D deficiency: histomorphometric analysis of iliac crest bone biopsies and circulating 25-hydroxyvitamin D in 675 patients. J Bone Miner Res 25:305–312
76. Feitosa Dda S, Bezerra Bde B, Ambrosano GM, Nociti FH, Casati MZ, Sallum EA, de Toledo S (2008) Thyroid hormones may influence cortical bone healing around titanium implants: a histometric study in rats. J Periodontol 79:881–887
77. Duarte PM, Cesar Neto JB, Goncalves PF, Sallum EA, Nociti J (2003) Estrogen deficiency affects bone healing around titanium implants: a histometric study in rats. Implant Dent 12:340–346
78. Zura R, Kaste SC, Heffernan MJ, Accousti WK, Gargiulo D, Wang Z, Steen RG (2018) Risk factors for nonunion of bone fracture in pediatric patients: an inception cohort study of 237,033 fractures. Medicine (Baltimore) 97:e11691
79. Yan W, Li X (2013) Impact of diabetes and its treatments on skeletal diseases. Front Med 7:81–90
80. Gong Z, Muzumdar RH (2012) Pancreatic function, type 2 diabetes, and metabolism in aging. Int J Endocrinol 2012:320482
81. Moseley KF (2012) Type 2 diabetes and bone fractures. Curr Opin Endocrinol Diabetes Obes 19:128–135
82. Donath MY, Shoelson SE (2011) Type 2 diabetes as an inflammatory disease. Nat Rev Immunol 11:98–107
83. Graves DT, Kayal RA (2008) Diabetic complications and dysregulated innate immunity. Front Biosci 13:1227–1239
84. Pacios S, Kang J, Galicia J, Gluck K, Patel H, Ovaydi-Mandel A, Petrov S, Alawi F, Graves DT (2012) Diabetes aggravates periodontitis by limiting repair through enhanced inflammation. FASEB J 26:1423–1430
85. Alblowi J, Tian C, Siqueira MF, Kayal RA, McKenzie E, Behl Y, Gerstenfeld L, Einhorn TA, Graves DT (2013) Chemokine expression is upregulated in chondrocytes in diabetic fracture healing. Bone 53:294–300
86. Hernandez RK, Do TP, Critchlow CW, Dent RE, Jick SS (2012) Patient-related risk factors for fracture-healing complications in the United Kingdom general practice research database. Acta Orthop 83:653–660
87. Blackwell KA, Raisz LG, Pilbeam CC (2010) Prostaglandins in bone: bad cop, good cop? Trends Endocrinol Metab 21:294–301
88. Nasrallah R, Laneuville O, Ferguson S, Hebert RL (2001) Effect of COX-2 inhibitor NS-398 on expression of PGE2 receptor subtypes in M-1 mouse CCD cells. Am J Physiol Renal Physiol 281:F123–F132
89. Welting TJ, Caron MM, Emans PJ, Janssen MP, Sanen K, Coolsen MM, Voss L, Surtel DA, Cremers A, Voncken JW, van Rhijn LW (2011) Inhibition of cyclooxygenase-2 impacts chondrocyte hypertrophic differentiation during endochondral ossification. Eur Cell Mater 22:420–436; discussion 436–7
90. Burd TA, Hughes MS, Anglen JO (2003) Heterotopic ossification prophylaxis with indomethacin increases the risk of long-bone nonunion. J Bone Joint Surg Br 85:700–705
91. Bhattacharyya T, Levin R, Vrahas MS, Solomon DH (2005) Nonsteroidal antiinflammatory drugs and nonunion of humeral shaft fractures. Arthritis Rheum 53:364–367

92. Giannoudis PV, MacDonald DA, Matthews SJ, Smith RM, Furlong AJ, De Boer P (2000) Nonunion of the femoral diaphysis. The influence of reaming and non-steroidal anti-inflammatory drugs. J Bone Joint Surg Br 82:655–658
93. Dodwell ER, Latorre JG, Parisini E, Zwettler E, Chandra D, Mulpuri K, Snyder B (2010) NSAID exposure and risk of nonunion: a meta-analysis of case-control and cohort studies. Calcif Tissue Int 87:193–202
94. Pearson RG, Clement RG, Edwards KL, Scammell BE (2016) Do smokers have greater risk of delayed and non-union after fracture, osteotomy and arthrodesis? A systematic review with meta-analysis. BMJ Open 6:e010303
95. Bornmyr S, Svensson H (1991) Thermography and laser-Doppler flowmetry for monitoring changes in finger skin blood flow upon cigarette smoking. Clin Physiol 11:135–141
96. Mosely LH, Finseth F (1977) Cigarette smoking: impairment of digital blood flow and wound healing in the hand. Hand 9:97–101
97. Sorensen LT, Jorgensen S, Petersen LJ, Hemmingsen U, Bulow J, Loft S, Gottrup F (2009) Acute effects of nicotine and smoking on blood flow, tissue oxygen, and aerobe metabolism of the skin and subcutis. J Surg Res 152:224–230
98. Fang MA, Frost PJ, Iida-Klein A, Hahn TJ (1991) Effects of nicotine on cellular function in UMR 106-01 osteoblast-like cells. Bone 12:283–286
99. Leow YH, Maibach HI (1998) Cigarette smoking, cutaneous vasculature, and tissue oxygen. Clin Dermatol 16:579–584
100. Glassman SD, Anagnost SC, Parker A, Burke D, Johnson JR, Dimar JR (2000) The effect of cigarette smoking and smoking cessation on spinal fusion. Spine (Phila Pa 1976) 25:2608–2615
101. Andersen T, Christensen FB, Laursen M, Hoy K, Hansen ES, Bunger C (2001) Smoking as a predictor of negative outcome in lumbar spinal fusion. Spine (Phila Pa 1976) 26:2623–2628
102. Robinson CM, Court-Brown CM, McQueen MM, Wakefield AE (2004) Estimating the risk of nonunion following nonoperative treatment of a clavicular fracture. J Bone Joint Surg Am 86-A:1359–1365
103. Rosenberg GA, Sferra JJ (2000) Treatment strategies for acute fractures and nonunions of the proximal fifth metatarsal. J Am Acad Orthop Surg 8:332–338
104. Kozin SH (2001) Incidence, mechanism, and natural history of scaphoid fractures. Hand Clin 17:515–524
105. DiGiovanni CW (2004) Fractures of the navicular. Foot Ankle Clin 9:25–63
106. Claes LE, Heigele CA, Neidlinger-Wilke C, Kaspar D, Seidl W, Margevicius KJ, Augat P (1998) Effects of mechanical factors on the fracture healing process. Clin Orthop Relat Res (355 Suppl):S132–47, 355S
107. Mouzopoulos G, Kanakaris NK, Kontakis G, Obakponovwe O, Townsend R, Giannoudis PV (2011) Management of bone infections in adults: the surgeon's and microbiologist's perspectives. Injury 42(Suppl 5):S18–S23
108. Lowy FD, Hammer SM (1983) Staphylococcus epidermidis infections. Ann Intern Med 99:834–839
109. Court-Brown CM, Keating JF, McQueen MM (1992) Infection after intramedullary nailing of the tibia. Incidence and protocol for management. J Bone Joint Surg Br 74:770–774
110. Roussignol X, Sigonney G, Potage D, Etienne M, Duparc F, Dujardin F (2015) Secondary nailing after external fixation for tibial shaft fracture: risk factors for union and infection. A 55 case series. Orthop Traumatol Surg Res 101:89–92
111. Chen AT, Vallier HA (2016) Noncontiguous and open fractures of the lower extremity: epidemiology, complications, and unplanned procedures. Injury 47:742–747
112. Kurtz SM, Lau E, Watson H, Schmier JK, Parvizi J (2012) Economic burden of periprosthetic joint infection in the United States. J Arthroplasty 27:61–5.e1
113. Bozic KJ, Kamath AF, Ong K, Lau E, Kurtz S, Chan V, Vail TP, Rubash H, Berry DJ (2015) Comparative epidemiology of revision arthroplasty: failed THA poses greater clinical and economic burdens than failed TKA. Clin Orthop Relat Res 473:2131–2138
114. Martinez-Pastor JC, Macule-Beneyto F, Suso-Vergara S (2013) Acute infection in total knee arthroplasty: diagnosis and treatment. Open Orthop J 7:197–204

115. Calhoun JH, Manring MM, Shirtliff M (2009) Osteomyelitis of the long bones. Semin Plast Surg 23:59–72
116. Oryan A, Alidadi S, Moshiri A, Maffulli N (2014) Bone regenerative medicine: classic options, novel strategies, and future directions. J Orthop Surg Res 9:18
117. Brady RA, Leid JG, Calhoun JH, Costerton JW, Shirtliff ME (2008) Osteomyelitis and the role of biofilms in chronic infection. FEMS Immunol Med Microbiol 52:13–22
118. Stewart PS, Costerton JW (2001) Antibiotic resistance of bacteria in biofilms. Lancet 358:135–138
119. Claro T, Widaa A, O'Seaghdha M, Miajlovic H, Foster TJ, O'Brien FJ, Kerrigan SW (2011) Staphylococcus aureus protein A binds to osteoblasts and triggers signals that weaken bone in osteomyelitis. PLoS One 6:e18748
120. Widaa A, Claro T, Foster TJ, O'Brien FJ, Kerrigan SW (2012) Staphylococcus aureus protein A plays a critical role in mediating bone destruction and bone loss in osteomyelitis. PLoS One 7:e40586
121. Mendoza Bertelli A, Delpino MV, Lattar S, Giai C, Llana MN, Sanjuan N, Cassat JE, Sordelli D, Gomez MI (2016) Staphylococcus aureus protein A enhances osteoclastogenesis via TNFR1 and EGFR signaling. Biochim Biophys Acta 1862:1975–1983
122. Shinji H, Yosizawa Y, Tajima A, Iwase T, Sugimoto S, Seki K, Mizunoe Y (2011) Role of fibronectin-binding proteins A and B in in vitro cellular infections and in vivo septic infections by Staphylococcus aureus. Infect Immun 79:2215–2223
123. Ahmed S, Meghji S, Williams RJ, Henderson B, Brock JH, Nair SP (2001) Staphylococcus aureus fibronectin binding proteins are essential for internalization by osteoblasts but do not account for differences in intracellular levels of bacteria. Infect Immun 69:2872–2877
124. Bost KL, Bento JL, Ellington JK, Marriott I, Hudson MC (2000) Induction of colony-stimulating factor expression following Staphylococcus or Salmonella interaction with mouse or human osteoblasts. Infect Immun 68:5075–5083
125. Bost KL, Ramp WK, Nicholson NC, Bento JL, Marriott I, Hudson MC (1999) Staphylococcus aureus infection of mouse or human osteoblasts induces high levels of interleukin-6 and interleukin-12 production. J Infect Dis 180:1912–1920
126. Mahalingam D, Szegezdi E, Keane M, de Jong S, Samali A (2009) TRAIL receptor signalling and modulation: are we on the right TRAIL? Cancer Treat Rev 35:280–288
127. Trouillet-Assant S, Lelievre L, Martins-Simoes P, Gonzaga L, Tasse J, Valour F, Rasigade JP, Vandenesch F, Muniz Guedes RL, Ribeiro de Vasconcelos AT, Caillon J, Lustig S, Ferry T, Jacqueline C, Loss de Morais G, Laurent F (2016) Adaptive processes of Staphylococcus aureus isolates during the progression from acute to chronic bone and joint infections in patients. Cell Microbiol 18:1405–1414
128. Trouillet-Assant S, Gallet M, Nauroy P, Rasigade JP, Flammier S, Parroche P, Marvel J, Ferry T, Vandenesch F, Jurdic P, Laurent F (2015) Dual impact of live Staphylococcus aureus on the osteoclast lineage, leading to increased bone resorption. J Infect Dis 211:571–581
129. Kavanagh N, Ryan EJ, Widaa A, Sexton G, Fennell J, O'Rourke S, Cahill KC, Kearney CJ, O'Brien FJ, Kerrigan SW (2018) Staphylococcal osteomyelitis: disease progression, treatment challenges, and future directions. Clin Microbiol Rev 31(2). https://doi.org/10.1128/CMR.00084-17. Print 2018 Apr
130. Flammier S, Rasigade JP, Badiou C, Henry T, Vandenesch F, Laurent F, Trouillet-Assant S (2016) Human monocyte-derived osteoclasts are targeted by staphylococcal pore-forming toxins and superantigens. PLoS One 11:e0150693
131. Jin T, Zhu YL, Li J, Shi J, He XQ, Ding J, Xu YQ (2013) Staphylococcal protein A, Panton-Valentine leukocidin and coagulase aggravate the bone loss and bone destruction in osteomyelitis. Cell Physiol Biochem 32:322–333
132. Badiou C, Dumitrescu O, George N, Forbes AR, Drougka E, Chan KS, Ramdani-Bouguessa N, Meugnier H, Bes M, Vandenesch F, Etienne J, Hsu LY, Tazir M, Spiliopoulou I, Nimmo GR, Hulten KG, Lina G (2010) Rapid detection of Staphylococcus aureus Panton-Valentine leukocidin in clinical specimens by enzyme-linked immunosorbent assay and immunochromatographic tests. J Clin Microbiol 48:1384–1390

Animal Models for Bone Tissue Engineering and Osteoinductive Biomaterial Research

Qifeng Lu, Xiao Lin, and Lei Yang

Abstract Animal bone defect model is the most important and widely used in vivo model for the study of osteoinductive biomaterials and bone tissue engineering. There are many different types of bone defect models at various anatomical sites, including skull and long bones, and using different animals, including mice, rats, rabbits, dogs, swine, and even nonhuman primates. Proper selection of animal model for a specific biomaterial or bone tissue engineering study is critical to obtain reasonable and reliable results. In this chapter, calvarial, weight-bearing long bone segmental defect models, metaphyseal defect models, and vertebral defect models are reviewed referring to several selection criteria of bone defect models. Several issues regarding model selection are discussed, including the characteristics of the model, the material being tested, and the experimental purpose. Considering the inconsistency between the current models and the real clinical conditions, we propose a suggestion for the future development of animal models for bone tissue engineering and osteoinductive biomaterial research.

Keywords Animal model · Bone defect · Fracture · Osteoinductive biomaterials
Bone tissue engineering

Qifeng Lu and Xiao Lin contributed equally to this work.

Q. Lu · X. Lin
Orthopaedic Institute and Department of Orthopaedics, The First Affiliated Hospital, Soochow University, Suzhou, Jiangsu, China

L. Yang (✉)
Orthopaedic Institute and Department of Orthopaedics, The First Affiliated Hospital, Soochow University, Suzhou, Jiangsu, China

Center for Health Science and Engineering (CHSE), School of Materials Science and Engineering, Hebei University of Technology, Tianjin, China
e-mail: ylei@hebut.edu.cn

B. Li et al. (eds.), *Racing for the Surface*,
https://doi.org/10.1007/978-3-030-34471-9_11

Introduction

Bone fracture is a common injury. Severe bone fractures are usually comminuted, which may cause bone defects. Nonunion would occur if the size of bone defect is beyond the self-healing ability of bone, which will cause serious pain and medical burden to patients. Furthermore, the repair of bone defects is difficult in clinical treatment due to limited autogenous bone supply.

The development of osteoinductive biomaterials and bone tissue engineering technology brings hope to the treatment of bone defects caused by fracture. For bone defects caused by tumors and osteomyelitis, biomaterials with only osteoinductive capability have limited treatment potential. Thus, this chapter focusses on defects caused by fracture. After development of bioactive bone substitutes, including osteoinductive biomaterials and bone tissue engineering products, evaluation of whether the materials meet the requirements of clinical treatment is necessary. Considering the complexity of the human body, results from in vitro tests cannot properly reflect the performance of biomaterials in vivo. Animal experiments are therefore essential for biomaterial evaluations by simulating the clinical human application.

However, the establishment of a universal animal model applicable to all studies for bone tissue engineering and osteoinductive biomaterial is impossible. A wide variety of bone defect models has been created for testing various materials. The repair of bone defect is affected by many factors, such as stability of the fractured ends of bone, blood supply, infection, and treatment methods, all of which affect the evaluation results of biomaterials. An incorrect selection of model or failure of detail control may seriously affect the results and even mislead the researchers. Therefore, proper animal models should simulate the real conditions of fracture and bone defect well and should be repeatable and standardized to obtain reliable results.

Here, we review the current bone defect models for the researches of osteoinductive biomaterials and bone tissue engineering and provide prospects for the development of defect models, which may serve as a reference for relevant researchers. This review includes three parts. The characteristics and analyses of the each bone defect model are introduced in the first part. The model selection criteria are described and suggested in the second part. Finally, the need for improvement and development of new bone defect models are proposed in the third part.

Animal Bone Defect Models

An ideal animal bone defect model used in osteoinductive biomaterials and bone tissue engineering research needs to meet the following criteria.

Clinical similarity: The animal model should have a similar in vivo environment simulating a specific human disease. For the bone fracture defect model, it should simulate bone defect of human when fracture occurs. Meanwhile, the animal model should be consistent with the disease severity. In terms of bone defect models, the size of bone defects should exceed the self-repair ability of the body where non-union occurs without immediate treatment. The smallest size causing nonunion is defined as the critical size defect (CSD) [1]. The CSD values vary with animal model and animal age, weight, defect location, and other pathological factors [2].

Repeatability: Standardized processes and operational methods should be used to produce animal bone defect model, which allows the model to be repeated under the same conditions, rendering the results from the bone defect model reliable and comparable.

There are many types of bone defect models used in the researches of osteoinductive biomaterials and bone tissue engineering. Based on the defect location, the bone defect models can be divided into calvarial bone defect model, long bone segmental defect model, metaphyseal defect model, vertebral body defect model, and long bone multiple defect model, etc. Moreover, each model can be created in different animal species. Within so many kinds of defect model, selecting a suitable model for research is a problem every researcher encounters. Before making choices, an in-depth understanding of these models is necessary.

Calvarial Bone Defect Models

Calvarial bone defects are usually created at the center of parietal bone by using a trephine with a specific diameter after cutting the skin and subcutaneous tissue [7]. This model is easy to create and the procedure is standardized to reach a high repeatability.

CSD in Calvarial Bone Defect Models

The CSD of a mouse skull is 4 mm in diameter for round defects [8]. In addition, bilateral skull defects with 3.5 mm in diameter [7] can be established in one mouse to obtain more reliable results. Table 1 summarizes CSD values for a number of different models.

Calvarial bone defect models are usually created in rats. However, the CSD of cranial bone defects is reported to be approximately 4–8 mm [3, 9–13] (Fig. 1a). Scholars initially recognized that the CSD was 8 mm in diameter [9], which is generally chosen in the level of the dura and sigmoid sinus [9]. However, these large-size cranial defects may damage the sagittal sinus, resulting in increasing the probability of bleeding. Therefore, scholars studied small-size skull defects. Sakata et al. created a 7-mm diameter cranial defect [13]. At 14 days post grafting, no calcified tissue was visible in the defect, and more calcification was seen after 35 days, but this did not fill the calvarial defects [13]. Subsequently, experiments of rat skull defects with diameters of 6, 5, and even 4 mm were reported [12–14]. These small-size bone defect models could make bilateral defects so that it might be used to

Table 1 Calvarial bone defect models for biomaterial evaluations and relevant references

Animal			Bone defect					
Species	Strain and age	Gender	Position	Number	Size	Fixation	Tested biomatcrials	Reference
Mouse	LMW transgenic mouse (2 months old)	Male	In central calvarial	Bilateral	3.5 mm in diameter	No	Scaffold (consisting of 70% type 1 collagen and 30% hydroxyapatite) with or without BMP2	L. Xiao [7]
	Nude mouse (8-week-old)	Female	Middle of parietal bone	One	4 mm in diameter (CSD)	No	hACSs co-transduced with Bac-flpo/ Bac-FCBW and Bac-mir148b which seeded into gelatin-coated PLGA scaffolds	Y. Liao [8]
Rat	Rats (mean age 7 months)	Male (365–480 g)	Skull	One	5 mm in diameter (CSD)	No	Mussel powder with or without additional bovine bone graft	D. Rizzo Trotta [3]
	SD rat (adult)	Male	To level of dura and sigmoid sinus	One	8 mm in diameter (CSD)	No	Injectable biopolymer of chitosan and inorganic phosphates was seeded with MSCs and BMP-2	J. Scott [9]
	Fisher-344 rat (12-week-old)	Male	In center of each parietal bone	Bilateral	5 mm in diameter	No	BMP-2 loaded CDHA porous scaffold with sulfated chitosan (SCS)	J. Zhao [10]
	Fischer-344 rat (2-month-old)	–	Skull	One	5 mm in diameter	No	BMP-2 mixed with gelatin gel was crosslinked using Tg	J. Fang [11]
	28-day-old CD strain	Male	Each parietal bone	Two	4 mm in diameter	No	Demineralized bone powder (DBP, 75–250 μm)	J. Glowacki [12]
	Sprague-Dawley (SD) rat (7-week-old males)	–	Skull	One	7 mm in diameter	No	Human mandibular periosteum Cells were cultured three-dimensionally on a collagen sponge	Y. Sakata [13]
	Immunodeficient CBH–ruul/_ nude (8 weeks of age)	Male	Skull	Bilateral	6 mm in diameter	No	CD14 mononuclear cells were induced to differentiate into osteoblast-like cells in combination with biodegradable polymer matrices made from poly-e-caprolactone	H. Chim [14]
	Skeletally mature Sprague-Dawley (7–8 week old)	Athymic, male rats	In the parietal bones	Bilateral	5 mm in diameter	No	Demineralized bone matrix (DBM)	A. Mhawi [15]

	Lewis rat (skeletally mature)	Male	Skull	One	5 mm in diameter	No	Polycaprolactone–tricalcium phosphate (PCL/TCP) scaffolds with two different fiber laydown patterns (coated with hydroxyapatite and gelatin)	A. Berner [17]
Rabbit	New Zealand white (NZW) rabbit (3 months old)	– (about 2.6 kg)	–	One	15 mm in diameter (CSD)	No	The cocultured cells (peripheral blood CD34+ cell) and MSC sheets were also composited with hydroxyapatite	G. Li [4]
	NZW rabbits (6 months old)	– (2.5–3 kg)	–	Four	8 mm in diameter	No	Stromal cell derived factor-1a	X. Liu [5]
	NZW rabbits (5 months old)	Female (3–4 kg)	In the parietal bone	Bilateral	8 mm in diameter (CSD)	No	ASCs were engineered to express the potent osteogenic (BMP-2) or chondrogenic (TGF-β3) factor, seeded into either apatite-coated PLGA or gelatin sponge scaffolds	C. Lin [18]
	NZW rabbits	–	Proximal to coronal suture in parietal bone	Bilateral∗2	8.0 mm in diameter	No	Particulate graft materials with and without P-15, Osteograf with CSP and Osteograf	N. Tovar [19]
Canine	Adult beagle dogs (12–24 months)	Average weight of 23.6 kg	At the parietal bone	Bilateral	20 mm ∗ 20 mm	No	Adipose-derived stem cells (ASCs) and coral scaffold	L. Cui [6]
	Beagle dogs (average 13.3 months)	Male	In bilateral parietal bone	Bilateral	20 mm in diameter (CSD)	No	Disk of OCP/Col (20 mm diameter, 2.5 mm thick) or commercially available sintered porous β-TCP	Y. Tanuma [20]
	Adult mongrel dogs	Male (15–25 kg)	On either side of the external sagittal crest	Bilateral	14 mm trephine defect	No	Implants of bovine bone morphogenetic protein (β-BMP) and a carrier consisting of matrix y-carboxyglutamic acid rich protein	K. Sato [21]
	Beagle dogs (12–13 months old)	Neutered male (average 9.8 kg, 8.6–11.1 kg)	From bregma posteriorly to interfrontal Suture anteriorly	One	3.5 ∗ 3.5 cm	No	0.2 mg/mL rhBMP-2 in absorbable collagen sponge, 0.2 mg/mL rhBMP-2 in absorbable collagen sponge with corticocancellous chips, 0.2 mg/mL rhBMP-2 in absorbable Collagen sponge with mastergraft granules, or 0.4 mg/mL rhBMP-2 in compression resistant matrix carrier	C.R. Kinsella [22]

compare different materials in one single animal, thus improving the reliability of the experiment. The most widely used model is the 5-mm defect model, which is considered to be CSD with bilateral [15, 16] or single defect [3, 11, 17]. With regard to the skull defect with 4-mm diameter, Glowacki et al. [12] found no obvious signs of bone healing even after 6 months of observation. It should be noted that the observation time period for the determination of CSD in the rat skull should be >8 weeks because skull repair is active within the first 4 weeks and essentially stops at the eighth week [15].

For the rabbit calvarial defect model, the CSD is generally considered as a full-thickness defect with a diameter of 15 mm [4] (Fig. 1b). Bilateral bone defects with smaller size were also established on rabbit skull. Lin et al. drilled an 8 mm-

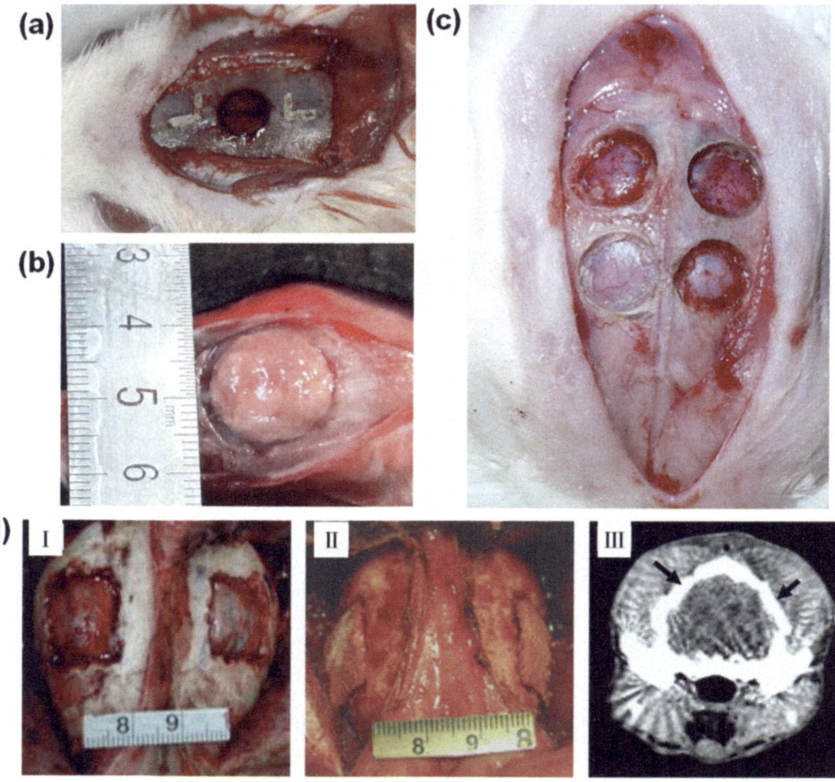

Fig. 1 Calvarial bone defect models: (**a**) a rat calvarial CSD of 5 mm in diameter [3]; (**b**) the composite hydroxyapatite/cell sheet transplanted into the calvarial defect of rabbit (15 mm in diameter) [4]; (**c**) four 8 mm-diameter defects were drilled into the rabbit calvarial bone with a trephine [5]; (**d**) canine calvarial defect model: (I) bilateral full-thickness bone defects with a size of 20 mm × 20 mm were created; (II) the experimental side (left) was repaired with a cell-coral construct and the control side (right) was filled with a coral alone; (III) coronal CT scan image of the same animal was taken immediately after surgery [6]

diameter bilateral full-thickness defect, which was used to compare the repair of bone defect with autologous mesenchymal stem cells (MSCs) [18]. Liu et al. created four 8 mm-diameter defects in the calvarium of rabbits using a trephine attached to a low-speed hand-piece [5] (Fig. 1c). Furthermore, Tovar et al. also drilled two bilateral 8.0 mm-diameter defects without dural injury in the proximal to the coronal suture of the parietal bone [19]. In this way, a variety of bone substitutes have be evaluated by the rabbit calvarial defect model (Table 1).

The CSD in canine skull defect model is usually 20 mm in diameter [20]. With regard to large animal models, making bilateral skull defects is convenient. Sato et al. manufactured bilateral 14-mm trephine defects in dog skull to evaluate the performance of β-sheep bone morphogenetic protein (β-BMP) and a carrier consisting of matrix y-carboxyglutamic acid rich protein [21]. Cui et al. created bilateral full-thickness bone defects with a size of 20 × 20 mm at the parietal bone [6] (Fig. 1d). In addition, a large skull defect area, measuring 35 × 35 mm, was also created in another study [22].

Comments on the Models

Some studies performed periosteal removal in the operative region [12, 23], which might have caused the failure of the skull repair. For example, the stripping of periosteum could even lead to the failure of 2-mm diameter defect healing in rats [24, 25]. Thus, the periosteal layer should be sutured to avoid the obstruction of blood supply when creating animal skull defects.

The operation procedure of the calvarial defect model seems simple, but many things needs special attention to create a successful, repeatable model. The skull is located under the skin; thus, direct skull exposure without intraoperative bleeding is simple. However, prevention of secondary damage is difficult. The surgeon should be careful not to damage the dura mater and the surrounding vessels intraoperatively, or it would cause uncontrollable bleeding. Also, the sagittal sinus may be injured during when creating large-size defects, resulting in obstruction of blood supply. In addition, bone marrow progenitor cells exist in the connective tissue near the sagittal suture, which might be involved in bone remodeling after injury. These factors may critically impact the repair of bone defects.

Segmental Defect Models of Weight-Bearing Long Bone

Weight-bearing bones refer to the bones that bear the main weight of the body in the extremities. The femur and tibia are the most common weight-bearing bones in animal models. The femur size is large, which is convenient for the establishment of bone defect model. And the infection resisting ability of femur model is believed to be strong. However, much muscle tissue around the femur can complicate bone exposure during operation. Because part of tibia is directly under the skin, tibial exposure is relatively simple, but infection is easy to occur. The selection of model should be based on the purpose of the study and the degree of familiarity of the anatomy.

Fixation Used in the Segmental Defect Model

After the long bone segmental defect forms, the body weight would cause the movement and displacement of the broken ends of the bone, which could affect bone healing and test results. Therefore, internal or external fixation must be used to ensure the stability of defect site.

Rats are often used to create the segmental defect models of weight-bearing long bone. It is however challenging to perform surgery on the slender bones. Kirschner wire is often used for fixing the tibia [30] and femur [31] of rats. However, Kirschner wire in bone marrow cavity occupies much space, causing inconvenience for implanting testing materials. In addition, Kirschner wire has no anti-rotation ability and disturbs the blood supply of defect ends. These shortcomings limit the application of Kirschner wire in the small animal model. Besides the Kirschner wire, external fixator can be used to secure the femoral defect [32, 33], which reveals good anti-rotation ability and preserves the blood supply of defect ends as well as adequate space for implantation. Moreover, the external fixator usually allows animals move freely without the joint being fixed. In recent years, plate fixation [26] (Fig. 2a) and a four-hole high-density polyethylene with stainless-steel screws [34] have been developed for rat femur models. Also, small steel plates and screws have been used to immobilize the femur [35] and tibia [27] (Fig. 2b) of rabbits (Table 2).

With regard to large animals, the weight of the goat and sheep is closer to that of the human body and the size of the bone is also similar to that of the human bones. So more fixation options are available for these animal models. The standard dynamic compression plate (DCP, ten-hole) [36, 37], external fixator [29, 38] (Fig. 2d), screws [28] (Fig. 2c), and internal fixation rods [39, 40] can be used as fixation devices for tibial defects. For femoral defects, internal fixation rods or interlocking nails can be used as fixation devices [39].

CSD in the Segmental Defect Models of Weight-Bearing Long Bone

The CSD of long bone is considered as 1.5–2.5 times the circumference of the diaphysis, or >1/10 of the length [1]. Different animals and bones have different CSDs.

The CSD of the tibia and femur in the rat model are 4 mm [30] and 5 mm [31–34], respectively. For the rabbit model, the CSD of the tibial defect model is determined as 5 mm [27]. However, there are different opinions on the femoral CSD in rabbit models. Fialkov et al. confirmed 12 mm as CSD for femoral diaphyseal defect model [41], but Fan [42] and Duan et al. [35] used a 15-mm defect as CSD.

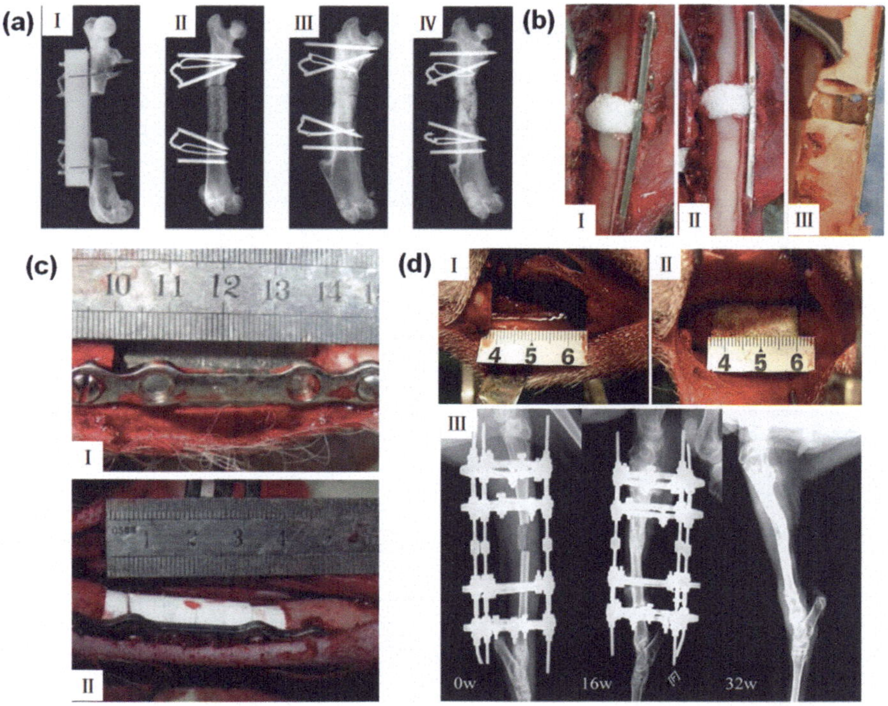

Fig. 2 Weight-bearing long bone segmental defect models with fixations: (**a**) Photo and radiographs of rat femoral defects with fixation (I) prior to implantation, (II) immediately post-implantation, (III) treated with HA/TCP loaded human MSCs at 12 weeks post-implantation, and (IV) treated with HA/TCP alone at 12 weeks post-implantation [26]. (**b**) A 5-mm bone defect was created in the middle of rabbit tibia and stabilized with a plate and screws. The defect was implanted with a complex of β-TCP granules and collagen, either with 200 mg of FGF-2 (I) or without FGF-2 (II). For control group, the bone defect was left empty (III) [27]. (**c**) The 40 mm defect of goat shank with fixation before (I) and after implantation of biomaterial (II) [28]. (**d**) Goat tibia defects model with external fixation. (I) A periosteal segmental defect of 26 mm length was created at the right tibia; (II) the defect was filled with BMSCs/β-TCP construct; (III) radiographs of goat tibial defects taken at different time points post-operation [29]

With regard to large animals, goats and sheep have relatively large limbs that are convenience for modeling surgery using internal fixators. Both the femur and tibia of goat may be used in bone defect models. The segmental defects in the femur is located in the middle of the shaft, and the length of the osteotomy defect could reach up to 25 mm [39]. The sizes of the tibial defect in goats are reported to be different (e.g., 26 mm [38] and 40 mm [28]). For sheep models, the tibia is the most widely used bone and the lengths of defect is either 3.0 cm [36, 37] or 3.5 cm [43], or even 5 cm [40].

Table 2 Weight-bearing long bone segmental defect models for biomaterial evaluations and relevant references

Animal			Bone defect					
Species	Strain and age	Gender and weight	Position	Number	Size	Fixation	Experimental biomaterials	Reference
Rat	Fischer 344 rat	–	Femoral diaphysis	One	8 mm long	Plate fixation	Mesenchymal stem cells (MSCs) combined with biphasic calcium phosphate ceramics	T. L. Livingston [26]
	Wistar albino rat	Female	Central diaphysis of tibia	One	4 mm long	Kirschner wire	RhBMP-2-containing absorbable collagen sponges	S. Sarban [30]
	Brown Norway rat (13 week old)	–	Femoral diaphysis	One	5 mm long	Kirschner wire	Biodegradable polypropylene Fumarate/tricalcium phosphate (PPF/TCP) with dicalcium phosphate dihydrate (DCPD) cement portals	R. Stewart [31]
	–	–	Midfemoral	One	5 mm long (CSD)	External fixator	11 mg of recombinant human BMP (rhBMP)-2 on a collagen sponge	V. Glatt [32]
	Wistar rat (9–11 weeks old)	– (350–440 g)	Mid-diaphyseal femur	One	5 mm long	External fixation	Combined poly (ethylene glycol) diacrylate (PEGDA) hydrogel with allogeneic "carrier" cells transduced with an adenovirus expressing BMP2	C. Sonnet [33]
	SD rat	Male (325–360 g)	Mid-diaphyseal femoral	One	5 mm long	Four-hole high-density polyethylene custom fabricated internal fixator	Recombinant human bone morphogenetic protein-2 (rhBMP-2) absorbable collagen Sponges carrying rhBMP-2	S.R. Angle [34]
Rabbit	NZW rabbit	Female (3.3–3.5 kg)	In middle of tibial diaphysis	Unilateral	5 mm long	A plate and four screws	A complex of β-tricalcium phosphate (β-TCP) granules, collagen, and fibroblast growth factor-2 (FGF-2)	H. Komaki 2006 [27]

	NZW rabbit (16–20 weeks old)		Femoral diaphysis	Unilateral	15 mm long	Five hole steel plates	BMP2-derived peptide Plus PLGA-[ASP-PEG] scaffold material	Z. Duan 2008 [35]
	NZW rabbit	Female (2.0–2.5 kg)	Femoral diaphysis	Unilateral	1.2 cm long	A 2.7-mm mandibular reconstruction plate fixated with three screws on either side of the osteotomy site	A biodegradable scaffold-poly(lactide-*co*-glycolide) (PLGA) foam-with similar porosity to human trabecular bone	J.A. Fialkov [41]
	New Zealand rabbit	Male (2–2.5 kg)	Femoral diaphysis between the second and the third holes of the plate	Unilateral	1.5 cm long	A four-hole steel plate of reconstruction	Tissue-engineered bone and implanted the sensory nerve tracts, blood vessel, or nothing	J.J. Fan [42]
Sheep	Merino sheep (7–8 years)	Average 43.3 kg	Mid-diaphyseal tibia	Unilateral	3 cm long (CSD)	A ten-hole, stainless steel, dynamic compression plate (DCP)	Two dosages of rhBMP-7, 3.5 mg and 1.75 mg, embedded in a slowly degradable medical grade poly(ε-caprolactone) (PCL) scaffold with β-tricalcium phosphate microparticles (Mpcl-TCP)	A. Cipitria [36]
	Merino sheep (6–7 years)	45 ± 2 kg	Mid-diaphyseal tibia	Unilateral	3 cm long (CSD)	A broad dynamic compression plate (DCP) (4.5 mm, ten holes, synthes)	Allogenic mpcs	A. Berner [37]
	Skeletally mature ewes	Female	Mid-diaphyseal tibia	Unilateral	5 cm long (CSD)	An intramedullary interlocking nail with two 4.5 mm cortical	The standard osteogenic protein-1 (OP-1/BMP-7) implant with carboxymethylcellulose (CMC)	G. E. Pluhar [40]

(continued)

Table 2 (continued)

Animal			Bone defect					
Species	Strain and age	Gender and weight	Position	Number	Size	Fixation	Experimental bio materials	Reference
	Sheep (2 years old)	Female	Mid-diaphyseal tibia	Unilateral	3.5 cm long	4.5 mm transcortical screws, four proximal screws, and four distal screws	Blood-derived progenitor cells that have endothelial characteristics (EPC)	N. Rozen [43]
Goat	Adult male goat	About 50 kg	Mid-diaphyseal tibia	Unilateral	40 mm long	A titanium screw	A novel nanohydroxyapatite/collagen/PLLA (nhacp) composite reinforced by chitin fibers	X. Li [28]
	Skeletally mature goat	22.3 ± 4.1 kg	Tibial diaphysis	One	26 mm long	External fixators	Bone marrow stromal cells and ß-tricalcium phosphate	G. Liu [29]
	Goat (1-year-old)	18.6–31.5 kg	Mid-diaphyseal tibia	Unilateral	2.6 cm long	Circular external fixators	Biphasic calcined bone (BCB) and autologous bone marrow-derived mesenchymal stem cells (BMSCs) transduced with human bone morphogenetic protein-2 (hBMP-2)	K. R. Dai 2005 [38]
	Adult goat (12–14 months)	16.5–23 kg (average of 19.6 kg)	Mid-diaphyseal femur	One	25 mm long	An internal fixation rod and interlocking nails	Coral cylinder alone, or filled with coral cylinder plus induced BMSCs	L. Zhu [39]
	Ewe (2-year-old)	–	Tibial diaphysis	One	35 mm long	External fixation	Large, cylindrical implants of a porous calcium phosphate ceramic; hydroxyapatite ceramic (HAC)	A. Boyde [44]

Comments on the Models

Segmental defect models of long bone model requires osteotomy to make bone defects and needs to fix the defect ends with fixator, which makes this type of model complicated. However, in real clinical practice, fractures and bone defects usually occur on the weight-bearing bones of the extremities and internal and external fixators are frequently used in treatment. Thus, the segmental defect models are close to clinical scenarios. Besides, if the design purpose of a novel bone substitute material is to repair a long bone defect of the limbs, the segmental defect model of weight-bearing bone is more suitable. In addition, real bone fractures are often accompanied by peripheral soft tissue injury and inadequate blood supply. The middle and lower segments of the tibia are in contact with subcutaneous tissue without muscle in between, rendering the tibia defect model similar to the clinical factures.

Segmental Defect Models of Non-weight Bearing Long Bone

The non-weight-bearing bones include ulnar, radius, and the fibula. When these bones are truncated, the stability of the broken ends may not be destroyed and the defect ends usually need no fixation, which obviously simplifies the preparation process of the model. To prevent the displacement of the testing material, cerclage wires can be used to prevent dislocation during modeling [46, 47, 49].

CSD in Segmental Defect Models of Non-weight Bearing Long Bone

For mice, the shaft of the radius is most commonly used and the lengths of defects are usually 1.5 and 2.5 mm [50, 51].

For rats, 5-mm osteotomy is usually created on the bilateral radial shaft [45, 52] (Fig. 3a). Although the segmental fibular defect model has two different defect sizes (2 mm and 4 mm) [53], 4 mm is generally considered to be the CSD for fibula [1, 54]. The segmental fibular defect model can also be performed in bilateral fibular bone [54] which can reduce the required number of animals. Besides, two different materials can be compared on one individual rat, which could reduce the experimental error and make the results more consistent.

In rabbit model radius is the most commonly used bone to generate a CSD. It is generally believed that the CSD value in long bone should be at least two times the diameter of the diaphysis [46, 47, 49] (Fig. 3b, c). The radius diameter of adult New Zealand white rabbit is approximately 5–6 mm. Therefore, some scholars have determined 10 mm-defect as the CSD [16, 55–58], although 15 mm-defect has also been used [59, 60]. Moreover, the osteotomy site of the defect is suggested to be selected at 2.5 cm above the radio-carpal joint. Besides the radius, the rabbit ulnar shaft can also be used. Two sizes, 15 mm [61] and 20 mm [62], are commonly adopted as the CSD values of the ulnar. Only a few scholars adopted nonstandardized

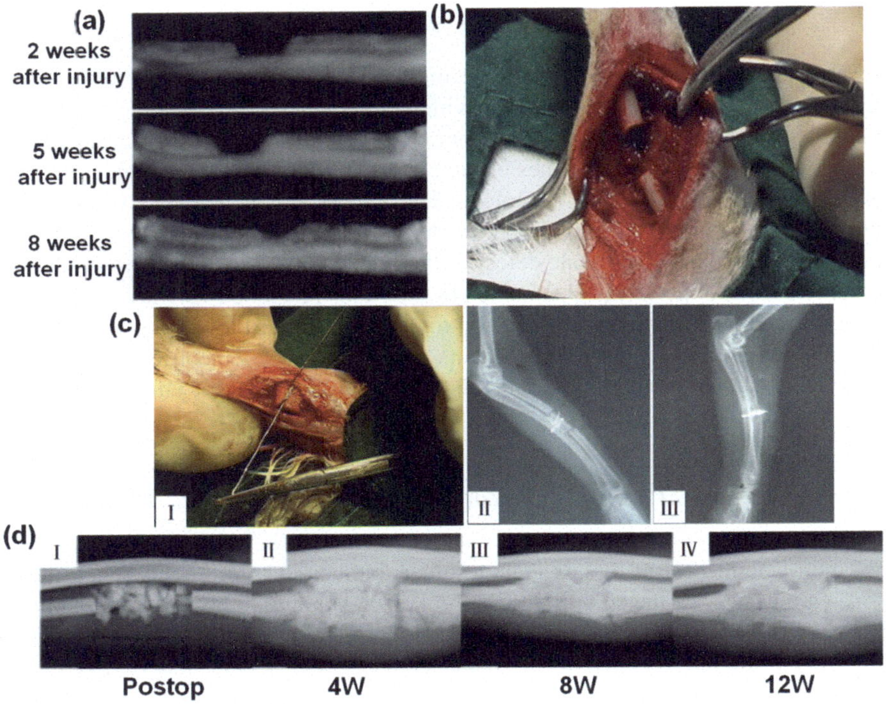

Fig. 3 Non-weight-bearing long bone segmental defect models: (**a**) Postoperative radiographs of chitosan (CS)-gelatin (Gel) scaffolds in rat radial bone defect (5 mm) at 2, 5, and 8 weeks [45]. (**b**) Radius of rabbit exposed by dissection of surrounding muscles and an osteoperiosteal segmental defect was created on the middle portion of each radius at least twice that of the diameter of the diaphysis [46]. (**c**) (I) Xenogenic growth plate grafting in radius defect of rabbit; (II–III) Radiographs of forelimb on 14th day post opertation (II-autograft, III-growth plate xenograft) [47]. (**d**) Radiographs of the postoperative changes in canine segmental ulnar defects filled with BMP and cells [48]

ulnar stem defects. For example, evaluate recombinant human bone morphogenetic protein-2 in promoting bone healing, Bouxsein et al. built a blade-width (0.5–1 mm) defect in the ulna [63].

Ulna defects are often made in dog models. The ulna is exposed through craniolateral approach [64], and the osteotomy site is located at the proximal 4 cm of the ulnar styloid process [65]. The length of the osteotomy has various sizes in literature: 20 mm [65], 22.5 mm [64], and 25 mm [48, 66, 67] (Fig. 3d). The radius of the dog can also be used to build segmental bone defects with a size of 10 mm [68].

Comments on the Models

Compared with the weight-bearing long bone defect models, the non-weight bearing long bone segmental defect models are relatively simple to create. However, the forearm defect model cannot avoid the interactions between radius and ulna, and the pathological and biomechanical changes in one could be compensated by another. Other shortcomings of these models include the great variations in the bone diameter and resultant sizes of defects created (Table 3).

Metaphyseal Defect Model

When preparing a metaphyseal defect model, a small drill is used to make a lateral cylindrical defect which is vertical to the axis of the femoral shaft [71–74], and then the bone substitute materials with proper diameters are implanted into the cylindrical defect.

Femur or tibia is commonly selected for this model due to their large sizes and good accessibility to their condyles. The metaphyseal bone is located subcutaneously and usually no extra fixation is needed after defect creation, which makes the model relatively simple to create with a high success rate [75].

CSD in Metaphyseal Defect Models

In mice metaphyseal defect model, defects with 1 mm in diameter were usually created at the proximal tibia [76] and distal femur [77, 78]. However, these small defects are difficult to evaluate bone substitutes. These defects are mainly used to test drug or investigate the mechanism of fracture healing.

For rats, bone defects with 3.5 mm in diameter at the proximal tibial diaphysis are often used for evaluating biomaterials [79]. Bone defects with different sizes, 5 mm in diameter and 5 mm in depth [80], 6 mm in diameter and 10 mm in depth [69, 73] (Fig. 4a), and 6.1 mm in diameter [70] (Fig. 4b) were fabricated on the metaphysis of rabbit femur. The size of defect in canine femoral condyles could reach 20 mm in depth and 8 mm in diameter [81] or 20 mm in depth and 10 mm in diameter [82].

Comment on the Models

The metaphyseal defect model only simulates a bone defect environment, not a long bone fracture or defect in clinical scenarios. Only a few categories of materials are suitable for evaluating by this type of models, including bone filler or substitutes for localized bone defects, granular or colloidal materials without structural support capabilities, and bone cement, etc. Besides simple modeling without the requirement for fixation, another advantage of the condyle defect model is that it can avoid the displacement of testing material (Table 4).

Table 3 Non-weight bearing long bone segmental defect models for biomaterial evalutions and relevant references

Animal			Bone defect					
Species	Strain and age	Gender and weight	Position	Number	Size	Fixation	Experimental biomaterials	Reference
Mice	C3H/hen mice (8–10 weeks old)	Female	Radial diaphysis	One	1.5 mm long	No	Electroporated with either BMP-9 plasmid, Luciferase plasmid or injected with BMP-9 plasmid but not electroporated	K.B. Nadav [50]
	C3H/hen mice (8–10 weeks old)	Female	Radial diaphysis	One	2.5 mm long	No	One million Tet-off BMP2 MSCs were suspended in 15 mL fibrin gel that was supplemented with PFTBA or not	K.B. Nadav [51]
Rat	Wistar rats	Weighing 250 ± 25 g	Radial diaphysis	Bilateral	5 mm long	No	Gelatin, chitosan, and their combination as tissue-engineered scaffolds	A. Oryan [45]
	SD rat (10-week-old)	Male	Radial diaphysis	Bilateral	5 mm long	No	Vitamin D-binding protein (DBP) (gelatin sponge)	J. Sun [52]
	Homozygous Athymic RNU Nude rat (adult)	– (200–300 g)	Fibula diaphysis	One	2–4 mm long	No	Adenovirus-transduced cells expressing bone morphogenetic protein 2	W. Zawaunyka [53]
	Rat	– (200–250 g and 260–290 g)	Fibula diaphysis	Bilateral	4 mm long	No	7-mm-long tubular specimen of demineralized bone matrix (DBM)	D. Chakkalakal [54]
Rabbit	New Zealand Albino rabbits (12 months old)	Male (3.0 ± 0.5 kg)	On middle portion of each radius	Bilateral	At least twice as long as the diameter of diaphysis	Cercelage wire	Fresh cortical autograft and allograft	Z. Shafiei [46]
	New Zealand Albino rabbits (12 months old)	Male (3.0 ± 0.5 kg)	On middle portion of each radius	Bilateral	At least twice as long as the diameter of diaphysis	Cercelage wire using stainless wire No. 01	Autograft and new xenograft (bovine fetal growth plate)	S.N. Dehghani [47]

	NZW rabbits (12 months old)	Male (3.0 ± 0.5 kg)	On middle portion of each radius	Bilateral	At least twice as long as the diameter of diaphysis	Cercelage wire	Autograft and new xenogenic bovine demineralized bone matrix (DBM)	A.S. Bigham [49]
	NZW rabbits (12 months old)	Both sexes (2.0 ± 0.5 kg)	Radial diaphysis	Bilateral	10 mm elongation	No	Combination of human platelet-rich plasma and coral	A.M. Parizi [16]
	NZW rabbits (12 months old)	Both sexes	Radial diaphysis	Unilateral	10 mm long (CSD)	No	Human platelet-rich plasma in combination with hydroxyapatite and coral on osteogenesis	S.S. Zahra [55]
	NZW rabbits (12 months old)	Both sexes (2.00 ± 0.50 k)	In mid portion of each radius	Bilateral	10 mm long (CSD)	No	Hydroxyapatite-human platelet-rich plasma (HPRP)	A. Oryan [56]
	NZW rabbits (12 months old)	Both sexes (2.0 ± 0.5 kg)	Radial diaphysis	Unilateral	10 mm long (CSD)	No	Hydroxyapatite with demineralized calf fetal growth plate (DCFGP)	B. Amin 2015 [57]
	NZW rabbits	Male (about 3.5 kg)	Radial diaphysis	Unilateral	10 mm long	No	Barrier membranes combination with recombinant human bone morphogenetic protein 2 (rhbmp-2)	G. Zellin 1997 [58]
	New Zealand rabbit	– (6 kg)	Radius diaphysis	One	15 mm long (CSD)	No	Rhbmp-2 combined either with chitosan hydrogel (rhbmp-2/CH) and chitosan hydrogel containing β-tricalcium phosphate (β-TCP) (rhbmp-2/CH/TCP)	L. Luca [59]
	NZW rabbits (3-month-old)	Male (2.0–2.5 kg)	In middle of radial shafts	One	1.5 cm long	No	Complexes of zein scaffolds and rabbit MSCs	J. Tu [60]

(continued)

Table 3 (continued)

Animal			Bone defect					
Species	Strain and age	Gender and weight	Position	Number	Size	Fixation	Experimental bio materials	Reference
	Japanese white rabbit (16–20 week old)	Male	Ulna diaphysis	Unilateral	1.5 cm long	No	Recombinant human bone morphogenetic protein (rhBMP)-2 and a novel carrier, PLGA-coated gelatin sponge (PGS)	S. Kokubo 2003 [61]
	NZW rabbits (6-month-old)	Male (3.5–4.0 kg)	Ulna (3.0 cm from ulnocarpal joint)	Bilateral	2.0 cm long	No	Implanted with a pastelike polylactic glycolic acid blood clot combination that was mixed with five different concentrations of rh-BMP2	M. Bostrom [62]
	NZW rabbits (8–9 months)	Male	Ulna (45–50 mm distal to olecranon process)	Bilateral	Blade-width defect (0.5–1 mm)	No	An absorbable collagen sponge containing rh-BMP2	M.L. Bouxsein 2001 [63]
Canine	Female beagle dogs (1 year old)	Female (9.4–11.0 kg)	Shaft of ulna	Unilateral	2.5 cm long (CSD)	No	Escherichia coli-derived recombinant human bone morphogenetic protein-2 combined with bone marrow-derived mesenchymal stromal cells	T. Itoi [48]
	Beagle dogs (1.5–0.27 years)	(13.7–2.1 kg)	Shaft of ulna	Bilateral	22.5 mm long (CSD)	No	Bone morphogenetic protein-2 and vascular endothelial growth factor	R. E. Geuze [64]
	Labrador retrievers (32 ± 5 months)	Both gender (31 ± 1 kg)	4 cm proximal of ulnar styloid process	Bilateral	20 mm long (CSD)	No	Local continuous growth hormone administration	L. Theyse [65]

	Mixed-breed hounds	– (22.56 ± 2.22 kg)	Mid-diaphyseal of ulnar	Bilateral	2.5 cm long (CSD)	No	Recombinant human bone morphogenetic protein-2 (rhbmp-2)/absorbable collagen sponge (ACS) with cancellous allograft bone or biphasic calcium phosphate ceramic granules	C.B. Jones [66]
	Male mongrel dogs	Male (35–45 kg)	Mid-ulna	Bilateral	2.5 cm long (about 2–2.5 times midshaft) (CSD)	No	Recombinant human bone morphogenetic protein-7 (recombinant human osteogenic Protein-1, rhop-11)	S. Cook [67]
	Adult mongrel dogs (3–4 years old)	Male (25.2 ± 3.5 kg)	Mid-diaphysis of radius	One	10 mm transverse bone defect	No	Greater omentum as a scaffold incorporation of ASCs (adipose-derived adult stem cells)	B.S. Amin [68]

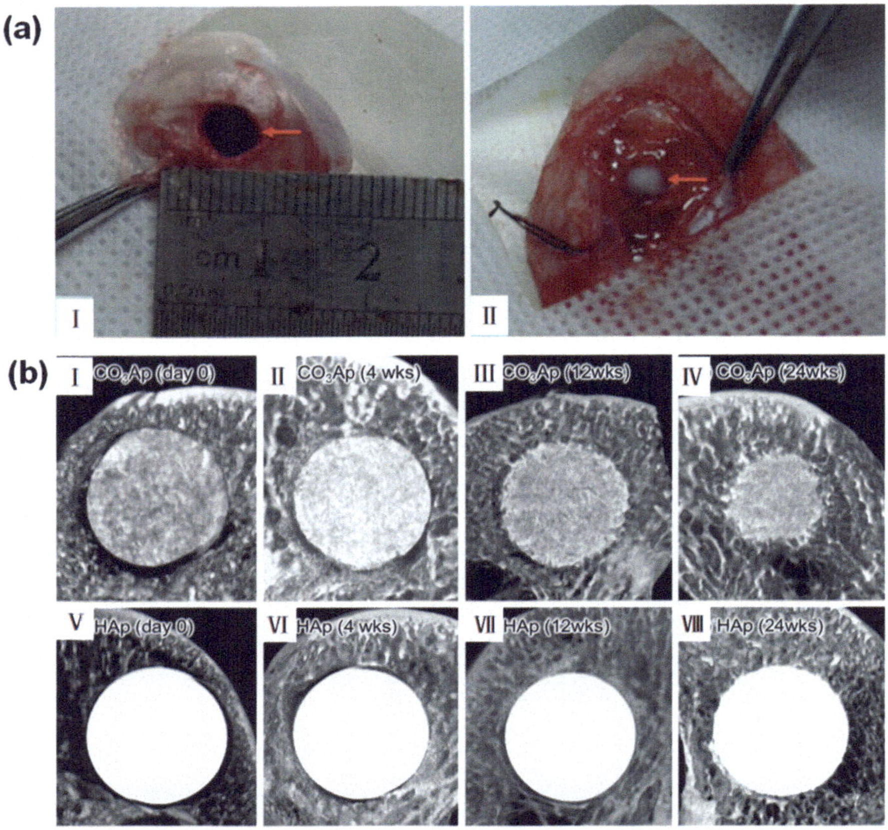

Fig. 4 Metaphyseal defect models: (**a**) Rabbit femoral condyle bone defects surgical procedure; (I) CSD (6 mm in diameter and 10 mm in length) was transversally created in the femoral condyles of rabbits; (II) the defect was filled with nHA/CS composite scaffolds [69]. (**b**) Micro-CT sagittal images of the rabbits' distal femurs from a top view of the CO_3Ap blocks (I–IV) and HAp blocks (V–VII) immediately after implantation (day 0) (I, V) and at 4 (II, VI), 12 (III, VII), and 24 (IV, VII) weeks after implantation in rabbits' distal femurs [70]

Vertebral Body Defect Models

In some studies, the injectable biomaterials were evaluated on long bone defect models [79]. However, the different parts of the bone have varied mechanical environment, which may lead to varied material degradation behavior and bone tissue response in vivo [83, 84]. To simulate the actual environment, materials specifically for vertebral repair or augmentation are strongly recommended to be evaluated by vertebral defect models (refer to [85] for illustrative images).

Table 4 Metaphyseal defect models for biomaterial evaluation and relevant references

Animal			Bone defect					
Species	Strain and age	Gender and weight	Position	Number	Size	Fixation	Experimental biomaterials	Reference
Mice	Osm and Osmr_/_ mice, backcrossed onto C57BL/6 (7–8 weeks old)	–	Proximal tibia	One	A hole with 1.0 mm in diameter	No	Mice were treated daily by intraperitoneal injection of 10 mL/g clodronate or control phosphate-buffered saline (PBS) liposomes	E. Wehrle [76]
	C57BL/six male mice	–	In the femur mid-diaphysis and the distal epimetaphyseal region	Two	0.9 mm in diameter for femur mid-diaphysis Holes with 0.9 mm in diameter, 1 mm deep for distal epimetaphyseal region	No	–	L. Monfoulet [77]
	C57BL/6N mice (7 weeks of age)	Male	At the anterior portion of the diaphysis of bilateral femurs	Two	1 mm in diameter	No	Bisphosphonate (YM529), a nitrogen-containing bisphosphonate	M. Nagashima [78]
Rat	Wistar rats	Male (body weight 350–450 g)	Proximal tibial diaphysis	One	3.5 mm in diameter	No	A full synthetic injectable bone substitute (SIBS) (like microparticles)	W. Xu [79]
Rabbit	New Zealand rabbits	2.0–2.5 kg body weight	Distal femur	One	6 mm in diameter and 10 mm in depth (CSD)	No	Injectable nanohydroxyapatite/chitosan composites scaffolds	X. Zhang [69]
	Japanese white male rabbits (19–20 weeks old)	3335 ± 322 g body weight	Distal femur	One	6.1 mm in diameter	No	Carbonate apatite blocks fabricated from dicalcium phosphate dihydrate blocks	M. Kanazawa [70]

(continued)

Table 4 (continued)

Animal			Bone defect					
Species	Strain and age	Gender and weight	Position	Number	Size	Fixation	Experimental biomaterials	Reference
	NZW rabbits (6-month-old)	Male	Distal femur	Bilateral	6 mm in diameter and 10 mm in depth (CSD)	No	β-TCP collagen composite bone substitute	H. Zheng [73]
	NZW rabbits (6-month-old)	–	Distal femur	Bilateral	5 mm in diameter and 4–5 mm in depth	No	Pthrp (107–111) (osteostatin) loaded onto silica-based ordered mesoporous SBA15 materials	C.G. Trejo [80]
Canine	Adult mongrel dogs (German shepherd type)	25–30 kg body weight	Distal epiphysis of the femur	Bilateral	8 mm in diameter and 20 mm in length	No	Skeletal fragments of different coral genera	G. Guillemin [81]
	Beagle dogs (aged 1–2 years)	Male (10–12 kg)	Femoral condyles	Bilateral	10 mm diameter (from medial to lateral)	No	Tetrapod-shaped granular artificial bone (Tetrabone®) or β-tricalcium phosphate granules (β-TCP)	S. Choi [82]

Small Animal Vertebral Body Defect Models

In small animals, rats are often used to fabricate vertebral body defect models. The CSD of rat vertebra was explored by Liang et al. [85]. Two different sizes of defect, 2 mm × 3 mm × 1.5 mm and 2 mm × 3 mm × 3 mm, were made in anterior part of L4, L5, and L6 vertebra, and the former bone defect was found to be the CSD. Besides, the bone defect model can also be constructed in the caudal vertebra of rats. For example, a cylindrical defect (1.8 mm in diameter and 2 mm in depth) in the center of the third caudal vertebra was created through a left posterolateral approach [86]. The advantages of caudal vertebra model include simple anatomy approach and free of the risk to important organs and vessels. Meanwhile, there is even simpler method of modeling, such as PMMA directly injected into the L1–L6 vertebral body of New Zealand rabbits [87]. However, the clinical significance of such modeling methods for biomaterial evalution is questionable.

Large Animal Vertebral Body Defect Models

Sheep vertebral body defect model is most frequently used. The similarity of body weight and bone size of sheep to those of humans allows testing biomaterials designed for clinical study. The L2-L5 vertebrae are usually used to create the model. Usually, a cylindrical hole with a certain diameter and depth is drilled in the center of vertebral body to simulate the space created by balloon in kyphoplasty (KP) and then the defect is filled with injectable biomaterials. The diameters of the hole include 4 mm [88], 6 mm [88, 89], 8 mm [90, 91], and 10 mm [92], while the depths of the hole are 9 mm [88], 10 mm [89], 15 mm [90–92], and even 20 mm [93].

In addition to these common cylindrical defects, the rectangular defect with 8 mm high, 10 mm deep, and 20 mm long, was used to test injectable calcium phosphate cement [94]. Verron et al. [94] built an osteoporosis vertebral bone defect model, which was in line with the fact that most of the patients receiving the treatment of bone filling material suffer from osteoporosis. However, the establishment of this osteoporotic model needs a long modeling time (4–6 months) and high costs.

Canines are less used in vertebral defect model than sheep. In the modeling process, the lateral side of the lumbar spine is shown through the dorsal to the transverse process. The central vertebral defect with the size of 18 mm × 5 mm × 22 mm was produced, and the researcher injected the biomaterial into the defect for testing [95].

The bone defect in swine vertebral body can be created by percutaneous puncture under fluoroscopic guidance [96]. The lumbar vertebrae of mini-swine were also used for preparing defect model (15 mm in depth and 4 mm in diameter) [97]. Compared with other species, swine models are noisier and more aggressive, and their daily management is more difficult [98] (Table 5).

Table 5 Vertebral body defect models for biomaterials evalutions and relevant references

Animal			Bone defect					
Species	Strain and age	Gender and weight	Position	Number	Size	Fixation	Experimental biomaterials	Reference
Rat	Fischer rats (6-months-old)	Male	In anterior part of VB (L4, L5, and L6)	Three	A large defect (2 mm deep × 3 mm wide × 3 mm long, CSD) and A small defect (2 mm deep × 3 mm wide × 1.5 mm long)	No	Sintered poly(lactic-*co*-glycolic acid) (PLGA) microsphere scaffolds loaded with Matrigel with or without recombinant human bone morphogenetic protein 2 (rhbmp2; 2.0 μg rhbmp2/10 μL Matrigel/scaffold)	H. Liang [85]
	Mature athymic nude rats	–	In center of third caudal vertebral body (Ca3)	One	1.8 mm in diameter and 2 mm-deep cylindrical defect	No	The ASCs were first labeled with reporter genes and then nucleofected with an rhbmp6-encoding plasmid. The ASC-BMP6 cells were suspended in fibrin gel (FG)	D. Sheyn [86]
	Wistar rats	Male (300–350 g)	L2 vertebra	One	$2 \times 5 \times 1.5$ mm^3	No	A composite bioartificial graft based on a hydroxyapatite bone scaffold (CEM-OSTETIC_) combined with human mesenchymal stem cells (MSCs)	V. Vaněček [99]
Rabbit	New Zealand rabbits (0.7–1.1 years, mean 1.0 year)	Female (weight: 2.3–2.8 kg)	L1–L6 vertebrae	Six	1 mL syringe injected 0.5–0.6 m PMMA into vertebrae	No	PMMA	R. Quan [87]
Canine	Skeletally mature large hounds	– (30–41 kg)	In the vertebral bodies of L1 and L3 just dorsal to the transverse process	Two	Central vertebral defect nominally measuring $18 \times 5 \times 22$ mm^3	No	Cap or PMMA cement	T.M. Turner [95]
	Sheep (2.5 years old ± 0.5)	40–50 kg	L2 to L5 vertebrae	Four	6 mm in diameter and 9 mm in depth	No	Material consists of calcium sulfate and β-tricalcium phosphate (β-TCP)	H.L. Yang [88]

Sheep	Adult female sheep (an average age of 2.5 ± 0.5 years)	40–50 kg	Center of the L2-L5 lumbar vertebra	Four	6 mm in diameter, 10 mm in depth	No	Calcium phosphate (cap), calcium sulfate (cas), and polymethylmethacrylate (PMMA)	X. Zhu [89]
	Female Rambouillet X Columbia Sheep (about 3.5 years old)	Mean 79.0 kg	L3, L4, and L5 vertebral bodies	Three	8 mm in diameter and depth of 15 mm	No	Calcium sodium phosphosilicate putty with or without autograft and NovaBone 45S5 Bioglass Particulate (Novabone, LLSSSSSC, Jacksonville, FL)	H. Kobayashi [90]
	Skeletally mature female Rambouillet X Columbian ewes	60–80 kg	In lateral cortex of L3 and L5	Two	8 mm in diameter and depth of 1 cm drilled hole	No	Calcium phosphate bone cement (Bone-Source1 (CPC); Stryker Orthopaedics, Mahwah, New Jersey) plus carboxymethyl cellulose (CMC)	H. Kobayashi [91]
	Male sheep (2.5 years old)	110–150 lb	L3, L4 and L5	Three	10 mm in diameter and 15 mm deep	No	Bioactive glass based composite putty (NovaBone Putty) compared to NovaBone (a bioactive glass particulate)	Z. Wang [92]
	Sheep (4-year-old sheep)	Mean 75 kg	Central portion of L2 and L4	Two	2.0 cm in diameter	No	Hyaluronic acid (HA) with hydroxyapatite-coated β-tricalcium phosphate (β-TCP), or material of rhnell-1 protein lyophilized onto β-TCP mixed with HA	A.W. James [93]
	Ovariectomized adult female sheep	–	L3 and L4 vertebral bodies	Two	8 mm high × 10 mm deep × 20 mm long,	No	Injectable alendronate-doped calcium phosphate cement	E. Verron [94]
Swine	Porcine species crossing Pietrain breed	Female	L3 vertebral body	One	An anterior defect of 1 cm^3	No	Tricalcium phosphate, tricalcium phosphate with bone morphogenetic protein (rhBMP-7) and autologous bone marrow aspirate with rhbmp-7	E. Manrique [96]
	Miniswines	–	Lumbar vertebra	One	15-mm in depth and 4-mm in diameter cylindrical bone defect	No	Mesenchymal stem cells overexpressing the BMP6 gene (MSC-BMP6)	G. Pelled [97]

Other Bone Defect Models

Femoral Wedge Bone Defect Model

Volker et al. produced femoral wedge bone defect model in rat to resemble clinically relevant situations of osteoporotic bone [100]. In Volker's research, a 5-mm wedge-shaped osteotomy as critical size defect was conducted at the distal metaphyseal area of rat femur and fixed using a T-shaped plate.

Multiple-Defect Model in Canine Femur and Tibia

Bone defects could be established at various sites on canine bones. The middle section of the femur can produce two bone defects with 4-mm in diameter for testing biomaterials [101]. The canine femoral multiple defect (CFMD) is in the lateral cortex of the upper segment of the femur. During modeling, a specific device is fixed to the lateral cortex of the proximal femur by two screws: one upper and one lower. The four consecutive fabricated bone defects are cylindrical with 10 mm in diameter and 15 mm in length [102–104]. In addition, each bone defect is at least 1.5 cm [102] or 1.2 cm [105] apart from the normal bone and bone marrow. This model is suitable for evaluating different materials in the same animal at the same time. Thus, accurate data comparison could be reached, and the number of animals could also be reduced. It is also an effective comparison tool for the evalution of cell delivery [103] and growth factor delivery [102, 105].

The canine tibia could also be used to prepare tibil a multiple-defect model (CTMD) [106]. In a previous study, the tibia was exposed via a medial approach, and then an electrical motor was used to create consecutive cylindrical bone defects with 4 mm in diameter, where different materials could be implanted at the same time [106].

Nonunion Models

The short stumpy limbs of swine might restrict the creation of standard CSD models. Thus, some special models, such as nonunion model, have emerged. Schubert et al. [107] created a critical bone defect with a size 1.5 times that of the diameter of the femoral shaft in miniature swine (aged not older than 6 months and weighing ≤50 kg) and used two 4.5-mm locking compression plates for stabilization. Canines can also be used to produce nonunion model. Saifzadeh et al. produced a 2-mm transverse defect in the right medial radial diaphysis [108], where the defect ends were not fixed to create a nonunion model for testing autogenous greater omentum (Table 6).

Table 6 Other bone defect models

Category	Animal information		Bone defect				Experimental biomaterials	Reference
	Strain and age	Gender and weight	Position	Number	Size	Fixation		
Rat femoral wedge bone defect model	Ovariectomized SD rat (10-week-old)	Female	Distal metaphyseal area of femur	One	3 or 5 mm wedge-shaped osteotomy	T-shaped mini-plate	–	V. Alt [100]
Canine femur diaphyseal local defect model	–	–	Mid-diaphyseal of femur	Two	4 mm in diameter	No	Calcium phosphate cement	H. Yuan [101]
Canine femoral multi-defect model (CFMD)	Coonhounds (mean 1.9 years, 1–3 years)	Female (31.6 ± 2.2 kg)	In metaphysis and proximal diaphysis of the femur	Four (outer edge of each defect was 1.5 cm from adjacent defect)	10 mm diameter * 15 mm long unicortical cylindrical defects	No	A scaffold comprising rhbmp-2/ACS in which the sponge wraps around tricalcium phosphate hydroxyapatite granules (rhBMP-2/ACS/TCP-HA)	V. Luangphakdy [102]
	Coonhounds (age 2–4)	Female (32.1 ± 1.8 kg)		Four	10 mm in diameter and 15 mm in length cylindrical defects	No	Mineralized cancellous allograft (MCA) used as an osteoconductive carrier scaffold for loading of HA-positive cells	T. Caralla [103]
	Coonhounds (age 2.5 years–01 months) (range 1–5 years)	Female and male (34.3 ± 4.2 kg)		Four	10-mm-diameter and 15-mm-long defects	No	Poly(L-lactide-*co*-glycolide) (PLGA), poly(L-lactide-*co*-e-caprolactone) (PLCL), tyrosine-derived polycarbonate (tyrpc), and poly(propylene fumarate) (PPF) Highly porous three-dimensional (3D) scaffolds	V. Luangphakdy [104]

Table 6 (continued)

Category	Animal information		Bone defect				Experimental biomaterials	Reference
	Strain and age	Gender and weight	Position	Number	Size	Fixation		
	Adult mongrel dogs	Male (30.0–35.8 kg)		Four (outer edge of each defect was 1.2 cm from adjacent defect)	1.0-cm diameter and 1.0-cm deep cylindrical defects	No	The OP-1 (3.5 mgrhop-1 in 1 g β sheep collagen I matrix) was implanted in each site combined with either clotted blood or aspirated bone marrow (BM)	H. Takigami [105]
Canine tibia multi-defect model (CTMD)	Mongrel dogs (2–3 years old)	Male (26.2 ± 2.5 kg)	From stifle to metatarsal region in tibia	Seven	4 mm in diameter of circular bone defect	No	Xenogenic bovine fetal demineralized bone matrix (DBM), commercial DBM, omentum, omentum-calf fetal DBM, cortical autograft, and xenogenic cartilage powder	B.S. Amin [106]
Nonunion models	MGH-miniature swine (<6 months)	<50 kg	In the mid-diaphysis of one femur	One	A critical bone defect of 1.5 times the diameter of the femoral shaft	Two 4.5-mm locking compression plates positioned on anterior and lateral sides	3D osteogenic differentiated AMSC (osteoblastic differentiated adipose mesenchymal stem cells) autograft	T. Schubert [107]
	Adult mongrel dogs	Male	Right medial radial diaphysis	One	2-mm transverse bone defect	No	Autogenous greater omentum as a free nonvascularized graft	S. Saifzadeh [108]

Selection of Bone Defect Model

A suitable bone defect model for biomaterials and tissue engineering study is critical. The selection of bone defect model involves the considerations of characteristics of animal models, research purposes, and biological characteristics and design specifications of bone substitute materials.

Age and Sex of Animal

The bone repair ability is very high in young animals. For rats, the speed of bone repair in immature bones was more than that of the mature ones [109, 110]. For example, the healing time of femoral fractures is only 4 weeks for 6-week-old rats, whereas the healing time is 26 weeks for 52 week-old rats [111]. The similar results have been found in rabbit models [112], which might be related to the varied secretion of bone morphogenetic protein (BMP) [113]. The ability of fast bone repair could affect the test results of bone substitute material. Therefore, adult rats were usually selected in the establishment of bone defect models [1, 109, 114, 115]. Similarly, rabbit bone growth stops within 19–32 weeks [116] and therefore 6–9 month-old male rabbits are usually selected for study [72, 73].

Estrogen cycle also has a significant effect on bone repair and turnover. In ovariectomized rats, particularly in elderly rats, the delayed union of the femoral fractures and reduction of bone mineral density (BMD) would occur [117]. Therefore, the ovariectomized rats were often used to establish osteoporotic fracture models. Studies on non-osteoporotic bones should select male animals to avoid this interference. But it is worthy mentioning that male animals usually have a strong sense of territory and the use of individual cages may be required to prevent fights, injuries, or even deaths among male animals. Thus, the age and sex of the animal should be well controlled for study and reported in details.

Comparison between Different Animal Species

In bone tissue engineering and osteoinductive biomaterial researches, the animal species used for model is also critical for reliable and consistent results. The animals of each species have their own unique skeletal characteristics including bone morphology, bone microstructure, bone turnover, and bone remodeling. The ideal animal for bone defect model should have bone characteristics highly similar to that of human, in order to guarantee that the results of animal experiments could be safely extrapolated to humans. However, the availability, costs, and ethical consideration will also be involved in the selection of animal species. A general summary of various commonly used species is provided below.

Rodents

There is a great discrepancy between the bone microstructures of rodents and human. Different from the osteon structure in adult humans, the long bone cortex of rats consisted mainly of primary and cancellous bone, which lacks a Haversian canal system [118].

Intramembrane and endochondral ossifications are two typical bone formation mechanisms during fracture healing [119]. In the fracture healing of rodents, endochondral ossification is more predominant although both kinds of fracture healing mechanisms are present [120, 121]. The bone healing capability of rodents is greater than that of human [1]. Unlike human bone that ceases to grow after sexual maturity, the bones of the sexually mature rodent would continue to grow for extended time [122].

Mice are inexpensive and widely available, easy to manage, and have pure strain and strong infection-resistance. The mouse genome is also thoroughly researched, and it means that bone substitute material can be studied at the molecular level. When the inbred line mice are used in the model, the differences between individuals are small, which makes the results more reliable. Furthermore, nude mice are well-developed animal model that has no immunological rejection for bone tissue engineering substitutes that incorporated with human cells. However, the mouse skeleton is too small to establish a large defect model. Moreover, for some bone implants, shrinking to the appropriate size of the defect to evaluated the clinical outcome is difficult [123]. In addition, compared with mouse model, the rat model has better practical value in biomechanical tests.

Rabbit

Both the macro- and microstructure of rabbit's bones are distinctly different from those of human [124]. Unlike the mature human bones with secondary hierarchical structure, rabbit skeletal bone has a primary vascular longitudinal tissue structure. It forms a vascular access to the Haversian canal that is wrapped around the medullary cavity and distributed on the periosteal surface [125]. The bones between these lamellar structures are composed of dense Haversian bone [125]. However, the bone structure of rabbit is still more representative for human than that of mouse [126].

Compared with primates and rodents, rabbits have faster bone remodeling, particularly in Haversian remodeling of cortical bones [127–129]. Thus, it is not advised to extrapolate from the rabbit's findings to possible human clinical responses [130].

Rabbits are easy to manage, and the cost is less than larger animals. Compared with rodents, the biggest advantage of rabbit is the large size of bones that benefits for bone defect creation. Also, the ear margin vein of rabbit is convenient for obtaining blood samples, which is beneficial for blood analysis.

Canine

Canine bone components are highly similar to those of humans, including hydroxyproline, extractable proteins, IGF-1 content [131], as well as water fraction, organic fraction, volatile inorganic fraction, and ash fraction [132]. Microscopically, canine bone has secondary osteonal structures, known as plexiform bone or laminar bone in the adjacent endosteum and periosteum [124]. The plexiform bone is found in the rapid growth of children, or in rapidly growing large animals [133]. Therefore, canine bone is considered as one of the best representatives of human bone structure [131]. Thus, dogs are suitable for animal models in biomaterials research and bone tissue engineering. However, canine bone defect models have not been widely used due to the high costs and difficult breeding as well as ethical limitations in some countries.

When using canine as animal model, it should be noted that bone remodeling process of dogs is quite different from that in humans. The mean bone turnover rate of young adult female beagles is much higher than that of humans [134, 135], which could partially explain the high fusion rates in canine lumbar fusion models and the low nonunion rates [98]. The rate of bone remodeling in different bones varies significantly for dogs. Considering adult male beagles, the bone turnover rate is close to 200% and 12% annually in the lumbar vertebra and talus, respectively. Although, the mean total turnover of trabecular bone is approximately 100% annually [134].

Goat and Sheep

Sheep and goats have the similar weight as humans, and their limbs can provide sufficient testing space for clinically sized implants [128, 136, 137]. The structure of sheep bone is remarkably different from that of human histologically. Compared with human bone that mainly consists of secondary bone structure [138], the sheep bone has mostly primary bone structure where osteons is <100 μm in diameter and contains at least two central blood vessels. Bone structure changes in sheep are related to age. Before the age of 3–4 years, the sheep's plexiform bone structure consists of a combination of woven and lamellar bone with sandwiched vascular plexuses [128]. When the sheep is 7–9 years old, Haversian remodeling occurs [128] and becomes more common as age increases [139]. Furthermore, the location of bone remodeling seemed to vary with bone types and the earliest indications of remodeling include the distal femur, radial shaft, and humerus [128]. Similar to sheep, the Haversian systems of goats are unevenly distributed in the whole skeleton, which mainly located in skull and the medial part of tibial shaft [140]. After comparing sheep with menopausal women in terms of bone mass, bone volume, and mineral deposition rate, Turner et al. concluded that old sheep were suitable as models for osteoporosis in the elderly [141].

Although the microscopic structure of sheep bone is different from humans, many studies have confirmed that bone turnover and remodeling activities in the sheep model are similar to those of humans [142], which makes the sheep as a valuable model animal for testing bone substitutes [136, 143]. Although the amount of bone ingrowth in sheep was greater than that in humans, Willie et al. found that sheep and humans had similar patterns in terms of bone ingrowth into porous implants [144]. Lamerigts et al. found similar sequences of events occurring in humans and goats for bone grafts [145]. Furthermore, Dai et al. found that the bone healing capacity and the blood supply of goat tibia were both similar to those of humans [38].

There are some differences between goats and sheep, including microscopic structure, bone metabolism, and remodeling. However, these two models were very similar in general. Therefore, choosing goats or sheep as animal models depends on availability and other personal preferences.

Swine

Compared with sheep model, swine model are closer to humans in bone morphology, bone metabolism, bone healing, and remodeling [98, 146, 147]. The cross-sectional diameter and area of femur are very similar to those of human [148]. In mature bone, the swine have well-developed Haversian systems [149] and a layered bone structure similar to human [150]. Furthermore, its bone density and bone mineral content are almost equal to those of human bones [131], although the trabecular meshwork of swine is denser than that of humans [151]. The bone growth rate in swine is 1.2–1.5 mm daily, which is comparable to that of human beings (1.0–1.5 mm daily). The processes of bone remodeling in swine, including trabecular and intra-cortical basic multicellular remodeling units (BMU) based remodeling, are all similar to those in humans [150, 151]. Moreover, mineralization rate of cortical bone was the same as that in humans [152]. Thus, swines are considered suitable models for bone-related studies [147].

Nevertheless, pigs weighing over 150 kg are considered not suitable to produce animal models for biomaterials and bone tissue engineering study. Even for miniature swine, the daily management is difficult, particularly because they are noisy and aggressive. Thus, in many cases sheep and dogs are preferred over swine as large animal models [128, 153].

Small animals, such as mice, rat, and rabbits account for >70% [154, 155] of all the animal models. In general, the advantages of small animals include good affordability, convenient management, and small individual difference. However, some problems exist when using small animal model to test the potential clinical products. Considering cell-seeded scaffolds for example, the viability and osteogenic potential of cells in small scaffolds are easy to maintain to repair bone defects in small animals but it becomes difficult in large scaffolds used for large bone defect repair. Besides, small animals have strong bone healing ability and weak immune response, which is different from the actual bone repair of the human body.

For large animal models, such as sheep, dogs, and swine, its large bones would allow the evaluation of implants with the size close to those used in humans. Moreover, compared with small animals, large animal bones have higher similarity to human bones. Although with many advantages, large animal models were not usually used due to the difficulties of accommodation, management, availability, and ethics.

Research Related Criteria

For experiments that are expected to get results in a short time, the small animal model (rodents and rabbits) is recommended due to their fast bone healing process. When long-term observation is required, large animal models are recommended, since the bone structure, bone repair mechanism, and immune response in large animals are closer to those in human body.

The study purpose of osteoinductive biomaterials and bone tissue engineering is also an important factor when choosing the model. If the experiment was to test the osteoinductive capability of the materials, skull defect models or long bone metaphyseal defect model could be used to simplify the experimental procedure.

However, more complicated models are needed for studies with specific research goals. For example, long bone segmental defect model should be utilized to test the substitute materials designed to repair long bone defects, in which fixation devices would be utilized to fix the ends of defects to simulate clinical practice. When bone substitute materials are used as bone cement for vertebroplasty (VP) and kyphoplasty (KP) surgery, the vertebral defect models are recommended.

The characteristics of materials also affect the selection of models. If the bone substitute materials were colloidal, shapeless granule, or powder, the long bone metaphyseal defect model or the multiple defect model might be a better choice, since it can avoid the dislocation and loss of biomaterials or scaffolds.

For materials containing human tissue cells, nude rodent model may be suitable models to prevent the immune system from interfering with cells. Nude rodent models are capable of ignoring the source of cells and the reaction between scaffold and the host; hence, it could be used to study the efficacy of engineered tissue products directly [156]. However, because nude rodents are immune-deficient animals and cannot reflect the normal immune functions, the data obtained from nude rodent models may need to be further confirmed by other animal models.

Bone Defect Models Simulating Clinical Scenarios

Actual Situation of Bone Defect in Clinical Settings

It should be recognized that most fractures were simple two-part fractures or minor comminuted fractures without bone defects. Bone defect is usually caused by comminuted broken ends of the fracture. In fact, serious violent injuries not only cause fracture end comminution, but also force the fractured fragment to move, rendering it impossible to maintain the position of the broken ends.

When the bone defect exceeds the self-repairing ability, nonunion would occur without external intervention therapy. However, the bone defect size is not the only factor that affects bone healing. The real bone defect is accompanied by severe damage of blood supply at the broken ends of the fracture, which would seriously hamper the fracture healing. Therefore, the occurrence of bone defect and fracture healing are not simple processes in clinical practice. Unfortunately, most of the current bone defect models did not well imitate the actual state of clinical bone defects.

Fracture Model and Nonunion Model

There are two models that simulate the actual state of a bone fracture: closed fracture model and nonunion model. The closed fracture model is the most widely used animal model for the study of fracture healing, which is caused by exerting strength damage without incision to completely simulate the clinical practice. To make consistent fracture models, Jackson made fracture of rat femur on a force-controllable blunt guillotine device after inserting intramedullary needle into marrow cavity [157]. Bonnarens and Einhorn [158] improved the device to make the whole modeling tool portable and easy to operate. Thus, the tools and concepts of closed fracture model making have been adopted by many scholars. Marturano et al. [159] made further improvements to the fracture device, making it suitable for preparing models on the femur of mice. The tibia of mouse may also be used to make a fracture model using similar methods [160–163]. Compared with femur, the location of tibia has less muscle attachment. However, the tibia is not an ideal model for fracture-related research due to the curved long axis, complicating the biomechanical test and the insufficient soft tissue around the bone [164, 165]. Furthermore, the fibula is accidentally fractured at a rate up to 30% during the modeling process, which can change the healing rate of the tibia [166].

For creating nonunion model, closed [167–169] or open [170] fractures are created first. Subsequently, the 2-mm periosteum of the two fracture ends are cauterized and corroded, which destroy the blood supply at the fracture ends, resulting in atrophic nonunion [167–170]. These nonunion models are obviously different from the nonunion caused by severe trauma in clinical practice, which causes not only the crushing of fracture ends, but also large segmental bone defects, and destroys the blood supply around the broken ends.

Fracture-Bone Defect Models

Closed fracture models can simulate the fractures seen in clinical practice. However, the closed fracture models are suitable for fracture healing research without space for implanting biomaterial or bone tissue engineering scaffolds. More importantly, these fracture models can self-heal without external intervention. Although current nonunion models cannot self-heal, they are not consistent with the clinical practice. Fracture defect models mimicking the clinical scenarios are still lacking.

For producing fracture-defect model, fracture is first produced. Blunt guillotine and three-point bending are classic methods of creating fractures, and the tools are simple and easy to operate. However, unlike aforementioned fracture models that have transverse fracture, fracture-defect models aim to create comminuted fracture. After creating the fracture, open operation can be used to expose the fracture ends and remove the broken pieces of bone to create a bone defect. The incision at the broken ends can be regarded as an open fracture, which is consistent with the open fractures in severe traumatic injury. Fracture-defect model can use tibia to prepare. This is because tibia has less muscle encircling, convenient positioning and is directly located in the subcutaneous tissue, which makes the removal of broken bones much easier. In clinical practice, the open comminuted fracture also occurs mostly in the tibia.

Challenges and Future Prospects

There is no standard answer for how much bone should be removed to reach the state of nonunion. In bone defect model, there is a concept of CSD and CSD values have been established in many animal models. While in fracture defect models, the factors affecting bone healing include not only the size of the defect, but also the degree of soft tissue injury at the fracture ends. Unfortunately, quantification of soft tissue damage in the process of modeling is difficult.

The present fracture creation tools are designed for small animal models. When produce similar models in large animals, redesign and development of new tools becomes necessary. However, it is not known whether the previous classical weight fall and the three-point bending principle are applicable to or effective on large animal models. In addition, small animals, such as mice and rats, used for fracture model have less individual difference and high repeatability. When fracture defect model is made using large animals, the individual difference between animals enlarges and the repeatability of the model decreases.

Summary

Compared with other animals, canine and swine are closer to humans in bone morphology, bone metabolism, fracture healing, and remodeling. However, the rat model is the most widely used bone defect model in the study of biomaterials and bone tissue engineering. It can be seen that in the actual selection of animal models, the factors that need to be considered are multifaceted. In the selection of animal species, researchers should consider the ultimate purpose of the biomaterial evalution, the material characteristics, their own facility capacity, costs and other resources.

In addition, it should be recognized that there is a significant difference between the animal bone defect models and clinical. In the future, the development of fracture defect models that are close to clinical reality may be come possible, making the evaluation results more reliable for the studies of biomaterials and bone tissue engineering.

Acknowledgements This work is supported by the National Natural Science Foundation of China (81622032, 51672184 and 81501858), Jiangsu Innovation and Entrepreneurship Program, National Basic Research Program of China (973 Program, 2014CB748600), Suzhou Science and Technology Project (SYS2019022), and the Priority Academic Program Development of Jiangsu High Education Institutions (PAPD).

References

1. Schmitz JP, Hollinger JO (1986) The critical size defect as an experimental model for cranio-maxillofacial nonunions. Clin Orthop Relat Res 205:299
2. Schmitz JP, Schwartz Z, Hollinger JO, Boyan BD (1990) Characterization of rat calvarial nonunion defects. Cells Tissues Organs 138:185–192
3. Trotta DR, Gorny C Jr, Zielak JC, Gonzaga CC, Giovanini AF, Deliberador TM (2014) Bone repair of critical size defects treated with mussel powder associated or not with bovine bone graft: histologic and histomorphometric study in rat calvaria. J Craniomaxillofac Surg 42:738–743
4. Guanghui L, Xi W, Jian C, Zhaoyu J, Dongyang M, Yanpu L et al (2014) Coculture of peripheral blood CD34+ cell and mesenchymal stem cell sheets increase the formation of bone in calvarial critical-size defects in rabbits. Br J Oral Maxillofac Surg 52:134–139
5. Liu X, Zhou S, Li Y, Yan J (2012) Stromal cell derived factor-1α enhances bone formation based on in situ recruitment: a histologic and histometric study in rabbit calvaria. Biotechnol Lett 34:387–395
6. Cui L, Liu B, Liu G, Zhang W, Cen L, Sun J et al (2007) Repair of cranial bone defects with adipose derived stem cells and coral scaffold in a canine model. Biomaterials 28:5477–5486
7. Liping X, Daisuke U, Sylvain C, Collin HB, Lyndon C, Liisa K et al (2014) Fibroblast growth factor-2 isoform (low molecular weight/18 kDa) overexpression in preosteoblast cells promotes bone regeneration in critical size calvarial defects in male mice. Endocrinology 155:965–974
8. Liao YH, Chang YH, Sung LY, Li KC, Yeh CL, Yen TC et al (2014) Osteogenic differentiation of adipose-derived stem cells and calvarial defect repair using baculovirus-mediated co-expression of BMP-2 and miR-148b. Biomaterials 35:4901–4910
9. Stephan SJ, Tholpady SS, Ce GBA, Botchway EA, Nair LS, Ogle RC et al (2010) Injectable tissue-engineered bone repair of a rat calvarial defect. Laryngoscope 120:895–901
10. Jun Z, Gang S, Changsheng L, Shaoyi W, Wenjie Z, Xiaochen Z et al (2012) Enhanced healing of rat calvarial defects with sulfated chitosan-coated calcium-deficient hydroxyapatite/bone morphogenetic protein 2 scaffolds. Tissue Eng Part A 18:185–197
11. Josephine F, Zhi Y, Shihjye T, Charisse T, Nimni ME, Mark U et al (2014) Injectable gel graft for bone defect repair. Regen Med 9:41–51
12. Glowacki J, Altobelli D, Mulliken JB (1981) Fate of mineralized and demineralized osseous implants in cranial defects. Calcif Tissue Int 33:71–76
13. Sakata Y, Ueno T, Kagawa T, Kanou M, Fujii T, Yamachika E et al (2006) Osteogenic potential of cultured human periosteum-derived cells—a pilot study of human cell transplantation into a rat calvarial defect model. J Craniomaxillofac Surg 34:461–465

14. Chim H, Schantz JT (2006) Human circulating peripheral blood mononuclear cells for calvarial bone tissue engineering. Plast Reconstr Surg 117:468–478
15. Mhawi AA, Peel SA, Fok TC, Clokie CM (2007) Bone regeneration in athymic calvarial defects with Accell DBM100. J Craniofac Surg 18:497–503
16. Parizi AM, Oryan A, Shafiei-Sarvestani Z, Bigham AS (2012) Human platelet rich plasma plus Persian Gulf coral effects on experimental bone healing in rabbit model: radiological, histological, macroscopical and biomechanical evaluation. J Mater Sci Mater Med 23:473–483
17. Berner A, Woodruff MA, Lam CXF, Arafat MT, Saifzadeh S, Steck R et al (2014) Effects of scaffold architecture on cranial bone healing. Int J Oral Maxillofac Surg 43:506–513
18. Lin CY, Chang YH, Li KC, Lu CH, Sung LY, Yeh CL et al (2013) The use of ASCs engineered to express BMP2 or TGF-β3 within scaffold constructs to promote calvarial bone repair. Biomaterials 34:9401–9412
19. Nick T, Ryo J, Riddhi G, Lukasz W, Fabio L, Charles M et al (2013) Modification of xenogeneic graft materials for improved release of P-15 peptides in a calvarium defect model. J Craniofac Surg 25:70–76
20. Tanuma Y, Matsui K, Kawai T, Matsui A, Suzuki O, Kamakura S et al (2013) Comparison of bone regeneration between octacalcium phosphate/collagen composite and β-tricalcium phosphate in canine calvarial defect. Oral Surg Oral Med Oral Pathol Oral Radiol 115:9–17
21. Sato K, Urist MR (2003) Induced regeneration of calvaria by bone morphogenetic protein (BMP) in dogs. Clin Orthop Relat Res 187:301
22. Kinsella CR, Bykowski MR, Lin AY, Cray JJ, Durham EL, Smith DM et al (2011) BMP-2-mediated regeneration of large-scale cranial defects in the canine: an examination of different carriers. Plast Reconstr Surg 127:1865
23. Mulliken JB, Glowacki J (1980) Induced osteogenesis for repair and construction in the craniofacial region. Plast Reconstr Surg 65:553–560
24. Freeman E, Turnbull RS (2010) The value of osseous coagulum as a graft material. J Periodontal Res 8:229–236
25. Turnbull RS, Freeman E (2010) Use of wounds in the parietal bone of the rat for evaluating bone marrow for grafting into periodontal defects. J Periodontal Res 9:39–43
26. Livingston TL, Gordon S, Archambault M, Kadiyala S, Mcintosh K, Smith A et al (2003) Mesenchymal stem cells combined with biphasic calcium phosphate ceramics promote bone regeneration. J Mater Sci Mater Med 14:211–218
27. Komaki H, Tanaka T, Chazono M, Kikuchi T (2006) Repair of segmental bone defects in rabbit tibiae using a complex of -tricalcium phosphate, type I collagen, and fibroblast growth factor-2. Biomaterials 27:5118–5126
28. Li X, Feng Q, Liu X, Dong W, Cui F (2006) Collagen-based implants reinforced by chitin fibres in a goat shank bone defect model. Biomaterials 27:1917–1923
29. Liu G, Zhao L, Zhang W, Cui L, Liu W, Cao Y (2008) Repair of goat tibial defects with bone marrow stromal cells and β-tricalcium phosphate. J Mater Sci Mater Med 19:2367–2376
30. Sarban S, Senkoylu A, Isikan UE, Korkusuz P, Korkusuz F (2009) Can rhBMP-2 containing collagen sponges enhance bone repair in ovariectomized rats?: a preliminary study. Clin Orthop Relat Res 467:3113
31. Rena S, Jessica G, Alan E, Chu TMG, Shawn G (2011) Increasing vascularity to improve healing of a segmental defect of the rat femur. J Orthop Trauma 25:472
32. Vaida G, Micah M, Alan I, Fangjun L, Nicola P, Damian G et al (2012) Improved healing of large segmental defects in the rat femur by reverse dynamization in the presence of bone morphogenetic protein-2. J Bone Joint Surg Am 94:2063–2073
33. Corinne S, Lashan SC, Olabisi RM, Kayleigh S, Zawaunyka L, Zbigniew G et al (2013) Rapid healing of femoral defects in rats with low dose sustained BMP2 expression from PEGDA hydrogel microspheres. J Orthop Res 31:1597–1604
34. Angle SR, Sena K, Sumner DR, Virkus WW, Virdi AS (2012) Healing of rat femoral segmental defect with bone morphogenetic protein-2: a dose response study. J Musculoskelet Neuronal Interact 12:28–37

35. Duan Z, Zheng Q, Guo X, Li C, Wu B, Wu W (2008) Repair of rabbit femoral defects with a novel BMP2-derived oligopeptide P24. J Huazhong Univ Sci Technol Med Sci 28:426–430
36. Amaia C, Reichert JC, Epari DR, Siamak S, Arne B, Hanna S et al (2013) Polycaprolactone scaffold and reduced rhBMP-7 dose for the regeneration of critical-sized defects in sheep tibiae. Biomaterials 34:9960–9968
37. Berner A, Reichert JC, Woodruff MA, Saifzadeh S, Morris AJ, Epari DR et al (2013) Autologous vs. allogenic mesenchymal progenitor cells for the reconstruction of critical sized segmental tibial bone defects in aged sheep. Acta Biomater 9:7874–7884
38. Dai KR, Xu XL, Tang TT, Zhu ZA, Yu CF, Lou JR et al (2005) Repairing of goat tibial bone defects with BMP-2 gene–modified tissue-engineered bone. Calcif Tissue Int 77:55–61
39. Zhu L, Liu W, Cui L, Cao Y (2006) Tissue-engineered bone repair of goat-femur defects with osteogenically induced bone marrow stromal cells. Tissue Eng 12:423
40. Pluhar GE, Turner AS, Pierce AR, Toth CA, Wheeler DL (2006) A comparison of two biomaterial carriers for osteogenic protein-1 (BMP-7) in an ovine critical defect model. J Bone Joint Surg Br 88:960–966
41. Fialkov JA, Holy CE, Shoichet MS, Davies JE (2003) In vivo bone engineering in a rabbit femur. J Craniofac Surg 14:324–332
42. Fan JJ, Mu TW, Qin JJ, Bi L, Pei GX (2015) Different effects of implanting sensory nerve or blood vessel on the vascularization, neurotization, and osteogenesis of tissue-engineered bone in vivo. Biomed Res Int 2014:412570
43. Nimrod R, Tova B, Alon B, Ben S, Michal ST, Yankel G et al (2009) Transplanted blood-derived endothelial progenitor cells (EPC) enhance bridging of sheep tibia critical size defects. Bone 45:918–924
44. Boyde A, Corsi A, Quarto R, Cancedda R, Bianco P (1999) Osteoconduction in large macroporous hydroxyapatite ceramic implants: evidence for a complementary integration and disintegration mechanism. Bone 24:579–589
45. Oryan A, Alidadi S, Bigham-Sadegh A, Moshiri A (2016) Comparative study on the role of gelatin, chitosan and their combination as tissue engineered scaffolds on healing and regeneration of critical sized bone defects: an in vivo study. J Mater Sci Mater Med 27:155
46. Shafiei Z, Bigham AS, Dehghani SN, Nezhad ST (2009) Fresh cortical autograft versus fresh cortical allograft effects on experimental bone healing in rabbits: radiological, histopathological and biomechanical evaluation. Cell Tissue Bank 10:19–26
47. Dehghani SN, Bigham AS, Nezhad ST, Shafiei Z (2008) Effect of bovine fetal growth plate as a new xenograft in experimental bone defect healing: radiological, histopathological and biomechanical evaluation. Cell Tissue Bank 9:91–99
48. Itoi T, Harada Y, Irie H, Sakamoto M, Tamura K, Yogo T et al (2016) Escherichia coli-derived recombinant human bone morphogenetic protein-2 combined with bone marrow-derived mesenchymal stromal cells improves bone regeneration in canine segmental ulnar defects. BMC Vet Res 12:201
49. Bigham AS, Dehghani SN, Shafiei Z, Nezhad ST (2008) Xenogenic demineralized bone matrix and fresh autogenous cortical bone effects on experimental bone healing: radiological, histopathological and biomechanical evaluation. J Orthop Traumatol 9:73–80
50. Kimelman-Bleich N, Pelled G, Zilberman Y, Kallai I, Mizrahi O, Tawackoli W et al (2011) Targeted gene-and-host progenitor cell therapy for nonunion bone fracture repair. Mol Ther 19:53
51. Kimelman BN, Pelled GD (2009) The use of a synthetic oxygen carrier-enriched hydrogel to enhance mesenchymal stem cell-based bone formation in vivo. Biomaterials 30:4639–4648
52. Sun JS, Chen PY, Tsuang YH, Chen MH, Chen PQ (2009) Vitamin-D binding protein does not enhance healing in rat bone defects: a pilot study. Clin Orthop Relat Res 467:3156–3164
53. Lazard ZW, Heggeness MH, Hipp JA, Corinne S, Fuentes AS, Nistal RP et al (2011) Cell-based gene therapy for repair of critical size defects in the rat fibula. J Cell Biochem 112:1563–1571
54. Chakkalakal D, Strates B, Garvin K, Novak J, Fritz E, Mollner T et al (2001) Demineralized bone matrix as a biological scaffold for bone repair. Tissue Eng 7:161–177

55. Shafiei-Sarvestani Z, Oryan A, Bigham AS, Meimandi-Parizi A (2012) The effect of hydroxyapatite-hPRP, and coral-hPRP on bone healing in rabbits: radiological, biomechanical, macroscopic and histopathologic evaluation. Int J Surg 10:96–101
56. Oryan A, Meimandi PA, Shafieisarvestani Z, Bigham AS (2012) Effects of combined hydroxyapatite and human platelet rich plasma on bone healing in rabbit model: radiological, macroscopical, histopathological and biomechanical evaluation. Cell Tissue Bank 13:639–651
57. Bigham-Sadegh A, Karimi I, Shadkhast M, Mahdavi MH (2015) Hydroxyapatite and demineralized calf fetal growth plate effects on bone healing in rabbit model. J Orthop Traumatol 16:141–149
58. Zellin G, Linde A (1997) Treatment of segmental defects in long bones using osteopromotive membranes and recombinant human bone morphogenetic protein-2: an experimental study in rabbits. Scand J Plast Reconstr Surg Hand Surg 31:97–104
59. Luca L, Rougemont AL, Walpoth BH, Boure L, Tami A, Anderson JM et al (2015) Injectable rhBMP-2-loaded chitosan hydrogel composite: osteoinduction at ectopic site and in segmental long bone defect. J Biomed Mater Res A 96A:66–74
60. Tu J, Wang H, Li H, Dai K, Wang J, Zhang X (2009) The in vivo bone formation by mesenchymal stem cells in zein scaffolds. Biomaterials 30:4369–4376
61. Satoshi K, Ryuhei F, Shoji Y, Shinji F, Kazutoshi N, Koichiro T et al (2003) Bone regeneration by recombinant human bone morphogenetic protein-2 and a novel biodegradable carrier in a rabbit ulnar defect model. Biomaterials 24:1643–1651
62. Bostrom M, Lane JM, Tomin E, Browne M, Berberian W, Turek T et al (1996) Use of bone morphogenetic protein-2 in the rabbit ulnar nonunion model. Clin Orthop Relat Res 327:272–282
63. Bouxsein ML, Turek TJ, Blake CA, D'Augusta D, Li X, Stevens M et al (2001) Recombinant human bone morphogenetic protein-2 accelerates healing in a rabbit ulnar osteotomy model. J Bone Joint Surg Am 83-A:1219
64. Geuze RE, Theyse LFH, Kempen DHR, Hazewinkel HAW, Kraak HYA, Oner FC et al (2012) A differential effect of bone morphogenetic protein-2 and vascular endothelial growth factor release timing on osteogenesis at ectopic and orthotopic sites in a large-animal model. Tissue Eng Part A 18:2052–2062
65. Theyse LF, Oosterlaken-Dijksterhuis MA, Van DJ, Dhert WJ, Hazewinkel HA (2006) Growth hormone stimulates bone healing in a critical-sized bone defect model. Clin Orthop Relat Res 446:259
66. Jones CB, Sabatino CT, Badura JM, Sietsema DL, Marotta JS (2008) Improved healing efficacy in canine ulnar segmental defects with increasing recombinant human bone morphogenetic protein-2/allograft ratios. J Orthop Trauma 22:550–559
67. Cook SD, Baffes GC, Wolfe MW, Sampath TK, Rueger DC (1994) Recombinant human bone morphogenetic protein-7 induces healing in a canine long-bone segmental defect model. Clin Orthop Relat Res 301:302
68. Bigham-Sadegh A, Mirshokraei P, Karimi I, Oryan A, Aparviz A, Shafiei-Sarvestani Z (2012) Effects of adipose tissue stem cell concurrent with greater omentum on experimental long-bone healing in dog. Connect Tissue Res 53:334–342
69. Zhang X, Zhu L, Cao Y, Liu Y, Xu Y, Ye W et al (2012) Repair of rabbit femoral condyle bone defects with injectable nanohydroxyapatite/chitosan composites. J Mater Sci Mater Med 23:1941–1949
70. Kanazawa M, Tsuru K, Fukuda N, Sakemi Y, Nakashima Y, Ishikawa K (2017) Evaluation of carbonate apatite blocks fabricated from dicalcium phosphate dihydrate blocks for reconstruction of rabbit femoral and tibial defects. J Mater Sci Mater Med 28:85
71. Betti LV, Bramante PCM, Cestari PTM, Granjeiro PJM, Garcia PRB (2011) Repair of rabbit femur defects with organic bovine bone cancellous block or cortical granules. Int J Oral Maxillofac Implants 26:1167
72. Gil-Albarova J, Vila M, Badiola-Vargas J, Sánchez-Salcedo S, Herrera A, Vallet-Regi M (2012) In vivo osteointegration of three-dimensional crosslinked gelatin-coated hydroxyapatite foams. Acta Biomater 8:3777–3783

73. Zheng H, Bai Y, Shih MS, Hoffmann C, Peters F, Waldner C et al (2014) Effect of a β-TCP collagen composite bone substitute on healing of drilled bone voids in the distal femoral condyle of rabbits. J Biomed Mater Res B Appl Biomater 102:376–383
74. Liu J, Mao K, Liu Z, Wang X, Cui F, Guo W et al (2013) Injectable biocomposites for bone healing in rabbit femoral condyle defects. PLoS One 8:e75668
75. Alireza RG, Lambers FM, Mehdi GR, Ralph M, Pioletti DP (2011) In vivo loading increases mechanical properties of scaffold by affecting bone formation and bone resorption rates. Bone 49:1357–1364
76. Guihard P, Boutet MA, Brounais-Le Royer B, Gamblin AL, Amiaud J, Renaud A et al (2015) Oncostatin m, an inflammatory cytokine produced by macrophages, supports intramembranous bone healing in a mouse model of tibia injury. Am J Pathol 185:765–775
77. Laurent M, Bénédicte R, Olivier C, Jean-Christophe F (2010) Drilled hole defects in mouse femur as models of intramembranous cortical and cancellous bone regeneration. Calcif Tissue Int 86:72–81
78. Nagashima M, Sakai A, Uchida S, Tanaka S, Tanaka M, Nakamura T (2005) Bisphosphonate (YM529) delays the repair of cortical bone defect after drill-hole injury by reducing terminal differentiation of osteoblasts in the mouse femur. Bone 36:502–511
79. Xu W, Ganz C, Weber U, Adam M, Holzhüter G, Wolter D et al (2011) Evaluation of injectable silica-embedded nanohydroxyapatite bone substitute in a rat tibia defect model. Int J Nanomedicine 2011:1543–1552
80. Trejo CG, Lozano D, Manzano M, Doadrio JC, Salinas AJ, Dapía S et al (2010) The osteoinductive properties of mesoporous silicate coated with osteostatin in a rabbit femur cavity defect model. Biomaterials 31:8564–8573
81. Guillemin G, Patat JL, Fournie J, Chetail M (2010) The use of coral as a bone graft substitute. J Biomed Mater Res A 21:557–567
82. Choi S, Liu IL, Yamamoto K, Honnami M, Sakai T, Ohba S et al (2014) Implantation of tetrapod-shaped granular artificial bones or Î²-tricalcium phosphate granules in a canine large bone-defect model. J Vet Med Sci 76:229–235
83. Smit TH (2002) The use of a quadruped as an in vivo model for the study of the spine—biomechanical considerations. Eur Spine J 11:137–144
84. Bloemers FW, Stahl JP, Sarkar MR, Linhart W, Rueckert U, Wippermann BW (2004) Bone substitution and augmentation in trauma surgery with a resorbable calcium phosphate bone cement. Eur J Trauma 30:17–22
85. Liang H, Wang K, Shimer AL, Li X, Balian G, Shen FH (2010) Use of a bioactive scaffold for the repair of bone defects in a novel reproducible vertebral body defect model. Bone 47:197–204
86. Dmitriy S, Ilan K, Wafa T, Doron CY, Anthony O, Susan S et al (2011) Gene-modified adult stem cells regenerate vertebral bone defect in a rat model. Mol Pharm 8:1592
87. Quan R, Ni Y, Zhang L, Xu J, Zheng X, Yang D (2014) Short- and long-term effects of vertebroplastic bone cement on cancellous bone. J Mech Behav Biomed Mater 35:102–110
88. Yang HL, Zhu XS, Chen L, Chen CM, Mangham DC, Coulton LA et al (2012) Bone healing response to a synthetic calcium sulfate/β-tricalcium phosphate graft material in a sheep vertebral body defect model. J Biomed Mater Res B Appl Biomater 100B:1911–1921
89. Zhu X, Chen X, Chen C, Wang G, Gu Y, Geng D et al (2012) Evaluation of calcium phosphate and calcium sulfate as injectable bone cements in sheep vertebrae. J Spinal Disord Tech 25:333
90. Kobayashi H, Turner AS, Kawamoto T, Bauer TW (2010) Evaluation of a silica-containing bone graft substitute in a vertebral defect model. J Biomed Mater Res A 92A:596–603
91. Kobayashi H, Fujishiro T, Belkoff SM, Kobayashi N, Turner AS, Seim HB et al (2010) Long-term evaluation of a calcium phosphate bone cement with carboxymethyl cellulose in a vertebral defect model. J Biomed Mater Res A 88A:880–888
92. Zhen W, Bin L, Lei C, Jiang C (2011) Evaluation of an osteostimulative putty in the sheep spine. J Mater Sci Mater Med 22:185–191

93. James AW, Chiang M, Asatrian G, Shen J, Goyal R, Chung CG et al (2016) Vertebral implantation of NELL-1 enhances bone formation in an osteoporotic sheep model. Tissue Eng Part A 22:840
94. Verron E, Pissonnier ML, Lesoeur J, Schnitzler V, Fellah BH, Pascal-Moussellard H et al (2014) Vertebroplasty using bisphosphonate-loaded calcium phosphate cement in a standardized vertebral body bone defect in an osteoporotic sheep model. Acta Biomater 10:4887–4895
95. Turner TM, Urban RM, Singh K, Hall DJ, Renner SM, Lim TH et al (2008) Vertebroplasty comparing injectable calcium phosphate cement compared with polymethylmethacrylate in a unique canine vertebral body large defect model. Spine J 8:482–487
96. Manrique E, Chaparro D, Cebrián JL, López-Durán L (2014) In vivo tricalcium phosphate, bone morphogenetic protein and autologous bone marrow biomechanical enhancement in vertebral fractures in a porcine model. Int Orthop 38:1993–1999
97. Pelled G, Sheyn D, Tawackoli W, Jun DS, Koh Y, Su S et al (2016) BMP6-engineered MSCs induce vertebral bone repair in a pig model: a pilot study. Stem Cells Int 2016:1–8
98. Reichert JC, Saifzadeh S, Wullschleger ME, Epari DR, Schutz MA, Duda GN et al (2009) The challenge of establishing preclinical models for segmental bone defect research. Biomaterials 30:2149–2163
99. Vaněček V, Klíma K, Kohout A, Foltán R, Jiroušek O, Šedý J et al (2013) The combination of mesenchymal stem cells and a bone scaffold in the treatment of vertebral body defects. Eur Spine J 22:2777–2786
100. Alt V, Thormann U, Ray S, Zahner D, Dürselen L, Lips K et al (2013) A new metaphyseal bone defect model in osteoporotic rats to study biomaterials for the enhancement of bone healing in osteoporotic fractures. Acta Biomater 9:7035–7042
101. Yuan H, Li Y, de Bruijn JD, de Groot K, Zhang X (2000) Tissue responses of calcium phosphate cement: a study in dogs. Biomaterials 21:1283–1290
102. Luangphakdy V, Shinohara K, Pan H, Boehm C, Samaranska A, Muschler GF (2015) Evaluation of rhBMP-2/collagen/TCP-HA bone graft with and without bone marrow cells in the canine femoral multi defect model. Eur Cell Mater 29:57–68
103. Caralla T, Joshi P, Fleury S, Luangphakdy V, Shinohara K, Pan H et al (2013) In vivo transplantation of autogenous marrow-derived cells following rapid intraoperative magnetic separation based on hyaluronan to augment bone regeneration. Tissue Eng Part A 19:125–134
104. Luangphakdy V, Walker E, Shinohara K, Pan H, Hefferan T, Bauer TW et al (2013) Evaluation of osteoconductive scaffolds in the canine femoral multi-defect model. Tissue Eng Part A 19:634–648
105. Takigami H, Kumagai K, Latson L, Togawa D, Bauer T, Powell K et al (2010) Bone formation following OP-1 implantation is improved by addition of autogenous bone marrow cells in a canine femur defect model. J Orthop Res 25:1333–1342
106. Bigham-Sadegh A, Karimi I, Alebouye M, Shafie-Sarvestani Z, Oryan A (2013) Evaluation of bone healing in canine tibial defects filled with cortical autograft, commercial-DBM, calf fetal DBM, omentum and omentum-calf fetal DBM. J Vet Sci 14:337
107. Schubert T, Lafont S, Beaurin G, Grisay G, Behets C, Gianello P et al (2013) Critical size bone defect reconstruction by an autologous 3D osteogenic-like tissue derived from differentiated adipose MSCs. Biomaterials 34:4428–4438
108. Saifzadeh S, Pourreza B, Hobbenaghi R, Naghadeh BD, Kazemi S (2009) Autogenous greater omentum, as a free nonvascularized graft, enhances bone healing: an experimental nonunion model. J Investig Surg 22:129–137
109. Aalami OO, Nacamuli RP, Lenton KA, Cowan CM, Fang TD, Fong KD et al (2004) Applications of a mouse model of calvarial healing: differences in regenerative abilities of juveniles and adults. Plast Reconstr Surg 114:713
110. Pritzker KP, Gay S, Jimenez SA, Ostergaard K, Pelletier JP, Revell PA et al (2006) Osteoarthritis cartilage histopathology: grading and staging. Osteoarthr Cartil 14:13–29
111. Meyer RA Jr, Tsahakis PJ, Martin DF, Banks DM, Harrow ME, Kiebzak GM (2010) Age and ovariectomy impair both the normalization of mechanical properties and the accretion of mineral by the fracture callus in rats. J Orthop Res 19:428–435

112. Hae-Ryong S, Ajay P, Jeong-Hee L, Hyung-Bin P, Do-Kyung R, Gon-Sup K et al (2002) Spontaneous bone regeneration in surgically induced bone defects in young rabbits. J Pediatr Orthop B 11:343–349
113. Nagai N, Qin CL, Nagatsuka H, Inoue M, Ishiwari Y, Nagai N et al (1999) Age effects on ectopic bone formation induced by purified bone morphogenetic protein. Int J Oral Maxillofac Surg 7:107–114
114. Bosch C, Melsen B, Vargervik K (1998) Importance of the critical-size bone defect in testing bone-regenerating materials. J Craniofac Surg 9:310–316
115. Takagi K, Urist MR (1982) The reaction of the dura to bone morphogenetic protein (BMP) in repair of skull defects. Ann Surg 196:100
116. Rivas R, Shapiro F (2002) Structural stages in the development of the long bones and epiphyses: a study in the New Zealand white rabbit. J Bone Joint Surg Am 84-A:85
117. Meyer RA Jr, Meyer MH, Tenholder M, Wondracek S, Wasserman R, Garges P (2003) Gene expression in older rats with delayed union of femoral fractures. J Bone Joint Surg Am 85:1243–1254
118. Holstein JH, Garcia P, Histing T, Kristen A, Scheuer C, Menger MD et al (2008) Advances in the establishment of defined mouse models for the study of fracture healing and bone regeneration. J Orthop Trauma 23:S31–S38
119. Batten RL (1982) Bone repair and fracture healing in man. Injury 13:532–533
120. Bostrom MP, Lane JM, Berberian WS, Missri AA, Tomin E, Weiland A et al (1995) Immunolocalization and expression of bone morphogenetic proteins 2 and 4 in fracture healing. J Orthop Res 13:357
121. Kawaguchi H, Kurokawa T, Hanada K, Hiyama Y, Tamura M, Ogata E et al (1994) Stimulation of fracture repair by recombinant human basic fibroblast growth factor in normal and streptozotocin-diabetic rats. Endocrinology 135:774–781
122. Kilborn SH, Trudel G, Uhthoff HK (2002) Review of growth plate closure compared with age at sexual maturity and lifespan in laboratory animals. Contemp Top Lab Anim Sci 41:21
123. Boutrand JP (2012) Chapter 12-Methods and interpretation of performance studies for bone implants. In: Boutrand J-P (ed) Woodhead publishing series in biomaterials, biocompatibility and performance of medical devices. Woodhead Publishing, Sawston, pp 271–307. ISBN 9780857090706
124. Wang X, Mabrey JD, Agrawal CM (1998) An interspecies comparison of bone fracture properties. Biomed Mater Eng 8:1–9
125. Martiniaková M, Omelka R, Chrenek P, Ryban L, Parkányi V, Grosskopf B et al (2005) Changes of femoral bone tissue microstructure in transgenic rabbits. Folia Biol 51:140–144
126. Muschler GF, Raut VP, Patterson TE, Wenke JC, Hollinger JO (2010) The design and use of animal models for translational research in bone tissue engineering and regenerative medicine. Tissue Eng Part B Rev 16:123–145
127. Castañeda S, Largo R, Calvo E, Rodríguez-Salvanés F, Marcos ME, Díaz-Curiel M et al (2006) Bone mineral measurements of subchondral and trabecular bone in healthy and osteoporotic rabbits. Skelet Radiol 35:34–41
128. Newman E, Turner AS, Wark JD (1995) The potential of sheep for the study of osteopenia: current status and comparison with other animal models. Bone 16:277S
129. Gilsanz V, Roe TF, Gibbens DT, Schulz EE, Carlson ME, Gonzalez O et al (1988) Effect of sex steroids on peak bone density of growing rabbits. Am J Physiol 255:E416–EE21
130. Viateau V, Guillemin G (2005) Experimental animal models for tissue-engineered bone regeneration In: Quarto R, Petite H (Eds) Engineered bone (pp 89–104). Austin: Landes Bioscience
131. Aerssens J, Boonen S, Lowet G, Dequeker J (1998) Interspecies differences in bone composition, density, and quality: potential implications for in vivo bone research. Endocrinology 139:663–670
132. Gong JK, Arnold JS, Cohn SH (1964) Composition of trabecular and cortical bone. Anat Rec 149:325–331
133. Stover BJ, Andersen AC (1971) The beagle as an experimental dog. Radiat Res 45:449

134. Kimmel DB, Jee WS (2010) A quantitative histologic study of bone turnover in young adult beagles. Anat Rec 203:31–45
135. Fernández-Tresguerres-Hernández-Gil I, Alobera-Gracia MA, Del-Canto-Pingarrón M, Blanco-Jerez L (2006) Physiological bases of bone regeneration I. Histology and physiology of bone tissue. Med Oral Patol Oral Cir Buca 11:E47–E51
136. Anderson M, Dhert BJ, Dalmeijer R, Leenders H, Van BC, Verbout A (1999) Critical size defect in the goat's os ilium. A model to evaluate bone grafts and substitutes. Clin Orthop Relat Res 364:231
137. Van Der Donk S, Buma P, Aspenberg P, Schreurs BW (2001) Similarity of bone ingrowth in rats and goats: a bone chamber study. Comp Med 51:336
138. Eitel F, Klapp F, Jacobson W, Schweiberer L (1981) Bone regeneration in animals and in man. A contribution to understanding the relative value of animal experiments to human pathophysiology. Arch Orthop Trauma Surg 99(1):59–64
139. Liebschner MAK (2004) Biomechanical considerations of animal models used in tissue engineering of bone. Biomaterials 25:1697–1714
140. Qin L, Mak AT, Cheng CW, Hung LK, Chan KM (2010) Histomorphological study on pattern of fluid movement in cortical bone in goats. Anat Rec Adv Integr Anat Evol Biol 255:380–387
141. Turner AS, Villanueva AR (1994) Static and dynamic histomorphometric data in 9- to 11-year-old ewes. Vet Comp Orthop Traumatol 07:101–109
142. Den Boer FC, Patka P, Bakker FC, Wippermann BW, Lingen A, Van VGQ et al (2010) New segmental long bone defect model in sheep: quantitative analysis of healing with dual energy x-ray absorptiometry. J Orthop Res 17:654–660
143. Spaargaren DH (1994) Metabolic rate and body size: a new view on the 'surface law' for basic metabolic rate. Acta Biotheor 42:263
144. Willie BM, Bloebaum RD, Bireley WR (2010) Determining relevance of a weight-bearing ovine model for bone ingrowth assessment. J Biomed Mater Res A 69(3):567–576
145. Lamerigts NM, Buma P, Huiskes R, Schreurs W, Gardeniers J, Slooff TJ (2000) Incorporation of morsellized bone graft under controlled loading conditions. A new animal model in the goat. Biomaterials 21:741–747
146. Raschke M, Kolbeck S, Bail H, Schmidmaier G, Flyvbjerg A, Lindner T et al (2001) Homologous growth hormone accelerates healing of segmental bone defects. Bone 29:368–373
147. Michael T, Stefan SM, Peter K, Joerg W, Karl Andreas S (2005) Bone regeneration in osseous defects using a resorbable nanoparticular hydroxyapatite. J Oral Maxillofac Surg 63:1626–1633
148. Raab DM, Crenshaw TD, Kimmel DB, Smith EL (2010) A histomorphometric study of cortical bone activity during increased weight-bearing exercise. J Bone Miner Res 6:741–749
149. Ermanno B, Paola B (2014) Osteoporosis-bone remodeling and animal models. Toxicol Pathol 42:957–969
150. Mosekilde L, Kragstrup J, Richards A (1987) Compressive strength, ash weight, and volume of vertebral trabecular bone in experimental fluorosis in pigs. Calcif Tissue Int 40:318–322
151. Li M, Weisbrode SE, Safron JA, Stills HG, Jankowsky ML, Ebert DC et al (1992) Calcium-restricted ovariectomized sinclair s-1 minipigs: an animal model of osteopenia and trabecular plate perforation. Bone 13:379
152. Kragstrup J, Richards A, Fejerskov O (1989) Effects of fluoride on cortical bone remodeling in the growing domestic pig. Bone 10:421–424
153. Swindle MM, Smith AC, Hepburn BJ (1988) Swine as models in experimental surgery. J Investig Surg 1:65–79
154. O'Loughlin PF, Morr S, Bogunovic L, Kim AD, Park B, Lane JM (2008) Selection and development of preclinical models in fracture-healing research. J Bone Joint Surg Am 90(Suppl 1):79–84
155. Martini L, Fini M, Giavaresi G, Giardino R (2001) Sheep model in orthopedic research: a literature review. Comp Med 51:292–299

156. Paige KT, Cima LG, Yaremchuk MJ, Vacanti JP, Vacanti CA (1995) Injectable cartilage. Plast Reconstr Surg 96:1390–1398
157. Jackson RW, Reed CA, Israel JA, Abou-Keer FK, Garside H (1970) Production of a standard experimental fracture. Can J Surg 13:415–420
158. Bonnarens F, Einhorn TA (1984) Production of a standard closed fracture in laboratory animal bone. J Orthop Res 2:97–101
159. Marturano J, Cleveland BC, Byrne MA, O'Connell S, Wixted J, Billiar K (2008) An improved murine femur fracture device for bone healing studies. J Biomech 41:1222–1228
160. Hebb JH, Ashley JW, Mcdaniel L, Lopas LA, Tobias J, Hankenson KD et al (2018) Bone healing in an aged murine fracture model is characterized by sustained callus inflammation and decreased cell proliferation. J Orthop Res 36(1):149–158
161. Lopas LA, Belkin NS, Mutyaba PL, Gray CF, Hankenson KD, Jaimo A (2014) Fractures in geriatric mice show decreased callus expansion and bone volume. Clin Orthop Relat Res 472:3523–3532
162. Dishowitz MI, Terkhorn SP, Bostic SA, Hankenson KD (2011) Notch signaling components are upregulated during both endochondral and intramembranous bone regeneration. J Orthop Res 30:296–303
163. Puolakkainen T, Rummukainen P, Lehto J, Ritvos O, Hiltunen A, Säämänen AM et al (2017) Soluble activin type IIB receptor improves fracture healing in a closed tibial fracture mouse model. PLoS One 12:e0180593
164. Holstein JH, Menger MD, Culemann U, Meier C, Pohlemann T (2007) Development of a locking femur nail for mice. J Biomech 40:215–219
165. Manigrasso MB, O'Connor JP (2004) Characterization of a closed femur fracture model in mice. J Orthop Trauma 18:687–695
166. Thompson Z, Miclau T, Hu D, Helms JA (2010) A model for intramembranous ossification during fracture healing. J Orthop Res 20:1091–1098
167. Makino T, Hak DJ, Hazelwood SJ, Curtiss S, Reddi AH (2010) Prevention of atrophic nonunion development by recombinant human bone morphogenetic protein-7. J Orthop Res 23:632–638
168. Kumabe Y, Sang YL, Waki T, Iwakura T, Takahara S, Arakura M et al (2017) Triweekly administration of parathyroid hormone (1–34) accelerates bone healing in a rat refractory fracture model. BMC Musculoskelet Disord 18:545
169. Kokubu T, Hak DJ, Hazelwood SJ, Reddi AH (2010) Development of an atrophic nonunion model and comparison to a closed healing fracture in rat femur. J Orthop Res 21:503–510
170. Hietaniemi K, Peltonen J, Paavolainen P (1995) An experimental model for non-union in rats. Injury 26:681–686

Biofabrication in Tissue Engineering

Guangyu Bao

Abstract Biofabrication has been extensively explored in tissue engineering over the past two decades. It uses bioactive materials and live cells as the building blocks to create spatially defined geometries. The goal of biofabrication is to create engineered tissue constructs to replace damaged or diseased human tissues with full functionality. The advantage is that it can rapidly fabricate tissue constructs to meet customized needs. The biomaterials used for biofabrication are called bioinks and usually comprise hydrogel precursor solutions or biocompatible thermal plastics. In this review, we review the commonly used biofabrication methods and critical aspects for creating scaffolds for tissue regeneration. We discuss the criteria for developing and selecting suitable biomaterials as the bioinks. Commonly used biomaterials and their applications are summarized to present the versatility of biofabrication. We also aim to highlight the challenges of this technology and initiate new ideas and opportunities in the future developments in the bioprinting approach and bioinks. The refinement in fabrication techniques, exploration in biology, and development in new bioinks are essential elements toward the advancement of biofabrication.

Keywords Biofabrication · 3D printing · Extrusion bioprinting · Tissue engineering
Bone engineering · Bioinks · Printability · Vascularization · Natural polymers
Synthetic polymers · Tunable mechanical properties

Introduction

Strategies for engineering scaffolds combining bioactive materials, cells, and growth factors have been extensively explored to restore damaged tissues and organs [1]. Traditionally, the tissue engineering (TE) ingredients are molded to form to a simple construct prior to implantation, or directly transferred to the host

G. Bao (✉)
Department of Mechanical Engineering, Macdonald Engineering Building Room 270,
Montreal, Quebec, Canada
e-mail: guangyu.bao@mail.mcgill.ca

B. Li et al. (eds.), *Racing for the Surface*,
https://doi.org/10.1007/978-3-030-34471-9_12

sites through injection [2]. These strategies allow the delivery of cells or drugs in promoting tissue regeneration while providing temporal physical support. However, the poor spatial control over the distribution of cells, pores, and biological and mechanical cues limits their ability in the creation of customized regenerative therapeutics and full-scale organs.

Biofabrication, also known as additive manufacturing, has drawn great attention over the past two decades in the field of TE [3–5]. Since the concept of 3D printing was introduced in 1986, this method has expanded printable materials from resins to solvent-free, aqueous, and cell-laden biological materials [6]. It has also been applied to areas including but not limited to TE [7], biomedical devices [8], drug delivery [9], and disease modeling [10].

In a typical process of biofabrication, biological materials are precisely positioned on a substrate in a layer-by-layer fashion. The precise spatial control allows for the creation of complex geometries and opens the arena for fabricating engineered tissues at a large scale [11]. The controlled deposition of multiple cell types may also facilitate the implants to regain their original biological functions [12]. The fast prototyping ability further enables the personalized tissue design. Although biofabrication shows promising results in TE, it is still at its infancy. Its ability is mainly constrained by feature size, resolution, biological and mechanical properties of biomaterials, and mild fabrication conditions.

In this review, methods for the biofabrication of tissue constructs are firstly introduced. Next, the criteria for selecting or developing biomaterials are discussed. Then, properties of commonly used biofabrication materials are described along with their applications. Possible solutions for overcoming the existing challenges and future perspectives are discussed to conclude this review.

Biofabrication Strategies

Precise deposition and patterning of biological materials for TE are generally realized through inkjet [13], laser-assisted [14] and extrusion-based bioprinting [15], or melt electrowriting (MEW) [16]. Schematics and examples of different biofabrication techniques are shown in Fig. 1. Selection of the biofabrication strategies should be based on the biological materials properties, required feature size, fabrication resolution, and whether live cells are involved during the fabrication process.

Inkjet Bioprinting

An inkjet printer is the most affordable type of printer for biological material patterning. It can be easily modified from commercially available 2D ink-based printers at a low cost [3]. During the printing process, the inkjet bioprinter can be driven by either thermal or acoustic forces to eject droplets of biomaterials, such as hydrogels

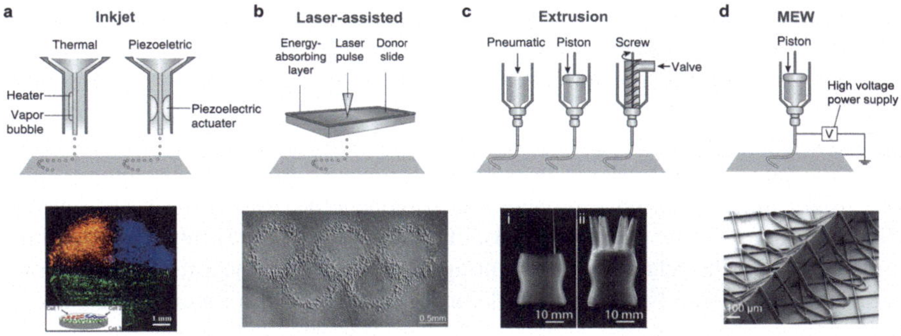

Fig. 1 Schematics and photos of constructs fabricated by (**a**) inkjet bioprinting, (**b**) laser-assisted bioprinting, (**c**) extrusion bioprinting, and (**d**) melt electrowriting (MEW). Images reproduced with permissions [3, 13, 16–18]

or a cell-laden medium [13]. For a thermal inkjet bioprinter, a microheater vaporizes the bioink and expels a small volume of bioink drop from the print head. The localized heat and pressure may damage cell viability, though cases with high viability and normal post-printing function of mammalian cells have been reported [19]. To avoid the potential damages to cell viability, piezoelectric crystals or an ultrasound field has been employed to replace the microheater. The mechanism is to create an acoustic wave at defined frequencies and break the bioink into droplets. Droplet size and ejection rate can be precisely controlled. The inkjet bioprinter offers position controls in the x, y, and z axes. The control in the z-direction is usually enabled by an electrical elevator stage.

One of the advantages of inkjet bioprinting is that it can create feature sizes on the order of several micrometers [20]. The fine feature size allows precise patterning of biological materials for biological research with high throughput. It can also print bioinks with low viscosity on the order of 10 mPa s [21]. Another advantage is that the high resolution provides the possibility to create gradients by spatially varying the amount of materials, cells, and growth factors [13]. However, the low print volume impedes the fabrication of 3D constructs with clinically relevant sizes and is the main drawback of this technique. Bioinks containing high cell density also tend to clog the printer head [6]. In general, inkjet bioprinting is more suitable to provide a simple and accessible patterning approach for biological studies than to create complex 3D scaffolds for TE.

Laser-assisted Bioprinting

In a process of printing using laser-assisted technology, a laser pulse focuses on an energy absorbing layer, with a donor-slide (bioink layer) adjacent underneath. A local evaporation from the energy absorbing layer transfers the high gas pressure to the bioink reservoir and propels a bioink droplet to a collector [14]. Similar to inkjet bioprinting, laser-assisted bioprinter provides position controls in x, y, and z axes with high printing resolution [18]. The nozzle-free printing process preserves high cell viability without affecting cellular functions.

Laser-assisted bioprinting can print bioinks with higher cell density and broader viscosity range (1–300 mPa s) in contrast to those from inkjet bioprinting [22]. Due to the nozzle-free characteristics, laser-assisted bioprinting is not affected by clogging issues. However, the high cost of the laser system prevents its acceptability in practice. The relatively low printing volume also restricts its application in building large scaffolds. In addition, laser-assisted bioprinting requires bioinks to have rapid gelation kinetics to obtain high shape fidelity, which limits the printable biomaterials.

Extrusion Bioprinting

Extrusion bioprinting, as the most commonly used biofabrication technology, offers a versatile printability for a wide range of materials and applications. During a typical extrusion bioprinting process, bioinks are extruded from a cartridge through a print nozzle and dispensed on a substrate. Cells can be encapsulated inside the bioinks before fabrication or seeded post-print [23–25]. The bioinks are deposited layer-by-layer with precise spatial control. The extrusion can be driven by pneumatic pressure or a piston/screw. For a pneumatic controlled extrusion bioprinter, the pressure applied on bioinks can be accurately controlled. This is particularly important for extruding bioinks containing live cells because cells are susceptive to both normal and shear stresses, which are related to the exerted pressure [6]. A mechanically driven extrusion bioprinter can precisely control the volumes of extruded bioinks by controlling the feeding rates of the piston or screw. It is commonly used to print highly viscous thermal plastics or control the printing volume of certain expensive biological reagents. A typical extrusion bioprinter provides a spatial control over all the three axes, with some robotic arm-based bioprinters offering six-axis control [26].

One of the advantages of the extrusion bioprinting is the wide range of printable biomaterials. Unlike the two bioprinting technologies discussed above, extrusion bioprinting can print materials with viscosity varying from 1 to greater than 10^7 mPa s, which spans the range of existing biomaterials [27–29]. In general, a bioink should have sufficient yield stress and viscosity to withstand its gravity and to maintain the shape fidelity [30]. Bioinks with high yield stress and viscosity can be extruded directly on a substrate; for materials with low yield stress and viscosity, several techniques have been explored to address this problem. For instance, low viscous bioinks can be extruded inside a supportive reservoir made of a Bingham fluid matrix [27]. A Bingham fluid matrix exhibits yield strength at a low shear rate and behaves like a fluid when the shear rate exceeds its critical value.

It can lock the extruded bioinks in place while self-healing after the retraction of a print needle [17]. This method is known as embedded bioprinting which enables the freeform modeling of complex biological and silicon structures. The low viscous bioink structures are crosslinked after printing. Another strategy to print low viscous materials is the implementation of in situ crosslinking during printing [28]. For UV-crosslinkable materials, a UV curing device can be mounted pointing to the print head. The extruded materials are gradually crosslinked while passing through the print needle. For crosslinker-required low viscous material extrusion, duo-extrusion, coaxial, and microfluidic extrusion techniques have been demonstrated to be effective [31]. Other advantages such as high printing volume also show the potential of extrusion bioprinting in fabricating scaffolds with organ-level sizes [32].

The main restriction on extrusion printing is that it has a relatively large limit for feature size. The feature size for extrusion bioprinting is determined by the applied pressure, print nozzle size, writing speed, distance between the printing nozzle and substrate, and the properties of the bioinks themselves [33–36]. Because most bioinks are composed of hydrogels and biological materials, it is difficult to analytically predict the feature size of the printed filament due to the instability and variations of hydrogels. Furthermore, although the majority of a bioink composition is water, the bioink is viscoelastic and slightly compressible due to the exitance of polymers. When the bioink is extruded through a nozzle, shear, extension, and compression forces are exerted on the bioink flow. Die-swelling effects exist at the outlet of the print nozzle, which renders a larger filament size than the analytical solution derived from the incompressibility assumption [37]. Therefore, further analytical and numerical modeling is required to refine the printability of extrusion bioprinting.

Melt Electrowriting (MEW)

MEW is an additive manufacturing method derived from electrospinning. During the MEW process, the polymer solution is charged with a typical voltage of 5–20 kV [38]. The electrified molten jet then deposits the polymer fibers with great consistency and minimal variation. MEW can fabricate feature sizes varying from 25 nm to 45 μm [39].

Traditionally, MEW is considered a non-biofabrication process given the high voltage and harsh organic solvent environment. However, combined with soft hydrogels after fabrication, the MEW-reinforced structures have achieved remarkable mechanical properties for bone and cartilage tissue engineering [16]. The soft hydrogel laden provides a suitable environment for cellular ingrowth while the MEW fibers provide a strong support against mechanical loadings. MEW has also been shown to improve the toughness of scaffolds [40] (Table 1).

Table 1 Comparison of commonly used biofabrication types

	Inkjet	Laser	Extrusion	MEW	Refs.
Resolution	50–75 μm	35–72 μm	5–10^3 μm	10^{-2} to 45 μm	[3, 39, 41]
Fabrication speed	–	–	+	+	[6]
Bioink viscosity	10^0–10^1 mPa s	10^0–10^2 mPa s	10^0–10^7 mPa s	10^2–10^3 mPa s	[21, 22, 28]
Gelation speed	Fast	Fast	Broad	Broad	[3, 6]
Cell density	<10^6 cells/mL	10^8 cells/mL	10^8 cells/mL	N/A	[3, 14, 18, 19]

Criteria for Biomaterial Design and Selection

Constructs produced from biofabrication are used as implant scaffolds or for other biomedical applications purposes. Therefore, the bioink materials should not only be printable and biocompatible, but also have the appropriate biological and mechanical properties to match those of real tissues [42]. Because scaffold materials ideally are expected to be replaced by the newly generated extracellular matrix overtime, the degradation ratio of the materials should be comparable to the tissue regeneration time [43]. In addition, the degradation byproducts should also be noncytotoxic [44]. The criteria for the design and selection of suitable biomaterials for biofabrication are discussed below.

Biocompatibility

Biofabrication often utilizes cell-laden bioinks as the building blocks. Therefore, the conditions for cell encapsulation are critical for supporting good cell viability. The factors determining the biocompatibility depend on the cytotoxicity, pH, osmosis, temperature, and the viscosity of the bioinks [6, 45]. For a nontoxic bioink, a physiological pH and osmosis condition are optimal for cell encapsulation. Cells have been demonstrated to withstand a low temperature encapsulation as low as 4 °C for 20 min [46]. The bioink viscosity affects the shear stress during cell encapsulation [47]. The high viscosity induces high shear stress during cell encapsulation. High shear stress could lead to damage to the cell membrane and affect both the viability and functionality during further culture.

Furthermore, the bioinks should actively support cellular functions post-print [48]. Most bioinks need to be crosslinked during or after fabrication. Thus, it is also necessary to evaluate the toxicity of different crosslinkers or crosslinking methods [49]. The main mechanisms for crosslinking bioinks include covalent, ionic, physical, and thermal crosslinking [6]. Covalent crosslinkers lead to the formation of a strong, stable, and irreversible hydrogel network and help maintain the post-print fidelity. However, they are usually composed of toxic chemicals or UV light exposure and introduce cytotoxicity at high doses [47]. Ionic crosslinking is one of the most popular crosslinking mechanisms used in biofabrication [49]. The electrostatic interactions form reversible bridging between opposite charged particles and polymeric chains. Ionic crosslinkers generally do not introduce cytotoxicity if biocompatible cations or anions are chosen. Physical and thermal crosslinking are also biocompatible. Nevertheless, physical crosslinks are weak against mechanical stimulations [50]. Some popular thermal sensitive polymers, such as poly(N-isopropylacryl amide) (PNIPAm), are not biodegradable [51]. Recent efforts such as using the Michael addition reaction, click chemistry, or enzymatic reactions [52–54] have been used to obtain high biocompatibility as well as achieving appropriate material properties.

Material Properties

Material properties can be divided into two aspects: biological and mechanical properties. In terms of biological properties, the material should provide cell adhesion, migration, and proliferation functions [55]. It has been established that the material's biological features affect the motility, orientation, proliferation, and cytoskeletal assembly of cells [3, 55, 56]. The fabricated 3D scaffolds should provide a biomimetic environment to support cell growth and tissue function.

In terms of mechanical properties of the bioinks, they are usually evaluated by their elastic and viscoelastic properties. Ultimately, the crosslinked constructs should have similar mechanical properties to match the targeted host tissue [48]. The mechanical properties of scaffolds have been found to contribute to cell fate [56]. For instance, human mesenchymal stem cells (hMSC) can be guided to differentiate into osteogenic or adipogenic cells by changing the stiffness of the extracellular matrix (ECM) when biological and chemical conditions are maintained [57]. The viscoelasticity of the fabricated matrix also plays an important role in facilitating cell growth, spreading, and differentiation. Cells tend to proliferate and spread quicker when the stress relaxation time of the matrix decreases [58]. The stress relaxation time as a viscoelastic property can be tuned by varying polymer molecular weight, or modifying crosslinking sites [59]; however, these treatments require complicated chemical modifications. Iterative trial-and-error is needed to achieve the desired properties. Fibrillar polymer reinforced interpenetrating networks can also be employed, but they are usually costly and may suffer from batch-to-batch variance [60]. A simple and robust approach for the tuning of the relaxation time without altering the matrix stiffness is to be discovered.

Structural Properties

Human tissues are generally highly porous to facilitate cell growth, nutrient/oxygen diffusion, waste removal, and to functionalize [42]. The pore sizes of scaffolds should resemble these hierarchical porous architectures. Additionally, it has been shown that tissue regeneration is highly dependent on pore structures of the implanted scaffolds. For example, pore sizes of 5–15 μm have been found desirable for fibroblast ingrowth, ~20 μm for hepatocyte ingrowth, 45–150 μm for liver tissue regeneration, 20–120 μm for effective wound healing, 100–300 μm for bladder smooth muscle cell ingrowth, and >90 μm for vascularization [61, 62]. However, the intrinsic pore sizes for most hydrogel networks are smaller than 100 nm. Such pore sizes are 100–10,000 fold less than the desirable sizes, which limits the ability of scaffolds to recruit native cells. Neither does it allow for the migration of loaded cells to remodel the matrix [63]. Conventional fabrication techniques, such as salt leaching, gas forming, phase separation, electrospinning, and freeze-drying have been explored to fabricate porous 3D constructs [42]. However, most of those techniques are not biocompatible. A recent effort has created micropores by printing a water-in-water emulsion bioink initiated with phase separation [64]. The emulsion printing method successfully introduced micropores while maintaining high cell viability. Yet, the separated polymers tended to fuse with their own phases and showed no emulsion after a certain time, with a limited printing window of around 30 min. Fabrication of pores on the order of 10 μm is the main challenge due to the restriction of printable feature size.

Degradation

Most biofabricated scaffolds are used as a temporal support for damaged tissue. Therefore, they are expected to degrade in a time matching the tissue regeneration speed. A degradation time from weeks to months are needed depending on the target host tissue [43]. The degradation is also a part of biocompatibility evaluation. The degraded byproducts should be nontoxic and rapidly metabolized from the human body without resulting in an immune response [49].

For a polymer network, degradation can happen at the polymeric backbone, side chains, crosslinks, or the combination of them [65]. Unlike biocompatibility which is an intrinsic property, the rate of degradation can be well engineered or chemically tuned [66]. For instance, partial oxidation and varying molecular weight distribution have been used to provide or accelerate the degradation speed for polymers with no/low degradability in vivo, such as alginate and chitosan [67, 68]. In addition, degradation can also be controlled with external stimuli, such as photo, thermal, magnetic, electrical, and mechanical triggers [66, 69]. The design of these triggering mechanisms allows simple control for applications with on-demand degradation requirement, for example, the delivery of drugs at pre-determined time points [70]. In principle, the degradation rate of polymer should match the ECM proteins production speed of the embedded and recruited cells without impeding the cells' ingrowth.

Printability

Most biofabrication techniques require the bioinks to pass through print nozzles with small inner diameters to create features at the microscopic level. The polymer chains inside a bioink form weak physical crosslinks [6]. When the bioink is subject to extrusion, the fragile network is not broken when the shear force is low and starts to flow when the shear exceeds a critical point. This critical value is called yield stress. The network reforms after the extrusion and the printed structure can retain its structure again. Yield stress is the most important factor in determining the printability of bioinks [10]. In general, the yield stress of a bioink should be high enough to maintain the shape of designed geometries but not exceeding the tolerance of cells if cells are encapsulated. Shear thinning behavior also facilitates the extrusion process. However, thixotropy behavior should be evaluated at the same time because certain bioinks do not or slowly resume their original yield stress when shear is removed [71]. Due to the viscoelastic behavior of bioinks, a die-swelling ratio greater than unity should also be taken into consideration during the prediction of feature size [37]. During a printing process, the distance and writing speed determine how the extruded material is deposited on a substrate. Situations of bioink filament can be accumulated, coiled, die-swelled, equi-dimensional, stretched, or discontinuous.

Embedded bioprinting methods have been developed for the extrusion of low viscous and low/no yield bioinks which cannot be collected directly on a substrate [27, 72]. For embedded bioprinting, the bioinks are extruded inside a supportive reservoir made of a Bingham fluid matrix. In the embedded printing cases—Oldroyd number, the ratio of the material yield stress to the viscous stresses in a flow, is the key factor to reduce the yield dimensions around the print nozzle [73]. The nozzle diameter and printing speed also contribute to the printing fidelity.

Commonly Used Biofabrication Materials

Materials currently used in the field of TE are mainly derived from natural polymers and synthetic materials. Naturally derived polymers, such as alginate, gelatin, gelatin methacryloyl (GelMA), hyaluronic acid (HA), chitosan, collagen, fibrin, and decellularized ECM, are the main players in the biofabrication field. The benefit of using natural polymers is that they are generally biocompatible, biodegradable, and support bioactivity. Synthetic polymers such as polyethylene glycol (PEG) and Pluronic are also popular due to their programmable mechanical properties, gelation kinetics, and biodegradation rate. However, most synthetic polymers do not offer cell-binding peptides and release toxic monomers during degradation. The details of several commonly used biofabrication materials are discussed below.

Naturally Derived Biomaterials

The most commonly used natural biomaterial for biofabrication is alginate [74, 75]. It has been widely used in biofabrication due to its biocompatibility, rapid gelation speed, abundance in nature, and tunable mechanical properties. Alginate is a linear block copolymer of D-mannuronic acid (M) and L-guluronic acid (G) residues. The negatively charged G block of alginate can be crosslinked by cations [1]. Calcium (Ca^{2+}) such as calcium chloride or calcium sulfate is commonly used as the ionic crosslinker due to its biocompatibility consideration. Alginate can be crosslinked by immersing the printed construct inside a crosslinker bath, or by coextruding with the crosslinker medium and then be crosslinked either through mixing or diffusion. Alginate has been demonstrated to stably fabricate feature sizes of as low as 75 μm [76]. Besides its excellent printability, its mechanical properties can also be easily tuned to a desirable target value [77]. Stiffness of alginate hydrogels can be tailored by varying polymer concentrations, crosslinking density, and molecular weight. Stress relaxation can be reduced by decreasing crosslinking density and the molecular weight of polymers. Although alginate does not have cell adhesion, Arg-Gly-Asp (RGD) peptides can be coupled to alginate chains to provide binding sites [58]. One drawback of alginate is that it cannot be degraded by the human body. Strategies such as partial oxidation have been applied to control the degradability of alginate [78].

Gelatin is also commonly used in biofabrication. Gelatin, as a fibrous protein, is derived from denatured collagen. It offers great biocompatibility and is nonimmunogenic [49]. Gelatin solution is solid-like at room temperature and becomes liquid at physiological temperature. The paste-like status between phase changing enables gelatin with an excellent printability and a generous printing time window [10]. However, physically crosslinked gelatin is not stable in vivo. Thus, it is generally used as a template blending material to improve biomaterials with poor printability [79–81]. For example, gelatin has been mixed with alginate and chitosan to improve the printability of bioinks. Methacrylate functionalized gelatin (GelMA) allows the free radical reaction to form a stable covalently crosslinked network [25]. The bioactive peptides promote cell adhesion and proliferation. GelMA has been demonstrated to support long-term cell culture for over 45 days [82].

Hyaluronic acid (HA) is high molecular weight glycosaminoglycan with repeating disaccharide units composed of (b-1,4)-linked D-glucuronic acid and (b-1,3)-linked *N*-acetyl-D-glucosamine [1]. It is presented in almost all connective tissues. Chondrocytes embedded inside HA bioinks have been shown to have high viability [83]. However, the poor mechanical properties and rapid degradation rate impede the use of unmodified HA in biofabrication [49]. To improve its stability in vivo, similar to gelatin, methacrylate can be conjugated to HA to improve both the mechanical properties and in vivo stability [28].

Chitosan is derived from the deacetylation of chitin and has been demonstrated to have excellent biodegradability and antibacterial and antifungal properties [84, 85]. It has been widely applied in various TE applications, such as cartilage tissue engineering and wound healing [86]. One advantage of chitosan is that it can be covalently, ionically, or physically crosslinked based on its application [50]. Chitosan has also been found to have thermal gelation behavior at 37 °C and thus is commonly employed as an injectable material [45, 87]. Another advantage is that the primary amine groups in chitosan chains provide strong mucosal adhesion through covalent binding with carboxylic acid groups on tissue surfaces, in contrast to most other natural polymers [88, 89]. However, chitosan is rarely used alone as a bioink due to its poor solubility in neutral pH solutions and slow gelation kinetics [90]. Researchers often blend chitosan with gelatin or agarose to improve its printability [80, 91].

Collagen, as the most abundant protein in a human body, has also been explored in the biofabrication field. Collagen can be physically crosslinked by raising both the pH and the temperature to physiological conditions [91, 92]. The crosslink mechanism is reversible. Collagen is often used as a cheap RGD source to supply other biomaterials which do not have cell adhesion. Extruding neutralized collagen at room temperature can result in the clogging of print nozzle. Furthermore, its low yield stress can hardly support its own weight. To address these problems, collagen is usually printed at 4°C using embedded bioprinting methods [27, 93]. The fabricated collagen scaffold has been shown to provide great support for high cell density extrusion (10^7 cells/mL). A collagen type-IV and laminin-rich Matrigel provides an irreversible thermal gelation network for 3D fabrication [94]. It has been demonstrated to support high viability for human epithelial cells and bone marrow stromal cells.

In general, most natural biomaterials lack a biomimetic environment for embedded cells to function. The lack of cell-specific ECM and network pores on the order of 10 μm limit cell spreading and migration and mass exchange. Therefore, most hydrogels do not support a high cell seeding density due to the unsuitable environment and competition in nutrition, which prevents cell–cell interactions and tissue-like functionalization. Bioinks derived from decellularized extracellular matrix (dECM) have been used as a strategy to target the cell-specific issue. Cartilage, heart, and adipose tissues have been demonstrated to provide crucial cues for cell engraftment, survival, and long-term function [95].

Synthetic Biomaterials

Poly(ethylene glycol) (PEG), as a synthetic polyether compound, has been widely used in biofabrication [96]. The PEG-based hydrogels are biocompatible, with highly controllable mechanical properties, gelation speed, and biodegradation rate. For instance, one of the PEG derivatives poly(ethylene glycol) diacrylate (PEGDA) undergoes chain-growth polymerization for gelation in the presence of light [28]. It can be crosslinked before, during, or after fabrication. PEG polymers with two reactive groups have also been used as first-stage crosslinkers to improve the printability of low viscous bioinks for multi-stage crosslinking systems [24].

Pluronic is a tri-block copolymer of poly(ethylene glycol)-block, poly(propylene glycol)-block, and poly(ethylene glycol) (PEO-PPO-PEO) sequences [49]. This temperature sensitive polymer is liquid at low temperature and solid at room temperature. This behavior is caused by the intermolecular interaction of PPO blocks, which leads to micelle formation when the temperature is above the critical micelle temperature. The sol-gel transition point depends on the concentration and molecular weight of the polymer. Due to its phase changing behavior and high yield stress at the solid state, Pluronic has been extensively used as sacrificial materials to form vasculatures or supportive layers for large and complex scaffold fabrication [46, 82, 97].

Some other popular polyesters such as poly(glycolic acid) (PGA) [98], poly(lactic acid) (PLA) [99], poly(ε-caprolactone) (PCL) [16], and poly(lactide-co-plycolide) (PLGA) [100] have also been used to fabricate scaffolds for bone, cartilage, muscle, marrow, tendon, and connective tissues. Due to their high mechanical strength and stiffness, they can withstand stresses in the host tissue environment and are generally used as the supporting structures for soft tissues or load bearing tissues. The excellent printability enables high-resolution biofabrication to create feature sizes in the macroscopic and microscopic regime. The hierarchical porous structures promote cellular ingrowth and vascularization [12] (Table 2).

Future Perspectives

Although biofabrication technologies have shown promising results for TE and biomedical applications, their development has been restricted by their feature size, building volume and speed, lack of vascularization, and limited selectable biomaterials. Their main challenges and future perspectives are discussed below to conclude this chapter.

Multi-scale Biofabrication

Currently, the dimensions of biofabricated scaffolds are mainly on the order of 10^2–10^4 μm. The scaffolds typically contain less than three different materials and lack tissue-like heterogeneity [100]. It is caused by the restrictions on current biofabrication technology and the printing costs. For instance, high printing speed yields high shear generated during extrusion, which leads to low viability for the encapsulated cells. The printed strands may also lose fidelity and fail to form the desirable morphology. Furthermore, biofabrication happens at ambient conditions. It may take hours to build a scaffold at the centimeter scale while the printed cells are exposed in an unfavorable environment. Besides the long fabrication time, the high expenses for bioprinters equipped with multiple bioink dispensers also affect

Table 2 Commonly used biofabrication materials and their properties

Biomaterial	Crosslinking mechanism	Biofabrication method	TE applications	Advantages	Disadvantages	Refs.
Agarose	Physical	Extrusion	Vascularization	Biocompatible, high mechanical strength, low cost, nonimmunogenic	Low cell adhesion, melting point higher than physiological temperature	[101–103]
Alginate	Ionic, covalent	Inkjet, laser, extrusion	Bone, cartilage, tumor, skin, cardiac, vessel	Biocompatible, rapid gelation speed, stable in vivo, tunable elasticity and viscoelasticity, low cost, nonimmunogenic	Low cell adhesion, nonporous, not biodegradable without modification	[31, 75, 104–106, 129, 133]
Chitosan	Covalent, physical, ionic	Extrusion, MEW	Bone, cartilage, skin, vocal fold	Biocompatible, stable in vivo, thermal gelation at 37°C, tunable elasticity and viscoelasticity, low cost, nonimmunogenic	Low solubility at neutral pH, slow gelation speed, low cell adhesion	[85, 103, 107, 108, 132]
Collagen	Physical, covalent	Extrusion, MEW	Bone, cartilage, skin, neural	Provide cell adhesion and promote proliferation, porous, gelation at 37°X, nonimmunogenic	Poor mechanical properties, slow gelation, high cost	[13, 93, 109, 110, 130, 131]
dECM	Physical	Extrusion	Cartilage, cardiac, adipose	Provide tissue-specific proteins, support long-term cell culture	Poor mechanical properties, slow gelation, long preparing time	[95, 111]
Gelatin	Physical, covalent, enzymatic	Extrusion, MEW	Bone, cartilage, aortic valve, skin, vascularization	Biocompatible, provide cell adhesion, good extrusion property, low cost, nonimmunogenic	Degrade fast in vivo, poor mechanical stability, do not form gel at physiological temperature	[64, 80, 112, 113, 129]
HA	Covalent	Laser, extrusion, MEW	Bone, cartilage, vessel	Biocompatible, good printability, nonimmunogenic	Fast degradation, low cell adhesion	[92, 109, 114, 115, 128]

(continued)

Table 2 (continued)

Biomaterial	Crosslinking mechanism	Biofabrication method	TE applications	Advantages	Disadvantages	Refs.
PCL	Physical	Extrusion, MEW	Bone, cartilage	Biocompatible, thermal responsive, promote cell adhesion and proliferation even in the absence of cell-binding moieties	Not compatible with cell encapsulation	[12, 116, 117, 127]
PEG	Covalent, ionic, physical	Extrusion, MEW	Bone, cartilage, skin, vessel	Biocompatible, soluble in water, nonimmunogenic	Low cell adhesion, require chemical modification	[92, 96, 118, 119, 127]
PLA	Physical	Extrusion, MEW	Bone, cartilage	High mechanical strength, biodegradable, biocompatible	Not compatible with cell encapsulation	[99, 120, 121]
PLGA	Physical	Extrusion, MEW	Bone, vessel	Mechanically strong and stable, rapid gelation upon extrusion	Not compatible with cell encapsulation	[117, 121, 122]
Pluronic	Thermal	Extrusion	Vascularization	High yield stress, great printability, fugitive, bioinert	Rapid degradation, poor structural properties, not suitable for implantation	[46, 82, 123]
Silk fibroin	Covalent, physical, enzymatic	Inkjet, extrusion, MEW	Bone, cartilage, soft tissues	Biocompatible, tunable mechanical properties	Clogging print head at high concentrations	[79, 118, 124, 125, 126]

the fabrication of complex scaffolds. In addition, print nozzles tend to get clogged while waiting for multi-material switching.

One approach to enabling large-scale biofabrication is to use fabricated scaffold modules and assemble them to form large tissues [3]. The scaffold modules can be fabricated in a manageable scale to ensure cell viability and functionality. A similar concept is currently used by scaffold-free TE, which is to employ the autonomous self-assembly of cellular spheroids that undergo fusion and cellular organization to mimic developing tissues [134]. This strategy requires an in-depth knowledge in manipulating the embryonic mechanism. An alternative approach is in situ bioprinting. Cells delivered using this method are directly deposited on the damaged tissue sites, where suitable humidity, temperature, and nutrition are supplied [49]. Another approach is to employ a multi-material coextrusion system to create a heterogeneous scaffold with spatially defined material gradients [31]. The preliminary work on developing such a system shows the fast fabrication ability with coextrusion of up to seven different bioinks as shown in Fig. 2a [32]. This strategy has been demonstrated to create constructs with gradients on cells and materials. Other approaches such as creating perfusion channels during scaffolds design have also shown

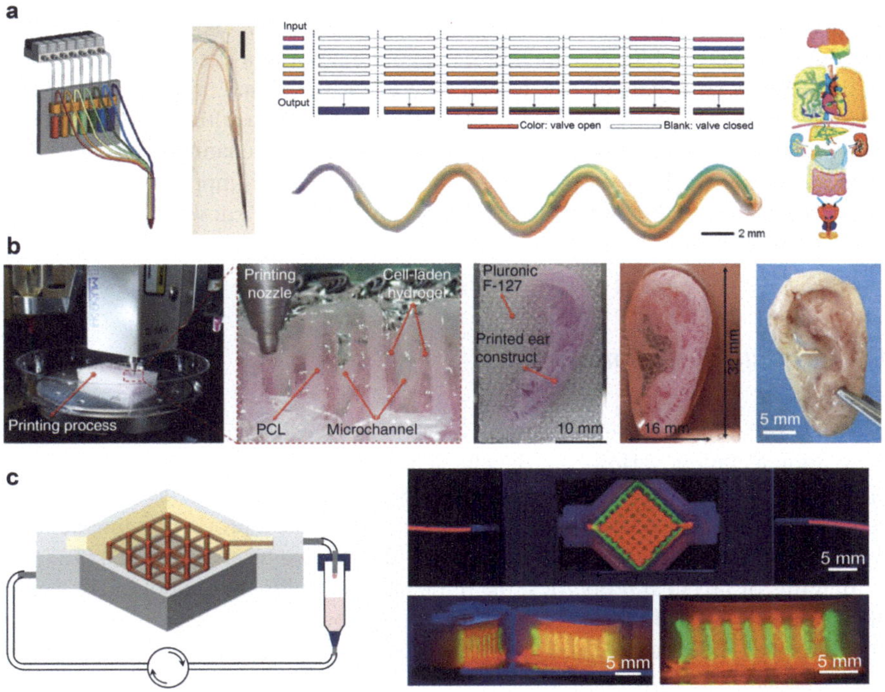

Fig. 2 Recent advances toward the fabrication of large-scale scaffolds. (**a**) Rapid and continuous extrusion of multiple bioinks. (**b**) Fabrication of an ear construct with mechanical strength similar to human ears. (**c**) Image showing biofabricated vasculature inside a hydrogel scaffold with 1 cm in thickness. Images reproduced with permissions [12, 32, 82]

promising result in large-scale scaffold fabrication. For instance, new cartilaginous matrix was generated inside fabricated ear construct to provide tissue-like resilience against mechanical deformation as demonstrated in Fig. 2b [12]. Those strategies may satisfy the requirements to fabricate scaffolds with clinically relevant sizes.

Vascularization

Creating a functional vasculature is the fundamental challenge that prevents the fabrication of organ-sized scaffolds. Cells need vessels and capillaries to deliver oxygen and nutrition and to remove waste [42]. Strategies of printing hollow features have been used to create engineered vasculatures. Currently, the engineered vasculature has been demonstrated to support scaffolds with up to 1 cm thickness [82]. The vasculature was created by flushing away sacrificial Pluronic F127 channels and seeding human umbilical vein endothelial cells (HUVECs) after scaffold fabrication. To date, no successful attempt has been reported to use engineered vasculature to support scaffolds thicker than 1 cm, as shown in Fig. 2c. This is due to the higher intraluminal pressure needed to drive cell culture medium to perfuse through the more complicated vascular structures in thicker scaffolds, which may burst the engineered blood vessels or damage the notorious shear-sensitive endothelial cells. A potential solution to combine the artificial and organic-grown vessels may solve the problem. Recent advances including coculturing of HUVECs and fibroblasts have created functional multi-scale vascular channels with perfused open lumens [135, 136]. Such strategies could be used to improve cell viability inside biofabricated thick scaffolds before we have a deeper knowledge of growing organic vascularity.

Biomaterial Development

A variety of biomaterials have been developed and investigated using different biofabrication techniques. However, the characterization usually only focuses on their printability, mechanical stiffness, and cytotoxicity. To ensure the in vivo stability and functionality of the biofabricated scaffolds, the physical properties, interactions with human body, and stability under loading should also be characterized [137].

In terms of physical properties, the implanted scaffolds should not only resemble the morphology of the native tissues or organs, but also have a similar pore size and porosity [42]. The pore size should also be customizable to fulfill the need for different cell types [61, 62]. Recent advances have been employing sacrificial surfactant particles or water-in-water emulsion bioinks to create porous scaffolds, but they are unable to generate pore gradients [64]. In addition, high porosity may impede the structural integrity of the scaffolds. Therefore, it is important to decouple the structural and material properties of bioink systems.

Interactions between the biofabricated scaffolds and human body are important to translate the biofabrication technologies to clinical use. The scaffolds should have a minimal immune response, which is a critical part in the evaluation of the biocompatibility of biomaterials [138]. Studies have shown that factors such as the size, surface roughness, sharpness, and shape contribute to immunoreactions in vivo even if the composition of the biomaterial is the same [139–141]. Also, biomaterials should actively recruit native cells to help remodel the implant matrix and deliver drugs and growth factors to promote tissue regeneration in a controllable manner. For instance, bone morphogenic proteins (BMPs) can promote osteoinduction in bone TE. However, a delay in BMP-2 administration from a biofabricated HA scaffold failed to enhance osteoinduction due to soft tissue formation [142]. It is necessary to regulate the body-scaffold interactions in both spatial and temporal dimensions.

Another challenge for the biofabricated scaffolds is that most hydrogels are neither tough nor anti-fatigue enough under a dynamic mechanical environment. According to the Lake-Thomas theory, the intrinsic fracture energy of a polymer network is determined by the number of chains across a unit crosslinked area [143]. Human tissues, such as cartilage, have a fracture energy on the order of 1000 J/m^2 [144]. In contrast, the intrinsic fracture energy of most hydrogels is ~10 J/m^2, which is much lower than the required toughness to function. A potential solution is to use bioinks that form a double-network (DN) after fabrication [145, 146]. A DN hydrogel consists of a densely crosslinked dissipative network to dissipate cracking energy and a loosely crosslinked elastic network to retain the original configurations after loading. The additions of polyacrylamide (PAAm) and PEGDA have been shown to significantly increase the fracture energy of alginate and chitosan hydrogels [147]. Besides toughness, the scaffolds should also be fatigue-resistant if the implant position is under cyclic loading. To design anti-fatigue hydrogels, the use of reversible chemical crosslinking mechanisms is necessary [148]. Commonly used reversible bonds include dynamic covalent bonds, ionic bonds, hydrogen bonds, hydrophobic interactions, dipole–dipole interactions, and host–guest interactions. Those bonds can reform after breaking upon resting, elevated temperature, pH change, presence of enzyme, or exposing under light. Developing materials to form a DN network with fatigue-resistant bonds and rapid crosslinking time will greatly facilitate the translation of biofabricated scaffolds to clinical use.

Toward the fabrication of functional large-scale tissue, biofabrication needs to address increasingly difficult challenges, including improving biofabrication resolution, advancing fundamental biology, and developing tunable, tough, anti-fatigue, and tissue-specific biomaterials. Collaboration in multidisciplinary work is encouraged to overcome the existing challenges and to bring personalized regenerative therapeutics into practice.

Acknowledgements I would like to thank Dr. Luc Mongeau and Dr. Jianyu Li (Department of Mechanical Engineering, McGill University) for inspiring my research on biofabrication for tissue engineering.

References

1. Slaughter BV, Khurshid SS, Fisher OZ, Khademhosseini A, Peppas NA (2009) Hydrogels in regenerative medicine. Adv Mater 21:3307–3329
2. Subia B, Kundu J, Kundu SC (2010) Biomaterial scaffold fabrication techniques for potential tissue engineering applications. Tissue Eng 3:141–159
3. Murphy SV, Atala A (2014) 3D bioprinting of tissues and organs. Nat Biotechnol 32(8):773–785
4. Zhu W, Ma X, Gou M, Mei D, Zhang K, Chen S (2016) 3D printing of functional biomaterials for tissue engineering. Curr Opin Biotechnol 40:103–112
5. Truby RL, Lewis JA (2016) Printing soft matter in three dimensions. Nature 540(7633):371–378
6. Malda J et al (2013) 25th anniversary article: engineering hydrogels for biofabrication. Adv Mater 25(36):5011–5028
7. Cui H, Nowicki M, Fisher JP, Zhang LG (2017) 3D bioprinting for organ regeneration. Adv Healthc Mater 6(1)
8. Liu X et al (2018) 3D printing of living responsive materials and devices. Adv Mater 30(4)
9. Yue K, Trujillo-de Santiago G, Alvarez MM, Tamayol A, Annabi N, Khademhosseini A (2015) Synthesis, properties, and biomedical applications of gelatin methacryloyl (GelMA) hydrogels. Biomaterials 73:254–271
10. Jiang T, Munguia-lopez JG, Flores-torres S, Grant J, De Leon-rodriguez A, Kinsella JM (2017) Directing the self-assembly of tumour spheroids by bioprinting cellular heterogeneous models within alginate / gelatin hydrogels. Sci Rep 7(1):4575
11. Pedde RD et al (2017) Emerging biofabrication strategies for engineering complex tissue constructs. Adv Mater 29(19):1–27
12. Kang H, Lee SJ, Ko IK, Kengla C, Yoo JJ, Atala A (2016) A 3D bioprinting system to produce human-scale tissue constructs with structural integrity. Nat Biotechnol 34(3):312–319
13. Xu T, Zhao W, Zhu JM, Albanna MZ, Yoo JJ, Atala A (2013) Complex heterogeneous tissue constructs containing multiple cell types prepared by inkjet printing technology. Biomaterials 34:130–139
14. Koch L, Gruene M, Unger C, Chichkov B (2013) Laser assisted cell printing. Curr Pharm Biotechnol 14:91–97
15. Visser J et al (2013) Biofabrication of multi-material anatomically shaped tissue constructs. Biofabrication 5(3):035007
16. de Ruijter M et al Out-of-plane 3D-printed microfibers improve the shear properties of hydrogel composites. Small 1702773:2017
17. O'Bryan CS et al (2017) Self-assembled micro-organogels for 3D printing silicone structures. Sci Adv 5:3
18. Guillotin B et al (2010) Laser assisted bioprinting of engineered tissue with high cell density and microscale organization. Biomaterials 31(28):7250–7256
19. Xu T et al (2006) Viability and electrophysiology of neural cell structures generated by the inkjet printing method. Biomaterials 27(18):3580–3588
20. Bracaglia LG, Smith BT, Watson E, Arumugasaamy N, Mikos AG, Fisher JP (2017) 3D printing for the design and fabrication of polymer-based gradient scaffolds. Acta Biomater 56:3–13
21. Kim JD, Choi JS, Kim BS, Choi YC, Cho YW (2010) Piezoelectric inkjet printing of polymers: stem cell patterning on polymer substrates. Polymer 51(10):2147–2154
22. Guillotin B, Guillemot F (2011) Cell patterning technologies for organotypic tissue fabrication. Trends Biotechnol 29(4):183–190
23. Abbadessa A et al (2016) A synthetic thermosensitive hydrogel for cartilage bioprinting and its biofunctionalization with polysaccharides. Biomacromolecules 17:2137–2147
24. Rutz AL, Hyland KE, Jakus AE, Burghardt WR, Shah RN (2015) A multimaterial bioink method for 3D printing tunable, cell-compatible hydrogels. Adv Mater 27(9):1607–1614

25. Klotz BJ, Gawlitta D, Rosenberg AJWP, Malda J, Melchels FPW (2016) Gelatin-methacryloyl hydrogels: towards biofabrication-based tissue repair. Trends Biotechnol 34(5):394–407
26. Hornick J (2017) 3D printing in healthcare. J 3D Print Med 1(1):13–17
27. Hinton TJ et al (2015) Three-dimensional printing of complex biological structures by freeform reversible embedding of suspended hydrogels. Sci Adv 1(9):e1500758
28. Ouyang L, Highley CB, Sun W, Burdick JA (2017) A generalizable strategy for the 3D bioprinting of hydrogels from nonviscous photo-crosslinkable inks. Adv Mater 29(8)
29. Bose S, Vahabzadeh S, Bandyopadhyay A (2013) Bone tissue engineering using 3D printing. Mater Today 16(12):496–504
30. Mouser VHM, Melchels FPW, Visser J, Dhert WJA, Gawlitta D, Malda J (2016) Yield stress determines bioprintability of hydrogels based on gelatin-methacryloyl and gellan gum for cartilage bioprinting. Biofabrication 8(3):035003
31. Colosi C et al (2016) Microfluidic bioprinting of heterogeneous 3d tissue constructs using low-viscosity bioink. Adv Mater 28:677–684
32. Liu W et al (2017) Rapid continuous multimaterial extrusion bioprinting. Adv Mater 29(3)
33. Ribeiro A et al (2018) Assessing bioink shape fidelity to aid material development in 3D bioprinting. Biofabrication 10(1):014102
34. Blaeser A, Duarte Campos DF, Puster U, Richtering W, Stevens MM, Fischer H (2016) Controlling shear stress in 3D bioprinting is a key factor to balance printing resolution and stem cell integrity. Adv Healthc Mater 5(3):326–333
35. Suntornnond R, Tan EYS, An J, Chua CK (2016) A mathematical model on the resolution of extrusion bioprinting for the development of new bioinks. Materials 9(9)
36. Jin Y, Chai W, Huang Y (2017) Printability study of hydrogel solution extrusion in nanoclay yield-stress bath during printing-then-gelation biofabrication. Mater Sci Eng C 80:313–325
37. Yuk H, Zhao X (2018) A new 3D printing strategy by harnessing deformation, instability, and fracture of viscoelastic inks. Adv Mater 30(6)
38. Huang Z, Zhang Y, Kotaki M, Ramakrishna S (2003) A review on polymer nanofibers by electrospinning and their applications in nanocomposites. Compos Sci Technol 63:2223–2253
39. Liu S, Reneker DH (2019) Droplet-jet shape parameters predict electrospun polymer nanofiber diameter. Polymer 168:155–158
40. Visser J et al (2015) Reinforcement of hydrogels using three-dimensionally printed microfibres. Nat Commun 6:1–10
41. Jiang T, Munguia-lopez JG, Flores-torres S, Kort-mascort J, Kinsella JM Extrusion bioprinting of soft materials: an emerging technique for biological model fabrication. Appl Phys Rev 6(1):011310, 2019
42. Loh QL, Choong C (2013) Three-dimensional scaffolds for tissue engineering applications: Role of porosity and pore size. Tissue Eng Part B Rev 19(6):485–502
43. Hubbell JA, West JL (1999) Polymeric biomaterials with degradation sites for proteases involved in cell migration. Macromolecules 32(1):241–244
44. Caliari SR, Burdick JA (2016) A practical guide to hydrogels for cell culture. Nat Methods 13(5):405–414
45. Monette A, Ceccaldi C, Assaad E, Lerouge S, Lapointe R (2016) Chitosan thermogels for local expansion and delivery of tumor-specific T lymphocytes towards enhanced cancer immunotherapies. Biomaterials 75:237–249
46. Kolesky DB, Truby RL, Gladman AS, Busbee TA, Homan KA, Lewis JA (2014) 3D bioprinting of vascularized, heterogeneous cell-laden tissue constructs. Adv Mater 26(19):3124–3130
47. Hospodiuk M, Dey M, Sosnoski D, Ozbolat IT (2017) The bioink: a comprehensive review on bioprintable materials. Biotechnol Adv 35:217–239
48. Burdick JA, Murphy WL (2012) Moving from static to dynamic complexity in hydrogel design. Nat Commun 3:1–8
49. Ozbolat IT, Hospodiuk M (2016) Current advances and future perspectives in extrusion-based bioprinting. Biomaterials 76:321–343
50. Berger J, Reist M, Mayer JM, Felt O, Peppas NA, Gurny R (2004) Structure and interactions in covalently and ionically crosslinked chitosan hydrogels for biomedical applications. Eur J Pharm Biopharm 57:19–34

51. Wang J, Lin L, Cheng Q, Jiang L (2012) A strong bio-inspired layered PNIPAM-clay nanocomposite hydrogel. Angew Chem Int Ed 51(19):4676–4680
52. Jin R et al (2010) Synthesis and characterization of hyaluronic acid–poly(ethylene glycol) hydrogels via Michael addition: an injectable biomaterial for cartilage repair. Acta Biomater 6(6):1968–1977
53. Chen D, et al. Drilling by light: ice-templated photo-patterning enabled by a dynamically crosslinked hydrogel. Materials Horizons; 2019.
54. Hou J, Li C, Guan Y, Zhang Y, Zhu XX (2015) Enzymatically crosslinked alginate hydrogels with improved adhesion properties. Polym Chem 6(12):2204–2213
55. Kular JK, Basu S, Sharma RI (2014) The extracellular matrix: structure, composition, age-related differences, tools for analysis and applications for tissue engineering. J Tissue Eng 5:1–17
56. Wen JH et al (2014) Interplay of matrix stiffness and protein tethering in stem cell differentiation. Nat Mater 13(10):979–987
57. Khetan S, Burdick JA (2010) Biomaterials patterning network structure to spatially control cellular remodeling and stem cell fate within 3-dimensional hydrogels. Biomaterials 31(32):8228–8234
58. Chaudhuri O et al (2016) Hydrogels with tunable stress relaxation regulate stem cell fate and activity. Nat Mater 15(3):326–334
59. Zhao X, Huebsch N, Mooney DJ, Suo Z (2010) Stress-relaxation behavior in gels with ionic and covalent crosslinks. J Appl Phys 107(6)
60. Lou J, Stowers R, Nam S, Xia Y, Chaudhuri O (2018) Stress relaxing hyaluronic acid-collagen hydrogels promote cell spreading, fiber remodeling, and focal adhesion formation in 3D cell culture. Biomaterials 154:213–222
61. Chiu YC et al (2011) The role of pore size on vascularization and tissue remodeling in PEG hydrogels. Biomaterials 32(26):6045–6051
62. Oh SH, Park IK, Kim JM, Lee JH (2007) In vitro and in vivo characteristics of PCL scaffolds with pore size gradient fabricated by a centrifugation method. Biomaterials 28(9):1664–1671
63. Yannas IV, Lee E, Orgill DP, Skrabut EM, Murphy GF (1989) Synthesis and characterization of a model extracellular matrix that induces partial regeneration of adult mammalian skin. Proc Natl Acad Sci 86(3):933–937
64. Ying G-L et al (2018) Aqueous two-phase emulsion bioink-enabled 3D bioprinting of porous hydrogels. Adv Mater 30(50):e1805460
65. Li J, Mooney DJ (2016) Designing hydrogels for controlled drug delivery. Nat Rev Mater 1(12):16071
66. Deshayes S, Kasko AM (2013) Polymeric biomaterials with engineered degradation. J Polym Sci Pt A Polym Chem 51(17):3531–3566
67. Vold IMN, Christensen BE (2005) Periodate oxidation of chitosans with different chemical compositions. Carbohydr Res 340(4):679–684
68. Boontheekul T, Kong HJ, Mooney DJ (2005) Controlling alginate gel degradation utilizing partial oxidation and bimodal molecular weight distribution. Biomaterials 26(15):2455–2465
69. Göpferich A (1996) Mechanisms of polymer degradation and erosion. Biomaterials 17(2):103–114
70. Kost J, Leong K, Langer R (1989) Ultrasound-enhanced polymer degradation and release of incorporated substances. Proc Natl Acad Sci 86:7663–7666
71. O'Bryan CS et al (2017) Three-dimensional printing with sacrificial materials for soft matter manufacturing. MRS Bull 42(8):571–577
72. Hinton TJ, Hudson A, Pusch K, Lee A, Feinberg AW (2016) 3D printing PDMS elastomer in a hydrophilic support bath via freeform reversible embedding. ACS Biomater Sci Eng 2(10):1781–1786
73. Grosskopf AK, Truby RL, Kim H, Perazzo A, Lewis JA, Stone HA (2018) Viscoplastic matrix materials for embedded 3D printing. ACS Appl Mater Interfaces 10(27):23353–23361
74. Hoesli CA et al (2012) Reversal of diabetes by βtC3 cells encapsulated in alginate beads generated by emulsion and internal gelation. J Biomed Mater Res - Part B Appl Biomater 100 B(4):1017–1028

75. Jia J et al (2014) Engineering alginate as bioink for bioprinting. Acta Biomater 10(10):4323–4331
76. Beyer ST, Bsoul A, Ahmadi A, Walus K (2013) 3D alginate constructs for tissue engineering printed using a coaxial flow focusing microfluidic device. In: 2013 transducers eurosensors XXVII 17th Int. Conf. solid-state sensors, actuators microsystems, pp 1206–1209
77. Wüst S, Godla ME, Müller R, Hofmann S (2014) Tunable hydrogel composite with two-step processing in combination with innovative hardware upgrade for cell-based three-dimensional bioprinting. Acta Biomater 10(2)
78. Jeon O, Alt DS, Ahmed SM, Alsberg E (2012) The effect of oxidation on the degradation of photocrosslinkable alginate hydrogels. Biomaterials 33(13):3503–3514
79. Das S et al (2015) Bioprintable, cell-laden silk fibroin-gelatin hydrogel supporting multilineage differentiation of stem cells for fabrication of three-dimensional tissue constructs. Acta Biomater 11:233–246
80. Ng WL, Yeong WY, Naing MW (2016) Polyelectrolyte gelatin-chitosan hydrogel optimized for 3D bioprinting in skin tissue engineering. Int J Bioprinting 2(1):53–62
81. Lim KS et al (2016) New visible-light photoinitiating system for improved print fidelity in gelatin-based bioinks. ACS Biomater Sci Eng 2(10):1752–1762
82. Kolesky DB, Homan KA, Skylar-scott MA, Lewis JA (2016) Three-dimensional bioprinting of thick vascularized tissues. PNAS 113(12):3179–3184
83. Park JY et al (2014) A comparative study on collagen type I and hyaluronic acid dependent cell behavior for osteochondral tissue bioprinting. Biofabrication 6(3)
84. Yuan Y, Chesnutt BM, Haggard WO, Bumgardner JD (2011) Deacetylation of chitosan: material characterization and in vitro evaluation via albumin adsorption and pre-osteoblastic cell cultures. Materials 4(8):1399–1416
85. Assaad E, Maire M, Lerouge S (2015) Injectable thermosensitive chitosan hydrogels with controlled gelation kinetics and enhanced mechanical resistance. Carbohydr Polym 130:87–96
86. Chenite A et al (2000) Novel injectable neutral solutions of chitosan form biodegradable gels in situ. Biomaterials 21:2155–2161
87. Li B et al (2015) Hydrosoluble, UV-crosslinkable and injectable chitosan for patterned cell-laden microgel and rapid transdermal curing hydrogel in vivo. Acta Biomater 22:59–69
88. Yang J, Bai R, Suo Z (2018) Topological adhesion of wet materials. Adv Mater 1800671:1–7
89. Li J et al (2017) Tough adhesives for diverse wet surfaces. Science 357(6349):378–381
90. Murphy SV, Skardal A, Atala A (2013) Evaluation of hydrogels for bio-printing applications. J Biomed Mater Res Pt A 101A(1):272–284
91. Gasperini L, Mano JF, Reis RL (2014) Natural polymers for the microencapsulation of cells. J R Soc Interface 11(100):20140817
92. Skardal A et al (2015) A hydrogel bioink toolkit for mimicking native tissue biochemical and mechanical properties in bioprinted tissue constructs. Acta Biomater 25:24–34
93. Lee Y-B et al (2010) Bio-printing of collagen and VEGF-releasing fibrin gel scaffolds for neural stem cell culture. Exp Neurol 223(2):645–652
94. Snyder JE et al (2011) Bioprinting cell-laden matrigel for radioprotection study of liver by pro-drug conversion in a dual-tissue microfluidic chip. Biofabrication 3(3):034112
95. Pati F et al (2014) Printing three-dimensional tissue analogues with decellularized extracellular matrix bioink. Nat Commun 5:3935
96. Wang Z, Abdulla R, Parker B, Samanipour R, Ghosh S, Kim K (2015) A simple and high-resolution stereolithography-based 3D bioprinting system using visible light crosslinkable bioinks. Biofabrication 7(4):045009
97. Bertassoni LE et al (2014) Hydrogel bioprinted microchannel networks for vascularization of tissue engineering constructs. Lab Chip 14:2202–2211
98. F. Y. Hsieh, H. H. Lin, and S. H. Hsu, "3D bioprinting of neural stem cell-laden thermo-responsive biodegradable polyurethane hydrogel and potential in central nervous system repair," Biomaterials, vol. 71:48–57, 2015.
99. Levato R, Visser J, Planell JA, Engel E, Malda J, Mateos-Timoneda MA (2014) Biofabrication of tissue constructs by 3D bioprinting of cell-laden microcarriers. Biofabrication 6:035020

100. Jung JW, Lee JS, Cho DW (2016) Computer-aided multiple-head 3D printing system for printing of heterogeneous organ/tissue constructs. Sci Rep 6(February):1–9
101. Righetti PG, Brost BCW, Snyder RS (1981) On the limiting pore size of hydrophilic gels for electrophoresis and isoelectric focussing. J Biochem Biophys Methods 4:347–363
102. Buckley CT, Thorpe SD, O'Brien FJ, Robinson AJ, Kelly DJ (2009) The effect of concentration, thermal history and cell seeding density on the initial mechanical properties of agarose hydrogels. J Mech Behav Biomed Mater 2(5):512–521
103. Duarte Campos DF et al (2015) The stiffness and structure of three-dimensional printed hydrogels direct the differentiation of mesenchymal stromal cells toward adipogenic and osteogenic lineages. Tissue Eng Pt A 21(3–4):740–756
104. Armstrong JPK, Burke M, Carter BM, Davis SA, Perriman AW (2016) 3D bioprinting using a templated porous bioink. Adv Healthc Mater 5(14):1724–1730
105. Gao Q, He Y, Fu JZ, Liu A, Ma L (2015) Coaxial nozzle-assisted 3D bioprinting with built-in microchannels for nutrients delivery. Biomaterials 61:203–215
106. Bencherif SA et al (2012) Injectable preformed scaffolds with shape-memory properties. Proc Natl Acad Sci 109(48):19590–19595
107. Wang L et al (2017) UV-crosslinkable and thermo-responsive chitosan hybrid hydrogel for NIR-triggered localized on-demand drug delivery. Carbohydr Polym 174(July):904–914
108. Wu Q, Therriault D, Heuzey MC (2018) Processing and properties of chitosan inks for 3D printing of hydrogel microstructures. ACS Biomater Sci Eng 4(7):2643–2652
109. Duan B, Kapetanovic E, Hockaday LA, Butcher JT (2014) Three-dimensional printed trileaflet valve conduits using biological hydrogels and human valve interstitial cells. Acta Biomater 10(5):1836–1846
110. Matthews JA, Wnek GE, Simpson DG, Bowlin GL (2002) Electrospinning of collagen nanofibers. Biomacromolecules 3(2):232–238
111. Jang J, Kim TG, Kim BS, Kim SW, Kwon SM, Cho DW (2016) Tailoring mechanical properties of decellularized extracellular matrix bioink by vitamin B2-induced photo-crosslinking. Acta Biomater 33:88–95
112. Wu Z, Su X, Xu Y, Kong B, Sun W, Mi S (2016) Bioprinting three-dimensional cell-laden tissue constructs with controllable degradation. Sci Rep 6:24474
113. Zhao Y, Li Y, Mao S, Sun W, Yao R (2015) The influence of printing parameters on cell survival rate and printability in microextrusion-based 3D cell printing technology. Biofabrication 7(4):045002
114. Müller M, Becher J, Schnabelrauch M, Zenobi-Wong M (2015) Nanostructured pluronic hydrogels as bioinks for 3D bioprinting. Biofabrication 7(3):035006
115. Highley CB, Song KH, Daly AC, Burdick JA (2019) Jammed microgel inks for 3D printing applications. Adv Sci 6(1):1801076
116. Lee JS, Hong JM, Jung JW, Shim JH, Oh JH, Cho DW (2014) 3D printing of composite tissue with complex shape applied to ear regeneration. Biofabrication 6(2):024103
117. Kundu J, Shim JH, Jang J, Kim SW, Cho DW (2015) An additive manufacturing-based PCL-alginate-chondrocyte bioprinted scaffold for cartilage tissue engineering. J Tissue Eng Regen Med 9(11):1286–1297
118. Zheng Z et al (2018) 3D bioprinting of self-standing silk-based bioink. Adv Healthc Mater 7(6):1–12
119. Wu D et al (2018) 3D bioprinting of gellan gum and poly (ethylene glycol) diacrylate based hydrogels to produce human-scale constructs with high-fidelity. Mater Des 160:486–495
120. Almeida CR, Serra T, Oliveira MI, Planell JA, Barbosa MA, Navarro M (2014) Acta Biomaterialia Impact of 3-D printed PLA- and chitosan-based scaffolds on human monocyte / macrophage responses : Unraveling the effect of 3-D structures on inflammation. Acta Biomater 10(2):613–622
121. Xiaohua L, Peter M (2004) Polymeric scaffolds for bone tissue engineering. Ann Biomed Eng 32(3):477–486
122. Karageorgiou V, Kaplan D (2005) Porosity of 3D biomaterial scaffolds and osteogenesis. Biomaterials 26(27):5474–5491

123. Homan KA et al (2016) Bioprinting of 3D convoluted renal proximal tubules on perfusable chips. Sci Rep 6:1–13
124. Sell SA, McClure MJ, Garg K, Wolfe PS, Bowlin GL (2009) Electrospinning of collagen/biopolymers for regenerative medicine and cardiovascular tissue engineering. Adv Drug Deliv Rev 61(12):1007–1019
125. Zhang H, Liu X, Yang M, Zhu L (2015) Silk fibroin/sodium alginate composite nano-fibrous scaffold prepared through thermally induced phase-separation (TIPS) method for biomedical applications. Mater Sci Eng C 55:8–13
126. Rodriguez MJ, Brown J, Giordano J, Lin SJ, Omenetto FG, Kaplan DL (2017) Silk based bioinks for soft tissue reconstruction using 3-dimensional (3D) printing with in vitro and in vivo assessments. Biomaterials 117:105–115
127. Shanjani Y, Pan CC, Elomaa L, Yang Y (2015) A novel bioprinting method and system for forming hybrid tissue engineering constructs A novel bioprinting method and system for forming hybrid tissue engineering constructs. Biofabrication 7(4):045008
128. Ouyang L, Highley CB, Rodell CB, Sun W, Burdick JA (2016) 3D printing of shear-thinning hyaluronic acid hydrogels with secondary cross-linking. ACS Biomater Sci Eng 2(10):1752–1762
129. Duan B, Hockaday LA, Kang KH, Butcher JT (2013) 3D bioprinting of heterogeneous aortic valve conduits with alginate/gelatin hydrogels. J Biomed Mater Res Pt A 101A(5):1255–1264
130. Rhee S, Puetzer JL, Mason BN, Reinhart-King CA, Bonassar LJ (2016) 3D bioprinting of spatially heterogeneous collagen constructs for cartilage tissue engineering. ACS Biomater Sci Eng 2(10):1800–1805
131. Lee V et al (2014) Design and fabrication of human skin by three-dimensional bioprinting. Tissue Eng Part C Methods 20(6):473–484
132. Ng WL, Yeong WY, Naing MW (2016) Development of polyelectrolyte chitosan-gelatin hydrogels for skin bioprinting. Proc CIRP 49:105–112
133. Martínez Ávila H, Schwarz S, Rotter N, Gatenholm P (2016) 3D bioprinting of human chondrocyte-laden nanocellulose hydrogels for patient-specific auricular cartilage regeneration. Bioprinting:1, 22–2, 35
134. Mironov V, Visconti RP, Kasyanov V, Forgacs G, Drake CJ, Markwald RR (2009) Biomaterials organ printing : Tissue spheroids as building blocks. Biomaterials 30(12):2164–2174
135. Lee VK et al (2014) Creating perfused functional vascular channels using 3D bio-printing technology. Biomaterials 35(28):8092–8102
136. Wüst S, Müller R, Hofmann S (2015) 3D Bioprinting of complex channels—effects of material, orientation, geometry, and cell embedding. J Biomed Mater Res Pt A 103(8):2558–2570
137. Ozbolat IT (2015) Bioprinting scale-up tissue and organ constructs for transplantation. Trends Biotechnol 33(7):395–400
138. Veiseh O et al (2015) Size- and shape-dependent foreign body immune response to materials implanted in rodents and non-human primates. Nat Mater 14(6):643–651
139. Ratner BD et al (2010) Proangiogenic scaffolds as functional templates for cardiac tissue engineering. Proc Natl Acad Sci 107(34):15211–15216
140. Kim YK, Que R, Wang S-W, Liu WF (2014) Modification of biomaterials with a self-protein inhibits the macrophage response. Adv Healthc Mater 3(7):989–994
141. Tsai RK, Rodriguez PL, Discher DE, Pantano DA, Harada T, Christian DA (2013) Minimal 'self' peptides that inhibit phagocytic clearance and enhance delivery of nanoparticles. Science 339(6122):971–975
142. Becker ST et al (2012) Endocultivation: the influence of delayed vs. simultaneous application of BMP-2 onto individually formed hydroxyapatite matrices for heterotopic bone induction. Int J Oral Maxillofac Surg 41(9):1153–1160
143. Lake GJ, Thomas AG (1967) The Strength of Highly Elastic Materials. Proc R Soc A Math Phys Eng Sci 300(1460):108–119
144. Zhao X (2014) Multi-scale multi-mechanism design of tough hydrogels: Building dissipation into stretchy networks. Soft Matter 10(5):672–687

145. Gong JP, Katsuyama Y, Kurokawa T, Osada Y (2003) Double-network hydrogels with extremely high mechanical strength. Adv Mater 15(14):1155–1158
146. Sun J et al (2012) Highly stretchable and tough hydrogels. Nature 489(7414):133–136
147. Yang Y, Wang X, Yang F, Shen H, Wu D (2016) A universal soaking strategy to convert composite hydrogels into extremely tough and rapidly recoverable double-network hydrogels. Adv Mater:7178–7184
148. Bai R, Yang J, Suo Z (2019) Fatigue of hydrogels. Eur J Mech A/Solids 74(2018):337–370

Additive Manufacturing of Bioscaffolds for Bone Regeneration

Osama Almayyahi, Irsalan Cockerill, Yufeng Zheng, and Donghui Zhu

Abstract As technology and computer applications reach new strides every day, the biomedical field is benefiting from this advancement and that includes the production of bioscaffolds for tissue engineering and bone regeneration. Even though the body is capable of healing minor injuries, some injuries might prove too challenging for the body to repair and that is where additive manufacturing (AM) comes in. This chapter discusses the primary materials as well as the major and mainstream methods used in AM of bioscaffolds for bone regeneration. In addition, new advancements in Computer-Aided Design (CAD) and three-Dimensional (3D) designs in AM are addressed. The benefits and drawbacks of different methods used in AM for bone scaffolds and their suitability for human trials and further applications on patients are also discussed.

Keywords Bioscaffold · Tissue engineering · Bone regeneration · Additive manufacturing · Bone graft

Introduction

According to the AAOS (American Academy of Orthopedic Surgeons), there are over six million fractures every year in America alone, and the cost associated with these injuries is approaching billions of dollars every year [1]. Throughout history, bone regeneration methods have been in development, but the standard for the past

O. Almayyahi · I. Cockerill (✉)
Department of Biomedical Engineering, University of North Texas, Denton, TX, USA
e-mail: IrsalanCockerill@my.unt.edu

Y. Zheng
Department of Materials Science and Engineering, Peking University, Beijing, China

D. Zhu
Department of Biomedical Engineering, Stony Brook University, Stony Brook, NY, USA
e-mail: donghui.zhu@stonybrook.edu

B. Li et al. (eds.), *Racing for the Surface*,
https://doi.org/10.1007/978-3-030-34471-9_13

few decades have been allograft or autograft bone transplants. However, the inherent limitations of this approach, such as availability, morbidity, and secondary surgeries—in the case of autografts—and disease transmission and immunological reactions—in the case of allografts—drove the biomedical and additive manufacturing industry to come together and come up with better approaches [2]. The fruits of this collaboration resulted in the emergence of engineered bone tissues made from bone cells and biocompatible materials [3]. These biomaterials are called bone bioscaffolds, and they are becoming the new standard replacing allografts and autografts. The field of bone regeneration using bioscaffolds is expanding, and many innovations are emerging to enhance the process of bone regeneration, inflicting as little pain as possible and reduced cost for patients' benefit. This desire also resulted in the innovation of biodegradable materials that degrade inside the body after serving their function, reducing the number of operations a patient will have to endure and, of course, reduce the cost related with multiple operations [4, 5]. Due to the fact foreign materials will be inserted inside the body, inherently, a lot of research will have to be conducted to investigate the response of the immune system and the toxicity of these materials.

Bone Structure and Properties

Bones form the skeletal system that in turn protects and supports many organs of the human body. Bones are also tasked with producing red and white blood cells, store minerals, and provide the structural support that enables us to walk, run, and carry items by supporting the entire musculoskeletal system [6]. Human bone is a dense, hard, connective tissue that consists of organic and inorganic components of about 10–20% content of water, 60–70% bone mineral, and the rest is contributed to collagen fibers. Additionally, bone consists of trace amounts of proteins and inorganic salts [6].

The bone structure is optimized to be as strong as possible and as lightweight as possible, relatively speaking. The center of the bone consists of bone marrow and is surrounded by two types of bone, the spongy bone, known as trabecular bone, and the compact bone, also known as cortical bone (Fig. 1a) [6]. The ratio of cortical tissue to trabecular tissue in one bone depends mainly on the function that bone serves. The cortical bone mainly consists of osteons that form cylindrical like shapes and contribute to the bone's resistance of bending [7]. Within the osteons, a network of osteocytes is found which are crucial to maintaining the structural integrity of the bone, as we will see in a later section in this chapter. The blood vessels in the cortical bone that facilitates exchanges between the osteocytes and the blood can be found in the center of the osteons, along with the nerves that run through the bone [8]. These vessels and nerves are found within a structure called haversian canals. The main function of trabecular bone is to resist compression where the osteocytes play an important role in sensing increased strain on the bone [9].

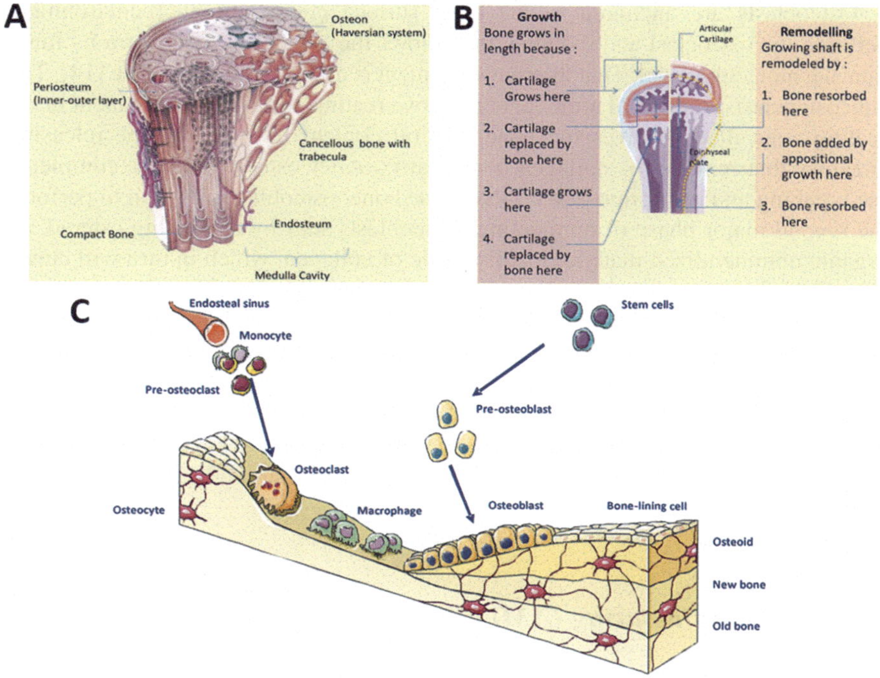

Fig. 1 (**A**) A human bone model showing cortical, hard bone, and trabecular, spongy bone; (**B**) A simplified illustration of the natural modeling and remodeling process of human bone and (**C**) the cellular mechanism underlying the phases of bone remodeling. (Permission obtained from © 2011 Stępien E. Published in [62] under CC BY-NC-SA 3.0 license. Available from: https://doi.org/10.5772/18940 and © 2018 Rosy S., Paulus R. Published in [63] under CC BY 3.0 license. Available from: https://doi.org/10.5772/intechopen.82452)

Bone Modeling and Remodeling

Contrary to what one might assume about bone, it is actually undergoing constant change in structure and goes through many growth cycles throughout a person's life (Fig. 1b) [10]. For example, a typical human is born with 270 bones but some of these bones are repurposed, merge with others, and the number of the bones actually decreases in a healthy human to 206. Another example is when a healthy human goes through loss of bone density that begins at early adulthood that would be noticeable in old age [11].

Osteoclasts, large multinucleate bone cells, are mainly responsible for controlling the density of bone by resorbing bones and are important for bone remodeling (Fig. 1c) [12]. Another type of cell that is important for bone remodeling is the osteoblasts which creates new bone in response to the osteoclasts activity [13].

Osteoclasts are anchored to the bone surface creating a microenvironment between the osteoclast and the bone that allows the osteoclast to perform its function at a molecular level. That microenvironment is called the sealing zone [14]. The way osteoclasts resorb and remodel bone is by creating an acidic environment in the sealing zone that dissolves the bone mineral content followed by the releasing enzymes that remove the collagen bone matrix. After osteoclasts have completed the breaking and resorption of a region in the bone, osteoblasts move in to perform the second major phase of remodeling. Osteoblasts start by depositing osteoid, an organic unmineralized material mainly made of collagen, which in turn will create a scaffold that promotes crystallization of phosphate and calcium. Some of these osteoblasts turn into osteocytes by merging with the scaffolds and the rest will become bone lining cells. That process is called bone remodeling. Alternatively, when the breaking down of bone and the building of bone by osteoblasts happen separately, the process is known as bone modeling. When osteoblasts build more bone mass without prior resorption of bone from osteoclasts, the result is an increase in bone mass and is important for maintaining bone strength and growth [15] (Fig. 1).

Additive Manufacturing (AM) in Bone Regeneration

Additive manufacturing (AM) is the process of which three-dimensional (3D) objects are made by adding a layer of material on top of another layer. The types of materials could be polymer-based, metal-based, ceramic-based, or composites, basically, any material that can be layered and hold its form with a certain level of detail [16]. AM has come a long way since its inception decades ago in the late 1980s with the emergence of stereolithography (STL) at 3D systems where they were able to successfully solidify a layer of liquid polymer using ultraviolet (UV) light, resulting in the production of the first commercially available AM machine, the SLA-1 [17].

The development of computer aided design (CAD) is probably one of the major contributors to the advancements of AM, its complexity, and its availability [18]. Today, many methods can be used in AM that are used as rapid prototyping or end-product manufacturing, namely, STL, selective laser sintering (SLS), directed energy deposition (DED), 3D modeling, and fused deposit modeling (FDM), and other commercial methods.

In a nutshell, the process of creating an object using AM starts with a CAD file that gets converted to STL file format. The STL then gets sliced according to manufacturing specifications and converted to G-code which holds instructions for making the part such as path to follow, speed, etc. The slices are then deposited using the AM machine, and usually this process is followed by a post processing procedure to clean up the model or to rid of support structures. An illustration of AM process is shown in Fig. 2.

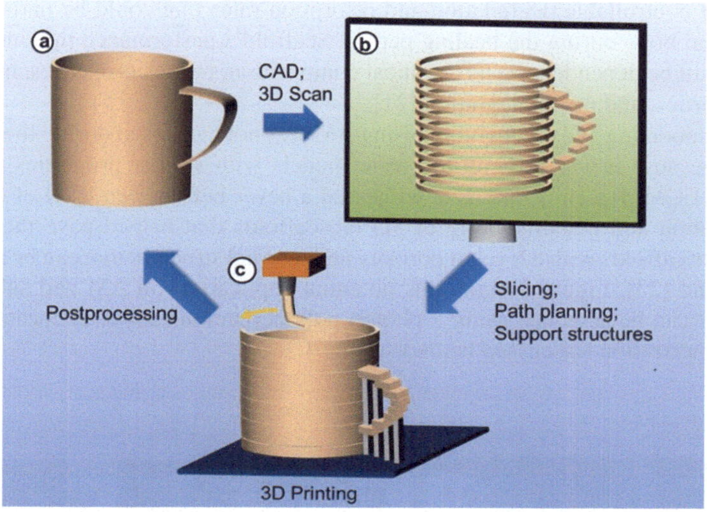

Fig. 2 An illustration of AM process from CAD file to the end-product 3D part. (**a**) Intended part to produce, (**b**) generation of the CAD file, and (**c**) the 3D-printed part which must get post-processed. (Reproduced from [64] https://pubs.acs.org/doi/10.1021/acs.chemrev.7b00074 with permission from ACS)

Although bone is characterized with self-healing properties through the remodeling process as discussed earlier, this process won't be sufficient to repair major defects in the bone and heal it completely, which would prompt external intervention. Generally, this intervention has been done using autografts (when one part of the patient is used to heal another part in the same patient) or allograft (when one part from a deceased donor is used to heal a living patient) [19]. The use of either of these methods can open the door for infection and allografts may compromise the immune system of the receiving patient. Moreover, the difficulty in finding a suitable donor makes either of them an inferior choice compared to bioscaffolds produced by means of AM, especially when factoring in the level of detail that can be achieved using AM to create custom bioscaffolds individualized for each patient. Therefore, due to the inherent limitations and risks associated with autografts and allografts, attention has been redirected toward AM to create the necessary bioscaffolds in bone tissue engineering to promote bone remodeling and regeneration and help it gain or promote its function, especially for large bone defects [20].

Bioscaffolds are 3D biocompatible and bioresorbable models or structures that are used to promote bone regeneration by mimicking the mechanical and extracellular matrix (ECM) of the natural bone. These scaffolds provide a structural template for bone cell attachment [21]. Manufactured scaffolds must contain pores with diameters exceeding 300 μm with interconnected channels to promote the exchange of nutrients and cell attachment to the scaffold [22]. Scaffolds must also have some degree of controllable degradation and resorption rates that could be harnessed to mimic real bone during the healing period. Scaffold's performance in bone regeneration will be dependent on its chemical composition, porous properties, mechanical properties, and osseointegration [21].

Bone modeling and regeneration using AM was not harnessed during the infancy of AM because it was difficult to create models with similar properties to those found in ECM. Recent years have witnessed a never before seen level of detail in manipulation and manufacturing of 3D bioscaffolds that helped pave the way to create bioscaffolds with a level of porosity and internal structure that can be adjusted on-demand [23]. Figure 3 highlights potential applications in AM and submicron scale that can be achieved using 2 photon polymerization (2PP), a subcategory of vat polymerization (discussed below).

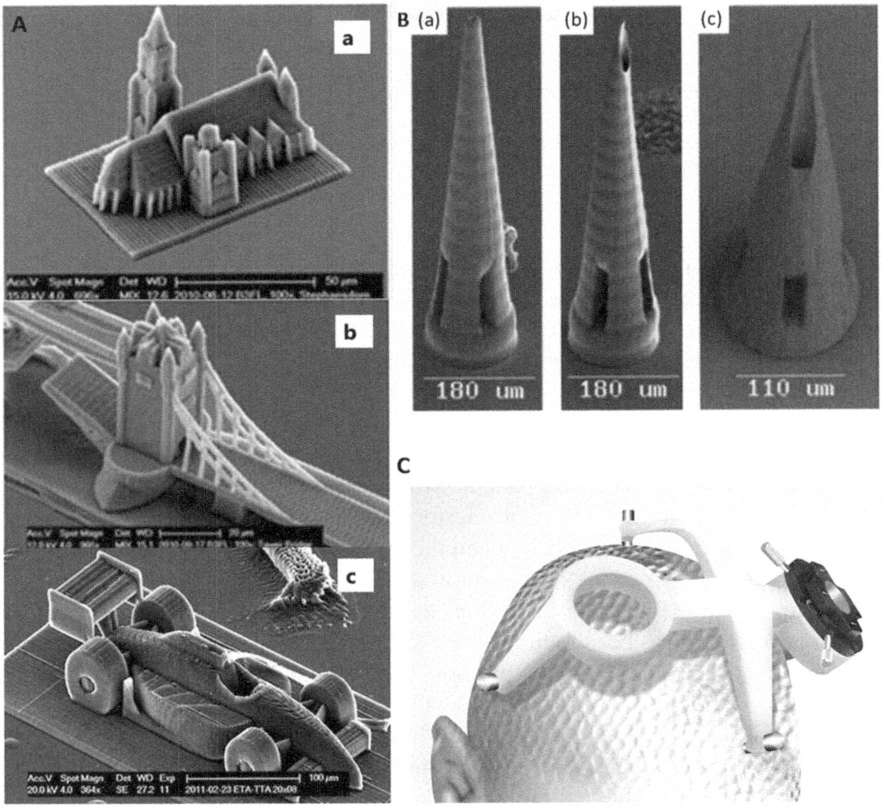

Fig. 3 Illustrations of dimensional resolution and applications of AM. (**A**) Real-world replicas [65, 66] and (**B**) microneedles [67] created by 2 photon polymerization (2PP), a high-resolution technique of SLA; (**C**) SLS-fabricated neurosurgical brain guide [68]. (Reproduced from [64] https://pubs.acs.org/doi/10.1021/acs.chemrev.7b00074 with permission from ACS)

Materials

Broad categorization of AM materials falls into any of these categories: metals, polymers, ceramics, and composites. Most are familiar with implant made of bio-ceramics and metals and their composite/alloys, such as those made for skull implant and hip joint replacements. Each type of material will have its own set of properties and characteristics with advantages and disadvantages that make it suitable for one type of application but not necessarily for others. Next, let's take a look at some of the popular materials used in creating bioscaffolds for bone regeneration.

Bio-ceramics

Bio-ceramics comes in many shapes and forms; they can be made very dense or very porous, in powder form, or could be even used to coat another type of material. Bio-ceramics range in biocompatibility from being inert or bioinert, to fully degradable, biodegradable, or bioactive in a biological environment [24]. Generally, a material qualifies as ceramic when it consists of metal-oxides, carbides, and nitrides and their combinations such as aluminum oxide and titanium oxide. Furthermore, they could be made with a combination of calcium phosphates, alumina, zirconia, carbon, and other types of materials [25].

One of the many characteristics to look for in a bioscaffold is the porosity. As bone comes with a range of porosity from micron and reaching several hundreds of microns [26], a bioscaffold would need to accommodate this property. We will next look into some of the bio-ceramics and how they would fulfill this need and other properties that should be found in the bioscaffold.

Calcium Phosphate

The abundant existence of calcium phosphate in bone and teeth has driven researchers to take a deep look into calcium phosphate as this material is inert, biocompatible, and somewhat degradable [27]. Furthermore, calcium phosphate was found to be osteoconductive, meaning it provides a template for new bone tissue formation and promotes osteoblast adhesion and propagation [28]. To create calcium phosphate bioscaffold, researchers had to go into the nanoscale to mineralize a scaffold using oxides, phosphates, and calcium and silicon combined with carbonates [29].

Ceramics are more brittle than desired so normally they are combined with polymers to achieve less brittleness. Hydroxyapatite and biphasic calcium phosphate are examples of popular types of bio-ceramics able to support bone in-growth and integration in bone tissue [25]. Calcium phosphate presents a major clinical advantage due to its osteoconductivity, and this characteristic was employed in conjunction with calcium aluminate to create a filler material for the bone that would promote new bone tissue growth, similar to real bone and cell proliferation and adhesion [30].

Bioglasses

Bioglass is a glass but it contains certain materials that include SiO_2, CaO, Na_2O, and P_2O_5. There is less silica in bioglass compared to regular glass found in everyday commercial merchandise. Bioglass contains a higher molar ratio of calcium to phosphorous, 5:1, and can promote apatite crystals, characterizing bioglass with good osteoconductivity and bioactivity. It is a good solution for biodegradable bone scaffolds and applications of tissue engineering. Bioglass can be mixed with varying amounts of elements such as strontium, zinc, or copper, which would promote the formation of healthy bone tissue [31]. Figure 4 shows several mesoporous bioactive glass (MBG) scaffolds of varying pore size and morphology created by extrusion-based 3D printing.

Bioglass® is a company that makes bioglass scaffolds and it highlights the bioactive behavior and advantages of using bioglass scaffolds. Their bioglass was proven to bind bone tissue by a function of formatted hydrocarbonate apatite (HCA) on the outer layer of the bioglass surface. Hydrocarbonate apatite is very similar to hydroxyapatite bone material as it promotes interaction with the damaged bone collagen fibrils. This results in enhanced proliferation of matrix mineralization of bone

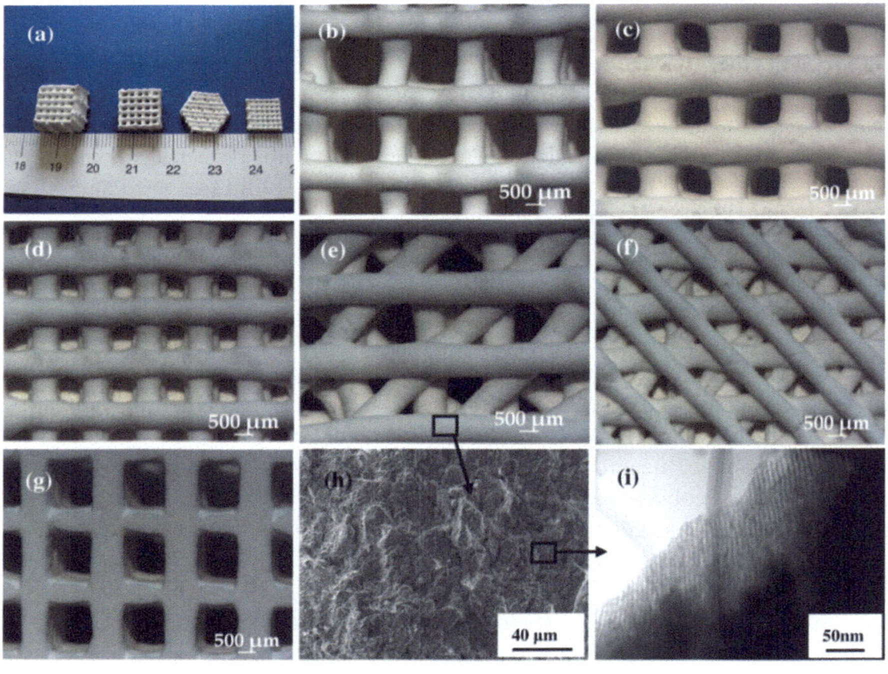

Fig. 4 3D printing of mesoporous bioactive glasses (MBG) [69, 70]. (**a**) Different MBG structures produced via extrusion-based 3D printing; (**b–g**) depicts changes in pore sizes (624–1307 μm) and pore morphology; (**h**) SEM of pore wall taken from (**h**); (**i**) TEM image of pore wall. (Reproduced with permission from [70])

ECM. It is worth noting that bioglass contains an elevated level of negative-surface charge density by silanols.

The degradability of bioglass will leave residual glass particles in living tissue, and the degradation of the bioglass as a whole would take up to 12 months or maybe longer. This degradation will result in the release of glass particles varying in size from below 300 μm, which will be taken care of by the osteoclasts, while particles above 300 μm may stay in the body for longer periods [32].

Because the bioglass surface is dense with a negative charge, it is a good factor to promote adsorption of serum proteins, making calcium phosphate-bioglass composite an excellent bio-ceramic scaffold material [33]. The main disadvantages of bioglass are its slow rate of degradation, difficulty creating models with specific shape and detail, and it causes increased concentrations of calcium ions and sodium ions in the scaffold site, which can be cytotoxic to cells as they must maintain ion concentrations within a given range.

Metals

Traditional Metals

The strong mechanical strength and toughness against fractures makes metal a good choice in dental and orthopedic bone healing applications. The most common metals are stainless steel (SS), cobalt-chromium (Co-Cr), and titanium (Ti) alloys. While these are not biodegradable, a brief discussion is warranted as they are the clinical benchmarks. A major disadvantage with using these materials is the release of metal ions which can accumulate in biological tissues, triggering an immune response and long-term chronic inflammation [34]. Additionally, due to the elastic moduli, these materials do not make good candidates for bone growth alone. With that said, advantages of using metal alloys are their ability to be used as a substitute for bone due to its porous structure, mechanical properties, biocompatibility, and shape memory effect, in the case of shape memory alloys [35]. Figure 5 shows a few examples of metal-based scaffolds fabricated by AM and their applications.

New advancements in AM have allowed researchers to adjust the level of porosity in metals such as Ti and NiTi alloys that could stimulate cell proliferation and attachment. Ti and TiO_2 scaffolds placed in rabbits showed good osteoconductivity and osseointegration [35]. W4-Mg and Fe-Mn alloys have also been in use in recent years to make metallic scaffolds.

Biodegradable Metals

Biodegradable metals would provide excellent support to the bone during the healing period and then degrade and disappear after the healing is completed [36]. Biodegradable metal scaffolds have significantly higher stiffness than polymer-based

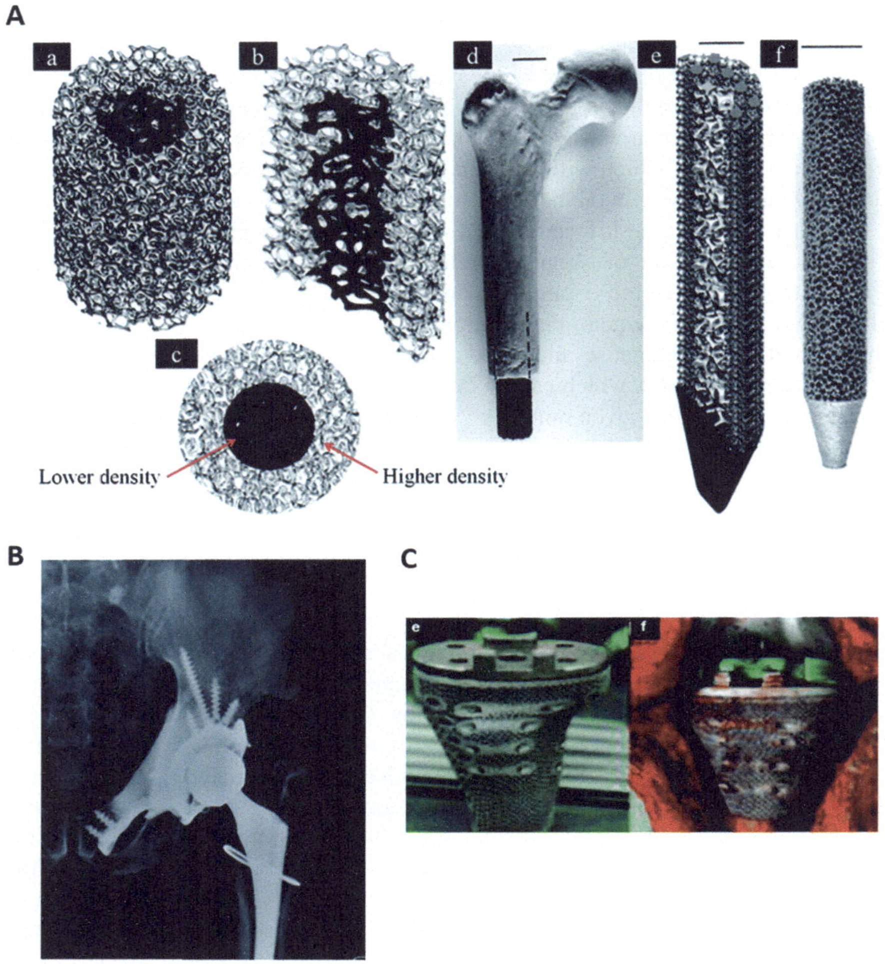

Fig. 5 Images of AM-produced bone scaffolds. (**A**) Ti6Al4V of varying inner and outer density femoral scaffolds produced via electron beam melting (EBM) [71]; (**B**) Implanted acetabular cage for a severe bone defect from a prototype [72]; (**C**) Proximal tibia prosthesis implanted after tumor resection [73]. (Reproduced from © 2018 Chaohua G. et al. Adapted from [74], originally published under CC BY 3.0 license)

scaffolds and are more suitable for situations where load-bearing is expected in the area around the injury [35]. Of course, a certain level of possible risk comes with deploying a biodegradable metal in vivo, so the metals or metal alloy to be used must not induce an adverse immune response, osteolysis, or inflammation. The biodegradability period is an important parameter to consider here. Magnesium biodegrades at a high rate in a chloride-rich biological environment, and the risk from the large amounts of hydrogen gas due to fast corrosion may take it out of the candidacy. Another option is iron-based biodegradable scaffolds [37], but the extremely slow degradation rate makes it unsuitable for this task. One way to slow down or speed up the degradation and manipulate the mechanical properties is to create alloys [35].

Magnesium alloys were developed to create degradable, biocompatible, and open-pore metal scaffolds that were found to promote bone formation [38]. Porous scaffolds of magnesium alloy made using AM were found to promote bone remodeling and regeneration without adverse effects in animal models. However, porous alloys of magnesium created more surface area and degraded at an accelerated pace. However, experiments using Mg-4 wt.% Y showed a slower degradation due to the presence of the element yttrium and is also biocompatible [39].

Polymers

Polymers, especially biodegradable polymers, are also widely used in bone and dental fixation and tissue engineering. Biodegradable polymer scaffolds should be biocompatible in a bio-environment to avoid triggering an immune response or cause any type of biological damage to the patient. Unlike permanent implants, biodegradable bioscaffolds eliminate the undesirable effects of conventional implants. Using biodegradable materials means less trauma a patient must endure, and less cost that is usually related to multiple surgeries [40].

Some of the biodegradable synthetic polymers that are most promising and used widely in biomedical research and applications are polyolefins, polyurethanes, polysiloxanes, and polyesters. Polyglycolic acid-polylactic acid (PGA-PLA) is a polyester that has seen increased attention in research for its good biocompatibility and properties to promote bone tissue regeneration [40, 41]. Another polyester with focused studies recently is polyorthoester as it is seen as a forerunner for its capabilities in promoting bone regeneration [40, 42]. Polycarbonate is also the focus of research as it demonstrated bone-like, load-bearing mechanical properties and excellent osteoconductive abilities when conjugated with an ethylester pendant group [43].

Composites

Using composites creates a level of diversity that cannot be found in one type of material in terms of mechanical strength characteristics, biological reactivity, bioactivity, and degradability. An example of a composite is Cerosium® which is very strong and almost similar to bone in strength, has the same modulus of elasticity as bone, and is made of a composite of alumina, calcium carbonate, silica, and magnesium carbonate. This composite comes with pores ranging in size 18 to 25 p [38].

Generally, composites come in combinations of polymer/metals, ceramics/polymer, and metals/ceramic. Calcium phosphate/polyester is an example of ceramic/polymer that has been extensively researched in recent years, showing a successful desired osteogenic differentiation and cell prefoliation [38].

Polymer/metals composites contain a metallic substance with a polymer coating and have an advantage over other types of composites because the metal part provides structural stability and load-bearing properties better than that found in ceramics and polymers. The polymer coating of the composite establishes a better medium for attachment with the surrounding organic material, creating a surface that allows for better protein adsorption and overall a better result for cell proliferation. Several examples of this include polymer coatings on titanium substrates [44–46]. Researchers noticed better cell proliferation and cell adhesion of osteoblasts and human mesenchymal stem cells (hMSCs) when using grafted titanium/polymer composites as compared to unmodified titanium.

Ceramic/metal composites are also a viable choice when mechanical properties and bioactivity are the main concern. A coating of β-TCP around magnesium showed good bioactivity and cell adhesion [47, 48]. Mechanical properties of this composite showed different results across different porosity settings [47, 49].

Fabrication Methods in Additive Manufacturing

Methods that preceded AM in creating scaffolds lacked consistency and repeatability which reduced their reliability. Today, AM comes with its wide range of methods to produce objects with desired mechanical properties to serve a multitude of functions. AM starts with a computer aided design (CAD) file which contains the information necessary to create a 3D model using vector information. Then, this CAD file is converted to an STL file which is regarded as the standard for most AM machines. Minor manipulations might be necessary at this time such as scaling, rotating, and aligning to the virtual platform. The model will be converted from one 3D model to multiple slices of cross-sectional 2D images. Machine preparation is necessary at this point to ensure no pausing during the fabrication process. At this time, prototyping is initiated and after it is done some post-processing is most likely necessary before the object is ready to be used in applications [50]. Next, we will discuss a few mainstreamed commercial methods to create scaffolds using AM, and the basic structure of different AM machinery is shown in Fig. 6.

Stereolithography (SLA)

This is arguably the first AM method that was introduced in the 1980s in which radiation from an ultraviolet (UV) light is used to solidify a radiation-sensitive, liquid risen photopolymer to create a 3D model that rests on a platform. After the

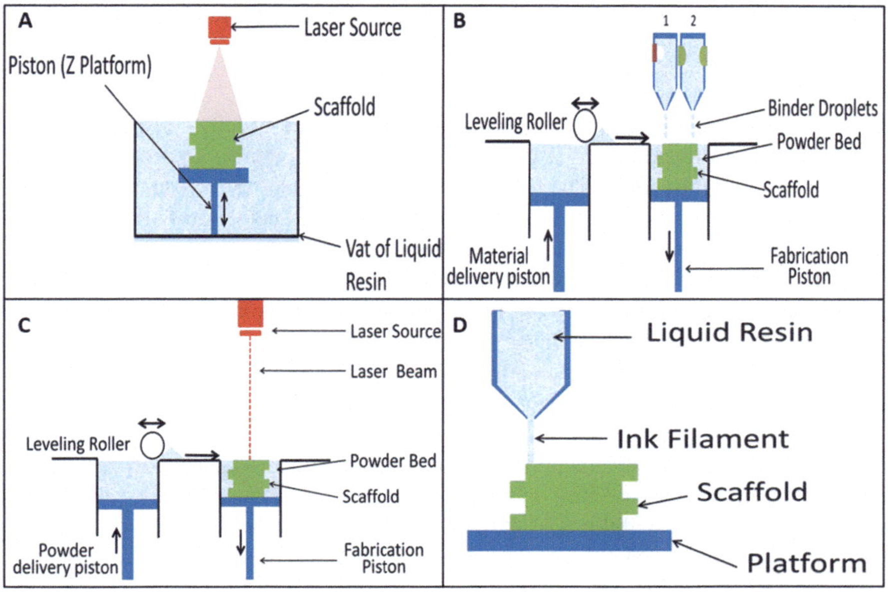

Fig. 6 Depiction of several AM processes. (**A**) An illustration of the basic premise of an SLA machine; (**B**) 3D ink-jet printing; AM can be either thermal-based (nozzle 1 in picture) or piezoelectric-based (nozzle 2 in figure); (**C**) Schematic showing the principle of SLS; (**D**) Premise of FDM highlighting the extrusion of polymer. (Permission granted from © 2015 Patrick R., et al. Published in [75] under CC BY 4.0 license. Available from: https://doi.org/10.3390/ijms19113308)

UV light solidifies one layer representing a cross-section of the final object, the platform is lowered, and a new layer of resin is applied on top of the model which will be cured with the UV light once again until the model is complete. The model could make use out of supporting structures to minimize the risk of falling to one side during the building process and keep its structural integrity. Figure 6a shows the basic parts of an SLA machine [51].

Aqueous poly(ethylene glycol) (PEG) hydrogel solution is one example of a photopolymer, or biopolymer, used in SLA that is capable of preventing dermal fibroblast cells, which are embedded in the hydrogel solution, from getting damaged during the SLA process [52]. This makes it possible to generate 3D models with biological characteristics embedded within them, which is a huge leap in the AM bio-advancements. Although most research is directed toward biopolymers to create these 3D biostructures, some researches head toward finding suitable bio-ceramics for this purpose.

The advantages of using SLA include high accuracy prototypes and easy post-processing that starts with the relatively easy task of removing the unused liquid photopolymer around the built object. Some disadvantages include the limited variety of material choices to fabricate with and high material cost [40].

Selective Laser Sintering (SLS) and Selective Laser Melting (SLM)

This is a process that takes advantage of the powder bed fusion (PBF) principle in which a laser is directed using a galvanometer to plot a path on a polymer bed that contains powdered polymer. The laser solidifies any particles that cross the laser path creating one layer by binding heated powder particles with neighboring particles [53]. In SLS, powder does not melt but fuses as a result of solid-state sintering. In SLM, powder is fully melted to generate fusion. The lifting mechanism on which the solidified powder rests on will be lowered one layer-width, and a roller pan will sweep the top of the platform with a new layer of particles covering the previously build layer. The laser is directed again on the path that corresponds with the current cross-section of the object. These steps are repeated until all cross-sections of the model are built. The powder that did not get solidified will surround the built object and act as support, removing the need to model supporting structures and thus reducing post-processing time. Although, metal-based models might need structural support since the heavy weight will surpass the abilities of the powder to hold it without tipping on one side [54]. Figure 6c shows a simplified representation of an SLS machine.

Some biocompatible and biodegradable polymer-based bioscaffolds manufactured using SLS include polycaprolactone (PCL), polylactide (PLA), and poly-L lactide (PLLA). For example, hydroxyapatite (HA)/PLAA composites were used to build bone scaffolds. The PLAA in this composite served as a binder because of its low melting temperature and fast degradation time. Results have shown that HA/PLAA composites resulted in somewhat similar bending strength and modulus of cancellous bone [55].

In one study [56], researchers used SLS to fabricate polycaprolactone (PCL) scaffolds, a bioresorbable polymer, used in healing of bone as well as cartilage. The created scaffolds showed good mechanical properties as analyzed by computational finite element analysis (FEA). They are similar to that of trabecular bone found in humans with yield strengths ranging within 2.0–3.2 MPa and compressive moduli ranging from 52—67 MPa, giving these scaffolds good load-bearing mechanical properties. To test the scaffolds–cell interaction, the scaffolds were implanted subcutaneously and exhibited enhanced tissue growth was observed in vivo [56].

In another example, researchers used SLS to create scaffolds using calcium phosphates/poly(hydroxybutyrate-co-hydroxyvalerate) composites and carbonated hydroxyapatite/poly(L-lactic acid) composites to provide structural and functional support for bone healing [57]. SLS enabled the researchers to control the porosity, numbers of pores, and the structural connectivity of the composites to fit its intended application. Cultured SaOS-2 osteoblast-like cells showed normal reactions, morphology, and high viability during the biological evaluation of the scaffolds in vitro after 3- and 7-days culture. Alkaline phosphatase activity and cell proliferation in these composites are attributed to the calcium phosphate which served as an osteoconductive agent [57].

The advantages of using SLS include a variety of material choices, excellent mechanical properties, decent accuracy, and low-cost polymers. Disadvantages include thermal degradation during building, high porosity, trapped powder, and high-cost machinery. Additionally, SLS has a rough surface finish compared to SLA [54].

3D Printing

This process makes use of ink-jet which places a binder material on a powder bed and creates a cross section of the final object by gluing particles together. Binder deposition happens by having the ink-jet head moving across the *x*-*y* plane and applying binder by dropping it where needed to create adhesion. After one layer is created, the platform is lowered, and a new layer of powder is applied and so on. Advantages of using 3D printing over other types include lower cost, fast process, multi-print heads, and a variety of material choices since, theoretically, any material that could be powdered is an acceptable candidate for 3D printing. Disadvantages include trapped powder and lower resolutions, and binding might not be sufficient to solidify the object [54]. Figure 6b shows a simplified representation of a 3D printer ink-jet machine.

In a study [22], researchers used rapid prototyping or 3D printing more specifically to create hydroxyapatite (HA) ceramic scaffolds for bone tissue engineering. 3D printed scaffolds were then tested for their biocompatibility using MC3T3-E1 cells within static and dynamic environments in vitro. The reason behind choosing HA was due to its chemical and relative structural similarity to the inorganic portion of real bone tissue. Researchers aimed to optimize the cell adhesion process and interaction with the organic environment by maximizing the surface area of the scaffolds. Results were very promising as the cell proliferation was very high within the HA granules [22].

Another study [58] focused on optimizing the binder to maximize biocompatibility and osteoconductivity with adequate mechanical properties using 3D printing of composites of calcium phosphate and collagen scaffolds for bone regeneration. To achieve maximum cytocompatibility and mechanical strength, the researchers incorporated an 8.75 wt.% phosphoric acid-based binder solution to improve printing, as well as polysorbate 80, a nonionic surfactant. Collagen 1–2 wt.% was also used to achieve better cell viability, flexural strength, and better formulation results. Although viscosity and surface tension could be problematic in the process of 3D printing, this effect was reduced through physiologic heat treatments and polysorbate 80. These scaffolds were tested using mice as a receiving host with 2 mm mid-femoral osteotomy defects, and the results showed osteoconductive behavior from the scaffold with a noticeable degradation of the scaffold as bone growth increased [58].

Fused Deposition Modeling (FDM)

This process uses a printing head which contains a melting chamber. A constant stream of polymer filament is liquefied due to heat transfer, and it is pushed out of a chamber by applying positive pressure. The extruded material will be applied on the build plate according to a set path to create the desired cross-section of the final object. Once the path for one layer is completed, the platform lowers a cross-section width and the process repeats. It is important to maintain constant temperature and filament feed as any interruption or changes in temperature will affect the amount of material extruded at one time, creating non-leveled layers. Complex geometries including highly detailed scaffolds can be obtained using this method combined with other deposition methods [59]. Figure 6d shows a simplified representation of an FDM machine.

This method is capable of creating honeycomb-like scaffolds with connected internal channels and networks and at the same time reserving the ability to enable the researchers to control the porosity of the scaffold and its channel size. In one study [59], researchers created a bioresorbable polymer poly(e-caprolactone) (PCL) filament made of aligned microfilaments. These filaments were used to make porous scaffolds for tissue engineering purposes. Using FDM, researchers were able to create pore sizes of 160–700 μm within the channels of the scaffold with 47–77% porosity. The scaffolds were comparable to porous metals in terms of stress-strain mechanical characteristics as stiffness ranged from 4 to 77 MPa [59]. Another study was conducted [60] to investigate the mechanical properties and cell cultural response of polycaprolactone scaffolds designed and fabricated via FDM. They used a honeycomb-like scaffold as well and showed better mechanical properties and improved biocompatibility with human fibroblasts [60].

A previous study [61] focused on the development of controlled porosity of polymer-ceramic composite scaffolds via FDM. The polymer/ceramic composite was produced using polypropylene (PP) polymer and tricalcium phosphate (TCP). Mechanical and structural properties of the scaffold were comparable to a dog bone tissue for reference. Average pore size of the scaffold was 160 μm with varying vol.% porosities (36%, 48%, and 52%). The best performing scaffolds in terms of compressive strength achieved 12.7 MPa with 36 vol.% porosity. In vitro testing showed outstanding human osteoblast cell (HOB) growth during the first 14 days and no toxic effects were observed during testing [61].

Advantages of using FDM include reduced waste material, no trapped material, and lower cost. Disadvantages include possible thermal degradation during building, and low resolution when not used in conjunction with other processes [54].

Conclusion and Prospective

The level of detail possible when using novel AM techniques is allowing researchers to mimic actual bone tissue never seen before. These techniques are in rapid development as the level of complexity that could be achieved is increasing exponentially, making it possible to create custom scaffolds at will with a great level of flexibility in terms of size, porosity, stress-strain, interconnectivity of scaffolds' internal channels, chemical composition, and more. These methods go beyond what was capable with autografting and allografting and inherently come free of their limitations.

With the rapid development of these methods and thorough scientific investigations, the way is being paved to go beyond in vitro and in vivo experiments with the goal of clinical implementation, supplying patients with implants specific to their body and functional requirements which will revolutionize the field of biomaterials in healthcare.

References

1. Laurencin C, Khan Y, El-Amin S (2006) Bone graft substitutes. Expert Rev Med Devices 3(1):49–57
2. Henkel J, Woodruff MA, Epari DR, Steck R, Glatt V, Dickinson IC, Choong PF, Schuetz MA, Hutmacher DWJBr (2013) Bone regeneration based on tissue engineering conceptions—a 21st century perspective. Bone Res 1(3):216–248
3. Greenwald AS, Boden SD, Goldberg VM, Khan Y, Laurencin CT, Rosier RNJJ (2001) Bone-graft substitutes: facts, fictions, and applications. J Bone Joint Surg Am 83:98–103
4. Porter B, Oldham J, He S, Zobitz M, Payne R, An K, Currier B, Mikos A, Yaszemski MJ (2000) Mechanical properties of a biodegradable bone regeneration scaffold. J Biomech Eng 122(3):286–288
5. Mikos AG, Temenoff JS (2000) Formation of highly porous biodegradable scaffolds for tissue engineering. Electron J Biotechnol 3(2):23–24
6. Marks SC Jr, Popoff SN (1988) Bone cell biology: the regulation of development, structure, and function in the skeleton. Am J Anat 183(1):1–44
7. Rho J-Y, Kuhn-Spearing L, Zioupos P (1998) Mechanical properties and the hierarchical structure of bone. Med Eng Phys 20(2):92–102
8. Sikavitsas VI, Temenoff JS, Mikos AG (2001) Biomaterials and bone mechanotransduction. Biomaterials 22(19):2581–2593
9. Goldstein SA (1987) The mechanical properties of trabecular bone: dependence on anatomic location and function. J Biomech 20(11–12):1055–1061
10. Mohan S, Baylink DJ (1991) Bone growth factors. Clin Orthop Relat Res (263):30–48
11. Jones G, Nguyen T, Sambrook P, Kelly P, Eisman JJB (1994) Progressive loss of bone in the femoral neck in elderly people: longitudinal findings from the Dubbo osteoporosis epidemiology study. BMJ 309(6956):691–695
12. Teitelbaum SL (2000) Bone resorption by osteoclasts. Science 289(5484):1504–1508
13. Rodan GA, Martin TJ (1981) Role of osteoblasts in hormonal control of bone resorption—a hypothesis. Calcif Tissue Int 33(4):349–351
14. Teitelbaum SL (2007) Osteoclasts: what do they do and how do they do it? Am J Pathol 170(2):427–435
15. Hadjidakis DJ, Androulakis II (2006) Bone remodeling. Ann N Y Acad Sci 1092(1):385–396
16. Mueller BJ (2012) Additive manufacturing technologies–rapid prototyping to direct digital manufacturing. Assemb Autom 32(2)
17. Wohlers T, Gornet T (2014) History of additive manufacturing. In: Wohlers report 2014—3D printing and additive manufacturing state of the industry, vol 24. Cambridge University Press, Cambridge, p 118
18. Wong KV, Hernandez A (2012) A review of additive manufacturing. ISRN Mech Eng 2012:208760

19. Delloye C, Cornu O, Druez V, Barbier O (2007) Bone allografts: what they can offer and what they cannot. J Bone Joint Surg Br 89(5):574–580
20. Melchels FP, Domingos MA, Klein TJ, Malda J, Bartolo PJ, Hutmacher DW (2012) Additive manufacturing of tissues and organs. Prog Polym Sci 37(8):1079–1104
21. Yavari SA, van der Stok J, Chai YC, Wauthle R, Birgani ZT, Habibovic P, Mulier M, Schrooten J, Weinans H, Zadpoor AAJB (2014) Bone regeneration performance of surface-treated porous titanium. Biomaterials 35(24):6172–6181
22. Leukers B, Gülkan H, Irsen SH, Milz S, Tille C, Schieker M, Seitz H (2005) Hydroxyapatite scaffolds for bone tissue engineering made by 3D printing. J Mater Sci Mater Med 16(12):1121–1124
23. Hu Q, Yang H, Yao Y (2007) A software method to model and fabricate the defective bone repair bioscaffold using in tissue engineering. In: International conference on life system modeling and simulation. Springer, New York, pp 445–452
24. Ducheyne P, Qiu QJB (1999) Bioactive ceramics: the effect of surface reactivity on bone formation and bone cell function. Biomaterials 20(23–24):2287–2303
25. Vallet-Regí M, Ruiz-Hernández EJAM (2011) Bioceramics: from bone regeneration to cancer nanomedicine. Adv Mater 23(44):5177–5218
26. Cooper D, Matyas J, Katzenberg M, Hallgrimsson B (2004) Comparison of microcomputed tomographic and microradiographic measurements of cortical bone porosity. Calcif Tissue Int 74(5):437–447
27. Ginebra M (2008) Calcium phosphate bone cements. Orthopaedic bone cements. Elsevier, Amsterdam, pp 206–230
28. Yuan H, Kurashina K, de Bruijn JD, Li Y, De Groot K, Zhang X (1999) A preliminary study on osteoinduction of two kinds of calcium phosphate ceramics. Biomaterials 20(19):1799–1806
29. Vallet-Regí M, González-Calbet JM (2004) Calcium phosphates as substitution of bone tissues. Prog Solid State Chem 32(1–2):1–31
30. Anselme KJB (2000) Osteoblast adhesion on biomaterials. Biomaterials 21(7):667–681
31. Hench LL, Splinter RJ, Allen W, Greenlee TK (1971) Bonding mechanisms at the interface of ceramic prosthetic materials. J Biomed Mater Res 5(6):117–141
32. Rahaman MN, Day DE, Bal BS, Fu Q, Jung SB, Bonewald LF, Tomsia AP (2011) Bioactive glass in tissue engineering. Acta Biomater 7(6):2355–2373
33. El-Ghannam A, Ducheyne P, Shapiro M (1999) Effect of serum proteins on osteoblast adhesion to surface-modified bioactive glass and hydroxyapatite. J Orthop Res 17(3):340–345
34. Gotman I (1997) Characteristics of metals used in implants. J Endourol 11(6):383–389
35. Wang X, Xu S, Zhou S, Xu W, Leary M, Choong P, Qian M, Brandt M, Xie YM (2016) Topological design and additive manufacturing of porous metals for bone scaffolds and orthopaedic implants: a review. Biomaterials 83:127–141
36. Hermawan H (2012) Biodegradable metals: state of the art. Biodegradable metals. Springer, Berlin, pp 13–22
37. Peuster M, Hesse C, Schloo T, Fink C, Beerbaum P, von Schnakenburg C (2006) Long-term biocompatibility of a corrodible peripheral iron stent in the porcine descending aorta. Biomaterials 27(28):4955–4962
38. Polo-Corrales L, Latorre-Esteves M, Ramirez-Vick JE (2014) Scaffold design for bone regeneration. J Nanosci Nanotechnol 14(1):15–56
39. Bobe K, Willbold E, Morgenthal I, Andersen O, Studnitzky T, Nellesen J, Tillmann W, Vogt C, Vano K, Witte F (2013) In vitro and in vivo evaluation of biodegradable, open-porous scaffolds made of sintered magnesium W4 short fibres. Acta Biomater 9(10):8611–8623
40. Gunatillake P, Mayadunne R, Adhikari R (2006) Recent developments in biodegradable synthetic polymers. Biotechnol Annu Rev 12:301–347
41. Hollinger JO (1983) Preliminary report on the osteogenic potential of a biodegradable copolymer of polyactide (PLA) and polyglycolide (PGA). J Biomed Mater Res 17(1):71–82
42. Andriano K, Tabata Y, Ikada Y, Heller J (1999) In vitro and in vivo comparison of bulk and surface hydrolysis in absorbable polymer scaffolds for tissue engineering. J Biomed Mater Res 48(5):602–612

43. Muggli DS, Burkoth AK, Keyser SA, Lee HR, Anseth KS (1998) Reaction behavior of biodegradable, photo-cross-linkable polyanhydrides. Macromolecules 31(13):4120–4125
44. Hélary G, Noirclère F, Mayingi J, Migonney V (2009) A new approach to graft bioactive polymer on titanium implants: improvement of MG 63 cell differentiation onto this coating. Acta Biomater 5(1):124–133
45. Oughlis S, Lessim S, Changotade S, Poirier F, Bollotte F, Peltzer J, Felgueiras H, Migonney V, Lataillade JJ, Lutomski D (2013) The osteogenic differentiation improvement of human mesenchymal stem cells on titanium grafted with polyNaSS bioactive polymer. J Biomed Mater Res A 101(2):582–589
46. Zorn G, Baio JE, Weidner T, Migonney V, Castner DG (2011) Characterization of poly(sodium styrene sulfonate) thin films grafted from functionalized titanium surfaces. Langmuir 27(21):13104–13112
47. Geng F, Tan LL, Jin XX, Yang JY, Yang K (2009) The preparation, cytocompatibility, and in vitro biodegradation study of pure β-TCP on magnesium. J Mater Sci Mater Med 20(5):1149–1157
48. Chen Z, Mao X, Tan L, Friis T, Wu C, Crawford R, Xiao Y (2014) Osteoimmunomodulatory properties of magnesium scaffolds coated with β-tricalcium phosphate. Biomaterials 35(30):8553–8565
49. Jasmawati N, Fatihhi S, Putra A, Syahrom A, Harun M, Öchsner A, Kadir MRA (2017) Mg-based porous metals as cancellous bone analogous material: a review. Proc Inst Mech Eng Part L: J Mater Des Appl 231(6):544–556
50. Mota C, Puppi D, Chiellini F, Chiellini E (2015) Additive manufacturing techniques for the production of tissue engineering constructs. J Tissue Eng Regen Med 9(3):174–190
51. Holck DE, Boyd EM Jr, Ng J, Mauffray RO (1999) Benefits of stereolithography in orbital reconstruction. Ophthalmology 106(6):1214–1218
52. Arcaute K, Mann B, Wicker RB (2006) Stereolithography of three-dimensional bioactive poly(ethylene glycol) constructs with encapsulated cells. Ann Biomed Eng 34(9):1429–1441
53. Shirazi SFS, Gharehkhani S, Mehrali M, Yarmand H, Metselaar HSC, Adib Kadri N, Osman NAA (2015) A review on powder-based additive manufacturing for tissue engineering: selective laser sintering and inkjet 3D printing. Sci Technol Adv Mater 16(3):033502
54. Stevens B, Yang Y, Mohandas A, Stucker B, Nguyen KT (2008) A review of materials, fabrication methods, and strategies used to enhance bone regeneration in engineered bone tissues. J Biomed Mater Res B Appl Biomater 85(2):573–582
55. Cruz F, Simoes J, Coole T, Bocking C (2005) Direct manufacture of hydroxyapatite based bone implants by selective laser sintering. In: Proceedings of the VRAP 2005, Leiria, Portugal, pp 119–126
56. Williams JM, Adewunmi A, Schek RM, Flanagan CL, Krebsbach PH, Feinberg SE, Hollister SJ, Das SJB (2005) Bone tissue engineering using polycaprolactone scaffolds fabricated via selective laser sintering. Biomaterials 26(23):4817–4827
57. Duan B, Wang M, Zhou WY, Cheung WL, Li ZY, Lu WW et al (2010) Acta Biomater 6(12):4495–4505
58. Inzana JA, Olvera D, Fuller SM, Kelly JP, Graeve OA, Schwarz EM, Kates SL, Awad HA (2014) 3D printing of composite calcium phosphate and collagen scaffolds for bone regeneration. Biomaterials 35(13):4026–4034
59. Zein I, Hutmacher DW, Tan KC, Teoh SH (2002) Fused deposition modeling of novel scaffold architectures for tissue engineering applications. Biomaterials 23(4):1169–1185
60. Hutmacher DW, Schantz T, Zein I, Ng KW, Teoh SH, Tan KC (2001) Mechanical properties and cell cultural response of polycaprolactone scaffolds designed and fabricated via fused deposition modeling. J Biomed Mater Res 55(2):203–216
61. Kalita SJ, Bose S, Hosick HL, Bandyopadhyay A (2003) Development of controlled porosity polymer-ceramic composite scaffolds via fused deposition modeling. Mater Sci Eng C 23(5):611–620
62. Stępien E (2011) Acceleration of new biomarkers development and discovery in synergistic diagnostics of coronary artery disease. In: Coronary angiography—advances in noninvasive imaging approach for evaluation of coronary artery disease. InTech, London

63. Rosy Setiawati PR (2018) Bone development and growth. IntechOpen, London
64. Ligon SC, Liska R, Stampfl J, Gurr M, Mülhaupt R (2017) Polymers for 3D printing and customized additive manufacturing. Chem Rev 117(15):10212–10290
65. Li Z, Siklos M, Pucher N, Cicha K, Ajami A, Husinsky W, Rosspeintner A, Vauthey E, Gescheidt G, Stampfl J, Liska R (2011) Synthesis and structure-activity relationship of several aromatic ketone-based two-photon initiators. J Polym Sci A 49(17):3688–3699
66. Torgersen J (2013) Novel Biocompatible Materials for in Vivo Two-Photon Polymerisation. Technische Universität Wien, Wien
67. Ovsianikov A, Chichkov B, Mente P, Monteiro-Riviere NA, Doraiswamy A, Narayan RJ (2007) Two photon polymerization of polymer–ceramic hybrid materials for transdermal drug delivery. Int J Appl Ceram Technol 4(1):22–29
68. Medical: FHC—EOS technology for manufacturing of stereotactic platforms for neurosurgery. Accessed 2 Mar 2019
69. Zhang J, Zhao S, Zhu Y, Huang Y, Zhu M, Tao C, Zhang C (2014) Three-dimensional printing of strontium-containing mesoporous bioactive glass scaffolds for bone regeneration. Acta Biomater 10(5):2269–2281
70. Baino F, Hamzehlou S, Kargozar S (2018) Bioactive glasses: where are we and where are we going? J Funct Biomater 9(1):25
71. Murr LE, Gaytan SM, Medina F, Lopez H, Martinez E, Machado BI, Hernandez DH, Martinez L, Lopez MI, Wicker RB, Bracke J (2010) Next-generation biomedical implants using additive manufacturing of complex, cellular and functional mesh arrays. Philos Trans A Math Phys Eng Sci 368(1917):1999–2032
72. Wong KC, Kumta SM, Geel NV, Demol J (2015) One-step reconstruction with a 3D-printed, biomechanically evaluated custom implant after complex pelvic tumor resection. Comput Aided Surg 20(1):14–23
73. enbin Luo LH, He L, Wenrui Q, Zhao X, Wang C, Li C, Tao Y, Han Q, Wang J, Qin Y (2017) Customized knee prosthesis in treatment of giant cell tumors of the proximal tibia: application of 3-dimensional printing technology in surgical design. Med Sci Monit 23:1691–1700
74. Gao C, Wang C, Jin H, Wang Z, Li Z, Shi C, Leng Y, Yang F, Liu H, Wang J (2018) Additive manufacturing techniques designed metallic porous implant for clinical application in orthopedics. RSC Adv 8(44)
75. Rider P, Kačarević ŽP, Alkildani S, Retnasingh S, Schnettler R, Barbeck M (2018) Additive manufacturing for guided bone regeneration: a perspective for alveolar ridge augmentation. Int J Mol Sci 19(11):3308

Anti-biofouling and Antimicrobial Biomaterials for Tissue Engineering

Yingnan Zhu, Jia Ke, and Lei Zhang

Abstract Biofouling from nonspecific protein adsorption and microorganism adhesion is a continuous challenge in numerous biomedical applications such as implants, biosensors, and tissue-engineered scaffolds. The bacteria attached to the biomaterial surface can encapsulate themselves within a protective extracellular polymeric layer, leading to the formation of biofilm that is difficult to combat or eliminate. A promising strategy to prevent device-related infections is the development of new biomaterials that are anti-biofouling and/or antimicrobial. In general, anti-biofouling materials exhibit low adhesion or resistance properties towards a variety of bacteria, while antimicrobial ones can kill microorganisms approaching the surfaces or in the surrounding areas. In this chapter, we briefly introduce the recent strategies in the design and applications of anti-biofouling and antimicrobial materials.

Keywords Biomaterials · Anti-biofouling · Anti-microbial · Biofilm · Infections
Polyethylene glycol (PEG) · Zwitterionic · Releasing-based · Contact-based
Medical implants · Tissue engineering

Y. Zhu · J. Ke · L. Zhang (✉)
Department of Biochemical Engineering, School of Chemical Engineering and Technology, Tianjin University, Tianjin, China

Frontier Science Center for Synthetic Biology and Key Laboratory of Systems Bioengineering (MOE), Tianjin University, Tianjin, China

Collaborative Innovation Center of Chemical Science and Engineering (Tianjin), Tianjin University, Tianjin, China
e-mail: zhuyn@tju.edu.cn; lei_zhang@tju.edu.cn

B. Li et al. (eds.), *Racing for the Surface*,
https://doi.org/10.1007/978-3-030-34471-9_14

Introduction and Principles of Anti-biofouling and Antimicrobial Biomaterials

Biofilm Formation and Associated Infections

In recent decades, biomaterials have been increasingly used in tissue engineering and many medical devices, involving wound dressings, orthopedic implants, vascular prostheses, urinary catheters, etc. Bacterial adhesion and biofilm formation on biomaterial surfaces have been considered as one of the major challenges that can lead to serious consequences, such as implant infections, the failure of medical devices, and associated health risks. Even under sterile surgical conditions, bacterial contamination of the implantation sites cannot be prevented. Hospitalized subjects are also at a high risk of acquiring device-related bacterial infections in some cases up to ~60% [1]. For example, bacteria attached on the urinary catheter can form a biofilm within 24 h [2]. The infection risk in patients with urinary catheter is ~50% after 10 days, and the subsequent treatment and replacement can cause considerable morbidity.

Biofilm formation by microorganisms is a complex and dynamic process, often involving more than one microbial species. The initial attachment of bacteria is reversible until the bacteria secrete adhesive proteins, extracellular polymeric substances (EPS), and then irreversibly bind to the material surface. The following step is bacterial colonization and proliferation inside the extracellular matrix. Communication occurs via quorum sensing among same [3, 4] or different bacterial species [5, 6] within the biofilm, which enables co-colonization within the same extracellular matrix. After rapid proliferation of bacteria, the biofilm grows to be mature and eventually the extracellular matrix ruptures, leading to the dispersal of planktonic bacteria and possible spread of the infection. Biofilm is an effective strategy to render bacteria highly tolerant to environmental stresses as well as strongly resistant to antibiotics [1]. Previous reports showed that an antibiotic dose up to 1000-fold higher is required to kill bacteria inside the biofilm [7]. Therefore, developing biomaterials with anti-biofouling and/or antimicrobial properties is in urgent demand as alternatives to antibiotics to fight against infections associated with medical devices.

A variety of biomaterials with or without intrinsic antimicrobial activity have been developed to combat bacterial biofilm and associated infections. They can be divided into either anti-biofouling or antimicrobial materials, the first one is able to prevent protein adhesion and bacterial attachment, while the other one can kill microorganisms by the biomaterial itself (e.g., a polymer) or by adding antimicrobial agents.

Anti-biofouling Biomaterials

Anti-biofouling biomaterials are designed to prevent the adhesion of microorganisms, proteins, and other biomolecules by minimizing interaction forces between the material surfaces and biological environments. Most anti-biofouling materials fall into three types: (1) PEG-based materials, (2) hydrophilic zwitterionic materials, or (3) superhydrophobic low surface energy materials.

PEG-Based Biomaterials

PEG is one of the frequently used polymers to endow biomaterial surfaces with protein-resisting properties owing to both the hydration effect and steric hindrance [8]. Considering that indirect bacterial attachment on surfaces can occur as a result of protein adsorption, a biomaterial that is able to repel protein adsorption can potentially also resist the contamination of bacteria [9]. Since PEG has been widely used against protein adhesion to substrates, many studies have focused on designing materials with PEG to resist bacterial adhesion [10–14]. The polymer chain can be hydrated with water molecules via hydrogen bonds, and the water layer acts as a barrier to impede the attachment of biomolecules and bacteria [1].

Park and coworkers [10] reported the preparation of PEG-based polyurethane substrates with terminal hydroxyl, amino, and sulfonate groups. *E. coli* and *S. epidermidis* were used to test the adhesion of bacteria in tryptic soy broth and human plasma-containing media. Results showed that the bacterial attachment was affected by both the PEG molecular weight and media. PEG of higher molecular weight showed better bacterial-resistant ability compared with the lower molecular weight equivalents.

Norde et al. [15] studied the influence of PEG brush length on the adhesion of different bacteria and yeasts. Two types of bacteria (*S. epidermidis* and *P. aeruginosa*) and two types of yeasts (*C. tropicalis* and *C. albicans*) were used in the tests. It was found that longer PEG brushes resulted in stronger resistance to bacteria and yeasts. In addition, more hydrophobic microorganisms (*P. aeruginosa* and *C. tropicalis*) were more prone to adhere onto the surface than the more hydrophilic ones (*S. epidermidis* and *C. albicans*), indicating that hydrophobic force was more favorable for the adhesion of microorganisms.

It is believed that the benefit of longer polymer chains is related to more efficient coverage of the material surface. Via self-assembled monolayers (SAMs), short PEG chains also showed anti-biofouling abilities. Prime et al. designed SAMs presenting oligo (ethylene glycol) groups to disturb bacterial attachment [16]. Cooper and coworkers researched SAMs with various terminal groups including $-CH_3$, -OH, -COOH, and $-(OCH_2CH_2)_3-OH$. It was found that $-(OCH_2CH_2)_3-OH$ SAMs displayed the lowest adhesion while $-CH_3$ surface have the highest fouling [17].

While PEG has often been termed the "gold standard" of the anti-biofouling field, it suffers several nonnegligible weaknesses in biomedical applications. It is prone to undergo oxidative damage and thus unstable in long-term applications. In addition, though PEG is generally considered as a biologically inert material with no immunogenicity or antigenicity, it has actually been demonstrated to provoke immune reaction in some conditions [18]. For example, PEG antibodies have been found in animal studies after immunization with PEG-modified proteins and nanoparticles, leading to the loss of therapeutic efficacy and related adverse effects [19–23].

Poly Zwitterionic-Based Biomaterials

Recently, zwitterionic polymers have been emerging as promising alternatives to PEG, which are electrically neutral with balanced positive and negative charges in one moiety. The charged pairs result in a stronger hydration via ionic effects than that of PEG formed by hydrogen bonds, which in turn can enhance the anti-biofouling ability of zwitterionic biomaterials [24].

Cheng et al. developed poly(sulfobetaine) (pSB) and poly(oligo ethylene glycol) (pOEG)-grafted glass surfaces via atom transfer radical polymerization (ATRP) and tested the adhesion of both *S. epidermidis* (Gram-positive) and *P. aeruginosa* (Gram-negative) strains. It was found that PSB-grafted surfaces showed reduced adhesion by 92% and 96% than bare glass in a short term (3 h). And PSB-grafted surface was more effective in resisting long-term bacterial adhesion and biofilm formation, while SAMs surface failed to achieve a significant effect (Fig. 1). This result was probably owing to the higher surface densities of polymer brushes grafted via ATRP compared with SAM method [25].

Cheng et al. also systematically studied the zwitterionic poly(carboxybetaine) (pCB) grafted from glass surfaces for their resistance to biofilm formation. Results showed that pCB coatings reduced long-term biofilm formation of *P. aeruginosa* up to 240 h by 95% at 25 °C while the unmodified glass was completely covered by bacterial biofilm. At the optimal growth temperature of 37 °C, the glass surface was completely covered in 15 h, while pCB-modified surface could inhibit 93% of *P. aeruginosa* accumulation for 64 h [26].

In addition to surface-modified materials, several hydrogels designed with zwitterionic polymers have attracted increasing attentions [27, 28]. For example, in 2013,

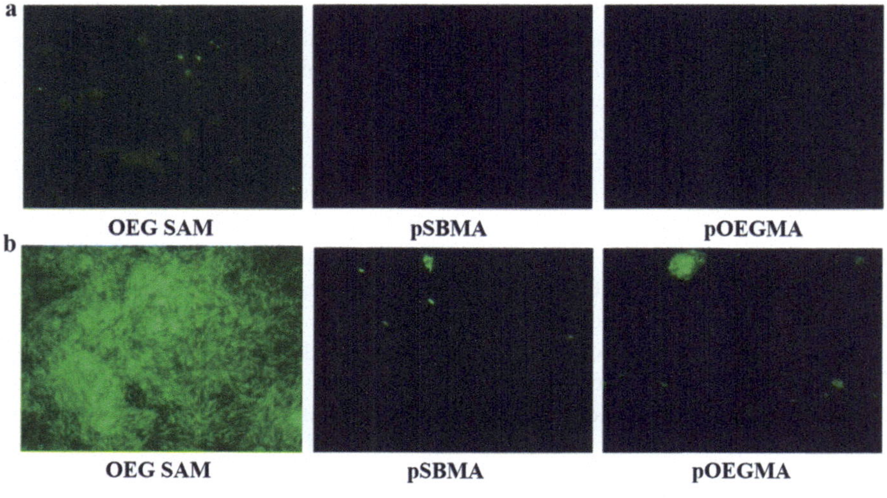

Fig. 1 (**a**) Fluorescence microscopy graphs of *S. epidermidis* attachment on various surfaces at 48 h. (**b**) Fluorescence microscopy graphs of *P. aeruginosa* attachment on various surfaces at 24 h. (Images reprinted with permission of Cheng et al. (2007). Copyright 2007 Elsevier Ltd. [25])

Zhang et al. developed a zwitterionic PCB hydrogel that could efficiently prevent foreign body capsule formation for 3 months and promote angiogenesis in the surrounding tissue when implanted subcutaneously in a mouse model. In the foreign body reaction, nonspecific protein adsorption is thought to be the first step to trigger the formation of a dense collagen layer. The collagen layer will isolate the implants from surrounding tissues, impeding mass transport and electrical communication between implants and the physiological environment [29]. The mechanism of pCB hydrogels is possibly due to the fact that the macrophage cells in anti-biofouling samples tend to differentiate to the pro-healing state.

Huang and coworkers synthesized zwitterionic pSB nanocomposite hydrogels as chronic wound dressings [30]. The prepared hydrogels displayed evident resistance to adsorption of bovine serum albumin (BSA), bacteria of Gram-positive *S. epidermidis*, and Gram-negative *P. aeruginosa* (Fig. 2). Zhang et al. proved that zwitterionic pCB hydrogel with high water content and excellent anti-biofouling properties could promote skin wound healing in a mouse model in comparison with pHEMA hydrogel and

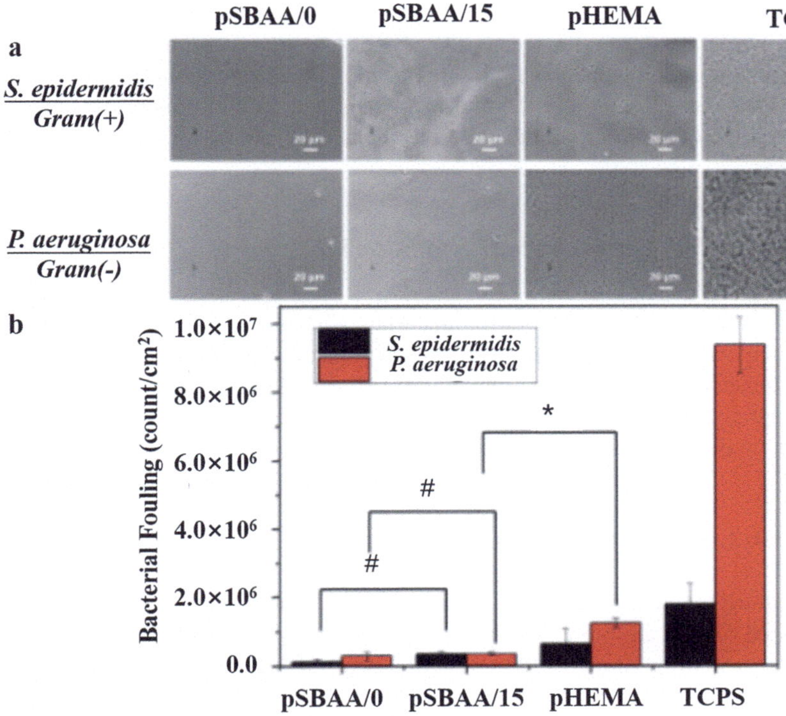

Fig. 2 (**a**) Bacterial fouling tests on pSBAA/0, pSBAA/15, pHEMA, and TCPS. *P. aeruginosa* and *S. epidermidis* were used in the tests and imaged using phase-contrast microscope. pSBAA/0 is hydrogels without adding any nanoclay and pSBAA/15 is hydrogels with 15% nanoclay. (**b**) The quantitative results for bacterial adsorption on all hydrogels. (Images reprinted with permission of Huang et al. (2016). Copyright 2016 The Royal Society of Chemistry [30])

the commercial product Duoderm [31]. A wound dressing is expected to be anti-biofouling also because traditional dressings can typically damage newly generated tissues upon removal and provide an opportunity for microorganism colonization.

Zwitterionic polymers have also successfully imparted anti-biofouling properties to various nanoparticles, such as gold nanoparticles [32–34], magnetic iron nanoparticles [35, 36], quantum dots [37, 38], and silica nanoparticles [39]. Jia covered silica nanoparticles with functional zwitterionic pCB layer via ATRP method and tested the stability of particles in protein-containing solutions. Results showed that the pCB layer is effective in protecting nanoparticles from nonspecific protein fouling [39].

Antimicrobial Biomaterials

In the past decade, the number of FDA-approved antimicrobial biomaterials has been continuously increasing, indicating the demand for alternatives to traditional antibiotics which often undergo drug resistance and difficulty to penetrate the biofilm [1]. Antimicrobial materials are designed to kill bacteria and prevent biofilm formation, while anti-biofouling materials are passive and vulnerable to microorganism invasion once their barriers are damaged. The antimicrobial biomaterials can be divided into those where the matrices integrated with antimicrobial agents that are released, or those where the materials themselves are active ingredients.

Releasing-Based Antimicrobial Biomaterials

An effective approach for imparting biomaterials with antimicrobial activity is to combine them with different releasing biocides/antibacterial agents, such as antibiotics, silver, quaternary ammonium compounds (QACs), and nitric oxide. These agents can be integrated with the biomaterials by suitable approaches, including physical adsorption, conjugation, or complexation.

Biomaterials Loaded with Antibiotics

The indwelling medical devices, for example, orthopedic implants and catheters, can be coated with an antibiotic-releasing layer to combat device-related infections. The greatest benefit by direct loading of antibiotics is that high systemic doses can be effectively avoided, preventing over-dosing problem and potential toxic side effects to other tissues in the body [40].

Antibiotics including vancomycin, cefamandole, gentamicin, cephalothin, carbenicillin, and amoxicillin have been widely used in controlled releasing devices [41]. Antibiotic-containing polymethyl methacrylate (PMMA) beads, which are fabricated by mixing the desired antibiotics with PMMA and forming into beads, have been clinically used as a kind of bone cement for about 30 years [42]. The primary advantage of PMMA beads is the clinical familiarity as well as the efficacy to eliminate acute infections in bone. However, since PMMA is non-biodegradable and hydrophobic, the incorporated antibiotics could not be totally released from the beads, thereby leading to a loss of 25–50% [40]. In addition, the non-biodegradable PMMA beads need to be removed by a second surgery if the antibiotic release has finished. Considering these issues, biodegradable materials have been developed as possible alternatives to bone cements, with the increasing popularity of cement-less prostheses in hip arthroplasty [43].

The use of antibiotic-containing biodegradable materials has the benefit of slow release of the antibiotic to the material–tissue interface, with the release of antibiotic following the kinetics of material degradation. For instance, poly(lactic acid) (PLA) and poly(lactic-co-glycolic acid) (PLGA) are effective biodegradable implant coatings that have been used to locally deliver antibiotics. Different from PMMA, these polymers can be used in several different forms, such as coatings, electrospun fibers, and microspheres [44]. Muller et al. combined fusidic acid and rifampicin on PLA to kill *S. aureus* both in vitro and in a rabbit infection model [45].

Clinical implant materials should be customizable to allow local antibiotic delivery to specific infection sites avoiding damaging bone cells. David et al. analyzed the inhibitory impact of PLA implants coated with single or double antibiotics (gentamicin, ciprofloxacin, colistin, daptomycin, or cefoxitin) on bacteria isolated from osteomyelitis. Results showed that all antibiotics, no matter alone or in combination, had a burst release and a dose-dependent antibacterial activity [46].

Recently, mesoporous materials such as hydroxyapatite (HA) has been loaded with different antibiotics (tobramycin, vancomycin, cephalothin, carbenicillin, and amoxicillin sodium salts). These antibiotic-containing HA materials have also been applied as practical methods for the decontamination of dental implants. All these studies have demonstrated that antibiotic-containing HA materials could fight against bacterial adhesion and impede biofilm formation as well as maintain a continuous agent release ability.

Biomaterials Loaded with Silver Nanoparticles (NPs)

Silver-loaded biomaterials have been used in medical implants due to the released silver ions being broad-spectrum against both Gram-positive and Gram-negative bacteria. Although the mechanism of their antimicrobial action is not yet completely understood, it is generally inferred that released silver ions are the primary molecular toxicant [47]. Silver ions released from silver-loaded materials destruct the bacterial membrane and damage the function of the enzymes and/or DNA of bacteria [48, 49]. Silver ions can react with the negatively charged groups in the cellular proteins and DNA, such as the carboxyl, phosphate, thiol, and amino groups [50]. It can also inactivate enzymes of the tricarboxylic acid (TCA) cycle, generating harmful hydroxyl radical. The needed concentration of silver for a required antibacterial effect ranges from 10 nM to 10 μM [51].

Silver NPs are facile to be incorporated into various materials for further applications, such as hydrogels, nanofibers, and films. Hydrogels formed by synthetic polymers such as poly(N-vinyl pyrrolidone) (PVP), poly(vinyl alcohol) (PVA), poly(acrylamide-co-acrylic acid), and natural polymers such as gelatin, chitosan, and alginate have been prepared to encapsulate silver NPs. Thomas and coworkers developed a technique called the breathing-in/breathing-out (BI-BO) method to load silver NPs. By exposing to solutions of different concentrations, hydrogels could sequentially swell and shrink, thus encapsulating silver NPs from the solution into the network. The antibacterial activity of hydrogel materials was influenced by the cycle numbers, and it was reported to be optimal after three cycles to kill *E. coli* [52].

Silver NP-loaded hydrogels can also be prepared by the formation of hydrogel and the encapsulation of NPs simultaneously [53]. Gonzalez et al. [54] prepared in situ silver NP-embedded matrix using $AgNO_3$ as the silver source and hydrogel polymer as the container and stabilizer. Hydroxyl ethyl methacrylate monomer (HEMA), cross-linking agents, and photoinitiator were added into the hydrogel synthesis system, while UV irradiation was used to reduce the silver ions and also form the HEMA hydrogel. Via the in situ synthesis and encapsulation methods, the aggregation and precipitation issues of silver NPs can be reduced. Zhang et al. reported a one-step in situ photo-polymerization reaction to simultaneously formed silver NPs and the PCB hydrogel. Results showed that silver NPs could be homo-dispersed in the hydrogel matrix without precipitation. In vitro tests proved that the resulting matrix could effectively kill both Gram-positive (*S. aureus*) and Gram-negative bacteria (*E. coli*) while resisting their adhesion [31].

Besides the antibacterial effects and material fabrication approach, silver-associated cytotoxicity must be considered. Eukaryotic cells have been shown to withstand 10 ppm exposure of silver [49]. High-level exposure of silver NPs could lead to nonnegligible toxicity to a variety of organs such as lung and liver.

Biomaterials Loaded with Quaternary Ammonium Compounds (QACs)

Unlike the release-based antibacterial silver ions, QAC-containing materials possess a long-term antibacterial mechanism [55]. Materials containing QAC have been proven to damage both Gram-positive and Gram-negative bacteria by disrupting the cellular membranes [56]. The positively charged ammonium groups interact with the negatively charged acidic phospholipid groups of the bacterial cellular membrane, disturbing the stability and integrity of the lipid bilayers. Further, the potassium ions release from the inner cytoplasm which in turn damage the original osmoregulation and other physiological functions of bacteria [57]. It was found that the antibacterial activity of QACs was relevant to the alkyl chain length. QACs possessing an alkyl chain of 12–14 carbons achieved an optimal activity against Gram-positive bacteria and yeast, whereas alkyl chain length of 14–16 carbons effectively resisted Gram-negative bacteria. QACs with alkyl chain lengths less than four or more than 18 were found to be virtually ineffective.

Biomaterials Loaded with Nitric Oxide (NO)

NO is a well-known factor to inhibit the platelet activation and adhesion. It has been used in many polymer-based materials, for example, silicone rubber, PVC, PVP, and PU, for medical applications in various blood-contacting devices to prevent thrombosis [58]. In recent years, NO has been found to resist biofilm formation, thus NO-loaded biomaterials have attracted increasing attentions to develop dual-functional (antithrombotic and antibacterial) biomaterials.

In 2005, an NO-stored sol-gel derived film was developed to coat silicone elastomer and subcutaneously implanted in a rat model to evaluate the anti-infection effect. After treated with NO-releasing coatings, the *S. aureus*-infected wounds showed an 82% reduction, indicating a promising application of NO-releasing biomaterials to treat *S. aureus* infections [59].

Anton et al. assessed possible benefits of a low-concentration NO-releasing carbon-based coating on monofilament polypropylene meshes in vitro and in vivo. NO-releasing coatings showed significant bactericidal effect on biofilms of *S. aureus*, *E. coli*, and *P. aeruginosa* in vitro. However, no obvious beneficial effects of this NO-releasing coating on subcutaneously in vivo implanted surgical meshes could be found [60].

Danie reported the synthesis of NO-modified xerogels using tertiary thiol-bearing silane to trigger the release of NO by photoactivation at physiological temperature. After exposing the NO-modified xerogels to visible irradiation, the bacterial adhesion (*P. aeruginosa*) was significantly reduced by 88% compared to TEOS xerogel controls [61].

Contact-Active Antibacterial Biomaterials

Another approach for the fabrication of antibacterial biomaterials is based on the non-releasing mechanism, where the polymers themselves are intrinsically antimicrobial and thus kill the bacteria in contact with the material surface. The polymers are often cationic and able to capture negatively charged bacterial cell envelop, interacting and further damaging the cell membrane to eventually kill the bacteria. Figure 3 showed the main classes of cationic natural and synthetic polymers possessing positive charge in the backbone or in the side chain [1].

Chitosan is an extensively studied, natural-derived cationic polymer, which is the N-deacetylated derivative of chitin. Chitosan-based materials, such as coatings and films, have been applied as wound dressings and scaffolds in tissue engineering [62]. Previous studies have demonstrated the ability of chitosan to inhibit the growth of a wide variety of bacteria, including *E. coli, P. fluorescens, S. aureus*, and *K. pneumoniae* [63]. It is reported to completely inactivate *E. coli* after a 2-day incubation with concentrations of 0.5–1% at pH 5.5. Meanwhile, only 0.1% concentration of chitosan was required to inhibit *E. coli* growth. Due to the different acetylation degree of chitosan, the antibacterial effect varied and displayed a higher sterilizing efficiency with 7.5% acetylation when compared with that of 15%.

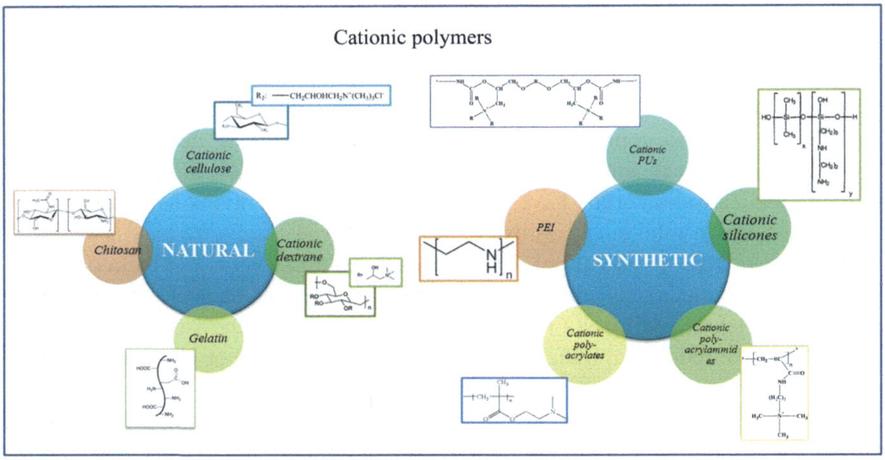

Fig. 3 The main classes of cationic natural and synthetic polymers. (Images reprinted with permission of Francolini et al. (2017). Copyright 2017 John Wiley & Sons Ltd. [1])

Chitosan has also been modified with other molecules or groups such as quaternary ammonium to augment its antimicrobial ability. The antibacterial activity of diethylaminoethyl chitosan and triethylaminoethyl chitin was evaluated against a number of bacterial species in vitro. The triethylaminoethyl chitin had a greater activity against *S. aureus* than against *E. coli*. And 500 ppm of triethylaminoethyl chitin was needed to completely eliminate *S. aureus* within 2 h.

Anton et al. immobilized chitosan via poly-acrylic acid (PAA) brushes and then grafted them on a polyethylene surface. *E. coli* and *S. aureus* were both used to test the samples by inhibition zone methods. After the treatment of chitosan, the polyethylene displayed clear inhibition zones of 35 mm^2 for *E. coli* and 275 mm^2 for *S. aureus* [64]. Chitosan and its derivatives have also been incorporated with other anionic polymers, including hyaluronic acid, alginate, carrageenan, heparin, and pectin.

Polyethyleneimine (PEI), a kind of synthetic hyperbranched polymer, is positively charged to serve as an antimicrobial agent against bacteria and fungi [65]. Compared with the unmodified PEI, low molecular weight counterparts with acid-labile imine linkers [66], disulfide bonds [67], or folate-PEG were designed with the aim to enhance biodegradability and biocompatibility.

Glass and metal surface have been coated with hydrophobic *N*,*N*-dodecyl methyl-PEI. The *E. coli* and *S. aureus* strains were 100% removed from the glass or polyethylene surface owing to the disruption of cell membrane and the leakage of cellular proteins [68, 69]. *N*,*N*-dodecyl methyl-PEI has also been used for the coating of orthopedic fracture-fixation hardware, which was made of titanium (Ti) and stainless steel. The treated surface was revealed to effectively prevent the biofilm formation of *S. aureus* both in vitro and in vivo [70].

Milovic used N-hexyl, methyl-PEI to covalently coat onto an amino-glass slide to combat *E. coli* and *S. aureus*, revealing a 10^9-fold reduction of live bacteria in the surface-exposed solutions and a 100% elimination of the surface-attached bacteria. In addition, the immobilized N-hexyl, methyl-PEI was proven to be harmless to monkey kidney cells while lethal to bacterial cells [71].

In 2015, Merve et al. coated brush-like polyethyleneimine (PEI) on polyurethane (PU) ureteral stents with the aim to develop permanent antibacterial surface since the biofilm formation on stents severely limited their long-term usage. PEI chains with different molecular weights (Mn: 1800 or 60,000 Da) were alkylated with bromohexane to break the bacterial membranes with increasing polycationic character. Both kinds of PEI brushes exhibited antibacterial activity by reducing the adhesion of *K. pneumonia, E. coli*, and *P. mirabilis* species to 10^2-fold, while no cytotoxicity was observed on L929 cells [72].

Besides PEI, cationic PU-based materials were also developed for contact-killing materials. Antibacterial QA compound-containing PU was coated to aluminum and PVC substrates, showing excellent biocompatibility and bacterial growth reduction to 83–100% against both *E. coli* and *S. aureus*. PU catheters were coated by a multistep process involving a vapor phase plasma-induced polymerization with acrylic acid and dimethyloctadecyl [3-(trimethoxysilyl) propyl] ammonium chloride. The coating was stable in aqueous media and uniformly dispersed on PU catheters, as well as displaying antimicrobial activity against *E. coli* strains in vitro [73].

Applications in Tissue Engineering

Wound Dressings

An ideal wound dressing is expected to provide a moist environment, protect the wound from microorganism invasion and infections, remove wound exudate, as well as promote wound healing. Materials such as hydrogel and hydrocolloid are suitable for the fabrication of wound dressings due to their hydrophilic properties. However, the moist environments are also prone to breed microbial infections, which will delay the wound healing process and induce other infection-associated complications. Therefore, wound dressings with antibacterial activity is of great necessity in clinical applications.

Fan designed a series of acrylic acid and N,N-methylene bisacrylamide hydrogels loaded with Ag/graphene composites of different mass ratios. The hydrogel with the optimal Ag to graphene mass ratio of 5:1 (Ag5G1) exhibited strong anti-infection abilities and excellent wound-healing performance (98% wound closure) within 2 weeks. The effect can be attributed to the antibacterial performance of Ag nanoparticles and the porous structure of graphene [74].

Chitosan itself has antibacterial properties owing to the cationic amino groups, thus chitosan-based wound dressings for anti-infection treatment have been developed recently. Nimal et al. prepared an injectable hydrogel composed of chitosan and tigecycline. Tigecycline can be released in a sustained manner to significantly inhibit bacterial growth, as well as to prevent skin infections [75]. Tetracycline hydrochloride

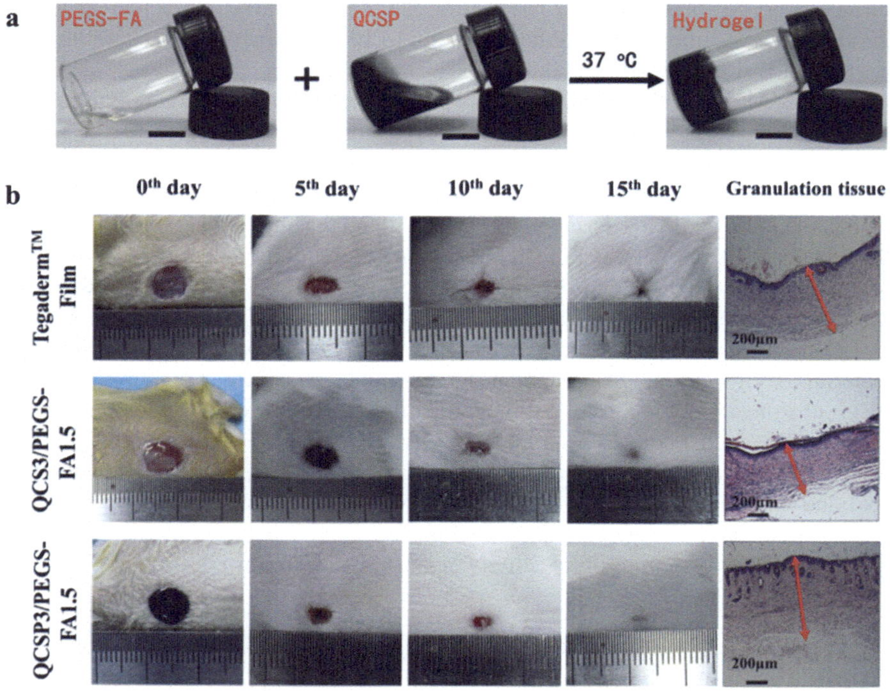

Fig. 4 (**a**) Photographs of PEGS-FA solution, QCSP solution, and hydrogel QCSP3/PEGS-FA1.5. (**b**) Photographs of wounds at 0th, 5th, 10th, and 15th day and granulation tissue at 15th day for commercial film dressing (Tegaderm™), hydrogel QCS3/PEGS-FA1.5, and hydrogel QCSP3/PEGS-FA1.5. (Images reprinted with permission of Zhao et al. (2017). Copyright 2017 Elsevier Ltd. [77])

was also incorporated into chitosan-PEG-PVP hydrogel as an antiseptic and scar preventive dressing. The prepared wound dressing promoted healing process with minimum scar formation and protected the open wound from bacterial invasions [76].

Zhao et al. developed a series of injectable conductive self-healed hydrogels based on quaternized chitosan-g-polyaniline (QCSP) and benzaldehyde group functionalized poly(ethylene glycol)-co-poly(glycerol sebacate) (PEGS-FA) as antibacterial and antioxidant wound dressing for cutaneous wound healing (Fig. 4). The antibacterial injectable hydrogel dressing prolonged the lifespan of dressing upon self-healing ability and promoted the in vivo wound healing process attributed to its multifunctional properties [77].

Nano metals such as silver, ZnO, and TiO_2 NPs have advantages of combating drug-resistant bacteria in infected wounds. A number of silver-containing wound dressings have been developed and approved by the FDA, including Tegaderm™, Duoderm®, Acticoat™, Fucidin®, 3M™, SilvaSorb®, PolyMem® Silver, etc. [78]. Moustafa et al. proposed an approach for the use of chitosan silver-based dressing for the control of diabetic foot infection with multidrug-resistant bacteria.

Chitosan hydrogel dressing loaded with the silver NPs showed promising antibacterial activities, as well as responsive healing properties for the wounds in both normal and diabetic rats [79].

In 2017, zwitterionic polycarboxybetaine (PCB) hydrogel and silver nanoparticles (AgNPs) were developed via a one-step method for the efficient treatment of infected wounds [31]. The PCB-AgNP hydrogel exhibited effective antibacterial ability against both Gram-positive bacteria (*S. aureus*) and Gram-negative bacteria (*E. coli*) in vitro and was able to efficiently treat *S. aureus* infections and accelerated cutaneous wound healing (Fig. 5a). Zwitterionic poly(sulfobetaine acrylamide) (pSBAA)-based hydrogel has also been impregnated with AgNPs and implemented to treat infected chronic wounds [30]. The AgNPs were grown within hydrogel networks by in situ free radical reduction and exhibited germicidal properties against Gram-positive *S. epidermidis* and Gram-negative *P. aeruginosa*. Results showed that pSBAA/Ag hydrogel was non-sticky to the new tissue and could accelerate the epithelialization in the infected diabetic wounds (Fig. 5b).

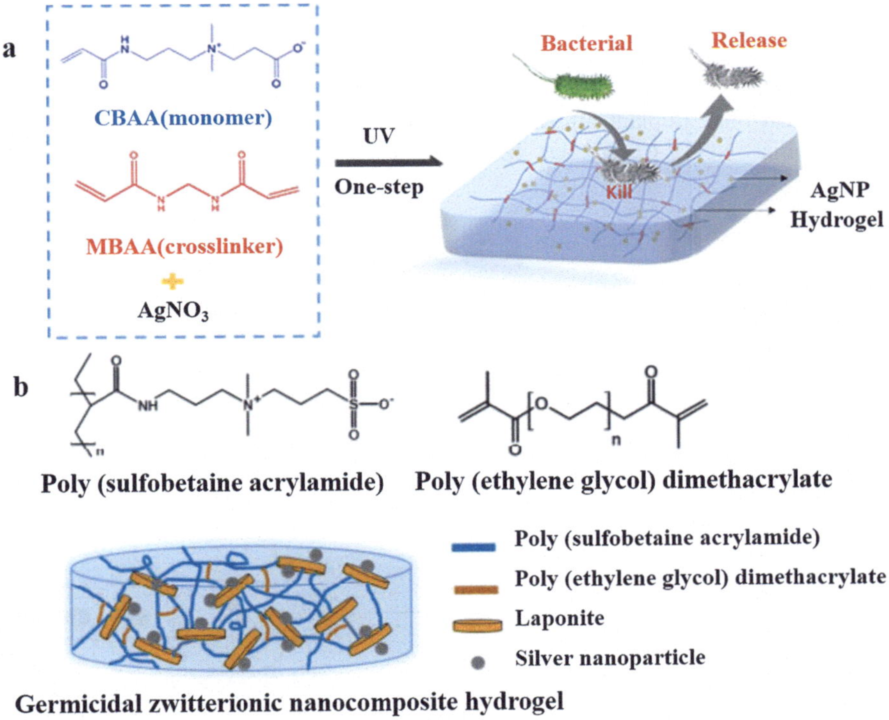

Fig. 5 (**a**) A schematic of the one-step synthetic route of the zwitterionic PCB-AgNP hydrogel. (Images reprinted with permission of Zhu et al. (2017). Copyright 2017 The Royal Society of Chemistry [31].) (**b**) Schematic illustration for structure of antimicrobial zwitterionic pSBAA/Ag composite hydrogel. (Images reprinted with permission of Huang et al. (2017). Copyright 2017 The Royal Society of Chemistry [30])

Orthopedic Implants

Orthopedic implant-related infection constitutes a major concern associated with high morbidity and health costs. There are a lot of new strategies to develop alternative antibacterial biomaterials to conventional antibiotics, such as zwitterionic modification, providing nanostructure-coated metal implants.

Recently, a surface-initiated atom transfer radical polymerization (SI-ATRP) strategy has been reported for surface zwitterionization of metal implants, such as commercial pure Ti (Fig. 6) [80] and biomedical grade 316L-type stainless steel (SUS 316L) [81]. Chang et al. presented a Ti surface with biocompatibility and antifouling properties grafting zwitterionic polySBMA using different anchoring agents of dopamine and silane. The resulting titanium surfaces grafted from dopamine- and silane-anchored polySBMA exhibited superlow fouling ability against the adhesion of proteins, human fibroblast cells (HT1080), *E. coli*, and *S. epidermidis*. Bacterial adhesion tests indicated that pristine metal surface was fully covered by *E. coli* and *S. epidermidis* after 24 h, whereas the SIATRP-treated Ti surfaces reduced 95% of bacterial adhesion relative to uncoated surfaces [80].

Bioceramics are excellent candidates to manufacture bone-like scaffolds which can load biologically active molecules to maintain, repair, or improve bone functions. Zwitterionization of bioceramics enables them to inhibit bacterial adhesion and prevent bone implant infections. SBA-15-type mesoporous material grafting zwitterionic $-NH3^{+}/-COO^{-}$ has been synthesized by the co-condensation of

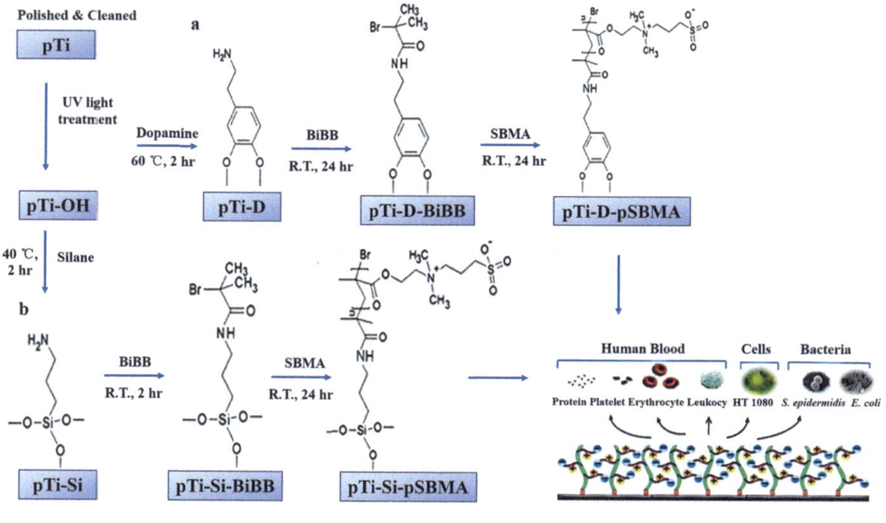

Fig. 6 Schematic illustration of the preparation process of zwitterionic pSBMA-grafted titanium disks via ATRP method with both (**a**) dopamine and (**b**) silane as respective anchoring agents. (Images reprinted with permission of Yu et al. (2014). Copyright 2014 American Chemical Society [80])

3-aminopropyltriethoxysilane (APTES) and carboxyethyl silanetriol sodium salt (CES). The water molecules above the zwitterionic surface would create a strong repulsive force to repel proteins from the surface, rendering the SBA-15 ultralow-fouling materials [82]. Furthermore, the ability of this material to inhibit bacterial adhesion was evaluated by simulating severe infection conditions. The in vitro adhesion assays showed that *E. coli* adhesion to zwitterionic SBA-15 was reduced by ~93% compared with the unmodified materials. After co-culturing with human Saos-2 osteoblasts to evaluate the biocompatibility at the physiological pH of 7.4, all materials exhibited good biocompatibility, with Saos-2 osteoblasts adhering, proliferating, and maintaining their initial morphology and function [83, 84].

Liu et al. reported that grafting pSBMA onto titanium alloy or dental implants led to promoted mineralization of the implant surface and increased osteointegration [85]. Ti6Al4V substrates were grafted with zwitterionic pSBMA brushes via SI-ATRP method, generating a stable super-hydrophilic and low-fouling surface without compromising mechanic property of the Ti6Al4V. The prepared surface was capable of attracting both cationic and anionic precursor ions during calcium phosphate apatite mineralization. The surface mineral coverage was enhanced from 32 to 71%, which significantly increased the attachment of the apatite crystals on the material surface.

Catheters

Catheters often need to be replaced at frequent intervals to prevent potential infections; however, this practice imposes considerable costs to the healthcare system. Imparting catheters with improved antibacterial ability can significantly reduce the frequency of implant-related infections [86]. An Ag alloy-coated latex-hydrogel catheter plus (Inc: Murray Hill, New Jersey, USA) was compared in vitro with a nitrofurazone-coated silicone catheter (Rochester Medical Group). Bacterial cells were detached from catheters by sonication and counted, and results showed that nitrofurazone-coated catheters performed better than Ag alloy-coated catheters [87]. Three kinds of catheters involving nitrofurazone-impregnated, Ag alloy-coated, and the standard polytetrafluoroethylene (PTFE) catheters were also compared within 6 weeks, the rate of symptomatic UTIs was 10.6% ($n = 2153$), 12.5% ($n = 2097$) and 12.6% ($n = 2144$), respectively [88].

In 2011, antibacterial coatings on catheters were obtained by an innovative and patented silver deposition technique based on the photo-reduction of the silver solution to form antibacterial silver NPs on the surface of the catheter (Fig. 7). The distribution, the size of clusters on the catheters surface, and the antibacterial capability of the devices against bacterial proliferation were evaluated. Inhibition zone tests performed against *E. coli* revealed a strong antibacterial activity of silver-treated catheters, as well as the main of antibacterial activity after soaking in high water flow for 30 days [89].

A strategy by combining the antibacterial effects of norfloxacin and silver NPs was used to resist bacterial adhesion and encrustation. The polymer films loaded with two

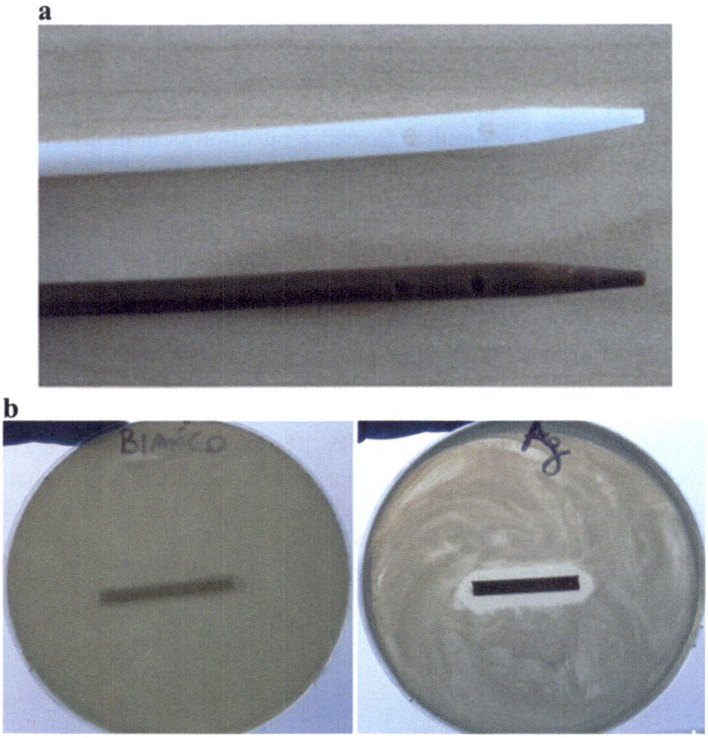

Fig. 7 (**a**) Visual comparison of untreated catheter and silver-treated catheter. (**b**) Test of *E. coli* growth on untreated samples (left) and silver-treated samples (right). (Images reprinted with permission of Pollini et al. (2011). Copyright 2011 Springer Science + Business Media, LLC [89])

antibacterial agents were applied on polyurethane (PUR) and silicon sheets and compared with commercially pure PUR and silicon. The coatings could resist the encrustation for at least 2 weeks in an in vitro encrustation model using artificial urine [90].

Conclusion

We have summarized the main approaches for designing biomaterials against bacterial infections, with special emphasis on advances developed in the past decades. Based on the different modes of action over bacteria, these biomaterials can be classified as anti-biofouling and antimicrobial to suit specific demands.

Anti-biofouling materials are expected to prevent the formation of biofilm on the surface by resisting bacterial attachment or adhesion. These materials usually impede bacteria/coating surface interactions by either exclusion steric repulsion, electrostatic repulsion or low surface energy. Major advances in recent years have been summarized in this chapter, including the functionalization of biomaterials with PEG, the development of alternatives to PEG such as zwitterionic polymers.

Antimicrobial materials exhibit a bactericidal activity that can kill bacteria by releasing antimicrobial agents or by contact-active mechanisms. Releasing-based agents including antibiotics, silver, quaternary ammonium salts, and nitric oxide have been widely explored to eliminate microbial contaminations. The widespread use of common antimicrobial agents, however, has accelerated the emergence of antibiotic resistance and raised concerns regarding potential toxicity of high-dose silver-containing compounds. Some new approaches involving the development of antimicrobial coatings based on AMPs, enzymes, and switchable cationic polymers have gained great promise recently.

Most of the biomaterials mentioned above have been tried in medical applications, such as wound dressings, orthopedic implants, vascular prostheses and urinary catheters. However, standardized methods to better support translation to the clinical level are still desirable. Meanwhile, the development of controllable antimicrobial biomaterials with actively responsible, switchable, or multifunctional properties is of great demand, to combat bacterial infections in tissue engineering.

Acknowledgments This chapter is partially supported by funds from the National Natural Science Funds for Innovation Research Groups 21621004, the Qingdao National Laboratory for Marine Science and Technology, QNLM2016ORP0407, National Natural Science Funds for Excellent Young Scholars 21422605, and Tianjin Natural Science Foundation 18JCYB- JC29500. The authors also gratefully acknowledge the helpful comments and suggestions of the reviewers, which have improved the presentation.

References

1. Francolini I, Vuotto C, Piozzi A, Donelli G (2017) Antifouling and antimicrobial biomaterials: an overview. APMIS 125(4):392–417
2. Riga EK, Vohringer M, Widyaya VT, Lienkamp K (2017) Polymer-based surfaces designed to reduce biofilm formation: from antimicrobial polymers to strategies for long-term applications. Macromol Rapid Commun 38(20)
3. Kim M, Lee S, Park HD, Choi SI, Hong S (2012) Biofouling control by quorum sensing inhibition and its dependence on membrane surface. Water Sci Technol 66(7):1424–1430
4. Schuster M, Sexton DJ, Diggle SP, Greenberg EP (2013) Acyl-homoserine lactone quorum sensing: from evolution to application. Annu Rev Microbiol 67(1):43–63
5. Abisado RG, Benomar S, Klaus JR, Dandekar AA, Chandler JR (2018) Bacterial quorum sensing and microbial community interactions. MBio 9(3)
6. Zhang YF, Shi WY, Song YQ, Wang JF (2019) Metatranscriptomic analysis of an in vitro biofilm model reveals strain-specific interactions among multiple bacterial species. J Oral Microbiol 11(1):1599670
7. Diaz C, Minan A, Schilardi PL, de Mele MFL (2012) Synergistic antimicrobial effect against early biofilm formation: micropatterned surface plus antibiotic treatment. Int J Antimicrob Agents 40(3):221–226
8. Chen SF, Li LY, Zhao C, Zheng J (2010) Surface hydration: principles and applications toward low-fouling/nonfouling biomaterials. Polymer 51(23):5283–5293

9. Ostuni E, Chapman RG, Liang MN, Meluleni G, Pier G, Ingber DE, Whitesides GM (2001) Self-assembled monolayers that resist the adsorption of proteins and the adhesion of bacterial and mammalian cells. Langmuir 17(20):6336–6343
10. Park KD, Kim YS, Han DK, Kim YH, Lee EHB, Suh H, Choi KS (1998) Bacterial adhesion on PEG modified polyurethane surfaces. Biomaterials 19(7–9):851–859
11. Kingshott P, Wei J, Bagge-Ravn D, Gadegaard N, Gram L (2003) Covalent attachment of poly(ethylene glycol) to surfaces, critical for reducing bacterial adhesion. Langmuir 19(17):6912–6921
12. Razatos A, Ong YL, Boulay F, Elbert DL, Hubbell JA, Sharma MM, Georgiou G (2000) Force measurements between bacteria and poly(ethylene glycol)-coated surfaces. Langmuir 16(24):9155–9158
13. Kenan DJ, Walsh EB, Meyers SR, O'Toole GA, Carruthers EG, Lee WK, Zauscher S, Prata CAH, Grinstaff MW (2006) Peptide-PEG amphiphiles as cytophobic coatings for mammalian and bacterial cells. Chem Biol 13(7):695–700
14. Fernandez ICS, van der Mei HC, Lochhead MJ, Grainger DW, Busscher HJ (2007) The inhibition of the adhesion of clinically isolated bacterial strains on multi-component cross-linked poly(ethylene glycol)-based polymer coatings. Biomaterials 28(28):4105–4112
15. Roosjen A, van der Mei HC, Busscher HJ, Norde W (2004) Microbial adhesion to poly(ethylene oxide) brushes: influence of polymer chain length and temperature. Langmuir 20(25):10949–10955
16. Prime KL, Whitesides GM (1991) Self-assembled organic monolayers: model systems for studying adsorption of proteins at surfaces. Science (New York, NY) 252(5009):1164–1167
17. Tegoulia VA, Cooper SL (2002) Staphylococcus aureus adhesion to self-assembled monolayers: effect of surface chemistry and fibrinogen presence. Colloids Surf B Biointerfaces 24(3):217–228
18. Zhang P, Sun F, Liu S, Jiang S (2016) Anti-PEG antibodies in the clinic: current issues and beyond PEGylation. J Control Release 244(Pt B):184–193
19. Richter AW, Akerblom E (1983) Antibodies against polyethylene glycol produced in animals by immunization with monomethoxy polyethylene glycol modified proteins. Int Arch Allergy Appl Immunol 70(2):124–131
20. Garay RP, El-Gewely R, Armstrong JK, Garratty G, Richette P (2012) Antibodies against polyethylene glycol in healthy subjects and in patients treated with PEG-conjugated agents. Expert Opin Drug Deliv 9(11):1319–1323
21. Armstrong JK, Hempel G, Koling S, Chan LS, Fisher T, Meiselman HJ, Garratty G (2007) Antibody against poly(ethylene glycol) adversely affects PEG-asparaginase therapy in acute lymphoblastic leukemia patients. Cancer 110(1):103–111
22. Hershfield MS, Ganson NJ, Kelly SJ, Scarlett EL, Jaggers DA, Sundy JS (2014) Induced and pre-existing anti-polyethylene glycol antibody in a trial of every 3-week dosing of pegloticase for refractory gout, including in organ transplant recipients. Arthritis Res Ther 16(2):R63
23. Longo N, Harding CO, Burton BK, Grange DK, Vockley J, Wasserstein M, Rice GM, Dorenbaum A, Neuenburg JK, Musson DG, Gu Z, Sile S (2014) Single-dose, subcutaneous recombinant phenylalanine ammonia lyase conjugated with polyethylene glycol in adult patients with phenylketonuria: an open-label, multicentre, phase 1 dose-escalation trial. Lancet 384(9937):37–44
24. Lowe S, O'Brien-Simpson NM, Connal LA (2015) Antibiofouling polymer interfaces: poly(ethylene glycol) and other promising candidates. Polym Chem 6(2):198–212
25. Cheng G, Zhang Z, Chen S, Bryers JD, Jiang S (2007) Inhibition of bacterial adhesion and biofilm formation on zwitterionic surfaces. Biomaterials 28(29):4192–4199
26. Cheng G, Li G, Xue H, Chen S, Bryers JD, Jiang S (2009) Zwitterionic carboxybetaine polymer surfaces and their resistance to long-term biofilm formation. Biomaterials 30(28):5234–5240
27. Cao B, Li L, Tang Q, Cheng G (2013) The impact of structure on elasticity, switchability, stability and functionality of an all-in-one carboxybetaine elastomer. Biomaterials 34(31):7592–7600

28. Shimizu T, Goda T, Minoura N, Takai M, Ishihara K (2010) Super-hydrophilic silicone hydrogels with interpenetrating poly(2-methacryloyloxyethyl phosphorylcholine) networks. Biomaterials 31(12):3274–3280
29. Zhang L, Cao Z, Bai T, Carr L, Ella-Menye J-R, Irvin C, Ratner BD, Jiang S (2013) Zwitterionic hydrogels implanted in mice resist the foreign-body reaction. Nat Biotechnol 31(6):553–556
30. Huang KT, Fang YL, Hsieh PS, Li CC, Dai NT, Huang CJ (2017) Non-sticky and antimicrobial zwitterionic nanocomposite dressings for infected chronic wounds. Biomater Sci 5(6):1072–1081
31. Zhu Y, Zhang J, Song J, Yang J, Xu T, Pan C, Zhang L (2017) One-step synthesis of an antibacterial and pro-healing wound dressing that can treat wound infections. J Mater Chem B 5(43):8451–8458
32. Yang W, Zhang L, Wang S, White AD, Jiang S (2009) Functionalizable and ultra stable nanoparticles coated with zwitterionic poly(carboxybetaine) in undiluted blood serum. Biomaterials 30(29):5617–5621
33. Moyano DF, Saha K, Prakash G, Yan B, Kong H, Yazdani M, Rotello VM (2014) Fabrication of corona-free nanoparticles with tunable hydrophobicity. ACS Nano 8(7):6748–6755
34. Yang W, Liu S, Bai T, Keefe AJ, Zhang L, Ella-Menye J-R, Li Y, Jiang S (2014) Poly(carboxybetaine) nanomaterials enable long circulation and prevent polymer-specific antibody production. Nano Today 9(1):10–16
35. Zhang L, Xue H, Gao C, Carr L, Wang J, Chu B, Jiang S (2010) Imaging and cell targeting characteristics of magnetic nanoparticles modified by a functionalizable zwitterionic polymer with adhesive 3,4-dihydroxyphenyl-L-alanine linkages. Biomaterials 31(25):6582–6588
36. Zhang X, Lin W, Chen S, Xu H, Gu H (2011) Development of a stable dual functional coating with low non-specific protein adsorption and high sensitivity for new superparamagnetic nanospheres. Langmuir 27(22):13669–13674
37. Giovanelli E, Muro E, Sitbon G, Hanafi M, Pons T, Dubertret B, Lequeux N (2012) Highly enhanced affinity of multidentate versus bidentate zwitterionic ligands for long-term quantum dot bioimaging. Langmuir 28(43):15177–15184
38. Muro E, Pons T, Lequeux N, Fragola A, Sanson N, Lenkei Z, Dubertret B (2010) Small and stable Sulfobetaine zwitterionic quantum dots for functional live-cell imaging. J Am Chem Soc 132(13):4556–4557
39. Jia G, Cao Z, Xue H, Xu Y, Jiang S (2009) Novel zwitterionic-polymer-coated silica nanoparticles. Langmuir 25(5):3196–3199
40. Shah SR, Kasper FK, Mikos AG (2013) Perspectives on the prevention and treatment of infection for orthopedic tissue engineering applications. Chin Sci Bull 58(35):4342–4348
41. Hetrick EM, Schoenfisch MH (2006) Reducing implant-related infections: active release strategies. Chem Soc Rev 35(9):780–789
42. Mi FL, Wu YB, Shyu SS, Schoung JY, Huang YB, Tsai YH, Hao JY (2002) Control of wound infections using a bilayer chitosan wound dressing with sustainable antibiotic delivery. J Biomed Mater Res 59(3):438–449
43. Norowski PA Jr, Bumgardner JD (2009) Biomaterial and antibiotic strategies for peri-implantitis. J Biomed Mater Res B-Appl Biomater 88B(2):530–543
44. Shi M, Kretlow JD, Nguyen A, Young S, Baggett LS, Wong ME, Kasper FK, Mikos AG (2010) Antibiotic-releasing porous polymethylmethacrylate constructs for osseous space maintenance and infection control. Biomaterials 31(14):4146–4156
45. Gong P, Li H, He X, Wang K, Hu J, Tan W, Zhang S, Yang X (2007) Preparation and antibacterial activity of Fe3O4@Ag nanoparticles. Nanotechnology 18(28):285604
46. Back DA, Bormann N, Calafi A, Zech J, Garbe LA, Muller M, Willy C, Schmidmaier G, Wildemann B (2016) Testing of antibiotic releasing implant coatings to fight bacteria in combat-associated osteomyelitis—an in-vitro study. Int Orthop 40(5):1039–1047
47. Xiu ZM, Zhang QB, Puppala HL, Colvin VL, Alvarez PJ (2012) Negligible particle-specific antibacterial activity of silver nanoparticles. Nano Lett 12(8):4271–4275
48. Gordon O, Slenters TV, Brunetto PS, Villaruz AE, Sturdevant DE, Otto M, Landmann R, Fromm KM (2010) Silver coordination polymers for prevention of implant infection: thiol

interaction, impact on respiratory chain enzymes, and hydroxyl radical induction. Antimicrob Agents Chemother 54(10):4208–4218
49. Schierholz JM, Lucas LJ, Rump A, Pulverer G (1998) Efficacy of silver-coated medical devices. J Hosp Infect 40(4):257–262
50. Knetsch MLW, Koole LH (2011) New strategies in the development of antimicrobial coatings: the example of increasing usage of silver and silver nanoparticles. Polymers 3(1):340–366
51. Kim JS, Kuk E, Yu KN, Kim J-H, Park SJ, Lee HJ, Kim SH, Park YK, Park YH, Hwang C-Y, Kim Y-K, Lee Y-S, Jeong DH, Cho M-H (2007) Antimicrobial effects of silver nanoparticles. Nanomedicine 3(1):95–101
52. Thomas V, Yallapu MM, Sreedhar B, Bajpai SK (2009) Breathing-in/breathing-out approach to preparing nanosilver-loaded hydrogels: highly efficient antibacterial nanocomposites. J Appl Polym Sci 111(2):934–944
53. Ho CH, Odermatt EK, Berndt I, Tiller JC (2013) Long-term active antimicrobial coatings for surgical sutures based on silver nanoparticles and hyperbranched polylysine. J Biomater Sci Polym Ed 24(13):1589–1600
54. Kumar R, Munstedt H (2005) Silver ion release from antimicrobial polyamide/silver composites. Biomaterials 26(14):2081–2088
55. Murata H, Koepsel RR, Matyjaszewski K, Russell AJ (2007) Permanent, non-leaching antibacterial surfaces—2: how high density cationic surfaces kill bacterial cells. Biomaterials 28(32):4870–4879
56. Tiller JC, Liao CJ, Lewis K, Klibanov AM (2001) Designing surfaces that kill bacteria on contact. Proc Natl Acad Sci U S A 98(11):5981–5985
57. Buffet-Bataillon S, Tattevin P, Bonnaure-Mallet M, Jolivet-Gougeon A (2012) Emergence of resistance to antibacterial agents: the role of quaternary ammonium compounds—a critical review. Int J Antimicrob Agents 39(5):381–389
58. Frost MC, Reynolds MM, Meyerhoff ME (2005) Polymers incorporating nitric oxide releasing/generating substances for improved biocompatibility of blood-contactincy medical devices. Biomaterials 26(14):1685–1693
59. Nablo BJ, Prichard HL, Butler RD, Klitzman B, Schoenfisch MH (2005) Inhibition of implant-associated infections via nitric oxide release. Biomaterials 26(34):6984–6990
60. Engelsman AF, Krom BP, Busscher HJ, van Dam GM, Ploeg RJ, van der Mei HC (2009) Antimicrobial effects of an NO-releasing poly(ethylene vinylacetate) coating on soft-tissue implants in vitro and in a murine model. Acta Biomater 5(6):1905–1910
61. Riccio DA, Coneski PN, Nichols SP, Broadnax AD, Schoenfisch MH (2012) Photoinitiated nitric oxide-releasing tertiary S-nitrosothiol-modified xerogels. ACS Appl Mater Interfaces 4(2):796–804
62. Liu XF, Guan YL, Yang DZ, Li Z, De Yao K (2001) Antibacterial action of chitosan and carboxymethylated chitosan. J Appl Polym Sci 79(7):1324–1335
63. Rabea EI, Badawy MET, Stevens CV, Smagghe G, Steurbaut W (2003) Chitosan as antimicrobial agent: applications and mode of action. Biomacromolecules 4(6):1457–1465
64. Popelka A, Novak I, Lehocky M, Junkar I, Mozetic M, Kleinova A, Janigova I, Slouf M, Bilek F, Chodak I (2012) A new route for chitosan immobilization onto polyethylene surface. Carbohydr Polym 90(4):1501–1508
65. Barros J, Dias A, Rodrigues MA, Pina-Vaz C, Lopes MA, Pina-Vaz I (2015) Antibiofilm and antimicrobial activity of polyethylenimine: an interesting compound for endodontic treatment. J Contemp Dent Pract 16(6):427–432
66. Kim YH, Park JH, Lee M, Kim YH, Park TG, Kim SW (2005) Polyethylenimine with acid-labile linkages as a biodegradable gene carrier. J Control Release 103(1):209–219
67. Lee Y, Mo H, Koo H, Park J-Y, Cho MY, Jin G-w, Park J-S (2007) Visualization of the degradation of a disulfide polymer, linear poly(ethylenimine sulfide), for gene delivery. Bioconjug Chem 18(1):13–18
68. Park D, Wang J, Klibanov AM (2006) One-step, painting-like coating procedures to make surfaces highly and permanently bactericidal. Biotechnol Prog 22(2):584–589

69. Hsu BB, Ouyang J, Wong SY, Hammond PT, Klibanov AM (2011) On structural damage incurred by bacteria upon exposure to hydrophobic polycationic coatings. Biotechnol Lett 33(2):411–416
70. Schaer TP, Stewart S, Hsu BB, Klibanov AM (2012) Hydrophobic polycationic coatings that inhibit biofilms and support bone healing during infection. Biomaterials 33(5):1245–1254
71. Milovic NM, Wang J, Lewis K, Klibanov AM (2005) Immobilized N-alkylated polyethylenimine avidly kills bacteria by rupturing cell membranes with no resistance developed. Biotechnol Bioeng 90(6):715–722
72. Gultekinoglu M, Tunc Sarisozen Y, Erdogdu C, Sagiroglu M, Aksoy EA, Oh YJ, Hinterdorfer P, Ulubayram K (2015) Designing of dynamic polyethyleneimine (PEI) brushes on polyurethane (PU) ureteral stents to prevent infections. Acta Biomater 21:44–54
73. Zanini S, Polissi A, Maccagni EA, Dell'Orto EC, Liberatore C, Riccardi C (2015) Development of antibacterial quaternary ammonium silane coatings on polyurethane catheters. J Colloid Interface Sci 451:78–84
74. Fan Z, Liu B, Wang J, Zhang S, Lin Q, Gong P, Ma L, Yang S (2014) A novel wound dressing based on Ag/Graphene polymer hydrogel: effectively kill bacteria and accelerate wound healing. Adv Funct Mater 24(25):3933–3943
75. Nimal TR, Baranwal G, Bavya MC, Biswas R, Jayakumar R (2016) Anti-staphylococcal activity of injectable nano Tigecycline/chitosan-PRP composite hydrogel using Drosophila melanogaster model for infectious wounds. ACS Appl Mater Interfaces 8(34):22074–22083
76. Anjum S, Arora A, Alam MS, Gupta B (2016) Development of antimicrobial and scar preventive chitosan hydrogel wound dressings. Int J Pharm 508(1–2):92–101
77. Zhao X, Wu H, Guo B, Dong R, Qiu Y, Ma PX (2017) Antibacterial anti-oxidant electroactive injectable hydrogel as self-healing wound dressing with hemostasis and adhesiveness for cutaneous wound healing. Biomaterials 122:34–47
78. Liu H, Wang C, Li C, Qin Y, Wang Z, Yang F, Li Z, Wang J (2018) A functional chitosan-based hydrogel as a wound dressing and drug delivery system in the treatment of wound healing. RSC Adv 8(14):7533–7549
79. El-Naggar MY, Gohar YM, Sorour MA, Waheeb MG (2016) Hydrogel dressing with a nano-formula against methicillin-resistant Staphylococcus aureus and Pseudomonas aeruginosa diabetic foot bacteria. J Microbiol Biotechnol 26(2):408–420
80. Yu BY, Zheng J, Chang Y, Sin MC, Chang CH, Higuchi A, Sun YM (2014) Surface zwitterionization of titanium for a general bio-inert control of plasma proteins, blood cells, tissue cells, and bacteria. Langmuir 30(25):7502–7012
81. Sin MC, Sun YM, Chang Y (2014) Zwitterionic-based stainless steel with well-defined polysulfobetaine brushes for general bioadhesive control. ACS Appl Mater Interfaces 6(2):861–873
82. Colilla M, Izquierdo-Barba I, Sánchez-Salcedo S, Fierro JLG, Hueso JL, Vallet-Regí MA (2010) Synthesis and characterization of zwitterionic SBA-15 nanostructured materials. Chem Mater 22(23):6459–6466
83. Colilla M, Martínez-Carmona M, Sánchez-Salcedo S, Ruiz-González ML, González-Calbet JM, Vallet-Regí M (2014) A novel zwitterionic bioceramic with dual antibacterial capability. J Mater Chem B 2(34):5639–5651
84. Izquierdo-Barba I, Sanchez-Salcedo S, Colilla M, Feito MJ, Ramirez-Santillan C, Portoles MT, Vallet-Regi M (2011) Inhibition of bacterial adhesion on biocompatible zwitterionic SBA-15 mesoporous materials. Acta Biomater 7(7):2977–2985
85. Liu P, Domingue E, Ayers DC, Song J (2014) Modification of Ti6Al4V substrates with well-defined zwitterionic polysulfobetaine brushes for improved surface mineralization. ACS Appl Mater Interfaces 6(10):7141–7152
86. Samuel U, Guggenbichler JP (2004) Prevention of catheter-related infections: the potential of a new nano-silver impregnated catheter. Int J Antimicrob Agents 23(supp-S1):75–78
87. Johnson JR, Johnston B, Kuskowski MA (2012) In vitro comparison of Nitrofurazone- and silver alloy-coated Foley catheters for contact-dependent and diffusible inhibition of urinary tract infection-associated microorganisms. Antimicrob Agents Chemother 56(9):4969–4972

88. Pickard R, Lam T, MacLennan G, Starr K, Kilonzo M, McPherson G, Gillies K, McDonald A, Walton K, Buckley B, Glazener C, Boachie C, Burr J, Norrie J, Vale L, Grant A, N'Dow J (2012) Types of urethral catheter for reducing symptomatic urinary tract infections in hospitalised adults requiring short-term catheterisation: multicentre randomised controlled trial and economic evaluation of antimicrobial- and antiseptic-impregnated urethral catheters (the CATHETER trial). Health Technol Assess 16(47):1–197
89. Pollini M, Paladini F, Catalano M, Taurino A, Licciulli A, Maffezzoli A, Sannino A (2011) Antibacterial coatings on haemodialysis catheters by photochemical deposition of silver nanoparticles. J Mater Sci Mater Med 22(9):2005–2012
90. Dayyoub E, Frant M, Pinnapireddy SR, Liefeith K, Bakowsky U (2017) Antibacterial and anti-encrustation biodegradable polymer coating for urinary catheter. Int J Pharm 531(1):205–214

Osteoinductive and Osteoconductive Biomaterials

Shreya Agrawal and Rohit Srivastava

Abstract With combinatorial approaches getting stronger to design materials with better functionalities and compatibility for restoring bone tissue, it is becoming important to understand the progress and evolution of existing and newly designed materials. For being clinically usable, they should have features that address the biomechanical, biochemical, and medical requirements.

Their various characteristics determine the cascade of events that take place at the site of bone healing. They should be selected based on the specific purpose with maximum benefit to the patient in long run. The current efforts in this domain are to render the orthopedic procedures minimally invasive and maximally effective.

This chapter encompasses the journey of classes of biomaterials used for their osteoinductive and osteoconductive properties and discusses the challenges for bringing them closer to fulfil the requisites.

Keywords Bone · Osteoblasts · Osteogenesis · Composites · Stem cells · Apatite Biomimicry · Implant · Bioactivity · Scaffold

Introduction

With an increased pace of life and increasing life expectancy, every year the number of bone graft procedures are increasing [1]. For bone repair, substitution, and augmentation, autografts have continued to be the gold standard till date, followed by allografts. Both of these grafts, in spite of having osteogenicity or osteoinductivity, have many shortcomings. Autografts incur high costs, are limited in their availability and cause additional trauma to the patient due to donor site morbidity. Allografts pose issues of potential viral transmission and immunogenicity apart from their limited supply and high cost. Their high cost and limited availability necessitate

S. Agrawal · R. Srivastava (✉)
Department of Biosciences and Bioengineering, Indian Institute of Technology Bombay, Bombay, Maharashtra, India
e-mail: rsrivasta@iitb.ac.in

B. Li et al. (eds.), *Racing for the Surface*,
https://doi.org/10.1007/978-3-030-34471-9_15

innovation in artificial novel biomaterials for bone substitution, repair, and augmentation.

To overcome this challenge of substituting bone tissue, researchers must have insight to reproduce the highly hierarchical composite microstructure of bone, which has not been possible yet. Any type of bone tissue engineering biomaterial should be mechanically and biologically compatible, osteoconductive, osteoinductive and allow integration with physicochemical characteristics irrespective of its source of origin [2]. With evolving technology in recent times, bone regenerative engineering is gradually becoming the solution for the new age with a bottom-up approach that incorporates stem cells, biomaterials, and growth factors as required, to regenerate bone tissue as compared to the conventional top-down approach commonly known as bone tissue engineering [3]. The three most commonly used terms pertaining to biologically relevant properties of biomaterials which are sometimes interchangeably used in regenerative medicine are as follows.

Osteoinduction is the regular phenomenon which induces osteogenesis during the bone healing process. The immature and pluripotent stem cells are stimulated to become preosteoblastic cells. Urist identified osteoinduction as the "the mechanism of cellular differentiation towards bone tissue due to the physicochemical effect or contact with another tissue" [4]. Osteoinduction is the major driving force in any bone healing situation as depicted in Fig. 1.

Osteoconduction is that attribute of a material that permits new bone to grow on the surface. So, any surface that allows bone growth either above it or inside it is an osteoconductive surface. In 1987, Wilson-Hench suggested that the process where the bone is aligned with a material's surface or contour is regarded as osteoconduction [5]. This property is indispensable for the success of bone implants. Few metals such as Ag, Cu as well as bone cement show little or negligible osteoconduction due to their poor biocompatibility [6].

Osseointegration can be realized as the apposition or interface of connecting bone and the implant. It is explained as "the direct attachment of an implant by the formation of bony tissue around the implant" (Dorland's illustrated medical dictionary). It can be realized as the "direct functional and structural bridge between an implant and bone" as bone deposition increases temporally.

Mechanism of Osteoinduction by Biomaterials

The real modus operandi of osteoinduction by biomaterials is still under exploration. The real question is whether there are any similarities between the mechanisms of osteoinduction by bone morphogenetic proteins (BMPs) and biomaterials. One of the possibilities is that after implantation, endogenous BMPs accumulate on the biomaterial surface and consequently induce bone formation [7–9] but there is no conclusive evidence for this hypothesis. Yuan and co-workers, suggested the role of BMPs in biomaterial-led-osteoinduction but their presence was not indispensable.

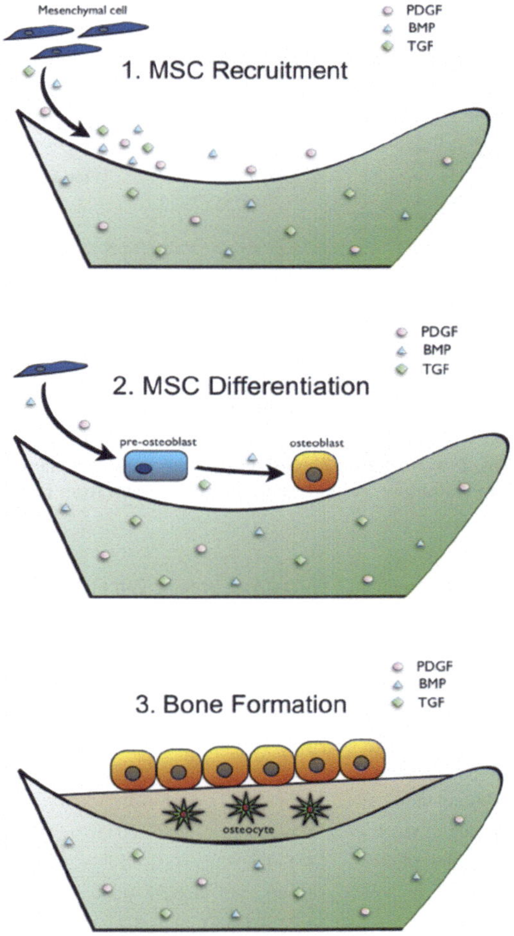

Fig. 1 Principles of osteoinductive materials. Principle 1: Osteoinductive materials recruit MSCs to bone graft surfaces by growth factor release. Principle 2: The material promotes MSC differentiation into osteoblasts. Principle 3: Osteoblasts forms ectopic bone in vivo [6]

A statistically lower number of animals under this pilot study makes this conclusions unreliable [7, 10]. There are some key contrasts in the osteoinduction mechanism by BMPs and biomaterials, such as: (a) BMPs generally induce bone through the endochondral pathway [11] while biomaterial-induced bone is always intramembranous [12, 13] ; (b) in smaller organisms like rats, bone is formed with much ease as in the case of BMPs as compared to biomaterials [14–19]; and (c) with biomaterials, bone formation happens inside the pores of biomaterials and not on the edge, while bone formation via BMPs happens on the external surface of the carrier, including the soft tissues that are located away from the implant surface [10, 20].

Although major aspects of the osteoinduction process by biomaterials are still unclear and under exploration, many studies in this field have proved its struc-

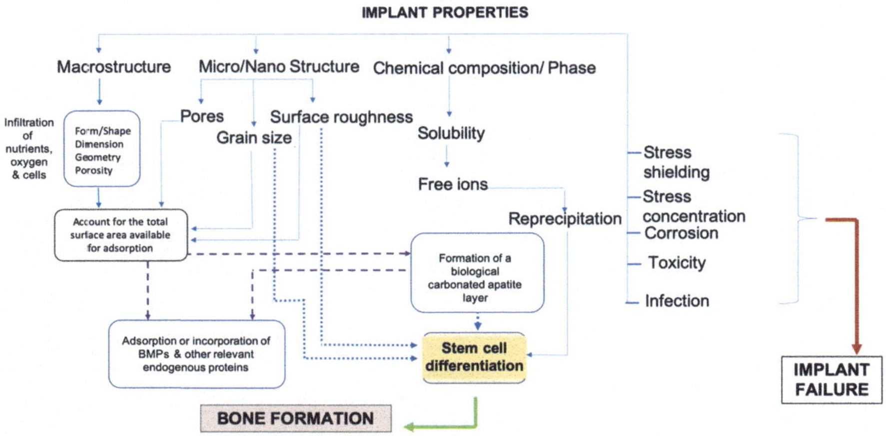

Fig. 2 Schematic summarizing hypothesized mechanisms of osteoinduction, bone formation, and implant failure. (Figure adapted and redrawn from Barradas et al. [21])

Table 1 Mechanical properties of bone and biomaterials [23, 24]

Material	Density (g/cm^3)	Elastic modulus (GPa)	Compressive strength (MPa)	Tensile strength (MPa)
Cortical bone	~2	17–24	130–180	50–151
Cancellous bone	~1	0.1–45	4–12	10–20
Hydroxyapatite (HA)	3.1	73–117	600	0.7
Magnesium	1.74–2.0	41–45	65–100	15–40
Stainless steel	8.1–8.9	185–205	170–310	50–200
Titanium alloy	4.4–4.5	110–117	758–1117	55–115
PCL	1.145	0.21–0.34	6.6–10.6	20.7–34.5
PLLA	1.210–1.430	2.4–4.2	18–93	55.2–82.7

ture (at the macro, micro, and nano-levels) along with its chemical components play the most crucial role in rendering a material osteoinductive. The physicochemical and structural characteristics of osteoinductive biomaterials determine the course and output of bone formation both directly and indirectly (Fig. 2). As their inherent mechanical properties vary a lot, it is quite complicated to fit it with human body requirements (Table 1). Properties like ultrastructure and surface topography can modulate the dynamics of interactions with BMPs, growth factors, and other endogenous proteins which initiate bone formation or result in implant failure if events like stress shielding or infection takes place [21].

Identification of Osteoinductive Materials [6]

To identify and select suitable osteoinductive biomaterials out of a plethora of materials, there is a specific sequence of biological activity characteristics which is as follows:

- Recruitment of mesenchymal-type osteoprogenitor cells;
- Transformation of mesenchymal stem cells (MSCs) into a mature, bone-forming lineage; and
- Induction of new bone formation when implanted intramuscularly.

Classes of Orthopedic Biomaterials

There are various classes of biomaterials that have been employed as osteoinductive and osteoconductive materials as listed in Table 2 [22].

Metals

Metals are the very first category of materials to be used as implants in ancient times as per reports from Egyptian era. Aluminum, lead, gold, and silver were among the first metals to be used while titanium and its alloys are the most widely used metallic biomaterials in modern times. Owing to their good biocompatibility, strength, non-toxicity, and corrosion resistance, metal alloys are used for replacing joints and fixing fractures but many times, due to their non-biodegradability, these implants need removal. Metallic biomaterials are not bioactive *per se*. They are generally applied in load-bearing implants which require high strength, sufficient biocompatibility, and low in vivo corrosion rates. Metallic implants also tend to loosen due to stress shielding-led-bone resorption, weak interfacial bonding with bone, and absence of a supporting structure for new tissue.

In recent times, there have been efforts to design and develop metal implants that can mimic the traits and microstructure of the trabecular bone for expanding the scope of the use of metals. Hence, these special materials are termed as open-cell porous metals, metallic scaffolds, metallic foams, or cellular metals with three dimensionally (3D) interconnected pores. The total porosity is 50–75% with pore sizes in the range 200–500 μm [23]. Another method to make multipurpose and bioactive metals consist of: (1) applying a surface coating over the implant with bioactive ceramics or (2) chemically modifying the surface of the metal to bring about the in vivo build-up of a bioactive ceramic, or (3) to trigger the adhesion of proteins and cells toward tissue–material interactions [24]. The major metals used as orthopedic biomaterials are discussed below.

Table 2 Classes of biomaterials used for orthopedic applications [22]

Type of material	Advantages	Disadvantages
Metals	Biocompatibility, non-toxicity and corrosion resistance	Not biodegradable
Bioceramics		
Bioactive glasses	Improve differentiation and osteogenesis	Low strength and brittleness
Hydroxyapatite	Bioactivity, biocompatibility, osteoconductivity, non-toxicity, and non-inflammatory	Brittle, very slow degradation
Tricalcium phosphate	Supports in vivo osteogenic differentiation	Slow degradation, incompressible nature
Natural polymers		
Collagen	Enzymatic biodegradability	Complexity of structure
Gelatin	Biocompatible, biodegradable	Poor mechanical properties
Chitosan	Support cell attachment, differentiation and migration, non-toxicity, non-allergenicity, mucoadhesivity, biocompatibility, biodegradability, and osteoconductivity	Poor mechanical strength
Hyaluronic acid	Biocompatibility, biodegradability, viscoelasticity, enzymatic biodegradability	Very rapid degradation and water solubility
Alginate	Biocompatibility, easy gelling, easy chemical modification	Nondegradable in mammals, sterilization causes degradation
Agarose	Wide range of gelling and melting temperatures, no need of cross-linking agents, little inflammatory response in vivo	Poor cell attachment
Synthetic polymers		
Poly (α-hydroxy acids)	Degradation products can be excluded from the body	Degradation by bulk erosion, relatively poor mechanical properties, hydrophobicity of polymer surface, acidic degradation byproducts
Poly (ε-caprolactone)	Biodegradable, non-toxic, a low processing temperature	Hydrophobicity, slow degradation
Polyurethanes	Excellent mechanical properties, good biocompatibility	Toxicity of degradation products (from aromatic diisocyanate component)

Magnesium (Mg)

The physical characteristics of magnesium are attractive, like high strength-to-mass ratios and an elasticity coefficient that matches with natural bone more as compared to other conventional metals. Apart from resemblance to human bone, they also have functional roles as natural ion content in human physiology and their biodegradation dynamics once inside the body [23]. Though they have been used for making short-term biodegradable implants with load-bearing purposes, their high corrosion rates and low bioactivity are some major challenging problems, which need to be resolved before tapping into their full potential in clinical applications. Various reports suggest the need to develop Mg alloys with modifiable in vivo decomposition rates and mechanical attributes similar to those of bone [25].

Endeavors to limit the corrosion dynamics of Mg have involved many approaches such as purification, alloying, and surface modification. Pure Mg also shows significantly reduced corrosion rates but owing to its low yield strength, it does not find widespread usage in orthopedics and other load-bearing applications. Surface modification is one of the very important strategies that reduces and controls the degradation behavior along with improving its surface biocompatibility. As compared to adjusting its bulk structure and composition, tuning the surface properties is a much simpler approach to adjust surface corrosion resistance. This helps in preserving favorable bulk attributes and is easily implemented on Mg alloys [25].

In one of the landmark studies with Mg alloys, Lee *et al.* systematically investigated the bone induction mechanism by using a Mg-5wt%Ca-1wt%Zn screw in an extensive clinical investigation. They employed this biodegradable implant and observed changes in elemental composition and crystallinity at the material interface. Controlled degradation of this alloy was followed by the growth of a biologically similar calcification matrix at the degrading surface which aids in initiating bone generation. This phenomenon facilitated faster healing with entirely replaced new bone at the place of the biodegradable implant within 1 year of implantation, as shown in an elaborated clinical study [26].

Despite alloying which is one of the major approaches for metals to tune their properties, its larger electronegative potential (−2.4 V) makes it harder to lower the degradation rate, only with the approach of alloying. Rather, the addition of a ceramic phase has proven to be a successful solution for optimizing Mg's attributes, such as mechanical properties and corrosion resistance. Witte *et al.* carried out some initial studies with Mg/hydroxyapatite (HA) composites. They reinforced 20% HA particles to Mg alloy AZ91D matrix and developed a metal matrix composite (MMC). The mechanical, corrosive, and compatibility properties of the composite were studied in vitro. It was revealed that adding HA particles lowered the pH, enhanced the resistance for corrosion in various fluids, and enhanced the cell viability for a variety of cell lines when compared with the AZ91D matrix [27].

It's important to be cautious while designing and fabricating such composites as these steps may have several hazards involved. One of the possible reactions is that of Mg with calcium phosphate which forms Mg phosphide that forms phosphine (a toxic gas) upon reaction with water. The possibility of this reaction increases with molten Mg and for amalgamations of HA due to hydroxyl groups in HA [28].

There are many commercially available Mg implants available today with various brand names. Cortical bone screws and pins from MAGNEZIX® are shown in Fig. 3. Various forms of Mg such as porous, fine grained, composite, and glassy structures, have been employed for implant fabrication purposes.

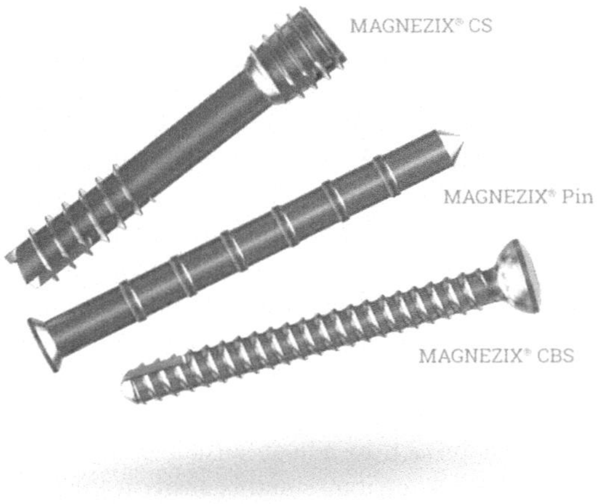

Fig. 3 Commercially available MAGNEZIX® cortical screw (CS), cortical bone screw (CBS), and pin (Image courtesy www.magnezix.de)

Steel

The first line of effective substitutive joint prosthesis came very late in the 1950s and it was a cemented prosthesis with a stainless steel stem developed by Charnley [29]. Stainless steel materials show resistance to a variety of harsh chemical agents due to their high chromium content (~12%). They also induce the synthesis of a firm coating of Cr_2O_3 which has self-HA healing and corrosion resistance properties. Austenitic stainless steel is the preferred form of steel among the available versions for implant manufacturing. To behave austenitically at ambient temperature, stainless steel requires austenite stabilizers like Ni or Mn. AISI 316L grade of stainless steel is generally employed for clinical applications and it consists of 17–20wt% Cr, 12–14wt% Ni, 0.03wt% C, and 2–3wt% Mo as well as traces of N, Mn, P, Si, and S [24].

One of the major problems found with medical-grade steel alloys is the release of nickel ions which pose negative effects for various organs [30]. Nickel is expensive as well as causes serious allergies to the human skin. These issues have compelled advances in creating better nickel-free stainless steels. Nitrogen has been found as one of the great substitutes for nickel for austenite stabilizing and strengthening. Also, there have been developments in bio-composites, such as those made up of nickel-free stainless steels and HA [31–33].

As per ASTM standards, two of such nitrogen-containing medical-grade stainless steels have been listed without the presence of nickel: ASTM ID: F2229 and ASTM ID: F2581. A number of research studies have been conducted on these alloys in the recent years, and showed better biocompatibility, osteointegration, and corrosion behavior compared to traditional steel compositions [30, 34–37].

Popkov *et al.* studied the impact of nanostructured calcium-phosphate coatings on steel and Ti. They implanted these in the tibial bone marrow of dogs. Stainless steel and Ti wires were administered for the first and second group, and stainless steel and Ti wires of the same diameter coated with HA were implanted for the third and fourth groups. It was found that the HA coated wire implants provided better bioactivity and osteointegration and the nanostructure of HA played a major role in bone formation. These wires may find applications in various orthopedic conditions like fractures, *osteogenesis imperfecta,* and rectification of bone deformities [38].

Various modifications and coatings over existing steel have been yielding great results. UNS S31254 grade steel is one such version with a higher nitrogen content and lower cost (close to Ti) and has been further improved by HA coating using pulsed laser deposition (PLD). In one of such research studies with this technology, HA columns were successfully coated on a UNS S31254 substrate. Surface roughness of the coating is a desirable property for increased cell growth and proliferation and was found to be increased by atomic force microscopy (AFM) and scanning electron microscopy (SEM). This material had superior antibacterial properties and bioactivity as well as better adhesive strength as compared to the standard. This research paves the way to an alternate route of improving the physiological behavior of conventional metals. The favorable outcomes of this study emphasize the obvious requirements for an improved orthopedic implant material [39].

Titanium

Pure Ti shows relatively poor strength despite having excellent biocompatibility. When it is alloyed, the strength quotient goes up, but the toxicity level also goes up. The superior compatibility of Ti implants is accredited to its surface events (Fig. 4) and the stable 3–10 nm layer of oxide that is spontaneously formed on its exposure to oxygen [40, 41]. This layer creates direct contact with bone while blocking the development of a fibrous capsule outside the implant. At the beginning of $Ti(O_2)$

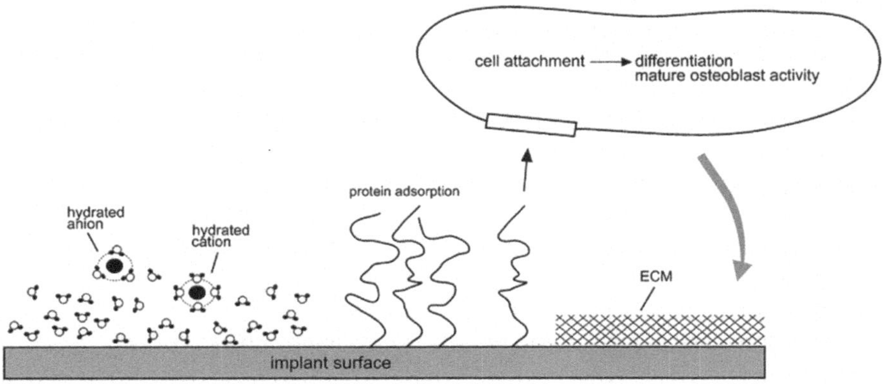

Fig. 4 Schematic representation of events consecutively taking place at the titanium surface after implantation into living bone tissue. Water binds to the surface, followed by the incorporation of hydrated ions, adsorption and desorption of proteins, eventually leading to cell attachment. After differentiation, mature osteoblasts produce the extracellular matrix (ECM) [42]

implantation, its non-physiological surface gets exposure of the physiological conditions and it becomes important to create a biomimicking coating onto the Ti surface. This helps in a gradual transformation between the Ti surface and the adjoining bone tissue and results in better bone cell adhesion, growth, and differentiation [42].

Apart from the above-mentioned surface changes, immobilization of proteins, enzymes, peptides, and other biological molecules on implants have been the latest interest in this domain [43–46]. As compared to the conventional method of inorganic calcium phosphate coatings, the newer approach for surface modification employs purely organic components of bone. The organic and natural mimicking coatings include: (1) extracellular matrix (ECM) proteins or peptide sequences from laminin, fibronectin, or heparin binding domain [47, 48]; (2) cell signalling agents (TGF, BMPs, FGF, IGF, PDGF, etc.) to trigger new bone formation [20, 49, 50], (3) DNA [46, 51, 52] , and (4) enzymes [53].

There have been studies on both organic and inorganic phases over the implant. Xia *et al.* applied HA and collagen (which constitute the bone) over implant surfaces to enrich the biological reactions at the tissue and implant interface. They successfully prepared a homogenous collagen/apatite coating using a biomimetic technique. These composite layers showed a homogeneous porous structure with fibrous collagen and crystalline apatite. The in vitro studies showed better osteoblast proliferation over a composite coating with higher collagen content than with the pure apatite coating. This kind of organic/inorganic bone mimicking molecular cocktail has a wide scope to be applied on implant surfaces and improve the osteointegration between the implant and adjacent bone [54].

There are certain other surface modification routes as well to simply deposit pure and firm HA over Ti alloy surfaces not including the mediation of organic molecules. The zwitterionic polymer-modified surfaces are nonreactive to biological moieties viz. proteins/cells and have the potential to induce mineralized cluster growth. In one such experiment by Nishida *et al.*, a monolayer of poly(ethylene glycol) methacrylate phosphate (Phosmer PE) was self-assembled on a Ti alloy surface [55] with a zwitterionic monomer (carboxymethyl betaine, CMB). This poly(CMB) -modified Ti alloy plate suppressed the adsorption of proteins and attachment of cells, and triggered deposition of around two times calcium (Ca^{2+}) as compared to the unaltered Ti alloy plate. In another experiment, the enzyme alkaline phosphatase (ALP), which plays the most crucial role in bone and cartilage mineralization process, was utilized to coat the Ti surface. It is believed to raise the concentration levels of inorganic phosphate as well as to lower the levels of extracellular pyrophosphate which impedes the process of mineralization. These coatings hastened mineralization over the Ti implant surface [53].

In one of the experiments to study surface topography effects, Yu *et al.* created two periodic microscale functionalized zones on Ti (MZT), i.e., nanoneedle zones and plain (buffer) zones. It was designed to relay spatially controlled topographical signals for better bone regeneration. The alternating buffer zones with no nanoneedle arrays were intercalated as the tips of the nanoneedles were too sharp to act as a contact area for the cells to effectively proliferate. In this way, the MZT displayed zone-wise apatite deposition and protein adsorption in N-zone and cell differentiation, mechanotransduction, and proliferation in the B and B-N zones. Enhanced osteoblast differentiation and nodule deposition was also observed with MZT. The constitution of the bone nodules on untreated Ti and MZT were also found to be different and the process of bone formation was enhanced by the MZT implant [56].

Tantalum

Ta is a transition metal and has recently emerged as a useful biomaterial for orthopedic applications due to its mechanical properties. This material has caught special attention due to properties such as greater volumetric porosity, friction coefficient, and modulus of elasticity that matches with natural compact and spongy bone. It has been successful in several orthopedic applications owing to its exceptional biocompatibility and favorable characteristics. It also possesses similarity with cancellous bone and is safe for in vivo use as verified by its clinical use in orthopedic surgery [57].

Despite its biocompatibility, inertness, and resistance to corrosion, the use of solid Ta in orthopedic implant devices has been restricted due difficulty in its manipulation. This issue led to the advent of designing porous Ta implants. As the technology of making implants with interconnected porosity has been here for a while, porous Ta trabecular metal (PTTM) also entered in the orthopedic domain in early 1990s [58]. "PTTM is commercially available as Trabecular Metal Material (Zimmer, Trabecular Metal Technology, Inc., Parsippany, NJ, USA) with an open-cell porous structure having 3D dodecahedron repeats similar to trabecular bone and the striking similarity can be seen" (Fig. 5). An initial foam-like vitreous carbon general scaffold is fabricated for these open-cell dodecahedron repeats and in due course it becomes the inner framework of the PTTM implant [59].

The pore dimension of PTTM is in the range of 300–600 μm with a porosity of 75–85%. For materials used in dental applications, like Ti and Ti alloys (with an elastic modulus of 106–115 GPa), PTTM is much more suitable due to its relevant properties. Porous Ta is also resistant to corrosion and is biocompatible. It can

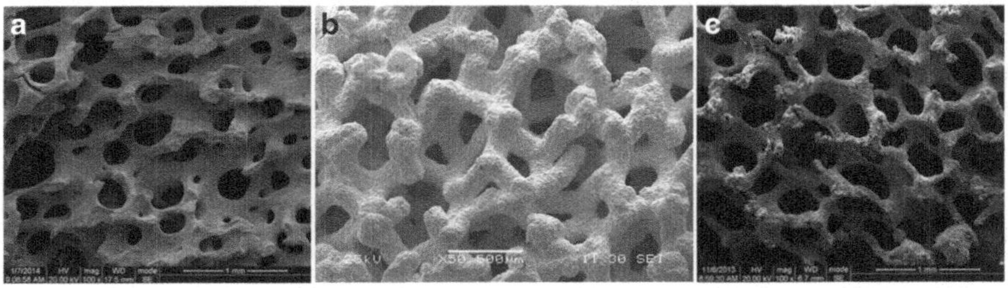

Fig. 5 Scanning electron micrograph of human trabecular bone (**a**), porous tantalum fabricated by the CVD/CVI (**b**) or PM (**c**) method [60]

extensively augment the proliferation and differentiation of primary osteoblasts extracted from aged patients than those on traditional solid Ti [60].

In one of the studies on novel porous Ta implant by Lu et al., its osteocompatibility and efficacy to achieve lumbar interbody fusion (LIF) was evaluated in a rabbit anterior lumbar fusion model. The fusion was accomplished 12 months postoperatively as confirmed both radiographically and histologically in the porous Ta group similar to autologous bone implanted at intervertebral spaces. Implant degradation, wear debris, and osteolysis were found to be nil and no significant local inflammation response was found inside or outside the implant. Even the composites and degradation products of the Ta implant were non-toxic and biocompatible [61].

In an attempt to compare orthopedic implant grade porous Ta (i.e., trabecular metal), the responses of human osteoblast cultures from two groups of females were examined (less than 45 years and more than 60 years) on a titanium fiber mesh (TFM) and tissue culture plastic. In relationship to the cells from older patients, the cell attachment, growth, and mineralization were better in cells derived from the patients of lower age as expected. Among all the three substrates, cell adhesion was not much different on porous Ta than TFM or tissue culture plastic but cell proliferation on porous Ta was found to be highly stimulated. The findings from this study substantiated that porous Ta can find many more clinical applications in degenerative skeletal conditions than titanium [62].

Another study by Lee *et al.*, compared Trabecular Metal™ Dental Implants (TM) and Tapered Screw-Vent® Dental Implants (TSV) for the potential of neo-osteogenesis and trabecular bone microarchitecture in fresh canine extraction sockets. It was revealed that there was more new bone in the TM implant than in the TSV at various healing time points. Histologically, trabecular metal implants exhibit higher amounts of bone with newly woven bone earlier than in the TSV implants [63].

NiTi Alloy: Apart from many of these traditionally used metals, a special alloy with a shape memory effect was identified by Buehler and Wang in 1967 [64] made up of Ni and Ti called as a NiTi alloy. The shape memory effect is a property that brings a material back to the old shape on rising its ambient temperature after it gets "plastically" deformed. Due to their elastic modulus and elastic recoverable strain, their use in load-bearing applications have been found to be more adequate than other metals. They exert compressive stress after the material has recovered from the pre-strain of heating and this makes them suitable as spinal correctors, staples for osteotomies, internal fixators for long bone shafts, fracture repair, vertebral spacers and anchoring of prostheses, etc. [65, 66]. Despite these advantages, NiTi alloys show problems of allergy and toxicity due to Ni ion release. The toxicity of Ni and its possible carcinogenicity constraints the use of NiTi alloys in many parts of the world [24].

Ceramics

Ceramics have a very old history of usage as biomaterials and they have evolved many generations from first, second, third to modern generation bioactive ceramics (Fig. 6). They possess high hardness, high melting temperatures, low conduction of

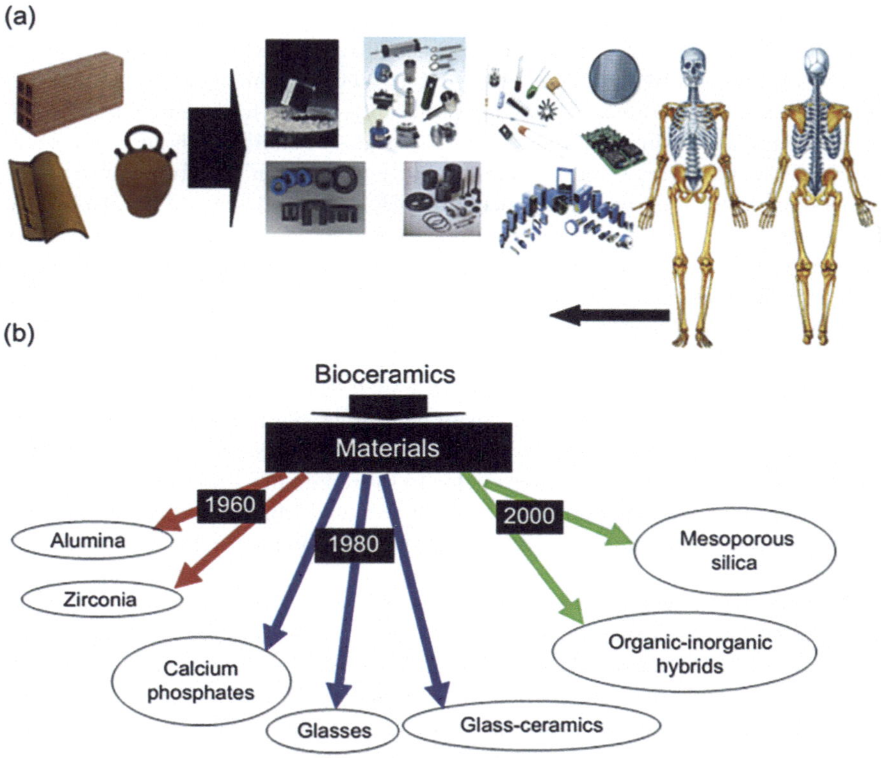

Fig. 6 Historical evolution of ceramics from routine objects to bioceramics. (**a**) Changing forms and factors of ceramics and (**b**) timeline of bioceramics [186]

electricity and heat along with biocompatibility, resistance to compression and corrosion. Their surface properties are also favorable such as high wetting degrees and surface tensions that facilitate the adhesion of proteins, cells, and other biological moieties. As biomaterials, they are suitable specially for the generation of hard tissue engineering, owing to their chemical and physical properties. However, these biomaterials have some disadvantages, such as brittleness and low strength [67, 68].

From bio-inert ceramics such as alumina, zirconia to bioactive bioglasses, mesoporous silica and organic-inorganic composites, bioceramics have come a long way. The major categories of ceramics utilized for bone tissue engineering are as follows.

Hydroxyapatite

Hydroxyapatite (HA) is a natural kind of calcium phosphate and is the largest depot of inorganic constituent of human bones. This is the reason why it is most widely used in bone regeneration [69]. The word apatite stems from the Greek word `απατaω' ("to deceive"), as it was not easy to be distinguished from other naturally occurring compounds such as aquamarine, amethyst, etc. The general formula for apatite is $Ca_5(PO_4)X$, where X can be any mono- and/or divalent anion such as fluoride, hydroxide, or carbonate. They are noticeably similar to the mineral phase of human bones and denture and thus are the molecules of choice for bone tissue engineering [70]. HA is very stable calcium phosphate and doesn't easily dissolve under in vivo conditions specified by temperature, pH, body fluids, etc. [71, 72] The HA surface acts as a site for nucleation of bone minerals [73, 74] and one of its best features is that it doesn't induce inflammatory reactions when used in clinical applications. Among apatites, carbonate apatite is the most abundant bioceramic phase of the human system. They can also be modified easily by ionic substitutions such as Ca^{2+}ions with Ba^{2+}, Sr^{2+}or Pb^{2+}, etc. [42]. Although HA has a natural occurrence and can be procured, due to its non-uniform and sometimes defective structures, HA for clinical applications and research needs to be synthesized in aqueous solution systems. HA possesses osteoconductive properties but not osteoinductive properties and the ionic substitution sometimes helps to overcome this drawback. The examples include fluoride ions for anionic substitution and Mg as cationic substitution which resulted in increased stability and favorable biological effects, respectively [74, 75].

Research has shown the potential and biocompatibility of HA with in vivo bone regeneration studies. This material enhances the differentiation and proliferation of MSCs by improving the attachment of osteoblasts [76, 77]. As it is quite brittle and hard, it is not usable for load bearing. It has been used for biomaterial surface coatings and as graft materials and also for bone regenerative applications in many forms, such as granules, cements, and pastes [78–80]. They have shown to improve many relevant aspects, such as osteoblast activity, implant contact area, and cellular responses of bone implants. Indirectly, they improve all the aspects of biomaterial performance [81, 82]. The HA coating methods can vary from spraying, sputtering, pulsed laser deposition or sol-gel techniques, etc. Their tunability and spectrum of usage can be further enhanced by combination with other flexible biomaterials like gels. Controlling their pore size and distribution, mechanical properties, biological activity, and user-friendliness can render them much more useful for bone regenerative applications [83–86].

Tricalcium Phosphate

Tricalcium phosphate (TCP; $Ca_3(PO_4)_2$) is among the most widely used calcium phosphates. It has two phases (viz. α and β). α-TCP has monoclinic crystal strucures while β-TCP displays rhombohedron structures [87, 88]. β-TCP displays more stability as it takes more time than α-TCP to degrade. All of these reasons make β-TCP more applicable for bone regeneration. Its resorption rate is better and thus is more widely employed to enhance the biocompatibility of other materials [89, 90]. β-TCP also helps the proliferation of osteoblasts and bone marrow stromal cells [91, 92]. The excellent biomineralization and cell adhesion activities are attributed to its nanoporous structure. Due to such favorable properties, β-TCP has actively been examined and used for bone regeneration purposes, such as in bone cements and for bone replacement [93, 94].

For blending the best properties of TCP and HA in one place, special materials have come into existence which are termed as biphasic materials and were first formulated in 1986. These are made in such a manner that their constituents do not become separated owing to homogeneous and intimate submicron level mixing. These biphasic ceramic blends exist as a mixture of HA (with more stability) and β-TCP (with more solubility) [95]. Their bioactivity, bioresorbability, and osteoinductivity have been under evaluation for use in orthopedics and dentistry [96–98]. Such biphasic mixtures have shown promising results for the osteogenic differentiation of MSCs, increased cell adhesion, and enhanced mechanical properties [99, 100]. A biodegradable blend of β-TCP matrix and HA nanofibers was designed by Ramay *et al.* They constructed these microporous nanocomposite scaffolds using gel-polymer methods. Apart from cues to enhance cell growth and neovascularization, they also possessed enhanced mechanical properties which made them apt for use in load-bearing applications for bone tissue engineering [101].

Whitlockite

Whitlockite (WH) is a CaP containing ceramic with Mg content and is represented as $Ca_9Mg(HPO_4)(PO_4)_6$. After HA, it is the second highest concentration of mineral in the human skeleton, with a Ca/P ratio of 1.43 and crystal structure with a rhombohedral space group [102, 103]. It has a negatively charged surface and shows good stability at low pH [104, 105]. It has higher solubility under physiological conditions which leads to the continuous release of ions. As compared to HA, it has a higher compressive strength but because of its difficult synthesis, its research and development has not progressed well [106, 107].

WH is formed in the presence of Mg ions and calcium phosphate under acidic solutions even at low temperatures. The same phenomenon happens in vivo when old bone is resorbed by osteoclasts under acidic conditions [108–110]. Jang *et al.* designed an easy process for the formation of a stable, high-purity WH without toxic by-products. WH has shown to induce elevated expression levels of osteogenic genes than other ceramics [105, 107]. WH with a composite hydrogel promoted better growth and osteogenic activity than HA as shown in a bone regeneration study in rat calvaria [106]. The property of WH to continuously release Mg and PO_4^{-3} seems to be the causal mechanism for osteogenic differentiation and bone growth as the Mg ions also decrease osteoclast activity [111].

Another form of calcium phosphate (viz. octacalcium phosphate (OCP)) that has a natural presence in human teeth is also considered important during the early phases of bone mineralization [112–115]. It has a very high biocompatibility and has been widely researched for bone implantation and coatings [116–120]. Even the amorphous form of calcium phosphate (ACP) transiently releases calcium and phosphate ions locally and has been utilized in several clinical applications [121–123].

The utilization of WH and HAP at a bone-like ratio (1:3) has recently been shown to exhibit remarkable osteogenic activity [124]. Such findings pave a hopeful way that understanding the generation and function of WH within local bone can guide in designing better calcium phosphate based materials. Because of certain mechanical disadvantages in clinical applications, research is being focused on employing calcium phosphate in combination with other materials [125].

In one comparative study with commercially available biomaterials, Ishikawa *et al.* studied three artificial bone substitutes with different ceramic constitutions viz. HA (Neobone®), carbonate apatite (CO_3Ap; Cytrans®), and β-tricalcium phosphate (β-TCP; Cerasorb®) [126]. The difference in their ultrastructure can be seen in Fig. 7. Their physicochemical responses along with their tissue responses to

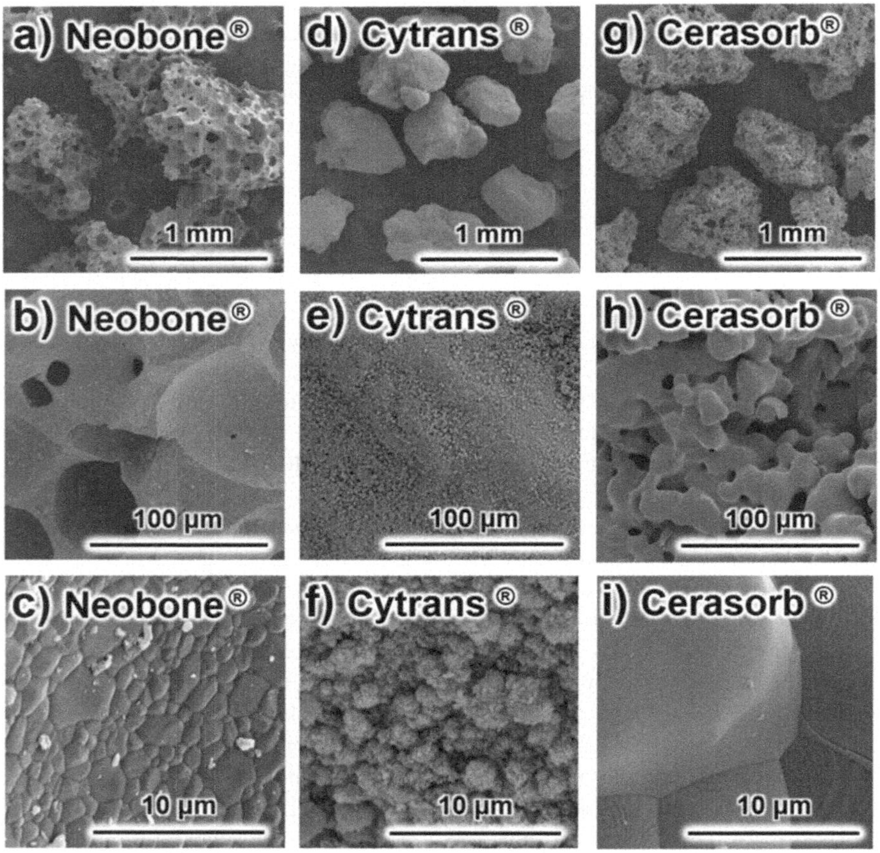

Fig. 7 Typical scanning electron microscopy images of Neobone® (HAp) (**a–c**), Cytrans® (CO_3Ap) (**d–f**), and Cerasorb® (β-TCP) (**g–i**). Both Neobone® and Cerasorb® displayed a porous structure, where Cerasorb® had much smaller pores compared to Neobone®. By contrast, Cytrans® was dense, and had no pores in the granules. Higher magnification showed that both Neobone® and Cerasorb® had smooth surfaces typical for sintering [126]

bone were studied in hybrid dogs. As per SEM investigations, Neobone® and Cerasorb® were found to be porous, whereas Cytrans® was relatively much dense. As the fabrication of Cytrans® bone substitutes was done through a dissolution–precipitation method, it had a greater specific surface area with smaller crystals when compared to the other two which were fabricated by sintering. After 12 weeks of implantation, CO_3Ap (Cytrans®) stimulated a higher new bone volume than HA and β-TCP. A bone-like composition and a higher specific surface area of CO_3Ap (Cytrans®) may have caused a better osteogenesis response.

Bioactive Glasses

Bioactive glasses (BGs) are surface active glass ceramics. "Bioactive" refers to "a material that elicits a specific biological response at the material surface which results in the formation of a bond between the tissues and the materials" [127]. They are one of the most favorable bioactive scaffold materials for bone regeneration. They have a specific advantage in bone tissue engineering applications due to their capacity to form HA, along with their established osteoconductivity and their strong bonding ability with bone and soft tissues [128]. Since its invention as silicate glass (45S5) by Hench *et al.* in 1971, it has been a favorite material for exploration in clinical applications. 45S5 Bioglass® has a specific composition of 45% silica (SiO_2), 24.5% calcium oxide (CaO), 24.5% sodium oxide (Na_2O), and 6% phosphorous pentoxide (P_2O_5). The University of Florida has the intellectual property rights of this name viz. Bioglass® and it denotes the original 45S5 composition. All the other glasses are termed as bioactive glasses [129]. There are many methods for fabricating BG scaffolds (Fig. 8) and the method directly influences their structural aspects such as porosity and thus functionalities [130]. BGs can be either silicate-, borate-, or phosphate-based systems but silicate-based systems have been most studied and also successfully translated to many commercially available products.

In a study by Bi *et al.*, the influence of the microstructure of BGs on bone regeneration capacity was examined. Three variants of microstructures of borate bioactive glass (1393B3) scaffolds (trabecular, fibrous, and oriented) were fabricated. They were then tested for their bone regeneration potential in a calvarial defect model of rats. The extent of formed and mineralized new bone and angiogenesis was assessed 12 weeks post-implantation by histomorphometry and SEM. HA formed at the site of scaffold at the end of study period had a ratio of Ca to P that matched with that of bone. Out of the three variants, the trabecular microstructure seemed to be the most promising as it showed better neo-osteogenesis, higher osteoinductive ability, and greater blood vessel infiltration than the other two microstructures. Thus, out of the three variants, trabecular microstructure may have the highest potential for bone regeneration using synthetic implants [128].

BGs can also be used to functionalize other scaffolds and improve their physicochemical and osteoinductive properties. Moses *et al.*, investigated copper doped BGs for functionalizing two silk scaffolds (*Bombyx mori* and *Antheraea assama*). The sol-gel coating of BG efficiently functionalized silk microfibers. An even,

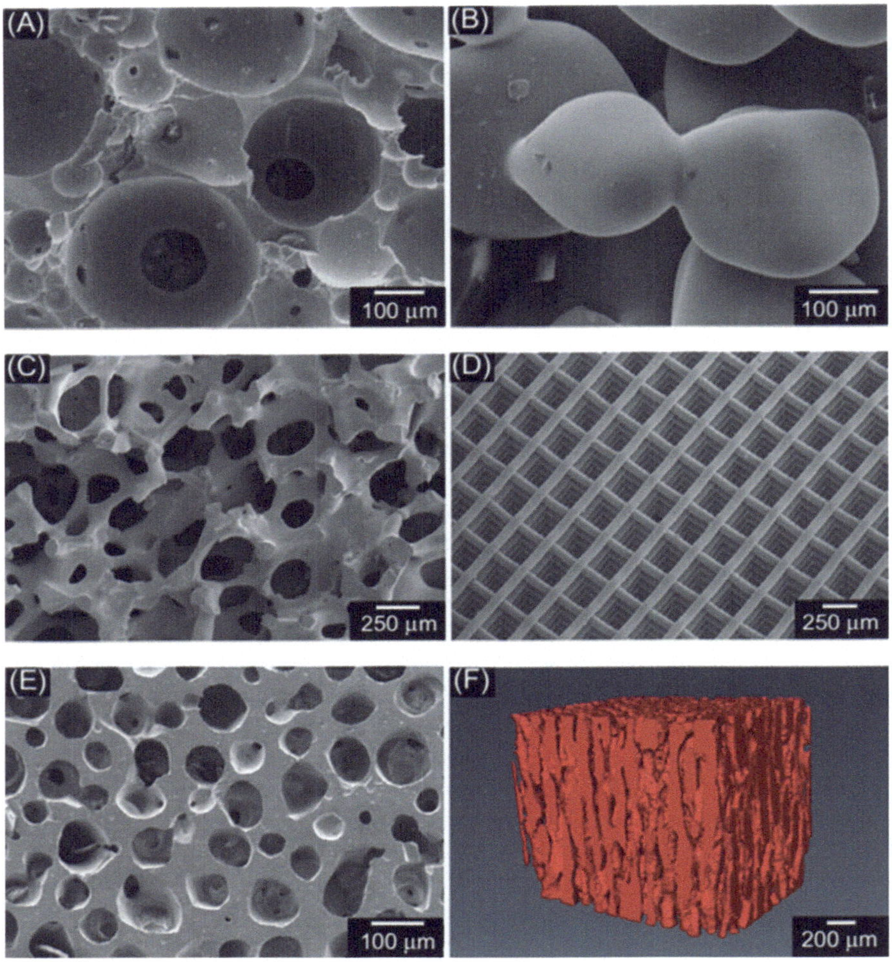

Fig. 8 Microstructures of bioactive glass scaffolds fabricated by different methods: (**a**) sol-gel; (**b**) thermal bonding (sintering) of particles (microspheres); (**c**) polymer foam replication technique to create "trabecular" scaffold; (**d**) grid-like microstructure prepared by robocasting; (**e**) oriented microstructure prepared by unidirectional freezing of suspensions (plane perpendicular to the orientation direction); and (**f**) micro-computed tomography image of the oriented scaffolds in (**e**) [130]

non-crystalline and nanoparticulate coating was formed over the silk microfibers. These functionalized silk microfibers had better physical characteristics like wettability, bulk density, stiffness, etc. These reinforced composite matrices were also better in the terms of their surface area along with open pores and a biomimetic microenvironment that allowed for cellular infiltration. The composite matrices got integrated and fully resorbed with new bone formation in rabbit femur defects. Such a kind of composite matrix opens interesting futuristic avenues to heal complex

Table 3 Commercial products made of bioactive glasses [132]

Glass name	Commercial products and notes	Manufacturer
45S5	NovaBone Putty and NovaBone Dental Putty: Moldable glass-based paste injectable into the bone/dental defect site by a syringe	NovaBone (USA); o-Sci Corp. (USA); Biomet 3i (USA); BioMin Technologies (UK)
	NovaBone particulate (90-710 μ) for orthopedics	
	NovaBone Morsels (porous granulate)	
	PerioGlas particulate (90-710 μ) for repairing jaw bone defects	
	NovaBone porous blocks	
	Biogran particulate (300-360 μ)	
	NovaMin particulate and BioMin for tooth enamel remineralization and dentinal hypersensitivity prevention	
45S5+ polyethylene	Porous composite orbital implant (Medpor-Plus)	Porex Surgical Inc (USA)
S53P4	Particulate and custom-made monolithic plates (repair of orbital floor fractures) (BoneAlive)	Abmin Technologies Ltd/ Vivoxid (Finland); Mo-Sci Corp. (USA)
13-93	Cast monolithic implants, quenched frit, rods, fibers, disks, granules and micro-sized powders for bone repair	Mo-Sci Corp. (USA)
S55F5	Cast monolithic implants, quenched frit, rods, fibers, disks, and micro-sized powders	Mo-Sci Corp. (USA)
13-93B3	Cotton-candy fibrous scaffolds (DermalFuse/ Mirragen) for wound healing	Mo-Sci Corp. (USA), Avalon Medical (USA)
70S30C	Sol-Gel powders for bone repair	MedCell (UK)
Y O-Al O -SiO glass	20- or 30 μ microspheres (TheraSphere) for the treatment of liver cancer by embolization combined with local radiotherapy	BTG International Ltd (USA)

bone defects [131]. The market of bioactive glasses is constantly emerging and some of the commercially available products are presented in Table 3 [132].

Studies show that the regeneration by BG scaffolds is dependent on many aspects such as its constitution, method of fabrication, ultrastructure, and pore characteristics. Even other factors like pretreatment of the scaffold and the presence of growth factors play a major role. Lastly, the animal model used for the experiment, size of the defect, and implantation time may alter the scaffold behavior and results [129, 133–135].

Natural Ceramic (Nacre)

Over the past two decades, there have been major efforts to design bone substitutes, which can be employed for easy and efficient clinical practice. Along with the quest for synthetic materials, the aspect of biomimicking and generating bone-like com-

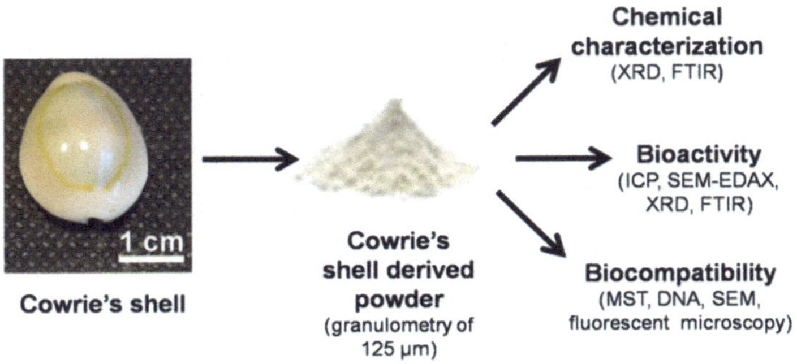

Fig. 9 Scheme of the experiment with cowrie shell derived powder [137]

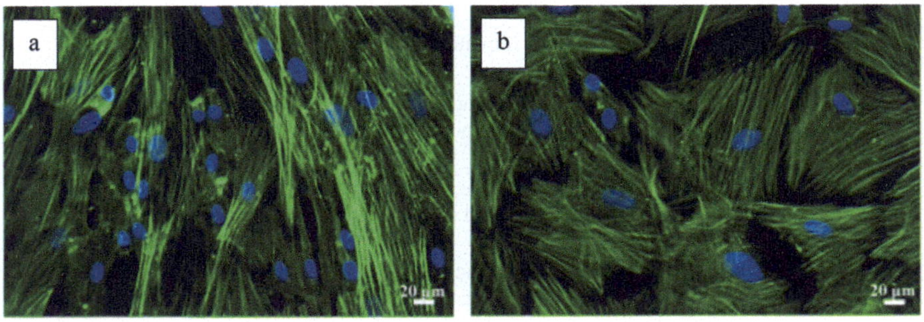

Fig. 10 Fluorescent microscopy views of human umbilical cord derived stem cells-cytoskeleton labeled cells after 7 days in presence of (**a**) CSDP and (**b**) NDP, respectively [137]

posites, and natural materials are also under constant exploration. Nacre is also called the mother-of-pearl and is one such material that has osteoinductive, osteoconductive, biocompatible, and biodegradable properties. Its usage in the human race as a medical entity dates back to historic times. Nacre is a composite material composed of an organic matrix and calcium carbonate produced by molluscs as the shell. In vivo and in vitro studies confirm its outstanding properties as a potential multipurpose biomaterial as a bone graft substitute [136].

Akilal et al. experimented on Cowrie's shell derived powder (CSDP) and Nacre derived powder (NDP) for bone regeneration (Fig. 9). Structural and physicochemical investigation of CSDP revealed its brick and mortar ultrastructure composed of aragonite crystals. Upon soaking at 37 °C in simulated body fluid (SBF) for a week, these crystals transformed to poorly crystalline B-type carbonate apatite, reflecting bioactive features. Upon culturing stem cells on both substrates, it was found that CSDP supported cell proliferation more than nacre derived powder (NDP) over the study time period.

The morphology of stem cells also seemed flattened over CSDP, indicating superior biocompatibility. The cytoskeletal labeling showed well elongation of fibers on CSDP and relatively flattened cells over NDP after 7 days (Fig. 10). After 21 days, the cells were confluent with highly elongated actin fibers. The study suggests that these naturally derived materials offer an economic and novel hope for bone regenerative medicine [137].

Polymers

As compared to metals and ceramics, polymers offer more flexibility, resilience, low cost, low conductivity, high durability, and moldability. Like all the biomaterial classes, polymers have also evolved in their subsequent generations. The first generation of polymer biomaterials included silicone rubber, polyethylene (PE), acrylic resins, polyurethanes, polypropylene (PP), and polymethylmethacrylate (PMMA). The major issue with first-generation biomaterials was the adsorption of various unspecific proteins on their surface post-implantation. The protein alignment in various conformations led to unspecific signaling pathways in the cellular microenvironment which further resulted in fibrous tissue growth which encapsulated the entire implant [138–140].

The second generation of polymer biomaterials was marked by the evolution of resorbable biomaterials that allowed for more control over their chemical breakdown and resorption. Biodegradable polymers of synthetic and natural origin such as polyglycolide (PGA), polylactide (PLA), poly(ϵ-caprolactone) (PCL), polydioxanone (PDS), polyhydroxybutyrate (PHB), chitosan (CS), polyorthoester, poly(2-hydroxyethyl-methacrylate) (PHEMA), polyethylene glycol (PEG), hyaluronic acid and other hydrogels have been researched and developed over this period. The hydrogel structure and 3D network help to hold large volumes of water and play multiple roles when used as an orthopedic biomaterial. Hydrogel polymers have shown potential for the repair of various tissues of the musculoskeletal system [141].

The most important properties for third generation biomaterials are bioactivity and biodegradability. The surfaces of these materials are also bioactivated with specific molecules to allow cell guidance and elicit particular responses. This class of biomaterials aim to closely match the ECM milieu and function by integrating specific cues. In short, they are able to modulate the important phases of cell behavior (viz. cell adhesion, migration, proliferation, and guided differentiation to the desired lineage) [24, 142, 143].

To develop new 3D scaffolds, natural as well as synthetic polymers have been employed for musculoskeletal regenerative medicine. Natural polymers have their own advantages but synthetic biodegradable polymers have drawn special interest because their physicochemical properties can be controlled in a better manner as demonstrated by their successful use in various medical purposes. Polymers that have been studied the most for bone tissue engineering include PLA, PGA, PCL, and PHB. Natural polymers like collagen and silk have been explored for applications in ligament tissue engineering [144, 145]. Composites of PCL and hyaluronic acid have yielded promising results and potential for meniscus tissue engineering [146]. "For applications such as cartilage and intervertebral disc tissue engineering, hyaluronic acid, polyglactin collagen, fibrin, alginates, chondroitin sulphate photo-crosslinked hydrogels and glycosaminoglycans have also been explored." [147–149].

A polymeric membrane of gelatin-CS with inclusions of HA and titania nanoparticles was fabricated with UV radiation as a safe cross-linking agent. It resulted in a homogeneous material with well-distributed nanoparticles. UV cross-linking yielded better biological and mechanical properties of the composite membrane. The osteogenic potential of the gelatin-based material was established by ALP assay with mouse embryonic fibroblasts (MEF) in vitro. This composite is a good substitute to the existing guided bone regeneration membranes [150]. Babitha et al. designed gelatin composites with natural polymer viz. zein but because of its poor osteoinductivity for human periodontal ligament stem cells (hPDLSCs), they added a ceramic phase to it. This resulted in the fabrication of novel zein/gelatin/nanoHA (zein/gelatin/nHAp) nanofibrous membranes. These nanofiber membranes showed better surface wettability and induced better cell proliferation and differentiation of hPDLSCs to osteogenic lineage. It was demonstrated that these composite membranes possess superior biocompatibility and osteoinductive ability for hPDLSCs [151].

The behavior of bone mesenchymal stromal cells (BMSCs) was examined on electrospun $CaCl_2$ treated poly(l-lactide) (PLLA) and gelatin composite fibers (PG-Ca). Mineralization ability for both fibers was also assessed in α minimum essential medium (αMEM) and it was found that the PG-Ca fibers strongly induced HA formation as compared to PG fibers. Apatite depositions were found after culturing BMSCs on both kinds of fibrous mats. Osteogenic differentiation of the proliferating BMSCs was also enhanced despite the absence of extra osteoinductive factors due to the continuous consumption of ions. PG-Ca fibrous mats showed better results than the control group. In essence, it was inferred that designing scaffolds which can infuse bone supportive ions such as Ca^{+2} and PO_4^{-3} around cells can be a great approach to facilitate bone formation [152].

In an attempt to mimic bone in 3D, Anada et al. used a two-step digital light processing technique and fabricated a spheroid structure with octacalcium phosphate (OCP), spheroids of human umbilical vein endothelial cells (HUVEC), and a gelatin methacrylate (GelMA) hydrogel. The whole construct was designed to mimic the inner architecture of bone wherein the peripheral hydrogel with OCP mimicked the cortical shell and the inner bone marrow-like space was created using HUVEC spheroids embedded in GelMA (Fig. 11). The results showed that evenly embedded OCP stimulated the osteoblastic differentiation of MSCs. The capillary-like structure formation from the spheroids was regulated by the concentration of GelMA. This biomimetic construct with a cell-loaded hydrogel base and dual ring structure seems to be an encouraging model for bone tissue engineering purposes [153].

Ingavle et al. studied the bone healing capacity of two natural polymer, i.e., alginate and hyaluronate, based biomineralized microspheres with entrapped MSCs. The polymers were altered with the adhesive tripeptide arginine-glycine-aspartic acid (RGD). Both the in vitro and in vivo studies (in Swiss alpine sheep) showed great results in terms of osteogenesis. When assessed against the untreated/acellular

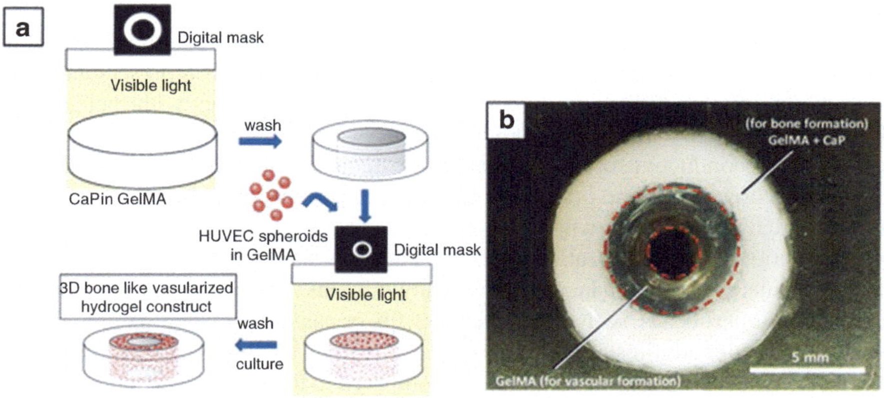

Fig. 11 (**a**) Diagrammatic representation of fabrication process for 3D hydrogel constructs. (**b**) A photograph of 3D hydrogel constructs for vascular and bone formation. Bar = 5 mm [153]

gels, the modified polymers had a substantial positive effect on parameters like blood vessel density and bone formation. These results indicated that hydrogels with stem cells can prove useful for bone regeneration in large animal bone defects [154].

Murahashi et al. loaded multi-layered PLLA nanosheets with recombinant human fibroblast growth factor-2 (rhFGF-2) for bone regeneration. They examined its effect in critical-sized mouse femoral defects and it was found that these nanosheets acted as a modified sustained-release carrier and efficiently induced bone regeneration. The nanosheets also induced FGFR1 gene activation and subsequent osteoblast differentiation [155].

Hydrogels are one of the very important classes of polymers that have immense potential as a biomaterial. Their optimization and design strategies must ensure features such as: (1) Non-immunogenicity and non-cytotoxicity, (2) osteoinductivity, osteoconductivity, osteogenicity, as well as osteocompatibility, (3) mimicry of the natural ECM to the extent possible, (4) degradability by endogenous enzymes or hydrolysis to synchronize with neo-osteogenesis, (5) mechanical and structural strength for treating load-bearing defects, (6) correct pore dimensions with interconnected porosity, and (7) injectability for patient compliance to enhance administration ease [156].

Due to their weaker mechanical properties, they are better suited for non-load-bearing applications. Various physical and chemical methods for attaching specific functional groups, hydrogen bonding, electrostatic interaction of the natural materials and inclusion of other phases, etc. are used to strengthen their mechanical traits and improve their bioactivity in order to augment their clinical value [144]. A recent development in the field of hydrogels is called nanogels which is a class of materials with spherical nanoparticles formed by cross-linking 3D polymer networks. They tend to expand in fluids by physical or chemical means. They show a range of hydrogel features like a great biocompatibility and tunable mechanical properties along with desirable dimensions of nanoparticles. Nanogels have great scope in the bone regeneration domain [157]. One more system of polymer hydrogels that can be used for bone regeneration includes microbeads. Microbeads represent an ideal system for entrapping live stem cells and therapeutic molecules with their cross-linking mechanisms. Still more studies are underway to develop perfectly biocompatible, osteoconductive, osteoinductive, and osteogenic microbead models [158].

Along with nanogel and microbead forms, a recent form of hydrogels that can be utilized are hydrogel fibers which may be responsive owing to their greater surface-volume ratio and immobilization capacity. They are made up of fibrous structures with a diameter of a few nanometers to many microns. The advantage of hydrogel fibers over microbeads is that the hydrogel fibers can be arranged axially in the syringe for better administration to the defect site. They can also be expected to stay at the implant site for an extended time as compared to the rounded structures of the microbeads [159].

Onat *et al.* reported the osteoinductivity of a polymer in their latest research. Most of the time, biodegradable polymers have been largely employed for complexing with and delivering osteoinductive moieties, but not by themselves as osteoinductive agents. This work reported the osteoinductive ability of poly(4-hydroxy-L-proline ester) (PHPE), which is a biodegradable cationic polymer with a specific property for cell penetration. The specific reactions cause PHPE to degrade in vivo and convert to trans 4-hydroxy-L-proline (trans-Hyp), a non-coded amino acid that plays crucial role in the formation and stability of collagen fibrils [160]. It was derived from this study that this polymer is nontoxic for the cells; cell exposure to PHPE induces higher COL1A1 expression leading to more synthesis of collagen, thus secretion in osteoblast-like cells. It also induces ECM mineralization in primary osteoblasts which further promotes ECM mineralization and bone tissue regeneration.

The effect of polymer molecular weight was demonstrated in one study by Davide et al. In bone tissue regeneration, polymers have mostly been used as various composites such as calcium-phosphate-based composites. They utilize the best features of both material classes and help to facilitate osteoinduction. This study examined the potential effects of the polymer phase as its molecular weight regulates fluid uptake, degradation dynamics, and the onset of surface reactions. They developed composites by the extrusion of two different molecular weight L/D, L-lactide copolymers with calcium phosphate apatite, namely M38 and M60. The M38 copolymer permitted higher fluid uptake leading to a better adsorptive ability for proteins in vitro. Figure 12 shows a higher amount of bone generated in the composite formed with lower molecular weight. The underlying reason may be its faster degradation and thus exposure of a rougher surface to trigger stem cells to differentiate osteogenically and cause bone formation [161].

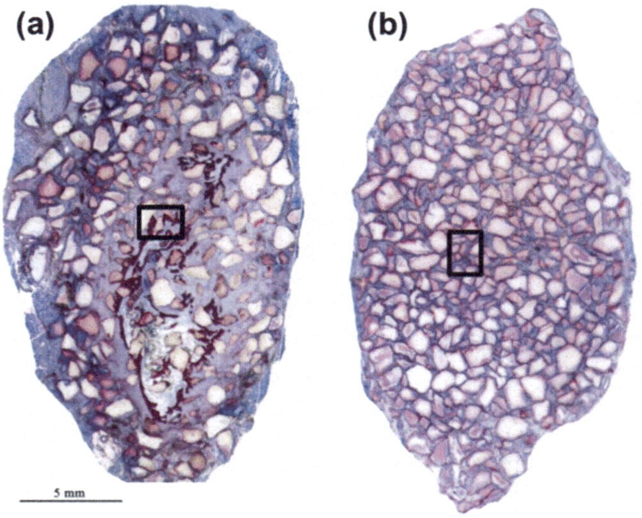

Fig. 12 Low magnification scans of composite explants. (**a**) Heterotopic bone formed in all 5 implants with M38. Bone is indicated by the dark pink areas among the M38 granules. (**b**) None of the M60 composites induced bone formation as no dark areas can be observed [161]

Composites

Significant efforts have been made to develop favorable bone replacement biomaterials for repairing larger bone defects. Still, the success in achieving perfect bone biomaterials with ideal physicochemical and osteoinductive properties seems far and that is where the role of composites comes in which can blend the best properties for various classes of materials. As the bone is also an organic–inorganic composite, this methodology can prove to be the best road to closely simulate bone. Some of the composite-based commercially available graft materials are listed in Table 4 [162].

Alidadi et al. evaluated the efficacy of xenogeneic demineralized bone matrix (DBM), with two polymers viz. CS, and PMMA for repairing critical-sized radial bone defects in rats. As compared to DBM, CS and PMMA were found to be slowly degrading, non-compatible polymers with slow biodegradation rates that impeded bone regeneration, though CS is not osteoconductive or osteoinductive alone [163].

Lai et al. fabricated a novel porous composite with a powdered form of Mg, poly (lactide-co-glycolide) (PLGA), and β-tricalcium phosphate (β-TCP). This PLGA/TCP/Mg (PTM) scaffold was evaluated against a PLGA/TCP composite (PT) for its osteogenic and angiogenic properties. New vessels and new bone formation were seen in the PTM group and the results were better than the PT group. These findings divulged the osteogenic and angiogenic abilities of the PTM scaffold due to the presence of three chief orthopedic biomaterials and their concerted synergy which

Table 4 Some commercially available bone graft materials [162]

Class	Commercial product	Composition	Claimed mechanism of action	Formulations
Allograft based	DBX® (Synthes)	DBM with sodium hyaluronate carrier	Osteoinduction, osteoconduction	Putty, paste, mix, injectable, strips
	Grafton® (Medtronic)	DBM fibers w/ w/o cancellous chips	Osteoinduction, osteoconduction, osteogenesis	Putty, strips, crunch gel, paste
	Orthoblast® (Citagenix)	DBM and cancellous chips in reverse phase medium	Osteoinduction, osteoconduction	Putty, paste, injectable paste
Ceramic based	Norian SRS® (Synthes)	Calcium phosphate	Osteoinduction, bioresorbable	
	Vitoss® (Stryker)	Beta-tricalcium phosphate and bioactive glass	Osteoinduction with bone marrow aspirate, bioresorbable	Putty, strips, injectables, morsels, shapes
	ProOsteon® (Biomet)	Corraline hydroxyapatite and calcium carbonate	Osteoconduction, bioresorbable	Injectable granules or block
	BonePlast® (Biomet)	Calcium sulfate	Osteoconduction, bioresorbable	Paste
	Osteoset® (Wright)	Calcium sulfate	Osteoconduction, bioresorbable	Pellets
Factor based	Infuse® (Medtronic)	rhBMP-2 on bovine collagen carrier	Osteoinduction	Bovine collagen carrier
	OP-1® (Stryker)	rhBMP-7 with type I collagen carrier	Osteoinduction	Putty (with carboxymethylcellulose addition), Paste
Polymer based	Healos® (Depuy)	Crosslinked collagen and HA	Osteoconduction, osteogenic with bone marrow	Strips
	Cortoss® (Stryker)	Non-resorbable polymer resin with ceramic particles	Osteoconduction	Injectable paste

is capable of clinical translation [164]. An experiment to compare the biological qualities and bone regenerative capacity of an uncoated porous PCL scaffold with $MgCO_3$-doped HA particles (PCL_MgCHA) and the same scaffold biomimetically coated with apatite-like crystals (PCL_MgCHAB) was carried out [165]. Both the scaffolds were found to be non-cytotoxic. PCL_MgCHAB displayed higher levels of ALP expression and collagen production. New bone trabeculae growth in PCL_MgCHAB was significantly higher as compared to PCL_MgCHA at the 4th and 12th weeks after implantation.

In a study, the osteogenicity and antimicrobial activity of HA and silver, respectively, were examined in a composite blend of HA/Ag. As the elevated concentrations of silver are known to trigger cytotoxicity, this study aimed to use silver nanoparticles so that even lower concentrations of silver can exhibit improved antimicrobial and anti-inflammatory effects. These particulate HA/Ag nanocomposites showed apatite formation in SBF and parameters like cell viability were found to be unaltered. This HA/Ag nanocomposite expressed bioactivity, and higher antimicrobial activity against *Escherichia coli, Staphylococcus aureus,* and *Candida albicans* even at lower concentrations of silver [166].

In one of the latest developments, a novel composite was designed with unique combinatorial chemistry. Strontium, which is known for bone strength enhancement and graphene which is known for its massive surface area, great specific conductance, high tensile modulus, and simple functional groups were amalgamated with the competence of nanotechnology. Chen *et al.* developed these innovative strontium-graphene oxide (Sr-GO) nanocomposites to release Sr ions in a sustained manner and were used as a reinforcement in collagen scaffolds. The Sr-GO-Col scaffold demonstrated better physicochemical and mechanical properties than the unmodified Col scaffolds. They also facilitated cell adhesion and osteogenic differentiation due to the MAPK signalling pathway stimulation. These Sr-GO-Col composites showed excellent bone regeneration, defect repairing, and angiogenesis results at 12 weeks in a critical-size calvarial bone defects in rats. The residual Sr-GO nanoparticles were found to be phagocytosed and degraded [167].

Apart from the above-mentioned materials, there have been many new explorations, such as with graphene for bone regenerative medicine *per se* with stem cells and with other molecules like PCL, alginate, gelatin, fibroin, etc. to produce composites and 3D scaffolds [168–171]. In one of the studies to combine the best of synthetic and naturally derived materials, a porous composite was prepared by PLA and DBM. This was implanted in a rabbit radius segmental bone defect to examine repair. The composite was compared with only PLA and only a bone autograft. The results from X-ray and histology confirmed that the effect of composite materials on bone repair was substantially higher than any of them acting alone [172].

A 3D scaffold with the combination of synthetic polymers and nanoceramics was explored for bone tissue engineering by Carrow et al. using additive manufacturing (AM). They reported the synthesis of a bioactive 3D scaffold nanocomposite from a poly(ethylene oxide terephthalate) (PEOT)/poly(butylene terephthalate) (PBT) (PEOT/PBT) copolymer and 2D nanosilicates. This particular combination was chosen with PEOT/PBT as they boost calcification and increase bone bonding ability, while 2D nanosilicates are known to lead hMSCS to an osteogenic lineage even without relying on osteoinductive agents. The stability and the bioactive properties of the composites were found to be increased. The osteo-related proteins were significantly upregulated along with the production of a mineralized matrix [173].

Role of Topography in Orthopedic Biomaterials

Bone as a Nanocomposite

The unique properties of natural bone are not derived just from its chemistry but also from its microarchitecture that span from nanoscale to macroscopic dimensions, with precise and meticulously engineered interfaces. The nanostructural composition of HA nanocrystals and collagen fibrils is what makes bone exceptional, a feature that could not be precisely reproduced artificially till date. In the stem cell microenvironment also, communications with ECM components exert indirect mechanical forces which influence stem cell behavior. The ECM has a unique composition with various types of polymers with different hierarchical dimensions such as collagen strands at the nanometer level to the micron range [174]. These symphonized spatiotemporal communications between the cells and their surroundings maintain and control their behaviors in the long term.

Cells keep receiving mechanotransducive cues in their natural microenvironment due to nanotopographical structures within the ECM around them. These cues influence their local migration, cell polarization, and other functions. This is the underlying reason how the nanoscale and microscale topographical features of a biomaterial can act as signalling modalities and can control cellular functions [175, 176].

Nanomaterials for Bone Regeneration and Repair

Nanomaterials and nanocomposites have been emerging as potential models for recreating the structure and function of bone and ECM. They serve as a structural framework for cells along with the regulation of key cell phases viz. growth, proliferation, differentiation, and migration that result in organized tissues. There can be various kinds of nanotopographical features on the biomaterial surface that influence cellular processes, some of which are represented in Fig. 13. They possess biomimicking and suitable traits such as high roughness and surface area that leads to higher protein adsorption than traditional biomaterials. Interactions at the interface of cells and a biomaterial are supposed to be moderated by integrin-led pathways that affect cell behavior [177].

Many research groups have been constantly striving to recreate the bone's mechanical properties (stiffness, strength, and toughness) as tissue engineered constructs through the inclusion of nanostructures in the base matrices to mimic bone's natural composition [178, 179].

A precise microscale shape pattern has been shown to better regulate the osteogenesis process and nodule deposition as it directly influences cellular differentiation. The periodicity and dimensions of nanotopographical features have been

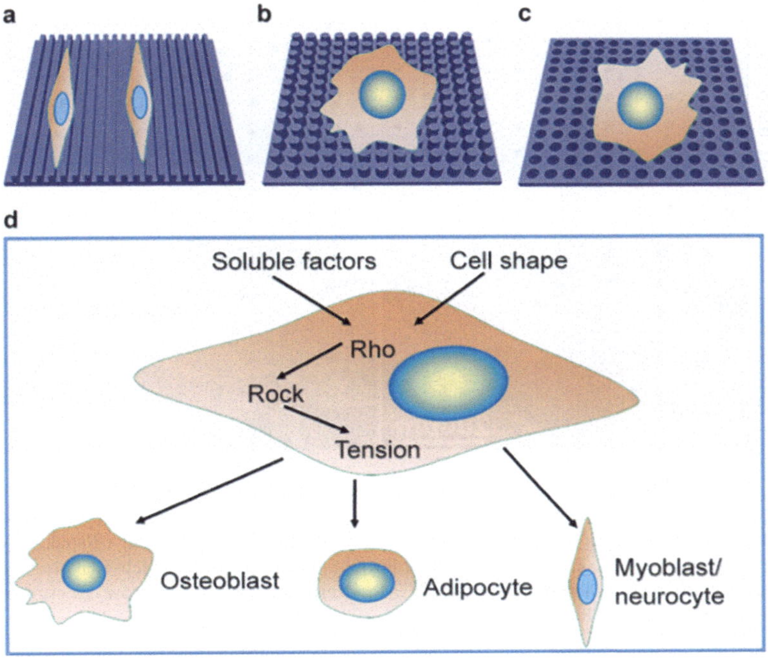

Fig. 13 Schematic depictions of representative nanotopography geometries. Three basic nanotopography geometries include nanogrooves (**a**), nanopost array (**b**), and nanopit array (**c**). The speculative pathways (**d**) for cell-shape-directed osteogenic and adipogenic differentiations of MSCs were examined in growth medium. *RhoA* Ras homolog gene family member A, *ROCK* Rho-associated protein kinase [185]

shown to affect cellular differentiation process in many studies [176, 180, 181]. The topographical cues also affect events like protein adsorption and apatite deposition, which subsequently influence bone-mineral deposition [182, 183].

The introduction of advanced small-scale technologies has facilitated the design and development of methods that allows us to closely fathom stem cell behavior and biomechanics. Even the spatial arrangement of nanosized features play a crucial role in determining cell behavior and differentiation. Irregularly placed nanopits were found to be more effective in eliciting hMSC differentiation as demonstrated by higher osteopontin expression (Fig. 14) [184]. Moreover, biologically inspired substrates with a tuned ultrastructure are evolving for better understanding and control of stem cell behavior [185]. In the advent of the third generation of biomaterials, the magnitude of importance of their physical properties has taken central stage along with their chemical composition.

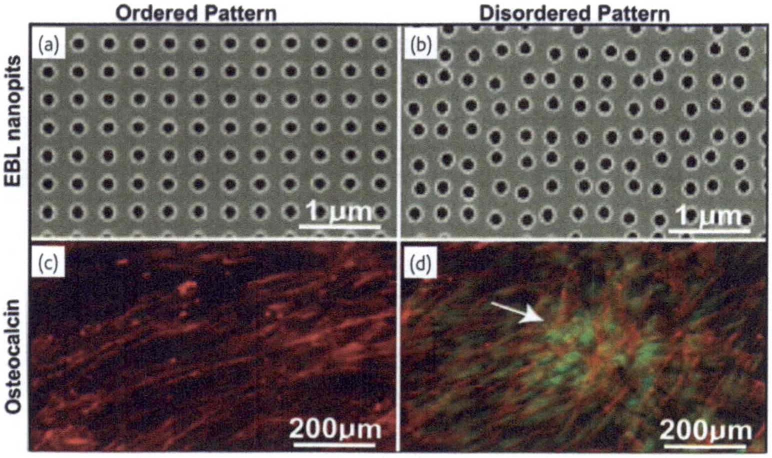

Fig. 14 Exploring the effect of spatially different nanotopographies on cell differentiation. (**a**, **b**) nanotopographies fabricated by electron beam lithography (EBL). The pits (120 nm diameter, 100 nm depth) (**a**) in a square arrangement and (**b**) with increasing disorder (displaced square ±50 nm from true center). The nanoscale disorder stimulates human mesenchymal stem cells to increase the expression of the bone-specific ECM protein osteopontin (**d**, arrow) as compared to the ordered structure (**c**) [184]

Conclusion

Apart from natural grafts, metals, ceramics, and polymers are the major classes which have been under investigation as budding candidates for orthopedic regenerative medicine. Each of them has advocated for their unique characteristics but the far superior and unique traits of bone necessitate better combinatorial chemistry from the various material genres. It's equally important that the material design schemes pay a high level of attention to minimize risks such as those caused by their degradation products to avoid various unforeseen hazards. As the in vitro results show inconsistency when tested in vivo, it's crucial to understand the host responses after implantation (Table 5).

The natural composition and strength of bone is so unique that despite innumerable scientific attempts, identical ultrastructure and mechanical properties could not be achieved for their replacement and repair. Moreover, bone is constantly remodelled by daily loaded actions, both in static and dynamic conditions and this makes it even more challenging to engineer bone-tissue similes.

The third and upcoming generation of bone replacement materials includes not only biocompatibility, safety, and inertness, but also the capacity to induce or stimulate healing. The ongoing research is incorporating an understanding of the molecular level mechanisms and adding further functionalities and activation strategies to enhance the active participation of biomaterials in osteogenesis and new bone for-

Table 5 Summary of various orthopedic biomaterials and their modification strategies

Base material	Modifying material	Key findings	References
Magnesium	Ca and Zn	Calcification matrix formation and complete bone healing	[26]
Magnesium	HA	Enhanced cell viability	[27]
Steel	HA	Superior antibacterial properties, bioactivity and adhesive strength	[39]
Titanium	HA and collagen	Higher osteoblast proliferation	[54]
Titanium	Alkaline phosphatase	Faster mineralization	[53]
BetaTCP	HA fibers	Enhanced cell growth, better mechanical properties	[101]
Bioglass	Silk microfibers	Aided in new bone formation	[131]
Gelatin	Zein and HA	Better proliferation of stem cells	[150]
Alginate-hyaluronate	Tripeptide RGD	Enhanced angiogenesis and osteogenesis	[153]
PLLA	Recombinant human fibroblast growth factor	Osteoinduction and osteoblast differentiation	[154]
HA	Silver	Better bioactivity and antimicrobial potential	[164]
Graphene oxide	Strontium	Bone regeneration, angiogenesis	[165]

mation. These include major considerations for macro, micro, and nanoscale architectures and geometries, surface topography, biomimicking coatings, compositions, microenvironment tailoring, hierarchically organized structures, enzyme or growth factor release patterns, etc. such that they can directly regulate cellular responses toward favorable outcomes.

The futuristic ideal and smart material for bone regenerative applications will not only need to be osteoinductive and osteoconductive but also "osteo-sensitive" to adapt as per the changing mechanical and biophysical needs. Along with the quest to invent new materials and relevant compounds, parallel learning from nature and integrating both of these approaches seems like the most plausible way to come up with closest possible bone composites and fast translation to the clinics.

References

1. Laurencin CT, Khan Y, Kofron M et al (2006) THE ABJS NICOLAS ANDRY AWARD: tissue engineering of bone and ligament. Clin Orthop Relat Res. 447:221–236. https://doi.org/10.1097/01.blo.0000194677.02506.45
2. El-Ghannam A (2005) Bone reconstruction: from bioceramics to tissue engineering. Expert Rev Med Devices. 2(1):87–101. https://doi.org/10.1586/17434440.2.1.87
3. Ozdemir T, Higgins AM, Brown JL (2013) Osteoinductive biomaterial geometries for bone regenerative engineering. Curr Pharm Des 19(19):3446–3455. http://www.ncbi.nlm.nih.gov/pubmed/23432675. Accessed 11 Apr 2019
4. Urist MR (1965) Bone: formation by autoinduction. Science (80-) 150(3698):893 LP–893899. http://science.sciencemag.org/content/150/3698/893.abstract

5. Wilson-Hench J (1987) Osteoinduction. Prog Biomed Eng 4:29
6. Miron RJ, Zhang YF (2012) Osteoinduction: a review of old concepts with new standrads. J Dent Res. 91(8):736–744. https://doi.org/10.1177/0022034511435260
7. Habibovic P, de Groot K (2007) Osteoinductive biomaterials—properties and relevance in bone repair. J Tissue Eng Regen Med. 1(1):25–32. https://doi.org/10.1002/term.5
8. Ripamonti U. The morphogenesis of bone in replicas of porous hydroxyapatite obtained from conversion of calcium carbonate exoskeletons of coral. 1991. http://wiredspace.wits.ac.za/handle/10539/19185. Accessed 27 Feb 2019.
9. Ripamonti U (1996) Osteoinduction in porous hydroxyapatite implanted in heterotopic sites of different animal models. Biomaterials. 17(1):31–35. https://doi.org/10.1016/0142-9612(96)80752-6
10. Yuan H, de Bruijn JD, Zhang X, van Blitterswijk CA, de Groot K (2001) Use of an osteoinductive biomaterial as a bone morphogenetic protein carrier. J Mater Sci Mater Med. 12(9):761–766. https://doi.org/10.1023/A:1013957431372
11. Reddi AH (1981) Cell biology and biochemistry of endochondral bone development. Coll Relat Res. 1(2):209–226. https://doi.org/10.1016/S0174-173X(81)80021-0
12. Yuan H, van den Doel M, Li S, van Blitterswijk CA, de Groot K, de Bruijn JD (2002) A comparison of the osteoinductive potential of two calcium phosphate ceramics implanted intramuscularly in goats. J Mater Sci Mater Med. 13(12):1271–1275. https://doi.org/10.1023/A:1021191432366
13. Ripamonti U (1991) The morphogenesis of bone in replicas of porous hydroxyapatite obtained from conversion of calcium carbonate exoskeletons of coral. J Bone Joint Surg Am. 73(5):692–703. http://www.ncbi.nlm.nih.gov/pubmed/1675218. Accessed 26 Feb 2019
14. Ohgushi H, Goldberg VM, Caplan AI (1989) Heterotopic osteogenesis in porous ceramics induced by marrow cells. J Orthop Res. 7(4):568–578. https://doi.org/10.1002/jor.1100070415
15. Goshima J, Goldberg VM, Caplan AI (1991) The osteogenic potential of culture-expanded rat marrow mesenchymal cells assayed in vivo in calcium phosphate ceramic blocks. Clin Orthop Relat Res. 262:298–311. http://www.ncbi.nlm.nih.gov/pubmed/1984928. Accessed 27 Feb 2019
16. Klein C, de Groot K, Chen W, Li Y, Zhang X (1994) Osseous substance formation induced in porous calcium phosphate ceramics in soft tissues. Biomaterials. 15(1):31–34. https://doi.org/10.1016/0142-9612(94)90193-7
17. Yang Z, Yuan H, Tong W, Zou P, Chen W, Zhang X (1996) Osteogenesis in extraskeletally implanted porous calcium phosphate ceramics: variability among different kinds of animals. Biomaterials. 17(22):2131–2137. http://www.ncbi.nlm.nih.gov/pubmed/8922598. Accessed 27 Feb 2019
18. Hari RA (1992) Regulation of cartilage and bone differentiation by bone morphogenetic proteins. Curr Opin Cell Biol. 4(5):850–855. https://doi.org/10.1016/0955-0674(92)90110-X
19. Wozney JM (1992) The bone morphogenetic protein family and osteogenesis. Mol Reprod Dev. 32(2):160–167. https://doi.org/10.1002/mrd.1080320212
20. Liu Y, de Groot K, Hunziker E (2005) BMP-2 liberated from biomimetic implant coatings induces and sustains direct ossification in an ectopic rat model. Bone. 36(5):745–757. https://doi.org/10.1016/j.bone.2005.02.005
21. Barradas AMC, Yuan H, Clemens A van B, Habibovic P (2011) Osteoinductive biomaterials: current knowledge of properties, experimental models and biological mechanisms. Eur Cells Mater. 21:407–429. https://doi.org/10.22203/eCM.v021a31
22. Chocholata P, Kulda V, Babuska V, Chocholata P, Kulda V, Babuska V (2019) Fabrication of scaffolds for bone-tissue regeneration. Materials (Basel). 12(4):568. https://doi.org/10.3390/ma12040568
23. Yazdimamaghani M, Razavi M, Vashaee D, Moharamzadeh K, Boccaccini AR, Tayebi L (2017) Porous magnesium-based scaffolds for tissue engineering. Mater Sci Eng C. 71:1253–1266. https://doi.org/10.1016/j.msec.2016.11.027
24. Navarro M, Michiardi A, Castaño O, Planell JA (2008) Biomaterials in orthopaedics. J R Soc Interface. 5(27):1137–1158. https://doi.org/10.1098/rsif.2008.0151

25. Kamrani S, Fleck C (2019) Biodegradable magnesium alloys as temporary orthopaedic implants: a review. BioMetals. 32(2):185–193. https://doi.org/10.1007/s10534-019-00170-y
26. Lee J-W, Han H-S, Han K-J et al (2016) Long-term clinical study and multiscale analysis of in vivo biodegradation mechanism of Mg alloy. Proc Natl Acad Sci U S A. 113(3):716–721. https://doi.org/10.1073/pnas.1518238113
27. Witte F, Feyerabend F, Maier P et al (2007) Biodegradable magnesium–hydroxyapatite metal matrix composites. Biomaterials. 28(13):2163–2174. https://doi.org/10.1016/J.BIOMATERIALS.2006.12.027
28. Courtenay J, Bryant M (2011) Bone ash replacement product is safer. Alum Times. 57
29. Charnley J (1960) Anchorage of the femoral head prosthesis to the shaft of the femur. J Bone Joint Surg Br. 42-B:28–30. http://www.ncbi.nlm.nih.gov/pubmed/13855642. Accessed 23 Apr 2019
30. Yang K, Ren Y (2010) Nickel-free austenitic stainless steels for medical applications. Sci Technol Adv Mater. 11(1):014105. https://doi.org/10.1088/1468-6996/11/1/014105
31. Younesi M, Bahrololoom ME, Fooladfar H (2010) Development of wear resistant NFSS–HA novel biocomposites and study of their tribological properties for orthopaedic applications. J Mech Behav Biomed Mater. 3(2):178–188. https://doi.org/10.1016/J.JMBBM.2009.08.003
32. Younesi M, Bahrololoom ME (2010) Formulation of the wear behaviour of nickel-free stainless-steel/hydroxyapatite bio-composites by response surface methodology. Proc Inst Mech Eng Part J J Eng Tribol. 224(11):1197–1207. https://doi.org/10.1243/13506501JET773
33. Younesi M, Bahrololoom ME, Ahmadzadeh M (2010) Prediction of wear behaviors of nickel free stainless steel–hydroxyapatite bio-composites using artificial neural network. Comput Mater Sci. 47(3):645–654. https://doi.org/10.1016/J.COMMATSCI.2009.09.019
34. Montanaro L, Cervellati M, Campoccia D, Arciola CR (2006) Promising in vitro performances of a new nickel-free stainless steel. J Mater Sci Mater Med. 17(3):267–275. https://doi.org/10.1007/s10856-006-7313-3
35. Torricelli P, Fini M, Borsari V et al (2003) Biomaterials in orthopedic surgery: effects of a nickel-reduced stainless steel on in vitro proliferation and activation of human osteoblasts. Int J Artif Organs. 26(10):952–957. https://doi.org/10.1177/039139880302601013
36. Thomann UI, Uggowitzer PJ (2000) Wear–corrosion behavior of biocompatible austenitic stainless steels. Wear. 239(1):48–58. https://doi.org/10.1016/S0043-1648(99)00372-5
37. Ren YB, Yang HJ, Yang K, Zhang BC (2007) In vitro biocompatibility of a new high nitrogen nickel free austenitic stainless steel. Key Eng Mater. 342-343:605–608. https://doi.org/10.4028/www.scientific.net/KEM.342-343.605
38. Popkov AV, Gorbach EN, Kononovich NA, Popkov DA, Tverdokhlebov SI, Shesterikov EV (2017) Bioactivity and osteointegration of hydroxyapatite-coated stainless steel and titanium wires used for intramedullary osteosynthesis. Strateg Trauma Limb Reconstr. 12(2):107–113. https://doi.org/10.1007/s11751-017-0282-x
39. Das A, Shukla M (2019) Surface morphology, bioactivity, and antibacterial studies of pulsed laser deposited hydroxyapatite coatings on Stainless Steel 254 for orthopedic implant applications. Proc Inst Mech Eng Part L J Mater Des Appl. 233(2):120–127. https://doi.org/10.1177/1464420716663029
40. Sul Y-T, Johansson CB, Petronis S et al (2002) Characteristics of the surface oxides on turned and electrochemically oxidized pure titanium implants up to dielectric breakdown:: the oxide thickness, micropore configurations, surface roughness, crystal structure and chemical composition. Biomaterials. 23(2):491–501. https://doi.org/10.1016/S0142-9612(01)00131-4
41. Raikar GN, Gregory JC, Ong JL et al (1995) Surface characterization of titanium implants. J Vac Sci Technol A Vacuum, Surfaces, Film 13(5):2633–2637. https://doi.org/10.1116/1.579462
42. de Jonge LT, Leeuwenburgh SCG, Wolke JGC, Jansen JA (2008) Organic–inorganic surface modifications for titanium implant surfaces. Pharm Res. 25(10):2357–2369. https://doi.org/10.1007/s11095-008-9617-0
43. Bierbaum S, Hempel U, Geißler U et al (2003) Modification of Ti6AL4V surfaces using collagen I, III, and fibronectin. II. Influence on osteoblast responses. J Biomed Mater Res Part A. 67A(2):431–438. https://doi.org/10.1002/jbm.a.10084

44. MacDonald DE, Markovic B, Allen M, Somasundaran P, Boskey AL (1998) Surface analysis of human plasma fibronectin adsorbed to commercially pure titanium materials. J Biomed Mater Res. 41(1):120–130. https://doi.org/10.1002/(SICI)1097-4636(199807)41:1<120::AID-JBM15>3.0.CO;2-R
45. Roessler S, Born R, Scharnweber D, Worch H, Sewing A, Dard M (2001) Biomimetic coatings functionalized with adhesion peptides for dental implants. J Mater Sci Mater Med. 12(10/12):871–877. https://doi.org/10.1023/A:1012807621414
46. van den Beucken JJJP, Vos MRJ, Thüne PC et al (2006) Fabrication, characterization, and biological assessment of multilayered DNA-coatings for biomaterial purposes. Biomaterials. 27(5):691–701. https://doi.org/10.1016/J.BIOMATERIALS.2005.06.015
47. Ferris D, Moodie G, Dimond P, Giorani CW, Ehrlich M, Valentini R (1999) RGD-coated titanium implants stimulate increased bone formation in vivo. Biomaterials. 20(23-24):2323–2331. https://doi.org/10.1016/S0142-9612(99)00161-1
48. Elmengaard B, Bechtold JE, Søballe K (2005) In vivo study of the effect of RGD treatment on bone ongrowth on press-fit titanium alloy implants. Biomaterials. 26(17):3521–3526. https://doi.org/10.1016/J.BIOMATERIALS.2004.09.039
49. Liu Y, Hunziker EB, Layrolle P, De Bruijn JD, De Groot K (2004) Bone morphogenetic protein 2 incorporated into biomimetic coatings retains its biological activity. Tissue Eng. 10(1-2):101–108. https://doi.org/10.1089/107632704322791745
50. Schmidmaier G, Wildemann B, Cromme F, Kandziora F, Haas NP, Raschke M (2002) Bone morphogenetic protein-2 coating of titanium implants increases biomechanical strength and accelerates bone remodeling in fracture treatment: a biomechanical and histological study in rats. Bone. 30(6):816–822. https://doi.org/10.1016/S8756-3282(02)00740-8
51. van den Beucken JJJP, Walboomers XF, Boerman OC et al (2006) Functionalization of multilayered DNA-coatings with bone morphogenetic protein 2. J Control Release 113(1):63–72. https://doi.org/10.1016/J.JCONREL.2006.03.016
52. van den Beucken JJJP, Walboomers XF, Leeuwenburgh SCG et al (2007) Multilayered DNA coatings: in vitro bioactivity studies and effects on osteoblast-like cell behavior. Acta Biomater. 3(4):587–596. https://doi.org/10.1016/J.ACTBIO.2006.12.007
53. De Jonge LT, Leeuwenburgh SCG, van de Beucken J, Wolke JGC, Jansen JA (2008) Electrosprayed enzyme coatings as bio-inspired alternatives to conventional bioceramic coatings for orthopedic and oral implants. Adv Funct Mater 19:755–762
54. Xia Z, Yu X, Wei M (2012) Biomimetic collagen/apatite coating formation on Ti6Al4V substrates. J Biomed Mater Res Part B Appl Biomater. 100B(3):871–881. https://doi.org/10.1002/jbm.b.31970
55. Nishida M, Nakaji-Hirabayashi T, Kitano H, Saruwatari Y, Matsuoka K (2017) Titanium alloy modified with anti-biofouling zwitterionic polymer to facilitate formation of biomineral layer. Colloids Surfaces B Biointerfaces. 152:302–310. https://doi.org/10.1016/j.colsurfb.2017.01.018
56. Yu P, Zhu X, Wang X et al (2016) Periodic nanoneedle and buffer zones constructed on a titanium surface promote osteogenic differentiation and bone calcification in vivo. Adv Healthc Mater. 5(3):364–372. https://doi.org/10.1002/adhm.201500461
57. George N, Nair AB (2018) Porous tantalum: a new biomaterial in orthopedic surgery. Fundam Biomater Met:243–268. https://doi.org/10.1016/B978-0-08-102205-4.00011-8
58. Edelmann AR, Patel D, Allen RK, Gibson CJ, Best AM, Bencharit S (2019) Retrospective analysis of porous tantalum trabecular metal–enhanced titanium dental implants. J Prosthet Dent. 121(3):404–410. https://doi.org/10.1016/j.prosdent.2018.04.022
59. Bencharit S, Byrd WC, Altarawneh S et al (2014) Development and applications of porous tantalum trabecular metal-enhanced titanium dental implants. Clin Implant Dent Relat Res. 16(6):817–826. https://doi.org/10.1111/cid.12059
60. Liu Y, Bao C, Wismeijer D, Wu G (2015) The physicochemical/biological properties of porous tantalum and the potential surface modification techniques to improve its clinical application in dental implantology. Mater Sci Eng C. 49:323–329. https://doi.org/10.1016/j.msec.2015.01.007

61. Lu M, Xu S, Lei Z-X et al (2019) Application of a novel porous tantalum implant in rabbit anterior lumbar spine fusion model. Chin Med J (Engl). 132(1):51–62. https://doi.org/10.1097/CM9.0000000000000030
62. Sagomonyants KB, Hakim-Zargar M, Jhaveri A, Aronow MS, Gronowicz G (2011) Porous tantalum stimulates the proliferation and osteogenesis of osteoblasts from elderly female patients. J Orthop Res. 29(4):609–616. https://doi.org/10.1002/jor.21251
63. Lee JW, Wen HB, Gubbi P, Romanos GE (2018) New bone formation and trabecular bone microarchitecture of highly porous tantalum compared to titanium implant threads: a pilot canine study. Clin Oral Implants Res. 29(2):164–174. https://doi.org/10.1111/clr.13074
64. Buehler WJ, Wang FE (1968) A summary of recent research on the nitinol alloys and their potential application in ocean engineering. Ocean Eng. 1(1):105–120. https://doi.org/10.1016/0029-8018(68)90019-X
65. Chu Y, Dai KR, Zhu M, Mi XJ (2000) Medical application of NiTi shape memory alloy in China. Mater Sci Forum. 327-328:55–62. https://doi.org/10.4028/www.scientific.net/MSF.327-328.55
66. Duerig TW, Pelton AR, Stöckel D (1996) The utility of superelasticity in medicine. Biomed Mater Eng. 6(4):255–266. https://doi.org/10.3233/BME-1996-6404
67. Vallet-Regí M, Salinas AJ (2019) Ceramics as bone repair materials. Bone Repair Biomater:141–178. https://doi.org/10.1016/B978-0-08-102451-5.00006-8
68. Arcos D, Izquierdo-Barba I, Vallet-Regí M (2009) Promising trends of bioceramics in the biomaterials field. J Mater Sci Mater Med. 20(2):447–455. https://doi.org/10.1007/s10856-008-3616-x
69. Yoshikawa H, Myoui A (2005) Bone tissue engineering with porous hydroxyapatite ceramics. J Artif Organs. 8(3):131–136. https://doi.org/10.1007/s10047-005-0292-1
70. Rey C (1990) Calcium phosphate biomaterials and bone mineral. Differences in composition, structures and properties. Biomaterials. 11:13–15. https://doi.org/10.1016/0142-9612(90)90045-R
71. Ramselaar MMA, Driessens FCM, Kalk W, De Wijn JR, Van Mullem PJ (1991) Biodegradation of four calcium phosphate ceramics;in vivo rates and tissue interactions. J Mater Sci Mater Med. 2(2):63–70. https://doi.org/10.1007/BF00703460
72. Rapacz-Kmita A, Ślósarczyk A, Paszkiewicz Z (2005) FTIR and XRD investigations on the thermal stability of hydroxyapatite during hot pressing and pressureless sintering processes. J Mol Struct. 744-747:653–656. https://doi.org/10.1016/J.MOLSTRUC.2004.11.070
73. Bohner M, Lemaitre J (2009) Can bioactivity be tested in vitro with SBF solution? Biomaterials. 30(12):2175–2179. https://doi.org/10.1016/J.BIOMATERIALS.2009.01.008
74. Samavedi S, Whittington AR, Goldstein AS (2013) Calcium phosphate ceramics in bone tissue engineering: a review of properties and their influence on cell behavior. Acta Biomater. 9(9):8037–8045. https://doi.org/10.1016/J.ACTBIO.2013.06.014
75. Ogata K, Imazato S, Ehara A et al (2005) Comparison of osteoblast responses to hydroxyapatite and hydroxyapatite/soluble calcium phosphate composites. J Biomed Mater Res Part A. 72A(2):127–135. https://doi.org/10.1002/jbm.a.30146
76. Douglas T, Pamula E, Hauk D et al (2009) Porous polymer/hydroxyapatite scaffolds: characterization and biocompatibility investigations. J Mater Sci Mater Med. 20(9):1909–1915. https://doi.org/10.1007/s10856-009-3756-7
77. Guo H, Su J, Wei J, Kong H, Liu C (2009) Biocompatibility and osteogenicity of degradable Ca-deficient hydroxyapatite scaffolds from calcium phosphate cement for bone tissue engineering. Acta Biomater. 5(1):268–278. https://doi.org/10.1016/J.ACTBIO.2008.07.018
78. Capilla MV, Olid MNR, Gaya MVO, Botella CR, Romera CZ (2007) Cylindrical dental implants with hydroxyapatite- and titanium plasma spray–coated surfaces: 5-year results. J Oral Implantol 33(2):59–68. https://doi.org/10.1563/1548-1336(2007)33[59:CDIWHA]2.0.CO;2
79. Zhou W, Liu Z, Xu S, Hao P, Xu F, Sun A (2011) Long-term survivability of hydroxyapatite-coated implants: a meta-analysis. Oral Surg. 4(1):2–7. https://doi.org/10.1111/j.1752-248X.2010.01112.x

80. Zhou M, Geng Y, Li S et al (2019) Nanocrystalline hydroxyapatite-based scaffold adsorbs and gives sustained release of osteoinductive growth factor and facilitates bone regeneration in mice ectopic model. J Nanomater. 2019:1–10. https://doi.org/10.1155/2019/1202159
81. Ramires PA, Wennerberg A, Johansson CB, Cosentino F, Tundo S, Milella E (2003) Biological behavior of sol-gel coated dental implants. J Mater Sci Mater Med. 14(6):539–545. https://doi.org/10.1023/A:1023412131314
82. Albrektsson T (1998) Hydroxyapatite-coated implants: a case against their use. J Oral Maxillofac Surg. 56(11):1312–1326. https://doi.org/10.1016/S0278-2391(98)90616-4
83. de Oliveira PT, Zalzal SF, Beloti MM, Rosa AL, Nanci A (2007) Enhancement ofin vitro osteogenesis on titanium by chemically produced nanotopography. J Biomed Mater Res Part A. 80A(3):554–564. https://doi.org/10.1002/jbm.a.30955
84. Göransson A, Arvidsson A, Currie F et al (2009) An in vitro comparison of possibly bioactive titanium implant surfaces. J Biomed Mater Res Part A. 88A(4):1037–1047. https://doi.org/10.1002/jbm.a.31911
85. Hwang NS, Varghese S, Lee HJ, Zhang Z, Elisseeff J (2013) Biomaterials directed in vivo osteogenic differentiation of mesenchymal cells derived from human embryonic stem cells. Tissue Eng Part A. 19(15-16):1723–1732. https://doi.org/10.1089/ten.tea.2013.0064
86. Dhivya S, Saravanan S, Sastry TP, Selvamurugan N (2015) Nanohydroxyapatite-reinforced chitosan composite hydrogel for bone tissue repair in vitro and in vivo. J Nanobiotechnology. 13(1):40. https://doi.org/10.1186/s12951-015-0099-z
87. Dickens B, Schroeder LW, Brown WE (1974) Crystallographic studies of the role of Mg as a stabilizing impurity in β-Ca3(PO4)2. The crystal structure of pure β-Ca3(PO4)2. J Solid State Chem. 10(3):232–248. https://doi.org/10.1016/0022-4596(74)90030-9
88. Mathew M, Schroeder LW, Dickens B, Brown WE (1977) The crystal structure of α-Ca3(PO4)2. Acta Crystallogr Sect B Struct Crystallogr Cryst Chem. 33(5):1325–1333. https://doi.org/10.1107/S0567740877006037
89. Horch H-H, Sader R, Pautke C, Neff A, Deppe H, Kolk A (2006) Synthetic, pure-phase beta-tricalcium phosphate ceramic granules (Cerasorb®) for bone regeneration in the reconstructive surgery of the jaws. Int J Oral Maxillofac Surg. 35(8):708–713. https://doi.org/10.1016/J.IJOM.2006.03.017
90. Yamada S, Heymann D, Bouler J-M, Daculsi G (1997) Osteoclastic resorption of calcium phosphate ceramics with different hydroxyapatite/β-tricalcium phosphate ratios. Biomaterials. 18(15):1037–1041. https://doi.org/10.1016/S0142-9612(97)00036-7
91. Yao C-H, Liu B-S, Hsu S-H, Chen Y-S, Tsai C-C (2004) Biocompatibility and biodegradation of a bone composite containing tricalcium phosphate and genipin crosslinked gelatin. J Biomed Mater Res. 69A(4):709–717. https://doi.org/10.1002/jbm.a.30045
92. Liu H, Cai Q, Lian P et al (2010) β-tricalcium phosphate nanoparticles adhered carbon nanofibrous membrane for human osteoblasts cell culture. Mater Lett. 64(6):725–728. https://doi.org/10.1016/J.MATLET.2009.12.050
93. Kamitakahara M, Ohtsuki C, Miyazaki T (2008) Review paper: behavior of ceramic biomaterials derived from tricalcium phosphate in physiological condition. J Biomater Appl. 23(3):197–212. https://doi.org/10.1177/0885328208096798
94. Bi L, Cheng W, Fan H, Pei G (2010) Reconstruction of goat tibial defects using an injectable tricalcium phosphate/chitosan in combination with autologous platelet-rich plasma. Biomaterials. 31(12):3201–3211. https://doi.org/10.1016/J.BIOMATERIALS.2010.01.038
95. Ellinger RF, Nery EB, Lynch KL (1986) Histological assessment of periodontal osseous defects following implantation of hydroxyapatite and biphasic calcium phosphate ceramics: a case report. Int J Periodontics Restorative Dent. 6(3):22–33. http://www.ncbi.nlm.nih.gov/pubmed/3015813. Accessed 28 Apr 2019
96. Daculsi G (1998) Biphasic calcium phosphate concept applied to artificial bone, implant coating and injectable bone substitute. Biomaterials. 19(16):1473–1478. https://doi.org/10.1016/S0142-9612(98)00061-1

97. Lobo SE, Livingston Arinzeh T, Lobo SE, Livingston AT (2010) Biphasic calcium phosphate ceramics for bone regeneration and tissue engineering applications. Materials (Basel). 3(2):815–826. https://doi.org/10.3390/ma3020815
98. Dorozhkin SV (2012) Biphasic, triphasic and multiphasic calcium orthophosphates. Acta Biomater. 8(3):963–977. https://doi.org/10.1016/J.ACTBIO.2011.09.003
99. Arinzeh TL, Tran T, Mcalary J, Daculsi G (2005) A comparative study of biphasic calcium phosphate ceramics for human mesenchymal stem-cell-induced bone formation. Biomaterials. 26(17):3631–3638. https://doi.org/10.1016/J.BIOMATERIALS.2004.09.035
100. Amirian J, Linh NTB, Min YK, Lee B-T (2015) Bone formation of a porous Gelatin-Pectin-biphasic calcium phosphate composite in presence of BMP-2 and VEGF. Int J Biol Macromol. 76:10–24. https://doi.org/10.1016/J.IJBIOMAC.2015.02.021
101. Ramay HR, Zhang M (2004) Biphasic calcium phosphate nanocomposite porous scaffolds for load-bearing bone tissue engineering. Biomaterials. 25(21):5171–5180. https://doi.org/10.1016/J.BIOMATERIALS.2003.12.023
102. Scotchford CA, Vickers M, Yousuf Ali S (1995) The isolation and characterization of magnesium whitlockite crystals from human articular cartilage. Osteoarthr Cartil. 3(2):79–94. https://doi.org/10.1016/S1063-4584(05)80041-X
103. Elliott JC (1994) Structure and chemistry of the apatites and other calcium orthophosphates. Elsevier, Amsterdam. https://www.sciencedirect.com/bookseries/studies-in-inorganic-chemistry/vol/18. Accessed 26 Jul 2019
104. Jang HL, Lee HK, Jin K, Ahn H-Y, Lee H-E, Nam KT (2015) Phase transformation from hydroxyapatite to the secondary bone mineral, whitlockite. J Mater Chem B. 3(7):1342–1349. https://doi.org/10.1039/C4TB01793E
105. Jang HL, Jin K, Lee J et al (2014) Revisiting whitlockite, the second most abundant biomineral in bone: nanocrystal synthesis in physiologically relevant conditions and biocompatibility evaluation. ACS Nano. 8(1):634–641. https://doi.org/10.1021/nn405246h
106. Kim HD, Jang HL, Ahn H-Y et al (2017) Biomimetic whitlockite inorganic nanoparticles-mediated in situ remodeling and rapid bone regeneration. Biomaterials. 112:31–43. https://doi.org/10.1016/J.BIOMATERIALS.2016.10.009
107. Jang HL, Bin ZG, Park J et al (2016) In vitro and in vivo evaluation of whitlockite biocompatibility: comparative study with hydroxyapatite and β -tricalcium phosphate. Adv Healthc Mater. 5(1):128–136. https://doi.org/10.1002/adhm.201400824
108. Cheng PT, Grabher JJ, LeGeros RZ (1988) Effects of magnesium on calcium phosphate formation. Magnesium. 7(3):123–132. http://www.ncbi.nlm.nih.gov/pubmed/2846970. Accessed 29 Apr 2019
109. Silver IA, Murrills RJ, Etherington DJ (1988) Microelectrode studies on the acid microenvironment beneath adherent macrophages and osteoclasts. Exp Cell Res. 175(2):266–276. https://doi.org/10.1016/0014-4827(88)90191-7
110. Teitelbaum SL (2000) Bone resorption by osteoclasts. Science. 289(5484):1504–1508. https://doi.org/10.1126/SCIENCE.289.5484.1504
111. Kim H-K, Han H-S, Lee K-S et al (2017) Comprehensive study on the roles of released ions from biodegradable Mg-5 wt% Ca-1 wt% Zn alloy in bone regeneration. J Tissue Eng Regen Med. 11(10):2710–2724. https://doi.org/10.1002/term.2166
112. Zapanta Le Geros R (1974) Variations in the crystalline components of human dental calculus: i. crystallographic and spectroscopic methods of analysis. J Dent Res. 53(1):45–50. https://doi.org/10.1177/00220345740530012801
113. Barrère F, van Blitterswijk CA, de Groot K (2006) Bone regeneration: molecular and cellular interactions with calcium phosphate ceramics. Int J Nanomedicine. 1(3):317–332. http://www.ncbi.nlm.nih.gov/pubmed/17717972. Accessed 29 Apr 2019
114. Bodier-Houllé P, Steuer P, Voegel J-C, Cuisinier FJG (1998) First experimental evidence for human dentine crystal formation involving conversion of octacalcium phosphate to hydroxyapatite. Acta Crystallogr Sect D Biol Crystallogr. 54(6):1377–1381. https://doi.org/10.1107/S0907444998005769

115. Suzuki O, Imaizumi H, Kamakura S, Katagiri T (2008) Bone regeneration by synthetic octacalcium phosphate and its role in biological mineralization. Curr Med Chem. 15(3):305–313. https://doi.org/10.2174/092986708783497283
116. Barrère F, Layrolle P, van Blitterswijk CA, de Groot K (2001) Biomimetic coatings on titanium: a crystal growth study of octacalcium phosphate. J Mater Sci Mater Med. 12(6):529–534. https://doi.org/10.1023/A:1011271713758
117. Socol G, Torricelli P, Bracci B et al (2004) Biocompatible nanocrystalline octacalcium phosphate thin films obtained by pulsed laser deposition. Biomaterials. 25(13):2539–2545. https://doi.org/10.1016/J.BIOMATERIALS.2003.09.044
118. Shelton RM, Liu Y, Cooper PR, Gbureck U, German MJ, Barralet JE (2006) Bone marrow cell gene expression and tissue construct assembly using octacalcium phosphate microscaffolds. Biomaterials. 27(14):2874–2881. https://doi.org/10.1016/J.BIOMATERIALS.2005.12.031
119. Kikawa T, Kashimoto O, Imaizumi H, Kokubun S, Suzuki O (2009) Intramembranous bone tissue response to biodegradable octacalcium phosphate implant. Acta Biomater. 5(5):1756–1766. https://doi.org/10.1016/J.ACTBIO.2008.12.008
120. Stefanic M, Krnel K, Pribosic I, Kosmac T (2012) Rapid biomimetic deposition of octacalcium phosphate coatings on zirconia ceramics (Y-TZP) for dental implant applications. Appl Surf Sci. 258(10):4649–4656. https://doi.org/10.1016/J.APSUSC.2012.01.048
121. ter Brugge PJ, Wolke JGC, Jansen JA (2003) Effect of calcium phosphate coating composition and crystallinity on the response of osteogenic cells in vitro. Clin Oral Implants Res. 14(4):472–480. https://doi.org/10.1034/j.1600-0501.2003.00886.x
122. Combes C, Rey C (2010) Amorphous calcium phosphates: synthesis, properties and uses in biomaterials. Acta Biomater. 6(9):3362–3378. https://doi.org/10.1016/J.ACTBIO.2010.02.017
123. Popp JR, Laflin KE, Love BJ, Goldstein AS (2012) Fabrication and characterization of poly(lactic-co-glycolic acid) microsphere/amorphous calcium phosphate scaffolds. J Tissue Eng Regen Med. 6(1):12–20. https://doi.org/10.1002/term.390
124. Cheng H, Chabok R, Guan X et al (2018) Synergistic interplay between the two major bone minerals, hydroxyapatite and whitlockite nanoparticles, for osteogenic differentiation of mesenchymal stem cells. Acta Biomater. 69:342–351. https://doi.org/10.1016/J.ACTBIO.2018.01.016
125. Jeong J, Kim JH, Shim JH, Hwang NS, Heo CY (2019) Bioactive calcium phosphate materials and applications in bone regeneration. Biomater Res. 23:4. https://doi.org/10.1186/s40824-018-0149-3
126. Ishikawa K, Miyamoto Y, Tsuchiya A et al (2018) Physical and histological comparison of hydroxyapatite, carbonate apatite, and β-tricalcium phosphate bone substitutes. Materials (Basel). 11(10):1993. https://doi.org/10.3390/ma11101993
127. Williams DF (2008) On the mechanisms of biocompatibility. Biomaterials. 29(20):2941–2953. https://doi.org/10.1016/J.BIOMATERIALS.2008.04.023
128. Bi L, Rahaman MN, Day DE et al (2013) Effect of bioactive borate glass microstructure on bone regeneration, angiogenesis, and hydroxyapatite conversion in a rat calvarial defect model. Acta Biomater. 9(8):8015–8026. https://doi.org/10.1016/j.actbio.2013.04.043
129. El-Rashidy AA, Roether JA, Harhaus L, Kneser U, Boccaccini AR (2017) Regenerating bone with bioactive glass scaffolds: a review of in vivo studies in bone defect models. Acta Biomater. 62:1–28. https://doi.org/10.1016/j.actbio.2017.08.030
130. Fu Q (2019) Bioactive glass scaffolds for bone tissue engineering. Biomed Ther Clin Appl Bioact Glas:417–442. https://doi.org/10.1016/B978-0-08-102196-5.00015-X
131. Moses JC, Nandi SK, Mandal BB (2018) Multifunctional cell instructive silk-bioactive glass composite reinforced scaffolds toward osteoinductive, proangiogenic, and resorbable bone grafts. Adv Healthc Mater. 7(10):1701418. https://doi.org/10.1002/adhm.201701418
132. Baino F (2018) Bioactive glasses—when glass science and technology meet regenerative medicine. Ceram Int. 44(13):14953–14966. https://doi.org/10.1016/j.ceramint.2018.05.180
133. Srinivasan S, Jayasree R, Chennazhi KP, Nair SV, Jayakumar R (2012) Biocompatible alginate/nano bioactive glass ceramic composite scaffolds for periodontal tissue regeneration. Carbohydr Polym. 87(1):274–283. https://doi.org/10.1016/j.carbpol.2011.07.058

134. Gkioni K, Leeuwenburgh SCG, Douglas TEL, Mikos AG, Jansen JA (2010) Mineralization of hydrogels for bone regeneration. Tissue Eng Part B Rev. 16(6):577–585. https://doi.org/10.1089/ten.TEB.2010.0462
135. Jones JR (2013) Review of bioactive glass: from Hench to hybrids. Acta Biomater. 9(1):4457–4486. https://doi.org/10.1016/j.actbio.2012.08.023
136. Zhang G, Brion A, Willemin A-S et al (2017) Nacre, a natural, multi-use, and timely biomaterial for bone graft substitution. J Biomed Mater Res Part A. 105(2):662–671. https://doi.org/10.1002/jbm.a.35939
137. Akilal N, Lemaire F, Bercu NB et al (2019) Cowries derived aragonite as raw biomaterials for bone regenerative medicine. Mater Sci Eng C. 94:894–900. https://doi.org/10.1016/j.msec.2018.10.039
138. Anderson JM, Rodriguez A, Chang DT (2008) Foreign body reaction to biomaterials. Semin Immunol. 20(2):86–100. https://doi.org/10.1016/j.smim.2007.11.004
139. Anderson JM (2001) Biological responses to materials. Annu Rev Mater Res. 31:81–110. http://www.bioen.utah.edu/faculty/pat/Courses/biomaterials/BiologicalResponse.pdf. Accessed 7 Mar 2017
140. Navarro M, Michiardi A, Castano OPA (2008) Biomaterials in orthopaedics. J R Soc Interface. 5:1137–1158
141. Ambrosio L, De Santis R, Nicolais L (1998) Composite hydrogels as intervertebral disc prostheses. In: Science and technology of polymers and advanced materials. Springer US, Boston, MA, pp 547–555. https://doi.org/10.1007/978-1-4899-0112-5_46
142. Hasan A, Byambaa B, Morshed M et al (2018) Advances in osteobiologic materials for bone substitutes. J Tissue Eng Regen Med. 12(6):1448–1468. https://doi.org/10.1002/term.2677
143. Campana V, Milano G, Pagano E et al (2014) Bone substitutes in orthopaedic surgery: from basic science to clinical practice. J Mater Sci Mater Med. 25(10):2445–2461. https://doi.org/10.1007/s10856-014-5240-2
144. Guarino V, Caputo T, Altobelli R, Ambrosio L (2015) Degradation properties and metabolic activity of alginate and chitosan polyelectrolytes for drug delivery and tissue engineering applications. AIMS Mater Sci. 2(4):497–502. https://doi.org/10.3934/matersci.2015.4.497
145. Guarino V, Causa F, Ambrosio L (2007) Bioactive scaffolds for bone and ligament tissue. Expert Rev Med Devices. 4(3):405–418. https://doi.org/10.1586/17434440.4.3.405
146. Chiari C, Koller U, Dorotka R et al (2006) A tissue engineering approach to meniscus regeneration in a sheep model. Osteoarthr Cartil. 14(10):1056–1065. https://doi.org/10.1016/J.JOCA.2006.04.007
147. Marijnissen WJC, van Osch GJV, Aigner J et al (2002) Alginate as a chondrocyte-delivery substance in combination with a non-woven scaffold for cartilage tissue engineering. Biomaterials. 23(6):1511–1517. https://doi.org/10.1016/S0142-9612(01)00281-2
148. Cancedda R, Dozin B, Giannoni P, Quarto R (2003) Tissue engineering and cell therapy of cartilage and bone. Matrix Biol. 22(1):81–91. https://doi.org/10.1016/S0945-053X(03)00012-X
149. Revell PA, Damien E, Di Silvio L, Gurav N, Longinotti C, Ambrosio L (2007) Tissue engineered intervertebral disc repair in the pig using injectable polymers. J Mater Sci Mater Med. 18(2):303–308. https://doi.org/10.1007/s10856-006-0693-6
150. Acevedo CA, Olguín Y, Briceño M et al (2019) Design of a biodegradable UV-irradiated gelatin-chitosan/nanocomposed membrane with osteogenic ability for application in bone regeneration. Mater Sci Eng C. 99:875–886. https://doi.org/10.1016/J.MSEC.2019.01.135
151. Ou Q, Miao Y, Yang F, Lin X, Zhang L-M, Wang Y (2019) Zein/gelatin/nanohydroxyapatite nanofibrous scaffolds are biocompatible and promote osteogenic differentiation of human periodontal ligament stem cells. Biomater Sci 7(5):1973–1983. https://doi.org/10.1039/c8bm01653d
152. Cao M, Zhou Y, Mao J et al (2019) Promoting osteogenic differentiation of BMSCs via mineralization of polylactide/gelatin composite fibers in cell culture medium. Mater Sci Eng C. 100:862–873. https://doi.org/10.1016/j.msec.2019.02.079

153. Anada T, Pan C-C, Stahl A et al (1096) Vascularized bone-mimetic hydrogel constructs by 3D bioprinting to promote osteogenesis and angiogenesis. Int J Mol Sci. 20(5):2019. https://doi.org/10.3390/ijms20051096
154. Ingavle GC, Gionet-Gonzales M, Vorwald CE et al (2019) Injectable mineralized microsphere-loaded composite hydrogels for bone repair in a sheep bone defect model. Biomaterials. 197:119–128. https://doi.org/10.1016/j.biomaterials.2019.01.005
155. Murahashi Y, Yano F, Nakamoto H et al (2019) Multi-layered PLLA-nanosheets loaded with FGF-2 induce robust bone regeneration with controlled release in critical-sized mouse femoral defects. Acta Biomater. 85:172–179. https://doi.org/10.1016/j.actbio.2018.12.031
156. Lee S-H, Shin H (2007) Matrices and scaffolds for delivery of bioactive molecules in bone and cartilage tissue engineering. Adv Drug Deliv Rev. 59(4-5):339–359. https://doi.org/10.1016/J.ADDR.2007.03.016
157. Vinogradov SV, Bronich TK, Kabanov AV (2002) Nanosized cationic hydrogels for drug delivery: preparation, properties and interactions with cells. Adv Drug Deliv Rev 54(1):135–147. https://doi.org/10.1016/S0169-409X(01)00245-9
158. Bai X, Gao M, Syed S, Zhuang J, Xu X, Zhang X-Q (2018) Bioactive hydrogels for bone regeneration. Bioact Mater. 3(4):401–417. https://doi.org/10.1016/j.bioactmat.2018.05.006
159. Heo YJ, Shibata H, Okitsu T, Kawanishi T, Takeuchi S (2011) Long-term in vivo glucose monitoring using fluorescent hydrogel fibers. Proc Natl Acad Sci. 108(33):13399–13403. https://doi.org/10.1073/pnas.1104954108
160. Onat B, Tuncer S, Ulusan S, Banerjee S, Erel GI (2019) Biodegradable polymer promotes osteogenic differentiation in immortalized and primary osteoblast-like cells. Biomed Mater. 14(4):045003. https://doi.org/10.1088/1748-605X/ab0e92
161. Barbieri D, Yuan H, Luo X, Farè S, Grijpma DW, de Bruijn JD (2013) Influence of polymer molecular weight in osteoinductive composites for bone tissue regeneration. Acta Biomater. 9(12):9401–9413. https://doi.org/10.1016/j.actbio.2013.07.026
162. Ricciardi BF, Bostrom MP (2013) Bone graft substitutes: claims and credibility. Semin Arthroplasty 24(2):119–123. https://doi.org/10.1053/j.sart.2013.07.002
163. Alidadi S, Oryan A, Bigham-Sadegh A, Moshiri A (2017) Comparative study on the healing potential of chitosan, polymethylmethacrylate, and demineralized bone matrix in radial bone defects of rat. Carbohydr Polym. 166:236–248. https://doi.org/10.1016/j.carbpol.2017.02.087
164. Lai Y, Li Y, Cao H et al (2019) Osteogenic magnesium incorporated into PLGA/TCP porous scaffold by 3D printing for repairing challenging bone defect. Biomaterials. 197:207–219. https://doi.org/10.1016/j.biomaterials.2019.01.013
165. Veronesi F, Giavaresi G, Guarino V et al (2015) Bioactivity and bone healing properties of biomimetic porous composite scaffold: in vitro and in vivo studies. J Biomed Mater Res Part A. 103(9):2932–2941. https://doi.org/10.1002/jbm.a.35433
166. Martínez-Sanmiguel JJ, Zarate-Triviño G, Hernandez-Delgadillo R et al (2019) Anti-inflammatory and antimicrobial activity of bioactive hydroxyapatite/silver nanocomposites. J Biomater Appl 33(10):1314–1326. https://doi.org/10.1177/0885328219835995
167. Chen Y, Zheng Z, Zhou R et al (2019) Developing a strontium-releasing graphene oxide/collagen-based organic-inorganic nanobiocomposite for large bone defect regeneration via MAPK signaling pathway. ACS Appl Mater Interfaces 11(17):15986–15997. https://doi.org/10.1021/acsami.8b22606
168. Arnold AM, Holt BD, Daneshmandi L, Laurencin CT, Sydlik SA (2019) Phosphate graphene as an intrinsically osteoinductive scaffold for stem cell-driven bone regeneration. Proc Natl Acad Sci U S A 116(11):4855–4860. https://doi.org/10.1073/pnas.1815434116
169. Narimani M, Teimouri A, Shahbazarab Z (2019) Synthesis, characterization and biocompatible properties of novel silk fibroin/graphene oxide nanocomposite scaffolds for bone tissue engineering application. Polym Bull. 76(2):725–745. https://doi.org/10.1007/s00289-018-2390-2
170. Wang W, Junior JRP, Nalesso PRL et al (2019) Engineered 3D printed poly(ε-caprolactone)/graphene scaffolds for bone tissue engineering. Mater Sci Eng C. 100:759–770. https://doi.org/10.1016/J.MSEC.2019.03.047

171. Purohit SD, Bhaskar R, Singh H, Yadav I, Gupta MK, Mishra NC (2019) Development of a nanocomposite scaffold of gelatin–alginate–graphene oxide for bone tissue engineering. Int J Biol Macromol. 133:592–602. https://doi.org/10.1016/J.IJBIOMAC.2019.04.113
172. Zhang Y, Wang J, Wang J et al (2015) Preparation of porous PLA/DBM composite biomaterials and experimental research of repair rabbit radius segmental bone defect. Cell Tissue Bank. 16(4):615–622. https://doi.org/10.1007/s10561-015-9510-0
173. Carrow JK, Di Luca A, Dolatshahi-Pirouz A, Moroni L, Gaharwar AK (2019) 3D-printed bioactive scaffolds from nanosilicates and PEOT/PBT for bone tissue engineering. Regen Biomater. 6(1):29–37. https://doi.org/10.1093/rb/rby024
174. Li X, Wang L, Fan Y, Feng Q, Cui F-Z, Watari F (2013) Nanostructured scaffolds for bone tissue engineering. J Biomed Mater Res Part A. 101A(8):2424–2435. https://doi.org/10.1002/jbm.a.34539
175. Bettinger CJ, Langer R, Borenstein JT (2009) Engineering substrate topography at the micro- and nanoscale to control cell function. Angew Chemie Int Ed. 48(30):5406–5415. https://doi.org/10.1002/anie.200805179
176. Peng R, Yao X, Ding J (2011) Effect of cell anisotropy on differentiation of stem cells on micropatterned surfaces through the controlled single cell adhesion. Biomaterials. 32(32):8048–8057. https://doi.org/10.1016/j.biomaterials.2011.07.035
177. McMahon RE, Wang L, Skoracki R, Mathur AB (2013) Development of nanomaterials for bone repair and regeneration. J Biomed Mater Res Part B Appl Biomater. 101B(2):387–397. https://doi.org/10.1002/jbm.b.32823
178. Pasqui D, Torricelli P, De Cagna M, Fini M, Barbucci R (2014) Carboxymethyl cellulose-hydroxyapatite hybrid hydrogel as a composite material for bone tissue engineering applications. J Biomed Mater Res Part A. 102(5):1568–1579. https://doi.org/10.1002/jbm.a.34810
179. Gelinsky M, Welzel PB, Simon P, Bernhardt A, König U (2008) Porous three-dimensional scaffolds made of mineralised collagen: Preparation and properties of a biomimetic nano-composite material for tissue engineering of bone. Chem Eng J. 137(1):84–96. https://doi.org/10.1016/J.CEJ.2007.09.029
180. Kilian KA, Bugarija B, Lahn BT, Mrksich M (2010) Geometric cues for directing the differentiation of mesenchymal stem cells. Proc Natl Acad Sci. 107(11):4872–4877. https://doi.org/10.1073/pnas.0903269107
181. Watari S, Hayashi K, Wood JA et al (2012) Modulation of osteogenic differentiation in hMSCs cells by submicron topographically-patterned ridges and grooves. Biomaterials. 33(1):128–136. https://doi.org/10.1016/j.biomaterials.2011.09.058
182. Cai K, Frant M, Bossert J, Hildebrand G, Liefeith K, Jandt KD (2006) Surface functionalized titanium thin films: zeta-potential, protein adsorption and cell proliferation. Colloids Surfaces B Biointerfaces. 50:1–8. https://doi.org/10.1016/j.colsurfb.2006.03.016
183. Palmer LC, Newcomb CJ, Kaltz SR, Spoerke ED, Stupp SI (2008) Biomimetic systems for hydroxyapatite mineralization inspired by bone and enamel. Chem Rev. 108(11):4754–4783. https://doi.org/10.1021/cr8004422
184. Dalby MJ, Gadegaard N, Tare R et al (2007) The control of human mesenchymal cell differentiation using nanoscale symmetry and disorder. Nat Mater 6(12):997–1003. https://doi.org/10.1038/nmat2013
185. Gong T, Xie J, Liao J, Zhang T, Lin S, Lin Y (2015) Nanomaterials and bone regeneration. Bone Res. 3:15029. https://doi.org/10.1038/boneres.2015.29
186. Vallet-Regí M, Salinas AJ (2009) Ceramics as bone repair materials. 2nd. Elsevier, Amsterdam. https://doi.org/10.1533/9781845696610.2.194

Bimetallic Nanoparticles for Biomedical Applications: A Review

David Medina-Cruz, Bahram Saleh, Ada Vernet-Crua, Alfonso Nieto-Argüello, Diana Lomelí-Marroquín, Lydia Yerid Vélez-Escamilla, Jorge L. Cholula-Díaz, José Miguel García-Martín, and Thomas Webster

Abstract Bimetallic nanoparticles, or BMNPs, are nanosized structures that are of growing interest in biomedical applications. Although their production shares aspects with physicochemical approaches for the synthesis of their monometallic counterparts, they can show a large variety of new properties and applications as a consequence of the synergetic effect between the two components. These applications can be as diverse as antibacterial treatments or anticancer or biological imaging approaches, as well as drug delivery. Nevertheless, utilization of BMNPs in such fields has received limited attention because of the severe lack of knowledge and concerns regarding the use of other nanomaterials, such as stability and biodegradability over time, tendency to form clusters, chemical reactivity, and biocompatibility. In this review, a close look at bimetallic systems is presented, focusing on their biomedical applications as antibacterial, anticancer, drug delivery, and imaging agents, showing significant enhancement of their features compared to their monometallic counterparts and other current used nanomaterials for biomedical applications.

Keywords Biomaterials · Anti-biofouling · Anti-microbial · Biofilm · Infections
Polyethylene glycol (PEG) · Zwitterionic · Releasing-based · Contact-based
Medical implants · Tissue engineering

David Medina and Bahram Saleh contributed equally to this work.

D. Medina-Cruz · B. Saleh · A. Vernet-Crua · T. Webster (✉)
Department of Chemical Engineering, Northeastern University, Boston, MA, USA

Nanomedicine Science and Technology Center, Northeastern University, Boston, MA, USA
e-mail: th.webster@neu.edu

A. Nieto-Argüello · D. Lomelí-Marroquín · L. Y. Vélez-Escamilla · J. L. Cholula-Díaz
School of Engineering and Sciences, Tecnologico de Monterrey,
Monterrey, Nuevo León, Mexico

J. M. García-Martín
Instituto de Micro y Nanotecnología, IMN-CNM, CSIC (CEI UAM + CSIC),
Tres Cantos, Spain

B. Li et al. (eds.), *Racing for the Surface*,
https://doi.org/10.1007/978-3-030-34471-9_16

Nanotechnology for Biomedical Applications

Nanotechnology and Nanomedicine: The Born of a New Era

In 1959, Richard P. Feynman gave his famous after-dinner lecture named "There's Plenty of Room at the Bottom" [1], setting the basis for the future of a field whose name was given 15 years later. The word *nanotechnology* was used for the first time by the Japanese investigator Norio Taniguchi in a 1974 paper on production technology involving the generation of objects and features in the nanometer scale [2]. Nowadays, nanotechnology is defined as the study and application of materials with sizes less than 100 nm. Since the beginning, it has been beneficial in a variety of fields like chemistry, physics, biology, biochemistry, as well as medicine [3–6].

Feynman suggested the use of nanotechnology to develop medicine to his colleague Albert R. Hibbs. Consequently, a new field was born from this interaction: nanomedicine that impacts the medical applications of nanomaterials and biological devices, nanoelectronic biosensors, and even possible future applications of molecular nanotechnologies, such as biological machines [7].

Nanostructures have been employed in a variety of biomedical applications, i.e., as antibacterial [8], anticancer [9], drug delivery [10], or imaging [11] agents. Thus, the number of nanotechnology publications has raised over time with an unexpected trend, with hopes that if we can master this technology, we will be able to improve not just medicine, but all aspects of the modern life.

The Use of Metallic Nanoparticles in Nanomedicine

The use of nanotechnology in the medical field has brought the utilization of different nanostructures to medical applications, such as carbon [12] or silica [13], among others. Nonetheless, metallic nanoparticles (NP) have been gaining popularity due to their large variety of chemical and physical properties, along with their tenability, which is hugely related to their performance, making them perfect materials for biomedical applications [14]. From noble metals like gold (Au) to metalloids, such as selenium (Se), or magnetic metals like iron (Fe), this diversity of materials has allowed broad applications in many areas of study such as antibacterial, anticancer, imaging, or drug delivery processes [15–17].

In this way, different metallic systems are implemented in the nanotechnological field, which can be classified into magnetic, pure metallic, metal oxide nanomaterials without magnetic behavior, and bimetallic nanoparticles.

Magnetic Nanoparticles

Magnetic nanoparticles can show different features when subjected to a magnetic field. These configurations are known to present either ferri- or ferromagnetic behavior. These magnetic properties are found in materials such as nickel (Ni), cobalt (Co), or iron (Fe). Among all the configurations, the most common ones are oxides (also known as ferrites). A further classification includes ferrites with a shell, pure metallic magnetic nanoparticles, and metallic with a shell [18].

Iron oxide nanomaterials are the most widely used for research along magnetic nanostructures, showing essential features for nanomedicine. For instance, the structures become superparamagnetic—magnetization can arbitrary change the direction under the effects of temperature—preventing aggregation, which is a significant concern in biomedical applications [18]. At the nanoscale regime, the surface chemistry of the structures becomes highly significant. However, the surface of ferrite nanostructures is relatively inert, because of the presence of a coating, mainly a layer of silica, among other materials, producing what is commonly called ferrites with a shell [19]. The use of surfacntants, polymers and processes such as oxidation can passivate pure metallic magnetic nanoparticles [18].

Notably, iron oxide nanomaterials are the most promising ones, due to their reduced toxicity in biological matrices, and their antibacterial and anticancer features [20]. As an example, Tokajuk et al. described how iron oxide nanoparticles may be utilized as a successful drug delivery agent for chlorhexidine, an active antimicrobial agent, against *Staphylococcus aureus* and *Candida albicans* [21].

Pure Metallic Nanoparticles Without Magnetic Behavior

Pure nanoparticles composed of single metals, for instance, gold (Au), silver (Ag), or platinum (Pt), have important biomedical applications. For instance, gold nanoparticles have been primarily used in hyperthermia treatments towards cancer—inducing apoptosis in cancerous cells—as a consequence of their optical properties [22]. Similarly, silver nanoparticles are known to present a powerful antibacterial activity, and therefore, they have been largely used to alter orthopedic implants with the aim to avoid infection [23]. However, not only noble metals can be used. For instance, bacteria-mediated selenium nanoparticles showed potent antibacterial activity against different bacteria in their antibiotic-resistant phenotypes, such as methicillin-resistant *S. aureus* (MRSA) or multidrug-resistant (MDR) *Escherichia coli* [24].

Metal Oxide Nanoparticles Without Magnetic Behavior

Metal oxide nanoparticles are mainly composed of transition metals combined with oxygen atoms, producing copper (Cu), zinc (Zn), or titanium (Ti) oxide formulations with relevant biomedical features. For instance, Amiri et al. described the antibacterial and antifungal effects of CuO nanoparticles [25], while Zhijun Jiao et al. explored a new drug delivery system utilizing titanium dioxide (TiO_2) nanoparticles due to its key function in improving the anticancer properties of doxorubicin as well as diminishing side effects [26].

Bimetallic Nanoparticles

Bimetallic nanomaterials are composed of a combination of two different metals. There is a significant variety of different compounds that can form bimetallic nanoparticles. Nevertheless, some common combinations include the use of transition metals such as gold and silver. These two metals can be combined either between them or with other metals such as Co or Fe in order to get different features including the enhancement of their biomedical applications [27]. This review will center its attention on the application of these bimetallic formulations as a powerful tool for the field of biomedicine.

Bimetallic Nanoparticles (BMNPs): A Step Further

Bimetallic nanoparticles have been widely used as biomedical agents for a long time. The reason behind this fact is related to the synergetic interaction created when two different metals are combined, leading to an enhancement of the features of the nanostructure in comparison with its monometallic configurations.

Due to the full range of different possible arrangement of the metals, such as alloy, core–shell, and cluster-in-cluster, new features arise. Therefore, the variations of the atomic structure in bimetallic nanoparticles present a significant advantage in comparison to monometallic nanoparticles, as they can only achieve a change in shape or size, leading to an improved optical or catalysis properties, as well as anticancer and antibacterial contributions. Nevertheless, and in comparison, with single-metal formulations, less research has been focused on bimetallic nanoparticles. Despite this, bimetallic structures have risen considerable interest because of their improved catalytic activity and selectivity compared to their monometallic counterparts, which have their origin on the synergetic effect between both the components [28–30].

For instance, Xu QH et al. demonstrated two-photon imaging of bacteria by Au–Ag NPs under near-infrared (NIR) femtosecond laser pulses which improved antibacterial activity by the use of the NIR two-photon photothermal effects [27]. Similarly, Cho et al. have developed a bimetallic hybrid Zn–Au nanomaterial with zinc cores and gold shells, with the aim to show their in vivo visibility by using positron emission tomography (PET) once the proton into the zinc core has been activated [31]. The structures were also able to produce radiosensitization when Au cover was irradiated by different sources and therefore emitted secondary electrons (SE).

Moreover, from the biomedical point of view, BMNPs could induce higher cytotoxicity when they are in contact with the body, as more than one metallic configuration would be reacting at a time, compromising the effectiveness of these systems, due to a more significant production of reactive oxygen species (ROS) and stimulation of metal-ion release [32, 33]. In an elegant study, T. Li et al. found that bimetallic nanostructures with a composition of 80% Ag and 20% Au had significantly lower toxicity compared to Ag NPs, whereas if the composition was changed to 20% Ag and 80% Au, the levels of toxicity increased unexpectedly [34, 35]. Their results confirmed that Ag NPs presented significant higher toxicity than Au NPs, and thus the presence of gold in silver structures potentially decreases their environmental repercussions by reducing the quantity of bioavailable Ag.

Synthesis of Bimetallic Nanoparticles

The production of nanoparticles, including bimetallic ones, can be split into two main divisions that depend on the approaches followed for the production of the structures: the top-down and the bottom-up methods [36]. The bottom-up method—also known as constructive process—uses atoms or molecules of the precursors as starting materials, which react to form clusters, followed by the formation of nanoparticles. This method includes the synthesis of nanoparticles by sol-gel, chemical vapor deposition (CVD), molecular self-assembly, atomic layer deposition, nanocluster sources, wet chemical reduction, and other approaches [37]. On the other hand, the top-down synthesis—also called destructive method—is characterized by the use of bulk materials as the starting point, which is transformed into a fine powder until nanometric particles are obtained or patterned into nanostructures. Some examples include mechanical milling and different nanolithography techniques, such as e-beam lithography, ultraviolet (UV) lithography, focused ion-beam lithography, and nanoimprint [38].

Recently, an alternative classification for the methods of the generation of metallic NPs has been proposed, based on the biocompatibility of the substances—except the metal precursor—used to produce the nanomaterials: the physicochemical synthesis (involving the top-down and bottom-up methodologies) and the green synthesis [39]. We will follow this classification, although devoting special attention to the nanocluster sources among the physicochemical methods due to their commercial availability [40, 41].

Physicochemical Approaches

The physicochemical approaches for the synthesis of nanomaterials are characterized for taking knowledge from both physics and chemistry in order to build efficient pathways for formulation. Therefore, there is a wide variety of these approaches to synthesize any nanomaterial, including BMNPs [42]. They rely upon the nature and composition of the targeted nanomaterial and are characterized by different reaction times, physical parameters, and chemical intermediates. Some examples of nanoparticles made by these methods are shown in Fig. 1.

One of the most important methods is laser ablation or pulsed laser deposition (PLD), where a specific spot on the surface is irradiated by a pulsed laser beam to

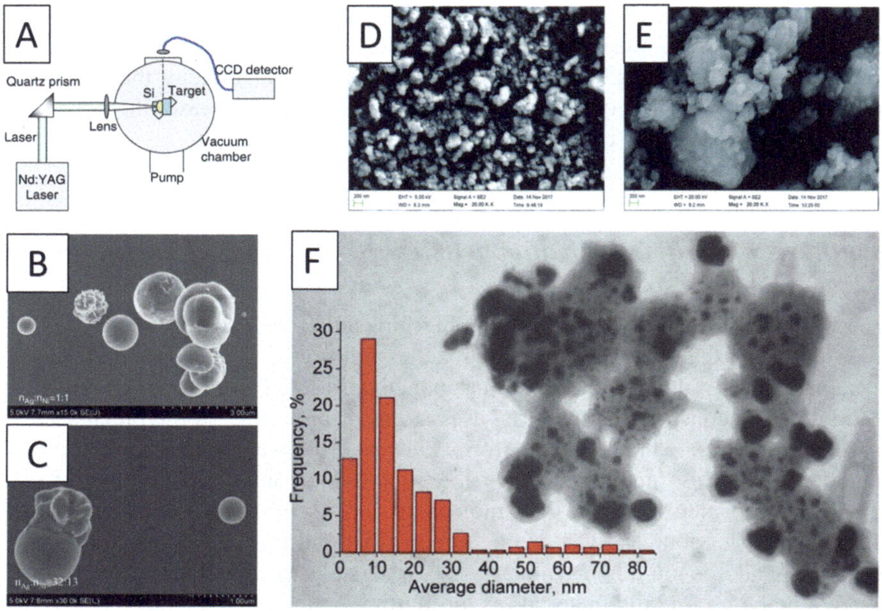

Fig. 1 The schematic diagram of the experimental setup for the generation of Ag–Ni bimetallic nanoparticles by laser-induced plasma (**A**) [43]; Scanning electron microscopy (SEM) images of nanoparticles with a mole ratio of silver and nickel nAg:nNi = 1:1 and 32:13, respectively (**B**, **C**) [43]; SEM images of Ti3+ and N self-doped SrTiOxNy samples warmed at different temperatures and atmospheres: (**D**) T-130-Vac and (**E**) T-200-Vac [44]; and transmission electron microscopy (TEM) image and size distribution of particles in Ag/Au/Trp colloids (**F**) [44]

the point of evaporation or sublimation [45]. Classified in a single-photon process—splitting chemical bonds—and multiphoton excitation—thermal evaporation—laser ablation is mostly used to produce semiconductor quantum dots (QD), carbon nanotubes (CNT), nanowires (NW), and core–shell NP [46]. When performed in a high vacuum chamber, the advantage of the laser ablation method is that the resulting products are highly pure, although there is low control in the size distribution, agglomeration, and crystal structures [47]. Another advantage is that the laser beam can irradiate a target dipped in a solution, which facilitates the functionalization of the obtained nanoparticles [48].

A variety of BMNPs may be synthesized following the physicochemical fashion, as can be seen in Table 1. For instance, Nakamura T. and Sato S. developed a solid solution of palladium–platinum (Pd–Pt) particles with total variable composition that was synthetized using high-intensity laser irradiation of liquid solution of palladium and platinum ions where no reducing agents or thermal processes were employed [54]. Similarly, a laser ablation deposition technique was utilized to generate silver–nickel (Ag–Ni) bimetallic nanoparticles [43], while Ag–Pd nanoparticles have been synthetized by colloids of Ag NPs using liquid methodology by pulsed laser ablation in aqueous media [55].

Table 1 Physicochemical methods of synthesis of bimetallic nanoparticles

Method of synthesis	Nanoparticle	Size (nm)	Morphology
Seed-mediated	Ag–Fe_3O_4 [49]	5–20 nm each nanoparticle	Two nanoparticles together
Chemical reduction	Au–Ni [50]	2.3–2.7 nm	Quasi-spherical
	Ag–Au [51]	5–15 nm depending on the ratio	Quasi-spherical
	Ag–Au [52]	15–30 nm depending on the metal ratio	Quasi-spherical
	Ag–Au [53]	20–27 nm depending on the metal ratio	Quasi-spherical
Chemical oxidation	$SrTiO_xNy$ [44]	50–200 depending on the temperature	Quasi-spherical
Laser ablation	Pd–Pt [54]	50–90 nm	Predominantly spherical
	Ag–Pd core–shell [55]	2.4–3.2 nm	Quasi-spherical
	Ag–Ni [43]	Varies	Irregular shape
	PEG–Fe–Au alloy [48]	15–40 nm	Spherical
Sol-gel	Ag–Pt-modified TiO_2 [56]	15–30 nm	Quasi-spherical
CVD	(Pd–Pt)/SiO_2 [57]	1–15 nm	Layers
	Pt–Co [58]	1.2–1.8 nm	Spheroids
	Ni–Pd @ MIL-101 [59]	2–3 nm	Spherical

Oxidation is a type of chemical reaction that involves a substance that gives away electrons, meaning that it is oxidized. This chemical principle can be used for the production of nanomaterials by oxidation of metal ions [60]. For instance, a facile in situ green approach for the generation of nanoparticles by process was elaborated by J. Liu et al. to prepare Ti^{3+} and N self-doped SrTiOxNy nanoparticles using TiN and H_2O_2 as precursors, with the resultant nanoparticles showing a pronounced absorption in a range between 400 and 800 nm, utilizing UV-visible (vis) diffuse reflectance spectroscopy (UV-vis DRS) [44].

On the other hand, reduction in a chemical reaction involves a substance gaining electrons by one of the atoms implicated in the chemical process. Due to the feasibility of the reaction and the disponibility of reagents, chemical reduction methods are the most widely utilized for the production of BMNPs [51]. For the chemical reduction to take place, a metal precursor, a reducing agent, and a stabilizing agent are needed [61]. The metal precursor is usually a metal salt in solution; the reducing agent may vary from organic compounds—such as ascorbic acid [62]—to inorganic salts—like sodium borohydride [50]. The stabilizing agents (such as polyvinylpyrrolidone (PVP), starch, and sodium carboxyl methyl cellulose) allow the dispersibility of the nanoparticles in common solvents [63]. The systematic modification of the experimental parameters controls the morphology and uniformity of the final product. As an example, hydrazine co-reduction of $[AuCl_4]^-$ and $[PdCl_4]^{2-}$ complex anions was used to synthetize colloidal suspensions of Au_xPd_{1-x} nanoalloys. Besides, polymeric compounds of high molecular weight, such as polyvinylpyrrolidone (PVP) or polyvinyl alcohol (PVA), were used as surface capping agents [64].

Peng and colleagues have elaborated a seed-mediated protocol in order to obtain hybrid Ag–FexOy nanoparticles that show plasmonic and magnetic characteristics [49]. The method is based on the chemical reduction of Ag+ ions when Fe NPs are present with noncrystalline FexOy surfaces previously produced by thermal decomposition.

Chemical vapor deposition (CVD) is a bottom-up technique used for the generation of a large number of nanomaterials under different atmospheres. In this process, a volatile precursor (or multiple precursors) reacts to produce the desired material [65]. Hierso et al. have proved the synthesis of bimetallic core–shell palladium–platinum NPs in a layered structure through the CVD method using a metal-organic precursor [57]. Alternatively, Choi et al. have defined the CVD technique as a fast method to produce monodisperse nanoparticles, which can be easily scaled to larger volumes, for example, highly monodisperse Pt–Co BMNPs [58]. Another example is the synthesis of Pt_3Co intermetallic nanoparticles supported on ceria by CVD [66].

A sol-gel process is a bottom-up method that consists of the synthesis of nanomaterials by hydrolysis and condensation reactions. The main benefits of sol-gel processing are the high purity and uniform nanostructures achievable at low temperatures. BMNPs made of Au–Pd, Au–Ag, and Au–Pt can be generated in a single step by a sol-gel methodology and stabilized in liquid and solid matrices [67].

Nanocluster sources deserve a special mention within the physicochemical approaches in terms of nanoparticle synthesis. BMNPs can be produced using gas-phase synthesis using the so-called nanocluster sources. Although there is a rich variety of types of NP sources that can be used, each one with advantages and drawbacks [68], the most popular ones are the magnetron-based cluster sources. This technique involves two chambers separated by a wall with a pinhole: the aggregation chamber where the NPs are formed and the deposition chamber where the NPs are collected. The sputtering process takes place in the first chamber, loaded with a low-pressure ideal gas (typically argon), and a negative voltage is administered to a metallic objective (the origin of atoms). The application of this voltage fractures the gas and constitutes a plasma formed by Ar^+ and Ar^{2+} ions (with ratio $Ar^{2+}/Ar^+ \ll 1$) and electrons. The ions are accelerated towards the target, and the atoms of the surface are ejected by momentum transfer, which aggregate in the aggregation zone to form NPs. The dimension of these NPs (typically 1–20 nm) is directed by the power applied to the target, the argon pressure, the position of the target in the aggregation zone, and the possible use of extra gas (usually helium). It is worth mentioning that the technique is called "magnetron sputtering" because magnets are placed behind the target, which helps to trap electrons over it and as a result increases the efficiency of the process, thus allowing for using lower pressure values and obtaining faster deposition rates. Finally, the second chamber is the collection chamber where only the NPs that go through the pinhole are collected on a substrate. This chamber is usually maintained in conditions of vacuum or ultra-high vacuum in order to guarantee the chemical purity of the NPs. Besides, the substrate can be heated to promote thermal diffusion effects, which can alter the structure of the NPs [69].

With a single magnetron-based source, BMNPs can be fabricated by using a bimetallic alloyed target. In general, the elements present in the NPs are similar or the exact ones on the target, but the internal structure of the compounds depends on the atomic diffusion of the two kinds of atoms. As an example, Oprea et al. showed the fabrication of Fe-B NPs from a $Fe_{50}B_{50}$ target, which was made of an internal Fe core while outside the shell with a combination of Fe and B oxides and oxynitrides [70]. They could be functionalized with polyethylene glycol (PEG) and exhibited superparamagnetic behavior at RT. Therefore, they may be tentatively used as contrast agents in magnetic resonance imaging [71].

By heating the substrate, one can use thermal-induced diffusion to modify the structure of the NPs, leading to the generation of perfectly defined core–shell structures. That was the approach used by Llamosa Pérez and co-workers: with a $Co_{95}Au_5$ target and by heating the substrate, significant structural changes were found: the deposition at low temperatures (close to RT) induced the synthesis of an alloyed fcc CoAu core with a non-complete cobalt oxide shell, whereas the use of elevate deposition temperatures (up to 770 K) led to the generation of a pure hcp Co core with an intermediate Au shell and a compact outer cobalt oxide shell [69].

Another possibility to obtain a wide diversity of crystalline and chemical combinations at atomic level from an alloyed magnetron target is to apply high-power impulses instead of a continuous DC voltage, in the so-called high-power impulse magnetron sputtering technique (HiPIMS) [72].

However, the most powerful way that permits for fine-tailoring of the chemical composition of the NPs is to use a cluster source with multiple targets [71, 73]: with the aim to produce BMNPs, two targets, A and B, are thus needed in this strategy. Different NPs are produced when the relative position of the targets in the aggregation zone is modified. If targets A and B are placed together, i.e., their distances to the pinhole are the same, NPs made of a homogeneous AB alloy are obtained [73]. In clear contrast, if target A is located behind target B, NPs are being made of an A core and a B shell, which is a route that can be used to obtain core–shell NPs beyond the limitations of thermal diffusion [74]. Martínez et al. used this route with an Au target behind a Ti one, and the obtained NPs were oxidized in-flight to have Au core–TiOx shell NPs at the end [75]. Finally, it is worth mentioning that this type of cluster source with multiple targets is commercially available [41].

Green Approaches

The application of green chemistry basics to nanotechnology supposed improvement in the synthesis of nanomaterials in the direction of sustainability. Green nanotechnology was born as the promise of using environmentally friendly, cost-effective, and safe methods for the generation of different nanomaterials [76]. Therefore, these processes can overcome the main restrictions of traditional physicochemical approaches—likewise, the generation of noxious by-products or the often uses of extreme reaction conditions—offering a suitable answer to environmental concerns [77].

Therefore, several green nanotechnological approaches have been reported for the generation of nanostructures. These methods involve the utilization of living entities—for instance, microorganisms, fungi, or plants—and biomaterials coming from different natural sources, like fruits, biopolymers, and organic waste materials. It is important to know that each of these alternatives to the traditional chemical methods offers a different mechanism for the generation of the nanomaterials. However, the mechanisms still have as their basis the use of a reducing agent as polysaccharide and a capping agent like protein or lipid.

The application of green methodologies for the production of bimetallic nanostructures is not widely reported as it could happen with other well-known nanoparticles, such as gold or silver. Nevertheless, some interesting examples can be found in the literature. In the next sections, the leading green synthetic approaches are explained together with some relevant examples, with some examples shown in Fig. 2.

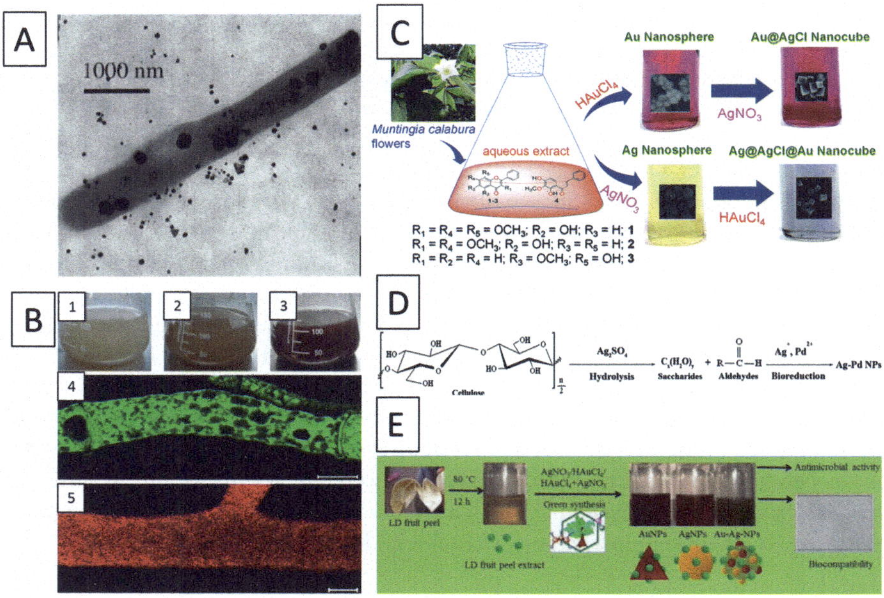

Fig. 2 Electron micrographs of the growth of gold crystals in *Lactobacillus strains* (**A**). Reprinted with permission from [78]. Copyright 2019 American Chemical Society; N. crassa biomass exposed to liquid solutions of $AgNO_3$ and $HAuCl_4$ (1) culture at time zero; (2) culture after 24 h in $AgNO_3$; (3) culture after 24 h in $HAuCl_4$; (4) hypha after 24 h in $AgNO_3$ scanned under confocal microscopy (Abs/Em 420/515–530 nm); (5) hypha after 24 h in $HAuCl_4$ scanned under confocal microscopy (Abs/Em 543/574–691 nm) (**B**) [79]; synthesis of metal nanoparticles using flower extract of *Muntingia calabura* (**C**) [80]; The main reaction mechanism for the biosynthesis of Ag–Pd bimetallic NPs (**D**) [80]; An economic and environmental friendly methodology created to produce colloidal solutions of gold–silver alloy (Au–Ag NPs) nanoparticles using *Lansium domesticum* (LD) fruit peel extract (**E**) [80]

Table 2 Generation of bimetallic nanoparticles mediated by bacteria

Bacteria	Method of synthesis	Type of NPs	Size/nm	Shape
Lactobacillus	Extracellular	Au–Ag [78]	100–500	Irregular
Phormidium tenue (marine cyanobacterium)	C-phycoerythrin pigment extracted from the marine cyanobacterium	Cd–S [82]	5.1 ± 0.2	Spherical
Deinococcus radiodurans	Protein extracts (Dps2)	Au–Ag [32, 33]	10–60	Spherical
Shewanella oneidensis	Intracellular and extracellular	Pd–Pt [83]	3–40	Spherical
Escherichia coli	Intracellular and extracellular	Au–Pd core–shell [84]	16	Spherical
Cupriavidus necator	Extracellular	Au–Pd core–shell [85]	10–50	Spherical
Spirulina platensis	Extracellular	Au–Ag [86]	17–25	Core–shell

Bacteria-Mediated Synthesis

The use of bacterial populations as natural biofactories was one of the first reported methods. Since then, both Gram-negative and -positive bacteria have been used for the successful generation of many hard systems with biomedical and technological applications [81]. It has been reported that these microorganisms can generate nanoparticles by converting metal ions into zero-valence metal nanoparticles, in a synthetic process that can be either intracellular or extracellular [24]. In Table 2, examples of bacteria-mediated bimetallic nanoparticle synthesis are presented.

For instance, *Lactobacillus* strains in buttermilk can support the process of growing of Au-Ag alloy crystals of submicron size as a result of exposing them to the precursor ions without affecting the bacterial survival by the crystal growth [78]. Similarly, Pd–Pt nanostructures were generated both inside and outside the cells of *Shewanella oneidensis*, a facultative bacterium, with the ability to grow and resist in both aerobic and anaerobic media [83].

Fungi-Mediated Synthesis

Fungi are eukaryotic organisms that include yeasts, molds, and mushrooms. In the past years, there has been a rising need in developing biogenic synthesis methods of metallic nanoparticles using fungi [87]. As it happens with bacteria, the reduction of metallic salts by fungi may occur both intra- and extracellularly, using enzymes and proteins that these microorganisms produce naturally.

Table 3 Synthesis of bimetallic nanoparticles mediated by fungi

Fungi	Nanoparticle		Size (nm)	Morphology
Fusarium oxysporum	CdS [91]		5–20	Spherical
Neurospora crassa	Au–Ag [79]	70/30	3–90	Spherical
		50/50	3–110	
		30/70	4–45	
Fusarium oxysporum	Au–Ag [92]		8–14	Spherical
Fusarium semitectum	Au–Ag [93]	50/50	10–35	Spherical
Volvariella volvacea	Au–Ag [94]		50–100	Predominantly spherical
Yeast (instant dry)	Au–Ag [95]		9–25	Quasi-spherical

Interestingly, up-scale manufacturing of nanostructures from different fungal strains has demonstrated possible for scale-up processing, since they can be grown even in vitro [88]. It is worth mentioning that for large-scale production of nanoparticles by biogenic synthesis, the use of fungi rather than bacteria is preferable since it is faster to produce fungus biomass than bacteria biomass [89]. Besides, the fungus mycelia have a large surface area for interaction [90]. Hence, the fungi can produce a high number of proteins, which may accelerate the process of the reduction of metallic salts. Some examples are given in Table 3, which shows previous studies into the synthesis of bimetallic nanoparticles utilizing fungi.

The filamentous fungus *Neurospora crassa*—a red bread mold—was analyzed and found to be propitious for the synthesis of bimetallic Au–Ag NPs [79], whereas crystallized and spherical-shaped Au–Ag NPs were generated and regulated using a fungus *Fusarium semitectum* in an aqueous media [93].

Plant-Mediated Synthesis

The use of plant extracts for the generation of nanomaterials offers suitable and easy-to-scale-up approaches. The chemical composition of the extracts is full of compounds with diverse functional groups, like aldehyde, hydroxyl, carboxyl, or amine groups. These groups are primarily responsible for the reduction of metal salts into zero-valence metals, conducive for the formation of nanoparticles with various shapes and dimensions [96]. The compounds found in plant extracts help in the stabilization of the colloidal NP, preventing particle aggregation and controlling their morphology over time [97, 98].

In general, plant-mediated synthesis includes the preparation of a liquid extract, which consists of cutting finely or milling distinct pieces of the plant for instance stems, leaves, flowers, or even roots until obtaining light and dried powder. Afterwards, this powder is added to a certain amount of water. The mixture is kept at a specific heat within a range of 25–100 °C for a period that could be from a few minutes to several hours. Then, the solution is decanted or filtered to remove insoluble solid materials. Subsequently, the essence is combined with a solution of the metal salts at a specific temperature and desired pH with or without agitation [99–102].

Table 4 Generation of bimetallic nanoparticles mediated by plant extracts

Plant	Type of alloy NPs		Size/nm	Shape
Cacumen platycladi	Ag–Pd [97]	3:1	11.9 ± 0.8	Mostly spherical
		1:1	9.1 ± 0.7	
		1:3	7.4 ± 0.4	
Cacumen platycladi	Au–Pd core–petals [103]		47.8	Flower-like
Cacumen platycladi	Au–Pd [104]		~7	Spherical
Antigonon leptopus	Au–Ag [105]		8–176	Spherical, triangular and irregular shapes
Dioscorea bulbifera	Pt–Pd [106]		20–25	Irregular shape
Ocimum basilicum	Au–Ag [107]		21 ± 11.53 (leaf extract) 25 ± 9.63 (flower extract)	Spherical
Azadirachta indica (neem)	Au–Ag core–shell [108]		50–100	Predominantly spherical
Agrimonia eupatoria	Cu–Pt core–shell [109]		30	Spherical
Muntingia calabura	Ag–AgCl–Au [80]		100–400	Cubical

In Table 4, different types of noble metal alloy NP using a variety of vegetable extracts made up of different parts of a living organism are shown. For instance, S. S. Shankar et al. synthetized Au–Ag core–shell extracellularly utilizing neem—*Azadirachta indica*—leaf solution [108]. When compared to previous studies that use microorganisms like fungi, it was stated that the rates of reduction of the metal ions utilizing this leaf extract were significantly faster. Alternatively, leaf extract of *Cacumen platycladi*—evergreen coniferous tree in the cypress family—was used to synthesize spherical Ag–Pd nanostructures [97] and Au–Pd nanoparticles [104], but also flower-shaped Au–Pd core–shell bimetallic nanostructures [103].

Similarly, Ganaei et al. demonstrated the efficient synthesis of bimetallic Au–Ag NP using a highly invasive soil weed coral vine (*Antigonon leptopus*) [105]. The liquid extracts of the weed were found to be capable of reducing the metal ions and therefore generate aggregates of nanometric size, which were finally stabilized to avoid further aggregation processes. Similarly, Chopade et al. used tubers from *Dioscorea bulbifera* to synthesize Pt–Pd NPs that were used as anticancer agents [106].

Biomolecule-Mediated Synthesis

Another green synthesis approach for the synthesis of bimetallic nanostructures is utilizing biopolymers such as chitosan, polyethylene glycol (PEG), poly(methyl vinyl ether-*co*-maleic anhydride) (PVM/MA), and starch, as well as other mole-

Table 5 Synthesis of bimetallic nanoparticles by biomolecules

Biomolecules	Type of NPs	Size/nm	Shape
Polyglycerol (HBP)	Au–Pt [34, 35]	8.0 ± 1.8	Spherical
	Au–Pd [34, 35]	7.8 ± 2.0	Spherical
	Au–Ru [34, 35]	7.1 ± 1.3	Spherical
Tryptophan	Au–Ag [52]	5–15	Quasi-spherical
Pueraria starch	Au–Ag [110]	32 ± 1.6	Spherical
Starch	Ag–Cu [111]	20 ± 1.6	Spherical
Cellulose	Ag–Pd [33]	59.7 ± 1.4	Quasi-spherical
Chitosan	Ag–Cu [62]	–	–
Gelatin	Au–Ag [112]	30	Spherical

cules, as can be seen in Table 5. These biopolymers serve as eco-friendly reducing and stabilizer agents in the synthesis reaction [113]. Haiqing Li et al. used the biocompatible hyperbranched polyglycerol (HBP) as an active reducing and stabilizing agent for the generation of bimetallic nanoparticles (Au–Pt, Au–Pd, and Au–Ru) with high solubility in aqueous media [34, 35]. The reports have shown the Au–Pt spherical particles with a size of 4.5 ± 1.7 nm, while on the other hand, Au–Pd and Au–Ru particles did not have a uniform size and they showed irregular morphologies.

In another study, the amino acid tryptophan was employed for the synthesis of Ag–Au NPs due to their ability to act as both reducing and capping agent. The Ag–Au NPs presented size of 5–15 nm with a quasi-spherical morphology [52]. Furthermore, a particular type of starch known as degraded pueraria starch (DPS) was used by Xia et al. to synthesize Au–Ag NPs with a diameter of around 32 nm and with a nearly spherical morphology [110].

X. Li et al. have recently reported an effective biological methodology related to the hydrolysis of cellulose with the aim to generate Ag–Pd alloy NP utilizing hydrothermal conditions [33]. Alternatively, a green approach to generate gelatin-capped bimetallic Au–Ag NP has been reported, with its principals on the interaction between silver nitrate or chloroauric acid with a 1.0 wt% liquid gelatin solution at 50 °C. The developed process is environmental friendly and reactive to chemiluminescence detection of a specific drug in its bulk powder, pharmaceutical injections, and biological samples [112].

Waste Material-Mediated Synthesis

On the one hand, waste materials coming from natural sources, such as agricultural wastes, have been recognized as an environmental burden for the society. On the other hand, they have stimulated new gateways for the production of renewable, low cost, and sustainable materials [114]. Nevertheless, few methodologies have been described for the generation of nanostructures using these raw materials with successful results.

The synthesis of BMNPs mediated by waste materials has been poorly explored, so it remains as an area of opportunity for future scientific research. S. Shankar et al. presented for the first time an economic and non-contaminant approach for the formation of a colloidal solution of Au–Ag NPs utilizing *Lansium domesticum* fruit peel essence as a mixed reducing and capping agent [115].

Bimetallic Nanoparticles as Biomedical Tools

In the next section, the successful application of bimetallic nanomaterials in the biomedical field will be explored as a potential solution towards cancer and bacterial infections, and as a tool to enhance biological imaging, drug delivery, and biosensing.

Antimicrobial Applications

The interaction between bacteria and humans has always been a complex and pivotal relationship with beneficial and undesired consequences. Living in a bacterial world as we do, pathogenic bacteria have learned how to infect humans. Fighting back was a laborious process until the discovery of penicillin by Sir Alex Fleming in 1932 [79]. The antibiotic era was born, providing society with a wide range of antibiotics to fight infections. However, bacteria fought back, and they developed what is known as human-induced antibiotic resistance—different from natural antibiotic resistance found in the interaction between living organisms, such as fungi, with bacteria. Different bacterial strains started developing resistance to the main antibiotics soon after their first use; hence, the development of new formulations became a race for pharmaceutical companies [116]. Therefore, humankind is approaching the post-antibiotic era, where infections that were easily treated in the past can now kill again [117].

Nanotechnology has emerged as a potential solution to overcome the main limitations of antibiotics. Although it has been reported, bacterial resistance to NP and other nanostructures is not prevalent among microorganism. For a long time, nanostructures composed of noble metals like gold and silver have been used as powerful antimicrobial agents, decreasing the growth of bacterial populations [118, 119]. Nevertheless, bacteria are starting to develop resistance to them employing natural defense mechanisms [120].

Therefore, new formulations are needed, enhancing antimicrobial approaches. The use of BMNPs as antimicrobial agents is not prominent in the biomedical field. A synergic antimicrobial effect can be achieved when different formulations of materials, such as silver nanostructures, are hybridized with other metal and oxide nanoparticles. For core–shell nanostructures, it is essential that one material acts as a shell, while the other will form the core of the material [121]. Often, an enhancement of the antimicrobial properties—among others, as it will be discussed—is

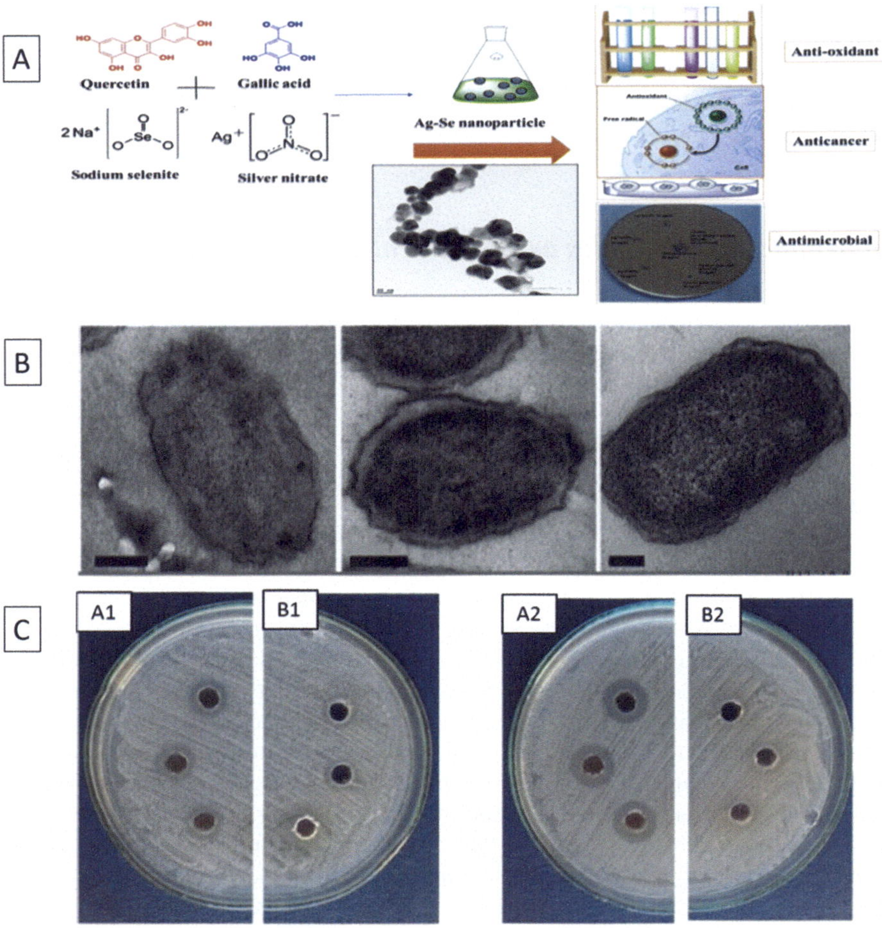

Fig. 3 Methodology for the synthesis of stable, mono-dispersed high-yielding bimetallic (Ag–Se) nanoparticles utilizing quercetin and gallic acid at room temperature (**A**) [122]; Morphological changes of *Escherichia coli* when subjected to Au–Pt NPs (40 μg mL^{-1}, 2 h) visualized with TEM; the NPs inferred blurring of the cytoplasm membrane, misplacing of the interior structure, and generation of a large-scale light area (**B**) [123]; Antimicrobial effect of Cu–Pt nanoparticles (A1, A2) and *Agrimoniae herba* extract (B1–B2) facing *Staphylococcus aureus* (A1, B1) and *Escherichia coli* (A2, B2) [109]

found compared to the monometallic formulation. Importantly, when talking about bimetallic nanoparticles—with some examples being shown in Fig. 3—as antimicrobial agents, they can be found in literature as the individual nanostructure, with the presence of a coating made of different molecules or combined with different drugs.

For instance, Sankar Ghosh et al. reported a natural synthesis using sodium citrate as a reducing agent for the development of bimetallic Au–Ag core–shell

NP. In the system, the Au NPs were utilized as seeds for constant discharging of silver atoms on its surface [124]. The structures presented antibacterial effects against Gram-negative (*Escherichia coli* and *Pseudomonas aeruginosa*) and the two Gram-positive (*Enterococcus faecalis* and *Pediococcus acidilactici*) bacteria when working with low concentrated silver shell, stronger efficiency was observed when treating Gram-negative bacteria. Y. Zhao's group showed that non-modified BMNPs of Au–Pt were powerful antibiotic materials, while pure Au NPs or pure Pt NPs did not display any antibacterial effect. A synthesis of the structures was achieved using a co-reduction of $HAuCl_4$ and K_2PtCl_4 with sodium borohydride in the vicinity of Tween 80 used as a stabilizer in water in an ice-water bath [123]. They have explained that the mechanism that triggers the antibacterial activity is related to the dissemination of membrane potential and the increase of adenosine triphosphate (ATP) levels by using Au–Pt nanoparticles.

Besides, the formulation of bimetallic nanoparticles allows the combination with other molecules of biological interest. Holden's group produced Ag–Au alloy BMNPs covered with glutathione through a synthetic process of galvanization replacement between maltose-coated Ag NPs and chloroauric acid ($HAuCl_4$) in an aqueous copolymer solution [125]. They successfully brought into comparison the antibacterial properties of the Ag–Au NPs to pure Ag NPs on *Porphyromonas gingivalis* W83, a pivotal pathogen in the development of periodontitis. They found that passivation of Ag NP with metals like gold can produce less toxicity in eukaryotes.

Alternatively, Antonoglou et al. employed a reproducible, easy hybrid polyol methodology employing non-extreme temperature and solvothermal conditions with the aim to isolate bimetallic Cu–Fe nanoparticles polluted with non-oxide compounds [126]. Besides, they studied how 1,2-propylene glycol (PG), tetraethylene glycol (TEG), and polyethylene glycol (PEG 8000), with distinct physicochemical properties, can control the dimensions, structure, composition, and the surface chemistry of the nanostructures.

Interestingly, the effect of antibiotics can be enhanced when combined with BMNPs. For instance, Naji M. et al. explored the profits of bimetallic Au–Ag NPs when used together with doxycycline for burn infections [127]. The bimetallic nanoparticles, produced by core–shell method, showed the presence of a synergistic antibacterial effects of doxycycline coupled with the bimetallic nanoparticles when treating *Pseudomonas aeruginosa*, *Escherichia coli*, *Staphylococcus aureus*, and *Micrococcus luteus*.

Belonging to the green chemistry approaches, some pathways have been reported for the production of BMNPs. For instance, R. Singh et al. showed the antimycobacterial properties of medicinal plants (*Barleria prionitis*, *Plumbago zeylanica*, and *Syzygium cumini*)—mediated synthesis of Au NPs, Ag NPs, and Au–Ag NPs, demonstrating strong effectivity, specificity, and selectivity to end with mycobacteria [100, 101]. In a similar research, Akinsiku et al. described the production and modeling of Ag NPs and silver/nickel-allied BMNPs (Ag–Ni NPs) using a plant-extract reduction method [128]. The antimicrobial properties of the nanoparticles were studied through the application of Ag NPs as antibacterial and antifungal agents. Dlugaszewska J. used the *Agrimoniae* herbal essence to get bimetallic core–shell Cu–Pt NP, showing a maximum performance when treating Gram-negative bacteria [109].

Anticancer Applications

Cancer is defined as an abnormal growth of tissue where cells divide uncontrollably, leading to an exponential cell population growth and the death of healthy cells surrounding the cancerous tissue [129]. In 2018, the global cancer burden rose to 18.1 million cases and 9.6 million cancer deaths. In the United States alone, there were approximately 1,735,350 cancer cases diagnosed and 609,640 cancer deaths in 2018 [130]. With the current cancer survivor rate, the associated cost of treatments will be 157.77 billion US dollars by 2020 [131].

Humankind started using surgery as a cancer treatment. Nowadays, we live in the era of chemotherapy and radiotherapy [132]. The first one involves chemical drugs that are introduced intravenously or orally and that, upon entering the bloodstream, attack cancer cells, inducing cell-cell death [133]. On the other hand, radiotherapy takes advantage of an ionic radiation dose in the tumor location, producing DNA damage that leads to cell failure [134]. Often, both techniques are used in combination with each other to improve the efficacy of treatment.

Nevertheless, they both include essential limitations, such as a different effect on each patient, and harmful side effects like anemia, organ damage, hair loss, and vomiting, among others [135]. As a consequence of the above-mentioned limitations in chemotherapy and radiation therapy, several new treatments have arisen in the last decades, such as immune therapy, hyperthermia, or gene therapy [136].

The use of nanotechnology for cancer treatment has gained influence, as it is possible to overcome some current limitations regarding current imaging techniques and drug delivery approaches [137]. Specifically, nanoparticles can be synthesized with the ability to differentiate cancer cells from healthy cells, which enables tumor targeting and is one of the potential advantages of nanotechnology for cancer treatment [138]. Consequently, NPs meet the essential cancer therapeutic requirements of efficacy and selective toxicity. One of the most encouraging nanomaterials for cancer treatments is gold nanoparticles [22] since they are significantly stable, sensitive, and can be manufactured with elevated levels of precision. Moreover, there has been continued research to use them as drug carriers, photothermal agents, contrast agents, and radiosensitizers. Besides gold, silver nanoparticles [139] have been proposed as potential anticancer agents, with unprecedent cytotoxic characteristics and the ability to provoke apoptosis in a range of cancer cells [140].

BMNPs can offer an enhancement in their properties compared to monometallic nanostructures due to a synergetic effect between the two metals employed. The major concerns related to physicochemical features in relation to the biological performance of metallic nanoparticle are primarily their dimensions, shape, composition, and charge among others; another key factor is the interaction with cancer cells. Besides, when combined with some techniques used in imaging or drug delivery for cancer, the improvement is typically related to this synergetic effect.

For instance, in an elegant study recently published, I. Shmarakov et al. proposed that the bimetallic Ag–Au composition could be utilized as improved anticancer materials because of their synergetic effect. Different colloidal solutions of Ag–Au NPs of alloy and core–shell type were analyzed to obtain the antitumor performance as a function of the molar ratio of the metal and the atomic distribution of Au and Ag within NPs. The nanoparticles were tested in vivo with mouse Lewis lung carcinoma (LLC). Their results suggested that there was a significant dependence between the in vivo antitumor performance and the gold and silver interaction coming from their ordered atomic distribution [141]. Consequently, NPs with Ag core surrounded by Au shell showed maximum effectivity in comparison with all the NPs analyzed towards LLC tumor growth and metastasizing inhibition.

Similarly, Kannan S. and Mishra S. K. synthesized a silver-neodymium bimetallic nanoparticle system that was generated by a microwave-assisted polyol synthetic procedure. The structures confirmed magnetic resonance imaging (MRI), computed tomography (CT), and NIR trimodal imaging capacity and stated worthy temperature response to irradiation when NIR laser was used. The nanoparticles were functionalized using chitosan on the surface, delivering suitable biocompatibility in addition to promoting the charging of paclitaxel, an anticancer drug [142]. Consequently, the bimetallic formulation supposes a combined therapeutic agent, giving the capacity to eliminate cancer cells in vitro at very low concentrations when compared to single therapy.

Hybrid Zn–Au NPs composed of zinc cores and gold shells were synthesized by Jongmin et al., with the objective of providing them with in vivo discernibility through positron emission tomography (PET) once the zinc core was activated and also to give them the capacity to produce radiosensitization by the use of secondary electrons generated on the gold shell after being irradiated with different radiation founts [31]. Quasi-spherical zinc NPs (<5 nm diameter) were produced and then recovered with a <4.25 nm gold layer to generate Zn–Au NPs. After several tests, they concluded that the Zn–Au NPs synthetized in this research have the ability to be utilized as PET-imageable radiosensitizers for applications such as radiotherapy and PET tracers for molecular imaging.

Alternatively, a methodology utilizing the reduction of silver nitrate and tetrachlorauratic acid at the same time and also employing tryptophan (Trp) as a reducing/stabilizing agent was developed. The obtained Ag–Au–Trp NPs, sized 5–15 nm, were capable of generating stable aggregates with dimensions ranging 370–450 nm and showed significantly less toxicity compared to Ag–Au NPs stabilized by sodium dodecyl sulfate (SDS), taking into account the estimation of biochemical parameters and oxidative damage produces considering a mouse model system. Ag–Au–Trp NPs demonstrated to have anticancer activity regarding a Lewis lung carcinoma model [52]. The results of the research reinforce the theory that the use of tryptophan in NP production successfully attenuates the potential hepato- and nephrotoxicity of NPs in vivo.

The bimetallic oxide nanostructures are known to enhance the carrier characteristics of the nanomaterials. For instance, Kumar and colleagues used ZnO–MgO BMNPs that were synthesized by precipitation process at low temperature. In order to elucidate the efficacy of the uptake and discharge of the anticancer drug, the adsorption and liberation of doxorubicin were analyzed, and also the kinetics of the procedure. They deduced that ZnO–MgO nanoparticles would potentially be a powerful drug carrier in a drug delivery system [143].

In a similar work, Alarfaj and El-Tohamy described the generation of gelatin-capped bimetallic Au–Ag NP with its principals on the interaction between silver nitrate and chloroauric acid with a 1.0 wt% liquid gelatin solution at 50 °C. The structures were utilized to improve a sensitive consecutive injection of the chemiluminescence luminol-potassium ferricyanide system in order to determine the anticancer drug raloxifene hydrochloride [112]. Therefore, they developed a method that is environmental friendly and sensitive for chemiluminescence detection of the desired drug when found in bulk powder form, pharmaceutical shots, and biological samples.

However, the nanoparticles show anticancer activity by themselves with no need for therapeutic drugs. As an example, a method of synthesis for the stable, monodispersed high-yielding bimetallic (Ag–Se) nanostructures utilizing quercetin and gallic acid at RT was studied. A variety of features, for instance, the amount of quercetin, gallic acid and Ag–Se salt, pH, temperature, and reaction time were investigated and improved to regulate the features of the nanoparticles [122]. The generated Ag–Se NPs were used as anticancer entities for Dalton lymphoma (DL) cell, and in addition, when used in vitro they showed a decrease of 80% of its growth at concentrations of 50 μg/mL. Besides, a Fe/Zn bimetallic nanoparticle using *Coriandrum sativum* leaf essence acting like reducing agent was prepared by an ultrasonic-assisted methodology. After generation, HeLa cancer cell line and normal cell line were tested in vitro to determine nanoparticles' cytotoxicity, showing 54.95% of cytotoxic effects at a concentration of 200 μg/mL when facing HeLa cancer cell line [144].

As mentioned in previous sections, over the last few years, green nanotechnology approaches have arisen as a novel solution for the production of bimetallic nanoparticles with biomedical applications. As an example, Chopade et al. demonstrated the quick and effective production of new Pt–Pd BMNPs by the use of a medicinal plant—*Dioscorea bulbifera*—acting as a reducing agent for the first time [106]. These nanostructures showed anticancer activity against HeLa cells. The bimetallic formulation exhibited more pronounced cell death—74.25%—in comparison to individual Pt NPs (12.6%) or Pd NPs (33.15%). Moreover, they reported an improved scavenging performance against 2,2-diphenyl-1-picrylhydrazyl, superoxide, nitric oxide, and hydroxyl radicals.

Imaging Applications

Imaging is largely applied in the biomedical field for diagnosis because of the ability of these techniques to provide an interface between vision and intuition [145]. Conjugation of NP with targeting ligands, for instance, peptides, small organic molecules, or antibodies, enables the development of targeted probes with elevated specificity [146]. The interaction of these modified-metallic nanoparticles within natural tissue results in real-time monitoring of molecules and single cells, tissues, and organs in vivo. Metallic nanoparticles are being used in different modalities for imaging, such as X-ray (CT), positron emission tomography (PET), single-photon emission CT (SPECT), ultrasound, and MRI [147], each of them has its strengths and limitations.

Nuclear imaging is a modality that generates data about biodistribution of radiolabeled drugs or other ligands; nonetheless, it particularly ends in noisy images which have low spatial resolution when compared with anatomic images. In this technique, emitted photons have to penetrate the massive lead parallel collimators in the first time which only permits photons emitted perpendicularly with respect to the camera face to prosperously interact with the imaging crystal [148]. Consequently, to improve the resolution of this imaging technique, single-photon emission tomography (SPECT) was developed, which obtaining volumetric images having a spatial resolution of ~7.5–10 mm. In the line of this research, bimetallic hybrid Zn–Au nanoparticles with zinc cores and gold shells were synthesized to allow them to be seen in vivo when the zinc cores are proton activated and also gave them the ability to provoke radiosensitization by using secondary electrons generated by the gold shell when irradiated by different radiation sources [31].

Another medical technique, PET helps reveal how biological tissues are functioning, using a radioactive drug, also called tracer, to show this activity. Radionuclides that decay with positron emissions, like fluorine-18, carbon-11, iodine-124, and oxygen-15, can be imaged with a PET scanner that gives a higher spatial resolution (2–5 mm for clinical scanners). The emitted positron easily finds an electron, ending in a reaction of annihilation where both photons present an energy around 511 keV and provokes trajectories around 180° in contrary directions [149]. Reduced-quality imaging of positrons can be achieved utilizing SPECT having ultra-high energy collimators; on the other hand, the PET camera was particularly designed to take profit of the physical features of positron emitters [150]. Bo Pang et al. developed an easy methodology for radiolabeled Pd–Cu–Au core–shell tripods to utilize them in PET and image-guided photothermal cancer treatment when radioactive 64Cu atoms were directly introduced inside the crystal lattice [151].

CT was created by Hounsfield et al. in 1967, in one side motivated for the enhancement in not only computer processing power but also the capacity of X-ray imaging. Utilizing accessible computing technology, CT produced high-resolution 3D modelings of patients' anatomy by the use of multiplane X-ray imaging and mathematical computer image regeneration algorithms [152]. CT is one of the convenient types of imaging; however, it is not considered a targeted imaging modality without using specific contrast agents [153]. Therefore, metallic nanoparticles have been used as a contrast agent for molecular and targeted imaging of tumors and cancers [153]. For example, B. Li et al. have developed a multifunctional copper–bismuth sulfide (Cu–BiS) nanocomposite as a contrast agent for CT, IR thermal imaging, and photothermal therapy [154, 155]. This enables them to mix the NIR results with nanostructure-improved imaging results, like MRI or X-ray CT ones.

A special mention should be made with magnetic NP, which have been utilized in a broad variety of applications due to their properties that are generated by their size, shape, composition, and the capacity of functionalization at the cellular and molecular levels. Notably, their unique magnetic properties make them an interesting system to be used as contrast entity MRI. When going to the bimetallic configuration, the structures can also exhibit super magnetic properties, which makes them great aspirants to work as MRI contrast agents.

For example, and as can be seen in Fig. 4, later progressions in the production and alteration of Fe–Pt nanoparticles, which have higher chemical stability compared to Fe and Fe–Co, have made them a viable option for MRI and CT [154, 155, 157]. These particles are formed through thermal annealing of the MgO-coated FePt–Fe_3O_4 NPs, after that an acid washing is done to eliminate MgO. Amendola et al. used laser ablation synthesis in solution to obtain PEG–FeAu NPs with excellent stability in air and in aqueous solutions, which exhibited good properties to act as harmful contrast agents for MRI [48]. In another work, J. Choi et al. enhanced imaging accuracy of Fe-based nanoparticle using a Gd-based coating [158]. In this work, a dual-mode nanoparticle contrast has been reported; the T_1 contrast material,

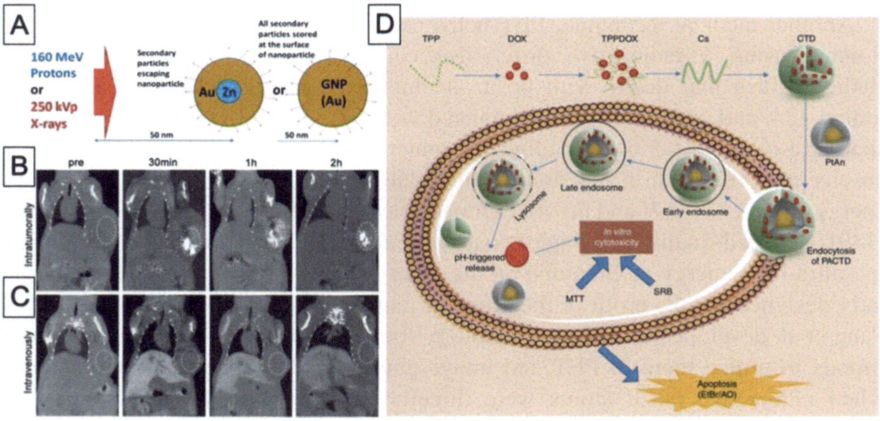

Fig. 4 (**A**) An incident proton or X-ray beam interacted with nonradioactive gold-coated zinc (Zn–Au NPs) or gold nanoparticles (GNPs) placed in vacuum, as well as with the secondary particles. All secondary particles scored at the surface of NP [31]; In vitro CT image of a intratumoral (**B**) and intravenous (**C**) injection of the solution of the Cu_3BiS_3 NPs. The position of tumor is labeled by the dotted circles [154, 155], and a schematic illustration of doxorubicin-encapsulated platinum–gold/chitosan BMNPs for sustained pH responsive release leading to site-specific in vitro biological activity and apoptosis (**D**) [156]

$Gd_2O(CO_3)_2$ (thickness = 1.5 nm), is found on the shell for high T_1 contrast characteristics, and the superparamagnetic T_2 contrast material, $MnFe_2O_4$ of 15 nm, is found at the core. Therefore, $Gd_2O(CO_3)_2$ shell come into direct contact with water molecules, and $MnFe_2O_4$ core maintains a long-range magnetic field for the relaxation of water molecules. Nevertheless, T_1 contrast agents are suitable for morphological assessment of the standard or pathological anatomy (e.g., for musculoskeletal applications), and T_2 contrast agent is suitable for the detection of blood and iron and calcification in many tissues [159]. Therefore, these nanoparticles can potentially be applied to a large variety of biotargets with improved diagnostic exactness.

Drug Delivery Applications

Nanostructures are often used in drug delivery applications. Properties of nanoparticles, such as the high surface-to-volume ratio and biocompatibility, enable an efficient loading and immobilization of a high amount of therapeutic agents including synthetic drug molecules, proteins and peptides, oligonucleotide and antibodies, either on or inside these nanostructures [160]. When talking about BMNPs (Fig. 4), tunable and adjustable physical property and critical parameters, which can happen by using a combination of elements, make these synergetic structures appealing for biomedical applications [161]. Targeted delivery of drugs is an essential biomedical application that aids in avoiding the systemic administration of drugs and increases the therapeutic efficiency of drugs.

For instance, Kumal et al. have recently reported a gold–silver–gold core–shell–shell (CSS) nanostructure to deliver oligonucleotide-like plasmid DNAs, siRNAs, and miRNAs—by the use of NIR irradiation cleavage, which enables targeted delivery of RNAs to target locations under spatial and temporal regulation [162]. They have explained that the photothermal liberation of siRNA from the surface of CSS nanoparticles is particularly higher than the one from Au NPs when comparable conditions were used. Regulating the dimensions and ratio of the core and shell sizes has helped them in the tuning and optimization of release efficiency.

Similarly, Taylor et al. have used a combination of iron, zinc, and silver with superparamagnetic iron oxide nanoparticles, also known as (SPIO) for the therapy of antimicrobial-resistant biofilms [163]. In this work, Fe, Zn, and Ag have been separately conjugated to SPION to couple the antibacterial features of these compounds to the superparamagnetic characteristics of SPION. It is reported that this could improve the antibacterial activity of these metal salts and enables targeting bacterial infections (i.e., *Staphylococcus aureus*) in a magnetic field using superior magnetic properties of SPION.

Moreover, bimetallic nanoparticles have also been used as a drug carrier for cancer therapy. For example, Au–Pt (Au core and Pt shell) bimetallic nanoparticles were synthesized by V. Maney et al. and encapsulated inside chitosan coating through ionic gelation with tripolyphosphate (TPP). Also, doxorubicin (DOX)—most potent anticancer drug available—was encapsulated inside chitosan layer and delivered to different cancerous cells (e.g., human embryonic kidney cells

(HEK293), breast adenocarcinoma (MCF-7), epithelial colorectal adenocarcinoma cells (Caco-2), and hepatocellular carcinoma cells (HepG2)) [156]. This novel combination has enabled them to encapsulate high amounts of DOX and further induce a pH-triggered release at intracellular acidic conditions of cancerous cells to bring about selective cancer targeting.

Similarly, D. Dutta et al. have developed cationic bovine serum albumin (BSA)-embedded (gold–silver) Au–Ag NPs to carry a suicide gene payload into the cancer cells 153 effectively. It has been explained that loaded plasmid DNA (pDNA, CD-UPRT) and Au–Ag NPs induce apoptosis in HeLa cells through two different mechanisms. Specifically, suicide gene (CD-UPRT) converts the non-toxic prodrug 5-FC into 5-FU and other toxic metabolites, while Au–Ag applies oxidative stress inside cells through ROS generation.

Photothermal Therapy Applications

Photothermal therapy (PTT) is a promising technique to fight against microbial infection and cancer [164]. Initially, this form of therapy is a variation of photothermal ablation (PTA) therapy that utilizes light (NIR wavelength range (700–2000 nm)) to burn tumors. However, PTA may destroy healthy cells and tissues due to unspecific targeting of the light. Recently, nanotechnology has been shown as a potential way to remarkably enhance PTA by applying a near-IR laser light absorbing material, such as metallic/bimetallic nanoparticles, to the target sight in the body [165]. Unique advantages appended to PTT by nanotechnologies, such as high specificity and minimal invasiveness, have made this technique an excellent potential for treating cancer metastasis and bacterial infections. Moreover, owing to using bimetallic and metal alloy nanoparticles, PTT can be guided with multimodal imaging or combined with the other available therapies to perform more effective targeting (Fig. 5).

For instance, Wang et al. have developed a dual-mode bimetallic nanoparticle for imaging and photothermal therapy of SK-BR-3 breast cancer cells [168]. In this work, the super magnetic properties of Fe_3O_4 and the photon luminescence effect of gold nanorods (Au_{rod}) make using a combination of diagnosis, imaging and therapy techniques possible. In another work, Fan et al. used an Au-coated Fe-based nanoparticle to target and destroy multidrug-resistant (MDR) bacteria (i.e., MDR *Salmonella* DT104) [169]. As reported, the plasmonic shell made of Au, in combination with the magnetic core of iron of developed nanoparticles, has been used for targeted separation, in addition to selective photothermal eradication of a MDR pathogen.

NP based on noble metals, like Au, Ag, and Pt, are mainly utilized for PTT because they present enhanced biocompatibility, decreased toxicity, and strong optical absorption in the near-IR (NIR, 700–1100 nm) [170]. For instance, Cheng et al. have used chitosan-capped seedless branched-bimetallic Au–Ag NPs against oral cancer cell line (SAS) [171]. This work has proposed a quick, facile, and one-pot approach for making spiky star-shaped Au–Ag NPs by adding aqueous $HAuCl_4$ to

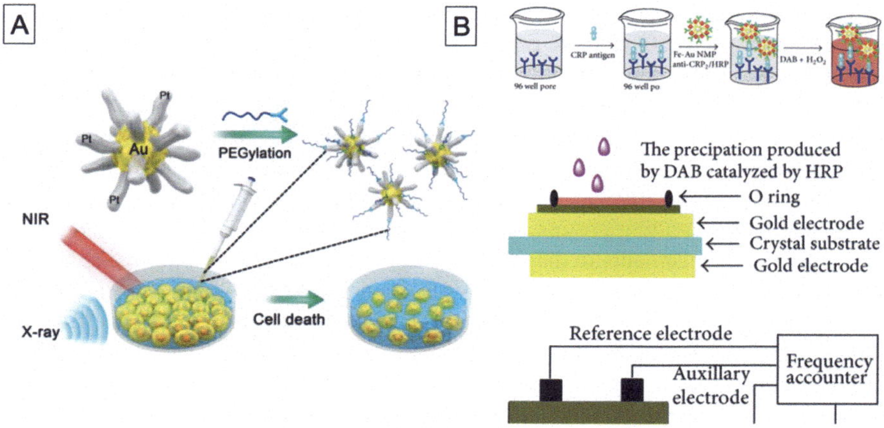

Fig. 5 The illustration of the PEGylation and photothermal/radiation synergistic therapy application of the Au–Pt nanoparticles. Reprinted with permission from [166] (**A**). Copyright 2019 American Chemical Society; Diagram of the generation of Fe–Au MNP-anti-CRP2/horseradish peroxidase (HRP) conjugation (**B**) and the immunoassay methodology with the employment of Fe–Au MNP-anti-CRP2/HRP as the signal tag (*top*) and the EQCM's measurement steps (*middle*) and detection system (*bottom*) [167] (**B**)

$AgNO_3$, and then mixing it with enough ascorbic acid acting as reducing agent. The morphology of the NPs is regulated by changing the ratio of gold to silver precursors.

Moreover, X. Liu et al. have developed Au–Pt nanostructures which are comprised of Au core and Pt nanobranches with a wide absorbance of NIR light and intense concentration of X-rays [166]. To improve the stability of AuPt nanostructures, they have been PEGylated and have been used against 4T1 cells under irradiation of an 808 nm laser.

Biosensing Applications

The development of nanoscience and nanotechnology revealed the critical role of bimetallic nanoparticles in areas such as nano-biosensors, immune labeling, and microelectronics because of their unique optical, electrical, and thermal characteristics [172].

The distinct properties of nanoparticles like the large surface-to-volume ratio, high electrical conductivity, favorable biocompatibility, excellent catalytic ability, and surface reaction activity provide metal nanoparticles enormous possibilities in improving sensor performance, a feature that can be extended to bimetallic configurations in the nanoscale. Therefore, they have been widely used to develop transducers for biosensors, based on the efficient combination of different components into a nanoscale hybrid structure, which enables to maintain the known characteristics of such components inasmuch to lead to novel features which are absent in each component individually.

For example, Gan et al. have developed an electrochemical immune sensor that can selectively sense C-reactive protein (CRP) using antigen–antibody reaction on a bimetallic nanoparticle [167]. CRP is primarily produced in the liver due to an acute inflammatory stimuli that its concentration in blood abruptly increases in the acute-phase response of inflammation [173]. They used a supermagnetic iron oxide nanoparticles (Fe_3O_4) with an Au coating, achieving a well-assembled, biocompatible, and environmentally safe metallic component with high affinity for connecting to the amine/thiol ending groups of organic molecules. Thus, they contributed significantly to advancements in diagnostics and biological detection [174, 175].

Alternatively, semiconductor nanocrystal quantum dots (QD) have been utilized for protease sensing using fluorescence resonance energy transfer (FRET) and bioluminescence by E. Chang et al., using the properties of these particles, since QDs are inorganic luminescent semiconductor structures [176]. They typically are composed of a Cd–Se core, a zinc sulfide (ZnS) shell [177]. QDs quenched by AuNP using different cleavable peptides have been used to investigate some proteases including chymotrypsin and collagenase [178], thrombin [179], and magnetic NP (Fig. 5).

In another work, Chiriac et al. have developed a giant magneto-impedance (GMI) using Co–Fe magnetic microparticles (0.9–2 μm) for the determination of the small concentration of biomolecules (such as RNA) [180]. It is explained that functionalized Co–Fe microparticles could successfully target biomolecules (i.e., ssDNA (single-stranded DNA), antibodies, proteins, enzymes) and form a dipole field under the external magnetic environment that was detected by the GMI sensor. Moreover, impedancemetric and conductometric are conventional approaches to develop immune sensors. For example, D. Lin et al. have developed an electrochemical impendence spectrum (EIS) in order to detect alpha-fetoprotein (AFP) by implementing Au–Pt nanodendrites [181]. AFP is a key tumor marker cancer diagnosis in liver, testis, and ovary. A sandwich-type electrochemical immunosensor was generated using mesoporous graphene-loaded Au–Pt nanodendrites, which are improved by polydopamine functionalized N-doped multiwalled carbon nanotube (PDA-N-MWCNT).

The Future of Bimetallic Nanoparticles

Despite the outstanding advances in the development of efficient bimetallic nanoparticles for biomedical applications, there are still several challenges for the use of these nanostructures. For instance, one of the top concerns that have to be addressed is the limited control over the resulting particle size that most of the synthetic methods offer. In the case of monometallic Au and Ag NPs, it has been reported that there is a particle size-dependent cytotoxic effect on several cancer cell types [182, 183]. For example, Pan et al. investigated the cytotoxicity of Au NPs in a particle size range of 0.5–15 nm [183]. They found that the tested cells were susceptible to relatively small AuNPs (1.4 nm). In contrast, 15 nm AuNPs were not cytotoxic at up to

60-fold higher concentrations. After extended research, it can be said that for noble metal-based BMNPs, there are no studies about the size-dependent cytotoxic effect of these nanostructures with a constant composition and atomic structure.

The next controversial issue is related to the mass transport across the original boundaries of the body, such as cellular membranes or intestinal barrier, where the biological affinity of the nanomaterials plays a pivotal role. Here, the use of BMNPs can enhance the biological affinity and mimic the conditions inside the body compared to other formulations. This behavior allows to efficiently induce cellular responses due to the nature of the different compounds employed in the nanosystem. For instance, N. D. Telling et al. evaluated the applicability of bacteria-mediated Zn- and Co-doped magnetite NP to use in biomedical fines, demonstrating that the magnetic particle response was modified once they interacted with cells producing a decline in their movement. This finding also shows that the medium-level dopant needs optimum magnetic characteristics in order to not modify their biological toxicity or influence osteogenic differentiation of the stem cells [184]. Therefore, although reported toxicity of Co and Zn ions, these data suggest that iron oxide nanoparticles can be tuned to adequately modify their magnetic features without jeopardizing cellular biocompatibility.

Another important milestone is related to the need for developing new mathematical and computer models that contribute to a greater knowledge and prediction of chemical, physical, and biological response to the presence of the nanostructures. These systems are challenging enough for the monometallic formulation, becoming even harder when two or more metals are incorporated within one single structure. However, several attempts have been made. For example, Akinsiku et al. have found that a theoretical model can clarify an experiment observation on the relationship between BMNP penetration through peptidoglycan layers and the activeness of microbial species, depending on the nature of the nanoparticles and pore size of the layer [128].

The nanotechnological inclusion in medical applications regarding bimetallic formulations is still a young field which needs more research, studies, and experiments.

Conclusion

Nanotechnology has undoubtedly risen as a powerful tool that benefits every aspect of society, from healthcare to environmental remediation, going through renewable technology and agriculture, with a meaningful economic impact that is expected to see exponential growth in time. Among all the potential nanotechnological applications, the utilization of nanoparticles in the biomedical area is leading to a novel horizon of techniques, developmental technologies, and improvements that are transforming the way that we fight diseases and pathogens. Despite this, the use of bimetallic formulations remains a field to be further explored, with plenty of unknowns and applications that need to be elucidated in order to build the better world that nanotechnology promises for humanity.

References

1. Olsman N, Goentoro L (2018) There's (still) plenty of room at the bottom. Curr Opin Biotechnol 54:72–79. https://doi.org/10.1016/j.copbio.2018.01.029
2. Taniguchi N (1974) On the basic concept of nano-technology. In: Proc. intl. conf. prod. London. https://ci.nii.ac.jp/naid/20000654683
3. Pearce JM (2012) Make nanotechnology research open-source. Nature 491(7425):519–521. https://doi.org/10.1038/491519a
4. Salata OV (2004) Applications of nanoparticles in biology and medicine. J Nanobiotechnol 2(1):3. https://doi.org/10.1186/1477-3155-2-3
5. Seeman NC (2003) Biochemistry and structural DNA nanotechnology: an evolving symbiotic relationship†. Biochemistry 42(24):7259–7269. https://doi.org/10.1021/bi030079v
6. Whitesides GM (2005) Nanoscience, nanotechnology, and chemistry. Small 1(2):172–179. https://doi.org/10.1002/smll.200400130
7. Andrew AM (2000) An introduction to support vector machines and other kernel-based learning methods by Nello Christianini and John Shawe-Taylor, Cambridge University Press, Cambridge, 2000, Xiii+189 pp., ISBN 0-521-78019-5. Robotica 18(6):687–689. https://doi.org/10.1017/S0263574700232827
8. Webster TJ, Seil I (2012) Antimicrobial applications of nanotechnology: methods and literature. Int J Nanomedicine 7:2767. https://doi.org/10.2147/IJN.S24805
9. Kamal MA, Jabir NR, Tabrez S, Ashraf GM, Shakil S, Damanhouri GA (2012) Nanotechnology-based approaches in anticancer research. Int J Nanomedicine 7:4391. https://doi.org/10.2147/IJN.S33838
10. Shi J, Votruba AR, Farokhzad OC, Langer R (2010) Nanotechnology in drug delivery and tissue engineering: from discovery to applications. Nano Lett 10(9):3223–3230. https://doi.org/10.1021/nl102184c
11. Cormode DP, Skajaa T, Fayad ZA, Mulder WJM (2009) Nanotechnology in medical imaging. Arterioscler Thromb Vasc Biol 29(7):992–1000. https://doi.org/10.1161/ATVBAHA.108.165506
12. Bekyarova E, Ni Y, Malarkey EB, Montana V, McWilliams JL, Haddon RC, Parpura V (2005) Applications of carbon nanotubes in biotechnology and biomedicine. J Biomed Nanotechnol 1(1):3–17. https://doi.org/10.1166/jbn.2005.004
13. Shi S, Chen F, Cai W (2013) Biomedical applications of functionalized hollow mesoporous silica nanoparticles: focusing on molecular imaging. Nanomedicine 8(12):2027–2039. https://doi.org/10.2217/nnm.13.177
14. Ramos AP, Cruz MAE, Tovani CB, Ciancaglini P (2017) Biomedical applications of nanotechnology. Biophys Rev 9(2):79–89. https://doi.org/10.1007/s12551-016-0246-2
15. Mody VV, Siwale R, Singh A, Mody HR (2010) Introduction to metallic nanoparticles. J Pharm Bioallied Sci 2(4):282–289. https://doi.org/10.4103/0975-7406.72127
16. Raliya R, Singh Chadha T, Haddad K, Biswas P (2016) Perspective on nanoparticle technology for biomedical use. Curr Pharm Des 22(17):2481–2490. http://www.ncbi.nlm.nih.gov/pubmed/26951098
17. Vernon RE (2013) Which elements are metalloids? J Chem Educ 90(12):1703–1707. https://doi.org/10.1021/ed3008457
18. Dutz S, Müller R, Eberbeck D, Hilger I, Zeisberger M (2015) Magnetic nanoparticles adapted for specific biomedical applications. Biomed Tech (Berl) 60(5):405–416. https://doi.org/10.1515/bmt-2015-0044
19. Kralj S, Makovec D, Čampelj S, Drofenik M (2010) Producing ultra-thin silica coatings on iron-oxide nanoparticles to improve their surface reactivity. J Magn Magn Mater 322(13):1847–1853. https://doi.org/10.1016/J.JMMM.2009.12.038
20. Cardoso VF, Francesko A, Ribeiro C, Bañobre-López M, Martins P, Lanceros-Mendez S (2018) Advances in magnetic nanoparticles for biomedical applications. Adv Healthc Mater 7(5):1700845. https://doi.org/10.1002/adhm.201700845

21. Tokajuk G, Niemirowicz K, Deptuła P, Piktel E, Cieśluk M, Wilczewska A, Dąbrowski J, Bucki R (2017) Use of magnetic nanoparticles as a drug delivery system to improve chlorhexidine antimicrobial activity. Int J Nanomedicine 12:7833–7846. https://doi.org/10.2147/IJN.S140661
22. Jain S, Hirst DG, O'Sullivan JM (2012) Gold nanoparticles as novel agents for cancer therapy. Br J Radiol 85(1010):101–113. https://doi.org/10.1259/bjr/59448833
23. Qing Y'a, Cheng L, Li R, Liu G, Zhang Y, Tang X, Wang J, Liu H, Qin Y (2018) Potential antibacterial mechanism of silver nanoparticles and the optimization of orthopedic implants by advanced modification technologies. Int J Nanomed 13:3311–3327. https://doi.org/10.2147/IJN.S165125
24. Medina Cruz D, Mi G, Webster TJ (2018) Synthesis and characterization of biogenic selenium nanoparticles with antimicrobial properties made by *Staphylococcus aureus*, methicillin-resistant *Staphylococcus aureus* (MRSA), *Escherichia coli,* and *Pseudomonas aeruginosa*. J Biomed Mater Res A 106(5):1400–1412. https://doi.org/10.1002/jbm.a.36347
25. Amiri M, Etemadifar Z, Daneshkazemi A, Nateghi M (2017) Antimicrobial effect of copper oxide nanoparticles on some oral bacteria and Candida species. J Dent Biomater 4(1):347–352. http://www.ncbi.nlm.nih.gov/pubmed/28959764
26. Chen Y, Wan Y, Wang Y, Zhang H, Jiao Z (2011) Anticancer efficacy enhancement and attenuation of side effects of doxorubicin with titanium dioxide nanoparticles. Int J Nanomedicine 6:2321–2326. https://doi.org/10.2147/IJN.S25460
27. Ding X, Yuan P, Gao N, Zhu H, Yang YY, Xu Q-H (2017) Au-Ag core-shell nanoparticles for simultaneous bacterial imaging and synergistic antibacterial activity. Nanomedicine 13(1):297–305. https://doi.org/10.1016/j.nano.2016.09.003
28. Allaedini G, Tasirin SM, Aminayi P (2016) The effects of cerium doping concentration on the properties and photocatalytic activity of bimetallic Mo/Ce catalyst. Russ J Phys Chem A 90(10):2080–2088. https://doi.org/10.1134/S0036024416080094
29. Jiang H-L, Xu Q (2011) Recent progress in synergistic catalysis over heterometallic nanoparticles. J Mater Chem 21(36):13705. https://doi.org/10.1039/c1jm12020d
30. Sun Y, Lei C (2009) Synthesis of out-of-substrate Au-Ag nanoplates with enhanced stability for catalysis. Angew Chem Int Ed 48(37):6824–6827. https://doi.org/10.1002/anie.200902305
31. Cho J, Wang M, Gonzalez-Lepera C, Mawlawi O, Cho SH (2016) Development of bimetallic (Zn@Au) nanoparticles as potential PET-imageable radiosensitizers. Med Phys 43(8 Part1):4775–4788. https://doi.org/10.1118/1.4958961
32. Li JL, Tian B, Li T, Dai S, Weng YL, Lu JJ, Xu XL, Jin Y, Pang RJ, Hua YJ (2018a) Biosynthesis of Au, Ag and Au–Ag bimetallic nanoparticles using protein extracts of Deinococcus radiodurans and evaluation of their cytotoxicity. Int J Nanomedicine 13:1411–1424. https://doi.org/10.2147/IJN.S149079
33. Li X, Odoom-Wubah T, Huang J (2018b) Biosynthesis of Ag–Pd bimetallic alloy nanoparticles through hydrolysis of cellulose triggered by silver sulfate. RSC Adv 8(53):30340–30345. https://doi.org/10.1039/C8RA04301A
34. Li H, Jo JK, Zhang LD, Ha C-S, Suh H, Kim I (2010a) Hyperbranched polyglycidol assisted green synthetic protocols for the preparation of multifunctional metal nanoparticles. Langmuir 26(23):18442–18453. https://doi.org/10.1021/la103483c
35. Li T, Albee B, Alemayehu M, Diaz R, Ingham L, Kamal S, Rodriguez M, Whaley Bishnoi S (2010b) Comparative toxicity study of Ag, Au, and Ag–Au bimetallic nanoparticles on Daphnia magna. Anal Bioanal Chem 398(2):689–700. https://doi.org/10.1007/s00216-010-3915-1
36. Anu Mary Ealia S, Saravanakumar MP (2017) A review on the classification, characterisation, synthesis of nanoparticles and their application. IOP Conf Ser Mater Sci Eng 263(3):032019. https://doi.org/10.1088/1757-899X/263/3/032019
37. Thiruvengadathan R, Korampally V, Ghosh A, Chanda N, Gangopadhyay K, Gangopadhyay S (2013) Nanomaterial processing using self-assembly-bottom-up chemical and biological approaches. Rep Prog Phys 76(6):066501. https://doi.org/10.1088/0034-4885/76/6/066501

38. Merkel TJ, Herlihy KP, Nunes J, Orgel RM, Rolland JP, DeSimone JM (2010) Scalable, shape-specific, top-down fabrication methods for the synthesis of engineered colloidal particles. Langmuir 26(16):13086–13096. https://doi.org/10.1021/la903890h
39. Lee Y-H, Chuang S-M, Huang S-C, Tan X, Liang R-Y, Yang GCC, Chueh PJ (2017) Biocompatibility assessment of nanomaterials for environmental safety screening. Environ Toxicol 32(4):1170–1182. https://doi.org/10.1002/tox.22313
40. Mantis Deposition Systems (2019). https://www.mantisdeposition.com/nanoparticlegenerators.html
41. Nanocluster Deposition Source (2019). http://www.oaresearch.co.uk/oaresearch/cluster/
42. Lin P-C, Lin S, Wang PC, Sridhar R (2014) Techniques for physicochemical characterization of nanomaterials. Biotechnol Adv 32(4):711–726. https://doi.org/10.1016/j.biotechadv.2013.11.006
43. Xiao Q, Yao Z, Liu J, Hai R, Oderji HY, Ding H (2011) Synthesis and characterization of Ag–Ni bimetallic nanoparticles by laser-induced plasma. Thin Solid Films 519(20):7116–7119. https://doi.org/10.1016/J.TSF.2011.04.201
44. Liu J, Ma X, Yang L, Liu X, Han A, Lv H, Zhang C, Xu S (2018) *In situ* green oxidation synthesis of Ti^{3+} and N self-doped $SrTiO_x N_y$ nanoparticles with enhanced photocatalytic activity under visible light. RSC Adv 8(13):7142–7151. https://doi.org/10.1039/C7RA13523H
45. Barnett GH, Chen CC, Gross RE, Sloan AE (2016) Introduction: laser ablation techniques. Neurosurg Focus 41(4):E1. https://doi.org/10.3171/2016.8.FOCUS16319
46. Tajdidzadeh M, Azmi BZ, Yunus WMM, Talib ZA, Sadrolhosseini AR, Karimzadeh K, Gene SA, Dorraj M (2014) Synthesis of silver nanoparticles dispersed in various aqueous media using laser ablation. ScientificWorldJournal 2014:324921. https://doi.org/10.1155/2014/324921
47. Sportelli MC, Izzi M, Volpe A, Clemente M, Picca RA, Ancona A, Lugarà PM, Palazzo G, Cioffi N (2018) The pros and cons of the use of laser ablation synthesis for the production of silver nano-antimicrobials. Antibiotics (Basel) 7(3):E67. https://doi.org/10.3390/antibiotics7030067
48. Amendola V, Meneghetti M, Bakr OM, Riello P, Polizzi S, Anjum DH, Fiameni S et al (2013) Coexistence of plasmonic and magnetic properties in Au89Fe11 nanoalloys. Nanoscale 5(12):5611. https://doi.org/10.1039/c3nr01119d
49. Peng S, Lei C, Ren Y, Cook RE, Sun Y (2011) Plasmonic/magnetic bifunctional nanoparticles. Angew Chem Int Ed 50(14):3158–3163. https://doi.org/10.1002/anie.201007794
50. Wang X, Sun S, Huang Z, Zhang H, Zhang S (2014) Preparation and catalytic activity of PVP-protected Au/Ni bimetallic nanoparticles for hydrogen generation from hydrolysis of basic NaBH4 solution. Int J Hydrog Energy 39(2):905–916. https://doi.org/10.1016/J.IJHYDENE.2013.10.122
51. Mukha I, Vityuk N, Grodzyuk G, Shcherbakov S, Lyberopoulou A, Efstathopoulos EP, Gazouli M (2017) Anticancer effect of Ag, Au, and Ag/Au bimetallic nanoparticles prepared in the presence of tryptophan. J Nanosci Nanotechnol 17(12):8987–8994. https://doi.org/10.1166/jnn.2017.14106
52. Shmarakov IO, Mukha IP, Karavan VV, Chunikhin OY, Marchenko MM, Smirnova NP, Eremenko AM (2014) Tryptophan-assisted synthesis reduces bimetallic gold/silver nanoparticle cytotoxicity and improves biological activity. Nanobiomedicine 1:6. https://doi.org/10.5772/59684
53. Pal A, Shah S, Devi S (2007) Preparation of silver, gold and silver–gold bimetallic nanoparticles in w/o microemulsion containing TritonX-100. Colloids Surf A Physicochem Eng Asp 302(1–3):483–487. https://doi.org/10.1016/j.colsurfa.2007.03.032
54. Nakamura T, Sato S (2015) Green and facile synthesis of Pd-Pt alloy nanoparticles by laser irradiation of aqueous solution. J Nanosci Nanotechnol 15(1):426–432. http://www.ncbi.nlm.nih.gov/pubmed/26328375
55. Mottaghi N, Ranjbar M, Farrokhpour H, Khoshouei M, Khoshouei A, Kameli P, Salamati H, Tabrizchi M, Jalilian-Nosrati M (2014) Ag/Pd core-shell nanoparticles by a successive method: pulsed laser ablation of Ag in water and reduction reaction of PdCl2. Appl Surf Sci 292:892–897. https://www.sciencedirect.com/science/article/pii/S0169433213023465

56. Zielińska-Jurek A, Zaleska A (2014) Ag/Pt-modified TiO2 nanoparticles for toluene photooxidation in the gas phase. Catal Today 230:104–111. https://www.sciencedirect.com/science/article/pii/S0920586113006494
57. Hierso J-C, Feurer R, Poujardieu J, Kihn Y, Kalck P (1998) Metal-organic chemical vapor deposition in a fluidized bed as a versatile method to prepare layered bimetallic nanoparticles. J Mol Catal A Chem 135(3):321–325. https://doi.org/10.1016/S1381-1169(98)00125-3
58. Choi DS, Robertson AW, Warner JH, Kim SO, Kim H (2016) Low-temperature chemical vapor deposition synthesis of Pt-Co alloyed nanoparticles with enhanced oxygen reduction reaction catalysis. Adv Mater 28(33):7115–7122. https://doi.org/10.1002/adma.201600469
59. Hermannsdörfer J, Friedrich M, Miyajima N, Albuquerque RQ, Kümmel S, Kempe R (2012) Ni/Pd@MIL-101: synergistic catalysis with cavity-conform Ni/Pd nanoparticles. Angew Chem Int Ed 51(46):11473–11477. https://doi.org/10.1002/anie.201205078
60. Lee Y-J, Barrera D, Luo K, Hsu JWP (2012) *In situ* chemical oxidation of ultrasmall MoOx nanoparticles in suspensions. J Nanotechnol 2012:1–5. https://doi.org/10.1155/2012/195761
61. Sharma VK, Yngard RA, Lin Y (2009) Silver nanoparticles: green synthesis and their antimicrobial activities. Adv Colloid Interf Sci 145(1–2):83–96. https://www.sciencedirect.com/science/article/pii/S0001868608001449
62. Zain NM, Stapley AGF, Shama G (2014) Green synthesis of silver and copper nanoparticles using ascorbic acid and chitosan for antimicrobial applications. Carbohydr Polym 112:195–202. https://doi.org/10.1016/J.CARBPOL.2014.05.081
63. Rac-Rumijowska O, Fiedot M, Suchorska-Wozniak P, Teterycz H (2017) Synthesis of gold nanoparticles with different kinds of stabilizing agents. In: 2017 40th international spring seminar on electronics technology (ISSE). IEEE, pp 1–6. https://doi.org/10.1109/ISSE.2017.8000972
64. Zaytsev SY, Plyusnin PE, Slavinskaya EM, Shubin YV (2017) Synthesis of bimetallic nanocompositions AuxPd1-x/γ-Al2O3 for catalytic CO oxidation. J Nanopart Res 19(11):367. https://doi.org/10.1007/s11051-017-4061-x
65. Yu J, Li J, Zhang W, Chang H (2015) Synthesis of high quality two-dimensional materials via chemical vapor deposition. Chem Sci 6(12):6705–6716. https://doi.org/10.1039/c5sc01941a
66. Saedy S, Palagin D, Safonova O, van Bokhoven JA, Khodadadi AA, Mortazavi Y (2017) Understanding the mechanism of synthesis of Pt_3 Co intermetallic nanoparticles *via* preferential chemical vapor deposition. J Mater Chem A 5(46):24396–24406. https://doi.org/10.1039/C7TA06737B
67. Devarajan S, Bera P, Sampath S (2005) Bimetallic nanoparticles: a single step synthesis, stabilization, and characterization of Au–Ag, Au–Pd, and Au–Pt in sol–gel derived silicates. J Colloid Interface Sci 290(1):117–129. https://doi.org/10.1016/J.JCIS.2005.04.034
68. Huttel Y (2017) Gas-phase synthesis of nanoparticles. Edited by Yves Huttel. ISBN: 978-3-527-34060-6. p.416
69. Llamosa Pérez D, Espinosa A, Martínez L, Román E, Ballesteros C, Mayoral A, García-Hernández M, Huttel Y (2013) Thermal diffusion at nanoscale: from CoAu alloy nanoparticles to Co@Au core/shell structures. J Phys Chem C 117(6):3101–3108. https://doi.org/10.1021/jp310971f
70. Oprea B, Martínez L, Román E, Vanea E, Simon S, Huttel Y (2015) Dispersion and functionalization of nanoparticles synthesized by gas aggregation source: opening new routes toward the fabrication of nanoparticles for biomedicine. Langmuir 31(51):13813–13820. https://doi.org/10.1021/acs.langmuir.5b03399
71. Oprea B, Martínez L, Román E, Espinosa A, Ruano M, Llamosa D, García-Hernández M, Ballesteros C, Huttel Y (2014) Growth and characterization of FeB nanoparticles for potential application as magnetic resonance imaging contrast agent. Mater Res Express 1(2):025008. https://doi.org/10.1088/2053-1591/1/2/025008
72. Mayoral A, Martínez L et al (2019) Tuning the size, composition and structure of Au and Co50 Au50 nanoparticles by high-power impulse magnetron sputtering in gas-phase synthesis. Nanotechnology 30(6):065606

73. Martínez L, Díaz M, Román E, Ruano M, Llamosa PD, Huttel Y (2012) Generation of nanoparticles with adjustable size and controlled stoichiometry: recent advances. Langmuir 28(30):11241–11249. https://doi.org/10.1021/la3022134
74. Llamosa D, Ruano M, Martínez L, Mayoral A, Roman E, García-Hernández M, Huttel Y (2014) The ultimate step towards a tailored engineering of core@shell and core@shell@shell nanoparticles. Nanoscale 6(22):13483–13486. https://doi.org/10.1039/c4nr02913e
75. Martínez L, Mayoral A, Espiñeira M, Roman E, Palomares FJ, Huttel Y (2017) Core@shell, Au@TiOx nanoparticles by gas phase synthesis. Nanoscale 9(19):6463–6470. https://doi.org/10.1039/c7nr01148b
76. Makarov VV, Love AJ, Sinitsyna OV, Makarova SS, Yaminsky IV, Taliansky ME, Kalinina NO (2014) "Green" nanotechnologies: synthesis of metal nanoparticles using plants. Acta Nat 6(1):35–44. http://www.ncbi.nlm.nih.gov/pubmed/24772325
77. Kadzinski M, Cinelli M, Ciomek K, Coles SR, Nadagouda MN, Varma RS, Kirwan K (2018) Co-constructive development of a green chemistry-based model for the assessment of nanoparticles synthesis. Eur J Oper Res 264(2):472–490. https://doi.org/10.1016/j.ejor.2016.10.019
78. Nair B, Pradeep T (2002) Coalescence of nanoclusters and formation of submicron crystallites assisted by Lactobacillus strains. Cryst Growth Des 2(4):293–298. https://pubs.acs.org/doi/10.1021/cg0255164
79. Castro-Longoria E, Vilchis-Nestor AR, Avalos-Borja M (2011) Biosynthesis of silver, gold and bimetallic nanoparticles using the filamentous fungus Neurospora crassa. Colloids Surf B: Biointerfaces 83(1):42–48. https://doi.org/10.1016/j.colsurfb.2010.10.035
80. Patra N, Taviti AC, Sahoo A, Pal A, Beuria TK, Behera A, Patra S (2017) Green synthesis of multi-metallic nanocubes. RSC Adv 7(56):35111–35118. https://doi.org/10.1039/C7RA05493A
81. Thakkar KN, Mhatre SS, Parikh RY (2010) Biological synthesis of metallic nanoparticles. Nanomedicine 6(2):257–262. https://doi.org/10.1016/J.NANO.2009.07.002
82. Mubarakali D, Gopinath V, Rameshbabu N, Thajuddin N (2012) Synthesis and characterization of CdS nanoparticles using C-phycoerythrin from the marine cyanobacteria. Mater Lett 74:8–11. https://www.sciencedirect.com/science/article/pii/S0167577X1200047X
83. Xu H, Xiao Y, Xu M, Cui H, Tan L, Feng N, Liu X, Qiu G, Dong H, Xie J (2019) Microbial synthesis of Pd–Pt alloy nanoparticles using *Shewanella oneidensis* MR-1 with enhanced catalytic activity for nitrophenol and azo dyes reduction. Nanotechnology 30(6):065607. https://doi.org/10.1088/1361-6528/aaf2a6
84. Deplanche K, Merroun ML, Casadesus M, Tran DT, Mikheenko IP, Bennett JA, Zhu J et al (2012) Microbial synthesis of core/shell gold/palladium nanoparticles for applications in green chemistry. J R Soc Interface 9(72):1705–1712. https://doi.org/10.1098/rsif.2012.0003
85. Hosseinkhani B, Søbjerg LS, Rotaru A-E, Emtiazi G, Skrydstrup T, Meyer RL (2012) Microbially supported synthesis of catalytically active bimetallic Pd-Au nanoparticles. Biotechnol Bioeng 109(1):45–52. https://doi.org/10.1002/bit.23293
86. Govindaraju K, Basha SK, Kumar VG, Singaravelu G (2008) Silver, gold and bimetallic nanoparticles production using single-cell protein (Spirulina platensis) Geitler. 43(15):5115–5122. https://doi.org/10.1007/s10853-008-2745-4
87. Zhao X, Zhou L, Rajoka MSR, Yan L, Jiang C, Shao D, Zhu J et al (2018) Fungal silver nanoparticles: synthesis, application and challenges. Crit Rev Biotechnol 38(6):817–835. https://doi.org/10.1080/07388551.2017.1414141
88. Siddiqi KS, Husen A (2016) Fabrication of metal nanoparticles from fungi and metal salts: scope and application. Nanoscale Res Lett 11(1):98. https://doi.org/10.1186/s11671-016-1311-2
89. Taherzadeh MJ, Fox M, Hjorth H, Edebo L (2003) Production of mycelium biomass and ethanol from paper pulp sulfite liquor by Rhizopus oryzae. Bioresour Technol 88(3):167–177. http://www.ncbi.nlm.nih.gov/pubmed/12618037
90. Pantidos N (2014) Biological synthesis of metallic nanoparticles by bacteria, fungi and plants. J Nanomed Nanotechnol 5(5). https://doi.org/10.4172/2157-7439.1000233

91. Ahmad A, Mukherjee P, Mandal D, Senapati S, Khan MI, Kumar R, Sastry M (2002) Enzyme mediated extracellular synthesis of CdS nanoparticles by the fungus, Fusarium oxysporum. J Am Chem Soc 124(41):12108–12109. https://doi.org/10.1021/JA027296O
92. Senapati S, Ahmad A, Khan MI, Sastry M, Kumar R (2005) Extracellular biosynthesis of bimetallic Au-Ag alloy nanoparticles. Small 1(5):517–520. https://doi.org/10.1002/smll.200400053
93. Dasaratrao Sawle B, Salimath B, Deshpande R, Dhondojirao Bedre M, Krishnamurthy Prabhakar B, Venkataraman A (2008) Biosynthesis and stabilization of Au and Au-Ag alloy nanoparticles by fungus, Fusarium semitectum. Sci Technol Adv Mater 9(3):035012. https://doi.org/10.1088/1468-6996/9/3/035012
94. Philip D (2009) Biosynthesis of Au, Ag and Au–Ag nanoparticles using edible mushroom extract. Spectrochim Acta A Mol Biomol Spectrosc 73(2):374–381. https://doi.org/10.1016/J.SAA.2009.02.037
95. Zheng D, Hu C, Gan T, Dang X, Hu S (2010) Preparation and application of a novel vanillin sensor based on biosynthesis of Au–Ag alloy nanoparticles. Sensors Actuators B Chem 148(1):247–252. https://doi.org/10.1016/J.SNB.2010.04.031
96. Shah M, Fawcett D, Sharma S, Tripathy SK, Poinern GEJ (2015) Green synthesis of metallic nanoparticles via biological entities. Materials (Basel) 8(11):7278–7308. https://doi.org/10.3390/ma8115377
97. Lu F, Sun D, Huang J, Du M, Yang F, Chen H, Hong Y, Li Q (2014) Plant-mediated synthesis of Ag–Pd alloy nanoparticles and their application as catalyst toward selective hydrogenation. ACS Sustain Chem Eng 2(5):1212–1218. https://doi.org/10.1021/sc500034r
98. Phan CM, Nguyen HM (2017) Role of capping agent in wet synthesis of nanoparticles. 121(17):3213–3219. https://doi.org/10.1021/acs.jpca.7b02186
99. Kuppusamy P, Yusoff MM, Maniam GP, Govindan N (2016) Biosynthesis of metallic nanoparticles using plant derivatives and their new avenues in pharmacological applications—an updated report. Saudi Pharm J 24(4):473–484. https://doi.org/10.1016/j.jsps.2014.11.013
100. Singh P, Kim Y-J, Zhang D, Yang D-C (2016a) Biological synthesis of nanoparticles from plants and microorganisms. Trends Biotechnol 34(7):588–599. https://doi.org/10.1016/j.tibtech.2016.02.006
101. Singh R, Nawale L, Arkile M, Wadhwani S, Shedbalkar U, Chopade S, Sarkar D, Chopade BA (2016b) Phytogenic silver, gold, and bimetallic nanoparticles as novel antitubercular agents. Int J Nanomedicine 11:1889–1897. https://doi.org/10.2147/IJN.S102488
102. Velusamy P, Kumar GV, Jeyanthi V, Das J, Pachaiappan R (2016) Bio-inspired green nanoparticles: synthesis, mechanism, and antibacterial application. Toxicol Res 32(2):95–102. https://doi.org/10.5487/TR.2016.32.2.095
103. Sun D, Zhang G, Huang J, Wang H, Li Q (2014) Plant-mediated fabrication and surface enhanced Raman property of flower-like Au@Pd nanoparticles. Materials 7(2):1360–1369. https://doi.org/10.3390/ma7021360
104. Zhan G, Huang J, Du M, Abdul-Rauf I, Ma Y, Li Q (2011) Green synthesis of Au–Pd bimetallic nanoparticles: single-step bioreduction method with plant extract. Mater Lett 65(19–20):2989–2991. https://doi.org/10.1016/J.MATLET.2011.06.079
105. Ganaie SU, Abbasi T, Abbasi SA (2016) Rapid and green synthesis of bimetallic Au–Ag nanoparticles using an otherwise worthless weed *Antigonon leptopus*. J Exp Nanosci 11(6):395–417. https://doi.org/10.1080/17458080.2015.1070311
106. Chopade B, Ghosh S, Nitnavare R, Dewle A, Tomar GB, Chippalkatti R, More P, Kitture R, Kale S, Bellare J (2015) Novel platinum-palladium bimetallic nanoparticles synthesized by Dioscorea Bulbifera: anticancer and antioxidant activities. Int J Nanomedicine 10:7477. https://doi.org/10.2147/IJN.S91579
107. Malapermal V, Mbatha JN, Gengan RM, Anand K (2015) Biosynthesis of bimetallic Au-Ag nanoparticles using Ocimum basilicum (L.) with antidiabetic and antimicrobial properties. Adv Mater Lett 6(12):1050–1057. https://doi.org/10.5185/amlett.2015.5997

108. Shankar SS, Rai A, Ahmad A, Sastry M (2004) Rapid synthesis of Au, Ag, and bimetallic Au Core–Ag Shell nanoparticles using Neem (Azadirachta indica) leaf broth. J Colloid Interface Sci 275(2):496–502. https://doi.org/10.1016/j.jcis.2004.03.003
109. Dobrucka R, Dlugaszewska J (2018) Antimicrobial activity of the biogenically synthesized core-shell Cu@Pt nanoparticles. Saudi Pharm J 26(5):643–650. https://doi.org/10.1016/J.JSPS.2018.02.028
110. Xia B, He F, Li L (2013) Preparation of bimetallic nanoparticles using a facile green synthesis method and their application. Langmuir 29(15):4901–4907. https://doi.org/10.1021/la400355u
111. Valodkar M, Modi S, Pal A, Thakore S (2011) Synthesis and anti-bacterial activity of Cu, Ag and Cu–Ag alloy nanoparticles: a green approach. Mater Res Bull 46(3):384–389. https://doi.org/10.1016/J.MATERRESBULL.2010.12.001
112. Alarfaj NA, El-Tohamy MF (2016) Eco-friendly synthesis of gelatin-capped bimetallic Au-Ag nanoparticles for chemiluminescence detection of anticancer raloxifene hydrochloride. Luminescence 31(6):1194–1200. https://doi.org/10.1002/bio.3089
113. Hebbalalu D, Lalley J, Nadagouda MN, Varma RS (2013) Greener techniques for the synthesis of silver nanoparticles using plant extracts, enzymes, bacteria, biodegradable polymers, and microwaves. ACS Sustain Chem Eng 1(7):703–712. https://doi.org/10.1021/sc4000362
114. Khatami M, Sharifi I, Nobre MAL, Zafarnia N, Aflatoonian MR (2018) Waste-grass-mediated green synthesis of silver nanoparticles and evaluation of their anticancer, antifungal and antibacterial activity. Green Chem Lett Rev 11(2):125–134. https://doi.org/10.1080/17518253.2018.1444797
115. Shankar S, Jaiswal L, Aparna RSL, Prasad RGSV (2014) Synthesis, characterization, in vitro biocompatibility, and antimicrobial activity of gold, silver and gold silver alloy nanoparticles prepared from Lansium domesticum fruit peel extract. Mater Lett 137:75–78. https://www.sciencedirect.com/science/article/abs/pii/S0167577X14015997
116. Ventola CL (2015) The antibiotic resistance crisis: part 1: causes and threats. P T 40(4):277–283. http://www.ncbi.nlm.nih.gov/pubmed/25859123
117. Zaman SB, Hussain MA, Nye R, Mehta V, Mamun KT, Hossain N (2017) A review on antibiotic resistance: alarm bells are ringing. Cureus 9(6):e1403. https://doi.org/10.7759/cureus.1403
118. Salomoni R, Léo P, Montemor A, Rinaldi B, Rodrigues M (2017) Antibacterial effect of silver nanoparticles in Pseudomonas aeruginosa. Nanotechnol Sci Appl 10:115–121. https://doi.org/10.2147/NSA.S133415
119. Shamaila S, Zafar N, Riaz S, Sharif R, Nazir J, Naseem S (2016) Gold nanoparticles: an efficient antimicrobial agent against enteric bacterial human pathogen. Nanomaterials (Basel) 6(4). https://doi.org/10.3390/nano6040071
120. Panáček A, Kvítek L, Smékalová M, Večeřová R, Kolář M, Röderová M, Dyčka F et al (2018) Bacterial resistance to silver nanoparticles and how to overcome it. Nat Nanotechnol 13(1):65–71. https://doi.org/10.1038/s41565-017-0013-y
121. Chou K-S, Chen C-C (2007) Fabrication and characterization of silver core and porous silica shell nanocomposite particles. Microporous Mesoporous Mater 98(1–3):208–213. https://doi.org/10.1016/J.MICROMESO.2006.09.006
122. Mittal AK, Kumar S, Banerjee UC (2014) Quercetin and gallic acid mediated synthesis of bimetallic (silver and selenium) nanoparticles and their antitumor and antimicrobial potential. J Colloid Interface Sci 431:194–199. http://www.ncbi.nlm.nih.gov/pubmed/25000181
123. Zhao Y, Ye C, Liu W, Chen R, Jiang X (2014) Tuning the composition of AuPt bimetallic nanoparticles for antibacterial application. Angew Chem Int Ed 53(31):8127–8131. https://doi.org/10.1002/anie.201401035
124. Banerjee M, Sharma S, Chattopadhyay A, Ghosh SS (2011) Enhanced antibacterial activity of bimetallic gold-silver core–shell nanoparticles at low silver concentration. Nanoscale 3(12):5120. https://doi.org/10.1039/c1nr10703h
125. Holden MS, Black J, Lewis A, Boutrin M-C, Walemba E, Sabir TS, Boskovic DS, Wilson A, Fletcher HM, Perry CC (2016) Antibacterial activity of partially oxidized Ag/Au nanopar-

ticles against the oral pathogen *Porphyromonas gingivalis* W83. J Nanomater 2016:1–11. https://doi.org/10.1155/2016/9605906
126. Antonoglou O, Giannousi K, Arvanitidis J, Mourdikoudis S, Pantazaki A, Dendrinou-Samara C (2017) Elucidation of one step synthesis of PEGylated CuFe bimetallic nanoparticles. Antimicrobial activity of CuFe@PEG vs Cu@PEG. J Inorg Biochem 177:159–170. https://doi.org/10.1016/j.jinorgbio.2017.09.014
127. Fakhri A, Tahami S, Naji M (2017) Synthesis and characterization of core-shell bimetallic nanoparticles for synergistic antimicrobial effect studies in combination with doxycycline on burn specific pathogens. J Photochem Photobiol B Biol 169:21–26. https://doi.org/10.1016/j.jphotobiol.2017.02.014
128. Akinsiku AA, Dare EO, Ajanaku KO, Ajani OO, Olugbuyiro JAO, Siyanbola TO, Ejilude O, Emetere ME (2018) Modeling and synthesis of Ag and Ag/Ni allied bimetallic nanoparticles by green method: optical and biological properties. Int J Biomater 2018:1–17. https://doi.org/10.1155/2018/9658080
129. Cooper GM (2000) The development and causes of cancer. https://www.ncbi.nlm.nih.gov/books/NBK9963/
130. Siegel RL, Miller KD, Jemal A (2018) Cancer statistics, 2018. CA Cancer J Clin 68(1):7–30. https://doi.org/10.3322/caac.21442
131. Mariotto AB, Robin Yabroff K, Shao Y, Feuer EJ, Brown ML (2011) Projections of the cost of cancer care in the United States: 2010-2020. J Natl Cancer Inst 103(2):117–128. https://doi.org/10.1093/jnci/djq495
132. Arruebo M, Vilaboa N, Sáez-Gutierrez B, Lambea J, Tres A, Valladares M, González-Fernández A (2011) Assessment of the evolution of cancer treatment therapies. Cancers 3(3):3279–3330. https://doi.org/10.3390/cancers3033279
133. Institute for Quality and Efficiency in Health Care: Executive Summaries (2005) Institute for Quality and Efficiency in Health Care: Executive. Institute for Quality and Efficiency in Health Care (IQWiG), Cologne. http://www.ncbi.nlm.nih.gov/pubmed/23101074
134. Gelband H, Jha P, Sankaranarayanan R, et al (2016) Cancer: disease control priorities, vol 3, 3rd edn. The International Bank for Reconstruction and Development/The World Bank, Washington, DC, p 2016. https://doi.org/10.1596/978-1-4648-0349-9
135. Ramirez LY, Huestis SE, Yap TY, Zyzanski S, Drotar D, Kodish E (2009) Potential chemotherapy side effects: what do oncologists tell parents? Pediatr Blood Cancer 52(4):497–502. https://doi.org/10.1002/pbc.21835
136. Alberts B, Johnson A, Lewis J, Raff M, Roberts K, Walter P (2002) Molecular biology of the cell. Garland Science, New York
137. Heath JR, Davis ME (2008) Nanotechnology and cancer. Undefined. https://www.semanticscholar.org/paper/Nanotechnology-and-cancer.-Heath-Davis/47006f38bd3a82be6d869ead7748f841a4184cf3
138. Gmeiner WH, Ghosh S (2015) Nanotechnology for cancer treatment. Nanotechnol Rev 3(2):111–122. https://doi.org/10.1515/ntrev-2013-0013
139. Yuan Y-G, Peng Q-L, Gurunathan S (2017) Silver nanoparticles enhance the apoptotic potential of gemcitabine in human ovarian cancer cells: combination therapy for effective cancer treatment. Int J Nanomed 12:6487–6502. https://doi.org/10.2147/IJN.S135482
140. De Matteis V, Cascione M, Toma C, Leporatti S (2018) Silver nanoparticles: synthetic routes, in vitro toxicity and theranostic applications for cancer disease. Nanomaterials 8(5):319. https://doi.org/10.3390/nano8050319
141. Shmarakov I, Mukha I, Vityuk N, Borschovetska V, Zhyshchynska N, Grodzyuk G, Eremenko A (2017) Antitumor activity of alloy and core-shell-type bimetallic AgAu nanoparticles. Nanoscale Res Lett 12(1):333. https://doi.org/10.1186/s11671-017-2112-y
142. Mishra SK, Kannan S (2017) A bimetallic silver–neodymium theranostic nanoparticle with multimodal NIR/MRI/CT imaging and combined chemo-photothermal therapy. Inorg Chem 56(19):12054–12066. https://doi.org/10.1021/acs.inorgchem.7b02103
143. Kumar R, Gokulakrishnan N, Kumar R, Krishna VM, Saravanan A, Supriya S, Somanathan T (2015) Can be a bimetal oxide ZnO-MgO nanoparticles anticancer drug carrier and deliver?

Doxorubicin adsorption/release study. J Nanosci Nanotechnol 15(2):1543–1553. http://www.ncbi.nlm.nih.gov/pubmed/26353689
144. Sathya K, Saravanathamizhan R, Baskar G (2018) Ultrasonic assisted green synthesis of Fe and Fe/Zn bimetallic nanoparticles for in vitro cytotoxicity study against HeLa cancer cell line. Mol Biol Rep 45(5):1397–1404. https://doi.org/10.1007/s11033-018-4302-9
145. Estelrich J, Sánchez-Martín MJ, Busquets MA (2015) Nanoparticles in magnetic resonance imaging: from simple to dual contrast agents. Int J Nanomedicine 10:1727–1741. https://doi.org/10.2147/IJN.S76501
146. Nune SK, Gunda P, Thallapally PK, Lin Y-Y, Laird Forrest M, Berkland CJ (2009) Nanoparticles for biomedical imaging. Expert Opin Drug Deliv 6(11):1175–1194. https://doi.org/10.1517/17425240903229031
147. Lindner JR, Link J (2018) Molecular imaging in drug discovery and development. Circ Cardiovasc Imaging 11(2):e005355. https://doi.org/10.1161/CIRCIMAGING.117.005355
148. Hacker M, Beyer T, Baum RP, Kalemis A, Lammertsma AA, Lewington V, Talbot J-N, Verzijlbergen F (2015) Nuclear medicine innovations help (drive) healthcare (benefits). Eur J Nucl Med Mol Imaging 42(2):173–175. https://doi.org/10.1007/s00259-014-2957-6
149. Shukla AK, Kumar U (2006) Positron emission tomography: an overview. J Med Phys 31(1):13–21. https://doi.org/10.4103/0971-6203.25665
150. Rahmim A, Zaidi H (2008) PET versus SPECT: strengths, limitations and challenges. Nucl Med Commun 29(3):193–207. https://doi.org/10.1097/MNM.0b013e3282f3a515
151. Pang B, Zhao Y, Luehmann H, Yang X, Detering L, You M, Zhang C et al (2016) 64Cu-Doped PdCu@Au Tripods: a multifunctional nanomaterial for positron emission tomography and image-guided photothermal cancer treatment. ACS Nano 10(3):3121–3131. https://doi.org/10.1021/ACSNANO.5B07968
152. Histed SN, Lindenberg ML, Mena E, Turkbey B, Choyke PL, Kurdziel KA (2012) Review of functional/anatomical imaging in oncology. Nucl Med Commun 33(4):349–361. https://doi.org/10.1097/MNM.0b013e32834ec8a5
153. Reuveni T, Motiei M, Romman Z, Popovtzer A, Popovtzer R (2011) Targeted gold nanoparticles enable molecular CT imaging of cancer: an in vivo study. Int J Nanomedicine 6:2859–2864. https://doi.org/10.2147/IJN.S25446
154. Li B, Ye K, Zhang Y, Qin J, Zou R, Xu K, Huang X et al (2015a) Photothermal theragnosis synergistic therapy based on bimetal sulphide nanocrystals rather than nanocomposites. Adv Mater 27(8):1339–1345. https://doi.org/10.1002/adma.201404257
155. Li Q, Wu L, Wu G, Su D, Lv H, Zhang S, Zhu W et al (2015b) New approach to fully ordered Fct-FePt nanoparticles for much enhanced electrocatalysis in acid. Nano Lett 15(4):2468–2473. https://doi.org/10.1021/acs.nanolett.5b00320
156. Maney V, Singh M (2017) An *in vitro* assessment of novel chitosan/bimetallic PtAu nanocomposites as delivery vehicles for doxorubicin. Nanomedicine 12(21):2625–2640. https://doi.org/10.2217/nnm-2017-0228
157. Senpan A, Caruthers SD, Rhee I, Mauro NA, Pan D, Hu G, Scott MJ et al (2009) Conquering the dark side: colloidal iron oxide nanoparticles. ACS Nano 3(12):3917–3926. https://doi.org/10.1021/nn900819y
158. Choi J-s, Lee J-H, Shin T-H, Song H-T, Kim EY, Cheon J (2010) Self-confirming 'AND' logic nanoparticles for fault-free MRI. J Am Chem Soc 132(32):11015–11017. https://doi.org/10.1021/ja104503g
159. Chavhan GB, Babyn PS, Thomas B, Shroff MM, Haacke EM (2009) Principles, techniques, and applications of T2∗-based MR imaging and its special applications. RadioGraphics 29(5):1433–1449. https://doi.org/10.1148/rg.295095034
160. McNamara K, Tofail SAM (2015) Nanosystems: the use of nanoalloys, metallic, bimetallic, and magnetic nanoparticles in biomedical applications. Phys Chem Chem Phys 17(42):27981–27995. https://doi.org/10.1039/C5CP00831J
161. Sun C, Lee J, Zhang M (2008) Magnetic nanoparticles in MR imaging and drug delivery☆. Adv Drug Deliv Rev 60(11):1252–1265. https://doi.org/10.1016/j.addr.2008.03.018

162. Kumal RR, Abu-Laban M, Hamal P, Kruger B, Smith HT, Hayes DJ, Haber LH (2018) Near-infrared photothermal release of SiRNA from the surface of colloidal gold–silver–gold core–shell–shell nanoparticles studied with second-harmonic generation. J Phys Chem C 122(34):19699–19704. https://doi.org/10.1021/acs.jpcc.8b06117
163. Taylor EN, Kummer KM, Durmus NG, Leuba K, Tarquinio KM, Webster TJ (2012) Superparamagnetic iron oxide nanoparticles (SPION) for the treatment of antibiotic-resistant biofilms. Small 8(19):3016–3027. https://doi.org/10.1002/smll.201200575
164. Rozanova N, Zhang JZ (2009) Photothermal ablation therapy for cancer based on metal nanostructures. Sci China Ser B Chem 52(10):1559–1575. https://doi.org/10.1007/s11426-009-0247-0
165. Sharma H, Mishra PK, Talegaonkar S, Vaidya B (2015) Metal nanoparticles: a theranostic nanotool against cancer. Drug Discov Today 20(9):1143–1151. https://doi.org/10.1016/j.drudis.2015.05.009
166. Liu X, Zhang X, Zhu M, Lin G, Liu J, Zhou Z, Tian X, Pan Y (2017) PEGylated Au@Pt nanodendrites as novel theranostic agents for computed tomography imaging and photothermal/radiation synergistic therapy. ACS Appl Mater Interfaces 9(1):279–285. https://doi.org/10.1021/acsami.6b15183
167. Gan N, Xiong P, Wang J, Li T, Hu F, Cao Y, Zheng L (2013) A novel signal-amplified immunoassay for the detection of C-reactive protein using HRP-doped magnetic nanoparticles as labels with the electrochemical quartz crystal microbalance as a detector. J Anal Methods Chem 2013:1–8. https://doi.org/10.1155/2013/482316
168. Wang C, Chen J, Talavage T, Irudayaraj J (2009) Gold nanorod/Fe $_3$ O $_4$ nanoparticle 'nano-pearl-necklaces' for simultaneous targeting, dual-mode imaging, and photothermal ablation of cancer cells. Angew Chem Int Ed 48(15):2759–2763. https://doi.org/10.1002/anie.200805282
169. Fan Z, Senapati D, Khan SA, Singh AK, Hamme A, Yust B, Sardar D, Ray PC (2013) Popcorn-shaped magnetic core-plasmonic shell multifunctional nanoparticles for the targeted magnetic separation and enrichment, label-free SERS imaging, and photothermal destruction of multidrug-resistant bacteria. Chem Eur J 19(8):2839–2847. https://doi.org/10.1002/chem.201202948
170. Yamada M, Foote M, Prow TW (2015) Therapeutic gold, silver, and platinum nanoparticles. Wiley Interdiscip Rev Nanomed Nanobiotechnol 7(3):428–445. https://doi.org/10.1002/wnan.1322
171. Cheng L-C, Huang J-H, Chen HM, Lai T-C, Yang K-Y, Liu R-S, Hsiao M, Chen C-H, Her L-J, Tsai DP (2012) Seedless, silver-induced synthesis of star-shaped gold/silver bimetallic nanoparticles as high efficiency photothermal therapy reagent. J Mater Chem 22(5):2244–2253. https://doi.org/10.1039/C1JM13937A
172. Yadi M, Mostafavi E, Saleh B, Davaran S, Aliyeva I, Khalilov R, Nikzamir M et al (2018) Current developments in green synthesis of metallic nanoparticles using plant extracts: a review. Artif Cells Nanomed Biotechnol 46(Suppl 3):S336–S343. https://doi.org/10.1080/21691401.2018.1492931
173. Salvo P, Dini V, Kirchhain A, Janowska A, Oranges T, Chiricozzi A, Lomonaco T, Di Francesco F, Romanelli M (2017) Sensors and biosensors for C-reactive protein, temperature and PH, and their applications for monitoring wound healing: a review. Sensors 17(12):2952. https://doi.org/10.3390/s17122952
174. Chin SF, Iyer KS, Raston CL (2010) Superparamagnetic core-shell nanoparticles for biomedical applications. In: 2010 international conference on enabling science and nanotechnology (ESciNano). IEEE, p 1. https://doi.org/10.1109/ESCINANO.2010.5700936
175. Zhou T, Wu B, Xing D (2012) Bio-modified Fe_3 O_4 core/Au shell nanoparticles for targeting and multimodal imaging of cancer cells. J Mater Chem 22(2):470–477. https://doi.org/10.1039/C1JM13692E
176. Chang E, Miller JS, Sun J, Yu WW, Colvin VL, Drezek R, West JL (2005) Protease-activated quantum dot probes. Biochem Biophys Res Commun 334(4):1317–1321. https://doi.org/10.1016/j.bbrc.2005.07.028

177. Welser K, Adsley R, Moore BM, Chan WC, Aylott JW (2011) Protease sensing with nanoparticle based platforms. Analyst 136(1):29–41. https://doi.org/10.1039/c0an00429d
178. Medintz IL, Clapp AR, Brunel FM, Tiefenbrunn T, Tetsuo Uyeda H, Chang EL, Deschamps JR, Dawson PE, Mattoussi H (2006) Proteolytic activity monitored by fluorescence resonance energy transfer through quantum-dot–peptide conjugates. Nat Mater 5(7):581–589. https://doi.org/10.1038/nmat1676
179. Choi JH, Chen KH, Strano MS (2006) Aptamer-capped nanocrystal quantum dots: a new method for label-free protein detection. J Am Chem Soc 128(49):15584–15585. https://doi.org/10.1021/JA066506K
180. Chiriac H, Tibu M, Moga A-E, Herea DD (2005) Magnetic GMI sensor for detection of biomolecules. J Magn Magn Mater 293(1):671–676. https://doi.org/10.1016/J.JMMM.2005.02.043
181. Lin D, Wu J, Wang M, Yan F, Ju H (2012) Triple signal amplification of graphene film, polybead carried gold nanoparticles as tracing tag and silver deposition for ultrasensitive electrochemical immunosensing. Anal Chem 84(8):3662–3668. https://doi.org/10.1021/ac3001435
182. Gliga AR, Skoglund S, Odnevall Wallinder I, Fadeel B, Karlsson HL (2014) Size-dependent cytotoxicity of silver nanoparticles in human lung cells: the role of cellular uptake, agglomeration and Ag release. Part Fibre Toxicol 11(1):11. https://doi.org/10.1186/1743-8977-11-11
183. Pan Y, Neuss S, Leifert A, Fischler M, Wen F, Simon U, Schmid G, Brandau W, Jahnen-Dechent W (2007) Size-dependent cytotoxicity of gold nanoparticles. Small 3(11):1941–1949. https://doi.org/10.1002/smll.200700378
184. Moise S, Céspedes E, Soukup D, Byrne JM, El Haj AJ, Telling ND (2017) The cellular magnetic response and biocompatibility of biogenic zinc- and cobalt-doped magnetite nanoparticles. Sci Rep 7(1):39922. https://doi.org/10.1038/srep39922

Peptide-mediated Bone Tissue Engineering

Abdullah Karadag, Hana'a Iqbal, and Hilal Yazici

Abstract Bone is a highly vascularized tissue and one of the most dynamic tissues in terms of self-renewal throughout one's life. It possesses a high regenerative capacity, which makes it possible that a majority of bone fractures will heal well without the need for major intervention. However, some large bone defects and fractures require medical intervention for bone repair and regeneration. Proteins, growth factors, and peptides have played a remarkable role in bone regeneration. However, the use of proteins and growth factors in tissue engineering has several limitations, such as cost and difficulty in production, immunogenicity, and a short half-life. In addition to these drawbacks, they have many active domains, which affect their functionality. Recently, an alternative to proteins and growth factors has emerged for the use in tissue engineering. This competent approach includes biomimetic peptides, which are amino acid sequences derived from the functional domains of soluble or extracellular matrix (ECM) proteins. Biological materials for tissue regeneration can be functionalized with these peptides to either mediate the adhesion of cells or to be released as soluble ligands. These short peptides are easy to design and synthesize, facilitating their use as cost-effective and efficient scaffolds for regenerative medicine. In this extensive chapter, several of the peptides that have potential for bone tissue engineering, including those that facilitate cell adhesion, prompt osteogenic differentiation of progenitor cells, or those that mediate angiogenesis which is a crucial requirement for proper bone regeneration will be discussed.

Keywords Peptides · Bone regeneration · Scaffold · Osteoinduction · Biomimetic

A. Karadag · H. Yazici (✉)
TUBITAK-Marmara Research Center, Genetic Engineering and Biotechnology Institute, Kocaeli, Turkey
e-mail: hilal.yazici@tubitak.gov.tr

H. Iqbal
TUBITAK-Marmara Research Center, Genetic Engineering and Biotechnology Institute, Kocaeli, Turkey

Dr. Panjwani Center for Molecular Medicine and Drug Research (PCMD), International Center for Chemical and Biological Sciences (ICCBS), University of Karachi, Karachi, Pakistan

B. Li et al. (eds.), *Racing for the Surface*,
https://doi.org/10.1007/978-3-030-34471-9_17

Introduction

General View of Tissue Engineering

Tissue and organ injury or failure in the human body results in a heavy economic burden on health care systems worldwide. Existing therapeutic options for tissue/organ loss include: drugs, replacements through synthetic materials, and organ/tissue transplantation which have significant limitations to prevent mortality or morbidity for many patients every year. These limitations led to the creation of a new research field, tissue engineering or regenerative medicine, which focuses on assembling functional constructs that can reinstate, maintain, or repair injured tissues or whole organs. To achieve the goal, new synthetic materials (metals, polymers, ceramic-based materials, etc.) and their functionalization with various proteins, combinations of cells, and cell-instructive peptide need to be designed and fabricated [1, 2].

In tissue engineering, there are three general approaches termed as conduction, induction, and cell transplantation. Conduction includes grafting a construct at the defect site in the tissue which allows cells to penetrate through the surrounding tissue for the regeneration of the tissue. This phenomenon forms the basis of guided tissue engineering intervention used clinically in orthopedics and dentistry for bone healing [3].

The induction approach is placing a peptide or protein-based growth factor using a scaffold as a carrier, at the site of tissue defect. The cells in the surroundings are recruited to the site of injury and start proliferation. This recruitment is initiated by the peptides and proteins that bind to cell surface receptors. This approach is more attractive than the conductive approach. Inductive approaches are currently in clinical use for bone formation and angiogenesis. The third option in tissue engineering is the transplantation of particular cell types in the damaged tissues. The population is generated and multiplied outside the body by taking small starting tissue at the defect site. Multiplied cells need to be transplanted to form a new tissue replacing the lost and defective one. This strategy is preferred when inductive molecules are unknown for the tissue of interest with two applications. One requires transplanting the target cells directly at the defect site to trigger tissue regeneration. The other one requires prefabrication of the cells in vitro that are then available to patients for implantation. The latter approach is currently utilized for cartilage and skin tissue replacements [1, 4, 5] (Fig. 1).

Bone Tissue Engineering

Bone is a very dynamic tissue with well-developed vascular beds that continue to self-renew throughout an individual's lifetime.

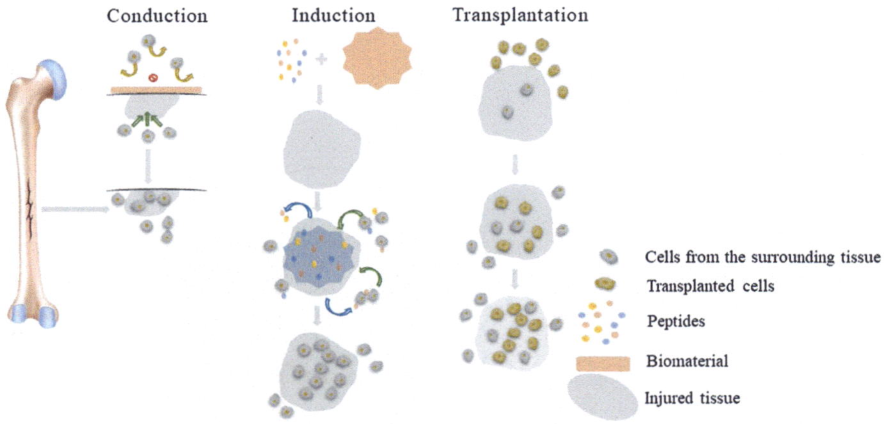

Fig. 1 Major strategies to engineer or regenerate tissues [1]

It is a composite material that is comprised of an organic matrix of type-I collagen and inorganic minerals of calcium phosphate. It plays an essential role in movement. The bone tissue also provides support to the skeleton which acts as a scaffold for the delicate soft tissue and internal organs with suitable load-bearing capacity. Besides these structural functions, bone has a storage capacity for Ca and P ions and a role in regulating the key electrolyte concentrations in the blood, making it crucial in homeostasis [5–7].

Bone Structure and Properties

The human skeleton consists of 206 bones of strikingly diverse sizes, shapes, and function. These include short bones in the ankle and writs, long bones of the limbs, irregular bones of the pelvis and vertebrae, and the flat bones in the skull and sternum. Bone tissue is composed of two layers: compact (cortical) and trabecular (cancellous) bones. Bone tissue, similar to all the organs in the human body, has a hierarchical organization spanning several folds of magnitude from the macroscale to nanoscale components (Fig. 2). For the nanoscale component, the extracellular matrix (ECM) comprises both a mineralized inorganic (carbonated apatite mineralites, 4 nm in thickness) and non-mineralized organic components (mainly collagen type I) [5].

Additionally, there are several noncollagenous matric proteins, such as sialoproteins, proteoglycans, and glycoproteins, which contribute to signaling in the immediate extracellular environment. The nanocomposite structure composed of flexible and tough collagen fibers strengthened by hydroxyapatite (HA) crystals, provide the necessary high fracture toughness and compressive strength of bone [3, 5].

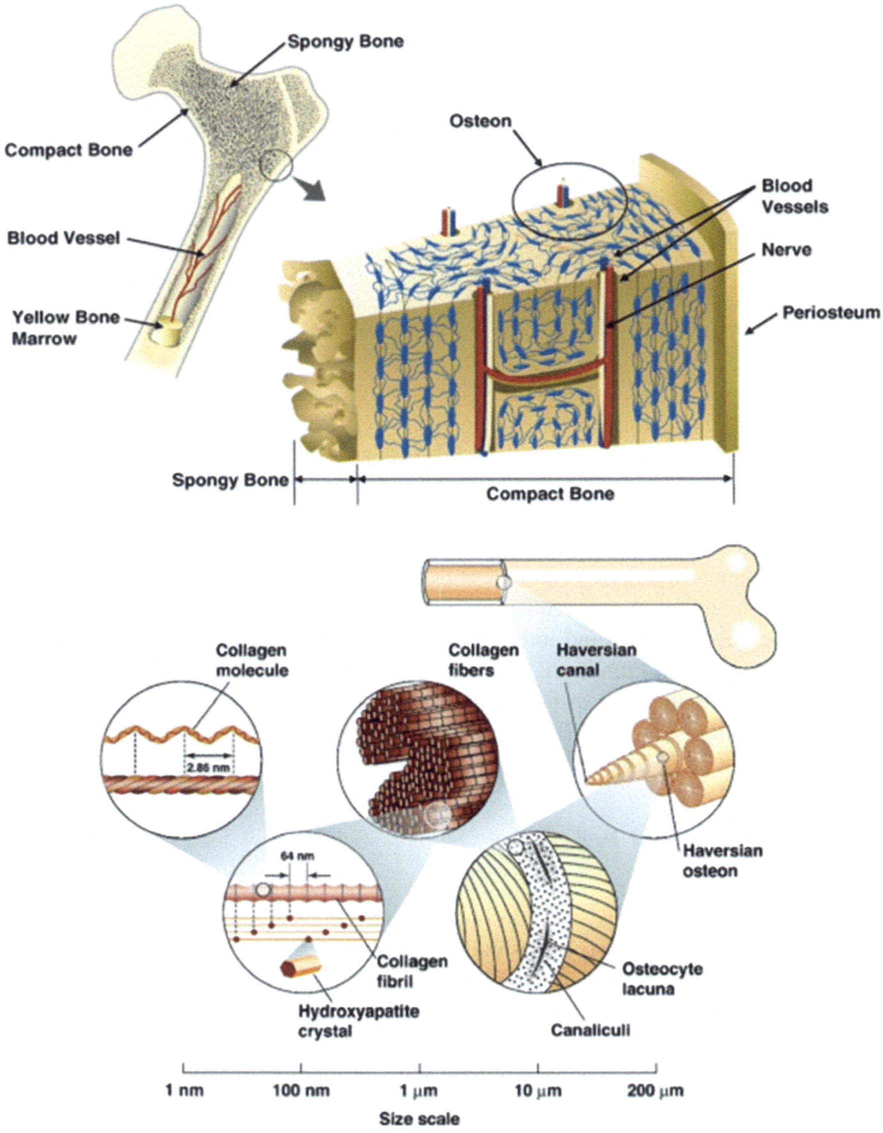

Fig. 2 The hierarchical organization of human cortical/compact bone [8]

Bone Healing

There are three major processes such as osteoconduction, osteoinduction, and osteogenesis that need to be considered in bone healing.

Osteoconduction is the process of bone formation which is marked by the passive growth of the resident cells, tissues, and blood vessels, supported by a grafted scaffold. In context to this, osteoconductive materials function as a scaffold for the response by host tissue to repair or form the new bone. Bone autograft, allograft, and inert filler structures such as calcium ceramics, and demineralized bone matrix (DBM) are some examples of osteoconductive materials. Osteoconductive scaffolds are usually composed of a structured matrix analogous to cancellous bone so as to assist the ingrowth of host cells and vasculature. Autografts comprised of cancellous bone, therefore, carry the utmost osteoconductive potential as compared to signaling molecules and growth factors, such as bone morphogenetic proteins (BMPs) when used alone, as these do not provide any specific physical supportive structure [9, 10].

Osteoinduction is bone repair and formation through the specific growth factors provided by a grafted scaffold, thus promoting the differentiation of mesenchymal stem cells (MSCs) to chondroblasts and osteoblasts. These growth factors can be of various types such as platelet-derived growth factors (PDGFs), BMPs, fibroblast growth factors (FGFs), and interleukins. Additionally, bone autografts, platelet-rich plasma (PRP), and bone marrow aspirate concentrates (BMAC) possess osteoinductive properties [4, 6] (Table 1).

The process of osteogenesis is hallmarked by new bone formation via specific cellular components within the graft. Autologous cancellous bone, as it has all the essential elements required for osteogenic stimulation, offers tremendous osteogenic potential. However, with an increase in our understanding of the molecular and cellular mechanisms involved in the bone repair and formation, alternative stimulators of osteogenesis have emerged, such as cell signaling proteins, growth factors, and cell-based treatments [4].

Table 1 Parts of bones and their specific osteoconductive, osteoinductive, and osteogenic properties [4]

	Osteoconductive	Osteoinductive	Osteogenic
Cortical autograft	+	+	+
Cancellous autograft	+++	+++	+++
Cortical allograft	+	+/−	−
Cancellous allograft	+	+/−	−
Demineralized bone matrix	+	++	−
Calcium ceramics	+	−	−
Bone marrow aspirate	−	++	+++
Bone morphogenetic protein	−	++	−
Platelet-rich plasma	−	+++	+

+ activity, − no activity, +/− activity depends on preparation process

The Role of Biomaterials in Bone Tissue Engineering

Due to the high regenerative capacity of bone, most fractures heal without the need of any surgery; however, large defects in bone require surgical intervention [5].

Currently, a bone graft is the gold standard treatment for large bone defects. Clinically, these bone grafts are classified on the basis of their origin into biological and synthetic grafts. The biological grafts comprise autografts, allografts, and xenografts. An autograft is the transplantation of bone taken from the patient's own body. This offers the greatest clinical outcome as it is compatible with the host bone and does not elicit any disease or immune-related complications. However, this approach is limited by donor site morbidity associated with the harvest of the graft bone which results in damage of the harvest site. Contrarily, an allograft is the transplantation of bone harvested from one individual and transplanted into the other individual within the same species. Allografts also have some limitations, such as the host immune response and potential transmission of pathogens. Xenografts, the bones harvested from one species and transplanted in another, have severe restrictions due to immunogenic barriers between the species. The limitations associated with biological bone transplants have directed the treatment paradigm toward the use of synthetic substitutes for bone repair, replacement, and augmentation [6, 11].

Osteoconductive Materials

The ability of the biological scaffolds to support bone formation on their surfaces, known as osteoconductivity, is one of the crucial prerequisites of biomaterials used for bone repair [12]. Materials with osteoconductive properties allow proliferation, migration, differentiation, and ECM deposition from osteoprogenitor cells in the bone defect, which are key initiators of new bone formation [13].

Biomaterial osteoconductivity during bone regeneration largely depends on their physicochemical characteristics, which include suitable chemical composition, architectural geometry, and surface properties [14]. Tricalcium phosphate (TCP) and hydroxyapatite (HA), both calcium phosphate (CaP) based ceramics, possess superior osteoconductive properties due to their similarity to natural bone mineral [15, 16]. Another osteoconductive material, bioglass, is capable of binding with the bone directly. Moreover, type-I collagen is an osteoconductive material owing to its structure and composition which is favorable for mineral deposition through noncollagenous matrix protein binding thus initiating and regulating mineralization [17].

Osteoinductive Materials

The capability of biomaterials to prompt the formation of bone at the site of the graft is known as osteoinductivity. In the past few decades, there have been tremendous advancements in revealing the role that biomaterials play in bone generation, although the precise mechanism of osteoinduction still remains to be explored [18].

Biological scaffolds possessing osteoinductive properties have been shown to influence the ectopic bone formation at many levels:

1. At the tissue level, the biomaterials actively facilitate the exchange of oxygen, nutrition, and waste between the material and the tissue. They also promote vascularization within the materials, essential for new tissue formation [19, 20].
2. At the cellular level, the differentiation of stem cells or the osteoprogenitor cells can be triggered toward the osteogenic lineage by the formation of a biological carbonated apatite layer [21, 22]. The liberated phosphate and calcium ions also serve as a potent chemotaxis for cell migration and direct cell growth at the graft site [23, 24].
3. At the molecular level, these biomaterials may play a role in enriching osteogenic proteins such as BMP-2 and BMP-7 owing to their high affinity to these biomolecules, thus promoting a series of cellular events on the surface of biomaterials [25].

Moreover, the released phosphate and calcium ions may accelerate the mineralization in the implanted site by reaching a supersaturation level in the graft void. The most extensively used osteoinductive materials are CaP-based biological ceramics. Osteoinduction has been shown on numerous types of CaP materials such as TCP, biphasic calcium phosphate, and HA. The fundamental element that confers the osteoinductive capacity to these CaP scaffolds is the presence of calcium and phosphate. Nevertheless, other materials that do not contain CaP, such as alumina ceramics, titanium, and poly (hydroxyethyl methacrylate) (PHEMA), have been found to be osteoinductive under certain conditions [17, 23].

Vascular Materials

Vascularization is a vital process during bone repair as any tissue with sizes beyond 200 μm requires formation of blood vessels for appropriate diffusion of oxygen through the tissue. Therefore, proper functional bone regeneration necessitates the close association with the vasculature, which must also properly integrate with the host blood vasculature [26].

Considering the critical requirement of establishing a functional vasculature during bone repair, biological materials that can promote various events of vascular network formation have been designed and extensively employed for bone tissue engineering. As the most used biomaterial formulation, hydrogels and scaffolds can be used as temporal matrix to facilitate progenitor cells and pericyte migration and to furnish mechanical support for vessel sprouting. Through tuning the architectural properties of scaffolds during fabrication, the influence of these on tissue-engineered constructs can be realized [27, 28].

Vascularization is also significantly influenced by the chemical composition of the biological scaffolds, as they interact with the endothelial cells directly during vessel formation [29, 30]. Though, most of the materials utilized for bone tissue engineering such as silk, PCL, collagen, and PLGA have been found to be compatible with endothelial cells, there are some which have been demonstrated as proangiogenic during bone repair [31]. One of such example is the hydrogel made of dextran, which has demonstrated a remarkable ability to stimulate neovascularization resulting in skin regeneration. Akermanite, a silicate bioceramic, has been shown to effectively induce angiogenesis during bone repair by supplying appropriate Si ion concentrations, stimulating cell proliferation and gene expression of human aortic endothelial cells [32]. Other biomaterials including heparin sulfate, fibrin, and hyaluronic acid (HA) also possess the ability to regulate blood vessel formation through their high affinity to angiogenic cytokines such as endothelial growth factor (EGD), vascular endothelial growth factor (VEGF), and basic fibroblast growth factor (bFGF) [29, 33]. These biomaterials enhance bone regeneration through vascularization by sequestering endogenous growth factors at the defect site [34].

Role of Biological Molecules in Bone Tissue Engineering

For the successful implantation of biomaterials, a favorable impact of cellular fate is a crucial requirement. There are factors including bacteria adhesion, nonspecific protein adsorption, and poor osseointegration that result in implant mobility, infection, and eventual failure of the implant. To facilitate various biological events, directed and selective cell adhesion is crucial after transplantation of the biomaterial. Therefore, for effective biomaterial implantation, new strategies are being developed to improve selective cell adhesion, recruitment of progenitor cells, and subsequent cellular differentiation. Moreover, the surfaces of the biomaterials are designed to reduce the nonspecific adhesion of the proteins. There are various bioactive molecules that have been utilized to resolve the aforementioned critical problems including: GAGs, proteins, and peptides (Fig. 3).

Proteoglycans

One of the major components of the ECM is proteoglycans, characterized by modifications with GAGs. GAGs mediate and regulate several key functions mainly through electrostatic interactions with ECM proteins. Therefore, they are a promising means to improve the biocompatibility of biomaterials. The osteogenic differentiation is influenced tremendously by the degree of sulfation of GAGs [28]. Additionally, the GAGs can regulate the functions of chemokines and growth factors by sequestering and releasing them [36]. Heparin has an anti-coagulative property, which is particularly valuable for vascular biomaterials [37]. Moreover, cell membrane proteoglycans interact synergistically with integrins, providing a potential target for multifunctional coatings. Besides these cell-attracting domains, molecules rendering cell-repelling properties to the biomaterial coatings are also incorporated.

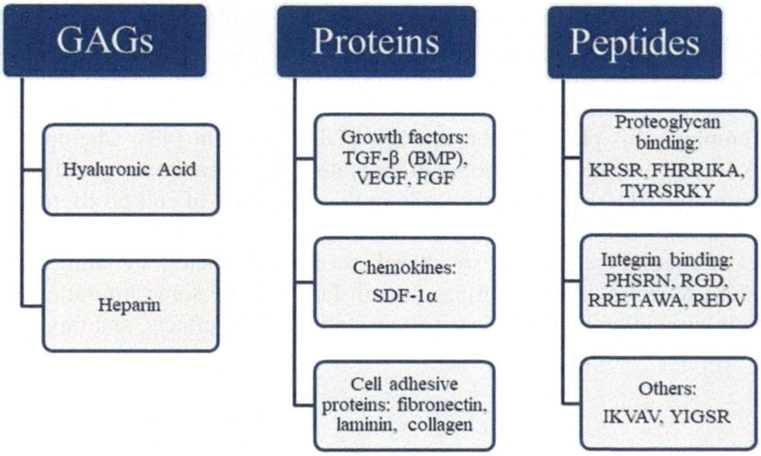

Fig. 3 Components of multifunctional coatings. Cellular function can be mediated by different glycosaminoglycans (GAGs), proteins: VEGF (vascular endothelial growth factor), BMP (bone morphogenetic protein), TGF-β (transforming growth factor), FGF (fibroblast growth factor), SDF-1 (stromal cell-derived factor 1, PDB: 1A15), or peptides (sequences are shown by the one letter code) [35]

One of the most commonly used components to prevent undesirable cell and protein adhesion is PEG [38].

Proteins

For bone tissue engineering, bone ECM proteins and peptides have been extensively used to address surface functionalization. Various proteins including vitronectin, collagen, laminin, and fibronectin are feasible mediators for cellular adhesion to biomaterials and regulate signaling events as they contain several moieties, such as proteoglycans or integrin-binding sites [39].

Moreover, growth factors are extensively used to stimulate cell proliferation and differentiation of primary and progenitor cells to osteoblasts or endothelial cells. Growth factors also induce other cellular events such as cell adhesion and proliferation or increase collagen synthesis, which favor implant integration and healing. Bone morphogenetic protein-2 (BMP-2), a growth factor of the transforming growth factor β (TGF-β) superfamily, has found its use in dental and orthopedic implants due to its potent osteoinductive effect [40]. Furthermore, the role of TGF-β1 has been implicated in the regulation of bone remodeling [41]. Vascular endothelial growth factors (VEGFs) stimulate angiogenesis and thus can be used for vascular materials and additionally for bone regeneration. The basic fibroblast growth factor (bFGF or FGF-2) and other FGFs are proangiogenic and stimulate cellular proliferation [35, 42]. A noncollagenous bone matrix protein, bone sialoprotein, also a member of SIBLING protein family, promotes migration of osteoprogenitor cells through bridging matrix metalloproteinase (MMP)-2 with integrin αvβ3 at the cell surface [40].

Chemokines orchestrate the migration of cells to the site of injury or the biomaterial surface. The recruitment of the MSCs and stimulation of osteogenic differentiation employs this phenomenon through SDF-1-gradient [43]. Chemokines are also involved in the immune response of biomaterials. Human parathyroid hormone-related protein (PTHrP) plays a key role in the regulation of cell proliferation, differentiation, and development of skeleton. The pivotal point of application of growth factors and chemokines is their steady release from biomaterial coatings. Besides successful use in tissue engineering, growth factors have some limitations due to their short circulating half-life, low protein stability, side effects, and rapid cellular internalization [44, 45].

Peptides

Peptides have emerged as useful alternatives to proteins for bone tissue repair as evidence suggests that employing smaller growth factor fragments or peptides prompts receptor-mediated signal transduction [46]. Moreover, these small biomolecules can be custom synthesized to induce directed immobilization and multifunctional properties. This can be achieved by introducing or ligating the immobilization anchors like dihydroxyphenylalanine (DOPA), basic amino acids, nucleic acids, thiols, and phosphonic acids to bioactive peptides. Moreover, cell-attracting or repelling moieties can be integrated [35].

Cell adhesion, migration, proliferation, and differentiation are favored by binding of the peptides to the receptors in the cell membrane (as illustrated in Fig. 4). The recent vigorous research in this field has led to the discovery of numerous peptides involved in the upregulation of bone repair response. The

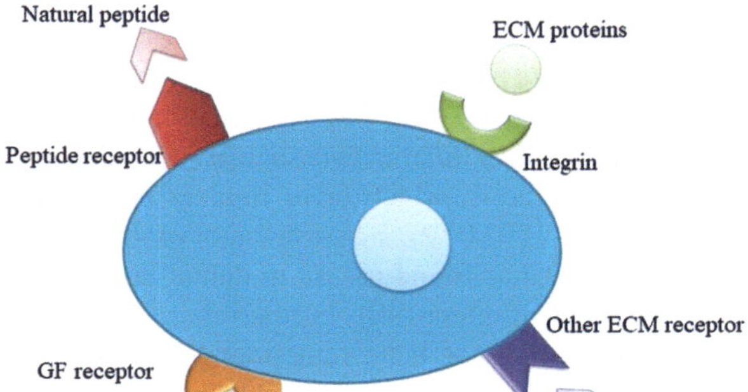

Fig. 4 Main sources of peptides for bone tissue engineering: derived from extracellular matrix proteins, soluble growth factors or are naturally occurring [47]

fragments of these peptides are typically derived from the ECM proteins, such as collagen (DGEA, GFOGER), fibronectin (RGD, PHSRN, LDV, REDV, KRSR), FGF-2 (TYRSRKY), bone sialoprotein (FHRRIKA, RGD), and laminin (YIGSR, IKVAV). The RRETAWA motif involved in selective endothelial cell binding was not derived from a protein associated with the ECM. One of the most frequently used peptides for the improvement of the tissue integration of biomaterials and to explore the integrin-mediated cell adhesion on the surfaces is the RGD motif [47].

Role of Peptides for Bone Tissue Engineering

Extensive research has established that modifying bone repair materials with suitable bioactive peptides could further improve their regenerative properties. It has been particularly noticed that peptide-modified biomaterials could stimulate new bone formation more efficiently as compared with non-modified materials.

While it has been thought earlier that biomaterials should possess a biotolerant surface so as to minimize the immune and fibrotic responses, increasingly evidence now suggests that interactive and biomimetic surfaces often demonstrate superior performances [48].

The efficacy of osseointegration relies on interactions between osteogenic macromolecules in the blood and implants. There are some biomaterials that do not readily adsorb plasma proteins to their surface, thus, these do not maintain bone-related cell activities competently and lead to inadequate bone formation [49].

Peptides Involved in Cell Adhesion

There are various insoluble proteins that reside in the ECM and form the scaffold for cells to live on. The types of biomolecules constituting the ECM and their spatial organization significantly govern cell behavior [50]. The organic fraction of the ECM is 90% type-I collagen and only 5% are noncollagenous proteins: such as osteopontin, osteocalcin, osteonectin, the adhesion proteins vitronectin and fibronectin, or the proteoglycans decorin, versican, or hyaluronan [51]. Most of these biomolecules mediate adhesive cell-ECM interactions and also play an active role in the regulation of osteoprogenitor and osteoblast proliferation, survival, and differentiation. These processes are mediated by ligand sequences in these proteins that bind to specific receptors on cell surfaces [47].

RGD Peptides

The Arg-Gly-Asp (RGD) tripeptide sequence present in fibronectin is a minimal cell adhesion peptide. It is the most studied and used peptide for biomaterial functionalization pertaining to its binding with multiple integrins stimulating in cell adhesion and differentiation. In addition to fibronectin, the RGD sequence is also present in many other ECM proteins, for instance, SIBLING proteins and vitronectin [52]. Several studies have reported enhanced osteoprogenitor cell attachment and/or differentiation with biomaterials coated with RGD peptides (Fig. 5) [53]. RGD peptides immobilized on diverse biomaterials such as alginate, titanium oxide nanotubes, or collagen sponges, have demonstrated increased cell adhesion and differentiation of MSCs or in vivo bone formation. Several variants of RGD have shown a similar osteogenesis promoting role [54].

Although categorized separately, many of the peptides discussed below contain an RGD sequence, which might be responsible partly or entirely for their biological functions.

Type-I Collagen-Derived Peptides

Osteoblastic cells consistently express major adhesion receptors for type-I collagen, the α2β1 integrins.

A synthetic triple helical peptide was engineered to contain the GFOGER sequence corresponds to 502–507 residues of the type-I collagen α1 chain. This engineered peptide selectively binds to the α2β1 integrins and promotes density-dependent cell adhesion and differentiation of MC3T3-E1 cells [55]. Coating of titanium surfaces with GFOGER increases the alkaline phosphatase (ALP) expres-

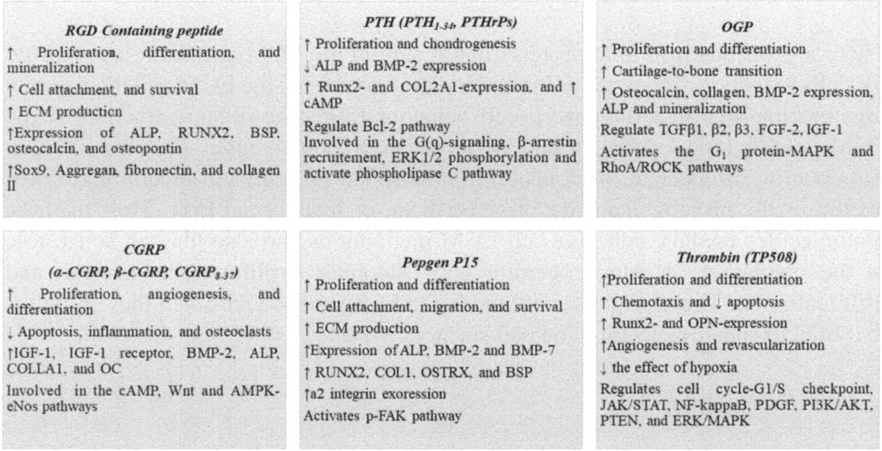

Fig. 5 Potential pathways and effect of peptides on osteoblastic cell lines [45]

sion, ECM mineralization, and the expression of osteocalcin, bone sialoprotein (BSP), and Runx2 in bone marrow stromal cells [56].

Similarly, the DGEA, a four-residue sequence initially suggested to be the recognition site of the α2β1 integrin in type-I collagen, has shown dose-dependent cell adhesion of murine MC3T3-E1 cells [57], improved attachment and spreading of hBMSCs [58], and an increased bone formation in rat tibial osteotomies when DGEA-immobilized HA was used.

P15 is another extensively studied peptide that is derived for type-I collagen. It is a 15 amino acid peptide identical to the cell-binding sequence of type-I collagen, which improves cell adhesion to bone substitutes and increases the production of ECM. P-15 significantly increases the transcript levels of BMP-2, BMP-7, and ALP when added to the scaffold (Fig. 5) [59]. It has also increased cell attachment and differentiation of human osteoblast-like HOS cells, and promoted MC3T3-E1 cell survival when adsorbed onto bovine bone-derived HA (anorganic bone matrix, ABM) [60].

PHSRN

Bone marrow stromal cells and osteoblasts express the α5β1 integrin during different stages of osteogenesis, which mediates cell adhesion to fibronectin and osteogenic differentiation. Even though only the RGD sequence can act as the ligand for α5β1, for a stable binding, both the RGD in the tenth type III repeat and the PHSRN sequence in the ninth type III repeat of fibronectin are required. A combination of RGD and PHSRN peptides conjugated to alginate hydrogels stimulated normal human osteoblasts (hOBs) for osteogenic differentiation and mineralization [61]. It is evident that the spatial configuration of the sequences in native fibronectin is crucial for assuring the correct binding as the two peptides interact with the same integrin.

FGF-2-Derived Peptides

Two peptides obtained from FGF-2 located in the cell-binding domain correspond to residue 36–41 (F36) and 77–83 (F77). When these peptides were immobilized on chitosan discs, they enhanced the cellular attachment and spreading of hBMSCs. Moreover, these peptides encourage a higher transcript level of ALP and promote ECM mineralization [54].

Laminin-Derived Peptides

Laminins are ECM proteins that bind to cell membranes through integrin receptors and other plasma membrane molecules. Peptides IKVAV and YIGSR, derived from the A and B1-chains of laminin, respectively, have shown a capability to promote

MC3T3-E1 attachment to plastic dishes coated with these peptides. YIGSR and IKVAV are peptides derived from the B1 and A-chains of laminin, respectively, have shown capability of promoting MC3T3-E1 attachment to peptide-coated plastic dishes.

The greatest osteogenic and adipogenic differentiation effects have been demonstrated by IKVAV as compared to the YIGSR peptide [62]. A more recently discovered laminin-derived peptide, Ln2-p3, has shown enhanced expression of several osteogenic markers and increased ALP activity of the cells when coated on titanium surfaces [63].

Osteopontin-Derived Peptide

Osteopontin (OPN) is a noncollagenous protein of ECM, which has demonstrated its role in bone mineralization and bone cell adhesion. The osteopontin-derived peptide (OPD) possessing RGD sequence and flanking sequences when conjugated to oligo(poly(ethylene glycol) fumarate) hydrogels showed a dose-dependent improvement of murine osteoblasts. Osteoblasts when cultured on the modified hydrogels demonstrated increased levels of secreted osteopontin, ALP activity, and ECM mineralization [64].

Heparin-Binding Peptides

Heparin-binding sequences in many ECM proteins interact with the transmembrane proteoglycans, which might have significant utility in the regulation of osteoprogenitor cell behavior.

Interactions between transmembrane proteoglycans and heparin-binding sequences found in many ECM proteins might also be of great importance to control the behavior of osteoprogenitor cells. Several studies have established that KSRS-functionalized surfaces increase the adhesion of human osteoblasts than those displaying RGD, and an increased adhesion of murine pre-osteoblasts and the expression of osteogenic markers [65].

RGD peptide and a heparin-binding domain (HBD), FHRRIKA, when functionalized on a scaffold revealed an increased ECM mineralization, and a higher degree of rat calvarial osteoblast cell and surface interactions [66].

A peptide with HBD in BMP-4 induced osteogenic differentiation of hMSCs through the ERK1/2 pathway activation. When immobilized in alginate gels, it has been demonstrated to induce a fourfold formation of new bone when grafted into a cranial defect model [67].

MEPE Peptide or AC-100

MEPE and its peptide motif, AC-100, possess an integrin-binding RGD and a consensus binding site for glycosaminoglycans SDGD. It has demonstrated significantly improved cell attachment and spreading, increased level of ECM mineralization, and superior cell differentiation with increased ALP expression in rat calvarial osteoblasts preincubated with the AC-100 fragment [68].

RRETAWA

Cyclic GA-CRRETAWAC-GA peptide-coated surfaces show an increased Runx2 and type-I collagen expression, and enhanced mineralization of ECM osteoprogenitors [69]. RRETAWA-conjugated PEG scaffolds with different stiffness profiles displayed improved MSC adhesion, higher ALP activity, and increased peptide densities. Moreover, it increased the expression of type-I collagen, osteopontin, and Runx2 [70].

Peptides Involved in Angiogenesis

During natural bone formation, vascularization is a vital process. After a bone injury, an inflammation process starts which is characterized by triggering of new blood vessel formation to recruit stem cells and soluble biomolecules that coordinate osteogenesis. In injuries where bone loss is critical, using only osteoconductive and/or osteoinductive strategies may result in graft failure due to the insufficient initial vascularization. Insufficient vascularization will lead to hypoxic conditions in the biomaterial, lacking essential osteoprogenitor cells and growth factors. Therefore, the discovery and design of proangiogenic factors and their use along with osteoinductive biomaterials are a quite appealing strategy for bone tissue engineering [26].

Osteopontin-Derived Peptide (OPD)

OPD mimics the OPN sites that are exposed during wound repair. OPD is capable of inducing the same number of newly formed blood vessels as VEGF [71]. When combined with a CO_3-apatite-collagen scaffold, OPD enhanced neovascularization in a tibial defect model [72].

Osteonectin-Derived Peptides

Osteonectin, also known as secreted protein acidic and rich in cysteine (SPARC), produces two fragments upon cleavage: SPARC113 and SPARC118. Both of these fragments contain the multifunctional tripeptide glycine-histidine-lysine (GHK) and have demonstrated potent angiogenic activity. In an in vivo model, these peptides incorporated into an MMP-degradable hydrogel prompted angiogenesis [73].

Exendin-4

Exendin-4 is an analog of glucagon-like peptide-1 (Glp-1) and it increases human umbilical vein endothelial cells (HUVECs) migration, sprouting, and tube formation in vitro and increases the sprout outgrowth from aortic rings [74].

TP508

TP508 is a 23-amino-acid peptide, commercialized as Chrysalin®. TP508, initially thought of enhancing skin wound repair when topically administered, corresponds to the receptor-binding domain of the human thrombin [75]. This peptide enhances the blood vessel formation and fracture healing. TP508 improves VEGF-stimulated angiogenesis and reduces the effects of chronic hypoxia (Fig. 5) [76].

QK Peptide

A VEGF-derived peptide, known as QK peptide, is the most studied peptide with proangiogenic properties. It binds with the Kdr and Flt-1 receptors, resulting in the stimulation of tube formation by HUVECs in Matrigel in vitro and neoangiogenesis in an ischemic hindlimb model [77].

RoY Peptide

In hypoxic conditions, the RoY peptide has demonstrated angiogenic activity by binding to a 78-KDa endoplasmic reticulum chaperone on the endothelial cell membrane, known as the glucose-regulated protein (GRP78). Moreover, a single local dose of RoY was found to normalize blood perfusion in a mouse hind limb ischemia model [78].

PBA2-1c

A multi-domain peptide, PBA2-1c, contains a heparin-binding domain and 159–163 amino acids from PDGF-BB. PBA2-1c binds with the α and β receptors of PDGF, eliciting cell proliferation, migration, and cell-induced collagen gel contraction, playing a role similar to that of which a recombinant PDGF molecule does [79].

Peptides Involved in Osteoinduction

BMPs are the main osteoinductive molecules in mammalian cells. BMPs belong to the TGF-β superfamily and interact with target cells by activating the intracellular Smad pathway directing gene expression. Out of all the identified 20 BMPs, BMP-2, BMP-4, BMP-6, BMP-7, and BMP-9 have consistently demonstrated the most osteogenic properties, being the most explored ones in the field of bone tissue engineering. A majority of the peptide molecules discussed in this section is derived from the BMPs, however, other sequences identified in the proteins of the ECM prompting osteodifferentiation have also been enlisted here.

BMP-Derived Peptides

BMPs have two sequences denoted as the "wrist" epitope, which binds to the BMP receptor type I, and the "knuckle" that binds to the BMP receptor type II. Contained within the knuckle epitope of BMP-2, a 20-mer sequence (NSVNSKIPKACCVPTELSAI) has osteogenic activity [80].

Recently, a triple functionalized substrate was generated by incorporating a BMP-2-derived peptide with an OPD angiogenic peptide on an RGD-conjugated hydrogel. This substrate induced enhanced mineralization in rat bone marrow stromal (BMS) cells, with increased transcription levels of the vasculogenic markers VE-cadherin and PECAM-1 in response to the OPD peptide. These properties made this triple functionalized substrate of particular interest for bone tissue engineering [81].

A slightly modified version of BMP-2 including the N-terminal phosphoserine and three C-terminal aspartates is P24 peptide. When this peptide was added to a porous nano-HA/collagen/poly-L-lactic acid (PLLA) scaffold with controlled release of the peptide through chitosan microspheres, it promoted the ALP activity in MSCs [82].

Another peptide of this family, BMP-9, also possess abundant osteogenic activity. A peptide derived from the knuckle epitope of the BMP-9, known as pBMP-9, has shown to elicit osteogenic marker mRNAs expression and ECM mineralization to a slight degree. These effects were not as pronounced as those of the growth factors at the equimolar concentration. A higher dose could, however, compensate its lowered activity in prompting transcription of certain markers but not all [83].

In a study, slightly different variants of BMP-2 (RKIPKASSVPTELSAISMLYL), BMP-9-derived peptide (RKVGKASSVPTKLSPISILYK), and BMP-7-derived peptide (RTVPKPSSAPTQLNAISTLYF) were implanted onto RGD-conjugated polyethylene terephthalate (PET) surfaces and compared. The BMP-2 showed the highest activity followed by the BMP-7 and the BMP-9-derived peptides in terms of the expression of osteogenic markers and the ECM thickness [84].

A BMP-2-derived peptide containing the DWIVA pentamer, which corresponds to the BMP receptors I and II binding site sequences, is known as the osteopromotive domain. This peptide has been shown to elicit the proliferation and differentiation of MC3T3-E1 cells in vitro when conjugated to titanium surfaces [85].

PTH_{1-34} or Teriparatide

Teriparatide is the commercialized N-terminal 34-residue fragment of the parathyroid hormone (PTH_{1-34}) which retains the major activities of PTH. Teriparatide plays a major role in the regulation of the mineral ion homeostasis. It also prompts osteoblast proliferation, differentiation, and inhibits apoptosis (Fig. 5). In situ bone growth was significantly enhanced by a synthetic scaffold made of polyethylene-glycol containing PTH_{1-34} [86].

Osteogenic Growth Peptide (OGP)

A naturally occurring 14-mer peptide, osteogenic growth peptide (OGP), is primarily found in serum and promotes bone anabolism, leading to increased bone formation and overall bone mass. OGP has a role in the regulation of osteoprogenitor cell proliferation, ALP activity, collagen production, differentiation, secretion of osteocalcin, and ECM mineralization (Fig. 5).

When dissociated, OGP is exposed to proteolytic cleavage, it produces a C-terminal pentapeptide YGFGG known as OGP_{10-14}. This pentapeptide activates the intracellular Gi-protein-MAP kinase pathway. Both OGP and OGP_{10-14} enhance the early expression of various markers related to osteogenesis, such as Runx2, ALP, osteopontin, osteoprotegrin, or osteocalcin [87].

Calcitonin Gene-Related Peptide (CGRP)

Calcitonin gene-related peptides (CGRP) have two forms: α and β. α-CGRP, consisting of 37 amino acids, is derived from the Calca gene. β-CGRP is derived from Calcb gene located in close proximity to the Calca gene. CGRP is found in the sensory nerve endings in the periosteum, metaphysis, and bone marrow. CGRP enhances the proliferation and differentiation, and inhibits the apoptosis of osteoprogenitor cells [88]. It also prompts the production of osteogenic molecules such as BMP-2 and IGF-1 (Fig. 5) [89].

Collagen-Binding (CB) Peptide

The collagen-binding (CB) peptide is a 28-residue peptide corresponding to the hydrophobic sequence of the BSP N-terminal. This peptide interacts with type-I collagen and stimulates the osteogenic differentiation of osteoblastic cells through Akt- and ERK-dependent signaling. This results in the increase in type-I collagen, osteopontin, osteocalcin, and ALP mRNA levels. Moreover, an increase in the ALP activity and matrix mineralization were observed [90].

Collagen-Binding Motif (CBM) Peptide

A 28-amino-acid fragment in osteopontin, called collagen-binding motif (CBM) was identified by Lee and coauthors, which is able to bind to type-I collagen. CBM enhanced hBMS cell differentiation as shown by the ECM mineralization and ALP expression. The high level of cellular phospho-Smads due to the peptide, was due to Smad pathway activation. When implanted as a CBM-collagen gel conjugate, accelerated calvarial defect repair was noted in rabbits [91].

Substance P

Substance P (SP) has been observed to enhance type-I collagen, Runx2, and osteocalcin mRNA expression, and matrix mineralization in murine calvarial osteoblasts [92]. Moreover, it prompted dose-dependent proliferation along with enhancing the expression of osteocalcin and ALP at low concentrations and inducing ECM mineralization at higher concentrations [93].

Endothelin-1

Endothelin-1 (ET-1) promotes the proliferation and differentiation of murine osteoprogenitor cells, enhanced formation of mineralized nodules and increased the activity of ALP in cultures [94]. ET-1 secreted by endothelial cells guides the MSC osteogenic and chondrogenic differentiation through AKT signaling [95]. It also plays a crucial role in the regulation of remodeling of the postnatal trabecular bone [96]. Despite its role in osteogenesis and bone remodeling, it is crucial to exercise caution when using it in bone tissue engineering as ET-1 also plays an important part in tumor progression.

BCSP™-1

A nine-amino acid synthetic fragment from human type-I collagen is known as the bone and cartilage stimulating peptide (BCSP™-1). BCSP™-1, when covalently immobilized on a commercial HA and tricalcium phosphate ceramic, can significantly enhance ALP expression by murine calvarial osteoprogenitor cells [97].

CTC Peptide

A peptide derived from the α-subunit of the collagen III C-terminal is known as the CTC peptide. It is a 12-mer cryptic peptide, which exerts a chemotactic effect on the perivascular stem cells (PSCs) and other cells. It also increases the transient expression of various osteogenic markers, such as osteopontin or type-I collagen. CTC has also been observed to stimulate ALP activity and accelerate matrix mineralization, without an increase in proliferation rate [98].

Cathelicidins

Cathelicidins are antimicrobial peptides found in cells of innate defense systems, various epithelial cells, bone marrow stroma, and MSCs. One of the cathelicidins-derived peptide, LL-37, has shown to induce monocyte differentiation to novel bone forming cells. These cells are known as monoosteophils and possess bone forming capabilities that could enhance the repair of femoral defects in mice [69].

Advantages of Peptides

The therapeutic use of peptides in tissue engineering is growing significantly as peptides offer several advantages over proteins (Fig. 6).

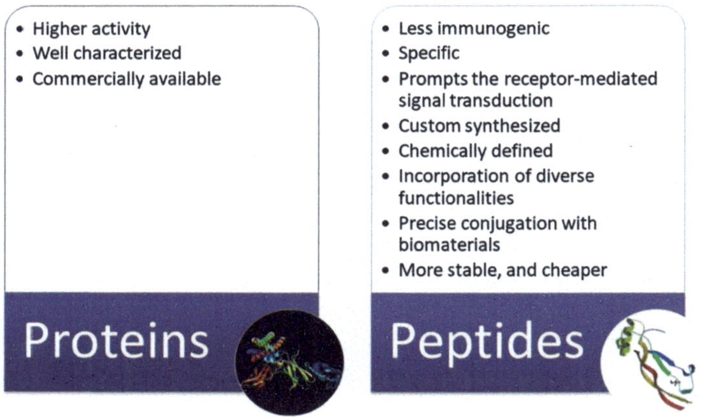

Fig. 6 Main advantages of proteins and peptides for tissue engineering [47]

Defined Chemical Properties of Peptides

Peptides are chemically defined, which is the most useful property of these biomolecules. This advantage consequently enables the refinement of their structures, corresponding experimental designs for discovering novel peptides or their combinations, and precise molecular manipulations necessary for mechanistic studies [99].

Incorporation of Non-native Chemistries and Functions into Peptides

Due to the possible synthetic methods of production, there are huge prospects of connecting moieties to peptides that are not routinely accomplished in expressed proteins or tissue-derived scaffolds. These functional groups can be natural amino acids apart from the genetically encoded ones, D-amino acids, amino acids with fluorinated side chains, polymer bioconjugates, fluorescent labels, chemical functionalities for cross-linking or polymerization, and posttranslational modifications. Chemically derived DOPA containing peptides, peptide-polymers, and peptidomimetics have been synthesized to coat and modify various synthetic biomaterials, thus, furnishing these synthetic scaffolds with adhesive properties [100].

Diverse Functions of Peptides

Peptides can possess a wide variety of functions besides cell binding, growth factor binding, surface binding, matrix-binding, self-assembly, and specific proteolytic susceptibility.

The possibility of peptide customization ensures that more efficient discovery and development of additional moieties is conceivable [99].

There are several functionalities that can be installed within a scaffold through peptides, though enzyme substrate sequences have been particularly studied, especially for hydrogels. Scaffolds can be designed that can degrade with controllable kinetics by various proteases including plasmin or matrix metalloproteinases, using peptide sequences of varying proteolytic susceptibility [101]. This enables the custom degradation of biomaterial or release of a matrix-tethered cargo through proteolysis [102].

Peptides can also be tailored to have growth factor binding abilities. This strategy has been utilized to identify VEGF-binding heparin-mimicking sulfated peptides by using bead-peptide libraries of sulfated peptides [103] or via rational design [104].

Peptides can also be engineered to self-assemble, a basic property that is native to ECM proteins. A number of peptides have been engineered and used in biomaterials for their self-assembling properties. These custom-made peptides include peptide amphiphiles [105], β-sheet fibrillizing peptides [33], short aromatic peptide derivatives [106], coiled coil peptides [107], β-hairpins [108], and others [109]. In engineering the custom peptides, design rules have been worked out to achieve predictable assembly into networks, fibers, gels, and tactics to decking these with functional peptide sequences which continue to be reported [28, 33].

Conjugation Capability of Peptides onto Biomaterials

Peptide conjugation onto or within synthetic scaffolds is achievable. This can be accomplished through specific chemistries, using a wide range of strategies such as the Michael addition of cysteine-containing peptides to vinyl sulfones, commercially available cross-linkers, UV-initiated cross-linking, amine/carboxylic acid coupling, acrylate or maleimide conjugation, and chemoselective chemistries such as "click" chemistries and native chemical ligation [99].

Enhancing Biofuntionality of Biomaterials Through Peptides

Materials from bioceramics and polymers to peptide-based scaffolds that enhance bone regeneration are applicable clinically for the treatment of diverse bone fractures and spinal fusions for different parts of the body. Current challenges to engineer materials for bone tissue engineering includes designing new materials mimicking the mechanical and biological context of bone tissue matrix supported with a vascularization system. Construction of new topographical features at nanoscale and functionalization of material surfaces with biological cues from the extracellular environment or any other biological environment are emerging approaches to design new biomimetic materials with desirable functions for bone tissue engineering [5].

Peptides as Coating Materials

Materials, which create a platform for tissue formation, play a significant role in nearly all tissue-engineering approaches. They are the skeleton of newly forming tissues with mechanical support. They also deliver inductive molecules or cells, which have the potential to control the structure and function of newly created tissue, to the site of interest. The interaction of biomaterials with biological systems is a very critical issue that needs to be considered in the design of any biomaterial. Features of biomaterials restricting the nonspecific adsorption of proteins on the surface result in weak interactions with cells for tissue formation. Thus, designing new materials which enhance cell attachment and adhesion through the presentation of peptides, proteins, and GAGs that bind to cell surface receptors and trigger a desired cell response is essential [1, 35].

In addition, multifunctional coatings with increased specificity and activity are also required to enhance the function of materials via various kinds of biomolecules. To maintain multifunctionality, a diverse bio-functional surface has the potential to control interactions between the surface and the surrounding biological environment via immobilizing different cell adhesive motifs or molecules [35].

As we mentioned above, biomoleclues especially peptides, are good candidates for bio-functionalization of materials with their diverse functions. Peptides and proteins, which we also listed above, are classified according to their source. Most peptides are derived from the ECM that surrounds and organizes cells into tissues or are derived from secreted proteins or peptides by cells into surrounding fluids. Proteins purified from a natural source and recombinantly manufactured ones allow researchers to identify the function and physicochemical properties of proteins for tissue regeneration regulations. Having knowledge about their critical domains through biological assays and bioinformatics tools created the possibility of utilizing synthetic peptides. Their presenting strategy can affect their functions. They can be presented either in an immobilized form from the surface of a material or releasable form for a material to interact with cells. All these surface functionalization strategies define the quality and functionality of biomaterials. In the following section, general techniques utilized for surface functionalization will be discussed [1].

Biomaterial Functionalization Strategies

There are various applicable strategies to functionalize biomaterials with biomolecules. The major principle of immobilization strategies is to increase the stability and functionality of biomolecules. However, the random orientation and structural deformation can occur during immobilization of molecules. It can cause to diminish activity of the biomolecules. This challenge becomes critical to differentiate immobilization strategies depending on binding strength, modularity, and complexity [110].

Especially, poor binding affinity, uncontrolled release, or strong attachment are all factors to determine the efficiency of immobilization strategies. For example, uncontrolled release of the cell adhesive motif RGD from the surface because of loose attachment can block integrin-mediated adhesion and thereby provoke apoptosis [111]. Conversely, strong anchoring of biomolecules can block biological signals that cause cell recruitment by chemokines. Hence, a feasible balance between controlled release and firm attachment is essential for efficient immobilization.

Physical and chemical immobilizations are two major techniques for the biomolecular functionalization of biomaterials. In the following sections, these techniques will be discussed in detail.

Physical Immobilization

Physical immobilization or adsorption is a very simple immobilization method performed under mild conditions. Surface functionalization with this methodology occurs by dipping biomaterials into a solution of proteins. Therefore, experimental conditions such as pH, temperature, and solvent define biomolecule-binding strength. These dependencies of experimental conditions determine whether there is stable attachment or not. However, this method is hardly disruptive to the biomolecules. Intermolecular forces, mainly ionic bonds and hydrophobic and polar interactions, specify biomolecular adsorption strength on the surface. Mostly, proteins can be attached on surfaces via electrostatic, and hydrophobic or hydrogen bond interactions. The forces involved in adsorption are weak which causes reorientation of molecules through conformational and condition dependent changes [35, 112]. On the other hand, the optimal conformation for each molecule, which minimizes the repulsive forces from the surface and previously attached proteins, can regulate to form a heterogeneous and randomly oriented biomolecule layer on the surface [113].

These conformational changes can be explained through the enthalpic or entropic state of the biomolecule. For example, the net free energy change must be negative during protein adsorption at a surface. The ordered structural content of many proteins decrease during the adsorption process through conformational changes. This fact yields an entropic gain and may act as an adsorption-driving force from an enthalpic point of view. The adsorption-driving force may originate from the interactions between protein and surface. The most important ones are van der Waals, hydrophobic, electrostatic interactions, and hydrogen bonding [114].

Surface properties of the inorganic materials lead to physical attachment of proteins. The surface hydrophobicity, charge, morphology, and topography are physical parameters which have control over a wide range of protein adsorption. For example, hydrophilic interactions are more favorable for protein adsorption on to the surface due to the nature of the protein. In terms of surface physical features, cells like rough surface topographies and morphologies for their attachment and adhesion [112, 114, 115].

Among all of the inter- and intramolecular forces we mentioned above, electrostatic interactions that are defined via the charge of biomolecules and the surfaces, can directly control adsorption as the most efficient forces to anchor positively charged molecules to negatively charged biomaterial surfaces or vice versa. Therefore, the affinity of electrostatic interactions can be tuned by raising the number of charged residues such as in poly-lysine or other charged amino acids. Other such cooperative effects could be coordinative and hydrogen bond interactions [35, 116] (Fig. 7).

Covalent Immobilization for Surface Functionalization

Covalent attachment is frequently utilized and is the most preferable for the immobilization of peptides, enzymes, and adhesive proteins onto material surfaces [1].

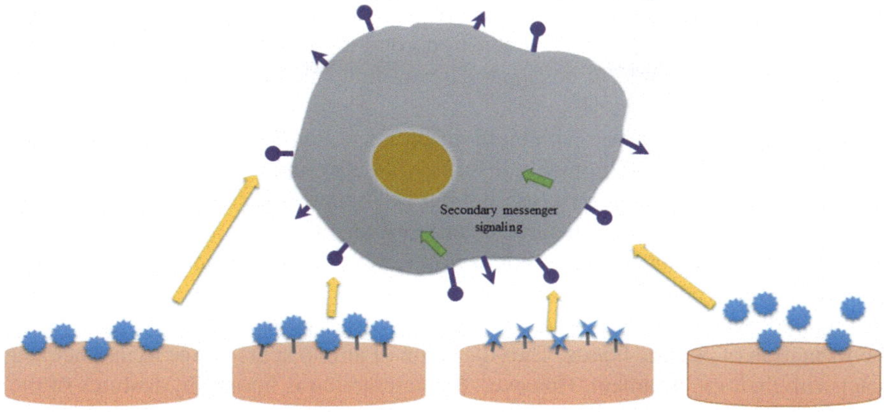

Fig. 7 Surface functionalization for biomolecules [1]

In this section, various chemistries and treatments have been applied so far in terms of covalent coupling. Alkanethiol, silane, carbodiimide, phenyl azides, acrylate, DOPA are well-known coupling agents used through chemical and photo immobilization, plasma treatment, and click chemistry as a treatment method and have been applied on biomaterial surfaces for the attachment of biomolecules [117].

Among them, carbodiimide-coupling chemistry is one of the most preferable approaches for conjugating proteins covalently to other molecules. 1-ethyl-3-(3-dimethylaminopropyl) carbodiimide hydrochloride (EDC) is used as a bioconjugation agent for the intermediate conjugation reaction to form amide bonds in between the carboxylic functional groups and the amino groups [117–119].

Silanization is another well-known strategy to introduce functional groups of molecules to the surface. A surface modified with an alkoxysilane such as (3-aminopropyl) triethoxysilane (APTES) covered by hydroxyl groups, which will be attractive for the coupling reaction with amino groups. Subsequently, the surface can be modified directly or thiol-containing biomolecules as a cross-linker can be used to enable the immobilization. This covalent strategy has disadvantages as a complex and sensitive surface treatment and potential hydrolysis of the siloxane bond, which would result in release of the bioactive moiety [35, 120].

Click chemistry is a new approach other than traditional esterification and amidation conjugation reactions. The basis of the click chemistry relies on regiospecificity. Small units together with heteroatom links were introduced to the surface through simple reaction conditions with high yield, and stereospecificity to produce newly designed materials. This chemistry triggers the reactions that can be selectively performed in the presence of natural occurring functional groups (bioorthogonal). The reaction proceeds mostly in water and it results in high yields [121].

Side and cross-reactions define whether the reactions are carried out simultaneously in one pot or stepwise. The reaction conditions and high yields make this chemistry highly suitable to covalently ligate molecules for biomaterial coatings [35, 117, 121].

To functionalize a surface with biomolecules is not only dependent on properties of the coupling agent and biomaterial surface but also functional groups in biomolecules that can be reacted by cross-linking or its derivative reagent. For peptide conjugation to any biomaterial surface, amino acid side chains are the most significant entity for covalent coupling. For example, peptides and proteins are composed of over 20 amino acids, which are identified by their side chain chemical structure, charge, hydrogen bonding, and reactivity properties, polymerized together through the formation of peptide bonds. The side chain of amino acids is free to interact and react with their environment. However, this interaction is limited by features of the side chains. For example, the aliphatic and aromatic residues are often located at the interior of the protein molecules due to their hydrophobic nature. The other amino acids (Asn: asparagine, Thr: Threonine, and Ser: Serine) are hydrophilic and contain relatively polar residues located near surfaces where they can be interacting with the surrounding aqueous environment. However, modifying Asn, Thr, and Ser, which are often posttranslationally modified with carbohydrates, with common reagent systems under aqueous conditions is difficult. The difficulty of this modification relies on the same nucleophilicity for hydroxyl and amide portions [1].

The other amino acids such as aspartic acid (Asp), glutamic acid (Glu), lysine (Lys), arginine (Arg), cysteine (Cys), histidine (His), and tyrosine (Tyr) have ionizable side chains that are suitable for covalent coupling. Each of these side chains can be in nucleophile to place in a reaction when they are in an unprotonated state. For example, Asp and Glu contain carboxylate groups that have ionization capability similar to C-terminal alpha carboxylate. Derivatization of carboxylate groups can be made through the use of amide bond forming agents or through active ester or reactive carbonyl intermediates. Lys, Arg, and His are good candidates for alkylation and acylation reactions due to their ionizable amine containing side chains. The imidazole ring of His makes it reactive species for electrophilic reactions. Cys is the only amino acid containing a sulfhydryl group, which gives a critical role to Cys in protein stabilization. The most important modification reaction of peptides and proteins is the derivatization of the side chain sulfhydryl of cysteine. The functional groups of side chains together with the N-terminal α-amino and the C-terminal α-carboxylate form the full complement of polypeptide reactivity [1].

These approaches are more complicated and time consuming than other immobilization methods. A major limitation of this methodology is the loss of protein mobility when they are immobilized on the surfaces, which is directly affected by possible representation of unfamiliar protein conformation on the surface. Remains of toxic monomer residues on the surface may cause biocompatibility problems in the area of implantation. Toxic monomer residues, which are directly caused by the instability of the molecules on the surface, are the most challenging issues in the task of chemical immobilization. Physical adsorption techniques can be addressed to reduce toxic residues. While it may help reduce the effect of toxic residues, its weak immobilization capability brings more problems to surface functionalization [122, 123].

Therefore, there is a need for alternative methodologies for surface functionalization in bone tissue engineering. In the next section, a new strategy based on peptide-based surface functionalization through material recognition capability will be discussed more in detail.

Material-Binding Peptides for Surface Functionalization

In recent years, material-binding peptides (MBPs) have shown remarkable potential in various application areas, taking advantage of molecular biomimetics. The idea behind this is to discover and design MBPs mainly inspired by nature and a biomimetic approach. This approach is revealed at the intersection of different disciplines such as material science and molecular biology. It mainly covers understanding the interactions between materials and biomolecules by learning from nature [124, 125].

All biological materials are highly organized often in a hierarchical manner from the molecular to the nanoscale, with complex nano-architectures [126, 127]. The contribution of biomacromolecules such as lipids, proteins, glycoproteins, and phosphoproteins, etc. is a key aspect of this architecture. For example, proteins control the formation of hard tissues like teeth, bone, and many other hard tissues occurring in different organisms. Due to the role of proteins for the fabrication of materials in nature, MBPs can be good candidates to fabricate, design, assemble, and functionalize many materials from various fields due to their molecular recognition and binding capabilities [123, 128–130].

Especially in bone tissue engineering, inorganic surface and protein/peptide or any biomolecule interactions is a key aspect due to bone structure which shows composite material features. The major concern in bone tissue engineering and implants is the uncontrolled interactions between synthetic materials and human tissues. The most successful approach to this issue is functionalization of the biomaterial surfaces with different molecules with desired functions including anti-fouling polymers or cell growth factors. To date, physical and covalent immobilization methods have been applied onto varieties of biomaterial surfaces. Covalent immobilization requires the presence of specific functional groups and synthetic pathways. Moreover, the functional groups used in these strategies have low selective properties, which restrict their use as we discussed in previous sections [110].

On the other hand, the behavior, stability, and cytotoxicity of the modified surfaces for all strategies under physiological conditions are not well understood. Therefore, material selectivity, coupling efficiency, and flexibility remain a challenge at the biomaterial interface. MBPs can provide a new platform for surface functionalization via immobilization of biomolecules with controlled attachment and assembly on solid surfaces [110, 123, 131].

During the last decade, MBPs have been selected using phage and cell surface display techniques possessing affinity and specificity to select inorganic surfaces [132, 133].

Their potential use was shown in many disciplines including surface functionalization, and biomineralization for tissue engineering and regenerative medicine [125]. There has been a great deal of interest in identifying and characterizing peptides that bind to various materials, such as TiO_2 [134], Au [133], SiO_2, and HAP utilized especially in tissue engineering and regenerative medicine [135].

In this section, two case studies will be demonstrated to give basic perspectives for the promising potential of MBPs in implantation and hard tissue engineering as molecular linker and material synthesizer.

In the first example, the capability of MBPs as molecular linkers will be given. Yazici et al. selected titanium binding peptides (TiBPs) through cell surface display and conjugated them with RGDS, which is a cell adhesive peptide to show their molecular linker capability for implant surface functionalization [110, 123].

These peptides can be conjugated with a variety of bioactive molecules to enhance cell attachment, cell proliferation, cellular spreading, and other cell behaviors or creating antimicrobial or anti-fouling surfaces. Therefore, these peptide-based molecular linkers provide a new platform to conjugate domains with different functionality. In this case study, TiBP was conjugated with biologically active signaling molecules (RGDS) while retaining their remarkable binding and selectivity to a solid substrate in the absence of cytotoxicity properties [123]. This study proved that a TiBP-RGDS bifunctional peptide enhanced osteoblast attachment and adhesion on a titanium implant surface. The proposed peptide-based surface coating can be applied on various materials to induce various desired biological activities on any biomaterial using an easily adaptable single-step biologically relevant set of conditions [110, 123, 128, 129].

In another example, the capability of MBPs as a synthesizer in the biomineralization process will be given. In nature, biomineralization, which occurs under mild conditions to form bone, teeth, sponge spicules, and similar tissues with nanoarchitecture in various organisms, has attracted attention in the field of bionanotechnology. Biological organisms have a capability to synthesize their inorganic structures or hard tissues with unique morphological, structural, and functional properties. This biological mineralization process usually involves a large number of proteins with various temporal and spatial distributions. To understand the exact role of proteins in the biomineralization process and biological material synthesis, the traditional approach, which involves extracting and purifying proteins from the organism of interest, has been utilized [136]. Although there are exciting examples for performing biomineralization using isolated proteins, this approach is limited because of the difficulties involved in the extraction and purification steps of these proteins from biological systems. De novo design is another approach that stems from the prediction of functional sequence of proteins using computational methods. Biomineralization through extracted proteins or de novo designed peptide sequences remains elusive due to impractical identification of all proteins and their sequences [137].

MBPs offer a unique and a more practical approach in the biomineralization process. MBPs have a recognition capability to solid materials; this recognition capability may influence the fabrication process of inorganic materials as well. To prove this, Gungormus et al. have explored the possibility of HA-binding peptides to regulate calcium phosphate formation in vitro. They found that the formation of calcium phosphate mineral could be controlled via a strong-binding peptide HABP1. The rate of mineralization decreased by the addition of HABP1, resulting in the formation of much larger plate-like particles compared to control samples. Meanwhile, the rate of transformation of the amorphous phase to the crystalline phase increased. The transition between the two phases can happen via interactions of HABP1 in the amorphous mineral surface. This interaction may stabilize the crystal structure by lowering surface energy, therefore, resulting in a growth-dominated mineralization pathway [137, 138].

As given in the examples above, MBPs have a great potential in biomaterial surface functionalization and the biomineralization process. Biofunctionalization of biomaterials through peptides with various methodologies are summarized in Table 2.

Peptides as Scaffold Materials

Self-Assembled Peptides

Self-assembling proteins and peptides have a remarkable potential due to their unique features as scaffolds for applications in tissue engineering. Their self-organization capability from basic building blocks to forming supramolecular structures and simulating the native ECM, make them as preferable scaffold materials for tissue engineering applications. Favorable properties of self-assembling peptides are mainly based on their modification capability at the sequence level. Moreover, their synthesis does not rely on difficulties of a recombinant protein expression and purification system. They can be easily produced through solid phase peptide synthesis. On the other hand, recombinant technologies can be the only alternative to solve any homogeneity and standardization issues necessary for applications [139, 140].

The unique structural properties of the self-assembled peptides are relevant with alternating hydrophilic and hydrophobic amino acid residues. Under physiological conditions or mild conditions, these residues spontaneously adopt a β-sheet structure when exposed to monovalent cation solutions. This process finalizes with the formation of self-assembled matrices with certain geometries. The RADA-16 series, which is commercially available (PuraMatrix), ELK, and EAK are well-known examples for self-assembled scaffolds in the literature. The RADA series is also the best example for the commercialization of self-assembled peptide-based scaffolds [141, 142].

Among self-assembling biomolecules, peptide amphiphiles are another class of peptide-based scaffolds that can enhance osteoprogenitor cells and prompt their differentiation. Many studies in the literature revealed the role of peptide amphiphiles through mineralized matrixes in promoting osteogenic differentiation of hMSCs

Table 2 Peptides and their assembling methodology and function [35, 110, 123]

Bioactive	Surface	Immobilization	Assembly	Favored cellular functions
BMP 4	Ti	Carbodiimide mediated	Covalent	Proliferation, mineralization
BMP-2	PLGA	Acylate-NHS-PEG	Covalent	Mineralization
BMP-2 derivative	Alginate	Carbodiimide mediated	Covalent	Mineralization
OGP	Si	Click Chemistry	Covalent	Mineralization
KRSR	Ti	Silanization	Covalent	Spreading, adhesion, mineralization
RGD + FHRRIKA	Si	Silanization	Covalent	Spreading, mineralization
RGD + PHSRN	Ti	Thiol	Covalent	Spreading, proliferation
RGD + BMP-2 + hydroxyapatite	Ti	DOPA	Covalent/ electrostatic	Adhesion, mineralization
RGD + bFGF	Si	Spin coating/ thermal annealing	Covalent/ electrostatic	Spreading, focal adhesion
OGP + fibronectin	Ti	Adsorption/ co-precipitation	Electrostatic	Adhesion, proliferation, differentiation
Heparin + 1 aminin + bFGF	PLLA	Covalent	Covalent/ electrostatic	Neurite outgrowth
Heparin + BMP-2	Ti	Silanization	Covalent/ electrostatic	Anti-inflammatory, proliferation, mineralization
Heparin + VEGF + fibronectin	Ti	Electrostatic	Layer by layer electrostatic	Anti-coagulative, adhesion, proliferation
Heparin + SDF-1	PGS	Electrostatic	Electrostatic	Progenitor cell recruitment
GFOGER	PEG	Covalent	Covalent	Bone regeneration, osseointegration
Chitosan + BMP-2	Ti	Covalent	Electrostatic	Differentiation
Hyaluronic acid + collagen	Ti	Silanization	Layer by layer covalent	Adhesion, proliferation, differentiation
Hyaluronic acid + collagen	Ti	Electrostatic	Electrostatic	Non-pathological smooth muscle cell phenotype
Collagen binding motif	Ti	Electrostatic		Osteogenic differentiation

(continued)

Table 2 (continued)

Bioactive	Surface	Immobilization	Assembly	Favored cellular functions
Collagen + lactoferrin	Ti	Electrostatic	Electrostatic	Adhesion, proliferation, differentiation
Collagen + CS + BMP-4	PLGA	Electrostatic	Electrostatic	Increase of bone-implant contact
Fibronectin-derived	Ti	Covalent, electrostatic	Covalent/ electrostatic	Cell spreading, adhesion

BMP bone morphogenetic proteins, *OGP* osteogenic growth peptide, *FGF* fibroblast growth factor, *VEGF* vascular endothelial growth factor, *SDF-1* stromal cell derived factor, *CS* chondroitin sulfate, *PEG* polyethylene glycol, *PGS* poly(glycerol sebacate), *PLGA* poly(lactic-*co*-glycolic acid), *PLLA* poly(lactic-*co*-glycolic acid)

and bone formation. On the other hand, peptide amphiphile matrices functionalized with MSCs and platelet-rich plasma were demonstrated to encourage bone formation and improve angiogenesis [105, 143, 144].

Peptide-Based Biomaterial Scaffolds

Synthetic biomaterials are a necessity for the controlled release of drugs, tissue restoration, and tissue engineering. One of the most common advantages shared by all synthetic scaffolds is reducing the possibility of carrying biological pathogens or contaminants [10]. Moreover, synthetic biomaterials can be engineered to meet particular needs with their promising in vivo biocompatibility. Recently, newly designed biomaterials have displayed a tremendous enhancement for in vivo biocompatibility.

Each synthetic material is composed of different substitutes, such as calcium phosphate and amino acids are substitutes for ceramics and for peptides, respectively. Some synthetic scaffolds are comprised of molecules that are not found in vivo such as ceramics or metal-based materials. However, they display desired features such as high tensile strength (e.g., bone tissue replacement materials). Other classes of materials that were discussed here are mainly peptide-based materials that are composed of spontaneously self-assembling oligopeptides and were discovered recently. One of the advantages of peptide-based scaffolds is their design flexibility that will allow us to conjugate them with various molecules that have different functions. For example, biological functional domains that enhance cell adhesion such as the cell attachment motif RGD, an integrin receptor-binding ligand or any peptide with mineralization and cell differentiation capacity, can be easily incorporated during synthesis of these peptides. This function allows researchers to gain various functions in one material. On the other hand, amino acids as a monomer of peptide-based scaffolds display outstanding physiological compatibility and minimal cytotoxicity. Having a biological substitute is an advantage for a scaffold design due to their breakdown products of biologically derived biomaterials and can be incorporated into synthesized biomolecules or metabolized in the host organism [142].

As all peptide-based scaffolds, the peptide amphiphiles are also composed of a unique sequence of amino acids which comprise of repeating units of positively charged (lysine or arginine) and negatively charged (aspartate or glutamate) amino acids with hydrophobic residues (alanine or leucine) in between. These self-complementary peptides consist of 50% charged amino acids. Therefore, their self-complementary properties rely on the type and sequence of their amino acid substitute [142, 145, 146].

Among peptide-based scaffolds, RAD16-I, which has the sequence AcN-RADARADARADARADA-CNH_2, and RAD16-II, which has the sequence AcN-RARADADARARADADA-CNH_2, are well-known examples in terms of their clinical use. Though both of these peptides have the same length and number of residues, RD16-I possesses a spacing modulus of one based on the formula (RADA)n, contrarily, RAD16-II has a spacing modulus of two based on the formula (RARADADA)n, where n denotes any number of repeats.

The self-assembly of peptides depends on various factors, such as peptide and salt concentration, which may determine the geometry and dimensions of the macroscopic matrices either as tapes, strings, or sheets. Circular dichroism (CD) is one common technique to define the structure of peptide-based scaffolds. It is a very important technique to define structure related parameters, which are very crucial techniques to design peptide-based scaffolds. CD spectroscopy revealed that RAD-, EAK-, and ELK-based peptides with their representative periodicities displayed strong β-sheet secondary structure in aqueous solutions. These secondary structures displayed two distinctive polar and nonpolar surfaces with simple rules of amino acid sequence and type, which give the structural property of the scaffold [147–149]. Having defined structural properties through this simple rule should be explained with some predictions or paradox. For example, the measured β-sheet secondary structure of RAD, EAK, and ELK peptides opposed anticipations based on the Chou-Fasman statistical predictions for protein helical preferences. Glutamate, leucine, and lysine all have high α-helical tendency in the Chou-Fasman model [13].

The secondary structure of self-assembling synthetic peptides is an outcome of when local and nonlocal intramolecular influences compete. Generally, local influences for defining secondary structures include the intrinsic helical propensity of amino acids. Nonlocal influences are illustrated by the periodicity and positioning of amino acids in the context of the peptide sequence—periodicity and amino acid positioning determined secondary structures for all synthetic peptides. Thus, nonlocal effects predominated over local effects [142, 149].

Interestingly, the role of local and nonlocal forces can be affected by environmental conditions in which the self-assembly occurs. Amphiphilic peptides, such as RAD16 and EAK16, can be solubilized at low millimolar concentrations in salt-free aqueous solutions. However, the amphiphilic peptides spontaneously form hydrogel-like matrices when the peptide solutions are exposed to salt solutions or physiological media. The ordered matrix formation is due to millimolar levels of monovalent cations. The ordered biomatrix comprises a hydrogel with a water content of >99%.

Contrary to the ordered biomaterial matrix that forms when exposed to monovalent cations, EAK16 and related peptides form highly disordered materials in the presence of millimolar levels of divalent cations. This demonstrates that the concentration of salt is critical to trigger a molecular switch to form a matrix. One explanation for this can be that the matrix formation is triggered due to the electrostatic interactions that occur through the salt between the negatively and positively charged amino acids of the adjacent peptides. Consequently, promoting the staggered configuration of the individual peptide. Alternatively, as these peptides are self-complementary in aqueous solutions, monomeric peptides might undergo folding to form intramolecular electrostatic interactions. However, if salt is added, it could disrupt the intramolecular electrostatic interactions, directing the peptides to adopt a configuration that favors intermolecular hydrophobic interactions between adjacent peptides.

Although salt concentration effects matrix formation, the length of the peptide and degree of hydrophobicity of the aliphatic amino acids are also critical. For example, amphiphilic peptides containing alanine (such as EAK16) form a salt-induced stable matrix when at least 16-mer peptides are present [9]. Amphiphilic peptides containing leucine (such as ELK8), by contrast, require eight-mer peptides to form salt-induced stable matrices. These observations show that increasing the hydrophobicity of the aliphatic residue contributes to matrix formation. A third hierarchal model includes features of both earlier models. The matrix could be stabilized as a result of the electrostatic intermolecular interactions between the charged amino acids of two adjacent peptides after the intermolecular hydrophobic interactions are formed. Moreover, other conditions of the process, such as the temperature and pH, can be adjusted to direct the resulting self-assembling matrix geometry. All of these models of peptide-based matrix formation and stabilization require further direct experimental confirmation to understand the function of each experimental condition [17, 150]. The sequence of the peptide may not have the only role for gel formation. For example, the RAD-based amphiphilic peptide sequence shares similarity to the ligand that binds with the cell adhesion receptor integrin RGD. Some of the ECM proteins contain RAD sequences that can bind to the isoforms of integrin [151]. The first hypothesis that was tested was if the cells can adhere and grow on peptide-based scaffolds in an integrin-dependent manner. Cell adhesion to RAD- and EAK-based scaffolds do not involve integrin binding [152], and both of these matrices promote cell adhesion and growth. Though the RAD sequence can bind to certain integrin receptors, the EAK does not bind to integrins. Moreover, high concentrations of RGD peptides do not affect cell attachment to RAD- and EAK-based matrices, which confirms that integrin-based attachment is not crucial for cell adhesion to these peptide-based matrices. The RAD- and EAK-peptide scaffolds support cell adhesion of diverse types of mammalian and avian primary and transformed cells [7, 17, 142, 150].

Conclusion

The expanding need of interventions for bone tissue regeneration can be met with the use of biomimetic peptides. These peptides have various advantages that render them useful for tissue engineering including their compact size. These small molecules with simple structures can be customized to include many properties such as directed immobilization. The peptides can also be used to functionalize the osteoinductive biomaterials which can enhance the cell attachment, differentiation, and phenotype development. These biomimetic peptides can confer bioactivity that synthetic scaffold lack, which can lead to better biomaterial–host interaction. Numerous peptides have been developed and explored for bone repair. However, inadequate verification from clinical trials limits the use of these peptides. These peptides have potential for tissue engineering, which needs to be explored more. Further investigations will result in biological molecules that can be utilized in the clinical settings for bone tissue engineering. The areas that need to be explored more in this regard include one of the most common problem of stability of the peptides resulting in low bioavailability and short duration of activity due to proteolysis. This can be achieved by improving the peptide designs incorporating cyclization, nonnatural amino acids, and stable peptide bonds. Moreover, polytherapy, i.e., combining several distinct peptides targeting specific bone repair phase or a specific population of cells involved in bone healing, should be explored further. This strategy can be used to couple osteoconductive peptide with another osteoinductive peptide or a vascularization inducing peptide to increase the bone healing. Another field to be expanded is the scaffolding technologies to incorporate controlled release of peptides to provide correct signal at the precise stage of the repair pathway. Therefore, further research in the peptide design and scaffolding technologies aiming for the upregulation of osteoinduction and osteoconduction is desirable to lead to future treatment involving biomimetic peptides in clinical setups.

Acknowledgments Dr. Hana'a Iqbal is jointly supported by institutional post-doctoral fellowship program, International Center for Chemical and Biological sciences (ICCBS) in Pakistan and the TUBITAK Marmara Research Center (MAM) in Turkey. The authors also thank TUBITAK-MAM for research funding (Project Code: 5183402).

References

1. Hirano Y, Mooney DJ (2004) Peptide and protein presenting materials for tissue engineering. Adv Mater 16(1):17–25. https://doi.org/10.1002/adma.200300383
2. Agarwal R, García AJ (2015) Biomaterial strategies for engineering implants for enhanced osseointegration and bone repair. Adv Drug Deliv Rev 94:53–62
3. Ho-Shui-Ling A, Bolander J, Rustom LE, Johnson AW, Luyten FP, Picart C (2018) Bone regeneration strategies: engineered scaffolds, bioactive molecules and stem cells current stage and future perspectives. Biomaterials 180:143–162
4. Calcei JG, Rodeo SA (2019) Orthobiologics for bone healing. Clin Sports Med 38(1):79–95

5. Stevens MM (2008) Biomaterials for bone tissue engineering. Mater Today 11(5):18–25
6. Roseti L, Parisi V, Petretta M, Cavallo C, Desando G, Bartolotti I, Grigolo B (2017) Scaffolds for bone tissue engineering: state of the art and new perspectives. Mater Sci Eng C 78:1246–1262
7. Chan K, Zhuo S, Ni M (2013) Natural and synthetic peptide-based biomaterials for bone tissue engineering. OA Tissue Eng 1(1):6
8. Nalla RK, Kruzic JJ, Kinney JH, Balooch M, Ager JW, Ritchie RO (2006) Role of microstructure in the aging-related deterioration of the toughness of human cortical bone. Mater Sci Eng C 26(8):1251–1260. https://doi.org/10.1016/j.msec.2005.08.021
9. Finkemeier CG (2002) Bone-grafting and bone-graft substitutes. JBJS 84(3):454–464
10. Giannoudis PV, Dinopoulos H, Tsiridis E (2005) Bone substitutes: an update. Injury 36(3):S20–S27
11. Pan P, Chen X, Metavarayuth K, Su J, Wang Q (2018) Self-assembled supramolecular systems for bone engineering applications. Curr Opin Colloid Interface Sci 35:104–111
12. Rezwan K, Chen Q, Blaker J, Boccaccini AR (2006) Biodegradable and bioactive porous polymer/inorganic composite scaffolds for bone tissue engineering. Biomaterials 27(18):3413–3431
13. LeGeros RZ (2008) Calcium phosphate-based osteoinductive materials. Chem Rev 108(11):4742–4753
14. Rodrigues C, Serricella P, Linhares A, Guerdes R, Borojevic R, Rossi M, Duarte M, Farina M (2003) Characterization of a bovine collagen–hydroxyapatite composite scaffold for bone tissue engineering. Biomaterials 24(27):4987–4997
15. Chen K-Y, Shyu P-C, Dong G-C, Chen Y-S, Kuo W-W, Yao C-H (2009) Reconstruction of calvarial defect using a tricalcium phosphate-oligomeric proanthocyanidins cross-linked gelatin composite. Biomaterials 30(9):1682–1688
16. Yu X, Tang X, Gohil SV, Laurencin CT (2015) Biomaterials for bone regenerative engineering. Adv Healthc Mater 4(9):1268–1285
17. Habibovic P, Yuan H, van der Valk CM, Meijer G, van Blitterswijk CA, de Groot K (2005) 3D microenvironment as essential element for osteoinduction by biomaterials. Biomaterials 26(17):3565–3575
18. Chai YC, Carlier A, Bolander J, Roberts SJ, Geris L, Schrooten J, Van Oosterwyck H, Luyten FP (2012) Current views on calcium phosphate osteogenicity and the translation into effective bone regeneration strategies. Acta Biomater 8(11):3876–3887
19. Li X, Liu H, Niu X, Fan Y, Feng Q, Cui F z, Watari F (2011) Osteogenic differentiation of human adipose-derived stem cells induced by osteoinductive calcium phosphate ceramics. J Biomed Mater Res B Appl Biomater 97(1):10–19
20. Ripamonti U (2006) Soluble osteogenic molecular signals and the induction of bone formation. Biomaterials 27(6):807–822
21. Barradas AM, Fernandes HA, Groen N, Chai YC, Schrooten J, van de Peppel J, van Leeuwen JP, van Blitterswijk CA, de Boer J (2012) A calcium-induced signaling cascade leading to osteogenic differentiation of human bone marrow-derived mesenchymal stromal cells. Biomaterials 33(11):3205–3215
22. Liu H, Yazici H, Ergun C, Webster TJ, Bermek H (2008) An in vitro evaluation of the Ca/P ratio for the cytocompatibility of nano-to-micron particulate calcium phosphates for bone regeneration. Acta Biomater 4(5):1472–1479
23. Breitwieser GE (2008) Extracellular calcium as an integrator of tissue function. Int J Biochem Cell Biol 40(8):1467–1480
24. Murphy WL, Peters MC, Kohn DH, Mooney DJ (2000) Sustained release of vascular endothelial growth factor from mineralized poly (lactide-co-glycolide) scaffolds for tissue engineering. Biomaterials 21(24):2521–2527
25. Krishnan L, Willett NJ, Guldberg RE (2014) Vascularization strategies for bone regeneration. Ann Biomed Eng 42(2):432–444
26. Kanczler JM, Oreffo RO (2008) Osteogenesis and angiogenesis: the potential for engineering bone. Eur Cell Mater 15(2):100–114

27. Salbach-Hirsch J, Ziegler N, Thiele S, Moeller S, Schnabelrauch M, Hintze V, Scharnweber D, Rauner M, Hofbauer LC (2014) Sulfated glycosaminoglycans support osteoblast functions and concurrently suppress osteoclasts. J Cell Biochem 115(6):1101–1111
28. Sun G, Zhang X, Shen Y-I, Sebastian R, Dickinson LE, Fox-Talbot K, Reinblatt M, Steenbergen C, Harmon JW, Gerecht S (2011) Dextran hydrogel scaffolds enhance angiogenic responses and promote complete skin regeneration during burn wound healing. Proc Natl Acad Sci 108(52):20976–20981
29. Jabbarzadeh E, Starnes T, Khan YM, Jiang T, Wirtel AJ, Deng M, Lv Q, Nair LS, Doty SB, Laurencin CT (2008) Induction of angiogenesis in tissue-engineered scaffolds designed for bone repair: a combined gene therapy–cell transplantation approach. Proc Natl Acad Sci 105(32):11099–11104. https://doi.org/10.1073/pnas.0800069105
30. Unger RE, Peters K, Huang Q, Funk A, Paul D, Kirkpatrick CJ (2005) Vascularization and gene regulation of human endothelial cells growing on porous polyethersulfone (PES) hollow fiber membranes. Biomaterials 26:3461–3469. https://doi.org/10.1016/J.BIOMATERIALS.2004.09.047
31. Madden LR, Mortisen DJ, Sussman EM, Dupras SK, Fugate JA, Cuy JL, Hauch KD, Laflamme MA, Murry CE, Ratner BD (2010) Proangiogenic scaffolds as functional templates for cardiac tissue engineering. Proc Natl Acad Sci 107(34):15211–15216. https://doi.org/10.1073/pnas.1006442107
32. Zhai W, Lu H, Chen L, Lin X, Huang Y, Dai K, Naoki K, Chen G, Chang J (2012) Silicate bioceramics induce angiogenesis during bone regeneration. Acta Biomater 8:341–349. https://doi.org/10.1016/j.actbio.2011.09.008
33. Rudra JS, Tian YF, Jung JP, Collier JH (2010) A self-assembling peptide acting as an immune adjuvant. Proc Natl Acad Sci 107(2):622–627
34. Hubbell JA, Chilkoti A (2012) Nanomaterials for drug delivery. Science 337(6092):303–305. https://doi.org/10.1126/science.1219657
35. Pagel M, Beck-Sickinger AG (2017) Multifunctional biomaterial coatings: synthetic challenges and biological activity. Biol Chem 398(1):3–22
36. Hintze V, Samsonov SA, Anselmi M, Moeller S, Becher J, Schnabelrauch M, Scharnweber D, Pisabarro MT (2014) Sulfated glycosaminoglycans exploit the conformational plasticity of bone morphogenetic protein-2 (BMP-2) and alter the interaction profile with its receptor. Biomacromolecules 15(8):3083–3092
37. Liu T, Zeng Z, Liu Y, Wang J, Maitz MF, Wang Y, Liu S, Chen J, Huang N (2014) Surface modification with dopamine and heparin/poly-L-lysine nanoparticles provides a favorable release behavior for the healing of vascular stent lesions. ACS Appl Mater Interfaces 6(11):8729–8743
38. Pagel M, Hassert R, John T, Braun K, Wießler M, Abel B, Beck-Sickinger AG (2016) Multifunctional coating improves cell adhesion on titanium by using cooperatively acting peptides. Angew Chem Int Ed 55(15):4826–4830
39. Siebers MC, ter Brugge PJ, Walboomers XF, Jansen JA (2005) Integrins as linker proteins between osteoblasts and bone replacing materials. A critical review. Biomaterials 26(2):137–146. https://doi.org/10.1016/j.biomaterials.2004.02.021
40. Karadag A, Fisher LW (2006) Bone sialoprotein enhances migration of bone marrow stromal cells through matrices by bridging MMP-2 to αvβ3-integrin. J Bone Miner Res 21(10):1627–1636
41. Tang Y, Wu X, Lei W, Pang L, Wan C, Shi Z, Zhao L, Nagy TR, Peng X, Hu J (2009) TGF-β1–induced migration of bone mesenchymal stem cells couples bone resorption with formation. Nat Med 15(7):757
42. Liu T, Liu S, Zhang K, Chen J, Huang N (2014) Endothelialization of implanted cardiovascular biomaterial surfaces: the development from in vitro to in vivo. J Biomed Mater Res A 102(10):3754–3772. https://doi.org/10.1002/jbm.a.35025
43. Eman RM, Öner FC, Kruyt MC, Dhert WJ, Alblas J (2013) Stromal cell-derived factor-1 stimulates cell recruitment, vascularization and osteogenic differentiation. Tissue Eng Part A 20(3–4):466–473

44. Guo J, Liu M, Yang D, Bouxsein ML, Thomas CC, Schipani E, Bringhurst FR, Kronenberg HM (2010) Phospholipase C signaling via the parathyroid hormone (PTH)/PTH-related peptide receptor is essential for normal bone responses to PTH. Endocrinology 151(8):3502–3513. https://doi.org/10.1210/en.2009-1494
45. Pountos I, Panteli M, Lampropoulos A, Jones E, Calori GM, Giannoudis PV (2016) The role of peptides in bone healing and regeneration: a systematic review. BMC Med 14(1):103
46. Mitchell AC, Briquez PS, Hubbell JA, Cochran JR (2016) Engineering growth factors for regenerative medicine applications. Acta Biomater 30:1–12
47. Visser R, Rico-Llanos GA, Pulkkinen H, Becerra J (2016) Peptides for bone tissue engineering. J Control Release 244:122–135
48. Yi J, Xiong F, Li B, Chen H, Yin Y, Dai H, Li S (2016) Degradation characteristics, cell viability and host tissue responses of PDLLA-based scaffold with PRGD and β-TCP nanoparticles incorporation. Regen Biomater 3(3):159–166
49. Punet X, Mauchauffé R, Rodríguez-Cabello JC, Alonso M, Engel E, Mateos-Timoneda MA (2015) Biomolecular functionalization for enhanced cell-material interactions of poly(methyl methacrylate) surfaces. Regen Biomater 2(3):167–175. https://doi.org/10.1093/rb/rbv014
50. Rosso F, Giordano A, Barbarisi M, Barbarisi A (2004) From cell-ECM interactions to tissue engineering. J Cell Physiol 199(2):174–180
51. Shekaran A, Garcia AJ (2011) Extracellular matrix-mimetic adhesive biomaterials for bone repair. J Biomed Mater Res A 96(1):261–272. https://doi.org/10.1002/jbm.a.32979
52. Fisher LW, Fedarko NS (2003) Six genes expressed in bones and teeth encode the current members of the SIBLING family of proteins. Connect Tissue Res 44(Suppl 1):33–40
53. Bilem I, Chevallier P, Plawinski L, Sone E, Durrieu M, Laroche G (2016) RGD and BMP-2 mimetic peptide crosstalk enhances osteogenic commitment of human bone marrow stem cells. Acta Biomater 36:132–142
54. Rezania A, Healy KE (2000) The effect of peptide surface density on mineralization of a matrix deposited by osteogenic cells. J Biomed Mater Res 52(4):595–600
55. Reyes CD, García AJ (2004) α2β1 integrin-specific collagen-mimetic surfaces supporting osteoblastic differentiation. J Biomed Mater Res A 69(4):591–600
56. Reyes CD, Petrie TA, Burns KL, Schwartz Z, García AJ (2007) Biomolecular surface coating to enhance orthopaedic tissue healing and integration. Biomaterials 28(21):3228–3235. https://doi.org/10.1016/j.biomaterials.2007.04.003
57. Gilbert M, Giachelli CM, Stayton PS (2003) Biomimetic peptides that engage specific integrin-dependent signaling pathways and bind to calcium phosphate surfaces. J Biomed Mater Res A 67(1):69–77
58. Hennessy KM, Pollot BE, Clem WC, Phipps MC, Sawyer AA, Culpepper BK, Bellis SL (2009) The effect of collagen I mimetic peptides on mesenchymal stem cell adhesion and differentiation, and on bone formation at hydroxyapatite surfaces. Biomaterials 30(10):1898–1909
59. Nguyen H, Qian JJ, Bhatnagar RS, Li S (2003) Enhanced cell attachment and osteoblastic activity by P-15 peptide-coated matrix in hydrogels. Biochem Biophys Res Commun 311(1):179–186
60. Hanks T, Atkinson BL (2004) Comparison of cell viability on anorganic bone matrix with or without P-15 cell binding peptide. Biomaterials 25(19):4831–4836
61. Nakaoka R, Hirano Y, Mooney DJ, Tsuchiya T, Matsuoka A (2013) Study on the potential of RGD-and PHSRN-modified alginates as artificial extracellular matrices for engineering bone. J Artif Organs 16(3):284–293
62. Frith JE, Mills RJ, Hudson JE, Cooper-White JJ (2012) Tailored integrin–extracellular matrix interactions to direct human mesenchymal stem cell differentiation. Stem Cells Dev 21(13):2442–2456
63. Min S-K, Kang HK, Jang DH, Jung SY, Kim OB, Min B-M, Yeo I-S (2013) Titanium surface coating with a laminin-derived functional peptide promotes bone cell adhesion. Biomed Res Int 2013:638348

64. Shin H, Zygourakis K, Farach-Carson MC, Yaszemski MJ, Mikos AG (2004) Modulation of differentiation and mineralization of marrow stromal cells cultured on biomimetic hydrogels modified with Arg-Gly-Asp containing peptides. J Biomed Mater Res A 69(3):535–543
65. Palchesko RN, Romeo JD, McGowan KA, Gawalt ES (2012) Increased osteoblast adhesion on physically optimized KRSR modified calcium aluminate. J Biomed Mater Res A 100(5):1229–1238
66. Rezania A, Healy KE (1999) Biomimetic peptide surfaces that regulate adhesion, spreading, cytoskeletal organization, and mineralization of the matrix deposited by osteoblast-like cells. Biotechnol Prog 15(1):19–32
67. Choi YJ, Lee JY, Park JH, Park JB, Suh JS, Choi YS, Lee SJ, Chung CP, Park YJ (2010) The identification of a heparin binding domain peptide from bone morphogenetic protein-4 and its role on osteogenesis. Biomaterials 31(28):7226–7238. https://doi.org/10.1016/j.biomaterials.2010.05.022
68. Sprowson AP, McCaskie AW, Birch MA (2008) ASARM-truncated MEPE and AC-100 enhance osteogenesis by promoting osteoprogenitor adhesion. J Orthop Res 26(9):1256–1262
69. Fromigué O, Brun J, Marty C, Da Nascimento S, Sonnet P, Marie PJ (2012) Peptide-based activation of alpha5 integrin for promoting osteogenesis. J Cell Biochem 113(9):3029–3038
70. Gandavarapu NR, Alge DL, Anseth KS (2014) Osteogenic differentiation of human mesenchymal stem cells on $\alpha5$ integrin binding peptide hydrogels is dependent on substrate elasticity. Biomater Sci 2(3):352–361
71. Hamada Y, Yuki K, Okazaki M, Fujitani W, Matsumoto T, Hashida MK, Harutsugu K, Nokihara K, Daito M, Matsuura N (2004) Osteopontin-derived peptide SVVYGLR induces angiogenesis in vivo. Dent Mater J 23(4):650–655
72. Hamada Y, Egusa H, Kaneda Y, Hirata I, Kawaguchi N, Hirao T, Matsumoto T, Yao M, Daito K, Suzuki M (2007) Synthetic osteopontin-derived peptide SVVYGLR can induce neovascularization in artificial bone marrow scaffold biomaterials. Dent Mater J 26(4):487–492
73. Van Hove AH, Burke K, Antonienko E, Brown E III, Benoit DS (2015) Enzymatically-responsive pro-angiogenic peptide-releasing poly (ethylene glycol) hydrogels promote vascularization in vivo. J Control Release 217:191–201
74. Kang H-M, Kang Y, Chun HJ, Jeong J-W, Park C (2013) Evaluation of the in vitro and in vivo angiogenic effects of exendin-4. Biochem Biophys Res Commun 434(1):150–154
75. Norfleet AM, Bergmann JS, Carney DH (2000) Thrombin peptide, TP508, stimulates angiogenic responses in animal models of dermal wound healing, in chick chorioallantoic membranes, and in cultured human aortic and microvascular endothelial cells. Gen Pharmacol 35(5):249–254
76. Olszewska-Pazdrak B, Carney DH (2013) Systemic administration of thrombin peptide TP508 enhances VEGF-stimulated angiogenesis and attenuates effects of chronic hypoxia. J Vasc Res 50(3):186–196
77. Santulli G, Ciccarelli M, Palumbo G, Campanile A, Galasso G, Ziaco B, Altobelli GG, Cimini V, Piscione F, D'Andrea LD (2009) In vivo properties of the proangiogenic peptide QK. J Transl Med 7(1):41
78. Hardy B, Battler A, Weiss C, Kudasi O, Raiter A (2008) Therapeutic angiogenesis of mouse hind limb ischemia by novel peptide activating GRP78 receptor on endothelial cells. Biochem Pharmacol 75(4):891–899
79. Lin X, Takahashi K, Liu Y, Derrien A, Zamora PO (2007) A synthetic, bioactive PDGF mimetic with binding to both α-PDGF and β-PDGF receptors. Growth Factors 25(2):87–93
80. Suzuki Y, Tanihara M, Suzuki K, Saitou A, Sufan W, Nishimura Y (2000) Alginate hydrogel linked with synthetic oligopeptide derived from BMP-2 allows ectopic osteoinduction in vivo. J Biomed Mater Res 50(3):405–409. https://doi.org/10.1002/(sici)1097-4636(20000605)50:3<405::Aid-jbm15>3.0.Co;2-z
81. He X, Yang X, Jabbari E (2012) Combined effect of osteopontin and BMP-2 derived peptides grafted to an adhesive hydrogel on osteogenic and vasculogenic differentiation of marrow stromal cells. Langmuir 28(12):5387–5397. https://doi.org/10.1021/la205005h

82. Niu X, Feng Q, Wang M, Guo X, Zheng Q (2009) Porous nano-HA/collagen/PLLA scaffold containing chitosan microspheres for controlled delivery of synthetic peptide derived from BMP-2. J Control Release 134(2):111–117
83. Bergeron E, Senta H, Mailloux A, Park H, Lord E, Faucheux N (2009) Murine preosteoblast differentiation induced by a peptide derived from bone morphogenetic proteins-9. Tissue Eng Part A 15(11):3341–3349
84. Zouani OF, Chollet C, Guillotin B, Durrieu M-C (2010) Differentiation of pre-osteoblast cells on poly (ethylene terephthalate) grafted with RGD and/or BMPs mimetic peptides. Biomaterials 31(32):8245–8253
85. Seol YJ, Park YJ, Lee SC, Kim KH, Lee JY, Kim TI, Lee YM, Ku Y, Rhyu IC, Han SB (2006) Enhanced osteogenic promotion around dental implants with synthetic binding motif mimicking bone morphogenetic protein (BMP)-2. J Biomed Mater Res A 77(3):599–607
86. Jung RE, Hämmerle CH, Kokovic V, Weber FE (2007) Bone regeneration using a synthetic matrix containing a parathyroid hormone peptide combined with a grafting material. Int J Oral Maxillofac Implants 22(2):258–266
87. Pigossi SC, de Oliveira GJ, Finoti LS, Nepomuceno R, Spolidorio LC, Rossa C Jr, Ribeiro SJ, Saska S, Scarel-Caminaga RM (2015) Bacterial cellulose-hydroxyapatite composites with osteogenic growth peptide (OGP) or pentapeptide OGP on bone regeneration in critical-size calvarial defect model. J Biomed Mater Res A 103(10):3397–3406
88. Xu G, Jiang D (2014) The role and mechanism of exogenous calcitonin gene-related peptide on mesenchymal stem cell proliferation and osteogenetic formation. Cell Biochem Biophys 69(2):369–378
89. Millet I, Vignery A (1997) The neuropeptide calcitonin gene-related peptide inhibits TNF-α but poorly induces IL-6 production by fetal rat osteoblasts. Cytokine 9(12):999–1007
90. Choi YJ, Lee JY, Chung CP, Park YJ (2013) Enhanced osteogenesis by collagen-binding peptide from bone sialoprotein in vitro and in vivo. J Biomed Mater Res A 101(2):547–554
91. Lee J-Y, Choo J-E, Choi Y-S, Park J-B, Min D-S, Lee S-J, Rhyu HK, Jo I-H, Chung C-P, Park Y-J (2007) Assembly of collagen-binding peptide with collagen as a bioactive scaffold for osteogenesis in vitro and in vivo. Biomaterials 28(29):4257–4267
92. Goto T, Nakao K, Gunjigake K, Kido M, Kobayashi S, Tanaka T (2007) Substance P stimulates late-stage rat osteoblastic bone formation through neurokinin-1 receptors. Neuropeptides 41(1):25–31
93. Wang L, Zhao R, Shi X, Wei T, Halloran BP, Clark DJ, Jacobs CR, Kingery WS (2009) Substance P stimulates bone marrow stromal cell osteogenic activity, osteoclast differentiation, and resorption activity in vitro. Bone 45(2):309–320
94. Von Schroeder H, Veillette C, Payandeh J, Qureshi A, Heersche J (2003) Endothelin-1 promotes osteoprogenitor proliferation and differentiation in fetal rat calvarial cell cultures. Bone 33(4):673–684
95. Tsai T-L, Wang B, Squire MW, Guo L-W, Li W-J (2015) Endothelial cells direct human mesenchymal stem cells for osteo- and chondro-lineage differentiation through endothelin-1 and AKT signaling. Stem Cell Res Ther 6(1):88
96. Clines GA, Mohammad KS, Grunda JM, Clines KL, Niewolna M, McKenna CR, McKibbin CR, Yanagisawa M, Suva LJ, Chirgwin JM (2011) Regulation of postnatal trabecular bone formation by the osteoblast endothelin A receptor. J Bone Miner Res 26(10):2523–2536
97. Wang V, Misra G, Amsden B (2008) Immobilization of a bone and cartilage stimulating peptide to a synthetic bone graft. J Mater Sci Mater Med 19(5):2145–2155
98. Agrawal V, Tottey S, Johnson SA, Freund JM, Siu BF, Badylak SF (2011) Recruitment of progenitor cells by an extracellular matrix cryptic peptide in a mouse model of digit amputation. Tissue Eng Part A 17(19–20):2435–2443
99. Collier JH, Segura T (2011) Evolving the use of peptides as components of biomaterials. Biomaterials 32(18):4198–4204
100. Lee H, Dellatore SM, Miller WM, Messersmith PB (2007) Mussel-inspired surface chemistry for multifunctional coatings. Science 318(5849):426–430. https://doi.org/10.1126/science.1147241

101. Patterson J, Hubbell JA (2011) SPARC-derived protease substrates to enhance the plasmin sensitivity of molecularly engineered PEG hydrogels. Biomaterials 32(5):1301–1310
102. Tokatlian T, Shrum CT, Kadoya WM, Segura T (2010) Protease degradable tethers for controlled and cell-mediated release of nanoparticles in 2- and 3-dimensions. Biomaterials 31(31):8072–8080
103. Maynard HD, Hubbell JA (2005) Discovery of a sulfated tetrapeptide that binds to vascular endothelial growth factor. Acta Biomater 1(4):451–459
104. Kim SH, Kiick KL (2007) Heparin-mimetic sulfated peptides with modulated affinities for heparin-binding peptides and growth factors. Peptides 28(11):2125–2136
105. Cui H, Webber MJ, Stupp SI (2010) Self-assembly of peptide amphiphiles: from molecules to nanostructures to biomaterials. Pept Sci 94(1):1–18
106. Zhou M, Smith AM, Das AK, Hodson NW, Collins RF et al (2009) Self-assembled peptide-based hydrogels as scaffolds for anchorage-dependent cells. Biomaterials 30(13):2523–2530
107. Woolfson DN, Mahmoud ZN (2010) More than just bare scaffolds: towards multi-component and decorated fibrous biomaterials. Chem Soc Rev 39(9):3464–3479
108. Rajagopal K, Lamm MS, Haines-Butterick LA, Pochan DJ, Schneider JP (2009) Tuning the pH responsiveness of β-hairpin peptide folding, self-assembly, and hydrogel material formation. Biomacromolecules 10(9):2619–2625
109. Jung JP, Gasiorowski JZ, Collier JH (2010) Fibrillar peptide gels in biotechnology and biomedicine. Pept Sci 94(1):49–59
110. Khatayevich D, Gungormus M, Yazici H, So C, Cetinel S, Ma H, Jen A, Tamerler C, Sarikaya M (2010) Biofunctionalization of materials for implants using engineered peptides. Acta Biomater 6(12):4634–4641
111. Mas-Moruno C, Rechenmacher F, Kessler H (2010) Cilengitide: the first anti-angiogenic small molecule drug candidate design, synthesis and clinical evaluation. Anticancer Agents Med Chem 10(10):753–768
112. Rabe M, Verdes D, Seeger S (2011) Understanding protein adsorption phenomena at solid surfaces. Adv Colloid Interf Sci 162(1–2):87–106
113. Rusmini F, Zhong Z, Feijen J (2007) Protein immobilization strategies for protein biochips. Biomacromolecules 8(6):1775–1789
114. Gray JJ (2004) The interaction of proteins with solid surfaces. Curr Opin Struct Biol 14(1):110–115
115. Nakanishi K, Sakiyama T, Imamura K (2001) On the adsorption of proteins on solid surfaces, a common but very complicated phenomenon. J Biosci Bioeng 91(3):233–244
116. Chen C, Zhang S-M, Lee I-S (2013) Immobilizing bioactive molecules onto titanium implants to improve osseointegration. Surf Coat Technol 228:S312–S317
117. Hajimiri M, Shahverdi S, Kamalinia G, Dinarvand R (2015) Growth factor conjugation: strategies and applications. J Biomed Mater Res A 103(2):819–838
118. Chiu LL, Weisel RD, Li RK, Radisic M (2011) Defining conditions for covalent immobilization of angiogenic growth factors onto scaffolds for tissue engineering. J Tissue Eng Regen Med 5(1):69–84
119. Liu L, Deng D, Xing Y, Li S, Yuan B, Chen J, Xia N (2013) Activity analysis of the carbodiimide-mediated amine coupling reaction on self-assembled monolayers by cyclic voltammetry. Electrochim Acta 89:616–622
120. Aissaoui N, Bergaoui L, Landoulsi J, Lambert J-F, Boujday S (2011) Silane layers on silicon surfaces: mechanism of interaction, stability, and influence on protein adsorption. Langmuir 28(1):656–665
121. Kolb HC, Finn MG, Sharpless KB (2001) Click chemistry: diverse chemical function from a few good reactions. Angew Chem Int Ed Engl 40(11):2004–2021
122. Tang L, Thevenot P, Hu W (2008) Surface chemistry influences implant biocompatibility. Curr Top Med Chem 8(4):270–280
123. Yazici H, Fong H, Wilson B, Oren E, Amos F, Zhang H, Evans J, Snead M, Sarikaya M, Tamerler C (2013) Biological response on a titanium implant-grade surface functionalized with modular peptides. Acta Biomater 9(2):5341–5352

124. Sarikaya M, Tamerler C, Jen AK-Y, Schulten K, Baneyx F (2003) Molecular biomimetics: nanotechnology through biology. Nat Mater 2(9):577
125. Sarikaya M, Tamerler C, Schwartz DT, Baneyx F (2004) Materials assembly and formation using engineered polypeptides. Annu Rev Mater Res 34:373–408
126. Mayer G, Sarikaya M (2002) Rigid biological composite materials: structural examples for biomimetic design. Exp Mech 42(4):395–403
127. Sakiyama-Elbert S, Hubbell J (2001) Functional biomaterials: design of novel biomaterials. Annu Rev Mater Res 31(1):183–201
128. Yazici H, Habib G, Boone K, Urgen M, Utku FS, Tamerler C (2019) Self-assembling antimicrobial peptides on nanotubular titanium surfaces coated with calcium phosphate for local therapy. Mater Sci Eng C 94:333–343
129. Yazici H, O'Neill MB, Kacar T, Wilson BR, Oren EE, Sarikaya M, Tamerler C (2016) Engineered chimeric peptides as antimicrobial surface coating agents toward infection-free implants. ACS Appl Mater Interfaces 8(8):5070–5081
130. So CR, Hayamizu Y, Yazici H, Gresswell C, Khatayevich D, Tamerler C, Sarikaya M (2012) Controlling self-assembly of engineered peptides on graphite by rational mutation. ACS Nano 6(2):1648–1656
131. Shiba K (2010) Exploitation of peptide motif sequences and their use in nanobiotechnology. Curr Opin Biotechnol 21(4):412–425
132. Donatan S, Yazici H, Bermek H, Sarikaya M, Tamerler C, Urgen M (2009) Physical elution in phage display selection of inorganic-binding peptides. Mater Sci Eng C 29(1):14–19
133. Hnilova M, Oren EE, Seker UO, Wilson BR, Collino S, Evans JS, Tamerler C, Sarikaya M (2008) Effect of molecular conformations on the adsorption behavior of gold-binding peptides. Langmuir 24(21):12440–12445
134. Sano K-I, Shiba K (2003) A hexapeptide motif that electrostatically binds to the surface of titanium. J Am Chem Soc 125(47):14234–14235
135. Naik RR, Brott LL, Clarson SJ, Stone MO (2002) Silica-precipitating peptides isolated from a combinatorial phage display peptide library. J Nanosci Nanotechnol 2(1):95–100
136. Ajikumar PK, Vivekanandan S, Lakshminarayanan R, Jois SD, Kini RM, Valiyaveettil S (2005) Mimicking the function of eggshell matrix proteins: the role of multiplets of charged amino acid residues and self-assembly of peptides in biomineralization. Angew Chem Int Ed 44(34):5476–5479
137. Gungormus M, Oren EE, Horst JA, Fong H, Hnilova M, Somerman MJ, Snead ML, Samudrala R, Tamerler C, Sarikaya M (2012) Cementomimetics—constructing a cementum-like biomineralized microlayer via amelogenin-derived peptides. Int J Oral Sci 4(2):69
138. Gungormus M, Fong H, Kim IW, Evans JS, Tamerler C, Sarikaya M (2008) Regulation of in vitro calcium phosphate mineralization by combinatorially selected hydroxyapatite-binding peptides. Biomacromolecules 9(3):966–973
139. Loo Y, Goktas M, Tekinay AB, Guler MO, Hauser CA, Mitraki A (2015) Self-assembled proteins and peptides as scaffolds for tissue regeneration. Adv Healthc Mater 4(16):2557–2586
140. Matson JB, Zha RH, Stupp SI (2011) Peptide self-assembly for crafting functional biological materials. Curr Opinion Solid State Mater Sci 15(6):225–235
141. He B, Zhao J, Ou Y, Jiang D (2018) Biofunctionalized peptide nanofiber-based composite scaffolds for bone regeneration. Mater Sci Eng C 90:728–738
142. Holmes TC (2002) Novel peptide-based biomaterial scaffolds for tissue engineering. Trends Biotechnol 20(1):16–21
143. Gelain F, Horii A, Zhang S (2007) Designer self-assembling peptide scaffolds for 3-D tissue cell cultures and regenerative medicine. Macromol Biosci 7(5):544–551
144. Stupp SI (2010) Self-assembly and biomaterials. Nano Lett 10(12):4783–4786
145. Wu EC, Zhang S, Hauser CAE (2012) Self-assembling peptides as cell-interactive scaffolds. Adv Funct Mater 22(3):456–468. https://doi.org/10.1002/adfm.201101905
146. Schachner M (2000) Neurobiology. Nervous engineering. Nature 405(6788):747–748. https://doi.org/10.1038/35015648

147. Aggeli A, Bell M, Boden N, Keen JN, Knowles PF, McLeish TC, Pitkeathly M, Radford SE (1997) Responsive gels formed by the spontaneous self-assembly of peptides into polymeric beta-sheet tapes. Nature 386(6622):259–262. https://doi.org/10.1038/386259a0
148. West MW, Wang W, Patterson J, Mancias JD, Beasley JR, Hecht MH (1999) De novo amyloid proteins from designed combinatorial libraries. Proc Natl Acad Sci U S A 96(20):11211–11216. https://doi.org/10.1073/pnas.96.20.11211
149. Xiong H, Buckwalter BL, Shieh HM, Hecht MH (1995) Periodicity of polar and nonpolar amino acids is the major determinant of secondary structure in self-assembling oligomeric peptides. Proc Natl Acad Sci U S A 92(14):6349–6353. https://doi.org/10.1073/pnas.92.14.6349
150. Lashuel HA, Labrenz SR, Woo L, Serpell LC, Kelly JW (2000) Protofilaments, filaments, ribbons, and fibrils from peptidomimetic self-assembly: implications for amyloid fibril formation and materials science. J Am Chem Soc 122(22):5262–5277. https://doi.org/10.1021/ja9937831
151. Prieto AL, Edelman GM, Crossin KL (1993) Multiple integrins mediate cell attachment to cytotactin/tenascin. Proc Natl Acad Sci U S A 90(21):10154–10158. https://doi.org/10.1073/pnas.90.21.10154
152. Zhang S, Holmes TC, DiPersio CM, Hynes RO, Su X, Rich A (1995) Self-complementary oligopeptide matrices support mammalian cell attachment. Biomaterials 16(18):1385–1393

Antibody Mediated Osseous Regeneration: A New Strategy for Bioengineering

Fernanda Coelho, Ticiana Sidorenko de Oliveira Capote, Marcell Costa de Medeiros, and Suzane Cristina Pigossi

Abstract This chapter provides a brief review of bone biology and metabolism, focusing on the regenerative potential of bone tissues. In this context, we discussed the main clinical approaches to enhance bone regeneration, concentrating on an innovative approach referred to as antibody-mediated osseous regeneration (AMOR). Bone morphogenetic proteins (BMPs) are some of the most relevant osteoinductive factors in the demineralized bone matrix. The main role of BMPs is the recruitment and differentiation of mesenchymal cells into an osteogenic lineage, resulting in new bone formation. As an alternative for the BMP-2 exogenous administration of an osteoinductive growth factor, the use of immobilized anti-BMP-2 antibodies in matrices has been proposed to capture the endogenous protein. The captured endogenous BMP-2 would be able to induce osteogenic differentiation of osteoprogenitor stem cells and improve the bone formation. In general, the association of anti-BMP-2 mAb with a scaffold has demonstrated success in new bone formation in different in vivo models with no evidence of adverse reactions.

Keywords Bone regeneration · Bone morphogenetic proteins · Antibodies · Stem cells · Osteoinductive · Osteoconductive · Osteogenic · Scaffold

Biology and Metabolism of Bone Tissue

Bone tissue has a high vascularization and its remodeling continues throughout one`s lifetime [1]. The diversity of the functions of bone is shown by its complex architecture [2]. The arrangement of the bone tissue occurs in a trabecular (cancellous bone) or compact pattern (cortical bone) [2, 3]. Interconnected trabeculae with free spaces filled by bone marrow compose the trabecular bone, and the cortical

F. Coelho (✉) · T. S. de Oliveira Capote · M. C. de Medeiros
Department of Morphology, School of Dentistry, São Paulo State University (UNESP), Araraquara, Sao Paulo, Brazil

S. C. Pigossi
Department of Clinics and Surgery, School of Dentistry, Alfenas Federal University (Unifal-MG), Alfenas, Minas Gerais, Brazil

B. Li et al. (eds.), *Racing for the Surface*,
https://doi.org/10.1007/978-3-030-34471-9_18

bone is constituted by repeating osteon units, which are composed of collagen fibers and calcium phosphate crystals [4].

Different proportions of those two architectural patterns are observed in the skeleton. Only 10% of the cortical bone is porous, being almost solid, whereas cancellous bone is more porous (50–90%) [3, 5]. According to Jimi et al. [6], the remodeling of the cancellous bone occurs at more than 30% per year and the cortical bone at approximately 3% per year. In 1 year, about 6% of all bones in the human body remodel.

Of the two components of the bone matrix: 65–70% are composed of a mineral part (hydroxyapatite), and 25–30% are made of an organic phase [3], which is composed of type I collagen (90–95%) [7].

Besides, the organic part of the bone matrix is also composed of several proteins (thrombospondin, osteonectin, osteocalcin, byglican, decorin, fibronectin, bone sialoprotein, osteopontin) with different functions [3]. The organic, non-mineralized part, presents an important role in the control of the growth and differentiation of osteocytes, osteoblasts, and osteoclasts in the bone remodeling process [2]. According to Stevens [1], over 200 different types of noncollagenous matrix proteins (sialoproteins, proteoglycans, glycoproteins, etc.) participate in a large number of signals in the immediate extracellular environment. The nanocomposite structure provides the requisite compressive strength and high fracture toughness of bone [1].

Bone tissue is composed of three cell types: osteoblasts, osteoclasts, and osteocytes. Osteoblasts are the most important cells for bone growth and metabolism; they are matrix producers [8]. They produce the extracellular matrix and regulate its mineralization [9]. Mature osteoblasts secrete and deposit most of the bone matrix proteins, besides regulating the formation of hydroxyapatite crystals in osteoids [6]. According to Jimi et al. [6], it is believed that the main functions of osteoblasts are related to the high activity of alkaline phosphatase in response to osteotropic hormones and cytokines, and the expression of a high number of extracellular matrix proteins related to bone tissue.

Osteocytes account for 90% of all cells in the adult skeleton [9]. They are derived from the osteoblasts that are included in the matrix [10]. They are mature cells found in bone lacunae, which communicate with other osteocytes through long cellular processes, and also participate in bone remodeling by sending signals resulting from mechanical stress [11]. Osteocytes may have an important role in initiating the bone cycle, perhaps by detecting microfractures or other perturbations in the bone structure and then signaling osteoclasts to those defects [7].

Osteoclasts, the tissue-resorbing and macrophage-like cells, degrade the bone structure through a combination of localized acidification, removing the mineral, and breaking down the matrix by protease secretion [11]. Osteoclasts are active at the beginning of the bone remodeling cycle and have the function of resorbing the existing bone. They attach their fenestrated membrane to the mineralized matrix on the surface of the bone, isolating a microenvironment, which will be a local site of bone resorption. There is a decrease in pH, and potent enzymes (β-glycerophosphatases, acid phosphatase, β-glucuronidases, aryl-sulfatases, cysteine-proteinases, metalloproteinases, etc.) are released. There is the formation of a depression in the bone due to the erosion, termed as lacunae [7].

Osteogenesis occurs not only during embryogenesis. It occurs throughout life and is related to the bone remodeling process in adults. It is a tightly regulated, active process initiated by stem cells for the purpose of forming a normal vascularized bone structure. The formation of bone tissue depends on the association of factors such as the expression of soluble molecules (hormones, growth factors, vitamins, cytokines, ions, etc.), specific cell types (osteoclasts and mesenchymal stem cells), scaffolds (extracellular matrix molecules, hydroxyapatite), and mechanical stimuli [12]. The bone tissue may be formed by intramembranous and/or endochondral ossification. Mesenchymal condensation nuclei are formed, where cells can differentiate directly into osteoblasts (intramembranous ossification), or into chondrocytes (endochondral ossification) [12].

Bone remodeling occurs by the balanced activity between osteoclasts and osteoblasts [10]. The beginning of the bone remodeling occurs in the quiescent phase. After the osteoclasts are attracted to the new site, they promote erosion of the bone matrix, forming lacunae with sizes of approximately a 50 μm depth and 100 μm in diameter. This process requires about 10 days. Resorption is discontinued and osteoblasts are attracted to bone remodeling site. Osteoblasts secrete an osteoid matrix composed primarily of type 1 collagen, filling the lacunae. This process requires about 80 days. The newly formed matrix is mineralized with hydroxyapatite. The remodeled area goes into the quiescent phase to complete the bone cycle of 60–120 days [7]. The process of bone remodeling is exemplified in Fig. 1.

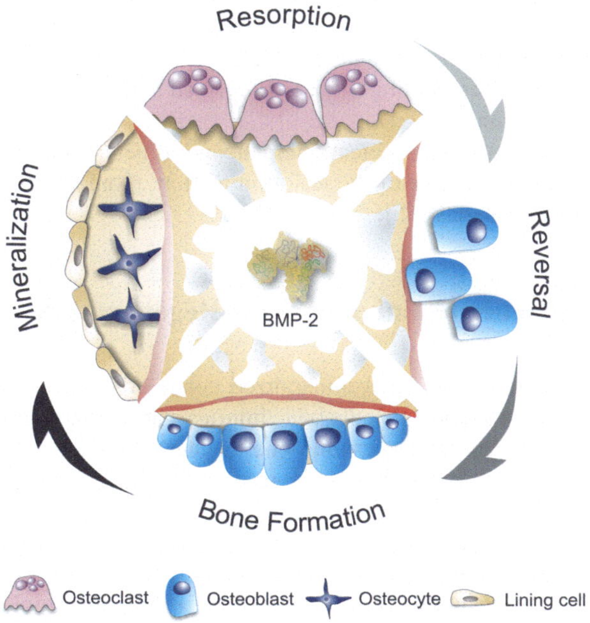

Fig. 1 The process of bone remodeling. (Source: Prepared by the authors)

Bone Regeneration

Unlike other tissues, bone can regenerate and repair itself in a process called bone regeneration. This process is a very efficient and is a rigorously regulated process where all components of bone tissue are involved to optimize the repair and restore skeletal function. It is characterized by a sequence of biological events of bone induction and bone conduction, which involves several types of cells, extracellular and intracellular molecular signaling pathways [11, 13].

Four components are related to the bone regeneration process: (a) morphogenetic signals, (b) response to the signal by the host cells, (c) a suitable carrier for the growth of the host cells, and (d) a well-vascularized and viable tissue of the host [4, 14]. Inflammatory cells, vascular cells, osteoclasts, and osteochondral progenitor cells are cells that are present in the repair process [15]. Increased expression of pro-inflammatory, osteogenic, and angiogenic growth factors released at the site of the bone lesion induces signaling cascades for tissue repair [16].

Several growth factors are present in the bone regeneration process, including: bone morphogenetic proteins (BMPs); fibroblast growth factor (FGF), insulin-like growth factors (IGFs), platelet-derived growth factor (PDGF); transforming growth factor-β (TGF-β), and vascular endothelial growth factor (VEGF) [17].

According to Schindeler et al. [15], bone healing occurs in a four-stage process: inflammation (formation of hematoma, infiltration of the hematoma by inflammatory cells, secretion of cytokines and growth factors, invasion of mesenchymal stem cells, formation of granulation tissue); formation of soft callus (formation of a cartilaginous callus and fibrocartilaginous tissue); formation of hard callus (removal of soft callus, revascularization, osteoblasts in high activity, formation of a mineralized bone matrix); and bone remodeling (the newly formed bone is remodeled into cortical and/or trabecular bone).

Physiological and pharmacological components influence the fracture healing, such as the location and extent of the lesion, infection, biomechanical forces, diseases, nutrition, and genetics [15].

Clinical Approaches to Enhance Bone Regeneration

As previously discussed, bone is a mineralized conjunctive tissue, with a special healing capacity which ensures that the bone injuries and fractures heal without scar formation [11]. However, the successful bone regeneration becomes more complicated depending on the size of the defect since large defects present greater difficulty of repair [18]. In this cases bone substitutes (autogenous, homogenous, heterogeneous, or synthetic) have been used to replace the missing bones.

The ideal bone substitute must present osteoconductive, osteogenic, and osteoinductive properties: osteoconductive bone substitute is able to stimulate osteogenic cells attachment, survival, and migration; osteoinductive bone substitutes induces stem cells differentiation toward an osteoblastic lineage by physical and biochemical factors, and osteogenic bone substitutes contains stem cells able of differentiating into osteogenic cells in the bone defect [4]. Based on that, the autogenous bone graft is therefore the gold standard since it has the osteoinductive factors and osteogenic cells required for bone regeneration [19]. Autogenous bone graft is removed from another part of the patient's body (tibia, iliac crest, mandible, or skullcap), which restricts its applications essentially due to the limited quantity of the autograft that can be achieved [20]. Moreover, the operating time required for harvesting autografts is expensive with substantial donor site injury often related with pain, infection, and hematoma [21].

Therefore, viable alternatives including allograft (from human donors/cadavers) and xenograft (from a nonhuman) bone are also regularly used for bone defect regeneration when autologous bone graft is not available [22]. However, although the biomechanical stability and elasticity are similar to autologous bone, the absence of osteogenicity associated with the lower rate of graft incorporation represents limitations to those bone grafts [2]. Furthermore, host rejection, disease transmission, and infection risk (even infrequent) have limited their uses [16].

Against all these limitations, synthetic bone substitutes have been developed as a safer, less expensive, and less invasive alternative compared to these bone implants. Those bone substitutes can be produced from biomaterials as hydroxyapatite (HA), bioactive glass, tricalcium phosphate (TCP), polymers, and ceramics [23]. The isolated use of those materials presents only an osteoconductive role, limiting their use in bone reconstruction [18]. Consequently, researchers have been proposed the association of diverse compounds as growth factors, hormones, and drugs to the synthetic bone substitutes to ensure osteoinductive properties and to improve the regenerative potential of those materials [18].

Synthetic Tissue Scaffolds for Bone Regeneration

Three-dimensional scaffolds are normally prepared with porous biodegradable materials which have the mechanical support and preserve the space necessary for cell growth and matrix production during new bone formation [4]. Moreover, the scaffold structure is able to transport diverse compounds in the target space in a high local concentration with the smallest side effects [18]. The ideal synthetic tissue scaffolds should be: (a) biocompatible without eliciting an immune response; (b) osteoconductive, osteoinductive, and osteogenic, promoting bone ingrowth; (c) absorbable in a predictable manner, with biocompatible components, and at the same time of bone growth; (d) easily adaptable to an irregular wound site; and (e) sterilizable without property modification. Additionally, the scaffolds need to present pore sizes with approximately 200–400 μm, correct mechanical and physical properties and not stimulate soft tissue growth at bone/implant interface [14].

The choice of the material to produce a scaffold is an essential stage since its properties will determine the mechanical and physical properties [3]. Several materials including biodegradable ceramics and polymers have been proposed. Ceramics can be from natural (e.g., coralline HA) or synthetic origin (β-TCP or synthetic HA) [24]. Although being osteoconductive and osteoinductive, those materials have some major drawbacks including small mechanical stability, which limits their applications for large bone defects regeneration. Likewise, their degradation/dissolution rates are difficult to estimate [3]. As an alternative, the natural (animal or vegetal source) and synthetic biodegradable polymers have been used. Those materials showed little immunogenic potential, bioactive behavior, chemical versatility, and capability of host's tissue interaction [3].

The synthetic tissue scaffolds also act as carriers to deliver different compounds to the bone defect area improving local protein retention and sustain a slow release, increasing the osteoinductive potential of the material [16]. According to Vo et al. [16] "the strategies for protein absorption into scaffolds involve either non-covalent (surface adsorption, physical entrapment, affinity binding, ionic complexation) or covalent immobilization on or into the delivery system (chemical conjugation)". The choice of the type of absorption is based on the material's physicochemical properties and interactions between the protein, defect type, and carrier [25]. Beyond the several hormones, growth factors and drugs, the antibody incorporation was proposed in an approach referred to as "antibody-mediated osseous regeneration (AMOR)" [26].

AMOR: Antibody Mediated Osseous Regeneration

Several therapies involving antibodies have been studied in genomic research based on its high degree of affinity and specificity to the antigen or target molecule that guarantee a high level of efficiency with fewer adverse events. The mechanism of action of the antibodies and the ability of targeting several molecules allows them to be applied to a wide range of therapeutic targets [27]. Currently, the number of monoclonal antibodies (mAbs) approved in clinical research is surprisingly growing in different therapeutic areas, including cancer treatment, organ transplantation, inflammatory disease, respiratory disease, cardiovascular disease, ophthalmologic disease, and infection [27, 28]. Based on these antibody therapies advantages, Freire et al. [26] proposed a strategy using BMP-2 specific immobilized antibodies (BMP-2 Abs) to promote bone regeneration. In this strategy, the BMP-2 specific immobilized antibodies can sequester endogenous BMP-2 and induce new bone formation. This new approach is referred to as AMOR.

Osteoblasts are responsible for the synthesis and secretion of BMPs. BMPs induce the differentiation of mesenchymal stem cells (MSCs), which stimulate the process of osteogenesis, allowing for healing and bone remodeling [29, 30]. Traditionally, pathways leading to differentiation of MSCs into osteoblasts are activated by BMPs binding to a membrane-specific ligand receptor named of BMP type 1 (BMPR1) and BMP type 2 receptors (BMPR2). Receptor binding stimulates signal transduction through the phosphorylation of various homologues of the Drosophila protein, mothers against decapentaplegic (Mad) and the Caenorhabditis elegans proteins (Sma) (SMAD) proteins and their nuclear translocation. The SMADs also function as transcription factors, controlling the expression of essential osteogenic genes involved in osteoblast proliferation (*Msx2*), matrix synthesis (*RUNX2*, osteopontin-*SPP1*, alkaline phosphatase-*ALPL*) and inhibition of osteoclast differentiation (*TNFRSF11B*- osteoprotegerin) [31]. According to [32]: "In addition to BMP/SMAD signaling, Mitogen Activated Protein Kinases (MAPK) cascades represent an alternative, non-canonical pathway for BMP-2 signal transduction. BMP-2 activates the signaling pathways p38, extracellular-signal-regulated kinase (ERK1/2) and c-Jun N-terminal kinase (JNK1/2) to induce the activation and expression of a specific transcription factor related to *RUNX2*. *RUNX2* plays an essential role in the osteoblastic differentiation of MSCs and directly stimulates the transcription of important downstream target genes, including those encoding osteocalcin (*Bglap*), collagen type 1 (*Col1A1*) and osteopontin (*Spp1*)". The mechanism of action of BMP-2 on the expression of osteogenic markers can be verified through the signaling cascade in Fig. 2.

Based on this high osteogenic activity, the use of recombinant human bone morphogenetic protein-2 (rhBMP-2) as a viable alternative to bone grafts has been approved by the US Food and Drug Administration (FDA) for clinical use and has been investigated in different bone cicatrization applications. Preclinical and clinical researches have demonstrated that an absorbable collagen sponge combined with rhBMP-2 can induce new bone formation with clinical and radiographic results equivalent to autogenous grafting [33–35]. However, according to [36]: "The clinical application of rhBMP-2 is associated with a number of biological and logistic drawbacks, including (1) requirement for administration of rhBMP-2 at superphysiological doses, (2) inability to sustain growth factor concentration over extended periods of time, (3) lower biological activity of rhBMP-2 relative to endogenous counterparts and (4) high costs of rhBMP-2". Therefore, the use of BMP-2 specific immobilized antibodies has been investigated to replace the exogenous pathway and avoid the adverse effects associated with rhBMP-2 use [26, 37].

This approach aims to capture endogenous BMP-2 using specific Abs immobilized in a solid scaffold. The binding scheme of the anti-BMP-2 antibodies in scaffolds for the capture of endogenous BMP-2 can be visualized in Fig. 3.

According to Freire et al. [26]: "To participate in AMOR, an Abs molecule must have the following properties: (1) high affinity for binding to endogenous BMP-2; (2) binding of BMP-2 epitopes remotely from the BMP-2 receptor-binding domains; (3) involve the BMP-2 cellular receptor on osteoprogenitor cells by the Ab-BMP-2 immune complex; (4) intracellular signal transduction by the Ab-BMP-2 immune complex; (5) absence of an adverse local or systemic immunological response in the host; and (6) mediation of osteogenic differentiation by Ab-BMP-2 immune complexes".

Freire et al. [26] performed in vitro and in vivo studies to evaluate the ability of the immobilized anti-BMP-2 Abs to capture the endogenous BMP-2 and mediate the formation of a new bone tissue. In these studies, antibodies were obtained from the immobilization of rhBMP-2 (Infuse®; Medtronic) in mice. Many clones were formed using the ClonaCell-HY hybridoma cloning kit (StemCell Technologies).

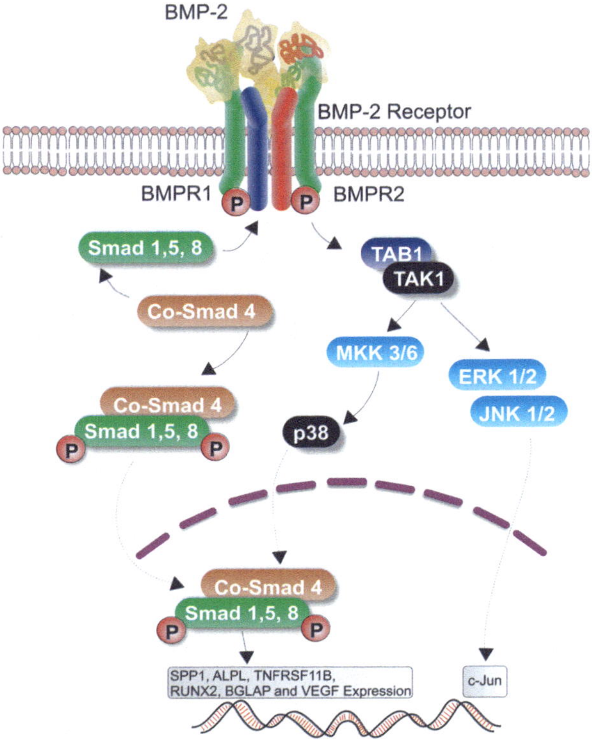

Fig. 2 BMP-2 signaling markers in the expression of osteogenic markers. (Source: Prepared by the authors)

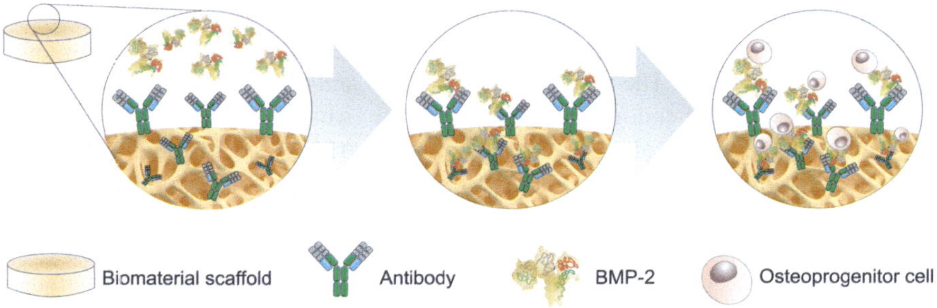

Fig. 3 Anti-BMP-2 immobilization scheme on membranes. Demonstration of capture of endogenous BMP-2 and osteoprogenitor stem cells. (1) Anti-BMP-2 mAb is withheld on membranes; (2) mAb binds endogenous BMP-2; and (3) specific receptor osteoprogenitor cells are attracted by endogenous BMP-2 and promote bone differentiation. (Source: Prepared by the authors)

Thereafter, the competence of the anti-BMP-2 Abs to link to BMP-2 and to permit BMP-2 in the immune complex to link to the BMP cell receptor was assessed by a flow cytometry assay. Through this assay it was observed that some mAbs could link to BMP-2 and permit BMP-2 to bind to cells, but most prevent the binding of BMP-2 to its cellular receptors. In the in vivo analysis, the BMP-2 antibody was immobilized on an absorbable collagen sponge and surgically placed in a rat calvarial defect. Micro-CT analysis demonstrated that bone regeneration was promoted by only a few anti-BMP-2 Abs clones immobilized on absorbable collagen sponge (ACS). The in situ expression of BMP-2 and osteocalcin was evaluated by immunohistochemistry. The analysis revealed higher expression of these markers in sites with greater bone regeneration. These studies demonstrated the competence of anti-BMP-2 Abs to link to endogenous BMP-2 and mediate the formation of a new bone in vivo, presenting a strategy for improved tissue engineering [26].

Investigation of the AMOR Approach in Animal Models

The AMOR approach may be used in clinics, and it is interesting to demonstrate its effectiveness in animal models. Besides the study [26], a variety of in vivo experimental approaches have been reported in the literature.

Ansari et al. [37] compared the effectiveness of a murine anti-BMP-2 mAb immobilization technique using diverse matrices including titanium microspheres, alginate hydrogel, and ACS. Those matrices were surgically grafted on rat critical-sized calvarial defects. After 8 weeks, the bone regeneration process was most effectively evidenced in the three types of scaffolds with immobilized anti-BMP-2 mAb compared to the isotype control mAb. The titanium scaffold presented greater bone formation followed by ACS. However, the titanium scaffold is not biodegradable, which limits its applications. In all the scaffolds used, the presence of the BMP-2, -4, and -7 antigens was identified through the immobilized anti-BMP-2 mAb, reinforcing the efficiency of the AMOR strategy for bone regeneration.

The wide range of applications of induced pluripotent stem cells (iPSCs) stimulated the curiosity of Wu et al. [38]. The iPSCs are somatic cells collected from patients, which could be transformed into pluripotent stem cells, using a suitable sequence of signaling proteins and growth factors, exhibiting the same pluripotency of embryonic stem cells (ESCs). In this context, Wu et al. [38] described an approach using BMP-2 Abs to guide osteogenic differentiation of iPSCs (derived from mesenchymal stromal cells or iMSCs) in an in vivo ectopic bone formation model. For this, subcutaneous injection of alginate microbeads with iMSCs and encapsulated anti-BMP-2 Abs were perfomed in 12 eight-week-old male mice. The presence of the anti-BMP-2 antibody was able to involve the BMP-2 recalls in the iMSCs. A subcutaneous implantation locus loaded with the iMSCs and anti-BMP-2 showed increased bone formation and vascularization in mice compared to exogenous BMP-2. The exogenous BMP-2 exhibited significantly lower dystrophic calcification and vascularization, which revealed that the anti-BMP-2 Ab/BMP2 immune complex was able to dictate the acquirement of the osteogenic phenotype of iMSCs and subsequent mineralization.

Guo et al. [39] investigated the AMOR approach for a nonunion tibia defect repair in a nonhuman primate model. Six animals of the species *Macaca fascicularis* were operated on for a 20 mm segmental osteotomy in their tibias. The isotype matched control mAb and the investigated material (absorbable collagen sponge incorporated with anti-BMP-2 mAb) were introduced into the defects created. After a surgical period of 12 weeks, histological, histomorphometric, and cone beam computed tomography (CBCT) analyses were performed. Quantitative 3D volumetric analysis by CBCT demonstrated the formation of a larger volume of mineralized tissue at sites that were implanted with an absorbable collagen sponge incorporated with anti-BMP-2 mAb compared to sites implanted with an isotype-matched control mAb; and the histological and histomorphometric analysis indicated that sites introduced with anti-BMP-2 showed new bone formation with a higher percentage of bone volume compared to the isotype matched control mAb.

Xie et al. [40] investigated the application of the AMOR approach in a mandibular continuity defect repair in nonhuman primates. Critical-sized mandibular continuity defects were formed in six adult male *Macaca fascicularis*. Collagen sponges (CS) incorporated with anti-BMP-2 mAbs were locally implanted. Three animals were designated to experimental (AMOR) and three to control (isotype-matched mAb) groups. 2D and 3D analysis of CBCT and histological examination demonstrated an increased bone density and volume in regions treated with anti-BMP-2 ACS-mAb compared to control in 6 and 12 weeks, postoperatively.

Khojasteh et al. [41] evaluated the AMOR strategy on the repair of a canine segmental mandibular continuity defect model. Consequently, a 15 mm unilateral segmental defect was created in the mandible and fixed with a titanium plate. Inorganic bovine bone mineral with 10% collagen (ABBM-C) was incorporated with anti-BMP-2 mAb or isotype-matched mAb. The rhBMP-2 served as a positive control. Morphometric analyses were observed by CBCT and histological images. Bone densities within healed defect sites at 12 weeks after surgery were 1360.81 ± 10.52 Hounsfield Unit (HU), 1044.27 ± 141.16 HU, and 839.45 ± 179.41 HU, in sites with implanted anti-BMP-2 mAb, rhBMP-2, and isotype mAb groups, respectively. Osteoid bone formation in anti-BMP-2 mAb (42.99% ± 8.67) and rhBMP-2 (48.97% ± 2.96) groups was not significantly different but was higher than in sites with an isotype control mAb (26.8% ± 5.35). In this way, the results of this study confirmed the feasibility of AMOR in a large clinically relevant animal model.

Conclusion

The efficiency of the AMOR approach in bone tissue engineering has been demonstrated in several in vivo studies. In general, the association of anti-BMP-2 mAb with a scaffold has demonstrated success in the new bone formation in different in vivo models with no evidence of adverse reaction. The effectiveness of this approach suggests that this strategy could be introduced into clinical use of tissue engineering with promising results.

References

1. Stevens MM (2008) Biomaterials for bone tissue engineering. Mater Today 11:18–25
2. Kneser U, Schaefer DJ, Polykandriotis E, Horch RE (2006) Tissue engineering of bone: the reconstructive surgeon's point of view. J Cell Mol Med 10:7–19
3. Salgado AJ, Coutinho OP, Reis RL (2004) Bone tissue engineering: state of the art and future trends. Macromol Biosci 4:743–765
4. Gong T, Xie J, Liao J, Zhang T, Lin S, Lin Y (2015) Nanomaterials and bone regeneration. Bone Res 3:15029
5. Sikavitsas VI, Temenoff JS, Mikos AG (2001) Biomaterials and bone mechanotransduction. Biomaterials 22:2581–2593
6. Jimi E, Hirata S, Osawa K, Terashita M, Kitamura C, Fukushima H (2012) The current and future therapies of bone regeneration to repair bone defects. Int J Dent 2012:148261
7. Christenson RH (1997) Biochemical markers of bone metabolism: an overview. Clin Biochem 30:573–593
8. Valerio P, Pereira MM, Goes AM, Leite MF (2004) The effect of ionic products from bioactive glass dissolution on osteoblast proliferation and collagen production. Biomaterials 25:2941–2948
9. Sommerfeldt DW, Rubin CT (2001) Biology of bone and how it orchestrates the form and function of the skeleton. Eur Spine J 10(Suppl 2):S86–S95
10. Williams PL, Warwick R, Dyson M, Bannister LH (1989) Gray's anatomy, 37th edn. Churchill Livingstone, Edinburgh
11. Wagh MR, Ravalia D (2015) Bone regeneration and repair: current and future aspects. Int J Sci Res 4:351–355
12. Deschaseaux F, Sensebe L, Heymann D (2009) Mechanisms of bone repair and regeneration. Trends Mol Med 15:417–429
13. Dimitriou R, Jones E, McGonagle D, Giannoudis PV (2011) Bone regeneration: current concepts and future directions. BMC Med 9:66
14. Burg KJ, Porter S, Kellam JF (2000) Biomaterial developments for bone tissue engineering. Biomaterials 21:2347–2359
15. Schindeler A, McDonald MM, Bokko P, Little DG (2008) Bone remodeling during fracture repair: the cellular picture. Semin Cell Dev Biol 19:459–466
16. Vo TN, Kasper FK, Mikos AG (2012) Strategies for controlled delivery of growth factors and cells for bone regeneration. Adv Drug Deliv Rev 64:1292–1309
17. Devescovi V, Leonardi E, Ciapetti G, Cenni E (2008) Growth factors in bone repair. Chir Organ Mov 92:161–168
18. Martin V, Bettencourt A (2018) Bone regeneration: biomaterials as local delivery systems with improved osteoinductive properties. Korean J Couns Psychother 82:363–371
19. Rose FR, Oreffo RO (2002) Bone tissue engineering: hope vs hype. Biochem Biophys Res Commun 292:1–7
20. Liedert A, Wagner L, Seefried L, Ebert R, Jakob F, Ignatius A (2010) Estrogen receptor and Wnt signaling interact to regulate early gene expression in response to mechanical strain in osteoblastic cells. Biochem Biophys Res Commun 394:755–759
21. Tiedeman JJ, Garvin KL, Kile TA, Connolly JF (1995) The role of a composite, demineralized bone matrix and bone marrow in the treatment of osseous defects. Orthopedics 18:1153–1158

22. Gazdag AR, Lane JM, Glaser D, Forster RA (1995) Alternatives to autogenous bone graft: efficacy and indications. J Am Acad Orthop Surg 3:1–8
23. Kumar P, Vinitha B, Fathima G (2013) Bone grafts in dentistry. J Pharm Bioallied Sci 5:S125–S127
24. LeGeros RZ (2002) Properties of osteoconductive biomaterials: calcium phosphates. Clin Orthop Relat Res:81–98
25. Schliephake H (2010) Application of bone growth factors—the potential of different carrier systems. Oral Maxillofac Surg 14:17–22
26. Freire MO, You HK, Kook JK, Choi JH, Zadeh HH (2011) Antibody-mediated osseous regeneration: a novel strategy for bioengineering bone by immobilized anti-bone morphogenetic protein-2 antibodies. Tissue Eng Part A 17:2911–2918
27. Suzuki M, Kato C, Kato A (2015) Therapeutic antibodies: their mechanisms of action and the pathological findings they induce in toxicity studies. J Toxicol Pathol 28:133–139
28. Viola M, Sequeira J, Seica R, Veiga F, Serra J, Santos AC, Ribeiro AJ (2018) Subcutaneous delivery of monoclonal antibodies: how do we get there? J Control Release 286:301–314
29. Bessa PC, Casal M, Reis RL (2008) Bone morphogenetic proteins in tissue engineering: the road from the laboratory to the clinic, part I (basic concepts). J Tissue Eng Regen Med 2:1–13
30. Kim NH, Lee SH, Ryu JJ, Choi KH, Huh JB (2015) Effects of rhBMP-2 on sandblasted and acid etched titanium implant surfaces on bone regeneration and osseointegration: spilt-mouth designed pilot study. Biomed Res Int 2015:459393
31. Rath B, Nam J, Deschner J, Schaumburger J, Tingart M, Grassel S, Grifka J, Agarwal S (2011) Biomechanical forces exert anabolic effects on osteoblasts by activation of SMAD 1/5/8 through type 1 BMP receptor. Biorheology 48:37–48
32. Huang J, Best SM, Bonfield W, Brooks RA, Rushton N, Jayasinghe SN, Edirisinghe MJ (2004) In vitro assessment of the biological response to nano-sized hydroxyapatite. J Mater Sci Mater Med 15:441–445
33. Alonso N, Tanikawa DY, Freitas Rda S, Canan L Jr, Ozawa TO, Rocha DL (2010) Evaluation of maxillary alveolar reconstruction using a resorbable collagen sponge with recombinant human bone morphogenetic protein-2 in cleft lip and palate patients. Tissue Eng Part C Methods 16:1183–1189
34. Carragee EJ, Hurwitz EL, Weiner BK (2011) A critical review of recombinant human bone morphogenetic protein-2 trials in spinal surgery: emerging safety concerns and lessons learned. Spine J 11:471–491
35. Herford AS, Boyne PJ, Rawson R, Williams RP (2007) Bone morphogenetic protein-induced repair of the premaxillary cleft. J Oral Maxillofac Surg 65:2136–2141
36. Epstein NE (2013) Complications due to the use of BMP/INFUSE in spine surgery: the evidence continues to mount. Surg Neurol Int 4:S343–S352
37. Ansari S, Freire MO, Pang EK, Abdelhamid AI, Almohaimeed M, Zadeh HH (2014) Immobilization of murine anti-BMP-2 monoclonal antibody on various biomaterials for bone tissue engineering. Biomed Res Int 2014:940860
38. Wu Q, Yang B, Cao C, Hu K, Wang P, Man Y (2018) Therapeutic antibody directed osteogenic differentiation of induced pluripotent stem cell derived MSCs. Acta Biomater 74:222–235
39. Guo L, Min S, Su Y, Tang J, Du J, Goh BT, Saigo L, Wang S, Ansari S, Moshaverinia A, Zadeh HH, Liu Y (2017) Collagen sponge functionalized with chimeric anti-BMP-2 monoclonal antibody mediates repair of nonunion tibia defects in a nonhuman primate model: an exploratory study. J Biomater Appl 32:425–432
40. Xie Y, Su Y, Min S, Tang J, Goh BT, Saigo L, Ansari S, Moshaverinia A, Zhang C, Wang J, Liu Y, Khojasteh A, Zadeh HH, Wang S (2017) Collagen sponge functionalized with chimeric anti-BMP-2 monoclonal antibody mediates repair of critical-size mandibular continuity defects in a nonhuman primate model. Biomed Res Int 2017:8094152
41. Khojasteh A, Hosseinpour S, Dehghan MM, Mashhadiabbas F, Rezai Rad M, Ansari S, Farzad Mohajeri S, Zadeh HH (2018) Antibody-mediated osseous regeneration for bone tissue engineering in canine segmental defects. Biomed Res Int 2018:9508721

Extracellular Matrix-based Materials for Bone Regeneration

Sheng Zhou, Shichao Zhang, and Qing Jiang

Abstract Decellularized extracellular matrix (ECM)-based scaffolds are rapidly expanding in regenerative medicine. The ECM is an intricate microenvironment with excellent biochemical, biophysical, and biomechanical properties, which can regulate cell adhesion, proliferation, migration, and differentiation, as well as drive tissue homeostasis and regeneration. Decellularized tissue-derived ECMs have been reported to be successful in clinical application of cardiovascular, respiratory, and gastrointestinal surgery. In bone tissue engineering, decellularized ECMs derived either from tissues such as bone, cartilage, and small intestinal submucosa or from cells such as stem cells, osteoblasts, and chondrocytes have shown promising results. We begin this chapter with a brief description of the composition of the ECM and its changes during osteogenesis in vivo and in vitro. Next, the decellularization methods are summarized, followed by the latest development in matrices from native tissues, or cultured cells and their application in bone tissue engineering. Finally, we investigated the different engineering strategies for the design of ECM-based scaffolds in bone regenerative medicine. With this information, we hope to better understand the ECM-based materials and to develop biomaterials more close to the clinical needs in bone tissue engineering.

Keywords Extracellular matrix · Decellularization · Tissue-derived extracellular matrix · Cell-derived extracellular matrix · Electrospinning · Three-dimensional printing · Hydrogel · Bone regeneration

S. Zhou
Department of Orthopaedics, School of Medicine, West Virginia University, Morgantown, WV, USA

Department of Sports Medicine and Adult Reconstructive Surgery, School of Medicine, Nanjing Drum Tower Hospital, Nanjing University, Nanjing, Jiangsu, People's Republic of China

S. Zhang · Q. Jiang (✉)
Department of Sports Medicine and Adult Reconstructive Surgery, School of Medicine, Nanjing Drum Tower Hospital, Nanjing University, Nanjing, Jiangsu, People's Republic of China
e-mail: qingj@nju.edu.cn

B. Li et al. (eds.), *Racing for the Surface*,
https://doi.org/10.1007/978-3-030-34471-9_19

Introduction

Bone-devastating defects due to trauma, tumor, infection, or congenital etiology lead to significant alternations in appearance and function, which have a significant influence on patients and society. It is estimated that the incidence of long bone fractures in the United States alone is about 1,500,000 annually [1]. More than 500,000 bone grafting operations are performed in the United States per year, and more than 2 millions in the world, to treat nonunions or large defects [2, 3]. A bone graft is the second most frequent transplanted tissue, coming right after blood transfusion [4, 5]. Autologous bone grafts are predominantly regarded as osteoconductive materials for bone replacement, with the ability to deliver a combination of differentiated osteoblasts, a mixture of bone growth factors, and the matrix of cancellous bone at a physiological level [6, 7]. Allogenic bone grafts are available from bone banks, which are cadaveric in origin and have been also widely used in bone reconstructive surgery [8]. Although allografts still have bone morphogenetic proteins (BMPs), they only possess osteoconductive and weakly osteoinductive capacities [9]. Autologous or allogeneic bone grafts are considered as the gold standard for bone defect reconstruction; however, their inherent insufficient tissue availability, risks of unknown disease transmission, severe donor site morbidity, and unpredictable late resorption of grafts render their clinical use [4].

Tissue engineering is becoming a promising therapeutic approach in treating bone diseases in light of a number of problems confronting currently available treatments [10, 11]. Three-dimensional (3D) scaffolds should offer initial support to seeded cells and provide biological and physical cues for proliferation, migration, and differentiation. This requires the development of scaffolds that have the same functions as in vivo extracellular matrix (ECM). Significant efforts have been made to generate synthetic substitutes (e.g., collagen, ceramic, and hydroxyapatite), which resemble the architecture and/or composition of native bone [12]. However, most of these materials do not display truly osteoinductive properties but merely have an osteoconductive ability [13]. Significant efforts have been made to increase their bioactivity by adding growth factors, ceramic particles, or ECM proteins or peptides [14–17]. However, quantities and release kinetics as well as optimal combinations for these factors are still not defined. These considerations prompt the generation of native ECM-based materials with osteoinductive properties for bone regeneration.

The ECM is a complex and essential meshwork between and around niche cells, consisting of glycoproteins, collagens, proteoglycans, and polysaccharides [18]. The ECM regulates cellular functions, such as cell attachment, survival, proliferation, migration, and differentiation with biochemical, mechanical, and biophysical properties [19]. Furthermore, the ECM holds great promise in tissue regeneration, as it has immune tolerance [20]. Unfortunately, ECM proteins and variances from different tissues are currently not completely understood. Thus, it is quite difficult to rebuild an ECM with appropriate architecture from synthetic materials or pure protein components.

The purpose of fabricating a native ECM-based scaffold is to mimic the target tissue ECM structure as much as possible. For example, lungs with acellular vasculature, airways, and alveoli have been reseeded with epithelial and endothelial cells to establish ventilation function [21]. It is exciting that such scaffolds based on natural ECM sources have been successful in the clinic for the regeneration of a number of tissues, such as heart valves [22] and the trachea [23]. Furthermore, several decellularized scaffolds have been commercialized and approved by the Food and Drug Administration (FDA) for use in humans, such as porcine urinary bladder (ACell), dermis tissue (Alloderm®; LifeCell), and porcine heart valves (Synergraft®; Cryolife) [24]. Unfortunately, due to the challenging regulatory- and cost-related issues and uncertain advantages from the standard-of-care, currently there is no decellularized ECM (dECM) for bone-specific clinical applications. Taking into account the impact of ECM-based biomaterials for bone regeneration on the quality of patients living, this technology needs to be continually evaluated.

This review first summarizes the composition of the ECM in bone and its changes during osteogenesis in vivo and in vitro, followed by a discussion of decellularization methods. The ECM from tissue and cells, and their influence on cell functions and bone regeneration are further discussed. We then focus on fabrication methods, such as electrospinning, 3D printing, and hydrogels, in generating ECM-based materials in bone regenerative medicine. Finally, we conclude with challenges and perspectives in the development of ECM-based materials for bone tissue engineering.

Characteristics of ECM from Bone Tissue and Changes During Osteogenesis

The specific composition and distribution of ECM constituents vary considerably depending on the tissue source [18]. In general, the ECM mainly consists of glycosaminoglycans (GAGs) (e.g., heparin sulfate, chondroitin sulfate, hyaluronan), fibrillar proteins (e.g., collagens, laminin, fibronectin), matricellular proteins (e.g., osteopontin, thrombospondin), and proteoglycans (e.g., decorin, versican, aggrecan). The bone ECM consists of organic (40%, mainly of type I collagen) and inorganic (60%, mainly of hydroxyapatite) components [25, 26]. Collagen, the main component of the ECM plays a vital role in bone formation and regeneration. Wang et al. showed that type I collagen could initiate and orientate carbonated apatite mineral growth without the aid of any other vertebrate ECM molecules of calcifying tissues [27]. Nudelman et al. demonstrated that collagen fibrils could control mineralization with hydroxyapatite nucleation inhibitors [28]. The bone ECM is also a reservoir for growth factors such as the transforming growth factor beta (TGFβ) family, pro-inflammatory cytokines, calcium phosphates (tricalcium phosphate or hydroxyapatite), and angiogenic growth factors like VEGF [29]. Protein components in the ECM connect cells through integrins, major cell surface transmembrane receptors, thereby mediating cell behavior [30]. In general, the ECM provides an excellent microenvironment for cell adhesion, migration, proliferation, and differentiation.

The distribution and composition of specific matrix components may be altered during bone developmental stage [31–33]. Fibronectin and versican can be observed during the mesenchymal condensation period while decorin exists in unmineralized bone but they all disappear in mineralized bone [31–33]. Biglycan is found in the bone marrow surrounding mesenchymal stem cells (MSCs) but disappears in unmineralized and mineralized bone matrices [32]. Changes in the ECM components during the osteogenesis of MSCs in vitro are similar to in vivo remodeling of the ECM [34]. Fibronectin, versican, and biglycan are abundant in the stem cell matrices while expression of versican and biglycan decreases significantly in the late stage matrices of osteogenesis [34]. Decorin is only strongly detected in the early stage of osteogenic matrices [34].

Biomaterials used in tissue engineering attempt to simulate the multifactorial aspects of ECM function. Unfortunately, synthetic growth factors and specific ECM proteins or peptides (fibrin, collagen, or hyaluronan) fail to achieve the organization and molecular complexity of native tissue matrices [14, 15, 35]. Moreover, the fact that decellularized allogenic bone grafts have a comparable clinical performance to autologous bone grafts highlights the important effect of the ECM in bone regeneration [36]. These reasons have motivated the utilization of the native ECM itself as a biomaterial source in bone regeneration.

Decellularization

Cellular components should be removed from tissues or cells to minimize the immune response of the ECM [37]. Decellularization is a process whereby all cellular and nuclear materials are eliminated from tissues or cultured cells while maintaining the composition and organization of an ECM [38]. A good decellularization requires a balance between preservation of bioactive cues in the ECM and reducing the risks of unknown disease transmission. To improve the efficiency of decellularization, a variety of decellularization methods have been developed and fall into four categories: physical, chemical, enzymatic, and biological decellularization methods (Fig. 1). The most effective and robust decellularization protocols are a combination of the above methods. A classic decellularization protocol generally utilizes physical or chemical treatments to lyse the cell membrane, followed by enzymatic methods to separate cellular components.

Physical Methods

In physical approaches, freezing and thawing, direct pressure, sonication, agitation, and osmosis are used to decellularize tissue or cell-derived ECM by destroying cellular membranes, with corresponding cell lysis [37]. A repeated freeze-thaw cycling is often used to form ice crystals inside cells and lyse the cells [37]. A physical

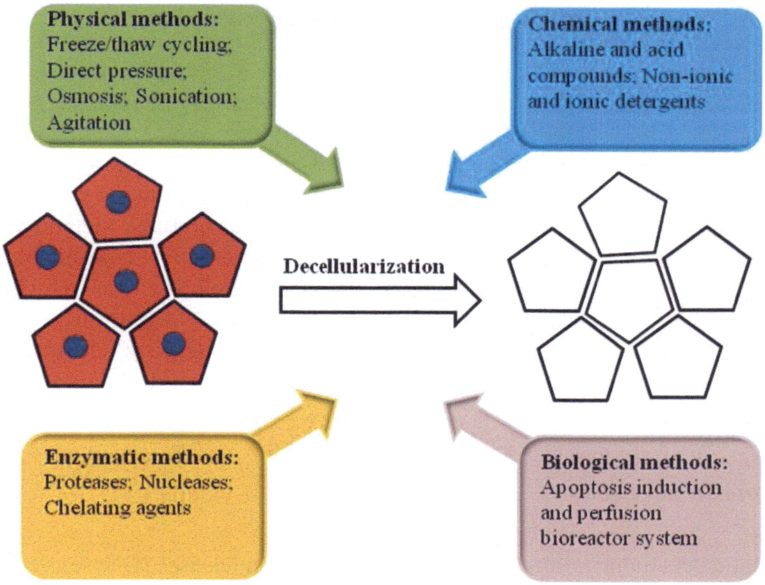

Fig. 1 Decellularization approaches can be classified into physical, chemical, enzymatic, and biological methods. Physical methods include freeze/thaw cycling, direct pressure, osmosis, sonication, and agitation. Alkaline and acid compounds, non-ionic and ionic detergents are chemical solutions used for decellularization. Enzymatic decellularization uses proteases, nucleases, and chelating agents. Biological decellularization is induced by apoptosis in a perfusion bioreactor system

method has the advantage of minimally altering the mechanical properties of the ECM and preventing the disruption of the ECM ultrastructure [39]. However, as this approach does result in incomplete removal of cellular debris, chemical or enzymatic methods may need to be added to obtain acellular tissues. For example, due to the density of compact bone, cold isostatic pressure at 30 °C for 10 min combined with deoxyribonuclease (DNase) treatment at 37 °C for 3 weeks has been utilized for decellularization [40].

Chemical Methods

Decellularization using a chemical solution can be classified into alkaline and acid compounds, and non-ionic and ionic detergents [41]. Alkaline bases can denature chromosomal and plasmid deoxyribonucleic acid (DNA), while acids can dissociate DNA from the ECM by solubilizing cytoplasmic components [39]. Alkaline bases used for decellularization include calcium hydroxide, sodium sulfide, sodium hydroxide, and ammonium hydroxide [37]. Commonly used acids include acetic acid and deoxycholic acid [37]. Alkaline and acid treatments are effective in disrupting cell membranes and intracellular organelles. However, they can cause damage and the removal of collagen or growth factors with a corresponding reduction in mechanical properties. Non-ionic detergents, such as Triton X-100, are generally considered to gently solubilize proteins while preserving protein structure and enzymatic activity [37]. However, effectiveness of non-ionic detergents selectively depends on the type of tissue. Ionic detergents, such as sodium dodecyl sulfate (SDS), can completely solubilize cytoplasmic and nuclear cellular membranes and fully denature proteins [42]. Moreover, SDS is difficult to be removed from the remaining matrix and may result in adverse cytocompatibility. Chemical treatments can remove all cellular materials, but can cause damage to collagen structures in the rest of ECM at the molecular and fibrillar levels [43]. Therefore, chemical treatments should be combined with other approaches for the least damage to ECM.

Enzymatic Methods

Enzymatic decellularization uses proteases (trypsin), nucleases (DNase, ribonuclease (RNase)), and chelating agents (ethylenediaminetetraacetic acid (EDTA)). Trypsin can detach cells from tissue surface by selectively cleaving cell adherent proteins. Trypsin is effective as a decellularizing adjuvant but can cause damage to a collagen matrix with long exposure times. Among these methods, nucleases (RNase and DNase solutions) are often used to degrade residual ribonucleic acid (RNA) or DNA, which are often added to detergent treatments when decellularization is not effective with detergents alone [44]. EDTA binds divalent metal cations at cell adhesion sites, which could cause cell dissociation from the ECM. EDTA is used with trypsin or detergents to achieve complete removal of cell nuclei [45, 46]. However, decellularization with EDTA may leave some cellular remnants [47]. In all, enzymatic methods can remove cellular contents and preserve most of the collagen components, but it has damage on the ECM structure and tensile strength [37].

Biological Methods

The above methods are efficient in decellularization, but result in damage to the remaining ECM and do not guarantee the preservation of the structural, biochemical, or biomechanical features of the ECM. Moreover, all of the above techniques rely on cell lysis, which lead to a paradoxical increase in immunogenicity caused by cell debris [48]. A novel method was developed by the specific activation of apoptosis [49, 50]. In this method, programmed cell death was activated for decellularization while preserving the integrity of the ECM. During apoptosis, cells lose contact with an ECM while cellular constituents are kept strictly within the apoptotic bodies and cell membranes [51, 52]. In this approach, cellular contents do not leak into the surrounding matrix, thus avoiding an unnecessary inflammatory reaction. This approach consists of apoptosis induction and a perfusion bioreactor system, which is helpful for removal of cellular material. Decellularization by apoptosis is an intriguing proposal that needs further investigation.

Application of Extracellular Matrix in Bone Tissue Engineering

A combination of decellularization methods is used to prepare a tissue-derived (e.g., bone, cartilage, or skin) ECM for bone regeneration. Utilization of specific decellularization procedures depends on the tissue type; for instance, harsh treatments are needed to decellularize compact bone and cartilage [40, 53]. Similar decellularization methods have also been applied in cell-derived ECMs, such as mild chemical agents and nucleases. As a cell-derived ECM is less dense than ones from native tissue, chemical agents combined with enzymatic methods are often utilized [34, 54]. Furthermore, the decellularization procedure for cell-derived ECM is more efficient and generally shorter, thus preventing the reduction of an aggregate modulus of dECM. Currently, there are no standard decellularization methods for tissues or cells. In principle, the optimal decellularization procedure should effectively remove cellular components while maintaining the ultrastructure, micromechanical, and bioactive molecules of the remaining ECM.

Tissue-Derived ECM

Tissue-derived ECM can be generated from allogenic (donor/cadaver) or xenogenic (animals) tissues or organs. As a tissue-derived ECM is directly derived from a mature organ, its structure and architecture meet biophysical requirements and offer an adequate template for host cells. Harsh decellularization treatments are required to reduce the risks of unknown disease transmission, which results in the impairment of bioactive cues in this type of ECM [8, 55]. To overcome this limitation, a tissue-derived ECM can be combined with high doses of growth factor cocktails, thereby losing the advantage of an *off-the-shelf* osteoinductive material. In the following part, we will discuss different types of tissue-derived ECM applied in bone regeneration, for instance, decellularized bone ECM, demineralized bone matrix (DBM), decellularized cartilage ECM, small intestinal submucosa (SIS)-ECM, and decellularized other tissue-derived ECM (Table 1).

Table 1 Application of tissue-derived ECM in bone regeneration

ECM type	Seeded cells	Scaffold formulation or other conditions	In vitro outcome	In vivo outcome	Reference(s)
Decellularized bone ECM					
Bovine trabecular bone	Human autologous bone marrow cells	–	–	Bone actively formed in human	[62]
	Human ESCs	Decellularized bone scaffolds with low (0.281 ± 0.018 mg/mm^3), medium (0.434 ± 0.015 mg/mm^3), and high (0.618 ± 0.027 mg/mm^3) density	Medium-density group yielding highest cell densities and newly bone matrix	–	[61]
Cow trabecular bone	Human ADSCs	–	Formation of compact and viable bone tissue constructs in perfusion culture	–	[60]
Porcine bone/bone marrow	Rat MSCs	–	Increased ALP activity	Observing clusters of the hematopoietic cells subcutaneously in rats	[40]
Demineralized bone matrix					
Bovine sponge bone DBM	HUVECs	Heparin cross-linked DBM pre-loaded with VEGF	Increased proliferation	Promoting cells and new microvessel invasion	[67]
DBM (Grafton®)	Human skin MSCs, human BMSCs, and human dental follicle MSCs	–	Having osteogenic capability in non-osteogenic induction media	Having bone formation in all three types of cells, with human dental follicle MSCs showing the best	[70]
	Porcine SDMSCs	Cells with fibrin glue injected into a DBM scaffold	Enhanced osteogenesis	Pronounced trabecular bone formation and osteocalcin expression in porcine maxillary sinus floor	[72]

Human DBM	Human ASPSCs and rat calvarial osteoblasts	–	ALP activity of human ASPSCs decreased over time while the activity of osteoblasts peaked on day 5; DBM did not increase more calcium deposition	More bone in rat ectopic pouch osteoinduction model	[74]
	Human BMSCs	The DBM putty with porous HAp granules (DBM/HAp putty)	Higher ALP activity	Newly generated mineralized tissues in rat abdominal muscle pouch model	[68]
Human bone matrix and DBM	Murine BMSCs	Incorporation of bone matrix/ DBM into PLGA films	Higher cell attachment	–	[69]
Porcine partially cancellous bone DBM	Human UCB-MSCs	Partially DBM scaffolds	Higher ALP activity and increased osteogenic markers	Rapid and more bone in rat critical-sized full-thickness circular defects	[71]
Rat cortical bone DBM	Murine or rat BMSCs	Cell/DBM/RTR polymers composites	–	More bone formation under kidney capsule or in rat calvarial defect	[75]
Decellularized cartilage ECM					
Equine joint cartilage	Human BMSCs	–	Forming cartilage matrix components, such as proteoglycans and collagen type II	Higher mineralization subcutaneously in rats	[85]
Human DBM and porcine DCC	–	PLGA microsphere-based scaffolds were first fabricated, with opposing gradients of DCC and DBM encapsulated	–	Not ideal	[84]

(continued)

Table 1 (continued)

ECM type	Seeded cells	Scaffold formulation or other conditions	In vitro outcome	In vivo outcome	Reference(s)
Porcine growth plate	Porcine BMSCs	Growth plate derived scaffolds	Successful osteogenic differentiation	Accelerating vessel in-growth, mineralization and de novo bone formation in rat critical-sized cranial defects	[86]
SIS-ECM					
Porcine mineralized SIS	Rabbit BMSCs	Heparinized mineralized SIS loaded with BMP2-related peptide P28 (mSIS/P28)	Promoting cell proliferation and viability, ALP activity and osteogenic markers	Stimulating bone regeneration in critical-sized mouse model	[91]
Porcine SIS	–	–	–	Not supporting new bone formation in rat femur critical defects	[94]
	MC3T3-E1; mouse BMSCs	–	Beneficial for cell attachment, proliferation, migration, and osteogenic differentiation	Facilitating bone regeneration in a calvarial defect mouse model	[89]
	Rat BMSCs	SIS sponge	–	Enhanced bone formation in full-thickness bone defects in the rat crania	[88]
		Cell/SIS composites	–	More efficient and rapid bone defect reparation in rabbit radial critical-sized defects; pure SIS cannot guide bone regeneration	[95]

Porcine SIS and MC3T3-E1	Human ADSCs	MC3T3-E1 were first seeded on SIS 4 weeks, then decellularized	Facilitating cell adhesion and proliferation; inducing osteogenesis even without osteogenic inductive factors	Enhanced bone formation in a mouse calvarial defect model	[92]
	MC3T3-E1	Cells first seeded on SIS in induction medium 2 weeks, then decellularized to get osteogenic and mineralized ECM construct (Os/M-SIS-ECM)	Promoting cell adhesion, proliferation, and osteogenic markers; enhanced osteogenic transdifferentiation of fibroblasts	Greatly enhanced bone formation in mouse calvarial defect model	[93]
Porcine SIS and rabbit BMSCs	Rabbit BMSCs	Depositing SIS on TBC, followed by mineralization treatment (mSIS/TBC)	Promoting cell proliferation and viability, ALP activity, and osteogenic markers	Both ECM types promoting bone formation with mSIS/TBC showing better in ovariectomized rats calvarial defects	[90]
Decellularized other tissue-derived ECM					
Human amnion membranes	Rat BMSCs	Lyophilized, multilayered	High cell density, good cell viability, and efficient osteogenic differentiation	Guiding bone regeneration in rat tibia defect model	[103]
Human dermis	Canine ADSCs	–	–	More rapid and greater bone formation in athymic murine calvarial bone defect	[101]
	Rat osteoblasts and murine BMSCs	–	Promoting cells attachment, proliferation, and differentiation	Bone formation in critical-sized mandibular defects in nude rats	[100]
Porcine liver, lung, spleen, cartilage, bone, and adipose tissue particles	Human ADSCs	GAGs-tissue ECM gel formation	Hydrogels containing bone ECM particles showing the best osteogenic differentiation	No significant difference	[105]

(continued)

Table 1 (continued)

ECM type	Seeded cells	Scaffold formulation or other conditions	In vitro outcome	In vivo outcome	Reference(s)
Porcine pericardium	–	–	–	Better bone formation in rabbit tibial defect model	[102]
Porcine periosteum	–	–	Initiating mineralization without aid of cells	Promoting regeneration in a critical bone defect rabbit model and subcutaneously in nude mouse	[98]
Porcine UBM	Human WJMSCs	ECM powders to form scaffold	Guiding adhesion, proliferation, and 3D colonization; positively affecting osteoblastic markers	–	[104]
Rat adipose tissue	Rat BMSCs	Mixed solution (chitosan, gelatin, and ECM suspension) to form scaffolds	Better osteogenic differentiation	–	[99]

Abbreviations: ALP alkaline phosphatase, *ASPSC* adipose-derived side population stem cell, *BMSCs* bone marrow mesenchymal stem cells, *DBM* demineralized bone matrices, *DCC* decellularized cartilage, *ECM* extracellular matrix, *ESCs* embryonic stem cells, *GAGs* glycosaminoglycans, *HAp* hydroxyapatite, *HUVECs* human umbilical vein endothelial cells, *MSC* mesenchymal stem cell, *RTR* reverse thermal responsive, *SDMSCs* skin-derived mesenchymal stem cell-like cells, *SIS* small intestinal submucosa, *TBC* true bone ceramic, *UBM* urinary bladder matrix, *UCB-MSCs* umbilical cord blood-derived mesenchymal stem cells, *VEGF* vascular endothelial growth factor, *WJMSCs* umbilical cord Wharton's Jelly

Decellularized Bone ECM

Decellularization of bone has been utilized to avoid anticipated host immune reactions [56] while maintaining ECM components [57, 58]. In fact, decellularized bone is a prototype ECM grafting material [59]. In vitro, decellularized bone increased the osteogenesis of human ASCs [60], human ESCs [61], and rat MSCs [40]. In vivo, a decellularized bone matrix combined with MSCs played a positive role in bone regeneration subcutaneously in rats [40]. Clinically, decellularized bovine trabecular bone discs combined with human autogenous bone marrow cells showed a beneficial effect for the reconstruction of segmental long bone defects [62].

DBM

DBM can be generated from bone ECM and retains the organic content of bone but lacks the mineral ones [63–65]. Commonly used in orthopedic applications, DBM has been studied for over 4 decades in bone grafting procedures [66] and currently, 25 different DBM products are available in the market [63]. After defatting human allograft cortical bone and formulated it through acidic washing, the DBM still has a similar microstructure to bone tissue and contains collagenous proteins and ECM-associated growth factors (e.g., TGFβ and BMPs), which are osteoinductive agents helpful for cell attachment, migration, and differentiation [67]. The resulting DBM has a gel-like consistency which can be further processed and stored as granules or powder. The biological activity of DBM is dependent on the sterilization method [65] or origin of the donor [64]. Due to demineralization, the mechanical performance of DBM is poor. To overcome these limitations, DBM has been blended with synthetic materials, such as hydroxyapatite (HAp) [68] and poly(lactic-*co*-glycolic acid) (PLGA) [69]. In vitro, DBM has the potential to enhance the proliferation, alkaline phosphatase (ALP) activity, or osteogenic markers of human bone marrow mesenchymal stem cells (BMSCs) [68, 70], human skin MSCs [70], human dental follicle MSCs [70], umbilical cord blood-derived mesenchymal stem cells (UCB-MSCs) [71], human umbilical vein endothelial cells (HUVECs) [67], murine BMSCs [69], porcine skin-derived mesenchymal stem cell-like cells (SDMSCs) [72], and mouse primary calvarial cells (mPCs) [73]. In vivo, a demineralized bone matrix with MSCs generated more mineralized bone matrix in a rat abdominal muscle pouch model [68, 74], rat critical-sized full-thickness circular defects [71, 75], and a porcine maxillary sinus floor [72].

Decellularized Cartilage ECM

The rudimentary formation or the growth of long bone length, and the healing of bone fractures all involve endochondral ossification [76–79]. During endochondral ossification, cartilage is present, followed by becoming hypertrophic, angiogenic, and forming mineralization, subsequently to be remodeled into bone [78]. This raises the interest of using the cartilage ECM for long bone defects, inspired from the endochondral ossification process [78]. Accordingly, this implies the use of decellularization of cartilage ECM for bone regeneration. However, autogenic cartilage grafts are limited by donor site morbidity [80, 81], while allogenic cartilage grafts are restricted by storability and immunogenicity [82].

Acellular cartilage is rich in aggrecan, type II collagen, and growth factors. A decellularized cartilage ECM can be obtained from a variety of cartilaginous sources, such as porcine articulating hyaline cartilage [83, 84], equine joint cartilage [85], and a porcine growth plate [86]. Colloidal nanoparticles of HAp, DBM, decellularized cartilage, and hyaluronic acid were developed to form colloidal gels with desirable rheological properties [83]. These decellularized cartilage ECM alone or combination with DBM showed enhanced osteogenesis in a rat critical-sized defect [83, 86] or subcutaneously in rats [85].

SIS-ECM

SIS, generated from the submucosal layer of porcine intestine, is composed of over 90% collagen by dry weight, with the most being type I and III collagen [38, 87, 88]. After decellularization, SIS contains GAGs and growth factors [38, 87]. The remaining aligned collagen fibers, GAGs, and growth factors in acellular SIS have sparked interest in bone tissue regeneration. Currently, SIS is FDA approved for urogenital clinical applications [87].

SIS scaffolds enhanced cell attachment, proliferation, migration, and osteogenesis of rat BMSCs [89–91], human adipose-derived stem cells (ADSCs) [92], and MC3T3-E1 [93] in vitro, and promoted bone formation in bone defects in the rat crania [88], a calvarial defect mouse model [89, 90, 92, 93], a rabbit radial critical-sized defect and ovariectomized rats calvarial defect [91]. However, pure SIS did not have the ability to support new bone formation in a rat femur critical defect [94] or in a rabbit radial critical-sized defect [95]. Therefore, biomineralization of SIS loaded with P28 (derived from the knuckle epitope of BMP2) was developed to synergistically enhance bone regeneration [91]. Osteogenic cell-derived ECM ornamented SIS provides a better osteogenic microenvironment [92, 93]. MC3T3-E1 cells were first seeded on SIS scaffolds for 4 weeks to get abundant ECM, followed by repeated freeze/thaw cycles for decellularization. After these steps, ECM ornamented SIS scaffolds could be obtained.

Decellularized Other Tissue-Derived ECM

Apart from the above tissue-derived ECM, other tissue-derived ECMs have also been developed to facilitate bone formation. The periosteum, covering the external surface of bone, is important in bone formation and regeneration [96, 97]. A decellularized natural periosteum, a 3D collagen network, can initiate mineralization in vitro even without cells [98]. As adipose tissue has the advantage of being abundant in the body and easy to obtain, Wang et al. used adipose-derived ECM for bone tissue engineering and found that it was beneficial for osteogenic differentiation in vitro [99]. Acellular dermal matrix, composed of collagen, elastin, and proteoglycan, maintains tensile properties and has favorable handling and is easy to implant [100]. Kim et al. found that acellular dermal matrix combined with canine ADSCs promoted a more rapid and greater new bone formation in athymic murine calvarial bone defect [101]. Human amnion membranes, porcine pericardium, and UBM were also subjected to decellularization and were found beneficial for osteogenesis in vitro or in vivo [102–104]. Although many other different tissue-derived ECM was used in bone regeneration, a bone-derived ECM showed the best potential in enhancing bone osteogenic differentiation [105].

Cell-Derived ECM

Tissue-derived ECMs have their own advantages, for example, some tissue ECMs, such as human DBM [87], decellularized skin [106], SIS [107], and amniotic membranes [108], have been already used in preclinical studies and clinical applications. Moreover, decellularized matrices from tissues can mimic the microstructure of native ECM. However, the availability of tissues for decellularization is problematic. Besides, inherent heterogeneity, potential pathogen transfer, uncontrollable degradation, and a limited ability for customization restrict their clinical applications. These considerations prompt efforts to generate cell-derived ECM-based materials in bone tissue engineering. Similar to a decellularized tissue-derived ECM, cell-derived matrices are biocompatible, and bioactive materials are made up of a complex assembly of fibrillar proteins, growth factors, and matrix macromolecules.

When applied to bone regeneration, a variety of cell types can be utilized to produce an ECM, such as human ADSCs [109], BMSCs [34, 110–119], amniotic fluid stems (AFS) [120], HUVEC [121], dermal fibroblasts [122, 123], lung fibroblasts [124], nasal inferior turbinate tissue-derived mesenchymal stromal cells (TMSCs) [125], embryonic stem cells [126], hair follicle keratinocytes (HFKTS) [126], murine BMSCs [54], MC3T3-E1 [127–133], L929 fibroblasts [130], rat BMSCs [134–136], primary rat osteoblasts [137], primary porcine chondrocytes [138], and mesenchymal sword of Damocles (MSOD) [139]. Human and rat BMSCs as well as MC3T3-E1 cells are used during in vitro tests the most to secrete an ECM for bone tissue engineering. Although different types of cells have been utilized to generate an ECM, the resulting ECM has different capabilities in promoting osteogenesis. Remarkably, early stage matrices of osteogenesis from human BMSCs enhance osteogenic differentiation more strongly compared with stem cell matrices, late stage matrices, and tissue culture polystyrene [34]. Chen et al. also reported that an undifferentiated murine BMSC-derived matrix suppressed spontaneous osteogenic differentiation in vitro [54]. Moreover, they showed that human BMSCs cultured on an MSC-derived ECM retained their "stemness" and osteogenic differentiation potential [111, 114]. However, Fu et al. reported that mineralization of an ECM was good for osteogenesis. They found that mineralized ECM/stem cell microspheroids (MECS) provided a cell-instructive structural framework which promoted osteogenesis, thereby achieving excellent bone formation outcomes in rat critical-sized defects [140].

As is well known, the functions of MSCs decline with age or passage number [141, 142]. Remarkably, the dECM produced by BMSCs from young mice (3 month-old) was capable of rejuvenating properties of BMSCs from aged mice (18 months old) [143]. Telomerase activity was significantly improved and ROS levels were reduced on a young mouse dECM [143]. Moreover, adult BMSCs cultured on dECM which were generated by fetal BMSCs had a smaller size, narrower size distribution, and greater proliferation rate, implying a delay in senescence [144]. Furthermore, these MSCs had an enhanced osteogenic differentiation capacity [144].

A cell-derived ECM has a greater ability for customization by choosing the types of cell generating the ECM, genetically modifying the source of cells to overexpress or knockdown the expression of target molecules, or modulating cell culture conditions (e.g., static versus perfusion; 2D versus 3D culture). A cell-derived ECM can be used to confer bioactivity to synthetic materials to generate ECM hybrid scaffolds. Moreover, scaffold-free tissue constructs have been developed to offer an alternative in tissue regeneration. However, the application of an ECM without scaffolds has not been extensively reported in bone tissue regeneration and is still in a relatively nascent stage of development. A transferable ECM, clumps of cells/ECM, and ECM sheet have been developed and offer tunability and flexibility to the desired applications.

ECM Hybrid Scaffolds

Tissue-engineered hybrid scaffolds can be generated by combining a cell-derived ECM in vitro with synthetic polymers or inorganic materials (Table 2). In this situation, an ECM could be obtained from autologous cells, thereby avoiding the shortcomings of an anti-host immune response or donor site morbidity. These ECM hybrid scaffolds can maintain the desired biomechanical characteristics, geometry, porosity, and the desired biological elasticity at the same time. Moreover, the ECM yield can be realized by anchoring on scaffolds. For instance, porcine gelatin foam is often selected due to its excellent cell adherence properties and common use in clinical applications [119, 145, 146].

3D cell-derived ECM hybrid scaffolds utilizing polymers, such as poly(sebacoyl diglyceride) (PSeD) [109], HAp [114, 116], titanium (Ti) [135, 136, 147], chitosan-alginate [110, 127], PLGA/poly(L-lactide) (PLA) [124], polycaprolactone (PCL) [120], tricalcium phosphate (TCP) [116, 121], porcine gelatin foam [119], and even calcined bovine bones [115], as templates, have been developed. These hybrid scaffolds possess stronger mechanical properties compared with an ECM without a scaffold, thereby increasing potential applications in bone tissue engineering.

Table 2 Applications of cell-derived ECM in bone regeneration

ECM type	Seeded cells	Scaffold formulation or other conditions	In vitro outcome	In vivo outcome	Reference(s)
ECM hybrid scaffolds					
Human ADSCs	Human BMSCs	PSed mesh scaffold (cells were first seeded on scaffold for 15 days, then decellularized)	Markedly promoting cell proliferation and exhibiting strongly osteogenic simulative effects	Enhanced reparative effects	[109]
Human BMSCs	Human BMSCs	Cells were embedded in collagen I and chitosan matrix in osteogenic medium for 14 days, then decellularized	Supporting osteogenic differentiation	–	[110]
		Cells seeded on plate or HAp-TCP scaffold in proliferation medium for 2 days and osteogenic medium for 11 days, then decellularized	Enhanced cell adhesion, proliferation, osteogenic differentiation, and mineralization	Promoting bone formation in mouse ectopic calcification model	[116]
		Using perfusion-flow (i.e., dynamic) combined with PILP method to fabricate a porous collagen/HAp composite (cells first seeded on construct, then decellularized)	Greater proliferation, retaining stemness and osteogenic differentiation capacity	Exhibiting less bone mineral formation subcutaneously in immunocompromised mice	[114]
		Cells first seeded on porous ceramic materials in proliferation medium for 1 week and osteogenic medium for 3 weeks in perfusion bioreactor, then decellularized	Higher deposited calcium	Inducing a mineralized matrix in nude mice	[115]

(continued)

Table 2 (continued)

ECM type	Seeded cells	Scaffold formulation or other conditions	In vitro outcome	In vivo outcome	Reference(s)
Human BMSCs and AFS cells	Human BMSCs	In 3D culture, cells were seeded on PCL scaffolds in osteogenic medium for 3 weeks, then decellularized	In 2D culture, increased calcium deposition; in 3D culture, both ECM types increased mineralized matrix with AFS cells inducing a greater	Both ECM types increased the rate of bridging, but did not have effect on the volume of mineralized matrix	[120]
Human dermal fibroblasts	Human BMSCs and ADSCs	SBC-ECM constructs (cells seeded on SBC scaffold for 21 days, then decellularized)	Enhancing osteogenic differentiation	Exhibiting excellent handling properties and high biocompatibility in rat critical-sized calvarial defect model	[122]
Human lung fibroblasts	Human PMSCs	PLGA/PLA mesh scaffolds immersed in ECM solution; heparin conjugated onto ECM via EDC chemistry; BMP2 immobilized via heparin and released at a controlled rate (BMP2-tethered ECM mesh scaffolds)	Enhanced ALP activity, mineralization, and osteogenic markers	More newly formed bone in mouse ectopic or rat calvarial bone defect model	[124]
HUVECs	Human BMSCs	ECM/TCP scaffolds (cells were seeded on scaffolds for 14 days, then decellularized)	Increasing ALP and upregulating osteogenic markers	–	[121]
MC3T3-E1	MC3T3-E1 and human MSCs	Reconstituted in chitosan-alginate scaffolds based on interfacial polyelectrolyte complexation	Cell adhesion	Inducing bone formation subcutaneously in SCID mice	[127]
MC3T3-L1(LMP-1 overexpressing)	–	Cells were first seeded on calcined bovine bone scaffolds for 7 days, then decellularized	–	Better regeneration in rabbit critical-sized bone defects	[129]

MSOD or overexpressing VEGF MSODs	MSOD	Cells first seeded on porous ceramic materials in proliferation medium for 4 days and osteogenic medium for 24 days in perfusion bioreactor, then decellularized	–	Enhanced bone formation and overexpressing VEGF resulting in selective ECM enrichment and superior vasculature recruitment in rat calvarial defects	[139]
Rat BMSCs	Rat BMSCs	Ti/ECM construct (cells were first seeded on Ti in osteogenic medium, then decellularized)	More calcium content, increase in mineralization	–	[134–136]
	–	Cells were seeded on titanium fiber mesh in a flow perfusion bioreactor, then decellularized	–	Not inducing bone formation, but enhancing vascularization intramuscularly in rats	[147]
Subcutaneous pockets in rats	MC3T3-E1	Incorporating ECM into HAp in vivo	Increase in cell proliferation	New bone formation at 8 weeks in rabbit radial defects	[150]
Transferable cell-derived ECM					
Human BMSCs	Human BMSCs	Mechanical homogenization in acetic acid	Enhancing osteogenesis	–	[153]
		ECM (mechanical homogenization in acetic acid) coated on 3D scaffolds	Enhancing osteogenesis	Higher blood vessel density, but no differences in mineral deposition subcutaneously in nude rats	[154]
		First urea extraction (u-ECM); remaining ECM pellet digested with pepsin	U-ECM enhancing cell proliferation, attachment, migration, chemotactic attraction, and osteogenesis	–	[152]
	PPARγ inhibitor-treated human BMSCs	ECM produced in osteogenic differentiation	Osteogenesis capability	ECM extending inhibitor-treated MSCs retention and resulting in complete repair of bone defects in mouse calvarial healing	[113]

(continued)

Table 2 (continued)

ECM type	Seeded cells	Scaffold formulation or other conditions	In vitro outcome	In vivo outcome	Reference(s)
Human placental MSCs (DMSC23 and CMSC29)	DMSC	ECM obtained from urea extraction (U-ECM), pepsin digestion (P-ECM), and mechanical homogenization in acetic acid (AA-ECM)	U-ECM promoting osteogenesis; P-ECM proliferation; AA-ECM no effect	–	[151]
Clumps of cells/ECM					
Human IFN-γ-treated BMSCs	IFN-γ-treated human BMSCs	IFN-γ-treated human BMSCs/ECM	–	Bone regeneration in mouse calvarial defect model	[118]
Primary porcine chondrocytes	Human fetal osteoblast and human MSCs	Construct made up of chondrocytes and cartilaginous ECMs, named LhCG	Increasing osteogenic markers	Bone formation subcutaneously in nude mice	[138]
Rat BMSCs	Rat BMSCs	Rat BMSCs/ECM	Increased expression of OPN, ALP activity, and calcium content	Bone formation in rat calvarial defect model	[158]
		Rat BMSCs/ECM	–	Successful bone regeneration in rat calvarial defects	[159]
	–	Cells were seeded in microwell-based agarose plates in regular or osteogenic differentiation medium for 5 days to harvest spheroids	Upregulated ALP and BMP2	Excellent bone regeneration in rat critical-sized defects	[140]
Extracellular matrix sheet					
Human dermal fibroblasts	Human BMSCs	–	Increasing ALP activity, bone-specific genes, and calcium deposition	–	[123]
Human BMSCs	HUVECs	Cell sheets wrapped onto TCP scaffold	–	Enhanced angiogenesis and functional anastomosis, and more bone matrix subcutaneously in mice	[165]

Leporine BMSCs	Leporine BMSCs	Cells cultured in osteogenic medium and lifted using a scraper as cell sheet	–	Enhanced bone formation subcutaneously in mice	[164]
Murine BMSCs	–	Cell sheets wrapped on devitalized mouse bone allografts	–	Enhanced bone formation in murine femoral defects	[168]
Rat BMSCs	–	Osteogenic ECM sheet combined with HAp construct	–	Enhanced bone formation in rat non-union model or subcutaneously in rats	[160]
	Rat BMSCs	Cells cultured in osteogenic medium and lifted using a scraper as cell sheet	–	Enhance bone formation in a rat nonunion model	[162]
		Cells cultured in dexamethasone and vitamin C and lifted using a scraper as cell sheet; HAp/sheet composites	–	New bone formation subcutaneously in rats	[163]
		Cells cultured in osteogenic medium and lifted using a scraper as cell sheet	–	New bone formation subcutaneously in rats	[161]
		Cells were seeded in plates in osteogenic medium for 3 weeks, then decellularized to get ECM; GO-collagen I scaffold wrapped with ECM	Higher proliferation rate and osteogenic ability	Better repair effects in rat cranial defect model	[166]

Abbreviations: 2D two-dimensional, *3D* three-dimensional, *ADSCs* adipose-derived stem cells, *AFS* amniotic fluid stem, *ALP* alkaline phosphatase, *BMSCs* bone marrow mesenchymal stem cells, *BMP* bone morphogenetic protein, *CMSC* fetal MSC isolated from the chorionic villi, *DMSC* maternal MSC isolated from the decidua basalis, *ECM* extracellular matrix, *EDC* 1-ethyl-3-(3-dimethylaminopropyl)-carbodiimide, *GO* grapheme oxide, *HAp* hydroxyapatite, *HUVECs* human umbilical vein endothelial cells, *IFN-γ* interferon-gamma, *LhCG* living hyaline cartilaginous graft, *LMP-1* LIM mineralization protein-1, *MSC* mesenchymal stem cell, *MSOD* mesenchymal sword of Damocles, *OPN* osteopontin, *PCL* polycaprolactone, *PILP* polymer-induced liquid-precursor, *PLA* poly(L-lactide), *PLGA* poly(lactic-*co*-glycolic acid), *PMSCs* placenta-derived mesenchymal stem cells, *PPARγ* peroxisome proliferator-activated receptor gamma, *PSed* poly(sebacoyl diglyceride), *SBC* synthetic biphasic calcium phosphate ceramic, *SCID* severe combined immunodeficiency, *TCP* tricalcium phosphate, *Ti* titanium, *VEGF* vascular endothelial growth factor

To prepare these ECM hybrid scaffolds, cells are first seeded directly onto the polymer surface, then they produce an ECM in proliferation or osteogenic differentiation culture medium to form a cell-ECM-template architecture [115, 135, 139]. Decellularization is then carried out by physical methods, such as freeze-dry cycling, in most studies. However, 0.5% Triton X-100 containing 20 mM NH_4OH [121, 122], or 1% SDS with a combination of DNase I and RNase A [114] have also been used in this step (Fig. 2A). A collagen/HAp composite with human BMSC-derived ECM exhibited less bone mineral formation subcutaneously in immunocompromised mice [114]. Moreover, PCL/ECM hybrid scaffolds increased the

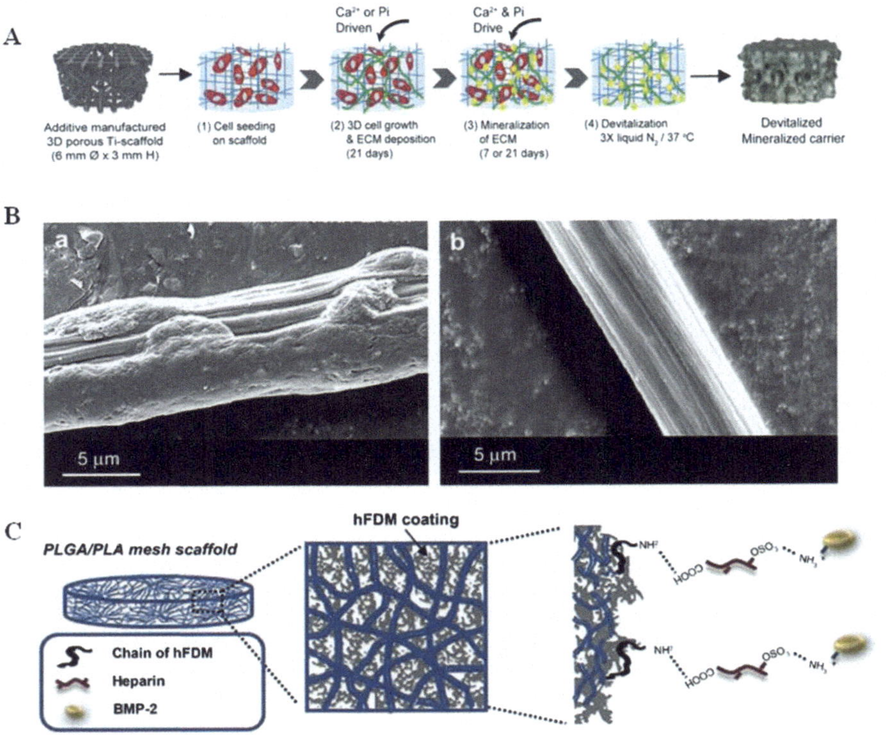

Fig. 2 Cell-derived extracellular matrix (ECM) hybrid materials. (**A**) Schematic figure showing steps of generating ECM-based materials: (1) cell seeding on a scaffold; (2) three-dimensional (3D) cell growth and ECM deposition for 21 days; (3) mineralization of ECM for 7 or 21 days; (4) devitalization by repeated freeze-thaw cycles. (**B**) ECM was coated by polyelectrolyte complexation. Scanning electron microscope (SEM) images of chitosan-alginate scaffolds with (left) and without (right) MC-3T3-E1 ECM. (**C**) Poly(lactic-*co*-glycolic acid)/PLA: poly(L-lactide) (PLGA/PLA) mesh scaffolds were immersed in a human lung fibroblast-derived matrix (hFDM) solution, followed by functionalized with heparin via 1-ethyl-3-(3-dimethylaminopropyl)-carbodiimide (EDC) chemistry, onto which bone morphogenetic protein-2 (BMP-2) is immobilized. Reprinted from [124, 127, 182] with permissions from Mary Ann Liebert, Inc. and Elsevier

bridging rate, but did not have an influence on the mineralized matrix volume in vivo [120]. It is possible that the lifespan of a cell-derived ECM is much shorter than the time scale needed for bone defect repair. Another possibility is the dominance of bulk scaffold effects in the in vivo environment. Although these studies show no promoting effects on osteogenic differentiation, most ECM hybrid scaffolds show an ability to promote cell proliferation, cell adhesion, osteogenic potential in vitro and enhance bone formation in vivo. Moreover, overexpressing VEGF MSODs resulted in superior vasculature recruitment [139] while overexpressing LMP-1 MC3T3-L1 upregulated the expression of growth factors stored in the cell-derived matrix [129], consolidating the perspective of generating ECM-based off-the-shelf materials. The use of engineered cells contributes to the development of improved cell-derived ECM, which are enriched by a defined factor and have enhanced abilities in osteoinduction (by BMPs), vascularization (by VEGF or PDGF-BB), or cell recruitment (by chemoattractants) [129, 139]. Clinically, a customized engineering ECM is urgently needed as atrophic nonunions requiring enhanced cell recruitment and vessels at the site [148], while bone defects due to surgical excision need an engineered ECM with enhanced osteoinductive properties without further angiogenic stimuli [149].

Another way to incorporate a cell-derived ECM into a scaffold is based on polyelectrolyte complexation [127]. In this way, Narayanan et al. first isolated an ECM from MC3T3-E1 cells, then reconstructed them into chitosan-alginate scaffolds. Retention of cell adhesion of the reconstituted ECM was proven by culturing MC3T3-E1 cells in vitro (Fig. 2A) and ECM-induced bone formation was seen subcutaneously in SCID mice [127]. Additionally, Kim et al. developed BMP2-tethered dECM derived from human lung fibroblast PLGA/PLA mesh scaffolds [124]. PLGA/PLA mesh scaffolds were first immersed in an ECM solution; heparin was then conjugated onto ECM via EDC (1-ethyl-3-(3-dimethylaminopropyl)-carbodiimide) chemistry, followed by BMP2 immobilized via heparin and released at a controlled rate [124] (Fig. 2B). This BMP2-tethered ECM/PLGA/PLA scaffold enhanced ALP activity, mineralization, and osteogenic markers in vitro, and generated more newly formed bone in a mouse ectopic and rat calvarial bone defect model [124].

By a variety of methods, the above studies attempted to coat a dECM on materials in vitro. Dolendo et al. deposited an ECM in vivo on HAp sponge scaffolds by implanting bare scaffolds subcutaneously in rats [150]. Decellularized scaffolds were obtained by soaking in a 1% SDS solution [150]. Decellularized scaffolds increased cell infiltration and new bone formation [150].

Current strategies to generate ECM hybrid scaffolds are mainly coating 3D materials with an ECM by placing matrix-depositing cells directly on the target substrate surface, which may result in the heterogeneous deposition of an ECM. Due to process heterogeneity and costs, this procedure may limit the large-scale production of coated substrates.

Transferable Cell-Derived ECM

To generate the transferable form of a cell-secreted ECM, cells are cultured on tissue culture plastic under a specific microenvironment, followed by decellularization, collection, and homogenization (Table 2). Transferable matrices can be generated by pepsin digestion [151, 152], urea extraction [151, 152], and mechanical homogenization in acetic acid [151, 153, 154] (Fig. 3A, B). Pepsin digestion is the most common method for solubilizing ECM, but the bioactivity of pepsin-digested ECM remains a debate [152, 155]. Recently, Kusuma et al. created transferable matrices from pepsin digestion which promoted proliferation while ones from urea extraction enhanced osteogenesis in vitro [151]. Additionally, these transferable ECM have the potential to coat 1.3–5.2 times the surface area covered by native ECM, aiding in scale-up of ECM technology. Other studies also showed that a transferable cell-derived ECM retained the ability to promote osteogenesis in vitro [152, 153]. Furthermore, Zeitouni et al. demonstrated that the transferable human BMSC-derived ECM had in vivo ability to enhance osteogenesis in mouse calvarial defects [113]. In addition to direct application of transferable matrices, Decaris et al. coated the transferable ECM on 3D polymeric constructs and found that they retained the capacity to promote MSC osteogenesis in vitro [154].

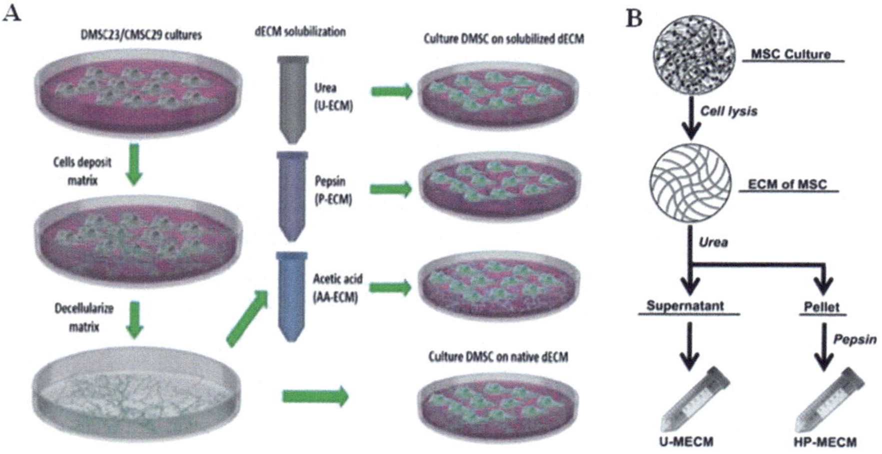

Fig. 3 Transferable cell-derived extracellular matrix (ECM). (**A**) Schematic figure showing steps of generating tissue culture polystyrene coated with native or transferable decellularized ECM. Transferable ECM can be obtained from urea extraction (U-ECM), pepsin digestion (P-ECM), and mechanical homogenization in acetic acid (AA-ECM); DMSC: maternal MSC isolated from the decidua basalis; CMSC: fetal MSC isolated from the chorionic villi. (**B**) Schematic figure showing the steps of mesenchymal stem cell (MSC) ECM extraction protocol. After decellularization, the ECM was dissolved in urea for 2 days, followed by centrifugation to pellet unextracted contents. The resulting supernatant was dialyzed against phosphate-buffered saline (PBS), and named as U-MECM. The remaining pellet was digested in pepsin, dialyzed against PBS, and named as HP-MECM. Reprinted from [151, 152] with permissions from American Chemical Society and Elsevier

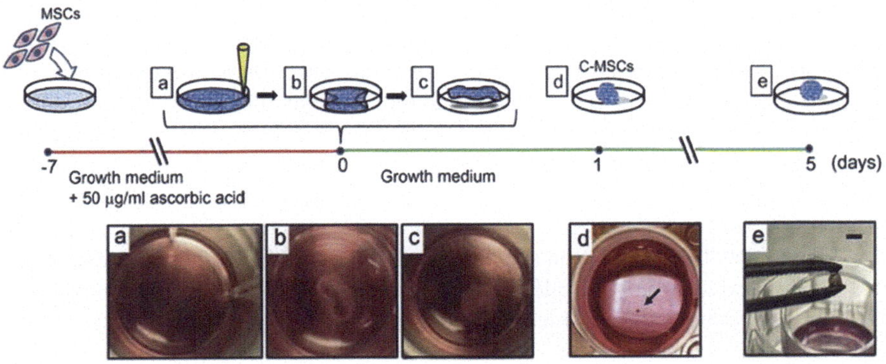

Fig. 4 Clumps of cell/extracellular matrix (ECM). Rat mesenchymal stem cells (MSCs)/ECM were generated in vitro. Rat MSCs were cultured in 50 µg/mL L-ascorbic acid for 7 days. (**a**) Using a micropipette tip to scratch confluent cells to tear off a cellular sheet (**b**), followed by detachment from the bottom of the plate (**c**). It rolled up to make clumps of cells/ECM after 24 h (**d**, arrow shows cell/ECM). (**e**) Maintaining cells/ECM in growth medium for 5 days. Reprinted from [158] with permissions from Elsevier

Clumps of Cells/ECM

Although synthetic scaffolds mimic some characteristics of the bone matrix, the effectiveness may vary due to biological incompatibility [156, 157]. Moreover, artificial scaffolds may bring about problems with host inflammatory and immunological reactions. Indeed, scaffold-free fabrication of MSCs and self-produced ECM for bone regeneration have been developed [118, 140, 158, 159] (Table 2) (Fig. 4). Clumps of rat BMSCs/ECM transplantation resulted in successful bone regeneration in a rat calvarial defect model [158]. However, xenotransplantation of clumps of human BMSCs/ECM into immunocompetent mice failed to induce bone formation owing to T cell infiltration [118]. IFN-γ-treated human BMSCs could eliminate these undesirable immune responses and result in successful bone formation [118]. Additionally, owing to type I collagen, cryopreserved BMSCs/ECM transplantation also caused successful bone regeneration, indicating a great advantage for clumps of MSCs/ECM in clinical applications [159]. Bioinspired from endochondral ossification process, living hyaline cartilaginous graft (LhCG), a 3D porous macroscopic construct, has been developed for enhancing bone formation in vitro and in vivo [138].

Extracellular Matrix Sheet

Complications from cell-derived ECM hybrid scaffolds, such as asynchronous scaffold absorption and not enough bone formation, and the production of harmful degradation products, lead to development of cellular sheets consisting of MSCs and

ECM [123, 160–166] (Table 2). Compared to hybrid scaffolds, a scaffold-free cell sheet has the advantage of more closely mimicking the natural growth microenvironment. Cell sheet engineering is to harvest confluent cells as contiguous, tissue-like cell sheets with intact cell junctions and their deposited ECM [167]. Cell sheets can be prepared using temperature-responsive culture dishes [168] or lifted using a scraper [161–164] (Fig. 5A) in bone regeneration. Scaffold-free MSC cell sheet transplantation could enhance bone formation subcutaneously in mice [164] or rats [163] or in a rat nonunion model [162]. Furthermore, injection of a scaffold-free cell sheet led to new bone formation subcutaneously in rats [161]. Combined with HAp or tubular coral scaffolds, a cell/ECM sheet showed the capacity to enhance bone formation subcutaneously in mice [164] or rats [160]. The periosteum plays a vital role in bone tissue engineering and synthetic tissue-engineered periostea utilizing cell sheet engineering have been developed [123, 165]. A human BMSCs/fibroblasts-derived ECM sheet could be a candidate for engineered periosteum for bone regeneration [123]. Moreover, a vascularized biomimetic periosteum consisting of premineralized and prevascularized MSC sheets was wrapped with β-TCP to improve vascularization and osteogenesis [165] (Fig. 5B).

Engineering of ECM-Based Materials

Engineering of ECM-based materials combines the advantages of control and precision of synthetic material manufacturing and the biocompatible and bioactive properties of an ECM. In this part, we discuss different engineering strategies, such as electrospinning, 3D printing, and hydrogels in the design of ECM-based materials for bone tissue engineering (Table 3).

Electrospinning

Electrospinning, a notable fiber-fabrication technique, is capable of producing nanofibers with fine flexibility and robust mechanical properties [169, 170]. These fibrous materials produced by electrospinning are similar to the hierarchical organized micro/nano scale fibrous networks in native bone ECM [171, 172]. An ECM-based solution [173] or synthetic polymers (e.g., poly(L-lactic acid) (PLLA) [131], gelatin [130], PCL [126, 174–177], PLGA [126]) can be forced through an electrostatic field to draw fibers. However, materials generated by electrospinning have limitations, such as poor cell growth, and differentiation, inadequate mechanical strength, and potential cytotoxicity [171]. A promising way for supporting cell viability is a combination of bioactive factors, such as ECM, with electrospinning technologies. A decellularized ECM is unstable and weak, which could not be transplanted. To overcome this drawback, electrospun nanofiber carriers were utilized for enhancing desirable mechanical properties [126]. Different approaches have been utilized to

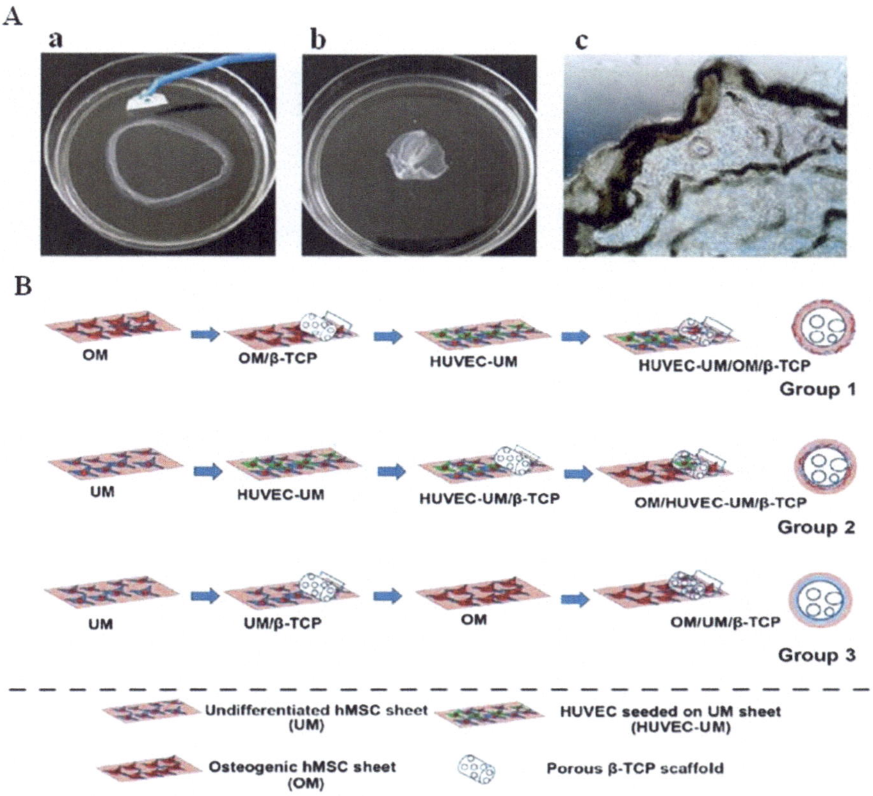

Fig. 5 Extracellular matrix sheet. (**A**) Macroscopic (**A**(a) and (b)) and microscopic (**A**(c)) appearance of a sheet from rat bone marrow mesenchymal stem cells (BMSCs). (**B**) Schematic figures showing the steps for preparing cell sheet/β-tricalcium phosphate (TCP) composite grafts (OM/UM/β-TCP, OM/HUVEC-UM/β-TCP, and HUVEC-UM/OM/β-TCP). Reprinted from [163, 165] with permissions from John Wiley and Sons and American Chemical Society

generate an ECM/electrospun scaffold for bone tissue engineering. Strategies for most of the studies in bone regeneration to anchor an ECM in electrospun nanofibers include first seeding cells onto the fibers, followed by decellularization to get ECM-decorated electrospun fibers [126, 130, 131, 175–177] (Fig. 6A). A different method has also been developed, in which an ECM was first produced, then lyophilized to get ECM powder, followed by being added to a polymer solution and electrospun to produce a hybrid electrospun nanofiber with an ECM [173, 174] (Fig. 6B). ECM-decorated electrospun nanofiber from either approach improved bone differentiation and formation in vitro and in vivo (Table 3).

ECM fibers and polymer nanofibers both play critical roles in the tissue regeneration process. Therefore, biofunctional nanofibers can be directly produced from natural ECM materials or generated from the mixture of synthetic polymers and a

Table 3 Fabrication methods of ECM-based materials in bone regeneration

ECM type	Seeded cells	Scaffold formulation or other conditions	In vitro outcome	In vivo outcome	Reference(s)
Electrospinning					
Human BMSCs and HUVECs	Human BMSCs	Lyophilized ECM incorporated into PCL solution, followed by electrospinning	Promoting cell proliferation, increased ALP activity and osteogenic genes	–	[173]
Human embryonic stem cells and human HFKTs	–	Hybrid electrospun NFL with ECM (cells were seeded on electrospun NFL in osteogenic medium for 2 weeks, then decellularized)	–	Bone regeneration rat subcutaneous transplantation model	[126]
MC3T3-E1	MC3T3-E1	Electrospun PCL fiber meshes (cells were seeded on meshes in proliferation medium for 4 weeks, then decellularized), followed by plasma treatment	Increased cell viability, calcium mineralization	–	[177]
	Mouse BMSCs	Electrospun PLLA nanofibers (cells were seeded on nanofibers in osteogenic medium for 2 weeks, then decellularized)	Enhanced cell adhesion, support cell proliferation and promoting early stage osteogenic differentiation	–	[131]
Rat BMSCs	Rat BMSCs	Electrospun PCL fiber meshes (cells were seeded on meshes in osteogenic medium for 12 or 16 days in a flow perfusion bioreactor, then decellularized)	Demineralization and devitalization resulting in decreased ALP activity and calcium content	–	[175]
		Electrospun PCL fiber meshes (cells were seeded on meshes in osteogenic medium for 12 days, then decellularized)	Osteogenic differentiation maintained even in the absence of dexamethasone	–	[176]

Rat BMSCs, MC3T3-E1, and L929 fibroblasts	Rat BMSCs	Fibrous meshes containing PLA/gelatin were electrospun. Cells were seeded on these meshes for 7 days, then decellularized to get ECM/mesh complexes	All three ECM types support cell proliferation; BMSC sheet promoting osteogenic differentiation most	BMSC sheet having the potential to induce ectopic osteogenesis subcutaneously in mice	[130]
Porcine bone, cartilage, fat, liver, spleen, lung derived ECM	Human ADSCs	Mixture (lyophilized, filtered ECM particles and PCL solution) electrospun	Scaffolds containing bone, cartilage, and fat promoting osteogenesis while ones from spleen and lung reducing.	–	[174]
3D printing					
Calves trabecular bone	Human ADSCs	3D printing hybrid scaffolds (PCL)	More cell adhesion and osteogenic markers	Greater bone regeneration in critical-sized murine calvarial defects	[180, 181]
Human PDCs	Human PDCs	3D porous, cylindrical Ti scaffolds (cells were seeded on scaffold, then decellularized)	BMP2 upregulated and ALP downregulated in a perfusion bioreactor culture setup	Augmentation bone formation subcutaneously in nude mice	[182]
Human nasal inferior turbinate TMSCs	Human nasal inferior turbinate TMSCs	3D printing PCL/PLGA and PCL/PLGA/β-TCP scaffold (cells first seeded on scaffolds in osteogenic medium, then decellularized)	Upregulating osteoblastic genes and increasing calcium deposition	Greater bone formation in rat ectopic and orthotopic models	[125]
MC3T3-E1	MC3T3-E1	3D printing HAp scaffold (cells were seeded on scaffold in osteogenic medium for 15 days, then decellularized)	Enhanced expression of actin and vinculin	–	[133]
		3D printing Ti-6Al-4V scaffold (cells were seeded on scaffold in osteogenic medium for 15 days, then decellularized)	Higher cell adhesion, proliferation, and growth	–	[132]
		Biomimetic 3D porous cell blocks combining collagen-ECM-based bioinks and alginate	Higher osteogenic activities	–	[128]

(continued)

Table 3 (continued)

ECM type	Seeded cells	Scaffold formulation or other conditions	In vitro outcome	In vivo outcome	Reference(s)
Porcine demineralization and decellularization bone	MC3T3-E1	3D printing PCL/β-TCP scaffolds were immersed in bone ECM solution	Excellent cell seeding efficiency, proliferation, and early and late osteogenic differentiation capacity	Outstanding bone regeneration in rabbit calvarial defect model	[183]
Hydrogels					
Bovine DBM and decellularized matrix material produced from DBM	mPCs	Pepsin digestion and solubilization to form hydrogels	ECM hydrogels showing the most enhanced cell proliferation	–	[73]
Human DBM and porcine DCC	–	Colloidal gels of HAp, and ECM	–	HAp-decellularized cartilage showing better osteogenesis in critical-sized rat calvarial defects	[83]

Abbreviations: 3D three-dimensional, *ADSCs* adipose-derived stem cells, *ALP* alkaline phosphatase, *BMP* bone morphogenetic protein, *BMSCs* bone marrow mesenchymal stem cells, *DBM* demineralized bone matrices, *DCC* decellularized cartilage, *ECM* extracellular matrix, *HAp* hydroxyapatite, *HFKTs* hair follicle keratinocytes, *HUVECs* human umbilical vein endothelial cells, *mPCs* mouse primary calvarial cells, *NFL* nanofiber layer, *PCL* polycaprolactone, *PDCs* periosteum-derived osteoprogenitor cells, *PLA* poly(L-lactide), *PLLA* poly (L-lactic acid), *PLGA* poly(lactic-*co*-glycolic acid), *TCP* tricalcium phosphate, *Ti* titanium, *Ti-6Al-4V* titanium alloy

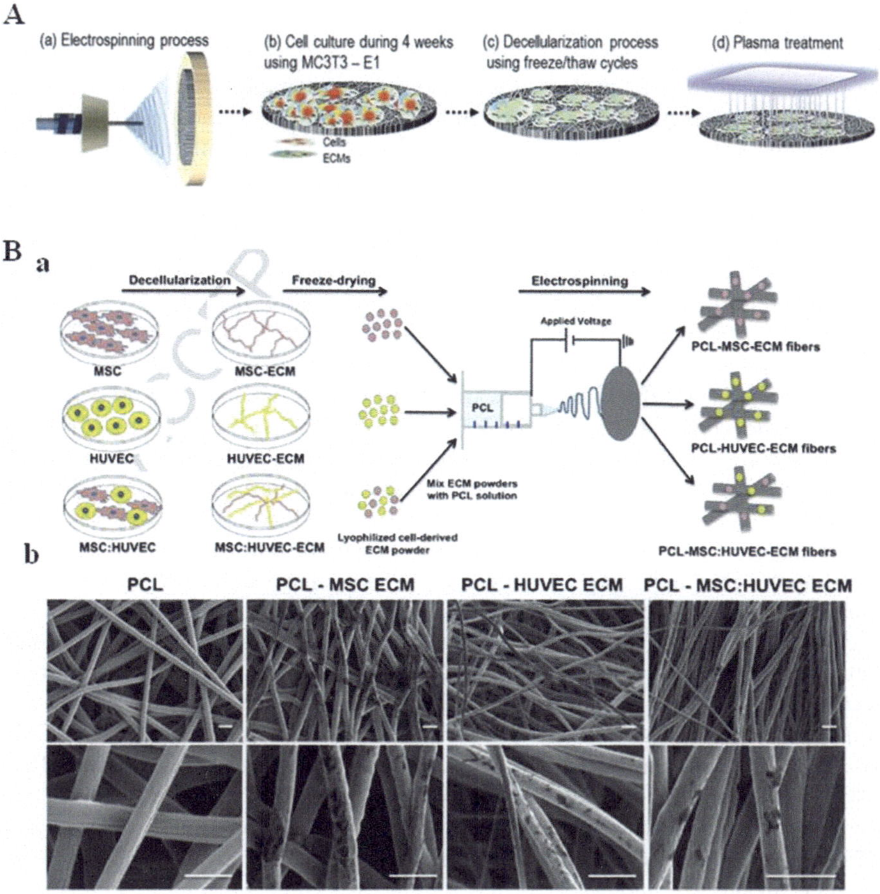

Fig. 6 Extracellular matrix (ECM)-based materials fabricating by electrospinning. (**A**) Schematic images showing the process of developing a decellularized ECM (dECM)/polycaprolactone (PCL) fibrous structure. (**A**(a)) Electrospinning and (**A**(b)) culturing of MC3T3-E1 cells on the micro/nanofibers for 4 weeks. (**A**(c)) Decellularization by freeze/thaw cycles to obtain a dECM and (**A**(d)) plasma treatment of the dECM/PCL fibrous structure. (**B**(a)) Schematic images showing the steps of fabrication of ECM microfibrous scaffolds. Lyophilized MSC, HUVEC, MSC:HUVEC derived ECM were mixed with a PCL solution respectively to make the corresponding scaffolds. (**B**(b)) Lower and higher magnification of PCL and different ECM PCL electrospun scaffolds. Scale bar = 5 μm. HUVEC: human umbilical vein endothelial cell. Reprinted from [173, 177] with permissions from Royal Society of Chemistry and Elsevier

natural ECM to form copolymer fibers. In this aspect, ECM-based materials fabricated by electrospinning technology would have enhanced physical properties owing to the synthetic materials and improved bioactivity due to the ECM component, which are promising for bone tissue regeneration.

3D Printing

3D printing utilizes patient data obtained from CT (computerized axial tomography) or MRI (magnetic resonance imaging) to fabricate an identical scaffold which matches the defects [178]. Building bone tissue via 3D printing technology is becoming increasingly popular due to its flexibility and versatility. Importantly, utilization of an ECM improves the biocompatibility of 3D printed scaffolds. Hard bone-like materials like PCL, ceramics (HAp/β-TCP), and bioglass-ceramic composites have been well utilized in bone tissue engineering [179]. To generate physiologically relevant bone, a dECM was combined in 3D printing [125, 133, 180–182]. Nyberg et al. fabricated 3D printed biocomposites of PCL and DBM through an extrusion system and showed that this scaffold has improved osteogenesis in vitro and in vivo [181]. Different methods for fabricating ECM-based 3D printed scaffolds were developed. Pati et al. first 3D-printed PL/PLGA and PCL/PLGA/β-TCP scaffolds, followed by human nasal inferior turbinate tissue-derived mesenchymal stromal cells (TMSCs) seeded on scaffolds and decellularized by the method of freeze/thaw cycles [125] (Fig. 7A). Kim et al. developed 3D-printed scaffolds immersed in a bone dECM solution, followed by lyophilization [183] (Fig. 7B). The above approaches may face a significant challenge that ECM composition in 3D geometrical scaffolds is not uniform [184].

3D bioprinting (the use of biological materials, such as cells, their extracellular matrix, and growth factors) represents a formidable technology in bone tissue engineering. The common bioprinting techniques are extrusion-based, laser-assisted, inkjet, and stereolithography [185, 186]. 3D bioprinting combined with a dECM-based bioink has been used to develop cartilage, heart, and adipose tissue [187, 188]. With regard to bone, Lee et al. developed a highly bioactive collagen/ECM/alginate-based bioinks construct showing enhanced osteogenic activities in vitro [128]. Bioprinted materials have weak mechanical properties and further efforts should be made to develop an ECM-based bioink with higher mechanical properties for bone regeneration.

Hydrogels

More recently, in addition to the direct usage of decellularized matrix scaffolds for clinical applications, they have also been reconstituted into an injectable hydrogels. Hydrogels, prepared from a natural ECM, can be applied alone or combined with other fabrication methods, such as electrospinning and 3D printing. When used alone, ECM hydrogels can be delivered via minimally invasive surgical techniques and theoretically fit to any 3D irregular spaces. Cell-derived ECM hydrogels, such as Matrigel, can support cell culture in a 3D microenvironment in vitro [189]. Sawkins et al. applied a stringent decellularization process to bovine DBM to obtain

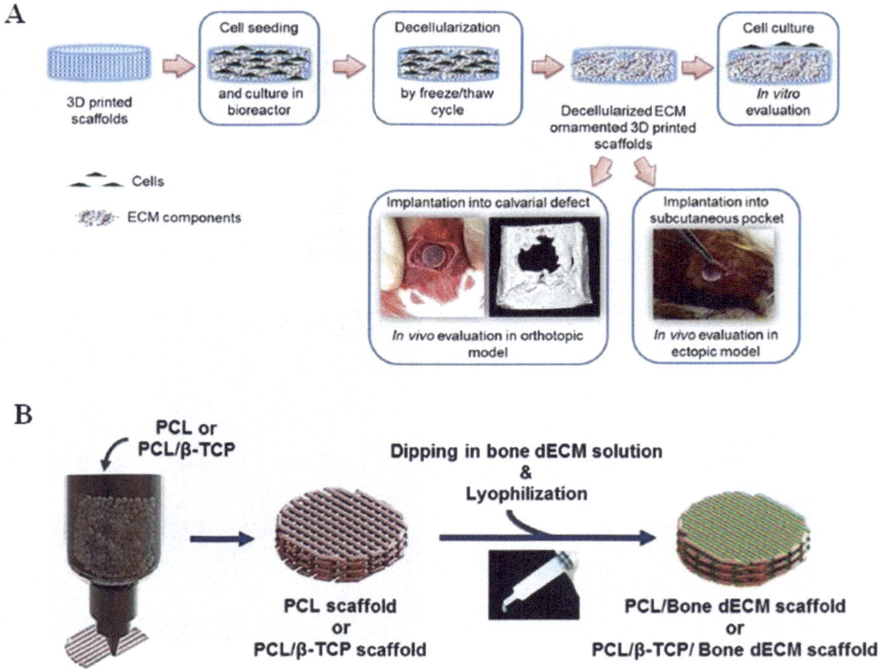

Fig. 7 Extracellular matrix (ECM)-based materials fabricated by three-dimensional (3D) printing. (**A**) Schematic figures showing the development of ECM-based 3D printed scaffolds. Human TMSCs were cultured on 3D printed scaffolds in a bioreactor for 14 days, followed by decellularization from freeze/thaw cycles. Then these scaffolds were evaluated in vitro and in vivo. (**B**) Schematic figures showing procedures of ECM-based 3D printed scaffold preparation. Polycaprolactone (PCL) or PCL/β-tricalcium phosphate (TCP) scaffolds were immersed in a porcine bone decellularized ECM (dECM) solution, followed by lyophilization for 48 h. Reprinted from [125, 183] with permissions from Elsevier and John Wiley and Sons

a bovine decellularized matrix (bECM) and produced a hydrogel made up of DBM or bECM (Fig. 8A). Mouse primary calvarial cells on bECM hydrogels demonstrated enhanced proliferation [73]. Hydrogel forms of DBM or bECM reduced the need for carrier liquids [73]. Moreover, colloidal nanoparticles of HAp, DBM, and DCC were combined to form hydrogels with desirable rheological properties ($\tau y \geq 100$ Pa) [83]. Remarkably, the addition of DCC, not DBM, increased the bone regenerative potential in rat critical-sized calvarial defects [83]. However, glycosaminoglycan composite hydrogels containing bone ECM particles showed better osteogenic differentiation capacity compared to ones containing from ECM from other tissues, such as fat, liver, and cartilage (Fig. 8B) [105]. Due to a lack of mechanical properties, hydrogel forms of an ECM are limited to clinical applications for bone graft substitutes.

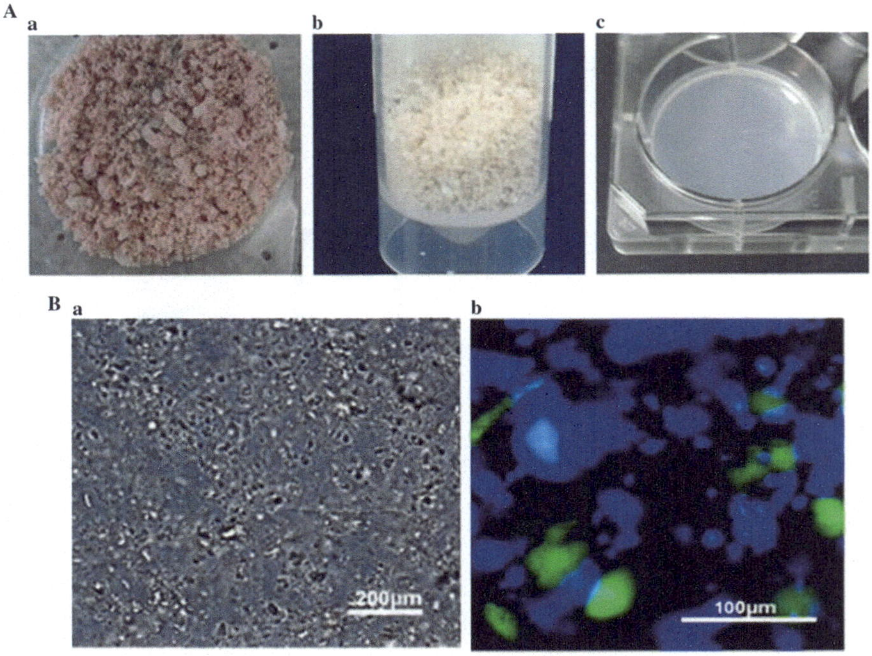

Fig. 8 Extracellular matrix (ECM)-based materials fabricated by hydrogels. (**A**(a)) Fragments from bovine tibiae; (**A**(b)) Decellularized bone obtained after mineral, lipid, and cell removal processes; (**A**(c)) An ECM hydrogel obtained after pepsin digestion and solubilization. Uniform porcine bone particle distribution was achieved. (**B**(a)) Without (bright field) and (**B**(b)) with encapsulated cells (fluorescence, blue indicating bone particles, and green indicating cells). Reprinted from [73, 105] with permissions from Elsevier and John Wiley and Sons

Conclusion and Perspective

Recently, much effort has been attempted to optimize and improve ECM-based materials. Bioactive molecules in the ECM drive tissue homeostasis and regeneration. The ECM functions not only as a blocking material but also as a regulator of cellular functions, such as cell attachment, proliferation, and differentiation. Furthermore, an appropriately fabricated ECM is biodegradable and does not give rise to adverse immune responses.

Tissue-derived and cell-derived ECM have their own advantages and disadvantages. For instance, decellularized tissue-derived ECM cannot mimic the development process of tissue while cell-formed decellularized matrices have the advantage of adaptability for different developmental stages [34]. Moreover, the availability of tissue-derived ECM is limited while a cell-derived ECM is not. However, ECMs from tissues mimic more the native ECM-like complicated architectures and composition than from cells.

To better generate ECM-based materials, a correct type of dECM should be selected. Moreover, although a novel decellularization method utilizing apoptosis has been developed, the process of decellularization should be optimized to improve the quantity, quality, and reproducibility of the ECM. Thanks to gene technology, customized engineering ECM can be realized by utilizing cells overexpressing specific components or growth factors [49]. These engineering ECM-based materials can adapt to different clinical needs, which have broad application prospects.

In general, ECM-based materials for bone regeneration are gaining clinical importance and market space, which deserve further investigation. Composition, microstructure, and biomechanical properties of the ECM should be considered when using them in tissue engineering. Development of methods to harness native regenerative capacities and enhance physicochemical properties of an ECM is key to providing clinical application for bone tissue regeneration.

Conflicts of Interest The authors declare that they have no competing financial interest.

References

1. Friedlaender GE et al (2001) Osteogenic protein-1 (bone morphogenetic protein-7) in the treatment of tibial nonunions. J Bone Joint Surg Am 83-A(Suppl 1):S151–S158
2. Bucholz RW (2002) Nonallograft osteoconductive bone graft substitutes. Clin Orthop Relat Res (395):44–52
3. Fernandez de Grado G et al (2018) Bone substitutes: a review of their characteristics, clinical use, and perspectives for large bone defects management. J Tissue Eng 9:2041731418776819. https://doi.org/10.1177/2041731418776819
4. Finkemeier CG (2002) Bone-grafting and bone-graft substitutes. J Bone Joint Surg Am 84-A:454–464
5. Van Heest A, Swiontkowski M (1999) Bone-graft substitutes. Lancet 353(Suppl 1):SI28–SI29
6. Kane R, Ma PX (2013) Mimicking the nanostructure of bone matrix to regenerate bone. Mater Today 16:418–423. https://doi.org/10.1016/j.mattod.2013.11.001
7. Pape HC, Evans A, Kobbe P (2010) Autologous bone graft: properties and techniques. J Orthop Trauma 24(Suppl 1):S36–S40. https://doi.org/10.1097/BOT.0b013e3181cec4a1
8. Zimmermann G, Moghaddam A (2011) Allograft bone matrix versus synthetic bone graft substitutes. Injury 42(Suppl 2):S16–S21. https://doi.org/10.1016/j.injury.2011.06.199
9. Habibovic P, de Groot K (2007) Osteoinductive biomaterials—properties and relevance in bone repair. J Tissue Eng Regen Med 1:25–32. https://doi.org/10.1002/term.5
10. Langer R, Vacanti JP (1993) Tissue engineering. Science 260:920–926
11. Griffith LG, Naughton G (2002) Tissue engineering—current challenges and expanding opportunities. Science 295:1009–1014. https://doi.org/10.1126/science.1069210
12. Yuan H et al (2010) Osteoinductive ceramics as a synthetic alternative to autologous bone grafting. Proc Natl Acad Sci U S A 107:13614–13619. https://doi.org/10.1073/pnas.1003600107
13. Parikh SN (2002) Bone graft substitutes in modern orthopedics. Orthopedics 25:1301–1309. ; quiz 1301–1310
14. Chan G, Mooney DJ (2008) New materials for tissue engineering: towards greater control over the biological response. Trends Biotechnol 26:382–392. https://doi.org/10.1016/j.tibtech.2008.03.011
15. Shekaran A, Garcia AJ (2011) Extracellular matrix-mimetic adhesive biomaterials for bone repair. J Biomed Mater Res A 96A:261–272. https://doi.org/10.1002/jbm.a.32979

16. Lind M et al (1996) Transforming growth factor-beta 1 stimulates bone ongrowth to weight-loaded tricalcium phosphate coated implants: an experimental study in dogs. J Bone Joint Surg Br 78:377–382
17. Inzana JA et al (2014) 3D printing of composite calcium phosphate and collagen scaffolds for bone regeneration. Biomaterials 35:4026–4034. https://doi.org/10.1016/j.biomaterials.2014.01.064
18. Hussey GS, Dziki JL, Badylak SF (2018) Extracellular matrix-based materials for regenerative medicine. Nat Rev Mater 3:159–173. https://doi.org/10.1038/s41578-018-0023-x
19. Hynes RO (2009) The extracellular matrix: not just pretty fibrils. Science 326:1216–1219. https://doi.org/10.1126/science.1176009
20. da Anunciacao A et al (2017) Extracellular matrix in epitheliochorial, endotheliochorial and haemochorial placentation and its potential application for regenerative medicine. Reprod Domest Anim 52:3–15. https://doi.org/10.1111/rda.12868
21. Ott HC et al (2010) Regeneration and orthotopic transplantation of a bioartificial lung. Nat Med 16:927–933. https://doi.org/10.1038/nm.2193
22. D'Onofrio A et al (2011) Clinical and hemodynamic outcomes after aortic valve replacement with stented and stentless pericardial xenografts: a propensity-matched analysis. J Heart Valve Dis 20:319–326
23. Macchiarini P et al (2008) Clinical transplantation of a tissue-engineered airway. Lancet 372:2023–2030. https://doi.org/10.1016/S0140-6736(08)61598-6
24. Cheng CW, Solorio LD, Alsberg E (2014) Decellularized tissue and cell-derived extracellular matrices as scaffolds for orthopaedic tissue engineering. Biotechnol Adv 32:462–484. https://doi.org/10.1016/j.biotechadv.2013.12.012
25. Clarke B (2008) Normal bone anatomy and physiology. Clin J Am Soc Nephrol 3(Suppl 3):S131–S139. https://doi.org/10.2215/CJN.04151206
26. Boskey AL (2013) Bone composition: relationship to bone fragility and antiosteoporotic drug effects. Bonekey Rep 2:447. https://doi.org/10.1038/bonekey.2013.181
27. Wang Y et al (2012) The predominant role of collagen in the nucleation, growth, structure and orientation of bone apatite. Nat Mater 11:724–733. https://doi.org/10.1038/nmat3362
28. Nudelman F et al (2010) The role of collagen in bone apatite formation in the presence of hydroxyapatite nucleation inhibitors. Nat Mater 9:1004–1009. https://doi.org/10.1038/nmat2875
29. Barradas AM, Yuan H, van Blitterswijk CA, Habibovic P (2011) Osteoinductive biomaterials: current knowledge of properties, experimental models and biological mechanisms. Eur Cell Mater 21:407–429. ; discussion 429
30. Legate KR, Wickstrom SA, Fassler R (2009) Genetic and cell biological analysis of integrin outside-in signaling. Genes Dev 23:397–418. https://doi.org/10.1101/gad.1758709
31. Sasano Y et al (2000) Immunohistochemical localization of type I collagen, fibronectin and tenascin C during embryonic osteogenesis in the dentary of mandibles and tibias in rats. Histochem J 32:591–598
32. Kamiya N, Shigemasa K, Takagi M (2001) Gene expression and immunohistochemical localization of decorin and biglycan in association with early bone formation in the developing mandible. J Oral Sci 43:179–188
33. Nakamura M et al (2005) Expression of versican and ADAMTS1, 4, and 5 during bone development in the rat mandible and hind limb. J Histochem Cytochem 53:1553–1562. https://doi.org/10.1369/jhc.5A6669.2005
34. Hoshiba T, Kawazoe N, Tateishi T, Chen G (2009) Development of stepwise osteogenesis-mimicking matrices for the regulation of mesenchymal stem cell functions. J Biol Chem 284:31164–31173. https://doi.org/10.1074/jbc.M109.054676
35. Papadimitropoulos A, Scotti C, Bourgine P, Scherberich A, Martin I (2015) Engineered decellularized matrices to instruct bone regeneration processes. Bone 70:66–72. https://doi.org/10.1016/j.bone.2014.09.007

36. Al-Abedalla K et al (2015) Bone augmented with allograft onlays for implant placement could be comparable with native bone. J Oral Maxillofac Surg 73:2108–2122. https://doi.org/10.1016/j.joms.2015.06.151
37. Gilbert TW, Sellaro TL, Badylak SF (2006) Decellularization of tissues and organs. Biomaterials 27:3675–3683. https://doi.org/10.1016/j.biomaterials.2006.02.014
38. Hoshiba T, Lu H, Kawazoe N, Chen G (2010) Decellularized matrices for tissue engineering. Expert Opin Biol Ther 10:1717–1728. https://doi.org/10.1517/14712598.2010.534079
39. Keane TJ, Swinehart IT, Badylak SF (2015) Methods of tissue decellularization used for preparation of biologic scaffolds and in vivo relevance. Methods 84:25–34. https://doi.org/10.1016/j.ymeth.2015.03.005
40. Hashimoto Y et al (2011) The effect of decellularized bone/bone marrow produced by high-hydrostatic pressurization on the osteogenic differentiation of mesenchymal stem cells. Biomaterials 32:7060–7067. https://doi.org/10.1016/j.biomaterials.2011.06.008
41. Kabirian F, Mozafari M (2019) Decellularized ECM-derived bioinks: prospects for the future. Methods. https://doi.org/10.1016/j.ymeth.2019.04.019
42. Seddon AM, Curnow P, Booth PJ (2004) Membrane proteins, lipids and detergents: not just a soap opera. Biochim Biophys Acta 1666:105–117. https://doi.org/10.1016/j.bbamem.2004.04.011
43. Hwang J et al (2017) Molecular assessment of collagen denaturation in decellularized tissues using a collagen hybridizing peptide. Acta Biomater 53:268–278. https://doi.org/10.1016/j.actbio.2017.01.079
44. Fitzpatrick JC, Clark PM, Capaldi FM (2010) Effect of decellularization protocol on the mechanical behavior of porcine descending aorta. Int J Biomater 2010:620503. https://doi.org/10.1155/2010/620503
45. He M, Callanan A (2013) Comparison of methods for whole-organ decellularization in tissue engineering of bioartificial organs. Tissue Eng B Rev 19:194–208. https://doi.org/10.1089/ten.TEB.2012.0340
46. Petersen TH et al (2010) Tissue-engineered lungs for in vivo implantation. Science 329:538–541. https://doi.org/10.1126/science.1189345
47. Woods T, Gratzer PF (2005) Effectiveness of three extraction techniques in the development of a decellularized bone-anterior cruciate ligament-bone graft. Biomaterials 26:7339–7349. https://doi.org/10.1016/j.biomaterials.2005.05.066
48. Boer U et al (2011) The effect of detergent-based decellularization procedures on cellular proteins and immunogenicity in equine carotid artery grafts. Biomaterials 32:9730–9737. https://doi.org/10.1016/j.biomaterials.2011.09.015
49. Bourgine PE, Pippenger BE, Todorov A Jr, Tchang L, Martin I (2013) Tissue decellularization by activation of programmed cell death. Biomaterials 34:6099–6108. https://doi.org/10.1016/j.biomaterials.2013.04.058
50. Bourgine PE et al (2014) Osteoinductivity of engineered cartilaginous templates devitalized by inducible apoptosis. Proc Natl Acad Sci U. S. A 111:17426–17431. https://doi.org/10.1073/pnas.1411975111
51. van Engeland M, Kuijpers HJ, Ramaekers FC, Reutelingsperger CP, Schutte B (1997) Plasma membrane alterations and cytoskeletal changes in apoptosis. Exp Cell Res 235:421–430. https://doi.org/10.1006/excr.1997.3738
52. Raff M (1998) Cell suicide for beginners. Nature 396:119–122. https://doi.org/10.1038/24055
53. Bruyneel AAN, Carr CA (2017) Ambiguity in the presentation of decellularized tissue composition: the need for standardized approaches. Artif Organs 41:778–784. https://doi.org/10.1111/aor.12838
54. Chen XD, Dusevich V, Feng JQ, Manolagas SC, Jilka RL (2007) Extracellular matrix made by bone marrow cells facilitates expansion of marrow-derived mesenchymal progenitor cells and prevents their differentiation into osteoblasts. J Bone Miner Res Off 22:1943–1956. https://doi.org/10.1359/jbmr.070725
55. Laurencin CT, Khan Y (2012) Regenerative engineering. Sci Transl Med 4:160–169. https://doi.org/10.1126/scitranslmed.3004467

56. Esses SI, Halloran PF (1983) Donor marrow-derived cells as immunogens and targets for the immune response to bone and skin allografts. Transplantation 35:169–174
57. Campana V et al (2014) Bone substitutes in orthopaedic surgery: from basic science to clinical practice. J Mater Sci Mater Med 25:2445–2461. https://doi.org/10.1007/s10856-014-5240-2
58. Rosenberg E, Rose LF (1998) Biologic and clinical considerations for autografts and allografts in periodontal regeneration therapy. Dent Clin N Am 42:467–490
59. Freiberg RA, Ray RD (1964) Studies of devitalized bone implants. Arch Surg 89:417–427
60. Frohlich M et al (2010) Bone grafts engineered from human adipose-derived stem cells in perfusion bioreactor culture. Tissue Eng PartA 16:179–189. https://doi.org/10.1089/ten.TEA.2009.0164
61. Marcos-Campos I et al (2012) Bone scaffold architecture modulates the development of mineralized bone matrix by human embryonic stem cells. Biomaterials 33:8329–8342. https://doi.org/10.1016/j.biomaterials.2012.08.013
62. Hesse E et al (2010) Repair of a segmental long bone defect in human by implantation of a novel multiple disc graft. Bone 46:1457–1463. https://doi.org/10.1016/j.bone.2010.02.011
63. Drosos GI, Kazakos KI, Kouzoumpasis P, Verettas DA (2007) Safety and efficacy of commercially available demineralised bone matrix preparations: a critical review of clinical studies. Injury 38(Suppl 4):S13–S21
64. Schwartz Z et al (1998) Ability of commercial demineralized freeze-dried bone allograft to induce new bone formation is dependent on donor age but not gender. J Periodontol 69:470–478. https://doi.org/10.1902/jop.1998.69.4.470
65. Munting E, Wilmart JF, Wijne A, Hennebert P, Delloye C (1988) Effect of sterilization on osteoinduction. Comparison of five methods in demineralized rat bone. Acta Orthop Scand 59:34–38
66. Iwata H, Sakano S, Itoh T, Bauer TW (2002) Demineralized bone matrix and native bone morphogenetic protein in orthopaedic surgery. Clin Orthop Relat Res (395):99–109
67. Chen L et al (2010) Loading of VEGF to the heparin cross-linked demineralized bone matrix improves vascularization of the scaffold. J Mater Sci Mater Med 21:309–317. https://doi.org/10.1007/s10856-009-3827-9
68. Lee JH et al (2011) Combined effects of porous hydroxyapatite and demineralized bone matrix on bone induction: in vitro and in vivo study using a nude rat model. Biomed Mater 6:015008. https://doi.org/10.1088/1748-6041/6/1/015008
69. Jayasuriya AC, Ebraheim NA (2009) Evaluation of bone matrix and demineralized bone matrix incorporated PLGA matrices for bone repair. J Mater Sci Mater Med 20:1637–1644. https://doi.org/10.1007/s10856-009-3738-9
70. Park BW et al (2012) In vitro and in vivo osteogenesis of human mesenchymal stem cells derived from skin, bone marrow and dental follicle tissues. Differentiation 83:249–259. https://doi.org/10.1016/j.diff.2012.02.008
71. Liu G et al (2010) In vitro and in vivo evaluation of osteogenesis of human umbilical cord blood-derived mesenchymal stem cells on partially demineralized bone matrix. Tissue Eng Part A 16:971–982. https://doi.org/10.1089/ten.TEA.2009.0516
72. Kang EJ et al (2010) In vitro and in vivo osteogenesis of porcine skin-derived mesenchymal stem cell-like cells with a demineralized bone and fibrin glue scaffold. Tissue Eng Part A 16:815–827. https://doi.org/10.1089/ten.TEA.2009.0439
73. Sawkins MJ et al (2013) Hydrogels derived from demineralized and decellularized bone extracellular matrix. Acta Biomater 9:7865–7873. https://doi.org/10.1016/j.actbio.2013.04.029
74. Supronowicz P et al (2011) Human adipose-derived side population stem cells cultured on demineralized bone matrix for bone tissue engineering. Tissue Eng Part A 17:789–798. https://doi.org/10.1089/ten.tea.2010.0357
75. Kurkalli BG, Gurevitch O, Sosnik A, Cohn D, Slavin S (2010) Repair of bone defect using bone marrow cells and demineralized bone matrix supplemented with polymeric materials. Curr Stem Cell Res Ther 5:49–56

76. Brighton CT, Sugioka Y, Hunt RM (1973) Cytoplasmic structures of epiphyseal plate chondrocytes. Quantitative evaluation using electron micrographs of rat costochondral junctions with special reference to the fate of hypertrophic cells. J Bone Joint Surg Am 55:771–784
77. Brighton CT, Hunt RM (1986) Histochemical localization of calcium in the fracture callus with potassium pyroantimonate. Possible role of chondrocyte mitochondrial calcium in callus calcification. J Bone Joint Surg Am 68:703–715
78. Gerber HP et al (1999) VEGF couples hypertrophic cartilage remodeling, ossification and angiogenesis during endochondral bone formation. Nat Med 5:623–628. https://doi.org/10.1038/9467
79. Kronenberg HM (2003) Developmental regulation of the growth plate. Nature 423:332–336. https://doi.org/10.1038/nature01657
80. Krych AJ, Harnly HW, Rodeo SA, Williams RJ 3rd (2012) Activity levels are higher after osteochondral autograft transfer mosaicplasty than after microfracture for articular cartilage defects of the knee: a retrospective comparative study. J Bone Joint Surg Am 94:971–978. https://doi.org/10.2106/JBJS.K.00815
81. Pallante AL et al (2012) Treatment of articular cartilage defects in the goat with frozen versus fresh osteochondral allografts: effects on cartilage stiffness, zonal composition, and structure at six months. J Bone Joint Surg Am 94:1984–1995. https://doi.org/10.2106/JBJS.K.00439
82. Pallante AL et al (2012) The in vivo performance of osteochondral allografts in the goat is diminished with extended storage and decreased cartilage cellularity. Am J Sports Med 40:1814–1823. https://doi.org/10.1177/0363546512449321
83. Townsend JM et al (2017) Colloidal gels with extracellular matrix particles and growth factors for bone regeneration in critical size rat calvarial defects. AAPS J 19:703–711. https://doi.org/10.1208/s12248-017-0045-0
84. Gupta V et al (2017) Microsphere-based osteochondral scaffolds carrying opposing gradients of decellularized cartilage and demineralized bone matrix. ACS Biomater Sci Eng 3:1955–1963. https://doi.org/10.1021/acsbiomaterials.6b00071
85. Gawlitta D et al (2015) Decellularized cartilage-derived matrix as substrate for endochondral bone regeneration. Tissue Eng Part A 21:694–703. https://doi.org/10.1089/ten.TEA.2014.0117
86. Cunniffe GM et al (2017) Growth plate extracellular matrix-derived scaffolds for large bone defect healing. Eur Cell Mater 33:130–142. https://doi.org/10.22203/eCM.v033a10
87. Badylak SF, Freytes DO, Gilbert TW (2009) Extracellular matrix as a biological scaffold material: structure and function. Acta Biomater 5:1–13. https://doi.org/10.1016/j.actbio.2008.09.013
88. Kim KS et al (2010) Small intestine submucosa sponge for in vivo support of tissue-engineered bone formation in the presence of rat bone marrow stem cells. Biomaterials 31:1104–1113. https://doi.org/10.1016/j.biomaterials.2009.10.020
89. Li M, Zhang C, Cheng M, Gu Q, Zhao J (2017) Small intestinal submucosa: a potential osteoconductive and osteoinductive biomaterial for bone tissue engineering. Mater Sci Eng C Mater Biol Appl 75:149–156. https://doi.org/10.1016/j.msec.2017.02.042
90. Sun T et al (2018) Composite scaffolds of mineralized natural extracellular matrix on true bone ceramic induce bone regeneration through Smad1/5/8 and ERK1/2 pathways. Tissue Eng Part A 24:502–515. https://doi.org/10.1089/ten.TEA.2017.0179
91. Sun TF et al (2018) Guided osteoporotic bone regeneration with composite scaffolds of mineralized ECM/heparin membrane loaded with BMP2-related peptide. Int J Nanomed 13:791–804. https://doi.org/10.2147/Ijn.S152698
92. Zhang C, Li M, Zhu J, Luo F, Zhao J (2017) Enhanced bone repair induced by human adipose-derived stem cells on osteogenic extracellular matrix ornamented small intestinal submucosa. Regen Med 12:541–552. https://doi.org/10.2217/rme-2017-0024
93. Li M, Zhang C, Mao YX, Zhong Y, Zhao JY (2018) A cell-engineered small intestinal submucosa-based bone mimetic construct for bone regeneration. Tissue Eng Part A 24:1099–1111. https://doi.org/10.1089/ten.tea.2017.0407

94. Moore DC, Pedrozo HA, Crisco JJ 3rd, Ehrlich MG (2004) Preformed grafts of porcine small intestine submucosa (SIS) for bridging segmental bone defects. J Biomed Mater Res A 69:259–266. https://doi.org/10.1002/jbm.a.20123
95. Zhao L, Zhao J, Wang S, Wang J, Liu J (2011) Comparative study between tissue-engineered periosteum and structural allograft in rabbit critical-sized radial defect model. J Biomed Mater Res B Appl Biomater 97:1–9. https://doi.org/10.1002/jbm.b.31768
96. Roberts SJ, van Gastel N, Carmeliet G, Luyten FP (2015) Uncovering the periosteum for skeletal regeneration: the stem cell that lies beneath. Bone 70:10–18. https://doi.org/10.1016/j.bone.2014.08.007
97. Zhang X et al (2005) Periosteal progenitor cell fate in segmental cortical bone graft transplantations: implications for functional tissue engineering. J Bone Miner Research 20:2124–2137. https://doi.org/10.1359/JBMR.050806
98. Lin XF et al (2018) Periosteum extracellular-matrix-mediated acellular mineralization during bone formation. Adv Healthc Mater 7:1700660. https://doi.org/10.1002/Adhm.201700660
99. Wang XY et al (2017) Preparation and characterization of a chitosan/gelatin/extracellular matrix scaffold and its application in tissue engineering. Tissue Eng Part C Methods 23:169–179. https://doi.org/10.1089/ten.tec.2016.0511
100. Schonmeyr B, Clavin N, Avraham T, Longo V, Mehrara BJ (2009) Synthesis of a tissue-engineered periosteum with acellular dermal matrix and cultured mesenchymal stem cells. Tissue Eng Part A 15:1833–1841. https://doi.org/10.1089/ten.tea.2008.0446
101. Kim HJ et al (2012) Effect of acellular dermal matrix as a delivery carrier of adipose-derived mesenchymal stem cells on bone regeneration. J Biomed Mater Res Part B Appl Biomater 100:1645–1653. https://doi.org/10.1002/jbm.b.32733
102. Hwang JW, Kim S, Kim SW, Lee JH (2016) Effect of extracellular matrix membrane on bone formation in a rabbit tibial defect model. Biomed Res Int 2016:6715295. https://doi.org/10.1155/2016/6715295
103. Li W et al (2015) Investigating the potential of amnion-based scaffolds as a barrier membrane for guided bone regeneration. Langmuir 31:8642–8653. https://doi.org/10.1021/acs.langmuir.5b02362
104. Penolazzi L et al (2012) Human mesenchymal stem cells seeded on extracellular matrix-scaffold: viability and osteogenic potential. J Cell Physiol 227:857–866. https://doi.org/10.1002/jcp.22983
105. Beachley V et al (2018) Extracellular matrix particle-glycosaminoglycan composite hydrogels for regenerative medicine applications. J Biomed Mater Res A 106:147–159. https://doi.org/10.1002/jbm.a.36218
106. Wainwright D et al (1996) Clinical evaluation of an acellular allograft dermal matrix in full-thickness burns. J Burn Care Rehabil 17:124–136
107. Badylak SF (2007) The extracellular matrix as a biologic scaffold material. Biomaterials 28:3587–3593. https://doi.org/10.1016/j.biomaterials.2007.04.043
108. Trelford JD, Trelford-Sauder M (1979) The amnion in surgery, past and present. Am J Obstet Gynecol 134:833–845
109. Wei W et al (2017) In vitro osteogenic induction of bone marrow mesenchymal stem cells with a decellularized matrix derived from human adipose stem cells and in vivo implantation for bone regeneration. J Mater Chem B 5:2468–2482. https://doi.org/10.1039/c6tb03150a
110. Ravindran S et al (2012) Biomimetic extracellular matrix-incorporated scaffold induces osteogenic gene expression in human marrow stromal cells. Tissue Eng Part A 18:295–309. https://doi.org/10.1089/ten.TEA.2011.0136
111. Lai Y et al (2010) Reconstitution of marrow-derived extracellular matrix ex vivo: a robust culture system for expanding large-scale highly functional human mesenchymal stem cells. Stem Cells Dev 19:1095–1107. https://doi.org/10.1089/scd.2009.0217
112. Zhang Z et al (2015) Bone marrow stromal cell-derived extracellular matrix promotes osteogenesis of adipose-derived stem cells. Cell Biol Int 39:291–299. https://doi.org/10.1002/cbin.10385

113. Zeitouni S et al (2012) Human mesenchymal stem cell-derived matrices for enhanced osteo-regeneration. Sci Transl Med 4:132–155. https://doi.org/10.1126/scitranslmed.3003396
114. Antebi B et al (2015) Stromal-cell-derived extracellular matrix promotes the proliferation and retains the osteogenic differentiation capacity of mesenchymal stem cells on three-dimensional scaffolds. Tissue Eng Part C Methods 21:171–181. https://doi.org/10.1089/ten.TEC.2014.0092
115. Sadr N et al (2012) Enhancing the biological performance of synthetic polymeric materials by decoration with engineered, decellularized extracellular matrix. Biomaterials 33:5085–5093. https://doi.org/10.1016/j.biomaterials.2012.03.082
116. Baroncelli M et al (2018) Human osteoblast-derived extracellular matrix with high homology to bone proteome is osteopromotive. Tissue Eng Part A 24:1377–1389. https://doi.org/10.1089/ten.tea.2017.0448
117. Cunniffe GM et al (2015) Porous decellularized tissue engineered hypertrophic cartilage as a scaffold for large bone defect healing. Acta Biomater 23:82–90. https://doi.org/10.1016/j.actbio.2015.05.031
118. Takeshita K et al (2017) Xenotransplantation of interferon-gamma-pretreated clumps of a human mesenchymal stem cell/extracellular matrix complex induces mouse calvarial bone regeneration. Stem Cell Res Ther 8:101. https://doi.org/10.1186/S13287-017-0550-1
119. Clough BH et al (2015) Bone regeneration with osteogenically enhanced mesenchymal stem cells and their extracellular matrix proteins. J Bone Miner Res 30:83–94. https://doi.org/10.1002/jbmr.2320
120. Deutsch ER, Guldberg RE (2010) Stem cell-synthesized extracellular matrix for bone repair. J Mater Chem 20:8942–8951
121. Kang YQ, Kim S, Bishop J, Khademhosseini A, Yang YZ (2012) The osteogenic differentiation of human bone marrow MSCs on HUVEC-derived ECM and beta-TCP scaffold. Biomaterials 33:6998–7007. https://doi.org/10.1016/j.biomaterials.2012.06.061
122. Tour G, Wendel M, Tcacencu I (2013) Human fibroblast-derived extracellular matrix constructs for bone tissue engineering applications. J Biomed Mater Res A 101:2826–2837. https://doi.org/10.1002/jbm.a.34600
123. Xing Q, Qian Z, Kannan B, Tahtinen M, Zhao F (2015) Osteogenic differentiation evaluation of an engineered extracellular matrix based tissue sheet for potential periosteum replacement. ACS Appl Mater Interfaces 7:23239–23247. https://doi.org/10.1021/acsami.5b07386
124. Kim IG et al (2015) Bioactive cell-derived matrices combined with polymer mesh scaffold for osteogenesis and bone healing. Biomaterials 50:75–86. https://doi.org/10.1016/j.biomaterials.2015.01.054
125. Pati F et al (2015) Ornamenting 3D printed scaffolds with cell-laid extracellular matrix for bone tissue regeneration. Biomaterials 37:230–241. https://doi.org/10.1016/j.biomaterials.2014.10.012
126. Shtrichman R et al (2014) The generation of hybrid electrospun nanofiber layer with extracellular matrix derived from human pluripotent stem cells, for regenerative medicine applications. Tissue Eng Part A 20:2756–2767. https://doi.org/10.1089/ten.TEA.2013.0705
127. Narayanan K, Leck KJ, Gao S, Wan AC (2009) Three-dimensional reconstituted extracellular matrix scaffolds for tissue engineering. Biomaterials 30:4309–4317. https://doi.org/10.1016/j.biomaterials.2009.04.049
128. Lee HJ et al (2015) A new approach for fabricating collagen/ECM-based bioinks using preosteoblasts and human adipose stem cells. Adv Healthcare Mater 4:1359–1368. https://doi.org/10.1002/adhm.201500193
129. Ma JX et al (2017) Biomimetic matrix fabricated by LMP-1 gene-transduced MC3T3-E1 cells for bone regeneration. Biofabrication 9:045010. https://doi.org/10.1088/1758-5090/Aa8dd1
130. Gao CY et al (2018) Directing osteogenic differentiation of BMSCs by cell-secreted decellularized extracellular matrixes from different cell types. J Mater Chem B 6:7471–7485. https://doi.org/10.1039/c8tb01785a

131. Fu Y, Liu LL, Cheng RY, Cui WG (2018) ECM decorated electrospun nanofiber for improving bone tissue regeneration. Polymers (Basel) 10:272. https://doi.org/10.3390/Polym10030272
132. Kumar A, Nune KC, Misra RDK (2016) Biological functionality and mechanistic contribution of extracellular matrix-ornamented three dimensional Ti-6Al-4V mesh scaffolds. J Biomed Mater Res A 104:2751–2763. https://doi.org/10.1002/jbm.a.35809
133. Kumar A, Nune KC, Misra RDK (2016) Biological functionality of extracellular matrix-ornamented three-dimensional printed hydroxyapatite scaffolds. J Biomed Mater Res A 104:1343–1351. https://doi.org/10.1002/jbm.a.35664
134. Pham QP et al (2008) The influence of an in vitro generated bone-like extracellular matrix on osteoblastic gene expression of marrow stromal cells. Biomaterials 29:2729–2739. https://doi.org/10.1016/j.biomaterials.2008.02.025
135. Datta N et al (2006) In vitro generated extracellular matrix and fluid shear stress synergistically enhance 3D osteoblastic differentiation. Proc Natl Acad Sci U S A 103:2488–2493. https://doi.org/10.1073/pnas.0505661103
136. Datta N, Holtorf HL, Sikavitsas VI, Jansen JA, Mikos AG (2005) Effect of bone extracellular matrix synthesized in vitro on the osteoblastic differentiation of marrow stromal cells. Biomaterials 26:971–977. https://doi.org/10.1016/j.biomaterials.2004.04.001
137. Kwon SH et al (2013) Modulation of BMP-2-induced chondrogenic versus osteogenic differentiation of human mesenchymal stem cells by cell-specific extracellular matrices. Tissue Eng Part A 19:49–58. https://doi.org/10.1089/ten.TEA.2012.0245
138. Lau TT, Lee LQP, Vo BN, Su K, Wang DA (2012) Inducing ossification in an engineered 3D scaffold-free living cartilage template. Biomaterials 33:8406–8417. https://doi.org/10.1016/j.biomaterials.2012.08.025
139. Bourgine PE et al (2017) Engineered extracellular matrices as biomaterials of tunable composition and function. Adv Funct Mater 27:1605486
140. Fu CC et al (2018) Embryonic-like mineralized extracellular matrix/stem cell microspheroids as a bone graft substitute. Adv Healthc Mater 7:1800705. https://doi.org/10.1002/Adhm.201800705
141. Choudhery MS, Badowski M, Muise A, Pierce J, Harris DT (2014) Donor age negatively impacts adipose tissue-derived mesenchymal stem cell expansion and differentiation. J Transl Med 12:8. https://doi.org/10.1186/1479-5876-12-8
142. Carlson ME, Conboy IM (2007) Loss of stem cell regenerative capacity within aged niches. Aging Cell 6:371–382. https://doi.org/10.1111/j.1474-9726.2007.00286.x
143. Sun Y et al (2011) Rescuing replication and osteogenesis of aged mesenchymal stem cells by exposure to a young extracellular matrix. FASEB J 25:1474–1485. https://doi.org/10.1096/fj.10-161497
144. Ng CP et al (2014) Enhanced ex vivo expansion of adult mesenchymal stem cells by fetal mesenchymal stem cell ECM. Biomaterials 35:4046–4057. https://doi.org/10.1016/j.biomaterials.2014.01.081
145. Bilousova G et al (2011) Osteoblasts derived from induced pluripotent stem cells form calcified structures in scaffolds both in vitro and in vivo. Stem Cells 29:206–216. https://doi.org/10.1002/stem.566
146. Rohanizadeh R, Swain MV, Mason RS (2008) Gelatin sponges (Gelfoam) as a scaffold for osteoblasts. J Mater Sci Mater Med 19:1173–1182. https://doi.org/10.1007/s10856-007-3154-y
147. Pham QP et al (2009) Analysis of the osteoinductive capacity and angiogenicity of an in vitro generated extracellular matrix. J Biomed Mater Res A 88:295–303. https://doi.org/10.1002/jbm.a.31875
148. Garcia P et al (2012) Temporal and spatial vascularization patterns of unions and nonunions: role of vascular endothelial growth factor and bone morphogenetic proteins. J Bone Joint Surg Am 94A:49–58. https://doi.org/10.2106/Jbjs.J.00795
149. Cricchio G, Lundgren S (2003) Donor site morbidity in two different approaches to anterior iliac crest bone harvesting. Clin Implant Dent Relat Res 5:161–169

150. Ventura RD, Padalhin AR, Min YK, Lee BT (2015) Bone regeneration using hydroxyapatite sponge scaffolds with in vivo deposited extracellular matrix. Tissue Eng Part A 21:2649–2661. https://doi.org/10.1089/ten.TEA.2015.0024
151. Kusuma GD et al (2018) Transferable matrixes produced from decellularized extracellular matrix promote proliferation and osteogenic differentiation of mesenchymal stem cells and facilitate scale-up. ACS Biomater Sci Eng 4:1760–1769. https://doi.org/10.1021/acsbiomaterials.7b00747
152. Lin H, Yang G, Tan J, Tuan RS (2012) Influence of decellularized matrix derived from human mesenchymal stem cells on their proliferation, migration and multi-lineage differentiation potential. Biomaterials 33:4480–4489. https://doi.org/10.1016/j.biomaterials.2012.03.012
153. Decaris ML, Mojadedi A, Bhat A, Leach JK (2012) Transferable cell-secreted extracellular matrices enhance osteogenic differentiation. Acta Biomater 8:744–752. https://doi.org/10.1016/j.actbio.2011.10.035
154. Decaris ML, Binder BY, Soicher MA, Bhat A, Leach JK (2012) Cell-derived matrix coatings for polymeric scaffolds. Tissue Eng Part A 18:2148–2157. https://doi.org/10.1089/ten.TEA.2011.0677
155. Keane TJ et al (2015) Tissue-specific effects of esophageal extracellular matrix. Tissue Eng Part A 21:2293–2300. https://doi.org/10.1089/ten.TEA.2015.0322
156. Shegarfi H, Reikeras O (2009) Review article: bone transplantation and immune response. J Orthop Surg 17:206–211. https://doi.org/10.1177/230949900901700218
157. Bose S, Roy M, Bandyopadhyay A (2012) Recent advances in bone tissue engineering scaffolds. Trends Biotechnol 30:546–554. https://doi.org/10.1016/j.tibtech.2012.07.005
158. Kittaka M et al (2015) Clumps of a mesenchymal stromal cell/extracellular matrix complex can be a novel tissue engineering therapy for bone regeneration. Cytotherapy 17:860–873. https://doi.org/10.1016/j.jcyt.2015.01.007
159. Motoike S et al (2018) Cryopreserved clumps of mesenchymal stem cell/extracellular matrix complexes retain osteogenic capacity and induce bone regeneration. Stem Cell Res Ther 9:73. https://doi.org/10.1186/s13287-018-0826-0
160. Onishi T et al (2018) Osteogenic extracellular matrix sheet for bone tissue regeneration. Eur Cell Mater 36:68–80. https://doi.org/10.22203/eCM.v036a06
161. Akahane M et al (2010) Scaffold-free cell sheet injection results in bone formation. J Tissue Eng Regen Med 4:404–411. https://doi.org/10.1002/term.259
162. Nakamura A et al (2010) Cell sheet transplantation of cultured mesenchymal stem cells enhances bone formation in a rat nonunion model. Bone 46:418–424. https://doi.org/10.1016/j.bone.2009.08.048
163. Akahane M et al (2008) Osteogenic matrix sheet-cell transplantation using osteoblastic cell sheet resulted in bone formation without scaffold at an ectopic site. J Tissue Eng Regen Med 2:196–201. https://doi.org/10.1002/term.81
164. Gao Z et al (2009) Vitalisation of tubular coral scaffolds with cell sheets for regeneration of long bones: a preliminary study in nude mice. Br J Oral Maxillofac Surg 47:116–122. https://doi.org/10.1016/j.bjoms.2008.07.199
165. Kang Y, Ren L, Yang Y (2014) Engineering vascularized bone grafts by integrating a biomimetic periosteum and beta-TCP scaffold. ACS Appl Mater Interfaces 6:9622–9633. https://doi.org/10.1021/am502056q
166. Liu SK et al (2018) Off-the-shelf biomimetic graphene oxide-collagen hybrid scaffolds wrapped with osteoinductive extracellular matrix for the repair of cranial defects in rats. ACS Appl Mater Inter 10:42948–42958. https://doi.org/10.1021/acsami.8b11071
167. Matsuda N, Shimizu T, Yamato M, Okano T (2007) Tissue engineering based on cell sheet technology. Adv Mater 19:3089–3099. https://doi.org/10.1002/adma.200701978
168. Long T et al (2014) The effect of mesenchymal stem cell sheets on structural allograft healing of critical sized femoral defects in mice. Biomaterials 35:2752–2759. https://doi.org/10.1016/j.biomaterials.2013.12.039

169. Zou B et al (2012) Electrospun fibrous scaffolds with continuous gradations in mineral contents and biological cues for manipulating cellular behaviors. Acta Biomater 8:1576–1585. https://doi.org/10.1016/j.actbio.2012.01.003
170. Zhang SC et al (2019) Direct electronetting of high-performance membranes based on self-assembled 2D nanoarchitectured networks. Nat Commun 10:1458. https://doi.org/10.1038/S41467-019-09444-Y
171. Khorshidi S et al (2016) A review of key challenges of electrospun scaffolds for tissue-engineering applications. J Tissue Eng Regen Med 10:715–738. https://doi.org/10.1002/term.1978
172. Reznikov N, Shahar R, Weiner S (2014) Bone hierarchical structure in three dimensions. Acta Biomater 10:3815–3826. https://doi.org/10.1016/j.actbio.2014.05.024
173. Carvalho MS et al (2019) Co-culture cell-derived extracellular matrix loaded electrospun microfibrous scaffolds for bone tissue engineering. Mater Sci Eng C Mater Biol Appl 99:479–490. https://doi.org/10.1016/j.msec.2019.01.127
174. Gibson M et al (2014) Tissue extracellular matrix nanoparticle presentation in electrospun nanofibers. Biomed Res Int 2014:469120. https://doi.org/10.1155/2014/469120
175. Thibault RA, Mikos AG, Kasper FK (2013) Winner of the 2013 young investigator award for the Society for Biomaterials annual meeting and exposition, April 10-13, 2013, Boston, Massachusetts Osteogenic differentiation of mesenchymal stem cells on demineralized and devitalized biodegradable polymer and extracellular matrix hybrid constructs. J Biomed Mater Res A 101:1225–1236. https://doi.org/10.1002/jbm.a.34610
176. Thibault RA, Scott Baggett L, Mikos AG, Kasper FK (2010) Osteogenic differentiation of mesenchymal stem cells on pregenerated extracellular matrix scaffolds in the absence of osteogenic cell culture supplements. Tissue Eng Part A 16:431–440. https://doi.org/10.1089/ten.TEA.2009.0583
177. Jeon H, Lee J, Lee H, Kim GH (2016) Nanostructured surface of electrospun PCL/dECM fibres treated with oxygen plasma for tissue engineering. RSC Adv 6:32887–32896. https://doi.org/10.1039/c6ra03840a
178. Jang J, Park JY, Gao G, Cho DW (2018) Biomaterials-based 3D cell printing for next-generation therapeutics and diagnostics. Biomaterials 156:88–106. https://doi.org/10.1016/j.biomaterials.2017.11.030
179. Bose S, Vahabzadeh S, Bandyopadhyay A (2013) Bone tissue engineering using 3D printing. Mater Today 16:496–504. https://doi.org/10.1016/j.mattod.2013.11.017
180. Hung BP et al (2016) Three-dimensional printing of bone extracellular matrix for craniofacial regeneration. ACS Biomater Sci Eng 2:1806–1816. https://doi.org/10.1021/acsbiomaterials.6b00101
181. Nyberg E, Rindone A, Dorafshar A, Grayson WL (2017) Comparison of 3D-printed polyvarepsilon-caprolactone scaffolds functionalized with tricalcium phosphate, hydroxyapatite, bio-oss, or decellularized bone matrix<sup/>. Tissue Eng Part A 23:503–514. https://doi.org/10.1089/ten.TEA.2016.0418
182. Chai YC et al (2017) Harnessing the osteogenicity of in vitro stem cell-derived mineralized extracellular matrix as 3D biotemplate to guide bone regeneration. Tissue Eng Part A 23:874–890. https://doi.org/10.1089/ten.tea.2016.0432
183. Kim JY et al (2018) Synergistic effects of beta tri-calcium phosphate and porcine-derived decellularized bone extracellular matrix in 3D-printed polycaprolactone scaffold on bone regeneration. Macromol Biosci 18:1800025. https://doi.org/10.1002/Mabi.201800025
184. Davis HE, Leach JK (2011) Designing bioactive delivery systems for tissue regeneration. Ann Biomed Eng 39:1–13. https://doi.org/10.1007/s10439-010-0135-y
185. Wang XH et al (2016) 3D bioprinting technologies for hard tissue and organ engineering. Materials (Basel) 9:802. https://doi.org/10.3390/Ma9100802
186. Mandrycky C, Wang ZJ, Kim K, Kim DH (2016) 3D bioprinting for engineering complex tissues. Biotechnol Adv 34:422–434. https://doi.org/10.1016/j.biotechadv.2015.12.011

187. Jang J et al (2016) Tailoring mechanical properties of decellularized extracellular matrix bioink by vitamin B2-induced photo-crosslinking. Acta Biomater 33:88–95. https://doi.org/10.1016/j.actbio.2016.01.013
188. Pati F et al (2014) Printing three-dimensional tissue analogues with decellularized extracellular matrix bioink. Nat Commun 5:3935. https://doi.org/10.1038/Ncomms4935
189. Hughes CS, Postovit LM, Lajoie GA (2010) Matrigel: a complex protein mixture required for optimal growth of cell culture. Proteomics 10:1886–1890. https://doi.org/10.1002/pmic.200900758

Calcium Phosphate Biomaterials for Bone Tissue Engineering: Properties and Relevance in Bone Repair

Kanchan Maji and Soumini Mondal

Abstract Bone defects are common and are associated with a significant burden of disease threatening the health of many people around the globe. Since the last decade, data obtained from case studies have demonstrated that 20% of patients who experience an osteoporotic hip break are unable to endure the primary year after medical treatment. Many similar cases suggest that there is a huge requirement for better treatment of unhealthy and broken bones. Human bone comprises of about 70% of calcium phosphate (CaP) mineral, therefore CaPs are possible alternative materials to fix a broken bone. CaP is broadly utilized for bone fixation because of its bioactive properties like osteoinductivity, osteoconductivity, and biodegradability. Therefore, examination of these properties and the impact of their different affecting factors are crucial for balancing CaP during the fabrication procedure to maximally fulfill required clinical prerequisites. The aim of this chapter is to highlight the systems behind the CaP-assisted bone development in the initial phase, specifically as a biocompatible bone graft substitute. In this study, the latest developments in the biological properties of CaP biomaterials, including hydroxyapatite (HA), tricalcium phosphate (TCP), and biphasic CaP (BCP), have been summarized. Moreover, recent advances on how their properties are altered by different factors are reviewed. Finally, perspectives regarding future developments of CaP materials are provided.

Keywords Calcium phosphate ceramic · Bone · Osteoconductivity
Osteoinductivity

Introduction

Medical advances have definitely paved a way in increasing our life span. However, increasing longevity raises new challenges. Age-related diseases result in significant reductions in the quality of our life. The loss of a skeletal tissue that accompanies trauma, injury, disease can result in significant morbidity as well as significant

K. Maji (✉) · S. Mondal
Department of Ceramic Engineering, NIT Rourkela, Rourkela, Odisha, India

B. Li et al. (eds.), *Racing for the Surface*,
https://doi.org/10.1007/978-3-030-34471-9_20

socioeconomic cost. It, thus, emphasizes the need for new and more reliable skeletal regeneration strategy. Today, hundreds of people across the globe are diagnosed with musculoskeletal diseases such as arthritis, osteoporosis, bone fractures, bone tumors, back ache, and other cerebrospinal disorders [1, 2]. To address this dire need of bone augmentation and tissue regeneration, regenerative medicine has come to the forefront in recent years with new advances in neoskeletal tissue formation. The successful outcome of this approach requires pluripotent stem cells, novel scaffolds, and growth factors that assure bone regeneration strategies for improved life quality. This chapter demonstrates various characteristic aspects of osteoconductive and osteoinductive biomaterials, and advances in their applications in the field of tissue engineering to address various bone defect problems. Damaged bone has to be regenerated naturally or it needs to be substituted with a prosthesis, or a bone material from another body part by surgery [3]. At present, demineralized bone, HA and other graft substitutes are developed and have been used to facilitate osteogenesis at damaged bony tissue parts. However, they have failed to bring about satisfactory results in the regeneration of tissues. Recently, growth factors, such as bone morphogenic factors (BMF) [4], platelet-derived growth factors (PDGF) [5], insulin-like growth factors (IGF) [6], and cytokines [7] have been reported to be very useful in the regeneration of bone tissues.

The application of bone graft substitutes is conducted by autograft and allograft methods. However, grafting methods suffer from several problems, such as limited availability of grafts [8]. Moreover, there always exists a chance of bacterial infection and blood loss during surgical processes. In addition, the areas from which the grafts are taken experience poor structural stability [9]. The autograft method has an advantage over allografts with respect to donor availability because they are obtained from allo-doners, but osteoinductive potential of allogenic bones are far inferior than autogenous, which makes it suitable for temporary support only. In order to circumvent these problems, active research has been directed to the development of bone graft substitutes which possess excellent biomechanical properties of metal grafts and biological properties of bone grafts.

CaP ceramics are the most widely used bone substitutes for clinical applications of bone grafting and orthopedics [10]. However, not a wide variety of CaP ceramics are responsible for influencing better biological performance in vivo [11–13]. On the other hand, most of the CaP ceramics are osteoconductive in nature, just a few specific groups are osteoinductive in nature [13]. These little differences in their intrinsic characteristics which in turn enhance osteoblast differentiation are identified with little difference in the physical and chemical properties of CaP ceramics. For instance, chemical properties, such as, surface chemistry and charge can impact biological phenomenon like protein adsorption [10], which can in this manner to effect osteoblastic differentiation by means of cell–extracellular matrix (ECM) interactions [14, 15]. In this manner, physical properties, for example, surface topography (roughness) can facilitate cell differentiation by helping cellular attachment on the material surface [16]. Moreover, some other surface characteristics of CaP ceramics can enhance the recruitment of important cell-attaching proteins and in this manner give conditions favorable to the development of fixed focal adhesive compounds [17]. Along these lines, understanding

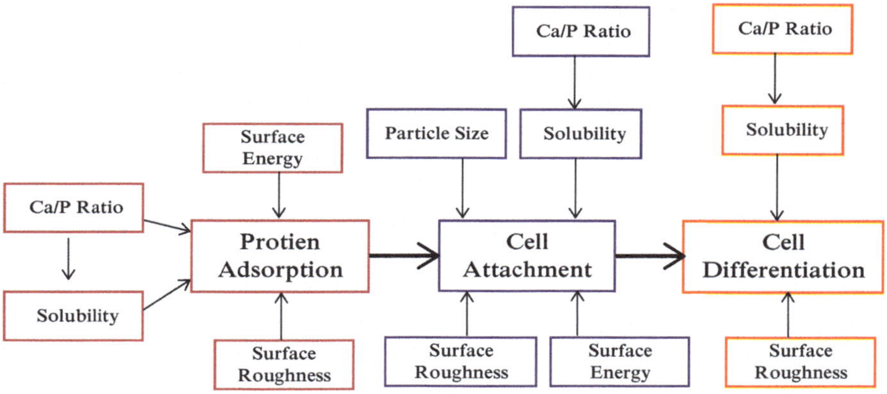

Fig. 1 Schematic of key properties of CaP ceramic materials that impact a series of biological events such as protein adsorption, cell attachment, and cell differentiation

the exact roles that material properties play for regulating the cell material interaction process is a primary step toward designing osteoinductive CaP materials. This chapter describes the physical as well as chemical characteristics of CaP ceramics and its influence with regards to bone tissue engineering (BTE). Specifically, it clarifies the variation in CaP ceramic properties like surface roughness, solubility and crystallinity, related to and contrasted with osteoconductivity and osteoinductivity (Fig. 1).

Bone and Its Properties

Bone is a mineralized connective tissue that shows four kinds of cells: osteoblasts, bone-covering cells, osteocytes, and osteoclasts [1, 2]. In spite of its inactive appearance, bone is an exceedingly powerful organ that is persistently resorbed by osteoclasts and reformed by osteoblasts. In this section, we address the present information about bone cell science, the bone network, and the variables that impact the bone rebuilding process.

Hierarchical Design of Bone

In order to investigate the mechanical properties of bone tissue, it is very much essential to have a fair understanding of the constituent mineral phases of bone, and the macrostructural co-relationship between them at different levels of hierarchical morphological arrangements [18–20]. These organizations are:

I. Basic macro-architecture: cancellous bone and cortical bone;
II. Basic micro-architecture (10–500 μm): haversian canals and osteons;

III. Sub-micro-architecture (1–10 μm): lamellae;
IV. Nanostructure (100 nm–1 μm): fibrillar collagen and embedded mineral; and
V. Sub-nanostructure (<100 nm): molecular organization of constituent mineral components, like collagen, and non-collagenous organic proteins.

This hierarchically arrangement of bone has an intricate, but optimized structural orientation of the components, making the bone material heterogeneous and anisotropic (Fig. 2).

Composition of Bone Materials

The CaP biomaterial is constantly talked about in connection with bone repair as CaP is the fundamental inorganic component of bone. In spite of the fact that the shape of bone changes in various pieces of the body, the physicochemical structure of bone for the different shape is biochemically similar. Bone tissue can be viewed as a composite material developed by a collagen biopolymer and CaP bioceramic. Normal bone comprises of 69% CaP bioceramic, considered as the standard bone material. The natural part (22%) built in by proteins, type I collagen (90%), and some other non-collagenous proteins, such as proteoglycans, lipids, and osteogenic stimulus (this is meant to be enhancement factors, such as bone morphogenetic proteins (BMPs) and vascular endothelial growth factors (VEGFs)) [22]. The other remaining 9% is filled up by water molecules. Table 1 [23] describes the characteristic properties of bone tissues and their capabilities in bone mineralization events.

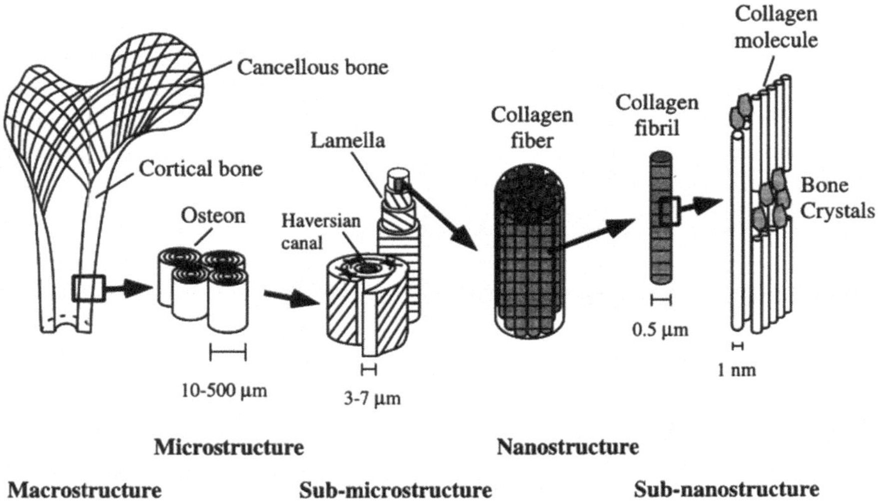

Fig. 2 Hierarchical architecture of bone [21]

Table 1 Organic composition of bone and their specific functions in bone maturation [14]

Name of organic component	Specific functions
Collagen	This protein is a structural protein and main constituent of bone tissue
ON-Osteonectin OP-Osteopontin	May be responsible for HA mineralization Deposition and mineralization of ECM
OC-Osteocalcin	Inhibit osteoclast activity
BSP-Bone sialoprotein	This protein helps in binding of calcium with the Arg-Gly-Asp (RGD) sequence

Bone Cells

Despite its strength and hardness, bone is a dynamic living tissue. Bone is composed of a series of complex events altogether arranged by different types of bone cells associated with each other and also with the ECM. The bone cells comprise of four types of cells namely (1) osteoblasts, (2) osteoclasts, (3) osteocytes, and (4) bone-covering cells. Osteoblasts are cells that are responsible for the creation and mineralization of the bone grid; whereas osteoclasts are accountable for bone resorption. Osteoblasts are derived from mesenchymal stem cells (MSCs). The dedication of MSCs toward the osteoprogenitor lineage requires the expression of specific genes, which follow modified steps, including the synthesis of BMPs and also members from the Wingless (Wnt) pathways. Run related translation factor 2 (Runx2) is the ace gene of osteoblast differentiation. Runx2 is also vital for osteoblast differentiation [22, 24]. Additionally, Runx2 has been shown to upregulate osteoblast-related qualities, for example, collagen ColIA1, alkaline phosphatase (ALP), bone sialoprotein (BSP), and osteocalcin (OCN).

Structure of Bone Grafts

Although bone has its own capacity to repair, the capacity declines with age, and is constrained to small defects. So far, grafts are important to help bone repair when bone loss is too enormous. A few grafts can be the choices in the clinic for bone repair. An autograft, which is gathered from the patient's own body, has no issue with biocompatibility and immune response [25]. Autografts might be cortical, cancellous, or cortico-cancellous (scanning ultra-micrographs of various bone grafts are introduced in Fig. 3). The cortical bone has higher mineral substance than the trabecular or cancellous bone [18]. The compressive solidness and quality of the cortical bone are a lot higher than those of the cancellous bone. In choosing a graft, the specialist must know about these major contrasts in bony structures [26, 27].

Cortical bone grafts are utilized for the most part for structural help and strength, and cancellous bone grafts for osteogenesis. Cancellous bone grafts are ordinarily

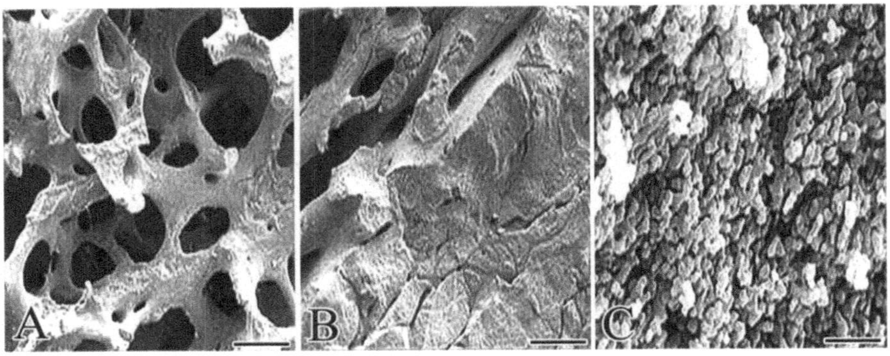

Fig. 3 SEM of ultra-micrographs of the microstructure of natural bone grafts. (**A**) Trabecular or cancellous bone graft. Note the porous honey comb-like microstructure of cancellous bone graft. (**B**) Cortico-cancellous bone graft. (**C**) Cortical or compact bone graft (scale bars: 100 μm)

utilized in break non-association, dental imperfections, maxillofacial deformities, spinal combination, and other little bone deformities [28, 29]. These grafts need mechanical stability; however, the permeable structure of cancellous bone grafts can upgrade bone cell ingrowth and improve the healing process, permitting quicker revascularization [30]. Cortical bone grafts are used less regularly, and they might be utilized as only as grafts [31]. An allograft, which is given by the donor, has osteogenic capacity yet the supply of allografts is constrained. The inadequacies of autografts and allografts legitimize the improvement of artificial bone joint biomaterials.

Bone Porosity

Interconnecting porosity is an important physicochemical property of bone. The size of pores and interconnection within bone determine the internal vascularization as well as tissue ingrowth [24, 32–35]. Bone pore sizes in a typical cortical bone territory vary from 1 to 100 μm while trabecular bone has pores extending from 200 to 400 μm [36]. The size range, degree, and interconnectivity of the pores are basic variables influencing dispersion of supplements, cell adhesion, migration and expression, and tissue ingrowth that are important for bone arrangement and repair or recovery [37].

Bone Strength

The high level of combination and introduction of the mineral and natural segments gives bone its mechanical strength. The property that is frequently used to describe the mechanical behavior of bone substitutes is their compressive strength. Since

Table 2 Summary of physical and mechanical properties of various implant materials in comparison to natural bone

Properties	Natural bone	Co-Cr alloy	Stainless steel	Synthetic HA
Density (g/cm^3)	1.8–2.1	8.3–9.2	7.9–8.1	3.1
Elastic modulus (GPa)	3–20	230	189–205	73–117
Compressive yield strength (MPa)	130–180	450–1000	170–310	600
Fracture toughness (MPam$^{1/2}$)	3–6	N/A	50–200	0.7

these materials are proposed to be utilized as bone substitutes, it is vital to remember that the compressive strength of human cortical bone ranges somewhere in the range of 90 and 230 MPa (with elastical strength from 90 to 190 MPa), while the compressive strength of cancellous bone ranges somewhere in the range of 2 and 45 MPa [38]. Table 2 describes the comparative mechanical strength of CaP with metals used as a biocompatible material.

Types of CaP Derivatives Present in the Body

The existence of CaP in the vertebrate has been reported to be in the form of apatite. The mineral present in teeth and bones is appreciated as calcium HA, $Ca_{10}(PO_4)_6(OH)_2$ [39] with minor segments of carbonate (CO_3), magnesium (Mg), and sodium (Na). The crystal structure of apatite in enamels is well understood among all of the apatites in our body. It was found to be platelet-like in shape (lengths and widths (30–45 nm) and thickness around 5 nm) embedded in collagen nanofibrils [40]. As shown in the XRD profile of enamel apatite [41] and a lot more extensive diffraction peaks of either bone or dentin apatite (Fig. 4), it is evident that apatite present in enamel possesses a bigger crystal size (around 2000 nm) as compared to any other apatite in our body.

Many other biological non-apatitic CaPs exist in our body, which are equally responsible for the regeneration of bone, as summarized in Table 3.

Categories of CaPs

Based on structural composition, CaPs for bone and teeth regeneration are classified as: (1) calcium-deficient apatite, CDA (i.e., Ca/P molar ratio less than the stoichiometric value of 1.67 for pure HA), (2) HA, $Ca_{10}(PO_4)_6(OH)_2$, (3) beta-tricalcium phosphate (β-TCP), $Ca_3(PO_4)_2$, and (4) biphasic CaP (BCP), an intimate mixture of HA and β-TCP of varying HA/β-TCP weight ratios, are available commercially (Table 4).

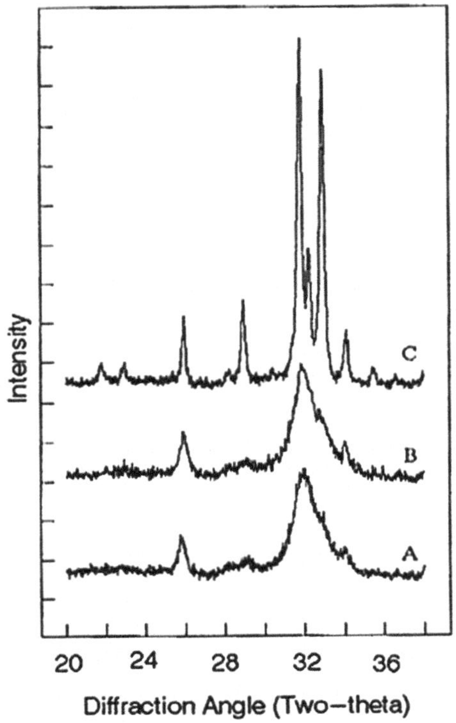

Fig. 4 X-ray diffraction profiles of biologic apatites from (**a**) bone, (**b**) dentin, and (**c**) enamel. The sharper diffraction peaks in **c** compared to either **b** or **a** indicates that enamel apatite crystals are much larger compared to either bone or dentin apatite crystals

Table 3 Occurrence of CaPs in human body [42]

Various calcium phosphate (CaP)	Location of occurrence
DCPD-Di calcium phosphate di-hydride	Presence in dental caries
TCMP-Mg-substitute tricalcium phosphate	Presence in mineralized soft tissue and dental caries
CFA-Carbonated fluroapatite	Presence in fish enamel
ACP-Amorphous CaP	Presence in mineralized soft tissue
CHA-Carbonated HA	Presence in mineralized dentin, urinary stone, and dental callus
CPPD-Calcium pyrophosphatedi-hydride	Presence in joints
OCP-Octacalciumphosphate	Presence in urinary stone

Hydroxyapatite (HA)

HA is broadly used as an alternative inorganic filler material in bone tissue engineering because of its compositional similarities with that of the inorganic counterpart of bone [43]. HA material is the most stable phase among various other forms

Table 4 Commercially available CaP-based biomaterials [52]

Composition	Property
CDA-Calcium-deficient apatite	Osteogenic agent Company-Impladent, NY
HA-hydroxyapatite	Ostegraf Company-Ceramed, CO
Hydroxyapatite derived from coral	ProOsteon Company-Interpore, CA
Hydroxyapatite/$CaSO_4$	Hapset Company-LifeCore, MN
Sintered bovine bone	Endobon Company-Merck, Germany
β-TCP-tricalcium phosphate	Vitoss Company-Orthovita, PA

of CaP, specifically it is the most steady in a dissolve stage [44, 45]. In spite of the fact that not exceedingly solvent, the surface of HA mineral is favorable as a nucleation site in culture medium (continuously soaked with calcium and phosphate particles) for the precipitation of apatite crystal [46]. In addition, stoichiometrically HA($Ca_5(PO_4)_3$)OH contain a Ca/P ratio of 1.67 and is believed to be as osteoconductive not osteoinductive [47].

Tricalcium Phosphate (TCP)

Proportion of Ca/P ratio in TCP is 1.5 and it is most likely to exit in two different phases, namely α-TCP phase and β-TCP phase; these two phases have indistinguishable different crystal structures [48]. The two phases are less steady than HA [49]. The α-TCP phase is believed to be osteoconductive as well as osteoinductive properties. It can encourage the formation of an apatite layer when incubated in biological fluid containing different ionic arrangements [50]. β-TCP is the more utilized form of TCP than α-TCP in bone regeneration.

Biphasic CaP (BCP)

This type of CaP belongs to a bone substitute group that consists of an intimate mixture of two ceramics with varying ratios. BCP powder is synthesized by the homogeneous mixing of HA and TCP powder by means of physical grinding, or by high temperature sintering of calcium-deficient HA (CDHA) above 700 °C, resulting in a composition of two individual phases [51]. The Ca/P ratio in BCPs mainly depends upon the calcium deficiency of CDHA and sintering temperature of CDHA, which generally falls in the range between pure β-TCP and HA. Furthermore, in

Table 5 Physicochemical characterization of various CaPs [59–63]

Calcium phosphate	Abbreviation	Chemical formula	Ca/P molar ratio	Solubility
Monocalcium phosphate monohydrate	MCPM	$Ca(H_2PO_4)_2.H_2O$	0.5	7.2×10^{-2}
Dicalcium phosphate dihydrate	DCPD	$CaHPO_4.2H_2O$	1.0	2.5×10^{-7}
Octacalcium phosphate	OCP	$Ca_8H_2(PO_4)_6.5H_2O$	1.33	2.51×10^{-97}
α-Tricalcium phosphate	a-TCP	a-$Ca_3(PO_4)_2$	1.5	3.16×10^{-26}
β-Tricalcium phosphate	b-TCP	b-$Ca_3(PO_4)_2$	1.5	1.25×10^{-29}
Amorphous CaP	ACP	$Ca_3(PO_4)_2.nH_2O$	1.2–2.2	–
Calcium-deficient hydroxyapatite	CDHA	$Ca_{10}x(HPO_4)$ $x(PO_4)_6x(OH)_2x$ $(0 < x < 1)$	1.5–1.67	–
Hydroxyapatite	HA	$Ca_{10}(PO_4)_6(OH)_2$	1.67	2.35×10^{-59}

biomedical applications, BCP, is known as a potential candidate for bone regeneration, drug delivery vehicle and carrier of growth factors.

Solubility of CaP

Cell-mediated biodegradation happens under acidic conditions [53]. Therefore, in vitro dissolution studies of CaP biomaterials may be predictive of their in vivo dissolution or biodegradation [25]. Monocalcium phosphate monohydrate (MCPM) is the most acidic and soluble CaP among all other CaPs. MCPM isn't biocompatible because of its very acidic nature and high solubility. Bone apatite is like CDHA apart from the presence of carbonate (CO_3^{-2}) and the other components, for example, Na^+, K^+, Mg^{2+}, Sr^{2+}, and Zn^{2+} [54–56]. HA and β-TCP are considered to be the most useful CaPs compared to other CaPs due to their osteogenic potential and also their ability to form reliable attachment with host bone tissues near the defect site. The dissolution property of β-TCP is greater than HA which makes it beneficial as a bioresorbable agent [57, 58]. The advancement of biphasic CaP (BCP)-based biomaterials comprising of HA and β-TCP [26–28] are likewise important to control degradation properties. Table 5 presents an overview of various CaPs and their properties.

Bioactivity and Resorbability of CaP Materials

The justification in the advanced applications of CaPs has been lying in their close proximity in composition and properties to that of bone, like osteoconductivity as well as bioactivity. To discover its huge potential as an artificial bone substitute, biomaterial researchers need an understanding of the basic properties of CaP material, like biological and mechanical properties. Some of the relevant properties of CaP biomaterials in terms of osteoinductivity, osteoconductivity, biodegradability, and the potential factors which influence these inherent properties of CaP biomaterials in hard tissue recovery utilizing tissue design are discussed in the following section.

Cell Signalling in CaP Mediated Osteoinductivity

As of now, most metallic implants do not possess osteogenetic characteristics, whereas some CaP ceramics used encourage osteogenesis without adding any more osteogenic agents from the outside. This type of characteristic is defined as osteoinduction. Osteoinduction implies the recruitment of immature cells and the stimulation of these cells to develop into a pre-osteoblasts lineage. Osteoconductivity of CaP has been studied in the literature but the mechanism behind has not been well explored [64]. Release of ions from CaP materials is thought to be the main contributor to this phenomenon. However, some osteogenic growth factors including TGF-β and BMPs appear to play an important role in the osteoinduction process via related signalling pathways [65]. Thus, to understand the detailed mechanism behind the osteoinduction process, researchers need to focus their study into the molecular level for a detailed osteoinduction signalling pathway. A schematic illustration of feasible signalling pathways can be created in Fig. 5.

Transforming growth factor-beta (TGF-β) is appearing to play a major role in the formation of bone cells from MSCs during mammalian development. In brief, the TGF-β superfamily is comprised of over forty members of proteins, such as TGF-βs, nodal, activin, and BMPs [66]. In this signalling event, first the signal was internalized across the plasma membrane through the formation of heteromeric complexes of specific type I and type II serine/threonine kinase receptors. After that, a specific type II receptor is activating via phosphorylation of type I receptors. Next, phosphorylation of some specific proteins, called Smad and R-mad initiate the signalling pathway by the help of the activated type I receptor. The transcriptional process starts inside nucleus as activated R-Smads translocate into the nucleolus by forming a complex with co-Smad and Smad4.

There are other signalling pathways, like BMPs, Wnt which are also capable of modulating new bone formation through this osteoinduction procedure [67]. This signalling route exhibits various regulatory functions in enhancing various processes during osteogenesis (like signal transduction, gene expression, and osteoblast differentiation).

Osteoconductibility of CaP Materials

Osteoconductivity is a process, where bioactive materials are implanted inside a defect bone, and consequently bone cells will adhere or attach on the material's outer surface and at a later time point, bone cells will invade inside the pore of the implanted material; this process is also defined as bone conduction [68]. These large

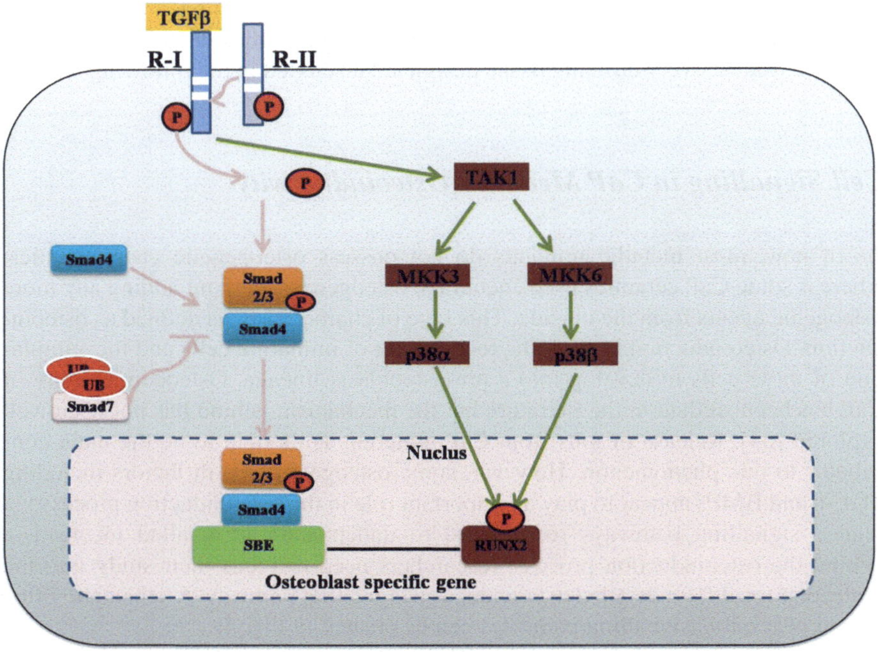

Fig. 5 TGF-β signalling pathway describing the differentiations of osteoblast during bone formation

amount of bone cells that occupy the implant material surface and internal pores, clearly indicate their osteoinductive property. Both material-dependent factors and defect sites are two decisive factors which influence the osteoconduction process during osteogenesis. Yu et al. showed in his work that material properties can be factor for inducing osteogenesis. The results clearly indicate that vascularization was different for different channel diameters in CaP scaffolds (Fig. 6), higher expression of the PLGF (placental growth factor), angiogenic factors HIF1 alpha (hypoxia-inducible factor 1), and migration factor CXCR4 (C-X-C chemokine receptor type 4), which are responsible for starting the development of micro vessels, was seen inside the CaP porous scaffolds with a channel diameter of 250 p.m. Whereas, the diameter of the 500 pm channel in the same CaP scaffold gradually increased the expression of VEGF-A (vascular endothelial growth factor A), which initiated the formation of macro vessels [69]. However, apart from the size of the interconnecting channels, macroporosity (pores >50 μm) and microporosity (porosity <50 μm) are thought to have a prominent function in cellular attachment on the material implant. For example, macroporosity influences cell adhesion and accordingly vascular growth as well as the development of bone tissue. On the other hand, the microporosity surface of bioceramic enhances protein adsorption, which in turn escalates cell attachment on the material.

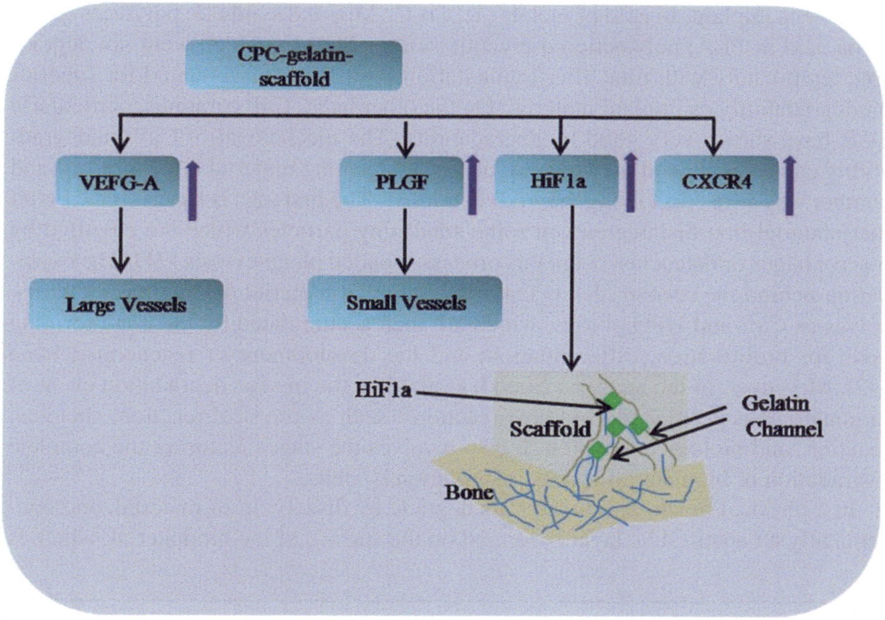

Fig. 6 Graphical representation of the angiogenesis strategy within a CaP scaffold internal channel pore: (**a**) different channel diameters influence different blood vessel formation and (**b**) the increased HIF1a expression in the internal pore channels influence the formation of blood vessels into its host [61]

Studies have also shown that BCP ceramics coated with HA and seeded with MSCs were augmented and have shown enhanced formation of new bone tissue in the BCP ceramics after 12 weeks of implantation inside rabbit tibia [70]. Based on the other literature references, it can be confirmed that CaP performs well in human patients. Still, not many studies have shown the osteoconductive properties of CaP ceramics in human patients. That is why researchers need to explore more about CaP osteoconductibility in humans with suitable approaches.

Biodegradability of CaP Materials

The essential properties of a perfect bioactive bone substitute is that substitute materials have to disintegrate at a similar rate at which the osteoblast cells start to develop into new bone cells on the material surfaces, until the substitute material is fully supplemented by new, active bone tissue, although, biomaterials are believed to exhibit identical biomechanical as well as biochemical characteristics and regenerate bone tissue in a similar fashion as autologous bone [59]. Till date, various biomaterials have been explored to determine their feasibility to be used as an absorbable implant. In case of metals (Ti, Co-Cr-Mo) and synthetic polymers (polylactic acid or PLA, polylactic-*co*-glycolic acid or PLGA), which were not degradable legitimately with time after implantation has not been accounted for superior biodegradability as implant material. On the other hand, CaP ceramics, particularly TCP, have shown very good biodegradability. The mechanism of CaP biodegradability can be explained in two ways, one is, "dispersing material into particles" and another way says "dissolving material into ions". The first idea is based on the belief that material first disintegrates in some small tiny particles which are engulfed by macrophages or osteoclast cells; this process is called phagocytosis [71]. The explanation behind the second idea is that the reinforced material disintegrates and dissolves as Ca^{2+} and HPO_4^{-2} ions, which are then accumulated by the bone forming cells for proliferation, differentiation, and the development of regenerated bone [72]. Moreover, in the study by Sheikh et al. [73], the in vivo degradation event of biomaterials is categorized by three reactions: such as physical reaction, chemical reaction, and biological reaction. It also involves the stages featuring the complete degradation of biomaterial and its assimilation by cells.

In a physical reaction, biomaterials degrade by dissolving of material, and consequently an apatite-like layer is formed on the surface of the biomaterial, which is

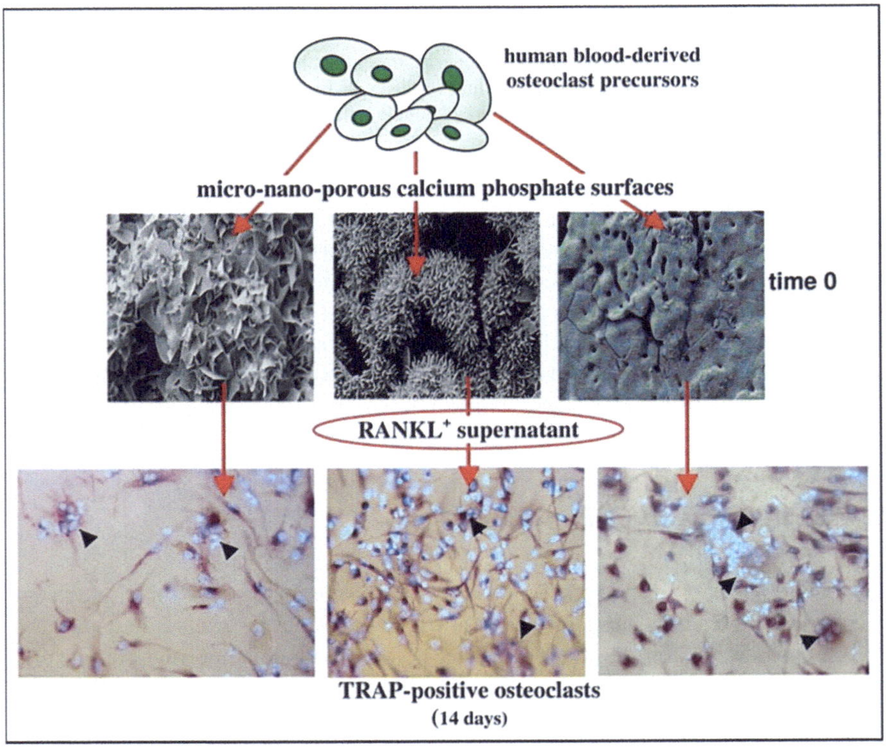

Fig. 7 Function of osteoclasts on the surface of CaP [77]

believed to be developed by dissolving, depositing, ion exchange events occurring on the material surface during the early degradation period until the material at last is crushed into tiny particles. Mechanical stability of biomaterials decreases rapidly during this time period. Biological reaction means degradation and adsorption or microscopic segregation of biomaterials in a biological fluid by the shared support of various bone forming cells including osteoclasts [74], osteoblasts, macrophages [75], fibroblasts, and multinucleated monster cells (Fig. 7) [76].

Recently, Wang et al. [78] demonstrated that biodegradation and osteoinductive ability of a BCP material were correlated to each other. Moreover, there are some specific mechanisms and much complexity in between both processes, which should be addressed in the future to develop promising biomaterials for bone repair and regeneration.

Characteristics of Osteoinductive Materials

Till date, among all materials that are currently used as bone grafts, CaP materials hold the most promise to be utilized in the clinic for bone tissue designing, because of its biocompatibility, osteoconductivity, and osteoinductivity. Additionally, alongside a 3D permeable structure and some specific basic intrinsic characteristics for CaP ceramic production, they are important to new bone regeneration. In the accompanying sections, we will talk in detail about the impact of different material attributes of CaP materials on osteoinductivity.

Effect of Crystallinity

A number of studies related to material characterization of CaP have suggested that crystallinity is an important factor in inducing bone formation. The concentration of ions in the culture medium and pH of the culture medium can be affected by crystallinity and solubility of CaP ceramics, which in turn are responsible for cell adhesion on a material surface. Hu et al. showed that BMSCs from rabbit adhered better on HA (higher crystallinity) and it was better than amorphous HA (lower crystallinity) [65]. In another study, Berube et al. showed that the adhesion of osteoblasts from rat calvarias, was better on higher crystalline HA surfaces than on lower crystalline HA surfaces [79]. In addition, Knabe et al. investigated the reasons behind the lower attachment of BMSC from rat bone onto $CaKPO_4$ pellet samples in comparison with α-TCP pellet samples, and explained that as the release of phosphate and potassium ion from $CaKPO_4$ samples decreased the concentration of calcium ion from the culture medium caused lower attachment of cells onto material surfaces [77]. Moreover, the authors suggested that the formation of an apatite layer on the material surface played an important role in influencing cell attachment and proliferation on the material surfaces.

Overall, the research on the crystallinity of materials indicates that crystalline and solubility of released ions from CaP ceramics may develop stable surfaces for cell adhesion and proliferation in physiological conditions.

Effect of Solubility

The adsorption of proteins on the surfaces of CaP ceramic materials depends on surface charge and solubility of the material, which in turn influences cell behavior by changing the concentration of ions in the solution [65]. In a recent investigation by Lee et al., it was observed that, CaP ceramics nanoparticles in a polypropylene fumarate [80] and poly(lactic-co-glycolic acid) matrix enhances the adsorption of proteins compared to without CaP ceramic samples [81]. Overall, these results indicated that higher ion concentrations and changes in pH near the surface of soluble CaP ceramics (e.g., HA, TCP, and BCP) promote cell adhesion by facilitating protein adsorption on the surfaces of the materials.

Effect of Surface Roughness

Although integrin binding and cell adhesion on the material surfaces can be influenced by material surfaces having nano and sub-micron level roughness properties [82]. As an example, Zhou et al. demonstrated that rabbit BMSCs attached on HA surfaces (with an Ra value of 11.9 nm) was more prominent than on surfaces having an Ra value of 54.2 nm, where the particle sizes were different for two nonidentical materials [83]. In addition, Dulgar-Tulloch et al. investigated these events: small grains in the range of ~50–100 nm influenced (decrease) the attachment of human BMSCs in comparison with large grain (200 nm) [84]. Altogether, these studies indicated that surface roughness (nano and micro) and crystal grain size less than 100 nm can promote protein adsorption as well as cell adhesion.

Effect of Surface Charge

Surface charge is another important factor along with crystallinity and solubility may influence protein adsorption as well as cellular attachment by significantly varying the charge concentration near material surfaces. In addition, cationic charged surfaces could have assumed a positive function in cell attachment by promoting the adsorption of proteins on the surface of the material. In a recent study by Feng et al., it has been shown that calcium-coated titanium implant surfaces increase osteoblast attachment compared to phosphate-coated titanium implant surfaces [85]. The authors further suggested that apatite-coated implants gave better cellular adhesion compared to calcium-coated implants. Overall, the results indicated that calcium-coated surface provided positively charged ions that appear to significantly increase the adsorption of negatively charged glycoproteins (e.g., fibronectin, vitronectin).

Expert Opinion and Five-Year View

An incredible test is to develop a biomaterial that carries on in a route identical to an autograft. The material of the scaffold needs to dissolve in a fashion, parallel to that of bone tissue regeneration. On the chance that resorption happens too rapidly, pseudoarthrosis may happen. On the other hand, if the rate of resorption happens very slowly, then bone ingrowth might be hindered and pushing again to pseudoarthrosis.

To these reasons, numerous design and fabrication processes have been adopted to modify and develop chemical and phase composition of bioactive CaP materials so that they should be ready to release of particular ions from the bone scaffold material into the surrounding space. This might influence the osteointegration process of the cell-scaffold construct. In addition, advancement in the design and fabrication of 3D porous bioactive ceramics is still needed to encourage control of graft material resorption and bone tissue regeneration in a desirable manner. Nanotechnology can give an alternative method for fabricating CaP bioceramics with increased mechanical properties and higher bioactivity, as well as resorbability. Nano-biotechnology has the capability to work with material parameters on a nuclear, subatomic, and supra-subatomic dimension. It is evident from the literature that grain size (in the nano level) of biomaterials might be a decisive factor for the enhancement of its mechanical performance. In addition, CaP bioceramics created with nanograin microstructures are characterized by prevalent bioactivity compared with traditional micrograin bioceramics.

In this manner, properties like dissolution and protein adsorption on CaP biomaterials, which are subject to surface science, energy, and roughness, can be improved to upregulate cell adhesion, proliferation, and differentiation, which in turn are collectively responsible for the osteoinductivity and osteoconductivity of CaP biomaterials.

Current Challenges and Future Directions

The mechanical stability and osteointegrity of scaffolds that must bear loads long-term are critical problems. Insufficient vascularization of the interior of thick bone substitutes, limiting cell ingrowth and survivability, is associated with poor osseointegration. Mechanical strength is heavily dependent on porosity and geometry of the scaffold, and pores and strut dimensions. Therefore, CaP biomaterials have to address this problem in a manner to solve the mechanical as well as biological issues related to scaffold tissue engineering.

CaP ceramics present a category that possesses osteoconductivity and osteoinductivity properties, making them ideal materials for bone regeneration process. The osteoinductive limit of CaP ceramics in vivo is influenced by the solubility of surface particles of CaP materials. In this perspective, both β-TCP and amorphous CaP put an impression of being osteoinductive and increase bone cell ingrowth quicker than a slowly dissolving HA.

Although the major advancement happened toward understanding the possible mechanism of CaP osteoinduction, still much work needs to be done within CaP materials toward inducing bone tissue regeneration. Primarily, tuning the physical properties of CaP free of its compositional chemistry. Secondarily, the morphological behavior of MSCs varies in the presence and absence of osteogenic supplements. CaP in the presence of osteogenic media hardly influences the osteogenic property of materials. In this regards, research involving the in vitro osteoinductive capacity of CaP without osteogenic media can give a thorough technique for investigation which assists in the interpretation of the in vivo nature of CaP materials.

In this manner, a lot of work needs to be done in understanding the adsorption of cell-glue proteins, for example, fibronectin onto CaP ceramic surfaces. Successful clinical applications of bone substitutes require an interplay among cells, biological signals, and biomaterials. Many unanswered questions and unexplored frontiers remain for the optimal use of nanostructured materials. Fundamental advances in life and materials sciences are required.

Acknowledgements The authors thank the Department of Ceramic Engineering and Biotechnology and Medical Engineering (NIT Rourkela, India), Dr. Sudip Dasgupta (Department of Ceramic Engineering, NIT Rourkela) for a critical review of the manuscript and for providing valuable suggestions.

References

1. Woolf AD, Pfleger B (2003) Burden of major musculoskeletal conditions. Bull World Health Organ 81:646–656
2. Brooks PM (2002) Impact of osteoarthritis on individuals and society: how much disability? Social consequences and health economic implications. Curr Opin Rheumatol 14(5):573–577
3. Mistry AS, Mikos AG (2005) Tissue engineering strategies for bone regeneration. In: Regenerative medicine II. Springer, Berlin, pp 1–22
4. Burt DW, Law AS (1994) Evolution of the transforming growth factor-beta superfamily. Prog Growth Factor Res 5(1):99–118
5. Ross R, Raines EW, Bowen-Pope DF (1986) The biology of platelet-derived growth factor. Cell 46(2):155–169
6. Humbel RE (1990) Insulin-like growth factors I and II. Eur J Biochem 190(3):445–462
7. Opal SM, Depalo VA (2000) Anti-inflammatory cytokines. Chest 117(4):1162–1172
8. Springer IN et al (2004) Particulated bone grafts–effectiveness of bone cell supply. Clin Oral Implants Res 15(2):205–212
9. Naran S, Menard RM (2015) Bone grafting: physiology and techniques. In: Ferraro's fundamentals of maxillofacial surgery. Springer, Berlin, pp 115–133

10. Giannoudis PV, Dinopoulos H, Tsiridis E (2005) Bone substitutes: an update. Injury 36(3):S20–S27
11. Radin S, Ducheyne P (1993) The effect of calcium phosphate ceramic composition and structure on in vitro behavior. II. Precipitation. J Biomed Mater Res 27(1):35–45
12. Nandi S et al (2010) Orthopaedic applications of bone graft & graft substitutes: a review. Indian J Med Res 132(1):15–30
13. Ferna E et al (1999) Calcium phosphate bone cements for clinical applications. Part I: solution chemistry. J Mater Sci Mater Med 10(3):169–176
14. Zeng H, Chittur KK, Lacefield WR (1999) Analysis of bovine serum albumin adsorption on calcium phosphate and titanium surfaces. Biomaterials 20(4):377–384
15. Zhu X et al (2010) Effect of phase composition and microstructure of calcium phosphate ceramic particles on protein adsorption. Acta Biomater 6(4):1536–1541
16. Boyan BD et al (1996) Role of material surfaces in regulating bone and cartilage cell response. Biomaterials 17(2):137–146
17. Wang C et al (2004) Phenotypic expression of bone-related genes in osteoblasts grown on calcium phosphate ceramics with different phase compositions. Biomaterials 25(13):2507–2514
18. Rho J-Y, Kuhn-Spearing L, Zioupos P (1998) Mechanical properties and the hierarchical structure of bone. Med Eng Phys 20(2):92–102
19. Gao H (2006) Application of fracture mechanics concepts to hierarchical biomechanics of bone and bone-like materials. Int J Fract 138(1-4):101
20. Huang S et al (2014) A novel model for porous scaffold to match the mechanical anisotropy and the hierarchical structure of bone. Mater Lett 122:315–319
21. Reznikov N, Shahar R, Weiner S (2014) Bone hierarchical structure in three dimensions. Acta Biomater 10(9):3815–3826
22. Lemons J (1993) Inorganic and organic composition for treatment of bone lesions. Google Patents
23. Szabo CM, Martin MB, Oldfield E (2002) An investigation of bone resorption and Dictyostelium discoideum growth inhibition by bisphosphonate drugs. J Med Chem 45(14):2894–2903
24. Inanç B, Elcin AE, Elcin YM (2007) Effect of osteogenic induction on the in vitro differentiation of human embryonic stem cells cocultured with periodontal ligament fibroblasts. Artif Organs 31(11):792–800
25. Xia Z et al (2006) In vitro biodegradation of three brushite calcium phosphate cements by a macrophage cell-line. Biomaterials 27(26):4557–4565
26. Choi K et al (1990) The elastic moduli of human subchondral, trabecular, and cortical bone tissue and the size-dependency of cortical bone modulus. J Biomech 23(11):1103–1113
27. Boivin G, Meunier PJ (2003) Methodological considerations in measurement of bone mineral content. Osteoporos Int 14(5):22–28
28. Fennis J, Stoelinga P, Jansen J (2004) Mandibular reconstruction: a histological and histomorphometric study on the use of autogenous scaffolds, particulate cortico-cancellous bone grafts and platelet rich plasma in goats. Int J Oral Maxillofac Surg 33(1):48–55
29. Silva R et al (2005) The use of hydroxyapatite and autogenous cancellous bone grafts to repair bone defects in rats. Int J Oral Maxillofac Surg 34(2):178–184
30. Moore WR, Graves SE, Bain GI (2001) Synthetic bone graft substitutes. ANZ J Surg 71(6):354–361
31. Finkemeier CG (2002) Bone-grafting and bone-graft substitutes. JBJS 84(3):454–464
32. Weiner S, Wagner HD (1998) The material bone: structure-mechanical function relations. Annu Rev Mater Sci 28(1):271–298
33. Nakase T et al (1994) Alterations in the expression of osteonectin, osteopontin and osteocalcin mRNAs during the development of skeletal tissues in vivo. Bone Miner 26(2):109–122
34. Florencio-Silva R et al (2015) Biology of bone tissue: structure, function, and factors that influence bone cells. Biomed Res Int 2015:421746
35. Ferro F et al (2010) Biochemical and biophysical analyses of tissue-engineered bone obtained from three-dimensional culture of a subset of bone marrow mesenchymal stem cells. Tissue Eng Part A 16(12):3657–3667

36. Klein M et al (2009) Pore characteristics of bone substitute materials assessed by microcomputed tomography. Clin Oral Implants Res 20(1):67–74
37. Karageorgiou V, Kaplan D (2005) Porosity of 3D biomaterial scaffolds and osteogenesis. Biomaterials 26(27):5474–5491
38. Hannink G, Arts JC (2011) Bioresorbability, porosity and mechanical strength of bone substitutes: what is optimal for bone regeneration? Injury 42:S22–S25
39. Vallet-Regi M, González-Calbet JM (2004) Calcium phosphates as substitution of bone tissues. Prog Solid State Chem 32(1-2):1–31
40. Wopenka B, Pasteris JD (2005) A mineralogical perspective on the apatite in bone. Mater Sci Eng C 25(2):131–143
41. Legeros RZ (1981) Apatites in biological systems. Progr Crystal Growth Character 4(1-2):1–45
42. Dorozhkin SV, Epple M (2002) Biological and medical significance of calcium phosphates. Angew Chem Int Ed 41(17):3130–3146
43. Sun F, Zhou H, Lee J (2011) Various preparation methods of highly porous hydroxyapatite/polymer nanoscale biocomposites for bone regeneration. Acta Biomater 7(11):3813–3828
44. Ferreira A, Oliveira C, Rocha F (2003) The different phases in the precipitation of dicalcium phosphate dihydrate. J Cryst Growth 252(4):599–611
45. Tang R et al (2003) Constant composition dissolution of mixed phases: II. Selective dissolution of calcium phosphates. J Colloid Interface Sci 260(2):379–384
46. Nancollas G, Tomazic B (1974) Growth of calcium phosphate on hydroxyapatite crystals. Effect of supersaturation and ionic medium. J Phys Chem 78(22):2218–2225
47. Zhao J et al (2014) Rietveld refinement of hydroxyapatite, tricalcium phosphate and biphasic materials prepared by solution combustion method. Ceram Int 40(2):3379–3388
48. Carrodeguas RG, De Aza S (2011) α-Tricalcium phosphate: synthesis, properties and biomedical applications. Acta Biomater 7(10):3536–3546
49. Kamitakahara M, Ohtsuki C, Miyazaki T (2008) Behavior of ceramic biomaterials derived from tricalcium phosphate in physiological condition. J Biomater Appl 23(3):197–212
50. Cicek G et al (2011) Alpha-tricalcium phosphate (α-TCP): solid state synthesis from different calcium precursors and the hydraulic reactivity. J Mater Sci Mater Med 22(4):809–817
51. Nilsson M et al (2002) Characterization of a novel calcium phosphate/sulphate bone cement. J Biomed Mater Res 61(4):600–607
52. LeGeros RZ (2008) Calcium phosphate-based osteoinductive materials. Chem Rev 108(11):4742–4753
53. Yamada S et al (1997) Osteoclastic resorption of biphasic calcium phosphate ceramic in vitro. J Biomed Mater Res 37(3):346–352
54. Berry E (1967) The structure and composition of some calcium-deficient apatites. J Inorg Nucl Chem 29(2):317–327
55. Hutchens SA et al (2006) Biomimetic synthesis of calcium-deficient hydroxyapatite in a natural hydrogel. Biomaterials 27(26):4661–4670
56. Ravi ND, Balu R, Sampath Kumar T (2012) Strontium-substituted calcium deficient hydroxyapatite nanoparticles: synthesis, characterization, and antibacterial properties. J Am Ceram Soc 95(9):2700–2708
57. Klein CP et al (1990) Studies of the solubility of different calcium phosphate ceramic particles in vitro. Biomaterials 11(7):509–512
58. Yamada S et al (1997) Osteoclastic resorption of calcium phosphate ceramics with different hydroxyapatite/β-tricalcium phosphate ratios. Biomaterials 18(15):1037–1041
59. Klein C et al (1983) Biodegradation behavior of various calcium phosphate materials in bone tissue. J Biomed Mater Res 17(5):769–784
60. Groot KD (1988) Effect of porosity and physicochemical properties on the stability, resorption, and strength of calcium phosphate ceramics. Ann N Y Acad Sci 523(1):227–233
61. de Groot K (2018) Ceramics of calcium phosphates: preparation and properties. In: Bioceramics calcium phosphate. CRC, Boca Raton, FL, pp 99–114
62. Sun L et al (2010) Preparation and properties of nanoparticles of calcium phosphates with various Ca/P ratios. J Res Natl Inst Standards Technol 115(4):243

63. ŚAlósarczyk A et al (1996) Calcium phosphate materials prepared from precipitates with various calcium: phosphorus molar ratios. J Am Ceram Soc 79(10):2539–2544
64. Eyckmans J et al (2010) A clinically relevant model of osteoinduction: a process requiring calcium phosphate and BMP/Wnt signalling. J Cell Mol Med 14(6b):1845–1856
65. Samavedi S, Whittington AR, Goldstein AS (2013) Calcium phosphate ceramics in bone tissue engineering: a review of properties and their influence on cell behavior. Acta Biomater 9(9):8037–8045
66. Zayzafoon M (2006) Calcium/calmodulin signalling controls osteoblast growth and differentiation. J Cell Biochem 97(1):56–70
67. Barradas AM et al (2012) A calcium-induced signalling cascade leading to osteogenic differentiation of human bone marrow-derived mesenchymal stromal cells. Biomaterials 33(11):3205–3215
68. LeGeros RZ (2002) Properties of osteoconductive biomaterials: calcium phosphates. Clin Orthop Relat Res 395:81–98
69. Lu J, Yu H, Chen C (2018) Biological properties of calcium phosphate biomaterials for bone repair: a review. RSC Adv 8(4):2015–2033
70. Li S et al (2003) Macroporous biphasic calcium phosphate scaffold with high permeability/porosity ratio. Tissue Eng 9(3):535–548
71. Kitsugi T et al (1993) Four calcium phosphate ceramics as bone substitutes for non-weight-bearing. Biomaterials 14(3):216–224
72. Ambard AJ, Mueninghoff L (2006) Calcium phosphate cement: review of mechanical and biological properties. J Prosthodont 15(5):321–328
73. Sheikh Z et al (2015) Mechanisms of in vivo degradation and resorption of calcium phosphate based biomaterials. Materials 8(11):7913–7925
74. Benahmed M et al (1996) Biodegradation of synthetic biphasic calcium phosphate by human monocytes in vitro: a morphological study. Biomaterials 17(22):2173–2178
75. Aparicio JL et al (2016) Effect of physicochemical properties of a cement based on silicocarnotite/calcium silicate on in vitro cell adhesion and in vivo cement degradation. Biomed Mater 11(4):045005
76. Heymann D, Pradal G, Benahmed M (1999) Cellular mechanisms of calcium phosphate ceramic degradation. Histol Histopathol 14(3):871–877
77. Sun H et al (2006) Proliferation and osteoblastic differentiation of human bone marrow-derived stromal cells on akermanite-bioactive ceramics. Biomaterials 27(33):5651–5657
78. Wang J et al (1998) Biological evaluation of biphasic calcium phosphate ceramic vertebral laminae. Biomaterials 19(15):1387–1392
79. Yang X (2017) Hydroxyapatite: design with nature. In: Orthopedic biomaterials. Springer, Berlin, pp 141–165
80. Venkatesan J et al (2015) Alginate composites for bone tissue engineering: a review. Int J Biol Macromol 72:269–281
81. Du C et al (2000) Formation of calcium phosphate/collagen composites through mineralization of collagen matrix. J Biomed Mater Res 50(4):518–527
82. Mendonça G et al (2008) Advancing dental implant surface technology–from micron-to nanotopography. Biomaterials 29(28):3822–3835
83. Hu Q et al (2007) Effect of crystallinity of calcium phosphate nanoparticles on adhesion, proliferation, and differentiation of bone marrow mesenchymal stem cells. J Mater Chem 17(44):4690–4698
84. Dulgar-Tulloch A, Bizios R, Siegel R (2009) Human mesenchymal stem cell adhesion and proliferation in response to ceramic chemistry and nanoscale topography. J Biomed Mater Res A 90(2):586–594
85. Xu B et al (2012) RhoA/ROCK, cytoskeletal dynamics, and focal adhesion kinase are required for mechanical stretch-induced tenogenic differentiation of human mesenchymal stem cells. J Cell Physiol 227(6):2722–2729

Bioactive Glasses in Orthopedic Applications

Jena Madison, Joy-anne N. Oliver, and Donghui Zhu

Abstract Bioactive glasses currently play a rather small role in orthopedic procedures when compared to metallic implants, despite its superior bioactivity within the body. Common metallic materials that are used to engineer and manufacture orthopedic implants generally consist of titanium and its alloys, cobalt-chromium alloys, and stainless steel 316L as they provide the best material properties based on their applications. However, these implants tend to carry the following risks: infection, surrounding tissue damage, and improper healing. Typically, it is not the implant that is directly responsible for the tissue damage or rejection, but the interaction between the bodily tissue and the implant. In some cases, when in contact with the physiological environment, a fibrous capsule of scar tissue has been known to form around the grafted implant, therefore interfering with the integration of the implant with the surrounding tissue. Bioactive glass presents as an alternative material that can be considered for surgical orthopedic applications. One major limitation that has prevented bioactive glasses dominance in the orthopedic surgical realm is its lack of ductility, strength, and ability to withstand load-bearing applications in supporting larger bones. This book chapter intends to showcase how bioactive glass is currently used in tissue engineering and its ability to improve and change current commonly used orthopedic implants.

Keywords Bioactive glasses · Tissue engineering · Bone regeneration Orthopedics

J. Madison
Department of Materials Science and Engineering, University of North Texas, Denton, TX, USA

J.-a. N. Oliver (✉)
Department of Materials Science and Engineering, University of North Texas, Denton, TX, USA

Department of Biomedical Engineering, University of North Texas, Denton, TX, USA
e-mail: Joy-anneoliver@my.unt.edu

D. Zhu (✉)
Department of Biomedical Engineering, Stony Brook University, Stony Brook, NY, USA
e-mail: Donghui.Zhu@stonybrook.edu

B. Li et al. (eds.), *Racing for the Surface*,
https://doi.org/10.1007/978-3-030-34471-9_21

Introduction

Biomaterials are man-made materials used within the body to repair or aid in function of an injured or damaged body part. Biomaterials must encompass several characteristics in order for them to be used within the body. Biocompatibility is a major requirement of biomaterials so that the biomaterial does not cause harm to the body. In order for biomaterials to be effective, they must not be toxic nor cause a severe reaction in the surrounding tissue [1]. The ability for biomaterials to form a bone-like hydroxyapatite layer on its surface provides a mineral phase that is present in natural bone tissue, and can therefore decrease the possibility of rejection [2]. This consideration improves on the fact that synthetic polymers and metals can easily be considered and treated as a foreign entity by the body due to their characterization of being chemically inert [2]. Biomaterials are most commonly used in orthopedic procedures to fix bone defects and injuries [1]. These materials can be bioactive as they can interact with the surroundings and illicit a biological reaction between the material and surrounding [3]. Not all biomaterials exhibit this bioactive property with its surrounding. Typically, there are three types of biomaterials: bioinert materials that are nontoxic and biologically inactive causing various thicknesses of fibrous tissues to surround, bioactive materials that is nontoxic and biologically active, thus forming interfacial bonds, and biodegradable or bioresorbable materials which are considered third-generation materials that can dissolve while being replaced with the surrounding tissue [4]. An example of a commonly used biomaterial is titanium that is used to make screws for dental implants. Titanium is a metallic implant that can be considered a biomaterial and can therefore interact with bone cells, also known asosseointegration [5]. Metallic implants are heavily utilized in orthopedic surgeries to repair injured and deformed bones.

Bioactive glasses are biomaterials that exhibit bioactivity when in the presence of a biological environment. They are nontoxic and exhibit bioactivity in orthopedics by interacting with the bodily fluid it is submerged in causing osteoconduction, and the ability to exhibit bone growth on the surface of a material. Bioinert materials such as stainless steel and cobalt-chromium alloys are more relied on currently for orthopedic surgeries due to their known success over a number of decades compared to their failures which include leaching by corrosion of the metal into the bone matrix due to drastic differences in elastic moduli between the materials and natural bone tissue [6], and possibly due to the lack of competitive materials. Orthopedic surgery treats disease of the musculoskeletal system that includes bones, joints, ligaments, muscles, nerves, and tendons. Therefore, bioactive glasses should be able to treat both soft and hard tissue diseases.

Current Orthopedic Application of Metallic Implants

Orthopedic surgeries are performed daily to repair bone tissue due to traumatic injuries, disease, and degeneration. The body is a powerful and restorative force; however, there are instances in which it is beyond self-repair and requires fixed orthopedic devices to aid in its recovery. Orthopedic procedures rely heavily on the use of an array of biomaterials to stabilize, replace, and repair bone tissue. Bone fractures and tissue injuries are traditionally repaired with screws, wires, and plates made mostly of different metallic materials. In order for a metal to be considered a biomaterial and be used in orthopedic procedures, it must be able to function within the body without causing significant harm or adverse reaction and be corrosion resistant. This metal must also be strong enough to support a substantial amount of weight and fluctuating forces.

Traditional orthopedic metallic materials include titanium and its alloys, stainless steel 316L, and cobalt-chromium alloys [7]. Metal orthopedic devices and their usage depend on the structure, i.e., body part that is being repaired, and the type of injury or deformity of the bone. Screws are implant structures that are either used alone or in conjunction with plates depending on the type of damage the bone has sustained and the location of the damage [8]. There are various forms of screws used in orthopedic procedures; however, most of them are composed of stainless steel. Stainless steel is a strong and resilient metal containing anticorrosive properties that withstand body pressure without deformation. The stainless steel 316L in orthopedic surgeries contains carbon, chromium, and nickel [9, 10]. Surgical stainless steel is the go-to material when it comes to traumatic orthopedic injuries. Surgical stainless steel is used to create a wide variety of different types of screws. Cortical anchor screws and cannulated screws are two major variations of screws that are used depending on the type of tissue that the screw will go through. Screws are inserted through the formation of a hole into the bone and then feed through the fragmented pieces [8]. Screws tend to weaken the bone through these holes that are formed leaving it vulnerable to further stress that can take place in the form of another fracture [8]. Screws range in size from a few millimeters when being inserted into a place like the wrist to several inches when applied into and down the shaft of long bones such as the femur for femoral repairs. Metallic orthopedic plates are used in conjunction with screws in order to stabilize bone tissue mostly through the use of compressive forces [8]. Metallic plates are often composed of either stainless steel or titanium. The combination of plates and screws apply a compressive force to either keep the bone from moving for proper healing in its correct alignment or to stabilize and bring together a bone that has fragmented into several pieces. Titanium carries anticorrosive properties and features better bone integration than stainless steel [11]. It has shown that although both material implants produce full functional recovery in similar time frames, titanium is shown to be more compatible with bone tissue than stainless steel as bone tissue easily bonds to the metal surface with significantly less periosteal reaction, soft tissue swelling, and callus formation [11]. Moreover, the tight integration between the titanium and bone helps to keep it in place preventing loosening of the implant. Medical surgical wire is another form of orthopedic device that is utilized in orthopedic surgeries to provide a more malleable option in setting fractures. Medical surgical wires can be made using any appropriate metal alloy such as cobalt-chromium. Depending on the diameter of the wire, it could be applied as a minimally invasive procedure [8]. Cobalt-chromium alloy is another material offering anticorrosive properties and high fatigue resistance. The high fatigue resistance gives the cobalt-chromium implant the ability to take the load of the bone allowing the bone to have better healing. These traditional metal fixtures for repairing diseased, arterially damaged and injured orthopedic ailments offer a huge array of choices. Fixtures include devices such as various screws, plates, and wires depending on which is most appropriate for the repair. Choosing the type of metal that is most compatible with the patient is another important factor that must be considered. Metal orthopedic devices are currently used and work well due to their strength and slow degradation within the body [7]. The requirement for device removal can be due to several factors, one being implant instability due to degradation that occurs over a long period of time because of the corrosion of the metal, allowing debris to enter the bone matrix and blood supply [5, 6]. The potential of medical equipment failing overuse can be due to load bearing, which increases instability [8]. The requirement for a second procedure amplifies the risk for the patient to occur further damage as well as infection and improper removal. Figure 1 displays examples of failures that have been observed with metal implants, therefore introducing the need for a follow-up surgery.

All metallic implants have the possibility to cause irritation to the surrounding tissue with some metals such as surgical stainless steel having a higher irritation rate than titanium [7]. Allergic reactions occur depending on the type of metal and patient sensitivity to the metal. Alongside irritation and allergic reaction, there is an infection, which is the most critical risk. Infection possesses the greatest threat to the patient's life and the associated body part. An infection starting off in the bone or surgical site can quickly lead to a systemic infection leading to the patient becom-

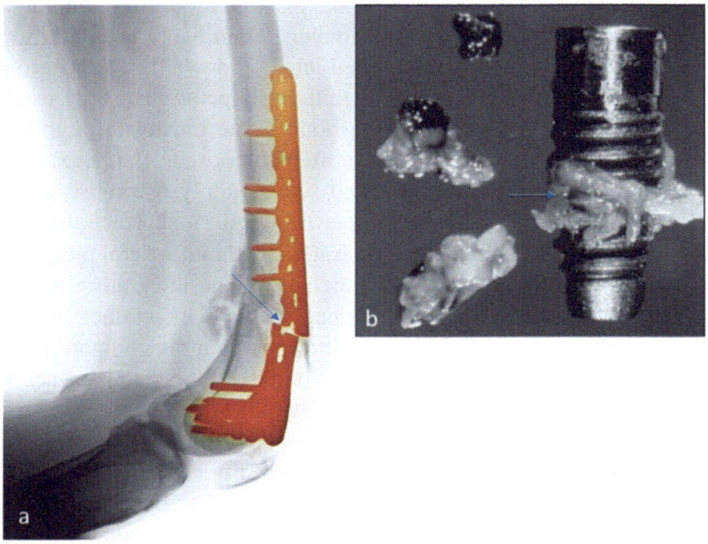

Fig. 1 Defective metal implants. (**a**) Fatigue failure in bone implant. Arrow indicates the location of failure [37]. (**b**) Failed titanium implant screw removed due to corrosion. Arrow indicates the presence of tissue that remained bonded to the implant [38]

ing septic. Sepsis is a major problem that can quickly lead to a patient fighting for their life due to the infection spreading throughout their body and to major organs causing them to shut down [8]. Additionally, localized infection can result in the loss of the implant or the limb that the implant supports. The risks associated with metal surgical devices have presented the question and investigation of other types of materials to fix and/or replace these materials. A major alternative to fixing these metals and making them more suitable for implantation is altogether replacing them with new materials that behave better biologically. Of the new types of materials that have been widely explored, bioactive glasses have surfaced. Bioactive glasses are currently being used as void fillers, or as the material of which implants are constructed and are showing success when being utilized in orthopedic procedures [12].

Bioactive Glass: Background and Future Perspective

Bioactive glasses are novel materials that are different from conventional glasses that have the ability to bond to tissue and are biocompatible [13]. In 1969, Larry Hench invented the third-generation material as his aim was to develop a material capable of bonding to bone, after having a conversation with a US Army colonel who had just returned from the Vietnam War and questioned if materials could be developed that could survive the harsh environment that the human body presents [14]. This was of concern because the present materials such as stainless steel that was in use could result in encapsulation around the implant [14]. Larry Hench then went ahead to develop the first novel material, Bioglass™ 45S5, a silicate-based glass-ceramic material that was trademarked by the University of Florida. The outstanding property of this material was that had the ability to facilitate bone growth away from the bone-implant interface and it was degradable [15]. This composition consisted of a Na_2O-CaO-SiO_2-P_2O_5 system that is high in calcium content [16]. This discovery set a strong foundation for the future developments and expectations in bioactive materials. Figure 2 provides a schematic of the composition requirements for Bioglass™ that would accomplish different glass characteristics.

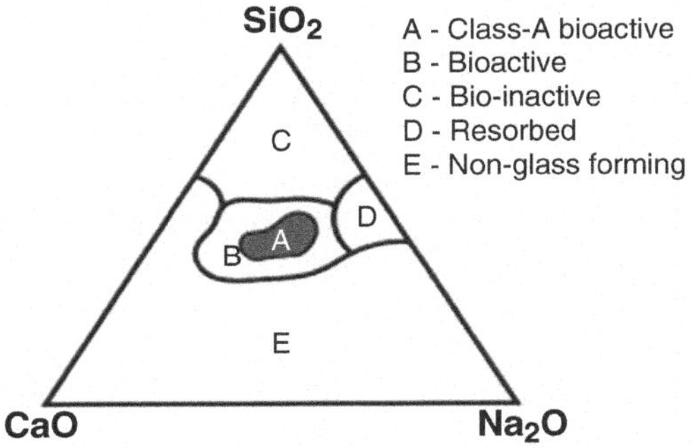

Fig. 2 Ternary phase diagram also known as the kinetic diagram of bioactivity on SiO-CaO-Na_2O system that is used to determine the mass compositions required to achieve different characteristics in glass [39]. Available via license CC by 2.0

This material is used in orthopedic procedures amongst other areas to help repair damages that have occurred due to fracture, degeneration, and/or deformation. Bioactive glasses inserted into damaged tissue interact with its surroundings to initiate healthy tissue growth. Ions like sodium and calcium are essential in creating bone tissue and bioactive glass brings these essential ions in to assist bone formation. Additionally, these materials can be deemed as safe since they are no stranger to the bodily environment. Bioactive glass presents as a healthier and more natural material for implantation versus the currently used metallic material [17]. The interaction of surroundings with the bioactive glass causes the glass to naturally degrade as new bone tissue is created. Although bioactive glass is a better biologic substance especially for orthopedic repair, they are not structurally sound enough to completely rely on. The continued advancements of bioactive glass can take care of this current issue by creating a stronger more stable material. The formation of a more structurally sound bioactive glass would result in bioactive glass being used in more and more orthopedic procedures, therefore replacing metallic implants little by little. This replacement or reduction of metallic implants will lead to better orthopedic procedure results. For now, the types of developed bioactive glasses may not have the required strength to serve as large hard tissue implants and high load bearing application but they can be used to develop smaller implanting materials such as screws and stent. A stent can be considered as an impression of a body part or cavity that can be used to maintain constant pressure and support healing in an area such as a skin graft. In other cases, it can be considered as a thin, tubular structure that provides temporary support in blood vessels. Additionally, bioactive glasses may also be applied as coatings to improve the bonding mechanism between metal implants and surrounding tissue [18]. Bioactive glass has the opportunity with advancements to completely change the orthopedic game with further advancements in not only bone tissue but also in the soft tissue structures that surround them.

Bioactive Glass Composition and Formation

Different glass compositions include oxide glasses that consist of silica-based glasses and nonsilica-based glasses, and non-oxide glasses that consists of the following types: heavy-metal fluoride glasses, glassy metals, semiconducting solids. Other types of glasses that exist are glass ceramics, glass composites, and natural glass. Bioactive glass is a silica-based compound that encompasses a solid and malleable form and can consist of different families of compositions depending on their properties. A typical phospho-silicate is composed of sodium and calcium ions. Sodium and calcium oxides are mixed within the phospho-silicate material to form the compound matrix Na_2O-CaO-SiO_2-P_2O_5 glass structure [19]. The glass is formed through melting of the compound at an extremely high temperature and then submerging it in water. The submerging of the substance in water causes crystallization in which it takes on a solid structure [19]. Although bioactive glass is formed in this fashion, it is not readily used in its crystalline solid structure for most applications.

Bioactive glass is utilized mostly in its putty form since it is easily molded to any shape, thus giving it the ability to fit into small crevices. Bioactive glass is also available in other forms apart from a putty-like texture; however, the ability for bioactive glasses to completely replace metal as implants is not yet possible due to vast differences in mechanical characteristics such as ductility since ceramics are inherently known to be more brittle than metals and can spontaneously fracture under load cycles of compressions in the application of large bone tissue implantation for instance. The flexibility of bioactive glass putty allows for it to fit in cracks to repair fractured bones [19]. Bioactive glasses phospho-silicate structure is composed of 45% silica, 25% calcium oxide, 24.5 sodium oxide, and 6% phosphorous pentoxide, all naturally occurring minerals inside of the body [3]. The fact that these minerals are already present within the body makes it easier for them to be integrated into the injured tissue and allow healing to naturally occur.

In order for a material to be considered when forming bioactive glasses, certain properties such as bioactivity must be present [3]. The higher the bioactivity the better the material integrates in the body when utilized in surgical procedures. Bioactivity is determined by the amount of silica present, the lower the concentration of silica the stronger the bioactive material. Bone tissue growth requires a threshold amount of silica present for bone formation. Areas that require repair with weaker frailer bones can have a higher concentration of silica because the strength of material is not as important. There are several different compositions of bioactive glasses each with different elemental concentrations, which determines how and where it can be applied based on the requirement of the repair area.

Bioactive Glasses Reaction Mechanism and Integration Inside the Body

Bioactive glass implantation causes a natural cascade of exchanges between the glass and its surroundings within the body. This cascade leads to degradation of bioactive glass while simultaneous forming new natural bone in its place. This breakdown of bioactive glass and its exchange with bone creates a less adverse effect compared to its metallic counterparts [20]. Metals do not degrade as do bioactive glasses since they are only known to interact with the surrounding tissue. Although bioactive glass offers a more natural and symbiotic relationship, it is has a few limitations that have permitted it from being solely and more commonly used in orthopedic application.

Bioactive glass contains many biologically occurring elements, specifically those that are responsible for bone tissue formation such as hydroxyapatite. Hydroxyapatite is an essential element of bioactive glass that binds to tissue and forms a surface where cells can adhere to and from new bone tissue [21]. Hydroxyapatite naturally occurs in bone tissue and elicits a chemical reaction within the body that attracts compounds and materials necessary to form new bone tissue [21]. Hydroxyapatite gives bone its strength in conjunction with calcium, therefore allowing the bone to not only regenerate but also give the bone its rigid structure while it is healing thereby protecting it [21]. When hydroxyapatite binds an ion exchange cascade occurs between the glass and the surrounding body fluids resulting in the natural formation of bone tissue and glass degradation simultaneously. The ion exchange occurring contributes to uptake of hydrogen ions, thereby increasing the concentration of hydroxide that drives up the localized pH [2]. The silica that is present in the glass is displaced with this pH change, causing an increase in the concentration of silica in its surrounding due to its release from the glass [2]. The increase in silica in conjunction with the influx of calcium and phosphate to the glass causes an amorphous calcium phosphate layer to build on the glass. This amorphous phosphate layer is responsible for bringing in bone growth factors that actually allow the natural production of bone to take place [2].

Osteoclast, osteocytes, and osteoblasts are bone cells that are targeted to and adhere to the surface of this amorphous layer [21]. Creating new bone tissue through the building of osteoblast takes a few weeks in which they are formed, activated and reabsorbed [10]. Once the work of laying down the foundation of the bone tissue has occurred via osteoclast, then osteoblasts come in and form the new bone tissue. Osteoblast forms bone tissue by laying down collagen fiber matrices where osteoclasts attach forming the bone cytoskeleton [21]. Osteoblasts form matrices and the matrices signal for various minerals such as calcium, phosphates, and magnesium to come together. The unity and conjunction of these cells are what give bones their sturdy structure [21]. Bioactive glasses are dissolvable within the human body and it continues to send signals with degradation. The signals that are sent out with degradation illicit an immune response, which continues to aid in bone tissue formation. The glass reacts initially with the body fluid, which results in the flow of ions into its surrounding environment, leading to the formation of a hydroxyapatite surface on top of the glass structure [21]. Bioactive glass can be a highly vascular structure due to its ability to form a porous structure, which allows ions and blood flow to enter resulting in an accelerated rate of bone formation [22]. Bioactive glass essentially acts as a stem cell for bones and bone formation [23]. The degradation process then takes shape in which dissolution begins to occur within the glass/crystal interface targeting the areas in which defects are present first [21]. Once the defects are targeted and broken down this is when the above-stated ion exchange begins to occur. This cascade of bodily fluid and glass exchange continue to occur until the glass is fully degraded [23]. As the bioactive glass and breaks down in the presence of the tissue envirenment as it interacts with the surrounding tissue, the needed elements for natural bone formation are laid down and bone formation begins layer-by-layer building on top of what was the bioactive glass as shown in Fig. 2.

Types of Bioactive Glasses

Bioactive glasses come in different forms, as they consist of different families, with each family consisting of different compositions, therefore expanding the structures and properties and in turn their applications. Currently, they are integrated in fixing bone tissue by replacing small bones and/or seal cracks in larger bones. Bioactive glass can also be used to form bone grafts, which are made through the combination of bioactive glass and polymers formed into matrices [21]. A commercially available bioactive glass, GlassBone, and its applications are presented in Fig. 3.

Bioglass™ 45S5 glass was the first bioactive glass ever used. Upon visual examination after repair via X-ray, the bioactive glass was undetectable from the bone tissue [12]. The Bioglass™ 45S5 is used in a variety of areas across the body encompassing the areas of oral and maxillofacial, otolaryngology, and orthopedic specialties. Bioglass™, however, has been highly used in oral and maxillofacial cases where jaw and tooth repair are necessary [12]. Bioglass™ 45S5 was further trademarked through the brand name NovaBone™. NovaBone™ is another bioactive glass that is highly utilized in dental and oral maxillofacial application [12]. It is used by extracting the patients' blood and mixing it with the NovaBone™ putty formula and is fixed in place either in conjunction with metallic implants or alone [24]. NovaBone™ as seen in Fig. 4 is a bioactive synthetic bone graft made of a composition of bioactive glass that is offered in a putty form and can be molded into any shape to be inserted in the respected place that is requiring repair. NovaBone™ has the ability to signal genetic pathways to accelerate natural bone growth and is primarily used within the field of orthopedics as a defect filler that promotes new bone formation. One specific feature about NovaBone™ is that it consists of a porous structure that is similar to cancellous bone. Therefore, it offers that open structure that facilitates rapid vascularization and mineralization that leads to complete device absorption and bone replacement. Additionally, NovaBone™ is very malleable and easy to mold and pack into any defect, it remains at the surgical site

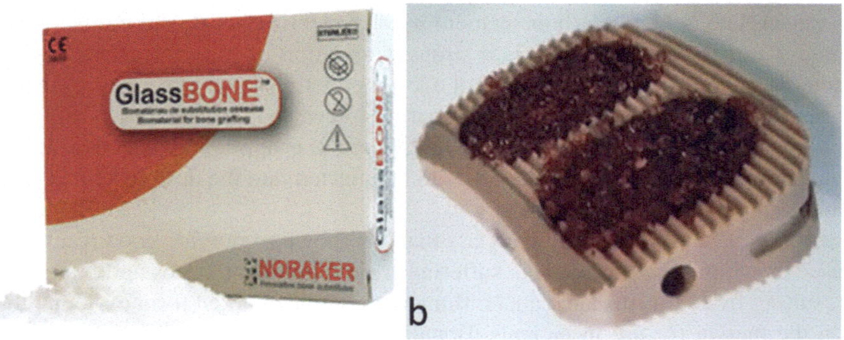

Fig. 3 (**a**) Commercially available GlassBone mixed with blood, (**b**) to make a putty and inserted in a spinal fusion cage. Image with permission courtesy of Noraker [40]

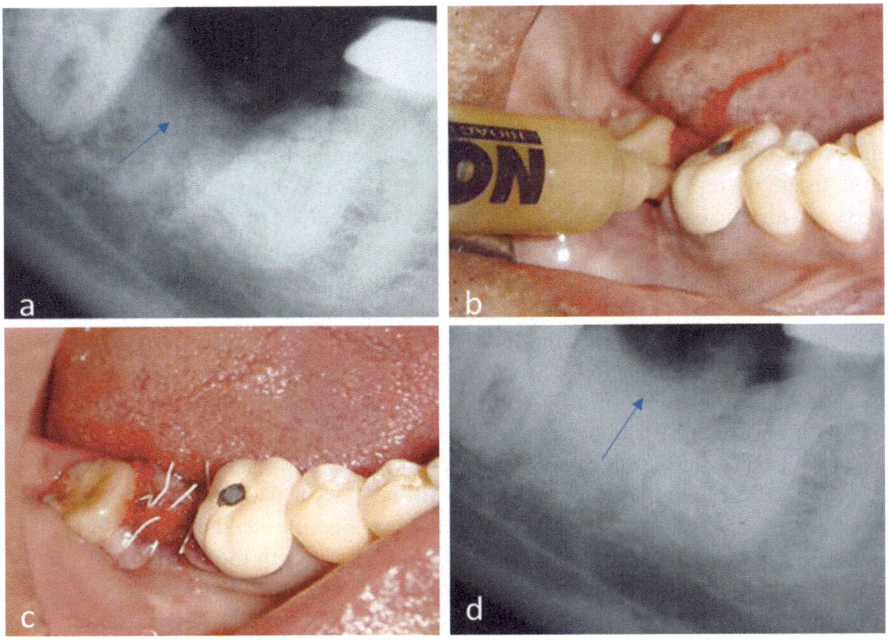

Fig. 4 (**a**) X-ray image of jawline proceeding tooth extraction. Arrow indicates area of bone loss experienced after extraction. (**b**) Application of NovaBone to surgical site for bone grafting. (**c**) Surgical site healing. (**d**) X-ray demonstrating complete healing of surgical site (shown by arrow). Images courtesy of [41]

without migrating, and it is ready to use out of the package. Other applications of NovaBone™ include cartridge delivery, osteoconductive bone grafting, providing osteoconductive matrix, and cellular and protein binding substrate in bone grafting [25]. NovaBone™ not only comes in a putty form that is delivered via syringe but also comes in forms of strips and scaffolds that can be molded and shaped [24].

Cortoss is another class of bioactive glass that is constructed of a polymethyl methacrylate, which is a bone cement used to stabilize weak and damaged vertebrae. Cortoss bioactive glass is a stronger variation of bioactive glass due to its cement property. The placement of Cortoss in the vertebrate requires strength because the spine undergoes weight-bearing forces at all times from as simple of a task as sitting up to standing and walking. Cortoss is applied to defective vertebrae via a syringe that allows the materials to be injected into the desired spot in a minimally invasive procedure [26].

S53P4 is a form of bioactive glass that has shown to provide good results when used as a bone graft in patients suffering from traumatic injuries and bone defects resulting from arthritis and cancer. Bonalive is a trademarked form of S53p4 that is on the market for use in patients. Bonalive putty is a form of bioactive glass that comes in a ready to use syringe, thus requiring no preparation shown below in Fig. 1 [27]. Unlike NovaBone™, Bonalive does not require mixing or any further preparation before it is ready to be used on a patient. The S53P4 bioactive glass is used in conjunction with polyethylene glycols and glycerol, which acts as a binder in the compound forming the putty-like consistency [12]. Bonalive not only comes in putty form but is also available in granule form. The granules form of Bonalive provides an antibacterial property by increasing the pH that gives the surroundings a more basic pH [27]. A basic pH decreases the growth of microbes which directly decrease and can eliminate the rate of infection in patients undergoing surgical repair [27]. Bonalive granules require preparation unlike its putty composition that comes in ready to use. The granules come in a syringe that requires the addition of saline and then is placed in its targeted site via a syringe containing a shoveled end. The shovel tip allows the granules to be guided directly into their targeted location, which is important because the granules must only be placed in their intended places. Any granules that are misplaced must be removed to prevent bone formation in an unnecessary location since these granules will cause bone formation wherever they are placed.

Bioactive Glass Integration with Metal Implants

Bioactive glass is not only being used independently in its putty form to repair defects and fractures but is also being integrated in metallic orthopedic appliances [28]. Metallic devices on their own are not naturally prone to bond and cause chemical reactions that drive bone reformation. Bioactive glasses have been integrated into traditional metallic medical equipment by giving metals a bioactive glass coating [29]. The plates used to piece together fractured bones are taken before being surgically implanted and are first coated with bioactive glass. The integration of the bioactive glass coated metallic implant allows for the metal to better stabilize and integrate into the bone allowing for better healing [28]. Metal, when implanted alone, tends to form fibrous tissue, which is a thickening of tissue that surrounds the implant. The thickening of the tissue is due to the rubbing of the metal and the nearby tissue, thus forming scar tissue. The coating of these metal implants with bioactive glass allows a reduction in the formation of scar tissue. The coating of metal implants with bioactive glass gives it better biological compatibility once inserted into the tissue [28]. Bioactive glass stimulates the creation of bone tissue through its release of ions that in turn bring in needed materials to create healthy new bone tissue in place of injured tissue. As shown below in Fig. 3, bioactive glass has been integrated into a metallic implant by applying it into the grooves of the screw. The addition of bioactive glass into the grooves of the screw allows it to directly contact the bone shaft and allowing it better integrate into the bone that it is surrounded by [28]. Bone formation will occur on the ridges that will engulf the screw into the bone tissue, thereby stabilizing the screw and the fracture or defect. Bioactive glass is heavily used in titanium implants through creating pores where bone can form. The occurrence of having natural forming bone rather than a purely metallic implant allows better mobility. Bioactive glass integration into ball and socket joint repairs is a great example of increasing mobility with bioactive glass as shown in Fig. 4. Ball and socket joints consist of hips and shoulder joints where a ball-like structure sits into a hollowed space with a long bone attached. In these ball and socket structures, the ball can utilize a bioactive glass coating on the ball or femoral head [28]. By applying a bioactive glass coating to the ball, it allows natural bone tissue to form, thus better integrating the metallic implant into the structure. A better integration of the ball and joint socket allows for natural tissue to form, which will reduce the chance of the metallic implant loosening and decrease in deterioration of the implant. The coating of the femoral head allows the implant to have a more customized and comfortable fit for each and every individual patient. A customized fit means better overall healing and result for each patient minimizing discomfort and complications [30].

Advantages of Bioactive Glass

Bioactive glass currently plays a major role in smaller bones such as in jaw and facial procedures where there is no weight-bearing capacity required. It is heavily utilized currently in these procedures due to bioactive glasses brittleness. Bioactive glasses fragility leaves it incapable of fulfilling and sustaining repair of larger bones where it will be subjected to weight bearing such as in the legs and hip. Bioactive glass provides a great alternative to metallic implants in smaller bones due to its ability to form tissue and degrade naturally. The degradation of the material gets rid of several of the risk factors associated with metallic implants such as allergic reaction, surrounding tissue irritation, and the possibility of a second surgery for removal.

Bioactive glass also works very well and better than metallic implants in areas that cannot tolerate metal such as the orbital. Repairing an orbital fracture with metallic implants is not optimal due to the orbital socket housing the eyeball, which cannot endure punctures. Placing metal screws and plates near the eye is not an option without risking significant consequences such as loss of vision. Bioactive glass in this scenario poses no threat to the eyeball or vision. Bioactive glass is not only safer, but also easy to use in making this repair as the putty is simply modeled to fill in and fix the fracture and deformity of the orbital. Metallic implants are harder to work with and are rigid in their structure, therefore, it cannot be easily fit into many spaces without alteration of the material. The ease of use and bioactive glass's ability to fix injuries without causing further damage to the surrounding tissue is not its only advantage it possesses over metallic implants.

A bioactive glass containing naturally occurring components is essential in creating bone grafting for fractures and deformities and can contribute to forming healthy bone tissue compared to alternative metallic structures [21]. It not only contains compounds and ions that participate in bone healing, but it also provides a stable surface for this bone formation to occur [21].

Metal implants when degraded illicit no benefit to its surrounding or to the healing process while bioactive glass tends to contribute to the healing and formation of bone through its entirety forming a symbiotic relationship with its environment.

Metal implants are susceptible to infection, which could compromise not only the bone tissue but the health of the patient. Bioactive glass contains a natural antibiotic property that allows healing to happen while killing off infection without any additional treatment. As previously stated, bone infection is a major complication of metallic medical device implantation. An infection starting in the bone not only puts the bone at risk, which could lead to further bone damage but also amputation. Amputation and removal of the limb can be a consequence of a spreading bone infection. Bone infection can also easily spread to other parts of the body like the heart and lungs. Infections of major organs can lead to sepsis and ultimately can take the patient's life [28]. Avoiding infection is a major concern but having a comfortable and functional joint is also key in successful surgery.

Bioactive glass as previously stated a more comfortable joint due to customization and provides a better integration into the bone than metal with its ability to chemically interact with the surrounding soft or hard tissue. Bioactive glass integrates into the structures versus sitting on top of the bone and being screwed into it and further damaging the bone like its metal counterparts. Integration is especially beneficial because a vast amount of metallic implants require a second surgery to remove them, which again presents patients with all of the initial risks they faced with the first surgery [30]. Risk factors of metallic implants include complications such as infection, re-fracture, blood clots, and nerve damage. Metallic structures can not only require additional operations, but in some cases, they corrode with continued contact with body fluids and those by-products created can cause harm and adverse reactions to the patients once the micro-particles enter the bloodstream. Bioactive glass as a third-generation implantation device can be considered to decrease the rate of implant rejection and reaction due to its ability to integrate into the bone and work with its natural properties instead of having its own properties that make it unsuitable inside of the body. A decrease in rejection rate can decrease the rate of complication and additional surgeries, which is sought after for both the patient and surgeon. Bioactive glass provides a more custom fit for not only each patient but for each injury type and body part. Having customization for structures to fix and stabilize fractures would mean a better result in healing and fixing structural issues. Metal implants are mass-produced and therefore come in one shape and size and do not always allow a proper fit for every patient. In order to get these metallic implants to properly fit in their designated location, it requires the surgeon to work harder to best replicate the shape of the injury. The surgeon's job becomes more challenging when they have to try to mold and customize metallic implants to properly fix and stabilize a patient's injury or defect.

Future Advancements in Bioactive Glass

Bioactive glasses have many issues that require working out but its potential to be a revolutionary game changer for orthopedic surgery is significant. One major factor that is holding bioactive glasses back from its full potential is its brittleness and high stress which affects its strength. The strength limitation in current bioactive glass cripples its ability to be used in larger weight-bearing bones that suffer injury. Larger weight-bearing bones endure compressive forces from walking, standing, and running [12]. Healing of larger bones such as femurs is contingent on gravitational forces and the pressure associated with compressive forces in order to properly heal and form bone tissue. Bioactive glass in its current formulation would shatter or deform if it were utilized solely in large bone repairs. Due to bioactive glasses' limited strength, it is better suited to be utilized in the smaller bones as is currently used; however, with advancements in bioactive glasses composition and strength it has great potential to increase its strength. The increase in bioactive glasses' strength gives it the possibility to be used in larger weight-bearing bones and the ability to withstand weight-bearing forces. The overall benefit that bioactive glass possesses completely outweighs those of metallic implants making it the superior material once it has weight-bearing capability. Research has shown this to be a very important aspect, and as much as porosity can improve bone regeneration, it negatively affects weight-bearing capability [31]. Bioactive glass being the superior material could lead to the reduction and possible elimination of metallic implants altogether, which would be a major advancement in the orthopedic surgical realm. The substitution of metal for glass can reduce complications associated with current orthopedic surgery procedures such as infection which is a major surgical problem occurring today. This does not discount the possibility of infection, since risk of infection is always possible once the skin is opened in a surgery. However, since the presence of a third-generation implant device interacts with the physiological environment in such a way that it facilitates bone growth and integration, the formation of a fibrous capsule around the implant is less likely to occur. The use of bioactive glass solely in things like bone fractures will result in increasing the rate of healing and patient success [30]. Bioactive glass is also being explored and looked into when it comes to creating 3D printed bone for patients [28]. Creating a 3D printed structure would be particularly useful when it comes to creating a surface for bone tissue to grow that may be too large of an area for a bioactive glass paste to be applied as shown in Fig. 5 [28]. A 3D printing option would be extremely beneficial in instances where a large amount of bone tissue is missing or is defective. This would provide a better option over its metallic counterpart in that customization would provide simplicity over molding a metal plate to fit a specific region. Fractures usually exhibit irregular shapes, making it difficult for the commonly implant metals to be manipulated to adhere to the required shape for the patient's body and injury.

Bioactive glass has potential to not only fix and repair bones when it comes to the orthopedic realm but is also being explored in the pharmaceutical realm. Bioactive glasses potential encompasses the prospect of being used in drug delivery systems to heal bone ailments [32]. Drug delivery systems through bioactive glass would occur naturally with the degradation that occurs within the bioactive glass [32]. Through the degradation of bioactive glass, the release of a variety of potential drugs can occur in the future. The drugs are to be integrated into the bioactive glass structure by being placed inside the deeper layers of the bioactive glass. The degradation of the glass would thereby allow continuous release of the drugs as those layers are exposed. The mechanism of drug release would be determined by the rate of degradation as well as the pH of the surrounding environment of the glass [32].

One of the potential drug classes that are already being sought after for this system is antibiotic drugs. Bioactive glasses are still undergoing testing to establish degradation rates and drug release mechanisms. Drug delivery systems are complex and contain numerous dynamics that are currently being investigated with concern to bioactive glasses as this area shows tremendous potential. The ability to directly target a specific area for treatment could be a game changer not just for antibiotics but also for potential cancer therapies [32]. Administering cancer therapy via bioactive glass has been an idea that would revolutionize the approach to cancer treatment that can allow treatment without subjecting patients to systemic chemotherapy treatments. Currently, major treatment for cancer therapy involves radiotherapy and chemotherapy. Another treatment includes external beam radiation. However, this application is limited because this high localized level of radiation can also kill surrounding tissues. These systemic drugs destroy not just the diseased tissue but the tissue surrounding it, leaving patients fragile and weak. Treating cancer with a targeted therapy would be revolutionary and would result in a decrease in side effects from treatment. A decrease in side effect could lead to a better rate and result of successful cancer treatment and remission for patients. This treatment would again play a major role in bone cancers, which can be extremely aggressive due to the bone marrow in the bone creating blood cells that help the body fight invaders and function properly. Instead of destroying numerous cells and circulating chemotherapy drugs through the entire body, the chemotherapy can be directly targeted to the cancerous cell within the bone tissue. Bioactive glass compositions including yttrium has been considered able to deliver high levels of radiation to kill cancer cells in organs such as the liver [31]. Bioactive glasses have been considered as being exceptional drug carrier materials because they possess the following properties that make them suitable options: they have the ability to enhance growth of new, healthy tissues after treatment with radionuclides, they have high biocompatibility and controlled biodegradability, which has all been well documented over the past years [31, 33]. Treating cancers through the use of bioactive glass can also be beneficial due to its continued long-term release of ongoing treatment without requiring multiple pills and doctor appointments. A reduction in doctor's appointments and treatments can also decline the associated cost of cancer, which is a tremendous financial burden for patients and their family. There are substantial numbers of doctor visits that are required to treat cancer, which also takes a huge toll on the patient and their families. This large number of doctor visits often times requires family members to take off multiple days from work as well as possibly requiring a leave of absence just to accommodate the necessary cancer treatments for their loved ones.

Bioactive glass is showing promise in the development of soft tissue, which can also result in better orthopedic surgical outcomes. Cartilage and ligaments are the main soft tissues that sustain injury and require orthopedic surgical repair [34]. Cartilage is one of the targeted structures in the potential of 3D printing of bioactive glass. The goal is to build cartilage via 3D printing in order to get a customized fit for each patient and therefore the ripped structures can be completely removed and repaired. 3D printing of cartilage and ligaments will also allow repair without damaging the surrounding structure or other structures via autograft of healthy tissue [28]. In creating 3D printed cartilage and ligaments, the problem lies in getting bioactive glass to replicate the properties of naturally occurring soft tissue to the point in which the body cannot distinguish it as being foreign such that bioactive glass does with bone regeneration. The 3D printing would provide an outstanding route in repairing cartilage and ligaments. Bioactive glass, when used to repair these soft tissues, would have to potentially prevent patients from having to graft their own tissue from another area or through a cadaver, which carries risk of infection and rejection. Cartilage and ligaments are only a few of the soft tissue engineering areas that are being sought after when it comes to the advancement of bioactive glass orthopedically.

The ability for bioactive glasses to bond to hard tissue has made it possible for them to be considered as a suitable application for soft tissue repair. The cellular processes involved in the early stages of bioactive glass bonding to bone are aspects to consider when investigating requirements for soft tissue bonding mechanisms. The ionic dissolution products of bioactive glasses as shown in hard tissue bonding in the early stages are the same important ions required for soft tissue bonding due to the high angiogenic potential that they produce [35]. Many soft tissues are being investigated that indirectly have to do with healing of orthopedic procedures. Neural and skin tissue require proper healing after orthopedic procedures and are being tested for advancement via bioactive glass [13]. One way in particular that bioactive glass is being explored for tissue engineering is in its ability to heal wounds by regenerating new skin tissue. These include soft tissue healing of gunshot wounds and cancer treatment surgeries [33, 35]. Being able to regenerate new skin would be effective and revolutionary not only for patients that need to heal their surgical wounds but also for patients who suffer from diseases such as diabetes who have trouble healing wounds. These patients sometimes require amputation to some of their body parts due to their decreased ability to heal wounds naturally. Having a leg amputated or even a toe amputated can cause a major inconvenience and lifestyle change to patients due to their now limited mobility. Patients with other skin diseases or injuries like patients that may have suffered from second and third degree skin burns may also substantially benefit from the healing effects of bioactive glass. Skin burns are hard to treat and require skin grafts and lots of treatment for proper healing but still face permanent damage. The idea of skin regeneration is being sought after in creating a skin-like material. Skin regeneration by using bioactive glass based cream/coating as a wound dressing has been accomplished by placing it on top of the damaged skin, creating a protective coating between the wound and the external environment and by default, protecting the wound from dangerous bacteria [13].

The investigation of bioactive glass for neural tissue repair is also important for healing orthopedic injuries due to the limitation that damaged neural tissue can have on the use of the associated limb. Despite boney repair neural damage can impact and reduce the ability of the patient to fully utilize and operate the repaired structure. Neural tissue currently is non-regenerative and therefore if it is not able to be successfully surgically repaired patients face suffering from improper use of limbs permanently. Therefore, the ability to regenerate neural tissue would significantly impact orthopedic-related injuries to muscle or boney tissues and enhance the overall use of the limb. Neural tissue is already showing great potential when collaborated with bioactive glass for neural tissue repair and regeneration. It has shown reactivity with neural tissue and currently elicits a chemical response biologically just as bone does. The ability to control the rate of exchange with the environment and its degradation present a major hurdle currently [13]. Research has already proven the possibility of bioactive glass participating in treatment for peripheral nerve injuries when Laura M. Marquardt et al. was able to show that bioactive glass can provide alignment to support directed axon growth [36].

Conclusion

Bioactive glasses have been studied extensively and have been around for several decades and have proven to be great biomaterials. Bioactive glass integrates biologically well once placed in the body and establishes great degradation. The degradation of the bioactive glass allows the complete formation and restoration of bone in place of the injured and defective tissue. Bioactive glasses are currently being used solely in smaller bones due to their low impact and weight-bearing capacities. Bioactive glass is heavily utilized in facial structures and has exhibited great results for healing the patient's bones and restoring their original functions. Bioactive glass has also played a smaller role in larger bones through its integration into metallic implants. Bioactive glass integration with metallic fixtures not only provides better bone integration of the metallic implants but also provides a naturally occurring antibacterial property. Bioactive glass has the potential to change the orthopedic realm by allowing proper healing and restoration of natural bone tissue with an enhancement in its strength.

Bioactive glass provides the potential to improve total healing from orthopedic injuries through its restoration of cartilage, ligament, skin and neural tissue. The healing of all these tissues affects the complete success of the orthopedic procedure and the proper use of the affected limb. Injury and defects to bone tissue are not the only ailments that affect the orthopedic world. Bone cancer is another major disease that impacts bone tissue and bioactive glass has shown potential to provide a targeted and thereby more effective chemotherapy treatment in the future. The continued study of an improvement of bioactive glasses is necessary due to its ability to cause major innovations in the medical world. These medical innovations can not only recover injury and diseases quicker but also allow better outcome. Bioactive glass creates a better outcome by decreasing the chance of complications, thereby not only saving tissue but most importantly saving lives.

References

1. Overview of biomaterials and their use in medical devices (2003)
2. Baino F, Hamzehlou S, Kargozar S (2018) Bioactive glasses: where are we and where are we going? J Funct Biomater 9(1):25. https://doi.org/10.3390/jfb9010025
3. Krishnan V, Lakshmi T (2013) Bioglass: a novel biocompatible innovation. J Adv Pharm Technol Res 4:78. https://doi.org/10.4103/2231-4040.111523
4. Chen Q, Thouas GA (2015) Metallic implant biomaterials. Mater Sci Eng R Rep 87:1–57. https://doi.org/10.1016/J.MSER.2014.10.001
5. Hosoki M, Nishigawa K, Tajima T, Ueda M, Matsuka Y (2018) Cross-sectional observational study exploring clinical risk of titanium allergy caused by dental implants. J Prosthodont Res 62:426–431. https://doi.org/10.1016/J.JPOR.2018.03.003
6. Hahn M, Buzanich G, Jähn K, Reinholz U, Radtke M (2019) Analysis of cobalt deposition in periprosthetic bone specimens by high-resolution synchroton XRF in undecalcified histological thin sections. Materialia 6:100290. https://doi.org/10.1016/J.MTLA.2019.100290
7. Hardware used in surgery (n.d.)
8. Orthopedic hardware (n.d.)
9. Navarro M, Michiardi A, Castaño O, Planell JA (2008) Biomaterials in orthopaedics. J R Soc Interface 5:1137–1158. https://doi.org/10.1098/rsif.2008.0151
10. Abad-Javier ME, Cajero-Juárez M, Contreras García ME (2016) 45S5 Bioglass porous scaffolds: structure, composition and bioactivity characterization. Epitoanyag J Silicate Based Comp Mater 68(4):124–128. https://doi.org/10.14382/epitoanyag-jsbcm.2016.22
11. Sahoo N, Anand S, Bhardwaj J, Sachdeva V, Sapru B (1994) Bone response to stainless steel and titanium bone plates: an experimental study on animals. Med J Armed Forces India 50:10. https://doi.org/10.1016/S0377-1237(17)31029-8
12. Jones JR, Brauer DS, Hupa L, Greenspan DC (2016) Bioglass and bioactive glasses and their impact on healthcare. Int J Appl Glas Sci 7:423–434. https://doi.org/10.1111/ijag.12252
13. Miguez-Pacheco V, Greenspan D, Hench LL, Boccaccini AR (2015) Bioactive glasses in soft tissue repair. Am Ceram Soc Bull:94, 27–31
14. Jones JR (2015) Reprint of: review of bioactive glass: from Hench to hybrids. Acta Biomater 23:S53–S82. https://doi.org/10.1016/J.ACTBIO.2015.07.019
15. Hench LL (2006) The story of bioglass®. J Mater Sci Mater Med 17:967–978. https://doi.org/10.1007/s10856-006-0432-z
16. Hench LL, Polak JM (2002) Third-generation biomedical materials. Science 295:1014–1017. https://doi.org/10.1126/science.1067404
17. Rahaman MN, Xiao W, Huang W (2017) Review—bioactive glass implants for potential application in structural bone repair. Biomed Glas 3:56–66. https://doi.org/10.1515/bglass-2017-0005
18. Sanyal S, Shukla M, Dandapat N, Ghosh S (2018) In vitro evaluation of bioactive glass ceramic coating for application on Ti6Al4V based biomedical implants. J Non-Cryst Solids 500:22–29. https://doi.org/10.1016/J.JNONCRYSOL.2018.04.043
19. Tilocca A (2007) Structure and dynamics of bioactive phosphosilicate glasses and melts from *ab initio* molecular dynamics simulations. Phys Rev B 76:224202. https://doi.org/10.1103/PhysRevB.76.224202
20. Woźniak MJ, Chlanda A, Oberbek P, Heljak M, Czarnecka K, Janeta M, John Ł (2019) Binary bioactive glass composite scaffolds for bone tissue engineering—structure and mechanical properties in micro and nano scale. A preliminary study. Micron 119:64–71. https://doi.org/10.1016/J.MICRON.2018.12.006
21. Bandyopadhyay-Ghosh S (2008) Bone as a collagen-hydroxyapatite composite and its repair. Trends Biomater Artif Organs 22(2):116–124
22. Nawaz Q, Ur Rehman MA, Roether JA, Yufei L, Grünewald A, Detsch R, Boccaccini AR (2019) Bioactive glass based scaffolds incorporating gelatin/manganese doped mesopo-

rous bioactive glass nanoparticle coating. Ceram Int 45:14608. https://doi.org/10.1016/J.CERAMINT.2019.04.179
23. Sanz-Herrera JA, Boccaccini AR (2011) Modelling bioactivity and degradation of bioactive glass based tissue engineering scaffolds. Int J Solids Struct 48:257–268. https://doi.org/10.1016/J.IJSOLSTR.2010.09.025
24. NovaBone® Products « Novabone (n.d.)
25. MTF Biologics, NovaBone (n.d.). https://www.mtfbiologics.org/our-products/detail/novabone. Accessed 24 May 2019
26. Jacobson RE, Granville M, Hatgis J, Berti A (2017) Low volume vertebral augmentation with Cortoss® cement for treatment of high degree vertebral compression fractures and vertebra plana. Cureus 9:e1058. https://doi.org/10.7759/cureus.1058
27. Osteostimulation bioactive bone bonding medical education spine & neurosurgery natural hydroxyapatite formation (n.d.)
28. Coating implants with bioactive glass to promote implant fixation (n.d.)
29. Lopez-Esteban S, Saiz E, Fujino S, Oku T, Suganuma K, Tomsia AP (2003) Bioactive glass coatings for orthopedic metallic implants. J Eur Ceram Soc 23:2921–2930. https://doi.org/10.1016/S0955-2219(03)00303-0
30. Drnovšek N, Novak S, Dragin U, Čeh M, Gorenšek M, Gradišar M (2012) Bioactive glass enhances bone ingrowth into the porous titanium coating on orthopaedic implants. Int Orthop 36:1739–1745. https://doi.org/10.1007/s00264-012-1520-y
31. Gerhardt L-C, Boccaccini AR (2010) Bioactive glass and glass-ceramic scaffolds for bone tissue engineering. Mater (Basel, Switzerland) 3:3867–3910. https://doi.org/10.3390/ma3073867
32. Soundrapandian C, Datta S, Kundu B, Basu D, Sa B (2010) Porous bioactive glass scaffolds for local drug delivery in osteomyelitis: development and in vitro characterization. AAPS PharmSciTech 11:1675–1683. https://doi.org/10.1208/s12249-010-9550-5
33. Lin Y, Mauro JC (2019) Bioactive glasses for Cancer therapy. Biomed Ther Clin Appl Bioact Glas 10(8):273–312. https://doi.org/10.1016/B978-0-08-102196-5.00010-0
34. Prasad K, Bazaka O, Chua M, Rochford M, Fedrick L, Spoor J, Symes R, Tieppo M, Collins C, Cao A, Markwell D, Ostrikov KK, Bazaka K (2017) Metallic biomaterials: current challenges and opportunities. Materials (Basel) 10(8):884. https://doi.org/10.3390/ma10080884
35. Miguez-Pacheco V, Greenspan D, Hench LL, Boccaccini AR (n.d.) Bioactive glasses in soft tissue repair. www.ceramics.org. Accessed 24 May 2019
36. Marquardt LM, Day D, Sakiyama-Elbert SE, Harkins AB (2014) Effects of borate-based bioactive glass on neuron viability and neurite extension. J Biomed Mater Res Part A 102:2767–2775. https://doi.org/10.1002/jbm.a.34944
37. Jonathan Cluett M (2018) Broken metal implants in your body. https://www.verywellhealth.com/broken-metal-implants-in-your-body-2549321. Accessed 6 June 2019
38. Olmedo DG, Tasat DR, Duffó GS, Guglielmotti MB, Cabrini RL (2009) The issue of corrosion in dental implants: a review. Undefined. https://www.semanticscholar.org/paper/The-issue-of-corrosion-in-dental-implants%3A-a-Olmedo-Tasat/baf287fd1db8cf45fbbc-93b5737e7b0005fd127e. Accessed 6 June 2019
39. Tilocca A (2010) Models of structure, dynamics and reactivity of bioglasses: a review. J Mater Chem 20:6848. https://doi.org/10.1039/c0jm01081b
40. Jones JR, Brauer DS, Hupa L, Greenspan DC (n.d.) Bioglass and bioactive glasses and their impact on healthcare. https://spiral.imperial.ac.uk/bitstream/10044/1/41893/2/glass age bioglass submitted endnote version spiral version.pdf. Accessed 6 June 2019
41. Mahesh L, Narayan TV, Bali P, Shukla S (2012) Socket preservation with alloplast: discussion and a descriptive case. J Contemp Dent Pract 13:934–937. http://www.ncbi.nlm.nih.gov/pubmed/23404031. Accessed 28 May 2019

Advances in Tissue Engineering and Regeneration

Krishanu Ghosal, Priyatosh Sarkar, Rima Saha, Santanu Ghosh, and Kishor Sarkar

Abstract Despite the advancement of medical research, damaged tissue through life threatening diseases or severe trauma or accident cannot be regenerated by the body itself due to large defect sizes or totally damaged tissue. Over the last few decades, tissue engineering (TE) has provided a revolutionary change in the biomedical field by which damaged tissues can be regenerated. Although, the regeneration is dependent on various factors such as the types of materials to be used as scaffolds, source of cells, and its stimulating factors (including growth factors, genes, or physical stimulus). In this chapter, we describe all parameters in detail for the success of tissue regeneration and we also cover all of the different types of tissue regeneration (such as bone, cartilage, neural, skin, cardiac, vascular, liver, and interfacial tissue).

Keywords Tissue engineering · Regenerative medicine · Polymeric scaffold · 3D scaffold · Bone tissue engineering · Biocompatible · Biodegradable · Stem cell Growth factor

Introduction

Due to the inherent capability of the human body, it has the ability to regenerate damaged tissue by itself, although the regeneration of the tissue is limited by several factors including the type of tissue, the size of damage, loss of function, and presence of multiple tissues. Presently, organ transplantation is the most acceptable clinical approach for the regeneration of such large or complex tissues but the shortage of donor tissues or organs is a major problem for this technique. Over the last few years, tissue engineering (TE) and regenerative medicine have shown remarkable promise to regenerate damaged tissues. TE is a multidisciplinary field having the principles of chemistry, engineering, and biology for the development of biological substitutes that can mimic normal tissue functions which are damaged

---

K. Ghosal · P. Sarkar · R. Saha · S. Ghosh · K. Sarkar (✉)
Gene Therapy and Tissue Engineering Lab, Department of Polymer Science and Technology, University of Calcutta, Kolkata, West Bengal, India
e-mail: kspoly@caluniv.ac.in; https:// www.kishorgttl.com

B. Li et al. (eds.), *Racing for the Surface*,
https://doi.org/10.1007/978-3-030-34471-9_22

through severe injuries, accidents, or by severe diseases. The basic concept of TE involves three main strategies consisting of: (1) induced pluripotent stem cells (iPSC) or adult stem cells or genetically modified cells such as CRISPR cas9 (productive), (2) biomaterial scaffold acts as cell support systems (conductive), and (3) cell signalling molecules such as growth factors, gene, or physical stimulus (inductive) to regenerate damaged tissue (Fig. 1). In the TE field, biomaterial scaffolds play an important role through creating a three-dimensional microenvironment for cellular attachment followed by propagation and differentiation. In addition to this, the scaffold also acts as the reservoir of signalling molecules which direct the cells to differentiate toward the specific tissue environment.

Currently, TE consists of two strategies such as cellular (in vitro) and acellular (in vivo) strategies. In the cellular strategy, cells are seeded on a scaffold containing cell signalling stimulus followed by culturing in vitro and subsequently the

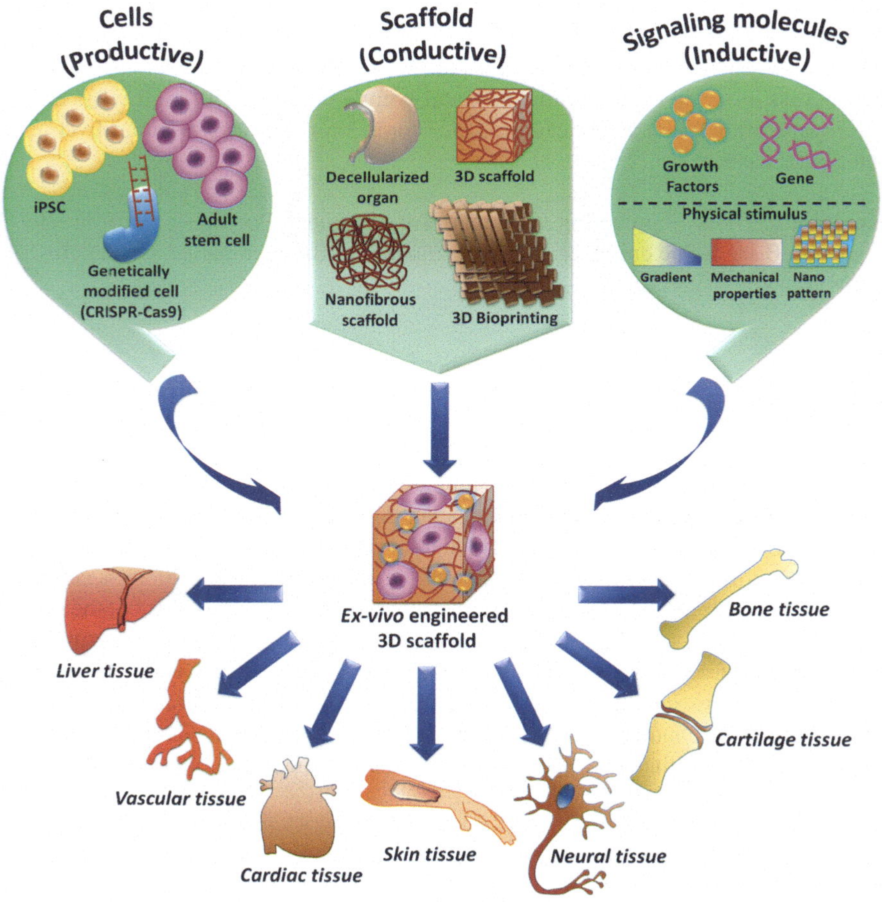

Fig. 1 Basic concept of tissue engineering and major barriers

engineered scaffold is implanted at the defect site in vivo for the generation of new tissue with the gradual degradation of the scaffold. On the other hand, the scaffold with signalling stimulus is implanted directly at the defect site in vivo without seeding any cells externally and the new tissue is regenerated through the infiltration of innate cells from either bone marrow or endothelial progenitors.

For successful TE, the selection of cell source is a very important factor especially when planned for clinical application. Cells derived from a patient's body or from their family members or relatives are ideal for the clinical application of TE but availability and accessibility of such cells become very difficult due to various reasons. Inherent unique properties of stem cells, which can differentiate to multiple functional tissue lineages may overcome the difficulty of cell availability. Embryonic stem cells (ESC) and adult stem cells (ASC) have been shown to be two different major sources of stem cell till date [1]. Despite the pluripotency of ESC, ethical controversy and teratomas formation of ESC restrict its clinical application of TE and regenerative medicine. Comparatively, ASC has shown to be a more attractive cell source in the TE field due to fewer problems than with ESC and poor rejection rate of regenerate tissue after implantation [2]. Adult stem cells are obtained from various mesenchymal tissues such as bone marrow, adipose tissue, umbilical cord, etc. and so it is also referred to as mesenchymal stem cells (MSC) [3–5]. Apart from this, induced pluripotent stem cells (iPSC) have gained tremendous attention in TE applications due to their preceding properties such as multipotent differentiation capability and robustness, autologous character and ease of reprograming technique [6, 7]. Although, unknown mechanisms behind the reprograming technique for iPSC limits its widespread clinical application in TE [8, 9]. Due to poor cell viability, vitality, and functionality of MSCs, genetically modified MSCs have shown promising results in TE and the regenerative medicine field owing to the improvement of viability, proliferative capability, and metabolic characteristics [10, 11].

In addition to the cell source, cell signalling molecules have great importance for the TE and regenerative medicine field [12–14]. In the presence of cell signalling biomolecules such as growth factors (GF) (e.g., bone morphogenetic protein-2 or 4 or 7 (BMP-2 or 4 or 7), transforming growth factor beta 1 (TGF-β1), insulin-like growth factor 1 (IGF-1), etc.); chemokines (e.g., CCL3, CCL27, CXCL12, etc.); cytokines (e.g., cardiotrophin-1 (CT-1), interleukin-6 (IL-6), etc.); and gene (BMP-2/4/7, SOX-6/9, etc.) or physical stimulus such as mechanical properties, nanopattern, etc., the stem cell differentiates toward respective functional tissues like bone, cartilage, cardiomyocytes, etc. The signalling molecules may be incorporated into the scaffold or added into the culture media during the in vitro culture of stem cells. Physical stimuli are created in scaffolds and the cells differentiate toward particular functional tissues depending upon the respective signals penetrating the scaffold.

Apart from the cell source and cell signalling molecules or stimuli, biomaterial-based scaffolds and their design have a big role to play for the success of TE. The scaffold not only functions as the extracellular matrix (ECM) which creates micro-environment for cellular attachment, proliferation, and differentiation but it also gives biophysical signalling to the cell for differentiation toward the specific type of functional tissue. By definition, biomaterials are such a type of implant material (which may be synthetic or natural) used for therapeutic purposes to replace or restore the function of body tissue and continuously or intermittently in contact with living tissue or body fluids [15]. Therefore, an ideal biomaterial in the TE field should possess several characteristics including nontoxic, biodegradable, biocompatible, non-immunogenic, ease of synthesis, and low production cost.

Despite the enormous advancement in the TE field, postsurgical infections remain a clinical threat globally. Biomaterial-related microbial infections are common problems in the TE field and there is a greater risk for failure in tissue regeneration due to this difficulty for curing infection [16, 17]. After implantation of the biomaterial at the defect site immediately is covered with proteins and protein components such as fibrin, fibrinogen, and fibronectin from inherent ECM to potentially form a thin film on the biomaterial and the protein film helps to anchoring free moving bacteria including both gram-positive and gram-negative as well as *Candida albicans* [18, 19].

Therefore, the main objective of this chapter is to discuss the advances along with their possible therapeutic approaches for future clinical applications.

Tissue Engineering for Different Tissue Regeneration

Bone Tissue Engineering

Currently, a huge number of patients in the US as well as in other countries all over the world suffer from bone-related disorders which demands a large number of functional bone grafts. A recent report stated that every year in the US more than half a million of patients pay more than 2.5 billion dollars to treat bone-related disorders [20]. In this regard, the treatment of severe bone defects using TE approaches is perceived as a good strategy because the regeneration process may involve the patient's own tissue for the repair process [21–23]. However, there are several studies which are focused on limitations and complications involved with current clinical treatment methods for bone regeneration (including allogenic and autologous transplantation) using allografts and autografts. Till now, autografts are recognized as a gold standard for treatment of bone defects as they are non-immunogenic, histocompatible, and offer the essential properties of a bone graft material. More precisely, autografts possess prerequisites which can start osteoinduction (i.e., growth factors, genes, proteins, etc.), and osteogenesis as well as osteoconduction.

However, there are certain limitations such as harvesting bone from the patient's iliac crest which needs a second surgery at the tissue harvest site, in addition, the expenses related to autograft transplantation, donor site morbidity, bleeding, and inflammation limit the wide spread application of autografts to treat bone defects [24–26]. Allografts are the second most popular bone grafting technique. Allografts are also histocompatible and can be obtained in several forms such as cancellous chips, demineralized bone matrices, cortical grafts, whole bone grafts, and osteochondral grafts depending on patient's requirements. Compared to autografts, allografts are more prone to infections and immune rejection. In addition to that, allografts have less osteoinductive properties and no cellular constituents because allografts from donors are decellularized via gamma irradiation or freeze-drying [27–29].

Other commonly used strategies for the bone repair process involve cement filler delivery of BMP, etc. However, clinical trials suggest that none of these strategies possess all of the ideal characteristics including high osteoinductive, osteoconductive, and angiogenic potentials, in vivo biocompatibility, storage, low patient morbidity, single surgical procedure, and long lifetimes at an economical cost for bone tissue regeneration.

The field of bone TE focuses on new alternative strategies with capabilities to eliminate the problems associated with conventional approaches as discussed above. Researchers, engineers, and clinicians have been working together to achieve an ideal bone graft material, which can enhance bone regeneration without significant complications at low costs. For successful bone TE, four criteria should be fulfilled: (1) the scaffold material should be biocompatible so that it can mimic the extracellular matrix of bone, (2) possess healthy osteogenic cells, (3) use growth factors which direct the cells to the phenotypically desired type through cell signalling, and (4) have enough vascularization so that nutrients can be supplied to the growing tissues. Although there has been notable progress in the field of bone TE, there are still lots of challenges that remain which should be addressed for real life clinical applications of bone TE. In this section, we will discuss such recent advances as well as remaining hurdles in the field of bone TE.

Different Types of Biomaterials for Bone TE Applications

Till now, a variety of biomaterials have been used for bone tissue engineering (BTE) applications including metals, ceramics, natural, and synthetic polymers as well as their composites. However, among these biomaterials, polymeric biomaterials and their composites are the most attractive for BTE, due to several advantages over metals and ceramics. These advantages include biodegradability, lower immune rejection compared to ceramic- and metal-based biomaterials, tunable mechanical properties, and easier processability. So, in this section, we mainly highlight various polymeric biomaterials and their composite scaffolds which are used in BTE applications.

A significant number of studies have been focused on natural polymers that are used for BTE applications. The main reason behind this is associated with the economic and environmental aspects of natural polymers. Natural polymers have several benefits including good biodegradability, low cytotoxicity, easy to manufacture, and zero disposable costs. Along with these, they also help in cell signaling and cell adhesion and differentiation. However, the inadequate mechanical properties limit their applications for BTE. So, in most cases, incorporation of a filler material is necessary to improve the physical properties of the overall system.

Among the natural polymers studied, chitosan is the most frequently used polymer for TE applications. Chitosan is a well-known biodegradable cationic polysaccharide derived from chitin which is the main element in the exoskeleton of crustaceans. Chitosan has a hydrophilic surface which helps to promote cell adhesion, proliferation as well as cell differentiation. In addition to that, the inherent antibacterial properties, very good biocompatibility, and minimal host response make them a suitable candidate for BTE. Along with these advantages, the molding capability of chitosan into a well-designed porous scaffold structure adds an extra advantage for its applicability in BTE applications. For example, Zhang and coworkers developed a nanofibrous hydroxyapatite/chitosan (nHAP/CTS) scaffold for BTE. They observed that nHAP/CTS induced a higher proliferation of stem cells (BMSCs) than membranous hydroxyapatite/chitosan (mHAP/CTS) and electrospun nanofibrous chitosan (nCTS). Here, nHAP acted as an osteoinductive agent and promoted osteogenic differentiation of stem cells. A further in vivo study confirmed that nHAP/CTS/BMSCs had a superior ability for bone reconstruction than other control groups for cranial bone defects by activating an integrin-BMP/Smad signaling pathway of BMSCs [30]. In another work, Guo and coworkers reported silver (Ag)-loaded strontium hydroxyapatite (SrHAP)/chitosan (CTS) porous scaffolds (Ag–SrHAP/CTS). Interestingly, the Sr and Ag ions released from the composite scaffold enhanced the osteoinductivity and antimicrobial activity of the scaffold whereas the presence of HAP within the scaffold increased the mechanical properties and osteoinduction properties of the composite system. The presence of Sr and Ag in Ag–SrHAP/CTS porous scaffolds increased the alkaline phosphatase (ALP) activity of stem cells, mineralization and upregulates expression of osteogenic related genes such as BMP-2 and COL-1 [31]. Saravanan et al. also demonstrated that chitosan/nano-hydroxyapatite/nano-silver particles (CTS/nHAP/nAg) can be a perfect candidate for BTE. The prepared scaffold demonstrated synergistic bacterial inhibition toward both gram-positive and gram-negative bacteria due to the presence of CTS and Ag nanoparticles along with very good cytocompatibility toward rat osteoprogenitor cells [32].

Alginate is another well-known natural polymer which has proved its presence in the field of BTE applications. Commercial alginates are produced from Algae named *Laminaria hyperborea*, *Ascophyllum nodosum*, and *Macrocystis pyrifera*. Alginates are linear unbranched polysaccharides consisting of different amounts of β-D-mannuronic acid and α-L-guluronic acid within the polymer chain. The main advantage of alginate is that it is biocompatible and it can easily form cross-linked hydrogels in the presence of multivalent cations. For example, Xie and his coworkers developed an injectable calcium phosphate–alginate–chitosan microcapsule for BTE. In detail, they studied the osteogenic potential of this injectable calcium phosphate–alginate–chitosan microcapsule with MC3T3-E1 cells in vivo. After 4 weeks of injection of these cell incorporated microcapsules into the dorsal subcutaneous area of nude mice, they observed lamellar-bone-like mineralization, collagen formation, and angiogenesis. A further study after 8 weeks of injection, confirmed expansion of newly formed collagen, and absorption of phosphate cement which suggested the applicability of alginate microcapsules for BTE [33].

Wang et al. also evaluated the applicability of sodium alginate for bone tissue engineering. They showed that, alginates with different compositions and purity had a significant effect on proliferation and differentiation of rat bone marrow cells. They found that, high purity and high G-type alginates had a similar ability for promoting cell proliferation. Not only that, but they also observed the presence of high guluronic acid into alginate helped the differentiation of stem cells into an osteogenic lineage which supported its use for BTE applications [34]. In another work, Guldberg and coworkers reported a peptide-modified alginate hydrogel hybrid system for the delivery of recombinant BMP-2 (rhBMP-2) to enhance the repair process of large bone defects [35]. Their result suggested that the hybrid system had a consistent ability of bone bridging on the defect site however in the absence of rhBMP-2 there was no substantial bone formation; which confirms that rhBMP-2 has a critical effect on bone regeneration. In addition to that, the alginate nanofiber mesh enhanced the rhBMP-2 mediated bone regeneration process by improving the infiltration of osteoprogenitor cells within the nanofiber mesh which confirmed that the hybrid alginate/nanofiber mesh could be a promising growth factor delivery system for the regeneration of large bone defects.

Unlike chitosan and alginate, collagen is a biological protein with a significant amount of glycine (nearly 30%) along its backbone. More specifically, glycine is present within collagen, in every third residue, which forms a $(Gly\text{-}Y\text{-}X)_n$ repeating unit. Due to the presence of amino acids, collagen has good biodegradability, low immunogenicity, and very good cell binding properties which make them a suitable candidate for TE applications [36, 37]. It is well established that collagen sponges can promote cell attachment and growth and enhance bone formation by promoting the differentiation of osteoblasts [38, 39]. However, the main disadvantage of collagen is its rapid degradation rate, which causes a rapid loss in mechanical properties of the scaffold. So, researchers are trying to develop new composite materials composed of collagen and other natural/synthetic biopolymers/nanofiller to decrease the biodegradation rate and increase the mechanical properties to make collagen a perfect candidate for TE applications [40].

As for example, Zhang and coworkers developed a biomimetic porous collagen/hydroxyapatite for BTE. Due to the presence of hydroxyapatite, the scaffold possesses good mechanical properties, in addition to that, hydroxyapatite increases osteointegration and osteoconductivity of the scaffold [41]. Murphy et al. studied the effect of pore size on cell attachment, proliferation, and migration in collagen–glycosaminoglycan scaffolds. They discovered that collagen–glycosaminoglycan scaffolds had very good cytocompatibility toward osteoblast cells. In addition to that, the pore size of the scaffolds played a crucial role in osteoblast proliferation and migration.

Silk fibroin is another well-known material for bone TE. Silk is a fibrous type of protein derived from the native silkworm and possesses unique chemical and physical properties. Silk fibers from cocoons (*Bombyx mori*) and are comprised of a structural protein fibroin covered with sericin, another protein which mainly acts as a glue to hold fibroin together [42]. Fibroin protein is made of multiple layers of antiparallel β sheets which form a crystalline section made of peptide chains and dominated by the hydrophobic sequence GAGAGSGAAG[SG(AG)2]8Y [43].

For couple of decades, silk has established its presence in the field of biomedical applications as sutures. However, researchers are now interested in its application for bone tissue regeneration as it possesses very good biocompatibility, slow biodegradability, high mechanical strength, tunable composition, and superior permeability. Other than these, their easy modification through amine and acidic side chains and easy processability of its aqueous solution into various forms such as gels, membranes, and sponges provide a perfect candidate for TE applications. The major drawback of silk is the need to remove sericin from it, as sericin can initiate a severe immune response at the implantation site.

As, for example, Kaplan and his coworkers used silk fibroin from *Bombyx mori* for BTE. Briefly, they deposited calcium phosphate on the silk fibroin scaffold and subsequently seeded bone marrow-derived stem cells and cultured it for 6 weeks in the presence and absence of BMP-2. They observed enhanced osteoconductivity in the presence of apatite and BMP-2 against control. The premineralization on the highly porous silk scaffold suggests that these scaffolds may be a suitable candidate for BTE [44]. In other work, they reported high strength silk protein scaffolds for BTE. The scaffold demonstrated a superior compressive strength of 13MPa which is suitable for load bearing bone grafts. The superior mechanical properties of the polymer were due to the protein interfacial bonding and micron-sized silk fiber. The combined effect of surface roughness, porosity, and scaffold stiffness resulted in human bone marrow-derived stem cells differentiating toward osteogenesis. To further confirm the effectiveness of these scaffold toward BTE, they performed in vivo implantation of the scaffolds to check their immunomodulatory responses which revealed no major immune responses to the implantation site [45]. Incorporation of hydroxyapatite into the scaffold can increase the mechanical properties of the overall system as well as it also enhances the osteoinduction and osteoconduction of the system. In this regard, Shao and coworkers reported coaxial electrospun aligned tussah silk fibroin fiber-hydroxyapatite composite scaffolds for bone TE. The composite scaffold demonstrated very high mechanical strength with an increase of 90-fold in the modulus and 2-fold increase in breaking strength than the bare silk fibroin scaffold. Upon culturing MG-63 cells on the scaffolds, they discovered that cells were well attached and proliferated on the scaffold surface. Additionally, ALP and biomineralization assays on the scaffold supported its potential application for BTE [46]. Other than these natural biopolymers, in recent years, synthetic biopolymers also received attention for BTE applications due to their improved mechanical properties and their tunable biodegradability. However, in the case of synthetic biopolymers, the major concern is their biocompatibility and their immunogenic responses. Although, some of the well-known synthetic biopolymers are already FDA approved, including poly(lactic-co-glycolic acid)(PLGA) and polylactic acid (PLA), the complicated synthesis procedure for large-scale production and cost of these biopolymers restrict their applications for many commercial biomedical applications.

Other than these synthetic biopolymers polyesters, polyester amides, polyurethanes, polyanhydrides, polycaprolactone, and polyhydroxysiloxanes have established their presence in the field of TE. However, for commercial clinical applications and to get FDA approval, it may take a long time. Researchers are also trying to develop new cost effective synthetic biopolymers based on recycling solid plastic waste and using other renewable low cost resources, such as soybean oil. For example, Sarkar et al. reported a novel biopolymer derived from commercial polyethylene terephthalate (PET) waste and other low cost renewable resources, such as citric acid, sebacic acid, and mannitol. Their synthesized polyester demonstrated very good biocompatibility both in vitro as well as in vivo. Upon osteoblast cell culture (MC3T3-E1 cells) on the 3D porous scaffolds, cells were well attached and proliferated within the scaffold as well as deposited minerals on the scaffold surface which confirmed the applicability of the scaffolds for bone TE applications [47]. Kolanthai et al. also reported another low cost synthetic biopolymer derived from soy oil by a simple melt condensation technique [48], and demonstrated very good osteogenesis from stem cells without the use of any growth factors. However, for real life TE applications of these polymers, a thorough long-term in vivo study should be performed. In the next section, we will discuss this current scenario and clinical trials in the field of TE.

Some of the natural and synthetic biopolymers for BTE applications are tabulated below (Table 1).

Current Scenario and Clinical Trials in the Field of Bone Tissue Engineering

Bone-related disorders have increased continuously and are expected to double by the year 2020, especially in those areas where people are older with increased obesity and poor physical activity [20]. In this context, engineered bone implants have been considered as one of the potential alternatives to conventional allograft transplantation or autologous bone graft transplantation due to their very limited supply [110]. Although there is significant improvement in the research of bone TE, still bone TE clinical practices have not progressed that much due to several challenges.

The success rate of any TE or regenerative medicine strategy is very much dependent upon its real life clinical applications. For successful bone TE, a clinical issue like tibial nonunions might require only stimulation of fracture healing however, other bone regeneration scenarios may require a physical mechanical support known as a scaffold. On the other hand, segmental bone defect repair may involve the need of vascularized bone with an ability to integrate to other neighboring host tissues, and this is a significant challenge [111]. In this section, we will highlight the current scenario and advances of the clinical translation of bone TE strategies for the management of various types of bone-related problems including fractures, arthrodesis, and segmental bone defects.

To treat fractures and arthrodesis, BMPs are usually delivered using a collagen sponge within a cage device or collagen putty to mediate spinal arthrodesis and

Table 1 Polymers used for bone TE applications

Material	Scaffold composition	Scaffold pattern	Fabrication technique	Signaling factor	In vitro/in vivo	References
Natural polymer						
Chitosan	CTS-alginate	3D porous scaffold	Freeze-drying	BMP-2	MSC/rat calvarial defect	[49]
	CTS-alginate/HAP	3D porous scaffold	In situ coprecipitation and freeze-drying	HAP	MG-63/mouse skull	[50]
	CTS-poly(butylene succinate)	3D porous scaffold	Melt-based compression molding followed by salt leaching	Self osteoconductivity	hBMSC/mice cranial defect	[51]
	CTS/nHAP	Nanofibrous	Electrospin	HAP	BMSC/rat cranial defect	[30]
	CTS/nHAP/PLLA	Microsphere loaded 3D porous	Emulsion-ionic cross-linking followed by freeze-drying	BMP-2	MC3T3-E1/rabbit femoral condyle defect	[52]
	CTS-collagen/rhBMP-2	Microsphere loaded 3D porous	Freeze-drying	rhBMP-2	Dog mandible defect	[53]
	CTS-collagen/β-TCP	3D porous scaffold	SFF printing process with the 3D-bioplotting technology	TCP	Merino sheep mandible defect	[54]
	CTS-collagen	Flat sponge	Freeze-drying	Self osteoconductivity	AMSC/miniature pig femur defect	[55]
	MPCS	Sponge	Freeze-drying	Self osteoconductivity	Human oral cavity	[56]
	CS-HTCC/GP	Injectable thermoresponsive hydrogel	Solution mixing	Self osteoconductivity	Mongrel dog periodontal defect	[57]

Material	Scaffold composition	Scaffold pattern	Fabrication technique	Signaling factor	In vitro/in vivo	References
	PH/rhBMP-2/NPs	Photopolymerizable hydrogel composite	Photochemical process	rhBMP-2	hMSC/mouse thigh and rabbit radius critical defect	[58]
Alginate	ALG coated by PLO/PLAR/DEAE-D	Core-shell microbead	Polyelectrolyte complexation technique	rhBMP-2	C_2C_{12}/rat posterolateral spinal fusion	[59]
	ALG-CTS/β-TCP/CPC	Microbead	Micro-encapsulation	TCP and CPC	MC3T3-E1/mice	[33]
	ALG/α-TCP	3D fibrous scaffold	Suspension injection followed by ionic cross-linking	TCP	MSC/rat calvarial defect	[60]
	ALG/*O*-CMC/CQ	Composite mat	Freeze-drying	CQ	hMSC	[61]
	PCL/BMP-2/ALG	3D porous scaffold	3D printing	BMP-2	MG-63	[62]
	ALG-CTS/rhBMP-2	Injectable hydrogel	Sonication	rhBMP-2	MSC/mice	[63]
	PCL/ALG-rhBMP-2	Nanofiber mesh	Electrospin	rhBMP-2	Rat femoral defect	[35]
	RGD–OSA/NSC	Composite hydrogels	Freeze-drying	Mechanical stimulation	hMSC	[64]
	ALG-*graft*-RGD	Microsphere scaffold	Electrostatic bead generator technique	NA	HOP and HUVEC/mice femur defect	[65]
Collagen	COL/HAP	3D porous scaffold	Freeze-drying	Stiffness	MSC/rat and rabbit radial defect	[66–68]
	COL (BIOGIDE®)	Layered membrane	NA	NA	Rat oral cavity	[69]
	FCOL/BCP	3D porous scaffold	NA	Self osteoconductivity	Canine spinal fusion defect	[70]

(continued)

Table 1 (continued)

Material	Scaffold composition	Scaffold pattern	Fabrication technique	Signaling factor	In vitro/in vivo	References
Collagen	COL/β-TCP/rhBMP-2	Sponge	Chemical cross-linking followed by freeze-drying	rhBMP-2	Rat	[71]
	COL/BGNF	Macro-porous/thin film	Evaporation followed by freeze-drying	Self osteoconductivity	MG-63	[72]
	COL/CTS/HAP/β-TCP/GCK	3D porous scaffold	Freeze-drying	CK	MG-63	[73]
	COL/Silica	Xerogel	Sol-gel technique	Self osteoconductivity	MC3T3/rat femoral and calvarial defect	[74, 75]
	PCL/MC	3D porous scaffold	Solution mixing followed by freeze-drying	MC	MC3T3-E1/sheep cranial bone defect	[76, 77]
Silk fibroin	SF/PLGA	Injectable hydrogel	Solution mixing	NA	MG-63/rabbit femoral condyle defect	[78]
	SF/rhVEGF$_{165}$/rhBMP-2	Injectable hydrogel	Sonication induced method	rhVEGF$_{165}$ and rhBMP-2	Rabbit nasofrontal bone defect	[79]
	SF	3D porous sponge	Salt leaching followed by freeze-drying	NA	hASC/sheep tibia and humerus bone defect	[80, 81]
	GM-SF	3D porous sponge	Salt leaching followed by freeze-drying	Self osteoconductivity	Rabbit femoral epi-condyle defect	[82]
	ESF	3D porous scaffold	Electrospin	NA	MC3T3-E1/rat calvarial defect	[83]
	SF-*graft*-GRGDS	3D porous scaffold	Salt leaching followed by freeze-drying	NA	hMSC/rat femoral defect	[84]
	SF/HAP	3D porous scaffold	3D printing	HAP	hMSC and hMMEC	[85]

Material	Scaffold composition	Scaffold pattern	Fabrication technique	Signaling factor	In vitro/in vivo	References
	Apatite coated SF	3D porous scaffold	Salt leaching followed by freeze-drying	Apatite	BMP-2 modified bMSC/rat mandibular defect	[86]
	SF/MBG	3D porous scaffold	Freeze-drying	MBG	Mice calvarial defect	[87]
Synthetic polymer						
Poly(ε-caprolactone)	PCL/$Ca_3(PO_4)_2$	3D porous scaffold	Fused deposition modeling	$Ca_3(PO_4)_2$	HBMC/mice	[88]
	PCL	3D porous scaffold	Selective laser sintering	NA	BMP-7 transduced HGF cell/mice	[89]
	PCL/β-TCP	3D porous scaffold	3D bioprinting	β-TCP	Beagle dog alveolar bone defects	[90]
	rhBMP2-coated PS/PCL	Macroporous fibrous scaffold	Electrospinning	rhBMP2	MG-63/rat calvarial defect	[91]
	PCL/PLGA/β-TCP/ECM	3D porous scaffold	3D printing	β-TCP	hTMSC/rat calvarial defect	[92]
Poly(lactic-co-glycolic acid)	PCL/PLGA/SIM	3D porous scaffold	3D printing	SIM	rBMSC/rat femoral defect	[93]
	PLGA/nHAP	Fibrous scaffolds	Electrospinning	nHAP	MC 3T3 E1	[94]
	PLGA/graphene	2D film	Solution casting	Graphene	rBMSC/rat furcation defect	[95]
	PLGA/TiO_2 nanotube	3D microsphere based porous scaffold	Emulsion and solvent evaporation method	TiO_2	G-292/rabbit calvarial defect	[96]
	PLGA/PRP	3D microsphere based porous scaffold	Emulsion and solvent evaporation method	PRP	FB and KL/mice forelimb defect	[97]

(continued)

Table 1 (continued)

Material	Scaffold composition	Scaffold pattern	Fabrication technique	Signaling factor	In vitro/in vivo	References
PGA	PGA/β-TCP	3D porous scaffold	Solvent casting and particulate leaching method	β-TCP	Bone defects in rat femoral medial-epicondyles	[98]
	PGA textile functionalized with RGD	3D fiber textile matt	By knitting resorbable threads made of monofilaments of PGA	Osteogenic inductor medium (OIM)	Human osteoblasts (hOB)	[99]
PLA	PLA coated with polydopamine and collagen I	3D porous scaffolds	3D printing	Polydopamine and collagen I	Porcine bone marrow stem cells (MSCs)	[100]
	PLA/HAP	Bioreactor	3D printing	HAP	Bone marrow stromal cells (BMSCs)	[101]
	PLA/nHAP	Electrospun fibers	Electrospinning	nHAP	Osteoblast cells	[102]
Polyester	Benzyloxymethyl glycolide, caprolactone	3D porous scaffolds	3D printing	None	Human MSCs	[103]
	BHET, CA, SA, MA	3D porous scaffold	Salt leaching	None	MC3T3-E1 cells	[47]
	SO, SA, CA, MA	3D porous scaffold	Salt leaching	None	Human MSCs	[48]
	Galactitol/adipic acid/Graphene oxide	2D film	None	Graphene oxide	MC3T3 cells	[104]
	Galactiol/adipic acid, galactitol/dodecanedioic acid/(Glutaric acid, malic acid, maleic acid, succinic acid, citric acid, turturic acid)	2D film/3D porous scaffold	Salt leaching	None	MC3T3-E1 cells	[105]
Polyester amide	l-Leucine (l-leu)/γ-benzyl-l-glutamate/tetraaniline	Thin films	Solvent evaporation	Camphorsulfonic acid	MC3T3-E1 cells	[106]
	SO/DETA/(SA,AA,SUA,DA)	2D films	None	None	MC3T3-E1 cells	[107]

Material	Scaffold composition	Scaffold pattern	Fabrication technique	Signaling factor	In vitro/in vivo	References
Polyurethane	PCL diol/PLLA/DMPA EDA/ SPIO NPs	3D scaffolds	3D printing	SPIO NPs	Human MSCs	[108]
	Dextran/poly(ethylene glycol)/ calcium stearate	3D foam	Foaming process	None	MG63 cells, BMSCs	[109]

CTS chitosan, *HAP* hydoxyapatite, *nHAP* nano-hydroxyapatite, *PLLA* poly(L-lactic acid), *β-TCP* β-tri calcium phosphate, *RhBMP-2* recombinant human bone morphogenetic protein-2, *SFF* solid freeform fabrication, *BMSC* bone marrow stem cell, *AMSC* allogenic mesenchymal stem cell, *MPCS* methylpyrrolidinone chitosan, *CS* chitosan sulfate, *HTCC* N-[(2-hydroxy-3-trimethylammonium)propyl]chitosan chloride, *GP* α,β-glycerophosphate, *PH* photopolymerizable gelatin, *NP* 2-N, 6-O-sulfated chitosan nanoparticle, *ALG* alginate, *PLO* poly-L-ornithine, *PLAR* poly-L-arginine, *DEAE-D* diethylaminoethyl dextran, C_2C_{12} myoblastic cells, *CPC* calcium phosphate cement, *MBG* mesoporous bioactive glass, *O-CMC* O-carboxymethyl chitosan, *CQ Cissus quadrangularis*, *PCL* polycaprolactone, *RGD* glycine-arginine-glycine-aspartic acid, *OSA* oxidized sodium alginate, *NSC* N-succinyl chitosan, *COL* collagen, *FCOL* fibrillar collagen, *BCP* biphasic calcium phosphate, *GCK Ginseng* compound K, *MC* mineralized collagen, *HOP* human osteoprogenitor cell, *HUVEC* human umbilical vein endothelial cell, *SF* silk fibroin, *VEGF* vascular endothelial growth factor, *hASC* human adipose-derived stem cell, *GM-SF* genetically modified silk fibroin, *GRGDS* glycine-arginine-glycine-aspartic acid-serine peptide, *hMMEC* human mammary microvascular endothelial cell, *MBG* mesoporous bioactive glass, *HBMC* human bone marrow cells, *HGF* human gingival fibroblasts, *PS* polystyrene, *PLGA* poly(lactic-co-glycolic acid), *ECM* extracellular matrix, *hTMSC* human nasal inferior turbinate tissue-derived mesenchymal stromal cell, *PRP* platelet-rich plasma, *FB* normal fibroblast, *KL* keloid fibroblasts, *PGA* polyglycolic acid, *PLA* polylactic acid, *BHET* Bis(2-hydroxyethyl) Terephthalate, *SA* sebacic acid, *CA* citric acid, *MA* mannitol, *SPIO* super magnetic iron oxide nanoparticles, *SO* soybean oil, *DETA* diethylenetriamine, *MC3T3-E1 cells* mouse pre-osteoblast cells, *AA* adipic acid, *SUA* suberic acid, *DA* dodecanedioic acid, *DMPA* 2,2-bis(hydroxymethyl)propionic acid, *EDA* ethylene diamine fracture healing. BMPs are growth factors which facilitate bone and cartilage formation as well as helping to repair bones [112]. Among the growth factors of the BMP family, BMP-2 and BMP-7 are the most popular among clinicians. In 2001, BMP-7 was FDA approved to cure tibial nonunions after clinical trials [113]. Later, in 2002, BMP-2 also became FDA approved for the treatment of acute open tibial fractures [114]. Although some of the initial studies were promising, later on, clinical reports revealed that high doses of BMP with poor accumulation in the body could cause osteolysis and heterotopic ossification [115]. Additionally, till now there has been a lot of debate on the clinical success of BMP for the cure of open tibial fractures [116]. Yamada et al. reported a hybrid bone graft (HBG) by combining bone marrow aspirates with porous β-tricalcium phosphate and trephine bone, for safe and effective autologous bone grafting. The HBG enhanced the posterolateral spinal fusion rate without any severe donor site morbidity [117]. However, for successful spinal fusions, the presence of an osteoinductive or osteoconductive material appears to be very essential. There are numerous reported data validating that calcium phosphate cement applications could produce similar results equivalent to autologous bone grafting for the management of tibial plateau fractures [118]. In addition, a series of case studies reported that, the use of a ceramic graft for impaction bone grafting of the acetabulum produced medium to good results. To support this phenomenon, Damron and his coworkers observed no significant differences between β-tricalcium phosphate or β-tricalcium phosphate-bone marrow aspirate groups in terms of radiographic parameter [119].

Segmental bone defects generally occur in the tibia and if the structural integrity cannot be maintained, it can result in amputation. For the treatment of segmental bone defects, the current strategies involve the use of: (a) vascularized autologous bone grafting, (b) isolated autologous bone grafting, (c) an induced membrane technique along with autologous bone grafting, and (d) bone transplantation. In this context, Karger et al. combined synthetic bone constructs with a membrane material with an autologous bone graft to treat posttraumatic bone defects. Reconstruction of large bone defects was found to be possible with the help of an induced membrane technique, although, sometimes it failed to repair soft collagenous bone tissue due to a complicated repairing process [120].

In another clinical study, Meinig et al. explored the application of polymeric membranes for the reconstruction of bone defects, and used PLA for the treatment of large bone defects. Upon implantation of this polymeric membrane along with autologous bone grafts into bone defects, successful treatment of segmental bone defects was observed [121]. Another novel strategy was procured by Golden and his coworkers for the reconstruction of diaphyseal tibial cortical defects. They used BMP-2 loaded allografts for the treatment of diaphyseal tibial fractures with cortical defects of 4 cm. The mean estimated blood loss was found to be considerably less in the case of BMP-2 loaded allograft groups compared to bare allografts. In addition, the BMP-2/allograft group didn't produce any undesirable antibodies to the BMP-2 protein, which suggested that BMP-2 loaded allograft was safe and effective for the treatment of tibial fractures related to extensive traumatic diaphyseal bone defects [122]. Recent clinical studies include stem cell-based therapies

[123] and an alteration in the Masquelet technique using a bioresorbable osteogenic/angiogenic membrane along with synthetic autograft materials [124]. Till now, fabrication of a vascularized mechanically robust bone construct which can provide an integrated solution to bone TE remains a great challenge for the clinical translation of bone TE.

Recently, the use of cell-based therapies for bone TE is an emerging area of research as well as clinical translation. However, current approaches have certain limitations, such as the origin of cells, number of cells to be used, and method of delivery. For example, in most of these cases, researchers and clinicians are using cells alone or in combination with a support material [111]. During the preliminary stages of tissue trauma, the damaged tissue site is rich in various cytokines such as TNF-α, interferons, interleukins as well as increased acute-phase serum protein concentrations. Usually, within the TE community, it has been assumed that the usefulness of stem and progenitor populations introduced into such a hostile inflammatory environment is detrimental to tissue regeneration. Although, some recent studies demonstrated that spatiotemporal manipulation of cell delivery to the injury site could be a promising strategy for the clinical advancement of segmental bone tissue engineering [125]. In this regard, Hutmacher and coworkers demonstrated that late injection of bone marrow stem cells to the injury site can regenerate large bone defects in a preclinical animal model with enhanced therapeutic efficacy compared to a "time of trauma" therapeutic strategy [125]. So it is very evident that the delivery time cells play is very crucial for successful tissue regeneration. Enhanced regeneration of tissue is noticed when cells are delivered to the injury site either intravenously or subcutaneously as well as direct injection to the wound site [126–128]. However, there is still a need for numerous preclinical studies with specific scaffolds and cell-based approaches for successful clinical translation in the field of bone TE. Figure 2 summarizes currently available different treatment strategies for bone TE.

Other than these two type of bone defects, osteochondral defects are also a major problem in the field of bone-related disorders. However, due to complex physiological properties, progressive alteration in mechanical properties of tissues and a lack of suitable biomaterials limit the translation of osteochondral TE into the clinical stage [129]. To repair osteochondral defects, current strategies involve microfracture surgery or autologous chondrocyte implantation which includes a two-step surgical procedure [130]. In this context, synthetic biphasic/triphasic scaffolds without any growth factors/cells have been employed for osteochondral TE, although their efficiency remains questionable [131]. Other than this, an alternative therapy was reported, where articular chondrocytes were collected and seeded on a scaffold material and implanted into the defect site in a single surgical procedure [132]. Initial results were quite promising however, the long-term effect of this single step surgical procedure still remains unknown. Researchers and clinicians are currently trying to develop new methods to treat osteochondral defects which include the use of chondroprogenitor cells from subcutaneous adipose tissue or retropatella loaded within scaffolds followed by implantation in a single surgical procedure [133].

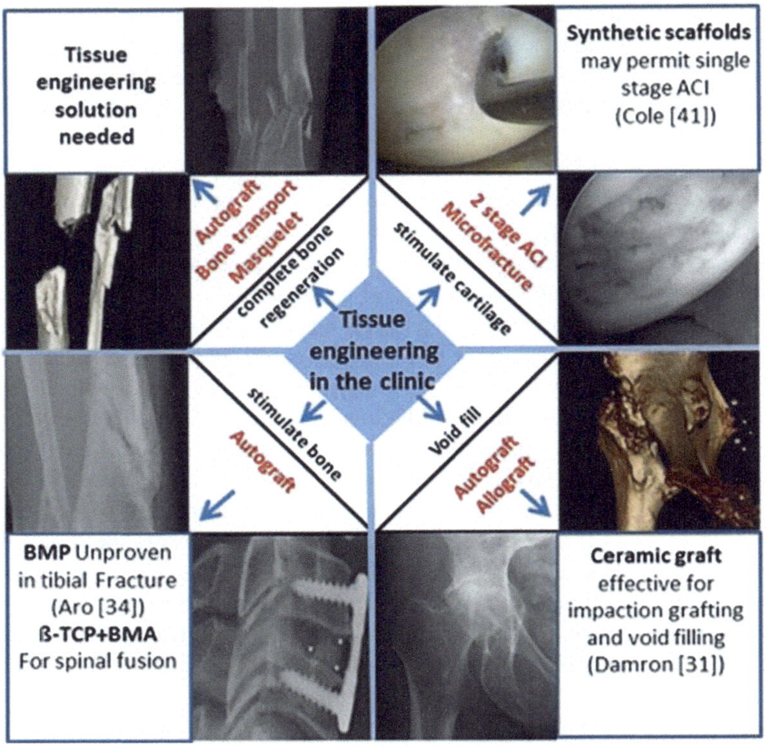

Fig. 2 Different clinical strategies for bone tissue engineering: (i) cartilage regeneration, (ii) bone void fillings, (iii) stimulation of fracture healing or arthrodesis, and (iv) reconstruction of segmental bone loss are represented in different quadrants. (i) In the top right quadrant are arthroscopic images of an osteochondral lesion (upper) and chondral lesion post microfracture (lower). (ii) In the lower right quadrant is a 3D reconstruction (upper) and radiograph (lower) of a patient with severe osteoarthritis and protrusio acetabuli. (iii) The lower left quadrant shows radiographs demonstrating fracture nonunions and spinal arthrodesis. (iv) The top left quadrant demonstrates a comminuted tibial fracture (upper) and segmental bone defect (lower). Current treatment strategies are detailed (in red) in corresponding triangles. Reproduced from permission [111]

Cartilage Tissue Engineering

According to Hunter, "From Hippocrates to the present age, it is universally allowed that ulcerated cartilage is a troublesome thing and that once destroyed, is not repaired" [134]. Over the past four decades, various researchers have proposed and undertaken a substantial amount of research to understand the biological repair process of articular cartilage. Previously conventional procedures were applied to repair cartilage defects such as micro fracture [135], autografts [136], and autologous chondrocyte implantations [137]. But such procedures have several limitations

Table 2 Various influential factors for cartilage regeneration

Factors	Desired value	References
Cell seeding density	Higher initial cell density (64 million cells/mL)	[144]
Mechanical loading (dynamic compression)	210% strain or 0.5–1.0 MPa at physiological frequency 0.01 to 1.0 Hz	[145, 146]
Osmolality	Physiological osmolality	[147]
Extracellular pH	7.2	[148]
Pore size	70–120 μm	[149]
Growth factors	PDGF, TGF-β, FGF, BMP, IGF	[150–154]

like inferior quality of the repaired tissue, autoimmune response, lack of integration, and lack of significant cell viability [138].

Regeneration of cartilage via tissue engineering was first approached by Green in 1977 [139]. The clinical application of cartilage tissue regeneration was first attempted by Mats Brittberg et al. in 1994 [140]. Chondrocytes were grown in *ex vivo* and transplanted into the defected area. Since then, cartilage tissue regeneration got a new direction by transplanting cells grown in an *ex vivo* environment. After the successful introduction of cartilage TE via cell-based regeneration process, the scenario of cartilage regeneration has been changed. This approach involves the isolation of chondrocyte cells, expansion by giving them an exact *ex vivo* environment, and then seeding them into a biocompatible ECM and implantation of the ECM into the damaged site. Viscoelastic properties of cartilage need to be retained from its key ECM components which is composed of water (70–80%), collagen (50–75%), and glycosaminoglycans (15–30%) [136]. So, the composition of the ECM plays a crucial role to provide proper compressiveness, tensile strength, and frictional properties to biomechanically enabled matrix within the arduous joint environment. Thereby, the choice of material is a critical key point to achieve success over TE in cartilage repair. Synthetic, semisynthetic, and naturally occurring biomaterials were also introduced for cartilage TE. Naturally occurring biomaterials include type-I and type-II collagen and their composites with other synthetic or natural biomaterials. Synthetic materials like PLA [141], PGA, [142] or their copolymers [143] are also used. Some other nonbiodegradable polymeric materials like polyethylmethacrylate and polytetrafluoroethylene, and hydroxyapatite composites were also investigated and found to be capable of articular surface restoration. Various influential factors which are responsible for cartilage regeneration are listed below (Table 2).

The ideal cell carrier matrix for cartilage repair should be capable of mimicking the natural environment of articular cartilage. Components of the cartilage ECM such as type-II collagen and glycosaminoglycans are the major differentiating element, they regulate the expression of the chondrocytic phenotype and therefore chondrogenesis takes place. Without proper phenotypic expression, chondrocytes undergo de-differentiation and form an inferior fibrinocartilaginous matrix composed of type-I collagen which fails to form hyaline cartilage.

Recent developments in cartilage regeneration are mainly focused on the preparation of an ECM based on biodegradable natural materials as well as synthetic materials which can give an appropriate environment to support cell differentiation to form cartilage. Natural materials like chitosan, alginate, hyaluronic acid, agarose, cellulose, collagen, chondroitin sulfate, fibrin glue, gelatin, hyaluronic acid, and silk fibroin are generally used as scaffold materials. Limitations are there which also restricts the use of natural biomaterials for cartilage regeneration, such as the autoimmune response which is a major drawback, enzymatic degradation and inferior mechanical properties, as well as disease transfer are major considerations. But synthetic materials are more predictable as the required properties can be tuned via modification. Tunable degradability and mechanical properties of synthetic materials are a key concern to form a scaffold architecture which can vary cellular behavior. Poly (α-hydroxy esters), PEG, poly(NiPAAm), poly(propylenefumarates), and polyurethanes are presently used as a scaffolding material for cartilage TE.

All of the above mentioned synthetic and natural polymers are used in the preparation of various physical forms like hydrogel, sponges, and fibrous meshes.

Hydrogels

Hydrogels are swollen networks capable of delivering bioactive molecules to the site of interest. Scaffold design variables are governed by physicochemical properties, mass transport property, and cellular interactions with the material for specific applications. Hydrogels designed for cartilage restoration should have high enough mechanical properties without damaging the cells, influencing cell adhesion properties, giving nontoxic materials to the surrounding tissues, and appropriate diffusion property. In cartilage tissue engineering, various biomaterials and methods have been applied to form hydrogel (Table 3).

Various critical factors are also involved to engineer cartilage TE constructs and they are listed in Table 4.

Many naturally occurring biopolymers exhibit the above mentioned properties and hydrogel scaffolds have been prepared using them. Polysaccharide-based hydrogels are promising materials for cartilage tissue regeneration. Their physicochemical structure also plays a critical role in cartilage tissue regeneration. Saccharide units regulating cell signalling, ease of synthesis of biologically active

Table 3 Various biomaterials and methods to form the hydrogel for cartilage TE

Biopolymer	Method of hydrogel formation	References
Collagen	Polymerize into a stable gel at neutral pH	[155, 156]
Alginate	Metal ion cross-linking	[157, 158]
Chitosan	Physical gelation by alcohol and ammonia vapor cross-linking	[159]
Silk	Ionic gelation	[160, 161]
Hyaluronic acid	Photo cross-linking of its acrylic derivatives	[162, 163]
Chondroitin sulfate	Photo cross-linking of its acrylic derivatives	[164]

Table 4 Factors involved for cartilage TE constructs

Desired biochemical parameters associated with ideal hydrogel constructs for cartilage TE
1. Maintenance of chondrocyte morphology, viability, and proliferation
2. Promoting chondrocyte differentiation
3. Preserving the cell phenotype, and upregulating the expression of: (a) Type-II collagen (b) Aggregan (c) The transcription factor Sox 9
4. Downregulating the expression of type-I collagen
5. Promoting the deposition of GAG
7. Integrating with host tissue
8. Biodegrading as the neo-cartilage formation proceeds

materials, and controllable biodegradation and biological activity make them more prominent for cartilage tissue regeneration [165]. Polysaccharides are able to form hydrogels by hydrogen bonding, ionic interactions, or chemical cross-linking and the methods employed to form such hydrogels are enlisted in Table 3. Gels based on collagen type-I containing chondrocytes were used to repair full-thickness defects and moderate regeneration of the articular surface was observed [166]. Denaturation of collagen forms gelatin which is also a promising material for cartilage regeneration. Genipin cross-linked gelatin showed collagen and GAG production in 9 days, and a tissue with a cell distribution resembling that of the native cartilage was developed after 30 days of cell culture [167]. Alginate is another biomaterial which shows favorable cellular responses and it can form a gel easily using various divalent cations such as calcium, magnesium, and barium. Chondrocyte embedded gelatin beads were reported to increase type-II collagen expression and this effect was significantly increased with the addition of BMP-2 [168]. Chitosan and its derivatives like dicarboxymethyl chitosan [169], N-acetyl chitosan [170], carboxymethyl chitosan [171], and quaternized chitosan [172] had also been used in cartilage TE. Recent report on chitosan and its hybrid material showed promising activity where they promoted the proliferation, differentiation, and viability of mesenchymal stem cells [173]. Recently, a silk biomaterial collected from insects or worm has gained attention due to its slower degradability, biocompatibility, and notable mechanical properties [174]. Meinel et al. prepared silk fibroin/collagen cross-linked scaffold and compared them over MSCs, where they found that stable and slow degradation rate over collagen and more cell density on silk fibroin scaffolds which also promotes the formation of cartilage-like ECM [174]. Chondroitin sulfate (CS) is a major matrix compound which is composed of repeated chains of alternating sugar units of N-acetyl galactosamine and glucuronic acid [175, 176]. Nishimoto et al. reported that CS upregulated the expression of collagen type-II mRNA, and thereby promoted cartilage formation [177]. Although, the cost and ineffectiveness of CS toward hydrogel formation have limited its application in cartilage TE.

Sponges

Sponges are porous scaffolds and their properties are dependent upon pore size and interconnectivity. A sponge's porous nature directs cell adhesion, where pore size and interconnectivity influence cell migration, matrix deposition and distribution, and nutrient and waste exchange [178]. Various methods have been applied to formulate a sponge material for cartilage TE-like porogen leaching [179–181], freeze-drying (10.1039/C1SM06179H, 10.1039/C4TA07057G), and 3D printing [182]. Different kinds of natural and degradable synthetic polymers and their hybrid materials were also applied to fabricate sponges for cartilage TE, including (polylactide (PLA) [141, 142, 183], poly(lactide-co-glycolide) [142, 181] (PLGA), PLGA/Collagen [184], PLA/PHBV [185], poly(ε-caprolactone) (PCL) [186], chitosan [187], PLA/chitosan [188], and cellulose systems [189]. PLA and a blend of PLA/PCL have been used to form ultraporous, wettable, and compressable sponges [190]. A hybrid of natural and synthetic material (PLA/gelatin fiber) based sponges with desired scaffolding properties such as elasticity, porosity, degradability, and biocompatibility have all been shown to be suitable for in vivo cartilage regeneration [191–193]. Blend of chitosan/gelatin sponge showed a significant increase in GAG content and the formation of type-II collagen with elastic fibers [194].

Meshes

Network structure of woven and nonwoven fibers is generally called meshes, where cell behavior can be dictated by void volume and the surface area. Differences between woven and nonwoven meshes are their construct strength where woven meshes possess greater strength over nonwoven meshes. Cells can be seeded onto these meshes in vitro and can be implanted in vivo for complete regeneration of defected cartilage [195]. The electrospinning technique is most commonly applied to fabricate meshes. Recently, this technique is gaining more interest as it can produce the fibers in a predetermined manner [196]. PGA, PLA, and their copolymer PLGA are the most prominent materials of choice to fabricate meshes for cartilage regeneration as they could be completely resorbed through metabolic pathways. Degradation of PLA is slower than PGA as PLA contains additional methyl group. . PLGA with a 50:50 composition degraded faster than 75:50 as PGA content increased [143]. Type-II collagen and hyaluronic acid immobilization greatly influenced the chondrocyte proliferation and GAG deposition [197, 198]. Another poly (α-hydroxy ester) is PCL with a slow degradation rate. PCL nanofiber scaffolds were applied to periosteal cells in vitro and in vivo and it was found that they could improve the formation of cartilage [199]. Hyaluronic acid based semisynthetic resorbable meshes (HYAFF® derivatives) have also been commercialized and could facilitate the formation of ECM-like cartilage [200].

Neural Tissue Engineering

The central nervous system is one of the vital components of the human body which plays a crucial role by transmitting signals between different body parts and the brain. At the cellular level, the nervous system consists of two different types of cells called neurons and glial cells. Neurons have a special type of structure which allows them to transmit signals in a form of electrochemical waves from one cell to another cell. Neurons withstand a small voltage difference between their membranes through concentrations of ions such as calcium, sodium, potassium, etc. Briefly, neurons transfer signals via a thin fiber known as an axon with the help of neurotransmitters such as dopamine, acetylcholine, serotonin, etc. at the junction of two different neurons known as synapses. Other than neurons, glial cells are also an important part of the central nervous system, and they provide structural and metabolic support to the nervous system. Other than the central nervous system, our body also consists of the peripheral nervous system. The peripheral nervous system connects the central nervous system with limbs and organs serving as a relay between the spinal cord, brain, and the other parts of the body.

With time, the need for neural TE constantly increases due to an increase in number of patients with neurodegenerative diseases and neural injuries and limited regeneration capability of nerve tissues. Currently, there are no well-established treatment methods which can completely repair damaged central nervous system tissue. Existing clinical strategies include stabilization of neural injuries and mitigation of symptoms using pharmacological agents, although the outcomes are not satisfying as most of the current pharmacological agents have many side effects [201, 202].

In this context, neural TE provides alternative strategies to regenerate damaged nerve tissues. Embryonic stem cells and neural stem cells are the most common cells which are used for neural TE. Transplanting stem cells into the brain with appropriate scaffold material enables them to differentiate into other cells, such as astrocytes and oligodendrocytes which help to improve the neurological functions of the brain. A scaffold acts as a support system for neural cells through which regenerating axons can proliferate from one side of the injury to the other side of nervous tissues. Scaffolds for neural TE should be biocompatible with ideal degradation rates, and possess enough mechanical properties so that cells can adhere to the scaffolds. In addition, presence of electroactive/electroconductive material within the scaffold can provide an added advantage for neural TE [203]. For example, Tu et al. reported a biomimetic graphene oxide (GO) composite composed of an acetylcholine-like unit (dimethylaminoethyl methacrylate, DMAEMA) or phosphorylcholine-like unit (2-methacryloyloxyethyl phosphorylcholine, MPC) for neurite overgrowth and branching. They used rat hippocampal neurons to check for neural cell adhesion, spreading, and proliferation on the scaffold surface. After 2 to 7 days of cell seeding, the number of neural cells and average length of the neural cells on GO–DMAEMA and GO–MPC composites were significantly enhanced compared to the control group. They proposed that, due to the presence of a neurotransmitter like unit (DMAEMAMPC) as well as the presence of electroconductive graphene, there was significant boost in neurite sprouting and outgrowth [204]. Schmidt and coworkers investigated electroconductive polymer polypyrrole (PPy)-coated electrospun PLGA nanofibers for neural TE. They studied the combined effect of nanofiber structures and electrical stimulation on neural tissue regeneration. Upon culturing rat pheochromocytoma 12 (PC12) cells and hippocampal neurons on the PPy-PLGA scaffold surface, cells were well adherent and differentiated on the scaffold surface compared to a control. Electrical stimulation studies revealed that when PC12 cells were stimulated with a 10mV/cm potential of PPy/PLGA scaffolds, they demonstrated 40–50% longer neurites and 40–90% greater neurite formation. Additional stimulation on the aligned nanofibers resulted in longer neurite outgrowth compared to random nanofibers [205]. Fan et al. discovered that polyaniline (PANI) could promote nerve tissue regeneration by enhancing brain-derived neurotrophic factor and ciliary neurotrophic factor through activation of the ERK1/2/MAPK signalling pathway [206]. Due to various fascinating properties of PANI, such as high conductivity, ease of synthesis, low cost, and suboptimal biocompatibility, it is often used for neural TE. However, due to slow biodegradability, PANI is often used with other degradable biopolymers for neural TE.

For example, Zhang and coworkers reported a micro nanostructured PANI-cellulose hydrogel scaffold for peripheral nerve regeneration. The resulting hydrogel demonstrated very good biocompatibility along with good mechanical properties as well as excellent guiding capacity for sciatic nerve regeneration within adult Sprague-Dawley rats. It is well established that cellulose hydrogels are inert materials in terms of neural tissue regeneration. PANI played a crucial role on nerve regeneration. The hierarchical microstructure and electrical conductivity of PANI remarkably induced the adhesion and differentiation of neurons both in vitro and in vivo which resulted in neural tissue regeneration [207]. In addition PANI and PPy, another conducting polymer named poly (3,4-ethylenedioxythiopene) (PEDOT) also established its presence for neural TE due to its optical transparency in a conductive state, low redox potential, and high stability. Pires et al. reported for the first time neural stem cell differentiation through stem cell differentiation using a cross-linked PEDOT substrate. They revealed that upon applying a pulsed laser current on the PEDOT substrate, resulted to neurite outgrowth and longer neurons [208]. In another similar type of study, Ostrakhovich and coworkers reported the differentiation of embryonic P19 cells and neural stem cells into a neural lineage in the presence of a PEDOT-PEG substrate. They observed that the ability of the substrate to induce differentiation of the stem cells toward a neural linage was directly associated with its conductivity [209].

Various synthetic nonconducting polymers were also studied, among them PLA, PGA, PLGA, PEG and poly(hydroxyethylmethacrylate) (PHEMA) are most often used for neural TE.

PLA scaffold provided support to neurilemma cells permitting elongation of axons as well as promoting vascular growth [210, 211].

PLGA has been extensively used for neural TE because of its versatile properties including permeability, swelling nature, deformation, and tunable degradation rate.

Kim and coworkers showed that multichannel PLGA scaffolds seeded with Schwann cells had synergistic effects on neural tissue regeneration. A further in vivo implantation of the multichannel scaffold along with Schwann cells revealed that the number of regenerated axons were significantly increased than a multichannel scaffold without Schwann cells [212]. Another big advantage of PLGA is that, PLGA can transport anti-tumor drugs [213] and glial-derived neurotrophic factor (GDNF) for the treatment of neurodegenerative diseases [214]. Scaffolds made with PEG was also extensively been used for nerve tissue regeneration. Improved neural cell growth, proliferation, and differentiation were observed on PEG surface [215–217]. Liu and coworkers reported enhanced neural stem cell growth, proliferation, and differentiation along with improved functional recovery in a rat model after transection of the spinal cord using PLGA/PEG scaffold [218].

Along with synthetic biopolymers, natural biopolymers were also extensively used for neural tissue regeneration. The main advantage of a natural biopolymer over synthetic polymers are their high biocompatibility and natural degradation kinetics along with tunable chemical properties. In the case of neural tissue regeneration, natural polymers can fulfill several important roles such as extracellular matrix formers, gelling agents, or drug release modifiers [218]. Till now, various natural polymers like chitosan, alginate, silk, and elastin collagen have been used for neural TE. However, collagen is the only biopolymer that was approved for clinical studies to regenerate neural tissues. For example, NeuraGen® demonstrated a high success rate in peripheral nerve regeneration in 43% of patients [219]. As another commercially accessible collagen nerve guide, Neuromaix® exhibited exceptional results for bridging long nerve gaps in its initial clinical trial [220]. In Table 5, we have enlisted a series of natural and synthetic biopolymers for neural TE.

Skin Tissue Engineering

Skin is the largest part of the body that primarily serves as a barrier between the environment and our body. It not only protects the underlying organs of our body, but it also acts as a shield against harmful thermal, mechanical, chemical, and microbial influences of nature. Loss of integrity or a large portion of damage to the skin may result in significant disability or even death.

Current regeneration approaches for skin TE are not that successful owing to complications, which arise due to wound erection, secretion, and development of an ECM. Currently autologous skin grafts, which have chances of donor site injury, are majorly used for the treatment. In order to improve the current scenario in the field of skin TE and wound healing, cell-derived ECM scaffolds are gaining much importance owing to their low risk of immune rejection and at the same time low risk for pathogen contamination [246]. Nevertheless, such scaffolds do not provide sufficient mechanical stiffness or rigidness compared to tissue-derived ECM scaffolds.

Form the viewpoint of skin TE, wound healing is the most important field of research, especially in the case of third degree burns [247]. In such cases, at the

Table 5 Polymers used for neural TE

Polymer used	Cells used to perform in vitro/animal used for in vivo study	Results	References
Collagen	Monkey model	Nerve regeneration was observed in presence of a collagen nerve guide and the final level of physiological recovery was comparable to direct suture repair of the median nerve	[221]
Chitosan–collagen hydrogel	Female Wistar Hannover rats	The presence of collagen and fibronectin increased neural tissue regeneration	[222]
Type-I collagen gel	Female CF1 mice	The collagen gel demonstrated very good biocompatibility along with controlled delivery of neurotrophic factors at the regeneration site. Furthermore the gel demonstrated enhanced peripheral nerve regeneration	[223]
Laminin-modified linear ordered collagen scaffolds	Rats	This biomaterial could guide axon growth, retain more ciliary neurotrophic factor on the scaffolds and enhance nerve regeneration as well as functional recovery	[224]
Magnetically aligned collagen gel	Schwann cell	The axial bias of neurite elongation became more pronounced with an increase in magnetic field strength, presumably due to a contact guidance response of growth cones at the neurite tips	[225]
Electrospun gelatin scaffold	Human chorion-derived MSCs	Upon culturing MSCs on the electrospun gelatin scaffold, cells demonstrated improved adhesion compared to conventional 2D plates along with enhanced differentiation toward motor neuron-like cells	[226]
Nanofibrous gelatin scaffold combined with growth factor loaded alginate microsphere	Rat pheochromocytoma (PC12) cells	The scaffold demonstrated a modulus value of 1.2 kPa which was similar to that of brain tissue (0.5–1) kPa. Additionally the controlled release of nerve growth factor induced neurite extension of PC12 cells	[227]
Electrospun PCL/gelatin nanofibrous scaffolds	Nerve stem cells (C17.2 cells)	PCL/gelatin (70:30) nanofibrous scaffolds enhanced nerve differentiation and proliferation compared to bare PCL nanofibrous scaffolds and served as a positive cue to support neurite outgrowth	[228]
3D multichannel bi-functionalized silk electrospun conduits	Dorsal root ganglia	The conduits provided sustained release of nerve growth factors. In addition, the mechanical behavior of the nerve graft was comparable to the rat sciatic nerves	[229]

Polymer used	Cells used to perform in vitro/animal used for in vivo study	Results	References
Hyaluronic acid doped-poly(3,4-ethylenedioxythiophene)/chitosan/gelatin gel	Rat pheochromocytoma (PC12) cells	8%PEDOT-HA/CS/Gel scaffold support cell adhesion, survival, proliferation, and synapse growth	[230]
Soft alginate hydrogel	Wistar rats	Alginate hydrogel demonstrated its capability in both preventing fibrous scars and providing a satisfactory functional recovery after a severe spinal cord injury	[231]
Injectable alginate hydrogel loaded with glial-cell-derived neurotrophic factor (GDNF)	Rat pheochromocytoma (PC12) cells, female Long–Evans rats	The local GDNF delivery from alginate: the hydrogel significantly increased the ingrowth of neurites in and around the site of a spinal cord hemisection. The rats treated with free GDNF-loaded hydrogel experienced superior functional recovery compared with the animals treated with GDNF microsphere-loaded hydrogels and non-treated animals	[232]
Polylysine-functionalized thermoresponsive chitosan hydrogel	Neural progenitor cells	The hydrogel demonstrated good cell adhesion properties with good cytocompatibility. Neuron survival was improved with the covalent attachment of polylysine to the chitosan hydrogel	[233]
Photo cross-linkable chitosan-based hydrogels	Human mesenchymal stem cells	Chitosan hydrogels facilitated enhanced neurite differentiation from primary cortical neurons and enhanced neurite extension from dorsal root ganglia as compared to agarose-based hydrogels with similar mechanical properties	[234]
Chitosan scaffold combined with neural growth factor (NGF)	Neural stem cells (NSCs)	Combination of chitosan scaffold and NGF exerts a synergistic effect on neuronal differentiation of NSCs	[235]
PLGA/chitosan scaffolds	Bone marrow stromal cells (BMSCs)	BMSCs cultivated in the PLGA/chitosan scaffolds could be guided in the presence of NGF toward mature neurons	[236]
Keratin hydrogel	RT4-D6P2T Schwann cells and male Swiss Webster mice	Keratins extracted from human hair enhanced the activity of Schwann cells by a chemotactic mechanism, increased their attachment and proliferation, and upregulated expression of important genes	[237]

(continued)

Table 5 (continued)

Polymer used	Cells used to perform in vitro/animal used for in vivo study	Results	References
Silk nanofiber hydrogel	Neural stem cells (NSCs)	Compared to non-annealed nanofibers, NSCs on annealed nanofibers with a stiffness similar to that of nerve tissues, differentiated into neurons without the presence of any specific differentiation biochemical factors	[238]
Silk hydrogels	Embryonic chick dorsal root ganglions (cDRGs)	Neurite extension of cDRGs cultured on 2 and 4% silk hydrogels displayed higher growth than softer or stiffer gels. Silk hydrogels released <5% of neurotrophin-3 (NT-3) over 2 weeks and showed preservation of growth factor bioactivity. Fibronectin- and NT-3-functionalized silk gels caused increased axonal bundling	[239]
Uniaxial multichannel silk fibroin scaffolds	Primary hippocampal neural cells	Multichannel silk fibroin scaffolds provided a favorable microenvironment for the growth of hippocampal neurons by guiding axonal elongation and cell migration	[240]
Chitosan/PLGA Scaffold	Bone marrow mesenchymal stem cells (MSCs), male Beagle dogs	Introduction of autologous bone marrow MSCs to a chitosan/PLGA scaffold improved the repair and rehabilitation of a large gap after peripheral nerve injury in dogs	[241]
PEG hydrogel	Neural precursor cells (NPCs)	3D PEG hydrogels led to better metabolic activity and lower apoptotic activity with substantial proliferation observed only in hydrogels that closely matched the stiffness of native brain tissue	[215]
PEG–heparin hydrogel	Neural stem cells (NSCs), Wistar rats	The hydrogel demonstrated excellent cytocompatibility toward NSCs. The effects of both structural/mechanical properties of the hybrid gel on neuroectodermal cells displayed the cell type-specific interplay of synergistic signaling events and highlighted the need for fine-tuned matrix characteristics	[216]

Polymer used	Cells used to perform in vitro/animal used for in vivo study	Results	References
Conductive polypyrrole-poly(lactide acid) fibers	Rat pheochromocytoma (PC12) cells	The results on cell adhesion rate, neurite-bearing cell rate and neurite alignment rate on an ECM coated with a conductive fiber-film (ECM-CFF) were ~95%, ~77%, ~70%, respectively, significantly larger than the corresponding values on bare CFF (17%, 29%, and 14%, respectively). The neurite length on ECM-CFF (~79 mm) was also larger than that on bare CFF (~25 mm)	[242]
Electrically conductive catechol functionalized HA hydrogels incorporated with single-walled carbon nanotubes (CNTs) and/or polypyrrole (PPy)	Human neural stem/progenitor cells (hNSPCs)	The HA–CA hydrogels containing CNT and/or PPy significantly promoted neural differentiation of human fetal neural stem cells (hfNSCs) and human induced pluripotent stem cell-derived neural progenitor cells (hiPSC-NPCs) with enhanced electrophysiological functionality	[243]
Poly(2-vinyl-4,6-diamino-1,3,5-triazine)/poly(1-vinylimidazole) copolymer P(VDT)-P(VI) hydrogels coated with polyaniline	Neural stem cells (NSCs)	The electroconducting scaffold promoted the adhesion, proliferation, and differentiation of NSCs to neurons	[244]
Polyaniline (PANI)- polyethyleneglycol diacrylate (PEGDA) macroporous hydrogels	Rat pheochromocytoma (PC12) cells, human mesenchymal stem cells	The presence of PANI effectively increased the electrical conductivity of the hybrid material from $1.1 \pm 0.5 \times 10^{-3}$ mS/cm with a PANI content of 3wt%. 3 wt% PANI also improved the biological response of PC12 and hMSC cells	[245]

primary stage of the healing process, a momentary repair is commonly achieved in the form of a blood clot. Immune cells, fibroblasts, and capillaries later invade the clot and form granulation tissue. However, this is a temporary solution and often it fails when a larger wound is concerned. To address this issue, researchers have been developing 3D scaffolds or 2D patches made of biodegradable biomaterials which can promote the wound healing process with minimal scar formation [248]. Till now, few biomaterials are used in the form of 3D scaffolds or 2D patches which are generally made of natural/synthetic polymers for wound healing.

For example, Tegaderm and Opsite are well-known commercial semipermeable polyurethanes used for wound healing. Their main purpose is to protect the skin from water loss and mechanical injury. Tegaderm is a commercially available vapor permeable wound dressing which supports cell adhesion and proliferation. Due to its optical transparency, it permits the observation of cell growth before re-epithelialization after grafting to the wound site. It is an inexpensive and cheap wound dressing for the cure of both severe burning and chronic wounds [249]. Opsite is also a commercially available elastic transparent self-adhesive polyurethane patches. The impermeability of Opsite toward bacteria provides an added advantage. In this context, a study reported by Gowland and coworkers tested, the in vitro killing ability of Tegaderm and Opsite toward pathogens. They found that Opsite was better than Tegaderm in terms of killing bacteria [250]. Other than these commercially available wound dressings, researchers and clinicians have been continuously trying to develop novel types of scaffold materials and wound dressings for skin tissue regeneration. For example, Yao and coworkers developed a Chitosan (CS)-Gelatin (Gel)-Hyaluronic Acid (HA) based artificial skin for skin tissue regeneration. The presence of HA within the scaffold enhanced water uptake, flexibility, and cytocompatibility of the scaffold. When culturing fibroblasts and keratinocytes on the scaffold, cells were well attached and proliferated [251]. In another similar type of work reported by Wang et al., they manufactured a skin substitute made of gelatin-chondroitin sulfate-hyaluronic acid (gelatin–C6S–HA). The gelatin–C6S–HA scaffold skin substitute showed positive effect on mice on promoting wound healing with a high rate of graft intake [252]. Hutmacher and coworkers reported 3D matrices composed of PLGA mesh and collagen–hyaluronic acid (CHA) sponge which was similar to bilayered skin. The bilayered skin supported both fibroblasts and keratinocytes without contraction over a period of 4 weeks. A further in vivo implantation of these scaffolds at the wound site presented a faster rate of wound contraction that was comparable to autografts [253].

Collagen is also one of the most widely used polymeric biomaterials for skin TE. For instance, Ma et al. fabricated a porous scaffold composed of collagen and chitosan. To improve the biostability of the scaffold, they used glutaraldehyde (GA) which enhanced the mechanical properties of the scaffold. In vitro culture of fibroblast cells on the GA-treated scaffold demonstrated that the scaffold maintained good cytocompatibility of collagen and accelerated cell infiltration and proliferation. A further in vivo study revealed that the scaffold could sufficiently support and accelerate fibroblast infiltration from the surrounding tissue at the wound site [254]. In another similar type of study, Dai et al. reported a composite membrane composed of collagen and PCL for skin tissue regeneration. Their result suggested that, the collagen-PCL composite films were favorable for fibroblast and keratinocyte growth and may be a promising candidate for skin repair (Fig. 3) [255].

Although there are some commercialized products such as Epicel®, Alloderm®, Dermagraft®, Orcel®, Opsite, etc. for skin tissue regeneration, there are still lots of challenges remaining in this field, which include generation of a complex dermo-epidermal substitute that can be securely and easily transplanted with minimal scarring in one single surgical procedure. Moreover, the development of novel ECM scaffolds and patches with minimal cost is still a main focus for skin TE.

Cardiac Tissue Engineering

Cardiovascular diseases (CVD) is one of the most epidemic diseases throughout the globe. It is the cause of 17 million deaths worldwide every year [256]. Cardiac tissue injury or cardiac tissue related diseases like myocardial infarction and cardiomyopathy are among the most common causes of illness and death all over the world [257]. The two major treatments for cardiac tissue repair are cardiomyoplasty and the implantation of cardiac grafts but unfortunately both of these methods are quite unsuccessful in clinical applications owing to the fact that cells have a very low electrical integrity with the host tissue and ingenuousness of the grafted tissue; so, for most of the cases, the damage is almost permanent [258]. To overcome this problem, researchers are now trying to develop new strategies with the help of TE to treat this type of critical heart disease. Due to its various fascinating properties (such as very good electrical conductivity and mechanical properties), graphene is one of the promising nanomaterials for cardiac TE applications. . Heart transplantation is currently the accepted longstanding approach for the administration of cardiac tissue injuries but at the same time there is a constant problem with lack of donors as the number of donors are far less than needed. Various approaches have been introduced for heart tissue regeneration which includes stem cell injections, introduction of various growth factors and different novel biomaterials into the myocardium, and designing various patches which can support damaged cardiac tissues [259]. But as discussed earlier, the success rate of these patches and scaffolds are very much limited and narrow till now [260]. Hence, designing a proper treatment for cardiac injury is a big challenge.

Myocardial infarction (MI), which is commonly known as a heart attack, is caused by the coronary occlusion of one or more blood vessels supplying blood to the heart. As cardiomyocytes are terminally differentiated cells, after myocardial infarction, adult mammalian heart tissue cannot efficiently regenerate new cardiac muscle cells, resulting into negative left ventricular remodeling and thus a heart attack. Here lies the importance of cardiac TE that is a vibrant area of research and its applications include cardiac patches, engineered blood vessels, and a vascular network. The major objective of cardiac TE is to restore heart function and regenerate

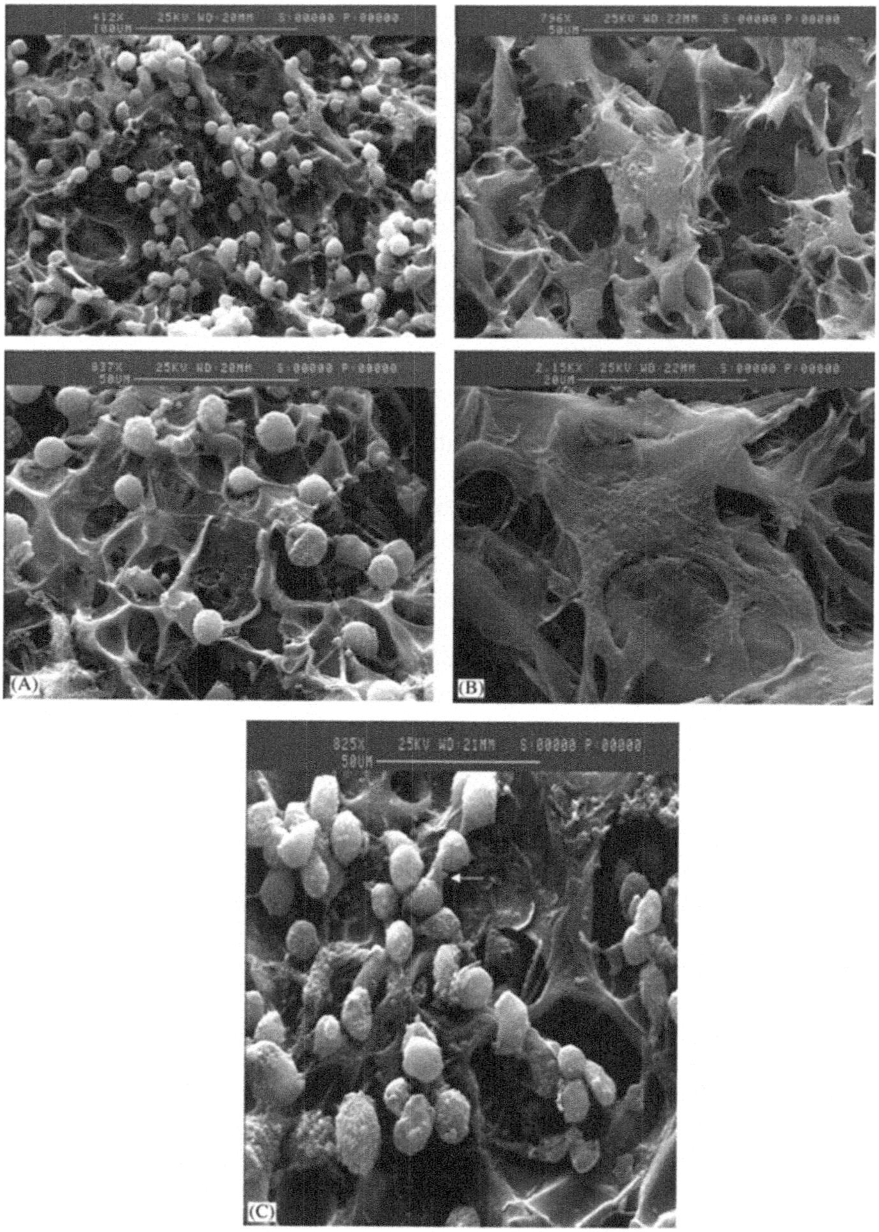

Fig. 3 The morphology of 3T3 fibroblasts following cell culture on collagen:PCL composites: (**A**) 1 day on 1:8 composite; (**B**) 3 days on 1:8 composite and (**C**) 1 day on 1:20 composite. Reproduced from permission [255]

cardiomyocytes. Three key requirements for cardiac tissue engineering are the implantation of biomaterials, cells, and cell-scaffold constructs.

The aim of this section is to give an overview about the synthetic and naturally occurring biomaterials, cells, and scaffolds that can be used for cardiac TE purposes.

Biomaterials in Cardiac Tissue Engineering

Biomaterials are commonly used to develop myocardial tissue engineering (MTE) constructs which encourage cardiomyocyte alignment and maturation in vitro before implantation. For successful cardiovascular TE, a biomaterial should withstand electrical integration with native tissue to allow for a continuous expanding/contracting motion of cardiomyocytes while the heart beats. In this context, among several biomaterials, biopolymers are the first choice for cardiovascular tissue engineering due to their tunable mechanical properties, easy processability and facile integration with other biomaterials to obtain better properties and degradation kinetics. Till now, both natural and synthetic polymers have been used in cardiac tissue engineering. However, they have their own pros and cons.

Natural Biopolymers Used in Cardiac Tissue Engineering

Among the natural biomaterials, collagen, gelatin, and alginate have been extensively investigated during the last few decades for cardiac tissue engineering applications (Table 6).

Synthetic Biopolymers in Cardiac Tissue Engineering

Synthetic biomaterials have suitable and reproducible chemical, mechanical, and physical properties (tensile strength, young's modulus, degradation rate) and have low immune responses and production costs.

Synthetic polymers used in cardiac TE have been depicted in Table 7 where applications of PLA, PGA, PCL, and other synthetic polymers have been explored as scaffold materials for cardiac TE constructs.

Injectable Biomaterials in Cardiac Tissue Engineering

Hydrogels designed for cardiac TE can be classified into three main groups. The first one can prevent adverse remodeling and recruitment of endogenous cells for repair; the second one can act as a temporary matrix for cell transplantation and exogenous repair; and the third may be designed to maintain left ventricle (LV) geometry for proper heart function.

Table 6 Natural biopolymers used in cardiac TE

Material	Cells used to perform in vitro study/animal used for in vivo study	Results	References
Chitosan/carbon nanofiber	Neonatal rat cardiomyocytes	The chitosan/carbon nanofiber scaffolds supported cultivation of cardiac cells and enhanced cardiogenic properties without exogenous electrical stimulation	[261]
Peptide-modified chitosan collagen hydrogel	Rat cardiomyocytes, adult male C57 black-6 mice	The scaffold provided the appropriate stiffness and mechanical properties to support cardiomyocyte survival both in vitro and in vivo	[262]
Graphene oxide-gold nanosheets containing chitosan scaffold	Rat smooth muscle cells (SMCs), mouse fibroblasts and human induced pluripotent stem (iPS) cells derived cardiomyocytes, adult male Wistar rats	The scaffold supported cell attachment and growth with no signs of cytotoxicity. In an in vivo animal model of MI, as well as in isolated heart, 5 weeks of implantation showed a significant enhancement of conduction velocity and contractility in the infarct zone	[263]
Chitosan-glycerol phosphate injectable hydrogel	Mouse embryonic stem cells (ESCs), female Sprague-Dawley (SD) rats	The injectable hydrogel effectively delivered stem cells to infarcted myocardium. The hydrogel also increased cell retention and graft size as well as improved cardiac function	[264]
Chitosan-polypyrrole scaffold	Neonatal rat cardiomyocytes, female Sprague-Dawley rats	The scaffold synchronized the contraction of physically isolated cardiomyocyte clusters without external electrical stimulation. The chitosan-polypyrrole scaffold can also improve electrical conduction across a fibrotic scar in the injured heart	[265]
RGD immobilized alginate scaffold	Neonatal rat cardiac cells, Sprague-Dawley rats	The scaffold promoted cell attachment to the matrix, enhanced cell survival and recovery, and induced the organization of cardiac muscle tissue	[266]
Alginate-gelatin blend film	C2C12 myoblast cells	Culturing C2C12 myoblast cells, the scaffold demonstrated good cell proliferation for all the blends containing more than 60% of gelatin, while the alginate/gelatin 20:80 showed the best response	[267]

(continued)

Table 6 (continued)

Material	Cells used to perform in vitro study/animal used for in vivo study	Results	References
Collagen cardiac patch incorporated with alginate microparticles	Cardiac stem cells (CSCs)	The resulting cardiac patch permitted the sustained release of hepatocyte growth factor and insulin-like growth factor-1 which enhanced cardiac stem cell migration and proliferation	[268]
In situ cross-linkable alginate-hyaluronic acid hydrogels	Neonatal rat heart cells (NRHC)	The hydrogel allowed generation of contractile bioartificial cardiac tissue from cardiomyocyte-enriched neonatal rat heart cells, which resembled the native myocardium	[269]
Peptide G4RGDY and heparin-binding peptide G4SPPRRARVTY (HBP) conjugated alginate scaffolds	Neonatal rat cardiac cells	The peptide bearing alginate macroporous scaffold exactly mimicked the microenvironment of the myocardial tissue. The cardiac tissue developed in this microenvironment revealed the striation and muscle fiber structure similar to that of a mature cardiac tissue	[270]
Micropatterned alginate substrate coated with fibronectin	Neonatal rat ventricular myocytes, human umbilical artery vascular smooth muscle cells	The heterogeneity in fibronectin pattern in microcontact printed substrates and heterogeneity in topography in micromolded substrates led to the formation of anisotropic cardiac and vascular smooth muscle tissues	[271]
Collagen type-I scaffold	Neonatal rat cardiac cells, male Wistar rats	After implantation in the infarcted heart of a rat model, engineered heart tissue showed undelayed electrical coupling to the native myocardium without evidence of arrhythmia induction	[272]
Collagen/elastin/PCL hybrid scaffold	Adipose-derived stem cells	The hybrid scaffold demonstrated favorable mechanical properties along with good cytocompatibility	[273]
Collagen type-I glycosaminoglycans (GAGs)-chondroitin sulfate (CS) hydrogel heart valve	Porcine mitral valve interstitial cells (VICs), endothelial cells (VECs)	Collagen gels could be used as matrices for the in vitro synthesis of tissue structures resembling mitral valve tissue. The addition of CS resulted in a more porous model which showed a positive influence on the bioactivity of seeded valve cells and tissue remodeling	[274]

(continued)

Table 6 (continued)

Material	Cells used to perform in vitro study/animal used for in vivo study	Results	References
Collagen scaffold along with bioglue	Human bone marrow-derived mesenchymal stem cells (hBMSCs), male juvenile Sprague-Dawley rats	The scaffolds were assembled into anatomically analogous 3D heart valve shapes using a novel protein-based glue, seeded with stem cells, conditioned in bioreactors, and induced stem cell differentiation	[275]
3D bioprinted gelatin patterned hydrogel	Human mesenchymal stem cell (hMSC), cardiomyocyte cells	The microchanneled hydrogel scaffold produced by 3D bioprinting induced myocardial differentiation of stem cells as well as supported cardiomyocyte growth and contractility	[276]
Electrospun gelatin nanofiber matrix	Neonatal rat cardiomyocyte cells	The nanofibrous scaffold exhibited similar modulus as native human myocardium tissue with fiber diameters of 200–600 nm and an average porosity of 49.9 ± 5.6%. Myoblasts showed good cell adhesion and proliferation	[277]
Gelatin-methacrylyol hydrogel (GelMa)	Human-induced pluripotent stem cells (hiPSCs)	The hydrogel demonstrated initial stiffness of approximately 220 Pa, supported tissue growth and dynamic remodeling, and facilitated high-efficiency cardiac differentiation (>70%) to produce spontaneous contracting GelMA human engineered cardiac tissues.	[278]
Anisotropic silk scaffold containing cardiac tissue-derived extracellular matrix (cECM)	Cardiac muscle cells (HL1 cells), human embryonic stem cell-derived cardiomyocytes, Sprague-Dawley rats	The silk-cECM scaffolds showed tunable architectures, degradation rates, and mechanical properties. Subcutaneous implantation in rats revealed that, the addition of cECM to aligned silk scaffolds led to 99% endogenous cell infiltration and promoted vascularization of a critically sized scaffold after 4 weeks in vivo. In vitro, silk-cECM scaffolds maintained the HL-1 atrial cardiomyocytes and human embryonic stem cell-derived cardiomyocytes and promoted a more functional phenotype in both cell types	[279]

(continued)

Table 6 (continued)

Material	Cells used to perform in vitro study/animal used for in vivo study	Results	References
Stacked silk-cell 3D construct	Primary rat cardiomyocytes and H9c2 cells, rat model	The presence of nanogrooves on silk films provided contact guidance to the growing of cardiomyocytes and allowed them to form unidirectionally aligned cell monolayers. Non-mulberry silk films exhibited significantly greater mechanical strength and lower immunogenicity in vitro. In vivo, supported better growth, proliferation, and maturation of both primary rat cardiomyocytes (PCMs) and H9c2 cells	[280]
Non-mulberry silk protein fibroin cardiac patch	Postnatal rat cardiomyocyte cells	The silk fibroin cardiac patch enabled efficient attachment of cardiomyocytes without affecting their response to extracellular stimuli. Cardiomyocytes which grew on the patch expressed connexin 43, exhibited aligned sarcomeres, and coupled electrically with each other resulting in synchronous beating	[281]

Table 7 Synthetic polymers used in cardiac tissue engineering

Material	Cells used to perform in vitro study/Animals used for in vivo study	Results	References
PGA coated with P4HB	Bone marrow-derived mono nuclear cell, nonhuman primates as in vivo model	Two decellularized engineered tubes and degradable suture pediatric pulmonary valves gave definite support for heart valve regeneration, immediately and up to 8 weeks after implantation. Deposition of elastin and collagen IV, suggested that invaded cells conferred true growth, not passive stretching of the engineered tube	[282]
PCL-U4U	BMSC, PBMCs	Synergistic upregulation of important trophic factors, bFGF and CXCL12 were observed	[283]

(continued)

Table 7 (continued)

Material	Cells used to perform in vitro study/Animals used for in vivo study	Results	References
Bovine pericardium	SDF1-α, fibronectin; sheep as in vivo model	Significant differences in the fraction of $CD90^+$, $CD34^+$, and $CD117^+$ cells were observed when comparing controls to impregnation coated FN, SDF-1α that help to develop native heart valve cells	[284]
Decellularized porcine valves	Fibronectin, HGF; dog	Fusion protein combined fibronectin and hepatocyte growth factor were used to develop a decellularized heart valve. One week after implantation, partial endothelization and vascularization were observed	[285]
Composite of poly(DL-lactide-co-caprolactone) (PCL-PLA)PLGA, and type-I collagen	Heart cells, 2-day-old neonatal Sprague-Dawley rats	PCL/PLA and collagen composite supported electromechanical coupling of cell and contractile function which was essential for mechanotransduction in cardiac constructs. Hydrophilic smooth and elastic surface of composite scaffolds enhanced cell density which promoted synthesis and assembly of cardiac protein Tn-I and Cx43. Moreover, mechanical intregrity of the composite scaffold kept scaffold pores open and interconnected that assisted in cardiac tissue regeneration	[286]
Polyurethane (PU) films	Cardiomyocytes	Cardiomyocytes on the PU surface remained adherent along protein patterned lanes. The elastomeric nature of PU allowed the contracting patterned cardiomyocytes to pull on the thin film as they beat. The alignment of the contracting cells manifested a linear force vector in order to achieve engineered cardiac tissue after myocardial infarction	[287]
PLGA electrospun fibers (ESFs)	Cardiomyocytes isolated from neonatal rats	Cardiomyocyte adhesion was enhanced. Cardiomyocyte contraction was faster and longer lasting on the laminin-coated ESFs and YIGSR-incorporated PLGA ESFs	[288]

(continued)

Table 7 (continued)

Material	Cells used to perform in vitro study/Animals used for in vivo study	Results	References
PLA-gelatin-PGA	Syngenic rat aortic smooth muscle cells, rats	PLA patches contributed to the mechanical intregrity by sustaining its physical dimension for at least 2 months; allowing better cellular penetration. Cell numbers and DNA content in the PLA patches increased promisingly during 3 weeks of culture. Cell seeded patches implanted into the surgically created defect in the right ventricular outflow tract of rats did not dilate or produce any inflammatory response	[289]
ε-Caprolactone-co-L-lactide (PCLA)	Vascular smooth muscle cells (SMCs) from rat aortas; syngenic rats	The porous PCLA patch increased cell colonization while the outer PLA layers preserved patch structure and dimensions. After 8 weeks of in vivo implantation, elastic tissue formation in the subendocardial layer was observed by seeding SMCs on a PCLA patch and heart systolic function was improved by echocardiography and LV distensibility was reduced as compared with unseeded graft repairs	[290]
PLA	Ventricular cardiomyocytes, neonatal rat	Nanofibers were fabricated from continuously aligned PLA in chloroform with diameters ranging from 50 to 3500 nm. Aligned nanofibers were utilized by myocytes to orient their contractile cytoskeleton and to organize themselves into a beating and multicellular tissue that could imitate the laminar, anisotropic architecture of cardiac tissue	[291]
PGA	Neonatal rat cells	After 1–2 weeks of cultivation, engineered cardiac constructs expressed cardiac specific proteins and ultrastructural features mimicking native cardiac tissue	[292]

(continued)

Table 7 (continued)

Material	Cells used to perform in vitro study/Animals used for in vivo study	Results	References
Camphorsulfonic acid doped polyaniline-poly(glycerol-sebacate)	Mouse myoblast cells (C2C12 cells)	The scaffold demonstrated good electrical conductivity along with favorable mechanical properties and biodegradation rates for cardiac tissue regeneration. Cell culture test results revealed good attachment, growth, and proliferation of C2C12 myoblasts and confirmed the biocompatibility of PANI containing composites	[293]
Aniline pentamer contained PU-polycaprolactone scaffold	Mouse L929 fibroblast cells, human umbilical vein endothelial cells (HUVECs)	The presence of aniline pentamer enhanced the electroconductivity of the polymer composite substantially. Cell culture studies demonstrated that the inherently electrical conductive scaffolds were nontoxic and supported cell proliferation and attachment combined with good antioxidant properties	[294]
Polyaniline films	Cardiac myoblast cells (H9c2 cells)	The electroactive polyaniline substrate supported the adhesion and proliferation of cardiac myoblasts	[295]

PLA poly-(lactic acid), *PEO* poly(ethylene oxide), *PAA* poly(acrylic acid), *PGA* polyglycolic acid, *P4HB* poly4-hydroxybutyrate, *PCL-U4U* polycaprolactone bisurea, *PBMC* peripheral blood mononuclear cells, *BMSC* bone marrow-derived stem cell, *bFGF* basic fibroblast growth factor, *CXCL12* C-X-C motif chemokine ligand 12, *PLGA* poly (L lactide-co-glycolide)

Hydrogels for Endogenous Repair and Cell Transplantation

Reis et al. [262] developed a thermoresponsive hydrogel made up of chitosan conjugated with the angiopoietin-1 derived peptide, QHREDGS, and mixed with collagen I suitable for the survival and maturation of transplanted cardiomyocytes. Conjugation of the QHREDGS peptide did not alter gelation property of the hydrogel but it contributed to attract more fibroblasts and improves the density of cardiomyocytes in a subcutaneous model. That hydrogel was used to successfully localize the site of infarction in a mouse MI model. They recovered some samples after 1 week of in vivo implantation (recovery rate was 50%). The recovered samples showed concentrated collagen deposited in a localized area compared to surrounding tissue which was indicating by the darker blue areas of Masson's trichrome-stained sections in Fig. 4a. Chitosan staining was done to confirm that in those localized areas, hydrogels were injected as shown in Fig. 4b. Penetration of SMA-positive and Factor-VIII-positive cells was observed in the injected hydrogels (Fig. 4c, d, respectively).

Dahlmann et al. developed an in situ hydrazone cross-linkable hydrogel based on alginate (Alg) and HA where mechanical and physical properties could be adjusted by varying the concentration, degree of derivatization, and composition of blends [269]. This hydrogel allowed formation of contractile cardiac tissue from CM-enriched neonatal rat heart cells, which mimics the native myocardium. Deng et al. fabricated alginate- and chitosan-based hydrogels, injected them in a rat MI model, could successfully heal scar thickness, sooth scar expansion, decrease scar fibrosis after 8 weeks, and maintain proper heart function by inducing endogenous cardiomyocyte proliferation at the site of infarction [296]. Huang et al. examined the angiogenic effect of an injectable biopolymer composed of fibrin, collagen I, and Matrigel in a rat MI model of left coronary artery occlusion followed by reperfusion [297]. They observed significantly higher levels of capillary formation and functional arterioles in the infarct zone which facilitated the rate of angiogenesis. Wang et al developed α-cyclodextrin/poly(ethylene glycol)–β- polycaprolactone (dodecanedioic acid)–polycaprolactone–poly(ethylene glycol) (MPEG–PCL–MPEG) hydrogels [298] for improved cell transplantation. After 4 weeks of postinjection in a rabbit model, increased cell retention and vessel density around the site of infarction was observed which minimized scar expansion and improved cardiac function compared to bone marrow-derived stem cells (BSCs) alone.

Bulk Material

Adverse remodeling of the heart after MI can be prevented by injecting a bulk material into the ventricular wall. Fujimoto et al. synthesized a novel biodegradable, thermoresponsive hydrogel based on N-isopropylacrylamide (NIPAAm), acrylic acid (AAc), and hydroxyethyl methacrylate-poly(trimethylene carbonate) (HEMAPTMC) [299]. The hydrogel was injected in the infarcted left ventricular (LV) wall of a rat model with phosphate buffered saline (PBS) as a control. After 8 weeks of postinjection with hydrogel, the LV cavity area decreased and contractility was preserved which was not the case with the PBS group. After the 8-week evaluation period, three of the eight animals in the PBS injection group had obvious ventricular aneurysm formation in the apex area (Fig. 5a) and others had well-defined scar areas. For the hydrogel injection group, no strong adhesion and no aneurysms were observed while the treated infarcts were covered with fat connective tissues (Fig. 5b).

Wu et al synthesized a temperature-sensitive, aliphatic polyester hydrogel (HG) conjugated with vascular endothelial growth factor (VEGF) which remained as an injectable liquid at room temperature and became a biodegradable solid at 37°C temperature [301]. In vivo studies revealed for up to 35 days after MI in a rat model, this hydrogel could restore ventricular function and preserved scar thickness. VEGF conjugated hydrogels improved blood vessel density at the site of infarction and promote angiogenesis and prevented adverse remodeling of the heart after MI.

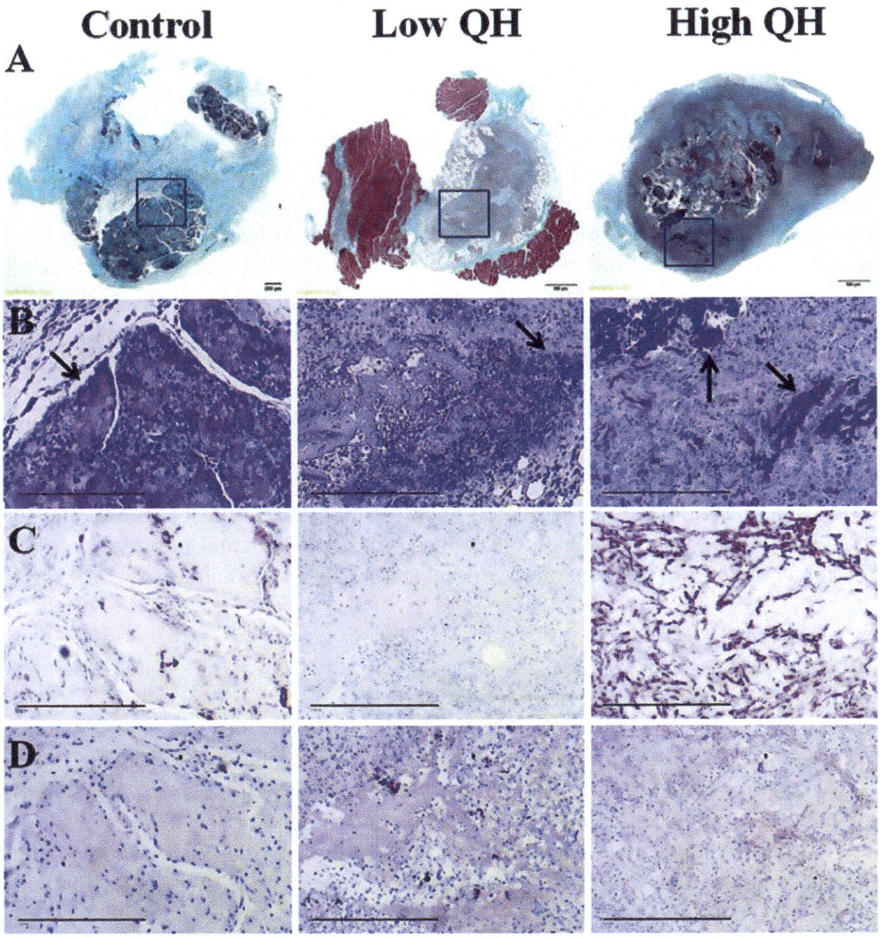

Fig. 4 Chitosan–collagen hydrogels with immobilized QHREDGS are suitable for subcutaneous cell injection. (**A–D**) Histological staining of subcutaneously injected hydrogel samples with encapsulated Lewis rat neonatal CMs, recovered 7 days postinjection. Full recovered nodules imaged with (**A**) Mason's trichrome stain with areas containing the recovered sample highlighted with a black box and confirmed with (**B**) chitosan staining (positive = dark pink/red), indicated with black arrows. (**C**) SMA staining indicates a higher expression in the High gel group compared to the Low and Control gel groups, while (**D**) Factor VIII staining shows no major differences between groups. Scale bars = 200 μm (A), 250 μm (**B–D**). Reproduced from permission [262]

Vascular Tissue Engineering

Presently, one of the leading causes of death is due to atherosclerotic cardiovascular disease (CVD) throughout the world [302, 303]. From 1950, expanded polytetrafluoroethylene (ePTFE, trade name "Dacron") has been used successfully to replace

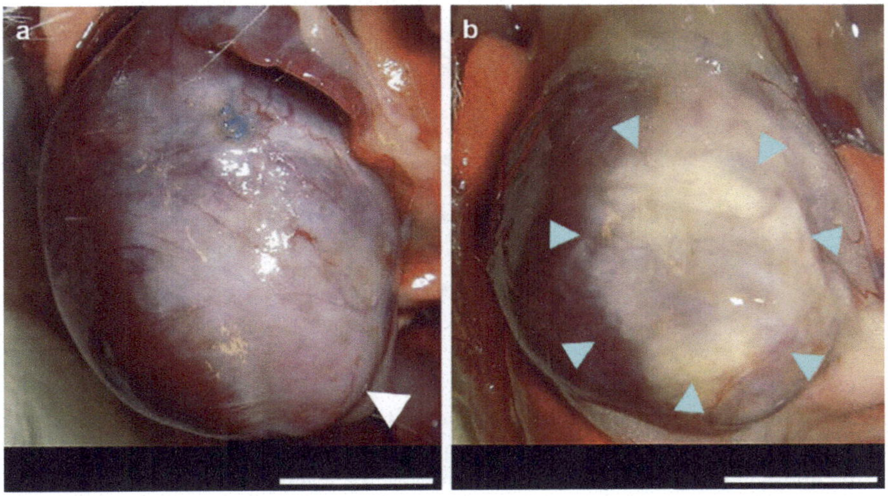

Fig. 5 Representative images at 8 weeks following the injection procedure of the anterior view of (**a**) PBS injected, and (**b**) poly(NIPAAm-co-AAc-co-HEMAPTMC) (86/4/10) injected hearts. White arrow shows an aneurysm formation in the apex area (**a**). Blue arrows indicate the injected hydrogel area (**b**). Reproduced from permission [299, 300]

larger diameter vessels of 6–10 mm [304]. But, the complication started for smaller diameter vessels, such as coronary vessels, where the diameter is around 3–4 mm [305]. To avoid this complication, tissue engineered vascular conduits (TEVC) have been developed to mimic the narrow article vascular conduit. Weinberg and Bell first developed TEVC by coating collagen on thin ePTFE mesh scaffolds followed by seeding bovine aortic endothelial cells (EC), smooth muscle cells (SMC) and adventitial fibroblast cells (AFC) as shown in Fig. 6 [37].

The nonbiodegradable nature of ePTFE creates several complications which limits its long-term clinical application. Later, researchers developed several techniques to improve biodegradability and structural integrity using collagen. But, the ideal vascular conduit should have high burst strength of 2000 and 3000 mmHg which are equal to those of human saphenous vein and internal mammary artery, respectively [306]. L'Heureux et al. fabricated collagen hydrogel and achieved maximum burst strength of reached <120 mmHg [307]. To overcome the problem associated with collagen, some other natural polymers such as elastin, fibronectin, etc. have been extensively studied but failed to achieve the high ideal burst strength with long duration. Therefore, researchers concentrated toward synthetic biocompatible, nonimmunogenic, and biodegradable polymers and in this context PCL, PGA, PLLA and their mixture or their combination with natural polymers were extensively studied. The Breuer group fabricated TEVC through coating of a PCL and PLA solution mixture on PGA nonwoven mesh tubes followed by mononuclear cells obtained from autologous bone marrow [308]. After implanting the TEVC in juvenile lambs, the volume of TEVC increased normally after 6 months of implantation and no

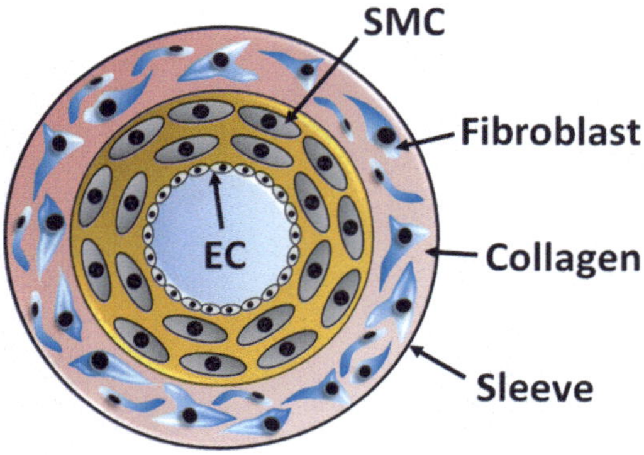

Fig. 6 Schematic representation of a collagen-based construct

evidence of aneurysmal dilation was observed. In a recent study, Shin'oka et al. prepared a vascular conduit having 10 mm of diameter, 20 mm of length, and 1 mm of thickness using PCL and PLA copolymer at a 1:1 weight ratio reinforced with woven PGA. After seeding the cells isolated from the explanted peripheral vein of a 4-year-old patient, the graft was implanted after 10 days of in vitro culturing. It was observed that the pulmonary artery of the patient was successfully regenerated without showing any postoperative complication even after 7 months of the operation [309]. After this successful vascular TE, a series of implantations were performed on 22 patients using the similar material composition [310]. But, the bone marrow cells (BMCs) of the patients were used instead of peripheral vein derived cells. After implanting the engineered vascular conduit, vascular tissues were successfully regenerated in all patients with no morbidity associated with stenosis, thrombosis, or aneurysmal dilation after 30 days of operation. But, thrombosis was found in one patient after 1 year of operation which might be caused by warfarin used as an anticoagulant. Therefore, more clinical trials are required based on this TEVC system to observe long-term postoperative effects on the patients.

Liver Tissue Engineering

Chronic liver disease is one of the major clinical issues for human health because of the progressive increase in liver-related deaths, especially by liver cirrhosis. It is anticipated that by 2020, cirrhosis of liver would be twelfth leading cause of death worldwide. In a recent study, more than 500 million people were affected by chronic liver diseases throughout the globe which included steatohepatitis, hepatocellular carcinoma, and nonalcoholic fatty liver disease and causes 2% of total death [311]. Other than these liver diseases, a large number of patients die every year because of hepatitis B or C infection, cholestasis, and metabolic syndrome which may result in multiorgan failure and chronic end stage liver diseases. Loss of liver functions and multiorgan failure are associated with high mortality rates [312]. Current clinical approach advices liver transplantation in case of acute liver failure, which is the only accepted efficacious treatment and advanced stages of liver therapy, but limited organ supplies and high cost effectiveness limits this therapy [313]. Thereby, liver TE shows a paradigm shift of transplantation to hepatocyte therapy via regenerative medicine strategies. Liver TE provides restoration of hepatocytes which is damaged due to disease or injury. Every TE platform requires three common components that are cells, scaffold materials, and growth factors. So, liver TE also requires cell sources which will differentiate into hepatocytes, cellular support where it will grow and expand, and growth factors which will help to differentiate the cells via phenotypic expression.

Cell sources for liver TE are a primary challenge due to the presence of different cells with different metabolic functions. Primary hepatocytes are the cell source studied to understand liver functions but their use is limited due to their shortage and they are mature cells with short telomere structures [314]. Other sources of hepatocyte cells include early fetal liver cells which consist of fetal hepatic stem cells and hepatoblasts [313].

Liver TE scaffold design and choice of biomaterials are difficult due to variations of cells with different functions. Scaffold design is an important parameter for TE as it can alter cell functions and the microenvironment. Different cellular functions were observed in the case of 2D and 3D scaffolds, where 3D culture systems show more benefits with respect to maturation and functions of hepatocytes [315, 316]. Biomaterials also play a crucial role to fabricate scaffolds which act as a cellular support system. In this context, animal-derived proteins are used to fabricate ECM scaffolds for liver TE , but due to a lack of appropriate mechanical strength and interbatch variation, they are not ideal. So, researchers and clinicians have now shifted toward chemically synthesized materials and materials derived from natural sources which possess properties needed to fabricate a cellular support system for liver TE such as good mechanical integrity, favorable biodegradability, etc. Synthetic polymers are most exploited for the fabrication of artificial liver matrices. Most commonly used polymers such as PLA, PGA, PCL, PLGA, and PEG are used to fabricate artificial liver matrices [317, 318]. Naturally occurring materials, such as collagen [319], chitosan [320], alginates [321], and hyaluronic acid [322] are most commonly used as an ECM material.

Interfacial Tissue Engineering

Osteoarthritis, rheumatoid arthritis (RA), and gout are the most well-known joint diseases that affect throughout the globe, especially in those areas where age is coupled with obesity and poor physical activity. In a recent survey, it was reported that, around 53 million people had arthritis in the US alone in the year 2011–2012 [323], which is expected to increase to 78 million by the end of 2040.

Arthritis mainly attacks load bearing joints such as the knee, elbow, finger, and hip joints and damages both subchondral bone as well as cartilage and results in severe pain. Current treatment methods for arthritis include knee replacement, hip replacement, allograft, microplasty, and autograft transplantation [324–326].

However, donor site morbidity, complex surgical procedures, immune rejection, degenerative alteration, and expensive cost of treatment have limited the effectiveness of the treatment [129]. So, researchers are now looking for alternative strategies to treat arthritis.

Interfacial TE has emerged due to its capability to regenerate complex multiphasic tissues which are present at the junction of bone–cartilage, ligament–bone, and muscle–tendon areas.

The main challenge for interfacial TE is the design of the scaffold. Different tissues at the interfacial tissue site possess different material composition and mechanical properties, and a single scaffold material may not mimic the properties of ECM at the interfacial site. For example, the osteochondral tissue interface is mainly consisted of chondrocytes, surrounded by an ECM which includes a major content of water, collagen, proteoglycan, and a minor amount of proteins. On the other hand, the boundary between subchondral bone and cartilage is separated by a calcified cartilage region. The subchondral bone portion at the interfacial tissue site is composed of collagen and hydroxyapatite, and the percentage of hydroxyapatite generally increases from the interfacial site upward to the bone section. So, from the structural point of view there is a progressive change of material composition at the interfacial site which results in changes in mechanical properties, stiffness, as well as cell types. In fact, the compressive strength of the interfacial region varies drastically with a value of 0.079, 2.1, and 320 MPa at the superficial, middle, and deep zone whereas the compressive modulus of the subchondral bone is 5.7 GPa [327, 328].

To mimic this complex tissue structure, researchers and clinicians take two different types of strategies to design scaffold for interfacial TE; the first one is a stratified scaffold design and the second one is a gradient scaffold design. The first generation of stratified scaffold for interfacial TE includes two well-distinguished cartilage and bone segments which are joined by glue or sutures [329]. For example, Schaefer et al. reported a biphasic scaffold based on PGA meshes and a blend of PLGA and PEG. They cultured chondrocytes on the PGA meshes and bovine calf periosteal cells on the blend of PLGA and PEG scaffolds and subsequently sutured them together. The resulting biphasic scaffold resulted in well-defined cartilaginous and bone tissues [330]. Mao and coworkers also designed a biphasic stratified scaffold composed of a PEG-diacrylate hydrogel, and the upper part of the scaffold consisted of MSC-derived chondrocyte cells and the bottom with MSC-derived osteoblast cells. Upon implantation at the dorsum of immunodeficient mice, two well-distinguished cartilaginous and osseous regions were observed after 12 weeks [331]. These preliminary studies on interfacial TE constructs demonstrated how we can engineer multiple tissues in a single stratified scaffold. However, due to the abrupt changes in mechanical properties, it is difficult to mimic the exact properties of interfacial tissues. To overcome this problem, researchers, are now shifting toward multiphasic or gradient scaffold designs where a smooth change in material composition can mimic the exact environment of the interfacial region. For example, Lu and coworkers developed a three-layer multiphasic scaffold consisting of gel only, gel/composite interface, and composite only region. Coculturing osteoblasts and chondrocytes led to development of three distinct but well-organized continuous sections of cartilage, calcified cartilage, and bone-like matrices. Additionally, the varied amount of calcium phosphate content enhanced the growth of multiple matrices: a GAG rich chondrocyte region, an interfacial site where both GAG and collagen is present, and a collagen-rich osteoblast region [332]. Harley et al. also designed a multiphasic scaffold for osteochondral TE. They used a new type of method "liquid phase co synthesis," which enabled the manufacture of a porous-layered scaffold that could mimic the interfacial region between articular cartilage and subchondral bone [333].

Baar and coworkers designed a stratified scaffold using a PEG-diacrylate-hydroxyapatite scaffold for the regeneration of the ligament to bone interface. Their engineered scaffold demonstrated obvious improvement compared to a commercially available single phase anterior cruciate ligament (ACL) treatment procedure [334]. Spalazzi et al. developed a triphasic scaffold which consists of polyglactin, PLGA, and bioactive glass. Upon coculturing of osteoblasts and fibroblasts on the scaffold, cells were well attached and proliferated [335].

Gradient scaffold design strategy is an emerging and promising approach to regenerate interfacial tissue. The advantage of gradient scaffolds over stratified scaffolds is that the gradient scaffold may possess a smooth change in either material composition/gene/growth factor that can mimic the interfacial tissue site more accurately than stratified scaffolds [129]. The gradient design strategy may lead to better transition of mechanical properties and provided better regional control over complex tissue interfaces. Till now, very limited studies have focused on the development of gradient scaffolds for interfacial TE. Among them Berkland and coworkers reported for the first time, a gradient scaffold fabrication technique using PLGA microspheres and a model dye [336]. Kalyon and coworkers designed a functionally graded scaffold made of nonwoven meshes of PCL incorporated with a gradient of tricalcium phosphate nanoparticles. Upon culturing osteoblast cells for a period of 4 weeks on the scaffold surface, cells formed a gradient in the ECM which is generally observed at the bone–cartilage interface [337]. Simon and coworkers reported a nanofiber scaffold with a gradient in amorphous calcium phosphate nanoparticles (nACP) for interfacial tissue regeneration. In detail, they simultaneously electrospan two types of nanofibers that were composed of PCL and PCL/amorphous calcium phosphate nanoparticles, creating a nonwoven mat of nanofibers with a composition gradient. When pre-osteoblast cells were cultured on the gradient scaffold, more proliferation of osteoblasts was observed with increasing percentage of calcium phosphate [338].

Scaffold Fabrication Techniques for Tissue Engineering Applications

Scaffolds play a very important role for TE as they provide the support to adhere, grow, differentiate, and proliferate cells to from new tissues/organs. Scaffolds could originate from natural, synthetic, or combined sources. They should be biocompatible to mimic ECM and 3D structure of native organs. Ideally, they should be biocompatible, biodegradable, and have optimal mechanical properties. The 3D structure of scaffold provides for a highly porous morphology which allows cells to spread into and provides effective transport of necessary nutrients, growth factors, and oxygen from the periphery to the inner portion of the scaffold. A scaffold should be degraded and only retain the shape of the final tissue structure [339–341]. Several fabrication techniques that have been practiced to fabricate 3D scaffolds including freeze-drying, salt leaching, phase separation, electrospinning, etc. In this section, we discuss the conventional to most advanced scaffold designing techniques employed in bone TE.

Freeze-Drying

Freeze-drying or lyophilization is the most widely used conventional technique for preparing 3D scaffolds. In this technique, firstly, the polymer sample in a solvent system has to be frozen followed by the process of sublimation under low pressure or in vacuum, where the water from the sample evaporates from its solid state without undergoing melting. Thus, this process creates 3D scaffolds by generating pores of different sizes and shapes. This process can generate up to 90% porosity in a scaffold with a pore size range from 20 to 200 μm [342]. There are lot of examples of research work using the freeze-drying technique to develop 3D scaffolds. Shahbazarab et al. [343] fabricated nanocomposite scaffolds containing zein (ZN), chitosan (CS), and nanohydroxyapatite (nHAp) for bone TE employing freeze-drying. Increases in the ZN and CS contents increased the percent porosity of the scaffold. The pore size could be tuned by controlling the rate of cooling and prefreezing temperature as the growth of ice crystals could influence their pore size [344, 345]. Tanir et al. [346] have prepared freeze-dried composite of chitosan, PLGA, and HAp having a pore size of about 100 μm. Elongated pores were observed, possibly [347] due to the formation of highly parallel ice crystals between the scaffold substrate layers,

Solvent Casting/Particle Leaching

Solvent casting or particle leaching is another conventional method for scaffold preparation. In this technique, the polymer is dissolved in an organic solvent along with porogens. Polymers get solidified by the evaporation of the organic solvent followed by the leaching out of porogens by selective dissolution resulting in porous scaffold. Porogens are generally calibrated minerals [348, 349] like, sodium chloride, citrate, and tartrate or organic particles like, sucrose. The technique could produce porous scaffolds having almost 90% porosity with about a 500-μm median pore diameter [348]. The pore size and porosity of the scaffold could be controlled by the selection of porogens as well as the amount of porogens used [350]. Use of a small amount of polymer could be an advantage of this technique but the major disadvantages that includes pore shape and inter-pore openings are not controllable [351, 352].

Phase Separation

The phase separation produces two phases: a polymer-rich phase and a polymer-lean phase. This could be done by baring the solution to an immiscible solvent system or by cooling down the solution under its solubility temperature [353]. In a typical phase diagram of temperature-composition for a polymer solvent system, there are two curves: one is a binodal curve and the other is spinodal curve [354]. The spinodal curve divides the system into unstable and metastable zones [355]. When the solution temperature moves above the binodal curve, the polymer solution becomes homogeneous. The critical point of the system represents the point where the two curves merge. The area between the binodal and spinodal curve is in a metastable zone. The separation of phases is initiated by decomposition of the spinodal curve to form a polymer-rich and polymer-lean phases [355, 356]. The pore size and interconnectivity can be controlled by several parameters which affect the early stage of phase separation, like, concentration and molecular weight of the polymer, route of quenching, type of solvent system, etc. In general, the phase separation process is used to construct porous membranes for separation and filtration purposes, but controlled phase separation processes like thermally induced phase separation can be used for scaffold formation.

Electrospinning

In the early 1930s, nonwoven fabric products for household or industrial purposes were fabricated by electrospinning process. Over the last few decades, the process has been revitalized to convert biocompatible as well as biodegradable polymer materials into fibrous-like structures having an average diameter of micro to nano-size range and which could be used as a scaffold for TE applications. In this electrospinning process, the polymer of interest passes through a capillary which will produce a polymer drop at the tip of the capillary. Between the tip of the capillary and the collection target, a voltage of high magnitude is applied. Once the strength of the electric field surmounts the surface tension of the polymer droplet, a polymer solution jet is introduced toward the collection target. During the travel of the polymer solution jet toward the collection area, the solvent evaporated and a fibrous-like structure is formed. A favored alignment could be achieved by introducing an electrically rotating drum as a collection point. It also may enable a range of fibrous material on the collection surface. This type of setup is favorable for mass production. The average diameter of the fiber-like filaments, and the inner-space between the fibers, could be modified by controlling several factors including the distance between the collection target and the capillary tip, the applied voltage, etc. Several natural as well as synthetic polymers have been used to form nonwoven fibrous-like scaffolds for bone TE applications. PCL, PLGA, and PGA are widely used synthetic polymers whereas collagen, fibrinogen, and silk are commonly used natural polymers are to prepare scaffolds for TE applications [357–360].

Gas Foaming

In the process of gas foaming, gas bubbles act as an internal phase which is dispersed through the continuous phase of the polymer [361]. The dispersed gas bubbles could be produced either by the addition of an inert gas at low pressure [362, 363], at high pressure [364] or by chemical reaction [365, 366]. Sodium bicarbonate is a commonly used foaming agent to produce an inert gas like CO_2. Once the dispersed bubbles are removed from the polymer continuous phase, a porous structure is generated. The major problem associated with this fabrication technique is the formation of porous polymer scaffold having a non-porous part at the bottom end and a highly porous part at the upper end. This is because of the fact that the gas tends to move up while the liquid phase intends to settle at the bottom of the system [363, 367]. To overcome this problem, surfactants like Pluronic F-108, are used to prevent the liquid to move downward [365]. There are several approaches used to minimize the polymer foams, i.e., (1) by increasing the viscosity of the polymer solution [365], (2) by selecting polymers which could endure rapid solidification with changes of temperature like gelatin [367], and (3) by using a cross-linker and initiator to the continuous phase to bring rapid polymerization after formation of gas bubbles [366]. Several polymers have been used to fabricate scaffolds for TE-like PEG-diacrylate (PEGDA) [366], PLA [368], alginate [365], and gelatin [367].

Future Perspectives and Conclusion

TE is a promising area of research, where researchers and clinicians are trying to endorse and enhance the capability to regenerate a damaged tissue, and to assist in recovering its shape as well as function where our body fails to regenerate by itself. Till now, a lot of studies have already focused on tissue regeneration in the areas of bone, heart, cartilage, neural, and skin TE. The first generation tissue regeneration involved 2D scaffolds. Next it was transformed into 3D scaffolds and now TE comes with 4D scaffolds and 3D biopriniting, where we can construct a fully functional organ outside of our body. However, a lot of research is still needed, particularly on the selection of biomaterials for different types of TE. Another big challenge is commercialization of tissue engineering products. Till now, a lot of research has reported on different types of TE in vitro. But very limited studies have focused on in vivo, especially on large animal models.

Acknowledgements This work was financially supported by the DST-SERB funding agency through two sanctioned projects EEQ/2016/000712 and ECR/2016/002018. K.S. also acknowledges UGC for a UGC-BSR Research Start-Up Grant (F.30-363/2017(BSR), Dt- 08/08/2017) and CU for a UPE-II Nanofabrication project fund (UGC/166/UPE-II, Dt- 03/04/2017) for financial support.

References

1. Bernstein HS, Srivastava D (2012) Stem cell therapy for cardiac disease. Pediatr Res 71(4-2):491
2. Naderi H, Matin MM, Bahrami AR (2011) Critical issues in tissue engineering: biomaterials, cell sources, angiogenesis, and drug delivery systems. J Biomater Appl 26(4):383–417
3. Clifford DM, Fisher SA, Brunskill SJ, Doree C, Mathur A, Clarke MJ, Watt SM, Martin-Rendon E (2012) Long-term effects of autologous bone marrow stem cell treatment in acute myocardial infarction: factors that may influence outcomes. PLoS One 7(5):e37373
4. Qayyum AA, Haack-Sørensen M, Mathiasen AB, Jørgensen E, Ekblond A, Kastrup J (2012) Adipose-derived mesenchymal stromal cells for chronic myocardial ischemia (MyStromalCell Trial): study design. Regen Med 7(3):421–428
5. Weber B, Zeisberger S, Hoerstrup S (2011) Prenatally harvested cells for cardiovascular tissue engineering: fabrication of autologous implants prior to birth. Placenta 32:S316–S319
6. Takahashi K, Yamanaka S (2006) Induction of pluripotent stem cells from mouse embryonic and adult fibroblast cultures by defined factors. Cell 126(4):663–676
7. Yu J, Vodyanik MA, Smuga-Otto K, Antosiewicz-Bourget J, Frane JL, Tian S, Nie J, Jonsdottir GA, Ruotti V, Stewart R (2007) Induced pluripotent stem cell lines derived from human somatic cells. Science 318(5858):1917–1920
8. Zhu Z, Huangfu D (2013) Human pluripotent stem cells: an emerging model in developmental biology. Development 140(4):705–717
9. Nishikawa S-i, Goldstein RA, Nierras CR (2008) The promise of human induced pluripotent stem cells for research and therapy. Nat Rev Mol Cell Biol 9:725
10. Dzau VJ, Gnecchi M, Pachori AS (2005) Enhancing stem cell therapy through genetic modification. J Am Coll Cardiol 46(7):1351–1353

11. Mastri M, Lin H, Lee T (2014) Enhancing the efficacy of mesenchymal stem cell therapy. World J Stem Cells 6(2):82–93
12. Brochhausen C, Lehmann M, Halstenberg S, Meurer A, Klaus G, Kirkpatrick CJ (2009) Signalling molecules and growth factors for tissue engineering of cartilage—what can we learn from the growth plate? J Tissue Eng Regen Med 3(6):416–429
13. Mieszawska AJ, Kaplan DL (2010) Smart biomaterials - regulating cell behavior through signaling molecules. BMC Biol 8(1):59
14. Cushnie EK, Ulery BD, Nelson SJ, Deng M, Sethuraman S, Doty SB, Lo KWH, Khan YM, Laurencin CT (2014) Simple signaling molecules for inductive bone regenerative engineering. PLoS One 9(7):e101627
15. Agrawal CM (1998) Reconstructing the human body using biomaterials. JOM 50(1):31–35
16. Auler ME, Morreira D, Rodrigues FFO, Abr Ão MS, Margarido PFR, Matsumoto FE, Silva EG, Silva BCM, Schneider RP, Paula CR (2010) Biofilm formation on intrauterine devices in patients with recurrent vulvovaginal candidiasis. Med Mycol 48(1):211–216
17. Gallo J, Holinka M, Moucha CS (2014) Antibacterial surface treatment for orthopaedic implants. Int J Mol Sci 15(8):13849
18. Lazar V (2011) Quorum sensing in biofilms—how to destroy the bacterial citadels or their cohesion/power? Anaerobe 17(6):280–285
19. Green M (2006) Biofilms, Infection, and Antimicrobial Therapy Edited by John L. Pace, Mark E. Rupp, and Roger G. Finch Boca Raton, FL: CRC Press, Taylor and Francis Group, 2006 512 pp., illustrated. $159.95 (cloth). Clin Infect Dis 43(12):1623–1623
20. Amini AR, Laurencin CT, Nukavarapu SP (2012) Bone tissue engineering: recent advances and challenges. Crit Rev Biomed Eng 40(5):363–408
21. Laurencin CT, Ambrosio A, Borden M, Cooper J Jr (1999) Tissue engineering: orthopedic applications. Annu Rev Biomed Eng 1(1):19–46
22. Laurencin C, Khan Y, El-Amin SF (2006) Bone graft substitutes. Expert Rev Med Devices 3(1):49–57
23. Nukavarapu S, Wallace J, Elgendy H, Lieberman J, Laurencin C (2011) An introduction to biomaterials and their applications. Bone Biomater 2:571–593
24. Banwart JC, Asher MA, Hassanein RS (1995) Iliac crest bone graft harvest donor site morbidity. A statistical evaluation. Spine 20(9):1055–1060
25. Ebraheim NA, Elgafy H, Xu R (2001) Bone-graft harvesting from iliac and fibular donor sites: techniques and complications. JAAOS 9(3):210–218
26. St TJ, Vaccaro AR, Sah AP, Schaefer M, Berta SC, Albert T, Hilibrand A (2003) Physical and monetary costs associated with autogenous bone graft harvesting. Am J Orthop 32(1):18–23
27. Delloye C, Cornu O, Druez V, Barbier O (2007) Bone allografts: what they can offer and what they cannot. J Bone Joint Surg 89(5):574–580
28. Lord C, Gebhardt M, Tomford W, Mankin H (1988) Infection in bone allografts. Incidence, nature, and treatment. J Bone Joint Surg Am 70(3):369–376
29. Tomford W, Starkweather R, Goldman M (1981) A study of the clinical incidence of infection in the use of banked allograft bone. J Bone Joint Surg Am 63(2):244–248
30. Liu H, Peng H, Wu Y, Zhang C, Cai Y, Xu G, Li Q, Chen X, Ji J, Zhang Y, OuYang HW (2013) The promotion of bone regeneration by nanofibrous hydroxyapatite/chitosan scaffolds by effects on integrin-BMP/Smad signaling pathway in BMSCs. Biomaterials 34(18):4404–4417
31. Xu Z-L, Lei Y, Yin W-J, Chen Y-X, Ke Q-F, Guo Y-P, Zhang C-Q (2016) Enhanced antibacterial activity and osteoinductivity of Ag-loaded strontium hydroxyapatite/chitosan porous scaffolds for bone tissue engineering. J Mater Chem B 4(48):7919–7928
32. Saravanan S, Nethala S, Pattnaik S, Tripathi A, Moorthi A, Selvamurugan N (2011) Preparation, characterization and antimicrobial activity of a bio-composite scaffold containing chitosan/nano-hydroxyapatite/nano-silver for bone tissue engineering. Int J Biol Macromol 49(2):188–193

33. Qiao P, Wang J, Xie Q, Li F, Dong L, Xu T (2013) Injectable calcium phosphate–alginate–chitosan microencapsulated MC3T3-E1 cell paste for bone tissue engineering in vivo. Mater Sci Eng C 33(8):4633–4639
34. Wang L, Shelton R, Cooper P, Lawson M, Triffitt J, Barralet J (2003) Evaluation of sodium alginate for bone marrow cell tissue engineering. Biomaterials 24(20):3475–3481
35. Kolambkar YM, Dupont KM, Boerckel JD, Huebsch N, Mooney DJ, Hutmacher DW, Guldberg RE (2011) An alginate-based hybrid system for growth factor delivery in the functional repair of large bone defects. Biomaterials 32(1):65–74
36. Lee CH, Singla A, Lee Y (2001) Biomedical applications of collagen. Int J Pharm 221(1-2):1–22
37. Weinberg CB, Bell E (1986) A blood vessel model constructed from collagen and cultured vascular cells. Science 231(4736):397–400
38. Freyman T, Yannas I, Gibson L (2001) Cellular materials as porous scaffolds for tissue engineering. Progr Mater Sci 46(3-4):273–282
39. O'Brien FJ, Harley BA, Yannas IV, Gibson LJ (2005) The effect of pore size on cell adhesion in collagen-GAG scaffolds. Biomaterials 26(4):433–441
40. Angele P, Abke J, Kujat R, Faltermeier H, Schumann D, Nerlich M, Kinner B, Englert C, Ruszczak Z, Mehrl R (2004) Influence of different collagen species on physico-chemical properties of crosslinked collagen matrices. Biomaterials 25(14):2831–2841
41. Chen L, Wu Z, Zhou Y, Li L, Wang Y, Wang Z, Chen Y, Zhang P (2017) Biomimetic porous collagen/hydroxyapatite scaffold for bone tissue engineering. J Appl Polym Sci 134(37):45271
42. Panilaitis B, Altman GH, Chen J, Jin H-J, Karageorgiou V, Kaplan DL (2003) Macrophage responses to silk. Biomaterials 24(18):3079–3085
43. Jin H-J, Kaplan DL (2003) Mechanism of silk processing in insects and spiders. Nature 424(6952):1057
44. Kim HJ, Kim U-J, Kim HS, Li C, Wada M, Leisk GG, Kaplan DL (2008) Bone tissue engineering with premineralized silk scaffolds. Bone 42(6):1226–1234
45. Mandal BB, Grinberg A, Seok Gil E, Panilaitis B, Kaplan DL (2012) High-strength silk protein scaffolds for bone repair. Proc Natl Acad Sci 109(20):7699–7704
46. Shao W, He J, Sang F, Ding B, Chen L, Cui S, Li K, Han Q, Tan W (2016) Coaxial electrospun aligned tussah silk fibroin nanostructured fiber scaffolds embedded with hydroxyapatite–tussah silk fibroin nanoparticles for bone tissue engineering. Mater Sci Eng C 58:342–351
47. Sarkar K, Meka SRK, Bagchi A, Krishna N, Ramachandra S, Madras G, Chatterjee K (2014) Polyester derived from recycled poly (ethylene terephthalate) waste for regenerative medicine. RSC Adv 4(102):58805–58815
48. Kolanthai E, Sarkar K, Meka SRK, Madras G, Chatterjee K (2015) Copolyesters from soybean oil for use as resorbable biomaterials. ACS Sustain Chem Eng 3(5):880–891
49. Florczyk SJ, Leung M, Li Z, Huang JI, Hopper RA, Zhang M (2013) Evaluation of three-dimensional porous chitosan–alginate scaffolds in rat calvarial defects for bone regeneration applications. J Biomed Mater Res A 101(10):2974–2983
50. Jin H-H, Kim D-H, Kim T-W, Shin K-K, Jung JS, Park H-C, Yoon S-Y (2012) In vivo evaluation of porous hydroxyapatite/chitosan–alginate composite scaffolds for bone tissue engineering. Int J Biol Macromol 51(5):1079–1085
51. Costa-Pinto AR, Correlo VM, Sol PC, Bhattacharya M, Srouji S, Livne E, Reis RL, Neves NM (2012) Chitosan–poly(butylene succinate) scaffolds and human bone marrow stromal cells induce bone repair in a mouse calvaria model. J Tissue Eng Regen Med 6(1):21–28
52. Niu X, Fan Y, Liu X, Li X, Li P, Wang J, Sha Z, Feng Q (2011) Repair of bone defect in femoral condyle using microencapsulated chitosan, nanohydroxyapatite/collagen and poly(L-lactide)-based microsphere-scaffold delivery system. Artif Organs 35(7):E119–E128
53. Shi S, Cheng X, Wang J, Zhang W, Peng L, Zhang Y (2009) RhBMP-2 microspheres-loaded chitosan/collagen scaffold enhanced osseointegration: an experiment in dog. J Biomater Appl 23(4):331–346

54. Haberstroh K, Ritter K, Kuschnierz J, Bormann K-H, Kaps C, Carvalho C, Mülhaupt R, Sittinger M, Gellrich N-C (2010) Bone repair by cell-seeded 3D-bioplotted composite scaffolds made of collagen treated tricalciumphosphate or tricalciumphosphate-chitosan-collagen hydrogel or PLGA in ovine critical-sized calvarial defects. J Biomed Mater Res B Appl Biomater 93B(2):520–530
55. Planka L, Necas A, Srnec R, Rauser P, Stary D, Jancar J, Amler E, Filova E, Hlucilova J, Kren L, Gal P (2009) Use of allogenic stem cells for the prevention of bone bridge formation in miniature pigs. Physiol Res 58(6):885–893
56. Muzzarelli RAA, Biagini G, Bellardini M, Simonelli L, Castaldini C, Fratto G (1993) Osteoconduction exerted by methylpyrrolidinone chitosan used in dental surgery. Biomaterials 14(1):39–43
57. Ji QX, Deng J, Xing XM, Yuan CQ, Yu XB, Xu QC, Yue J (2010) Biocompatibility of a chitosan-based injectable thermosensitive hydrogel and its effects on dog periodontal tissue regeneration. Carbohydr Polym 82(4):1153–1160
58. Cao L, Werkmeister JA, Wang J, Glattauer V, McLean KM, Liu C (2014) Bone regeneration using photocrosslinked hydrogel incorporating rhBMP-2 loaded 2-N, 6-O-sulfated chitosan nanoparticles. Biomaterials 35(9):2730–2742
59. Abbah S-A, Liu J, Lam RWM, Goh JCH, Wong H-K (2012) In vivo bioactivity of rhBMP-2 delivered with novel polyelectrolyte complexation shells assembled on an alginate microbead core template. J Control Release 162(2):364–372
60. Lee G-S, Park J-H, Shin US, Kim H-W (2011) Direct deposited porous scaffolds of calcium phosphate cement with alginate for drug delivery and bone tissue engineering. Acta Biomater 7(8):3178–3186
61. Soumya S, Sajesh KM, Jayakumar R, Nair SV, Chennazhi KP (2012) Development of a phytochemical scaffold for bone tissue engineering using Cissus quadrangularis extract. Carbohydr Polym 87(2):1787–1795
62. Kim M, Jung W-K, Kim G (2013) Bio-composites composed of a solid free-form fabricated polycaprolactone and alginate-releasing bone morphogenic protein and bone formation peptide for bone tissue regeneration. Bioprocess Biosyst Eng 36(11):1725–1734
63. Park D-J, Choi B-H, Zhu S-J, Huh J-Y, Kim B-Y, Lee S-H (2005) Injectable bone using chitosan-alginate gel/mesenchymal stem cells/BMP-2 composites. J Craniomaxillofac Surg 33(1):50–54
64. Wang Y, Peng W, Liu X, Zhu M, Sun T, Peng Q, Zeng Y, Feng B, Zhi W, Weng J, Wang J (2014) Study of bilineage differentiation of human-bone-marrow-derived mesenchymal stem cells in oxidized sodium alginate/N-succinyl chitosan hydrogels and synergistic effects of RGD modification and low-intensity pulsed ultrasound. Acta Biomater 10(6):2518–2528
65. Grellier M, Granja PL, Fricain J-C, Bidarra SJ, Renard M, Bareille R, Bourget C, Amédée J, Barbosa MA (2009) The effect of the co-immobilization of human osteoprogenitors and endothelial cells within alginate microspheres on mineralization in a bone defect. Biomaterials 30(19):3271–3278
66. Chen G, Lv Y, Dong C, Yang L (2015) Effect of internal structure of collagen/hydroxyapatite scaffold on the osteogenic differentiation of mesenchymal stem cells. Curr Stem Cell Res Ther 10(2):99–108
67. Chen G, Dong C, Yang L, Lv Y (2015) 3D scaffolds with different stiffness but the same microstructure for bone tissue engineering. ACS Appl Mater Interfaces 7(29):15790–15802
68. Chen G, Yang L, Lv Y (2016) Cell-free scaffolds with different stiffness but same microstructure promote bone regeneration in rabbit large bone defect model. J Biomed Mater Res A 104(4):833–841
69. Taguchi Y, Amizuka N, Nakadate M, Ohnishi H, Fujii N, Oda K, Nomura S, Maeda T (2005) A histological evaluation for guided bone regeneration induced by a collagenous membrane. Biomaterials 26(31):6158–6166
70. Zerwekh JE, Kourosh S, Scheinberg R, Kitano T, Edwards ML, Shin D, Selby DK (1992) Fibrillar collagen-biphasic calcium phosphate composite as a bone graft substitute for spinal fusion. J Orthop Res 10(4):562–572

71. Takahashi Y, Yamamoto M, Tabata Y (2005) Enhanced osteoinduction by controlled release of bone morphogenetic protein-2 from biodegradable sponge composed of gelatin and β-tricalcium phosphate. Biomaterials 26(23):4856–4865
72. Kim H-W, Song J-H, Kim H-E (2006) Bioactive glass nanofiber–collagen nanocomposite as a novel bone regeneration matrix. J Biomed Mater Res A 79A(3):698–705
73. Muthukumar T, Aravinthan A, Sharmila J, Kim NS, Kim J-H (2016) Collagen/chitosan porous bone tissue engineering composite scaffold incorporated with Ginseng compound K. Carbohydr Polym 152:566–574
74. Alt V, Kögelmaier DV, Lips KS, Witt V, Pacholke S, Heiss C, Kampschulte M, Heinemann S, Hanke T, Schnettler R, Langheinrich AC (2011) Assessment of angiogenesis in osseointegration of a silica–collagen biomaterial using 3D-nano-CT. Acta Biomater 7(10):3773–3779
75. Lee E-J, Jun S-H, Kim H-E, Koh Y-H (2012) Collagen–silica xerogel nanohybrid membrane for guided bone regeneration. J Biomed Mater Res A 100A(4):841–847
76. Wang S, Yang Y, Zhao Z, Wang X, Mikos AG, Qiu Z, Song T, Sun X, Zhao L, Zhang C, Cui F (2017) Mineralized collagen-based composite bone materials for cranial bone regeneration in developing sheep. ACS Biomater Sci Eng 3(6):1092–1099
77. Cui F, Wang S, Wang X, Yang Y, Zhao Z, Zhang C, Mikos AG, Song T, Qiu Z (2018) A high-strength mineralized collagen bone scaffold for large-sized cranial bone defect repair in sheep. Regener Biomater 5(5):283–292
78. Fini M, Motta A, Torricelli P, Giavaresi G, Nicoli Aldini N, Tschon M, Giardino R, Migliaresi C (2005) The healing of confined critical size cancellous defects in the presence of silk fibroin hydrogel. Biomaterials 26(17):3527–3536
79. Zhang W, Wang X, Wang S, Zhao J, Xu L, Zhu C, Zeng D, Chen J, Zhang Z, Kaplan DL, Jiang X (2011) The use of injectable sonication-induced silk hydrogel for VEGF165 and BMP-2 delivery for elevation of the maxillary sinus floor. Biomaterials 32(35):9415–9424
80. Correia C, Bhumiratana S, Yan L-P, Oliveira AL, Gimble JM, Rockwood D, Kaplan DL, Sousa RA, Reis RL, Vunjak-Novakovic G (2012) Development of silk-based scaffolds for tissue engineering of bone from human adipose-derived stem cells. Acta Biomater 8(7):2483–2492
81. Uebersax L, Apfel T, Nuss KMR, Vogt R, Kim HY, Meinel L, Kaplan DL, Auer JA, Merkle HP, von Rechenberg B (2013) Biocompatibility and osteoconduction of macroporous silk fibroin implants in cortical defects in sheep. Eur J Pharm Biopharm 85(1):107–118
82. Nagano A, Tanioka Y, Sakurai N, Sezutsu H, Kuboyama N, Kiba H, Tanimoto Y, Nishiyama N, Asakura T (2011) Regeneration of the femoral epicondyle on calcium-binding silk scaffolds developed using transgenic silk fibroin produced by transgenic silkworm. Acta Biomater 7(3):1192–1201
83. Park SY, Ki CS, Park YH, Jung HM, Woo KM, Kim HJ (2010) Electrospun silk fibroin scaffolds with macropores for bone regeneration: an in vitro and in vivo study. Tissue Eng Part A 16(4):1271–1279
84. Meinel L, Betz O, Fajardo R, Hofmann S, Nazarian A, Cory E, Hilbe M, McCool J, Langer R, Vunjak-Novakovic G, Merkle HP, Rechenberg B, Kaplan DL, Kirker-Head C (2006) Silk based biomaterials to heal critical sized femur defects. Bone 39(4):922–931
85. Sun L, Parker ST, Syoji D, Wang X, Lewis JA, Kaplan DL (2012) Direct-write assembly of 3D silk/hydroxyapatite scaffolds for bone co-cultures. Adv Healthc Mater 1(6):729–735
86. Jiang X, Zhao J, Wang S, Sun X, Zhang X, Chen J, Kaplan DL, Zhang Z (2009) Mandibular repair in rats with premineralized silk scaffolds and BMP-2-modified bMSCs. Biomaterials 30(27):4522–4532
87. Wu C, Zhang Y, Zhou Y, Fan W, Xiao Y (2011) A comparative study of mesoporous glass/silk and non-mesoporous glass/silk scaffolds: physiochemistry and in vivo osteogenesis. Acta Biomater 7(5):2229–2236
88. Choong C, Triffitt JT, Cui ZF (2004) Polycaprolactone scaffolds for bone tissue engineering: effects of a calcium phosphate coating layer on osteogenic cells. Food Bioprod Process 82(2):117–125
89. Williams JM, Adewunmi A, Schek RM, Flanagan CL, Krebsbach PH, Feinberg SE, Hollister SJ, Das S (2005) Bone tissue engineering using polycaprolactone scaffolds fabricated via selective laser sintering. Biomaterials 26(23):4817–4827

90. Park SA, Lee H-J, Kim K-S, Lee SJ, Lee J-T, Kim S-Y, Chang N-H, Park S-Y (2018) In vivo evaluation of 3D-printed polycaprolactone scaffold implantation combined with β-TCP powder for alveolar bone augmentation in a beagle defect model. Materials 11(2):238
91. Nguyen T-H, Lee B-T (2013) In vitro and in vivo studies of rhBMP2-coated PS/PCL fibrous scaffolds for bone regeneration. J Biomed Mater Res A 101A(3):797–808
92. Pati F, Song T-H, Rijal G, Jang J, Kim SW, Cho D-W (2015) Ornamenting 3D printed scaffolds with cell-laid extracellular matrix for bone tissue regeneration. Biomaterials 37:230–241
93. Zhang ZZ, Zhang HZ, Zhang ZY (2019) 3D printed poly (ε-caprolactone) scaffolds function with simvastatin-loaded poly (lactic-co-glycolic acid) microspheres to repair load-bearing segmental bone defects. Exp Ther Med 17(1):79–90
94. Li M, Liu W, Sun J, Xianyu Y, Wang J, Zhang W, Zheng W, Huang D, Di S, Long Y-Z, Jiang X (2013) Culturing primary human osteoblasts on electrospun poly(lactic-co-glycolic acid) and poly(lactic-co-glycolic acid)/nanohydroxyapatite scaffolds for bone tissue engineering. ACS Appl Mater Interfaces 5(13):5921–5926
95. Wu X, Zheng S, Ye Y, Wu Y, Lin K, Su J (2018) Enhanced osteogenic differentiation and bone regeneration of poly(lactic-co-glycolic acid) by graphene via activation of PI3K/Akt/GSK--3β/β-catenin signal circuit. Biomater Sci 6(5):1147–1158
96. Eslami H, Azimi Lisar H, Jafarzadeh Kashi TS, Tahriri M, Ansari M, Rafiei T, Bastami F, Shahin-Shamsabadi A, Mashhadi Abbas F, Tayebi L (2018) Poly(lactic-co-glycolic acid) (PLGA)/TiO2 nanotube bioactive composite as a novel scaffold for bone tissue engineering: In vitro and in vivo studies. Biologicals 53:51–62
97. Chen C, Wang H, Zhu G, Sun Z, Xu X, Li F, Luo S (2018) Three-dimensional poly lactic-co-glycolic acid scaffold containing autologous platelet-rich plasma supports keloid fibroblast growth and contributes to keloid formation in a nude mouse model. J Dermatol Sci 89(1):67–76
98. Cao H, Kuboyama N (2010) A biodegradable porous composite scaffold of PGA/β-TCP for bone tissue engineering. Bone 46(2):386–395
99. Ortiz M, Escobar-Garcia DM, Álvarez-Pérez MA, Pozos-Guillén A, Grandfils C, Flores H (2017) Evaluation of the osteoblast behavior to PGA textile functionalized with RGD as a scaffold for bone regeneration. J Nanomater 2017:4852190
100. Teixeira BN, Aprile P, Mendonça RH, Kelly DJ, Thiré RM (2019) Evaluation of bone marrow stem cell response to PLA scaffolds manufactured by 3D printing and coated with polydopamine and type I collagen. J Biomed Mater Res B Appl Biomater 107(1):37–49
101. Zhang H, Mao X, Zhao D, Jiang W, Du Z, Li Q, Jiang C, Han D (2017) Three dimensional printed polylactic acid-hydroxyapatite composite scaffolds for prefabricating vascularized tissue engineered bone: an in vivo bioreactor model. Sci Rep 7(1):15255
102. Apalangya VA, Rangari VK, Tiimob BJ, Jeelani S, Samuel T (2019) Eggshell based nano-engineered hydroxyapatite and poly (lactic) acid electrospun fibers as potential tissue scaffold. Int J Biomater 2019:6762575
103. Seyednejad H, Gawlitta D, Dhert WJ, Van Nostrum CF, Vermonden T, Hennink WE (2011) Preparation and characterization of a three-dimensional printed scaffold based on a functionalized polyester for bone tissue engineering applications. Acta Biomater 7(5):1999–2006
104. Natarajan J, Madras G, Chatterjee K (2017) Development of graphene oxide-/galactitol polyester-based biodegradable composites for biomedical applications. ACS Omega 2(9):5545–5556
105. Natarajan J, Movva S, Madras G, Chatterjee K (2017) Biodegradable galactitol based crosslinked polyesters for controlled release and bone tissue engineering. Mater Sci Eng C 77:534–547
106. Cui H, Liu Y, Deng M, Pang X, Zhang P, Wang X, Chen X, Wei Y (2012) Synthesis of biodegradable and electroactive tetraaniline grafted poly (ester amide) copolymers for bone tissue engineering. Biomacromolecules 13(9):2881–2889
107. Natarajan J, Dasgupta Q, Shetty SN, Sarkar K, Madras G, Chatterjee K (2016) Poly (ester amide) s from soybean oil for modulated release and bone regeneration. ACS Appl Mater Interfaces 8(38):25170–25184

108. Wang Y-J, Jeng U-S (2018) Hsu, S.-h., Biodegradable water-based polyurethane shape memory elastomers for bone tissue engineering. ACS Biomater Sci Eng 4(4):1397–1406
109. Gerges I, Tamplenizza M, Lopa S, Recordati C, Martello F, Tocchio A, Ricotti L, Arrigoni C, Milani P, Moretti M (2016) Creep-resistant dextran-based polyurethane foam as a candidate scaffold for bone tissue engineering: Synthesis, chemico-physical characterization, and in vitro and in vivo biocompatibility. Int J Polym Mater Polym Biomater 65(14):729–740
110. Pneumaticos SG, Triantafyllopoulos GK, Basdra EK, Papavassiliou AG (2010) Segmental bone defects: from cellular and molecular pathways to the development of novel biological treatments. J Cell Mol Med 14(11):2561–2569
111. Black CR, Goriainov V, Gibbs D, Kanczler J, Tare RS, Oreffo RO (2015) Bone tissue engineering. Curr Mol Biol Rep 1(3):132–140
112. Roddy E, DeBaun MR, Daoud-Gray A, Yang YP, Gardner MJ (2018) Treatment of critical-sized bone defects: clinical and tissue engineering perspectives. Eur J Orthop Surg Traumatol 28(3):351–362
113. Friedlaender GE, Perry CR, Cole JD, Cook SD, Cierny G, Muschler GF, Zych GA, Calhoun JH, LaForte AJ, Yin S (2001) Osteogenic protein-1 (bone morphogenetic protein-7) in the treatment of tibial nonunions: a prospective, randomized clinical trial comparing rhOP-1 with fresh bone autograft. The. J Bone Joint Surg 83(Pt 2):S151
114. Govender S, Csimma C, Genant HK, Valentin-Opran A (2002) Recombinant human bone morphogenetic protein-2 for treatment of open tibial fractures: a prospective, controlled, randomized study of four hundred and fifty patients. JBJS 84(12):2123–2134
115. Chrastil J, Low JB, Whang PG, Patel AA (2013) Complications associated with the use of the recombinant human bone morphogenetic proteins for posterior interbody fusions of the lumbar spine. Spine 38(16):E1020–E1027
116. Aro HT, Govender S, Patel AD, Hernigou P, de Gregorio AP, Popescu GI, Golden JD, Christensen J, Valentin A (2011) Recombinant human bone morphogenetic protein-2: a randomized trial in open tibial fractures treated with reamed nail fixation. JBJS 93(9):801–808
117. Yamada T, Yoshii T, Sotome S, Yuasa M, Kato T, Arai Y, Kawabata S, Tomizawa S, Sakaki K, Hirai T (2012) Hybrid grafting using bone marrow aspirate combined with porous β-tricalcium phosphate and trephine bone for lumbar posterolateral spinal fusion: a prospective, comparative study versus local bone grafting. Spine 37(3):E174–E179
118. Russell TA, Leighton RK (2008) Comparison of autogenous bone graft and endothermic calcium phosphate cement for defect augmentation in tibial plateau fractures: a multicenter, prospective, randomized study. JBJS 90(10):2057–2061
119. Damron TA, Lisle J, Craig T, Wade M, Silbert W, Cohen H (2013) Ultraporous β-tricalcium phosphate alone or combined with bone marrow aspirate for benign cavitary lesions: comparison in a prospective randomized clinical trial. JBJS 95(2):158–166
120. Karger C, Kishi T, Schneider L, Fitoussi F, Masquelet A-C (2012) Treatment of posttraumatic bone defects by the induced membrane technique. Orthop Traumatol Surg Res 98(1):97–102
121. Meinig RP (2010) Clinical use of resorbable polymeric membranes in the treatment of bone defects. Orthoped Clin 41(1):39–47
122. Jones AL, Bucholz RW, Bosse MJ, Mirza SK, Lyon TR, Webb LX, Pollak AN, Golden JD, Valentin-Opran A (2006) Recombinant human BMP-2 and allograft compared with autogenous bone graft for reconstruction of diaphyseal tibial fractures with cortical defects: a randomized, controlled trial. JBJS 88(7):1431–1441
123. Reichert JC, Cipitria A, Epari DR, Saifzadeh S, Krishnakanth P, Berner A, Woodruff MA, Schell H, Mehta M, Schuetz MA (2012) A tissue engineering solution for segmental defect regeneration in load-bearing long bones. Sci Transl Med 4(141):141ra93–141ra93
124. Ren L, Kang Y, Browne C, Bishop J, Yang Y (2014) Fabrication, vascularization and osteogenic properties of a novel synthetic biomimetic induced membrane for the treatment of large bone defects. Bone 64:173–182
125. Berner A, Henkel J, Woodruff MA, Steck R, Nerlich M, Schuetz MA, Hutmacher DW (2015) Delayed minimally invasive injection of allogenic bone marrow stromal cell sheets

regenerates large bone defects in an ovine preclinical animal model. Stem Cells Transl Med 4(5):503–512
126. Orlic D, Kajstura J, Chimenti S, Jakoniuk I, Anderson SM, Li B, Pickel J, McKay R, Nadal-Ginard B, Bodine DM (2001) Bone marrow cells regenerate infarcted myocardium. Nature 410(6829):701
127. Cameron SH, Alwakeel AJ, Goddard L, Hobbs CE, Gowing EK, Barnett ER, Kohe SE, Sizemore RJ, Oorschot DE (2015) Delayed post-treatment with bone marrow-derived mesenchymal stem cells is neurorestorative of striatal medium-spiny projection neurons and improves motor function after neonatal rat hypoxia–ischemia. Mol Cell Neurosci 68:56–72
128. Dreger T, Watson JT, Walter Akers D, Molligan J, Achilefu S, Schon LC, Zhang Z (2014) Intravenous application of CD271-selected mesenchymal stem cells during fracture healing. J Orthop Trauma 28(1):S15
129. Ghosal K, Khanna R, Sarkar K (2018) Biopolymer based interfacial tissue engineering for arthritis. In: Orthopedic biomaterials. Springer, Berlin, pp 67–88
130. Perera J, Gikas P, Bentley G (2012) The present state of treatments for articular cartilage defects in the knee. Ann R Coll Surg Engl 94(6):381–387
131. Verhaegen J, Clockaerts S, Van Osch G, Somville J, Verdonk P, Mertens P (2015) TruFit plug for repair of osteochondral defects—where is the evidence? Systematic review of literature. Cartilage 6(1):12–19
132. Cole BJ, Farr J, Winalski C, Hosea T, Richmond J, Mandelbaum B, De Deyne PG (2011) Outcomes after a single-stage procedure for cell-based cartilage repair: a prospective clinical safety trial with 2-year follow-up. Am J Sports Med 39(6):1170–1179
133. Gibbs DM, Vaezi M, Yang S, Oreffo RO (2014) Hope versus hype: what can additive manufacturing realistically offer trauma and orthopedic surgery? Regen Med 9(4):535–549
134. Hunter W (1743) Of the structure and disease of articular cartilages. Philos Trans Lond 42:514–521
135. Mithoefer K, McAdams T, Williams RJ, Kreuz PC, Mandelbaum BR (2009) Clinical efficacy of the microfracture technique for articular cartilage repair in the knee: an evidence-based systematic analysis. Am J Sports Med 37(10):2053–2063
136. Athanasiou KA, Darling EM, Hu JC (2009) Articular cartilage tissue engineering. Synth Lect Tissue Eng 1(1):1–182
137. Roberts S, Menage J, Sandell L, Evans E, Richardson J (2009) Immunohistochemical study of collagen types I and II and procollagen IIA in human cartilage repair tissue following autologous chondrocyte implantation. Knee 16(5):398–404
138. Pelttari K, Wixmerten A, Do we really need cartilage tissue engineering? Swiss Med Wkly. 2009;139(4142).
139. Green JW (1977) Articular cartilage repair. Behavior of rabbit chondrocytes during tissue culture and subsequent allografting. Clin Orthop Relat Res 124:237–250
140. Brittberg M, Lindahl A, Nilsson A, Ohlsson C, Isaksson O, Peterson L (1994) Treatment of deep cartilage defects in the knee with autologous chondrocyte transplantation. N Engl J Med 331(14):889–895
141. Zhou Q, Gong Y, Gao C (2005) Microstructure and mechanical properties of poly (L-lactide) scaffolds fabricated by gelatin particle leaching method. J Appl Polym Sci 98(3):1373–1379
142. He X, Kawazoe N, Chen G (2014) Preparation of cylinder-shaped porous sponges of poly (L-lactic acid), poly (DL-lactic-co-glycolic acid), and poly (-caprolactone). Biomed Res Int 2014:106082
143. Shin HJ, Lee CH, Cho IH, Kim Y-J, Lee Y-J, Kim IA, Park K-D, Yui N, Shin J-W (2006) Electrospun PLGA nanofiber scaffolds for articular cartilage reconstruction: mechanical stability, degradation and cellular responses under mechanical stimulation in vitro. J Biomater Sci Polym Ed 17(1-2):103–119
144. Williams GM, Klein TJ, Sah RL (2005) Cell density alters matrix accumulation in two distinct fractions and the mechanical integrity of alginate–chondrocyte constructs. Acta Biomater 1(6):625–633

145. Steinmeyer J, Ackermann B, Raiss RX (1997) Intermittent cyclic loading of cartilage explants modulates fibronectin metabolism. Osteoarthr Cartil 5(5):331–341
146. Parkkinen J, Ikonen J, Lammi M, Laakkonen J, Tammi M, Helminen H (1993) Effects of cyclic hydrostatic pressure on proteoglycan synthesis in cultured chondrocytes and articular cartilage explants. Arch Biochem Biophys 300(1):458–465
147. Hall AC (1999) Differential effects of hydrostatic pressure on cation transport pathways of isolated articular chondrocytes. J Cell Physiol 178(2):197–204
148. Wu MH, Urban JP, Cui ZF, Cui Z, Xu X (2007) Effect of extracellular pH on matrix synthesis by chondrocytes in 3D agarose gel. Biotechnol Prog 23(2):430–434
149. Griffon DJ, Sedighi MR, Schaeffer DV, Eurell JA, Johnson AL (2006) Chitosan scaffolds: interconnective pore size and cartilage engineering. Acta Biomater 2(3):313–320
150. Lammi MJ (2007) Cellular signaling in cartilage tissue engineering. Curr Signal Transduct Ther 2(1):41–48
151. Schmidt M, Chen E, Lynch S (2006) A review of the effects of insulin-like growth factor and platelet derived growth factor on in vivo cartilage healing and repair. Osteoarthr Cartil 14(5):403–412
152. Awad HA, Halvòrsen Y-DC, Gimble JM, Guilak F (2003) Effects of transforming growth factor β 1 and dexamethasone on the growth and chondrogenic differentiation of adipose-derived stromal cells. Tissue Eng 9(6):1301–1312
153. Gooch K, Blunk T, Courter D, Sieminski A, Vunjak-Novakovic G, Freed L (2002) Bone morphogenetic proteins-2,-12, and-13 modulate in vitro development of engineered cartilage. Tissue Eng 8(4):591–601
154. Mauck RL, Nicoll SB, Seyhan SL, Ateshian GA, Hung CT (2003) Synergistic action of growth factors and dynamic loading for articular cartilage tissue engineering. Tissue Eng 9(4):597–611
155. Aigner J, Tegeler J, Hutzler P, Campoccia D, Pavesio A, Hammer C, Kastenbauer E, Naumann A (1998) Cartilage tissue engineering with novel nonwoven structured biomaterial based on hyaluronic acid benzyl ester. J Biomed Mater Res 42(2):172–181
156. Nehrer S, Domayer S, Dorotka R, Schatz K, Bindreiter U, Kotz R (2006) Three-year clinical outcome after chondrocyte transplantation using a hyaluronan matrix for cartilage repair. Eur J Radiol 57(1):3–8
157. Bosnakovski D, Mizuno M, Kim G, Takagi S, Okumura M, Fujinaga T (2006) Chondrogenic differentiation of bovine bone marrow mesenchymal stem cells (MSCs) in different hydrogels: influence of collagen type II extracellular matrix on MSC chondrogenesis. Biotechnol Bioeng 93(6):1152–1163
158. Caterson EJ, Nesti LJ, Li WJ, Danielson KG, Albert TJ, Vaccaro AR, Tuan RS (2001) Three-dimensional cartilage formation by bone marrow-derived cells seeded in polylactide/alginate amalgam. J Biomed Mater Res 57(3):394–403
159. Montembault A, Tahiri K, Korwin-Zmijowska C, Chevalier X, Corvol M-T, Domard A (2006) A material decoy of biological media based on chitosan physical hydrogels: application to cartilage tissue engineering. Biochimie 88(5):551–564
160. Yoo M-K, Kweon HY, Lee K-G, Lee H-C, Cho C-S (2004) Preparation of semi-interpenetrating polymer networks composed of silk fibroin and poloxamer macromer. Int J Biol Macromol 34(4):263–270
161. Kim U-J, Park J, Li C, Jin H-J, Valluzzi R, Kaplan DL (2004) Structure and properties of silk hydrogels. Biomacromolecules 5(3):786–792
162. Burdick JA, Chung C, Jia X, Randolph MA, Langer R (2005) Controlled degradation and mechanical behavior of photopolymerized hyaluronic acid networks. Biomacromolecules 6(1):386–391
163. Sahoo S, Chung C, Khetan S, Burdick JA (2008) Hydrolytically degradable hyaluronic acid hydrogels with controlled temporal structures. Biomacromolecules 9(4):1088–1092
164. Li Q, Williams CG, Sun DD, Wang J, Leong K, Elisseeff JH (2004) Photocrosslinkable polysaccharides based on chondroitin sulfate. J Biomed Mater Res 68(1):28–33

165. Suh J-KF, Matthew HW (2000) Application of chitosan-based polysaccharide biomaterials in cartilage tissue engineering: a review. Biomaterials 21(24):2589–2598
166. Nehrer S, Breinan H, Ramappa A, Hsu H, Minas T, Shortkroff S, Sledge C, Yannas I, Spector M (1998) Chondrocyte-seeded collagen matrices implanted in a chondral defect in a canine model. Biomaterials 19(24):2313–2328
167. Lien S-M, Li W-T, Huang T-J (2008) Genipin-crosslinked gelatin scaffolds for articular cartilage tissue engineering with a novel crosslinking method. Mater Sci Eng C 28(1):36–43
168. Gründer T, Gaissmaier C, Fritz J, Stoop R, Hortschansky P, Mollenhauer J, Aicher WK (2004) Bone morphogenetic protein (BMP)-2 enhances the expression of type II collagen and aggrecan in chondrocytes embedded in alginate beads. Osteoarthr Cartil 12(7):559–567
169. Mattioli-Belmonte M, Gigante A, Muzzarelli R, Politano R, De Benedittis A, Specchia N, Buffa A, Biagini G, Greco F (1999) N, N-dicarboxymethyl chitosan as delivery agent for bone morphogenetic protein in the repair of articular cartilage. Med Biol Eng Comput 37(1):130–134
170. Kubota N, Tatsumoto N, Sano T, Toya K (2000) A simple preparation of half N-acetylated chitosan highly soluble in water and aqueous organic solvents. Carbohydr Res 324(4):268–274
171. Muzzarelli RA, Tanfani F, Emanuelli M, Mariotti S (1982) N-(carboxymethylidene) chitosans and N-(carboxymethyl) chitosans: Novel chelating polyampholytes obtained from chitosan glyoxylate. Carbohydr Res 107(2):199–214
172. Murata J-i, Ohya Y, Ouchi T (1996) Possibility of application of quaternary chitosan having pendant galactose residues as gene delivery tool. Carbohydr Polym 29(1):69–74
173. Medrado GCB, Machado CB, Valerio P, Sanches MD, Goes AM (2006) The effect of a chitosan–gelatin matrix and dexamethasone on the behavior of rabbit mesenchymal stem cells. Biomed Mater 1(3):155–161
174. Meinel L, Hofmann S, Karageorgiou V, Zichner L, Langer R, Kaplan D, Vunjak-Novakovic G (2004) Engineering cartilage-like tissue using human mesenchymal stem cells and silk protein scaffolds. Biotechnol Bioeng 88(3):379–391
175. Yeh MK, Cheng KM, Hu CS, Huang YC, Young JJ (2011) Novel protein-loaded chondroitin sulfate–chitosan nanoparticles: preparation and characterization. Acta Biomater 7(10):3804–3812
176. Jo S, Kim S, Noh I (2012) Synthesis of in situ chondroitin sulfate hydrogel through phosphine-mediated Michael type addition reaction. Macromol Res 20(9):968–976
177. Nishimoto S, Takagi M, Wakitani S, Nihira T, Yoshida T (2005) Effect of chondroitin sulfate and hyaluronic acid on gene expression in a three-dimensional culture of chondrocytes. J Biosci Bioeng 100(1):123–126
178. Bhardwaj T, Pilliar RM, Grynpas MD, Kandel RA (2001) Effect of material geometry on cartilagenous tissue formation in vitro. J Biomed Mater Res 57(2):190–199
179. He X, Kawazoe N, Chen G (2014) Preparation of cylinder-shaped porous sponges of poly(L-lactic acid), poly(DL-lactic-co-glycolic acid), and poly(-caprolactone). Biomed Res Int 2014:8
180. Pan Z, Ding J (2012) Poly(lactide-co-glycolide) porous scaffolds for tissue engineering and regenerative medicine. Interface Focus 2(3):366–377
181. Wang W, Li B, Li Y, Jiang Y, Ouyang H, Gao C (2010) In vivo restoration of full-thickness cartilage defects by poly(lactide-co-glycolide) sponges filled with fibrin gel, bone marrow mesenchymal stem cells and DNA complexes. Biomaterials 31(23):5953–5965
182. Murphy SV, Atala A (2014) 3D bioprinting of tissues and organs. Nat Biotechnol 32(8):773
183. Ma PX, Choi J-W (2001) Biodegradable polymer scaffolds with well-defined interconnected spherical pore network. Tissue Eng 7(1):23–33
184. Chen G, Sato T, Ushida T, Ochiai N, Tateishi T (2004) Tissue engineering of cartilage using a hybrid scaffold of synthetic polymer and collagen. Tissue Eng 10(3-4):323–330
185. Richards E, Rizvi R, Chow A, Naguib H (2008) Biodegradable composite foams of PLA and PHBV using subcritical CO2. J Polym Environ 16(4):258–266
186. Calimeri T, Battista E, Conforti F, Neri P, Di Martino M, Rossi M, Foresta U, Piro E, Ferrara F, Amorosi A (2011) A unique three-dimensional SCID-polymeric scaffold (SCID-synth-hu) model for in vivo expansion of human primary multiple myeloma cells. Leukemia 25(4):707

187. Croisier F, Jérôme C (2013) Chitosan-based biomaterials for tissue engineering. Eur Polym J 49(4):780–792
188. Lim JI, Park H-K (2012) Fabrication of macroporous chitosan/poly (l-lactide) hybrid scaffolds by sodium acetate particulate-leaching method. J Porous Mater 19(3):383–387
189. Moss T, Paulus IE, Raps D, Altstädt V, Greiner A (2017) Ultralight sponges of poly (para-xylylene) by template-assisted chemical vapour deposition. e-Polymers 17(4):255–261
190. Mader M, Jérôme V r, Freitag R, Agarwal S, Greiner A (2018) Ultraporous, compressible, wettable polylactide/polycaprolactone sponges for tissue engineering. Biomacromolecules 19(5):1663–1673
191. Chen W, Chen S, Morsi Y, El-Hamshary H, El-Newhy M, Fan C, Mo X (2016) Superabsorbent 3D scaffold based on electrospun nanofibers for cartilage tissue engineering. ACS Appl Mater Interfaces 8(37):24415–24425
192. Chen W, Ma J, Zhu L, Morsi Y, Al-Deyab SS, Mo X (2016) Superelastic, superabsorbent and 3D nanofiber-assembled scaffold for tissue engineering. Colloids Surf B Biointerfaces 142:165–172
193. Nettles DL, Elder SH, Gilbert JA (2002) Potential use of chitosan as a cell scaffold material for cartilage tissue engineering. Tissue Eng 8(6):1009–1016
194. Xia W, Liu W, Cui L, Liu Y, Zhong W, Liu D, Wu J, Chua K, Cao Y (2004) Tissue engineering of cartilage with the use of chitosan-gelatin complex scaffolds. J Biomed Mater Res B Appl Biomater 71(2):373–380
195. Chung C, Burdick JA (2008) Engineering cartilage tissue. Adv Drug Deliv Rev 60(2):243–262
196. Li WJ, Danielson KG, Alexander PG, Tuan RS (2003) Biological response of chondrocytes cultured in three-dimensional nanofibrous poly (ϵ-caprolactone) scaffolds. J Biomed Mater Res A 67(4):1105–1114
197. Hsu SH, Chang SH, Yen HJ, Whu SW, Tsai CL, Chen DC (2006) Evaluation of biodegradable polyesters modified by type II collagen and Arg-Gly-Asp as tissue engineering scaffolding materials for cartilage regeneration. Artif Organs 30(1):42–55
198. Yoo HS, Lee EA, Yoon JJ, Park TG (2005) Hyaluronic acid modified biodegradable scaffolds for cartilage tissue engineering. Biomaterials 26(14):1925–1933
199. Casper M, Fitzsimmons J, Stone J, Meza A, Huang Y, Ruesink T, O'Driscoll S, Reinholz G (2010) Tissue engineering of cartilage using poly-ε-caprolactone nanofiber scaffolds seeded in vivo with periosteal cells. Osteoarthr Cartil 18(7):981–991
200. Radice M, Brun P, Cortivo R, Scapinelli R, Battaliard C, Abatangelo G (2000) Hyaluronan-based biopolymers as delivery vehicles for bone-marrow-derived mesenchymal progenitors. J Biomed Mater Res 50(2):101–109
201. Shoichet MS, Tate CC, Baumann MD, LaPlaca MC (2008) Strategies for regeneration and repair in the injured central nervous system. In: Indwelling neural implants: strategies for contending with the in vivo environment. CRC/Taylor and Francis, Boca Raton, FL
202. Reichert WM (2007) Indwelling neural implants: strategies for contending with the in vivo environment. CRC, Boca Raton, FL
203. Samadikuchaksaraei A (2007) An overview of tissue engineering approaches for management of spinal cord injuries. J Neuroeng Rehabil 4(1):15
204. Tu Q, Pang L, Wang L, Zhang Y, Zhang R, Wang J (2013) Biomimetic choline-like graphene oxide composites for neurite sprouting and outgrowth. ACS Appl Mater Interfaces 5(24):13188–13197
205. Lee JY, Bashur CA, Goldstein AS, Schmidt CE (2009) Polypyrrole-coated electrospun PLGA nanofibers for neural tissue applications. Biomaterials 30(26):4325–4335
206. Fan L, Xiong Y, Fu Z, Xu D, Wang L, Chen Y, Xia H, Peng N, Ye S, Wang Y (2017) Polyaniline promotes peripheral nerve regeneration by enhancement of the brain-derived neurotrophic factor and ciliary neurotrophic factor expression and activation of the ERK1/2/MAPK signaling pathway. Mol Med Rep 16(5):7534–7540
207. Xu D, Fan L, Gao L, Xiong Y, Wang Y, Ye Q, Yu A, Dai H, Yin Y, Cai J (2016) Micro-nanostructured polyaniline assembled in cellulose matrix via interfacial polymerization for applications in nerve regeneration. ACS Appl Mater Interfaces 8(27):17090–17097

208. Pires F, Ferreira Q, Rodrigues CA, Morgado J, Ferreira FC (2015) Neural stem cell differentiation by electrical stimulation using a cross-linked PEDOT substrate: expanding the use of biocompatible conjugated conductive polymers for neural tissue engineering. Biochim Biophys Acta 1850(6):1158–1168
209. Ostrakhovitch E, Byers J, O'Neil K, Semenikhin O (2012) Directed differentiation of embryonic P19 cells and neural stem cells into neural lineage on conducting PEDOT–PEG and ITO glass substrates. Arch Biochem Biophys 528(1):21–31
210. Farzamfar S, Esmailpour F, Rahmati M, Vaez A, Mirzaii M, Garmabi B, Shayannia A, Ebrahimi E, Vahedi H, Salehi M. Poly-lactic acid/gelatin nanofiber (PLA/GTNF) conduits containing platelet-rich plasma for peripheral nerve regeneration. Int J Health Stud. 2017;3(2).
211. Boni R, Ali A, Shavandi A, Clarkson AN (2018) Current and novel polymeric biomaterials for neural tissue engineering. J Biomed Sci 25(1):90
212. Kim M, Kim J, Hyun J (2017) Development of Schwann cell-seeded multichannel scaffolds for peripheral nerve regeneration. J Neurol Sci 381:612–613
213. Mir M, Ahmed N, Rehman A (2017) Recent applications of PLGA based nanostructures in drug delivery. Colloids Surf B Biointerfaces 159:217–231
214. Garbayo E, Montero-Menei C, Ansorena E, Lanciego JL, Aymerich MS, Blanco-Prieto MJ (2009) Effective GDNF brain delivery using microspheres—a promising strategy for Parkinson's disease. J Control Release 135(2):119–126
215. Lampe KJ, Mooney RG, Bjugstad KB, Mahoney MJ (2010) Effect of macromer weight percent on neural cell growth in 2D and 3D nondegradable PEG hydrogel culture. J Biomed Mater Res A 94(4):1162–1171
216. Freudenberg U, Hermann A, Welzel PB, Stirl K, Schwarz SC, Grimmer M, Zieris A, Panyanuwat W, Zschoche S, Meinhold D (2009) A star-PEG–heparin hydrogel platform to aid cell replacement therapies for neurodegenerative diseases. Biomaterials 30(28):5049–5060
217. Mahoney MJ, Anseth KS (2006) Three-dimensional growth and function of neural tissue in degradable polyethylene glycol hydrogels. Biomaterials 27(10):2265–2274
218. Liu C, Huang Y, Pang M, Yang Y, Li S, Liu L, Shu T, Zhou W, Wang X, Rong L (2015) Tissue-engineered regeneration of completely transected spinal cord using induced neural stem cells and gelatin-electrospun poly (lactide-co-glycolide)/polyethylene glycol scaffolds. PLoS One 10(3):e0117709
219. Wangensteen KJ, Kalliainen LK (2010) Collagen tube conduits in peripheral nerve repair: a retrospective analysis. Hand 5(3):273–277
220. Bozkurt A, Claeys KG, Schrading S, Rödler JV, Altinova H, Schulz JB, Weis J, Pallua N, van Neerven SG (2017) Clinical and biometrical 12-month follow-up in patients after reconstruction of the sural nerve biopsy defect by the collagen-based nerve guide Neuromaix. Eur J Med Res 22(1):34
221. Archibald S, Shefner J, Krarup C, Madison R (1995) Monkey median nerve repaired by nerve graft or collagen nerve guide tube. J Neurosci 15(5):4109–4123
222. Gonzalez-Perez F, Cobianchi S, Heimann C, Phillips JB, Udina E, Navarro X (2017) Stabilization, rolling, and addition of other extracellular matrix proteins to collagen hydrogels improve regeneration in chitosan guides for long peripheral nerve gaps in rats. Neurosurgery 80(3):465–474·
223. Ceballos D, Navarro X, Dubey N, Wendelschafer-Crabb G, Kennedy WR, Tranquillo RT (1999) Magnetically aligned collagen gel filling a collagen nerve guide improves peripheral nerve regeneration. Exp Neurol 158(2):290–300
224. Cao J, Sun C, Zhao H, Xiao Z, Chen B, Gao J, Zheng T, Wu W, Wu S, Wang J (2011) The use of laminin modified linear ordered collagen scaffolds loaded with laminin-binding ciliary neurotrophic factor for sciatic nerve regeneration in rats. Biomaterials 32(16):3939–3948
225. Dubey N, Letourneau P, Tranquillo R (1999) Guided neurite elongation and Schwann cell invasion into magnetically aligned collagen in simulated peripheral nerve regeneration. Exp Neurol 158(2):338–350

226. Faghihi F, Mirzaei E, Ai J, Lotfi A, Sayahpour FA, Barough SE, Joghataei MT (2016) Differentiation potential of human chorion-derived mesenchymal stem cells into motor neuron-like cells in two-and three-dimensional culture systems. Mol Neurobiol 53(3):1862–1872
227. Büyüköz M, Erdal E, Alsoy Altinkaya S (2018) Nanofibrous gelatine scaffolds integrated with nerve growth factor-loaded alginate microspheres for brain tissue engineering. J Tissue Eng Regen Med 12(2):e707–e719
228. Ghasemi-Mobarakeh L, Prabhakaran MP, Morshed M, Nasr-Esfahani M-H, Ramakrishna S (2008) Electrospun poly (ε-caprolactone)/gelatin nanofibrous scaffolds for nerve tissue engineering. Biomaterials 29(34):4532–4539
229. Dinis T, Elia R, Vidal G, Dermigny Q, Denoeud C, Kaplan D, Egles C, Marin F (2015) 3D multi-channel bi-functionalized silk electrospun conduits for peripheral nerve regeneration. J Mech Behav Biomed Mater 41:43–55
230. Wang S, Guan S, Zhu Z, Li W, Liu T, Ma X (2017) Hyaluronic acid doped-poly(3,4-ethylenedioxythiophene)/chitosan/gelatin (PEDOT-HA/Cs/Gel) porous conductive scaffold for nerve regeneration. Mater Sci Eng C 71:308–316
231. Steel DA, Basu S (2017) Does trajectory matter? A study looking into the relationship of trajectory with target engagement and error accommodation in subthalamic nucleus deep brain stimulation. Acta Neurochir 159(7):1335–1340
232. Ansorena E, De Berdt P, Ucakar B, Simón-Yarza T, Jacobs D, Schakman O, Jankovski A, Deumens R, Blanco-Prieto MJ, Préat V (2013) Injectable alginate hydrogel loaded with GDNF promotes functional recovery in a hemisection model of spinal cord injury. Int J Pharm 455(1-2):148–158
233. Crompton K, Goud J, Bellamkonda R, Gengenbach T, Finkelstein D, Horne M, Forsythe J (2007) Polylysine-functionalised thermoresponsive chitosan hydrogel for neural tissue engineering. Biomaterials 28(3):441–449
234. Valmikinathan CM, Mukhatyar VJ, Jain A, Karumbaiah L, Dasari M, Bellamkonda RV (2012) Photocrosslinkable chitosan based hydrogels for neural tissue engineering. Soft Matter 8(6):1964–1976
235. Yi X, Jin G, Tian M, Mao W, Qin J (2011) Porous chitosan scaffold and ngf promote neuronal differentiation of neural stem cells in vitro. Neuro Endocrinol Lett 32(5):705–710
236. Kuo Y-C, Yeh C-F, Yang J-T (2009) Differentiation of bone marrow stromal cells in poly (lactide-co-glycolide)/chitosan scaffolds. Biomaterials 30(34):6604–6613
237. Sierpinski P, Garrett J, Ma J, Apel P, Klorig D, Smith T, Koman LA, Atala A, Van Dyke M (2008) The use of keratin biomaterials derived from human hair for the promotion of rapid regeneration of peripheral nerves. Biomaterials 29(1):118–128
238. Bai S, Zhang W, Lu Q, Ma Q, Kaplan DL, Zhu H (2014) Silk nanofiber hydrogels with tunable modulus to regulate nerve stem cell fate. J Mater Chem B 2(38):6590–6600
239. Hopkins AM, De Laporte L, Tortelli F, Spedden E, Staii C, Atherton TJ, Hubbell JA, Kaplan DL (2013) Silk hydrogels as soft substrates for neural tissue engineering. Adv Funct Mater 23(41):5140–5149
240. Zhang Q, Zhao Y, Yan S, Yang Y, Zhao H, Li M, Lu S, Kaplan DL (2012) Preparation of uniaxial multichannel silk fibroin scaffolds for guiding primary neurons. Acta Biomater 8(7):2628–2638
241. Xue C, Hu N, Gu Y, Yang Y, Liu Y, Liu J, Ding F, Gu X (2012) Joint use of a chitosan/PLGA scaffold and MSCs to bridge an extra large gap in dog sciatic nerve. Neurorehabil Neural Repair 26(1):96–106
242. Zhou X, Yang A, Huang Z, Yin G, Pu X, Jin J (2017) Enhancement of neurite adhesion, alignment and elongation on conductive polypyrrole-poly (lactide acid) fibers with cell-derived extracellular matrix. Colloids Surf B Biointerfaces 149:217–225
243. Shin J, Choi EJ, Cho JH, Cho A-N, Jin Y, Yang K, Song C, Cho S-W (2017) Three-dimensional electroconductive hyaluronic acid hydrogels incorporated with carbon nanotubes and polypyrrole by catechol-mediated dispersion enhance neurogenesis of human neural stem cells. Biomacromolecules 18(10):3060–3072

244. Xu B, Bai T, Sinclair A, Wang W, Wu Q, Gao F, Jia H, Jiang S, Liu W. Directed neural stem cell differentiation on polyaniline-coated high strength hydrogels. 2016;1.
245. Guarino V, Alvarez-Perez MA, Borriello A, Napolitano T, Ambrosio L (2013) Conductive PANi/PEGDA macroporous hydrogels for nerve regeneration. Adv Healthc Mater 2(1):218–227
246. Nyambat B, Chen C-H, Wong P-C, Chiang C-W, Satapathy MK, Chuang E-Y (2018) Genipin-crosslinked adipose stem cell derived extracellular matrix-nano graphene oxide composite sponge for skin tissue engineering. J Mater Chem B 6(6):979–990
247. Boucard N, Viton C, Agay D, Mari E, Roger T, Chancerelle Y, Domard A (2007) The use of physical hydrogels of chitosan for skin regeneration following third-degree burns. Biomaterials 28(24):3478–3488
248. Atiyeh BS, Gunn SW, Hayek SN (2005) State of the art in burn treatment. World J Surg 29(2):131–148
249. Phan T, Lim I, Tan E, Bay B, Lee S (2005) Evaluation of cell culture on the polyurethane-based membrane (Tegaderm TM): implication for tissue engineering of skin. Cell Tissue Bank 6(2):91–97
250. Holland K, Davis W, Ingham E, Gowland G (1984) A comparison of the in-vitro antibacterial and complement activating effect of 'OpSite'and 'Tegaderm' dressings. J Hosp Infect 5(3):323–328
251. Liu H, Yin Y, Yao K (2007) Construction of chitosan—gelatin—hyaluronic acid artificial skin in vitro. J Biomater Appl 21(4):413–430
252. Wang T-W, Sun J-S, Wu H-C, Tsuang Y-H, Wang W-H, Lin F-H (2006) The effect of gelatin–chondroitin sulfate–hyaluronic acid skin substitute on wound healing in SCID mice. Biomaterials 27(33):5689–5697
253. Ng KW, Hutmacher DW (2006) Reduced contraction of skin equivalent engineered using cell sheets cultured in 3D matrices. Biomaterials 27(26):4591–4598
254. Ma L, Gao C, Mao Z, Zhou J, Shen J, Hu X, Han C (2003) Collagen/chitosan porous scaffolds with improved biostability for skin tissue engineering. Biomaterials 24(26):4833–4841
255. Dai N-T, Williamson M, Khammo N, Adams E, Coombes A (2004) Composite cell support membranes based on collagen and polycaprolactone for tissue engineering of skin. Biomaterials 25(18):4263–4271
256. Mozaffarian D, Benjamin EJ, Go AS, Arnett DK, Blaha MJ, Cushman M, Das SR, De Ferranti S, Després J-P, Fullerton HJ (2016) Executive summary: heart disease and stroke statistics—2016 update: a report from the American Heart Association. Circulation 133(4):447–454
257. Shin SR, Li Y-C, Jang HL, Khoshakhlagh P, Akbari M, Nasajpour A, Zhang YS, Tamayol A, Khademhosseini A (2016) Graphene-based materials for tissue engineering. Adv Drug Deliv Rev 105:255–274
258. Morgan KY, Black LD III (2014) It's all in the timing: Modeling isovolumic contraction through development and disease with a dynamic dual electromechanical bioreactor system. Organogenesis 10(3):317–322
259. Zhang D, Shadrin IY, Lam J, Xian H-Q, Snodgrass HR, Bursac N (2013) Tissue-engineered cardiac patch for advanced functional maturation of human ESC-derived cardiomyocytes. Biomaterials 34(23):5813–5820
260. Lakshmanan R, Krishnan UM, Sethuraman S (2012) Living cardiac patch: the elixir for cardiac regeneration. Expert Opin Biol Ther 12(12):1623–1640
261. Martins AM, Eng G, Caridade SG, Mano JOF, Reis RL, Vunjak-Novakovic G (2014) Electrically conductive chitosan/carbon scaffolds for cardiac tissue engineering. Biomacromolecules 15(2):635–643
262. Reis LA, Chiu LL, Liang Y, Hyunh K, Momen A, Radisic M (2012) A peptide-modified chitosan–collagen hydrogel for cardiac cell culture and delivery. Acta Biomater 8(3):1022–1036
263. Saravanan S, Sareen N, Abu-El-Rub E, Ashour H, Sequiera GL, Ammar HI, Gopinath V, Shamaa AA, Sayed SSE, Moudgil M, Vadivelu J, Dhingra S (2018) Graphene oxide-gold

nanosheets containing chitosan scaffold improves ventricular contractility and function after implantation into infarcted heart. Sci Rep 8(1):15069
264. Lu W-N, Lü S-H, Wang H-B, Li D-X, Duan C-M, Liu Z-Q, Hao T, He W-J, Xu B, Fu Q, Song YC, Xie X-H, Wang C-Y (2008) Functional improvement of infarcted heart by co-injection of embryonic stem cells with temperature-responsive chitosan hydrogel. Tissue Eng Part A 15(6):1437–1447
265. Cui Z, Ni NC, Wu J, Du G-Q, He S, Yau TM, Weisel RD, Sung H-W, Li R-K (2018) Polypyrrole-chitosan conductive biomaterial synchronizes cardiomyocyte contraction and improves myocardial electrical impulse propagation. Theranostics 8(10):2752
266. Shachar M, Tsur-Gang O, Dvir T, Leor J, Cohen S (2011) The effect of immobilized RGD peptide in alginate scaffolds on cardiac tissue engineering. Acta Biomater 7(1):152–162
267. Rosellini E, Cristallini C, Barbani N, Vozzi G, Giusti P (2009) Preparation and characterization of alginate/gelatin blend films for cardiac tissue engineering. J Biomed Mater Res A 91(2):447–453
268. O'Neill HS, O'Sullivan J, Porteous N, Ruiz Hernandez E, Kelly HM, O'Brien F, Duffy GP. A collagen cardiac patch incorporating alginate microparticles permits the controlled release of HGF and IGF-1 to enhance cardiac stem cell migration and proliferation. 2016.
269. Dahlmann J, Krause A, Möller L, Kensah G, Möwes M, Diekmann A, Martin U, Kirschning A, Gruh I, Dräger G (2013) Fully defined in situ cross-linkable alginate and hyaluronic acid hydrogels for myocardial tissue engineering. Biomaterials 34(4):940–951
270. Sapir Y, Kryukov O, Cohen S (2011) Integration of multiple cell-matrix interactions into alginate scaffolds for promoting cardiac tissue regeneration. Biomaterials 32(7):1838–1847
271. Agarwal A, Farouz Y, Nesmith AP, Deravi LF, McCain ML, Parker KK (2013) Micropatterning alginate substrates for in vitro cardiovascular muscle on a chip. Adv Funct Mater 23(30):3738–3746
272. Zimmermann W-H, Melnychenko I, Wasmeier G, Didié M, Naito H, Nixdorff U, Hess A, Budinsky L, Brune K, Michaelis B (2006) Engineered heart tissue grafts improve systolic and diastolic function in infarcted rat hearts. Nat Med 12(4):452
273. Heydarkhan-Hagvall S, Schenke-Layland K, Dhanasopon AP, Rofail F, Smith H, Wu BM, Shemin R, Beygui RE, MacLellan WR (2008) Three-dimensional electrospun ECM-based hybrid scaffolds for cardiovascular tissue engineering. Biomaterials 29(19):2907–2914
274. Flanagan TC, Wilkins B, Black A, Jockenhoevel S, Smith TJ, Pandit AS (2006) A collagen-glycosaminoglycan co-culture model for heart valve tissue engineering applications. Biomaterials 27(10):2233–2246
275. Tedder ME, Simionescu A, Chen J, Liao J, Simionescu DT (2010) Assembly and testing of stem cell-seeded layered collagen constructs for heart valve tissue engineering. Tissue Eng Part A 17(1-2):25–36
276. Tijore A, Irvine SA, Sarig U, Mhaisalkar P, Baisane V, Venkatraman S (2018) Contact guidance for cardiac tissue engineering using 3D bioprinted gelatin patterned hydrogel. Biofabrication 10(2):025003
277. Elamparithi A, Punnoose AM, Paul SF, Kuruvilla S (2017) Gelatin electrospun nanofibrous matrices for cardiac tissue engineering applications. Int J Polym Mater Polym Biomater 66(1):20–27
278. Kerscher P, Kaczmarek JA, Head SE, Ellis ME, Seeto WJ, Kim J, Bhattacharya S, Suppiramaniam V, Lipke EA (2016) Direct production of human cardiac tissues by pluripotent stem cell encapsulation in gelatin methacryloyl. ACS Biomater Sci Eng 3(8):1499–1509
279. Stoppel WL, Hu D, Domian IJ, Kaplan DL, Black LD III (2015) Anisotropic silk biomaterials containing cardiac extracellular matrix for cardiac tissue engineering. Biomed Mater 10(3):034105
280. Mehrotra S, Nandi SK, Mandal BB (2017) Stacked silk-cell monolayers as a biomimetic three dimensional construct for cardiac tissue reconstruction. J Mater Chem B 5(31):6325–6338
281. Patra C, Talukdar S, Novoyatleva T, Velagala SR, Mühlfeld C, Kundu B, Kundu SC, Engel FB (2012) Silk protein fibroin from Antheraea mylitta for cardiac tissue engineering. Biomaterials 33(9):2673–2680

282. Reimer J, Syedain Z, Haynie B, Lahti M, Berry J, Tranquillo R (2017) Implantation of a tissue-engineered tubular heart valve in growing lambs. Ann Biomed Eng 45(2):439–451
283. Ballotta V, Smits AI, Driessen-Mol A, Bouten CV, Baaijens FP (2014) Synergistic protein secretion by mesenchymal stromal cells seeded in 3D scaffolds and circulating leukocytes in physiological flow. Biomaterials 35(33):9100–9113
284. De Visscher G, Lebacq A, Mesure L, Blockx H, Vranken I, Plusquin R, Meuris B, Herregods M-C, Van Oosterwyck H, Flameng W (2010) The remodeling of cardiovascular bioprostheses under influence of stem cell homing signal pathways. Biomaterials 31(1):20–28
285. Ota T, Sawa Y, Iwai S, Kitajima T, Ueda Y, Coppin C, Matsuda H, Okita Y (2005) Fibronectin-hepatocyte growth factor enhances reendothelialization in tissue-engineered heart valve. Ann Thorac Surg 80(5):1794–1801
286. Park H, Radisic M, Lim JO, Chang BH, Vunjak-Novakovic G (2005) A novel composite scaffold for cardiac tissue engineering. In Vitro Cell Dev Biol Animal 41(7):188–196
287. McDevitt TC, Woodhouse KA, Hauschka SD, Murry CE, Stayton PS (2003) Spatially organized layers of cardiomyocytes on biodegradable polyurethane films for myocardial repair. J Biomed Mater Res A 66(3):586–595
288. Yu J, Lee A-R, Lin W-H, Lin C-W, Wu Y-K, Tsai W-B (2014) Electrospun PLGA fibers incorporated with functionalized biomolecules for cardiac tissue engineering. Tissue Eng Part A 20(13-14):1896–1907
289. Ozawa T, Mickle DA, Weisel RD, Koyama N, Ozawa S, Li R-K (2002) Optimal biomaterial for creation of autologous cardiac grafts. Circulation 106(12_suppl_1):I-176–I-182
290. Matsubayashi K, Fedak PW, Mickle DA, Weisel RD, Ozawa T, Li R-K (2003) Improved left ventricular aneurysm repair with bioengineered vascular smooth muscle grafts. Circulation 108(10_suppl_1):II-219–II-225
291. Badrossamay MR, McIlwee HA, Goss JA, Parker KK (2010) Nanofiber assembly by rotary jet-spinning. Nano Lett 10(6):2257–2261
292. Xu J, Zhou X, Ge H, Yang D, Guo T (2006) Machine vision and feedback control system allow the precise control of vascular deformation in vitro. Rev Sci Instrum 77(6):064304
293. Qazi TH, Rai R, Dippold D, Roether JE, Schubert DW, Rosellini E, Barbani N, Boccaccini AR (2014) Development and characterization of novel electrically conductive PANI–PGS composites for cardiac tissue engineering applications. Acta Biomater 10(6):2434–2445
294. Baheiraei N, Yeganeh H, Ai J, Gharibi R, Azami M, Faghihi F (2014) Synthesis, characterization and antioxidant activity of a novel electroactive and biodegradable polyurethane for cardiac tissue engineering application. Mater Sci Eng C 44:24–37
295. Bidez PR, Li S, MacDiarmid AG, Venancio EC, Wei Y, Lelkes PI (2006) Polyaniline, an electroactive polymer, supports adhesion and proliferation of cardiac myoblasts. J Biomater Sci Polym Ed 17(1-2):199–212
296. Deng B, Shen L, Wu Y, Shen Y, Ding X, Lu S, Jia J, Qian J, Ge J (2015) Delivery of alginate-chitosan hydrogel promotes endogenous repair and preserves cardiac function in rats with myocardial infarction. J Biomed Mater Res A 103(3):907–918
297. Huang NF, Yu J, Sievers R, Li S, Lee RJ (2005) Injectable biopolymers enhance angiogenesis after myocardial infarction. Tissue Eng 11(11-12):1860–1866
298. Wang T, Jiang X-J, Tang Q-Z, Li X-Y, Lin T, Wu D-Q, Zhang X-Z, Okello E (2009) Bone marrow stem cells implantation with α-cyclodextrin/MPEG–PCL–MPEG hydrogel improves cardiac function after myocardial infarction. Acta Biomater 5(8):2939–2944
299. Fujimoto KL, Ma Z, Nelson DM, Hashizume R, Guan J, Tobita K, Wagner WR (2009) Synthesis, characterization and therapeutic efficacy of a biodegradable, thermoresponsive hydrogel designed for application in chronic infarcted myocardium. Biomaterials 30(26):4357–4368
300. Caplan AI (2005) Mesenchymal stem cells: cell–based reconstructive therapy in orthopedics. Tissue Eng 11(7-8):1198–1211
301. Wu J, Zeng F, Huang X-P, Chung JC-Y, Konecny F, Weisel RD, Li R-K (2011) Infarct stabilization and cardiac repair with a VEGF-conjugated, injectable hydrogel. Biomaterials 32(2):579–586

302. Mortality G Collaborators C. O. D. Global, regional, and national life expectancy, all-cause-specific mortality for 249 causes of death, 1980-2015: a systematic analysis for the Global Burden of Disease Study 2015. Lancet. 2016;388(10053).
303. Mozaffarian D, Benjamin EJ, Go AS, Arnett DK, Blaha MJ, Cushman M, Das SR, Ferranti SD, Després J-P, Fullerton HJ, Howard VJ, Huffman MD, Isasi CR, Jiménez MC, Judd SE, Kissela BM, Lichtman JH, Lisabeth LD, Liu S, Mackey RH, Magid DJ, McGuire DK, Mohler ER, Moy CS, Muntner P, Mussolino ME, Nasir K, Neumar RW, Nichol G, Palaniappan L, Pandey DK, Reeves MJ, Rodriguez CJ, Rosamond W, Sorlie PD, Stein J, Towfighi A, Turan TN, Virani SS, Woo D, Yeh RW, Turner MB (2016) Executive summary: heart disease and stroke statistics—2016 update. Circulation 133(4):447–454
304. Soyer T, Lempinen M, Cooper P, Norton L, Eiseman B (1972) A new venous prosthesis. Surgery 72(6):864–872
305. DeBakey ME, Crawford ES, Garrett HE, Beall AC Jr, Howell J (1965) Surgical considerations in the treatment of aneurysms of the thoraco-abdominal aorta. Ann Surg 162(4):650
306. Konig G, McAllister TN, Dusserre N, Garrido SA, Iyican C, Marini A, Fiorillo A, Avila H, Wystrychowski W, Zagalski K, Maruszewski M, Jones AL, Cierpka L, de la Fuente LM, L'Heureux N (2009) Mechanical properties of completely autologous human tissue engineered blood vessels compared to human saphenous vein and mammary artery. Biomaterials 30(8):1542–1550
307. L'Heureux N, Germain L, Labbé R, Auger FA (1993) In vitro construction of a human blood vessel from cultured vascular cells: a morphologic study. J Vasc Surg 17(3):499–509
308. Brennan MP, Dardik A, Hibino N, Roh JD, Nelson GN, Papademitris X, Shinoka T, Breuer CK (2008) Tissue engineered vascular grafts demonstrate evidence of growth and development when implanted in a juvenile animal model. Ann Surg 248(3):370
309. Shin'oka T, Imai Y, Ikada Y (2001) Transplantation of a tissue-engineered pulmonary artery. N Engl J Med 344(7):532–533
310. Matsumura G, Hibino N, Ikada Y, Kurosawa H, Shin'oka T (2003) Successful application of tissue engineered vascular autografts: clinical experience. Biomaterials 24(13):2303–2308
311. Mazza G, Al-Akkad W, Rombouts K, Pinzani M (2018) Liver tissue engineering: from implantable tissue to whole organ engineering. Hepatol Commun 2(2):131–141
312. Shi X-L, Zhang Y, Gu J-Y, Ding Y-T (2009) Coencapsulation of hepatocytes with bone marrow mesenchymal stem cells improves hepatocyte-specific functions. Transplantation 88(10):1178–1185
313. Li Y-S, Harn H-J, Hsieh D-K, Wen T-C, Subeq Y-M, Sun L-Y, Lin S-Z, Chiou T-W (2013) Cells and materials for liver tissue engineering. Cell Transplant 22(4):685–700
314. Zheng M-H, Ye C, Braddock M, Chen Y-P (2010) Liver tissue engineering: promises and prospects of new technology. Cytotherapy 12(3):349–360
315. Levenberg S, Burdick JA, Kraehenbuehl T, Langer R (2005) Neurotrophin-induced differentiation of human embryonic stem cells on three-dimensional polymeric scaffolds. Tissue Eng 11(3-4):506–512
316. Levenberg S, Huang NF, Lavik E, Rogers AB, Itskovitz-Eldor J, Langer R (2003) Differentiation of human embryonic stem cells on three-dimensional polymer scaffolds. Proc Natl Acad Sci 100(22):12741–12746
317. Rimann M, Graf-Hausner U (2012) Synthetic 3D multicellular systems for drug development. Curr Opin Biotechnol 23(5):803–809
318. Tsang VL, Chen AA, Cho LM, Jadin KD, Sah RL, DeLong S, West JL, Bhatia SN (2007) Fabrication of 3D hepatic tissues by additive photopatterning of cellular hydrogels. FASEB J 21(3):790–801
319. Lee H, Han W, Kim H, Ha D-H, Jang J, Kim BS, Cho D-W (2017) Development of liver decellularized extracellular matrix bioink for three-dimensional cell printing-based liver tissue engineering. Biomacromolecules 18(4):1229–1237
320. Detzel CJ, Kim Y, Rajagopalan P (2010) Engineered three-dimensional liver mimics recapitulate critical rat-specific bile acid pathways. Tissue Eng Part A 17(5-6):677–689

321. Miranda JP, Rodrigues A, Tostoes RM, Leite S, Zimmerman H, Carrondo MJ, Alves PM (2010) Extending hepatocyte functionality for drug-testing applications using high-viscosity alginate–encapsulated three-dimensional cultures in bioreactors. Tissue Eng Part C Methods 16(6):1223–1232
322. Katsuda T, Teratani T, Ochiya T, Sakai Y (2010) Transplantation of a fetal liver cell-loaded hyaluronic acid sponge onto the mesentery recovers a Wilson's disease model rat. J Biochem 148(3):281–288
323. Barbour KE, Helmick CG, Theis KA, Murphy LB, Hootman JM, Brady TJ, Cheng YJ (2013) Prevalence of doctor-diagnosed arthritis and arthritis-attributable activity limitation—United States, 2010–2012. MMWR Morb Mortal Wkly Rep 62(44):869
324. Hangody L, Kish G, Karpati Z, Szerb I, Udvarhelyi I (1997) Arthroscopic autogenous osteochondral mosaicplasty for the treatment of femoral condylar articular defects. A preliminary report. Knee Surg Sports Traumatol Arthrosc 5(4):262–267
325. Carranza-Bencano A, García-Paino L, Padron JA, Dominguez AC. Neochondrogenesis in repair of full-thickness articular cartilage defects using free autogenous periosteal grafts in the rabbit. A follow-up in six months. Osteoarthritis Cartilage. 2000;8(5):351–8.
326. Sledge SL (2001) Microfracture techniques in the treatment of osteochondral injuries. Clin Sports Med 20(2):365–378
327. Yang PJ, Temenoff JS (2009) Engineering orthopedic tissue interfaces. Tissue Eng Part B Rev 15(2):127–141
328. Keeney M, Pandit A (2009) The osteochondral junction and its repair via bi-phasic tissue engineering scaffolds. Tissue Eng Part B Rev 15(1):55–73
329. Gao J, Dennis JE, Solchaga LA, Awadallah AS, Goldberg VM, Caplan AI (2001) Tissue-engineered fabrication of an osteochondral composite graft using rat bone marrow-derived mesenchymal stem cells. Tissue Eng 7(4):363–371
330. Schaefer D, Martin I, Shastri P, Padera R, Langer R, Freed L, Vunjak-Novakovic G (2000) In vitro generation of osteochondral composites. Biomaterials 21(24):2599–2606
331. Alhadlaq A, Mao JJ (2005) Tissue-engineered osteochondral constructs in the shape of an articular condyle. JBJS 87(5):936–944
332. Jiang J, Tang A, Ateshian GA, Guo XE, Hung CT, Lu HH (2010) Bioactive stratified polymer ceramic-hydrogel scaffold for integrative osteochondral repair. Ann Biomed Eng 38(6):2183–2196
333. Harley BA, Lynn AK, Wissner-Gross Z, Bonfield W, Yannas IV, Gibson LJ (2010) Design of a multiphase osteochondral scaffold III: Fabrication of layered scaffolds with continuous interfaces. J Biomed Mater Res A 92(3):1078–1093
334. Paxton JZ, Donnelly K, Keatch RP, Baar K (2008) Engineering the bone–ligament interface using polyethylene glycol diacrylate incorporated with hydroxyapatite. Tissue Eng Part A 15(6):1201–1209
335. Spalazzi JP, Doty SB, Moffat KL, Levine WN, Lu HH (2006) Development of controlled matrix heterogeneity on a triphasic scaffold for orthopedic interface tissue engineering. Tissue Eng 12(12):3497–3508
336. Singh M, Morris CP, Ellis RJ, Detamore MS, Berkland C (2008) Microsphere-based seamless scaffolds containing macroscopic gradients of encapsulated factors for tissue engineering. Tissue Eng Part C Methods 14(4):299–309
337. Erisken C, Kalyon DM, Wang H (2008) Functionally graded electrospun polycaprolactone and β-tricalcium phosphate nanocomposites for tissue engineering applications. Biomaterials 29(30):4065–4073
338. Ramalingam M, Young MF, Thomas V, Sun L, Chow LC, Tison CK, Chatterjee K, Miles WC, Simon CG Jr (2013) Nanofiber scaffold gradients for interfacial tissue engineering. J Biomater Appl 27(6):695–705
339. Ferroni L, Gardin C, Sivolella S, Brunello G, Berengo M, Piattelli A, Bressan E, Zavan B (2015) A hyaluronan-based scaffold for the in vitro construction of dental pulp-like tissue. Int J Mol Sci 16(3):4666–4681

340. Barbara Z, Eriberto B, Stefano S, Giulia B, Chiara G, Ferrarese N, Letizia F, Edoardo S. Dental pulp stem cells and tissue engineering strategies for clinical application on odontoiatric field. In: Biomaterials science and engineering. IntechOpen; 2011.
341. Bose S, Vahabzadeh S, Bandyopadhyay A (2013) Bone tissue engineering using 3D printing. Mater Today 16(12):496–504
342. Mallick S, Tripathi S, Srivastava P (2015) Advancement in scaffolds for bone tissue engineering: a review. J Pharm Biol Sci 10:37–54
343. Shahbazarab Z, Teimouri A, Chermahini AN, Azadi M (2018) Fabrication and characterization of nanobiocomposite scaffold of zein/chitosan/nanohydroxyapatite prepared by freeze-drying method for bone tissue engineering. Int J Biol Macromol 108:1017–1027
344. Ghorbanian L, Emadi R, Razavi SM, Shin H, Teimouri A (2013) Fabrication and characterization of novel diopside/silk fibroin nanocomposite scaffolds for potential application in maxillofacial bone regeneration. Int J Biol Macromol 58:275–280
345. Santo VE, Rodrigues MT, Gomes ME (2013) Contributions and future perspectives on the use of magnetic nanoparticles as diagnostic and therapeutic tools in the field of regenerative medicine. Expert Rev Mol Diagn 13(6):553–566
346. Endogan Tanir T, Hasirci V, Hasirci N (2015) Preparation and characterization of Chitosan and PLGA-based scaffolds for tissue engineering applications. Polym Compos 36(10):1917–1930
347. Nwe N, Furuike T, Tamura H (2009) The mechanical and biological properties of chitosan scaffolds for tissue regeneration templates are significantly enhanced by chitosan from gongronella butleri. Materials 2(2):374–398
348. Mikos AG, Thorsen AJ, Czerwonka LA, Bao Y, Langer R, Winslow DN, Vacanti JP (1994) Preparation and characterization of poly (L-lactic acid) foams. Polymer 35(5):1068–1077
349. Mikos AG, Sarakinos G, Vacanti JP, Langer RS, Cima LG. Biocompatible polymer membranes and methods of preparation of three dimensional membrane structures. Google Patents; 1996.
350. Ma PX (2004) Scaffolds for tissue fabrication. Mater Today 7(5):30–40
351. Johnson T, Bahrampourian R, Patel A, Mequanint K (2010) Fabrication of highly porous tissue-engineering scaffolds using selective spherical porogens. Biomed Mater Eng 20(2):107–118
352. Liao CJ, Chen CF, Chen JH, Chiang SF, Lin YJ, Chang KY (2002) Fabrication of porous biodegradable polymer scaffolds using a solvent merging/particulate leaching method. J Biomed Mater Res 59(4):676–681
353. Nam YS, Park TG (1999) Porous biodegradable polymeric scaffolds prepared by thermally induced phase separation. J Biomed Mater Res 47(1):8–17
354. Akbarzadeh R, Yousefi AM (2014) Effects of processing parameters in thermally induced phase separation technique on porous architecture of scaffolds for bone tissue engineering. J Biomed Mater Res B Appl Biomater 102(6):1304–1315
355. Martínez-Pérez CA, Olivas-Armendariz I, Castro-Carmona JS, García-Casillas PE. Scaffolds for tissue engineering via thermally induced phase separation. In: Advances in regenerative medicine. IntechOpen; 2011.
356. Nam YS, Park TG (1999) Biodegradable polymeric microcellular foams by modified thermally induced phase separation method. Biomaterials 20(19):1783–1790
357. Boland ED, Wnek GE, Simpson DG, Pawlowski KJ, Bowlin GL (2001) Tailoring tissue engineering scaffolds using electrostatic processing techniques: a study of poly (glycolic acid) electrospinning. J Macromol Sci Pt A 38(12):1231–1243
358. Li WJ, Laurencin CT, Caterson EJ, Tuan RS, Ko FK (2002) Electrospun nanofibrous structure: a novel scaffold for tissue engineering. J Biomed Mater Res 60(4):613–621
359. Yoshimoto H, Shin Y, Terai H, Vacanti J (2003) A biodegradable nanofiber scaffold by electrospinning and its potential for bone tissue engineering. Biomaterials 24(12):2077–2082
360. Wnek GE, Carr ME, Simpson DG, Bowlin GL (2003) Electrospinning of nanofiber fibrinogen structures. Nano Lett 3(2):213–216

361. Lips P, Velthoen I, Dijkstra PJ, Wessling M, Feijen J (2005) Gas foaming of segmented poly (ester amide) films. Polymer 46(22):9396–9403
362. Barbetta A, Carrino A, Costantini M, Dentini M (2010) Polysaccharide based scaffolds obtained by freezing the external phase of gas-in-liquid foams. Soft Matter 6(20):5213–5224
363. Barbetta A, Rizzitelli G, Bedini R, Pecci R, Dentini M (2010) Porous gelatin hydrogels by gas-in-liquid foam templating. Soft Matter 6(8):1785–1792
364. Barry JJ, Silva MM, Popov VK, Shakesheff KM, Howdle SM (1838) Supercritical carbon dioxide: putting the fizz into biomaterials. Philos Trans R Soc A Math Phys Eng Sci 2005(364):249–261
365. Barbetta A, Barigelli E, Dentini M (2009) Porous alginate hydrogels: synthetic methods for tailoring the porous texture. Biomacromolecules 10(8):2328–2337
366. Keskar V, Marion NW, Mao JJ, Gemeinhart RA (2009) In vitro evaluation of macroporous hydrogels to facilitate stem cell infiltration, growth, and mineralization. Tissue Eng Part A 15(7):1695–1707
367. Barbetta A, Gumiero A, Pecci R, Bedini R, Dentini M (2009) Gas-in-liquid foam templating as a method for the production of highly porous scaffolds. Biomacromolecules 10(12):3188–3192
368. Ju YM, Park K, Son JS, Kim JJ, Rhie JW, Han DK (2008) Beneficial effect of hydrophilized porous polymer scaffolds in tissue-engineered cartilage formation. J Biomed Mater Res B Appl Biomater 85(1):252–260

Scaffolds for Tissue Engineering: A State-of-the-Art Review Concerning Types, Properties, Materials, Processing, and Characterization

Andréa Arruda Martins Shimojo, Isabella Caroline Pereira Rodrigues, Amanda Gomes Marcelino Perez, Eliana Maria Barbosa Souto, Laís Pellizzer Gabriel, and Thomas Webster

Abstract Given the constant lack of donors for organ transplantation, tissue engineering has been considered a very important tool for regenerative medicine to overcome the limitations of conventional treatments. Tissue engineering is mainly based on obtaining biodegradable three-dimensional (3D) scaffolds. Based on a bibliometric study covering the last three decades of scientific research in scaffolds, this review will address the existing types of scaffolds (solid and fluid); the necessary scaffold properties for adequate tissue regeneration, such as biocompatibility and adequate mechanical properties; the materials that can be used to manufacture the scaffold, from metals to natural and synthetic polymers; scaffold fabrication techniques, considering their advantages and disadvantages and which are the main selection criteria; and finally, the methods of scaffold characterization, such as chemical, morphological, mechanical, and biological.

Keywords Scaffold · Tissue engineering · Biomaterial · Biopolymer Biodegradable · Tissue regeneration

A. A. M. Shimojo · I. C. P. Rodrigues · L. P. Gabriel (✉)
School of Applied Sciences, University of Campinas, Limeira, SP, Brazil
e-mail: lgabriel@unicamp.br

A. G. M. Perez
School of Chemical Engineering, University of Campinas, Campinas, SP, Brazil

E. M. B. Souto
Department of Pharmaceutical Technology, Faculty of Pharmacy, University of Coimbra, Coimbra, Portugal

T. Webster
Department of Chemical Engineering, Northeastern University, Boston, MA, USA
e-mail: th.webster@neu.edu

B. Li et al. (eds.), *Racing for the Surface*,
https://doi.org/10.1007/978-3-030-34471-9_23

Introduction

Tissue engineering (TE) is a relatively new research line within the field of regenerative medicine, which has the aim of restoring, keeping, or improving the function of a tissue or group of organs through a specific combination of cells, scaffolds, and bioactive factors, such as growth factors and cytokines [1, 2]. The main goal of TE is to overcome the limitations of conventional treatments based on organ transplants.

Currently, the major obstacle for the clinical transplant of organs is the lack of donors. Based on OPTN (Organ Procurement and Transplantation Network) data, in 2018, only in the United States of America, 36,529 people received organ transplants, while more than 113,000 are still on the waiting list.

Furthermore, the increase of life expectancy and the malfunction and/or loss of tissue caused by injury or diseases have led to the reduction of the quality of life for many patients and an increase in socioeconomic costs associated with improving health around the world. In this context, TE has become a promising and important research field, once it can offer viable and less invasive alternatives for the repair and regeneration of tissues and damaged organs.

TE is based on obtaining three-dimensional (3D) biodegradable scaffolds where specific cells can proliferate and differentiate in a structure similar to tissues or organs.

Scaffolds are temporary 3D matrices that work as an extracellular matrix, organizing cells three-dimensionally and stimulating the growth and formation of the desired tissue. Besides allowing for the adherence and migration of cells inside the scaffold and promoting cell proliferation and differentiation, the scaffold must provide an environment where the cells can keep their phenotype and synthesize proteins and/or other necessary biomolecules. A scaffold must yet promote vascularization and nutrient migration, and possess degradation rates and mechanical properties suitable to support new tissue formation [1]. The scaffolds can also work as carriers of cells, growth factors, and/or other bioactive molecules [3].

In regenerative medicine, scaffolds represent the conductive capacity inside the cell proliferation triangle—which also includes undifferentiated cells and growth factors or other bioactive molecules (Fig. 1)—and can be used to carry cells before

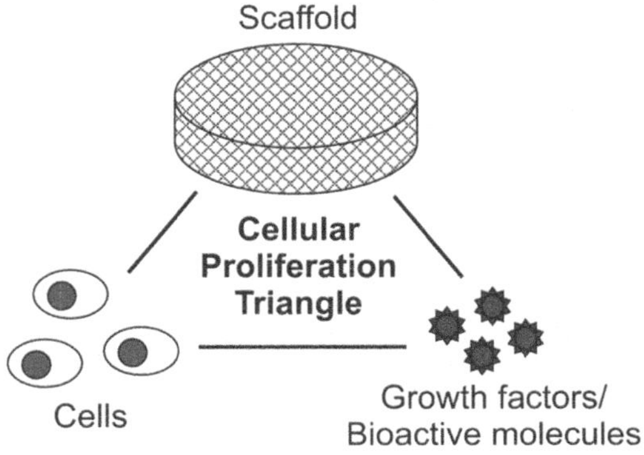

Fig. 1 Cell proliferation triangle

their in vivo implantation, or work only as a bioactive material attracting cells on the tissue where they are implanted [2, 4].

Scaffolds from different materials, manufactured by different technologies, have been used for hard and soft tissues regeneration, such as bones, cartilage, tendons, ligaments, skin, blood vessels, and muscles [5].

Many 3D matrices have been used as scaffolds to promote the proliferation and differentiation of various progenitor cells, including adult mesenchymal stem cells. However, most of these matrices do not provide a suitable biological environment so that cells can proliferate and differentiate in the same way as in the in vivo systems.

Therefore, the development of scaffolds that specifically address cell culture and can mimic the extracellular matrix and be mechanically stable, biocompatible, and biodegradable, still represents a big challenge to TE. In this review, a bibliometric analysis of the last 30 years was done to summarize the current state of the art in TE scaffold design, manufacturing, and use, as well as the advances made to overcome limitations of traditional techniques. Posteriorly, the main topics related to TE and scaffolds are extensively reviewed.

Methods

Database and Search Strategy

The data for this study were collected from the SciVerse Scopus database on May 17, 2019. The search term selected was "scaffold AND technique." The duration of this study was set from 1990 to 2019. The types of documents were limited to "articles" and "reviews." Documents within the subject areas of "arts and humanities," "social sciences," "psychology," "business, management and accounting," "decision sciences," and "economics, econometrics and finance" were excluded.

Bibliometric Mapping

To verify the trends in this research field over time, data downloaded from Scopus were imported into VOSViewer software (Leiden University, Leiden, Netherlands). This software can be used to create networks based on keywords extracted from publications [6]. A minimum of two occurrences was set to filter the keywords and the most relevant ones were extracted by the VOSViewer built-in mining text function [6]. All of the terms extracted and presented by the software were filtered manually to 30 relevant keywords (see Appendix), chosen to represent the scaffolds types, materials, properties, and fabrication techniques, and are reviewed and discussed throughout this work. These scaffold-related terms were then selected to generate the co-occurrence map. This map represents the frequency of occurrence of each keyword in the retrieved documents using the size of the circles and portrays its co-occurrences through colors (clusters). For the keyword analysis, the maps were divided in three parts, each one representing 10 years from the last three decades, to study the historical developments on scaffolds for TE.

Results

The total retrieved publications, using the described search methodology, were 19,934 studies from 1990 to 2019. The growth in scientific studies involving scaffold techniques over the years was verified through the annual publications obtained on Scopus until 2018 (Fig. 2). The results show an increase in publications over the last decade which was most expressive, consisting of 71.6% (13,426 papers) of all publications, although there was a decrease in publications in 2018, probably because of the emphasis on commercialization, considering that, on 2018 only at the United States, there was 49 public companies operating in TE and regenerative medicine sector, undergoing clinical trials or commercial stages [7].

Three network maps were created (Fig. 3), each one representing 10 years of scientific production related to scaffolds over the last three decades: from 1990 to 1999 (Fig. 3a), from 2000 to 2009 (Fig. 3b), and from 2010 to 2019 (Fig. 3c).

Comparing the three scientific landscapes, until 1999, the terms "scaffolds" and "tissue engineering" were new and beginning to show some occurrence and links with different applications (for example, cartilage, bone, and nerve regeneration) and with important scaffold properties (for example, biodegradation, mor-

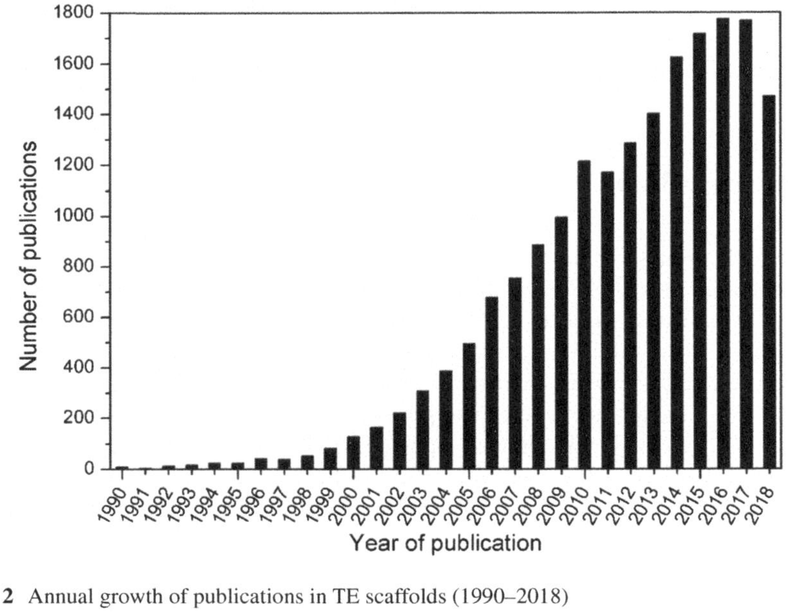

Fig. 2 Annual growth of publications in TE scaffolds (1990–2018)

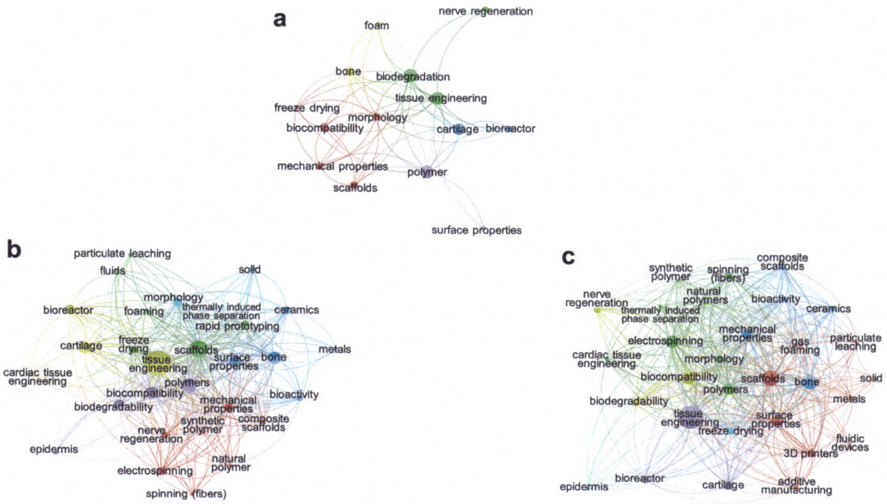

Fig. 3 Network visualization maps of keywords in scaffolds from the last three decades. (**a**) 1990–1999; (**b**) 2000–2009; (**c**) 2010–2019

phology, biocompatibility, mechanical properties, and surface property). However, only polymers, as materials, and a few fabrication techniques, such as freeze-drying, foaming, and bioreactors, significantly appeared during this period (Fig. 3a).

From 2000 to 2009, this scenario changed (Fig. 3b), with an increase of 9000% occurrences of the term "scaffolds" when compared to the previous decade, and with more links with many clusters. In this period, the studies gained force for many previous applications (for example, bone, cartilage, and nerve regeneration) and new applications (for example, cardiac and epidermis). In addition to the properties already studied, bioactivity also began to be explored. Besides that, scaffolds were now being produced with numerous different materials (natural and synthetic polymers, ceramics, metals, and composites) and fabrication techniques (electrospinning, particulate leaching, rapid prototyping, spinning, and thermally induced phase separation). The material with more occurrences remained to be polymers and the most cited techniques in this period were electrospinning, rapid prototyping, and freeze-drying.

Furthermore, on the last decade, the network map (Fig. 3c) shows a higher density of links between all the clusters. All of the applications, materials, techniques, and properties of scaffolds remained the same as the previous decade, but the occurrence of the terms increased considerably. The term "rapid prototyping" begin to change to "additive manufacturing" or "3D printing." Electrospinning became the technique with more occurrences, followed by additive manufacturing (or 3D printing or rapid prototyping). Although polymers still occurred more than other materials, a growth was observed for metals, ceramics, and composites. Moreover, a tendency in the evaluation of the biocompatibility and mechanical properties of scaffolds was observed during this time period.

Discussion and Literature Review

Scaffolds and tissue engineering are relatively new terms to the scientific community and started to appear in the early 1990s. As can be seen from the annual growth of publications graph, only a few works about scaffolds were published to 1999 and then, the research on this field began to grow exponentially (Fig. 2). The same can be verified on the network visualization maps of each decade (Fig. 3), in which the terms related to scaffolds increased as well as the occurrences and the amount of links between them. The terms presented on the network maps will be fully discussed on the following sections, separated into types, materials, properties, fabrication techniques, and characterization of scaffolds.

Types of Scaffolds

Scaffolds are divided into solids and fluids (injectable) and can be manufactured into several shapes (Fig. 4), as sponges, hydrogels, fibers, membranes, micro- and nanoparticles, tubes, and spheres [8–14],which depends on their desired application and the fabrication process used.

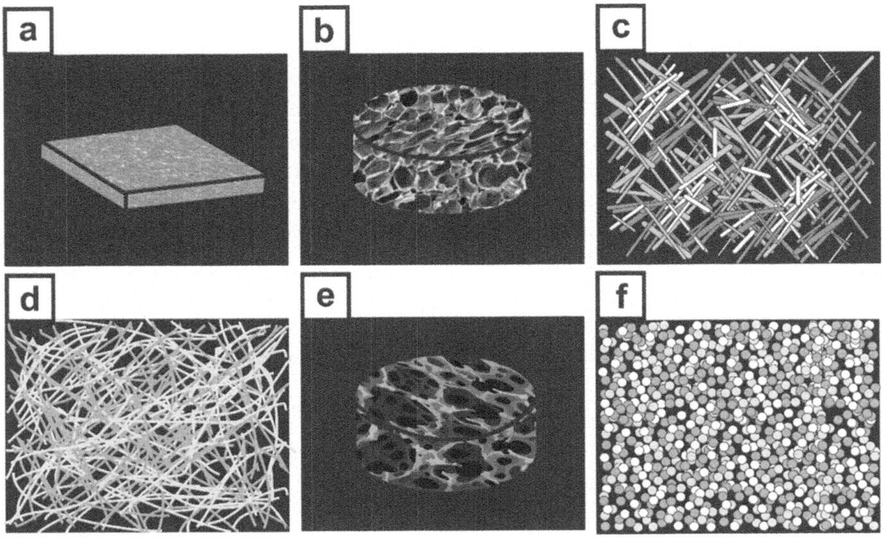

Fig. 4 Illustrative figure of different scaffold types: (**a**) membranes; (**b**) sponge; (**c**) tubes; (**d**) fibers; (**e**) foam; and (**f**) microparticles

Solid Scaffolds

In TE, solid scaffolds include sponges, foams, fibers, membranes, and tubes. These scaffolds present a stable and well-defined 3D porous structure. However, their application for the regeneration of different tissues is limited to morphology, pore dimensions along the structure, and mechanical properties of the scaffolds.

The materials used to fabricate these scaffolds must be able to create structures that will not collapse under the conditions of in vitro cell cultivation (inside an aqueous environment) or when implanted in vivo.

Although these scaffolds can be manufactured with a high control of its architecture, suitable nutrient transportation and effective adherence and cellular migration, the main disadvantage of conventional scaffolds is the need of surgical intervention (or high invasiveness) for their implantation.

Solid scaffolds can be used for various applications, especially those requiring a structural base capable of supporting their in vivo application, such as for the regeneration of bones, muscles, ligaments, and other tissues and organs [15, 16].

Fluid Scaffolds

Fluid scaffolds, in the hydrogel form, have been considered promising in the drug delivery area, as well as in TE, mostly due to their minimally invasive application [17]. Fluid scaffolds are usually flat hydrogels, micro- or nanoparticle hydrogels, or are formed by spheres.

From the clinical point of view, using fluid scaffolds is very interesting, because it minimizes patient discomfort, risk of infection, formation of scars, and treatment cost [16].

The fluid material can homogeneously fill the defect or the point of repair, incorporate many therapeutic agents and does not demand highly invasive surgical procedures for implantation. In addition, the high hydration of the hydrogels mimics the extracellular matrix, consequently being ideal for cell proliferation and differentiation.

Hydrogel provides an initial structural support that retains the cells on the damaged area for cellular growth and the synthesis of a new extracellular matrix, and is easily degradable when cells secrete the extracellular matrix. This strategy allows for cell transplantation and the combination with hydrogels with growth factors in a minimally invasive way.

Usually, the cells are isolated through a small biopsy, expanded in vitro, and encapsulated in the hydrogel precursors, for in situ solidification, or in the hydrogel already formed. Subsequently, these materials are transplanted to the patient by injection, using appropriate needles.

Fluid scaffolds have been widely used, mainly in wound healing, treatment of cartilage lesions, regeneration of soft tissues and in drug delivery [18].

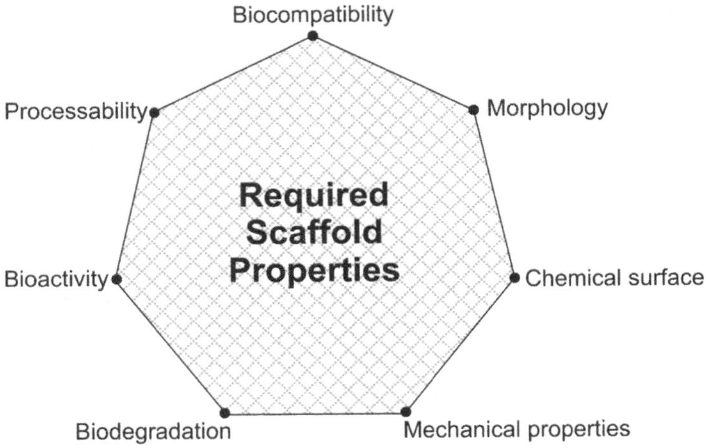

Fig. 5 Required scaffold properties

Required Scaffold Properties

Since the main function of scaffolds is to provide a suitable temporary support that allows for the cellular processes necessary for tissue regeneration, many requirements depending on the cell type and the tissue to be regenerated must be considered in the development of TE scaffolds. Among these requirements, the following can be highlighted (Fig. 5). The detailed description of each property is presented in Table 1.

Materials

Material selection for scaffold production is one of the most important steps in TE. Several different materials have been proposed for their manufacturing, among them metals, ceramic materials, natural and/or synthetic polymers, and composites. One of the main motivations for the study of biodegradable materials is the fact that these materials present some degree of degradation when exposed to physiological environments. This behavior has the advantages of the absence of a new surgical procedure for the removal of implanted materials [22].

Inorganic materials have still been used mostly for the production of scaffolds for bone TE and other mineralized tissues [23]. Currently, the main metallic materials used as scaffolds for implants are stainless steel and cobalt and titanium alloys. Among the ceramic materials, alumina, zirconia, hydroxyapatite (HA), calcium phosphate, and bioglass can be highlighted. However, these are not biodegradable and their processability is very limited. Besides that, their application is invasive, requiring surgical intervention.

Table 1 Detailed description of each scaffold's properties

Scaffold desirable properties	Description
Biocompatibility	Scaffolds must be biocompatible and demonstrate satisfactory performance in order to produce adequate response to the host tissue without producing cytotoxic or immune response. No by-product of its degradation can cause inflammatory or toxic reactions
Biodegradability	Scaffolds must have degradation rates compatible with the new tissue formation. The degradation of the scaffolds may occur by mechanisms involving physical (dissolution) or chemical (hydrolysis) processes and/or biological processes, such as enzymatic cleavage
Pore morphology	Scaffolds should exhibit high porosity with cell–scaffold interactions, in order to control the adequate diffusion of nutrients and oxygen to cells, metabolite dispersal, local pH stability, and cell signaling
Pore size	Pore size is an important feature due to cell penetration and tissue vascularization. The scaffolds must satisfy the condition of providing an empty volume of pores; according to Liu and Ma (2004), scaffolds with high porosity (>90%) allow the effective release of bioactive molecules and are appropriate substrates for nutrient exchange [19]
Chemicalproperties	A scaffold surface can control the effect of cell adhesion and proliferation, due to being the primary site of interaction between the cell and scaffold. The scaffold surface must present properties that allow cellular adhesion and promote proliferation and differentiation. These properties primarily comprise the chemical composition of the material surface with suitable functional groups, which influence its hydrophobicity and charge [20]
Mechanical resistance	Scaffolds must have adequate mechanical properties for manipulation in vitro and in vivo. Scaffolds directed to the regeneration of hard tissues must have a compression modulus around 10–1500 MPa, while scaffolds directed to soft tissues must have a modulus around 0.4–350 MPa [21]
Bioactivity	Scaffolds can be used as carriers or reservoirs for bioactive and/or signaling molecules that can accelerate tissue regeneration
Processability	In order to become clinically and commercially viable, scaffolds must be easily processed in a variety of shapes and sizes and present low fabrication costs. Besides, the fabrication process must be reproducible and scalable, and easily sterilizable and storable

Magnesium and its alloys represent promising solutions in the field of biodegradable metal biomaterials and are being widely investigated for orthopedic applications [24]. From the physiological point of view, magnesium is an essential mineral for human nutrition and crucial for bone health [25]. Furthermore, the mechanical properties of magnesium alloys are similar to those found in human bone (elastic modulus 40 GPa) [22].

Although magnesium alloys have potential for biomedical applications, their processing is extremely challenging. Magnesium has a low boiling point, is combustible in the form of billets or plates and has an increased risk of explosion due to the increased surface area when presented as a powder.

Therefore, over the last few years, these non-degradable materials have been replaced by a variety of natural and/or synthetic polymeric materials manufactured with different microstructures, which mainly include hydrogels, porous matrices, and fibrous matrices [19, 26].

Synthetic polymers have the advantage of presenting a reproducible production process under large scales and higher control of mechanical properties, degradation rate, and microstructure.

Thus, biodegradable synthetic polymers including linear aliphatic polyesters, polyglycolic acid (PGA), polylactic acid (PLA), and polylactic-*co*-glycolic acid (PLGA) copolymers have been widely used as vehicles for cell transplantation and scaffolds for the engineering of different tissues mainly due to its relatively hydrophilic nature [27].

Beyond that, these polyesters have in vitro and in vivo controllable degradation rates and are among the synthetic polymers approved by the Food and Drug Administration (FDA) for certain human clinical applications.

Other linear aliphatic polyesters such as poly(ε-caprolactone) (PCL) have also been investigated in TE mostly for long-term implants because of its significantly slower degradation rate than PLA, PGA, and PLGA [28].

Polyethylene glycol (PEG) has been used as fluid scaffolds in the form of hydrogels. Nevertheless, the toxicity and the low degradability of this material limit its application considerably, requiring prior modification.

Over the past few years, polyurethanes (PU) have also been widely used mainly because of the easiness of controlling its mechanical and morphological properties [29].

Despite the advantages of biodegradable synthetic polymers, natural polymers have been considered attractive materials for scaffold manufacturing, due to their similarities with the extracellular matrix, chemical versatility, biological performance, and specific cellular interactions.

Besides, they are susceptible to the enzymes of the organism, being inherently biodegradable. However, they are often immunogenic, may contain pathological impurities, have laborious manipulation and/or processing and exhibit variability from batch to batch. The most commonly used natural polymers in TE include fibrin, collagen, gelatin, chitosan, alginate, and hyaluronic acid.

Fibrin, one of the major constituents of blood clots, has been used in mixtures with thrombin to produce fluid scaffolds composed of fibrin gels (mesh) [30]. Since it is an autologous product, fibrin is completely biocompatible, thus it has desirable non-immunogenic responses. Furthermore, it is completely biodegradable and can be applied in the injured site using a non-invasive procedure. Fibrin scaffolds can be formed in situ or used as cell carriers associated to scaffolds of other materials. However, inadequate mechanical properties limit its application in TE considerably, especially in hard tissues [31].

Type I collagen extracted from animal tissues and gelatin prepared from collagen denaturation have been widely used as scaffolds for the regeneration of various tissues, especially soft tissues. However, these biomaterials can potentially transmit pathogens and immunological reactions, also showing handling difficulties and inadequate mechanical properties [31].

Silk fibroin has also been used for porous scaffold manufacturing mainly because of its excellent mechanical properties. However, its degradation rate is considerably slow and there is some concern about its cytotoxicity [28].

Properties such as biodegradability, biocompatibility, adhesiveness, foldability in different forms, and chemical modification versatility make chitosan (a cationic derivative from chitin) a promising biomaterial for several applications in TE [32]. In addition to these properties, chitosan can form hydrogels in situ and carry growth factors and adhesion proteins.

Alginate crosslinked with Ca^{2+} has been used in TE as a cell carrier in vivo and as fluid scaffolds, in the form of particulate hydrogels. Nevertheless, using $CaSO_4$ to prepare these hydrogels hinders the control of the gelation process, resulting in non-uniform structures that directly affect the cellular response.

The functions and applications of hyaluronic acid in TE are basically associated to its structural characteristics and possible chemical modifications of the polymer, which determine its rheological, solubility, hydration and specific cell recognition properties. Hyaluronic acid is non-immunogenic, biocompatible, and biodegradable; however, its application during scaffold preparation requires prior modification of the polymer, since the native hyaluronic acid has limited mechanical properties and low residence time in vivo. Considering that it can be obtained in different forms, solid or fluid scaffolds, hyaluronic acid has been successfully used for the regeneration of hard and soft tissues.

Over the last few years, in order to have better control of the biodegradability and mainly to improve the mechanical properties of scaffolds, efforts in TE have been directed to obtain composite scaffolds that can mimic in vivo systems.

Beyond crosslinking, chemical modification, addition of additives and reinforcing agents (fibers and particles), several studies have been combining biocompatible polymers, which have limited mechanical properties, with different inorganic materials. The addition of these materials, especially ceramics, can improve the mechanical properties of the new scaffold, provide essential osteoconductivity for the regeneration of tissues and enable the mineralization of bone tissues.

Biodegradable polymers have also been combined with bioactive molecules to improve biological properties of new materials, accelerate cellular processes involved in tissue regeneration, as well as promote specific cellular recognition. Some studies have also associated platelet-rich plasma with natural and synthetic polymers, aimed to improve the properties of the fibrin network and to enable a controlled release of growth factors and cytokines, which accelerate regeneration and healing of tissues [33–35].

Therefore, the material choice for scaffold production must consider the advantages and disadvantages of these materials, as well as the intended application, and it is still a major challenge in TE.

Fabrication Techniques

Several technologies have been used to fabricate different types of scaffolds (Fig. 6), among those which stand out are: solvent casting/particulate leaching, gas foaming, freeze-drying, thermally induced phase separation, electrospinning, rotary jet spinning, rapid prototyping, and bioreactors [36–41].

Despite advances in TE, scaffold manufacturing processes are still limited. Conventional technologies usually depend on time-consuming, inconsistent, inflexible and laborious manual processes, which use toxic organic solvents, porogenic materials with difficult size control and removal of pores and have format limitations [42].

Therefore, the chosen processing technique must generally comply with the following criteria:

- The process or production shall not affect material properties, such as its biocompatibility or physicochemical properties.
- The technique must allow for a control of porosity, size, distribution, and interconnectivity of pores.
- Different groups of matrices must exhibit minimal variations in their properties, when processed under the same conditions.

The main features that distinguish many selected technologies are the desired application and/or the use of solvents, heat, pressure, or additives responsible for pore generation (Table 2) [3].

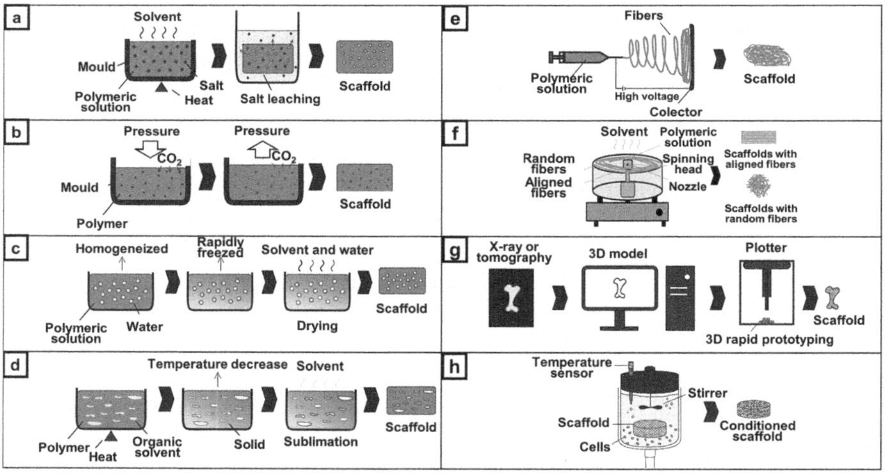

Fig. 6 Schematic illustration of the fabrication process of scaffolds: (**a**) solvent casting/particulate leaching; (**b**) gas foaming; (**c**) emulsion freeze-drying; (**d**) thermally induced phase separation; (**e**) electrospinning; (**f**) rotary spinning; (**g**) 3D printing; and (**h**) bioreactor

Table 2 Parameters of scaffold fabrication processes

Fabrication method	Heat	Solvent	Pressure	Electric field	Productivity	Process control	Porosity	Pore interconnectivity
Solvent casting/particulate leaching	Yes	Yes	No	No	Low	Yes	High	Low
Gas foaming	No	No	Yes	No	Low	Yes	High	Low
Emulsion freeze-drying	No	Yes	Yes	No	Low	Yes	High	Low
Thermally induced phase separation	Yes	Yes	No	No	Low	Yes	High	Low
Electrospinning	No	Yes	No	Yes	Low	Yes	High	High
Rotary jet spinning	No	Yes	No	No	High	Yes	High	High
Additive manufacturing	No	No	No	No	High	Yes	High	High

Solvent Casting/Particulate Leaching

This method involves mixing particles of a water-soluble salt (sodium chloride, sodium citrate) in the solution of a biodegradable polymer. The mixture is placed on a mold with the desired shape and the solvent is then removed by evaporation or freeze-drying (Fig. 6a). To obtain a porous structure, the salt particles are leached. This method, besides simple, allows for adequate control of the pore size and porosity, which can be obtained by the salt/polymer ratio and the particle size of the salt added. However, the geometric shape of the pore is limited to the shape of the cubic crystals of the salt and the removal of soluble particles from the interior of the polymeric matrix becomes difficult for thick scaffolds, limiting their thickness between 0.5 and 2 mm. In addition, the limited interconnectivity of the pores prevents uniform cell inoculation and tissue growth.

In order to overcome these issues, the group of Ma and Choi (2001) has developed scaffolds of biodegradable polymers with spherical pores and controlled interconnectivity, using paraffin beads to generate the pores. The main advantage of this method is that it can provide a porous network completely interconnected. Besides, paraffin is insoluble in water and some water-soluble polymers can be used in the scaffold production by this technique [43].

Gas Foaming

Gas foaming can be used to fabricate highly porous polymeric foams without using organic solvents. In this technique, a gas such as carbon dioxide (CO_2) is applied using high pressure, enabling the formation of a single polymer/gas phase. Subsequently, the pressure is reduced to create a thermodynamic instability of the dissolved CO_2, resulting in nucleation and pore growth and enabling for foam formation (Fig. 6b). The advantages of this method are the absence of organic solvents and the possibility of manufacturing polymeric foams with high porosity [44]. Besides that, processes that do not include heating allow for the incorporation of temperature-sensitive biomolecules. The disadvantage of this method is the production of structures with closed pores without sufficient interconnected pores [19].

Freeze-Drying

This method consists of creating an emulsion, by homogenizing a solution of polymer (in an organic solvent) and water, rapidly cooling the emulsion to trap the liquid inside the structure and removing the solvent and water by freeze-drying (Fig. 6c). Scaffolds with high porosity and large pore sizes can be manufactured by this method [45, 46]. Nevertheless, the pores obtained do not show high interconnectivity. Besides, porosity and pore size are affected by parameters such as freezing temperature and cooling rate [47]. This technique has been widely applied to fabricate scaffolds mainly used in soft TE.

Thermally Induced Phase Separation

In this technique, the polymer is primarily dissolved in an organic solvent at high temperature, and the phase separation (liquid–liquid or solid–liquid) is induced by a temperature decrease of the solution. Subsequently, the removal of the solidified solvent is accomplished by sublimation, generating a porous polymeric scaffold (Fig. 6d). The pore morphology of the polymer depends on the solvent, the concentration of the polymer solution, and phase separation temperature.

An advantage of this method is that the scaffolds obtained generally have good mechanical properties. However, this method normally results in scaffolds with pore sizes between 20 and 500 μm, but mainly smaller than 100 μm, which are not ideal for the regeneration of many tissues [48].

Electrospinning

Electrospinning is a process of fibrous scaffold preparation that employs an electric field to control the formation and deposition of polymeric fibers on a given substrate (Fig. 6e). The geometry of these fibers is mainly influenced by parameters such as viscosity, electrical conductivity, and surface tension of the polymeric solution.

This technique can fabricate scaffolds with fiber diameters ranging from micrometers to several hundred nanometers [49, 50]. A wide variety of polymeric blends has been electrospun for the formation of scaffolds with high surface area, high porosity, and low density, used mostly for fibrous TE [51]. However, the productivity of this process is low and high electrical fields are necessary for fiber formation.

Rotary Jet Spinning

The rotary jet spinning technique consists on inducing the formation of fibers from a polymeric solution in an organic solvent, through the action of a centrifugal force, which ejects the solution, generally through nozzles at the head spinning, also evaporating the solvent (Fig. 6f).

This technique presents many advantages such as high efficiency and productivity, good process control, and fabrication of highly aligned and porous scaffolds [52, 53]. However, it is necessary to evaluate if the organic solvent was totally evaporated throughout the process, since it can cause scaffold toxicity.

Additive Manufacturing

Additive manufacturing (AM) is a technology based on the advances of computer science that has emerged for the production of custom models, such as layer by layer (Fig. 6g). Specifically, in TE, this technique combines knowledge of computed tomography, magnetic resonance imaging, and computational models (CAD) to construct 3D scaffolds.

The main advantages of this technology are the possibility of manufacturing scaffolds with customized geometries, the fabrication of scaffolds with anisotropic structures, and total control of the manufacturing process by computers.

The most commonly used AM methods in TE are the powder-based technologies of selective laser sintering (SLS) and electron beam melting (EBM). From these technologies, custom models of hard body parts can be produced with precise control of morphology. However, the porous structures produced by these techniques affect mechanical properties of the scaffolds, and reduce the scaffold integrity [54].

Both processes fuse powder in a specific geometry of the model to be printed, SLS fuses or sintered the powder through a carbon dioxide laser, and EBM melts the powder with an electron laser beam.

Li et al. (2018) and Salmoria et al. (2018) produced scaffolds by an SLS process from a commercial magnesium alloy and a composite of poly(L-*co*-D,L) lactic acid (PLDLA) and bioglass, respectively, and the reported results showed biocompatibility and mechanical properties appropriate to bone repair [55, 56]. Yan et al. (2018) produced titanium mandibular scaffolds by EBM and reported that the mandibular defect was completely recovered after 2 months of in vivo implantation using 12 animal models [57].

Another promising AM technique that has emerged recently is 3D cell printing (or 3D bioprinting), which enables the fabrication of cell-embedded scaffolds using a one-step fabrication process in order to mimic complex structures. The challenge of this technology is to develop appropriate bioinks containing living cells in conjunction with microfluid systems capable of supporting cells and to present properties adequate to printability [58].

Choi et al. (2016) developed a bioink using decellularized skeletal muscle and applied the cell-printed technology to produce functional muscle embedded with myoblast cells mimicking the structure and function of skeletal muscle [59]. Ahn et al. (2017) noted that cell viability and printability are closely related [60].

This AM technology still presents many challenges, including the development of bioinks, long fabrication time, and limited thickness of the scaffolds.

Bioreactors

Scaffold conditioning using bioreactors has become a new and interesting approach for TE applications (Fig. 6h). Bioreactors are systems with adjustable parameters capable of stimulating biotransformation or cell expansion using whole cells or its components [61]. These systems can be used for three main applications on TE: cell expansion in vitro, cell viability maintenance during cultivation on scaffolds and validation of the scaffold function and cell differentiation [62].

Although TE seeks to create 3D scaffolds capable to regenerate or even replace tissues and organs, most research is limited to thin layered structures because of the poor diffusion of nutrients and oxygen through thicker scaffolds to the cells on static cultures. Besides that, cells are not able to proliferate and uniformly distribute on the scaffolds, when cultured in vitro on static setups [63, 64].

Santoro et al. (2015) studied the influence of perfusion flow on tumor cells, comparing dynamic culture using bioreactors with static cultures. The application of a bioreactor with perfusion flow improved cell distribution and proliferation on electrospun PCL scaffolds [65].

Scaffolds must withstand the same conditions of the tissue where it will be applied. Moreover, the cells seeded on the scaffold must be able to differentiate into the desired cells. Therefore, bioreactors must simulate in vivo conditions for scaffolds, controlling mechanical, electrical and physicochemical parameters, such as temperature, pH, flow, oxygen, nutrient, and shear stress [66].

Lee et al. (2008) produced composite scaffolds based on PCL and type I collagen and evaluated its response at conditions of high pressure and flow, similar to physiological vascular conditions. The scaffolds presented good stability and biomechanical properties and were able to support cell adhesion and the proliferation of endothelial and smooth muscle cells [67]. Shepherd et al. (2018) produced scaffolds of type I collagen using freeze-drying and evaluated the effect of combining a flow bioreactor with megakaryocytes to the production of platelets. The system was capable of retaining the cells and effectively releasing platelets [68].

In addition to the culture and conditioning applications of bioreactors, they have been widely used in the new scaffold production technique of decellularization and recellularization. Decellularization consists of using chemical, enzymatic, and/or physical agents to remove cellular components of tissues and organs, leaving just the structure of a biological scaffold. Through this process, the decellularized extracellular matrix (dECM) has its structural integrity preserved and presents similar properties of the native tissues or organs, without being immunogenic [69]. After the decellularization, the material obtained (dECM) can be used for whole organ recellularization [70–72] or to create scaffolds using other fabrication techniques [73–76], such as AM [77].

Nichols et al. (2018) produced porcine decellularized lung scaffolds and used a bioreactor for its recellularization with different cells able to promote lung regeneration. The bioengineered (recellularized) lungs were then transplanted to pigs and did not indicate any rejection [78].

Jang et al. (2017) used 3D cell printing of heart tissue-derived decellularized extracellular matrix to create scaffolds for cardiac repair. A pre-vascularized patch was developed showing therapeutic efficiency [79].

Applications

Scaffolds have been studied for numerous applications on TE, considering all types of tissues, such as bone, cartilage, and skin, among others. In the last few years, different types of scaffolds, materials, and fabrication techniques have been combined, seeking an ideal scaffold for clinical applications. Some of these recent combinations for various scaffold applications are presented in Table 3.

Table 3 Application of scaffolds in different tissues

Tissue application	Material	Scaffold type	Fabrication technique	Reference
Bone	Hydroxyapatite	Microparticles	3D printing	[80]
	PDLLA-*co*-TMC	Fibrous	Electrospinning	[81]
	Chitosan/alginate	Solid	Freeze-drying	[82]
	PU/hydroxyapatite	Fibrous	Electrospinning	[83]
	Chitosan/gelatin/alginate/hydroxyapatite	Beads	Foaming	[84]
	Strontium hydroxyapatite/chitosan	Membrane	Freeze-drying	[85]
	PLGA	Membrane	Solvent casting/particulate leaching	[86]
	Bone (ECM)	Solid	Decellularization	[87]
Cartilage	Keratin	Fibrous	Electrospinning	[88]
	Cartilage (ECM)	Solid	Decellularization	[89]
	Cellulose/alginate	Solid	3D printing	[90]
	Chitosan/PVA/$CaCO_3$	Fibrous	Electrospinning	[91]
	PBLF	Microspheres	Emulsion	[92]
	PU/hydroxyapatite and PU/PEO	Solid	3D printing	[93]
	Cellulose	Fibrous	Freeze-drying	[94]
	PEG/heparin	Fibrous	3D printing/electrospinning	[95]
	PLGA/dECM	Solid	3D printing	[96]
	Hyaluronic acid	Hydrogel	–	[97]
Cardiac	Cardiac tissue (ECM)	Solid	Decellularization	[98]
	PANI-PGS	Membrane	Solvent casting/particulate leaching	[99]
	Gelatin/hyaluronic acid	Solid	3D printing	[100]
	PU-siloxane	Membrane	Thermally induced phase separation	[101]
	PLCL	Fibrous	Electrospinning	[102]
	PLA/chitosan	Fibrous	Electrospinning	[103]
	Collagen-alginate	Membrane	Freeze-drying	[104]
	Protein/polysaccharide	Sponge	Freeze-drying	[105]
	PGS/PBS-DLA	Membrane	Solvent casting/particulate leaching	[106]
Neural	PU	Solid	3D printing	[30]
	Nerve (ECM)	Solid	Decellularization	[107]
	PCL/gelatin	Fibrous	Electrospinning	[108]
	Gelatin methacrylate/PEGDA	Solid	3D printing	[109]

(continued)

Table 3 (continued)

Tissue application	Material	Scaffold type	Fabrication technique	Reference
Epidermal	Chitosan–agarose	Hydrogel	–	[110]
	PCL/gelatin/collagen type I	Fibrous	Electrospinning	[111]
	Silk fibroin-keratin	Membrane	Freeze-drying	[112]
	Fibrin	Membrane	3D printing	[113]
	Skin (ECM)	Solid	Decellularization	[114]
Other tissues	WSC/galactose and WSC/collagen	Hydrogel	–	[115]
	Silk fibroin	Solid	Freeze-drying	[116]
	PLGA/ECM	Membrane	Freeze-drying	[117]
	PPY/PDLLA	Membrane	Emulsion	[118]

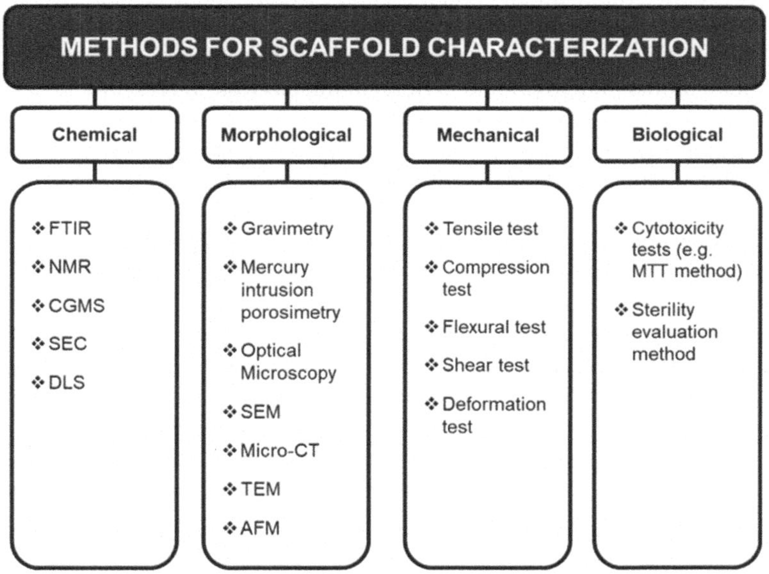

Fig. 7 Methods for scaffold characterization

Methods for Scaffold Characterization

Several analytical methods have been used to characterize the physical-chemical, mechanical, and biological properties of the scaffolds in their different formats (Fig. 7). These methods are usually based on international standards (American Society for Testing and Materials—ASTM) and include known analytical techniques [119].

Scaffolds are primarily characterized by their chemical properties, which include the chemical composition, impurities contained, and chemical nature of the surface groups.

These parameters can be identified and determined quantitatively by techniques such as Fourier Transform Infrared (FTIR) or Nuclear Magnetic Resonance (NMR ^{1}H, ^{13}C or ^{31}P) combined or not with Gas Chromatography (CG) coupled with Mass Spectrometry (MS) or other appropriate analytical methods.

Furthermore, the size of the molecules, or their molar mass, is usually evaluated. The techniques used in this case include size exclusion chromatography (SEC) and dynamic light scattering (DLS). In all cases, complete solubilization of the material in an appropriate solvent is required. SEC is indicated for the determination of the molar mass of linear polymers, while DLS can be used for both linear and branched polymers. These methods are usually comparative; thus, the results must include the solvent used, the temperature in which the measurements were taken, the standard used as reference, and the concentration of the solutions analyzed.

In order to determine the scaffold morphology (porosity, pore size, interconnectivity, tortuosity, roughness, and topography), a variety of equipment and software have been used. The most common methods include gravimetry, mercury intrusion porosimetry, optical microscopy (OM), scanning electron microscopy (SEM), and micro-computed tomography (micro-CT) [120].

The gravimetric method determines, in a fast and simple way, the total porosity of the material; however, the measurement accuracy is limited.

Mercury intrusion porosimetry determines total volume, average diameter, and the size distribution of pores, but has the disadvantages of high toxicity, high cost of mercury, and the possibility of scaffold collapse with the high pressures required by this method.

Optical microscopy is mainly employed in preliminary observations of scaffolds. Although it has a number of advantages including simple preparation of samples and low cost of analysis, the OM resolution is limited mainly to sizes around 200 nm, preventing a detailed characterization of the structures.

Through the association of micrographs obtained by SEM and computer software, it is possible to determine the average pore diameter and porosity, in addition to obtain an estimated interconnectivity and pore wall thickness. However, to ensure accurate measurements, samples must be carefully sectioned to avoid changes in the porous structure. In addition, the sample sensitive to the high vacuum required by this technique must be properly fixed to prevent its collapse.

Micro-CT accurately provides all information about the 3D morphology of the scaffolds and has the advantage of being non-destructive and not requiring pretreatment of the sample with toxic chemical compounds. However, this technique still presents a high cost and is not suitable for scaffolds containing metals.

Besides these methods, the average nanoparticle diameter and surface (topography and roughness) of the scaffolds can be observed by transmission electron microscopy (TEM) and atomic force microscopy (AFM), respectively.

The mechanical properties evaluated in the scaffolds usually involve tests related to stress and strain or that show the response of these scaffolds to the application of a physical force.

The scaffolds must be evaluated under conditions that mimic the intended application. Besides, a special assembly of the evaluated specimens may be required depending on the size and format of the scaffold and the equipment used.

Mechanical tests include compression, tensile, flexural, shear, and deformation performed on servo-hydraulic equipment and can be performed on dry or swollen scaffolds to mimic conditions in vivo. The parameters generally measured are the modulus of elasticity (Young's modulus) and the shear modulus that indicates the scaffold stiffness.

The swelling and degradation profiles of scaffolds are also important parameters and are usually determined by a gravimetric method in medium, which mimics conditions in vivo.

The swelling profile of the scaffolds is generally obtained at 37 °C in phosphate buffered saline (PBS) at pH between 7.2 and 7.4 or in Dulbecco's Modified Eagle's Medium (DMEM). Degradation tests may also be carried out under these conditions or in the presence of suitable enzymes.

From the biological point of view, scaffolds are characterized primarily by the in vitro biocompatibility of the materials used in their fabrication.

This evaluation is performed through cytotoxicity tests, especially by cell viability, which consists of placing the scaffold directly or indirectly in contact with a culture of animal cells and verifying the cellular changes that resulted by different mechanisms, including the incorporation of vital dyes or the inhibition of the formation of cellular colonies. The most used cell viability methods are the neutral red incorporation method and the MTT method (3-(4,5-dimethylthiazol-2-yl)-2,5-diphenyl tetrazolium bromide).

Furthermore, the sterility of scaffolds is also evaluated by counting the total bacteria and fungi (molds and yeasts), using the pour-plate method, surface-spread method, and serial dilution method.

Conclusions and Future Perspectives

During the last few years, important advances have been made in regenerative medicine, especially in the TE field. According to the bibliometric analysis of this work, the last two decades presented increasing scientific production and new techniques related to scaffolds. However, existing therapies still have several limitations. In fact, no combination between cells, scaffolds, and bioactive molecules have fulfilled all the necessary criteria to mimic the conditions in vivo and effectively promote the regeneration of different tissues.

A specific combination between cell type, culture regime, and scaffold must be carefully selected, since it has been demonstrated that the physicochemical characteristics of the scaffolds directly affect the cellular behavior and, consequently, the process of tissue regeneration. In addition, the incorporation of bioactive molecules in this system has been shown to contribute to accelerate these processes.

Therefore, an ideal combination of these parameters for the effective regeneration of different tissues is still a challenge, and researchers have been increasingly directing their work to biodegradable and biocompatible materials, undifferentiated cells, and autologous bioactive molecules.

The studies developed so far indicate that future advances in TE depend on new systems that can modulate cellular behavior and result in functional and effective tissues.

Many challenges are still limited to the multidisciplinarity of this area and the complexity of the biological systems involved in the regeneration of different tissues. Its success depends on combined efforts of researchers to understand, modulate, and optimize the results of basic sciences and clinical applications.

Acknowledgements This work was supported by FAPESP, grant #2017/13273-6, São Paulo Research Foundation and Coordenação de Aperfeiçoamento de Pessoal de Nível Superior—Brasil (CAPES)—Finance Code 001.

Appendix

Keywords	No. of occurrences (1990–1999)	No. of occurrences (2000–2009)	No. of occurrences (2010–2019)
3D printers	–	–	369
Additive manufacturing (or rapid prototyping)	–	107	112
Bioactivity	–	65	207
Biocompatibility	11	429	1562
Biodegradability (or biodegradation)	36	244	287
Bioreactors	6	225	428
Bone	13	427	1444
Cardiac tissue engineering	–	13	67
Cartilage	27	350	633
Ceramics	–	128	181
Composite scaffolds	–	32	183
Electrospinning	–	161	1221
Epidermis	2	27	47
Fluidic devices (or fluids)	–	11	34
Freeze-drying	2	106	284
Gas foaming (or foaming or foam)	3	20	34
Mechanical properties	8	177	765
Metals	–	18	81
Morphology	9	180	360

Keywords	No. of occurrences (1990–1999)	No. of occurrences (2000–2009)	No. of occurrences (2010–2019)
Natural polymers	–	17	56
Nerve regeneration	6	81	251
Particulate leaching	–	10	31
Polymer	33	601	941
Scaffolds	10	944	2174
Solid	–	54	20
Spinning (fibers)	–	26	231
Surface property	4	364	769
Synthetic polymers	–	13	38
Thermally induced phase separation	–	9	54
Tissue engineering	33	2564	5748

References

1. Langer R, Vacanti JP (1993) Tissue engineering. Science 260(5110):920–926. ISSN 0036–8075. Disponível em: < <Go to ISI>://WOS:A1993LB79100031 >
2. Crane D, Everts P (2008) Platelet rich plasma (PRP) matrix grafts. Pract Pain Manage 8(1):11–26
3. Agrawal CM, Ray RB (2001) Biodegradable polymeric scaffolds for musculoskeletal tissue engineering. J Biomed Mater Res 55(2):141–150. ISSN 0021–9304. Disponível em: < <Go to ISI>://WOS:000167221200001 >
4. Barnett JR, Pomeroy GC (2007) Use of platelet-rich plasma and bone marrow-derived mesenchymal stem cells in foot and ankle surgery. Tech Foot Ankle Surg 6(2):89–94. ISSN 1536-0644
5. Cross LM et al (2016) Nanoengineered biomaterials for repair and regeneration of orthopedic tissue interfaces. Acta Biomater 42:2–17. ISSN 1742–7061. Disponível em: < <Go to ISI>://WOS:000383292700001 >
6. van Eck NJ, Waltman L (2009) VOSviewer: a computer program for bibliometric mapping. In: 12th international conference of the international-society-for-scientometrics-and-informetrics, Rio de Janeiro, Brazil, 14–17 July 2009, pp 886–897
7. Kim YS et al (2019) An overview of the tissue engineering market in the United States from 2011 to 2018. Tissue Eng Part A 25(1–2):1–8. ISSN 1937–3341. Disponível em: < <Go to ISI>://WOS:000463809500001 >
8. Poursamar SA et al (2016) The effects of crosslinkers on physical, mechanical, and cytotoxic properties of gelatin sponge prepared via in-situ gas foaming method as a tissue engineering scaffold. Mater Sci Eng C 63:1–9. ISSN 0928–4931. Disponível em: < <Go to ISI>://WOS:000374916800001 >
9. Naahidi S et al (2017) Biocompatibility of hydrogel-based scaffolds for tissue engineering applications. Biotechnol Adv 35(5):530–544. ISSN 0734–9750. Disponível em: < <Go to ISI>://WOS:000405767000002 >
10. Jakobsson A et al (2017) Three-dimensional functional human neuronal networks in uncompressed low-density electrospun fiber scaffolds. NanomedNanotechnol Biol Med 13(4):1563–1573. ISSN 1549–9634. Disponível em: < <Go to ISI>://WOS:000402678800022 >

11. Lee WD et al (2017) Sol gel-derived hydroxyapatite films over porous calcium polyphosphate substrates for improved tissue engineering of osteochondral-like constructs. Acta Biomater 62:352–361. ISSN 1742–7061. Disponível em: < <Go to ISI>://WOS:000413175200028 >
12. Neufurth M et al (2017) 3D printing of hybrid biomaterials for bone tissue engineering: calcium-polyphosphate microparticles encapsulated by polycaprolactone. Acta Biomater 64:377–388. ISSN 1742–7061. Disponível em: < <Go to ISI>://WOS:000416498200033 >
13. Singh N et al (2016) Chitin and carbon nanotube composites as biocompatible scaffolds for neuron growth. Nanoscale 8(15):8288–8299. ISSN 2040–3364. Disponível em: < <Go to ISI>://WOS:000374159600057 >
14. Lachman N et al (2017) Synthesis of polymer bead nano-necklaces on aligned carbon nanotube scaffolds. Nanotechnology 28(24):6. ISSN 0957–4484. Disponível em: < <Go to ISI>://WOS:000402514600001 >
15. Dutta RC, Dutta AK (2009) Cell-interactive 3D-scaffold; advances and applications. Biotechnol Adv 27(4):334–339. ISSN 0734–9750. Disponível em: < <Go to ISI>://WOS:000267478100003 >
16. Zhao X et al (2017) Antibacterial anti-oxidant electroactive injectable hydrogel as self-healing wound dressing with hemostasis and adhesiveness for cutaneous wound healing. Biomaterials 122:34–47. ISSN 0142–9612. Disponível em: < <Go to ISI>://WOS:000394472500004 >
17. Xing RT et al (2016) An injectable self-assembling collagen-gold hybrid hydrogel for combinatorial antitumor photothermal/photodynamic therapy. Adv Mater 28(19):3669–3676. ISSN 0935–9648. Disponível em: < <Go to ISI>://WOS:000376480500005 >
18. Dhandayuthapani B et al (2011) Polymeric scaffolds in tissue engineering application: a review. Int J Polym Sci 2011:19. ISSN 1687–9422. Disponível em: < <Go to ISI>://WOS:000307633400011 >
19. Liu XH, Ma PX (2004) Polymeric scaffolds for bone tissue engineering. Ann Biomed Eng 32(3):477–486. ISSN 0090–6964. Disponível em: < <Go to ISI>://WOS:000222465100019 >
20. Chang HI, Wang Y (2011) Cell responses to surface and architecture of tissue engineering scaffolds. In: Regenerative medicine and tissue engineering-cells and biomaterials. InTech
21. Hollister SJ (2005) Porous scaffold design for tissue engineering. Nat Mater 4(7):518–524. ISSN 1476–1122. Disponível em: < <Go to ISI>://WOS:000230190900013 >
22. Staiger MP et al (2006) Magnesium and its alloys as orthopedic biomaterials: a review. Biomaterials 27(9):1728–1734. ISSN 0142–9612. Disponível em: < <Go to ISI>://WOS:000234962500007 >
23. Ran JB et al (2017) Constructing multi-component organic/inorganic composite bacterial cellulose-gelatin/hydroxyapatite double-network scaffold platform for stem cell-mediated bone tissue engineering. Mater Sci Eng C 78:130–140. ISSN 0928–4931. Disponível em: < <Go to ISI>://WOS:000404944700016 >
24. Witte F (2010) The history of biodegradable magnesium implants: a review. Acta Biomat 6(5):1680–1692. ISSN 1742–7061. Disponível em: < <Go to ISI>://WOS:000277847500002 >
25. Palacios C (2006) The role of nutrients in bone health, from A to Z. Crit Rev Food Sci Nutr 46(8):621–628. ISSN 1040–8398. Disponível em: < <Go to ISI>://WOS:000241365800002 >
26. Geesala R et al (2016) Porous polymer scaffold for on-site delivery of stem cells—protects from oxidative stress and potentiates wound tissue repair. Biomaterials 77:1–13. ISSN 0142–9612. Disponível em: < <Go to ISI>://WOS:000367118200001 >
27. Oh SH, Lee JH (2013) Hydrophilization of synthetic biodegradable polymer scaffolds for improved cell/tissue compatibility. Biomed Mater 8(1):16. ISSN 1748–6041. Disponível em: < <Go to ISI>://WOS:000314115100002 >
28. Yang SF et al (2001) The design of scaffolds for use in tissue engineering. Part 1. Traditional factors. Tissue Eng 7(6):679–689. ISSN 1076–3279. Disponível em: < <Go to ISI>://WOS:000172903100001 >
29. Yu JH et al (2017) Fabrication and characterization of shape memory polyurethane porous scaffold for bone tissue engineering. J Biomed Mater Res A 105(4):1132–1137. ISSN 1549–3296. Disponível em: < <Go to ISI>://WOS:000395008300018 >

30. Hsieh FY, LIN HH, Hsu SH (2015) 3D bioprinting of neural stern cell-laden thermo-responsive biodegradable polyurethane hydrogel and potential in central nervous system repair. Biomaterials 71:48–57. ISSN 0142–9612. Disponível em: < <Go to ISI>://WOS:000362612800005 >
31. Malafaya PB, Silva GA, Reis RL (2007) Natural-origin polymers as carriers and scaffolds for biomolecules and cell delivery in tissue engineering applications. Adv Drug Deliv Rev 59(4–5):207–233. ISSN 0169-409X. Disponível em: < <Go to ISI>://WOS:000247714800003 >
32. Trinca RB et al (2017) Electrospun multilayer chitosan scaffolds as potential wound dressings for skin lesions. Eur Polym J 88:161–170. ISSN 0014–3057. Disponível em: < <Go to ISI>://WOS:000396952500014 >
33. Shimojo AAM et al (2015) Performance of PRP associated with porous chitosan as a composite scaffold for regenerative medicine. Sci World J 2015:396131. ISSN 2356–6140
34. Shimojo AAM et al (2016) In vitro performance of injectable chitosan-tripolyphosphate scaffolds combined with platelet-rich plasma. Tissue Eng Regen Med 13(1):21–30. ISSN 1738-2696
35. Shimojo AAM et al (2016) Stabilization of porous chitosan improves the performance of its association with platelet-rich plasma as a composite scaffold. Mater Sci Eng C Mater Biol Appl 60:538–546
36. Moghadam MZ et al (2017) Formation of porous HPCL/LPCL/HA scaffolds with supercritical CO2 gas foaming method. J Mech Behav Biomed Mater 69:115–127. ISSN 1751–6161. Disponível em: < <Go to ISI>://WOS:000400199600012 >
37. Deng Y et al (2017) A novel akermanite/poly (lactic-co-glycolic acid) porous composite scaffold fabricated via a solvent casting-particulate leaching method improved by solvent self-proliferating process. Regenerat Biomater 4(4):233–242. ISSN 2056–3418. Disponível em: < <Go to ISI>://WOS:000409116500004 >
38. Repanas A, Andriopoulou S, Glasmacher B (2016) The significance of electrospinning as a method to create fibrous scaffolds for biomedical engineering and drug delivery applications. J Drug Deliv Sci Tech 31:137–146. ISSN 1773–2247. Disponível em: < <Go to ISI>://WOS:000370905200015 >
39. Liu W et al (2017) Low-temperature deposition manufacturing: a novel and promising rapid prototyping technology for the fabrication of tissue-engineered scaffold. Mater Sci Eng C Mater Biol Appl 70:976–982. ISSN 0928–4931. Disponível em: < <Go to ISI>://WOS:000387625700007 >
40. Guo J et al (2017) Novel porous poly(propylene fumarate-co-caprolactone) scaffolds fabricated by thermally induced phase separation. J Biomed Mater Res A 105(1):226–235. ISSN 1549–3296. Disponível em: < <Go to ISI>://WOS:000389145400024 >
41. Janik H, Marzec M (2015) A review: fabrication of porous polyurethane scaffolds. Mater Sci Eng CMater Biol Appl 48:586–591. ISSN 0928–4931. Disponível em: < <Go to ISI>://WOS:000348749200073 >
42. Leong KF, Cheah CM, Chua CK (2003) Solid freeform fabrication of three-dimensional scaffolds for engineering replacement tissues and organs. Biomaterials 24(13):2363–2378. ISSN 0142–9612. Disponível em: < <Go to ISI>://WOS:000182280400027 >
43. Ma PX, Choi JW (2001) Biodegradable polymer scaffolds with well-defined interconnected spherical pore network. Tissue Eng 7(1):23–33. ISSN 2152–4947. Disponível em: < <Go to ISI>://WOS:000167235200003 >
44. Poursamar SA et al (2015) Gelatin porous scaffolds fabricated using a modified gas foaming technique: characterisation and cytotoxicity assessment. Mater Sci Eng C Mater Biol Appl 48:63–70. ISSN 0928–4931. Disponível em: < <Go to ISI>://WOS:000348749200009 >
45. Zhang J et al (2015) Pore architecture and cell viability on freeze dried 3D recombinant human collagen-peptide (RHC)-chitosan scaffolds. Mater Sci Eng C Mater Biol Appl 49:174–182. ISSN 0928–4931. Disponível em: < <Go to ISI>://WOS:000350514100018 >
46. Vishwanath V, Pramanik K, Biswas A (2016) Optimization and evaluation of silk fibroin-chitosan freeze-dried porous scaffolds for cartilage tissue engineering application. J

Biomater Sci Polym Ed 27(7):657–674. ISSN 0920–5063. Disponível em: < <Go to ISI>://WOS:000373015000008 >
47. Perez-Puyana V et al (2019) Influence of the processing variables on the microstructure and properties of gelatin-based scaffolds by freeze-drying. J Appl Polym Sci 136(25):8. ISSN 0021–8995. Disponível em: < <Go to ISI>://WOS:000462061700025 >
48. Conoscenti G et al (2017) PLLA scaffolds produced by thermally induced phase separation (TIPS) allow human chondrocyte growth and extracellular matrix formation dependent on pore size. Mater Sci Eng C Mater Biol Appl 80:449–459. ISSN 0928–4931. Disponível em: < <Go to ISI>://WOS:000410254400053 >
49. Bazbouz MB et al (2019) Dry-jet wet electrospinning of native cellulose microfibers with macroporous structures from ionic liquids. J Appl Poly Sci 136(10):15. ISSN 0021–8995. Disponível em: < <Go to ISI>://WOS:000453915300014 >
50. Shamsabadi AS et al (2019) Electrospinning of gold nanoparticles incorporated PAN nanofibers via in-situ laser ablation of gold in electrospinning solution. Mater Res Exp 6(5):12. ISSN 2053–1591. Disponível em: < <Go to ISI>://WOS:000459994100003 >
51. Wang WY et al (2018) Electrospinning preparation of a large surface area, hierarchically porous, and interconnected carbon nanofibrous network using polysulfone as a sacrificial polymer for high performance supercapacitors. RSC Adv 8(50):28480–28486. ISSN 2046–2069. Disponível em: < <Go to ISI>://WOS:000442616800023 >
52. Hou T et al (2017) Highly porous fibers prepared by centrifugal spinning. Mater Des 114:303–311. ISSN 0264–1275. Disponível em: < <Go to ISI>://WOS:000390650800038 >
53. Rogalski JJ, Bastiaansen CWM, Peijs T (2017) Rotary jet spinning review—a potential high yield future for polymer nanofibers. Nanocomposites 3(4):97–121. ISSN 2055–0324. Disponível em: < <Go to ISI>://WOS:000424579900001 >
54. Gleadall A et al (2018) Review of additive manufactured tissue engineering scaffolds: relationship between geometry and performance. Burns Trauma 6:16. ISSN 2321–3868. Disponível em: < <Go to ISI>://WOS:000437332900001 >
55. Li Y et al (2018) Additively manufactured biodegradable porous magnesium. Acta Biomater 67:378–392. ISSN 1742–7061. Disponível em: < <Go to ISI>://WOS:000424853600033 >
56. Salmoria GV et al (2018) Properties of PLDLA/bioglass scaffolds produced by selective laser sintering. Polym Bull 75(3):1299–1309. ISSN 0170–0839. Disponível em: < <Go to ISI>://WOS:000425107700025 >
57. Yan RZ et al (2018) Electron beam melting in the fabrication of three-dimensional mesh titanium mandibular prosthesis scaffold. Sci Rep 8:10. ISSN 2045–2322. Disponível em: < <Go to ISI>://WOS:000422637200009 >
58. Jang J et al (2018) Biomaterials-based 3D cell printing for next-generation therapeutics and diagnostics. Biomaterials 156:88–106. ISSN 0142–9612. Disponível em: < <Go to ISI>://WOS:000419539100008 >
59. Choi YJ et al (2016) 3D cell printing of functional skeletal muscle constructs using skeletal muscle-derived bioink. Adv Healthc Mater 5(20):2636–2645. ISSN 2192–2640. Disponível em: < <Go to ISI>://WOS:000387158900005 >
60. Ahn G et al (2017) Precise stacking of decellularized extracellular matrix based 3D cell-laden constructs by a 3D cell printing system equipped with heating modules. Sci Rep 7:11. ISSN 2045–2322. Disponível em: < <Go to ISI>://WOS:000407864400005 >
61. Eibl D, Eibl R (2009) Bioreactors for mammalian cells: general overview. In: Cell and tissue reaction engineering: with a contribution by Martin Fussenegger and Wilfried Weber. Springer, Berlin, p 55–82. ISBN 978-3-540-68182-3
62. Gelinsky M, Bernhardt A, Milan F (2015) Bioreactors in tissue engineering: advances in stem cell culture and three-dimensional tissue constructs. Eng Life Sci 15(7):670–677. ISSN 1618–0240. Disponível em: < <Go to ISI>://WOS:000363416600002 >
63. Ravichandran A, Liu YC, Teoh SH (2018) Review: bioreactor design towards generation of relevant engineered tissues: focus on clinical translation. J Tissue Eng Regen Med 12(1):E7–E22. ISSN 1932–6254. Disponível em: < <Go to ISI>://WOS:000423431200002 >

64. Vunjak-Novakovic G et al (1998) Dynamic cell seeding of polymer scaffolds for cartilage tissue engineering. Biotechnol Prog 14(2):193–202. ISSN 8756–7938. Disponível em: < <Go to ISI>://WOS:000073011600003 >
65. Santoro M et al (2015) Flow perfusion effects on three-dimensional culture and drug sensitivity of Ewing sarcoma. Proc Natl Acad Sci U S A 112(33):10304–10309. ISSN 0027–8424. Disponível em: < <Go to ISI>://WOS:000359738300057 >
66. Ahmed S et al (2019) New generation of bioreactors that advance extracellular matrix modelling and tissue engineering. Biotechnol Lett 41(1):1–25. ISSN 0141–5492. Disponível em: < <Go to ISI>://WOS:000454783700001 >
67. Lee SJ et al (2008) Development of a composite vascular scaffolding system that withstands physiological vascular conditions. Biomaterials 29(19):2891–2898. ISSN 0142–9612. Disponível em: < <Go to ISI>://WOS:000256144900008 >
68. Shepherd JH et al (2018) Structurally graduated collagen scaffolds applied to the ex vivo generation of platelets from human pluripotent stem cell-derived megakaryocytes: enhancing production and purity. Biomaterials 182:135–144. ISSN 0142–9612. Disponível em: < <Go to ISI>://WOS:000444928200013 >
69. Taylor DA et al (2018) Decellularized matrices in regenerative medicine. Acta Biomater 74:74–89. ISSN 1742–7061. Disponível em: < <Go to ISI>://WOS:000437998200005 >
70. Sabetkish S et al (2015) Whole-organ tissue engineering: decellularization and recellularization of three-dimensional matrix liver scaffolds. J Biomed Mater Res A 103(4):1498–1508. ISSN 1549–3296. Disponível em: < <Go to ISI>://WOS:000350395300020 >
71. Taylor DA et al (2018) Building a total bioartificial heart: harnessing nature to overcome the current hurdles. Artif Organs 42(10):970–982. ISSN 0160-564X. Disponível em: < <Go to ISI>://WOS:000449690800009 >
72. Petersen TH et al (2010) Tissue-engineered lungs for in vivo implantation. Science 329(5991):538–541. ISSN 0036–8075. Disponível em: < <Go to ISI>://WOS:000280483500028 >
73. Vo TN et al (2015) In vitro and in vivo evaluation of self-mineralization and biocompatibility of injectable, dual-gelling hydrogels for bone tissue engineering. J Control Release 205:25–34. ISSN 0168–3659. Disponível em: < <Go to ISI>://WOS:000352966200005 >
74. Gong WH et al (2016) Hybrid small-diameter vascular grafts: anti-expansion effect of electrospun poly epsilon-caprolactone on heparin-coated decellularized matrices. Biomaterials 76:359–370. ISSN 0142–9612. Disponível em: < <Go to ISI>://WOS:000366961100030 >
75. Cunniffe GM et al (2015) Porous decellularized tissue engineered hypertrophic cartilage as a scaffold for large bone defect healing. Acta Biomater 23:82–90. ISSN 1742–7061. Disponível em: < <Go to ISI>://WOS:000359964000009 >
76. Yang Q et al (2017) Silk fibroin/cartilage extracellular matrix scaffolds with sequential delivery of TGF-beta 3 for chondrogenic differentiation of adipose-derived stem cells. Int J Nanomed 12:6721–6733. ISSN 1178–2013. Disponível em: < <Go to ISI>://WOS:000410234600001 >
77. Kabirian F, Mozafari M (2019) Decellularized ECM-derived bioinks: prospects for the future. Methods. ISSN 1046-2023
78. Nichols JE et al (2018) Production and transplantation of bioengineered lung into a large-animal model. Sci Transl Med 10(452):12. ISSN 1946–6234. Disponível em: < <Go to ISI>://WOS:000440494900002 >
79. Jang J et al (2017) 3D printed complex tissue construct using stem cell-laden decellularized extracellular matrix bioinks for cardiac repair. Biomaterials 112:264–274. ISSN 0142–9612. Disponível em: < <Go to ISI>://WOS:000389166700023 >
80. Cox SC et al (2015) 3D printing of porous hydroxyapatite scaffolds intended for use in bone tissue engineering applications. Mater Sci Eng C Mater Biol Appl 47:237–247. ISSN 0928–4931. Disponível em: < <Go to ISI>://WOS:000347581600029 >
81. Bao M et al (2014) Electrospun biomimetic fibrous scaffold from shape memory polymer of PDLLA-co-TMC for bone tissue engineering. ACS Appl Mater Interfaces 6(4):2611–2621. ISSN 1944–8244. Disponível em: < <Go to ISI>://WOS:000332144600055 >

82. Venkatesan J, Bhatnagar I, Kim SK (2014) Chitosan-alginate biocomposite containing fucoidan for bone tissue engineering. Mar Drugs 12(1):300–316. ISSN 1660–3397. Disponível em: < <Go to ISI>://WOS:000336087500018 >
83. Tetteh G et al (2014) Electrospun polyurethane/hydroxyapatite bioactive Scaffolds for bone tissue engineering: the role of solvent and hydroxyapatite particles. J Mech Behav Biomed Mater 39:95–110. ISSN 1751–6161. Disponível em: < <Go to ISI>://WOS:000343338800010 >
84. Sharma C et al (2016) Fabrication and characterization of novel nano-biocomposite scaffold of chitosan-gelatin-alginate-hydroxyapatite for bone tissue engineering. Mater Sci Eng C Mater Biol Appl 64:416–427. ISSN 0928–4931. Disponível em: < <Go to ISI>://WOS:000376547700051 >
85. Lei Y et al (2017) Strontium hydroxyapatite/chitosan nanohybrid scaffolds with enhanced osteoinductivity for bone tissue engineering. Mater Sci Eng C Mater Biol Appl 72:134–142. ISSN 0928–4931. Disponível em: < <Go to ISI>://WOS:000392165600017 >
86. Rashkow JT, Lalwani G, Sitharaman B (2018) In vitro bioactivity of one- and two-dimensional nanoparticle-incorporated bone tissue engineering scaffolds. Tissue Eng Part A 24(7–8):641–652. ISSN 1937–3341. Disponível em: < <Go to ISI>://WOS:000429016300011 >
87. Ling Y et al (2018) In vivo immunogenicity of bovine bone removed by a novel decellularization protocol based on supercritical carbon dioxide. Artif Cells Nanomed Biotechnol 46:334–344. ISSN 2169–1401. Disponível em: < <Go to ISI>://WOS:000459181400033 >
88. Xu HL et al (2014) Water-stable three-dimensional ultrafine fibrous scaffolds from keratin for cartilage tissue engineering. Langmuir 30(28):8461–8470. ISSN 0743–7463. Disponível em: < <Go to ISI>://WOS:000339463000027 >
89. Rahman S et al (2018) Optimising the decellularization of human elastic cartilage with trypsin for future use in ear reconstruction. Sci Rep 8:11. ISSN 2045–2322. Disponível em: < <Go to ISI>://WOS:000425190500001 >
90. Markstedt K et al (2015) 3D bioprinting human chondrocytes with nanocellulose-alginate bioink for cartilage tissue engineering applications. Biomacromolecules 16(5):1489–1496. ISSN 1525–7797. Disponível em: < <Go to ISI>://WOS:000354503700005 >
91. Sambudi NS et al (2015) Electrospun chitosan/poly(vinyl alcohol) reinforced with CaCO3 nanoparticles with enhanced mechanical properties and biocompatibility for cartilage tissue engineering. Compos Sci Tech 106:76–84. ISSN 0266–3538. Disponível em: < <Go to ISI>://WOS:000347868200007 >
92. Fang JJ et al (2015) Novel injectable porous poly(gamma-benzyl-L-glutamate) microspheres for cartilage tissue engineering: preparation and evaluation. J Mater Chem B 3(6):1020–1031. ISSN 2050-750X. Disponível em: < <Go to ISI>://WOS:000349146700010 >
93. Hung KC et al (2016) Water-based polyurethane 3D printed scaffolds with controlled release function for customized cartilage tissue engineering. Biomaterials 83:156–168. ISSN 0142–9612. Disponível em: < <Go to ISI>://WOS:000371651700013 >
94. Naseri N et al (2016) 3-Dimensional porous nanocomposite scaffolds based on cellulose nanofibers for cartilage tissue engineering: tailoring of porosity and mechanical performance. RSC Adv 6(8):5999–6007. ISSN 2046–2069. Disponível em: < <Go to ISI>://WOS:000368858000002 >
95. Bas O et al (2017) Biofabricated soft network composites for cartilage tissue engineering. Biofabrication 9(2):15. ISSN 1758–5082. Disponível em: < <Go to ISI>://WOS:000401338900001 >
96. Xu YY et al (2018) Construction of bionic tissue engineering cartilage scaffold based on three-dimensional printing and oriented frozen technology. J Biomed Mater Res A 106(6):1664–1676. ISSN 1549–3296. Disponível em: < <Go to ISI>://WOS:000431004500020 >
97. Han SS et al (2018) In situ cross-linkable hyaluronic acid hydrogels using copper free click chemistry for cartilage tissue engineering. Polym Chem 9(1):20–27. ISSN 1759–9954. Disponível em: < <Go to ISI>://WOS:000418370400003 >
98. Silva AC et al (2019) Comparable decellularization of fetal and adult cardiac tissue explants as 3D-like platforms for in vitro studies. J Vis Exp (145):8. ISSN 1940-087X. Disponível em: < <Go to ISI>://WOS:000462909500001 >

99. Qazi TH et al (2014) Development and characterization of novel electrically conductive PANI-PGS composites for cardiac tissue engineering applications. Acta Biomater 10(6):2434–2445. ISSN 1742–7061. Disponível em: < <Go to ISI>://WOS:000336345900008 >
100. Gaetani R et al (2015) Epicardial application of cardiac progenitor cells in a 3D-printed gelatin/hyaluronic acid patch preserves cardiac function after myocardial infarction. Biomaterials 61:339–348. ISSN 0142–9612. Disponível em: < <Go to ISI>://WOS:000357229900032 >
101. Baheiraei N et al (2016) Electroactive polyurethane/siloxane derived from castor oil as a versatile cardiac patch, part I: synthesis, characterization, and myoblast proliferation and differentiation. J Biomed Mater Res A 104(3):775–787. ISSN 1549–3296. Disponível em: < <Go to ISI>://WOS:000369160800022 >
102. Lakshmanan R et al (2016) Engineering a growth factor embedded nanofiber matrix niche to promote vascularization for functional cardiac regeneration. Biomaterials 97:176–195. ISSN 0142–9612. Disponível em: < <Go to ISI>://WOS:000377735800015 >
103. Liu YW, Wang SY, Zhang R (2017) Composite poly(lactic acid)/chitosan nanofibrous scaffolds for cardiac tissue engineering. Int J Biol Macromol 103:1130–1137. ISSN 0141–8130. Disponível em: < <Go to ISI>://WOS:000408286400128 >
104. O'neill HS et al (2018) A collagen cardiac patch incorporating alginate microparticles permits the controlled release of hepatocyte growth factor and insulin-like growth factor-1 to enhance cardiac stem cell migration and proliferation. J Tissue Eng Regenerat Med 12(1):E384–E394. ISSN 1932–6254. Disponível em: < <Go to ISI>://WOS:000423431200036 >
105. Rosellini E et al (2018) Protein/polysaccharide-based scaffolds mimicking native extracellular matrix for cardiac tissue engineering applications. J Biomed Mater Res A 106(3):769–781. ISSN 1549–3296. Disponível em: < <Go to ISI>://WOS:000423354200015 >
106. Merle B et al (2018) Dynamic mechanical characterization of poly(glycerol sebacate)/poly (butylene succinate-butylene dilinoleate) blends for cardiac tissue engineering by flat punch nanoindentation. Mater Lett 221:115–118. ISSN 0167-577X. Disponível em: < <Go to ISI>://WOS:000430446700031 >
107. Ediriwickrema LS et al (2017) Decellularization of porcine and primate optic nerve lamina towards cell culture with neural progenitor cells. Invest Ophthalmol Vis Sci 58(8):2. ISSN 0146–0404. Disponível em: < <Go to ISI>://WOS:000432176300317 >
108. Vatankhah E et al (2014) Artificial neural network for modeling the elastic modulus of electrospun polycaprolactone/gelatin scaffolds. Acta Biomater 10(2):709–721. ISSN 1742–7061. Disponível em: < <Go to ISI>://WOS:000330921700015 >
109. Zhu W et al (2017) 3D printing scaffold coupled with low level light therapy for neural tissue regeneration. Biofabrication 9(2):10. ISSN 1758–5082. Disponível em: < <Go to ISI>://WOS:000399408400002 >
110. Miguel SP et al (2014) Thermoresponsive chitosan-agarose hydrogel for skin regeneration. Carbohydr Polym 111:366–373. ISSN 0144–8617. Disponível em: < <Go to ISI>://WOS:000340302100042 >
111. Gautam S et al (2014) Surface modification of nanofibrous polycaprolactone/gelatin composite scaffold by collagen type I grafting for skin tissue engineering. Mater Sci Eng C Mater Biol Appl 34:402–409. ISSN 0928–4931. Disponível em: < <Go to ISI>://WOS:000330489500050 >
112. Bhardwaj N et al (2015) Silk fibroin-keratin based 3D scaffolds as a dermal substitute for skin tissue engineering. Integr Biol 7(1):53–63. ISSN 1757–9694. Disponível em: < <Go to ISI>://WOS:000347724900005 >
113. Cubo N et al (2017) 3D bioprinting of functional human skin: production and in vivo analysis. Biofabrication 9(1):12. ISSN 1758–5082. Disponível em: < <Go to ISI>://WOS:000390344900004 >
114. Farrokhi A et al (2018) Evaluation of detergent-free and detergent-based methods for decellularization of murine skin. Tissue Eng Part A 24(11–12):955–967. ISSN 1937–3341. Disponível em: < <Go to ISI>://WOS:000430057100001 >

115. Du C et al (2014) Induced pluripotent stem cell-derived hepatocytes and endothelial cells in multi-component hydrogel fibers for liver tissue engineering. Biomaterials 35(23):6006–6014. ISSN 0142–9612. Disponível em: < <Go to ISI>://WOS:000337212200003 >
116. Zhu MF et al (2014) Fabrication of highly interconnected porous silk fibroin scaffolds for potential use as vascular grafts. Acta Biomater 10(5):2014–2023. ISSN 1742–7061. Disponível em: < <Go to ISI>://WOS:000335095300023 >
117. Lih E et al (2016) Biomimetic porous PLGA scaffolds incorporating decellularized extracellular matrix for kidney tissue regeneration. ACS Appl Mater Interf 8(33):21145–21154. ISSN 1944–8244. Disponível em: < <Go to ISI>://WOS:000382179400004 >
118. Xu HX et al (2014) Conductive PPY/PDLLA conduit for peripheral nerve regeneration. Biomaterials 35(1):225–235. ISSN 0142–9612. Disponível em: < <Go to ISI>://WOS:000328006100022 >
119. Simon CG et al (2015) ASTM international workshop on standards and measurements for tissue engineering scaffolds. J Biomed Mater Res B Appl Biomater 103(5):949–959. ISSN 1552–4973. Disponível em: < <Go to ISI>://WOS:000356671800001 >
120. Loh QL, Choong C (2013) Three-dimensional scaffolds for tissue engineering applications: role of porosity and pore size. Tissue Eng B Rev 19(6):485–502. ISSN 1937–3368. Disponível em: < <Go to ISI>://WOS:000326962100003 >

Recent Developments of Zn-based Medical Implants

Qichan Hu, Yingchao Su, and Donghui Zhu

Abstract After decades of developing strategies to employ biodegradable metals in medical devices, there is an increasing interest to use zinc (Zn) and Zn-based alloys as novel and promising alternatives to magnesium and iron. Over the last decade, extensive research has been done on Zn regarding its mechanical properties, degradation behavior, and biocompatibility. This chapter summarizes the recent progress in improving the properties of pure Zn as well as Zn alloys to make them appropriate for medical applications, especially for orthopedic implantation.

Keywords Zinc · Biodegradable metal · Degradation · Biocompatibility Orthopedic implant

Introduction

It is well known that great progress has been made in the field of biomaterials and their clinical applications. The evolution of biomedical implants is firmly related to the development of biomaterials. Today biomedical implants are mainly used in the fields of nervous system [1, 2], cardiovascular system [3], skin system [4], cosmetic implants [5], and skeletal and dental systems [6, 7]. Currently, various bioimplants are employed in the form of ceramics, glasses, polymers, composites, glass-ceramics, and metal alloys. Metals are stiff and have superior mechanical strength compared with polymers and ceramic materials. They have been widely used as orthopedic implants, cardiovascular interventional devices, and tissue engineering

Q. Hu
Department of Biomedical Engineering, University of North Texas, Denton, TX, USA

Y. Su (✉) · D. Zhu (✉)
Department of Biomedical Engineering, Stony Brook University, Stony Brook, NY, USA
e-mail: Yingchao.Su@stonybrook.edu; Donghui.Zhu@stonybrook.edu

B. Li et al. (eds.), *Racing for the Surface*,
https://doi.org/10.1007/978-3-030-34471-9_24

scaffolds [8, 9]. Permanent implants for patients are commonly made of traditional metallic biomaterials like titanium alloys, stainless steel, cobalt-chromium alloys, and tantalum because of their characteristics of corrosion resistance [8–10]. However, the permanent residence of the implant in the body can lead to chronic deleterious effects and the function of the implant is not needed after the treatment site is fully restored. For example, a traditional coronary stent must remain inert in the human body for many years, but sometimes its effectiveness ends after serious side effects including chronic inflammation [11], late-stage thrombosis [12], and stent-strut disruption (fracture) [13]. In this case, it is necessary to perform a second surgery to pull out the implant, resulting in additional injury and expense. To reduce the long-term side effects involved with traditional metallic biomaterials, the next generation of so-called "biodegradable" metals is in the process of development [3, 14]. It is expected that biodegradable metals could corrode gradually and harmlessly in the body, maintain mechanical integrity during the critical tissue healing stages, and then dissolve completely upon completion of their task [15]. For instance, special stitches may be absorbed after wound healing and coronary stents may degrade completely after fulfilling their tasks as vascular scaffolds [16].

The development of biodegradable metal implants has attracted much attention over the last two decades. Researches have mainly concentrated on iron (Fe) [17–19] and magnesium (Mg) [14, 20] and their alloys, which have been widely studied as potential biodegradable metals for medical applications [21–23]. Nevertheless, the extensive experience of these material systems has shown critical limitations in their applicability to clinical applications [10, 24–27]. For example, although Fe and its alloys display superior mechanical properties, the degradation rates make them hard to meet clinical needs because of their corrosion products, which seem to be excreted or metabolized at an unsatisfactory rate, accumulate and repel adjacent cells and biological matrices, and do not allow cells to integrate around and within the original footprint of the degrading implant [28]. Mg-based biodegradable implants are attractive primarily because of their excellent biocompatibility, but pure Mg and its alloys display low mechanical strength in addition to a fast corrosion rate, followed by the production of hydrogen gas, increased pH values, and loss of mechanical integrity [29, 30].

Zn and Zn-based alloys have been suggested as new additions to the list of biodegradable metals and as optimistic substitutes to magnesium and iron in medical applications [31–33]. The chemical activity of Zn is exhibited by an electrode potential (−0.762 V) which is between that of magnesium (−2.372 V) and iron (−0.444 V) [34–36]. Zn also displays moderate corrosion rate that falls between the slowly degrading Fe and the rapidly degrading Mg [8]. In addition, Zn is easier to cast and process because of their relatively low melting points (Tm = 420 °C) (Mg Tm = 650 °C, Fe Tm = 1538 °C) and good machinability [35, 37, 38]. Thus, Zn is a promising alternative candidate among the new generation of biodegradable metals, and much effort has been made to develop metallic Zn as biodegradable implants since the early attention given to Mg and Fe.

Biological Significance

Zn has been found to be a completely intracellular element with 40% in the nucleus, 50% in the cytoplasm, organelles and specialized vesicles, and the rest in the cell membrane [39]. As far as the position in the whole body is concerned, 85% of Zn is present in muscles and bones, 11% is present in the skin and liver, and the rest distributes throughout the other tissues [40]. It has been determined that the biological half-life of Zn is between 162 and 500 days [41], and the suggested dietary allowance for Zn is 10–15 mg/day [41], which is well below the median lethal dose (LD50) of 27 g/day. Dietary Zn is absorbed through the small intestine in the form of Zn ions and amino acid complexes and it is regulated by metallothionein, which binds or unbinds the free Zn ions at picomolar to nanomolar levels because they cannot go through the cell membrane by means of active diffusion due to their hydrophilic property [42]. At an early stage, Zn is transported to organs like pancreas, liver, kidneys, and spleen to play its role [43]. However, in the long term, 90% of the absorbed Zn is deposited in the muscular and skeletal system [44].

As an essential trace element in the human body, Zn plays an indispensable role in human health [45]. Once Zn is transported into the cytoplasm via channels within the cell wall, it plays many different roles as shown in Fig. 1 [46]. Zn has a place in

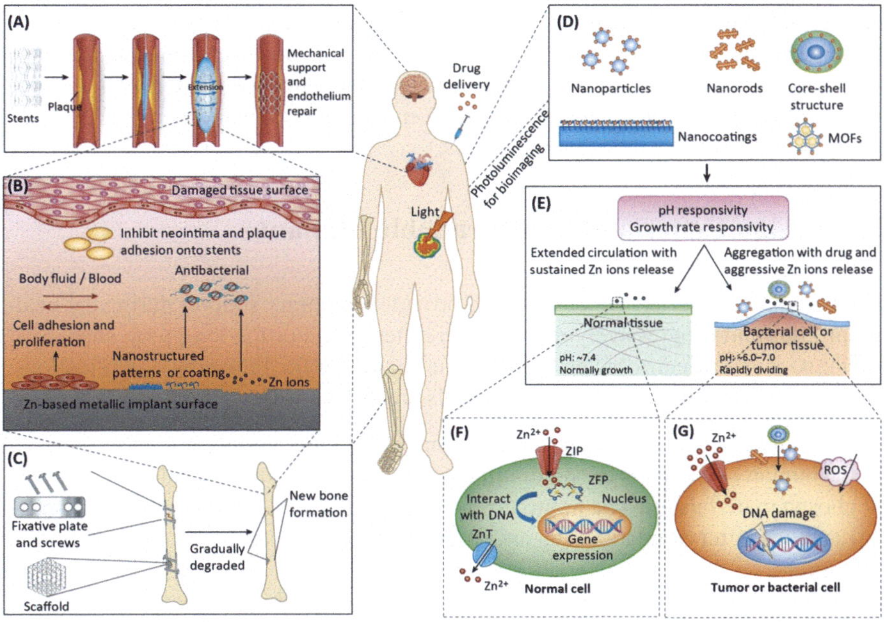

Fig. 1 Biomedical applications of biomaterials in human body, including (**A–C**) biodegradable metallic Zn for stent and orthopedic applications; (**D**) Zn-based ceramic nanomaterials and (**E–G**) their applications for antibacterial and cancer treatments [46]

the function of more than 300 enzymes and hence it is a necessary element for the catalysis and co-catalysis of these enzymes involved with wound healing, brain development, and membrane stability [47]. Zn is critical for the structural integrity of many proteins, especially metallic proteins and membrane proteins [39]. It is shown that Zn has a direct effect on different cellular signal transduction, closely associated with gene expression, RNA transcription, DNA replication and repair [48]. Zn has been found to act indirectly as an antioxidant within the cells [39, 49] and exhibit antiatherogenic effects because of its membrane stabilizing abilities [50]. Zn also plays a crucial role in bone formation, and mineralization, and can be found in the bone extracellular matrix, where it is co-deposited with calcium hydroxyapatite [51]. In addition, Zn-based nanostructured biomaterials can be used for drug delivery and bioimaging and have sensitive responses to pH which is related to the growth of bacteria, and normal or tumor tissues [52, 53].

Since Zn acts as a crucial trace element in the human body, its deficiency can give rise to many problems like retarded growth, impaired parturition (dystocia), neuropathy, decreased food intake, diarrhea, dermatitis, hair loss, bleeding tendencies, hypotension, and hypothermia [54]. The reason for Zn deficiency is usually insufficient dietary intake, or caused by malabsorption and chronic illnesses, such as diabetes, malignancy, liver disease, and sickle cell disease [42, 55]. However, excessive amounts of Zn^{2+} may be harmful to vital organs like the kidney, liver, spleen, brain, and heart [26]. For instance, the result of Zn overdose is copper deficiency, hypocupremia, anemia, leucopenia, neutropenia, and impairment of the Cu–Zn–superoxide dismutase antioxidant enzyme [31, 56]. However, some studies have shown that high concentrations of Zn could prevent osteoporosis through the promotion of osteoblastogenesis and suppression of osteoclastogenesis [57, 58].

The Design Criteria of Biodegradable Implants

Biodegradable materials are designed to provide temporary support during the healing process and progressively degrade thereafter [59]. For the long-term success of implants, it is critical to select appropriate materials for biomedical applications. The requirements for accepted biomaterials are as follows:

1. Biocompatibility. It is related to the behavior of biomaterials in various parts of the human body and is the predominant factor for implant material selection. The chosen materials must be nontoxic and have no induced inflammatory or allergic reactions after implantation, and perfect integration into the tissues around implants.
2. Degradation. Biodegradable metals must degrade in the complex physiological environment of the human body with matching degradation kinetics to the healing period. The degradation products should be transported and eliminated from the body without causing local or systematic accumulation.

3. Mechanical properties. Biodegradable metals should provide adequate mechanical support during the healing process throughout the implantation period. It is still difficult to clarify the exact requirements for specific clinical events such as narrowed artery or fractured bone, but it is reasonable to design depending on the intended applications.

Undoubtedly, the specific design and selection criteria of biodegradable materials depend on the intended applications. Future works need to focus on the development of more improved properties in Zn-based alloys by different methods of processing, coating, and alloying with elements that are functional in the human body, such as Ca, Mg, Fe, Zr, Sn, and Sr.

Biocompatibility

The success of Zn-based biomaterials will depend to a large extent on whether they cause a biocompatible response of blood and tissue-specific cells in the vicinity of the implant. So far, it has not been clarified how the biological reactions proceed between the corrosion products of Zn and Zn-based implants and the surrounding cells and tissues along with numerous proteins. Although the mechanism is still not clear, some studies have been conducted on several cell lines. For instance, the results of direct contact culture with human aortic endothelial cells (HAEC), human dermal fibroblasts (HDF), and human aortic smooth muscle cells (AoSMC) [60] revealed great tolerance for Zn^{2+} (LD50 265 mM), but the viabilities of HDF (LD50 50 mM) and AoSMC (LD50 70 mM) decreased obviously with an increasing concentration of Zn^{2+} ions. However, the three human vascular cell types could adhere and proliferate on the Zn surface coated with collagen-based gelatin ahead of direct to culture on the Zn substrate. Ma et al. [25] evaluated the short-term cellular behavior of primary human coronary artery endothelial cells (HCECs) exposed to a concentration gradient (0–140 μM) of extracellular Zn^{2+}. As a result, low concentrations of Zn^{2+} improved cell viability, while high Zn^{2+} concentrations displayed a deleterious effect on HCECs behavior. It was concluded from the above results that the controlled release of Zn^{2+} is one of the most significant considerations for Zn-based alloy design when used as biodegradable implants.

Li et al. [54] revealed that the addition of other elements like magnesium, calcium, and strontium to Zn is helpful for their hemocompatibility and cytocompatibility. Liu et al. [61] found that the hemolysis rate of the rolled Zn–1Mg– 0.1Mn alloy was extremely low, indicating good biocompatibility according to ASTM-F756-00. Besides, there is no evidence of thrombogenicity in alloys with acceptable blood compatibility. Drelich et al. [62] reported good biocompatibility of Zn by implanting Zn wires into the murine artery. It was found that inflammatory reactions diminished gradually between 10 and 20 months, and no local toxicity was observed. Nevertheless, these tests are too preliminary to validate biological behavior, many more in vitro and in vivo experiments are needed to be carried out.

Corrosion Properties

As a kind of chemically reactive metal, pure Zn dissolves in aqueous solutions via the electrochemical reactions as follows [63, 64]:

$$Zn \rightarrow Zn^{2+} + 2e^{-} - 0.7618\,V \tag{1}$$

$$O + 2HO + 4e^{-} \rightarrow 4OH^{-} + 0.4010\,V \tag{2}$$

Among them, (1) and (2) represent the anode and cathode reactions, respectively. Some of the major products of Zn cations dissolved in aqueous solution are produced through the following reactions:

$$Zn^{2+} + 2OH^{-} \rightarrow ZnO + H_2O \tag{3}$$

$$Zn^{2+} + 4OH^{-} \rightarrow ZnO_2^{-} + H_2O \tag{4}$$

$$Zn^{2+} + 2OH^{-} \rightarrow Zn(OH)_2 \tag{5}$$

$$Zn^{2+} + 4OH^{-} \rightarrow Zn(OH)_4^{2-} \tag{6}$$

Based on the above reactions, there is no gas production during the formation of $Zn(OH)_2$ and ZnO. It is expected that hydrogen release is negligible during Zn corrosion in contrast to the case of Mg. The decrease of the hydrogen reduction rate is commonly due to the presence of $Zn(OH)_4^{2-}$ ions in aqueous solutions [65].

It has been presented in several recent studies that the degradation mechanism relies on minor changes in the electrolyte pH, temperature, composition, and different reaction schemes. Thomas et al. [66] studied the pH-related degradation of Zn that can correlate corrosion kinetics and thermodynamics in chloride solutions. At pH $\approx$ 7.3, several species participate in the corrosion reaction, including Zn^{2+}, $ZnCl^{+}$, and $ZnOH^{+}$. It was concluded that the acidification from Zn anodic dissolution at pH $\approx$ 7–10 may interrupt passive surface layers previously formed. In another study, Thomas et al. [67] investigated Zn in both active and passive states by measuring the dissolved O_2 content adjacent to the metallic surface. It is indicated that the corrosion rate strongly depends on the dissolved O_2 concentration. Yang et al. [68] explained the degradation mechanism of Zn stents implanted into the abdominal aorta of rabbits for 12 months. They found that relative uniform corrosion was dominated by dynamic blood before endothelialization, but the later degradation was dependent on the diffusion of water molecules, hydrophilic solutes and ions

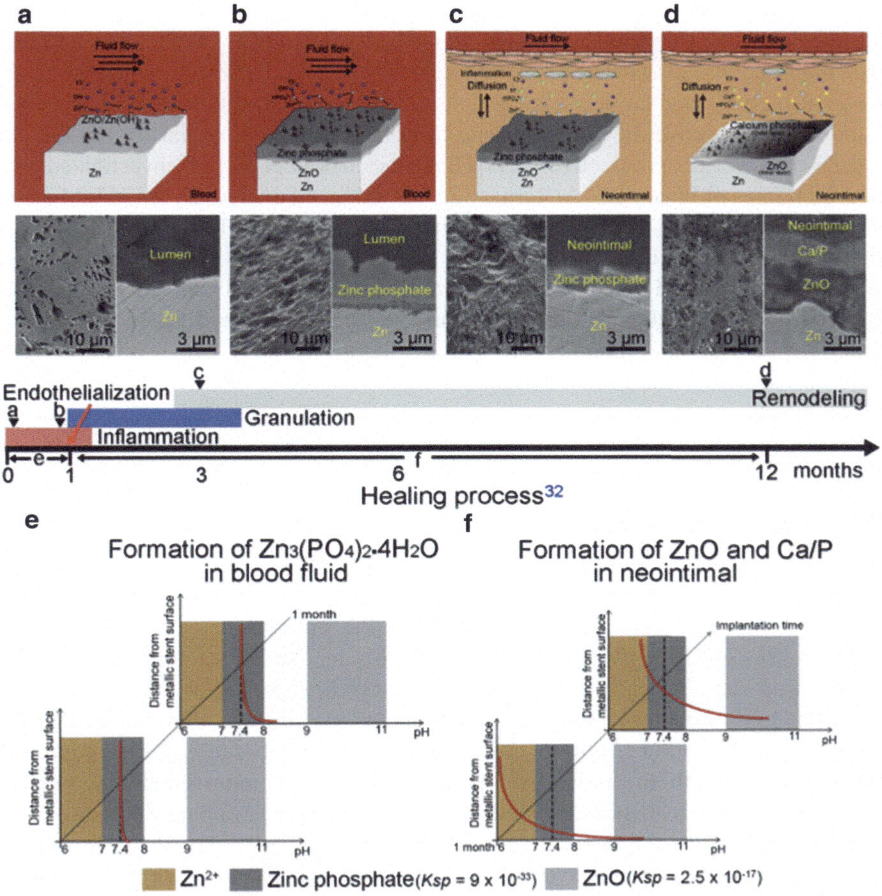

Fig. 2 Different stages and mechanism of Zn degradation when implanting in vivo for 12 months. (**a–d**) The surface morphology changes and formation of the different degradation products, (**e**, **f**) the mechanism of the degradation products formation in different microenvironment [68]

which led to localized corrosion. It was also revealed that phosphate generated in the blood flow transformed into Zn oxide and small amounts of calcium phosphate during the conversion of the degradation microenvironment (Fig. 2) [68].

The addition of alloying elements is one of the most common ways to further modulate the corrosion properties of Zn-based biodegradable metals. The corrosion of pure Zn depends more on its own properties, but Zn-based alloys corrosion is greatly adjusted by the different components of second phases such as Mg_2Zn_{11}, $CaZn_{13}$, and $SrZn_{13}$ [21, 69, 70]. Even small changes in the composition of alloying elements could greatly change the degradation mode in comparison with that of pure Zn [71]. Li et al. [70] reported that the corrosion rate of Zn–1.0Ca and Zn–1.0Sr alloys increased from 0.08 of pure Zn to 0.09 and 0.95 mm/year, respectively, by adding Ca and Sr to pure Zn through the method of weight loss test and the same growing trend was also measured by a potentiodynamic polarization test. The same group also performed another experiment indicating that the addition of Ca and Sr could further improve the corrosion rate from 0.09 to 0.11 mm/year for Zn–1.0Ca and Zn–1.0Ca–1.0Sr, respectively [54]. Bakhsheshi-Rad et al. [72] investigated the corrosion rates of binary Zn–Al, ternary Zn–Al–Mg, and quaternary Zn–Al–Mg–Bi alloys, and the results showed that the corrosion rate of Zn–Al–Mg–Bi was a little higher than those of the Zn–Al–Mg and Zn–Al alloys due to the formation of a secondary phase (α-Mg_3Bi_2) which induced more extensive galvanic corrosion. The result of corrosion behavior observed in in vivo experiment performed by Bowen et al. [73] showed intergranular corrosion for the Zn–Al alloy which was different from pure Zn. Such kind of a corrosion is mainly caused by the distributed Al precipitates in the grain boundaries.

Mechanical Properties

The physical and mechanical properties of metallic Zn are shown as follows: density = 7.14 g/cm^3; Young's modulus = 70 GPa; and ultimate tensile strength (UTS) = 126–246 MPa [74]. The addition of alloying elements and proper thermal deformation can efficiently enhance the mechanical properties of the as-cast pure Zn. A few binary and ternary Zn alloys (containing magnesium, aluminum, lithium, calcium, copper, and/or strontium) have been investigated because of their superior mechanical strength in comparison to pure Zn. It is demonstrated that binary Zn-based alloys show better mechanical properties compared to pure Zn. Especially under compression, the as-extruded Zn-1X (Ca, Mg, Sr) alloy exhibits superplasticity that makes them a favorable candidate for medical applications. Zn alloys display a broad range of ultimate tensile strengths (87–399 MPa) and elongations (0.9–170%). It is reported that even a small percentage of alloying elements can greatly enhance mechanical properties. Take adding 0.15% Mg to pure Zn as an example, the ultimate tensile strength and the elongation fraction are improved from 18 to 250 MPa and from 0.32 to 22%, respectively [21, 54]. Furthermore, the strength and ductility of Zn alloys could be improved by hot rolling and hot extrusion [8, 21, 54, 75]. For instance, the yield strength (YS), UTS, and elongation of as-cast Zn–1Mg–1Ca are 80 MPa, 130 MPa, and 1%, respectively, while the YS, UTS, and elongation properties increase to 138 MPa, 197 MPa, and 8.5% after hot rolling and 205 MPa, 250 MPa, and 5.2% after hot extrusion, respectively [54]. Hence, conventional metallurgy is also a possible way to meet the strength requirements for Zn alloys for most applications.

Animal Testing of Zn and Zn-Based Biodegradable Metal Implants

Cardiovascular Implantation

Zn is believed to be a possible vascular scaffold material since it has the properties of anti-inflammatory and anti-proliferative, stabilizes the membrane of endothelial cells [76] and reduces the risk of atherosclerosis [50] and in-stent restenosis, which is one of the most common causes of implant failure.

Bowen et al. studied the in vivo performance of Zn wires (99.99%) over 6 months when implanted into the abdominal aorta of adult rats [76, 77]. They found uniform corrosion in the first 1.5 and 3 months and relative severe corrosion after 4.5 and 6 months. The corrosion rate after 1.5 months was below the degradable stent benchmark (0.02 mm/year) and increased to 0.05 mm/year (~0.4 and ~0.97 mg/day, respectively) after 6 months, both of which are far below the recommended daily dose of Zn (15 mg/day) [76], indicating little or no systemic toxicity concerns of Zn stent. Also, a histological examination of Zn wires exhibited excellent biocompatibility with the arterial tissue, and inflammatory response, localized necrosis, and progressive intimal hyperplasia were not observed. It has also been shown that cell density in the neointimal tissue is low, and smooth muscle cells adjacent to the implant interface are clearly lacking, indicating that the Zn implant can inhibit the restenosis pathway [77]. In a study [62], pure Zn stents were implanted into the abdominal aorta of rabbits for 20 months. The authors found that the Zn wires exhibited a steady and linear corrosion without local toxicity for up to at least 20 months post-implantation, despite the formation of a passivating corrosion product layer around the wire. Although the long-term presence of the Zn implant elicited an inflammatory response, it was observed that chronic inflammation diminished between 10 and 20 months as suppressed by the fibrotic encapsulation of the implants.

Orthopedic Implantation

Zn, as an essential element with osteogenic potential, has attracted considerable attention in recent years. It has been demonstrated that Zn could both induce bone formation by activating aminoacyl-tRNA synthetase and the gene expression of the transcription factors runt-related transcription factor 2 (Runx2), stimulating cellular protein synthesis and osteoblastic differentiation and inhibiting osteoblastic activity including suppressing osteoclast-like cell formation, stimulating cellular apoptosis of mature osteoclasts and a suppressive effect on the receptor activator of nuclear factor (NF)-kB ligand (RANKL)-induced osteoclastogenesis [78]. Moreover, Zn deficiency leads to delayed bone growth development, postmenopausal osteoporosis, and osteopenia [79, 80]. Therefore, it has been reported that the incorporation of

Zn into bioceramic, bioglass, bone cement, and implant coatings enhances their mechanical properties and promotes adhesion, proliferation, and differentiation of osteoblasts [81, 82].

Zn implants have recently been explored for orthopedic applications. Unfortunately, it is hard to meet the clinical requirements with pure Zn due to their properties of softness, fragileness, and low mechanical strength in the practice. Therefore, pure Zn should be modified to meet the desired mechanical properties. In recent years, Zn alloys have been developed and designed for orthopedic applications. Li et al. [70] employed Zn-1X binary alloys (Zn–1Mg, Zn–1Ca, and Zn–1Sr) to systematically investigate their mechanical properties, degradation, and in vitro and in vivo biocompatibility. By alloying with Mg, Ca, and Sr elements, the microhardness, yield strength, UTS, and elongation have been enhanced significantly compared with that of pure Zn. It is also shown that the corrosion rate of Zn-1X alloys is significantly higher than that of pure Zn. Although the tensile strength of Zn-1X alloys decreases a little after 8 weeks of degradation in Hank's solution simulated body fluid solution, it remains at a high level to meet the requirements for fracture repair, which usually takes 3–4 months to maintain mechanical properties, such as from callus formation to mineralization and remodeling. In vitro studies showed reduced adhesive platelet numbers and improved cell viabilities in comparison to the pure Zn group. The aforementioned results are further confirmed by implantation of Zn-1X alloy pins in the mouse femur, which demonstrated thicker cortical bone and higher bone volume compared to the sham control group and no inflammation around the implantation site and no death observed after operation (Fig. 3a). In another study [83], a metallic matrix composite was chosen as a novel strategy to solve the problems of pure Zn for biomedical applications. The authors prepared Zn-HA composites by adding hydroxyapatite (HA: $Ca_{10}(PO_4)_6(OH)_2$) into a Zn matrix via spark plasma sintering. It was found that the degradation rates of

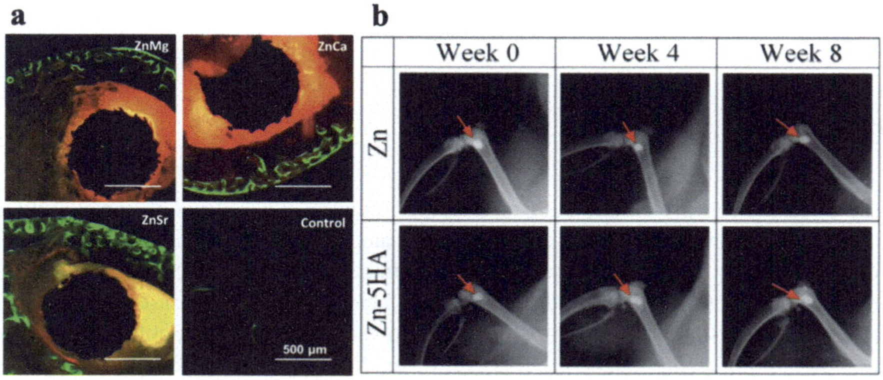

Fig. 3 (**a**) Representative histology of the cross sections of mouse distal femoral shaft under fluorescent microscopy at week 8. Green fluorescence indicates cortical bone thicker than that of sham control group [72]. (**b**) Micro-CT analysis of femoral condyle with pure Zn and Zn-5HA composite implants (red arrow) [83]

Zn-HA composites were adjustable due to the biphasic effects of HA, which decreased corrosion rate with a 1 wt.% addition of HA which accelerated the corrosion rate obviously with the addition of HA higher than 5 wt.%. In order to further investigate the degradation behavior of Zn-HA composites, they implanted pure Zn and a Zn-5HA composite into the femoral condyle of rats. The in vivo results showed that both pure Zn and Zn-HA composites were able to maintain mechanical integrity and excellent biocompatibility without inflammation or other adverse effects around the implants (Fig. 3b), but the Zn-5HA composite showed a more obvious effect in stimulating new bone formation at week 8 in contrast to pure Zn. However, benchmark behavior for Zn in orthopedic applications has not been studied comprehensively. Therefore, it is difficult to verify the results for Zn-based orthopedic implants at present.

Summary and Future Challenges

An ideal biodegradable material should possess a suitable corrosion rate, superior mechanical properties, and favorable biocompatibility that match the tissue healing process. However, in spite of these advantages, certain challenges, like cellular toxicity and inflammatory reactions, must be overcome to make Zn and Zn-based alloys clinically feasible for particular applications in the future.

Despite adding alloying elements to Zn has been reported to increase its ultimate tensile, yield strengths, elongation rate, and so on, careful selection of alloying elements is necessary to meet clinical requirements and ensure biosafety in addition to the mechanical concerns. Additional studies also need to be performed to investigate approaches to control the corrosion rates and the concentration of the degraded products from Zn-based alloys to eliminate toxic levels of metal concentrations because the released metal ions could cause systematic toxicity to the human body along with local toxicity to the peri-implant cells. Considering these side effects, surface modification, composite, or other methods should be utilized to develop a new class of Zn-based material. Besides, clarifying the molecular mechanism that how different alloying ions interact with surrounding cells is helpful for subtle control of the degradation rates.

In general, Zn alloys are still relatively new as a class of biodegradable metallic materials with a large data set obtained from basic research. Hence, further investigations are needed to solve the remaining challenges so as to translate the promising results into clinical application to benefit patients rapidly.

References

1. Takmakov P, Ruda K, Scott Phillips K, Isayeva IS, Krauthamer V, Welle CG (2015) Rapid evaluation of the durability of cortical neural implants using accelerated aging with reactive oxygen species. J Neural Eng 12(2):026003

2. Huang Y, van Dessel J, Martens W, Lambrichts I, Zhong W-J, Ma G-W et al (2015) Sensory innervation around immediately vs. delayed loaded implants: a pilot study. Int J Oral Sci 7(1):49–55
3. Waksman R, Pakala R (2010) Biodegradable and bioabsorbable stents. Curr Pharm Des 16(36):4041–4051
4. Guo R, Merkel AR, Sterling JA, Davidson JM, Guelcher SA (2015) Substrate modulus of 3D-printed scaffolds regulates the regenerative response in subcutaneous implants through the macrophage phenotype and Wnt signaling. Biomaterials 73:85–95
5. Yahyavi-Firouz-Abadi N, Menias CO, Bhalla S, Siegel C, Gayer G, Katz DS (2015) Imaging of cosmetic plastic procedures and implants in the body and their potential complications. AJR Am J Roentgenol 204(4):707–715
6. Kurtz SM, Devine JN (2007) PEEK biomaterials in trauma, orthopedic, and spinal implants. Biomaterials 28(32):4845–4869
7. Klein MO, Schiegnitz E, Al-Nawas B (2014) Systematic review on success of narrow-diameter dental implants. Int J Oral Maxillofac Implants 29(Suppl):43–54
8. Shen C, Liu X, Fan B, Lan P, Zhou F, Li X et al (2016) Mechanical properties, in vitro degradation behavior, hemocompatibility and cytotoxicity evaluation of Zn–1.2Mg alloy for biodegradable implants. RSC Adv 6(89):86410–86419
9. Niinomi M (2008) Metallic biomaterials. J Artif Organs 11(3):105–110
10. Katarivas Levy G, Ventura Y, Goldman J, Vago R, Aghion E (2016) Cytotoxic characteristics of biodegradable EW10X04 Mg alloy after Nd coating and subsequent heat treatment. Mater Sci Eng C Mater Biol Appl 62:752–761
11. Farb A, Weber DK, Kolodgie FD, Burke AP, Virmani R (2002) Morphological predictors of restenosis after coronary stenting in humans. Circulation 105(25):2974–2980
12. Cook S, Wenaweser P, Togni M, Billinger M, Morger C, Seiler C et al (2007) Incomplete stent apposition and very late stent thrombosis after drug-eluting stent implantation. Circulation 115(18):2426–2434
13. Chung W-S, Park C-S, Seung K-B, Kim P-J, Lee J-M, Koo B-K et al (2008) The incidence and clinical impact of stent strut fractures developed after drug-eluting stent implantation. Int J Cardiol 125(3):325–331
14. Witte F, Hort N, Vogt C, Cohen S, Kainer KU, Willumeit R et al (2008) Degradable biomaterials based on magnesium corrosion. Curr Opin Solid State Mater Sci 12(5–6):63–72
15. Li H, Zheng Y, Qin L (2014) Progress of biodegradable metals. Prog Nat Sci Mater Int 24(5):414–422
16. Ramcharitar S, Serruys PW (2008) Fully biodegradable coronary stents. Am J Cardiovasc Drugs 8(5):305–314
17. Mueller PP, Arnold S, Badar M, Bormann D, Bach F-W, Drynda A et al (2012) Histological and molecular evaluation of iron as degradable medical implant material in a murine animal model. J Biomed Mater Res A 100A(11):2881–2889
18. Hermawan H, Alamdari H, Mantovani D, Dubé D (2008) Iron–manganese: new class of metallic degradable biomaterials prepared by powder metallurgy. Powder Metall 51(1):38–45
19. Hermawan H, Purnama A, Dube D, Couet J, Mantovani D (2010) Fe-Mn alloys for metallic biodegradable stents: degradation and cell viability studies. Acta Biomater 6(5):1852–1860
20. Heublein B, Rohde R, Kaese V, Niemeyer M, Hartung W, Haverich A (2003) Biocorrosion of magnesium alloys: a new principle in cardiovascular implant technology? Heart 89(6):651–656
21. Mostaed E, Sikora-Jasinska M, Mostaed A, Loffredo S, Demir AG, Previtali B et al (2016) Novel Zn-based alloys for biodegradable stent applications: design, development and in vitro degradation. J Mech Behav Biomed Mater 60:581–602
22. Gu X-N, Zheng Y-F (2010) A review on magnesium alloys as biodegradable materials. Front Mater Sci China 4(2):111–115
23. Wang C, Yu Z, Cui Y, Zhang Y, Yu S, Qu G et al (2016) Processing of a novel Zn alloy micro-tube for biodegradable vascular stent application. J Mater Sci Technol 32(9):925–929

24. Guillory RJ, Bowen PK, Hopkins SP, Shearier ER, Earley EJ, Gillette AA et al (2016) Corrosion characteristics dictate the long-term inflammatory profile of degradable zinc arterial implants. ACS Biomater Sci Eng 2(12):2355–2364
25. Ma J, Zhao N, Zhu D (2015) Endothelial cellular responses to biodegradable metal zinc. ACS Biomater Sci Eng 1(11):1174–1182
26. Murni NS, Dambatta MS, Yeap SK, Froemming GRA, Hermawan H (2015) Cytotoxicity evaluation of biodegradable Zn-3Mg alloy toward normal human osteoblast cells. Mater Sci Eng C Mater Biol Appl 49:560–566
27. Aghion E, Levy G (2010) The effect of Ca on the in vitro corrosion performance of biodegradable Mg–Nd–Y–Zr alloy. J Mater Sci 45(11):3096–3101
28. Zhao S, McNamara CT, Bowen PK, Verhun N, Braykovich JP, Goldman J et al (2017) Structural characteristics and in vitro biodegradation of a novel Zn-Li alloy prepared by induction melting and hot rolling. Metall Mater Trans A 48(3):1204–1215
29. Kim S-M, Jo J-H, Lee S-M, Kang M-H, Kim H-E, Estrin Y et al (2014) Hydroxyapatite-coated magnesium implants with improved in vitro and in vivo biocorrosion, biocompatibility, and bone response. J Biomed Mater Res A 102(2):429–441
30. Wong HM, Yeung KWK, Lam KO, Tam V, Chu PK, Luk KDK et al (2010) A biodegradable polymer-based coating to control the performance of magnesium alloy orthopaedic implants. Biomaterials 31(8):2084–2096
31. Seitz J-M, Durisin M, Goldman J, Drelich JW (2015) Recent advances in biodegradable metals for medical sutures: a critical review. Adv Healthc Mater 4(13):1915–1936
32. Wang C, Yang HT, Li X, Zheng YF (2016) In vitro evaluation of the feasibility of commercial Zn alloys as biodegradable metals. J Mater Sci Technol 32(9):909–918
33. Törne K, Larsson M, Norlin A, Weissenrieder J (2016) Degradation of zinc in saline solutions, plasma, and whole blood. J Biomed Mater Res B Appl Biomater 104(6):1141–1151
34. Liu X, Sun J, Qiu K, Yang Y, Pu Z, Li L et al (2016) Effects of alloying elements (Ca and Sr) on microstructure, mechanical property and in vitro corrosion behavior of biodegradable Zn–1.5Mg alloy. J Alloys Compd 664:444–452
35. Niu J, Tang Z, Huang H, Pei J, Zhang H, Yuan G et al (2016) Research on a Zn-Cu alloy as a biodegradable material for potential vascular stents application. Mater Sci Eng C Mater Biol Appl 69:407–413
36. Huang T, Zheng Y, Han Y (2016) Accelerating degradation rate of pure iron by zinc ion implantation. Regen Biomater 3(4):205–215
37. Vojtěch D, Kubásek J, Serák J, Novák P (2011) Mechanical and corrosion properties of newly developed biodegradable Zn-based alloys for bone fixation. Acta Biomater 7(9):3515–3522
38. Guleryuz LF, Ipek R, Arıtman I, Karaoglu S (2017) Microstructure and mechanical properties of Zn-Mg alloys as implant materials manufactured by powder metallurgy method. AIP Conf Proc 1809(1):020020
39. Tapiero H, Tew KD (2003) Trace elements in human physiology and pathology: zinc and metallothioneins. Biomed Pharmacother 57(9):399–411
40. Calesnick B, Dinan AM (1988) Zinc deficiency and zinc toxicity. Am Fam Physician 37(4):267–270
41. Trumbo P, Yates AA, Schlicker S, Poos M (2001) Dietary reference intakes: vitamin A, vitamin K, arsenic, boron, chromium, copper, iodine, iron, manganese, molybdenum, nickel, silicon, vanadium, and zinc. J Am Diet Assoc 101(3):294–301
42. Plum LM, Rink L, Haase H (2010) The essential toxin: impact of zinc on human health. Int J Environ Res Public Health 7(4):1342–1365
43. Choi J, Kim H, Kim P, Jo E, Kim H-M, Lee M-Y et al (2015) Toxicity of zinc oxide nanoparticles in rats treated by two different routes: single intravenous injection and single oral administration. J Toxicol Environ Health A 78(4):226–243
44. Wastney ME, Aamodt RL, Rumble WF, Henkin RI (1986) Kinetic analysis of zinc metabolism and its regulation in normal humans. Am J Physiol 251(2):R398–R408

45. Vallee BL, Falchuk KH (1993) The biochemical basis of zinc physiology. Physiol Rev 73(1):79–118
46. Su Y, Cockerill I, Wang Y, Qin Y-X, Chang L, Zheng Y et al (2019) Zinc-based biomaterials for regeneration and therapy. Trends Biotechnol 37(4):428–441
47. Mocchegiani E, Muzzioli M, Giacconi R (2000) Zinc, metallothioneins, immune responses, survival and ageing. Biogerontology 1(2):133–143
48. Falchuk KH (1998) The molecular basis for the role of zinc in developmental biology. Mol Cell Biochem 188(1–2):41–48
49. Cousins RJ (1998) A role of zinc in the regulation of gene expression. Proc Nutr Soc 57(2):307–311
50. Hennig B, Toborek M, Mcclain CJ (1996) Antiatherogenic properties of zinc: implications in endothelial cell metabolism. Nutrition 12(10):711–717
51. Katarivas Levy G, Goldman J, Aghion E, Katarivas Levy G, Goldman J, Aghion E (2017) The prospects of zinc as a structural material for biodegradable implants—a review paper. Metals 7(10):402
52. Yang J, Gao F, Han D, Yang L, Kong X, Wei M et al (2018) Multifunctional zinc-based hollow nanoplatforms as a smart pH-responsive drug delivery system to enhance in vivo tumor-inhibition efficacy. Mater Des 139:172–180
53. Kang Y, Wu Y-Z, Hu X, Xu X, Sun J, Geng R et al (2017) Multicolor bioimaging with biosynthetic zinc nanoparticles and their application in tumor detection. Sci Rep 7(1):45313
54. Li H, Yang H, Zheng Y, Zhou F, Qiu K, Wang X (2015) Design and characterizations of novel biodegradable ternary Zn-based alloys with IIA nutrient alloying elements Mg, Ca and Sr. Mater Des 83:95–102
55. Roohani N, Hurrell R, Kelishadi R, Schulin R (2013) Zinc and its importance for human health: an integrative review. J Res Med Sci 18(2):144–157
56. Bowen PK, Shearier ER, Zhao S, Guillory RJ, Zhao F, Goldman J et al (2016) Biodegradable metals for cardiovascular stents: from clinical concerns to recent Zn-alloys. Adv Healthc Mater 5(10):1121–1140
57. Brzóska MM, Rogalska J (2013) Protective effect of zinc supplementation against cadmium-induced oxidative stress and the RANK/RANKL/OPG system imbalance in the bone tissue of rats. Toxicol Appl Pharmacol 272(1):208–220
58. Luo X, Barbieri D, Davison N, Yan Y, de Bruijn JD, Yuan H (2014) Zinc in calcium phosphate mediates bone induction: in vitro and in vivo model. Acta Biomater 10(1):477–485
59. Prakasam M, Locs J, Salma-Ancane K, Loca D, Largeteau A, Berzina-Cimdina L (2017) Biodegradable materials and metallic implants—a review. J Funct Biomater 8(4):44
60. Shearier ER, Bowen PK, He W, Drelich A, Drelich J, Goldman J et al (2016) In vitro cytotoxicity, adhesion, and proliferation of human vascular cells exposed to zinc. ACS Biomater Sci Eng 2(4):634–642
61. Liu X, Sun J, Zhou F, Yang Y, Chang R, Qiu K et al (2016) Corrigendum to "Micro-alloying with Mn in Zn–Mg alloy for future biodegradable metals application" [Mater. Des. 94 (2016) 95–104]. Mater Des 96:377
62. Drelich AJ, Zhao S, Guillory RJ, Drelich JW, Goldman J (2017) Long-term surveillance of zinc implant in murine artery: surprisingly steady biocorrosion rate. Acta Biomater 58:539–549
63. Qu Q, Li L, Bai W, Yan C, Cao C (2005) Effects of NaCl and NH4Cl on the initial atmospheric corrosion of zinc. Corros Sci 47(11):2832–2840
64. Mouanga M, Berçot P, Rauch JY (2010) Comparison of corrosion behaviour of zinc in NaCl and in NaOH solutions. Part I: corrosion layer characterization. Corros Sci 52(12):3984–3992
65. Zhang XG (2013) Corrosion and electrochemistry of zinc. Springer Science & Business Media, New York, 481 p
66. Thomas S, Birbilis N, Venkatraman MS, Cole IS (2012) Corrosion of zinc as a function of pH. Corrosion 68(1):015009-1–015009-9
67. Thomas S, Cole IS, Gonzalez-Garcia Y, Chen M, Musameh M, Mol JMC et al (2014) Oxygen consumption upon electrochemically polarised zinc. J Appl Electrochem 44(6):747–757

68. Yang H, Wang C, Liu C, Chen H, Wu Y, Han J et al (2017) Evolution of the degradation mechanism of pure zinc stent in the one-year study of rabbit abdominal aorta model. Biomaterials 145:92–105
69. Sikora-Jasinska M, Mostaed E, Mostaed A, Beanland R, Mantovani D, Vedani M (2017) Fabrication, mechanical properties and in vitro degradation behavior of newly developed ZnAg alloys for degradable implant applications. Mater Sci Eng C Mater Biol Appl 77:1170–1181
70. Li HF, Xie XH, Zheng YF, Cong Y, Zhou FY, Qiu KJ et al (2015) Development of biodegradable Zn-1X binary alloys with nutrient alloying elements Mg, Ca and Sr. Sci Rep 5:10719
71. Zhu D, Cockerill I, Su Y, Zhang Z, Fu J, Lee K-W et al (2019) Mechanical strength, biodegradation, and in vitro and in vivo biocompatibility of Zn biomaterials. ACS Appl Mater Interfaces 11(7):6809–6819
72. Bakhsheshi-Rad HR, Hamzah E, Low HT, Kasiri-Asgarani M, Farahany S, Akbari E et al (2017) Fabrication of biodegradable Zn-Al-Mg alloy: mechanical properties, corrosion behavior, cytotoxicity and antibacterial activities. Mater Sci Eng C Mater Biol Appl 73:215–219
73. Bowen PK, Seitz J-M, Guillory RJ, Braykovich JP, Zhao S, Goldman J et al (2018) Evaluation of wrought Zn-Al alloys (1, 3, and 5 wt % Al) through mechanical and in vivo testing for stent applications. J Biomed Mater Res B Appl Biomater 106(1):245–258
74. Port F (1991) Zinc handbook: properties, processing, and use in design. CRC Press, Boca Raton
75. Gong H, Wang K, Strich R, Zhou JG (2015) In vitro biodegradation behavior, mechanical properties, and cytotoxicity of biodegradable Zn-Mg alloy. J Biomed Mater Res B Appl Biomater 103(8):1632–1640
76. Bowen PK, Drelich J, Goldman J (2013) Zinc exhibits ideal physiological corrosion behavior for bioabsorbable stents. Adv Mater 25(18):2577–2582
77. Bowen PK, Guillory RJ, Shearier ER, Seitz J-M, Drelich J, Bocks M et al (2015) Metallic zinc exhibits optimal biocompatibility for bioabsorbable endovascular stents. Mater Sci Eng C Mater Biol Appl 56:467–472
78. Yamaguchi M (2010) Role of nutritional zinc in the prevention of osteoporosis. Mol Cell Biochem 338(1):241–254
79. Mahdavi-Roshan M, Ebrahimi M, Ebrahimi A (2015) Copper, magnesium, zinc and calcium status in osteopenic and osteoporotic post-menopausal women. Clin Cases Miner Bone Metab 12(1):18–21
80. Baltaci AK, Sunar F, Mogulkoc R, Acar M, Toy H (2014) The effect of zinc deficiency and zinc supplementation on element levels in the bone tissue of ovariectomized rats: histopathologic changes. Arch Physiol Biochem 120(2):80–85
81. Saino E, Grandi S, Quartarone E, Maliardi V, Galli D, Bloise N et al (2011) In vitro calcified matrix deposition by human osteoblasts onto a zinc-containing bioactive glass. Eur Cell Mater 21:59–72; discussion 72
82. Zreiqat H, Ramaswamy Y, Wu C, Paschalidis A, Lu Z, James B et al (2010) The incorporation of strontium and zinc into a calcium–silicon ceramic for bone tissue engineering. Biomaterials 31(12):3175–3184
83. Yang H, Qu X, Lin W, Wang C, Zhu D, Dai K et al (2018) In vitro and in vivo studies on zinc-hydroxyapatite composites as novel biodegradable metal matrix composite for orthopedic applications. Acta Biomater 71:200–214

Recent Physical Interaction-based Bioadhesives

Kaige Xu, Qiang Chang, Yuqing Liu, and Malcolm Xing

Abstract For centuries, research scientists and surgeons have been searching for the best way to close a wound from soft tissue to hard tissue. A number of wound closure techniques have been developed from simple primary wound closure to more complexed and sophisticated ones. Among them, covalent bonding and non-covalent bonding derived tissue bioadhesives have been given enormous attention due to their unique advantages, including convenient operation, minimum invasion, versatile application, etc. Compared to covalent bonding derived adhesives, non-covalent bonding or physical interaction-based adhesives are particularly outstanding because of their reversible and repeatable adhesion, non-chemical crosslinkers or initiators introducing, relative mild adhesion, and they are bioinspired to nature. This review explores and assesses the range of innovative techniques that have been researched to close different types of wounds. We highlight the research on non-covalent interactions derived and mechanical structure based bioadhesives, including the list of adhesion mechanisms, materials and compositions, adhesion strength, and the potential applications of different adhesives in recent research. In addition, further considerations for the next generation of bioadhesive will also be discussed at the end of this review.

Keywords Tissue bioadhesive · Wound closure · Non-covalent interaction Electrostatic interaction · Hydrogen bonding · Hydrophobic force · van der Waals force · Mechanical structure based adhesive

Introduction

Nowadays, several thousands of surgical operating procedures are performed around the world. Wound healing is one of the most significant aspects of surgical operation because surgeons rely on it to combine separated tissue together and control bleeding [1–6]. Suture is also a common method to gather injured tissues because of its simplicity and efficiency to close a wound [7, 8]. Another technique

K. Xu · Q. Chang · Y. Liu · M. Xing (✉)
Department of Mechanical Engineering, University of Manitoba, Winnipeg, MB, Canada
e-mail: Malcolm.Xing@umanitoba.ca

B. Li et al. (eds.), *Racing for the Surface*,
https://doi.org/10.1007/978-3-030-34471-9_25

frequently used in wound healing is staples. Staples are quicker, easier to control, and reduce the wound infection rate more effectively compared to sutures [9–12]. Although both of these wound healing methods have their advantages, they are invasive to surrounding healthy tissues and occasionally cause tissue trauma and scarring. In addition, these procedures oftentimes need another removal step after use, being costly and requiring skilled surgeons to perform these procedures [13–16]. Besides, sutures and staples are not suitable for repairing bone fractures or viscera that contains liquid.

A preferable option to overcome these limitations and moderate the application of these aggressive techniques is to use tissue adhesives. Tissue adhesives could be simply spread over the entire fracture or injured area as it can hold the tissues together even with light pressure [17–19]. In this way, the stress applied on the injured tissue by adhesion is much lower compared to sutures or staples and the load is only applied on the contact surface [20, 21]. Also, adhesives could be used to glue fractured bones, and viscera by using water-resistant adhesives [22–27]. The most important advantage of applying adhesives is that they lead to minimal or no tissue scars, since there are no further invasions on normal tissues like sutures or staples during the application process.

Recently, more and more tissue adhesives are being researched, and some of them have already been used as commercial medical adhesive glues for a long time, such as fibrin glue which was first introduced in the 1940s [28, 29], and cyanoacrylate adhesives which were first applied in the 1980s [30–32]. An ideal tissue adhesive is one which should be able to perform strong adhesion both in dry and wet conditions, be stable under the general physiological environment, exhibit rapid curing or crosslinking ability under room temperature or moderate environment, and have good biocompatibility, biodegradability, and nontoxicity [33–37]. Sometimes, extremely high adhesion strength can be harmful too, since it may cause unwanted adhesion between surgical instruments and tissues, even injured tissues; an optimum balance should be maintained in terms of adhesion strength. Water resistance should also be a useful and necessary characteristic of adhesives, which means adhesives should keep an adhesion ability even in aqueous conditions, as the environment around human tissues is mostly in an aqueous state. On account of adhesion mechanisms, tissue adhesives can be divided into: non-covalent interactions derived adhesives, mechanical structure based adhesives, and covalent chemical bonded adhesives [15, 38, 39].

The research and utilization of covalent chemical bonded adhesives have been reviewed comprehensively [40–42]. Hence, this current paper emphatically reviews recently developed non-covalent interactions derived bioadhesives, and mechanical structure based bioadhesives. Covalent chemical bonded adhesives are known for their strong adhesion and long-lasting adhesion period; in contrast, physical interaction-based adhesives have gained unique feats including rapid reversible and repeatable adhesion just like insects or geckos climbing walls, stimulus responsive adhesion, relatively mild adhesion strength that avoids non-necessary adhesion sometimes even on normal tissues, and no additional chemicals introduced due to non-chemical reactions occurring during adhesion [43]. Furthermore, some of these bioadhesives are inspired from natural biological entities, for example, an octopus' sucker can adhere on rocks even in mazy ocean salt aqueous environment, mainly functionalized with their mechanical structures [44]. This review mainly describes the physical interaction-based adhesives involving their materials and compositions, adhesion strength, and potential applications of different adhesives. The discussion of adhesion strength involves adhesion substrates, adhesion strength in dry and wet conditions, and animal models; in addition, the advantages and disadvantages of each adhesion research example are also mentioned following the adhesive strength analysis.

Non-covalent Interaction Derived Bioadhesives

Non-covalent interactions broadly distribute in nature, for example, water molecules in liquid water are attracted and attached to each other, geckos facilely climb vertical walls without creeping down, and it even contributes to maintaining and underpinning the three-dimensional structure of DNA [45–47]. Inspired from the wonderfulness of nature, versatile non-covalent interactions have been identified and defined, which include electrostatic interactions, hydrogen bonding, hydrophobic interactions, van der Waals forces, cation-π complexation, metal coordination, and π-π interactions [48–50]; diverse types of tissue adhesives have been researched and produced via these non-covalent interactions [51–53]. Notably, some of these non-covalent interactions are always synergetic in an adhesion system, no single intermolecular interaction is able to independently support all adhesion strength; for example, van der Waals forces could be detected almost in all adhesion systems; and even the catechol moieties are proven insufficient to ensure proper underwater adhesion alone, but researchers find that non-covalent interactions are involved [54, 55].

Electrostatic Interaction

The electrostatic interaction is defined as the attractive interaction between functional groups having opposite electric charges; the electrostatic interactions occur rifely when opposite charges (positive and negative) are separated within a limited distance because of ionization or attachment of ionic categories [56–59]. Adhesives that obtain adhesion abilities between adhesive biomaterials and tissue surfaces or substrates via electrostatic interactions are categorized as electrostatic-interactions functionalized adhesives [60–62]. Proteins, polypeptides, and polysaccharides contain a large amount of amine, hydroxyl, and carboxylic acid functionalities groups; these positively or negatively charged groups play crucial roles in raising the surface affinity via electrostatic effects and could be used as an adhesive glue after electrical adsorption with charged ions [63–65]. Occasionally, oppositely charged polysaccharides, like positively charged chitosan and negatively charged alginate could form an adhesion nanosheet attached on charged surfaces for wound repair. As the most significant aspects of these biomaterials, they are all highly adaptable, biocompatible, and biodegradable due to their biological origin. In addition to this, electrostatic interaction functionalized adhesives are generally produced with a relative simple design, easy operation procedures, and straightforward techniques, like self-assembling freestanding design, layer-by-layer (LBL) method, spin-coating techniques, etc. [66–73]. Concomitantly, uncomplicated concepts have been proposed for obtaining ordinary applications and low costs, but combined with disadvantages of relatively weak adhesion and limited application.

As one type of electrostatic interaction functionalized adhesives, polysaccharides and proteins such as chitosan [17, 78–82], and gelatin [83–90], possess positively charged primary amine groups in physiological criteria; they further can build blocks and tough matrix to strongly adhere to negatively charged tissue surfaces (such as porcine skin [91, 92], cartilage [93–100], heart, artery, and liver) via electrostatic interactions [42, 101–107]. In addition, the dissipated energy by hysteresis between tissue surfaces or negative charged substrates and positive charged polysaccharides also plays a critical role, which enhances the adhesive strength [108–110]. However, the main limitation of this type of electrostatic interaction is that the target tissue surface has to contain negatively charged groups; for example, a chitosan derived tissue adhesive was researched as a long-lasting mucoadhesive, but the cohered surface should contain negative carboxylic acid groups [111].

Beyond that, positively charged metal ions including Fe^{3+}, Cu^{2+}, Zn^{2+}, and Ca^{2+} are usually chelated with negative polysaccharides, and polypeptides or proteins [112–122]. For instance, the dopamine-modified gelatin material was developed as a water-resistant adhesive by simulating the mussels' adhesive proteins, then by adding $FeCl_3$ to form hexavalent Fe compounds, and effectively bonding both strands and creating a metallo-adhesive with tissue proteins, and levels of adhesion can be adjusted with Fe^{3+} concentrations [123–130]. By applying the same concept, cupric ions can also form a strong interaction and bind as a formation of catechol-Cu^{2+} with the catechol group contained mussel-inspired dopamine-conjugated gelatin macromer, but since Cu^{2+} is relatively cytotoxic, it can usually be replaced with other health friendly metal ions. The zinc ion is one of the metal-binding candidates since Zn^{2+} is significantly less toxic than Ni^{2+} and Cu^{2+}. In another study, DOPA functionalized PAA polymer was chosen to efficiently bind with Zn^{2+} ions and form durable sticky hydrogel adhesives relying on the electrostatic interactions between negatively charged carboxylic groups that are contained in PAA and metal cations (Zn^{2+}); in addition, Zn^{2+} could also create a divalent metal-chelation complex with catechol groups from DOPA, which further offers underwater adhesion (Fig. 1a) [74, 131, 132]. Alginate is a natural polysaccharide extracted from marine algae [133–135], it can form a hydrogel quickly with even low concentration of a Ca^{2+} solution, a soft tissue adhesive was made from sodium alginate and gelatin, which is similar to the extracellular matrices of tissue comprised of various sugars and amino-based macromolecules, and then electrostatic bonded with Ca^{2+} to form stable materials; in addition, positively charged amino groups could electrostatically attract with negatively charged carboxyl groups [136]. Beyond that, a poly(acrylamide)-alginate adhesive hydrogel system could create a strong wet adhesion strength since there are physically bridged linkages between Ca^{2+} and the alginate via electrostatic interactions entangled with chemical bonding which could effectively dissipate energy after peeling-off stretching from the substrates (Fig. 1d) [77]. Although these biomaterials have great biocompatibility, high concentrations of metal ions are relatively toxic to human body; on the other hand, the light low metal ion' concentration would weaken the electrostatic interactions, thereby reducing the adhesive strength.

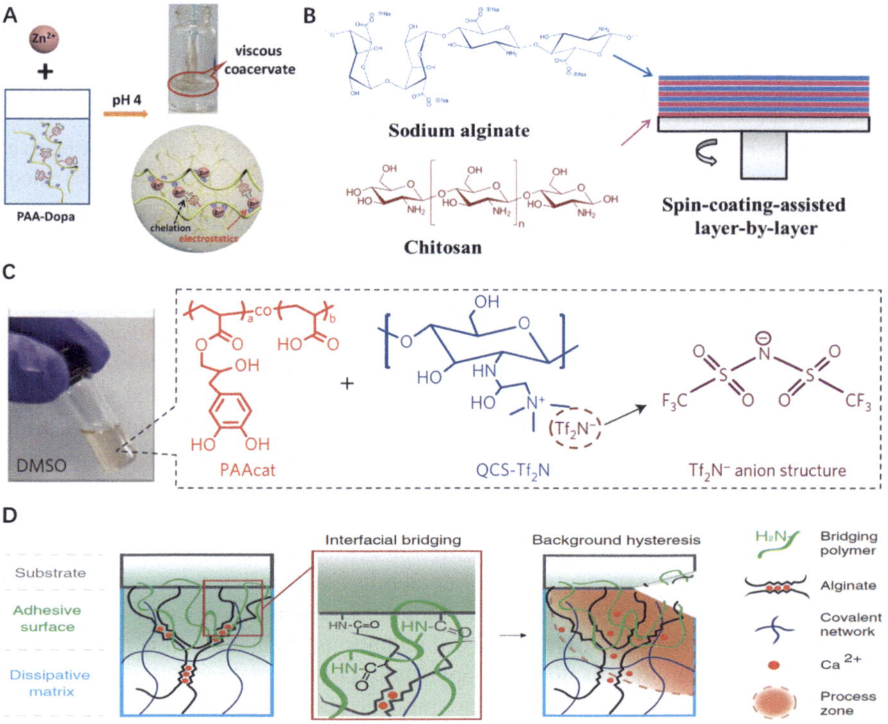

Fig. 1 Electrostatic interaction functionalized adhesives: (**A**) positively charged Zn^{3+} ions electrostatically interact with negatively charged carboxylic groups from poly(acrylic acid) (PAA) chains (Reprinted from [74] with permission from the Royal Society of Chemistry); (**B**) positively charged chitosan and negatively charged sodium alginate formed LBL polysaccharide nanosheet via a spin-coating method (Reprinted from [75] with permission from John Wiley & Sons, Inc.); (**C**) PAA based adhesive functionalized with catechols and quaternized chitosan that is ion-paired with Tf_2N^- (Reprinted from [76] with permission from Springer Nature Publishing AG); (**D**) dissipated energy by hysteresis between negatively charged substrates and positively charged polysaccharides. (Reprinted from [77] with permission from the American Association for the Advancement of Science)

The oppositely charged polypeptide functionalized adhesives were able to solve this problem, as both parts are peptides that contain a lot of lysine and arginine [137], which have great biocompatibility. Inspired by sandcastle worms, a polyanionic peptides/polycationic peptides adhesive upon electrostatic complexation was researched, among them, polyanionic peptides contain vast amount of *O*-phosphoserine and cationic peptides involve large amount of lysine and arginine [76]. In order to enhance the underwater adhesion strength, a catechol-decorated polyanion was premixed with a polycation in dimethyl sulfoxide (DMSO), then phase inversed when it touches water and activates via water-DMSO exchange at the same time. Furthermore, they developed duo polyelectrolytes meeting both electrostatic interaction and solvent exchange to prove this strategy; so, the first part is catechols functionalized PAA, and the second part is quadruplicate chitosan which is ion-conjugated with bis(trifluoromethane-sulfonyl)imide (Fig. 1c) [138]. No matter which polypeptide or chitosan, they are nontoxic and have excellent biocompatibility, and this adhesive is water-resistant and has satisfied adhesion characteristics; however, the relatively cytotoxic DMSO is the only defective point [139].

In another tissue adhesive research, the duo polysaccharide composited adhesive nanosheet containing nontoxic organic solvent or metal ions and was fabricated through a spin-coating method with LBL concept (Fig. 1b), using alternative deposition of oppositely charged polyelectrolytes through electrostatic interactions as well as van der Waals forces, which is a technique for macromolecular organization without requiring any chemical crosslinkers. The LBL assembly skill, which is utilized to fabricate a freestanding adhesive film, is simple and straightforward and works like an adhesive plaster; briefly, assembling chitosan and alginate by electrostatic interactions, which contain positively charged amine groups and negatively charged carboxylic acid groups, respectively [140–143]. This LBL adhesive film has much better biocompatibility, and the manufacturing method is simple and straightforward, but the assembled material itself is weak since its thickness is in the micrometer level, and, thus, it has to be supported with extra polyvinyl alcohol or a PDMS film layer, and in addition, the adhesion is limited [75].

Concluding the above, electrostatic interaction functionalized adhesives mainly refer to positively charged polysaccharides and proteins bonded to negatively charged tissue surfaces, positively charged metal ions chelated with negative polysaccharides, polypeptides or proteins, oppositely charged polypeptides functionalized adhesives, and duo polysaccharide composite adhesives.

Hydrogen Bonding

Hydrogen bonding, as a significant interaction, is playing a vital role in physical, chemical, and even biological processes. Generally, hydrogen bonding is defined as that when a hydrogen atom in a polar molecule (water molecule) is attracted by a small negatively charged atom (like an oxygen atom, nitrogen atom, etc.) under certain conditions, it would form strong forces between them, and this strong force is called hydrogen bonding; as well, it could be regarded as acting as a bond between these two atoms, and this bond is called a hydrogen bond; hydrogen bonding is also a physical and non-covalent bonding interaction [144]. Among the non-covalent interactions, hydrogen bonding interactions play a significant role in the construction of secondary and tertiary structures of biosystems [145]. To date, hydrogen bonding units have been widely incorporated into polymeric structures or polymerizable monomers to synthesize repeatable and tunable adhesives because of their reversible bonding and debonding nature, and it is responsive for all kinds of external stimuli, like solvent, pH, and temperature [146, 147]. In addition, hydrogen bonds could coexist with multiple other functional groups and enhance original interactions (Fig. 2a) [148]. As one of the earliest work in this area, self-supplementary quadruple hydrogen bonds formed ureido-4-pyrimidinone (UPy) which was produced to synthesize a methacrylic monomer, and further copolymerized with butylacrylate to make reversible adhesive polymers via different UPy content (Fig. 2b) [149]. This adhesion polymer presented the increasing adhesion strength with increasing UPy contents due to strong hydrogen bonding interactions between UPy and glass substrates and a mica surface in later research since the existence of UPy groups could form many hydrogen bonds at the touching interfaces, but the UPy dimers are fully dissociated above 80 °C [150]. Furthermore, researchers produced an reversible adhesion polymer, poly(dimethylsiloxane) (PDMS) modified with UPy end-groups via similar approach; in this study, the glass substrates could be bonded with this supramolecular polymer by heating to 120 °C first and then cooling to 20 °C, specifically, a reversible adhesion behavior was demonstrated as the polymer could bond glasses by heating to 120 °C after forcing the two glass slides to be separated [151]. This temperature-responsive feature offers novel repeatable adhesion, but limits the application, as well, since the adhesion tests were only made on glass.

Beyond that, the UV light was introduced as a remote stimulus to replace the direct-temperature control, and prepared the reversibly detached supramolecular adhesives; this adhesion system relies on the UPy-terminated poly(ethylene-*co*-butylene) (PEB) supramolecular structure and could adhere on steel and glass surfaces also through strong hydrogen bonding interaction between UPy and the substrates (Fig. 2c) [152]. Basically, this UV stimulus is achieved via a light-heat conversion procedure, where the UV-absorbing groups attract UV light and induce the temperature difference.

Furthermore, a pH responsive adhesive was prepared by a grafting PAA thin film onto a variety of substrates through a plasma polymerization process and potentially used to glue fat tissues. In this research, the PAA chains were grafted onto glass plates after plasma polymerization, and then showed oil holding at low pH values since the carboxylic acid pendant groups of PAA accept protons and the formation of PAA becomes tangled; at the same time, low pH value mainly leads to the building of intramolecular hydrogen bonds among PAA layers instead of the intermolecular hydrogen bonds that are created between PAA and circled water molecules

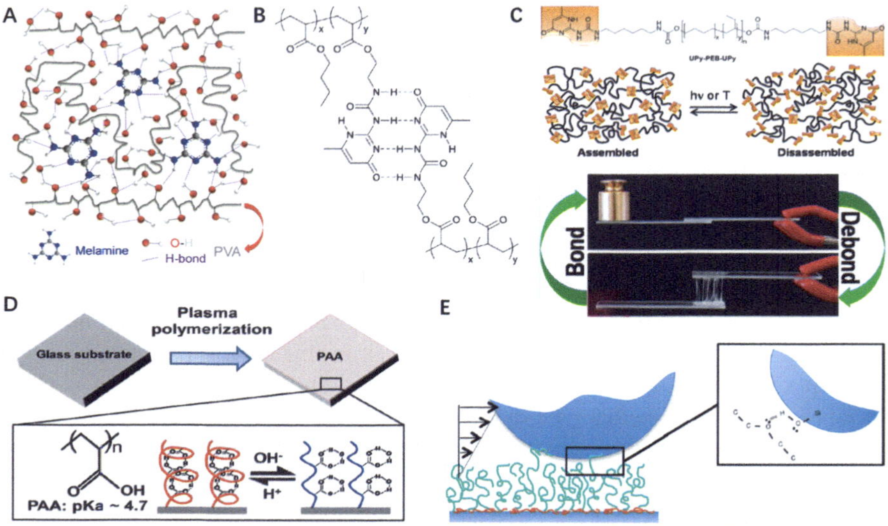

Fig. 2 Hydrogen bonding interactions in functionalized composites: (**A**) physical crosslink network based on hydrogen bonds assembly between melamine and poly(vinyl alcohol) (PVA) (Reprinted from [148] with permission from the American Chemical Society.); (**B**) self-complementary quadruple hydrogen bonding formed UPy derived polyacrylates (Reprinted from [40] with permission from the Royal Society of Chemistry.); (**C**) UV responsive UPy-PEB-UPy based reversible adhesion polymer (Reprinted from [152] with permission from the American Chemical Society.); (**D**) pH induced hydrogen bonding interactions switched between intramolecular and intermolecular hydrogen bonds (Reprinted from [153] with permission from the American Association for the Advancement of Science.); (**E**) PEG-silica adhesion brush via hydrogen bonding between them. (Reprinted from [154] with permission from the Royal Society of Chemistry)

(Fig. 2d), therefore, it adheres oil in underwater conditions at low pH value, which may apply in closing fat tissues [153].

Due to the presence of silanol assembled on the surface of silica nanoparticles, silica can easily form hydrogen bonding interactions with hydroxyl-groups-abundant PVA under acidic conditions to form a PVA-silica adhesion patch, or hydrated hydrogen-bond-acceptor poly(ethylene glycol) (PEG) to produce a PEG-silica adhesion brush (Fig. 2e), or etc. [154] For example, at a lower pH value, silanol groups on a silica surface grow during the hydrolysis process in underwater conditions which can supply a large amount of hydroxyl groups; also, when presenting in a PVA solution, silica particles break the intramolecular and intermolecular hydrogen bonds in PVA and create new hydrogen bonds with hydroxyl groups of PVA through silanol; also, silica particles can adhere to glass substrates via interacting with hydroxyl groups through the same mechanism, and form a PVA-silica adhesion film that exhibits great adhesion strength on glass as the same as any substrates that contain hydroxyl groups [155].

van der Waals Force

The van der Waals force is the attractive force acting between two neutral atoms or molecules when they are separated in a larger distance than their own dimensions, and this force is decreasing as the distance is increasing; as well, the van der Waals force is a non-covalent bonding interaction [156]. Natural creatures have evolved many kinds of mechanisms to glue on and climb surfaces. As one of the most common and important type of non-covalent interactions, the van der Waals force is found in many natural adhesive systems, among them, gecko footpads always come first to mind [157]. Geckos can climb and adhere on vertical and inverted surfaces relying on its complex hierarchical adhesion system, which uses fibrillar arrays covering footpads to produce and maximize interfacial adhesion to the surface through van der Waals forces and freely attach or detach by adjusting the loading angle (Fig. 3a) [158]. Geckos may be professional climbers on a dry surface, but not on wet ones, because the geckos' invertible dry adhesion relies on van der Waals forces between an absorbing structure and target substrates, but the reversible wet adhesion is generated mainly by adding viscosity and surface tension around the contact interface, called capillary interactions belonging to the non-covalent interaction category [158, 159]. Since the discovery of this mechanism of the adhesion in geckos, a variety of synthetic adhesives have been reported. For instance, an elastomeric and biocompatible gecko-inspired bioadhesive was researched based on poly(glycerol sebacate acrylate) (PGSA); in this study, PGSA was poured into a negative nanoporous silicon template and cured by UV light to obtain a PGSA pattern with nail-like nanopillars (Fig. 3b) [160]. The adhesion strength on stainless steel, glass, and porcine skin surfaces decreased with the growing ratio of a tip diameter to base diameter of the nanopillars; besides, it showed better adhesion on a dry surface than a wet surface, but they sought the help of covalent crosslinking with porcine skin by coating oxidized dextran on nanopillar surfaces. This limitation restricts the gecko-inspired adhesives to only function well on dry surfaces.

The gecko is not the only one that makes use of van der Waals forces, both plants like ivy and sundew, and marine organisms like mussel also get the utmost out of this force. Ivy or sundew is also known for their natural ability to climb and strongly absorb on many solid substrates, this is because not only their aerial rootlets secrete an adhesive made of polysaccharide that can chemically cure on the substrates, but also their footpad rootlet contains abundant spherical nanoparticles-like nanofibers that would build intermolecular interactions with solid surfaces through van der Waals force [161]. Then, researchers produced adhesive hydrogels for wound healing with polysaccharides sodium alginate which electrostatically interacted with calcium ions, and did not forget to mimic the nanofiber footpad of these plants to form van der Waals forces with target surfaces [162]. Mussel foot proteins express wonder adhering to mineral surfaces like mica and TiO_2 and this is usually attributed to catecholic ligands of DOPA by forming stable bidentate modes on surfaces, but the truth of hydrophobic methyl-terminated (CH_3-) and an alcohol-terminated (OH-) mussel decorated surfaces adsorption on mica substrates via van der Waals interactions cannot be neglected [163].

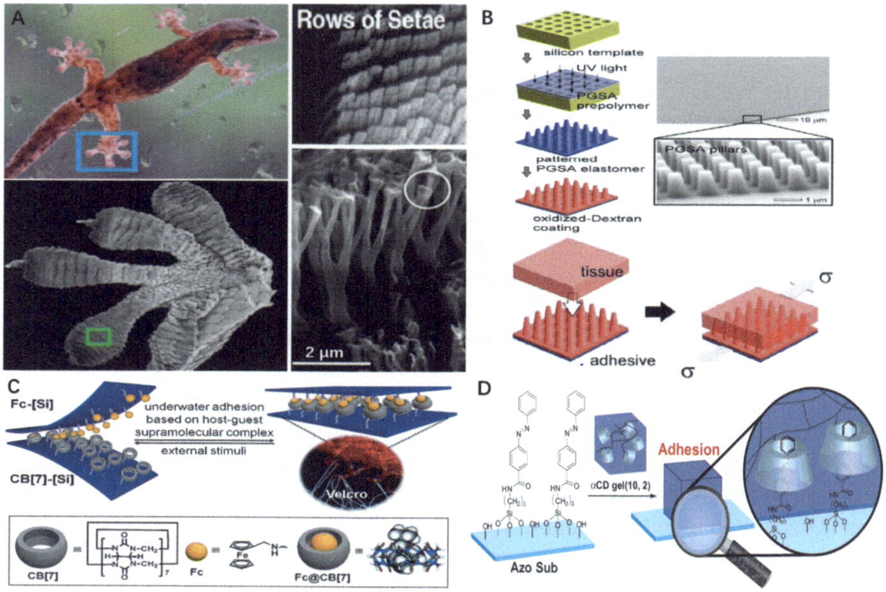

Fig. 3 Biomimicking van der Waals forces and hydrophobic interactions derived adhesives: (**A**) Gecko's footpad and its fibrillar arrays setae (Reprinted from [157] with permission from the American Chemical Society.); (**B**) gecko-inspired PGSA pattern with nail-like nanopillars showing tissue adhesion, made from a negative nanoporous silicon template (Reprinted from [160] with permission from the National Academy of Sciences.); (**C**) Fc-guest modified plate gained underwater adhesion onto a CB[7]-host grafted plate (Reprinted from [164] with permission from John Wiley & Sons, Inc.); (**D**) αCD gel glue onto an Azo modified surface via host-guest hydrophobic interaction. (Reprinted from [165] with permission from John Wiley & Sons, Inc.)

Hydrophobic Interactions

As one type of non-covalent bonding interactions, hydrophobic interactions are described as nonpolar groups or molecules that have the tendency to clump up together rather than distributing themselves in aqueous solution, usually in a water medium, since this leads to minimizing contact of these molecules with water; notably, hydrophobic interactions between hydrophobes are spontaneous [166]. In a dry environment, hydrophobic interactions would be undemanding applied on substrates and sufficient binding would occur among some nonpolar molecules, like silane groups [167]. However, some adhesion bindings are disrupted in the presence of water because water molecules could provide hydrogen bonding interactions that could substantially lack the adhesive strength; hydrophobic interactions may then be introduced into these adhesion systems to solve this issue. For example, the nature of popular host-guest supramolecular interactions include hydrophobic interactions since the macrocyclic host contains a hydrophobic cavity and adsorbing the hydrophobic guest inside, and researchers have already explored the potential of this interaction to achieve underwater adhesion [168]. The cucurbit[n]uril (CB[n]) macrocycles involve a hydrophobic hole and two carbonyl-fringed groups, and then these two carbonyl groups are modified with polymer-grafted silicon surfaces as the negative CB[n] host substrates. On the other hand, aminomethylferrocene (Fc) is also functionally grafted onto silicon surfaces to form a positive guest substrate, in this way, adhesion of both substrates is easily promoted by manually pressing the surfaces together and as a result, glass substrates are well glued underwater (Fig. 3c) [164]. The αCD or βCD is another widely used host supramolecular containing hydrophobic cavity and could offer host-guest interactions for hard surface adhesion for surfaces functionalized with hydrophobic units including adamantane, ferrocene, or azobenzene. In this research, an αCD gel is firstly prepared by homogenous radical copolymerization of acrylamide-CDs with a chemical crosslinker; glass plates are silanized with (3-aminopropyl)triethoxysilane (APTES), then functionally grafted with adamantane or azobenzene to build guest substrates, and then, the host gel adheres on the guest substrates; this bonding adhesion is reversible by regulating UV intensity since the irradiation of UV light could suppress the αCD-azobenzene or αCD-adamantane interactions (Fig. 3d) [165].

DNA contains hydrophobic nucleosides that are selective in pairing with each other via hydrophobic interactions attributed to highly stabilized DNA duplexes [169]. Inspired from this, scientists successfully produced an adhesive polyacrylamide chemical crosslinking hydrogel functionalized by a nucleobase (including adenine, thymine, guanine, cytosine, and uracil) from DNA or RNA, and this hydrogel gained great adhesion and could adhere to super hydrophobic polytetrafluoroethylene (PTFE) surfaces due to strong hydrophobic interactions between hydrogel and target substrate surfaces [170]. Hydrophobic interactions could offer considerable adhesion even in wetting conditions, while either host-guest hydrophobic interaction or DNA inspired hydrophobic interactions between hydrophobic groups require relatively complex synthetic processes.

Other Non-covalent Interactions

There are still some other non-covalent interactions, such as cation-π complexation, metal coordination, π-π interactions, etc. Several of these physical interactions are usually involved in one adhesion system and synergistically work with one or some electrostatic interactions, hydrogen bonding, van der Waals forces, hydrophobic interactions, or even covalent chemical bonding based adhesives (Fig. 4a) [171, 172]. As one example, the cation-π complexation is defined as the non-covalent interaction between the electron-rich π orbitals (benzene, ethylene, etc.) and cations (Li^+, K^+, Na^+, etc.) [173]. Cation-π binding would be particularly strong as cations act with the delocalized π orbitals perpendicular to the plane of aromatic rings, and this interaction surpasses hydrogen bonding in aqueous solutions [174]. In one research, lysine, leucine, or tyrosine is first coated onto mica substrate surface because strong binding of primary amines onto mica via ions exchange with the K^+

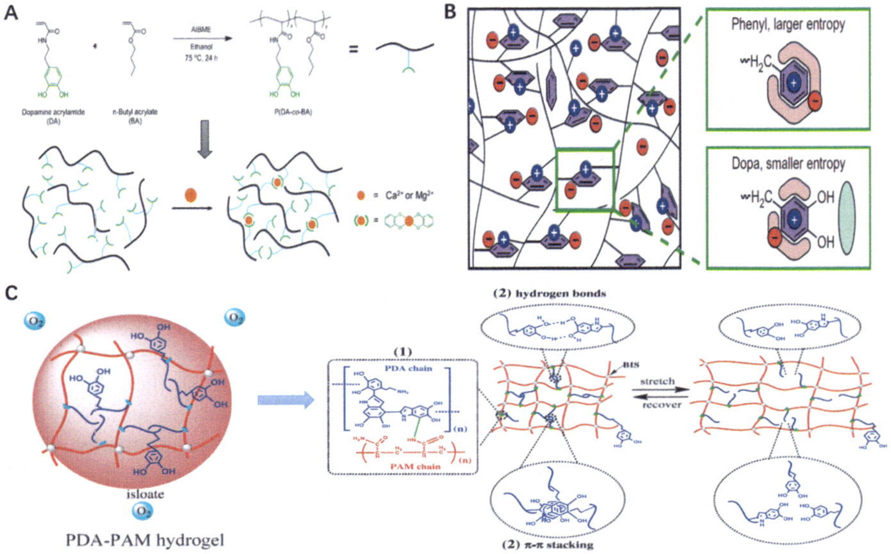

Fig. 4 Other non-covalent interactions synergistically functionalized in an adhesion system: (**A**) a physically crosslinked catechol polymer with metal cations via metal coordination interactions (Reprinted from [171] with permission from the American Chemical Society.); (**B**) cation-π binding mechanism between aromatic or peptides and cations (Reprinted from [175] with permission from Elsevier B.V.); (**C**) reversible duo non-covalent bonding interactions (π-π stacking and hydrogen bonding) occurring in a PDA-PAM adhesion hydrogel system under stretching and releasing states. (Reprinted from [176] with permission from Springer Nature Publishing AG)

ions exists on a single-crystalline mica surface. Next, two amino acid-modified mica plates are compressed together in a high salt buffer surrounding environment, two mica plates are found attached tightly and need to further be forced to separate each other. This is because of the cation-π binding between the cation ion in the surrounding salt solution and the aromatic functional groups (Fig. 4b) [175].

As a special category of dispersion forces, π-π interactions are a subtype of dispersion forces defined between unsaturated (poly)cyclic molecules, and commonly existed between graphene layers [177]. Han and co-workers designed a tough tissue adhesion polydopamine-polyacrylamide (PDA-PAM) hydrogel (free of catechol in PDA), providing reversible non-covalent bonds in a hydrogel system through the π-π stacking among catechol groups, as well as hydrogen bonding to make PDA chains entangled in a PAM network, but these invertible bonds also contribute to the toughness and self-healing ability of this adhesion hydrogel system (Fig. 4c) [176]. In another study, a physically linked 3D hierarchical functionalized-boron nitride nanosheet (f-BNNS)/clay/poly(*N*-isopropylacrylamide) (pNIPAM) ternary network adhesion hydrogel is synthesized through generating several non-covalent interactions; hydrogen bonds abundant in f-BNNS and pNIPAM chains at the two outer layers bonded via π-π and hydrogen bonding interactions, and this hydrogel could repeatably adhere on a variety of surfaces including tissue, aluminum, copper, plastics, etc. depending on hydrogen bonding, metal coordination, π-π complexation, and cation-π interactions [178].

Mechanical Structure Based Bioadhesives

As one particular division of adhesives, mechanical structure based bioadhesives usually adhere target substrates without either non-covalent interactions or covalent chemical bonding, but with a biomimetic physical structure-based absorption, mainly including octopus arm' suckers [179, 180]. Mechanical structure based adhesives would be defined as mimicking the natural hollow acetabulum, an octopus' sucker for example, to achieve the adhesion ability on any foreign target surfaces via the negative pressure between the acetabulum and surfaces; in addition, this adhesion behavior is repeatable [181].

In common with mussels underwater, the octopus can not only adhere to slippery, rough, and abnormal surfaces in the ocean, but the octopus' suckers provide several functions, such as movement, anchoring their bodies to the base, catching and holding objectives, and chemo tactile recognition [182, 183]. All of those behaviors are possible because octopus' suckers exhibit great controllable wet adhesion despite the target surface materials by controlling the negative cavity pressure and thus convincing adhesion on the touching surfaces via utilizing the muscle actuation of their suckers [184]. Inspired from this periodic infundibulum shaped sucker' adhesion behavior, many artificial suckers as the type of mechanical structure based adhesives have been researched to achieve adhesion capabilities [185, 186]. Generally, the researched octopus' suckers inspired adhesive systems induced by a pressure difference would be divided into three types, nanosucker-modified panel with negative pressure produced via solvent treatment during the adhesion process, suction cup adhesive achieving the negative pressure via the physically external preload, and the octopus-inspired thermosensitive smart adhesive pads [187–192].

For example, researchers produced nanosuckers from protruding nanoballs via a solvent treatment technique. The UV resin covered polystyrene (PS) nanosphere assembled layer was immersed into highly polar solvents (like acetonitrile, nitromethane, and propylene carbonate), then the solvent would swell the outer layer of interior nanospheres but not etching any polymer parts to form the cavity nanosuckers, which could adhere the resin layer onto target substrates via the negative pressure between resin's nanosuckers and surfaces (Fig. 5c) [187]. In this study, although the great adhesion ability gained via the negative pressure by mimicking the octopus' physical cavity sucker structure, toxic solvent has to be applied in order to obtain such adhesives.

To further study the sucker morphology, scientists have produced octopus-inspired micrometer-scale dome-like protuberance filled panel adhesives (Fig. 5a). Firstly, polyurethane-acrylate (s-PUA) based liquid precursor was filled into the micrometer-scale holes of a negative silicon mold with specific calculated designed structures and was crosslinked via the precursor using UV light, then an octopus sucker architecture panel was obtained, such perforated-cylinder structures could trap air bubbles inside and produce negative pressure between the suckers and target surfaces when pressing the panel and extruding the air bubbles [193]. Since the

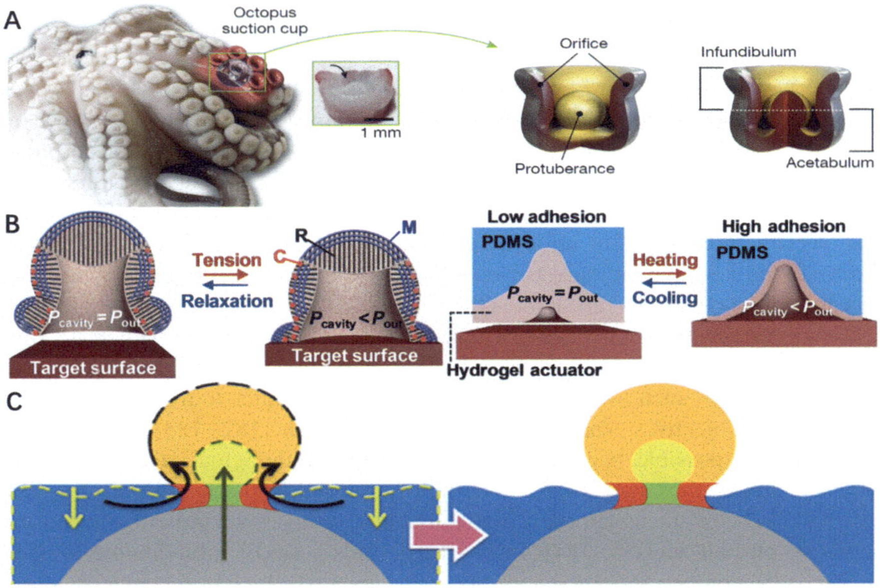

Fig. 5 Physical structure characterized adhesive devices: (**A**) octopus suction cup anatomical architecture (Reprinted from [193] with permission from Springer Nature Publishing AG.); (**B**) the adhesion can be tuned via a pressure difference, and thermosensitive hydrogel (Reprinted from [195] with permission from John Wiley & Sons, Inc.); (**C**) solvent exchange caused pressure difference between the inside and outside of a suction cup. (Reprinted from [187] with permission from the American Chemical Society)

adhesion behavior is mainly based on the physical structure, this adhesive system exhibits reversible and repeatable adhesion in both wet and dry conditions. The most important point is that there is no complicated chemical solvent treatment during the adhesion, and it is more related on the physical microstructure of the suckers. In another research, nanosilica crystals and ethoxylated trimethylolpropane triacrylate (ETPTA) were added on a silicon wafer first by a spin-coating technique, and the mixed solution of PVA and hydrochloric acid was then cast upon the top surface of the silicon wafer silica particles uniformly embedded in a PVA film via covalent bonding on a PVA film through a hydroxyl condensation reaction between the PVA macromolecules and silica particles containing hydroxyl groups to form the positive mold. Then, the PDMS solution was coated and cured on a PVA mold to form the PDMS adhesive pad with nanosucker arrays on the surface. Based on the same mechanism, this would produce negative pressure between the nanosuckers and target surfaces when pressing the pad, and the advantages of this method are gaining more uniform arrays of nanosuckers by utilizing the PVA mold [194].

Researchers never stop maximizing nature's gift and its rich imaginations. In another study, a smart octopus-inspired adhesive system could provide a temperature sensitive on/off adhesion behavior (Fig. 5b). A hole-patterned PDMS film has been made by curing on a silicon mold, and a thermo-responsive hydrogel pNIPAM was then covered on PDMS for the building of a thermo-responsive actuator fence into the PDMS cavity. The coated pNIPAM wall layer could then function as the octopus' muscle to momentarily adjust the cavity volume within the adhesive substrate and produce negative pressure; since pNIPAM hydrogel experiences a phase transition below the lower critical solution temperature (32 °C), which results in changing of the volume, it would go an opposite way when above 32 °C [195]. The temperature-controlled pressure-induced smart adhesive system does not even need manual pressing for producing negative pressure, which can also be achieved with the goal of remote control.

Adhesives Strength and Applications

Except for the primary considerations (biocompatibility and biodegradability of bioadhesives), the adhesive strength is another significant property that needs to be studied in adhesive research. The most important evaluation criteria of adhesion strength is firstly to glue adhesives on different substrates, including hydrophilic substrates like glasses, stainless steel metals, zinc metals, etc., and hydrophobic substrates such as wood, silicons (such as (poly(dimethylsiloxane)) (PDMS)) or fluorine based materials, etc. [196–198]; further incorporating animal tissues or organs like porcine skin, bones, lungs, livers, etc. [199–201]; and then testing with lap-shear or lap-peeling methods, normal direction tensile tests, or friction tests (Fig. 6).

However, animal in vivo assessments are performed in the most straightforward and efficient manner to assess the adhesion strength, as well as biocompatibility and biodegradability of bioadhesives. Animal in vivo tests mainly deal with closuring the wounds or bonding the separated skin tissue together, and sealing or patching the defects and trauma on the viscera (Fig. 7) [205]. Sometimes, bioadhesives that are used to close skin tissue wounds may not be highly required of significant great water-resistance ability of adhesives since there is relatively less liquid in the surroundings, but an excellent waterproof property is more necessary and requisite to adhere defects on viscera because almost all viscera organs such as lungs, heart, stomach, liver, or even the uterus, etc. are appearing in an internal aqueous environment, and much higher adhesion strength is also useful due to resistant high pressure like the lungs [200, 206, 207].

Adhesion Strength of Electrostatic Interaction-Based Adhesives

Many researchers have investigated the electrostatic interaction-based adhesives with successful excellent adhesion strength. For example, Zhang and co-workers designed an adhesion film through a LBL technique; two glass substrates were firstly treated

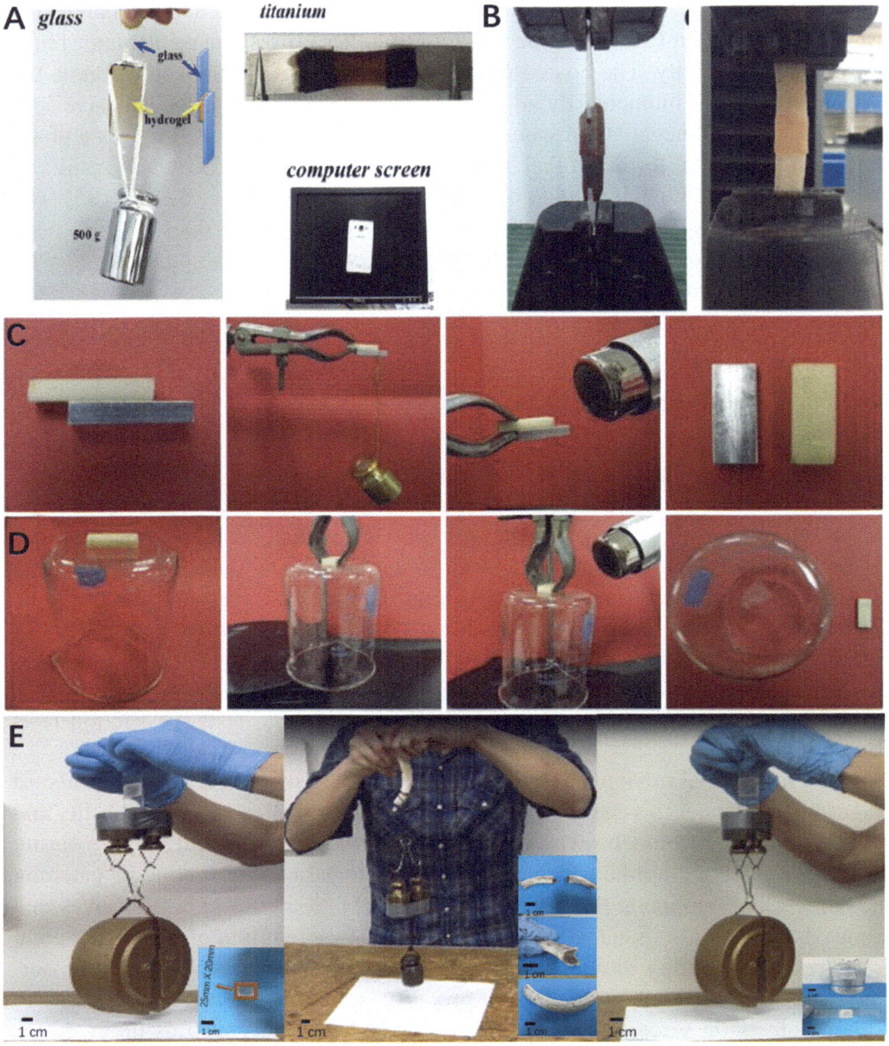

Fig. 6 Evaluating adhesive strength on different substrates: (**A**) glasses, titanium, and plastic substrates (Reprinted from [202] with permission from the American Chemical Society.); (**B**) shear tests on bones and skin tissues (Reprinted from [203] with permission from the American Chemical Society.); (**C**) aluminum, stainless steel, and (**D**) glass substrates (Reprinted from [198] with permission from Elsevier B.V.); (**E**) egg albumen adhesive showed excellent adhesion strength on glass slides (left), adhere fractured bones together (middle), great water-resistance adhesion ability (right). (Reprinted from [204] with permission from John Wiley & Sons, Inc.)

with a piranha solution, and then immersed in a poly(allylamine hydrochloride)-dextran (PAH-D) solution and a hyaluronic acid (HA) solution for 15 min; then, a positively charged amine-containing PAH-D and negatively charged glucuronic acid-containing HA were self-assembled as films via electrostatic interactions

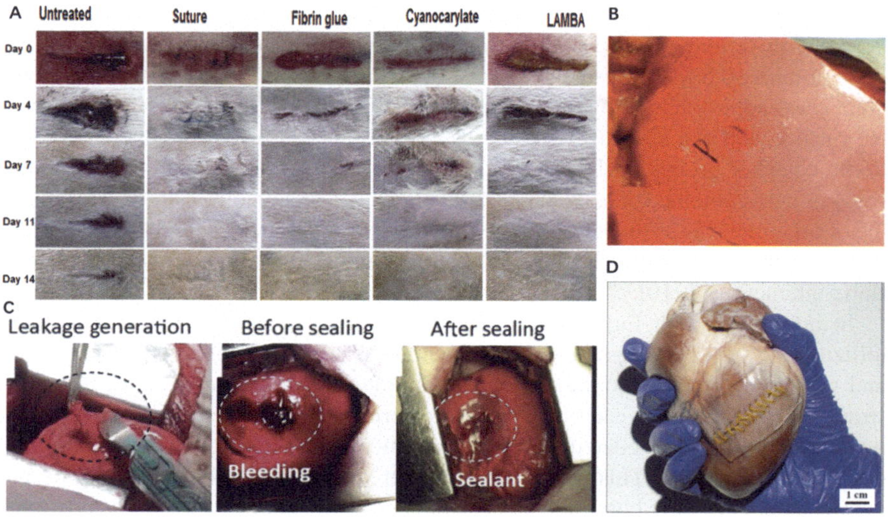

Fig. 7 Adhesion tests on tissue: (**A**) DOPA modified protein adhesive glue rat skin incision and compared to suture, commercial fibrin glue, cyanoacrylate adhesive; (Reprinted from [208] with permission from Elsevier B.V.); (**B**) layer-by-layer assembled polysaccharide nanosheet adhesive sealing the visceral pleural defect of a beagle dog (Reprinted from [75] with permission from John Wiley & Sons, Inc.); (**C**) chemical crosslinked gelatin methacryloyl fixing lung leakage (Reprinted from [207] with permission from Elsevier B.V.); (**D**) octopus-inspired nanosucker arrays adhesive patch adhering to a pig heart. (Reprinted from [194] with permission from the American Chemical Society)

between each layer on each glass substrate, so two glass substrates could be electrostatically glued together by pressing them to hold either a positive or negative layer on the top surface [140]. The lap shear strength could be as high as 7.87 MPa between the two glued glass substrates under lap shear tests after 30 deposition cycles; as well, a glass substrate deposited with 30 cycles of PAH-D/HA film could glue on a bare glass substrate with lap shear value of 4.83 MPa; in addition, 30 cycles of freestanding adhesion films could adhere two pieces of periostea together with lap shear strength of 3.61 MPa, so it may have a potential application on fractured bone gluing and repairing. However, this adhesive film was only glued on hydrophilic substrates under dry conditions, it may not be waterproof since the film is hydrogel based, which would swell and disassemble in an aqueous wet environment [209].

Wang and co-workers developed a Zn^{2+} bridged DOPA functionalized PAA adhesive hydrogel, where the positively charged Zn^{2+} could be gathered by negatively charged carboxylic groups on PAA via electrostatic interactions to form the coacervate in the absence of an acid environment; beyond that, Zn^{2+} could also chelate with catechol groups on DOPA and obtain a highly adhesion hydrogel [74]. This adhesive could strongly glue on both hydrophilic and hydrophobic materials: aluminum, stainless steel, polyethylene (PE), and Teflon under both dry and wet conditions (incubated underwater for 24 h at room temperature), and lap shear adhesion tests were conducted to assess the adhesion value. In dry condition, the shear adhesive strength could reach 6 MPa on aluminum, 3.5 MPa on steel, 1 MPa on PE, and 0.5 MPa on Teflon; in wet condition, the shear adhesion value decreases to 1.6 MPa on aluminum, 0.5 MPa on steel, 0.4 MPa on PE, and 0.3 MPa on Teflon. It showed great shear adhesion binding on aluminum surface and could adhere on a hydrophobic surface, and had a relatively strong water-resistance ability which increased due to the catechol groups on DOPA; so, it may be used for skin wound closure, and some viscera contained or surrounded by a relatively less aqueous solution. However, the curing time was as long as 24 h for both dry and wet adhesion, and the decreased wet adhesive strength was because of the reversibility of the non-covalent electrostatic interactions [210].

In another research, Fujie et al. produced adhesion polysaccharide nanosheets by utilizing the spin coating-assisted LBL method where positively charged chitosan and negatively charged sodium alginate solution were spincoated on a SiO_2 substrate, respectively, and each layer could adhere together via electrostatic interaction; PVA solution was coated on top as a sacrificial layer to help peel the nanosheet off the substrate [75]. This adhesive nanosheet was designed to seal a lung tissue defect, then a pressure burst test was used to assess the adhesion strength and compared it to a fibrin sheet; the bursting pressure of polysaccharide nanosheet was 31.7 cm H_2O, which was lower than the fibrin sheet (45.0 cm H_2O) after 5 min post-operation, and the bursting pressure increased to 56.7 cm H_2O at 3 and 24 h after repair, which was equal to the pressure of the fibrin sheet. Although, the adhesion strength is not extremely high since all adhesion was based on non-covalent interactions; the polysaccharide nanosheet still could be applied as sealant for lung defect repair because it worked well as commercial fibrin adhesive sheet [211].

Adhesion Strength of Hydrogen Bonding Based Adhesives

One of the most popular and earliest hydrogen bonding based adhesives is Upy supramolecular derived polymers; for instance, Heinzmann and co-workers introduced UPy functionalized telechelic poly(ethylene-cobutylene) (PEB) adhesive polymers (UPy-PEB-UPy); this adhesive polymer showed excellent adhesion strength on quartz glass, regular glass, and stainless steel, and its adhesive strength is evaluated by shear tests [152]. On the quartz slides, the shear strength value was 0.9 MPa, and the shear value was 1.2 MPa on stainless steel substrate; while this adhesion was reversible, the shear force was 32 N in a shear test of rebonded failed samples on quartz substrates with bond areas of 10 mm × 12.5 mm. Another silicon based UPy derived adhesive UPy-polymer (UPy-aminopropyl polydimethylsiloxane) also possessed good adhesion on glass where two UPy-polymer glued glass substrates could carry a load of 1 kg for more than 24 h, and even hold a 2 kg load for 2 min with adhesion areas of 26 mm × 20 mm [151]. Although these adhesives obtained great adhesion strength, such UPy modified polymers needed to be heated to 120 °C in order to gain adhesion, which limited their utilization. Faghihnejad et al. developed another UPy-poly(butyl acrylate) (PBA) adhesive copolymer, and the adhesion force reached 57 mN at 23 °C and increased to 65 mN at 40 °C with a circle contact area of 120 μm in diameter [48]. However, those UPy based adhesives were synthesized via relative complicated reactions.

Furthermore, Xu and co-workers developed an egg albumen based strong adhesive glue by a simple and straight method. Moreover, they evaluated the shear adhesive strength on both hydrophilic and hydrophobic substrates, as well as porcine skin tissue and an in vivo skin wound closure model. The shear strength was 5.3 kPa on hydrophobic PDMS substrates, but is as high as 216 kPa on hydrophilic glass substrates, and 56.2 kPa on porcine skin tissues [199]. It is important to mention that all of those values are higher than commercial medical fibrin glue and cyanoacrylate adhesives. Also, they evaluated the adhesive ability and biocompatibility by in vivo animal tests, and the skin wound site was glued together after 5 min and repaired after 5 days post-operation. There were no chemicals introduced to egg adhesives since no chemical reaction existed for producing and applying this adhesive, but it should be discrete while this egg adhesive is potentially to be applied to human body because eggs cause allergic impacts in 1–2% children worldwide [212].

In addition, a silica-PVA adhesive hybrid film was made by Hu and co-workers based on a mechanism that hydroxyl groups in PVA can bind to silanol groups of silica particles via hydrogen bonding interactions, and they also glued glass substrates with the same concept [155]. Two glass substrates were glued with silica/PVA adhesive films, and assessed the adhesive strength of the overlap joint with conducting lap sheer technique, the measured adhesion value was 2.29 MPa, which was higher than the pure PVA adhesive film (1.52 MPa). Compared to a UPy based adhesive polymer, this hybrid adhesive film was made without complex chemical reactions, but this hybrid adhesive film only adheres on glass substrates that may be because of the limitation of hydrogen bonding interactions between hydroxyl and silanol groups, and it may potentially be applied on the skin tissue for wound closure [213].

Adhesion Strength of van der Waals Force Derived Adhesives

Scientists usually research the van der Waals force derived adhesives inspired by nature and bio-mimic their (gecko, ivy, sundew etc.) physical functional surface structures. For example, Mahdavi and co-workers designed a biodegradable PGSA polymer adhesive with aligned nanopillars inspired from the nanotopography of gecko feet pad surface [160]. Then, the gecko feet-like nanopillars decorated PGSA polymer pattern adhesive was glued on porcine tissue slides and the adhesion strength was evaluated by a shear adhesion test; this adhesive was mainly based on surface characteristic non-covalent van der Waals forces, so it showed a weak adhesion strength in underwater condition, then they modified the gecko feet-like polymer pattern surface with oxidized dextran aldehyde functionalities (DXTA) to provide covalent crosslinking with target substrates since aldehyde has been proved to effectively minimize host inflammation as well [214]. In this way, the DXTA modified pattern showed a 2.5×10^4 N/m^2 maximum shear adhesion force without nanopillars on top, but gained a great adhesion force of 4.8×10^4 N/m^2 when mimicking gecko feet-like surface characteristics on porcine tissue slides; the in vivo adhesion studies are used to evaluate the adhesive strength by gluing fascial flaps with PGSA patterns; the shear adhesion strength value of the gecko feet-like patterned polymer was 0.7 N/cm^2 with DXTA modification and lower than 0.3 N/cm^2 without a DXTA coating after 48 h implantation. Therefore, this adhesive has potential applications for both skin wound closure and viscera injury reparation. Otherwise, based on this research, it proved that van der Waals force could certainly provide and enhance the adhesive but if weak in a wet condition, then it has to seek the assistance of covalent bonding, which makes the design more complicated.

In another research, Sun and co-workers developed a polysaccharides assembled adhesive hydrogel involving sodium alginate and gum arabic inspired from the leaves of sundew which consists of a surface network of nanofibers and nanoparticles [162]. The maximum lap shear adhesion force reached up to 176.0 nN on a 1 in. × 1 in. adhesion area on piranha solution treated glass substrates (24 h soaked treatment), but a pressing force of 25 N was applied in order to adhere the glass slides. In addition, Yu et al. measured the adhesive interaction energy value between a hydrophobic methyl-terminated mussel foot protein monolayer and a mica surface, which was 9.1 mJ/m^2; and this adhesion energy was contributed by relatively weak van der Waals interactions and other types of physical adhesion interactions (hydrophobic interactions, hydrogen bonding, etc.), but there were still other covalent binding like a catechol group and thiol linkage [215].

Adhesion Strength of Hydrophobic Interaction-Based Adhesives

As one of the most important adhesion interactions, hydrophobic interactions cannot be neglected. Ahn and co-workers introduced a reversible silicon based adhesive with waterproof ability via the host-guest concept. They modified silicon wafers with cucurbit[7]uril (CB[7]) and aminomethylferrocene (Fc), then pressed and glued both silicon substrates together, and assessed the underwater adhesion strength by lap shear tests [164]. After 13 h adhesion dispose, the maximum shear adhesion strength reached to 1.18 MPa, which was higher than commercial 3 M double-side tape (0.76 MPa); also, the pull apart force for the substrates pair was as high as 110 N on an adhesion area of 1 × 1 cm^2, but the number was only around 70 N of 3 M double-side tape. Even better, the adhesion strength could reach 0.69 MPa after 5 repeated adhesion cycles. This adhesive could be potentially applied on skin wounds for closure, but the processing time is relatively long, and they only completed the adhesion tests on silicon substrates because the host and guest groups would be first modified on a silicon wafer surface.

In another study, Takashima and co-workers synthesized a cyclodextrin-related supramolecular adhesive which can adhere on hard surfaces through hydrophobic interaction of host-guest recognition [165]. An acrylamide-cyclodextrins gel was first glued on guest group (adamantane, azobenzene, and ferrocene) modified glass substrates, and then the adhesive strength was evaluated with a friction test system (8 mN vertical load, 5 × 5 mm^2 adhesion area); the maximum initial friction force was 65 mN between the adhesive gel and the azobenzene grafted glass slide, where the friction force of the ferrocene modified glass substrate was 60 mN. This adhesive concept gives a hint of application on bone hard tissue adhesion; however, they would need to overcome the disadvantage that the guest coated glass slides should be stored in a fridge for 24 h.

Except for host-guest hydrophobic interaction-based adhesives, there is another bioinspired adhesive hydrogel nucleobase via hydrophobic interactions that has been studied. Liu et al. created a nucleobase-tackified adhesive hydrogel by introducing independent nucleobases into polyacrylamide (PAAm). This species adhesive hydrogel glued many types of surfaces including PTFE, plastic, glass, rubber, steel, and wood; as well as many kinds of tissues or organs such as the, heart, liver, spleen, lung, kidney, bone, and muscle [170]. The adhesive strength was evaluated by 90° peeling test, and the maximum peeling adhesive strength is 780 N/m which was contributed by an adenine-tackified PAAm hydrogel on an aluminum alloy substrate; while, even the weakest thymine-tackified PAAm hydrogel offered a 166 N/m peeling adhesive strength, which was stronger than a pure PAAm hydrogel (40 N/m). The 10 min adhesion period was not that long, but the possible adhesion interactions may cover hydrophobic interactions, hydrogen bondings, π-π stacking, cation-π interaction, and so on.

Adhesion Strength of Mechanical Structure Based Adhesives

As a special type of bioadhesive, the octopus' suction cup inspired adhesives could adhere on substrates which rely on their mechanical hollow structures via the pressure difference between inside and outside of the suction cups or negative pressure produced inside the cavate suctions. Many researchers of octopus suctions inspired adhesives have tested and proved that they are functioning well on both wet and dry surfaces.

For example, Chang and co-workers developed an octopus' suction inspired polystyrene (PS) nanospheres modified UV resin based adhesive pad where the PS nanospheres were dissolved in highly polar solvents to form cavity nanosuckers on the resin surface which would produce negative pressure to glue on target surfaces [187]. The lap-shear method was generated to measure the shear adhesion strength on a strip, which was as high as 75.2 N on an area of 100 mm^2 with a 0.3 kg preload in dry condition. This adhesive could also easily glue on a glass and silicon wafer and lift them up with the same adhesion area. This adhesive could generate extremely high shear adhesion under dry condition, but the organic solvent and UV source have to be applied, which may cause a risk of solvent residual and complicate the manufacturing procedure. Also, the preload should be used before adhesion even if the load is relatively small, which may limit its potential medical application.

Therefore, this octopus' suctions inspired adhesive may be used to close the skin wounds potentially if the UV crosslink source and organic solvent could be controlled at a safe level.

As another instance, Baik and co-workers researched octopi's suction cups inspired polyurethane-acrylate copolymer (s-PUA) adhesive patch that is based on mechanical interlocking on surfaces via a pressure difference of the inside and outside of the suction, and they tested the adhesion strength with a pull-off direction tensile test, and a peel-off shearing test [193]. With the application of 10–35 kPa preload, the normal adhesions of the octopus-inspired structure' patch on a silicon substrate (with an adhesion area of 1 × 1 cm^2) were measured as 120, 42, 39, and 26 kPa under the conditions of oil, water, moist (50% relative humidity), and dry ambient environment, respectively, which were much higher than other designed architecture patches (perforated cylinder, cylindrical pillar, and cylindrical hole). In addition, the maximum normal adhesion strength on glass substrates was 40 kPa under water, and 26 kPa in dry condition, which is close to those on a silicon surface. Furthermore, the highest peel-off strength was 0.25 N on a pigskin surface with the same adhesion area. This adhesive could glue on different substrates and even tissues with considerable adhesion strength, but the relative high preload produced a negative pressure inside the suctions, which limits their application. This octopi-inspired adhesive patch could be potentially used on health tissues for wearable devices adhesion, but it is too hard to apply it for wound closuring or defects fixing since the high preload would cause secondary damages.

Chen and co-workers designed octopus-inspired nanoscale silica colloidal crystals coated poly(dimethyl siloxane) (PDMS) nanosuckers adhesives, and tested the adhesion strength via both a normal direction tensile test and shear direction peel-off shearing test on glass surfaces under dry and wet conditions [194]. Among them, the normal direction adhesion force reached 3.0 N under dry conditions, and 2.8 N in wet conditions both on glass substrates with an adhesion area of 1 cm^2, whereas the shear forces were 1.3 and 1.2 N on dry and wet glass slides, respectively. In addition, this PDMS nanosucker adhesive could glue a porcine heart with an area of 10 cm^2. The normal and shear adhesion forces are comparatively high; however, this adhesive also needs to be pressed before adhesion to produce negative air pressure on the nanosuckers, but they did not mention the value of this preload. It may glue the wounds on heart tissue and further in vivo evaluation may be needed; in addition, it could glue and fix the defects on skin tissues potentially relying on its adhesion strength and materials.

Lee and co-workers fabricated another octopus-inspired thermo-responsive adhesive pad with the formation of pNIPAM hydrogel deposited nanoarrays patterned on a PDMS film, and assessed the adhesion strength via normal adhesion force measurement with an adhesion tester [195]. At 35 °C, the normal adhesion strength could get to 46 kPa on a Si surface, and it could reach 81 kPa at 47 °C and 94 kPa at 61 °C around 30 min, and the adhesion force would be down to 0.3 kPa at 22 °C, which means releasing from target surfaces. This smart adhesive relies on temperature to produce negative pressure via structure changes and does not need the exterior preload before adhering, which may widely apply on skin wound closures; but they have not completed enough wet adhesion tests, so it may not be considered for gluing viscera inside the body. Also, there is no shear adhesion test in this research, and the normal adhesion force increased with temperature, among them, 47 and 61 °C are too high for biological uses.

Conclusion

Each non-covalent interaction derived bioadhesive has certain advantages and disadvantages, and these merits and demerit have been analyzed in detail after presenting the adhesive strength of every research example, as well as potential applications. As a conclusion, electrostatic interaction-based adhesives may have no waterproof ability even with significant adhesive strength, some of them express water resistance and great hydrophobic surface adsorption, but the underwater adhesion strength is relative low because of the reversible non-covalent interactions; then, this bug is fixed with the help of DOPA mussel, which results in a long adhesion curing time. Hydrogen bonding derived adhesives gained higher adhesion strength but the synthesis procedure is complicated, some simple fabricated bioadhesives may need to be assisted by other types of non-covalent interactions or may reveal the limited applications for tissue wounds adhesion. In addition, van der Waals forces based adhesives have weak wet adhesion behavior or even dry adhesive strength, and need to receive the help from other kinds of physical interactions or covalent binding. Hydrophobic interaction-based adhesives reveal a versatile surface-adhesion ability and produce even higher adhesive strength than 3 M tapes, but the most severe issue is that the processing time is too long. Furthermore, as one special bioadhesive, mechanical structure based adhesives have excellent normal direction adhesion force, but they may not be ready to be widely applied on tissue wound closures or defect fixations due to the particular adhesion mechanism.

This review summarized the concepts of non-covalent interactions based adhesives that have been researched in biomaterial and biomedical areas. Different adhesion mechanisms, and composite materials for each type of bioadhesive are presented; concurrently, the adhesion strength analysis, advantages and disadvantages, and potential applications of each particular adhesive example are also fully investigated in detail.

Future Work

Concluding all the characteristics above, future work in the area of medical and biological adhesives is expected to develop non-harmful and degradable adhesives that exhibit better strength both in dry and wet conditions; and would not cause a strong tissue inflammatory response. Furthermore, those adhesives should glue well both on hydrophilic and hydrophobic surfaces. Extra features such as antibacterial, scarless, accelerating wound healing, and adhesion strength can be introduced to the adhesive materials to enhance their performance for different biomedical uses. In addition, adhesives can combine all mechanism categories, including electrostatic interactions, hydrogen bonding interactions, van der Waals force, hydrophobic interactions, and negative pressure inspired suction cups, together to make super adhesive glues for both soft and hard tissue repair.

References

1. Armitage J, Lockwood S (2011) Surgery (Oxford) 29:496–501
2. Spotnitz WD, Burks S (2012) Transfusion 52:2243–2255
3. Cooney G, Kiernan A, Winter D, Simms C (2018) Br J Surg 105:395–400
4. Dattilo PP Jr, King MW, Cassill NL, Leung JC (2002) J Text Apparel Technol Manag 2:1–5
5. Gouttefangeas C, Eberle M, Ruck P, Stark M, Müller J, Becker HD, Rammensee HG, Pinocy J (2001) Clin Exp Immunol 124:398–405
6. Izuta Y, Yasumoto M, Yoshikawa M, Niitani M, Hamada N, Sugita T (2017) Adv Mod Med 1:317–324
7. Zuhr O, Akakpo DL, Huerzeler M (2017) Quintessence Int 48:647–660
8. Pai D, Shenoy R, Chethan K (2018) Int Surg J 5:1690–1696
9. Quinn JV (2005) Tissue adhesives in clinical medicine, PMPH-USA. BC Decker, Hamilton
10. Takayama S, Yamamoto T, Tsuchiya C, Noguchi H, Sato J, Ishii Y (2017) Eur J Orthop Surg Traumatol 27:113–118
11. Kochar MP, Singh SP (2016) Int Surg J 2:369–372
12. Yuenyongviwat V, Iamthanaporn K, Hongnaparak T, Tangtrakulwanich B (2016) Bone Joint Res 5:185–190
13. Khan R, Fick D, Yao F, Tang K, Hurworth M, Nivbrant B, Wood D (2006) Bone Joint J 88:238–242
14. Siemer S, Lahme S, Altziebler S, Machtens S, Strohmaier W, Wechsel H-W, Goebell P, Schmeller N, Oberneder R, Stolzenburg J-U (2007) Eur Urol 52:1156–1163
15. Mehdizadeh M, Yang J (2013) Macromol Biosci 13:271–288
16. Meddahi-Pellé A, Legrand A, Marcellan A, Louedec L, Letourneur D, Leibler L (2014) Angew Chem Int Ed 53:6369–6373
17. Balakrishnan B, Soman D, Payanam U, Laurent A, Labarre D, Jayakrishnan A (2017) Acta Biomater 53:343–354
18. Shin K, Choi JW, Ko G, Baik S, Kim D, Park OK, Lee K, Cho HR, Han SI, Lee SH (2017) Nat Commun 8:15807
19. Fan C, Fu J, Zhu W, Wang D-A (2016) Acta Biomater 33:51–63
20. Schreader KJ, Bayer IS, Milner DJ, Loth E, Jasiuk I (2013) J Appl Polym Sci 127:4974–4982
21. Wistlich L, Rücker A, Schamel M, Kübler AC, Gbureck U, Groll J (2017) Adv Healthc Mater 6:1600902
22. Linderman SW, Golman M, Gardner TR, Birman V, Levine WN, Genin GM, Thomopoulos S (2018) Acta Biomater 70:165–176
23. Han L, Wang M, Li P, Gan D, Yan L, Xu J, Wang K, Fang L, Chan CW, Zhang H (2018) ACS Appl Mater Interfaces 10:28015–28026
24. Ho SS, Keown AT, Addison B, Leach JK (2017) Biomacromolecules 18:4331–4340
25. Hu W, Lu S, Ma Y, Ren P, Ma X, Zhou N, Zhang T, Ji Z (2017) J Mater Chem B 5:575–585
26. Hu W, Zhang Z, Lu S, Zhang T, Zhou N, Ren P, Wang F, Yang Y, Ji Z (2018) Biomater Sci 6:3030–3041
27. Ebrahimian S, Tahmasbi S, Bananzadeh AM, Esfahani FN, Nadri S, Abbaszadeh A (2018) Entomol Appl Sci Lett 5:43–46

28. Cronkite EP, Lozner EL, Deaver JM (1944) J Am Med Assoc 124:976–978
29. Coey J, Whittaker P, Williams G, Ikram U, Page O (2019) Rhinology 57(1):21–31
30. Bré LP, Zheng Y, Pêgo AP, Wang W (2013) Biomater Sci 1:239–253
31. Howell JM, Bresnahan KA, Stair TO, Dhindsa HS, Edwards BA (1995) Antimicrob Agents Chemother 39:559–560
32. Bishara SE, Ajlouni R, Laffoon JF (2003) Am J Orthod Dentofac Orthop 123:21–24
33. Maier GP, Bernt CM, Butler A (2018) Biomater Sci 6:332–339
34. Wang X, Shi J, Wang H (2018) J Alloys Compd
35. Chen Y, Cheng W, Teng L, Jin M, Lu B, Ren L, Wang Y (2018) Macromol Mater Eng 303:1700660
36. Wang B, Lee JS, Jeon YS, Kim J, Kim JH (2018) Polym Int 67:557–565
37. Devaud YR, Züger S, Zimmermann R, Ehrbar M, Ochsenbein-Kölble N (2018) Fetal Diagn Ther 45:1–9
38. Modaresifar K, Azizian S, Hadjizadeh A (2016) Polym Rev 56:329–361
39. Hofman AH, van Hees IA, Yang J, Kamperman M (2018) Adv Mater 30:1704640
40. Heinzmann C, Weder C, de Espinosa LM (2016) Chem Soc Rev 45:342–358
41. Tallawi M, Rosellini E, Barbani N, Cascone MG, Rai R, Saint-Pierre G, Boccaccini AR (2015) J R Soc Interface 12:20150254
42. Ryu JH, Hong S, Lee H (2015) Acta Biomater 27:101–115
43. Favi PM, Yi S, Lenaghan SC, Xia L, Zhang M (2014) J Adhes Sci Technol 28:290–319
44. Chun S, Kim DW, Baik S, Lee HJ, Lee JH, Bhang SH, Pang C (2018) Adv Funct Mater 28:1870372
45. Lin I-C, Seitsonen AP, Coutinho-Neto MD, Tavernelli I, Rothlisberger U (2009) J Phys Chem B 113:1127–1131
46. Savage N (2015) Nature 519:S7
47. Tjong V, Tang L, Zauscher S, Chilkoti A (2014) Chem Soc Rev 43:1612–1626
48. Faghihnejad A, Feldman KE, Yu J, Tirrell MV, Israelachvili JN, Hawker CJ, Kramer EJ, Zeng H (2014) Adv Funct Mater 24:2322–2333
49. Kanduč M, Schlaich A, Schneck E, Netz RR (2016) Langmuir 32:8767–8782
50. Courtois J, Baroudi I, Nouvel N, Degrandi E, Pensec S, Ducouret G, Chanéac C, Bouteiller L, Creton C (2010) Adv Funct Mater 20:1803–1811
51. Iturri J, Xue L, Kappl M, García-Fernández L, Barnes WJP, Butt HJ, del Campo A (2015) Adv Funct Mater 25:1499–1505
52. Yang SY, O'Cearbhaill ED, Sisk GC, Park KM, Cho WK, Villiger M, Bouma BE, Pomahac B, Karp JM (2013) Nat Commun 4:1702
53. Liu K, Yao X, Jiang L (2010) Chem Soc Rev 39:3240–3255
54. DelRio FW, de Boer MP, Knapp JA, Reedy ED Jr, Clews PJ, Dunn ML (2005) Nat Mater 4:629
55. Waite JH (2017) J Exp Biol 220:517–530
56. Sharp KA, Honig B (1990) Annu Rev Biophys Biophys Chem 19:301–332
57. Smith A, Robinson T, Salt M, Hamilton K, Silvia B, Blasiak R (2009) Comp Biochem Physiol B: Biochem Mol Biol 152:110–117
58. Baer A, Hänsch S, Mayer G, Harrington MJ, Schmidt S (2018) Biomacromolecules 19(10):4034–4043
59. Wang L, Wu Y, Li J, Qiao H, Di L (2018) Int J Biol Macromol 120(Pt A):529–536
60. Pinnaratip R, Meng H, Rajachar RM, Lee BP (2018) Biomed Mater 13:025003
61. Xiao S, Ren B, Huang L, Shen M, Zhang Y, Zhong M, Yang J, Zheng J (2018) Curr Opin Chem Eng 19:86–93
62. Wang C, Li Z, Chen J, Yin Y, Wu H (2018) Appl Surf Sci 427:1092–1098
63. Shao H, Stewart RJ (2010) Adv Mater 22:729–733
64. Hwang DS, Zeng H, Srivastava A, Krogstad DV, Tirrell M, Israelachvili JN, Waite JH (2010) Soft Matter 6:3232–3236
65. Zhao B, Zeng H, Tian Y, Israelachvili J (2006) Proc Natl Acad Sci 103:19624–19629
66. Sousa MP, Mano JF (2017) Biomimetics (Basel) 2:19

67. Faley SL, Baer BB, Larsen TS, Bellan LM (2015) Biomicrofluidics 9:036501
68. Kim JH, Han MJ, Seo S (2015) J Polym Sci B Polym Phys 53:453–460
69. Hogrebe NJ, Reinhardt JW, Tram NK, Debski AC, Agarwal G, Reilly MA, Gooch KJ (2018) Acta Biomater 70:110–119
70. Koutsopoulos S (2018) Peptide applications in biomedicine, biotechnology and bioengineering. Elsevier, Amsterdam, pp 387–408
71. Wang D, He Y, Zhang J, Li W, Fu X, Tian M, Zhou Y, Yao Z (2018) Int J Adhes Adhes
72. Wang D, Zhang J, Zhong Y, Chu M, Chang W, Yao Z (2018) RSC Adv 8:18904–18912
73. Liu T, Wang Y, Zhong W, Li B, Mequanint K, Luo G, Xing M (2019) Adv Healthc Mater 8:1800939
74. Wang W, Xu Y, Li A, Li T, Liu M, von Klitzing R, Ober CK, Kayitmazer AB, Li L, Guo X (2015) RSC Adv 5:66871–66878
75. Fujie T, Matsutani N, Kinoshita M, Okamura Y, Saito A, Takeoka S (2009) Adv Funct Mater 19:2560–2568
76. Zhao Q, Lee DW, Ahn BK, Seo S, Kaufman Y, Israelachvili JN, Waite JH (2016) Nat Mater 15:407
77. Li J, Celiz A, Yang J, Yang Q, Wamala I, Whyte W, Seo B, Vasilyev N, Vlassak J, Suo Z (2017) Science 357:378–381
78. Ono K, Saito Y, Yura H, Ishikawa K, Kurita A, Akaike T, Ishihara M (2000) J Biomed Mater Res A 49:289–295:
79. Ishihara M, Ono K, Saito Y, Yura H, Hattori H, Matsui T, Kurita A (2001) International Congress Series 1223:251–257
80. Ono K, Ishihara M, Ozeki Y, Deguchi H, Sato M, Saito Y, Yura H, Sato M, Kikuchi M, Kurita A (2001) Surgery 130:844–850
81. Ishihara M, Ono K, Sato M, Nakanishi K, Saito Y, Yura H, Matsui T, Hattori H, Fujita M, Kikuchi M (2001) Wound Repair Regen 9:513–521
82. Nie W, Yuan X, Zhao J, Zhou Y, Bao H (2013) Carbohydr Polym 96:342–348
83. Cohen B, Pinkas O, Foox M, Zilberman M (2013) Acta Biomater 9:9004–9011
84. Shefy-Peleg A, Foox M, Cohen B, Zilberman M (2014) Int J Polym Mater Polym Biomater 63:699–707
85. Cohen B, Shefy-Peleg A, Zilberman M (2014) J Biomater Sci Polym Ed 25:224–240
86. Foox M, Raz-Pasteur A, Berdicevsky I, Krivoy N, Zilberman M (2014) Polym Adv Technol 25:516–524
87. Pinkas O, Zilberman M (2014) J Biomater Sci Polym Ed 25:555–573
88. Cohen B, Panker M, Zuckerman E, Foox M, Zilberman M (2014) J Biomater Appl 28:1366–1375
89. Aoki H, Taguchi T, Saito H, Kobayashi H, Kataoka K, Tanaka J (2004) Mater Sci Eng C 24:787–790
90. Saito H, Taguchi T, Kobayashi H, Kataoka K, Tanaka J, Murabayashi S, Mitamura Y (2004) Mater Sci Eng C 24:781–785
91. Taguchi T, Saito H, Uchida Y, Sakane M, Kobayashi H, Kataoka K, Tanaka J (2004) Mater Sci Eng C 24:775–780
92. Taguchi T, Saito H, Aoki H, Uchida Y, Sakane M, Kobayashi H, Tanaka J (2006) Mater Sci Eng C 26:9–13
93. Nakajima N, Sugai H, Tsutsumi S, Hyon SH (2007) Key Engineering Materials 342:713–716
94. Araki M, Tao H, Nakajima N, Sugai H, Sato T, Hyon S-H, Nagayasu T, Nakamura T (2007) J Thorac Cardiovasc Surg 134:1241–1248
95. Naitoh Y, Kawauchi A, Kamoi K, Soh J, Okihara K, Hyon S-H, Miki T (2013) Urology 81:1095–1100
96. Hyon SH, Nakajima N, Sugai H, Matsumura K (2014) J Biomed Mater Res A 102:2511–2520
97. Matsumura K, Nakajima N, Sugai H, Hyon S-H (2014) Carbohydr Polym 113:32–38
98. Tsujita H, Brennan AB, Plummer CE, Nakajima N, Hyon S-H, Barrie KP, Sapp B, Jackson D, Brooks DE (2012) Curr Eye Res 37:372–380

99. Kazusa H, Nakasa T, Shibuya H, Ohkawa S, Kamei G, Adachi N, Deie M, Nakajima N, Hyon S-H, Ochi M (2013) J Appl Biomater Funct Mater 11:e180–e186
100. Kamitani T, Masumoto H, Kotani H, Ikeda T, Hyon S-H, Sakata R (2013) J Thorac Cardiovasc Surg 146:1232–1238
101. Johansson E, Blomberg E, Wågberg L (2009) Langmuir 25:2887–2894
102. Kreke MR, Badami AS, Brady JB, Akers RM, Goldstein AS (2005) Biomaterials 26:2975–2981
103. Dyer MA, Ainslie KM, Pishko MV (2007) Langmuir 23:7018–7023
104. Liu Y, Li K (2006) J Adhes 82:593–605
105. Li S, Dong W, Zong Y, Yin W, Jin G, Hu Q, Huang X, Jiang W, Hua Z-C (2007) Mol Ther 15:515–523
106. Mo X, Iwata H, Ikada Y (2010) J Biomed Mater Res A 94:326–332
107. Yamada K, Aoki T, Ikeda N, Hirata M (2007) J Appl Polym Sci 104:1818–1827
108. Smith BL, Schaffer TE, Viani M, Thompson JB (1999) Nature 399:761
109. Baljon AR, Robbins MO (1996) Science 271:482
110. Kim S, Sitti M (2006) Appl Phys Lett 89:261911
111. Kim K, Kim K, Ryu JH, Lee H (2015) Biomaterials 52:161–170
112. Robbins KC (1944) Am J Physiol--Legacy Content 142:581–588
113. Shen L, Hermans J, McDonagh J, McDonagh R, Carr M (1975) Thromb Res 6:255–265
114. Boyer MH, Shainoff JR, Ratnoff OD (1972) Blood 39:382–387
115. Laudano AP, Doolittle RF (1981) Science 212:457–459
116. O'Dell BL (1992) Nutr Rev 50:48–50
117. Oh DX, Kim S, Lee D, Hwang DS (2015) Acta Biomater 20:104–112
118. Bitton R, Berglin M, Elwing H, Colin C, Delage L, Potin P, Bianco-Peled H (2007) Macromol Biosci 7:1280–1289
119. Bitton R, Bianco-Peled H (2008) Macromol Biosci 8:393–400
120. Bitton R, Josef E, Shimshelashvili I, Shapira K, Seliktar D, Bianco-Peled H (2009) Acta Biomater 5:1582–1587
121. Gao S, Zhu Y, Wang J, Zhang F, Li J, Jin J (2018) Adv Funct Mater 28:1801944
122. Kim S, Huang J, Lee Y, Dutta S, Yoo HY, Jung YM, Jho Y, Zeng H, Hwang DS (2016) Proc Natl Acad Sci 113:E847–E853
123. Hong S, Pirovich D, Kilcoyne A, Huang CH, Lee H, Weissleder R (2016) Adv Mater 28:8675–8680
124. Choi YC, Choi JS, Jung YJ, Cho YW (2014) J Mater Chem B 2:201–209
125. Goldmann AS, Schödel C, Walther A, Yuan J, Loos K, Müller AH (2010) Macromol Rapid Commun 31:1608–1615
126. Karabulut E, Pettersson T r, Ankerfors M, Wågberg L (2012) ACS Nano 6:4731–4739
127. Yan Y, Zheng Z, Deng C, Zhang X, Yang P (2013) Chem Commun 49:5055–5057
128. Kord Forooshani P, Lee BP (2017) J Polym Sci A Polym Chem 55:9–33
129. Guo J, Kim GB, Shan D, Kim JP, Hu J, Wang W, Hamad FG, Qian G, Rizk EB, Yang J (2017) Biomaterials 112:275–286
130. Zeng H, Hwang DS, Israelachvili JN, Waite JH (2010) Proc Natl Acad Sci 107:12850–12853
131. Lansdown AB, Mirastschijski U, Stubbs N, Scanlon E, Ågren MS (2007) Wound Repair Regen 15:2–16
132. Sudheesh Kumar P, Lakshmanan V-K, Anilkumar T, Ramya C, Reshmi P, Unnikrishnan A, Nair SV, Jayakumar R (2012) ACS Appl Mater Interfaces 4:2618–2629
133. Hong SH, Kim S, Park JP, Shin M, Kim K, Ryu JH, Lee H (2018) Biomacromolecules 19(6):2053–2061
134. Attalla R, Ling CS, Selvaganapathy PR (2018) Adv Healthc Mater 7:1701385
135. Hwang I, Kim HN, Seong M, Lee SH, Kang M, Yi H, Bae WG, Kwak MK, Jeong HE (2018) Adv Healthc Mater 7:1800275
136. Yuan L, Wu Y, Fang J, Wei X, Gu Q, El-Hamshary H, Al-Deyab SS, Morsi Y, Mo X (2017) Artif Cells Nanomed Biotechnol 45:76–83

137. Zhao H, Sun C, Stewart RJ, Waite JH (2005) J Biol Chem 280:42938–42944
138. Ono T, Sugimoto T, Shinkai S, Sada K (2007) Nat Mater 6:429–433
139. Da Violante G, Zerrouk N, Richard I, Provot G, Chaumeil JC, Arnaud P (2002) Biol Pharm Bull 25:1600–1603
140. Zhang J, Chen D, Li Y, Sun J (2013) Polymer 54:4220–4226
141. Lee H, Lee Y, Statz AR, Rho J, Park TG, Messersmith PB (2008) Adv Mater 20:1619–1623
142. Richert L, Lavalle P, Payan E, Shu XZ, Prestwich GD, Stoltz J-F, Schaaf P, Voegel J-C, Picart C (2004) Langmuir 20:448–458
143. Kong Y, Xu R, Darabi MA, Zhong W, Luo G, Xing MM, Wu J (2016) Int J Nanomedicine 11:2543
144. Grabowski SJ (2011) Chem Rev 111:2597–2625
145. Li J, Wang Z, Wen L, Nie J, Yang S, Xu J, Cheng SZ (2016) ACS Macro Lett 5:814–818
146. Cordier P, Tournilhac F, Soulié-Ziakovic C, Leibler L (2008) Nature 451:977
147. Brunsveld L, Folmer B, Meijer EW, Sijbesma R (2001) Chem Rev 101:4071–4098
148. Song P a, Xu Z, Guo Q (2013) ACS Macro Lett 2:1100–1104
149. Sijbesma RP, Beijer FH, Brunsveld L, Folmer BJ, Hirschberg JK, Lange RF, Lowe JK, Meijer E (1997) Science 278:1601–1604
150. Yamauchi K, Lizotte JR, Long TE (2003) Macromolecules 36:1083–1088
151. van Gemert GM, Peeters JW, Söntjens SH, Janssen HM, Bosman AW (2012) Macromol Chem Phys 213:234–242
152. Heinzmann C, Coulibaly S, Roulin A, Fiore GL, Weder C (2014) ACS Appl Mater Interfaces 6:4713–4719
153. Cheng Q, Li M, Yang F, Liu M, Li L, Wang S, Jiang L (2012) Soft Matter 8:6740–6743
154. Kalasin S, Santore MM (2015) Macromolecules 49:334–343
155. Pingan H, Mengjun J, Yanyan Z, Ling H (2017) RSC Adv 7:2450–2459
156. Dzyaloshinskii IE, Lifshitz E, Pitaevskii LP, Priestley M (1992) Perspectives in theoretical physics. Elsevier, Amsterdam, pp 443–492
157. Gong G, Zhou C, Wu J, Jin X, Jiang L (2015) ACS Nano 9:3721–3727
158. Autumn K, Sitti M, Liang YA, Peattie AM, Hansen WR, Sponberg S, Kenny TW, Fearing R, Israelachvili JN, Full RJ (2002) Proc Natl Acad Sci 99:12252–12256
159. Qian J, Gao H (2006) Acta Biomater 2:51–58
160. Mahdavi A, Ferreira L, Sundback C, Nichol JW, Chan EP, Carter DJ, Bettinger CJ, Patanavanich S, Chignozha L, Ben-Joseph E (2008) Proc Natl Acad Sci 105:2307–2312
161. Xia L, Lenaghan SC, Zhang M, Wu Y, Zhao X, Burris JN, Stewart CN (2011) J Nanopart Res 13:1029–1037
162. Sun L, Huang Y, Bian Z, Petrosino J, Fan Z, Wang Y, Park KH, Yue T, Schmidt M, Galster S (2016) ACS Appl Mater Interfaces 8:2423–2434
163. Yu J, Kan Y, Rapp M, Danner E, Wei W, Das S, Miller DR, Chen Y, Waite JH, Israelachvili JN (2013) Proc Natl Acad Sci 2013:15015
164. Ahn Y, Jang Y, Selvapalam N, Yun G, Kim K (2013) Angew Chem 125:3222–3226
165. Takashima Y, Sahara T, Sekine T, Kakuta T, Nakahata M, Otsubo M, Kobayashi Y, Harada A (2014) Macromol Rapid Commun 35:1646–1652
166. Meyer EE, Rosenberg KJ, Israelachvili J (2006) Proc Natl Acad Sci 103:15739–15746
167. Yuan Y, Zhang Y, Fu X, Kong W, Liu Z, Hu K, Jiang L, Lei J (2017) J Appl Polym Sci 134:45292
168. Nakahata M, Takashima Y, Harada A (2016) Macromol Rapid Commun 37:86–92
169. Schweitzer BA, Kool ET (1995) J Am Chem Soc 117:1863–1872
170. Liu X, Zhang Q, Gao G (2017) Adv Funct Mater 27:1703132
171. Li J, Ejima H, Yoshie N (2016) ACS Appl Mater Interfaces 8:19047–19053
172. Zhu W, Peck Y, Iqbal J, Wang D-A (2017) Biomaterials 147:99–115
173. Ma JC, Dougherty DA (1997) Chem Rev 97:1303–1324
174. Meot-Ner M, Deakyne CA (1985) J Am Chem Soc 107:474–479
175. Gebbie MA, Wei W, Schrader AM, Cristiani TR, Dobbs HA, Idso M, Chmelka BF, Waite JH, Israelachvili JN (2017) Nat Chem 9:473
176. Han L, Yan L, Wang K, Fang L, Zhang H, Tang Y, Ding Y, Weng L-T, Xu J, Weng J (2017) NPG Asia Mater 9:e372

177. Pérez EM, Martín N (2015) Chem Soc Rev 44:6425–6433
178. Tong X, Du L, Xu Q (2018) J Mater Chem A 6:3091–3099
179. Baik S, Kim J, Lee HJ, Lee TH, Pang C (2018) Adv Sci (Weinh) 5:1800100
180. Chun S, Kim DW, Baik S, Lee HJ, Lee JH, Bhang SH, Pang C (2018) Adv Funct Mater 28(52):1805224
181. Ko H, Javey A (2017) Acc Chem Res 50:691–702
182. Wilker JJ (2017) Nature 546:358
183. Chen R, Fu Q, Liu Z, Hu X, Liu M, Song R (2017) Robotics and Biomimetics (ROBIO), 2017 IEEE International Conference on 1002–1007
184. Kier WM, Smith AM (2002) Integr Comp Biol 42:1146–1153
185. Hou J, Wright E, Bonser RH, Jeronimidis G (2012) J Bionic Eng 9:484–493
186. Varenberg M, Gorb SN (2009) Adv Mater 21:483–486
187. Chang W-Y, Wu Y, Chung Y-C (2014) Nano Lett 14:1546–1550
188. Wolf MP, Salieb-Beugelaar GB, Hunziker P (2018) Prog Polym Sci 83:97–134
189. Niu X, Mo Z, Yang X, Sun M, Zhao P, Li Z, Ouyang M, Liu Z, Gao H, Guo R (2018) Microchim Acta 185:328
190. Tramacere F, Beccai L, Mattioli F, Sinibaldi E, Mazzolai B (2012) Robotics and Automation (ICRA), 2012 IEEE International Conference on 3846–3851
191. Tomokazu T, Kikuchi S, Suzuki M, Aoyagi S (2015) Intelligent Robots and Systems (IROS), 2015 IEEE/RSJ International Conference on 2929–2936
192. Follador M, Tramacere F, Mazzolai B (2014) Bioinspir Biomim 9:046002
193. Baik S, Park Y, Lee T-J, Bhang SH, Pang C (2017) Nature 546:396
194. Chen Y-C, Yang H (2017) ACS Nano 11:5332–5338
195. Lee H, Um DS, Lee Y, Lim S, Kim HJ, Ko H (2016) Adv Mater 28:7457–7465
196. Zhou J, Xu T, Wang X, Liu C, Liao X, Huang X, Shi B (2017) RSC Adv 7:4024–4029
197. Lim C, Lee DW, Israelachvili JN, Jho Y, Hwang DS (2015) Carbohydr Polym 117:887–894
198. Luo X, Lauber KE, Mather PT (2010) Polymer 51:1169–1175
199. Xu K, Liu Y, Bu S, Wu T, Chang Q, Singh G, Cao X, Deng C, Li B, Luo G (2017) Adv Healthc Mater
200. Gao Y, Han Y, Cui M, Tey HL, Wang L-H, Xu C (2017) J Mater Chem B 5(23):4535–4541
201. Zhang H, Bré LP, Zhao T, Zheng Y, Newland B, Wang W (2014) Biomaterials 35:711–719
202. Han L, Lu X, Liu K, Wang K, Fang L, Weng L-T, Zhang H, Tang Y, Ren F, Zhao C (2017) ACS Nano 11:2561–2574
203. Lu D, Wang H, Li T, Li Y, Wang X, Niu P, Guo H, Sun S, Wang X, Guan X (2017) Chem Mater 29:5493–5503
204. Xu K, Liu Y, Bu S, Wu T, Chang Q, Singh G, Cao X, Deng C, Li B, Luo G (2017) Adv Healthc Mater 6:1700132
205. Zhang YN, Avery RK, Vallmajo-Martin Q, Assmann A, Vegh A, Memic A, Olsen BD, Annabi N, Khademhosseini A (2015) Adv Funct Mater 25:4814–4826
206. Jiang J, Wan W, Ge L, Bu S, Zhong W, Xing M (2015) Chem Commun 51:8695–8698
207. Assmann A, Vegh A, Ghasemi-Rad M, Bagherifard S, Cheng G, Sani ES, Ruiz-Esparza GU, Noshadi I, Lassaletta AD, Gangadharan S (2017) Biomaterials 140:115–127
208. Jeon EY, Hwang BH, Yang YJ, Kim BJ, Choi B-H, Jung GY, Cha HJ (2015) Biomaterials 67:11–19
209. Annabi N, Rana D, Sani ES, Portillo-Lara R, Gifford JL, Fares MM, Mithieux SM, Weiss AS (2017) Biomaterials 139:229–243
210. Gu Z, Li S, Zhang F, Wang S (2016) Adv Sci 3:1500327
211. Takazawa R, Yamato M, Kageyama Y, Okano T, Kihara K (2005) Tissue Eng 11:618–625
212. Savage JH, Matsui EC, Skripak JM, Wood RA (2007) J Allergy Clin Immunol 120:1413–1417
213. Kohl JG, Malicky DM, Jones AM (2017) Prog Org Coat 107:1–4
214. Wang D-A, Varghese S, Sharma B, Strehin I, Fermanian S, Gorham J, Fairbrother DH, Cascio B, Elisseeff JH (2007) Nat Mater 6:385
215. Yu J, Kan Y, Rapp M, Danner E, Wei W, Das S, Miller DR, Chen Y, Waite JH, Israelachvili JN (2013) Proc Natl Acad Sci 110:15680–15685

Tellurium, the Forgotten Element: A Review of the Properties, Processes, and Biomedical Applications of the Bulk and Nanoscale Metalloid

David Medina-Cruz, William Tien-Street, Ada Vernet-Crua, Bohan Zhang, Xinjing Huang, Athma Murali, Junjiang Chen, Yang Liu, Jose Miguel Garcia-Martin, Jorge L. Cholula-Díaz, and Thomas Webster

Abstract Tellurium (Te) is a brittle, mildly toxic, and rare metalloid with an extremely low abundance in the planet. The element has been used in both its bulk and nanoscale forms for several applications in solar cell industry, semiconductors, catalysis, or heavy metal removal, among others. The end of the last century witnessed an explosion in new strategies for synthesizing different Te nanostructures with controlled compositions, sizes, morphologies, and properties, which allow these structures to enhance their impact in numerous applications. Nanomedicine has recently taken advantage of the metalloid in its nanoscale, showing promising applications as antibacterial, anticancer, and imaging agents. Nevertheless, the biological role of Te within living organisms remains mostly unknown, and just a few reports appear working on this matter. In this chapter, the forgotten elements are extensively studied in terms of its chemical, physical, and geological properties, and its main applications are summarized and studied for both bulk and nanosized tellurium. At the end, tellurium's most important biomedical applications are presented with the aim to establish a general concept of the metalloid as a powerful biomedical tool with a bright future yet to be discovered.

D. Medina-Cruz · A. Vernet-Crua · B. Zhang · X. Huang · A. Murali · J. Chen
Y. Liu · T. Webster (✉)
Department of Chemical Engineering, Northeastern University, Boston, MA, USA
e-mail: th.webster@neu.edu

W. Tien-Street
Department of Chemical Engineering, Northeastern University, Boston, MA, USA

Department of Bioengineering, Northeastern University, Boston, MA, USA

J. M. Garcia-Martin
Instituto de Micro y Nanotecnología, IMN-CNM, CSIC (CEI UAM+CSIC), Tres Cantos, Spain

J. L. Cholula-Díaz
School of Engineering and Sciences, Tecnologico de Monterrey, Monterrey, NL, Mexico

B. Li et al. (eds.), *Racing for the Surface*,
https://doi.org/10.1007/978-3-030-34471-9_26

Keywords Tellurium · Nanomaterial · Properties · Applications · Biomedical Antibacterial

Tellurium, the Last Member of the Chalcogen Family

Tellurium (Te) is a metalloid element whose name is originated from "*tellus*" which means earth, whose properties are in between metals and nonmetals. It is a grayish white substance in appearance, with metallic brightness, that can conduct electricity like a metal. However, chemically it behaves as a nonmetal, with intermediate ionization energy and with oxides that are amphoteric—able to react both as a base and as an acid—or weakly acid [1].

Tellurium has 52 protons in its nucleus, and an atomic weight of 127.6 grams per mole, becoming the fourth element in Group 16, the chalcogen family, also known as the oxygen (O) family, with this element, sulfur (S), selenium (Se), tellurium (Te), and the radioactive polonium (Po) belonging to it. Livermorium (Lv), chemically uncharacterized, is supposed to be a chalcogen as well.

Oxygen, sulfur, selenium, and tellurium are all essential compounds on earth, elemental forms that have existed in their current state since before the planet was formed, being present in the interstellar medium from which the solar system was generated. On the other hand, polonium is naturally generated after the decay of the other chalcogens, while livermorium does not occur naturally at all [2].

In terms of abundance, oxygen makes up 21% of the atmosphere, 89% of the water, and 46% of the earth's crust, while sulfur makes up 0.035% of the planet's crust and 0.25% of the human body. Selenium is the 67th most abundant element in the planet, while tellurium is found in amounts of 5 ppb in the earth's crust and 15 ppb in seawater, being one of the least abundant elements in the whole planet. On the other hand, polonium appears in trace amounts on earth, during processes of radioactive decay of uranium (U) and thorium (Th), whereas livermorium is always produced artificially in particle accelerators [1].

Chemically speaking, oxygen is often excluded from the chalcogenides as a consequence of its distinct chemical behavior from the rest, while tellurium is chemically related to selenium and sulfur. All the chalcogens have six valence electrons, with −2, +2, +4, and +6 as common oxidation states. Overall, they have relatively low atomic radii, with the lighter chalcogens—O and S—being typically nontoxic in their elemental form and ligated to all life forms. However, the heavier ones—Se, Te, and Po—are toxic. Despite this, selenium is an essential nutrient and, although chemically related, tellurium often has unpleasant effects for all living organisms [3, 4].

In terms of variability in their crystallographic form, S, O, Se, and Po have 20, 9, 5, and 2 known allotropes, respectively, while tellurium has only one crystal structure with hexagonal crystallography. It is well known that the feasibility of an element to bond to carbon will determine its bioavailability: starting from oxygen, the tendency of forming compounds with carbon decreases, with less number of them belonging to organotellurium compounds [5].

Together with the abundance, feasibility to form compounds, and chemistry, it is possible to consider tellurium one of the least known elements, not only by researchers but also to all known life forms, hence becoming the forgotten element.

Discovery: A Difficult Task

Franz-Joseph Müller von Reichenstein was born the first on June 1, 1740, in the city of Vienna (Austria), where his father worked as *Thesauriatsrath*–treasurer. Müller finished his law and philosophy degrees in Vienna and went to the School of Mines at Schemnitz (Slovakia). Back then, he started to become actively engaged in processes involving mining, mineralogy, chemistry, and mechanics. Soon after, he was promoted as a surveyor for geological monitoring in Hungary and, under the command of the king Joseph II, he became chief inspector of all the mines, smelters, and salt-works in Transylvania (Romania).

One of Müller's tasks was to analyze ore samples and occurrences of rock or sediments that contained economically important elements, typically metals that could be extracted from the deposit using different mechanical techniques. Mainly, he was working with gold ores from Kleinschlatten (today Zlatna, Romania). Müller described the samples as "*Faczebajer weißes blättriges Golderz*"—white leafy gold ore from Faczebaja—or *antimonalischer Goldkies*—antimonic gold pyrite. One of Müller's duties involved quantifying and determining the composition of the ores. After some experiments, he concluded that these particular sediments did not contain antimony as he guessed. However, bismuth sulfide could be present.

The following year, after the conclusion of some additional tests, Müller reported that his guessing was erroneous and that the ore contained a high concentration of gold and a non-identified metal, analogous to antimony. He wrote a paper entitled "*An Experiment with the Regulus Thought to Be Metallic Antimony Occurring in the Mariahilf Mine on Mt. Fazebay near Salatna*" [6], reporting the discovery and some of the experiments and findings developed.

The publication set the beginning of extensive research that brought Müller to conduct more than 50 tests to different samples of that particular ore. He determined the specific gravity of the mineral and recorded three properties of this compound. In his own words, the unknown element had a white smoke with a radish-like odor that suddenly passed off when heated. Besides, it was able to impart a red color when mixed with sulfuric acid (H_2SO_4), and the appearance of a black precipitate was noticed when mixed and diluted with water. Despite his enthusiasm and perseverance, Müller was not able to identify this metal. He tried to give it a name, and this was *aurum paradoxium* or *metallum problematicum*.

Consequently, Müller started sending samples of this *metallum problematicum* to different chemists, mineralogists, and scientists all over Europe, seeking answers. A small sample arrived at the hands of the Swedish chemist Torbern Olaf Bergman, who was already considered one of the most important chemists in Europe due to his publication in 1775 of the *Dissertation on Elective Attractions*, containing the most extensive chemical affinity tables ever published. However, with such a small sample, Bergman could only prove that it was not antimony, and no further investigations were made.

Müller's discovery was overlooked for a long time, until on January 25, 1798, a German chemist, Martin Heinrich Klaproth, received a sample from Müller. Klaproth's work was aimed to improve the contribution of analytical chemistry and mineralogy in Central Europe. Therefore, he is known as the discoverer of uranium and zirconium, though he did not obtain any of them in the pure metallic state.

Once the *metallum problematicum* sample arrived at Klaproth, he started to isolate the new element. Fortunately, the process that he followed was reported in his notes, which have survived to the present time. Briefly, the ore was pulverized with aqua regia, and the subsequent residue was filtered and slightly diluted with water. Then an alkaline solution made of potassium hydroxide (KOH) was incorporated to the solution and as a result a white precipitate was found. Nonetheless, the excess of alkali media dissolved it and finally left a brown flocculent deposit, which contained both gold and hydrous ferric oxide. Once the precipitate was removed by filtration, a solution of hydrochloric acid (HCl) was added until neutrality, then a copious precipitate appeared. Afterwards, the precipitate was washed and dried and finally stirred up with oil. The resultant paste was introduced into a glass retort, which was gradually heated until a red color appeared. When the apparatus was cooled down to room temperature, metallic globules of the element in the receiver and retort were found. The solid was named tellurium [7, 8].

In 1789, a Hungarian scientist, Pál Kitaibel (1757–1817), discovered the element independently in ore from his hometown Deutsch-Pilsen (Hungary). He sent his research reports to Klaproth for comment, getting a favorable answer. Unfortunately for him, his work was not mentioned in Klaproth's later report, which claimed that existence of a new metal, since the discovery was attributed to Müller who found it 7 years previously [9].

Years later, S, Se, and Te were involved in the discovery of periodicity, as they were selected by Johann Wolfgang Döbereiner as a trio of elements belonging to the same group and sharing similar properties. John Newlands published several papers around 1865 where the elements were posted on a list ordered by increasing atomic weight and similar physical and chemical properties, which had a recurrence in intervals of eight. Newlands's version of the periodic table contained a "group b" composed of oxygen, sulfur, selenium, tellurium, and osmium [10].

It was not until 1869 when Dmitri Mendeleev proposed the known periodic table, in which one could find "group VI" formed by—from top to bottom—oxygen, sulfur, selenium, and tellurium. Other elements such as chromium, molybdenum, tungsten, and uranium were firstly part of the group, nonetheless they were regrouped and separately named "group VIB." Some changes were made until 1988 when uranium was moved into the actinide series, and, therefore, group VIA comprised of oxygen, sulfur, selenium, tellurium, and polonium [5, 10].

Presence in the Universe and on the Earth: Occurrence and Sources

As it was stated before, the four lightest chalcogens—O, S, Se, and Te—are all essential elements. In our planet, while S and O occur as a constituent of copper ores, Se and Te can be found in small traces, actually so small that there are only 5 parts per billion of Te in the earth's crust and 15 parts per billion of the metalloid in seawater. Therefore, it is ranked as the 75th crust element by content, which shows that it is an extremely rare compound that represents 9 parts per billion of the universe by weight [11, 12].

The conditions present during the formation of the Earth are thought to be one of the causes of the rarity of this metalloid on the planet. During those times, there was an absence of water and oxygen, and therefore, the stable form of elements was controlled by the reductive power of free hydrogen. Under this scenario, certain elements like tellurium were found as volatile hydrides, which were severely depleted through an evaporation process.

However, it is possible to find the metalloid in many minerals in nature; some of them are listed in Table 1.

Tellurium is sometimes found in its native or elemental form; however, it is more commonly found forming tellurides of gold such as calaverite and krennerite, petzite, and sylvanite. Interestingly, gold itself is usually an element which does not like to combine with anything—due to its behavior—but when it is present in forming a chemical compound, tellurium is the most common combination [14, 15].

Table 1 Tellurium ore found in nature

Compound	Formula	Structure	Occurrence
Altaite	$PbTe$	Isometric crystal	Central/East Asia
Hessite	Ag_2Te	Monoclinic crystal	USA
Sylvanite	$AuAgTe_4$	Monoclinic crystal	Romania, Australia, Canada, USA
Calaverite	$AuTe_2$	Monoclinic crystalline	Romania, Australia, USA
Tetradymite	Bi_2Te_2S	Trigonal crystal	Slovakia
Coloradoite	$HgTe$	Cubic crystal	Australia
Empressite	$AgTe$	Orthorhombic crystal	USA
Kostovite	$AuCuTe_4$	Orthorhombic crystal	USA, China, Russia, Bulgaria
Krennerite	$AuTe_2/Au_3AgTe_8$	Orthorhombic crystal	Romania
Melonite	$NiTe_2$	Trigonal crystal	USA
Petzite	Ag_3AuTe_2	Isometric crystal	Romania
Rickardite	Cu_7Te_5	Orthorhombic crystal	USA
Stützite	$Ag_{5-X}Te_3$	Hexagonal crystal	Romania
Tellurobismuthite	Bi_2Te_3	Trigonal crystal	USA
Temagamite	Pd_3HgTe_3	Orthorhombic crystal	USA
Vulcanite	$CuTe$	Orthorhombic crystal	Japan, Russia, USA, Norway
Weissite	$Cu_{2-X}Te$	Hexagonal crystal	USA, Australia, Sweden

Source: From [13]

The metalloid is also found combined as tellurides of more common metals. Natural tellurite and tellurate minerals are formed by oxidation of tellurides near the Earth's surface. Because of the differences in their chemical behavior, tellurium does not normally substitute sulfur in minerals as selenium does. Consequently, a variety of common sulfide minerals contain high amounts of selenium, with only traces of tellurium [15].

Because of its low abundance and rare presence of the planet, tellurium is mostly produced as a by-product of milled copper, iron, and other base-metal rich ores. These mineral deposits contain trace amounts of tellurium, a reason why its production is ligated to indirect mining. Therefore, the metalloid mainly comes from copper refining as a by-product with the concentrations of 1–4%.

Large-tonnage, low-grade ores from copper and copper–gold porphyry-type deposits are the reported primary sources. In the Ural Mountains of Russia, massive ores with high concentrations of iron and copper have been found, proving that seafloor volcanogenic massive sulfide (VMS) deposits—a type of metal sulfide ore deposits created by volcanic-associated hydrothermal events—have the ability to produce extended quantities of tellurium as a by-product. Besides, tellurium is also dispersed as a trace element or as micro-inclusions of its minerals in pyrite ores, which can also yield copper and lead.

The most important sources of telluride minerals have been found around creeks in Colorado Springs (US), several dispersed locations in Western Australia, a few places in southwestern, such as Dashuigou—in the Sichuan province—and the northwest of China. Other pivotal locations for recovery of tellurium are associated with VMS, like the ones in the Kankverg gold–tellurium mine, in the Skellefte district (Sweden), or a gold mining reservoir in the northern of Sonora (Mexico). All these sources provide important extraction points associated with gold deposits. However, the yield of tellurium extractions from these ores is low in comparison to other methodologies, such as anode sludge—a type of electrolytic metal recovery—from milling of large base metal deposits.

China and Sweden are the two major tellurium producers of the world, rendering about 15% of global production. Globally speaking, 400–500 tons were produced in 2017 from the sources mentioned above. Other countries, like Japan or Peru, are also significant tellurium producers. Europe was the dominating continent in the production of tellurium, reaching about 30% of the whole production (Fig. 1) [16]. In terms of the effect in the market, those countries that are producers have a significant impact, with China, Japan, and Peru leading the market.

In the United States, Te demand is mainly focused on domestic supply and importation. Almost all domestic production is provided by ASARCO copper refinery located in Amarillo, Texas. At this facility, tellurium is extracted as a by-product of copper anode slimes and skimming of lead refinery ores mined from deposits in the western United States. In the west of the country, some additional recovery spots supply tellurium from the skimming of lead. Nevertheless, supplies are also imported from China and Canada (about 75%), as well as the Philippines and Belgium [16] (Table 2).

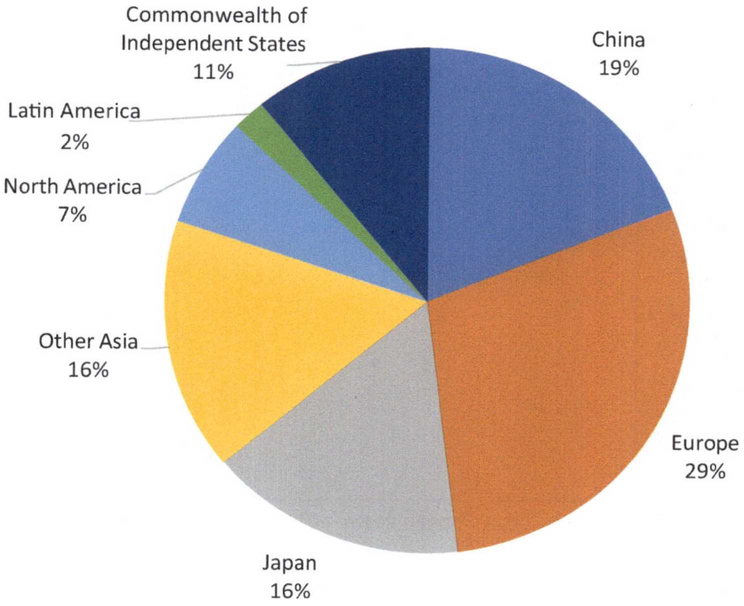

Fig. 1 Percentage of estimated world tellurium refinery production

Table 2 World refinery production of tellurium

Year/Country	2010 (tons)	2011	2012	2013	2014	2015	2016	2017
China	–	–	–	–	–	–	280	280
Canada	8	10	11	10	9	10	18	20
Japan	51	40	45	45	32	35	33	38
Peru	30	30	–	–	36	36	–	–
Russia	34	34	35	40	32	35	35	35
Sweden	–	–	7	–	31	40	39	40
USA	W	W	W	W	W	W	W	W

W avoid giving company proprietary data
Source: Minerals Yearbook 2011–2018

The metalloid value shows a progressive and descended trend (Fig. 2), reaching 40$ per kilogram in 2017. Some factors that affect this trend are related to the current production methods and the improvement that they are experiencing, leading to high efficiency in recovery. However, in future a more critical enhancement of this method is expected with the exploitation of subaquatic deposits. Figure 3 shows the variability of tellurium value. The price of the metalloid has fluctuated due to some external factors; however, it is generally going downwards. The continuous downward trend of prices is due to the new mine, improved refining technology, and improved recovery efficiency.

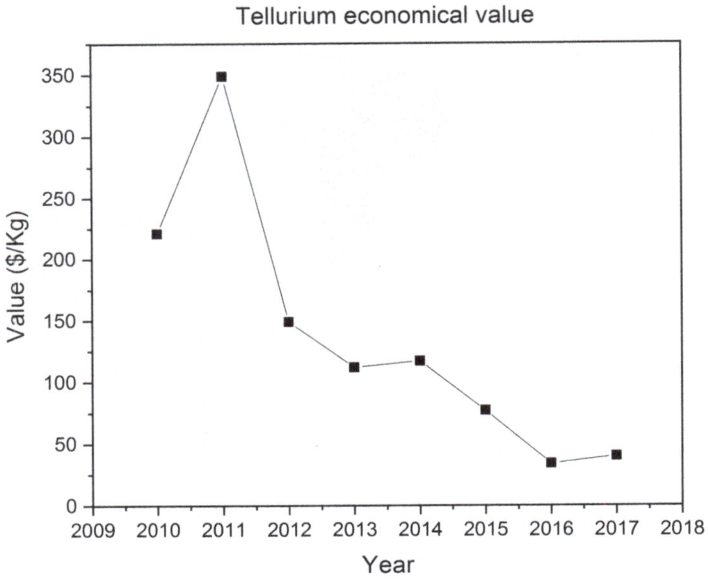

Fig. 2 Economic value of tellurium over time

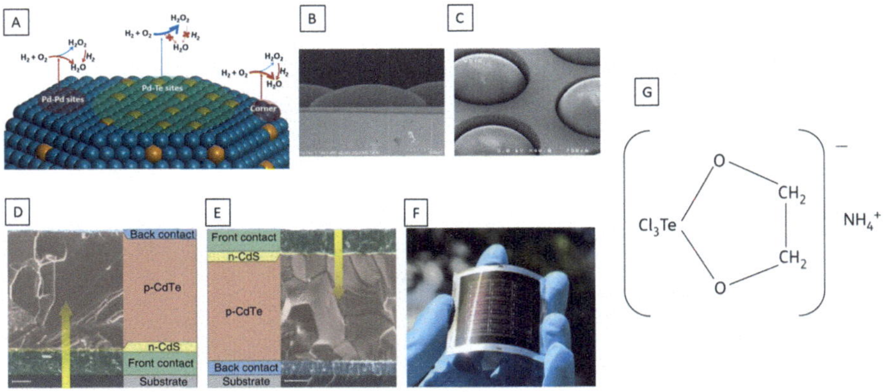

Fig. 3 (**A**) Different applications of tellurium. The proposed mechanism for the direct production of H_2O_2 over Pd–Te/TiO_2 catalysts (in blue and brown, spheres of Pd and Te atoms, respectively) [17]; Scanning electron micrograph (SEM) of chalcogenide microlens array, with the cross-sectional image of the molded chalcogenide microlens array (**B**) and the appearance of molded chalcogenide microlens array (**C**) [18]; SEM image and schematic of the cross-section of a CdTe solar cell in the conventional superstrate configuration (**D**) and the substrate configuration (**E**) which permits the use of opaque substrates such as metal foils. The scale bars represent 1 μm. The yellow arrows show the direction of illumination; (**F**) Photograph of a sample with several CdTe solar cells on the flexible metal foil [19]; and the chemical structure of the tellurium-based compound AS101 (**G**) [20]

Physicochemical Properties of Tellurium

Physical Properties

Tellurium can be found in two forms in nature: a crystalline structure and an amorphous powder—this later is not considered an allotrope for many authors. Amorphous tellurium is a brownish-black powder usually obtained by precipitation methodologies—for instance, the reduction of a solution of tellurium dioxide (TeO_2) with sulfurous acid (H_2SO_3). When heated, it is thermodynamically prompt to transform into crystalline tellurium, releasing around 2500 calories of heat. The transformation is undergone with a strong influence of the features of the element, such as the finesses of division when powdered [21].

On the other hand, crystalline tellurium is a brittle, silvery crystalline mass that is facile to convert into powder. This structure can be obtained by sublimation of tellurium or by its slow formation using a decomposition reaction—such as the one observed in hydrogen telluride (H_2Te). Moreover, crystalline tellurium can be originated by oxidation of an aqueous solution of an alkali telluride at atmospheric pressure. The shape of the crystals obtained is found to be prismatic in most of the cases [22, 23], and regarding the crystalline structure, some sources say hexagonal [24, 25], some others orthorhombic [26] or even trigonal (rhombohedral) [25].

In general, tellurium can be defined with normal physical parameters, and some of them are reviewed in Table 3—it is important to mention that there is an influence of the metalloid allotrope structure on these parameters, with variations if one allotrope is studied compared to other one.

The density of crystalline tellurium shows variability depending on the structure and the production method: 6.0 g/cm^3 for the amorphous, whereas for the crystal ranges from 6.24 g/cm^3 after fusion to 6.15 g/cm^3 after precipitation. Therefore, this density is altered under the influence of heat and related to the presence of the two dynamic forms [27]. Tellurium shows a low resistance to scratch in the Mohs scale, with a value close to calcite. Besides, it shows poor elastic behavior when placed under pressure.

Tellurium melts at 449.5 °C in the amorphous phase or at 452 °C in the crystalline one, and boils at nearly 990 °C for the amorphous but at 1390 °C for the crystal under atmospheric pressure. However, it volatilizes at temperatures close to 430 °C in a cathode ray vacuum producing a yellowish vapor [28]. Similar to density, the specific heat of the solid is variable. It can take values from 0.0475 cal/(g.K) for the distilled element to 0.0524 cal/(g.K) for the precipitated amorphous substance. It has been reported that it can increase by around 8% due to the exposure to X-rays, consequently inducing a possible change of the internal structure [29].

In solid form, tellurium is a deficient heat and electricity conductor. Although the metalloid has many metal-like properties, it breaks apart rather quickly and does not conduct electric current very well. However, depending on the specific direction of the atomic alignment of its structure, it can be considered an excellent electrical semiconductor. This conductivity may be increased slightly when exposed to light,

Table 3 Tellurium physical properties and values *I*

Physical parameter	Value
Density (when solid at RT)	6.0 (amorphous) to 6.25 (crystal) g/cm^3
Density (when liquid)	5.70 g/cm^3
Molar volume	2.05×10^{-5}
Mohs hardness	2.25
Bulk modulus	64 GPa
Shear modulus	16 GPa
Young modulus	43 GPa
Refractive index	1.002495 at $\lambda = 589$ nm (vapor) 1.26 at 260 nm–4.5 at 720 nm (thin film)
Melting point	449.5 °C (amorphous)–452 °C (crystal)
Boiling point	988 °C (amorphous) 1390 °C (crystalline)
Heat of fusion	17.5 kJ/mol
Heat of vaporization	48 kJ/mol
Specific heat	199–219 J/(kg. K)
Thermal conductivity	1–3.4 W/(m·K) in single crystal; polycrystalline: about 6 W/m·K
Electrical Resistivity	1–50 mΩ·m
Magnetic susceptibility	Between -4×10^{-8} m^3/kg and -3×10^{-10} m^3/kg
Atomic radius	123 pm
Covalent radius	138 pm
Van der Waals radius	206 pm
Crystal structure	Trigonal or hexagonal or orthorhombic

showing a photoconductor behavior that will be used for some of its most important applications as will be discussed later [30]. At elevate temperatures, higher than 360 °C, there is a negative dependence between the specific resistance and temperature, the higher the temperature the lower the specific resistance [31]. Nevertheless, tellurium in liquid state is considered a relatively good electrical conductor. Also, at the freezing point, the specific conductance is about 15 times that of the solid, while at 500 °C it is equal to 1/6 of that of mercury (Hg) at the ordinary temperature [32].

It has been observed that tellurium presents less resistance toward direct current than to an alternating current of high frequency. As it was stated before, a small increase in electrical conductivity can be observed when the material is exposed to light, this effect is significantly more visible in other chalcogens such as selenium. Therefore, tellurium can be considered as a photothermal material [33].

The "Hall effect"—the shifting of the equipotential lines that takes place when the current flows along a thin strip of metal located between the poles of a magnetic field—is more significant for tellurium than any other metal. Therefore, this metalloid is diamagnetic with a susceptibility which varies slightly with the temperature, however, have a sharp decrease at the melting point [33].

The linear coefficient of thermal expansion of tellurium is 17×10^{-6} °C^{-1}, so at room temperature (RT) the linear expansion is 3.4×10^{-4} [34]. For the range of wavelengths between 3000 and 5000 Å and for different positions relative to the plane of incidence, various optical constants including the refractive index, reflecting power and absorption constant, have been determined for the isolated crystals. Light absorption by this metal in vapor phase approaches a maximum at 1200 °C [35].

Vapor density determinations enable to see that the rare element at elevate temperatures (around 1500 °C) can be found in diatomic form, while at temperatures of 2100 °C the dissociation into single atoms starts to occur. However, some researchers showed that tellurium element is monatomic at 357 °C. Similar to sulfur and selenium, at lower temperatures, complex formulations of the molecule can be observed [36, 37].

When crystalline, tellurium shows a crystal structure made up of spiral chains of bonded atoms packed in a hexagonal array. The crystalline structure leads to a high grade of symmetry which allows the presence of enantiomorphic forms—a couple of configurations related to each other as left and right hand, therefore considered as mirror images that can never be identical although any reorientation is achieved—containing spirals of opposite handedness. The standard processes of crystal structure determination do not have the ability to discriminate between the enantiomorphs, nor is this feasible using anomalous dispersion unless there is sufficient sphericity in the tellurium electron density due to bonding. Besides, this sphericity allows the observation of tiny but measurable dissimilarities from unity in the flipping ratios for polarized neutron scattering due to the polarization dependence of the Schwinger scattering—proportional to the difference between the nuclear charge and the X-ray scattering factor [38, 39].

Besides crystalline and amorphous tellurium, the colloidal version of the metalloid is also known. On reduction of dilute solutions of telluric acid (TeO_2) using an active reducing agent—such as hydrazine, hydroxylamine, sulfurous acid, or salts of these compounds—brown or blue colloidal solutions can be obtained, with negatively charged particles [40, 41].

Different methodologies for preparing colloidal tellurium have been reported. Tellurium can be obtained in a colloidal solution when a brief amount of the metalloid is dissolved in a boiling solution of KOH and then—with energetic stirring—poured into cold water.

Other processes to obtain colloidal aqueous suspensions are by the cathodic pulverization of the element underwater and by the electrolysis of an aqueous solution of telluric acid containing either potassium cyanide or ammonium oxalate [42, 43]. When added to glass, colloidal tellurium produces a change in coloration, from blue to brown depending on the size of the particles: the blue glass containing more massive particles, while the brown formed in the presence of small ones. Polytellurides can be added to glass, and when they are present, color can be imparted, which is red or violet-red [44].

Chemical Properties

According to the aforementioned physical properties, Te has semiconductor behavior tendencies, and its chemical properties are considered similar to sulfur and selenium. Nonetheless, the compounds formed are less stable than sulfur and selenium analogs [45], which explains why the number of known compounds with tellurium in nature is smaller than those for its family members on top of the periodic table.

In this section, some of the most stable, abundant, and natural forms, in which tellurium can be found are explained, with some of the primary reactions that they can undergo, giving rise to other different compounds. Before starting, it is recommendable to state the difference between telluride and tellurite/tellurate. While the first one is referred to the anion Te^{2-} and its derivatives, tellurite/tellurate is referred to the anions $TeO_3{}^{2-}$ and $TeO_4{}^{-}$ and their derivatives.

Tellurides Tellurium shows the ability to combine with other metals to form tellurides, chemical species with properties similar to sulfides and with a metallic appearance. When alkali metals—lithium, sodium, potassium, rubidium, and cesium—are found together with the metalloid, they have a high solubility in water, rendering Te^{2-} and HTe^{-} ions that are dissolved in the aqueous media.

As it was stated before, one of the most common ways of finding tellurium in nature is in association with gold, and remarkable examples of these tellurides are found in minerals such as krennerite ($AuTe_2$) and sylvanite ($AgAuTe_4$), among others explained in a previous section, which shows a covalent bond. Nevertheless, most of these minerals do not have a great economic impact in industry, except combination with cadmium and lead, which are widely used in industry as thermoelectric materials [46].

Sodium and potassium combine with tellurium with huge heat production. The reaction might be controlled by dissolving the alkali metal in liquid ammonia (NH_3), producing Na_2Te, K_2Te, and the polytelluride Na_4Te_3. These compounds are easily oxidized, and they have to be protected from atmospheric oxygen [47]. Moreover, concentrated solutions of alkali hydroxides—such as NaOH and KOH—may be used to produce red-colored solutions of a mixture of telluride and tellurite. If water is added, the mixture undergoes decomposition, releasing tellurium. However, if sodium hyposulfite ($Na_2S_2O_3$) is used, crystalline Na_2Te will be obtained as a unique product [48].

Noble metals other than gold, such as silver (Ag) and, to less extent, copper (Cu), are displaced by tellurium from solutions of their salts, showing a metal-like behavior. However, its true metalloid nature is shown when, in the form of electrodes embedded in an alkaline solution, it is dissolved at both electrodes, as positive Te^{4+} ions in the anode and as negative Te^{-} ions in the cathode. If any of the electrodes containing tellurium is replaced by one made of platinum (Pt), the metalloid will be separated at the platinum electrode [49].

Hydrogen telluride Tellurium can react with hydrogen at high temperatures to form hydrogen telluride (H_2Te). The action of strong acids on several tellurides will derive in the generation of this compound, a gas at room temperature, with a strong smell. Hydrogen telluride behaves as hydrogen compounds of sulfur and selenium, decomposing by an oxygenated atmosphere in aqueous solution [50].

$$2\mathrm{H}_2Te + \mathrm{O}_2 \rightarrow 2\mathrm{H}_2\mathrm{O} + 2Te$$

At room temperature, it degrades into their constituent elements in a photocatalytic reaction.

$$H_2Te \rightarrow H_2 + Te$$

It also reacts with many metals, producing tellurides. If finely divided, the element is not affected by oxygen in the atmosphere; however, it burns with a blue flame producing tellurium dioxide (TeO_2) at temperatures higher than its melting point, even in the driest oxygen atmosphere [51, 52].

Tellurium dioxide When tellurium is heated in an open atmosphere, it burns producing TeO_2, a white compound that is volatile if heated. This oxide is found in two different structures in nature: β-TeO_2, with orthorhombic crystallographic structure—three unequal axes at right angles—and yellow color, and α-TeO_2, with tetragonal structure—three axes at right angles, two of them equal. α-TeO_2 will slowly convert to β-TeO_2 under pressure. Both the structures contain four coordinated Te atoms with the oxygen atom at the corners of a trigonal bipyramid. Mainly, in the α-form, all the vertices are shared to give a rutile-like structure—tetragonal unit cell—where the O-Te-O bond angle is 140° [50].

The dioxide shows poor solubility in water, while readily soluble in strong acids and alkali metal hydroxides, such as NaOH. Amphoteric in nature, it can react with acids to produce tellurium salts, and with bases to give rise to some tellurides. Besides, it can be easily oxidized to telluric acid ($Te(OH)_6$) [53].

Tellurium acids Tellurous acid, H_2TeO_3, is produced by the oxidation of the metalloid with nitric acid (HNO_3). White in appearance, it is slightly soluble in water, and it shows poor acid properties. In the presence of strong acids, it will behave as a base, generating extremely unstable salts with water. As it could not be easily expected, it behaves different from selenous acid (H_2SeO_3), being a metastable compound with also no similitudes with sulfuric acid (H_2SO_4). If mixed with water, crystals with an octahedral structure are formed, with a defined composition, $Te(OH)_6$ that, after heating, will form telluric acid, H_2TeO_4, to finally produce tellurium trioxide (TeO_3) in anhydrous form, with indifference toward water and a yellowish color [50].

Telluric acid may be generated from the oxidation of Te or TeO_2 with strong oxidizing agents such as hydrogen peroxide (H_2O_2) or chromium trioxide (CrO_3).

$$TeO_2 + H_2O_2 + 2H_2O \rightarrow Te(OH)_6$$

If crystallized, telluric acid solutions give $Te(OH)_6{\cdot}4H_2O$ at temperatures below or close to 10 °C. Although technically being oxidizing compounds, it is slow in reactions. When anhydrous in the form of TeO_3, the acid is stable in air at high temperatures; however, above 100 °C it will generate polymetatelluric acid [$(H_2TeO_4)_{10}$], a white powder with an amorphous structure, and allotelluric acid, with unknown structure. If heated above 300 °C, it will produce the α-TeO_3, with a closed crystalline configuration [54].

Te chemistry in water It is well known that, as occurs with many other elements, water has a pivotal role in the tellurium chemistry. Te combined with water (H_2O) and ozone (O_3) will produce telluric acid at RT, while peroxides will react with the element, with a strong dependence on the physical state of the element; colloidal tellurium is readily oxidized in a short time in the presence of water, while crystalline tellurium is not readily attacked [55].

Tellurium nitrides Tellurium is oxidized by HNO_3 to produce TeO_2. However, under certain conditions, and using a lot of materials and a high column of liquid, tellurium nitride might be produced as a flesh-colored precipitate that can be dried without decomposition. This compound has been isolated. However, tellurium nitrides are instable. For instance, tetratellurium tetranitride (Te_4N_4) has an unconfirmed structure, while a well-defined tellurium nitride [$Te_6N_8(TeCl_2)_4(THF)_4$] can be obtained when $TeCl_4$ is reacted with a THF solution of $N(SiMe_3)_3$ [56, 57].

Tellurium sulfates The metalloid is easily dissolved in H_2SO_4 to produce a red solution, with a subsequent generation of tellurium and sulfur dioxide (SO_2). If extremely hot and concentrated, the sulfuric acid will produce the same red solution. However, if boiled, white rhombic crystals of pyrotelluryl sulfate, $TeO_2{\cdot}SO_3$ will be produced, a basic tellurium sulfate. The crystals decompose by heating in the presence of water and are soluble in HCl [58, 59].

Tellurium halides Crystalline tellurium can be inflamed by fluorine and chlorine, producing the consequent halogen species. Under carefully controlled conditions in a sealed tube, tellurium reacts with chloride to form Te_3Cl_2. With bromine, the element will produce dibromide, while iodine reacts only at high temperatures, producing a tetraiodide compound. Nevertheless, HCl does not have any effect on the metalloid [60].

Tellurium iodine (TeI) is a gray solid formed by the hydrothermal reaction of Te and I in hydroiodic acid (HI). When this reaction is conducted near 270 °C, it gives the α-TeI, with a triclinic phase. When the same mixture is heated to 150 °C, it is possible to obtain the metastable monoclinic phase β-TeI [61]. Nevertheless, the diiodine formulation (TeI_2) has not been isolated yet [62].

Tellurium tetrabromide ($TeBr_4$) has a similar tetrameric structure to $TeCl_4$. The bromide is a good conductor [63], while the chloride is a volatile compound that can sublimate at 200 °C, being prepared merely by the chlorination of tellurium powder by the addition of heat, isolating the product through distillation. The chloride is especially useful in organic synthesis since when added to alkenes, it can give Cl-C-C-$TeCl_3$ derivatives [64]. Tellurium tetrafluoride (TeF_4) is a stable and hygroscopic crystalline solid that can give rise to tellurium oxide after reaction with water or silica [65].

Organic compounds of tellurium Tellurium forms organic compounds, but with less efficiency and number that selenium and sulfur, due to its metalloid nature. However, it may exert both bi- and tetra-valency in these compounds. The tellurium analogs of common organosulfur functional groups are known, with diorganomono- and ditellurides being the most common organotellurium compounds.

Nonetheless, they do not have many applications. For instance, dimethyl telluride ($Te(CH_3)_2$) is used in the metalorganic vapor phase epitaxy as a volatile source of the metalloid, being the unique organotellurium compounds that have been quantified in environmental samples. Also, diphenyl ditelluride ($Te(Ph)_2$) is used as a source of $PhTe^-$ groups in organic synthesis—particularly in the reduction of aldehydes, alkenes, alkynes, or nitro compounds. Furthermore, certain telluroxides can give rise to alkenes upon heating for the formation of olefins [66, 67].

Despite these few applications, some exciting research is being conducted. For instance, Pradeep Mathur et al. reported an unusual shortening of metal–tellurium bonds and inter- and intramolecular X→Te secondary interactions in the coordination patterns of the aryltellurium halides (ArTeX; X = Br, I) in transition metal carbonyl complexes (Fe, Co, Mn, Re, Ru), which can have significant applications in catalysis [68]. Besides, and as it will be discussed in the next sections, organotellurium compounds can show anticancer effect. Powis et al. reported water-soluble organotellurium compounds of the diaryl telluride, alkyl aryl telluride, and dialkyl telluride type, carrying sulfopropyl groups, were reported as the most efficient tellurium-based inhibitors of thioredoxin reductase ever tested [69]. The thioredoxin system (NADPH, thioredoxin reductase/thioredoxin) is essential for cancer cell growth and inhibition of apoptosis process; therefore, it is a suitable target for anticancer drug development.

Isotopes of Tellurium

Tellurium has 8 stable (or nearly) isotopes, 31 unstable ones, and 17 isomers—the most important are summarized in Table 4. While both selenium and tellurium have similar decay tendencies, their isotopes do not produce proton emission. Indeed, some of the most neutron-starved isotopes of tellurium undergo alpha decay—the atomic nucleus emits an alpha particle and thereby transforms or 'decays' into a different atomic nucleus [70].

Table 4 The most important tellurium isotopes

Isotope	Natural abundance (%)	Chemical form	Half-life
Te-120	0.09	Elemental	Stable
Te-122	2.55	Elemental, oxide	Stable
Te-123	0.89	Elemental, oxide	9.2×10^6 years
Te-124	4.74	Elemental, oxide	Stable
Te-125	7.07	Elemental	Stable
Te-126	18.84	Elemental, oxide	Stable
Te-128	31.74	Elemental, oxide	2.41×10^{24} years
Te-130	34.08	Elemental, oxide	3.0×10^{24} years

Tellurium naturally occurring on earth consists of eight nearly stable isotopes. Two of them are stable: ^{128}Te and ^{130}Te. They undergo double beta decay with half-lives of, respectively, 2.2×10^{24} years—the most extended half-life of all radioactive nuclides—and 8.2×10^{20} years. The longest-lived artificial radioisotope of tellurium is ^{121}Te with a half-life of nearly 19 days. Many nuclear isomers have longer half-lives [71].

^{124}Te can be used as a starting material in the production of radionuclides by a cyclotron or other particle accelerators. Except beryllium (Be), Te is found to be the lightest material observed to usually show alpha decay, commonly with isotopes ^{104}Te to ^{110}Te [70].

While ^{120}Te is used for the production of I-120, which has an application as a PET and beta emitting isotope, ^{122}Te is used in the production of the radioisotope I-122, which is used in gamma imaging. ^{123}Te is used for the production of radioactive I-123, which is used in thyroid imaging, and ^{124}Te is used for the synthesis of both I-123 and the PET isotope I-124. Finally, ^{130}Te is used in the research into double beta decay [72].

Bulk Tellurium: Applications

All the features of tellurium are applied for the development of a multitude of applications in the fields of metallurgy, semiconductors, biological, and other developing technologies. Currently, this rare metal is mostly involved in the field of metallurgy, for the production of useful alloys, and used in solar cells for photovoltaic efficiency. Some biological applications are studied, but not extensively [73]. In this section, some of the most common applications are reviewed here, from metallurgy to biological activities that are present in the bulk form of the metalloid.

Tellurium in Metallurgy

Metallurgy has been using tellurium for a long time as an additive, with the aim to improve several mechanical properties of materials with a high impact in society, such as steel or other ferrous alloys. Besides, tellurium can be mixed with copper and lead, with useful applications, and it can also be found as an additive for welding works.

Approximately 55% of the demand in the market for tellurium is for alloys, combined with other materials. Generally, in steel, as a free machining additive, in copper, to improve machinability—with no reduction of conductivity—in lead, to enhance resistance to vibration and fatigue, and in cast iron—iron–carbon alloys with more than 2% carbon—to help control the depth of chill. In addition, it is widely used in malleable iron as a carbide stabilizer: the addition of 0.04% Te to steel improves mechanical properties, such as bending, cutting, shaping, and turning that impact the final application of the materials. The metalloid is also added to lead to enhance its strength and durability, while decreasing the corrosiveness of acids [74].

Despite the use of tellurium as an alloying agent, the field of metallurgy involves many other applications of the metalloid. As described before, tellurium is a relatively rare element [75]. Consequently, a considerable percentage of metallurgic activities are oriented to the recovery of the element, through several processes that will be described below [76].

Anode slimes comprise nearly 90% of tellurium production. Mainly pyrometallurgical and hydrometallurgical operations are used to remove precious metals and recover tellurium by subsequent electrolysis. However, another important source of the metalloid can be found in the lead smelting processes, usually from metallurgical dust and alkaline skimming. The reductive smelting of tellurium-bearing materials can convert tellurium to alkaline tellurides, such as Na_2Te, which is readily amenable. However, it is also highly reactive at the open atmosphere [75].

Despite the aforementioned processes, the most popular approach and, probably, the most extended, for the recovery of tellurium, involve a hydrometallurgical method from electrolytic copper anode slimes, a process that is quite involved. Depending on the employed telluride—some of the most common ones are associated with gold and silver—phase and chemical composition, and associated metal content of the slime, different processes for tellurium recovery are used. These copper slimes are hydrometallurgically treated through direct leaching of the slimes via sulfuric acid, pressure, aeration, or pressurized leaching in alkaline solution [77].

The most common method of leaching for decopperization in slime treatments is acid pressure leaching, which involves diluting sulfuric acid and leach, under oxygen partial pressure conditions in an autoclave reactor. This acid pressure process leads to the dissolution of all copper and typically up to 80% of the tellurium that can be subsequently recovered [78, 79].

Alternatively, alkaline pressure leaching is primarily used for dissolving slimes that are not easily decopperized and contain high selenium and arsenic content. After dissolving Se and As, tellurium in the leach residue is solubilized with copper in an adapted sulfuric acid leach and recovered via a tellurium circuit. Following the leaching, precipitation of Te involves several steps for recovering the element and a subsequent filtering process that will render the metalloid [80].

Tellurium in Catalysis

The use of tellurium as a chemical and catalyst makes around a 25% of the world market for the metalloid, commonly being employed as vulcanizing agent and accelerator in the synthesis of rubber, as well as a component of catalysts for synthetic fiber fabrication. Heat resistance in rubbers can also be improved upon addition of a low concentration of the rare element. Tellurium catalysts are also utilized particularly for the oxidation of organic compounds. However, one of the most common catalytic applications is related to chlorination, halogenation, and hydrogenation reactions [17].

Giles et al. synthesized various structurally related organo-tellurium agents and proved that a mixture of electrochemical methods, in vitro assays, and cell culture tests can be utilized to rationalize the antioxidant activity of these catalytic agents [81].

Tellurium in Chalcogenide Glasses

The metalloid is used in the formation of chalcogenide glasses, with applications in lenses or fibers. These glasses contain one or more chalcogens—not O or Po—in a covalent network of solids. These glasses behave differently from oxides, lowering the band gaps, contributing to unique electrical and optical properties. Tellurium has the lowest glass-forming ability compared to the other chalcogens since it has the highest molecular weight [18], but it is found to be useful as well due to its characteristics.

Consequently, materials such as AgInSbTe and GeSbTe are chalcogenide glass formers that are applied in rewritable optical disks and phase-change memory devices, as will be discussed in the next section. Nevertheless, they are fragile glass formers: by regulation of the temperature—heating and annealing (cooling)—they can change from amorphous (glassy) to crystalline state and vice versa, thereby switching their optical and electrical properties and making them suitable for the storage of information. Besides, the glasses are used in infrared detectors, moldable infrared optics lenses, and infrared optical fibers, transmitting across an extensive range of the infrared electromagnetic spectrum [82, 83]. Although not all chalcogenide compositions existed in glassy form, some materials can be used to produce alloys in order to form glasses.

Tellurium in Electronic Applications

High-purity tellurium is used in electronics applications, such as thermoelectric and photoelectric devices. Alloys such as Hg-Cd in telluride form can be used as thermal imaging, where this compound allows the conversion from raw image into a crisp picture on the screen [27].

Bismuth telluride (Bi_2Te_3) is a gray semiconductor powder that can be alloyed with antimony or selenium to work as an efficient thermoelectric material. Therefore, semiconductor materials made of bismuth telluride are widely known, mainly used as thermoelectric cooling devices in fields such as electronics or consumer products. These devices consist of a series of pairs of semiconducting materials that are connected to a direct current that provokes that one side of the thermoelement cools, whereas the other produces heat. The application in consumer products are limited, nonetheless, they are mainly utilized in military applications—such as the cooling of infrared detectors, integrated circuits, laser diodes, and medical instrumentation. However, they can be found in some sophisticated portable food coolers as well [84].

On the other hand, TeO_2 is used in the media layer of different rewritable optical discs, including compact discs (CD-RW) and digital video discs (DVD-RW), although their use is starting to be limited. Cadmium zinc telluride (CdZnTe) or CZT, a direct bandgap semiconductor, is employed in solid-state X-ray detectors, semiconductor radiation detectors, photoreactive grafting, electro-optic modulators, and solar cells, among others, while mercury cadmium telluride (HgCdTe) or CMT is utilized as an infrared-sensitive semiconductor material [85]—the quantity of Cd in the compound can be tuned in order to modify the optical absorption of the material so as to get the desired infrared wavelength.

The future of electronic applications of tellurium could be related to the next-generation computer chips, known as phase-change memory (PCM), that are based on the metalloid. PCM is a non-volatile random-access memory that exploited the unique behavior of chalcogenide glasses. The technological development of these devices has split in two directions. The first one is devoted to the enhancement of $Ge_2Sb_2Te_5$ or GST, an interesting phase-change material, with a high heat of fusion that can melt and solidify at a certain temperature, being able to store and release large amounts of energy. This tellurium-based material can recrystallize in 20 ns, allowing bit rates—the number of bits that are processed per unit of time—of up to 35 Mbit/s to be written and direct overwrite capability up to 10^6 cycles [86]. The second current of research is aimed to find alternatives to GST, working on the use of a GeTe-Sb_2Te_3 superlattice to accomplish nonthermal phase changes by slightly modifying—with the use of a laser pulse—the coordination state of the germanium atoms.

Nevertheless, the most well-known application of the element is related to solar cells. Tellurium has led to the development of some of the most efficient solar cells in our society, as part of the synthesis of cadmium telluride (CdTe) semiconductors for solar panels. This alloy gives the highest electrical generation efficiency for photovoltaic solar cells and allows for low-cost power generation [27, 87]. Therefore, CdTe photovoltaics (PV) is the unique thin-film technology with less costs than conventional solar cells, composed of crystalline silicon. Considering one lifecycle, and in comparison, to all solar cell technologies, CdTe PV has the smallest carbon footprint and water usage as well as the lowest energy payback time—time that a PV system has to work to generate the same amount of energy that was utilized for its manufacturing and installation. Besides, it can be recycled, therefore attenuating the growing concern of environmental toxicity of Cd [27].

The most prominent manufacturer of CdTe PV, First Solar, manufactures over 3 gigawatts in 2016, from just 25 megawatts in 2005. This massive commercial success has driven the demand of tellurium to be dominated by the solar cell industry and could help for the environmental issue of renewable energy. Its low-cost technology has made CdTe PV a success, due to the combination of adequate efficiency with lower module area costs. Costs of CdTe PV modules can be divided into direct costs, which reached \$0.57 per watt in 2013, [88] and capital cost per new watt of capacity which is nearly \$0.9 per watt—including land and buildings.

This, however, has put a limit on other researchable applications for such a rare metal [89], leading to many unknowns that should be addressed and whose applications will be discussed in the next section.

Tellurium in Biological Applications

Sharing little with tellurium semiconductor technology, the organic applications of the metalloid are not fully defined and investigated, with many unknowns that should be addressed, especially related to the interaction between the element and some biological systems. What is known so far is that Te has exhibited a host of antibiotic properties and possible applications for anticancer drug design. Surprisingly, although tellurium belongs to a group of elements omnipresent in all life, its role in biology is limited. The metalloid has no apparent function in both prokaryotic and eukaryotic organisms [11].

Synthetic organo-tellurium compounds have demonstrated effectiveness as antibacterial agents. The non-toxic immunomodulator ammonium trichloro(dioxyethylene-O,O′)tellurate, or AS101, has been demonstrated to be effective in septic mice and goldfish infection models, used for its beneficial immunomodulating effects, and also exhibiting indirect antibacterial effects [20]. Nevertheless, the metalloid exhibits a certain degree of toxicity. Even in contact or taken internally, it can cause nausea, vomiting, and damage to the central nervous system. Humans metabolize tellurium into some dimethyl telluride $Te(CH_3)_2$, a gas that can be exhaled with an odor similar to odor in people exposed to or poisoned with tellurium [90].

Tellurium in Other Applications

Tellurium and its compounds are used in many other applications beyond semiconductors and in the field of biology, but to a lesser extent. For instance, organic tellurides can be applied as initiators for living radical polymerization, and electron-rich mono- and di-tellurides are utilized as antioxidants in the polymer and plastics industry. Besides, it is also used for achieving low-temperature growth of CdHgTe by metalorganic vapor phase epitaxy processes (MOVPE) which in turn is a significant process in the generation of optoelectronics [91].

Alternative and complex formulations containing the metalloid have also been studied. For instance, Shieh et al. prepared ternary Te-Fe-Cu chain polymers from the self-assembly of the precursors in tetrahydrofuran (THF) [92]. The chain polymers showed semiconducting behaviors with low band gaps of 0.59 and 0.41 eV, respectively. Similarly, an efficient methodology for the production of 2′-tellurium-modified phosphoramidite was explored [93]. The 2′-Te-nucleosides were transformed into 3′-phosphoramidites, which are the building blocks for DNA/RNA synthesis.

Still, the potential applications of bulk tellurium are not fully developed, since just metallurgy and solar cells' technology occupied the vast majority of the production and activity of the element.

Synthesis of Tellurium Nanostructures

Traditional Synthesis of Nanomaterials

The synthesis of nanomaterials is a pivotal point in nanotechnology. Several concerns have been appropriately addressed by researchers in order to elucidate and explain synthetic routes for a smooth, reproducible, and cost-effective production of these valuable materials. Physicochemical approaches—or traditional ways—for the synthesis of nanomaterials have been the unique answer for a long time, taking knowledge from both physics and chemistry for the development of efficient synthetic mechanisms. Methods such as reduction-oxidation (redox) reactions, chemical vapor deposition, or sputtering techniques have been widely used for precise control of the nanomaterial characteristics, such as size, shape, and dimensionality [94]. Such characteristics define the optical, chemical, physical, electromagnetic, or mechanical properties of the nanostructures and, hence, their potential applications [95]. Traditional synthetic approaches are able to strictly control these features, the reason why they are consolidated in research. Also, the feasibility to scale up the processes allow them to have a place in the industry.

Synthesis of tellurium nanomaterials has been taking knowledge of these processes, as other nanosystems do. The synthesis and features' controls of Te nanomaterials have been extensively reported all over the years, establishing protocols for quick synthesis of nanomaterials with a large range of applications. Tellurium ions are easily reduced elemental tellurium structures showing different sizes, shapes, and dimensionality depending on the synthetic routes. The reducing agent, as well as the tellurium precursor, strongly influence the nucleation and posterior nanoparticle generation [96].

In this section, the primary methods for the synthesis of tellurium nanomaterials will be revised in terms of the dimensionality of the newly generated products.

Zero-Dimensional (O-D) Tellurium Nanostructures

Zero-dimensional nanomaterials have all the dimensions at the nanoscale. The most common representatives of these materials are nanoparticles that can be either amorphous or crystalline. If belonging to the latter group, nanoparticles can be made of single crystals or show a polycrystalline structure. Besides, these nanostructures can be composed of single- or multi-chemical elements exhibiting various shapes and forms besides the conventional spherical shape. These nanoparticles (NP) are commonly found either individually, monodispersed in a solution, or incorporated in a matrix.

Nanoparticles have extraordinary potential as biomedical and industrial agents, due to the presence of properties that are not in the bulk materials and their relatively small size. The synthesis of these materials is facile compared to other more complex structures; hence, many research articles reported the synthesis of tellurium nanoparticles [97, 98].

As an example, Tsai et al. reported an easy and quick synthesis of single-crystalline tellurium nanoparticles (TeNPs) with uniform size using H_2Te gas as a precursor that was bubbled through a cylindrical polymer brush into an aqueous solution at standard conditions [99]. Alternatively, Weidong He's group developed an easy approach for the synthesis of colloidal Te nanocrystals with a binary uniform size distribution at room temperature by using Na_2Te as a precursor and oleic acid as an oxidizing agent through a one-step procedure, reporting a new platform for efficient synthesis and manipulation of Te nanostructures [97].

In an elegant study, Jiang's group reported the synthesis of tellurium colloidal NP through laser ablation technique employing various protic and aprotic solvents, with the aim to understand the thermodynamic behavior and the kinetics related to the growth of colloidal nanoparticles, a pivotal milestone for understanding single-nanoparticle synthesis [100]. They showed that the unique growth kinetics related to the formation of nanochains that is strongly dependent on the polarity and dielectric constant of solvent molecules. A size and structure dependency on the ability of chemical reduction was observed on the synthesized nanomaterials.

Ultrasonic irradiation in organic solvent was reported as an efficient way to generate pure Te and TeO_2 nanoparticles [101], carrying thoughtful research of the dependence of the consequences of ultrasonic power, irradiation time, solvent, and surfactant on the morphology and particle dimensions of the nanostructures. Another sonochemical process was used by Mousavi-Kamazani et al. when they reported the generation of 0-D Te nanosystems with different morphologies such as spherical NP or rice-like structures with dimensions of 15–40 nm [102]. The efficiency and performance of a variety of tellurium nanosystems in quantum dot-sensitized solar cells (QDSSCs) were also evaluated.

One-Dimensional (1D) Tellurium Nanostructures: Nanowires, Nanotubes, Nanorods, and Nanobelts

Nanomaterials with one dimension outside the nanoscale are considered one-dimensional materials, showing needle, rod, tube, or wire-like-shaped structures. As with zero-dimensional materials, these structures can be either amorphous or crystalline (single crystalline or polycrystalline). Once synthesized, they can be stand-alone materials or appear embedded within another medium or matrix. Nanowires

(NW), nanotubes (NT), nanorods (NR), and nanobelts (NB) are typically considered as one-dimensional structures, although other alternatives morphologies can be included in this classification [103].

Tellurium Nanowires

Nanowires are structures with a width and depth of a few nanometers or less, but a much longer length. Electron movements are different depending on the direction, while there is free motion in the direction along the wire, the other two dimensions are confined by quantum mechanics, radically changing the properties of the material. Therefore, researchers are able to synthesize nanowires that have an outstanding length-to-width ratio.

For instance, Xia et al. synthesized TeNW of the trigonal crystalline structure by a solution-phase, self-seeding method, which resembling the Ostwald repining process. The product was formed by reducing TeO_2 with hydrazine at different reflux temperatures. The homogeneous, relatively monodisperse one-dimensional nanostructures could form stable dispersions in ethylene glycol or water and be utilized as building blocks or templates to produce more sophisticated nanostructures [104].

Alternatively, Qian et al. prepared uniform TeNW with 4–9 nm in diameter by a hydrothermal method assisted by polyvinyl alcohol pyrrolidone (PVP), which can achieve large-scale selective synthesis. It was observed a strong dependence on the formation of the nanowires with the temperature, the amount of PVP, and the reaction time. The results concluded that the key to the selective synthesis of Te nanowires is to control the growth rates of (100), (101), and (110) planes and precisely control the reaction kinetics. This method provided a simple and feasible way for the preparation of high-quality tellurium nanostructures and had excellent optical properties [105].

Tellurium Nanorods

Nanorods have a typical size between 1 and 100 nm with standard aspect ratios—length divided by width—of 3–5. Synthetic pathways for the synthesis of these structures allow them to grow at different rates, producing an elongate nanomaterial with outstanding properties and different applications, such as display technologies or theragnostic. According to Komarneni's group report, tellurium nanorods can be easily synthesized by a hydrothermal process using the reducing property of glucose and Na_2TeO_3 as precursor. The glucose acted as both reducing agent and directional template [106].

Furthermore, Zhaoping Liu's group described a surfactant-assisted approach for the synthesis of uniform NR of trigonal tellurium with size control. These NR were synthetized from an original colloidal suspension of amorphous tellurium and TeNPs at RT—first formed through the reduction of $(NH_4)_2TeS_4$ by Na_2SO_3 in aqueous solution [43]. They demonstrated that by utilizing different surfactants during the manufacturing, nanorods with well-controlled diameters and lengths could be reproducibly synthesized. W. Huang et al., in a similar study, reported an easy one-pot synthesis of tellurium nanorods functionalized with a polysaccharide-protein complex by using a hydrothermal approach [107]. The NR stayed stable in aqueous media and in phosphate-buffered saline, they also showed high hemocompatibility.

Tellurium Nanotubes

Nanotubes are nanomaterials that have a tube-like structure. Although the most famous ones are made of carbon, other materials are gaining interested in the field due to their properties. For example, Na_2TeO_4 and formamide were used to prepare TeNT via a simple hydrothermal reduction approach. The diameters of the nanotubes could variate from 200 to 600 nm and lengths from 4 to 15 nm [108]. A simple strategy of selective synthesis of tellurium nanotubes with an inclined and hexagonal section by surfactant-assisted solvent, thermal method under mild conditions was also presented by Song's group. The NT were synthesized by ethanol reduction in the presence of cetyltrimethylammonium bromide (CTAB) with TeO_2 as the precursor. They also found when the surfactant CTAB is replaced by cellulose acetate (CA), TeNT with hexagonal cross-sections can be generated as well [109].

In an independent study, T. Liu et al. developed an easy microwave-assisted monosaccharide-reducing approach for the preparation of nanotubes, showing the vital role of the reductant agent in the synthesis as a modulator of the final shape [110]. Similarly, single crystalline TeNT with triangular cross sections can be prepared through a simple approach of vaporizing and condensation of the metallic tellurium in an inert atmosphere onto a substrate, with diverse morphologies being produced depending on the synthetic parameters [22]. The simple approach might provide some new applications and stimulate theoretical studies related to the stability of the high-energy configuration.

Tellurium Nanobelts

Tellurium belt-shaped structures can be readily synthesized using traditional approaches. For example, Wang et al. prepared high-quality TeNB by thermal evaporation and deposition method in a vacuum system. According to the experimental data, the formation of TeNB followed the gas-solid mechanism mediated by agglomeration. In addition, it has been proved that TeNB evolve into helical bands when subjected to electron beam (EB) action [111].

Wan et al. group first reported the synthesis of lead tellurite nanobelts using composite molten salts (KNO_3/$LiNO_3$) approach, which is an economic, one-step, easy to control, realized in conditions of low-temperature and ambient atmosphere [112]. A vacuum vapor deposition technique was followed by Qun Wang's and colleagues to prepare high-quality ultrawide TeNB, reporting that helical structures were achieved when TeNB were in contact with electron beams [111] (Fig. 4).

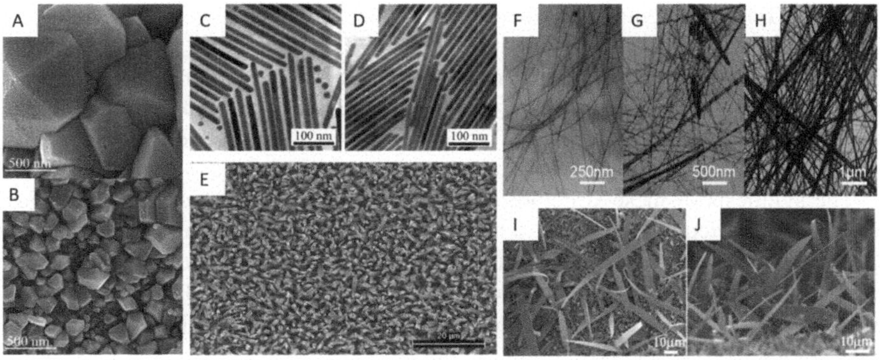

Fig. 4 Different tellurium nanostructures. SEM images of the synthesized TeO_2 in the (**A**) absence of ultrasonic irradiation and (**B**) presence of ultrasonic irradiation [101]; TEM images of Te nanorods from a reaction mixture after an aging time of (**C**) 15 h, and (**D**) 20 h at room temperature [43]; SEM images of tellurium nanotubes synthesized on a Si (100) substrate (**E**) [22]; TEM images of the morphological evolution of the tellurium nanowires at 4 h (**F**), 6 h (**G**), and 12 h (**H**) [105]; and SEM images of Te nanobelts deposited on Si substrates at different magnifications (**I, J**) [105]

Two-Dimensional Te Nanostructures: Tellurene

Since the discovery of graphene in 2004 [113], there has been a tremendous interest in the research of elementary two-dimensional (2D) materials, which often present differentiated physicochemical properties with respect to their bulk counterparts [114]. Examples of synthesized elementary 2D materials beyond graphene are silicene [115], germanene [116], stanene [117], borophene [118], antimonene [119], and black/blue phosphorene [120, 121]. The characteristic feature of these materials is that they belong to the groups III-V. More recently, theoretical and experimental investigations on group VI elemental 2D materials, such as selenene and tellurene has opened up the possibility to synthesize novel topological insulators [122], thermoelectric materials [123], and photodetectors [124]. For instance, Huang et al. grew mono- and few-layer Te films on graphene by molecular beam epitaxy (MBE) [125]. They found that the band gap increases monotonically with decreasing Te film thickness, reaching the energy gap ($E_g = 1.5$ eV, near-infrared) for the monolayer Te nanostructure. Moreover, Wang et al. synthesized 2D Te using a hydrothermal method, in which Na_2TeO_3 was reduced by hydrazine in alkaline solution at 160–180 °C [126]. The solution-grown Te nanomaterial was used to fabricate a tellurene-based field effect transistor that exhibited on/off ratios on the order of 10^6, field-effect mobilities of about 700 $cm^2\ V^{-1}s^{-1}$ and stability at normal conditions for over 2 months.

Complex Tellurium Nanostructures

Complex tellurium nanostructures often possess a nanocrystalline structure and involved the presence of features at the nanoscale, and are composed of multiple arrangements of nanosize crystals. They can also contain dispersions of nanoparticles, bundles or nanowires, and nanotubes and multinanolayers.

The synthesis of these structures is not an easy task. However, some methods have been reported. For instance, a two-step hydrothermal method was studied for the preparation of a novel gold-modified tellurium hybrid with a trip-shaped planar microstructure [79]. Alternatively, the synthesis of a self-standing Bi_2Te_3 network of interconnected nanowires was reported [127]. The reaction was achieved in a three-dimensional porous anodic alumina template. They showed how the crystalline structure and composition of the 3D Bi–Te nanowire network were modified by changing the applied voltage and the relaxation time off without applying current density during the deposition. They observed as well that the templates in which the complex Te structures were grown could be dissolved and the network of interconnected nanowires is self-standing without affecting its composition and orientation properties.

Chiral Tellurium Nanostructures

Chiral structures are the ones that are non-superimposable with their mirror image. Chiral Te nanostructures can be synthesized by using chiral biomolecules as initiators. For example, hydrazine can reduce the metalloid precursors in the presence of a large number of thio-chiral biomolecules. Consequently, Te nanostructures with different chiral shapes, such as small nanocrystals, large crystals, broken hexagonal tubes, and long hollow tubes, can be generated under different synthetic conditions. These different chiral nanostructures are generally characterized by high optical activity [128].

Once generated, they are useful in research, since semiconductors with chiral geometries at the nanoscale and mesoscale can provide a powerful material substrate for polarization optics, photocatalysis, and biomimetics. The relationship between the geometry of the chiral semiconductors and their chiroptical properties is not clearly expressed, in comparison with metallic and organic optical materials. W. Feng et al. reported that semiconductor helices can be prepared with an absolute yield of ~0.1% and enantiomeric excess (e.e.) of 98% or above from cysteine-stabilized cadmium telluride nanoparticles dispersed in methanol [129].

Tellurium-Based Alloys and Hetero-Nanostructures

Alloys and hetero-structural nanomaterials have been widely used in nanotechnology. For instance, designing a large number of nanomaterials with heterojunctions at the interface between two different crystal regions is key in the development of advanced materials for emerging applications.

As a consequence of its easily controlled morphology and bandgap, successfully synthesized tellurium-based alloys and heterostructures have attracted widespread attention. Single crystal $Se_{0.5}Te_{0.5}$ alloy nanorods can be synthesized in solution in dimensions of 50–250 nm. The morphology of the alloy was determined by the helical chains of Se or Te, because they have a similar lattice structure [130].

These alloys can have an important impact on the main applications of nanoscale tellurium. For instance, Y. Yang et al. noticed the challenge in the design and production of one-dimensional metal chalcogenide nanostructured materials with controllable components and properties [131]. Therefore, they reported a usual chemical transformation process for the synthesis of more than 45 types of one-dimensional nanostructures of alloyed or hybrid metal chalcogenide forms, all coming from a common template Te_XSe_Y–Se core–shell nanowires with variable compositions. Nine different kinds of alloys—including just one chalcogen in the structure—NW (such as AgSeTe, HgSeTe, CuSeTe, BiSeTe, PbSeTe, CdSeTe, SbSeTe, NiSeTe, and CoSeTe) were synthesized, providing a new general route for the controllable synthesis of a new generation of one-dimensional metal chalcogenide nanostructures.

Large-Scale Production of Tellurium Nanostructures

The scale-up of processes for the synthesis of nanomaterials is a challenge for research, due to problems related to reproducibility and feasibility for adaptation to industrial scale. However, some examples are gaining importance over time. Tellurium is an example of one of these materials whose scale-up deserves the investment, due to the increase of its use in the industrial sector.

A large-scale synthesis of tellurium nanoribbons in aquatic media was reported [132]. The super-long single-crystal nanoribbons were produced in tetraethylene pentamine aqueous solution at a temperature of 80 °C, the resulting structures presented a width of 200–300 nm and length of hundreds of micrometers. In a similar study, H. Zhu et al. reported in large scale utilizing an easy hydrothermal approach to synthesize various one-dimensional tellurium nanostructures such as nanotubes, nanowires, and nanorods on by the use of an NaOH solution [133]. The impact of reaction conditions like pH value, temperature, reducing agent, and reactant concentration over the dimensions and morphology of the nanostructures was studied as well as the relationship between the growth rate and structural modification on the systems.

Kim's group successfully reported the large-scale synthesis of single crystalline TeNW by the process of vaporization and later deposition of the tellurium metal and above a Si substrate using an inert atmosphere [21]. The resulting structures presented a high degree of purity due to the unique use of tellurium metal in the vaporization process. The industrial application of this material is easily facilitated due to the low temperature and high yield of the synthetic approach (Fig. 5).

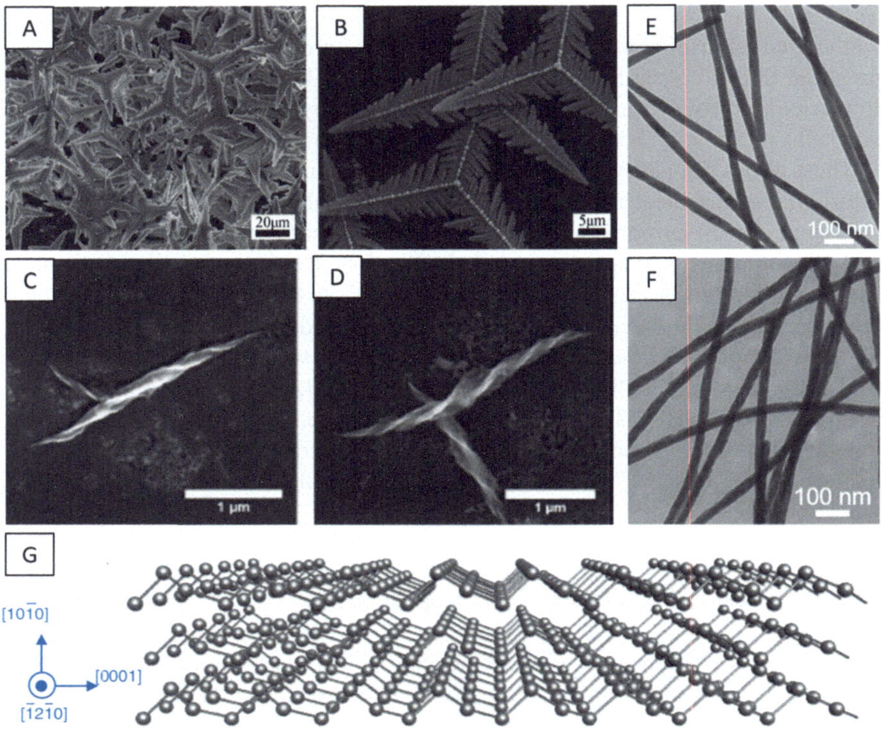

Fig. 5 Different 3D tellurium nanomaterials. Low-magnification SEM image of Au-decorated tripod-shaped tellurium hybrids (**A**, **B**) [134]; STEM tomography of R-helix assembled from D-Cys CdTe nanoparticles (**C**) and L-helix assembled from L-Cys CdTe nanoparticles (**D**) [129]; and TEM image of AgSeTe nanowires (**E**) and HgSeTe nanowires (**F**) [131]. 3D illustration of the structure of tellurene (**G**) [126]

Green Synthesis of Tellurium Nanomaterials

All the methodologies explained before for the synthesis of different dimensions are based on physicochemical approaches, known as traditional chemical or physical processes. These approaches work pretty well, with established and easy-to-follow methods for a quick and reproducible generation of nanomaterials. Nevertheless, they are not free of drawbacks, such as the use of hazardous materials, production of toxic by-products, often use of harsh conditions—such as very basic or acid conditions or high temperatures—and expensive instruments.

Therefore, new approaches for the generation of nanomaterials are needed, and nature may have the answer. Green nanotechnology was born for the use of environmentally friendly, cost-effective, and green approaches for the generation of nanomaterials. These green synthesis approaches involve the synthesis of nanomaterials using living organisms and biological compounds from different sources, such as bacteria, yeast, food, plants, and waste materials.

Various applications take profit of the advantages of biological mediated synthesis over chemical or physical nanoparticles syntheses, including lower capital and operating costs. In addition, they showed enhanced biocompatibility and stability of the nanomaterials, mainly due to the application of biosurfactants or capping agents on their surfaces. The size, morphology, and properties of nanoparticles can be controlled by modifying parameters such as temperature, pH, reaction time, metal ion concentration, or quantity of organic matter present in the reaction.

Many microorganisms have shown the ability to reduce metallic ions to elemental nanoparticles as part of their natural detoxification processes. For instance, Zonaro et al. reported the use of the tellurite-reducing bacterial strain *Ochrobactrum* sp. MPV1 isolated from polluted sites [135]. They discovered that by regulating the culture conditions and exposure time to the tellurite oxyanions, differently sized zero-valent Te nanoparticles were produced. Besides, these nanoparticles showed antimicrobial and biofilm eradication activity against different bacterial strains. Similarly, Ramos-Ruiz et al. focused on the recovery of elemental tellurium (Te^0) from aqueous streams with the presence of water-soluble oxyanions such as tellurate (Te^{6+}) and tellurite (Te^{4+}) utilizing microorganisms like bacteria [77]. The study showed how a microbial culture composed of an anaerobic mix—present in methanogenic granular sludge—had the ability to catalyze the reduction of the two oxyanions to generate Te^0 nanoparticles (NPs) in a medium free of sulfur. They also found that the redox mediators and electron donors had an effect on the morphologies and locations of Te^0 NPs, suggesting that NP production can be specifically designed for a particular application.

Fungi are another kind of microorganisms that are able to show resistance to tellurium, whose behavior can be used by researchers to develop alternative synthetic ways. For instance, Abo Elsoud et al. used six fungal isolates with the ability to reduce potassium tellurite (K_2TeO_3) into elemental tellurium nanoparticles with potential biomedical applications [136].

Hydrothermal methods are also considered green pathways for the generation of nanoparticles, allowing a clean and water-based reaction. In this line of research, Medina Cruz et al. developed an environmentally friendly approach for coating nanocolumnar titanium with Te nanorods with enhanced antibacterial properties against Gram-positive and -negative bacteria [137].

Despite these examples, many other green nanotechnologies, such as the use of waste material or natural biomolecules, should be tested as potential pathways for the synthesis of tellurium nanostructures, overcoming the main limitations of traditional physicochemical approaches.

Nanoscale Tellurium: Applications Beyond Biomedicine

As it happens with many other metallic elements, the properties of tellurium nanomaterials changed and showed differences with the ones found in a bulk form. Tellurium nanomaterials, due to their large specific surface area and the effect of quantum-confinement, exhibit specific physical and chemical properties, with a large array of applications. Therefore, in this section, some of the main applications of tellurium in nanomaterial form are presented and discussed, whereas its role in biology and medicine will be discussed in the following two sections.

Tellurium Nanostructures as a Photoconductive Conversion Material

Several physicochemical studies show that nonmetal elements with a relatively high refractive index should be photoconductive materials [138]. Therefore, many elements have been checked for this property, with some successful discoveries. Intensive research has been done over the years in terms of electrical properties of tellurium by a group at Purdue University [139]. They discovered a powerful photoconductive effect in tellurium films, while the conductivity of the metal was found unaltered by radiation [140].

After the discovery, one of the most intriguing potential applications for TeNP happens to be in renewable energy—in the use of nanophotonic materials for solar energy harvesting and photothermal conversion. Ma et al. found that the photothermal conversion utilizing TeNP surpassed that of plasmonic or all-dielectric nanoparticles previously known, indicating a substantial potential utility for tellurium nanoparticles as an improved photothermal conversion material for solar-enabled water evaporation. This advanced photothermal conversion property of tellurium nanoparticles may provide utility in desalination plants and in hydroelectric power generation as less energy is required to convert water into steam for use in these applications [41].

Besides nanoparticles, other structures showed photoelectric properties, such as Te nanowires that are synthesized by layer-by-layer (LBL) strategy [141]. Carotenuto et al. prepared a material that was composed of hexagonal tellurium and α-phase of tellurium oxide, whose electrical properties were studied, demonstrating the linear functionality of the photoconductivity of the film—in sandwich contact configuration—in relation to the light power density [74]. Besides, ultrathin TeNW can be synthetized by the Langmuir–Blodgett technique, showing reversibly switched photoelectric properties in a range from the higher- and lower-conductivity states if function of the light was on or off respectively, and the influence on the photocurrent was due to the light intensity and the amount of monolayers of the nanowire. In a similar study, Hackney et al. studied the photoconductive and polarization properties of single CdTe nanowires [142]. These NW showed a power conversion efficiency of 0.56%, a higher value in comparison with similar but more complex nanomaterials.

Nanoscale Tellurium as a Catalyst

Tellurium can be used as a catalyst as well when brought to the nanoscale, although its use is not really extended over the scientific community. Nevertheless, some examples can be found in the literature. For instance, with the aim to enhance the catalytic performance of platinum-based compounds in formic acid electro-oxidation, an electrochemical leaching process was used to prepare dealloyed Pt_XTe_Y/C catalysts [143]. Zhou et al. conducted studies, in which the Te-based catalyst showed increased performance and stability in the electro-oxidation of formic acid with a mass activity at 0.25 V being 10.6 times higher and at 0.4 V being 16.5 times higher than the commercially available Pt/C catalysts [143]. The improved electro-oxidation of formic acid may prove useful in the generation of formic acid fuel cells.

Alternatively, Tian et al. achieved a highly selective hydrogen peroxide synthesis directly from hydrogen and oxygen using a highly efficient palladium-tellurium ($Pd\text{-}Te/TiO_2$) catalyst that showed selectivity of nearly 100% toward H_2O_2 under mild conditions [17]. They demonstrated that the Te-modified palladium surface could significantly weaken the dissociative activation of O_2, leading to the non-dissociative hydrogenation of the molecule.

Nanoscale Tellurium as a Chemical Transformation Template and Building Blocks

The use of prefabricated and nanostructured materials for influencing the placement of building blocks is the basis for the so-called templating techniques. These techniques offer the possibility to prepare nanostructured materials with several applications. Depending on particular techniques, the templating material should have suitable properties. For instance, in a solution-based template synthesis, a proper template should have not only a uniform dispersion but also a high reactivity when used as a chemical template [144, 145].

Tellurium, in its nanostructure form, is a suitable material for templates. For instance, Te nanowires with a high surface-to-volume ratio are often used as templates in the synthesis of 1D functional nanowires, together with other metals, such as silver, cadmium, or lead. Various methodologies have been developed for the facile synthesis of different metal telluride nanowires by using a template made of Te nanowires [146]. Another example was reported by Samal and Pradeep, who reported a facile production of platinum telluride nanoparticles (Pt_3Te_4 NPs) in a solution phase at room temperature by the use of a template-assisted methodology [147].

It is well established that elemental selenium and tellurium have similar anisotropic and isomorphous crystal properties, leading to an epitaxial growth process that can be designed to achieve the production of complex and heterogeneous Te–Se nanostructures. A method for the generation of Te_xSe_y and Se core–shell NW by epitaxial growth was recently reported. These nanostructures are used as templates for the formation of some kinds of monometallic chalcogenide alloys (MSeTe, M = Ag, Hg, Cu, Bi, Pb, Cd) with nanowire-like structure [131]. Alternatively, these hybrid metal chalcogenide nanowires could also be synthetized only by mixing two or more types of metal precursors in one batch, thus showing a quick and facile production.

Besides, tellurium compounds can be used as mediators for the synthesis of other nanomaterials. For instance, Fernández-Lodeiro et al. confronted the challenge of preserving the properties of gold nanoparticles during a long time period in both solution and dry powder form [148]. By using organotellurium derivatives, they were able to overcome this challenge when adding them as reducing and stabilizing agents in the process of developing the gold nanoparticles. Diphenyl ditelluride (Ph_2Te_2), which had never been exploited in the synthesis of gold nanoparticles, was the key component in the synthetic approach due to its photochemical and oxidative properties. Moreover, the same group reported the synthesis of novel $PtTe_2$ through an annealing process using new nanostructured Pt–Te organometallic NPs as a single-source precursor, the resulting nanoparticles were multi-crystallite and in various sizes [149].

Nanoscale Tellurium as a Piezoelectric Energy Harvester

A few tellurium nanostructures have been reported to show both a high work function and a narrow bandgap energy. These can be used as a single-component metallic bonding-based piezoelectric materials [150]. Wang's group developed an easy fabrication method for a high-power density piezoelectric energy device based on tellurium [151]. They used a trigonal Te nanowire that shows an asymmetric crystal structure in its radial direction. Therefore, the structure has a strong potential to serve as a raw material for ultrathin nanogenerator (NG).

New advances for the piezoelectric properties of tellurium have been recently developed. They introduced the fabrication of a flexible strain sensor based on a single tellurium wire [152]. These Te nanowires were synthesized using a simple solvothermal process, rendering ultralong wires with a diameter of 40 nm and length of about 3 mm [152]. Polydimethylsiloxane (PDMS) was used to cover the tellurium wire—strain sensor—with the aim to improve the adhesion and prevent the Te wire from oxidization [152].

Alternatively, in similar work, Wen He's group reported a novel strategy for the synthesis of Te nanoflakes via a hydrothermal process at low temperature [153]. Once created, the Te nanoflakes were used for the building of a nanogenerator device. This system was composed of a sandwich-like structure with PDMS-coated Te structures in the center. The device exhibits fully flexible mechanical performance, reaching an open-circuit voltage and a closed-circuit current of 3 V and 290 nA during periodic bending tests.

Nanoscale Tellurium in Ion Detection and Removal

Contamination of heavy metals is a serious concern that environment researchers are trying to solve. From all the heavy metals, probably mercury is the most dangerous element to human health. Different concentrations of this inorganic element can be found in rivers, potable water, and industrial effluents, leading to different degrees of toxicity [154]. Te and Hg have a strong hybridization and can undergo a galvanic replacement reaction that induces the dissolution and aggregation of Te in the presence of mercury ions. In line with this fact, Huang et al. reported a pure gel-based membrane composed of TeNT and agarose gel for the effective detection and removal of mercury ions [155]. They took profit of the high surface area of the Te nanostructure and the strong hybridization between both elements [155]. A detection limit of about 10 nM for the Hg^{2+} ion was found [155], which is the maximum allowable level of Hg in potable water. Moreover, compared with other metal ions, the Te-based membrane responds selectively toward Hg^{2+} by a factor of 100 [155, 156].

Nanoscale Tellurium in Batteries

Rechargeable Li-ion batteries consist of lithium ions going in the direction from negative to positive electrode during discharge and reverse when charging. They have applications in modern consumer electronics and even in electric vehicles. However, the quickly increasing demand and the need of advancement in the field has been shifting research efforts towards low cost energy storage cells for batteries, with enhanced energy/power density and a superior cycling with trustful stability [157].

Lithium–chalcogen batteries with high theoretical specific energy are treated as the most promising candidates, but they still face challenges, such as the low electrochemical performance and poor cycling stability. Compared with S or Se, Te has a higher material density (6.24 g cm^{-3}) and higher rate capability due to its high electronic conductivity (200 S m^{-1}), which turned it into a suitable candidate.

In 2017, He et al. used a nanoporous structure based on a metal-organic framework (MOF) composed of cobalt and nitrogen codoped carbon polyhedra (C-Co-N) which was generated and applied as tellurium host for Li–Te batteries [158]. Results showed increasing cycling stability with superior capacity retention of 93.6% and ~99% coulombic efficiency after 800 cycles.

Besides, it is known that pure Te is reported to decay quickly without carbon support with a significant volume change [159]. The combination of Te nanostructures with carbon nanomaterials increases the reaction area due to the high aspect ratio. A superior mechanically reduced Te/C nanocomposite electrode material was created with elevated energy density, excellent cyclability, and quick rate capability. The nanocomposite electrodes could be properly used as both cathode or anode in

Li–Te secondary batteries or in rechargeable Li-ion batteries, respectively [160]. Besides, binder/collector free Te cathodes for high-energy Li–Te batteries have been prepared via a hydrothermal carbonization process [161].

Nanoscale Tellurium for Gas Sensing

Elemental Te shows a significant gas response at room temperature, especially for nitrogen dioxide (NO_2) [162], carbon monoxide (CO) [163], ammonia (NH_3) [164], and hydrogen sulfide (H_2S) [165], among others, which allows the fabrication of gas-sensitive devices of small size but with high sensitivity. The sensitivity to these gases depends mainly on the microstructure of tellurium, film thickness, and the environmental conditions.

Tsiulyanu's group reported that the sensitivity of Te films strongly increases with a decrease in film thickness [166]. They also showed that there exists an inverse relationship between conductivity and film thickness, especially noticeable when the thickness is on the nanometer scale. The effect of the surface grain boundary at low thickness is the main reason for this behavior. As a consequence, they reported that when the thickness of the film is decreased, the film conductivity is decreased as well. Therefore, there will be a stronger sensitivity.

Park et al. reported a method of synthesis with controlled size, morphology, and crystallinity for large and hollow tellurium nanofibers, showing an outstanding performance as NO_2 sensor at room temperature [167]. Similarly, Kumar's group reported the use of different tellurium nanostructures for NO gas sensing. TeNT on Ag templates were prepared, showing an enhanced detection of NO compared to other gases like H_2S or NH_3. There was a significant improvement in terms of selectivity response to NO in comparison with previously known Te thin film sensors [168].

Sen's group also reported the response of Te films toward H_2S gas, which showed sensitivity toward 0.1 ppm of the gas at RT [165]. It was shown that the response time decreased with an increase in gas concentration and took nearly 5 min at a concentration of 1 ppm, while the recovery time showed the contrary tendency, taking nearly 20 min under the same conditions. This response allows for the potential development of H_2S gas sensor. Sen's group also reported a sensor based on Te film, produced through a vacuum evaporation process, which showed high sensitivity for NH_3 [169]. The study showed an increase on the resistance of the Te films due to the exposure to NH_3 and, at concentrations lower than 100 ppm, the response of the film is shown to be linear.

Nanoscale Tellurium as a Doping Agent

Element doping is a technique in which an intentional introduction of impurities is done into another material of high purity with the purpose to modulate several of its properties. It is a usual technique in semiconductor production, with the aim to change electrical, optical, and structural properties of the material [170]. Te has exhibited wide applications in the metallurgy industry, while nearly 80% of the metalloid was consumed for industrial use [171]. As it was discussed in a previous section, the addition of Te into steel and Cu produces an alloy with better machinability and enhanced mechanical properties, while in Pb alloy, the addition of the metalloid improves its strength and durability, decreasing the corrosive action acids such as H_2SO_4 [172].

Therefore, tellurium is often selected as the dopant for the production of multicomponent materials for optimization properties. The reason behind it is that the metalloid has a large atomic radius, a heavy atomic mass, and a narrow band gap structure. As a consequence, the doping of Te mainly shows a great synergistic effect in applications in thermoelectric devices [173], photoconductive systems [174], and in the enhancement of conductive properties [175].

Zhang et al. showed the synthesis of single crystals of tellurium nanostructures doped with black phosphorus (Te-doped BP) with superior crystalline quality, generated by a process of chemical vapor transport [176]. They studied for the first time the properties of electrocatalytic oxygen evolution reaction (OER) of few-layer nanosheet made of Te-doped BP and synthetized by a process of liquid exfoliation.

The metalloid has also been used as doping in copper-silver-selenium-tellurium alloys for higher thermoelectric properties [177], while Te doping in amorphous selenium for increased photo-generation efficiency has been reported [178]. Tao et al. reported the use of tellurium as doping palladium; small concentration of the metalloid could selectively convert CO_2 to CO with a low overpotential, improving the catalytic properties of the material [179] (Fig. 6).

The Biological Role of Tellurium

Despite the previously described diversity and reactivity of tellurium compounds, the rare metalloid is having trouble to find its position as an essential agent in biology. Living organisms and biomolecules try to avoid the presence of the metalloid in their mechanisms and actions with feasibility. However, a few examples of the interaction of the element and different organisms can be found in nature.

Tellurium in Bacteria

Some microorganisms have the ability to proliferate in the presence of the main tellurium salts in nature—tellurate and tellurite. These living microorganisms can generate elemental tellurium, insoluble Te^0, which does not exhibit any risk for the biological activity of the organisms by a process of reduction of both anions [180]. Therefore, tellurium resistance can be commonly related to reductive processes. Interestingly, these metabolic transformations are related to existing mechanism

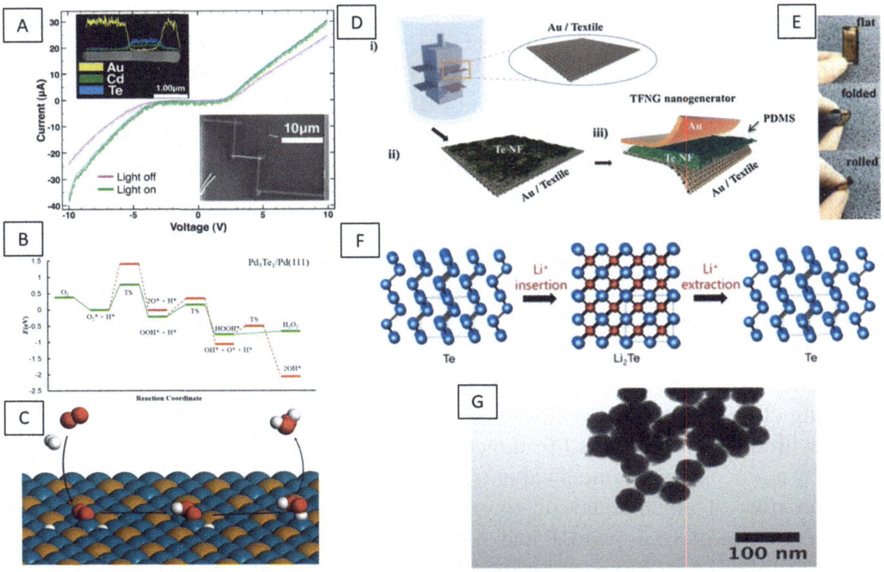

Fig. 6 (**A**) Comparative relation between dark and illuminated (150 W halogen lamp) I–V curves. Top inset: SEM image of an Au-Cd0.42Te0.58-Au nanowire, EDS data showed the multicomponent structure of the nanowires. Bottom inset: SEM image of a single Au-Cd0.42Te0.58-Au nanowire connected to microelectrodes via Pt-deposited films [142]; Energy profiles for the synthesis of H_2O_2 on Pd3Te1/Pd(111) surface (**B**) and representation of the lower (green) pathway (**C**). Blue, brown, red, and white spheres are Pd, Te, O, and H atoms, respectively [17]; The schematic diagrams of the fabrication process and general characterization of the TFNG device (**D**) and schematic of the process for fabricating the TFNG devices (**E**) [153]; (**F**) Schematic representation of crystallographic transformation during cycling (blue: Te atoms, red: Li atoms) [153]; TEM images of multi-crystallite PtTe2 NPs generated after the annealing process (**G**) [149]

based on selenium metabolism. A remarkable example is H_2Se, a compound that is naturally generated from SeO_3^{2-} via selenodiglutathione (GSSeSG) enzyme. The hypothesis now came into action: there is an existing possibility that tellurium "sky-jacks" the metabolic route related to selenium, and therefore, H_2Te is produced in a similar manner. As a consequence, tellurodiglutathione (GSTeSG) might be found there, nonetheless has not yet been reported in the metalloid metabolism [181]. In line with this, Bajaj and Winter reported that selenite reducing heterotrophic non-halophilic aerobic bacteria could successfully reduce tellurite anion and, therefore, produce the elemental form, which do not present any toxicity associated, hence generate extracellular nanospheres during the detoxification process [182]. They also showed that small quantities of selenite in the medium, which has less toxicity than tellurite, favor the bioreduction of tellurite leading to the generation of extracellular SeTe nanospheres.

Tellurate and tellurite anions play a key role in the metabolism of some microorganisms, as they promote the growth of some specific bacteria by acting as electron acceptors in the respiratory chain. Therefore, some of these bacterial species have been reported, including *Bacillus selenitireducens*, *Sulfurospirillum barnesii*, and *Bacillus beveridgei* sp. nov. [183], commonly extracted from deep ocean hydrothermal vent worms. The microorganisms showed the capacity to generate Te^0 nanoparticles with a uniform distribution of size, as a result of growing in a medium with the presence of hen grown with tellurate or tellurite anions, acting as terminal electron acceptors.

Ramos-Ruiz et al. reported how a mixture of anaerobic microbes present in a methanogenic granular sludge had the ability to catalyze the reduction of both Te oxyanions with the aim to synthetize Te^0 nanoparticles without the presence of sulfur in the media [77]. They discovered that the redox mediators and electron donors play a key role in the morphologies and location of the metalloid NPs; hence, the synthesis of these structures can be modified for any particular application. Alternatively, Presentato et al. reported the ability of the Gram-positive bacteria *Rhodococcus aetherivorans* BCP1 cells showed the capacity to generate Te nanostructures as NR or NP through the bioconversion of TeO_3^{2-} based on the oxyanion initial concentration and time of cellular incubation [184]. Pugin and colleagues used a common biochemical strategy to look for an innovative telluride reductase present on an Antarctic bacteria *Pseudomonas* sp. strain BNF22 with the aim to generate tellurium nanosystems [42]. They identified a new tellurate reductase as glutathione reductase, which was consequently produced in higher quantities by *Escherichia coli*. The characterization of this enzyme demonstrated that the tellurite reductase was NADPH-dependent.

Tellurium in Fungi

In the case of fungi, biovolatisation—a process where a dissolved sample is vaporized by a living organism—has a pivotal function in the elimination of tellurium. Analogous to selenium compounds present in nature, specific fungal species are capable of digesting tellurium structures by biomethylation, which results in $(CH_3)_2Te$, a highly toxic and volatile compound which is continuously eliminated from the system [185].

Recently, Abo Elsoud et al. studied tellurium myconanoparticles. Six fungal isolates were analyzed for their capacity of reduction of tellurium from K_2TeO_3 to elemental NP [136]. They reported that the most likely fungal isolate was *Aspergillus welwitschiae*. Special conditions such as free-sulfur medium containing sodium tellurite deposited in the soil, tellurium-containing amino acids (tellurocysteine, tellurocystine, and telluromethionine) and also proteins are generated. These pathways are largely known for selenium, but the ability of incorporation of the metalloid to these mechanisms has just started to be documented. For instance, Ramadan and colleagues reported that *Aspergillus fumigatus, Aspergillus terreus,* and *Penicillium chrysogenum* can grow on a medium free of sulfur mended with 0.2% (w/v) tellurite [186]. Tellurium was incorporated into different kinds of proteins with low and high molecular weight. These novel detected tellurium-containing proteins had a remarkably elevated level of tellurium, as well as telluro-cysteine, -cystine, -methionine, and -serine.

Tellurium in Plants

For instance, plants show high variability in the levels of the metalloid that they can stand, depending on factors such as the presence of the element in the soil or the surrounding environment. Samples all over the world have shown that there is an extremely low abundance of the metal in the surface of the planet, estimated around 0.027 ppm [11]. Cowgill et al. reported, after extensive research with more than a thousand samples from different locations in the USA, that plants that are recognized to collect selenium were capable of doing the same with tellurium up to concentrations of nearly 1 ppm [187].

Anan et al. wanted to show the metabolism of Te in plants, and they decided to choose garlic as a plant model due to its recognized Se accumulation. Garlic was grown in aquatic media and exposed to sodium tellurate [188]. The use of HPLC coupled with inductively coupled plasma mass spectrometry (ICP-MS) allowed the identification of metabolites that contained tellurium in the aqueous extracts of garlic leaves. They discovered two metabolites: Te-methyltellurocysteine oxide (MeTeCysO) and cysteine S-methyltellurosulfide. Therefore, it was shown that telluroamino acid is synthesized in a higher plant for the first time.

Therefore, it was hypothesized that despite the unknown role of tellurium in the biochemistry and biology of these species, it could be detoxified using the normal routes of selenium, with the aim to remove it from polluted areas, allowing a proper development of the plant. These tellurium phytoremediation was stated to happen via different routes, such as phytoextraction, rhizofiltration, phytodegradation, phytostabilization, or phytovolatilization [189].

Tellurium in Human Biology

Tellurium-related metabolism and toxicity in humans has not been extensively studied; hence, there remain many unknowns that should be addressed. Probably, the lack of information and research around human toxicity is because tellurium has hardly been used on an industrial scale, hence with a low impact profile in society. Therefore, its toxicity and related concerns are related to the academic environment. Nevertheless, this is about to change, and rapidly. Tellurium is now present in daily goods, through both intentional application and contamination. Consequently, society is becoming more and more exposed to the element.

A critical remark should be stated from the beginning: tellurium and its compounds are rather known as toxic; therefore, they have largely been treated as not suitable for drug development due to this early relationship with toxicity and different undesired effects on humans. Certainly, researches until the nineteenth century showed that the ingestion of tellurium-containing compounds, such as TeO_2 or tellurite, generated breath with a "disagreeable garlic-like odor" associated with both humans and animals. More severe clinical manifestations may appear including a metallic flavor, sickness, and vomiting [90]. Clinical reports of two children with tellurium-containing solution ingestion were reported. Clinical features included vomiting, black discoloration of the oral mucosa, and a garlic odor breath. No sequelae were found on the patients, which is common on tellurium-associated toxicity.

No toxicity has been generally applied to tellurium. Nonetheless, there are effects specifically related to some tellurium compounds depending on the chemical structure in which the metalloid is present. For instance, inorganic and organic tellurium compounds do not behave the same inside the body. Besides, the oxidation state has a strong dependence on its biochemistry. Consequently, organotellurium compounds with moderate degree stability on the Te-C bonds are usually treated as less toxic in comparison to inorganic Te compounds [190].

How can tellurium enter in the biology of a living human being? Direct oral intake of tellurium compounds has been reported to cause acute or chronic poisoning. Another way to consider is that the inhalation of Te-dust can penetrate the body through the lungs. Once tellurium compounds enter into the body, they can generate toxicity in different ways, mainly by strong interaction with proteins and enzymes that contain cysteine. Various in vivo studies performed in rats and mice utilizing compounds such as tellurite showed, for example, that intake of TeO_3^{2-} provokes transient demyelination of peripheral nerves as a consequence of the repression of squalene epoxidase (squalene monooxygenase), an enzyme that uses NADPH and molecular oxygen to oxidize squalene to 2,3-oxidosqualene (squalene epoxide). Nogueira and colleagues realized that diphenyl ditelluride presented neurotoxicity in mice, which was in part related to the interaction of diphenyl ditelluride with the thiol groups of cysteine-containing proteins and enzyme [191]. The use of diphenyl ditelluride (DPDT) and tellurium tetrachloride ($TeCl_4$) for toxicity in transformed (HT-29, Caco-2) and non-transformed colon cells (CCD-18Co) was evaluated as well [192]. Notable rise in caspase 3/7 and 9 performance was detected with DPDT leading to apoptosis. No meaningful increases in caspases were seen with $TeCl_4$ leading to necrosis. These caspases, cysteine-aspartic proteases, are a family of protease enzymes that play a pivotal role in programmed cell death and inflammation.

The chemistry of sulfur and tellurium might have an explanation for biological toxicity related to tellurium materials. Nonetheless, the high affinity of tellurium to selenium is related to some biochemical activities of the metalloid inside the body. Therefore, another motive of tellurium toxicity can be associated with the binding between tellurium compounds and selenium that can be found in some selenium proteins and enzymes, causing significant toxicity for human cells. As a consequence, it has been hypothesized that tellurium compounds are more likely to bind with selenium than sulfur since it exhibits some specificity for cellular Se proteins [193]. Additionally, undesired interactions between Te compounds and S- and

Se-containing biomolecules may have severe consequences. Interestingly, tellurium compounds can weaken the cell's antioxidant defense at the same time that actively produce reactive oxygen species (ROS). These ROS are related to different molecules such as super-oxides, hydrogen peroxide, or ozone [194]. Tellurium—as similar metalloids and metals—induces the overproduction of these species that generate mitochondrial dysfunction, and as a consequence, the cells die due to a process of apoptosis—self-programmed death [195, 196]. For instance, the antibacterial activity presented in tellurium nanowires was explained due to the generation of ROS in a dose-dependent amount [197].

Despite all the mechanisms of tellurium toxicity, the human body is capable of digesting and eliminating tellurium. However, the specific metabolic pathway is not totally understood yet. Again, it seems to resemble selenium's pathways. As previously stated, tellurium might "highjack" the metabolic routes that are commonly used for the other chalcogens. After the intake of tellurite and tellurate, the anions are transformed to telluride, probably via GSTeSG and Te(0), generally considering chemical processes rather than enzymatic ones. Once it reaches the liver, telluride is involved in a process of methylation, causing the generation of two compounds: dimethyltellurium ($(CH_3)_2Te$) and trimethyltellurium ($(CH3)_3Te^+$). These methylated species are possibly the most abundant forms of tellurium in circulation, generally found in the kidney and then recirculated to the spleen and the lungs. Finally, tellurium is eliminated through the urine (mostly as ionic, polar $(CH_3)_3Te^+$) and also via the breath, mostly as volatile (CH3)2Te [198].

In line with these findings, Kron, Hansen, and Werner decided to investigate the metabolic behavior of tellurium in humans [199]. Therefore, tellurium in various structures was administered perorally to healthy male human volunteers in the form of sodium tellurate, sodium tellurite, metallic colloid, and intrinsically bound in cress. After the administration, the urinary excretion of tellurium was determined. From the cumulative tellurium elimination in the first 4 days after the administration, a percentage intestinal absorption of $25 \pm 10\%$ for soluble tellurium salts was calculated. They reported that renal tellurium excretion is faster after administration of hexavalent tellurium than after the tetravalent form. This can explain the higher toxicity of the tetravalent tellurium compounds found in animal experiments. Other species such as dimethyltellurium can also accumulate in red blood cells. This circumstance has been researched in rats, where $(CH_3)_2Te$ bound to hemoglobin has been observed. Elevated concentrations in the spleen have been related to the accumulation of such tellurium compounds in red blood cells.

Although different research into tellurium nanostructures have demonstrated the reduction of toxicity in comparison to the bulk material, these compounds can still have some problems associated to their nanometric size as they tend to accumulate in the body and cause cytotoxic effects that may compromise their effectivity as biomedical agents. Therefore, there is still a need to find a complete explanation of the different mechanisms and applications. This particular element may offer in the biological field.

Tellurium Nanomaterials for Biomedical Applications

As primarily known, and previously commented, the biological role of Te bulk compounds is relatively limited due to the high toxicity involved and the lack of study on the matter [200]. Nevertheless, in its nanoscale form, Te has shown promising applications in the biomedical field, such as antibacterial [29], antifungal [135], anticancer [201], and imaging [202] applications. Therefore, tellurium nanomaterials are presented as a new field of investigation for their use in biomedicine and as an alternative of other well-known metallic NP that have been present in the field for a long time, like Au or Ag [11].

Nanoscale Tellurium as an Antimicrobial Agent

Tellurium has been largely known for its cytotoxic properties. Indeed, toxic tellurium agents were already known at the beginning of the twentieth century. In the pre-antibiotic era, tellurite was used as an antibiotic, inhibiting the growth of many microorganisms [203]. Tellurite was used in the research field as an antibiotic nonetheless not used to treat infections in humans. It was on 1932, when Sir Alexander Fleming brought into comparison the antibacterial properties of penicillin and tellurite and could be generally observed that penicillin-insensitive bacteria were also tellurite-sensitive and in the other way around. Fleming reported the antibacterial capacity of tellurite ions [204], and since then TeO_3^{2-} has been utilized constantly to isolate tellurite-resistant strains as *Escherichia coli* O157 [205]. Nevertheless, in those days, issues like tellurium toxicity and possible selectivity were not contemplated.

With the rise of the antibiotic era, tellurite was not the most promising candidate, a trend that has persisted to today. Different synthetic tellurides have been found to be toxic when in contact with cells or animal models. Recently, antimicrobial resistance (AMR) has become one of the main concerns in the healthcare system [206], leading us to the threat of a post-antibiotic era. New cases arise every day, as data from the Centers of Disease Control and Prevention (CDC) showed that there are more than 23,000 deaths/year as a consequence of antimicrobial resistance infections, thus situating AMR as one of the leading causes of death in the USA. Different metals, such as silver or zinc, have been largely recognized as powerful antibacterial agents in bulk toward both Gram-negative and -positive bacteria [207]. Moreover, and thanks to the use of nanotechnology—and the high surface-to-volume ratio that nanomaterials present—new materials present antimicrobial activity and therefore can be used as potential solutions to the AMR crisis [208].

Some tellurium bulk compounds showed antibacterial properties in a range of bacterial strains, for instance, Tellurium AS101—an organocompound—showed potential antibacterial activity against the species *Enterobacter cloacae* [20]. In addition, many other tellurite compounds were used during the twentieth century as bactericidal agents. Nevertheless, the toxicity associated with metallic tellurium discourage their use. Notwithstanding that, as said beforehand, the application of nanotechnology allowed the employment of this material as a novel antimicrobial agent with low cytotoxicity associated [209].

Distinct structures—from nanocrystals to nanowires and nanoparticles—have been tested showing a common antibacterial behavior against a broad spectrum of bacteria [210]. For instance, tellurium nanostructures were synthesized using enzymes showing antibacterial activity against *S. aureus* and *E. coli* [42]. In a similar study, Zare's group reported how *Bacillus* sp. BZ, extracted from the Caspian Sea in northern Iran, showed the capacity to generate elemental tellurium nanoparticles. The nanorods demonstrated bactericidal effect over different bacteria such as *Salmonella typhi* and *Pseudomonas aeruginosa* [211]. Besides, TeNP produced by physicochemical approaches also presented antibacterial activity on *E. coli*, *P. aeruginosa*, and *Acinetobacter baumannii* [212]. Similarly, TeNR obtained through an environmentally friendly hydrothermal approach showed antibacterial activity against antibiotic-resistant bacteria such as methicillin-resistant *Staphylococcus aureus* (MRSA) and multi-drug-resistant (MDR) *Escherichia coli* in a large variety of concentrations [201].

Although there is no strictly predetermined mechanism, it has been primarily hypothesized that the antimicrobial properties may be mainly related to the production of ROS or metal—ion release mechanisms [213]. The metal—ion release mechanism is applied when nanoparticles are in solution, as it is known they tend to dissolve and release ions; these ions can interact with the bacteria and penetrate the walls, generating toxicity that ends in cell death [213, 214].

In summary, tellurium nanomaterials have proven to be powerful antimicrobial agents and an alternative solution to the widespread use of antibiotics in a number of different structures and mechanisms of synthesis. Nonetheless, the future prospect of tellurium nanostructures as antimicrobial agents is largely undiscovered but is most likely to be focused on finding novel structures with low cytotoxic effects associated and establishing a common mechanism for the antibacterial behavior [208].

Nanoscale Tellurium as an Anticancer Agent

The American Cancer Society defines cancer as a conjunction of more than 100 diseases that are characterized by the abnormal and overgrowth of cells in different parts of the body such as lungs, breasts, or blood. The number of cases reported each year lead to an ever-growing concern over this disease. Different studies suggest that in 2018, more than 20 million people would suffer some variant of cancer and, as a consequence, approximately ten million people would die [215]. Due to poor lifestyles and environmental quality, cancer is estimated to be a rising problem in the future years [216].

Nanomaterials have been suggested as an alternative to the well-established treatments for cancer such as chemotherapy [217], radiotherapy [218], and surgery [219] as a potential solution to overcome some significant drawbacks [220, 221] and the increasing resistance to chemotherapy drugs [222]. As a result, these traditional treatments would no longer be effective and will end in an increasing death in patients [223].

A variety of tellurium organo-compounds—such as AS101 or octa-O-bis-(R,R)-tartrate ditellurane (SAS)—have been reported as potential anticancer agents [20, 223] by their capability of inactivating enzymes that compromise tumor surveillance [224]. In addition, and in order to enhance the performance of tellurium compounds as anticancer agents, new nanomaterials have recently emerged in different sizes and morphologies, although still in a reduced number. In one such study, tellurium nanorods synthesized using hydrothermal processes and coated with PVP showed in vitro anticancer properties on melanoma cells with low cytotoxicity for human dermal fibroblast cells [201].

Similar to the level of understanding existing of these nanostructures' antimicrobial properties, the mechanisms of action of these nanostructures are not clearly defined or fully researched. At first sight, the process of oxidation and later repression of cysteine proteins and enzymes can be treated as a drawback, and it can present diverse profits. The oxidation of cysteine residues in proteins is considered an effective and therefore selective mechanism, by which it can target successfully the majority of the reactive residues. This knowledge is the foundation for an innovative field in anticancer investigation of tellurium compounds. Redox modulators, also known as "sensor/effector" agents [225], are based on the mixture between high activity and chemical selectivity with the aim to detect cancerous cells above healthy ones and therefore and selectively end with these cells without causing severe damage to healthy ones. Investigations done along the last two decades demonstrated that a high number of cancerous cells reproduce in conditions of oxidative stress (OS), i.e., in the presence of elevated levels of ROS and an impaired antioxidant defense. By further controlling this disturbed redox balance, the possibility to reach a critical level known as "redox threshold" exists, which unleashes an apoptotic signal cascade that finally ends with cancer cell death. The concentrations of ROS (and other stressors) are generally low in healthy cells, and they are less affected by these mechanisms [226]. Hence, it seems that the catalysts in question "identifies" or "sense" a particular "biochemical signature" of OS in cancer cells and generate their effects in consequence.

Therefore, tellurium-based catalysts with glutathione peroxidase (GPx)-like activity have a significant impact on the anticancer field. As an example, a fairly selective activity of 2,3-bis(phenyltellanyl)naphthoquinone studied in different cancer cell lines, such as HT29 and CT26 human colon cancer cells, that are particularly more sensitive to tellurium compound in comparison with normal cultured NIH 3T3 fibroblast cells. Comparable effects were observed in a model of human chronic lymphocytic leukemia (CLL). Significant reduction of cell proliferation was observed when CLL B-cells were isolated from patient blood and were treated with 2,3-bis(phenyltellanyl)naphthoquinone. On the contrary, healthy B-cells extracted from identical patients and control peripheral blood mononuclear cells (PBMC) were considerably less affected. More detailed studies exposed that redox modulation is the explanation for the observed results. As a consequence, compounds such as 2,3-bis(phenyltellanyl)naphthoquinone have the ability to rise OS, with three possible mechanisms by producing ROS, by transforming less reactive ROS into more active species or by catalyzing the ROS-driven oxidation of proteins and enzymes [227, 228]. The catalysis can be performed on either tellurium site or using other sites, like the redox-active quinone.

With independence of the chemical and biochemical processes involving Te compounds, their cytotoxicity is considered more complicated and elaborated than thought beforehand. Late investigations suggest the feasibility of tellurium-agents to control the activity of particular apoptosis-inducing proteins in cancer cells and therefore trigger an antioxidant response in normal cells. The compound AS101 shows the biochemical complexity related to different tellurium agents. This compound particularly inactivates cysteine proteases by binding with and lately oxidizing the catalytic thiol to a disulfide. In addition, it participates in the inhibition of caspases and therefore down-regulating caspase-1 inflammatory products, such as interleukin-18 (IL-18) and IL-1b. The up-regulation of glial cell line-derived neurotrophic factor (GDNF) is produced by direct inhibition of anti-inflammatory cytokine IL10, which at the same time provokes the associated cell survival pathways [229].

With time, different vinyl tellurium compounds have been applied as therapeutic agents, which in turn can inhibit cysteine proteases, alike AS101 [230]. The compound RT-04, as an example, inhibits principally cathepsin B, and hence, it is capable of inducing apoptosis in HL60 cells without toxic effects observed in normal bone marrow cells. Materials surrounding these and comparable compounds offers as a huge field for future investigations so as to the production of further complex—tellurium agents.

Besides those mechanisms, hyperthermia—which describes the self-destruction of the cell if temperatures higher than 43 °C are applied—is also considered for developing powerful cancer treatment [231, 232]. For example, due to their optical properties, tellurium nanoparticles can be excited using near-infrared light with high throughput [41]. Thus they can be used as a target to induce hyperthermia in tumors. As an example, tellurium nanorods were tested on a variety of cancer lines—such as hepatocarcinoma and melanoma—as well as healthy vascular epithelial cells; the results showed significant anticancer effect with low cytotoxicity and no modification on the morphology and size of the nanostructures after treatment [233].

Future studies on anti-cancerous properties of tellurium nanoparticles are necessary and should strive to elucidate the unknown mechanisms of cell interaction of these novel nanoparticles. Moreover, further research including the optimization of the sizes and morphologies of various created tellurium nanostructures is required for discovering potential applications. In addition, extensive in-depth cytotoxicity studies would allow us to comprehend the vast utility of these compounds in human therapeutic applications [234].

Nanoscale Tellurium as an Imaging Agent and a Biological Marker

The imaging of tissues and organs have become a vital tool for accurate medical diagnostics and research in the medical field [235]. Imaging comprehends different techniques such as computed tomography (CT) [236]—based on the use of X-ray—and magnetic resonance imaging (MRI) [237]—based on the phenomenon of nuclear magnetic resonance (NMR)—both allowing for the visualization of components in the body for further diagnosis of diseases. Even though these techniques have been well established, they suppose a high economic impact [238], and new materials for screening diseases have been found in the last years coming from the use of nanotechnology [239].

One of the advantages of using nanomaterials is that they could be novel contrast agents for the use in imaging techniques, as they are designed to interact with parts of the body selectively; consequently, they give accurate responses and improved images [240]. Their unique physical properties allow for tuning their characteristics and for adapting to the desired configuration in order to obtain a high throughput [241]. For instance, they can be used as multi-model imaging tools [242]—for doing more than one imaging technique at a time—or to target specific compounds such as macrophages [243]. Another important key to be used in vivo is their biodistribution, related to the low cytotoxicity to human cells and the ability to be eliminated using renal or hepatic ways [244]. Consequently, there is a vast variety of materials that provide imaging response [245], such as iron oxide nanoparticles (MIONs) used in MRI as powerful magnetic nanoparticles [246] or gold nanoparticles for CT specifically for tumor-targeted imaging [247].

The different physical, chemical, and especially spectroscopic characteristics of the metalloid have brought into attention the chance to use it as a useful biological marker. It is well known that the chemistry of the element somewhat resembles that of sulfur, the reason why it might be possible to include it into amino acids like cysteine and methionine [248]. Intriguingly, the presence of compounds such as telluromethionine in the medium have resulted in the natural addition of tellurium into methionine and consequently into proteins and enzymes. The modification of light S for a heavier element such as Te have potential benefits, as an example in terms of protein structural studies using X-ray crystallography. It has been largely reported that selenium can be introduced biologically in the form of selenomethionine nonetheless, it is only able to contribute to certain protein structure information by the multi-wavelength anomalous dispersion (MAD) methodology. On the other hand, as it is heavier than the other elements, tellurium is able to provide clear signals. The increasing benefits of using Te as labeling agent cover the suitableness of the labeling methodologies, selective labeling at the methionine sites, the stability of telluromethionine, elevated isomorphism with the parent molecule, and—from a crystallographic point of view—high phasing power, relative abundance, and mobility of target sites. Further investigations on the labeling field can lead to surprising discoveries.

At the beginning of the 1980s, Knapp, Kirsch and collaborators produced a variety of fatty acids that contained different forms of tellurium, along with the radioactive 123mTe isotope. These compounds were largely investigated in rats and dogs. The presence of tellurium in the fatty acids provoked their difficult metabolization, and the results showed they were "stopped" in the myocardium (heart muscle) [249]. Some possible applications of this technology can be in the field of nuclear imaging or medicine as the Te-containing fatty acids can accumulate in particular organs and therefore be used for diagnosis of heart and pancreatic diseases [250].

Nonetheless, quantum dots (QDs) are considered a reference for biomedical applications of nanomaterials. QDs are fluorescent semiconductor nanoparticles with size-dependent emission of light, with a broad absorption and narrow emission spectra band [251]. These novel materials are employed in bioimaging: they present a core and a shell structure composed by semiconductor materials, therefore, in here is where tellurium plays its major role [252]. Here, different fluorescent tellurium particles are formulated not only as CdTe but also like CdSeTe, CdHgTe, and CdTe/ZnTe, each of them showing its own biological profile [253, 254].

P. Xu et al. researched innovative fluorescent nanocomposites based on gambogic acid (GA) and cadmium–tellurium (CdTe) QDs. The incorporation of cysteamine in the structure was designed with the aim of using them for cancer labeling and combined treatment [255]. Furthermore, other structures of CdTe QDs allowed the direct imaging of human serum proteins. Alternatively, Peng and colleagues developed an easy one-pot approach for the synthesis of Gd-doped CdTe QDs in aquatic media as fluorescence and magnetic resonance imaging dual-modal agent, showing a close size distribution and average dimensions of about 5 nm [256]. Mason et al. configured and prepared red CdTe and NIR CdHgTe QDs for fluorescent imaging [257]. They demonstrated the fluorescent imaging by utilizing CdTe and CdHgTe QDs as fluorescent probes both in vitro and in vivo. Their results gave sensitive detection over background autofluorescence in tissue biopsies and live mice, making them suitable probes for in vivo imaging for deep tissues or whole animals.

There is no report on the literature demonstrating TeNP with imaging properties in spite of the QDs structure. However, there is a strong belief that this metalloid can be used alone as a contrast agent due to the few reports on bulk tellurium materials such as Te-123m-labeled 23-(isopropyl telluro)-24-nor-5 alpha-cholan-3 beta-on [249] or tellurium-123m-labeled-9-telluraheptadecanoic acid [250] for adrenal gland and cardiac imaging agent, respectively (Fig. 7).

Conclusions

Tellurium has been known since the eighteenth century as a rare element of severe isolation. Despite its properties, which allow finding it useful for photoconductive, photothermal, and electronic applications, among others, many of its potential applications remain surrounded by uncertainty, especially in terms of bio-interactions. Widely used in the industry for enhancement of the properties of other materials, the

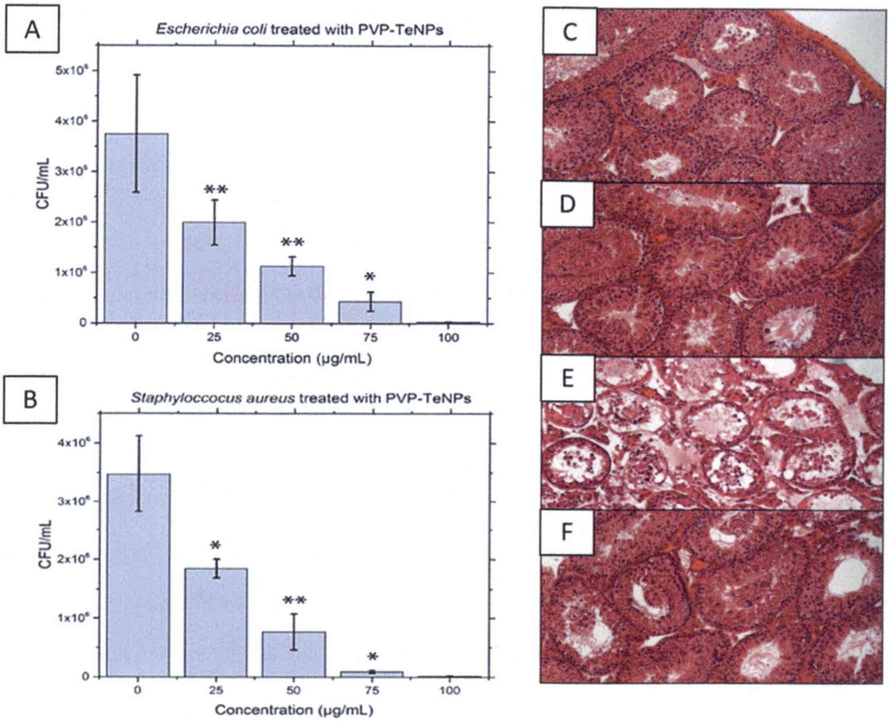

Fig. 7 Colony-forming assay of (**A**) *Escherichia coli* and (**B**) *f* after 8 h treatment with PVP-coated tellurium nanorods. $N = 3$. $*p < 0.01$ in comparison to control (0 µg/mL concentration), $**p < 0.005$ in comparison to control (0 µg/mL concentration) [201]; Effect of AS101 treatment on the testicular damage induced by 5 weekly injections of 200 mg/kg cyclophosphamide (Cy) ($n = 5$). Histological sections of testis from mice injected with (**C**) phosphate-buffered saline, (**D**) AS101, (**E**) Cy or (**F**) Cy + AS101. The cross-sections of Cy-treated mice were characterized by empty and atrophic seminiferous tubules, compared with the standard cellular content of the controls. Cy + AS101 group showed many tubules with undamaged spermatogenesis therefore with less severe damage associated [229]

research associated with the development of alternative ways to either produce or use it remains in a low profile. However, it seems like the future opens a broad range of possibilities to tellurium that, nowadays, remain as the forgotten element.

References

1. CRC handbook of chemistry and physics (1977) CRC Press, Cleaveland
2. Bouroushian M (2010) Electrochemistry of the chalcogens. Springer, Berlin, pp 57–75. https://doi.org/10.1007/978-3-642-03967-6_2
3. Chivers T, Laitinen RS (2015) Tellurium: a maverick among the chalcogens. Chem Soc Rev 44(7):1725–1739. https://doi.org/10.1039/c4cs00434e

4. Frieden E (1972) The chemical elements of life. Sci Am 227(1):52–60. http://www.ncbi.nlm.nih.gov/pubmed/5044408
5. Dobbin L (1900) A handbook of physics and chemistry. Edinb Med J 7(1):67. https://www.ncbi.nlm.nih.gov/pmc/articles/PMC5260252/
6. Szabadváry F (1997) Neue Deutsche Biographie, Band 18. https://www.deutsche-biographie.de/pnd134201868.html#ndbcontent
7. Cambridge University Press (1909) Encyclopædia Britannica, 11th edn. https://www.britannica.com/topic/Encyclopaedia-Britannica-English-language-reference-work/Eleventh-edition-and-its-supplements
8. Nicholson W (ed.) (1802) Journal of Natural Philosophy, Chemistry and the Arts. In: vol. III. https://books.google.com/books?id=aAgAAAAAMAAJ&pg=PR15&lpg=PR15&dq=%22Abstract+of+a+Memoir+of+Klaproth+on+a+New+Metal+Denominated+Tellurium&source=bl&ots=b3BmzfsbN3&sig=ACfU3U2Bh_9bv-skhWWaOjaJPuAK4YMSHg&hl=en&sa=X&ved=2ahUKEwirwbXRiZ7gAhWEd98KHYBwCLgQ6AE
9. Lukács D (1977) [Pál Kitaibel]. Orv Hetil 118(44):2660–62. http://www.ncbi.nlm.nih.gov/pubmed/335328
10. Rouvray DH (2004) Elements in the history of the periodic table. Endeavour 28(2):69–74. https://doi.org/10.1016/j.endeavour.2004.04.006
11. Ba LA, Döring M, Jamier V, Jacob C (2010) Tellurium: an element with great biological potency and potential. Org Biomol Chem 8(19):4203–4216. https://doi.org/10.1039/c0Ob00086h
12. Woollins JD, Laitinen R (eds) (2011) Selenium and tellurium chemistry. Springer, Berlin. https://doi.org/10.1007/978-3-642-20699-3
13. Zajacite-(Ce) mineral data. 2019. http://webmineral.com/data/Zajacite-(Ce).shtml#.XGLr1TNKiUk
14. Minerals Information Center, National (2017) Mineral commodity summaries. https://minerals.usgs.gov/minerals/pubs/mcs/2017/mcs2017.pdf. Accessed 2 Feb 2019
15. Rosing MT (2008) On the evolution of minerals. Nature 456(7221):456–458. https://doi.org/10.1038/456456a
16. Survey, U.S. Geological (2018) Minerals yearbook. In: Minerals yearbook, vol III. https://doi.org/10.3133/MYBVIII
17. Tian P, Xu X, Ao C, Ding D, Li W, Si R, Tu W, Xu J, Han Y-F (2017) Direct and selective synthesis of hydrogen peroxide over palladium-tellurium catalysts at ambient pressure. ChemSusChem 10(17):3342–3346. https://doi.org/10.1002/cssc.201701238
18. Zhou T, Zhu Z, Liu X, Liang Z, Wang X (2018) A review of the precision glass molding of chalcogenide glass (ChG) for infrared optics. Micromachines 9(7):337. https://doi.org/10.3390/mi9070337
19. Kranz L, Gretener C, Perrenoud J, Schmitt R, Pianezzi F, La Mattina F, Blösch P et al (2013) Doping of polycrystalline CdTe for high-efficiency solar cells on flexible metal foil. Nat Commun 4(1):2306. https://doi.org/10.1038/ncomms3306
20. Daniel-Hoffmann M, Sredni B, Nitzan Y (2012) Bactericidal activity of the organo-tellurium compound AS101 against Enterobacter Cloacae. J Antimicrob Chemother 67(9):2165–2172. https://doi.org/10.1093/jac/dks185
21. Mohanty P, Park J, Kim B (2006a) Large scale synthesis of highly pure single crystalline tellurium nanowires by thermal evaporation method. J Nanosci Nanotechnol 6(11):3380–3383. http://www.ncbi.nlm.nih.gov/pubmed/17252770
22. Mohanty P, Kang T, Kim B, Park J (2006b) Synthesis of single crystalline tellurium nanotubes with triangular and hexagonal cross sections. J Phys Chem B 110(2):791–795. https://doi.org/10.1021/JP0551364
23. Zhu Y-J, Wang W-W, Qi R-J, Hu X-L (2004) Microwave-assisted synthesis of single-crystalline tellurium nanorods and nanowires in ionic liquids. Angew Chem Int Ed 43(11):1410–1414. https://doi.org/10.1002/anie.200353101
24. ESPI Metals (2019) Espimetals—tellurium. http://www.espimetals.com/index.php/technical-data/253-tellurium

25. The Merck index online—chemicals, drugs and biologicals. 2019. https://www.rsc.org/merck-index
26. Johnstone AH (2007) CRC handbook of chemistry and physics-69th edition editor in chief R. C. Weast, CRC Press Inc., Boca Raton, Florida, 1988, Pp. 2400, Price £57.50. ISBN 0-8493-0369-5. J Chem Technol Biotechnol 50(2):294–295. https://doi.org/10.1002/jctb.280500215
27. Lin S, Li W, Chen Z, Shen J, Ge B, Pei Y (2016) Tellurium as a high-performance elemental thermoelectric. Nat Commun 7(1):10287. https://doi.org/10.1038/ncomms10287.
28. Zhan L, Xu Z (2014) State-of-the-art of recycling E-wastes by vacuum metallurgy separation. Environ Sci Technol 48(24):14092–14102. https://doi.org/10.1021/es5030383
29. Ollivier PRL, Bahrou AS, Marcus S, Cox T, Church TM, Hanson TE (2008) Volatilization and precipitation of tellurium by aerobic, tellurite-resistant marine microbes. Appl Environ Microbiol 74(23):7163–7173. https://doi.org/10.1128/AEM.00733-08
30. Dirmyer MR, Martin J, Nolas GS, Sen A, Badding JV (2009) Thermal and electrical conductivity of size-tuned bismuth telluride nanoparticles. Small 5(8):933–937. https://doi.org/10.1002/smll.200801206
31. Nyk J, Onderka B (2012) Thermodynamics of oxygen in dilute liquid silver–tellurium alloys. Monatsh Chem 143(9):1219–1224. https://doi.org/10.1007/s00706-012-0771-z
32. Yang T, Ke H, Wang Q, Tang Y'a, Deng Y, Yang H, Yang X et al (2017a) Bifunctional tellurium nanodots for photo-induced synergistic cancer therapy. ACS Nano 11(10):10012–10024. https://doi.org/10.1021/acsnano.7b04230
33. Fritzsche H (1952) Interpretation of the double reversal of the hall effect in tellurium. Science 115(2995):571–572. https://doi.org/10.1126/science.115.2995.571.
34. Otjacques C, Raty J-Y, Coulet M-V, Johnson M, Schober H, Bichara C, Gaspard J-P (2009) Dynamics of the negative thermal expansion in tellurium based liquid alloys. Phys Rev Lett 103(24):245901. https://doi.org/10.1103/PhysRevLett.103.245901
35. Churchill HOH, Salamo GJ, Yu S-Q, Hironaka T, Hu X, Stacy J, Shih I (2017) Toward single atom chains with exfoliated tellurium. Nanoscale Res Lett 12(1):488. https://doi.org/10.1186/s11671-017-2255-x
36. Bijelic A, Rompel A (2017) Ten good reasons for the use of the tellurium-centered Anderson–Evans polyoxotungstate in protein crystallography. Acc Chem Res 50(6):1441–1448. https://doi.org/10.1021/acs.accounts.7b00109
37. Harrison WTA, Johnston MG, IUCr (2014) Crystal structure of ammonium divanadium (IV,V) tellurium(IV) heptaoxide. Acta Cryst 70(7):27–30. https://doi.org/10.1107/S1600536814011015
38. Brown PJ, Forsyth JB, IUCr (1996) The crystal structure and optical activity of tellurium. Acta Cryst 52(3):408–412. https://doi.org/10.1107/S0108767395017144
39. Myers JP, Fronczek FR, Junk T (2016) The first crystal structures of six- and seven-membered tellurium- and nitrogen-containing (Te—N) heterocycles: 2 *H* -1,4-benzotellurazin-3(4 *H*)-one and 2,3-dihydro-1,5-benzotellurazepin-4(5 *H*)-one. Acta Cryst 72(1):1–5. https://doi.org/10.1107/S2053229615022378
40. Bloomer WD, McLaughlin WH, Neirinckx RD, Adelstein SJ, Gordon PR, Ruth TJ, Wolf AP (1981) Astatine-211—tellurium radiocolloid cures experimental malignant ascites. Science 212(4492):340–341. http://www.ncbi.nlm.nih.gov/pubmed/7209534.
41. Ma C, Yan J, Huang Y, Wang C, Yang G (2018) The optical duality of tellurium nanoparticles for broadband solar energy harvesting and efficient photothermal conversion. Sci Adv 4(8):eaas9894. https://doi.org/10.1126/sciadv.aas9894
42. Pugin B, Cornejo FA, Muñoz-Díaz P, Muñoz-Villagrán CM, Vargas-Pérez JI, Arenas FA, Vásquez CC (2014) Glutathione reductase-mediated synthesis of tellurium-containing nanostructures exhibiting antibacterial properties. Appl Environ Microbiol 80(22):7061–7070. https://doi.org/10.1128/AEM.02207-14
43. Liu Z, Hu Z, Liang J, Li S, Yang Y, Peng S, Qian Y (2004) Size-controlled synthesis and growth mechanism of monodisperse tellurium nanorods by a surfactant-assisted method. Langmuir 20(1):214–218. https://doi.org/10.1021/LA035160D

44. Graf C, Assoud A, Mayasree O, Kleinke H (2009) Solid state polyselenides and polytellurides: a large variety of Se–Se and Te–Te interactions. Molecules 14(9):3115–3131. https://doi.org/10.3390/molecules14093115
45. Ogra Y, Kobayashi R, Ishiwata K, Suzuki KT (2008) Comparison of distribution and metabolism between tellurium and selenium in rats. J Inorg Biochem 102(7):1507–1513. https://doi.org/10.1016/j.jinorgbio.2008.01.012
46. Cabri LJ (1965) Phase relations in the Au-Ag-Te systems and their mineralogical significance. Econ Geol 60(8):1569–1606. https://doi.org/10.2113/gsecongeo.60.8.1569
47. Zhang Q, Malliakas CD, Kanatzidis MG (2009) $\{[Ga(En)_3]_2(Ge_2Te_{15})\}_n$: a polymeric semiconducting polytelluride with boat-shaped Te_8^{4-} rings and cross-shaped Te_5^{6-} units. Inorg Chem 48(23):10910–10912. https://doi.org/10.1021/ic9019074
48. Dana JD, Dana ES, Gaines RV, Dana JD (1997) Dana's new mineralogy : the system of mineralogy of James Dwight Dana and Edward Salisbury Dana. Wiley, Hoboken. http://webmineral.com/danaclass.shtml#.XFYjLlxKiUk
49. Getman FH (1933) A study of the tellurium electrode. Trans Electrochem Soc 64(1):201. https://doi.org/10.1149/1.3504515
50. Lid DR (2006) CRC handbook of chemistry and physics. American Chemical Society, Boca Raton. https://doi.org/10.1021/JA069813Z
51. Bruère MA (1891) Direct action of hydrogen sulphide, hydrogen selenide, and hydrogen telluride on haemoglobin. J Anat Physiol 26(Pt 1):62–75. http://www.ncbi.nlm.nih.gov/pubmed/17231959.
52. Patnaik P (2003) Handbook of inorganic chemicals. McGraw-Hill, New York. https://books.google.com/books/about/Handbook_of_Inorganic_Chemicals.html?id=Xqj-TTzkvTEC
53. Greenwood NN, Earnshaw A (1997) Chemistry of the elements. Butterworth-Heinemann, Oxford
54. Cotton FA, Wilkinson G, Murillo CA, Bochmann M (n.d.) Advanced inorganic chemistry. Wiley, New York
55. Mikhaylov AA, Medvedev AG, Churakov AV, Grishanov DA, Prikhodchenko PV, Lev O (2016) Peroxide coordination of tellurium in aqueous solutions. Chem Eur J 22(9):2980–2986. https://doi.org/10.1002/chem.201503614
56. Laitinen RS, Maaninen A, Pietikäinen J (1998) Selenium- and tellurium-containing chalcogen nitrides. Phosphorus Sulfur Silicon Relat Elem 136(1):397–412. https://doi.org/10.1080/10426509808545966
57. Massa W, Lau C, Möhlen M, Neumüller B, Dehnicke K (1998) [Te6N8(TeCl4)4]—tellurium nitride stabilized by tellurium tetrachloride. Angew Chem Int Ed 37(20):2840–2842. https://doi.org/10.1002/(SICI)1521-3773(19981102)37:20<2840::AID-ANIE2840>3.0.CO;2-N
58. Eagleson M (1994) Concise encyclopedia chemistry. Walter de Gruyter, Berlin. https://books.google.com/books/about/Concise_Encyclopedia_Chemistry.html?id=Owuv-c9L_IMC
59. Laitinen RS, Oilunkaniemi R (2011) Tellurium: inorganic chemistry based in part on the article tellurium: inorganic chemistry by William R. McWhinnie which appeared in the Encyclopedia of Inorganic Chemistry, First Edition. In: Encyclopedia of Inorganic and Bioinorganic Chemistry. Wiley, Chichester. https://doi.org/10.1002/9781119951438.eibc0222
60. Wiberg E, Wiberg N, Holleman AF (2001) Inorganic chemistry. Academic Press, San Diego. https://northeastern.on.worldcat.org/search?queryString=no%3A+48056955#/oclc/48056955
61. Devillanova F, Du Mont W-W (2013) Handbook of chalcogen chemistry, vol 1. Royal Society of Chemistry, Cambridge. https://doi.org/10.1039/9781849737456
62. Kniep R, Mootz D, Rabenau A (1976) Zur Kenntnis Der Subhalogenide Des Tellurs. Zeitschrift Fr Anorganische Und Allgemeine Chemie 422(1):17–38. https://doi.org/10.1002/zaac.19764220103
63. Binnewies M, Milke E (2002) Thermochemical data of elements and compounds. Wiley, Hoboken

64. Petragnani N, Comasseto JV (1991) Tellurium reagents in organic synthesis; recent advances. Part 1. Synthesis 1991(10):793–817. https://doi.org/10.1055/s-1991-26577
65. King RB (1977) Inorganic chemistry of the main-group elements. In: Addison CC (ed) Inorganic chemistry of the main-group elements, vol 4. Royal Society of Chemistry, Cambridge. https://doi.org/10.1039/9781847556400
66. Petragnani N (2007) Tellurium in organic synthesis. Academic Press, London
67. Sadekov ID, Zakharov AV (1999) Stable tellurols and their metal derivatives. Russ Chem Rev 68(11):909–923. https://doi.org/RC990909
68. Torubaev Y, Pasynskii A, Mathur P (2012) Organotellurium halides: new ligands for transition metal complexes. Coord Chem Rev 256(5–8):709–721. https://doi.org/10.1016/J.CCR.2011.11.011
69. Engman L, Kandra T, Gallegos A, Williams R, Powis G (2000) Water-soluble organotellurium compounds inhibit thioredoxin reductase and the growth of human cancer cells. Anticancer Drug Des 15(5):323–330. http://www.ncbi.nlm.nih.gov/pubmed/11354308
70. Alessandrello A, Arnaboldi C, Brofferio C, Capelli S, Cremonesi O, Fiorini E, Nucciotti A, et al (2002) New limits on naturally occurring electron capture of 123Te. https://doi.org/10.1103/PhysRevC.67.014323
71. Meija J, Coplen TB, Berglund M, Brand WA, De Bièvre P, Gröning M, Holden NE et al (2016) Atomic weights of the elements 2013 (IUPAC technical report). Pure Appl Chem 88(3):265–291. https://doi.org/10.1515/pac-2015-0305
72. Magill J (2003) The universal nuclide chart. In: Nuclides.Net. Springer, Berlin, pp 197–207. https://doi.org/10.1007/978-3-642-55764-4_9.
73. Kim S, Thiessen PA, Bolton EE, Chen J, Gang F, Gindulyte A, Han L et al (2016) PubChem substance and compound databases. Nucleic Acids Res 44(D1):D1202–D1213. https://doi.org/10.1093/nar/gkv951
74. Carotenuto G, Palomba M, De Nicola S, Ambrosone G, Coscia U (2015) Structural and photoconductivity properties of tellurium/PMMA films. Nanoscale Res Lett 10(1):313. https://doi.org/10.1186/s11671-015-1007-z.
75. Makuei FM, Senanayake G (2018) Extraction of tellurium from lead and copper bearing feed materials and interim metallurgical products—a short review. Miner Eng 115:79–87. https://doi.org/10.1016/J.MINENG.2017.10.013.
76. Wang S (2011) Tellurium, its resourcefulness and recovery. JOM 63(8):90–93. https://doi.org/10.1007/s11837-011-0146-7
77. Ramos-Ruiz A, Field JA, Wilkening JV, Sierra-Alvarez R (2016) Recovery of elemental tellurium nanoparticles by the reduction of tellurium oxyanions in a methanogenic microbial consortium. Environ Sci Technol 50(3):1492–1500. https://doi.org/10.1021/acs.est.5b04074
78. Hait J, Jana RK, Kumar V, Sanyal SK (2002) Some studies on sulfuric acid leaching of anode slime with additives. Ind Eng Chem Res 41(25):6593–6599. https://doi.org/10.1021/IE020239J
79. Wang S, Cui W, Zhang G, Zhang L, Peng J (2017b) Ultra fast ultrasound-assisted decopperization from copper anode slime. Ultrason Sonochem 36:20–26. https://doi.org/10.1016/J.ULTSONCH.2016.11.013
80. Yang T, Zhu P, Liu W, Chen L, Zhang D (2017b) Recovery of tin from metal powders of waste printed circuit boards. Waste Manag 68:449–457. https://doi.org/10.1016/j.wasman.2017.06.019
81. Giles GI, Fry FH, Tasker KM, Holme AL, Peers C, Green KN, Klotz L-O, Sies H, Jacob C (2003a) Evaluation of sulfur, selenium and tellurium catalysts with antioxidant potential. Org Biomol Chem 1(23):4317. https://doi.org/10.1039/b308117f
82. Anne M-L, Keirsse J, Nazabal V, Hyodo K, Inoue S, Boussard-Pledel C, Lhermite H et al (2009) Chalcogenide glass optical waveguides for infrared biosensing. Sensors 9(9):7398–7411. https://doi.org/10.3390/s90907398
83. Li P, Zhang Y, Chen Z, Gao P, Wu T, Wang L-M (2017) Relaxation dynamics in the strong chalcogenide glass-former of Ge22Se78. Sci Rep 7(1):40547. https://doi.org/10.1038/srep40547

84. Pei Y, Wang H, Snyder GJ (2012) Band engineering of thermoelectric materials. Adv Mater 24(46):6125–6135. https://doi.org/10.1002/adma.201202919
85. Ramos-Ruiz A, Wilkening JV, Field JA, Sierra-Alvarez R (2017) Leaching of cadmium and tellurium from cadmium telluride (CdTe) thin-film solar panels under simulated landfill conditions. J Hazard Mater 336:57–64. https://doi.org/10.1016/j.jhazmat.2017.04.052
86. Simpson RE, Fons P, Kolobov AV, Fukaya T, Krbal M, Yagi T, Tominaga J (2011) Interfacial phase-change memory. Nat Nanotechnol 6(8):501–505. https://doi.org/10.1038/nnano.2011.96
87. Yuan QL, Yin HY, Nie QL (2013) Nanostructured tellurium semiconductor: from nanoparticles to nanorods. J Exp Nanosci 8(7–8):931–936. https://doi.org/10.1080/17458080.2011.620021
88. Ayre J (2013) First solar reports largest quarterly decline in CdTe module cost per-watt since 2007. Solar Love. https://cleantechnica.com/2013/11/07/first-solar-reports-largest-quarterly-decline-cdte-module-cost-per-watt-since-2007/
89. Todorov TK, Singh S, Bishop DM, Gunawan O, Lee YS, Gershon TS, Brew KW, Antunez PD, Haight R (2017) Ultrathin high band gap solar cells with improved efficiencies from the world's oldest photovoltaic material. Nat Commun 8(1):682. https://doi.org/10.1038/s41467-017-00582-9
90. Yarema MC, Curry SC (2005) Acute tellurium toxicity from ingestion of metal-oxidizing solutions. Pediatrics 116(2):e319–e321. https://doi.org/10.1542/peds.2005-0172
91. Yan Y, Zhang J, Ren L, Tang C (2016) Metal-containing and related polymers for biomedical applications. Chem Soc Rev 45(19):5232–5263. https://doi.org/10.1039/c6cs00026f
92. Shieh M, Ho C-H, Sheu W-S, Chen B-G, Chu Y-Y, Miu C-Y, Liu H-L, Shen C-C (2008) Semiconducting tellurium–iron–copper carbonyl polymers. J Am Chem Soc 130(43):14114–14116. https://doi.org/10.1021/ja8065623
93. Jiang S, Sheng J, Huang Z (2011) Synthesis of the tellurium-derivatized phosphoramidites and their incorporation into DNA oligonucleotides. Curr Protoc Nucleic Acid Chem 1:1.25.1–1.25.16. Hoboken: John Wiley & Sons, Inc. https://doi.org/10.1002/0471142700.nc0125s47.
94. Maurya D, Sardarinejad A, Alameh K, Maurya DK, Sardarinejad A, Alameh K (2014) Recent developments in R.F. magnetron sputtered thin films for PH sensing applications—an overview. Coatings 4(4):756–771. https://doi.org/10.3390/coatings4040756
95. Gurunathan S, Kim J-H (2016) Synthesis, toxicity, biocompatibility, and biomedical applications of graphene and graphene-related materials. Int J Nanomedicine 11:1927–1945. https://doi.org/10.2147/IJN.S105264
96. Liu J-W, Xu J, Hu W, Yang J-L, Yu S-H (2016) Systematic synthesis of tellurium nanostructures and their optical properties: from nanoparticles to nanorods, nanowires, and nanotubes. ChemNanoMat 2(3):167–170. https://doi.org/10.1002/cnma.201500206
97. He W, Krejci A, Lin J, Osmulski ME, Dickerson JH (2011) A facile synthesis of Te nanoparticles with binary size distribution by green chemistry. Nanoscale 3(4):1523. https://doi.org/10.1039/c1nr10025d
98. Taylor R, Coulombe S, Otanicar T, Phelan P, Gunawan A, Lv W, Rosengarten G, Prasher R, Tyagi H (2013) Small particles, big impacts: a review of the diverse applications of nanofluids. J Appl Phys 113(1):011301. https://doi.org/10.1063/1.4754271
99. Tsai H-W, Yaghoubi A, Chan T-C, Wang C-C, Liu W-T, Liao C-N, Lu S-Y, Chen L-J, Chueh Y-L (2015) Electrochemical synthesis of ultrafast and gram-scale surfactant-free tellurium nanowires by gas–solid transformation and their applications as supercapacitor electrodes for p-Doping of graphene transistors. Nanoscale 7(17):7535–7539. https://doi.org/10.1039/C5NR00876J
100. Jiang Z-Y, Xie Z-X, Zhang X-H, Xie S-Y, Huang R-B, Zheng L-S (2004) Synthesis of α-tellurium dioxide nanorods from elemental tellurium by laser ablation. Inorg Chem Commun 7(2):179–181. https://doi.org/10.1016/J.INOCHE.2003.10.037

101. Arab F, Mousavi-Kamazani M, Salavati-Niasari M (2017) Facile sonochemical synthesis of tellurium and tellurium dioxide nanoparticles: reducing Te(IV) to Te via ultrasonic irradiation in methanol. Ultrason Sonochem 37:335–343. https://doi.org/10.1016/j.ultsonch.2017.01.026
102. Mousavi-Kamazani M, Rahmatolahzadeh R, Shobeiri SA, Beshkar F (2017) Sonochemical synthesis, formation mechanism, and solar cell application of tellurium nanoparticles. Ultrason Sonochem 39:233–239. https://doi.org/10.1016/J.ULTSONCH.2017.04.031
103. Zhang A, Zheng G, Lieber CM (2016a) Nanowires. Building blocks for nanoscience and nanotechnology. Springer, Basel. http://www.springer.com/series/3705
104. Mayers B, Xia Y (2002) One-dimensional nanostructures of trigonal tellurium with various morphologies can be synthesized using a solution-phase approach. J Mater Chem 12(6):1875–1881. https://doi.org/10.1039/b201058e
105. Qian H-S, Yu S-H, Gong J-Y, Luo L-B, Fei L-f (2006) High-quality luminescent tellurium nanowires of several nanometers in diameter and high aspect ratio synthesized by a poly (vinyl pyrrolidone)-assisted hydrothermal process. Langmuir 22(8):3830–3835. https://doi.org/10.1021/la053021l
106. Lu Q, Gao F, Komarneni S (2004) Biomolecule-assisted reduction in the synthesis of single-crystalline tellurium nanowires. Adv Mater 16(18):1629–1632. https://doi.org/10.1002/adma.200400319
107. Huang W, Wu H, Li X, Chen T (2016) Facile one-pot synthesis of tellurium nanorods as antioxidant and anticancer agents. Chem Asian J 11(16):2301–2311. https://doi.org/10.1002/asia.201600757
108. Xi G, Peng Y, Yu W, Qian Y (2005) Synthesis, characterization, and growth mechanism of tellurium nanotubes. Cryst Growth Des 5(1):325–328. https://doi.org/10.1021/CG049867P
109. Song J-M, Lin Y-Z, Zhan Y-J, Tian Y-C, Liu G, Yu S-H (2008) Superlong high-quality tellurium nanotubes: synthesis, characterization, and optical property. Cryst Growth Des 8(6):1902–1908. https://doi.org/10.1021/cg701125k
110. Liu T, Zhang G, Su X, Chen X, Wang D, Qin J (2007) Tellurium nanotubes synthesized with microwave-assisted monosaccharide reduction method. J Nanosci Nanotechnol 7(7):2500–2505. http://www.ncbi.nlm.nih.gov/pubmed/17663271
111. Qun W, Li G-D, Liu Y-L, Xu S, Wang K-J, Chen J-S (2007) Fabrication and growth mechanism of selenium and tellurium nanobelts through a vacuum vapor deposition route. J Phys Chem C 111(35):12926–12932. https://doi.org/10.1021/JP073902W
112. Wan B, Hu C, Liu H, Chen X, Xi Y, He X (2010) Glassy state lead tellurite nanobelts: synthesis and properties. Nanoscale Res Lett 5(8):1344–1350. https://doi.org/10.1007/s11671-010-9651-9
113. Novoselov KS, Geim AK, Morozov SV, Jiang D, Zhang Y, Dubonos SV, Grigorieva IV, Firsov AA (2004) Electric field effect in atomically thin carbon films. Science 306(5696):666–669. https://doi.org/10.1126/science.1102896.
114. Chen Y, Fan Z, Zhang Z, Niu W, Li C, Yang N, Chen B, Zhang H (2018) Two-dimensional metal nanomaterials: synthesis, properties, and applications. Chem Rev 118(13):6409–6455. https://doi.org/10.1021/acs.chemrev.7b00727
115. Feng B, Ding Z, Meng S, Yao Y, He X, Cheng P, Chen L, Wu K (2012) Evidence of silicene in honeycomb structures of silicon on Ag(111). Nano Lett 12(7):3507–3511. https://doi.org/10.1021/nl301047g
116. Derivaz M, Dentel D, Stephan R, Hanf M-C, Mehdaoui A, Sonnet P, Pirri C (2015) Continuous Germanene layer on Al(111). Nano Lett 15(4):2510–2516. https://doi.org/10.1021/acs.nanolett.5b00085
117. Yuhara J, Fujii Y, Nishino K, Isobe N, Nakatake M, Xian L, Rubio A, Le Lay G (2018) Large area planar stanene epitaxially grown on Ag(1 1 1). 2D Mater 5(2):025002. https://doi.org/10.1088/2053-1583/aa9ea0
118. Mannix AJ, Zhou X-F, Kiraly B, Wood JD, Alducin D, Myers BD, Liu X et al (2015) Synthesis of borophenes: anisotropic, two-dimensional boron polymorphs. Science 350(6267):1513–1516. https://doi.org/10.1126/science.aad1080

119. Martínez-Periñán E, Down MP, Gibaja C, Lorenzo E, Zamora F, Banks CE (2018) Antimonene: a novel 2D nanomaterial for supercapacitor applications. Adv Energy Mater 8(11):1702606. https://doi.org/10.1002/aenm.201702606
120. Liu H, Neal AT, Zhu Z, Luo Z, Xu X, Tománek D, Ye PD (2014) Phosphorene: an unexplored 2D semiconductor with a high hole mobility. ACS Nano 8(4):4033–4041. https://doi.org/10.1021/nn501226z
121. Zhang JL, Zhao S, Han C, Wang Z, Zhong S, Sun S, Guo R et al (2016b) Epitaxial growth of single layer blue phosphorus: a new phase of two-dimensional phosphorus. Nano Lett 16(8):4903–4908. https://doi.org/10.1021/acs.nanolett.6b01459
122. Xian L, Pérez Paz A, Bianco E, Ajayan PM, Rubio A (2017) Square selenene and tellurene: novel group VI elemental 2D materials with nontrivial topological properties. 2D Mater 4(4):041003. https://doi.org/10.1088/2053-1583/aa8418
123. Sharma S, Singh N, Schwingenschlögl U (2018) Two-dimensional tellurene as excellent thermoelectric material. ACS Appl Energ Mater 1(5):1950–1954. https://doi.org/10.1021/acsaem.8b00032
124. Wang Q, Safdar M, Xu K, Mirza M, Wang Z, He J (2014) Van Der Waals epitaxy and photoresponse of hexagonal tellurium nanoplates on flexible mica sheets. ACS Nano 8(7):7497–7505. https://doi.org/10.1021/nn5028104
125. Huang X, Guan J, Lin Z, Liu B, Xing S, Wang W, Guo J (2017b) Epitaxial growth and band structure of Te film on graphene. Nano Lett 17(8):4619–4623. https://doi.org/10.1021/acs.nanolett.7b01029
126. Wang Y, Qiu G, Wang R, Huang S, Wang Q, Liu Y, Du Y et al (2018) Field-effect transistors made from solution-grown two-dimensional tellurene. Nature Electron 1(4):228–236. https://doi.org/10.1038/s41928-018-0058-4
127. Ruiz-Clavijo A, Caballero-Calero O, Martín-González M (2018) Three-dimensional Bi_2Te_3 networks of interconnected nanowires: synthesis and optimization. Nanomaterials 8(5). https://doi.org/10.3390/nano8050345
128. Ben-Moshe A, Wolf SG, Sadan MB, Houben L, Fan Z, Govorov AO, Markovich G (2014) Enantioselective control of lattice and shape chirality in inorganic nanostructures using chiral biomolecules. Nat Commun 5(1):4302. https://doi.org/10.1038/ncomms5302
129. Feng W, Kim J-Y, Wang X, Calcaterra HA, Qu Z, Meshi L, Kotov NA (2017) Assembly of mesoscale helices with near-unity enantiomeric excess and light-matter interactions for chiral semiconductors. Sci Adv 3(3):e1601159. https://doi.org/10.1126/sciadv.1601159
130. Mayers B, Gates B, Yin Y, Xia Y (2001) Large-scale synthesis of monodisperse nanorods of Se/Te alloys through a homogeneous nucleation and solution growth process. Adv Mater 13(18):1380–1384. https://doi.org/10.1002/1521-4095(200109)13:18<1380::AID-ADMA1380>3.0.CO;2-W
131. Yang Y, Wang K, Liang H-W, Liu G-Q, Feng M, Xu L, Liu J-W, Wang J-L, Yu S-H (2015) A new generation of alloyed/multimetal chalcogenide nanowires by chemical transformation. Sci Adv 1(10):e1500714. https://doi.org/10.1126/sciadv.1500714
132. He Z, Shu-Hong Y (2005) Large scale synthesis of tellurium nanoribbons in tetraethylene pentamine aqueous solution and the stability of tellurium nanoribbons in ethanol and water. J Phys Chem B 109(48):22740–22745. https://doi.org/10.1021/JP0544484
133. Zhu H, Zhang H, Liang J, Rao G, Li J, Liu G, Du Z, Fan H, Luo J (2011) Controlled synthesis of tellurium nanostructures from nanotubes to nanorods and nanowires and their template applications. J Phys Chem C 115(14):6375–6380. https://doi.org/10.1021/jp200316y
134. Wang D, Zhao Y, Jin H, Zhuang J, Zhang W, Wang S, Wang J (2013b) Synthesis of Au-decorated tripod-shaped Te hybrids for applications in the ultrasensitive detection of arsenic. ACS Appl Mater Interfaces 5(12):5733–5740. https://doi.org/10.1021/am401205w
135. Zonaro E, Lampis S, Turner RJ, Junaid S, Qazi S, Vallini G (2015) Biogenic selenium and tellurium nanoparticles synthesized by environmental microbial isolates efficaciously inhibit bacterial planktonic cultures and biofilms. Front Microbiol 6:584. https://doi.org/10.3389/fmicb.2015.00584

136. Abo Elsoud MM, Al-Hagar OEA, Abdelkhalek ES, Sidkey NM (2018) Synthesis and investigations on tellurium myconanoparticles. Biotechnol Rep 18:e00247. https://linkinghub.elsevier.com/retrieve/pii/S2215017X17303454
137. Medina-Cruz D, González MU, Tien-Street W, Castro MF, Crua AV, Fernández I, Martínez L, Huttel Y, Webster TJ, García-Martín JM (2019) Synergic antibacterial coatings combining titanium nanocolumns and tellurium nanorods. Nanomedicine 17:36–46. https://doi.org/10.1016/J.NANO.2018.12.009
138. Palik ED (1998) Handbook of optical constants of solids. Academic Press, Cambridge. https://www.sciencedirect.com/book/9780125444156/handbook-of-optical-constants-of-solids
139. Bottom VE (1952) The hall effect and electrical resistivity of tellurium. Science 115(2995):570–571. https://doi.org/10.1126/science.115.2995.570.
140. Blackband WT (1951) A photo-conductive effect in tellurium film. Nature 168(4277):704. https://doi.org/10.1038/168704a0
141. Liu J-W, Zhu J-H, Zhang C-L, Liang H-W, Yu S-H (2010) Mesostructured assemblies of ultrathin superlong tellurium nanowires and their photoconductivity. J Am Chem Soc 132(26):8945–8952. https://doi.org/10.1021/ja910871s
142. Hackney Z, Mair L, Skinner K, Washburn S (2010) Photoconductive and polarization properties of individual CdTe nanowires. Mater Lett 64(18):2016–2018. https://doi.org/10.1016/j.matlet.2010.06.032
143. Zhou F, Chen J, Wang Y, Zhang J, Wei X, Luo R, Wang G, Wang R (2017) Remarkable catalytic activity of electrochemically dealloyed platinum–tellurium nanoparticles towards formic acid electro-oxidation. Int J Hydrog Energy 42(26):16489–16494. https://doi.org/10.1016/J.IJHYDENE.2017.05.162
144. Huczko A (2000) Template-based synthesis of nanomaterials. Appl Phys A Mater Sci Process 70(4):365–376. https://doi.org/10.1007/s003390051050
145. Sotiropoulou S, Sierra-Sastre Y, Mark SS, Batt CA (2008) Biotemplated nanostructured materials. Chem Mater 20(3):821–834. https://doi.org/10.1021/cm702152a
146. Yang H, Finefrock SW, Albarracin Caballero JD, Wu Y (2014) Environmentally benign synthesis of ultrathin metal telluride nanowires. J Am Chem Soc 136(29):10242–10245. https://doi.org/10.1021/ja505304v
147. Samal AK, Pradeep T (2010) Pt_3Te_4 nanoparticles from tellurium nanowires. Langmuir 26(24):19136–19141. https://doi.org/10.1021/la103466j
148. Fernández-Lodeiro J, Rodríguez-González B, Santos HM, Bertolo E, Luis Capelo J, Santos AAD, Lodeiro C (2016) Unraveling the organotellurium chemistry applied to the synthesis of gold nanomaterials. ACS Omega 1(6):1314–1325. https://doi.org/10.1021/acsomega.6b00309
149. Fernández-Lodeiro J, Rodríguez-Gónzalez B, Novio F, Fernández-Lodeiro A, Ruiz-Molina D, Capelo JL, dos Santos AA, Lodeiro C (2017) Synthesis and characterization of PtTe2 multi-crystallite nanoparticles using organotellurium nanocomposites. Sci Rep 7(1):9889. https://doi.org/10.1038/s41598-017-10239-8
150. Royer D, Dieulesaint E (1979) Elastic and piezoelectric constants of trigonal selenium and tellurium crystals. J Appl Phys 50(6):4042–4045. https://doi.org/10.1063/1.326485
151. Lee TI, Lee S, Lee E, Sohn S, Lee Y, Lee S, Moon G et al (2013) High-power density piezoelectric energy harvesting using radially strained ultrathin trigonal tellurium nanowire assembly. Adv Mater 25(21):2920–2925. https://doi.org/10.1002/adma.201300657
152. Liang T, Zha J-W, Wang D-r, Dang Z-M (2014b) Remarkable piezoresistance effect on the flexible strain sensor based on a single ultralong tellurium micrometre wire. J Phys D Appl Phys 47(50):505103. https://doi.org/10.1088/0022-3727/47/50/505103
153. He W, Van Ngoc H, Qian YT, Hwang JS, Yan YP, Choi H, Kang DJ (2017b) Synthesis of ultra-thin tellurium nanoflakes on textiles for high-performance flexible and wearable nanogenerators. Appl Surf Sci 392:1055–1061. https://doi.org/10.1016/J.APSUSC.2016.09.157
154. Nolan EM, Lippard SJ (2008) Tools and tactics for the optical detection of mercuric ion. Chem Rev 108(9):3443–3480. https://doi.org/10.1021/cr068000q

155. Wang C-W, Lin Z-H, Roy P, Chang H-T (2013a) Detection of mercury ions using silver telluride nanoparticles as a substrate and recognition element through surface-enhanced raman scattering. Front Chem 1:20. https://doi.org/10.3389/fchem.2013.00020
156. Khayat A, Dencker L (1984) Interactions between tellurium and mercury in murine lung and other organs after metallic mercury inhalation: a comparison with selenium. Chem Biol Interact 50(2):123–133. http://www.ncbi.nlm.nih.gov/pubmed/6744461
157. Zhang Q, Uchaker E, Candelaria SL, Cao G (2013) Nanomaterials for energy conversion and storage. Chem Soc Rev 42(7):3127. https://doi.org/10.1039/c3cs00009e
158. He J, Lv W, Chen Y, Wen K, Xu C, Zhang W, Li Y, Qin W, He W (2017a) Tellurium-impregnated porous cobalt-doped carbon polyhedra as superior cathodes for lithium–tellurium batteries. ACS Nano 11(8):8144–8152. https://doi.org/10.1021/acsnano.7b03057.
159. Zhong B, Zhang Y, Li W, Chen Z, Cui J, Li W, Xie Y, Hao Q, He Q (2014) High superionic conduction arising from aligned large lamellae and large figure of merit in bulk $Cu_{1.94}Al_{0.02}Se$. Appl Phys Lett 105(12):123902. https://doi.org/10.1063/1.4896520
160. Seo J-U, Seong G-K, Park C-M (2015) Te/C nanocomposites for Li-Te secondary batteries. Sci Rep 5(1):7969. https://doi.org/10.1038/srep07969
161. Xu J, Xin S, Liu J-W, Wang J-L, Lei Y, Yu S-H (2016) Elastic carbon nanotube aerogel meets tellurium nanowires: a binder- and collector-free electrode for Li-Te batteries. Adv Funct Mater 26(21):3580–3588. https://doi.org/10.1002/adfm.201600640
162. Zhang M, Su HC, Rheem Y, Hangarter CM, Myung NV (2012) A rapid room-temperature NO_2 sensor based on tellurium–SWNT hybrid nanostructures. J Phys Chem C 116(37):20067–20074. https://doi.org/10.1021/jp305393c
163. Tsiulyanu D, Marian S, Liess H-D (2002) Sensing properties of tellurium based thin films to propylamine and carbon oxide. Sensors Actuators B Chem 85(3):232–238. https://doi.org/10.1016/S0925-4005(02)00113-2
164. Becher C, Maurel L, Aschauer U, Lilienblum M, Magén C, Meier D, Langenberg E et al (2015) Strain-induced coupling of electrical polarization and structural defects in SrMnO3 films. Nat Nanotechnol 10(8):661–665. https://doi.org/10.1038/nnano.2015.108
165. Sen S, Bhandarkar V, Muthe KP, Roy M, Deshpande SK, Aiyer RC, Gupta SK, Yakhmi JV, Sahni VC (2006) Highly sensitive hydrogen sulphide sensors operable at room temperature. Sensors Actuators B Chem 115(1):270–275. https://doi.org/10.1016/J.SNB.2005.09.013
166. Tsiulyanu D, Tsiulyanu A, Liess H-D, Eisele I (2005) Characterization of tellurium-based films for NO2 detection. Thin Solid Films 485(1–2):252–256. https://doi.org/10.1016/J.TSF.2005.03.045
167. Park H, Jung H, Zhang M, Chang CH, Ndifor-Angwafor NG, Choa Y, Myung NV (2013) Branched tellurium hollow nanofibers by galvanic displacement reaction and their sensing performance toward nitrogen dioxide. Nanoscale 5(7):3058. https://doi.org/10.1039/c3nr00060e
168. Kumar V, Shashwati S, Sharma M, Muthe KP, Jagannath N, Gaur K, Gupta SK (2009) Tellurium nano-structure based NO gas sensor. J Nanosci Nanotechnol 9(9):5278–5282. http://www.ncbi.nlm.nih.gov/pubmed/19928213
169. Sen S, Muthe KP, Niraj J, Gadkari SC, Gupta SK, Jagannath, Roy M, Deshpande SK, Yakhmi JV (2004) Room temperature operating ammonia sensor based on tellurium thin films. Sensors Actuators B Chem 98(2–3):154–159. https://doi.org/10.1016/J.SNB.2003.10.004
170. Chen X, Lou Y, Dayal S, Qiu X, Krolicki R, Burda C, Zhao C, Becker J (2005) Doped semiconductor nanomaterials. J Nanosci Nanotechnol 5(9):1408–1420. http://www.ncbi.nlm.nih.gov/pubmed/16193954
171. Sznopek JL (2006) Drivers of U.S. mineral demand. http://www.usgs.gov/pubprod
172. Guo WX, Shu D, Chen HY, Li AJ, Wang H, Xiao GM, Dou CL et al (2009) Study on the structure and property of lead tellurium alloy as the positive grid of lead-acid batteries. J Alloys Compd 475(1–2):102–109. https://doi.org/10.1016/J.JALLCOM.2008.08.011
173. Liang T, Su X, Yan Y, Zheng G, Zhang Q, Chi H, Tang X, Uher C (2014a) Ultra-fast synthesis and thermoelectric properties of Te doped skutterudites. J Mater Chem A 2(42):17914–17918. https://doi.org/10.1039/C4TA02780A

174. Avidan A, Oron D (2008) Large blue shift of the biexciton state in tellurium doped CdSe colloidal quantum dots. Nano Lett 8(8):2384–2387. https://doi.org/10.1021/nl801241m
175. Tang K, Gu S, Wu K, Zhu S, Ye J, Zhang R, Zheng Y (2010) Tellurium assisted realization of P-type N-doped ZnO. Appl Phys Lett 96(24):242101. https://doi.org/10.1063/1.3453658
176. Zhang Z, Khurram M, Sun Z, Yan Q (2018) Uniform tellurium doping in black phosphorus single crystals by chemical vapor transport. Inorg Chem 57(7):4098–4103. https://doi.org/10.1021/acs.inorgchem.8b00278
177. Qiu PF, Wang XB, Zhang TS, Shi X, Chen LD (2015) Thermoelectric properties of Te-doped ternary CuAgSe compounds. J Mater Chem A 3(44):22454–22461. https://doi.org/10.1039/C5TA06780D
178. Park W-D, Tanioka K (2014) Tellurium Doping effect in avalanche-mode amorphous selenium photoconductive film. Appl Phys Lett 105(19):192106. https://doi.org/10.1063/1.4902011
179. Tao H, Sun X, Back S, Han Z, Zhu Q, Robertson AW, Ma T et al (2018) Doping palladium with tellurium for the highly selective electrocatalytic reduction of aqueous CO $_2$ to CO. Chem Sci 9(2):483–487. https://doi.org/10.1039/C7SC03018E
180. Ogra Y (2009) Toxicometallomics for research on the toxicology of exotic metalloids based on speciation studies. Anal Sci 25(10):1189–1195. https://doi.org/10.2116/analsci.25.1189
181. Olm E, Fernandes AP, Hebert C, Rundlöf A-K, Larsen EH, Danielsson O, Björnstedt M (2009) Extracellular thiol-assisted selenium uptake dependent on the x(c)- cystine transporter explains the cancer-specific cytotoxicity of selenite. Proc Natl Acad Sci U S A 106(27):11400–11405. https://doi.org/10.1073/pnas.0902204106
182. Bajaj M, Winter J (2014) Se (IV) triggers faster Te (IV) reduction by soil isolates of heterotrophic aerobic bacteria: formation of extracellular SeTe nanospheres. Microb Cell Factories 13:168. https://doi.org/10.1186/s12934-014-0168-2.
183. Baesman SM, Bullen TD, Dewald J, Zhang D, Curran S, Islam FS, Beveridge TJ, Oremland RS (2007) Formation of tellurium nanocrystals during anaerobic growth of bacteria that use te oxyanions as respiratory electron acceptors. Appl Environ Microbiol 73(7):2135–2143. https://doi.org/10.1128/AEM.02558-06
184. Presentato A, Piacenza E, Darbandi A, Anikovskiy M, Cappelletti M, Zannoni D, Turner RJ (2018) Assembly, growth and conductive properties of tellurium nanorods produced by rhodococcus aetherivorans BCP1. Sci Rep 8(1):3923. https://doi.org/10.1038/s41598-018-22320-x
185. Chasteen TG, Bentley R (2003) Biomethylation of selenium and tellurium: microorganisms and plants. Chem Rev 103(1):1–25. https://doi.org/10.1021/cr010210+
186. Ramadan SE, Razak AA, Ragab AM, el-Meleigy M (1989) Incorporation of tellurium into amino acids and proteins in a tellurium-tolerant fungi. Biol Trace Elem Res 20(3):225–232. http://www.ncbi.nlm.nih.gov/pubmed/2484755
187. Cowgill UM (1988) The tellurium content of vegetation. Biol Trace Elem Res 17:43–67. http://www.ncbi.nlm.nih.gov/pubmed/2484368
188. Anan Y, Yoshida M, Hasegawa S, Katai R, Tokumoto M, Ouerdane L, Łobiński R, Ogra Y (2013) Speciation and identification of tellurium-containing metabolites in garlic, Allium Sativum. Metallomics 5(9):1215. https://doi.org/10.1039/c3mt00108c
189. Babula P, Adam V, Opatrilova R, Zehnalek J, Havel L, Kizek R (2008) Uncommon heavy metals, metalloids and their plant toxicity: a review. Environ Chem Lett 6(4):189–213. https://doi.org/10.1007/s10311-008-0159-9
190. Nogueira CW, Zeni G, Rocha JBT (2004) Organoselenium and organotellurium compounds: toxicology and pharmacology. Chem Rev 104(12):6255–6286. https://doi.org/10.1021/cr0406559
191. Nogueira CW, Rotta LN, Perry ML, Souza DO, Teixeira da Rocha JB (2001) Diphenyl diselenide and diphenyl ditelluride affect the rat glutamatergic system in vitro and in vivo. Brain Res 906(1–2):157–163. https://doi.org/10.1016/S0006-8993(01)02165-5
192. Vij P, Hardej D (2012) Evaluation of tellurium toxicity in transformed and non-transformed human colon cells. Environ Toxicol Pharmacol 34(3):768–782. https://doi.org/10.1016/j.etap.2012.09.009

193. Garberg P, Engman L, Tolmachev V, Lundqvist H, Gerdes RG, Cotgreave IA (1999) Binding of tellurium to hepatocellular selenoproteins during incubation with inorganic tellurite: consequences for the activity of selenium-dependent glutathione peroxidase. Int J Biochem Cell Biol 31(2):291–301. https://doi.org/10.1016/S1357-2725(98)00113-7
194. Nathan C, Cunningham-Bussel A (2013) Beyond oxidative stress: an immunologist's guide to reactive oxygen species. Nat Rev Immunol 13(5):349–361. https://doi.org/10.1038/nri3423
195. Brenneisen P, Reichert A (2018) Nanotherapy and reactive oxygen species (ROS) in cancer: a novel perspective. Antioxidants 7(2):31. https://doi.org/10.3390/antiox7020031
196. Kim KS, Lee D, Song CG, Kang PM (2015) Reactive oxygen species-activated nanomaterials as theranostic agents. Nanomedicine 10(17):2709–2723. https://doi.org/10.2217/nnm.15.108
197. Chou T-M, Ke Y-Y, Tsao Y-H, Li Y-C, Lin Z-H (2016) Fabrication of Te and Te-Au nanowires-based carbon fiber fabrics for antibacterial applications. Int J Environ Res Public Health 13(2):202. https://doi.org/10.3390/ijerph13020202
198. Feldmann J, Hirner AV (1995) Occurrence of volatile metal and metalloid species in landfill and sewage gases. Int J Environ Anal Chem 60(2–4):339–359. https://doi.org/10.1080/03067319508042888
199. Kron T, Hansen C, Werner E (1991) Renal excretion of tellurium after peroral administration of tellurium in different forms to healthy human volunteers. J Trace Elem Electrolytes Health Dis 5(4):239–244. http://www.ncbi.nlm.nih.gov/pubmed/1822332
200. Taylor A (1996) Biochemistry of tellurium. Biol Trace Elem Res 55(3):231–239. https://doi.org/10.1007/BF02785282
201. Brown CD, Cruz DM, Roy AK, Webster TJ (2018) Synthesis and characterization of PVP-coated tellurium nanorods and their antibacterial and anticancer properties. J Nanopart Res 20(9):254. https://doi.org/10.1007/s11051-018-4354-8
202. Na N, Liu L, Taes YEC, Zhang C, Huang B, Liu Y, Ma L, Ouyang J (2010) Direct CdTe quantum-dot-based fluorescence imaging of human serum proteins. Small 6(15):1589–1592. https://doi.org/10.1002/smll.201000684
203. Turner RJ, Weiner JH, Taylor DE (1999) Tellurite-mediated thiol oxidation in Escherichia Coli. Microbiology 145(Pt 9):2549–2557. www.microbiologyresearch.org
204. Fleming A (1932) On the specific antibacterial properties of penicillin and potassium tellurite. Incorporating a method of demonstrating some bacterial antagonisms. J Pathol Bacteriol 35(6):831–842. https://doi.org/10.1002/path.1700350603
205. Chapman PA, Siddons CA, Zadik PM, Jewes L (1991) An improved selective medium for the isolation of Escherichia Coli O 157. J Med Microbiol 35(2):107–110. https://doi.org/10.1099/00222615-35-2-107
206. Ventola CL (2015) The antibiotic resistance crisis: part 1: causes and threats. P T 40(4):277–283. http://www.ncbi.nlm.nih.gov/pubmed/25859123
207. Webster TJ, Seil I (2012) Antimicrobial applications of nanotechnology: methods and literature. Int J Nanomedicine 7:2767. https://doi.org/10.2147/IJN.S24805.
208. Wang L, Hu C, Shao L (2017a) The antimicrobial activity of nanoparticles: present situation and prospects for the future. Int J Nanomedicine 12:1227–1249. https://doi.org/10.2147/IJN.S121956.
209. Figueroa M, Fernandez V, Arenas-Salinas M, Ahumada D, Muñoz-Villagrán C, Cornejo F, Vargas E et al (2018) Synthesis and antibacterial activity of metal(loid) nanostructures by environmental multi-metal(loid) resistant bacteria and metal(loid)-reducing flavoproteins. Front Microbiol 9:959. https://doi.org/10.3389/fmicb.2018.00959
210. Lin Z-H, Lee C-H, Chang H-Y, Chang H-T (2012) Antibacterial activities of tellurium nanomaterials. Chem Asian J 7(5):930–934. https://doi.org/10.1002/asia.201101006
211. Zare B, Faramarzi MA, Sepehrizadeh Z, Shakibaie M, Rezaie S, Shahverdi AR (2012) Biosynthesis and recovery of rod-shaped tellurium nanoparticles and their bactericidal activities. Mater Res Bull 47(11):3719–3725. https://doi.org/10.1016/J.MATERRESBULL.2012.06.034

212. Jassim AMN, Farhan SA, Salman JAS, Khalaf KJ, Al Marjani MF, Mohammed M (2015) Study the antibacterial effect of bismuth oxide and tellurium nanoparticles. Int J Chem Biomol Sci 1(3):81–84. https://www.semanticscholar.org/paper/Study-the-Antibacterial-Effect-of-Bismuth-Oxide-and-Jassim-Farhan/327caffd1131ea7c8c69c3f0ebf36df4ef493137
213. Slavin YN, Asnis J, Häfeli UO, Bach H (2017) Metal nanoparticles: understanding the mechanisms behind antibacterial activity. J Nanobiotechnol 15(1):65. https://doi.org/10.1186/s12951-017-0308-z
214. Dutta P, Harrison A, Sabbani S, Munson RS, Dutta PK, Waldman WJ (2011) Silver nanoparticles embedded in zeolite membranes: release of silver ions and mechanism of antibacterial action. Int J Nanomedicine 6:1833. https://doi.org/10.2147/IJN.S24019.
215. Bray F, Ferlay J, Soerjomataram I, Siegel RL, Torre LA, Jemal A (2018) Global cancer statistics 2018: GLOBOCAN estimates of incidence and mortality worldwide for 36 cancers in 185 countries. CA Cancer J Clin 68(6):394–424. https://doi.org/10.3322/caac.21492
216. Boini S, Briançon S, Guillemin F, Galan P, Hercberg S (2004) Impact of cancer occurrence on health-related quality of life: a longitudinal pre-post assessment. Health Qual Life Outcomes 2:4. https://doi.org/10.1186/1477-7525-2-4.
217. Shewach DS, Kuchta RD (2009) Introduction to cancer chemotherapeutics. Chem Rev 109(7):2859–2861. https://doi.org/10.1021/cr900208x
218. Baskar R, Lee KA, Yeo R, Yeoh K-W (2012) Cancer and radiation therapy: current advances and future directions. Int J Med Sci 9(3):193–199. https://doi.org/10.7150/ijms.3635
219. Benjamin DJ (2014) The efficacy of surgical treatment of cancer—20 years later. Med Hypotheses 82(4):412–420. https://doi.org/10.1016/j.mehy.2014.01.004
220. Nakamura S (2018) [Radiotherapy and new cancer drugs—new side effects?]. Gan to Kagaku Ryoho 45(3):424–27. http://www.ncbi.nlm.nih.gov/pubmed/29650897
221. Ramirez LY, Huestis SE, Yap TY, Zyzanski S, Drotar D, Kodish E (2009) Potential chemotherapy side effects: what do oncologists tell parents? Pediatr Blood Cancer 52(4):497–502. https://doi.org/10.1002/pbc.21835
222. Housman G, Byler S, Heerboth S, Lapinska K, Longacre M, Snyder N, Sarkar S (2014) Drug resistance in cancer: an overview. Cancers 6(3):1769–1792. https://doi.org/10.3390/cancers6031769
223. Sredni B (2012) Immunomodulating tellurium compounds as anti-cancer agents. Semin Cancer Biol 22(1):60–69. https://doi.org/10.1016/j.semcancer.2011.12.003
224. Silberman A, Kalechman Y, Hirsch S, Erlich Z, Sredni B, Albeck A (2016) The anticancer activity of organotelluranes: potential role in integrin inactivation. Chembiochem 17(10):918–927. https://doi.org/10.1002/cbic.201500614
225. Fry F, Jacob C (2006) Sensor/effector drug design with potential relevance to cancer. Curr Pharm Des 12(34):4479–4499. https://doi.org/10.2174/138161206779010512
226. Zhang X-F, Liu Z-G, Shen W, Gurunathan S (2016c) Silver nanoparticles: synthesis, characterization, properties, applications, and therapeutic approaches. Int J Mol Sci 17(9):1534. https://doi.org/10.3390/ijms17091534
227. Giles NM, Gutowski NJ, Giles GI, Jacob C (2003b) Redox catalysts as sensitisers towards oxidative stress. FEBS Lett 535(1–3):179–182. https://doi.org/10.1016/S0014-5793(02)03890-5
228. Fry FH, Holme AL, Giles NM, Giles GI, Collins C, Holt K, Pariagh S et al (2005) Multifunctional redox catalysts as selective enhancers of oxidative stress. Org Biomol Chem 3(14):2579. https://doi.org/10.1039/b502197a
229. Carmely A, Meirow D, Peretz A, Albeck M, Bartoov B, Sredni B (2009) Protective effect of the immunomodulator AS101 against cyclophosphamide-induced testicular damage in mice. Hum Reprod 24(6):1322–1329. https://doi.org/10.1093/humrep/den481
230. Cunha RL, Gouvêa IE, Feitosa GPV, Alves MFM, Brömme D, Comasseto JV, Tersariol ILS, Juliano L (2009) Irreversible inhibition of human cathepsins B, L, S and K by hypervalent tellurium compounds. Biol Chem 390(11):1205–1212. https://doi.org/10.1515/BC.2009.125

231. Ahmed K, Zaidi SF (2013) Treating cancer with heat: hyperthermia as promising strategy to enhance apoptosis. J Pak Med Assoc 63(4):504–508. http://www.ncbi.nlm.nih.gov/pubmed/23905451
232. Luk KH, Hulse RM, Phillips TL (1980) Hyperthermia in cancer therapy. West J Med 132(3):179–185. http://www.ncbi.nlm.nih.gov/pubmed/7376656
233. Huang W, Huang Y, You Y, Nie T, Chen T (2017a) High-yield synthesis of multifunctional tellurium nanorods to achieve simultaneous chemo-photothermal combination cancer therapy. Adv Funct Mater 27(33):1701388. https://doi.org/10.1002/adfm.201701388
234. Huang Y, Fan C-Q, Dong H, Wang S-M, Yang X-C, Yang S-M (2017c) Current applications and future prospects of nanomaterials in tumor therapy. Int J Nanomedicine 12:1815–1825. https://doi.org/10.2147/IJN.S127349
235. Iglehart JK (2006) The new era of medical imaging—progress and pitfalls. N Engl J Med 354(26):2822–2828. https://doi.org/10.1056/NEJMhpr061219
236. Garvey CJ, Hanlon R (2002) Computed tomography in clinical practice. BMJ 324(7345):1077–1080. http://www.ncbi.nlm.nih.gov/pubmed/11991915
237. Grover VPB, Tognarelli JM, Crossey MME, Cox IJ, Taylor-Robinson SD, McPhail MJW (2015) Magnetic resonance imaging: principles and techniques: lessons for clinicians. J Clin Exp Hepatol 5(3):246–255. https://doi.org/10.1016/j.jceh.2015.08.001
238. Picano E (2005) Economic and biological costs of cardiac imaging. Cardiovasc Ultrasound 3:13. https://doi.org/10.1186/1476-7120-3-13
239. Toy R, Bauer L, Hoimes C, Ghaghada KB, Karathanasis E (2014) Targeted nanotechnology for cancer imaging. Adv Drug Deliv Rev 76:79–97. https://doi.org/10.1016/j.addr.2014.08.002
240. Murthy SK (2007) Nanoparticles in modern medicine: state of the art and future challenges. Int J Nanomedicine 2(2):129–141. http://www.ncbi.nlm.nih.gov/pubmed/17722542
241. Smith BR, Gambhir SS (2017) Nanomaterials for in vivo imaging. Chem Rev 117(3):901–986. https://doi.org/10.1021/acs.chemrev.6b00073
242. Leary J, Key J (2014) Nanoparticles for multimodal in vivo imaging in nanomedicine. Int J Nanomedicine 9:711. https://doi.org/10.2147/IJN.S53717
243. Weissleder R, Nahrendorf M, Pittet MJ (2014) Imaging macrophages with nanoparticles. Nat Mater 13(2):125–138. https://doi.org/10.1038/nmat3780
244. Choi HS, Frangioni JV (2010) Nanoparticles for biomedical imaging: fundamentals of clinical translation. Mol Imaging 9(6):291–310. http://www.ncbi.nlm.nih.gov/pubmed/21084027
245. Nune SK, Gunda P, Thallapally PK, Lin Y-Y, Laird Forrest M, Berkland CJ (2009) Nanoparticles for biomedical imaging. Expert Opin Drug Deliv 6(11):1175–1194. https://doi.org/10.1517/17425240903229031
246. Shen Z, Wu A, Chen X (2017) Iron oxide nanoparticle based contrast agents for magnetic resonance imaging. Mol Pharm 14(5):1352–1364. https://doi.org/10.1021/acs.molpharmaceut.6b00839
247. Popovtzer R, Agrawal A, Kotov NA, Popovtzer A, Balter J, Carey TE, Kopelman R (2008) Targeted gold nanoparticles enable molecular CT imaging of cancer. Nano Lett 8(12):4593–4596. http://www.ncbi.nlm.nih.gov/pubmed/19367807
248. Liu X, Silks LA, Liu C, Ollivault-Shiflett M, Huang X, Li J, Luo G, Hou Y-M, Liu J, Shen J (2009) Incorporation of tellurocysteine into glutathione transferase generates high glutathione peroxidase efficiency. Angew Chem Int Ed 48(11):2020–2023. https://doi.org/10.1002/anie.200805365
249. Knapp FF, Ambrose KR (1980) Tellurium-123m-labeled 23-(lsopropyl telluro)-24-Nor-5a-Cholan-3f3-Ol: a new potential adrenal imaging agent. J Nucl Med 21:251–257. http://jnm.snmjournals.org/content/21/3/251.full.pdf
250. Okada RD, Knapp FF, Elmaleh DR, Yasuda T, Boucher CA, Strauss HW (1982) Tellurium-123m-labeled-9-telluraheptadecanoic acid: a possible cardiac imaging agent. Circulation 65(2):305–310. https://doi.org/10.1161/01.CIR.65.2.305
251. Valizadeh A, Mikaeili H, Samiei M, Farkhani S, Zarghami N, kouhi M, Akbarzadeh A, Davaran S (2012) Quantum dots: synthesis, bioapplications, and toxicity. Nanoscale Res Lett 7(1):480. https://doi.org/10.1186/1556-276X-7-480

252. Barroso MM (2011) Quantum dots in cell biology. J Histochem Cytochem 59(3):237–251. https://doi.org/10.1369/0022155411398487
253. Bailey RE, Nie S (2003) Alloyed semiconductor quantum dots: tuning the optical properties without changing the particle size. J Am Chem Soc 125(23):7100–7106. https://doi.org/10.1021/JA035000O
254. Chen L, Chen C, Li R, Li Y, Liu S (2009) CdTe quantum dot functionalized silica nanosphere labels for ultrasensitive detection of biomarker. Chem Commun (19):2670. https://doi.org/10.1039/b900319c
255. Xu P, Li J, Shi L, Selke M, Chen B, Wang X (2013) Synergetic effect of functional cadmium-tellurium quantum dots conjugated with gambogic acid for HepG2 cell-labeling and proliferation inhibition. Int J Nanomedicine 8:3729. https://doi.org/10.2147/IJN.S51622
256. Jiang C, Shen Z, Luo C, Lin H, Huang R, Wang Y, Peng H (2016) One-pot aqueous synthesis of gadolinium doped CdTe quantum dots with dual imaging modalities. Talanta 155:14–20. https://doi.org/10.1016/j.talanta.2016.04.021.
257. Zhang J, Su J, Liu L, Huang Y, Mason RP (2008) Evaluation of red CdTe and near infrared CdHgTe quantum dots by fluorescent imaging. J Nanosci Nanotechnol 8(3):1155–1159. http://www.ncbi.nlm.nih.gov/pubmed/18468115

Index

B. Li et al. (eds.), *Racing for the Surface*,
https://doi.org/10.1007/978-3-030-34471-9

H

I

Q

R

S

The manufacturer's authorised representative in the EU is Springer Nature Customer Service Centre GmbH, Europaplatz 3, 69115 Heidelberg, Germany. If you have any concerns regarding our products, please contact ProductSafety@springernature.com

Printed and bound by CPI Group (UK) Ltd, Croydon, CR0 4YY
30/09/2026
02225449-0020